2nd International Conference on Chemistry and Material Science (IC2MS 2019)

IOP Conference Series: Materials Science and Engineering
Volume 833

Malang, Indonesia
2 - 3 November 2019

Editors:

Ibrahim bin Jantan
Kenji Mochizuki
Apinpus Rujiwatra
Akhmad Sabarudin
David R. Turner
Hendra Gunosewoyo
Lukman Hakim
Yuniar Ponco Prananto

ISBN: 978-1-7138-4726-7
ISSN: 1757-8981

Printed from e-media with permission by:

Curran Associates, Inc.
57 Morehouse Lane
Red Hook, NY 12571

Some format issues inherent in the e-media version may also appear in this print version.

Institute of Physics
Dirac House, Temple Back
Bristol BS1 6BE UK

Phone: 44 1 17 929 7481
Fax: 44 1 17 920 0979

techtracking@iop.org

Additional copies of this publication are available from:

Curran Associates, Inc.
57 Morehouse Lane
Red Hook, NY 12571 USA
Phone: 845-758-0400
Fax: 845-758-2633
Email: curran@proceedings.com
Web: www.proceedings.com

TABLE OF CONTENTS

IOP Conf. Series: Materials Science and Engineering **833** (2020) 011001 doi:10.1088/1757-899X/833/1/011001

2-3 November 2019
Malang, Indonesia

IC2MS

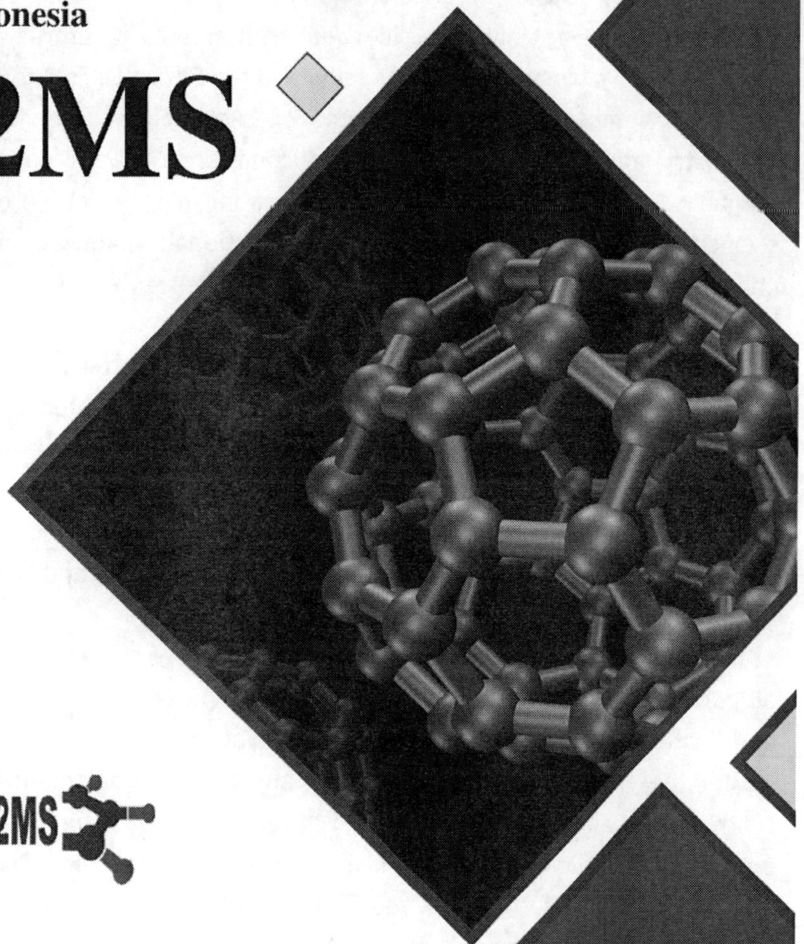

IC2MS

The 2nd International Conference on Chemistry and Material Science

IOP Conf. Series: Materials Science and Engineering **833** (2020) 011001 doi:10.1088/1757-899X/833/1/011001

PREFACE

Chemistry is often regarded as the central of science, while material science is considered as the mother of technology. The pair holds an important role in advancing the quality of life, environment, and welfare. The International Conference on Chemistry and Material Science (IC2MS) is organized to provide an international platform for promoting mutual exchange between scientists, discussing innovative ideas in scientific research, and as a building block for academicians to expand their global perspective and to initiate collaboration, and for government and industrial sectors to update their information with the newest research findings for their future research-based plans and policies.

This year, the conference is organized by the Department of Chemistry, Faculty of Science, Brawijaya University, in collaboration with Chemistry Departments of State University of Malang, Jember University, and Ma Chung University. The 2nd IC2MS was held on 2-3 November 2019 with a theme of "Advancing to the Frontier of Innovation in Chemistry and Material Science". The conference covers not only in chemistry area (Organic, Inorganic, Physical, Analytical, Biochemistry), but also promote papers in material science such as molecular science, functional material, material synthesis and characterisation, nanomaterials and nanodevices, as well as renewable energy and fuel cell.

This conference presents 4 plenary lectures by the keynote speakers, 16 parallel sessions (including invited speakers), and two poster sessions. In total, 220 titles were presented during this conference. The organising committee was also had to rejected some submitted abstracts due to the limited time and venue capacity. Participants were coming from seven different countries and more than 40 national and international universities/institutions (Figure 1). There are also 10 best and favourite (student) posters awarded in this event.

The organising committee of the 2nd IC2MS are greatly appreciating and thanking to all people involve in this event for their active participation and supports. To all keynote and invited speakers, thank you for sharing your research achievements. To the PPIKID of UB for the great financial commitment, and to the Department of Chemistry, Faculty of Mathematics and Natural Sciences (FMIPA), Brawijaya University, for every support given during this event. We also would like to acknowledge all co-host universities and the Savana Hotel of Malang, thank you for the cooperation and assistance. Thank you and may this event give benefits to all. See you in the 3rd IC2MS 2020.

Best Regards,

Dr. Yuniar Ponco Prananto, MSc
Chairman of the 2nd IC2MS 2019

IOP Conf. Series: Materials Science and Engineering **833** (2020) 011001 doi:10.1088/1757-899X/833/1/011001

EDITORIAL BOARD

SCIENTIFIC COMMITTEE

IOP Conf. Series: Materials Science and Engineering **833** (2020) 011001 doi:10.1088/1757-899X/833/1/011001

ORGANIZING COMMITTEE

Dr. Yuniar Ponco Prananto, MSc
Chairman

Dr. Elvina D. Iftitah, MSi
Secretary

Ellya Indahyanti, M.Eng
Treasurer

Anna Safitri, MSc., PhD
Secretariat Office Coordinator

Dr. Arie Srihardyastuti, M.Kes
Conference Program Coordinator

Prof. Aulaniam, DES
Institution Collaboration Coordinator

Moh. Farid Rahman, MSi
Funding Coordinator

Dra. Sri Wardhani, MSi
Conference Catering Coordinator

Drs. Suratmo, MSc
Transportation Coordinator

Drs. Danar Purwonugroho, MSi
Logistic and Facility Coordinator

PROCEEDINGS LAYOUTS AND TECHNICAL TEAM

Irwansyah Putra Pradana, MSi.
Ravi Mahesta, SSi.
Fath Dwisari, MSi.
Angelina Rosmawati, MSi.
Majida Ramadhan, MSi.

==

CONTACT ADDRESS:
Jl. Veteran, Malang, East Java, Indonesia 65145.
Phone: +62-341-575835; Fax: +62-341-575839;
Email: ic2ms@ub.ac.id (cc-ed to prananto@ub.ac.id).
Official site: http://ic2ms.ub.ac.id

==

IOP Conf. Series: Materials Science and Engineering **833** (2020) 011001 doi:10.1088/1757-899X/833/1/011001

List of Keynote Speakers

Keynote speaker 1 : Prof. Ibrahim bin Jantan (*Scopus docs = 176; h-index = 25*)
Taylor's University, Malaysia
Moderator: Dr. Elvina D. Iftitah, MSi

Keynote speaker 2 : Dr. David R. Turner (*Scopus docs = 119; h-index = 29*)
Monash University, Australia
Moderator: Dr. Yuniar Ponco Prananto, MSc

Keynote speaker 3 : Assoc. Prof. Kenji Mochizuki (*Scopus docs = 21; h-index = 9*)
Shinshu University, Japan
Moderator: Assoc. Prof. akhmad Sabarudin, Dr.Sc

Keynote speaker 4 : Akhmad Sabarudin, Dr.Sc (*Scopus docs = 59; h-index = 17*)
Brawijaya University, Indonesia
Moderator: Barlah Rumhayati, PhD

List of Invited Speakers

Invited speaker 1 : Dr. Hendra Gunosewoyo (*Scopus docs = 26; h-index = 11*)
Curtin Unversity, Australia

Invited speaker 2 : Dr. Mohamad Rafi (*Scopus docs = 19; h-index = 4*)
IPB University, Indonesia

Invited speaker 3 : Assoc. Prof. Lee Siew Ling (*Scopus docs = 91; h-index = 13*)
Universiti Teknologi Malaysia, Malaysia

Invited speaker 4 : Dr.rer.nat. Witri Wahyu Lestari (*Scopus docs = 27; h-index = 5*)
University of Sebelas Maret, Indonesia

Invited speaker 5 : Anugrah Ricky Wijaya, DSc (*Scopus docs = 13; h-index = 5*)
State University of Malang, Indonesia

Invited speaker 6 : Dr.rer.nat. Rachmat T. Tjahjanto (*Scopus docs = 12; h-index = 2*)
Brawijaya University, Indonesia

Link for Presented Papers

Oral presenters : 116 presenters
http://ic2ms.ub.ac.id/wp-content/uploads/2019/10/Oral-Session-Schedule.pdf

Poster presenters : 104 presenters
http://ic2ms.ub.ac.id/wp-content/uploads/2019/10/Poster-Session-Schedule.pdf

IOP Conf. Series: Materials Science and Engineering **833** (2020) 011001 doi:10.1088/1757-899X/833/1/011001

SELECTED PICTURES TAKEN FROM THE IC2MS-2019

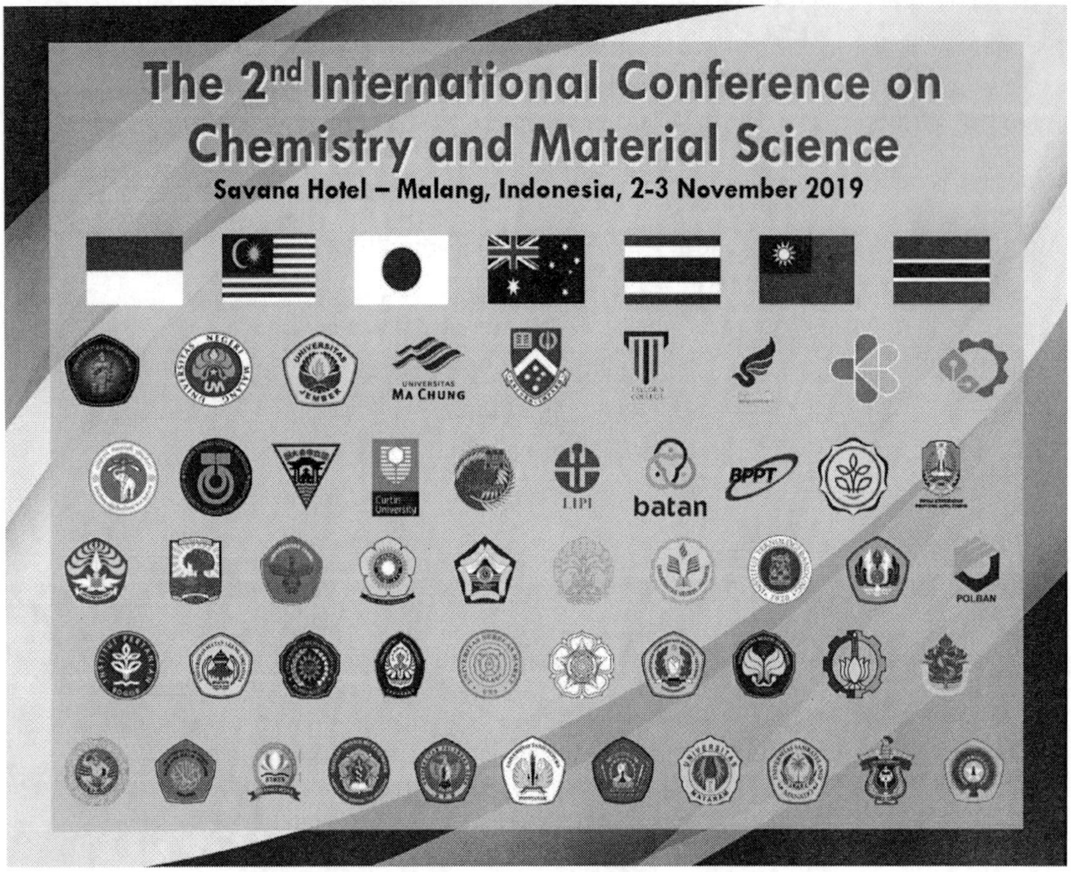

Picture 1. Flags of the participants' country and logos of the participants' institution.

Picture 2. Opening remarks by the chairman of the 2nd IC2MS

IOP Conf. Series: Materials Science and Engineering **833** (2020) 011001 doi:10.1088/1757-899X/833/1/011001

Picture 3. Poster session and poster award winners.

Picture 4. One of the invited speaker's talk.

Peer review statement

All papers published in this volume of *IOP Conference Series: Materials Science and Engineering* have been peer reviewed through processes administered by the proceedings Editors. Reviews were conducted by expert referees to the professional and scientific standards expected of a proceedings journal published by IOP Publishing.

IOP Conf. Series: Materials Science and Engineering **833** (2020) 012001 doi:10.1088/1757-899X/833/1/012001

Sonication-Assisted Pine Cone Flower Cellulose Hydrolysis Using Formic Acid

'Urfa Zakiyya 'Uyunin, Masruri MASRURI*, Zubaidah Ningsih, Arie Srihardyastutie

Chemistry Department, Faculty of Mathematics and Natural Science, Brawijaya University, Indonesia

*Corresponding author : masruri@ub.ac.id

Abstract. Nanocellulose has many applications in industrial sectors, such as in pulp and papermaking, production of synthetic textile fibers, dye, ink, and coating materials. The cellulose itself can be isolated from the waste of pine cone flower. This paper reports formic acid in the different concentrations (10%, 30%, and 60%) for hydrolyzing of cellulose using sonication technique (48 kHz) at 45 °C for 60 mins. The hydrolyzed cellulose was analyzed using FTIR and XRD spectrometry. It was found that the hydrolyzed-cellulose isolated was 92.4%, 94.6%, and 89.6%, respectively. The FTIR spectra provided the band for O-H (3435 cm^{-1}) and C-O-C (1180-1060 cm^{-1}) functional groups. However, the spectra also showed the C=C band for lignin impurities at 1661 cm^{-1}. Furthermore, The XRD data gave similar 2-theta values for all hydrolyzed cellulose at 16°, 22°, and 34° respectively. The crystallite size was 18.34, 15.09, and 15.07 nm. Meanwhile, the crystallinity index was 50.50, 52.70, and 51.60% respectively.

1. Introduction

The waste of pine cone flower has not attracted public attention. In fact, it contains some important chemicals such as secondary metabolites, volatile chemicals, and cellulose. Some secondary metabolite were reported previously as reagent for green synthesis of nanoparticles [1-2]. The volatile components were investigated as starting materials for industrial fine chemicals [3-4]. Meanwhile, the cellulose part is undergoing investigation.

Cellulose is the linear polysaccharide constructed by β-D-glucopyranose units and linked with the β-(1,4)-glycoside bond. The length of the linkage increases the size of the polymer and also the size of the cellulose particle. Hydrolysis toward these glycoside-bonds reduces the polymer chain and also decrease the particle size of cellulose. However, cellulose chains also consist of amorphous and crystalline domains. Nanocellulose is largely used in papermaking, food, cosmetic, biomedical and biocomposite packaging material for drug delivery [5,6]. Nanocellulose composite is also used for antibacterial and antifungal.

The general methods for hydrolysis are using sulphuric acid, hydrochloric acid, and phosphoric acid [5]. However, strong acid reagent causes equipment corrosion, over-decomposition, and requires a lot of water for neutralization. Previously, Garcı´a et al. [7] reported the hydrolysis of pine cone cellulose using sulfuric acid 64% at 45°C. Three different hydrolysis times (30, 45, and 90 min) were conducted and the yield reported was 32-37% and crystallinity of nanocellulose of 80.8-88.5%. In this paper, we report a greener hydrolysis under sonication condition by applying formic acid as a reagent for hydrolysis. The effect of formic acid concentration variation to the cellulose properties was studied. The

FTIR and XRD analysis were applied for functional groups characterization, crystallinity index, and crystallite size measurement.

2. Methods

2.1. Material and Chemicals
Pine cone flowers were collected from the pine forest in Coban Rais, Batu city, Indonesia. The chemicals used for research as bought from the manufacturer include formic acid (Smart Lab), sodium hypochlorite (Smart Lab), and sodium hydroxide (Merck).

2.2. Instruments
Some instruments operated for this research includes batch sonicator (DELTA D150H 48 kHz), FTIR spectrophotometer (Shimadzu FTIR 8400s), XRD spectrometer (PanAnalytical E'xpert Pro).

2.3. Cellulose preparation
Dried pine cone flower powder (50 g) in beaker glass was added with 500 mL of sodium hydroxide solution 6%. This mixture was heated at 70°C and magnetically stirred at 300 rpm for 4 h. The result was washed with distilled water to a neutral pH, and bleached with sodium hypochlorite 6%. Then, the resulted cellulose was washed with distilled water until neutral pH and prepared as starting material.

2.4. Hydrolysis procedure
Dried cellulose powder (5.0 g) was added with 10% formic acid solution (1:20 w/v ratio) and sonicated for 60 mins. Cold water was added to stop the reaction, and left overnight. It was further separated and centrifugated. The supernatant was collected as cellulose nanoparticle (FA 10). The obtained nanocellulose were characterized by FTIR and XRD. A similar separated procedure was performed by using 30% (FA 30) and 60% (FA 60) of formic acid solution.

2.5. Crystallite dimension measurement
Crystallite size (D_{hkl}) was calculated using the modified Debye Scherrer equation (1) [8]

$$D_{hkl} = \frac{0.9\lambda}{\beta_1 \cos\theta} \quad (1)$$

Where D_{hkl} is the crystal dimension upright to the diffracting planes, $\lambda = 1.5406$ Å. $\beta_{1/2}$ is the full width at half-maximum of diffraction peak. θ is the correlation Bragg angle.

3. Result and discussion

3.1. Cellulose preparation
According to Rambabu et al. [9], the composition of the pine cone flower was cellulose (44 wt%), hemicellulose (27 wt%), and lignin (27 wt%). Besides, other secondary components such as waxes, fats, tannins, gums and other phenolic and terpenoid compounds were also reported [9]. To obtain a pure and high strength of cellulose, several treatments have to be undertaken. Hot water treatment generally removes the waxes, fats, tannins, and gums from cellulose. The alkaline treatment using sodium hydroxide solution displaces the hemicellulose and pectin. Moreover, the mixture of cellulose with sodium hypochlorite can reduce lignin attached in cellulose. Cellulose produced from the waste of pine cone flower is depicted in Figure 1. Alkali treatment of cellulose using sodium hydroxide provides slightly brown cellulose (b), and after the bleaching process, it gives a light yellow cellulose (c).

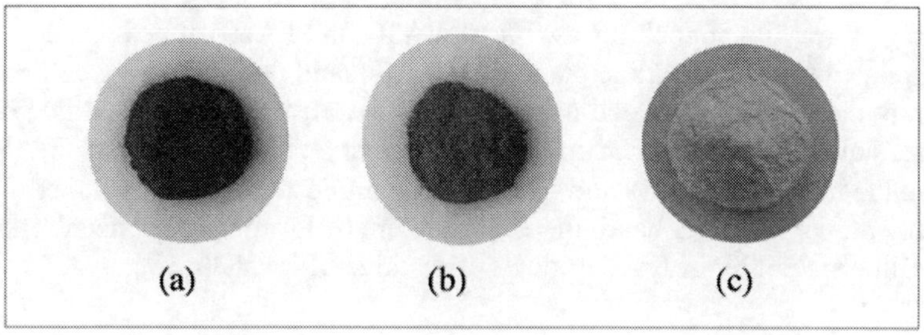

Figure 1. Preparation cellulose from the waste of pine cone flower: waste powder (a), after alkali treatment (b), and after bleaching (c).

3.2. Hydrolysis using formic acid

Cellulose hydrolysis is more affected by acidity of the reagent, due to acidic provided by proton to hydrolyze the glycoside bond in cellulose. Sonication can accelerate the process [10]. Table 1 summarized the data obtained from the hydrolysis process of cellulose in three different concentrations of formic acid. The recorded data correlated the yield of cellulose during hydrolysis, crystallinity index, and crystallite size achieved from XRD analysis.

Table 1. Yield, crystallinity index, and crystallite size of nanocellulose.

Sample	Yield (%)	Crystallinity index (%)	Crystallite size (nm)
FA 10	92.4	50.50	18.34
FA 30	94.6	52.70	15.09
FA 60	89.6	51.60	15.07

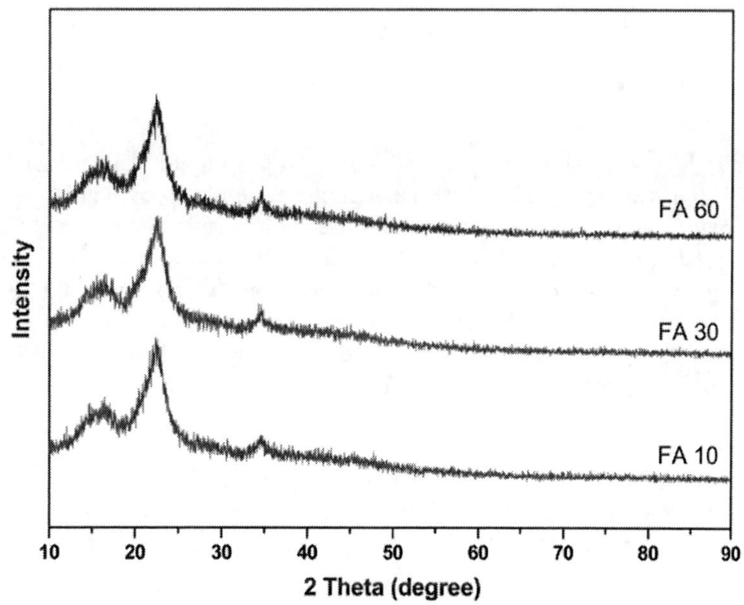

Figure 2. XRD pattern of three samples hydrolysis result.

The yield of nanocellulose obtained between 89.6% and 94.6%. The highest yield of nanocellulose is produced from 30% of formic acid hydrolysis (FA 30), and the lowest yield is provided using 60% of formic acid. In addition, the XRD analysis provided diffractogram (Figure 2). Three peaks diffractogram showed similar region for all nanocelluloses (FA10, FA30, and FA60) at peak (2θ) = 16°, 22°, and 34°. These values correlated to the crystalline plane of 110, 200, and 004 indicating cellulose [11].

The crystallinity index was calculated using Origin Pro software and crystallite size was calculated using the Scherrer equation as tabulated in Table 1. The crystallinity index of the three samples was obtained between the range of 50.50% and 52.70%. The highest crystallinity index was shown at 10% formic acid treatment. On the other hand, the enhancement of formic acid concentration showed a little decrease in crystallite size of hydrolysis products. The value of crystallite size FA 30 was similar to FA 60.

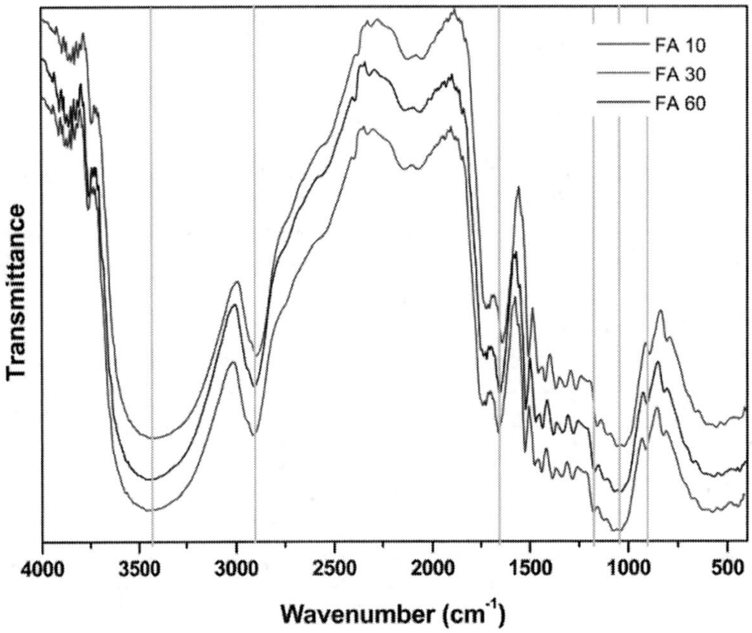

Figure 3. FTIR spectra of three nanocellulose samples.

Figure 3 shows the FT-IR spectra of nanocellulose achieved from hydrolysis using three concentrations of formic acid (FA10, FA30, and FA60). The nanocellulose samples provide identical absorption band spectra. The hydroxyl group stretching of nanocellulose is observed at 3435 cm^{-1}. The C-H symmetric and asymmetric stretching bands are shown at 2914 cm^{-1}. Moreover, the absorption band at 1180 and 1060 cm^{-1} correlated to C-O-C pyranose ring bending vibration, and also peak at 905 cm^{-1} is a band for glycosidic-bond bending vibration. However, remanence of phenolic or tannin compounds are still presence in the nanocellulose produced. It is indicated by the observed band at 1661 cm^{-1}. This finding is in agreement with that reported by Kian et al. [12].

4. Conclusion

Nanocellulose can be prepared with sonication-assisted formic acid hydrolysis. The optimum result is obtained using formic acid of 30%. It gives a higher yield of nanocellulose with the highest crystallinity index and a lower of the crystallite size. This finding lead in further application of nanocellulose for nanohydrogel-based nanocellulose.

Acknowledgment

This research is supported by Penelitian Dasar Unggulan Perguruan Tinggi (2019), DRPM of Ministry for Research, Technology and Higher Education of Indonesia. The authors gratefully acknowledge the contributions of Mr. Widji Sulistjo and Mr. Hadi Kurniawan.

IOP Conf. Series: Materials Science and Engineering **833** (2020) 012001 doi:10.1088/1757-899X/833/1/012001

References

[1] Azkiya N I, Masruri M and Ulfa S M 2018 *IOP Conf. Ser. Mater. Sci. Eng.* **299** 012070
[2] Masruri M, Pangestin D N, Ulfa S M, Riyanto S, Srihardyastutie A and Rahman M F 2018 *IOP Conf. Ser. Mater. Sci. Eng.* **299** 012072
[3] Masruri M and Pamungkas KK 2019 *IOP Conf. Ser. Mater. Sci. Eng.* **546** 042023
[4] Masruri M, Amini R W and Rahman M F 2016 *Indones. J. Chem.* **16** 59-64
[5] Li B, Xu W, Kronlund D, Määttänen A, Liu J, Smått J, Peltonen J, Willför S, Mu X and Xu C 2015 *Carbohyd. Polym.* **133** 605-612
[6] Melikoğlu A Y, Bilek A E and Cesur S 2019 *Carbohyd. Polym.* **215** 330-337
[7] Garcı´a D G, Balart R, Martinez J L, Monica Ek and Moriana R 2018 *Cellulose* **25** 2925-2938
[8] Segal L, Creely J J, Martin A E, Jr and Conrad C M 1959 *Text. Res. J.* **29** 786–794.
[9] Rambabu N, Panthapulakkal S, Sain M and Dalai A K 2015 *Ind. Crop. Prod.* **10** 1-9
[10] Hamid S B A, Zain S K, Das R and Centi G 2016 *Carbohyd. Polym.* **138** 349-355
[11] Rosa M F, Medeiros E S, Malmonge J A, Gregorsk K S, Wood D F, Mattoso L H C, Glenn G, Orts W J and Imam S H 2010 *Carbohyd. Polym.* **81** 83-92
[12] Kian L K, Jawaid M, Ariffin H and Karim Z 2018 *Int. J. Biol. Macromol.* **114** 54-63

Synthesis and molecular docking study of 6-chloropyrazine-2-carboxylic acid derivatives

Nur Pasca Aijijiyah[1], Muhammad Riza Ghulam Fahmi[2], Sri Fatmawati[1], Mardi Santoso[1,*]

[1]Department of Chemistry, Faculty of Science, Institut Teknologi Sepuluh Nopember, Kampus ITS Sukolilo Surabaya 60111, Indonesia

[2]Ma Chung Research Center for Photosynthetic Pigments, Universitas Ma Chung, Villa Puncak Tidar N-01, Malang 65151,Indonesia

*Corresponding author : tsv09@chem.its.ac.id

Abstract. One of the most lethal and frequent infectious diseases worldwide is tuberculosis. Multi and extensively tuberculosis drug-resistant constitutes a serious problem and emphasizes the need for novel anti-tubercular agents. Accordingly, various pyrazine-2-carboxamides were synthesized and evaluated as potential anti-tuberculosis agents. The synthesis involved reaction of pyrazinoic acids with thionyl chloride to yield acyl chlorides which on treatment with various anilines gave various pyrazine-2-carboxamides. Based on structure-activity relationships extracted from previously published, this paper reported the synthesis and molecular docking study of 6-chloropyrazine-2-carboxamides. Synthesis involved reaction of 6-chloropyrazinoic acid with 2,4,6-trichlorobenzoyl chloride instead of thionyl chloride which listed under the Chemical Weapons Convention as it may use for the production of chemical weapons. Structure identification of 6-chloropyrazine-2-carboxamides was carried out by ^{1}H NMR, ^{13}C NMR, FTIR, and high-resolution mass spectroscopy. It is predicted that 6-chloro-*N*-octylpyrazine-2-carboxamide has better bioactivity against *Mycobacterium tuberculosis*, based on molecular docking study.

1. Introduction

Mycobacterium tuberculosis (MTB) bacteria leads to tuberculosis disease wich primarily attacks the lungs [1]. In 2018, the World Health Organization reported that Indonesia is a country with a high TB burden [2]. The Ministry of Health Indonesia (2019) reported that there were 511,873 cases of TB in 2018 [3]. The genetic mutation of the TB bacteria causes resistance against first-line TB drugs called multi-drug resistant TB. This disease becomes a more serious problem with the formation of extensively drug-resistant TB wich bacteria MTB resistance against first-line and several second-line TB drugs [4]. Nowadays, TB treatment uses first-line drugs (isoniazid and rifampicin) and second-line drugs such as amikacin (injectable drugs) and levoflaxin (fluoroquinolones derivatives) [5]. However, this treatment is not enough to treat TB patients with resistance to TB drugs. [6]. The Food and Drug Administration has assigned two kinds of antituberculosis drugs for the medication of the patient with drug-resistant TB, namely bedaquiline (TMC207) and nitroimidazole bicyclic (PA-824), but both of them are reported to have a side effect that causes abnormal heart rhythms and sudden cardiac death [7]. Accordingly, the development of new antituberculosis drugs are needed.

As one of the first-line TB drugs, pyrazinamide (pyrazine-2-carboxamide) considered having an important role in the short course therapy of TB that can synergize with rifampicin and has a good sterilizing effect[8]. However, genetic mutations in TB bacteria cause resistance to this drug [9]. In the past 10 years, various pyrazine-2-carboxamide derivatives were prepared and evaluated their role as anti *Mycobacterium tuberculosis* $H_{37}Rv$. The structure-activity relationship (SAR) of the pyrazine-2-carboxamide derivatives shows that substitution with phenyl and alkyl group at the nitrogen of amide group, and chloro at C6 of pyrazine ring increased the antituberculosis activity against *Mycobacterium tuberculosis* H37Rv [10–14]. Pyrazine-2-carboxamides as amides can be synthesized through reactions involving several reagents such as thionyl chloride [15-16], 1,1'-carbonyldiimidazole (CDI) [14] and hydroxybenzotriazole (HOBt) [7]. Nevertheless, these reagents have several disadvantages. Thionyl chloride is a chemical listed under the Chemical Weapons Convention (CWC) as it may use for the production of chemical weapons [17]. Hydroxybenzoyltriazole (HOBt) has explosive property [18], while the reaction of carboxylic acid and 1,1'-carbonyldiimidazole (CDI) give an intermediate which less reactive than the acyl chloride [19]. 2,4,6-Trichlobenzoyl chloride (TCBC) is a reagent used in the synthesis of esters and thioesters compound [20–23]. The reaction of TCBC and carboxylic acid produce the mixed anhydride which has a sterically hindered aromatic moieties that can regioselectively give the corresponding ester [24]. Herewith, we report the synthesis of 6-chloropyrazine-2-carboxylic acid derivatives (**1a-c**), and their prediction as anti-tuberculosis agent based on molecular docking study.

2. Experimental Section

2.1. General information

6-Chloropyrazine-2-carboxylic acid, *p*-ethylaniline, cycloheptylamine, and *n*-octylamine were purchased from Tokyo Chemical Industry. Triethylamine (TEA), tetrahydrofuran (THF), sodium hydroxide, sodium carbonate, and magnesium sulfate were purchased from Merck. 2,4,6-Trichlorobenzoyl chloride (TCBC) and *p*-dimethylaminopyridine (DMAP) were recieved from Sigma-Aldrich. The chromatography analysis and purification were carried out using Merck with 60 GF_{254} silica gel plates type for thin-layer chromatography (TLC) and 60 PF_{254} silica gel containing gypsum type for radial chromatography. The compounds were detected with ultraviolet lights at 245 and 366 nm. Melting temperature measurement was performed on a Fisher John apparatus. The proton and carbon NMR were analyzed using Jeol resonance at 400 MHz, using tetramethylsilane and $CDCl_3$ as standard and solvent respectively. The high resolution mass data (HRMS) were obtained using Waters-Xevo QTof instrument, while Infrared data were recorded using FTIR Shimadzu 8400S.

2.2. General method for synthesis of 6-chloropyrazine-2-carboxylic acid derivatives (1a-c)

In the round-bottom flask, the 6-Chloropyrazine-2-carboxylic acid (1 eqiv), TCBC (1 eqiv), and TEA (1 eqiv) were mixtured in THF (80 mL) solvent and stirred for 20 mins at room temperature. DMAP (1 eqiv) was added to the mixture while the corresponding amine (0.25 eqiv) was added droplet. After that, the mixture was further stirred with heating at 50-55 °C for 1 hour and was allowed to stand at rt after the completion of the reaction and then filtrated off. The cold water was added to the obtained filtrate and extracted several times using dichloromethane. The organic layer was combined and washed subsequently with solutions of 5% HCl, 5% NaOH, 5% Na_2CO_3 and water. After that, the organic layer was dried over $MgSO_4$ and evaporated under vacuum. The crude product was purified using chromatogram with n-hexane:ethyl acetate (10:1 or 20:1) eluent to produce 6-chloropyrazine-2-carboxylic acid derivatives (**1a-c**).

2.3. Molecular docking studies

Molecular docking study of 6-chloropyrazine-2-carboxylic acid derivatives (**1a-c**) was performed on MVD using MolDock SE Algorithm.The geometry of the compound (**1a-c**) structures was optimized using Semi-empiris (AM1) technique on Hyperchem 8.0.7 software. The protein used for docking was *M. Tuberculosis* InhA (PDB ID: 4DRE) where the three-dimensional (3D) structure was taken from the

RCSB Protein Databank. The protein and reference ligand (NAI) were prepared using Molegro Virtual Docking (MVD) 5.0. The preferred docking pose of each compound was performed by 50 runs of Moldock search by the Lamarckian genetic algorithm (LGA) and the results were visualized using MVD.

3. Result and Discussion

3.1. Chemistry

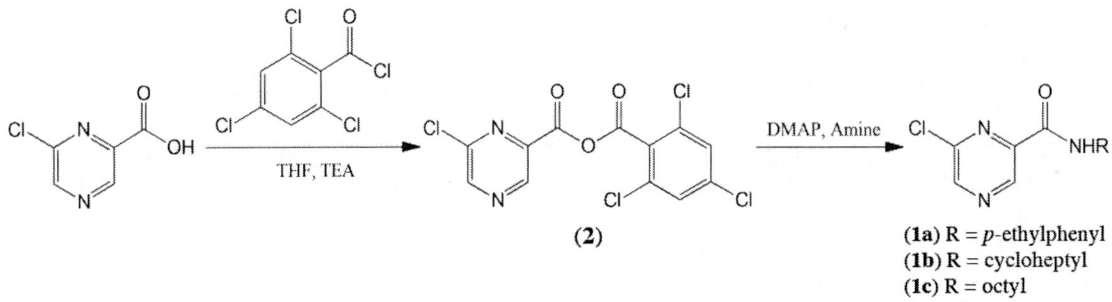

Scheme **1.** Synthesis of 6-chloropyrazine-2-carboxylic acid derivatives (**1a-c**).

6-Chloropyrazine-2-carboxylic acid derivatives (**1a-c**) were synthesized based on the synthesis of 2-phenylethyl esters [23]. The reaction of 6-chloropyrazine-2-carboxylic acid with TEA and TCBC in THF at room temperature produced a mixed anhydride (**2**), then treated with DMAP and the corresponding amine for 1 hour at 50-55 °C (Scheme 1). The mixture was filtered after cooling, extracted with dichloromethane, washed subsequently with solutions of 5% hydrochloric acid, 5% sodium hydroxide, 5% sodium carbonate, water, and then being evaporated. After that, the residue was purified to produce 6-chloropyrazine-2-carboxylicacid derivatives (**1a-c**) in 71-75% yields.

^{1}H NMR spectra of synthesis compounds (**1a-c**) showed broad singlet signals of NH protons at 7.55-9.34 ppm, which confirmed that carboxylic acid derivatives (**1a-c**) have formed. The aromatic proton shows six protons at 7.23-9.39 ppm for **1a** compounds, two protons at 8.72-9.27 ppm for **1b**, and at two protons at 8.73-9.28 ppm for **1c** compound. The characteristic of ethyl proton (**1a**), *N*-cycloheptyl protons (**1b**), and *N*-octyl proton (**1c**) appear at 1.24-2.65 ppm with 5 protons, 13 protons at1.54-4.16 ppm, and 17 protons at 0.86-3.46 ppm, respectively. ^{13}C NMR spectra showed carbon signals of carbonyl groups at 159.31-161.66 ppm. It was supported by the IR spectra which showed sharp absorption bands around 1653-1693 cm^{-1} for the carbonyl groups and 3273-3369 cm^{-1} for the NH groups. The HRMS data shows *m/z* at 262.0606 (^{35}Cl) and 264.0656 (^{37}Cl) ([M+H]$^+$) for compound **1a**, *m/z* at 254.0967 (^{35}Cl) and 256.0997 (^{37}Cl) ([M+H]$^+$) for compound **1b**, and *m/z* at 270.1252 (^{35}Cl) and 272.1357 (^{37}Cl) ([M+H]$^+$) for compound **1c**.

3.2. Molecular docking studies

Molecular docking or computational studies has an important role in the development of new drugs through the design of structure-based drugs. The docking approach can help the researchers to find out the behavior of a drug in the active site of the proteins through the interaction between them at the atomic level [25]. Interaction analysis of the synthesis compounds (**1a-c**) at the InhA protein (PDB ID: 4DRE) binding site of *M. tuberculosis* was carried out with MVD using MolDock SE Algorithm. The result revealed that compound (**1c**) has the lowest rerank score which means that it has the best binding affinity. Compound (**1a**) and (**1b**) have the hydrogen bonding and π-π interactions, while compound (**1c**) only has the π-π interactions with Gly14, Phe41, Leu63, and Gly40 which reported in Tabel 1 and can be seen in Figure 1.

Table 1. Rerank score and interaction of compound (**1a-c**) within the InhA protein binding site of *M.* tuberculosis.

Compound	Rerank score	Interaction
(**1a**)	-79.8858	H-Bond: Val65 π-π:Ile16, Ile122, Asp64, and Val65
(**1b**)	-79.1557	H-Bond: Gly96 π-π: Gly14, Val65, Phe41, Ile95, and Gly95
(**1c**)	-85.7435	H-Bond: - π-π:Gly14, Phe41, Leu63, and Gly40

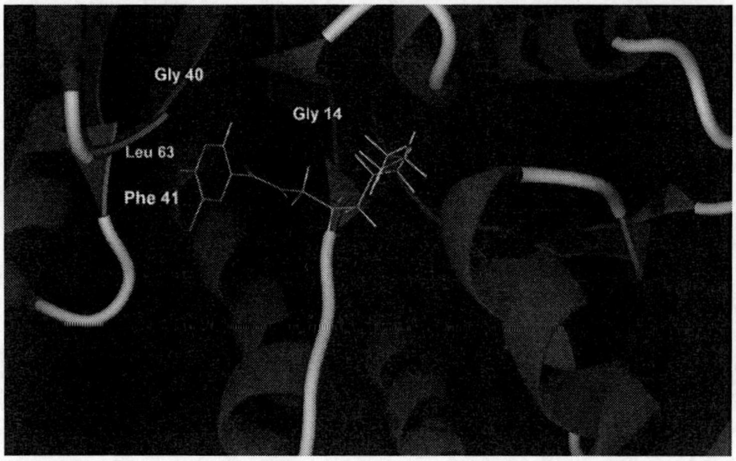

Figure 1. The docking pose of compound (**1c**) at the InHA protein binding site of *M. tuberculosis.*

4. Conclusion

The synthesis of 6-chloropyrazine-2-carboxylic acid derivatives (**1a-c**) has been successfully carried out with the presence of TCBC. Amidation of 6-chloropyrazine-2-carboxylic acid gave compounds (**1a-c**) in 71-75% yields. The molecular docking studies result revealed that 6-chloro-*N*-octylpyrazine-2-carboxamide (**1c**) has the best activity against *Mycobacterium tuberculosis* $H_{37}Rv$ strain.

Acknowledgments

The authors are very grateful to the Indonesian Government for the PenelitianTesis Master Grant No. 796/PKS/ITS/2019.

References

[1] Garcia-monco J C 2014 *Handbook of Clinical Neurology* vol 121, ed J Biller and J M Ferro (Amsterdam: Elevier BV) 1485–99
[2] World Health Organisation 2018 *Global Health TB Report 2018* (Genewa: WHO)
[3] Ministry of Health Indonesia 2019 *Profil Kesehatan Indonesia 2018* (Jakarta: Kemenkes RI)
[4] Bouz G, Juhás M, Niklová P, Jand'ourek O, Paterová P, Ek J J, Tůmová L, Kovalíková Z, Kastner P, Dolezal M and Zitko J 2017 *Molecules* **22** 1–19
[5] Unissa A N, Hanna L E and Swaminathan S 2016 *Chem. Biol. Drug Des.* **87** 537–50
[6] Hopewell P C, Kato-maeda M and Ernst J D 2016 *Murray and Nadel's Textbook of Respiratory Medicine* vol 10, ed V C Broaddus *et al* (Philadelphia: Elsevier Saunders) 593–628
[7] Onajole O K, Pieroni M, Tipparaju S K, Lun S, Stec J, Chen G, Gunosewoyo H, Guo H, Ammerman N C, Bishai W R and Kozikowski A P 2013 *J. Med. Chem.* **56** 4093–103
[8] Ugwu D I, Ezema B E, Eze F U and Ugwuja D I 2014 *Int. J. Med. Chem.* **2014** 1–18
[9] Njire M, Tan Y, Mugweru J, Wang C, Guo J, Yew W W, Tan S and Zhang T 2016 *Adv. Med. Sci.*

61 63–71
[10] Dolezal M, Cmedlova P, Palek L, Vinsova J, Kunes J, Buchta V, Jampilek J and Kralova K 2008 *Eur. J. Med. Chem.* **43** 1105–13
[11] Servusová B, Eibinová D, Doležal M, Kubíček V, Paterová P, Peško M and Král'ová K 2012 *Molecules* **17** 13183–98
[12] Zitko J, Servusová B, Paterová P, Mandíková J, Kubíček V, Kučera R, Hrabcová V, Kuneš J, Soukup O and Doležal M 2013 *Molecules* **18** 14807–25
[13] Zitko J, Barbora S V, Paterova P, Navratilova L, Trejtnar F, Kunes J and Dolezal M 2016 *Chem. Pap.***70** 649–57
[14] Bouz G, Semelková L, Jand'ourek O, Konečná K, Paterová P, Navrátilová L, Kubíček V, Kuneš J, Doležal M and Zitko J 2019 *Molecules* **24** 1–19
[15] Sarmah K K, Rajbongshi T, Bhowmick S and Thakuria R 2017 *Acta Crystallogr. Sect. B Struct. Sci. Cryst. Eng. Mater.* **73** 1007–16
[16] Servusová B, Vobicková J, Paterová P, Kubíček V, Kuneš J, Doležal M and Zitko J 2013 *Bioorganic Med. Chem. Lett.* **23** 3589–91
[17] Kostiainen O 2000 *Handbook of Analytical Separations* vol 2, ed M J Bogusz (Amsterdam: Elsevier BV) 405–35
[18] Wehrstedt K D, Wandrey P A and Heitkamp D 2005 *J. Hazard. Mater.* **126** 1–7
[19] Zhao W 2017 *Handbook for Chemical Process Research Development* (Boca Raton: CRC Press Taylor & Francis Group)
[20] Ewin R A and Pertusati F 2013 *Encycl. Reag. Org. Chem.* 2–6
[21] Tsakos M, Schaffert E S, Clement L L, Villadsen N L and Poulsen T B 2014 *Nat. Prod. Rep.* **1** 25
[22] Kamadatu L and Santoso M 2015 *Proc. 1st Int. Seminar on Science and Technology August 5th 2015, Surabaya* vol 1 pp J633-193–4
[23] Maharani N and Santoso M 2015 *Adv. Environ. Biol.* **9** 214–8
[24] Dhimitruka I and SantaLucia J 2006 *Org. Lett.* **8** 47–50
[25] Meng X-Y, Zhang H-X, Mezei M and Cui M 2011 *NIH Public Access* **7** 146–57

Crystallinity of nanocellulose isolated from the flower waste of pine tree (*Pinus merkusii*)

Mahrullina Mahirotul Aisiyah, Masruri MASRURI*, Arie Srihardyastutie

Chemistry Department, Faculty of Mathematics and Natural Science, Brawijaya University, Indonesia

*Corresponding author: masruri@ub.ac.id

Abstract. Pine flower is an agricultural waste that has high cellulose content. Cellulose is a major material for making nanocellulose. Nanocellulose has been isolated from the flower waste of the pine tree (*Pinus merkusii*). The process was initiated by delignification and bleaching process with sodium hydroxide 6% and sodium hypochlorite 6% followed by hydrolysis with acetic acid under stirring at 45°C for 1 hour. Three different concentrations of acetic acid (10%, 30%, and 60%) were studied toward nanocellulose crystallinity. Nanocellulose was characterized by FTIR, XRD spectroscopy, and TEM analysis. The result shows that the % yield of nanocellulose was 87.4%, 94.2%, and 91.8% respectively. Nanocellulose has high crystallinity as indicated by the crystallinity index. Besides, the nanocellulose width is 12.82 – 21.36 nm and the length is 267.12 – 326.87 nm.

1. Introduction

Pine flower is an agricultural and industrial waste that is rarely used [1]. Pine flower contain cellulose and secondary metabolites. The previous study shows secondary metabolites of pine flower can be extracted in water and used as a reagent for synthesis copper-nanoparticles [2]. On the other hand, the main content of pine flower is cellulose (±43%) [3]. Cellulose is a linear polymer consisting of β-D-glucose monomers. Each monomer linked with the β-(1,4)-glycosidic bond and the degree of polymerization is more than 20000 units [4].

Cellulose is a precursor to produce nanocellulose. Nanocellulose is a promising material as a polymer matrix nanocomposite produced from a natural biopolymer [5]. Nanocellulose has unique structural and remarkable physicochemical properties such as biocompatibility, biodegradability, renewability, adaptable surface chemistry, high surface area, and many hydroxyl groups contents for modification, also environmentally friendly [6-7]. Nanocellulose can be applied in many sectors such as in biomedical, pharmaceuticals, electronics, barrier films, nanocomposites, membranes, texturing agents in cosmetics, paper, and supercapacitors [8-9].

Nanocellulose can be obtained with various methods, such as hydrolysis with acid, oxidative hydrolysis, and hydrolysis with sonication assisted [10-11]. The most acid widely used for hydrolyzing cellulose is sulphuric acid, hydrochloric acid, and phosphoric acid [12]. Previously, Lusiana et al. [13] reported nanocellulose can be produced by hydrolyzing cellulose from pine flower using different concentrations of sulfuric acid at 45°C for 60 min. The sulfuric acid concentration affects the crystallinity and the particle size of nanocellulose. Sulfuric acid 10% and 30% obtain high crystallinity nanocellulose and the particle size is 15 nm wide and 185 nm length. However, sulfuric

IOP Conf. Series: Materials Science and Engineering **833** (2020) 012003 doi:10.1088/1757-899X/833/1/012003

acid 60% obtain amorphous nanocellulose. The strong acid is corrosive, toxic, non-environmentally friendly, and cause decomposition of cellulose. In this paper, we report hydrolysis cellulose with weaker acid (acetic acid) by conventional stirring to prevent cellulose decomposition and produced high crystallinity of nanocellulose. The purpose of this study is to determine the effect of cellulose hydrolysis from pine flower (*Pinus merkusii* Jungh et De Vrise) with acetic acid in various concentrations on its crystallinity.

2. Methods

2.1. Material and chemicals

Pine flower is (*Pinus merkusii* Jungh et De Vrise) from Batu, East Java, Indonesia. Chemicals used in this research are sodium hydroxide (Merck), sodium hypochlorite 12% w/v (technical grade), and acetic acid 99,7% (SMART-LAB).

2.2. Method

2.2.1. Sample preparation

Pine flower was washed thoroughly with water then soaked in hot water for 2 h before drying in an oven. The dried pine flower was ground using grinding mill. The powder of pine flower obtained was soaked in hot water for 4 h to remove the sap and other contaminants [3].

2.2.2. Cellulose isolation

Cellulose isolation followed the reported procedure [3, 10] with modification. The powder of pine flower was added with sodium hydroxide 6% at 70 °C for 4 h. The residue was washed until neutral pH with distilled water. The neutral residue was bleached with sodium hypochlorite 6 % at 70 °C for 2 h. The bleaching process was repeated until white cellulose obtained then, the cellulose washed until neutral pH with distilled water and dried in an oven.

2.2.3. Nanocellulose production

Nanocellulose production followed the reported procedure [3] with modification. Cellulose produced from step 2.2.2 was hydrolyzed with acetic acid 10%, 30% and 60% at 45°C for 60 min. The ratio of cellulose and acetic acid was 1:10 w/v. The cold water was added to the mixture to stop the reaction. Then, it was centrifugated to separate the acid and product. The % yield of nanocellulose was calculated by the equation below.

$$\%\text{Yield} = \frac{\text{Sample weight after hydrolysis (g)}}{\text{Sample weight before hydrolysis (g)}} \times 100 \quad (1)$$

2.3. Characterization

2.3.1. Fourier Transform Infrared (FTIR) analysis

The FTIR spectra of pine flower and nanocellulose were recorded with Fourier Transform Infrared Spectrometer SHIMADZU 8400s. The sample was crushed with potassium bromide then pressed until thin pellets were formed and measured in the wavenumber of range 4000-400 cm^{-1}.

2.3.2. X-Ray Diffraction (XRD) analysis

The X-ray diffractogram of nanocellulose was collected by XRD PanAnalytical type E'xpert Pro at room temperature and ranging from 10° to 90° with Cu Kα radiation (1.5406 Å). The crystallinity index (CI) of nanocellulose was determined by Seghal's method [14] as indicated below.

$$\mathrm{CI} = [(I_{002} - I_{amorph}) / I_{002}] \times 100\% \tag{2}$$

In equation 2, CI is the crystallinity index, I_{002} is the intensity of the 0 0 2 lattice diffraction at 2θ = 22.8°, and I_{amorph} is the intensity of diffraction at 2θ = 18°. I_{002} represents both crystalline and amorphous regions, while I_{amorph} represents the amorphous region.

2.3.3. TEM Analysis

The morphology and particle size of nanocellulose analyzed using a transmission electron microscope (TEM) HT7700 operated in 120 V. The sample is coated with carbon before analysis.

3. Result and discussion

Cellulose in the cell walls of pine flower was covered by lignin, hemicellulose, and other components such as waxes, pectin, and secondary metabolites [12]. The hot water pretreatment removes the dirt, waxes, and secondary metabolite in pine flower. The alkaline treatment with sodium hydroxide 6% can completely remove hemicellulose and pectin, while the bleaching process with sodium hypochlorite 6% can reduce lignin to a very small amount. The several treatments make the brown pine powder turn to yellowish-white, indicated more pure cellulose has been successfully isolated [3,10].

Cellulose isolated from pine flower was hydrolyzed with 10%, 30%, and 60% acetic acid, produced yellowish-white powder labeled AA 10, AA 30 and AA 60 showed in Figure 1. The % yield of AA 10, AA 30, and AA 60 is 87.4%, 94.2%, and 91.8% respectively (calculated from equation 1). The highest % yield was obtained from hydrolysis with acetic acid 30% (AA 30).

(a) (b) (c)

Figure 1. The product of cellulose hydrolysis with acetic acid. AA 10 (a), AA 30 (b), and AA 60 (c).

Cellulose hydrolysis is affected by acid strength. Acetic acid as a Brownsted acid produced hydronium ion in aqueous media, it acts as a catalyst to hydrolyze the glycosidic bond in cellulose. Acetic acid can hydrolize cellulose to nanocellulose. The particle size (witdh and length) of nanocellulose is determined with the direct measurement from the TEM image (Figure 2). The width of three nanocellulose sample are around 12.82 – 21.36 nm and the length range are 267.12 – 326.87 nm (shown in Table 1).

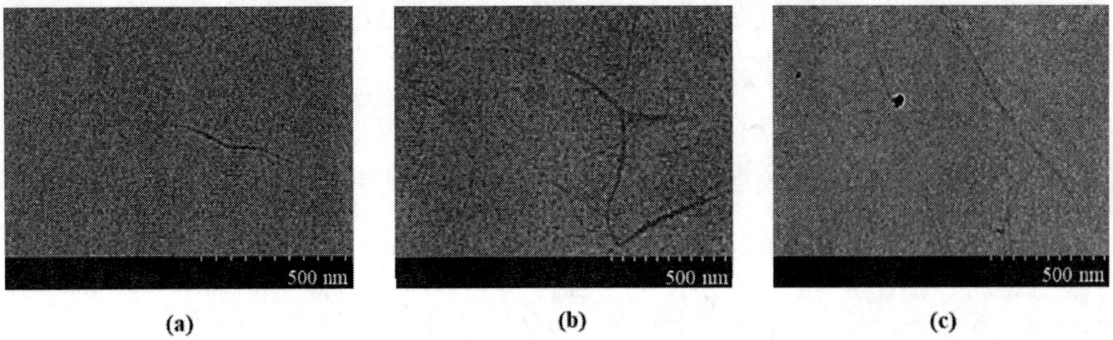

Figure 2. The transmission electron microscope image of AA 10 (a), AA 30 (b), and AA 60 (c).

Figure 3 shows the infrared spectra of pine flower and all products of nanocellulose. All samples have similar peak patterns. The peak at 3100 - 3700 cm^{-1} was identified as stretching vibration of O-H groups [12]. Stretching vibration of C-H cellulose was indicated by the peak at 2894 cm^{-1} [3]. The peak at 1643 cm^{-1} shows stretching vibration of C=C aromatics, indicated a small amount of lignin remained in nanocellulose sample [10]. The pine flower has a peak at 1700 cm^{-1} indicating C=O stretching from hemicellulose [10]. After hot water pretreatment, alkaline treatment, bleaching and followed by hydrolysis with acetic acid, hemicellulose was not identified in AA 10, AA 30 and AA 60. Pyranose rings in the cellulose are shown by the peak at 1160-1027 cm^{-1} (bending vibration of C-O-C) and the glycosidic bonds by the peak at 896 cm^{-1} [3].

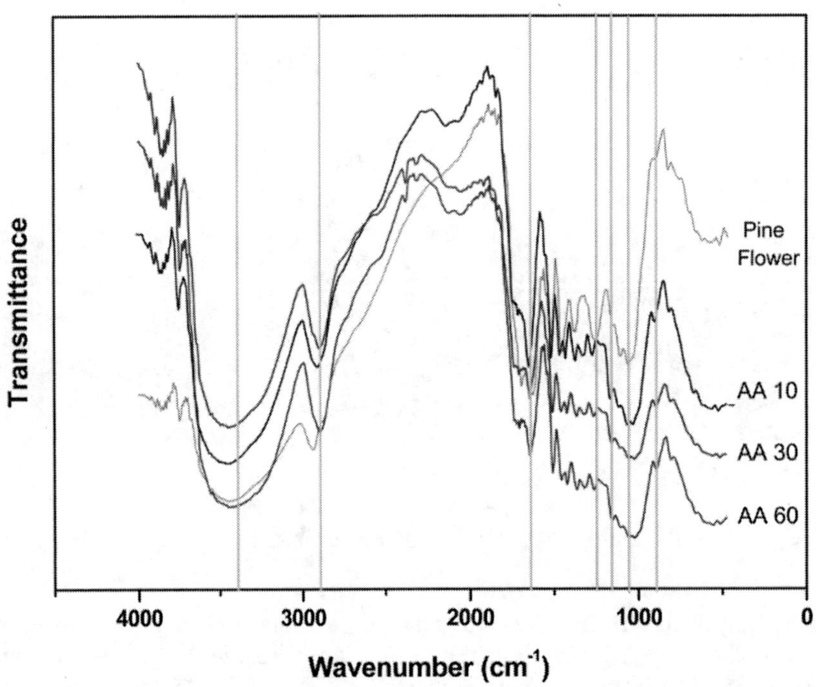

Figure 3. Fourier Transform Infrared (FTIR) spectra of pine flower, AA 10, AA 30, and AA 60.

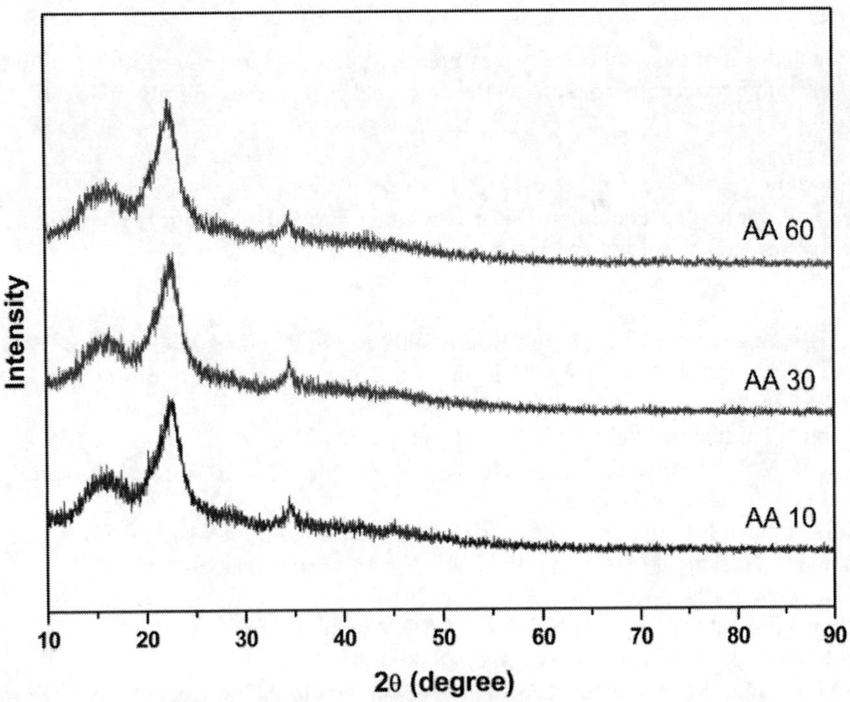

Figure 4. X-Ray Diffraction Pattern of AA 10, AA 30 and AA 60.

The X-ray diffraction pattern of AA 10, AA 30 and AA 60 are shown in Figure 3. AA 10, AA 30, and AA 60 have similar diffractogram profiles. Three peaks at 2θ = 16° (plane 101), 22° (plane 002), 34° (plane 004) indicated cellulose [15,16].

Table 1. Crystallinity index, width and length of nanocellulose.

Sample	Peak (2θ)	Crystallinity index (%)	Width (nm)	Length (nm)
AA 10	22.61	55.60	12.82	290.82
AA 30	22.69	51.00	21.36	267.12
AA 60	22.45	50.90	17.85	326.87

The crystallinity index of AA 10, AA 30, and AA 60 were calculated using Seghal's equation (equation 2). The crystallinity and crystallite size of AA 10, AA 30, and AA 60 are showed in Table 1. The highest crystallinity index obtained from AA 10 (cellulose hydrolyze with acetic acid 10%), whereas the lowest crystallinity index is AA 60 (cellulose hydrolyze with acetic acid 60%). The data show that the crystallinity index decreased while the concentration of acetic acid for hydrolyzed the cellulose increased.

The high crystallinity index of nanocellulose indicated that the amorphous region in cellulose has been reduced by hydrolysis procedures. The amorphous region is less stable than the crystalline regions, it can be dissolved after hydrolysis with acid [6]. Hydrolyzed with strong acid at a high concentration not only remove the amorphous region but also break the crystalline region on cellulose. The crystallinity index of cellulose isolated from pine flower hydrolyzed with acetic acid 60% (AA 60) is higher than cellulose hydrolyze with sulfuric acid 60%. Lusiana et al. reported nanocellulose produced from hydrolysis using 60% of sulfuric acid is amorphous because sulfuric acid makes the cellulose from pine flower decomposed [13].

4. Conclusion

This study concludes that the nanocellulose sample with AA 10 gave the best crystallinity index (CI) of 55.60%. The high concentration of acetic acid makes the crystallinity index of nanocellulose decrease.

Acknowledgment

This research is supported by Penelitian Dasar Unggulan Perguruan Tinggi (PDUPT) 2019, DRPM of Ministry for Research, Technology and Higher Education of Indonesia.

References

[1] Ayrilmis A Buyuksari U Avci E and Koc E 2009 *Forest Ecol. Manag.* **259** 65–70
[2] Masruri M Pangestin D N Ulfa S M Riyanto S Srihardyastutie A and Rahman M F 2018 *IOP Conf. Ser.: Mater. Sci. Eng.* **299** 012072
[3] Rambabu N Panthapulakkal S Sain M and Dalai A K 2015 *Ind. Crop. Prod.* **10** 1-9
[4] Malladi R Nagalakshmaiah N Robert M and Elkoun S 2018 *ACS. Sus. Chem. Eng.* DOI: 10.1021/acssuschemeng.7b03437
[5] Brinchi L Cotana F Fortunati E and Kenny J M 2013 *Carb. Pol.* **94** 154-169
[6] Trache D Hussin M H Haafiz M K M and Thakur V K 2017 *Nanoscale* DOI:10.1039/c6nr09494e
[7] Phanthong P Reubroycharoen P Hao X Xu G Abudula A and Guan G 2018 *Carbon Resour. Convers* DOI: doi.org/10.1016/j.crcon.2018.05.004
[8] Islam M T Alam M M Torino P Patrucco A Montarsolo A and Zoccola M 2014 *AATCC J. of Research.* **1** 5 17-23
[9] Li J Cha R Mou K Zhao K Long K Luo H Zhou F and Jiang X 2018 *Advanced Healthcare Materials* 71800334 Jhon Wiley and Sons Canada
[10] Liu C Lia B Duc H Lva D Zhanga Y Yua G Mua X and Penga H 2016 *Carb. Pol* **151** 716–724
[11] Hamid S B A, Zain S K, Das R and Centi G 2016 *Carb. Pol* **138** 349-355
[12] Zhang P P Tong D S Lin C X, Yang H M Zhong Z K Yu W H Wang H and Zhou C H 2014 *Asia-Pac. J. Chem. Eng.* DOI: 10.1002/apj.1812
[13] Lusiana S E Srihardyastutie A Masruri M 2019 *J. Phys.: Conf. Ser.* **1374** 012023
[14] Segal L Creely J J Martin A E Jr and Conrad C M 1959 *Tex. Res. J.* **29** 786–794
[15] Neto W P F Silvério H A Dantasb N O and Pasquinia D 2013 *Ind. Crop. Prod.* **42** 480– 488
[16] Henrique M A Silvério H A Neto W P F and Pasquini D 2013 *J. Env.Man.* **121** 202-209

Isolation, Identification, and Inhibition of Saponin Isolates from Pineapple (*Ananas comosus* L.) and Candlenut (*Aleurites moluccanus* L.) against Xanthine Oxidase by In Vitro Assay

Mimma Amalia[1], Mieke Alvionita[1], Subandi[1*], Evi Susanti[1], and Dian Nugraheni[1]

Department of Chemistry, Faculty of Mathematics and Sciences, Universitas Negeri Malang, Jalan Semarang 5, Malang 65145, Indonesia

*Corresponding author: subandi.fmipa@um.ac.id

Abstract. The exploration of gout treatment agent based on the natural product gains high attention in recent years. Although some current drugs effectively treat gout, fewer unbeneficial effects cannot be avoided. Thus, searching for new alternatives with fewer side effects is needed. This present work was aimed at investigating the potency of saponins in pineapple (*Ananas comosus* L.) and candlenut (*Aleurites moluccanus* L.) extract as an inhibitor of xanthine oxidase, an enzyme that responsible for gout disease, by in vitro study. The stage of this work was divided into six stages, including (1) sample preparation, (2) extraction of saponins using 70% ethanol, (3) isolation of saponins by thin-layer chromatography, (4) phytochemical test of saponins, (5) inhibitory activity testing of saponins using in vitro analysis, and (6) determination the types of saponins by UV-Vis, FT-IR and LC-MS/MS spectroscopy. The result from the preparative TLC of pineapple extract, there were three spots with Rf value of 0.95 (pi1), 0.65 (pi2), and 0.38 (pi3), while from candlenut extract there were two saponin isolates with Rf value of 0.95 (ci1) and 0.79 (ci2). The inhibition activity of each isolate against xanthine oxidase was 103.41% (pi1), 269.73% (pi2), 84.16% (pi3), 222.90% (ci1), and 157.05% (ci2) relative to Allopurinol 100 ppm. Based on spectroscopic analysis results, isolate with the greatest inhibition activity was likely a diosgenin type ($C_{27}H_{42}O_3$) of saponin.

1. Introduction

According to the WHO survey in 2015, Indonesia is the fourth largest country in the world with people who suffer from gout. Gout occurs every 840 per 100,000 people and represents about 5% of total arthritis [1]. Gout is caused by high levels of uric acid in the blood, due to the high intake of purines and excessive synthesis of uric acid [2].

The drug that is commonly used to treat gout is Allopurinol which reduces the activity of uric acid synthesizing enzymes, namely xanthine oxidase. However, long term use of Allopurinol can cause some side effects and even more serious diseases such as inflammation, kidney failure, and chronic heart [3].

Some traditional medicinal plants contain secondary metabolites that can inhibit the xanthine oxidase [4]. Some of them that have potential as xanthine oxidase inhibitors have been investigated for instance srikaya fruit and sendok leaf extract [5], rambutan bark extract [6], salam leaf extract [7], and mlinjo bark extract [8]. Besides that, based on testimonies from the community, pineapple and candlenut juice mixes are also effective for treating gout.

Extract pineapple and candlenut contains diosgenin saponin compounds that have the potential as xanthine oxidase inhibitors based on in silico study [9]. Therefore, in this present study, the isolation, identification, and test of saponin isolate activity from pineapple (*Ananas comosus* l.) and candlenut (*Aleurites moluccanus* l.) on xanthine oxidase was carried out by in vitro assay.

2. Methods

2.1. Materials and Instruments
This study used pineapple (*Ananas comosul* L.) and candlenut (*Aleurites molucannus* L) flesh as samples. Some chemicals used in this research were ethanol 70%, acetone p.a, Allopurinol tablet (100 mg), phosphate buffer 0.05 M (pH of 7.5), n-hexane p.a, acetic acid p.a, iod powder, aquadest, HCl 0.1 N, gel silica plat G60F254 (MERCK), xanthine (0Sigma) and xanthine oxidase from bovine (Sigma). UV-VIS spectrophotometer (Genesys 10S UV-Vis) and LC-MS/MS (MassLynx 4.1 SCN 884 Xevo G2-S QTof/Xevo G2-S Tof) were utilized to characterize the structure of bioactive compound in both samples.

2.2. Procedures
2.2.1. Sample Preparation. The pineapple was sliced thinly followed by drying process under the sunlight. The pineapple then was dried again in the oven at the temperature of 55-65 °C for four days until it reached the constant weight. The grinding process was carried out to obtained the powder form of dried pineapple. The next step was sifting and measuring the weight of smooth powder. Whereas the candlenut was ground directly using mortal and pastel then be weighed.

2.2.2. Sample Extraction. The powder of both samples was macerated using 600 mL of ethanol 70% p.a for 72 hours at a temperature of 30 °C. Then the filtration was conducted using Buchner funnel. Rotatory vacuum evaporator was used to obtain the thick extract of samples. This extract was weighed to calculate the rendement.

2.2.3. Isolation of Bioactive Compound. The bioactive compound contained in the sample was isolated by qualitative and preparative thin-layer chromatography (TLC). The first step was preparing the concentrated samples that were previously dissolved in ethanol 70% p.a. The gel silica plat containing the sample was put in the chamber filled by the different eluents with varied compositions. TLC of pineapple extract was chloroform: ethanol: acetic acid (1: 2: 1) and candlenut extract n-hexane: ethyl acetate was (1:4). Iodine powder was utilized as a stain viewer to sharpen the spot formed. The spot was exposed under the UV light at a wavelength of 254 nm and 365 nm. The appropriate eluents of qualitative TLC in this step were used for the next preparative TLC. The stained spot was scraped from the plates, weighed and dissolved in methanol, homogenized using vortex apparatus and then centrifuged. The filtrate obtained was then air dried, weighed, and dissolved in ethanol. The following stages were structural analyzing using UV-Vis, FT-IR, and LC-MS / MS spectrophotometry and inhibition activity testing by in vitro analysis.

2.2.4. Phytochemical Characterization. The type of phytochemical characterization used in this study was saponin analysis. The purpose of this test was to know the saponin content in both samples. 0.1 grams of concentrated sample was added by 5 mL hot aquadest and shaked. The positive result was shown by the formation of foam. One to two drops of HCl 0.1 N was added to maintain the foam produced.

2.2.5. Inhibitory Activity Testing towards XO. Both isolated sample were examined their inhibitory activity towards XO. Allopurinol was used as a positive control of this testing. The activity of XO was examined based on the absorbance in the UV spectrophotometer at a wavelength of 284 nm. The greatest inhibition activity was used for the following structural characterization i.e. UV-Vis, FT-IR, and LC-MS/MS spectrophotometer. Table 1 presents the reagent and volume used in this analysis.

Table 1. Volume and Reagent Used in the Inhibitory Activity Testing

Reagent	Volume (mL)													
	wi	**Cwi**	**pi1**	**Cpi1**	**pi2**	**Cpi2**	**pi3**	**Cpi3**	**ci1**	**Cci1**	**ci2**	**Cci2**	**A**	**CA**
Inhibitor	-	-	1.0	1.0	1.0	1.0	1.0	1.0	1.0	1.0	1.0	1.0	1.0	1.0
Phosphate buffer pH 7.5	3.9	4.0	3.9	4.0	3.9	4.0	3.9	4.0	3.9	4.0	3.9	4.0	3.9	4.0
Xanthine 0.15 mM	2.0	2.0	2.0	2.0	2.0	2.0	2.0	2.0	2.0	2.0	2.0	2.0	2.0	2.0
Pra incubation at temperature of 37 °C for 8 minutes														
Xanthine oxidase 0.05 U/mL	0.1	-	0.1	-	0.1	-	0.1	-	0.1	-	0.1	-	0.1	-
HCl 1.0 N	-	1.0	-	1.0	-	1.0	-	1.0	-	1.0	-	1.0	-	1.0
Incubation at temperature of 37 °C for 20 minutes														
HCl 1.0 N	1.0	-	1.0	-	1.0	-	1.0	-	1.0	-	1.0	-	1.0	-

Note:
wi : without inhibitor
Cwi : control without inhibitor
pi1 : pineapple isolates 1
Cpi1 : control pineapple isolates 1
pi2 : pineapple isolates 2
Cpi2 : control pineapple isolates 2
pi3 : pineapple isolates 3
Cpi3 : control pineapple isolates 3
ci1 : candlenut isolates 1
Cci1 : control candlenut isolates 1
ci2 : candlenut isolates 2
Cci2 : control candlenut isolates 2
A : allopurinol
CA : control allopurinol

The xanthine oxidase activity can be calculated using the following equation,

$$\text{Activity} = \frac{(\text{absorbance E}-\text{absorbance EK}) \times \text{vol} \times \text{df}}{\varepsilon \times 0.1} \quad (1)$$

where E is absorbance of sample at 284 nm, EK is absorbance of sample control at 284 nm, vol is total volume when measuring (mL), df is dilution factor, ε is uric acid coefficient (12.2 mM), and 0,1 showed the volume of xanthine oxidase used (mL).

The amount of uric acid produced can be calculated using Lambert-Beer law using the following equation,

$$A = \varepsilon \,.\, b \,.\, C \quad (2)$$

where A is absorbance at 284 nm (sample absorbance - absorbance control), b is cuvette width (1 cm), and C is gout concentration (mM)

The inhibition activity towards xanthine oxidase can be calculated using this equation,

$$\% \text{ Inhibition} = 1 - \frac{B}{A} \times 100\ \% \quad (3)$$

where A indicates the absorbance of sample without inhibitor while B indicates the absorbance of sample with inhibitor

The percentage of inhibition activity of the sample relative to Allopurinol (% DIA) can be calculated by this formula:

$$\%\text{DIA} = \frac{\text{inhibition activity of isolate}}{\text{inhibition activity of Allopurinol}} \times 100\% \quad (4)$$

2.2.6. Structural Characterization of Saponins

2.2.6.1. UV-Vis Spectroscopy. Isolates obtained from preparative TLC were dissolved in methanol p.a and analyzed using a UV-Vis spectrophotometer at a wavelength of 200-400 nm.

2.2.6.2. FT-IR Spectroscopy. This process was carried out to analyze the functional group of bioactive compounds. The sample from preparative TLC was dissolved in ethanol 70%. The solvent was evaporated then the product was mixed with KBr until homogenous. This result was then analyzed using FT-IR spectrophotometer at a wavelength number of 400-4000 cm^{-1}.

2.2.6.3. LC-MS/ MS Spectroscopy. This step was purposed to identify the saponins compound based on the m/z result. It was about 0.5 of both isolate volume from preparative TLC that was used to this structural characterization.

3. Results and Discussion

3.1. Pineapple and Candlenut Flesh Extract
The concentrated extract obtained from 613.6 grams of pineapple flesh was 17.45 grams while the candlenut flesh extract with the weight of 31.29 grams yielded 5.09 grams of viscous concentrated extract.

3.2. Pineapple and Candlenut Flesh Extract
Qualitative TLC is used to determine the type of eluents that are suitable for separating compounds in a sample while preparative TLC aims to isolate the bioactive compound. Based on preliminary experiments, the eluent that can separate well in TLC of pineapple extract was chloroform: ethanol: acetic acid (1: 2: 1) followed by exposing under UV_{254} and UV_{365}. The stain results of qualitative and preparative TLC of pineapple extract under UV detection are shown in Figure 1. Based on those profiles, there are 3 spots found in qualitative TLC. These three spots have different Rf values, i.e 0.95, 0.65, and 0.38. These three spots indicate three isolates (pi1, pi2, and pi3) which also have different mass, i.e 1.165, 1.134, and 1.092 grams, respectively.

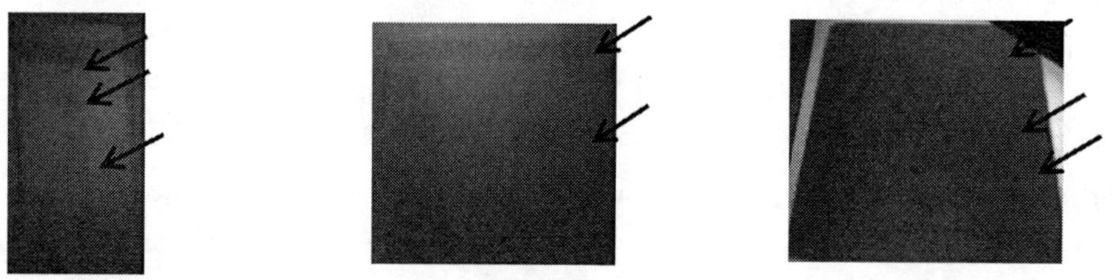

(a) **(b)** **(c)**

Figure 1. (a) Qualitative TLC of Pineapple Extract (b) Preparative TLC of Pineapple Extract Detected by UV_{254} (2 isolates) and (c) UV_{365} (3 isolates)

Meanwhile, the TLC result of candlenut presents in Figure 2. The mixture of n-hexane: ethyl acetate (1:4) was used as a mobile phase of this TLC. The results obtained from qualitative and preparative candlenut TLC are different. Qualitative TLC showed one stain with Rf (0.95). However, for preparative TLC produced 2 stains with Rf (0.93 and 0.87). Figure 2 indicates that there are two isolates obtained from preparative TLC of candlenut extracts. Those two spots were taken and labeled as ci1 and ci2 with an isolate mass of 1.195 and 1.155 grams, respectively.

3.3. Phytochemical Result
Phytochemical analysis of all samples from ethanol extract, ethanol viscous extract, and all isolates in pineapple and candlenut positively contain saponins which are indicated by the formation of foam. The foam produced is due to a combination of non-polar sapogenin chains and water-soluble polar side chains. The more foam produced, the greater the concentration of saponin content in the sample.

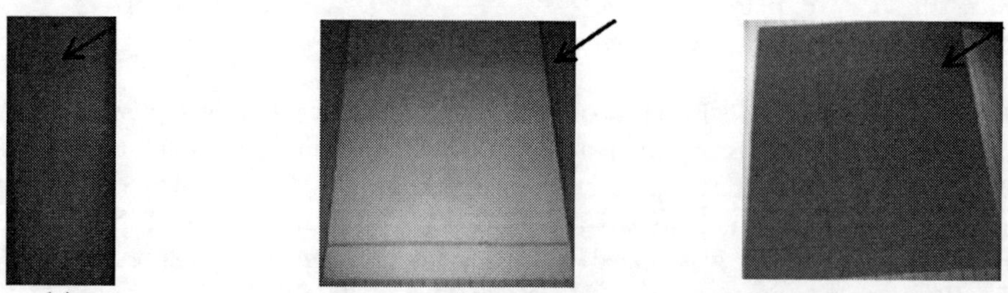

(a) **(b)** **(c)**

Figure 2. (a) Qualitative TLC of Candlenut Extract (1 isolate) (b) Preparative TLC of Candlenut Extract detected by UV_{254} (1 isolate) (c) UV_{365} (1 isolate)

3.4. The Inhibition Activity of Sample Isolates

Enzyme activity test was set by in vitro analysis using inhibitors and without inhibitors. Each isolate of both samples obtained from preparative TLC was tested their inhibition power toward xanthine oxidase (XO). The activity of XO was examined based on the absorbance at a wavelength of 284 nm using UV spectrophotometer. This test also used Allopurinol, a commercial drug for gout treatment, as a positive control. Based on the results of pineapple isolates, each isolate has a different inhibition activity against XO. The greatest inhibition power was shown by isolate 2 of pineapple (pi2) which more potential than Allopurinol. In addition, the candlenut isolates also show great inhibition activity. The inhibition percentage of isolate 1 of candlenut (ci1) was the greatest compared with isolate 3 and Allopurinol activity. The calculation results of this step were presented in Tables 2 and 3.

Table 2. Inhibition Activity of Pineapple Isolates and Allopurinol

Samples	Measurement	Absorbance		Average XO Activity (U/mL)	Average Concentration of Uric Acid (mM)	Inhibition Power (%)	Relative inhibition power to Allopurinol (%)
		Control	Sample				
Without Inhibitor	1	0.006	0.086	3.72	0.0066	0%	-
	2	0.005	0.086				
	3	0.005	0.087				
Allopurinol 100 ppm	1	0.001	0.087	4.49	0.0070	23.49%	100%
	2	0.001	0.087				
	3	0.001	0.086				
Isolate 1 Pineapple (52 mg)	1	0.023	0.081	3.22	0.0050	25.43%	103.41%
	2	0.020	0.083				
	3	0.019	0.082				
Isolate 2 Pineapple (71 mg)	1	0.048	0.078	1.56	0.0024	63.36%	269.73%
	2	0.048	0.079				
	3	0.050	0.078				
Isolate 3 Pineapple (70 mg)	1	0.024	0.087	3.41	0.0053	19.77%	84.16%
	2	0.024	0.089				
	3	0.023	0.090				

3.5. UV-Vis Analysis Result

The profile of UV-Vis spectrum containing saponin in pineapple sample is shown in Figure 3. Meanwhile, the UV-Vis spectrum of candlenut isolates is shown in Figure 4. Based on the UV-Vis spectrum profile, the three saponin isolates of pineapple and the two saponin isolates of candlenut contain the n → σ* electronic transition indicating OH groups and the n → π * electronic transition, which means that there are conjugated double bonds (C=C) in the isolates. The presence of the -OH group is one of the characteristics of the saponin structure.

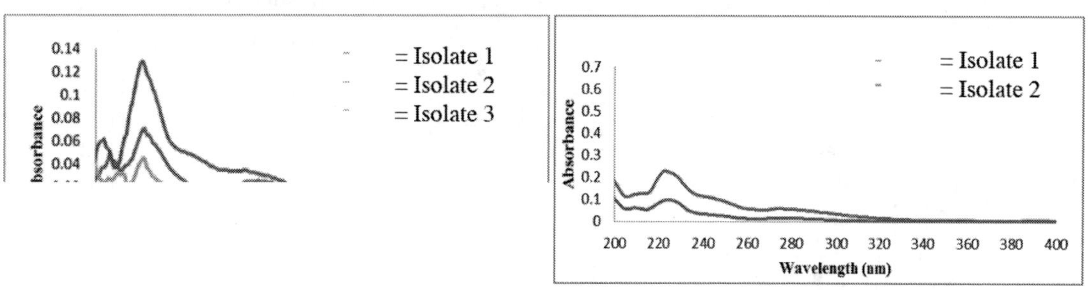

Figure 3. UV-Vis Spectrum of (a) the Three Pineapple Saponin Isolates and (b) Candlenut Saponin Isolates

Table 3. Inhibition Activity of Candlenut Isolates and Allopurinol

Samples	Measurement	Absorbance		Average XO Activity (U/mL)	Average Concentration of Uric Acid (mM)	Power Inhibition (%)	Relative inhibitory power to Allopurinol (%)
		Control	Sample				
Without Inhibitor	1 2 3	0.011 0.01 0.011	0.116 0.117 0.116	4.85	0.0087	0%	-
Allopurinol 100 ppm	1 2 3	0.001 0.001 0.001	0.087 0.087 0.086	4.49	0.0070	23.49%	100%
Isolate 1 candlenut (50 mg)	1 2 3	0.031 0.031 0.03	0.08 0.081 0.082	2.81	0.0041	52.36%	222.90%
Isolate 2 candlenut (116 mg)	1 2 3	0.056 0.055 0.057	0.115 0.115 0.116	3.11	0.0049	43.85%	157.05%

3.6. FT-IR Analysis Results

Based on the results of FT-IR data interpretation, it is known that pineapple second isolate and candlenut first isolate contain -OH because in the region around 3500-3200 cm^{-1} has a strong and wide intensity peak. Absorbtion bands around 3300-2700 cm^{-1} indicate the presence of aliphatic C-H groups, bands in the region of 1100-990 cm^{-1} indicate the presence of C-O groups and reinforced with bands with wave numbers 1000-630 cm^{-1} indicate aromatic C-H groups. The resulting clusters are characteristic of saponin compounds.

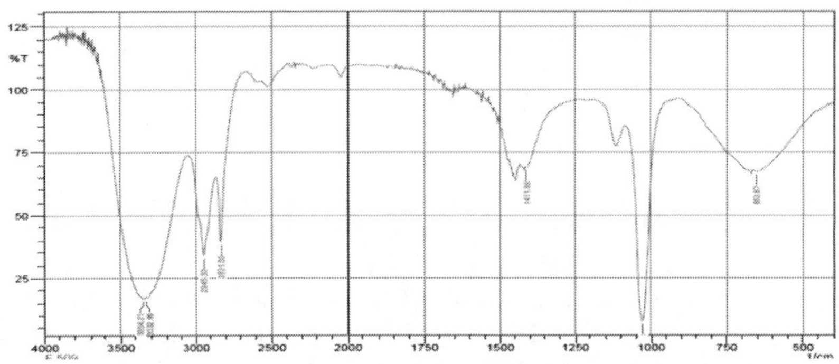

Figure 4. FT-IR Spectrum of Pineapple Second Isolates

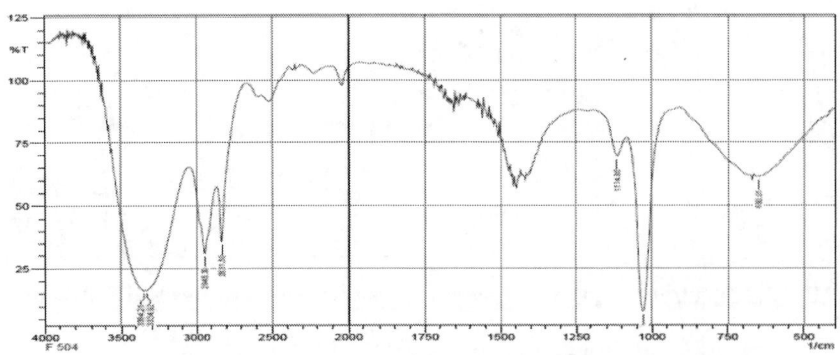

Figure 5. FT-IR Spectrum of Candlenut First Isolates

IOP Conf. Series: Materials Science and Engineering **833** (2020) 012004 doi:10.1088/1757-899X/833/1/012004

3.7. LC-MS / MS Analysis Results

Analysis using LC-MS/MS spectrophotometer aims to determine the type of saponins found in pineapple and candlenut isolates. LC analysis of pineapple isolates 2 produces 17 chromatograms, which showed that the isolates tested were not pure (Figure 7.a). Saponin compounds from pineapple fruit isolates are found at the seventh peak at 11.56 retention times (Figure 6a). In Figure 6b there are compounds that are thought to be saponins at m / z 415.2110 with the molecular formula $C_{27}H_{43}O_3$.

Although there are many higher peaks on the LC chromatogram compared to peaks with a retention time of 11.56 (Figure 6a), it turns out that these compounds are not saponin compounds. This has been analyzed based on elemental composition that there are no saponin-containing compounds, other than those whose retention time is 11.56, because the characteristics of steroid and triterpenoid compounds are C, H, and O atoms. In addition, steroid saponin compounds contain 30 C atoms and triterpenoid saponins contain atoms C as many as 27. The peak with a retention time of 16.19 at m / z 313.2734 with the molecular formula $C_{19}H_{37}O_3$ contains C, H, and O atoms, but does not contain as many as 27 or 30 C atoms so that the compounds are not classified as saponin.

That the compound with a retention time of 11.56 minutes is saponin, it can be seen in PubChem by reducing the mass of 1 sma H atom at m/z value 415.2110, because the compound is positively ionized, so the molecular formula $C_{27}H_{43}O_3$ with m/z 414.2110 is obtained, which based on literature (PubChem) is a triterpenoid saponin compound, Diosgenin. The fragmentation of diosgenin compounds produced by LC-MS / MS can be seen in Figure 7.

LC analysis of candlenut 1 isolate yielded 22 chromatograms, which showed that the isolate tested was not pure (Figure 6a). Saponin compounds from isolate 1 candlenut were found at the 10th peak LC at 11.72 minutes of retention time (Figure 8a). In Figure 8b there are compounds that are thought to be saponins at m / z 415.2109 with the molecular formula $C_{27}H_{43}O_3$.

Other compounds in candlenut isolates other than those whose retention time at LC is 11.72, are not saponin compounds which can be proven based on the results of elemental composition analysis. Compounds containing saponins that contain C, H, and O atoms. Besides steroid saponin compounds it contains 30 C atoms and triterpenoid saponins contain 27 C atoms. The peak with a retention time of 11.02 at m / z 273.1479 with molecular formula contains C, H, and O atoms, but does not contain 27 or 30 C atoms.

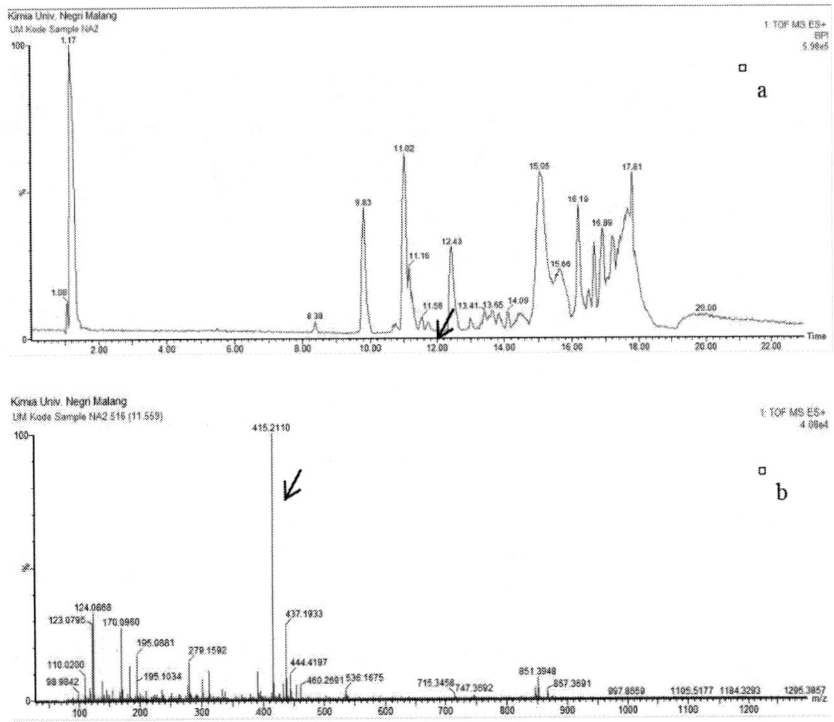

Figure 6. (a) LC Spectrum of Pineapple Isolate 2 and (b) MS after enlarged at retention time 11.56

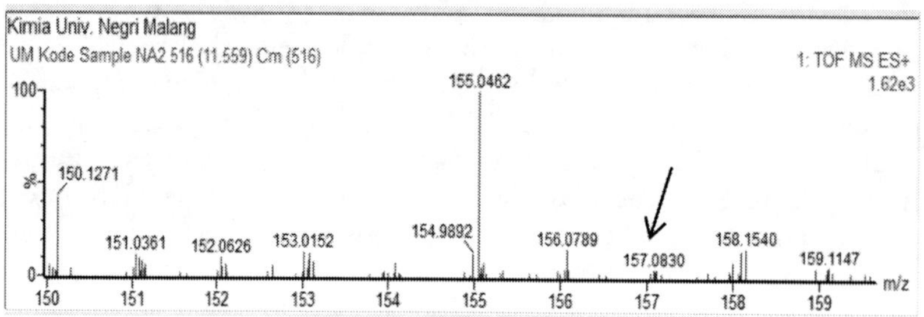

(a)

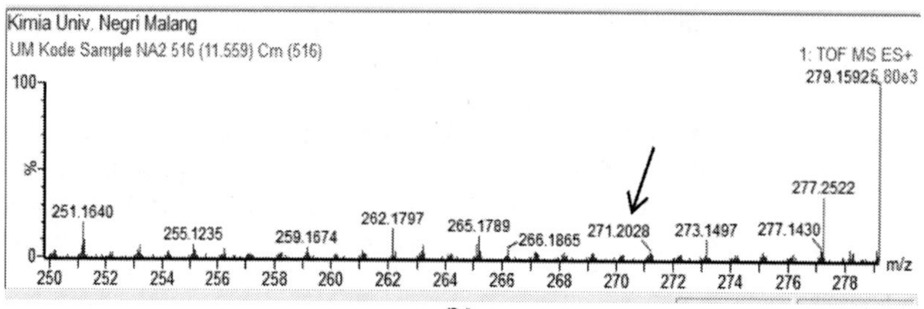

(b)

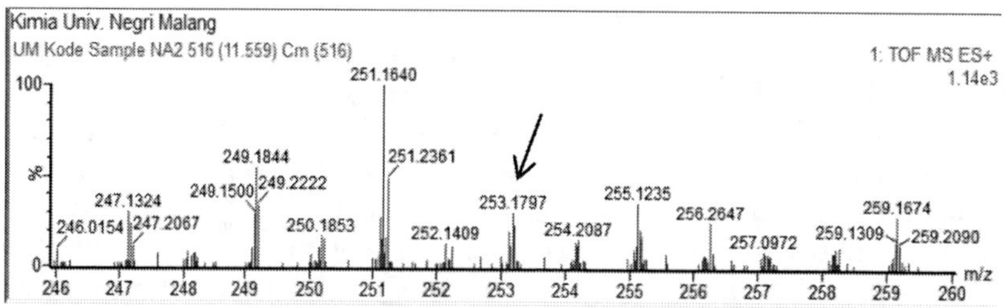

(c)

Figure 7. a-c MS Spectrum Fragmentation of Diosgenin Compounds in Pineapple

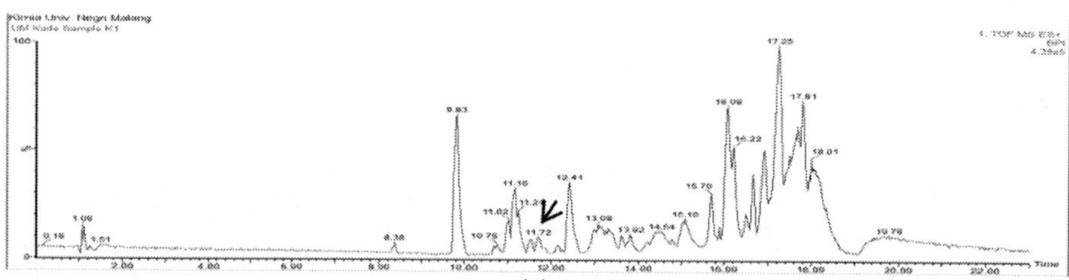

(a)

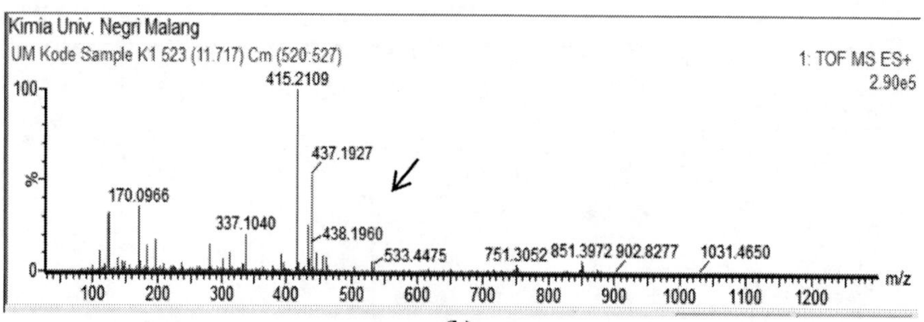

(b)

Figure 8. (a) LC Spectrum isolates 1 of Candlenut and (b) MS after enlarged at retention time 11.72

The saponin compound at 11.72 minutes at LC with m / z 415.2109 is Diosgenin with the molecular formula $C_{17}H_{21}O_3$ then it can be seen in Pubchem by reducing the mass of 1 sma H atom at m / z value because the compound is positively ionized. Then the fragmentation results of LC-MS / MS analysis of compounds with 11.76 minutes retention time (Figure 9) are matched with the diosgenin fragmentation patterns (PubChem Figure 7). In Figure 8, fragments of Diosgenin compound with molecular weight of 415.2109 were fragmented at m / z 157.0828; 253.1444; and 271.1666 obtained from PubChem. The results of MS fragmentation of candlenut isolates also have a spectrum pattern with m / z which is almost the same as the results of the fragmentation spectrum in Figure 7, namely at m / z 157.0828; 271.1666; and 253.1444 which can be seen in Figure 9.

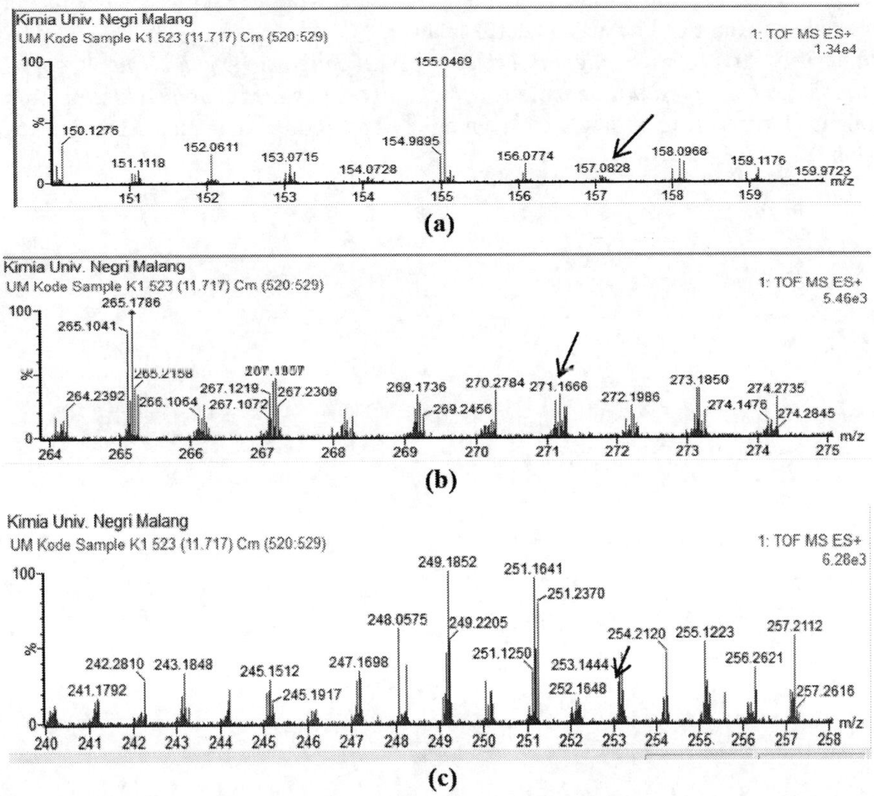

Figure 9. a-c MS Spectrum Fragmentation of Diosgenin Compounds in Candlenut

4. Conclusions

Based on this present work, it can be concluded that from the preparative TLC of ethanol extract of pineapple flesh, three isolates of saponins were obtained while from candlenut extract, there are two isolates. The greatest inhibition activity of pineapple was shown by isolate 2 while from candlenut is isolate 1. Based on spectroscopic analysis, those both isolates may be predicted as triterpenoid saponins, Diosgenin ($C_{27}H_{42}O_3$).

Acknowledgment

This study was fully funded by PNBP Universitas Negeri Malang 2019 with the contract number of 19.3.55/UN32/LT/2019.

References

[1] Sustrani L, Alam S and Hadibroto I 2004 *Asam Urat* (Jakarta: PT Gramedia Pustaka Utama)
[2] Misnadiarly 2007 *Rematik: Asam Urat-Hiperusisemia dan Arthritis Gout* (Jakarta: Pustaka Obor Populer)
[3] Pacher P, Alex N and Csaba S 2006 *J. Pharmacol.* **58** 87-144
[4] Kong L D, Cai Y, Cheng C and Tan R X 2000 *J. Ethnopharmacol.* **73** 199-207

[5] Surahman A 2013 *Uji Fitokimia dan Daya Inhibisi Ekstrak Daun Sendok (Plantago major* L.*) dan Buah Srikaya (Annona squamosal* L.*) terhadap Aktivitas Xantin Oksidase* (Malang: Jurusan Kimia Fakultas Matematika dan Ilmu Pengetahuan Alam Universitas Negeri Malang)
[6] Putri N E, Rissiyelli and Mauldina M G 2016 *Jurnal Farmasi* **3** 12-20
[7] Puspitasari A 2018 *Karakterisasi dan Identifikasi Kandungan Kimia Daun Salam Serta Uji Efek Penghambatan Enzim Xantin Oksidase Ekstrak Etanol Daun Salam (Eugenia polyantha* Wight.*)* (Yogyakarta: Jurusan Farmasi Universitas Sanata Dharma)
[8] Wulandari S 2012 *Inhibisi Xantin Oksidase oleh Ekstrak Etanol Kulit Melinjo (Gnetum gnemon) Relatif Terhadap Allopurinol* (Malang: Jurusan Kimia Fakultas Matematika dan Ilmu Pengetahuan Alam Universitas Negeri Malang)
[9] Arti E T B 2018 *Isolasi, Identifikasi dan Uji In Silico Senyawa Saponin dari Buah Nanas (Ananas comosus L.) dan Kemiri (Aleurites moluccanus L.) sebagai Inhibitor Xantin Oksidase* (Malang: Jurusan Kimia Fakultas Matematika dan Ilmu Pengetahuan Alam Universitas Negeri Malang)

IOP Conf. Series: Materials Science and Engineering **833** (2020) 012005 doi:10.1088/1757-899X/833/1/012005

The potency of *Polylathia longifolia* from Indonesia and the Philippines as therapeutic agents on inflammatory bowel disease (IBD) in Rats *(Rattus norvegicus)* induced by Indomethacin

A Aulanni'am[1,2*], T Z Anita[1], D S Nahari[1], I A Aluka[2], E I Agustine[2], T Novita[2], A A Pentaloka[2], D K Wuragil[2], W Riawan[3], and M A G Beltran[4]

[1]Chemistry Department, Brawijiya University, Malang, 65145, Indonesia
[2]Faculty of Veterinary Medicine, Brawijaya University, Malang, 65145, Indonesia
[3]Faculty of Medicine, Brawijiya University, 65145, Indonesia
[4]College of Veterinary Medicine, Tarlac Agricultural University, Camiling, Tarlac, 2306, Philippines

*Corresponding author: aulani@ub.ac.id

Abstract. Herbal medicines have the potential to be used as therapeutic agents. *Polyalthia longifolia* is widely grown in both Indonesia and the Philippines, but it has not commonly used for its therapeutic purposes. Some studies reported that *P. longifolia* leaves extract has anti-inflammatory activity. In this study, the potential of *Polyalthia longifolia* leaves extracts for inflammatory bowel disease (IBD) in Indometachin-induced rats was tested. Four groups of rats were used for this research, were control (T1), rats with IBD with 10 mg/kg BW of sulfasalazine therapy (T2),) rats with IBD treated with 300 mg/kg BW of Indonesia *P. longifolia* leaves extract therapy (T3) and rats IBD treated with 300mg/kg BW of The Philippines *P. longifolia* leaves extract therapy (T4). Histopathology of gastric, duodenum, jejunum, and colon were analyzed, with protein profile and pro-inflammatory cytokines expressions. The results showed that *P. longifolia* leaves extract therapy origin from Indonesia and the Philippines were potent as anti-inflammatory agents comparable to commercially available drugs against IBD. This works proposed the use of *P. longifolia* leaves as IBD therapy.

1. Introduction

Herbal medicines are a prominent part of healthcare whole world [1]. The use of herbal medicine has enhanced rapidly over the past few decades. Therapy with herbal medicine is the primary healthcare for 80% population in the world [2], especially in developing countries, as well as the increasing use of herbal medicine in developed countries [3]. Therapy with herbal medicine has fewer side effects when compared with synthetic drugs. Natural herbs contain phytochemicals that can use to treat various diseases such as inflammatory bowel disease (IBD). One of the plants that apply for this treatment is *Polyalthia longifolia*. This plant belongs to *the Annonacea* family which commonly found in tropical countries [4] such as Indonesia and the Philippines. This plant is widely used as an ornamental plant, reducing sound pollution and reducing fever and tonic. This plant was used to treat fever, diabetes, hypertension. Some studies indicated that this plant has various classes of compounds terpenoids, alkaloids, and flavonoids. The content of alkaloids and flavonoids has anti-inflammatory activities, including inhibiting infiltration, reducing inflammatory mediators, and reducing oxidative stress [5,6]. Based on the therapeutic effect of the active compounds *P. longifolia* on the inflammatory response, it has potential in inflammatory regulation of intestinal inflammation. Inflammatory bowel disease is a gastrointestinal inflammation that has developed into a global condition with increasing incidents in developed countries and industries in Asia. Long-term treatment with synthetic drugs such as sulfasalazine will cause resistance, aggravate bleeding, male infertility, pulmonary toxicity, and nephrotoxicity [7]. Therefore, it was necessary to expand the knowledge of the use of herbal medicines such as *P. longifolia* leaves as IBD medicine. This study evaluates *P. longifolia* leaf extract used as an IBD drug induced by indomethacin and compared with sulfasalazine as gold standard.

2. Material and Methods

2.1. Plant Material
Leaves of *Polyalthia longifolia* were obtained from Indonesia and the Philippines. Indonesia leaves of *P. longifoia* were obtained from Malang City, while samples from the Philippines from Camiling, Tarlac and were brought to Biosains Institute, Brawijaya University. The taxonomy for plants was identified and authenticated in plant taxonomy laboratory, Brawijaya University.

2.2. Animals
Male Wistar rats aged two months and weighed around 150-200g each. Rats were maintained under standard laboratory conditions of temperature, humidity and 12 h light and dark. Animals have free access to feed and water *(ad libitum)*. Before treatments, rats were acclimatized for a week. The experimental protocol was approved by the Ethic Commission of Brawijaya University (No. 1035-KEP-UB).

2.3. Experimental Design
The animals were in four groups: (T1) negative control, (T2) IBD with Sulfasalazine, (T3) IBD that is treated using *P. longifoia* Indonesian leaves extract, and (T4) treated *P. longifoia* Philippines leaves extract. Inflammatory bowel disease in rats were induced with indomethacin, with a dose of 15mg/kg orally (p.o). Treatment 2 was given 10 mg/kg BW of sulfasalazine as gold standard of IBD. T3 and T4 were given 300mg/kg BW of ethanolic extract of *P. longifoia* leaves from Indonesia and The Philippines, respectively. The therapies were given for 7 days orally. After therapy, rats were sacrificed and gastrointestinal contents from the gastric, duodenum, jejunum, and colon were collected, and organs were preserving in 10% neutral-buffer formalin and others in phosphate buffer saline for further use.

2.4. Histopathological Observations
The gastric, duodenum, jejunum and colon were collected and fixed in 10% neutral-buffered formalin, embedded in paraffin wax and were cut into sections of 3-5μmm thickness. The sections were stained using haematoxylin and eosin for histopathological observation. The effect on the different treatments on inflammation were observed especially the infiltration of the inflammatory cells, desquamation, erosion, and damage of tissue structures microscopically.

2.5. Immunohistochemistry
The paraffin section of the organs also was deparaffinized with xylol and multi-concentrations of alcohol for 15 min. Furthermore, slides were left overnight at 4°C and were washed with distilled water and PBS solution. Slides were incubated with 3% hydrogen peroxide for 40 min then washed for 5 min with PBS solution. Blocking stages were conducted with 1%BSA in PBS for overnight then washed again with PBS. Primary antibody for COX-2 and SMAD3 (Santa Cruz Biotechnology with ratio 1:50000) were added, left overnight and washed with PBS. After this, the secondary antibodies (anti-rabbit biotin conjugated) were added for 1h and washed again with PBS. Drops of *Strepta Avidin-Horseradish Peroxide* (SA-HRP) were made and left for 40 min and washed again with PBS. Drops of DAB was done and left for 10 min and washed with distilled water. Mayer hematoxylin was dropped on the slides, washed with distilled water, dried and then mounted. Slides were observed microscopically.

IOP Conf. Series: Materials Science and Engineering **833** (2020) 012005 doi:10.1088/1757-899X/833/1/012005

2.6. *Protein Profile Using SDS-PAGE*

0.1g of duodenum and colon were crushed and PMSF + PBS-T were added to the solution. The solution was homogenated and sonicated for 10 min and was centrifuged at 6000 rpm at 4 ° C for 15 min. The supernatant was collected, and cold ethanol was added to the solution (1:1), stored at 4 ° C for 12 h. Samples were centrifuged at a 6000 rpm at 4 ° C for 15 min. The pellets were dried until ethanol was lost. Subsequently, Tris-HCl pH 6.8 (1: 1) was added, and stored at -20 ° C.

3. Result and Discussion

3.1. *Histopathology*

Histopathological analysis of the small intestine was conducted by the Haematoxylin-Eosin (HE) method on all treated groups. Based on previous research, induction of indomethacin caused necrosis in the small intestine, so that the goblet cell and mucus layer were lost, and villi damage (desquamations and rupture) were observed, which was also confirmed in this study. Observation showed the desquamation of epithelia, erosion, congestion, and infiltration of inflammatory cells and villi damages were seen.

3.1.1. Gastric histopathology. The therapy of sulfasalazine as a gold standard has maintained IBD remission [10]. It was observed that the leaf whether from Indonesia and the Philippines have potency as anti-inflammatory [11]. It also reduced the inflammatory symptoms by inhibiting the product in PGE2. As earlier studied [12,13], leaves extract from P. longifolia also had antioxidant activity. One of the compounds was liriodenine [14]. Liriodenine has anti inflammatory and antioxidant activities that can reduce inflammation and repair tissue damage, as seen in the histopathological picture.

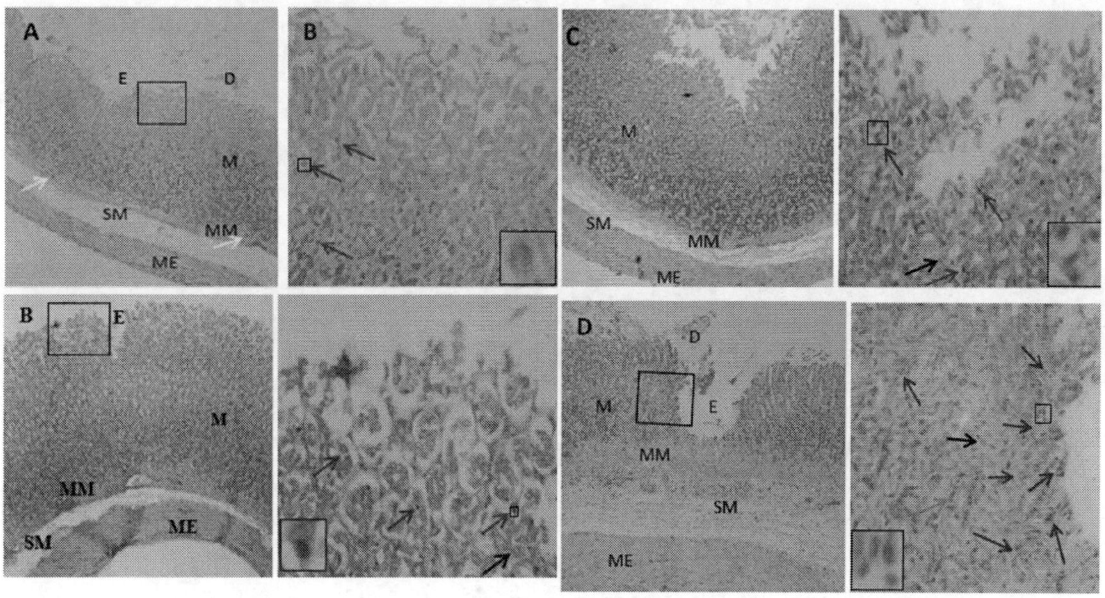

Figure 1. Histopathology of gastric mucosa. (A). Gold Standard group; (B). *P. longfolia* extract from Indonesia therapy; (C). *P. longifolia* extract from The Philippines, and (D) IBD condition.

Legend:

M : Mucosa
MM : Muscularis mucosa
SM : Submucosa
ME : Muscularis external
D : Desquamation of the epithelia
E : Erosion

→ : Congestion
→ : Infiltration
→ : Parietal cells
→ : Chief cells

3.1.2. Duodenum histopathology. Indomethacin induction in rat caused intestine tissue necrosis so the duodenal villi were eroded, and goblet cells disappeared caused by the villi damages. During inflammatory conditions, indomethacin caused villi desquamation resulted in the loss of goblet cells [15]. Based on Figure 2, the histopathology of the duodenum, group treated with sulfasalazine showed an improvement in the intestinal epithelial structure after treatment, as well as no erosion in the epithelial layer, without the infiltration of inflammatory cells and measured villous forms. The use of leaves extracts from Indonesia group (T3) showed that villous improvement, evidenced by the visible arrangement of columnar epithelial cells arranged by the regular villi and the appearance of goblet cells as mucus secretor for the protection of the villi layer in the duodenum. Histopathological features of the duodenum of rats treated with the Philippine extract showed the repairment of duodenal villi, which was followed by the appearance of goblet cells and the epithelial constituents of the villi.

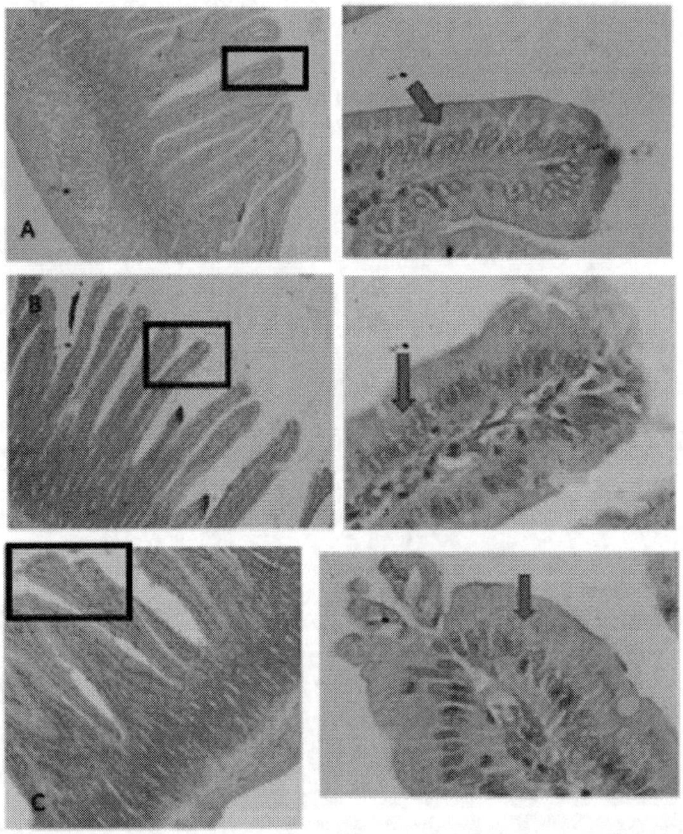

Figure 2. Histopathology of duodenum. (A). Gold Standard group; (B). P. longifolia extract from Indonesia therapy; (C). P. longifolia extract from Philippines.

Legend:

: Columnar epithelial tissue   : Goblet cell

3.1.3. Jejunum histopathology. Thick walls and villi characterized the jejunum histology because they contained many blood vessels. Like duodenum, jejunum suffered damage when exposed to toxic substances such as indomethacin. Figure 3 showed the histopathological features of jejunum in the IBD group treated with sulfasalazine (Figure 3A), showed the improvements in villous shape, neat columnar epithelial cell structure and the appearance of goblet cells. However, inflammation cells were still found coming out of the lymphatic channels and spaces in the Crypt of Lieberkühn, which indicated improvement was not yet completed. The histopathology of jejunum with IBD treated with the leaves extract from Indonesia and the Philippines showed improvement as evidenced by the reduction in inflammatory cell infiltration, neat layer of the columnar epithelial cell structure, goblet cells and no desquamation compared to the control group (Figure 3D).

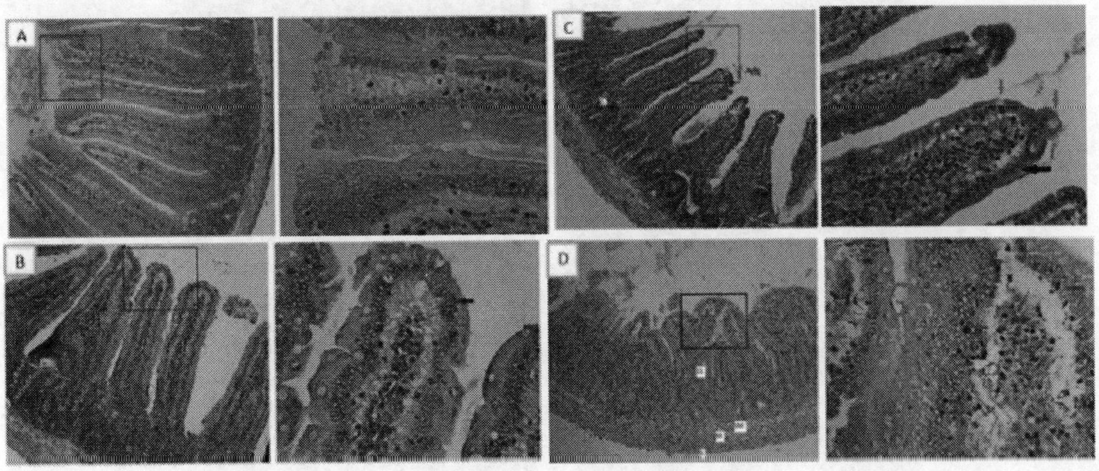

Figure 3. Histopathology of jejunum; (A). Gold Standard group; (B). *P. longifolia* extract from Indonesia therapy; (C). *P. longifolia* extract from Philippines, and (D) IBD condition.

Legend:

S : Serosa
M : Mucosa
SM : Submucosa
CL : Crypt of Lieberkuhn

→ : Cell infiltration
→ : Columnar epithelium tissues
→ : Goblet cells

3.1.4. Colon histopathology. Histopathology analysis showed that the sulfasalazine therapy group has undergone tissue repair in the colonic mucosa marked by the appearance of goblet cells in the mucosal lining of the colonic epithelium. Goblet cells in the colon function as a barrier in the colonic mucosa by removing mucin compounds. As for the colonic histopathology of the leaves extract therapies group from both countries, indicated tissue repairs. However, inflammation cell infiltration was still found, and the surface structure of the epithelium were not yet intact.

The goblet cells in the process of repairing the digestive tract were observed due to sulfasalazine, which has the ability to suppress the production of free radicals thus accelerating the time of inflammation and increasing the work of TGF-β as an anti-inflammatory cytokine for regenerating intestinal stem cells and differentiating into several types of cells namely enterocytes, goblet cells, and cells Paneth [16]. In addition, the content of secondary metabolites in the ethanol extract of *P. longifolia* leaves has quercetin and rutin, which can inhibit the formation of free radicals, reduce pro-inflammatory mediators, improving the protective function of the epithelium in the intestine organ. Flavonoids were known to increase colonic permeability [17].

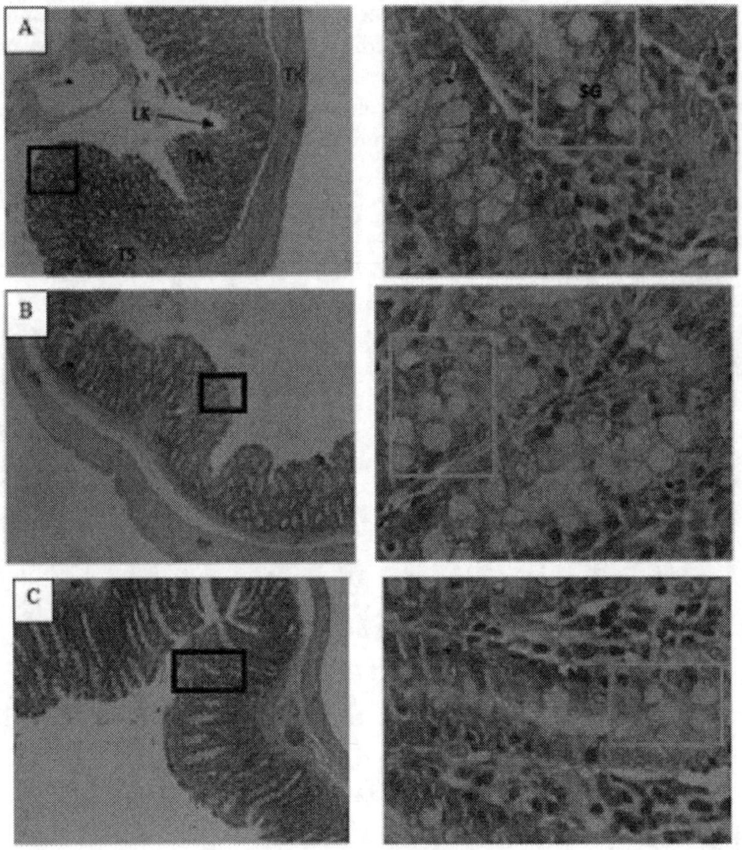

Figure 4. Histopathology of colon. A. Gold Standard group; B. EEPL from Indonesia group; C. EEPL from The Philippines group.

Legend:

 = repairing = inflammation cells infiltration

3.2. *Immunohistochemistry of COX-2 and Smad3*

Inflammatory bowel disease (IBD) is a chronic inflammatory condition that occurs in the gastrointestinal tract. Characterization of the inflammatory process is with the presence of COX-2 expression which increases rapidly and decreases anti-inflammatory cytokines such as Smad3. COX-2 and Smad3 expression were observed by immunohistochemistry methods and analyzed with one-way ANOVA statistics. The results showed that the treatment affected COX-2 and Smad3 expression ($p < 0.05$).

Table 1. Expression of COX-2 on gastric and Smad3 on duodenum in rats' model IBD.

Groups	**COX-2**	**Smad3**
T1 (control)	0.6533 ± 0.05465^{a}	14.8000 ± 0.33466^{a}
T2 (Sulfasalazine	1.2433 ± 0.11130^{b}	12.6000 ± 0.56569^{b}
Indonesia	1.3267 ± 0.05164^{b}	12.2000 ± 0.43818^{b}
Philippines	1.5467 ± 0.08262^{c}	8.0333±0.52789c

Note: The notations a, b and c show a significant difference between treatment groups (significance <0.05). While the same notation shows no significant difference between treatment groups.

Table 1 shows the number of COX-2 and Smad3 expressions when Sulfasalazine, leaves extracts from Indonesia and The Philippines as compared to the adverse treatment. Results showed improvement of IBD by decreasing COX-2 expression and decreasing Smad3 compared with the untreated group. The leaves extract from the Philippines showed significant differences ($P < 0.05$) in treating IBD. The leaves extract from Indonesia has a comparable result as with the gold standard treatment prescribed with sulfasalazine and showed significant differences between the treated and untreated groups. The leaves extract from the Philippines was found better because of increased COX-2 expression and decreased Smad3; this result concurred with the studies of Peng *et al.* [5,17].

Therapy of *P. longifolia* leaves extract from the Philippines and Indonesia has the potential of accelerating healing and cell regeneration in IBD rats. Flavonoid and alkaloid compounds in the leaves of *P. longifolia* can act as antioxidants and anti-inflammatory agents. They play a role in exhibiting COX-2 so that there were limited number of inflammatory cells that migrate to the wound tissue. Furthermore, the inflammatory reaction was shortened and the proliferative ability of Smad3 was not inhibited [16]. Flavonoid and alkaloid compounds stabilize the reactive oxygen species (ROS) by reacting with the compounds from free radicals so that these compounds become inactive [18]. This was evidenced by the histopathological picture of the duodenum which looked normal and has improved.

3.3. Protein profile of duodenum and colon

Table 2. The result of protein profile of duodenum analysis based on SDS-PAGE.

Group	Molecule weight (kDa)							
	289.7	**201.4**	**140.7**	**87.3**	**59.5**	**45.6**	**35.8**	**21.2**
Gold standard sulfasalazine (P1)	√	√	√	-	√	√	-	√
Polyathia longifolia from Indonesia (P2)	√	√	√	-	-	√	√	√
Polyathia longifolia from the Philippines (P3)	√	-	-	-	√	√	√	√

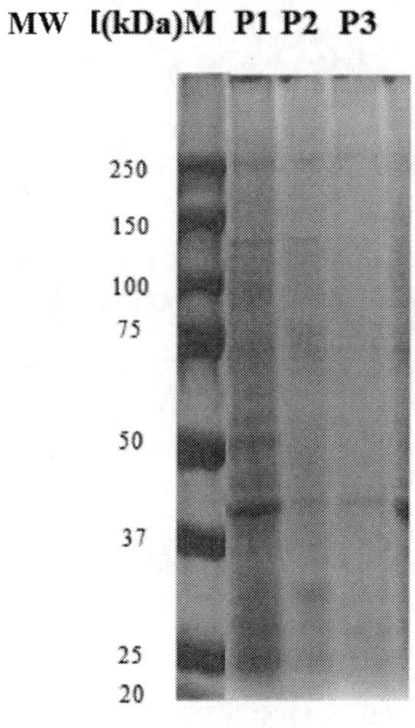

Figure 5. Protein bands of duodenum (SDS PAGE 12%).
Legend: M = marker; P1 = Sulfasalazine group; P2 = EEPL from Indonesia group; P3 = EEPL from Philippines

The 40 kDa protein appeared in sulfasalazine therapy and leaves extract from the Philippines. This protein was thought as Vasoactive Intestinal Polypeptide (VIP) which was secreted by cells in the intestinal mucosa. VIP was produced by inflammatory tissue; VIP increased the permeability of duodenal tissue stimulated the secretion of fluid and electrolytes from duodenal tissue which triggered watery diarrhea and dehydration [19].

Table 3. The result of protein profile of colon analysis based on SDS-PAGE.

Group	Molecular weight of Protein (kDa)						
	288.2	198.3	167.4	114.6	43.5	37.1	22.0
Negative control (K-)	√	√	√	√	√	√	√
Gold standard sulfasalazine (K1)	√	√	√	-	√	√	√
Polyathia longifolia from Indonesia (K2)	√	√	√	-	√	√	√
Polyathia longifolia from the Philippines (K3)	√	√	√	-	√	√	√

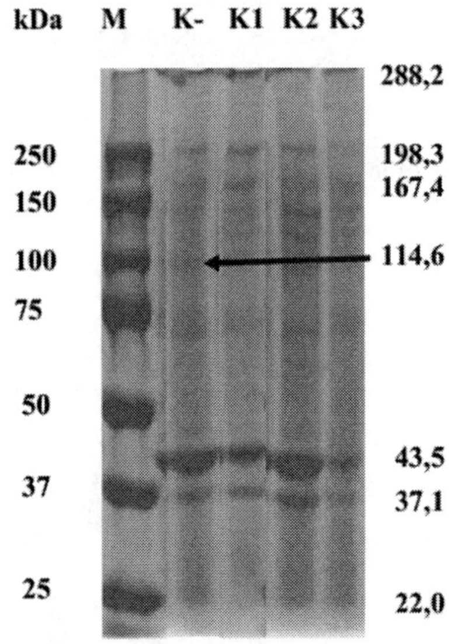

Figure 6. Protein bands of colon (SDS PAGE 12%).
Legend: M = marker; K- = negative control; K1 = Sulfasalazine group; K2 = EEPL from Indonesia group; K3 = EEPL from Philippines

Protein with molecular weights 114.6 kDa were not synthesized in all treatment groups, but it was synthesized in healthy conditioned rats. A protein with molecular weight of 114 kDa is a protein that binds to glycocalyx in blood vessels, namely the β-galactosidase [20]. This β-galactosidase was enzyme that very important to convert carbohydrates into disorders in IBD conditions. This enzyme was located at the peak of the villi to hydrolyze lactose to glucose and galactose [21]

4. Conclusion

Polyalthia longifolia has been proven to improve gastrointestinal tissue in rats with infectious bowel disease-induced indomethacin. It was proven by the histopathological profiles, showed decreasing pro-inflammatory signs, enhanced of anti-inflammatory expression, and improved protein profiles.

References

[1] Jo R Am L and Ceballos R. 2013 *Altern Integr Med* **2** 1–6.
[2] Ekor M. 2014 *Front Pharmacol* **4** 1–10
[3] Msomi N Z and Simelane M B 2019 *Herbal Medicine* **11** 215–271
[4] Sampath M and Vasanthi M 2013 *Int J Pharm Pharm Sci* **5** 336–41
[5] Peng J, Zheng T, Li X, Liang Y, Wang L, and Huang Y 2019 *Front Pharmacol* **10** 1–15.

[6] Vezza T, Rodríguez-nogales A, Algieri F, Utrilla MP, Rodriguez-cabezas ME, and Galvez J 2016 *Nutrients* **8** 1–22
[7] Aulanni'am, Rosdiana A, and Rahmah NL 2011 *J Life Sciences* **4** 57–64
[8] Shofia V, Aulanni'am A, and Prasetyawan S 2015 *Int J ChemTech Res* **8** 640–644
[9] Nagarjun S, Dhadde SB, Veerapur VP, Thippeswamy BS, and Chandakavathe BN 2017 *Biomed Pharmacother Elsevier Masson* **89** 1061–6
[10] Indonesian Endocrinology Association (PERKENI). 2011. National Consensus for IBD therapy in Indonesia. PERKENI. pp.1-42 (Write in Bahasa Indonesia)
[11] Sharma R, Mandal S, Gp R, Gupta N, and Dp S 2016 *International Journal of Drug Development & Research* **3** 351–359
[12] Sampath M 2013 *J Appl Pharm Sci* **3** 148–52
[13] Sashidhara K V., Singh SP, Srivastava A, and Puri A 2011 *Nat Prod Res* **25** 918–26
[14] Obaid AK, Salbiah S, Abdul S, Bakri YM, Nafiah MA, Awang K 2018 *J Pharm Sci Res* **10** 2718–21
[15] Balqis U, Hanafiah M, Januari C, Salim MN, Aisyah S, Fahrimal Y 2015 *Med Vet* **9** 1999–2002
[16] Craven A, Derubertis R, December R, October A 2008 *Gastroenterology* **92** 1998–2008
[17] Saxena A, Kaur K, Hegde S, Kalekhan FM, Baliga MS, Fayad R 2014 *J Tradit Complement Med* **4** 203–217
[18] Lichtenberger LM, JJ Romero, and EJ Dial 2007 *Br J Pharmacol* **150** 913–919
[19] Triantafilidis, J.K.A., C. Triantafyllidi and A. Papalios 2016 *Annals of Gastroenterology* **22** 268-281
[20] Lahav, J 1993 *Thrombospondin* (CRC Press-London)
[21] Nivetha A and V. Mohanasrinivisan 2018, *Research Journal of Pharmacy and Technology*, **11** 1778-1783

IOP Conf. Series: Materials Science and Engineering **833** (2020) 012006 doi:10.1088/1757-899X/833/1/012006

A molecular docking study of dehydroevodiamine as an inhibitor of epstein-barr virus protease

R N Azizah[1], Suharti[*1] and Yahmin[1]

[1]Chemistry Department, State University of Malang, Jalan Semarang No. 5 Malang, Indonesia

*Corresponding author: suharti.fmipa@um.ac.id

Abstract. Epstein-Barr Virus (EBV) is a type of γ-herpes virus which cause kissing disease. The virus induces cancer and causes latent infection. EBV protease is one of the constituent capsid proteins that play an important role in assembling virions on nucleus and spreading them. Therefore, this enzyme potentially became one of inhibition target which have impact on the termination EBV life cycle. During this time, drugs to inhibit this enzyme had not been studied. This study aimed to examine dehydroevodiamine as a potential inhibitor EBV protease by molecular docking method. The docking was done through both blind and specific docking techniques and the K_i values were calculated using docking approach when RMSD is 0 Å. Molecule visualization was done using PyMol and dehydroevodiamine profile identification was done on Ro5. The results showed that dehydroevodiamine has binding affinity of -9.8 kcal/mol and -7.3 kcal/mol; predicted K_i (STP) of $1{,}426729 \times 10^{-8}$ and $1{,}431479 \times 10^{-6}$ for blind and specific docking, respectively. Dehydroevodiamine profiles does not violate Ro5. These values indicated the potential of dehydroevodiamine as an oral drug candidate for kissing disease. This finding opens possibility to do further work on wet-lab-levels.

1. Introduction

Technology bring many benefits in science. One of them is supporting drug design process, especially in drug discovery step. Drug design is very important because many disease vectors exist around human. The purpose of drug design is to find chemical compounds that can enter certain cavities on target proteins both geometrically and chemically [1]. Drug design process using computer is also known as Computer-Aided Drug Design (CADD). In CADD, there are terms Ligand-Based Drug Design (LBDD) and Structure-Based Drug Design (SBDD). LBDD is a process designing drugs based on knowledge of molecules bound by biological targets, while SBDD is based on knowledge of 3D structure from biological targets.

In drug design, drug discovery step is often done using wet lab approach. However, this approach is not economic and effective. That approach needs to involve trials and errors. The estimated total incurred and capitalized R&D costs per new drug to be $1395 million and $2558 million, respectively in 2013. Furthermore, the increased out-of-pocket cost per approved drug to $1861 million and capitalized cost to $2870 million [2]. On the other hand, drug design is multidisciplinary and time consuming. Therefore, in silico approach can be considered to find potential medicines for diseases in drug discovery step. The method used for drug discovery is molecular docking. Molecular docking has become an increasingly important tool for drug discovery [3]. Molecular docking is a computational method for placing ligands which are often small molecules into receptor binding sites.[4] Molecular docking aims to predict position, orientation, and conformation of ligand to the binding site of target macromolecule.[5] In general, molecular docking underlies the discovery position that best suits with pattern of flexible ligands and proteins as rigid receptors. Molecular docking method is used to find compounds which have potentially became inhibitor for receptor, especially receptor that initiate diseases. One disease which does not have clear treatment is disease caused by Epstein-Barr Virus (EBV).

EBV was first observed at Burkitt Lymphoma (BL) cells in 1964. EBV was the first human tumour virus to be discovered.[6] EBV is present in all populations and infects over 90% of individuals within the first 10 years of life.[7] EBV vaccine to prevent primary infection or disease, or therapeutic vaccines to treat EBV malignancies have not been licensed.[8] EBV is dangerous because it is latent and can induce other diseases, especially cancer. Primary EBV infection occurs through saliva contact with patients EBV. Therefore, disease caused by EBV is commonly referred to as kissing disease. EBV can also spread due to sexual contact because it can be found in genetal fluid secretions. EBV is associated with infectious mononucleosis (IM) disease. The IM is characterized by fever, sore throat, cervical and general lymphadenopathy, hepatosplenomegaly, somatic complaints of fatigue and malaise.[9] Many neurologic manifestations of EBV infection have been documented, including encephalitis, aseptic meningitis, transverse myelitis, and Guillain-Barré syndrome.[10] Other diseases that may arise due to EBV include epithelial malignancies such as gastric carcinoma, mesenchymal malignancies such as follicular dendritic cell sarcoma and lymphoid malignancies such as pyothorax-associated lymphoma, Burkitt lymphoma, Hodgkin lymphoma and non-Hodgkin lymphoma.

Diseases often arise due to disorders of biomolecules, such as DNA and protein. EBV capsid proteins which form protease enzyme play an important role in the process of assembling virions at nucleus and spreading them from one cell to another. This enzyme catalyzes division of assembled proteins after formation procapsids. These cleavage sites are found in the bonds of -Ala-Ser-, -Ala-Ala- and Ala-Thr. Based on structure, EBV protease has an active site which shaped a very shallow pocket located on top of strands β5 and β6 of the middle β-barrel. The 3D structure of EBV protease was obtained from experiment using X-ray crystallography method with a resolution of 2.3 Å after that enzyme was inhibited with diisopropyl-fluorophosphate (DFP).[11] This enzyme has potentially became a biological target to be inhibited. EBV protease inhibition aims to fully inhibit enzyme work, thereby impacting in the termination of EBV life cycle.

Drugs as inhibitors most often are small molecules and produce therapeutic benefits for patients. EBV protease does not have a standard inhibitor as a drug. Kissing disease had been treated with drugs that did not have selectivity and specific toxicity to EBV. Therefore, this study tries to examine the ability of dehydroevodiamine as a potential inhibitor of EBV protease which is able to work with selectivity and specific toxicity. Dehydroevodiamine was chosen because it belongs to the type of alkaloid derivative. Alkaloids are often used as samples of ligands because they have good biological activity. Alkaloids had been the subject of many research related to their biological activities, such as anti-tumour[12], antitussive and expectorant[13], anti-alzheimer and antioxidant[14], analgesic activity[15] and antiviral[16]. This compound also has N atoms which is expected to donate electrons to the active site enzyme. The reason underlies that dehydroevodiamine has potentially became used as an inhibitor EBV protease, all at once as a drug for kissing disease. The purpose of this study was to examine dehydroevodiamine as an inhibitor of EBV protease with molecular docking method and its possibility to be implemented as an oral kissing disease drug. This study also will open possibility to do further work on wet-lab-levels.

2. Experiment

2.1. Research object

The research objects are 3D structure EBV protease and dehydroevodiamine. The device used are hardware consisting of a set computer with specifications: Processor Intel(R) Core(TM)2 Quad CPU

Q8300 @2.50GHz (4 CPU), 4GB RAM and 500GB hard disk and software consisting of Windows 7 64-bit Ultimate (6.1, Build 7601), Edu PyMOL v.1.7.4.5., PyRx 0.9.5 and LIGPLOT$^+$.

2.2. Procedure

2.2.1. Molecular docking and visualization result of molecular docking
The structure of dehydroevodiamine ligand is downloaded at https://pubchem.ncbi.nlm.nih.gov in 3D with (*.sdf) format. The structure is minimized using Open Babel which is integrated in PyRx 0.9.5 software. Insert new item is selected and dehydroevodiamine ligand is opened. Minimization is carried out by right clicking and choosing to minimize selected. Ligand is changed to pdbqt format, then save is selected.

The 3D structure of EBV protease is downloaded at https://www.rcsb.org in (*.pdb) format. This enzyme is opened in the Edu PyMOL software. In the names panel, letter "A" (action) selected, then the polar hydrogen atom is added by selecting hydrogens and clicking add. Enzyme sterilization is carried out by removing ligands and water. The procedure is done by selecting show sequence "S". Ligand and water sequences are selected, then right clicked and selected remove. Sterile enzyme are stored by clicking save molecule.

Molecular docking process between EBV protease and dehydroevodiamine is carried out using Vina Wizard program integrated in PyRx 0.9.5 software. The start button is clicked, add ligand and add macromolecule panels will be appeared. Ligand preparations and enzyme sterile are added. The molecular docking process is continued by clicking on forward. The binding site is determined through grid box position. In blind docking technique, the grid box used is Center X: 5,934, Center Y: 9,336, Center Z: 28,411 and Dimensions (Å) X: 67,785, Dimensions (Å) Y: 57,655, Dimensions (Å) Z: 94,953. In specific docking technique, grid box is adjusted to position catalytic triad at the active site enzyme (His48-Ser116-His139). These residues of triad catalytic are chosen with selected molecules menu in the navigator. The "+" button next to 1O6E sterile enzyme are clicked. The toggle selection spheres is also clicked, then catalytic triad is selected. Full screen is selected, and grid box is organized to limit catalytic triad area. Forward is chosen to continue molecular docking process. The results of molecular docking are enzyme-ligand complex, binding affinity value enzyme-ligand and Root Means Square Deviation (RMSD).

The binding affinity value has a relationship with Inhibition Constant (K_i). K_i values are predicted under Standard Temperature and Pressure (STP) conditions. The calculation of predicted K_i is at Equation 1.

Equilibrium : $E + I \leftrightarrow EI$
Binding : $E + I \rightarrow EI\ (K_b)$
Dissociation : $EI \rightarrow E + I\ (K_d)$
Therefore,

$$K_b = \frac{1}{K_d}$$
$$\ln K_b = -\ln K_d$$

K_i = dissociation constant enzyme-inhibitor complex (~K_d)

$$\ln K_b = -\ln K_i$$
$$\Delta G_{inhibition} = RxTx \ln K_i$$
$$\ln K_i = \frac{\Delta G}{RxT}$$
$$K_i = e^{\frac{\Delta G}{RxT}} \quad (1)$$

Visualization of enzyme-ligand complex is carried out using Edu PyMOL (3D) and LIGPLOT$^+$ (2D) software. In Edu PyMOL software, the results of docking and sterile enzymes are opened in PDB file format at the same time. In names panel section, letter "A" is clicked, then preset-ligand site is selected. The desired form of visualization is clicked. Visualization with LIGPLOT$^+$ software is carried out by selecting menu file and open PDB file. Complex is entered by clicking browse. In the select ligand to plot section, ligands in the complex are selected.

2.2.2. Rule of five (Ro5)

Lipinski rule also known as Rule of Five (Ro5) consists of (1) molecular mass <500 Dalton, (2) lipophilicity level (log P) <5, (3) molar refraction must be in the range of 40-130, (4) hydrogen bond donors <5 and (5) hydrogen bond acceptors <10. Analysis Profile analysis of drug candidates with Ro5 is performed using SCF Bio IITD webserver. File in PDB format is uploaded and submitted.

3. Results and Discussion

3.1. Molecular docking and visualization result of molecular docking

EBV protease has a 3D structure composed of A chain and B chain. These chains are presented in Table 1. These chains form the structure of EBV protease which can be observed in Figure 1. The active site of this enzyme is composed of catalytic triad His48-Ser116-His139, so this enzyme can be categorized as serine protease. Histidine has an imidazole group, making it possible for proton transfer process. Serine can act as a covalent binding on the acyl group. ISP (propan-2-yl dihydrogen phosphate) ligand has been co-crystallized with EBV protease and it bound at its active site suggesting that ISP ligand acts as an inhibitor. ISP is a general inhibitor for serine protease. The enzyme binds ISP ligand at its active site through A chain and B chain. The distance between ISP binding to EBV protease in A chain and B chain can be observed in Figure 2(A) and 2(B).

Table 1. A chain and B chain sequences of EBV protease

A Chain
MVQAPSVYVCGFVERPDAPPKDACLHLDPLTVKSQLPLKKPLPLTVEHLPDAPVGSVFGLYQ SSAGLFSAASITSGDFLS LLDSIYHDCDIAQSQRLPLPREPKVEALHAWLPSLSLASLHPDIPQTTADGGKLSFFDHVSICAL GRRRGTTAVYGTDLA WVLKHFSDLEPSIAAQIENDANAAKRESGCPEDHPLPLTKLIAKAIDAGFLRNRVETLRQDRG VANIPAESYLKA
B Chain
MVQAPSVYVCGFVERPDAPPKDACLHLDPLTVKSQLPLKKPLPLTVEHLPDAPVGSVFGLYQ SSAGLFSAASITSGDFLS LLDSIYHDCDIAQSQRLPLPREPKVEALHAWLPSLSLASLHPDIPQTTADGGKLSFFDHVSICAL GRRRGTTAVYGTDLA WVLKHFSDLEPSIAAQIENDANAAKRESGCPEDHPLPLTKLIAKAIDAGFLRNRVETLRQDRG VANIPAESYLKA

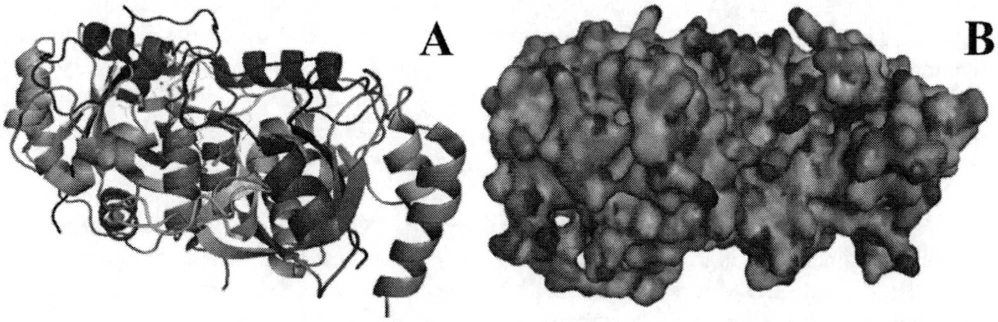

Figure 1. Visualization of EBV protease structure using Edu PyMOL (A) cartoon and (B) surface

IOP Conf. Series: Materials Science and Engineering **833** (2020) 012006 doi:10.1088/1757-899X/833/1/012006

ISP ligand binds to the enzyme using noncovalent and covalent interactions. Covalent interaction indicate that ISP inhibits protease EBV irreversibly. ISP has corrosive properties to metal and tissue. Thus, ISP might not be used as inhibitors which could be consumed by human. When ISP enters the body, enzymes that contain serine in its active site will become permanently inactive. This is very dangerous. One of the enzymes in the body that contain serine in the active site is acetylcholinesterase. Acetylcholinesterase works on the nervous system in human body. Paralysis in some organs can occur if this enzyme is inhibited. Therefore, ISP in this study was substituted using dehydroevodiamine. Dehydroevodiamina is classified as a secondary metabolite compound alkaloid derivative. Dehydroevodiamine have heterocyclic nitrogen atoms which are expected to donate electrons to bind in the active site of enzyme and contribute to the formation of hydrogen bonds. Many hydrogen bonds will make the binding affinity value more negative.

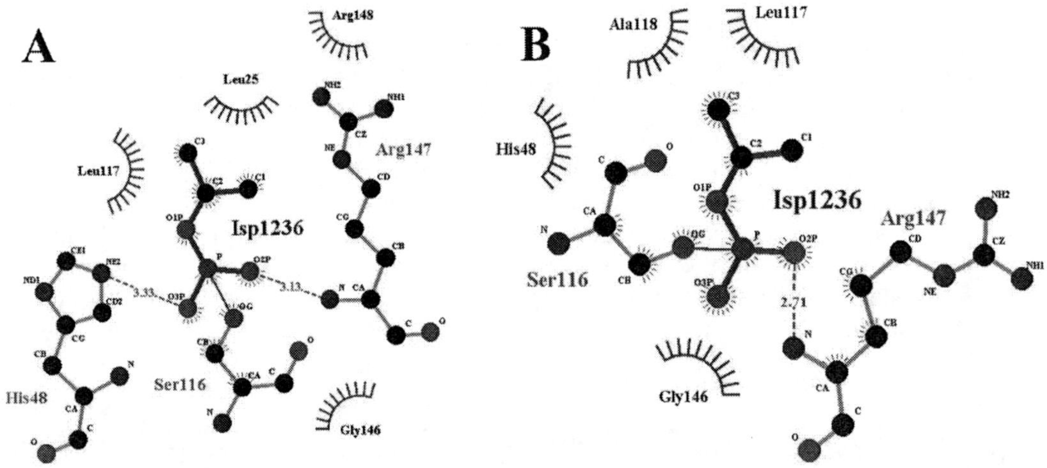

Figure 2. The position of ISP binding in protease EBV at (A) A chain and (B) B chain

Binding of drugs at active site enzyme will produce a competitive inhibition mechanism, whereas binding which does not occur at the active site enzyme will produce a noncompetitive inhibition mechanism. Implementation of blind docking technique will generally obtain drug candidates that work with noncompetitive inhibition mechanisms. This is due to the active site enzyme located on the inside, so it takes a much higher energy to arrive at that place. This technique will bind ligand in any binding site of enzyme that produces the most negative binding affinity value. Conversely, implementation of specific docking technique will obtain drug candidates which work with competitive inhibition mechanism.

Based on calculation by PyRx 0.9.5 software, each ligand has binding affinity value varies in the number of certain modes. The difference in these modes lies in the position, orientation and conformation when ligand is bound to enzyme. The scoring taken is the most negative of the calculated modes. A negative scoring value refers to an increasingly stable interaction between enzyme and ligand. Another parameter that can be observed is RMSD. The RMSD value is calculated relatively in terms of the best mode by only utilizing movable heavy atoms. RMSD affects the orientation of ligand position when interacting with target enzyme. The prioritized RMSD from these results of molecular docking is valued at 0 Å which means identical. RMSD value less than 2 Å means good.

The results blind docking and specific docking of EBV protease and dehydroevodiamine, respectively, obtained binding affinity values were -9.8 kcal/mol and -7.3 kcal/mol, when RMSD value 0 Å. Binding affinity values are obtained in kcal/mol. This value indicates total bonding energy from molecular docking process, so it can be referred as Gibbs free energy ($\Delta G_{binding}$). $\Delta G_{binding}$ expresses the spontaneity of chemical reactions.[17] $\Delta G_{binding}$ is a thermodynamic benchmark that

IOP Conf. Series: Materials Science and Engineering **833** (2020) 012006 doi:10.1088/1757-899X/833/1/012006

shows stability of ligand conformation when bound to a biological target. The ligand conformation to bind with receptor will be more stable if binding affinity value is getting negative.

$\Delta G_{binding}$ has a relationship with K_i. In this study, the value of binding affinity is converted to K_i. K_i values are prediction for reference when laboratory experiments are performed. The predicted K_i values from calculation based on blind docking and specific docking techniques are 1.42679×10^{-8} and $1.431479\text{x}10^{-6}$, respectively. The smaller binding affinity value, the smaller predicted K_i value. Binding affinity and predicted K_i values show if dehydroevodiamine has low energy when bind with EBV protease. It means that dehydroevodiamine can bind to biological target easily. Thus, dehydroevodiamine was proposed as drug candidate for kissing disease.

Visualization blind docking with Edu PyMOL software showed residues EBV protease that interact directly with dehydroevodiamine to form hydrogen bonds are Gln35 and Lys40. The distance hydrogen bond between dehydroevodiamine-GLN35 is 3.2 Å and dehydroevodiamine-LYS40 is 2.3 Å. Visualization is strengthened using LIGPLOT$^+$ software to see hydrophobic interactions. Based on this visualization result, a hydrogen bond is formed between dehydroevodiamine and Lys40 residue. This hydrogen bond distance is 3.07 Å. Some residues that carry out hydrophobic interactions with dehydroevodiamine are Phe136, Pro41, Phe137, Gln35, Leu38, Leu42, Leu60, Gly59, Val57 and Pro37. The results of blind docking visualization using Edu PyMOL software can be observed in Figure 3(A) and 3(B), while visualization with LIGPLOT$^+$ software can be observed in Figure 3(C).

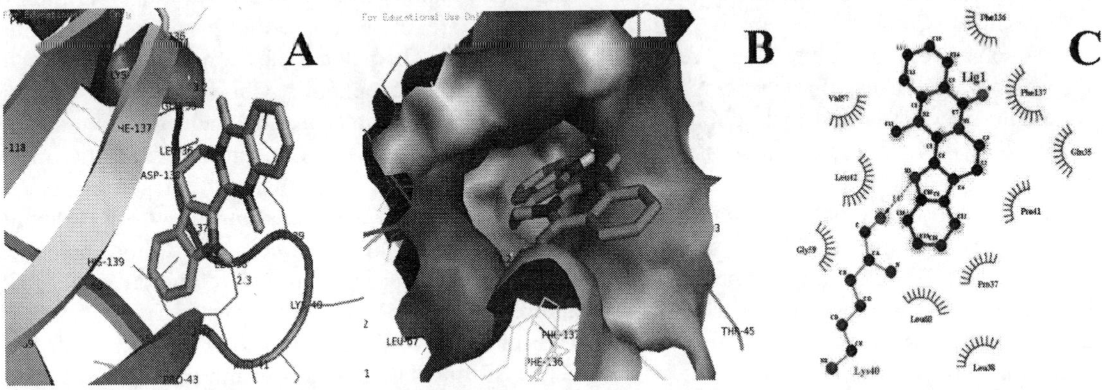

Figure 3. Visualization blind docking dehydroevodiamine to EBV protease (A) cartoon (PyMOL), (B) surface (PyMOL) and (C) 2D diagram (LIGPLOT$^+$)

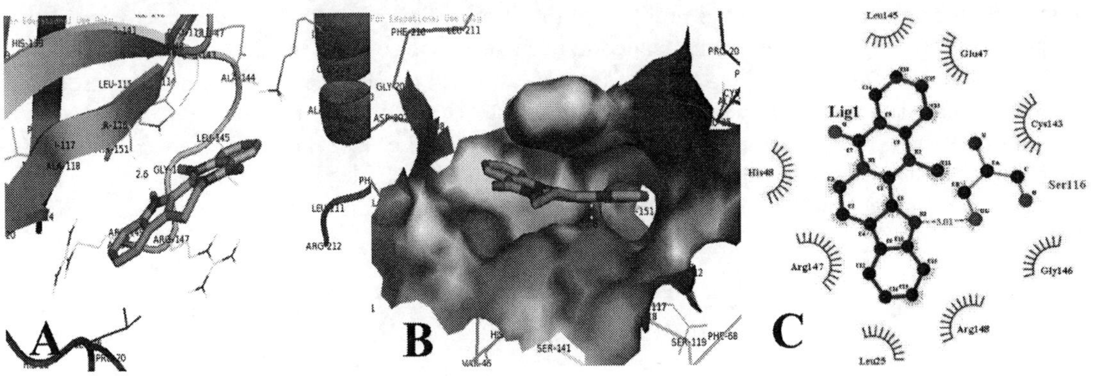

Figure 4. Visualization spesific docking dehydroevodiamine to EBV protease (A) cartoon (PyMOL), (B) surface (PyMOL) and (C) 2D diagram (LIGPLOT$^+$)

Visualization specific docking with Edu PyMOL software showed residue that interact directly with dehydroevodiamine is SER116 with hydrogen bond distance 2.6 Å. The dehydroevodiamine is precisely attached to the binding site ISP ligand. Visualization specific docking with Edu PyMOL software is strengthened using LIGPLOT^{+} software to see hydrophobic interactions. Based on the results, hydrogen bonds are formed between dehydroevodiamine with Ser116 residue. This hydrogen bond distance is 3.01 Å. Other residues that contribute in hydrophobic interactions for stability enzyme binding with ligand are Leu145, Glu47, Cys143, Gly146, Arg148, Leu25, Arg147 and His48. The results of specific docking visualization using Edu PyMOL software can be observed in Figure 4(A) and 4(B), while visualization with LIGPLOT^{+} software can be observed in Figure 4(C).

3.2. Ligand profile based on rule of five (Ro5)

Medicines taken orally will pass through the body with possibilities that can prevent it from reaching target. Factors which influence the body's treatment of drugs, namely absorption, distribution, metabolism and excretion (ADME) are better known as pharmacokinetics. Pharmacokinetics of drugs need to be considered when designing a drug, so the drug can effectively reach target.

The Lipinski Rule also known as Rule of Five (Ro5) provides guidance in the process of finding a drug which is good for oral consumption. Ro5 is based on a distribution of calculated properties among several thousand drugs [18]. Compounds that have high molecular weight, are generally large. The greater size of compound will more difficult to penetrate cell membrane (lipid bilayer). Lipophilicity is also known as the logarithmic value of partition coefficient (P). Partition coefficient is ratio of solubility compounds in a water-fat system. The main factors which contributes partition of solutes in a two-phase system, in this case water-fat, are molecular volume and intermolecular interaction of solution-solute. Log P related with hydrophobic characteristic of compounds. The higher log P value indicates that the compound is increasingly hydrophobic. Compound used as drug candidate may not be too hydrophobic, because it will block the rate of movement to arrive at lipid bilayer. It will affect the distribution of drug in body in a long time. As a result, the selectivity of the bond between drug candidate and target enzyme is reduced. Log P value that is too negative makes it difficult for drug candidate to enter lipid bilayer. Molar refraction has a relationship which is directly proportional with polarisability. The higher molar refraction value make compound more polar and will be penetrated difficulty to the lipid bilayer. Conversely, the lower molar refraction value make compound more nonpolar and will be difficulty to reach lipid bilayer. The value of H-donor and H-acceptor supported by molecular weight play important role in determining permeability of drug to pass lipid bilayer. Lipinski rules states if molecules profile fall outside its limits of the rules, the molecule will be difficult to absorb orally. Lipinski rules refers to candidate drug's properties as a suitable or unsuitable molecule as an oral drug. In practice, minimum of two Lipinski Rules must be achieved. Violations exceeding two rules are predicted to have a high failure rate for drugs to reach the target receptor and this is a potential problem in the drug discovery process. In general, drugs that do not suit with Lipinski Rules are recommended to be made as intravenous injection types.

The dehydroevodiamine profile in this study was adjusted to Ro5, because it would be used as an oral drug candidate. Dehydroevodiamine profile test with Ro5 parameters was carried out by Supercomputing Facility for Bioinformatics & Computational Biology, IIT Delhi Webserver. Ro5 data for dehydroevodiamine are presented in Table 2. Based on Table 2, dehydroevodiamine can be used as a potential oral drug candidate for kissing disease, because its profile does not violate the Lipinski Rules.

Table 2. Dehydroevodiamine profile based on Ro5

No.	Rule of 5	Spesification	Dehydroevodiamine Profile
1.	Molecular weight	<500 Dalton	301 Dalton
2.	Log P	<5	2,167700
3.	H-bond donors	<5	0
4.	H-bond acceptors	<10	4
5.	Molar refraction	(40-130)	89,9544750

IOP Conf. Series: Materials Science and Engineering **833** (2020) 012006 doi:10.1088/1757-899X/833/1/012006

4. Conclusion

Dehydroevodiamine compound has potential to be an EBV protease inhibitor with carried out hydrophobic interactions and hydrogen bonds. This compound is also potentially used as an oral kissing disease drug.

References

[1] Ashish B 2010 *Systematic Reviews in Pharmacy* **1**(1) 99-105.
[2] Joseph A D, Henry G G and Ronald W H 2016 *Journal of Health Economics* **47** 20–33.
[3] Xuan-Yu M, Hong-Xing Z, Mihaly M and Meng C 2011 *Current Computer Aided-Drug Design* **7**(2) 146–157.
[4] Elizabeth Y, Jessica H and Paul A R 2015 *Journal of Molecular Recognition* **28**(10) 581–604.
[5] Vincent Z, Aurélien G and Olivier M 2009 *Journal of Cellular and Molecular Medicine* **13**(2) 238–248.
[6] Lawrence S Y, Lee F Y and Paul G M 2016 *Nature Reviews Cancer* **16**(12) 789–802.
[7] Alberto A and Kassandra L M 2010 *Journal of Neuroimmune Pharmacology* **5**(3) 271–277.
[8] Jeffrey I C 2015 *Clinical & Translational Immunology* **4**(1) 1-6.
[9] Hal B J 2011 *Pediatrics in Review* **32**(9) 375-384.
[10] Asif D, Ari B, Elizabeth L F J, Susan R, Raymond T, Martin P, Helen H and Daune M 2006 *Journal of Child Neurology* **21**(5) 384–391.
[11] Marlyse B, Jean-Francois H, David L, Guy S, Eric F, Gérard A, Jean-Marie S, Rob W H R and Wim P B 2002 *Journal of Molecular Biology* **324**(1), 89–103.
[12] Ze L, Chang-Fa H, Xiao-Shan L and Jikai J 2010 *Basic & Clinical Pharmacology & Toxicology* **108** 304–309.
[13] Dongdong W, Shu W, Xiaong C, Xiaolong X, Jingyi Z, Lihuan N and Xia L 2012 *Journal of Ethnopharmacology* **139**(1) 189–193.
[14] Hyun A J, Byung-Sun M, Takako Y, Je-Hyun L, Yeong S K and Jae S C 2009 *Biological & Pharmaceutical Bulletin* **32**(8) 1433–1438.
[15] Jian-Hua S, Xiang-Hai C, Tao F, Yun-Li Z, Jing-Kun W, Lu-Yong Z, Ming Y, and Xiao-Dong L 2010 *Journal of Ethnopharmacology* **129**(2) 174–181.
[16] Kailiang W, Bo S, Ziwen W, Meng W, Zheng L, Yanna H, Zhijin F, Na M and Qingmin W 2010 *Journal of Agricultural and Food Chemistry* **58**(5) 2703–2709.
[17] Raymond C and Jason O 2011 *General Chemistry The Essential Concepts 6th Edition* (New York: McGraw-Hill Companies).
[18] Christopher A L, Franco L, Beryl W D and Paul J F 1997 *Advanced Drug Delivery Reviews* **23** 3-25.

Acknowledgement

This work was supported by PNBP funding addressed to Dr. Suharti and Team with the contract number of 20.3.116/UN32.14.1/LT/2019, Department of Chemistry, Universitas Negeri Malang.

The Potency of Antioxidant Perfume of Essential Oils to Reduce Free Radical Content in Air

Selena B Deshinta[1*], F A D Cahyo[1], G D Aggreini[1], Edi P Utomo[1], I Tazi[2]

[1]Department of Chemistry, Faculty of Science, Brawijaya University, Indonesia
[2]Department of Physics, Faculty of Science, Universitas Islam Negeri Maulana Malik Ibrahim Malang, Indonesia

*Corresponding author: sbungadeshinta@gmail.com, epu@ub.ac.id

Abstract. Free radical contamination is very dangerous for health. Perfume contains essential oils whose components are potential as antiradical. Antioxidant perfume is a perfume made from essential oil with a certain concentration which is very effective to reduce free radicals in the air. This study aims to determine the effectiveness of essential oils as antiradical compounds using laboratory experiments in the reduction of free radical concentration of DPPH with various concentrations of fruit extract which containing terpenoid component. DPPH concentration changes in the air followed by using electronic nose (E-Nose) equipped with multisensory gas. The experimental results showed that the critical concentration of fruit extract which effectively decreased free radical i.e. apple, orange, grape, melon, and lemon were 7.47%, 6.21%, 15.61%, 7.58%, and 6.22%. The greater concentration of these critical concentrations of fruit extracts is potentially as prooxidants.

1. Introduction

Free radicals are potentially damaging to the environment and human health. Because when free radicals enter the body, it will rapidly attract biological macromolecules electrons that surround proteins, nucleic acids, and deoxyribonucleic acids (DNA) which then interfere metabolic processes in the body [1]. It is now known that various types of volatile non-methane organic compounds from different vegetation have been emitted into the atmosphere every day. These organic compounds include isoprene, monoterpene $C_{10}H_{16}$, sesquiterpene $C_{15}H_{24}$, as well as the number of oxygenated compounds such as methanol, hexene, 2-methyl-3-buten-2-ol, and 6-methyl-5-hepten-2. In the troposphere layer, the organic compound with free radical is hydroxyl (OH), nitrate (NO_3) and ozone (O_3) which the reaction reacts to the lower layer of the troposphere [2]. Different half-life times, for example linalool a monoterpene often found in perfume products, has a half-life of 52 minutes with OH radicals, 55 minutes with ozone and 6 minutes with NO_x. Limonene has a citrus scent having a half-life of 49 minutes (OH), 2.0 hours (O_3) and 5 minutes (NO_x) [3].

Although terpenoids can react with these radicals to reduce the reactive effects of free radicals, but some researchers have found that their reaction products actually produce aerosol particles in the troposphere layer [4]. The termination reactions of these radicals produce stable products even in aerosol form. The reaction products include formaldehyde, methyl vinyl ketone, methacroline, organonitrate and others [5]. The products of perfume, air freshener or cosmetics usually contain terpenoid compounds as their aroma components. That causes when evaporated or sprayed into the air, the terpenoid components will help to reduce the presence of free radicals in the atmosphere with different half-lives depending on the type of terpenoid component [6].

Antioxidant perfume is a perfume made from essential oil with a certain concentration which is very effective to reduce the free radical in the air because there are antiradical compounds in essential oils. There are some natural ingredients that are potential as antioxidants and can be formulated as perfumes, including essential oils [7]. Some essential oils have antioxidant activity, such as phenol group compounds, such as eugenol. Eugenol can lower free radicals through the abstraction of hydrogen atoms [8], while terpenoids through a chain-based reaction mechanism carrying HOO radicals, which react rapidly with terpenoid radicals to halt chain reactions [9]. Measurement of antioxidants in the air requires instruments such as electronic nose (E-nose). E-nose is an instrument designed to mimic the mammalian olfactory system [10]. In this study, there will be held an experiment decrease free radical activity that comes from compound 2,2-diphenyl-1-picrylhydrazyl (DPPH) in the presence of perfume material derived from various extracts of aromatic fruits. The changes in free radical activity in the air are monitored using electronic nose (E-Nose).

2. Research Methods

2.1. Chemicals and Instrumentation

Essential oils were obtained from extraction of fruit by maceration using propylene glycol solvent. DPPH or 1,1-diphenyl-2-picrihidazyl was used as free radical source.

The extract was characterized using gas chromatography - mass spectrometer (Shimadzu QP-2010S equipped with RTX-5MS capillary column (30 m x 0.25 mm)). The gas carrier was set up at 50ml/min, from 60-300°C with 10°/min speed, injector temperature of 250 °C, interface temperature of 250 °C, and the spectrum library from Wiley 9. Electronic nose or e-nose instrumentation was prepared and assembled as shown in Figure 1. Inside the sensor room, there are 10 sensors (arrays) of various types and sensitivities as shown in Table 1. These sensors consist of semiconducting SnO_2 metal sensitive to particles in the air. If the air is clean, the sensor shows a low conductivity. It is expected that these sensors give respond to the component in the fruit extract. The sensor conductivity increases as the particle concentration in the air increases.

Table 1. Sensor Sensitivity

Type of Sensor	Sensitive to gas
MQ-2	LPG, iso-butane, propane, methane, alcohol, H_2, and cigarette smoke
MQ-3	Benzene
MQ-4	Methane and benzene
MQ-5	LPG, natural gas and urban gas
MQ-6	LPG, iso-butane, propane
MQ-7	CO
MQ-8	H_2
MQ-131	CO_2
MQ-135	Acid cigarettes, CO_2, NH_3, NOx and benzene

2.2. Reaction Procedure

The fruit samples used were oranges, grapes, lemons, melons, apples and each of them were destroyed and weighed until 100 g, and then soaked in propylene glycol by ratio of (1: 3) for 3 x 24 hours. The extract product was concentrated using a rotary evaporator vacuum. Next, a solution of fruit extract as a source of aroma with concentration of 5, 10, 15, 20, and 25% (v/v) in propylene glycol solvent was prepared. DPPH was dissolved in propylene glycol to obtain a 0.2 mMol liquid DPPH and stored in dark coloured bottles, before further use.

In the e-nose instrumentation, two hose were installed for the vapor of DPPH solution and a solution of fruit perfume extract. The first stage of the vapour was the DPPH flowing into the sensor room for data collection of signals and then blowing out, each of extraction took 30 seconds for a total 300 seconds. Then, the extract steam for each concentration and type of fruit extract was also performed in the same way. For each extract replacement, the instrument was flushed using propylene glycol vapor for 2 minutes. Observation of the effect of fruit extract on the change of DPPH free radical concentration was done by blowing the solution of DPPH and fruit extract simultaneously with the same treatment as described above for 300 seconds. Gas sensor response data was collected in the form file.xls which further processed to obtain the relationship of concentration with the magnitude of each sensor response.

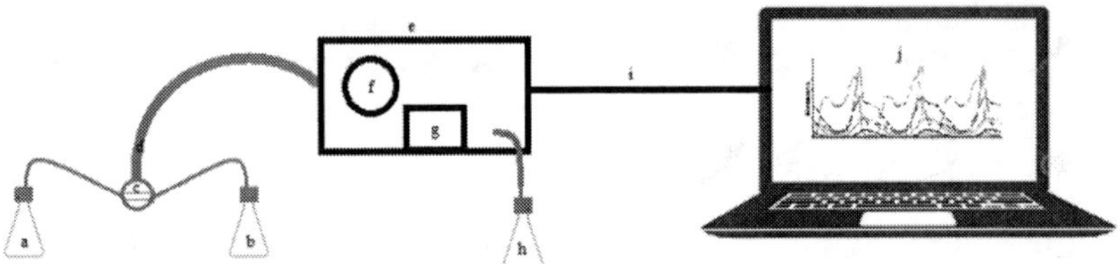

Figure 1. Trial series using e-nose. Description: (a) The sample bottle contains the DPPH, (b) the sample bottle containing the volatile, (c) valves, (d) the steam connecting hose, (e) E-nose instrumentation; (f) the sensor room; (g). power supply, (h) vapour trap bottle containing propylene glycol, (i) electrical cable, (j) computer.

3. Result and Discussion

3.1. GC-MS Result

Characterization of fruit extract as a source of perfume using gas chromatography-mass spectrometer provides information on the types of compounds contained in the extract, as shown in Table 2.

Table 2. Components in fruit extract GC-MS analysis result

Fruit Extract	Major Components	Minor Components
Lime *(Citrus hystrix)*	linalool, citronellal; ß-phelandrene, isopulegol, ß-citronelol	sabinene; thujene; terpinen-4-ol; cephrol
Melon (*Cucumis melo*)	myrsenol; lomine; jasminaldehyde; amilcinnamaldehyde; amil cinnamic aldehyde; alpha hexyl cinnamic aldehyde	caproil alcohol; penticarbinol; gardeniol; acetic styralyl; aceticneryl; 2-decalactone
Lemon (*Citrus limon*)	d1-Limonene; cinene; nesol; 2-methyl-4-isoprophenilcyclohexa-1-ene	
Grapes (*Vitis vinifera L)*	d1-Limonene; sinene; nesol; 4-methyl-3,6,9-trioxadecan-1-ol	Ethyl butanoic; ethyl butyrate acid; linalool l; benzylcarbinol; geraniol; aceticneryl; acetic geranyl; methyl dihydrojasmonate; cis-2-hexenol,
Apeple (*Malus sylvestris)*	heptanoic allyl, citronellal	tert-Butyl acetoacetate, gamma-undecanalactone, trans-alfa ionone, Benzyldimethylcarbinol acetate, 2-tertiobutylcyclohexyl acetate, o-tert-Butylcyclohexanol

3.2. E-nose Responses

The molecules in the fruit extract do not contain small size compounds that can be addressed by all types of sensors in Table 1. However, Figure 2 shows that the presence of molecular parts similar to the molecule responded, causing the sensors to respond to the vapor extract of fruit and DPPH vapor with different sensitivity.

The e-nose detector response curve to the DPPH vapor of both single vapor and steam mixture with fruit extract shows significant signal intensity change. E-nose response to the highest DPPH radical vapor on sensor 6, then sensor 2 and sensor 1. The response of air sensor containing a mixture of DPPH radical vapours and fruit extracts showed a decrease in the response intensity to a minimum concentration of fruit extracts which tended to increase again (Figure 3). This proves that in the sensor space that is analogous to the air space, there has been an interaction between DPPH molecules and scented molecules of fruit extract. This interaction results in a decrease of sensor signals conductivity.

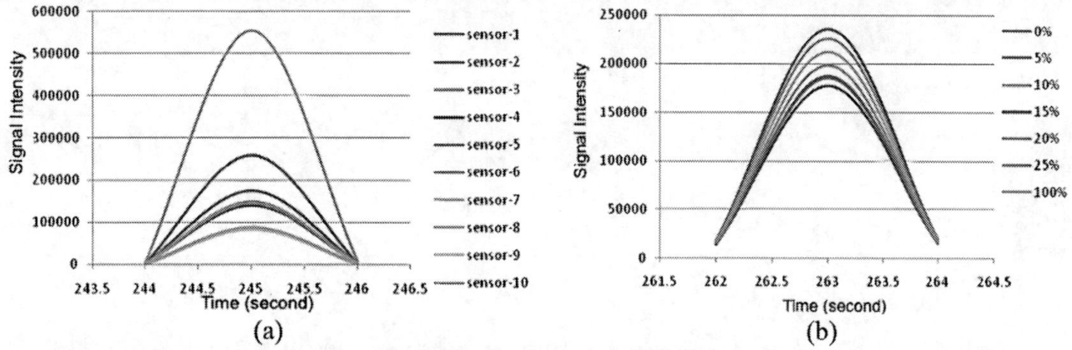

Figure 2. E-nose responses to (a) steam of DPPH solution and (b) steam of 20% lime extract.

3.3. Critical Concentration of Fruit Extract

The free radical intensity of DPPH can be optimally optimized by the effect of a particular concentration of fruit extract [11]. E-nose instrumentation will detect using appropriate sensor to produce an average graph which is the representative image of the free radical response of DPPH + fruit extract. The valley peak is a critical point indicating the optimum concentration of fruit extracts that can be used as antioxidant free radical antidote. As depicted in Figure 3 and summarised in Table 3, if the concentration exceeds at the critical point, then the fruit extract will increase and strengthen the intensity of DPPH free radical which means to be pro-radical [12].

Table 3. Sensor sensitivity of DPPH vapour and fruit extracts

No.	Extracts	Sensors									
		1	2	3	4	5	6	7	8	9	10
1	Apple	-	-	v	v	-	-	-	v	v	-
2	Lime	-	-	-	v	v	-	v	v	v	v
3	Grape	-	-	v	v	-	-	-	-	-	-
4	Melon	-	-	v	v	v	v	v	v	v	v
5	Lemon	v	v	-	-	-	-	-	-	-	-

The sensitivity of fruit extract and DPPH free radicals to the sensors in e-noses varies depending on the type of fruit extract. Regression analysis of the sensor response curve to the DPPH mixed vapor and fruit extract showed the quadratic equation where $R^2 \approx 1$. By solving the quadratic equation, each curve obtained the optimum average concentration of each fruit extract, referred as the critical concentration, as shown in Table 4. The decrease in intensity of the sensor response is predicted due to a decrease in sensor conductivity caused by its interaction with DPPH radicals. The DPPH vapor response numbers decrease as the concentration of extract increases, reaching a critical state after the critical value of the curve tends to increase again. This phenomenon showed that there has been an interaction between the DPPH radical and the chemical components of the fruit extract vapor inside the electronic nose. The reduction of the response can be identified as the curve of antioxidant activity of fruit extract, while the rebound curve can be said as an increasing of the radical concentration or can be called pro-oxidant [12].

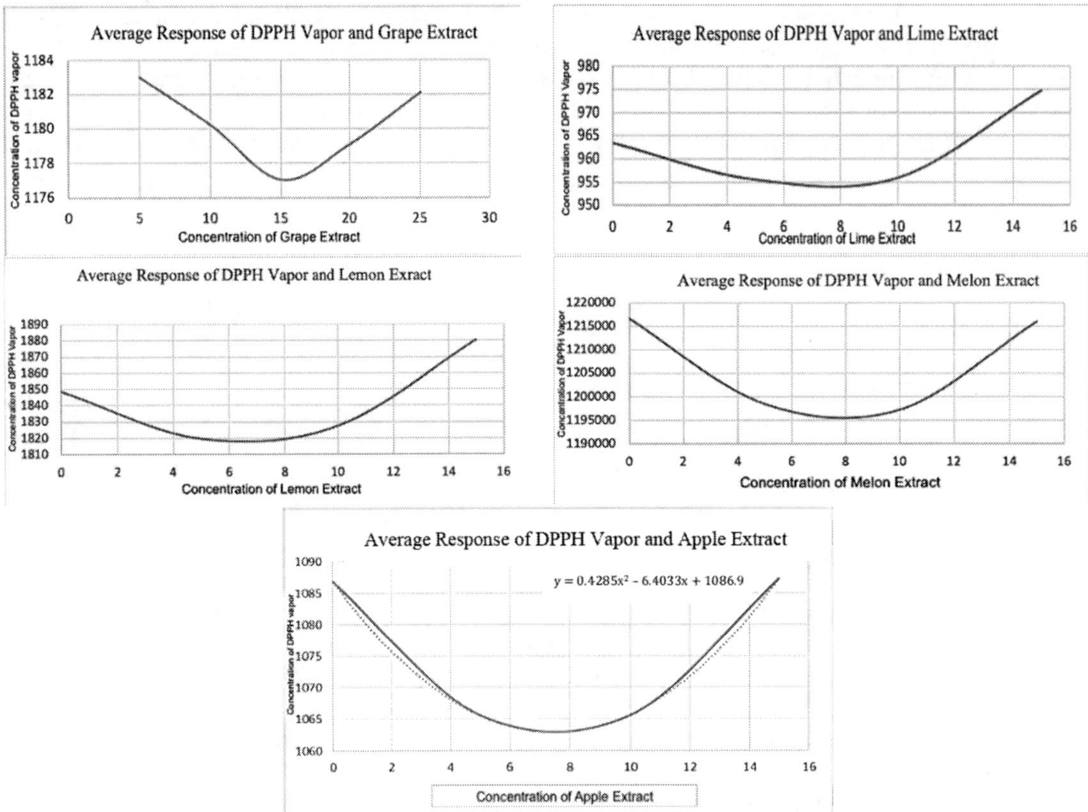

Figure 3. E-nose response to DPPH vapor in the presence of fruit extracts.

Table 4. The optimum concentration of each fruit extract in reducing DPPH free radical level

No.	Extracts	Critical Concentration (%)
1	Apple	7.47 ± 0.55
2	Lime	6.22 ± 0.07
3	Grape	15.61 ± 0.07
4	Melon	7.59 ± 0.04
5	Lemon	6.23 ± 0.81

It takes a relatively small concentration of fruit extract to be able optimally become antioxidant antidote to free radicals. Referring to Table 2 and Table 4, the main components of lime that have the greatest ability to reduce DPPH levels require 6.218% as the optimum concentration in antioxidant perfume formulations. This suggests that the main component of the orange linalool, which is a monoterpene derivative, can reduce radicals better than the minor components in other fruit extracts. The optimum concentration of limonene which is the main compound of lemon in antioxidant perfumes shows unrelated figures of 6.232%. Limonene is a monoterpene compound that has capability of reducing radical by conjugation system [13]. Components of apples serving as DPPH reducers are allyl hexanoic and citronellal. Melons have a major component of mirsenol, which is a terpenoid compound. This compound has a double bond conjugated, so it is strongly suspected to act as an antioxidant in melon extract. Although the wine-scented component is derived from a group of ester compounds, but the presence of a major component of limonene, and geraniol allows the grape extract to be antioxidant. Although it has the same main components, the wine has many minor components while the lemon does not have a minor compound. This suggests that in lemon there is more limonene than wine, so in compare with lemon, the more optimum concentration of wine is needed to reduce DPPH.

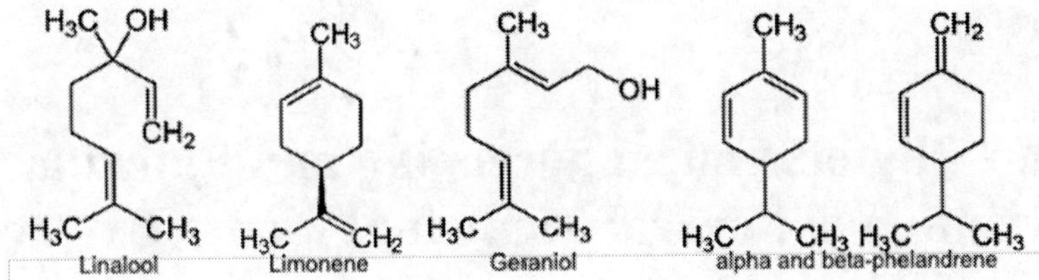

Figure 4. Terpenoid components in fruit extract that can be able to act as an antioxidant [9].

4. Conclusion
Experimental results of extra fruit effect on DPPH free radical decline showed that all fruits were able to decrease free radical of DPPH in the following order of abilities: lime> lemon> apple> melon> grape. Linalool in oranges is a monoterpene derivative that most potentially reduces free radicals and is followed by limonene in lemons. Diterpenoid compounds that are widely found in fruits have different ability to reduce radical. This confirms that the optimum concentration is important to know because the concentration of these compounds will become pro-radical. Fruit extracts that contain essential components potentially as perfume materials that can reduce free radicals in the air. Furthermore, E-nose can be used for tracing and simultaneously determining the activity of free radicals in the air in the gas phase.

Acknowledgment
Authors thanks to the support from the 2nd IC2MS organising committee and acknowledge the Department of Chemistry – Brawijaya University, Malang.

References
[1] Noneng Z, Dwina R, and Muhayatun S 2013 *J. Sains Dan Teknol. Nukl. Indones.* **10** 1
[2] Abid L, Shaiq G, Rayees A, Hilal A, Tauseef A, and Imtiyaz A 2013 *Afr. J. Pure Appl. Chem.* **7** 3 91–113
[3] Lien Ai PH, Hua H, and Chuomg PH 2008 *Int J Biomed Sci* **4** 2 89– 96
[4] Devasagayam T, Tilak J, Boloor K, Sane K, Ghaskadbi S, and Lele R 2004 *Japi* **52** 10 794–804
[5] Mariah U, Praptining R, Lussana D, and J. S. T. No 2013 Konsep Pengetahuan Lingkungan Green Chemistry Pada Program Studi Pendidikan Biologi," in *Prosiding Seminar Biologi* **10**
[6] Sandro S, Shoaib M, Gilles K, Gregory V, Julia H, and Stefan F 2017 *Toxicol. In Vitro* **42** 263–272
[7] Hatice Z and Ayse B 2014 *Molecules* **19** 11 17773–17798
[8] M Hossain, N Brunton, C Barry-Ryan, A B Martin-Diana, and M Wilkinson 2008 *Rasayan Journal of Chemistry* **1** 4 751-756
[9] J. Gramann 2005 *Vitamins & Hormones* **72** 505–535
[10] P Kalita, M P Saikia, and N H Singh 2015 *IJIREEICE* 123–128
[11] Tisserand R and Balacs T 1996 *Int J Aromather*, **3** 28-32
[12] Zhu B, Zhu J, Fan R and Mao L 2011 *Molecular Toxicology* **5** 1-43
[13] Hartsel J, Eades J, Hickory B, and Makriyannis A 2016 *Nutraceuticals* 735-754

Preliminary Phytochemical Screening and Fluorescence Characterization of Several Medicinal Plants Extract from East Java Indonesia

Q Fardiyah[1,2], Suprapto[1], F Kurniawan[1,3,*], T Ersam[4], A Slamet[5], Suyanta[6]

[1]Instrumentation and Analytical Sciences Laboratory, Department of Chemistry, Institut Teknologi Sepuluh Nopember, Surabaya, Indonesia
[2]Analytical Chemistry Laboratory, Department of Chemistry, Faculty of Mathematic and Natural Science, Brawijaya University, Malang, Indonesia
[3]ITS Halal Center, Institute of Research and Community Service, Institut Teknologi Sepuluh Nopember, Surabaya, Indonesia
[4]Natural Product Chemistry Laboratory, Department of Chemistry, Institut Teknologi Sepuluh Nopember, Surabaya, Indonesia
[5]Department of Environmental Engineering, Institut Teknologi Sepuluh Nopember, Surabaya, Indonesia
[6]Department of Chemistry Education, Faculty of Mathematics and Natural Science, Yogyakarta State University, Yogyakarta, Indonesia

*Corresponding author: fredy@chem.its.ac.id

Abstract. The preliminary phytochemical screening and fluorescence characterization of several medicinal plant extracts from East Java Indonesia have been carried out. Some medicinal plants used in this study are *Annona muricata, Strobilanthes crispus, Piper ornatum and Andrographis paniculata*. The objective of this research is to find out the result of phytochemical screening and characterization of several medicine plants by fluorescence spectroscopy. The plant part in the form of leaves is extracted using maceration method. The extraction process was carried out for 3 days with ethanol solvent in room temperature. The result of preliminary phytochemical screening showed the existence of secondary metabolite compounds in several medicinal plant extracts, among others, alkaloid, saponin, terpenoid, steroid, tannin and flavonoid. The fluorescence characterization spectra obtained shows that several medicinal plant extracts have optimum absorption at wavelength of excitation of 323 – 342 nm and wavelength of emission of 662 – 685 nm.

1. Introduction

Indonesia is famous for its biodiversity, and one of which is medicinal plants. Since long time ago, Indonesian people have known and used it as another treatment to overcome health problems. The use and utilization of medicinal plants has increased significantly in recent years. The public is increasingly aware of the importance of medicinal plants for health. Some of the advantages of using medicinal plants include abundant availability, easily obtained, inexpensive, harmless, effective and does not cause many side effects for users. The use of medicinal plants is broad enough to treat several types of diseases therapeutically[1]. Types of diseases that can be treated using medicinal plants include diabetes mellitus, liver, heart, tumors, cancer and others. The choice of medicinal plants in treating several dangerous diseases has been proven over the past few years so that it is used as a basis for finding new therapeutically effective drugs such as anticancer, antitumor, antimicrobial, antioxidant and antihepatotoxic compounds[2–5].

According to data from the World Health Organization (WHO), most people from developing countries choose traditional methods of treatment that come from medicinal plants. The bioactive compounds contained will be the best source for getting various types of drugs. However, this must be further investigated to discuss further, protection, and effectiveness[6]. Medicinal plants have some bioactive components that deliver physical activity in our body. Some organic compounds contained in medicinal plants include, alkaloids, saponins, terpenoids, steroids, tannins and flavonoids[7]. These compounds have many benefits for human body. Primary and secondary living metabolism have a role to produce that bioactive component. Secondary metabolites compounds are very diverse chemical for therapeutic treatment in humans, medicine, farming and other utilization[8].

In this study four different types of medicinal plants were used from Indonesian East Java. The four plants are *Annona muricata*, *Strobilanthes crispus*, *Piper ornatum* and *Andrographis paniculata*. The choice of medicinal plants is based on several properties and benefits of these medicinal plants. In addition, the existence of these four types of plants is often found around and its availability is quite abundant.

Phytochemical screening is a method of bioactive compounds identification that is unknown in plant extracts through qualitative analysis[3]. Phytochemical screening is a preliminary stage in a phytochemical study that aims to provide an overview of the class of compounds contained in plants that are being studied. Phytochemical screening method is done by looking at the color testing reaction using a color reagent[9]. Knowledge of the chemical components contained in medicinal plants needs to be studied. This information will be significant for the synthesis of complex active components of chemical compounds contained in medicinal plants. Phytochemical screening in medicinal plants, in addition to being used to identify active compounds that are beneficial to the body's health (positive effects of herbal medicines) can also be used to identify active compounds that cause toxins (negative effects of herbal medicines). This causes the phytochemical screening process of medicinal plants to be important before conducting further analysis[10–12].

To determine the character of the bioactive compounds contained in plant extracts, one of the methods that can be used is the fluorescence spectrophotometry method[13]. Characterization using the fluorescence spectrophotometric method can be used to determine the ability of some extracts of medicinal plants in absorbing and releasing ultraviolet radiation. The principle of the analysis of this method is by measuring the absorption of the sample of light energy entering the wavelength of its excitation and releasing the energy of the light at its emission wavelength[14].

In this research, preliminary phytochemical screening and fluorescence characterization have been carried out on several medicinal plants such as *Annona muricata, Stronbilanthes crispus, Piper ornatum and Andrographis paniculata* from the East Java Indonesia.

2. Material and methods

2.1. Material

The materials used in this study include the leaves of several types of medicinal plants which were macerated using ethanol pa (Merck) solvent. Preliminary phytochemical screening is done using the following ingredients: aquadest, chloroform (Merck), anhydrous acetic acid (Merck), H_2SO_4, HCl (Merck), $FeCl_3$ 1%, Mg powder, (Merck), Dragendorf, Mayer, and Liebermann-Burchard reagents and others.

Plants materials consist of fresh leaves, *Annona muricata*, *Stronbilanthes crispus*, *Piper ornatum* and *Andrographis paniculata* were obtained from Batu, East Java, Indonesia.

Figure 1. Several medicinal plants.

Some types of plant material collected by the leaves. Leaf samples were air dried and protected from direct sunlight for less than 5 days. The dried leaf samples are mashed using a blender. Furthermore, dry powder or simplicia is stored in a closed container at room temperature.

Medicinal plant extracts were extracted using maceration methods. A total of 50 gram of simplicia was macerated with 500 ml of ethanol solvent in the maceration vessel. Maceration process is carried out for 3 days and stirred for a certain time interval. The maceration results were filtered and then evaporated at 40 °C. Extracts of medicinal plants obtained were ready for phytochemical screening and characterized by fluorescence spectrophotometry.

2.2. Phytochemical testing of several medicinal plants extracts

Ethanol extracts from several medicinal plants were screened for the following phytochemical constituents: alkaloids, saponins, terpenoids, steroids, tannins and flavonoids. Standard methods were used in testing the presence of bioactive compounds in medicinal plant extracts[15,16].

2.2.1 Alkaloids test. 2 mL of plant extract was dissolved in 8 mL of distilled water. Solution is then filtered and added with 1% HCl then put into two different test tubes. In the first test tube, 2-3 drops of Dragrendroff reagent are added. To the second test tube 2-3 drops of Mayer's reagent are added. Plant extracts contain alkaloid compounds if the test result produces a brown, red or orange red precipitate (Dragendorff reagent) and white or yellowish precipitate (Mayer reagent).

2.2.2 Saponins test. 2 mL of plant extract is dissolved in 8 mL of distilled water and shaken until homogeneous. Solution is then filtered and placed into a test tube. After the solution is cold, it is shaken vigorously. Plant extracts contain saponin compounds if the test result produces a stable foam.

2.2.3 Terpenoids test. 2 mL of plant extract was dissolved in 8 mL of distilled water. Solution is then filtered and added with 2 mL chloroform then placed into a test tube. 2 mL H_2SO_4 is added through the test tube wall. Plant extracts contain terpenoid compounds if the test result produces brown or orange rings in a layer of two solvents.

2.2.4 Steroid test. 2 mL of plant extract was dissolved in 8 mL of distilled water. Solution is then filtered and added with 2mL chloroform and anhydrous acetic acid then placed in a test tube. 2 mL H_2SO_4 was added through the test tube wall. Plant extracts contain steroid compounds if the test result produces a blue or green color.

2.2.5 Tannins test. 2 mL of plant extract was dissolved in 8 mL of distilled water. Solution is then filtered and added with 2-3 drops of 1% $FeCl_3$ then placed into a test tube. Plant extract contain tannin compounds if the test result produces blue or blackish green color.

2.2.6 Flavonoids test. 2 ml of plant extract was dissolved in 8 mL of distilled water. Solution is then filtered and added with Mg powder and placed into a test tube. 4-5 drops of concentrated HCl was added through the test tube. Plant extracts contain flavonoid compounds if the test result produces a pink scarlet color.

IOP Conf. Series: Materials Science and Engineering **833** (2020) 012008 doi:10.1088/1757-899X/833/1/012008

2.3. Fluorescence Characterization of Several Medicinal Plant Extracts
20 µL plant extracts were dissolved in 100 mL of demineralized water in a measuring flask. 10 µL of the solution is taken and dissolved in 50 mL of demineralized water. The sample solution is taken sufficiently then placed into a cuvette and ready to be measured. Sample measurements were carried out using the Perkin Elmer LS 55 fluorescence spectrophotometer. This instrumentation was linked with Winlab FL software. The sample measurements were carried out at the maximum wavelength obtained from the results of the previous measurements[14].

3. Result
Phytochemical test resulted from ethanol extracts from several medicinal plant extracts is presented in Table 1. This table shows the presence of active phytochemical constituent compounds which are organic compounds contained in several extracts of medicinal plants that have been studied.

Table 1. Phytochemical component of several medicinal plants.

No	Type of plant	Alkaloid	Saponin	Terpenoid	Steroid	Tannins	Flavonoid
1.	*A. muricata*	+	+	+	-	+	+
2.	*S. Crispus*	+	+	+	-	+	+
3.	*P. ornatum*	-	-	+	+	+	+
4.	*A. paniculata*	+	-	+	-	+	+

Phytochemical test results showed that all types of plants studied contained terpenoids, tannins, and flavonoids. Saponin compounds were not found in *Annona muricata* and *Piper ornatum* plants. Only *Piper ornatum* plants contain steroid compounds and do not contain alkaloid compounds.

Table 2. Fluorescence Characterization of several medicinal plants extract.

No.	Plants	Fluorescence characterization			
		Excitation		Emission	
		Wavelength (nm)	Fluorescence Intensity (%)	Wavelength (nm)	Fluorescence Intensity (%)
1	*A. muricata*	342	573,65	682	567.71
2	*S. Crispus*	328	638.74	662	618.02
3	*P. ornatum*	330	648.12	683	473.03
4	*A. paniculata*	323	654.67	685	511.96

Fluorescence characterization of several medicinal plant extracts is presented in Table 2. In this table, fluorescence characterization is based on two processes namely the excitation process and the emission process. Each process has its maximum wavelength determined based on its fluorescent intensity value.

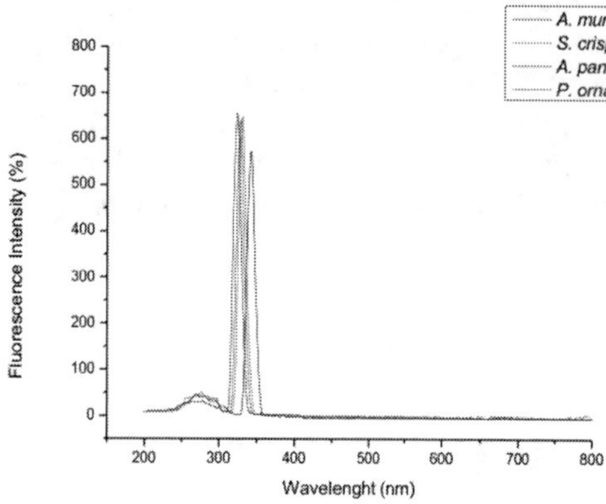

Figure 2. Fluorescence spectrum of excitation.

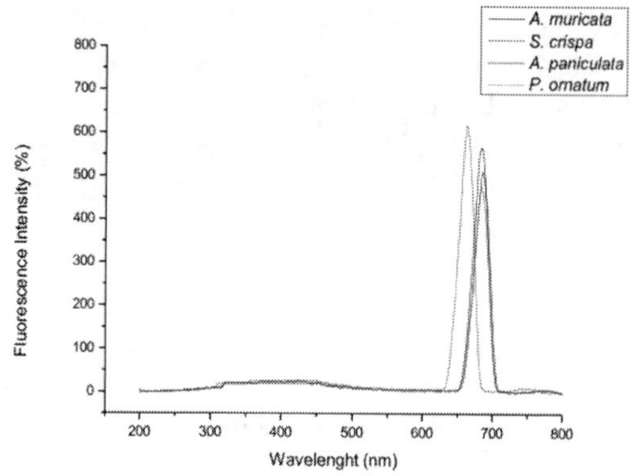

Figure 3. Fluorescence spectrum of emission.

Figures 2 and 3 represent the spectrum of fluorescence characterization of several medicinal plant extracts. The fluorescence spectrum results from two processes. The first is the excitation process which is the process of absorption of ultraviolet radiation by phytochemical components in extracts of several medicinal plants. The second is the emission process, which is a process of releasing energy when it returns to the ground state. Fluorescence spectrum resulted from several extracts of medicinal plants are presented in Figure 4 below:

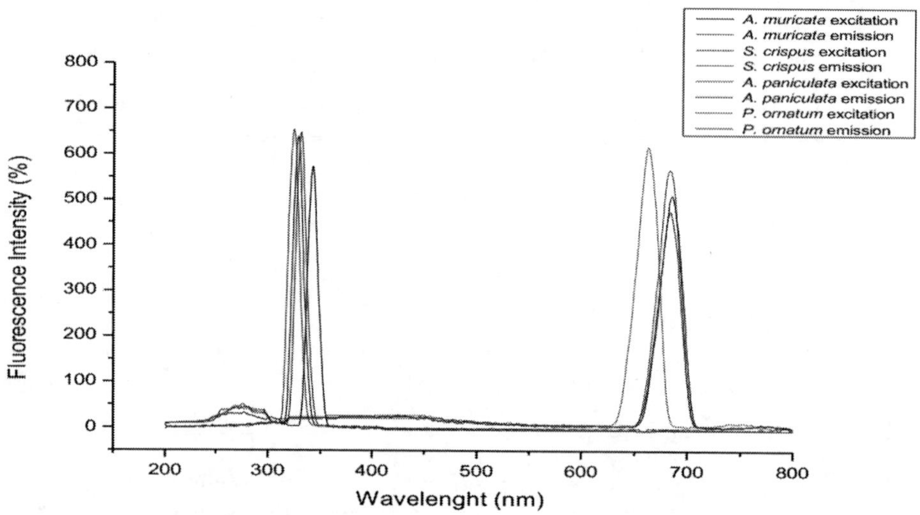

Figure 4. Fluorescence spectrum of several medicinal plants extracts.

In Figure 4, there is an overall fluorescence spectrum of two processes, namely the energy absorption process (the excitation process) and the release of energy (the emission process) of ultraviolet radiation. In this spectrum, there are differences in the characteristics of each medicinal plants extracts. The difference can be seen from the maximum wavelength and fluorescence intensity. In this study, four types of medicinal plants were used with different phytochemical components.

4. Discussion

Preliminary phytochemical screening of ethanol extracts from several medicinal plants has been carried out. The results of phytochemical analysis indicated the presence of phytochemical constituents contained in extracts of medicinal plants. The secondary metabolites component contained in the extracts of medicinal plants included the following compounds: alkaloids, saponins, terpenoids, steroids, tannins and flavonoids.

Almost all extracts of medicinal plants studied contain alkaloids, except for extracts of medicinal plants of *Piper ornatum*. In contrast, steroid compounds were only found in *Piper ornatum* medicinal plant extracts but were not found in other medicinal plant extracts. Saponin compounds were found in the extracts of medicinal plants *Strobilanthes crispus* and *Piper ornatum* and were not found in the extracts of medicinal plants *Andrographis paniculata* and *Annona muricata.*

Results of fluorescence characterization studies of several medicinal plant extracts showed the absorption of ultraviolet radiation at the excitation wavelength. In the excitation process, the maximum wavelength is generated in the ultraviolet area (200-400 nm). The four types of medicinal plants studied gave the results of excitation spectra that had different maximum wavelengths. This difference is caused by the content of different compounds in each of these medicinal plant extracts. In the excitation spectrum the shortest wavelength (323nm) is produced by the extract of the medicinal plant *Andrographis paniculata* and the furthest wavelength (342nm) is produced by the extract of the medicinal plant *Annona muricata*. This is influenced by the different compound components of the two medicinal plant extracts. From the results of phytochemical screening tests, it is known that the two extracts of these medicinal plants have the same compound content in them, namely the presence of alkaloid compounds, terpenoids, tannins and flavonoids as well as not containing steroid compounds. The difference is that there is no saponin content in *Andrographis paniculata* extracts. The presence of saponin compounds in the *Annona muricata* extract is possible to cause a shift in wavelength. Saponin compounds have two basic chemical structures, glycone and aglycone. The combination of these two structures allows the saponin compound to absorb ultraviolet radiation at longer wavelengths so that this plant extract produces the furthest excitation wavelength compared to the three other medicinal plant extracts. This process continues with the process of re-emitting ultraviolet radiation by releasing a certain amount of energy. This process was carried out to return to a stable state that was ground state. This process was known as the emission process. In the emission process, the maximum wavelength is generated in the visible light area (400-800 nm). The four types of medicinal plants studied gave different maximum wavelength emission spectra results. This difference is caused by the content of different compounds in each of these medicinal plant extracts. In the shortest wavelength emission spectra (662nm) produced by the extract of the medicinal plant *Strobilanthes cryspus* and the farthest wavelength (685nm) produced by the extract of the medicinal plant *Andrographis paniculata*. This is influenced by the different compound components of the two medicinal plant extracts.

Another parameter that needed to be considered was the value of the intensity of the fluorescence produced in each fluorescence process. The intensity of fluorescence indicates the number of compound components contained in extracts of medicinal plants that can absorb or emit ultraviolet radiation. The difference in the value of fluorescence intensity during the excitation and emission processes shows the ability of each medicinal plant extract to absorb ultraviolet radiation. Of the four types of medicinal plants studied, *Piper ornatum* extracted medicinal plants had the highest absorption ability of ultraviolet radiation indicated by a difference in the value of fluorescence intensity of 175.09%. The lowest absorption ability of ultraviolet radiation produced by the extract of the medicinal plant *Annona muricata* with a difference in the value of fluorescence intensity of 5.94%. The phytochemical component of the flavonoid and tannin compounds contained in the extracts of medicinal plants made the extract to absorb ultraviolet radiation. The area of absorption of ultraviolet radiation was broad enough to cover the regions of UV A and UV B. This can be seen in the fluorescence spectrum produced by some extracts of medicinal plants. At the same time, the maximum intensity of extract absorption at its wavelength can be known.

The results of fluorescence characterization indicated that some medicinal plants have a specific fluorescence spectrum both when absorbing (excitation) and releasing (emitting) radiation of ultraviolet. In the excitation process, the highest fluorescence intensity was produced by the extract of the medicinal plant *Andrographis paniculata* with a wavelength of 323 nm. While in the emission process, the highest fluorescence intensity was produced by the extract of the medicinal plant *Strobilanthes crispus* with a wavelength of 662 nm. In this study, fluorescence characterization showed that several extracts of medicinal plants can absorb ultraviolet radiation specifically at wavelength 323 - 342 nm and release ultraviolet radiation at wavelength 662-685 nm.

5. Conclusion

The results concluded that the preliminary screening of phytochemicals from some medicinal plants in general contained terpenoid compounds, tannins and flavonoids. Almost all extracts of medicinal plants studied contained alkaloids, except for extracts of medicinal plants *Piper ornatum*. In contrast, steroid compounds were only found in extracts of medicinal plants *Piper ornatum* but not found in extracts of other medicinal plants studied in this study. In the extracts of medicinal plants *Strobilanthes.crispus* and *Piper ornatum,* phytochemical components of the compound saponins were found, but not in the extracts of medicinal plants *Andrographis paniculata* and *Annona muricata*. Fluorescence characterization showed that several extracts of medicinal plants can absorb ultraviolet radiation specifically at wavelength 323 - 342 nm and release ultraviolet radiation at wavelength 662-685 nm.

References

[1] Sofowora A, Ogunbodede E and Onayade A 2013 *African Journal of Traditional, Complementary and Alternative Medicines* **10**
[2] Adesegun S, Ayoola G, Coker H, Adepoju-Bello A, Obaweya K, Ezennia E and Atangbayila T 2008 *Tropical Journal of Pharmaceutical Research* **7**
[3] Y. Gavamukulya, F. Abou-Elella, F. Wamunyokoli, H. AEl-Shemy 2014 *Asian Pac. J. Trop. Med.* **7** S355–S363
[4] J. Philipson, C. Wright, 1996, Plants with Antiprotozoal Activity: Tease and Evans, Pharmacognosy, 14th ed., WB Saunders Company
[5] P. Dewick 2002 Tumor Inhibitors from Plants *Pharmacogn. J* p.394–406.
[6] S. Arunkumar, Muthuselvam 2009 *World J. Agric. Sci.* **5** 572–576.
[7] H. Endoga, D. Okwu, B. Mbaebie 2005 *Afr. J. Biotechnol.* **4** 685–688.
[8] K. Vasu, J. Goud, A. Suryam, Singara, M. Chary 2009 *Afr. J. Microbiol. Res.* **3** 418–421.
[9] M. Cowan, 1999, *Clin. Microbiol. Rev.* **12** 564–582.
[10] J. Parekh, S. Chanda 2007 *Afr. J. Biomed Res.* **10** 175–181.
[11] F. Mojab, M. Kamalinejad, N. Ghaderi, H. Vanidipour 2003 *Iran J. Pharm Res.* **3** 77–82.
[12] J. Parekh, S. Chanda 2008 *Plant Arch, Plant Arch.* **8** 657–662.
[13] R.S. Nanna, M. Banala, A. Pamulaparthi, A. Kurra, S. Kagithoju, Evaluation of Phytochemicals and Fluorescent Analysis of Seed and Leaf Extracts of Cajanus cajan L., (n.d.) 8.
[14] Q. Fardiyah, Suprapto, F. Kurniawan 2018 *AIP Conference Proceedings* **2049** 020017.
[15] A. Sofowora, Phytochemical Screening of Medicinal Plants and Traditional Medicine in Africa, Spectrum Book, Ltd., n.d.
[16] G. Trease, W. Evans, 1989, Pharmacognosy, 11th ed., Bailliere Tindall and Co Publishers.

Acknowledgments

The author would like to thank the Minister of Research, Technology, and Higher Education of the Republic of Indonesia (RISTEKDIKTI) for research grants through graduate education scholarships in the country.

Physicochemical Properties and Antibacterial Activity of Castor Oil and Its Derivatives

M I Fitranda[1], Sutrisno[1*], S Marfu'ah[1]

[1]Chemistry Department, Faculty of Mathematics and Natural Sciences, State University of Malang, Jl. Semarang No. 5 Malang

[*]Corresponding author: sutrisno.kimia@um.ac.id

Abstract. Castor oil is vegetable oil sourced from castor seeds (*Ricinus communis* Linn). The main content of fatty acids in castor oil are ricinoleic acid (92%), oleic acid (3.53%), linoleic acid (2.90%), stearic acid (1.02%), and myristic acid (0.55%). Research on the antibacterial activity of castor oil and ricinoleic fatty acid has been carried out but for the K-soap and fatty acids methyl esters of castor oil have not been conducted. This research aims to produce castor oil derivatives, namely K-soap, free fatty acids (FFAs) and fatty acids methyl esters of (FAMEs) and evaluate their antibacterial activity. The results of the study included (1) K-soap (solid, white, melting point 168–175°C), (2) free fatty acids (liquid, yellow, boiling point 210°C , density 0.98 g.mL^{-1}, refractive index 1.46, viscosity 693.22 cSt, and the value of acids, saponification, and esters are 145.88, 294.52, 148.64), (3) fatty acids methyl esters (liquid, yellow, boiling point 170°C, density 0.98 g.mL^{-1}, refractive index 1.46, viscosity 27.31 cSt, and the value of acids, saponification and esters are 0.33, 392.7, 392.37). K-soap, free fatty acids, and methyl esters from castor oil have antibacterial activity against *Escherichia coli* and *Staphylococcus aureus* bacteria.

1. Introduction

According to renewal, natural resources can be classified into two, namely renewable resources and unrenewable resources. Unrenewable resources, namely oil, natural gas, and coal are very broad but still limited. The source of these chemicals will become drained with sustainable exploitation [1]. Conversely, renewable natural resources are reliable resources for fuels and chemicals in the long run [2]. This natural resource has many forms, namely carbohydrates (starch, cellulose), lignin, and triglycerides [3].

Triglycerides or oils have high potential to be developed, one of which is castor oil. Castor oil (*Ricinus communis*) is one of the vegetable oils with a specific fatty acid content, which contains hydroxyl groups and unsaturated chains [4]. The presence of hydroxyl groups (-OH) attached to the hydrocarbon chain in ricinoleic acid makes castor oil chemically different from other oils, especially its high viscosity and polarity. These properties make it very important for industrial production of lubrication [5], coatings, plastics, and cosmetics [6].

Castor oil has the potential as an antibacterial or antimicrobial agent [7][8]. The antibacterial activity of castor oil can be related to the content of fatty acids in it, ricinoleic fatty acid. Ricinoleic acid inhibits the growth of many viruses, bacteria, yeast, and mold, such as undecylenic acid derivatives [9][10]. The broad spectrum and strong biological activity of free fatty acids (FFAs), is its ability to kill or inhibit bacterial growth. The main target of FFAs action is cell membrane, which interferes with the electron transport chain and oxidative phosphorylation thereby disrupting cellular energy production [11].

FFAs is a compound produced from chemical or enzymatic hydrolysis of oil or fat. The study of the antibacterial activity of a variety of pure fatty acids is carried out by taking into account the variable length of the chain and the presence or absence of a double bond C=C in the fatty acid acyl groups against Gram-positive and Gram-negative bacteria. In the context of its antibacterial activity, there are several different results between oil and free fatty acids. Tamarin oil (*Tamarindus indica*) is not antibacterial active [12], while free fatty acids are antibacterial against *Escherichia coli* and *Staphylococcus aureus* [13].

FFAs are already known as antibacterial and antifungal substances. In general, free fatty acids function as anionic surface agents, and the anionic surfactants are less potent at physiological logical values. The antibacterial activity of free fatty acids has a relationship with its structure. Changing the COOH group to $CONMe_2$ a increased the activity [14]. Besides FFA as an oil derivative, K-soap also has antibacterial activity against *Staphylococcus aureus* bacteria [15].

In the current research, there have been no reported of how antibacterial activity is based on polarity or ability to produce cell membrane liabilities. So, it is necessary to assess the Gram-positive and Gram-negative antibacterial activity of the oil, its free fatty acids, its k-soap, and its fatty acid methyl ester from vegetable oil. The broad spectrum of antibacterial activity, modes of action, and non-specific safety make it attractive as an antibacterial agent for various applications in medicine, agriculture and food preservation, especially where the use of conventional antibiotics is undesirable or prohibited [11]. This research aims to produce castor oil derivatives (K-soap, FFAs, and FAMEs), evaluate their antibacterial activity against Gram-negative bacteria (*Escherichia coli*) and Gram-positive (*Staphylococcus aureus*), and examine the relationship of polarity with their antibacterial activity.

2. Experimental Procedure

2.1. Materials, Equipments and Intrumentations

2.1.1. Material. Castor oil is obtained from an online store in Kedungkandang, Malang. The reagents and solvents used are potassium hydroxide, hydrochloric acid solution, oxalic acid solution, crystalline sodium chloride, methanol, ethanol, hexane, glycerol, acetone and chloroform, each of which has a p.a quality.

2.1.2. Equipment and Instrumentation. A set of glassware, magnetic stirrer, stative, dropper pipette, stirring rod, filter paper, burette, analytical balance, thermometer, hot plate, set of melting point apparatus, autoclave, microtube, ose needle, Laminar Air Flow (LAF), bunsen, calipers, vortex, magnetic stirer, incubator, refractometer Abbe, Capillary-viscosimeter, a set of titration tools, a set of Shimadzu 8400S FT-IR spectrophotometers, and a set of GC-MS Shimadzu QP2010S .

2.2. Synthesis K-soap from Castor Oil

Castor oil (20.00 g) and 130 mL of 3 M potassium hydroxide solution were refluxed at 80 °C for 3 h and stirred with a magnetic stirrer to form two layers and there was no change anymore. This mixture is then allowed to stand and separate. Soft potassium soap is added 50 mL saturated sodium chloride solution and stirred with a magnetic stirrer until all soap is precipitated. Potassium soap deposits are washed with 20 mL distilled water and filtered with filter paper. Washed potassium soap is heated in an oven (60 °C, 2 h). Then, mashed with mortar and pastle. Potassium soap powder is characterized and identified.

2.3. Synthesis Free Fatty Acids (FFAs) from K-soap
Potassium soap (10.00 g) and 10 mL of distilled water are put into a 100 mL beaker glass then stirred with a magnetic stirrer until the K-soap is emulsified. After that, it is added dropwise 1 M hydrochloric acid solution accompanied by stirring with a magnetic stirrer until it is completely lumpy. This lump is filtered with filter paper and washed with water until the washing water is neutral. The residue obtained is left at room temperature until it melts. The liquid obtained was centrifuged at a speed of 3000 rpm for 20 minutes to separate the liquid from the remaining water and obtained residues (free fatty acids) and centrates. Free fatty acids are characterized and identified.

2.4. Synthesis Fatty Acids Methyl Esters (FAMEs) from Castor Oil
Castor oil (20.00 g), 15 mL methanol and 0.2 g of potassium hydroxide are refluxed at 60 °C for 4 h while stirring with a magnetic stirrer. The process is stopped if the mixture does not change and two layers are formed. The top layer is washed with warm water until the used washing water is neutral. The obtained mixture is then added to anhydrous magnesium sulfate to bind water. Then the mixture is filtered with filter paper to obtain water-free methyl ester filtrate. The methyl ester filtrate was characterized and identified.

2.5. Physicochemical Properties of Castor Oil and Its Derivatives
Physicochemical properties of the components are obtained by analyzing the state and color, boiling and melting points, density, refractive index, viscosity, solubility, acid value, saponification value and ester value.

2.5.1. State and Color. State and color are observed visually.

2.5.2. Boiling Point. The sample is put into a test tube that contains boiling stones and perculators. The test tube is heated to boiling and the boiling point value is constant.

2.5.3. Melting Point. The sample is placed on the glass preparation and inserted into the object into the melting point apparatus. The device is turned on and the product is observed when it begins to fuse until the product is completely fused.

2.5.4. Density. The remaining samples attached to the side of the pycnometer are cleaned. Next, the mass of the pycnometer is weighed with the sample. The mass is recorded and the density of the sample is calculated.

2.5.5. Refractive Index. Samples were dropped on the surface of the glass preparation, closed, and read the refractive index on the refractometer Abbe.

2.5.6. Viscosity. The sample is placed in the Capillary-viscosimeter and pumped with filler until it exceeds the boundary mark. The sample flow time is calculated from the upper boundary mark to the lower boundary mark that approaches the capillary tube.

2.5.7. Solubility. 3 mL of solvent is put into a test tube. The sample is added dropwise up to 0.2 mL into the test tube and shake it strongly. Observe the solubility and record the solubility of the sample in the solvent used.

2.5.8. Acid Value. 1.00 g of sample, 2.5 mL of hexane, and 4 drops of the phenolphtalein indicator were put into a 100 mL Erlenmeyer. The mixture is shaken to form a solution (homogeneous mixture). Then the mixture is titrated with 0.1 M KOH solution in water to a constant pink color solution for 20-30 seconds.

2.5.9. Saponification Value. 1.00 g of sample and 25 mL of 0.5 M KOH solution were put into a 250 mL three neck flask. The mixture is heated in a series of reflux apparatus at 60 °C for 1 h while stirring. Then it is cooled to room temperature and put in a 100 mL Erlenmeyer and 5 drops of phenolphtalein indicator are added. Then it is titrated with 0.5 M HCl solution until the pink turns yellow. Titration is also done on blanks.

2.6. Identification of Castor Oil and Its Derivatives

2.6.1. Analysis of Infra-Red Spectrum. Samples are mixed with KBr with a sample-KBr ratio (1: 1), pressed in a vacuum pump to form chips or pellets. Then the pellets were put into FT-IR spectrophotometry and analyzed with a frequency of 4000-400 cm^{-1}. The resulting IR spectrum is then analyzed based on a typical band at a frequency of 4000-400 cm^{-1} which can be used to identify the functional groups contained in the sample.

2.6.2. Analysis of Gas Chromatography-Mass Spectrometer. The sample was injected on a GC-MS Shimadzu QP2010S device. Operational parameters are adjusted to the conditions in GC-MS. Then the chromatogram produced by the recorder and the mass spectra of each compound was observed.

2.7. Antibacterial Activity

2.7.1. Media preparation. 5 grams of instant Nutrient Agar (NA) is dissolved in 250 mL of distilled water. Then this mixture is heated while stirring until the mixture is homogeneous. 10 mL of homogeneous mixture was poured into a petri dish. Subsequently the mixture was sterilized by autoclaving at 121°C for 15 minutes. Then the temperature and pressure of the autoclave are lowered, then allowed to reach ambient conditions. NA media are removed from the autoclave and left until the media solidifies.

2.7.2. Suspension of Escherichia coli and Staphylococcus aureus Bacteria. 10 mL Nutrient Broth (NB) was put into two test tubes. A total of 1 *E. coli* bacterial ose needles were inserted aseptically in the first tube and *S. aureus* bacteria was inserted in the second tube. Next, each mixture in the tube is rotated with vortex until a homogeneous mixture is obtained.

2.7.3. Inhibition Test. 6 mm diameter of the wellbore made on the media. Gram-negative (*Escherichia coli*) and Gram-positive (*Staphylococcus aureus*) bacteria are inoculated into the media. Then, 20 μL of the sample is put into the wellbore. Media that have been planted with bacteria and samples incubated at 37 ° C for 24 h. The diameter of the clear zone produced is measured by the calipers.

3. Result and Discussion

3.1. Physicochemical Properties of Castor Oil and Its Derivatives

Castor oil is liquid and yellow. Soap that is synthesized from oil is solid and has a white color. Changes in state during the derivatization reaction from oil (liquid, yellow) to solid and white in the saponification reaction indicate the formation of potassium soap (K-soap). Likewise, the change in state and color from the result of the acidification reaction of K-soap (solid, white) to liquid and yellow, indicates the formation of free fatty acids (FFAs). While indications of the formation of fatty acid methyl esters (FAMEs) are characterized by viscosity observed visually thinner than castor oil. The performance states and colors of castor oil and their derivatives are shown in **Figure 1**.

Physical properties can be influenced by intermolecular forces and molecular weights of compounds. Based on the results of physicochemical properties can be seen that castor oil has a different boiling point with fatty acids and methyl esters. Fatty acids have the highest boiling point than both because they are able to form hydrogen bonds between molecules so that the density of fatty acids is higher and requires high energy to break bonds between molecules. The difference in boiling points shows that fatty acids and methyl esters have been successfully synthesized from oil.

(a) (b) (c) (d)

Figure 1. Castor oil and its derivative (a) castor oil, (b) K-soap, (c) FFAs, (d) FAMEs

Table 1. Physicochemical Properties Results of Castor Oil and Its Derivatives

Properties	Castor oil	K-soap	FFAs	FAMEs
State	Liquid	Solid	Liquid	Liquid
Color	Yellowish	White	Yellow	Yellowish
Boiling point (°C)	191	-	210	170
Melting point (°C)	-	168-175	-	-
Density (g/cm^3)	0.96	-	0.98	0.92
Refractive Index (25 °C)	1.477	-	1.462	1.461
Viscosity (cSt)	520.52	-	693.22	27.31
Solubility in				
- water	insoluble	emulsified	insoluble	insoluble
- methanol	insoluble	slightly	soluble	soluble
- ethanol	insoluble	slightly	soluble	soluble
- chloroform	soluble	insoluble	soluble	soluble
- n-hexane	insoluble	insoluble	insoluble	soluble
Acid value, AV (mg KOH/g sample)	1.23	-	145.88	0.33
Saponification value, SV (mg KOH/g sample)	406.72	-	294.52	392.70
Ester value (SV – AV) (mg KOH/g sample)	405.49	-	148.64	392.37

Based on the results of physicochemical properties can be seen that castor oil has a different viscosity with fatty acids and methyl esters. Fatty acids have the highest viscosity than both because fatty acids have the highest molecular weight than oils and methyl esters so that the viscosity is greater. The difference in viscosity shows that fatty acids and methyl esters have been successfully synthesized from oil.

Based on the results of physicochemical properties can be seen that castor oil has a different density with fatty acids and methyl esters. Fatty acids have the highest density than both because they are able to form hydrogen bonds between molecules so that the density of fatty acids is higher. The difference in density shows that fatty acids and methyl esters have been successfully synthesized from oil.

The solubility of a compound is influenced by the rules of like dissolves like [16]. Based on the results of the solubility test, showed that the sequence of the nature of polarity from low to high, namely castor oil, methyl esters, fatty acids and potassium soap. The difference in solubility shows that potassium soap, fatty acids, and methyl esters have been successfully synthesized from oil.

Based on the definition of acid numbers, the higher the acid numbers, the higher the amount of free fatty acids in oil or fat. Based on the results of the characterization, the fatty acids synthesized had higher acidity than oil and methyl esters. This shows that fatty acids were successfully synthesized from oil. Based on the results of synthesis, it can be seen that castor oil has the highest saponification rate. This shows that the triglyceride content in castor oil is the highest compared to fatty acids and their methyl esters because in the oil structure they have a structure with moles greater than the fatty acids and methyl esters.

Esters also show the amount of esters in oil or fat. Based on the results of physicochemical properties, it can be seen that castor oil has the highest ester number. This shows that castor oil contains quite high esters. Differences in acid numbers, saponification rates and esters indicate that fatty acids and methyl esters have been successfully synthesized from oil.

3.2. Identification of Castor Oil and Its Derivatives

3.2.1 Infra Red Spectrophotometer Analysis

Other supporting data on the success of synthesis of castor oil derivatives is the result of IR spectrophotometer identification, with IR spectra as shown in **Figure 2- Figure 5**.

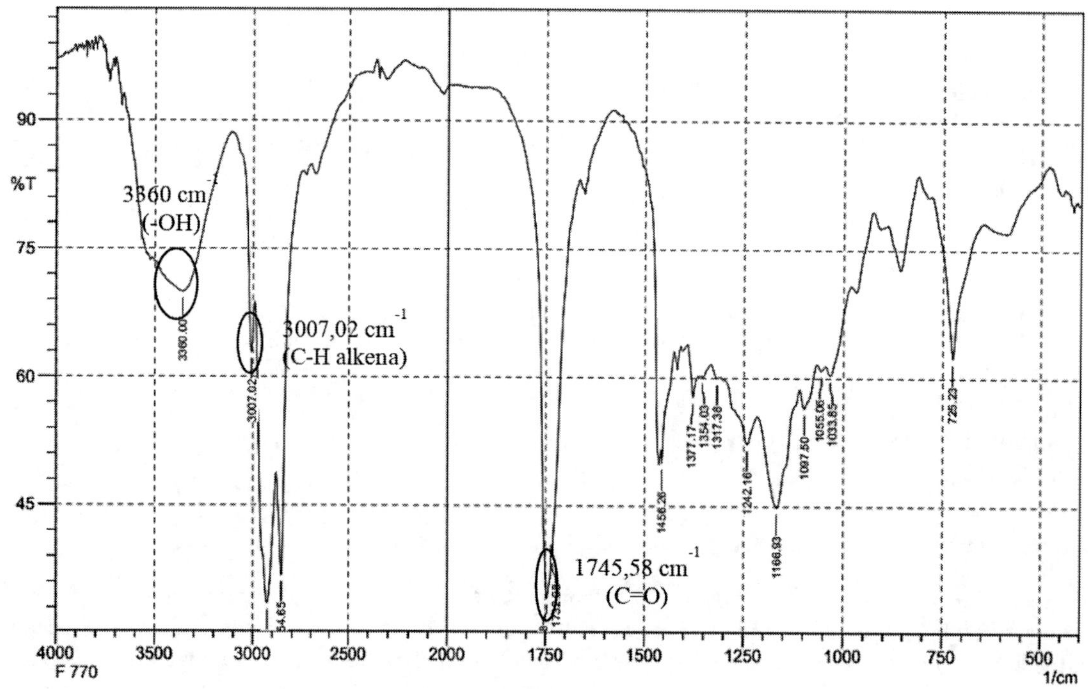

Figure 2. IR spectrum of castor oil

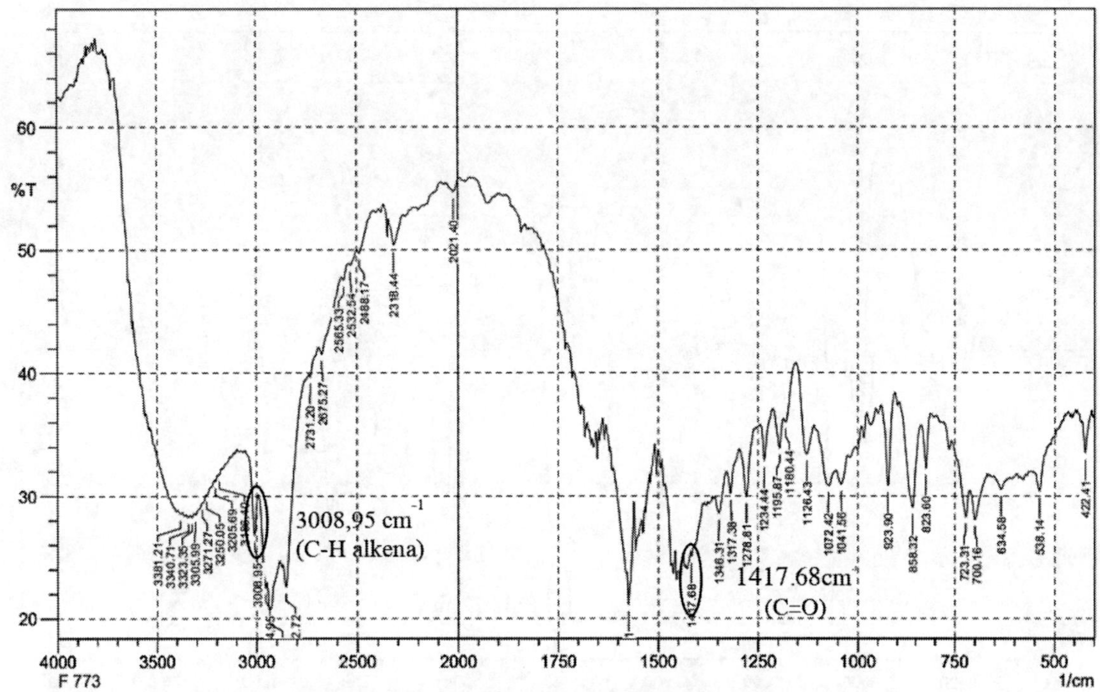

Figure 3. IR spectrum of K-soap from castor oil

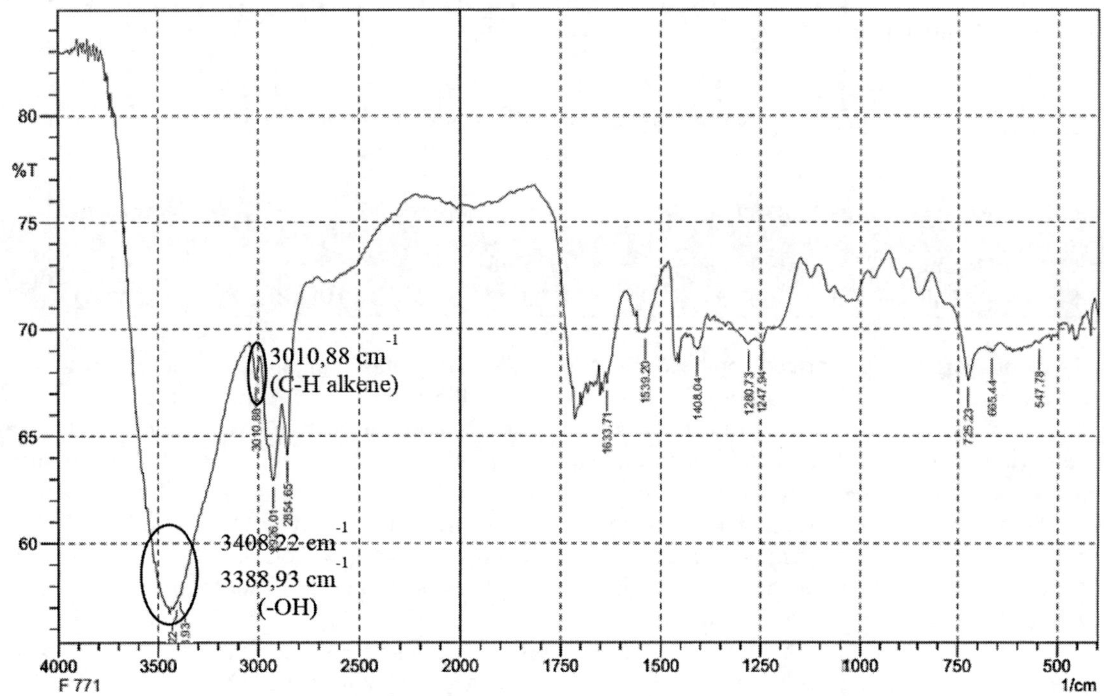

Figure 4. IR spectrum of FFAs from castor oil

IOP Conf. Series: Materials Science and Engineering **833** (2020) 012009 doi:10.1088/1757-899X/833/1/012009

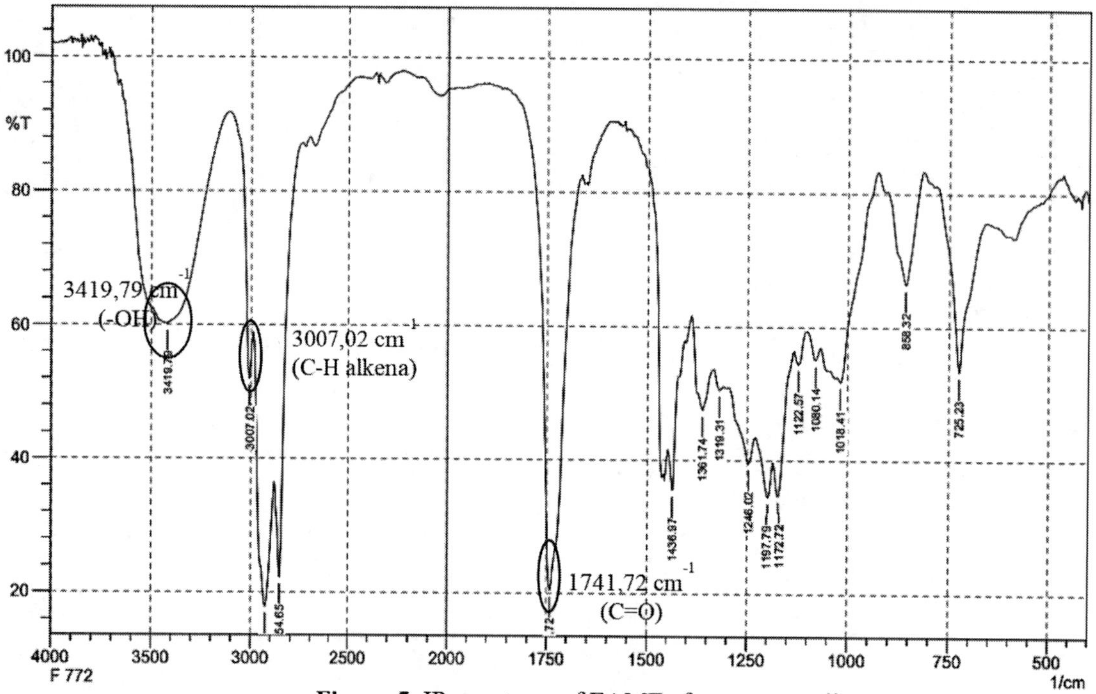

Figure 5. IR spectrum of FAMEs from castor oil

Based on the interpretation of IR spectra in **Figure 2-Figure 5**, typical bands that distinguish the four substances summarized in **Table 2**.

Table 2. IR Spectrum Interpretation of Castor Oil and Its Derivatives.

Vibration	Castor oil	K-soap	Fatty acids	Methyl esters
O-H stretching	3360.00	-	3408.22 3388.93	3419.79
C=O stretching	1745.58	1417.68	-	1741.72
Stretch C-H alkene	3007.02	3008.95	3010.88	3007.02

IR analysis was performed to determine the functional groups contained in castor oil and its derivatives. Based on the results of the interpretation of the IR spectra (**Table 2**), there are several typical absorption bands. Castor oil has a wave number 3360.00 cm^{-1} with a weak intensity indicating the existence of O-H bond stretch vibrations and wave number 1745.58 cm^{-1} with strong and sharp intensities indicating the existence of stretching vibrations of the C=O bond as esters.

The synthesized potassium soap has a wave number of 1417.68 cm^{-1} with a strong and sharp intensity indicating the stretching vibration of the C=O bond as a carboxylic salt [17]. The existence of stretching vibration C=O bonds as carboxylic salts indicates potassium soap successfully synthesized from castor oil.

Synthesized fatty acids have wave numbers 3408.22 cm^{-1} and 3388.93 cm^{-1} appear with weak intensity and width indicate the existence of O-H bonding stretch vibrations. The synthesized methyl ester has a wave number of 1741.72 cm^{-1} with a strong and sharp intensity indicating the existence of a stretching vibration C=O as an ester and a wave number 3419.79 cm^{-1} with a weak intensity indicating the existence of a stretching vibration O-H bond as an acyl group. Based on the results of the interpretation of the IR spectrum, the difference in the intensity of the O-H bond stretching vibrations between oil, fatty acids and methyl esters shows that fatty acids and methyl esters have been successfully synthesized from castor oil.

3.2.2 Gas Chromatography-Mass Spectrometer Analysis

GC-MS analysis was performed to determine the content of fatty acid methyl esters synthesized. Based on GC- MS results (**Table 3**), the synthesized methyl esters are composed of 5 fatty acids which are indicated by the appearance of 5 peaks on the chromatogram. The abundance percentage of fatty acid content based on its fatty acid methyl ester was obtained from % Area chromatogram data. The chromatogram of FAMEs are shown in **Figure 6**.

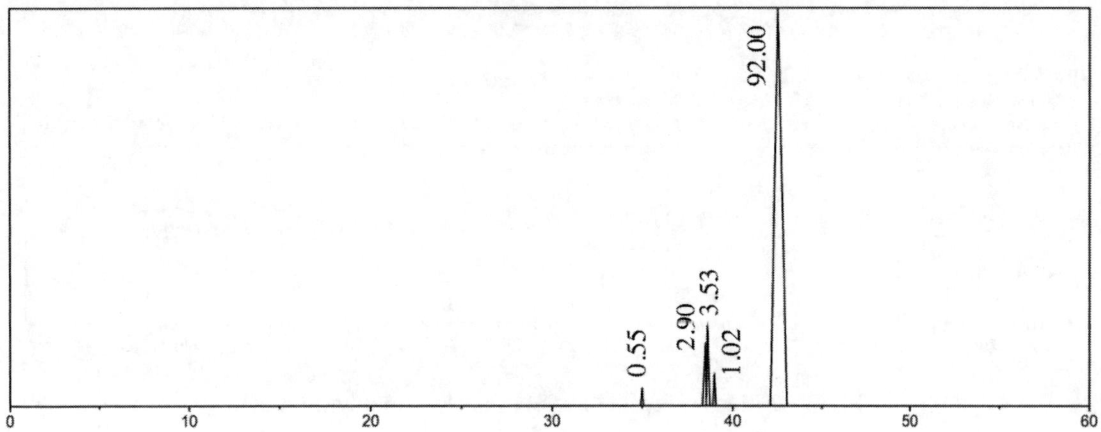

Figure 6. Chromatogram of FAMEs from castor oil

Type of fatty acids based on its fatty acid methyl ester was obtained from mass-spectrometry data for each peak on a chromatogram. Analysis of mass spectrometry by EI-MS (electron impact mass spectrometry) for each peak of the five peaks of GC results obtained by mass spectra as shown in **Figure 7** to **Figure 11.** The appropriate content of fatty acids from their fatty acid methyl esters as listed in **Table 3**, obtained from interpretation of each mass spectrum from nine peaks, and subsequently refered to WILEY229.LIB.

Table 3. Fatty Acid Content as Methyl Esters from GC-MS Analysis

Peak	Retention time	Relative percentage(%Area)	Fatty acid as its FAMEs	Fatty acid symbol
1	34.997	0.55	Myristic acid	C 14:0
2	38.516	2.90	Linoleic acid	C 18:2 9*c*12*c*
3	38.622	3.53	Oleic acid	C 18:1 9*c*
4	39.006	1.02	Stearic acid	C 18:0
5	42.521	92.00	Ricinoleic acid	C 18:1 9*c* 12-OH

IOP Conf. Series: Materials Science and Engineering **833** (2020) 012009 doi:10.1088/1757-899X/833/1/012009

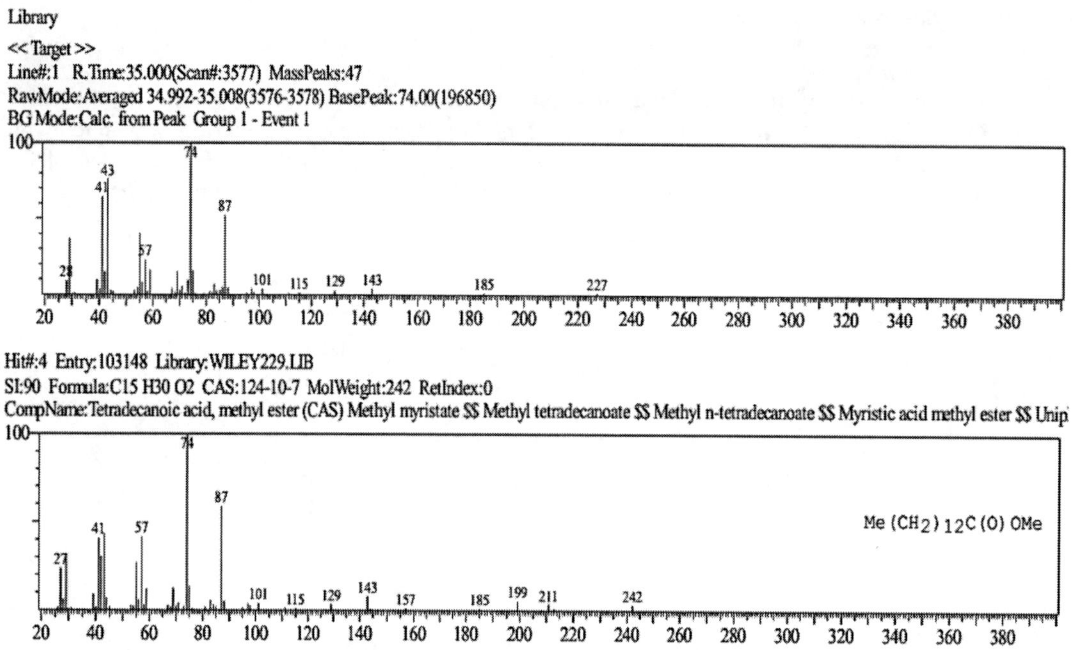

Figure 7 Mass spectra (EI-MS) of peak 1 (methyl myristate)

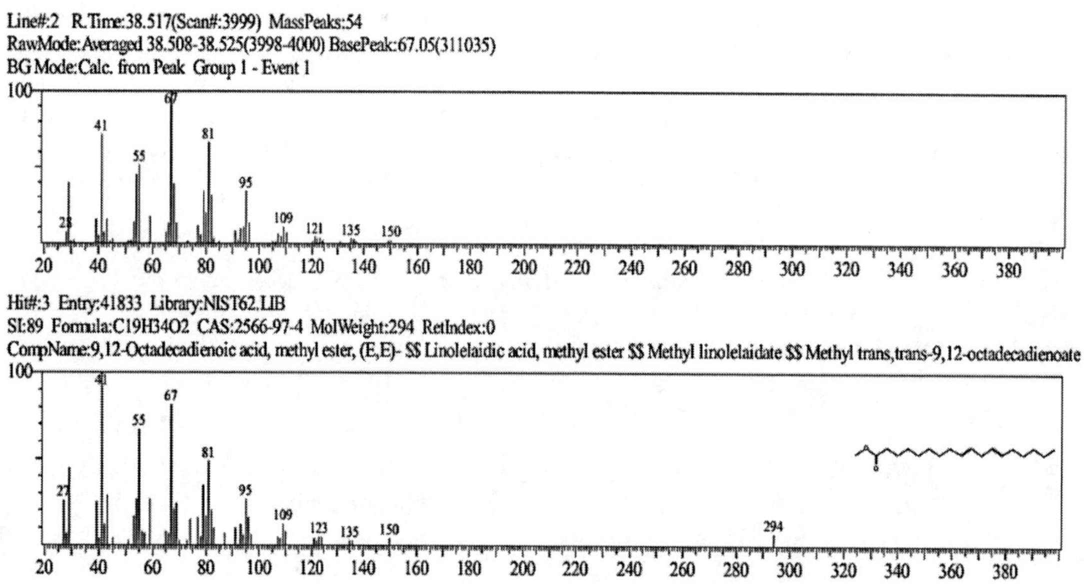

Figure 8 Mass spectra (EI-MS) of peak 2 (methyl linoleate)

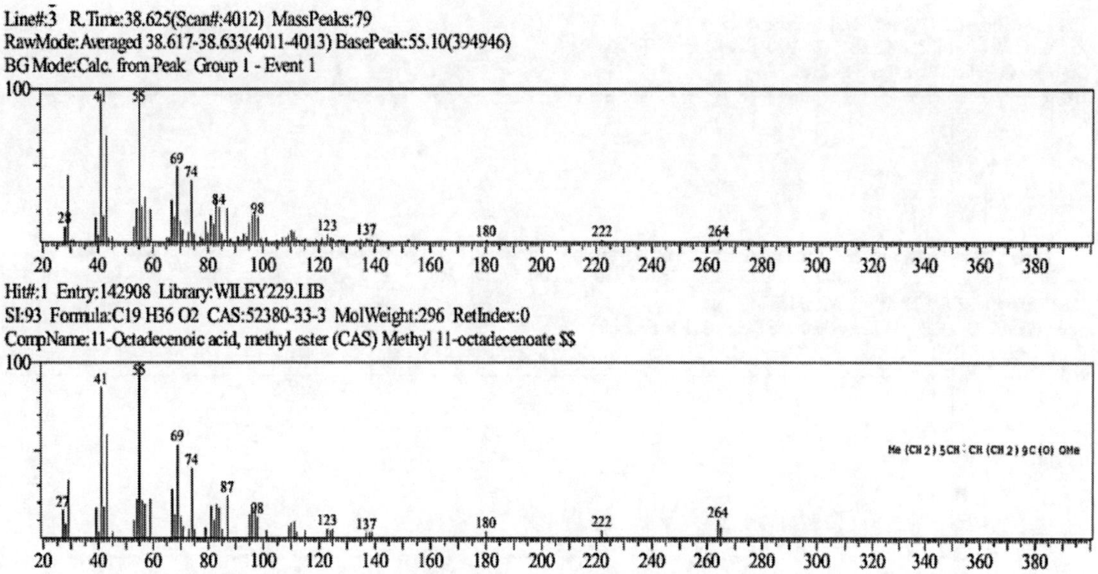

Figure 9 Mass spectra (EI MS) of peak 2 (methyl oleate)

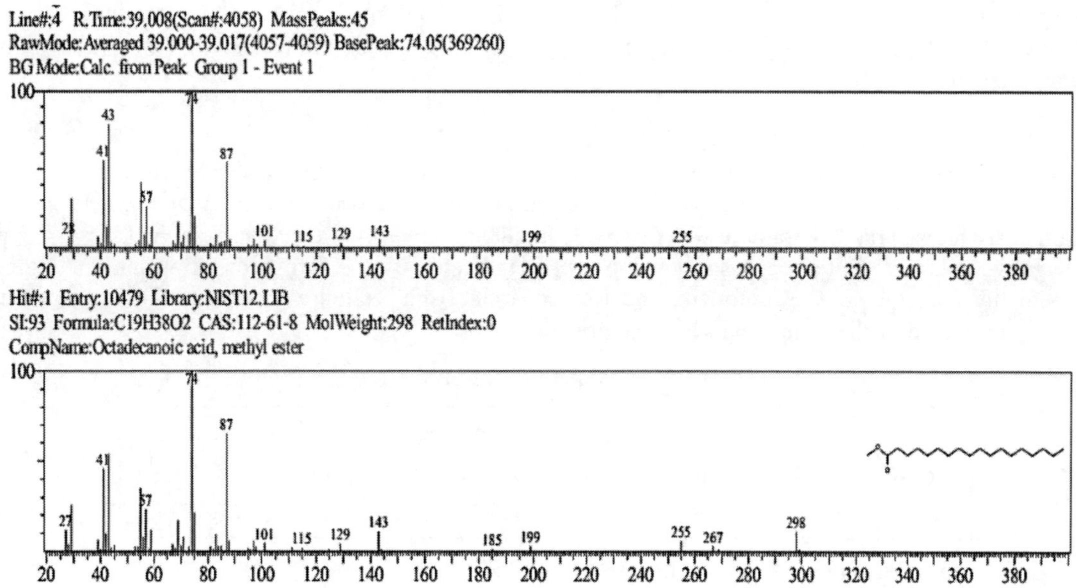

Figure 10 Mass spectra (EI-MS) of peak 2 (methyl stearate)

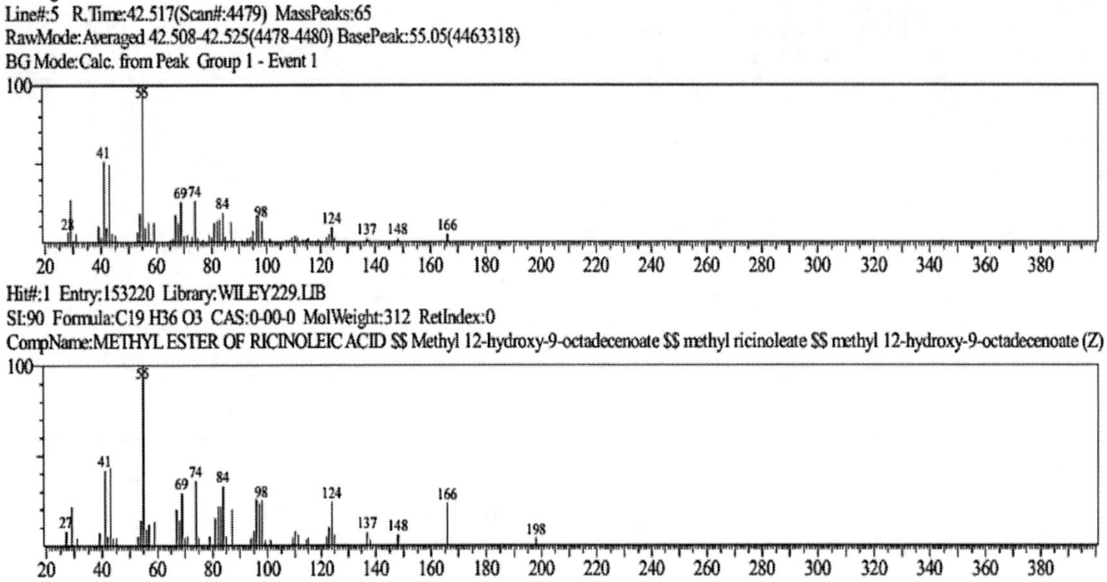

Figure 11 Mass spectra (EI-MS) of peak 2 (methyl ricioleate)

Based on the results of EI-MS, it can be seen that the synthesized fatty acids methyl esters are composed of 0.55% myristic acid, 2.90% linoleic acid, 3.53% oleic acid, 1.02% stearic acid, and 92.00% ricinoleic acid. The retention time, type of fatty acids, %area and fatty acid symbol summarized in **Table 3.**

3.3. Antibacterial Activity of Castor Oil Derivatives

Potassium soap, free fatty acids, and methyl esters which were synthesized were tested for their antibacterial activity against *Eschericia coli* as gram negative bacteria and *Staphylococcus aureus* as gram positive bacteria. The sample was dissolved in ethanol 60% and this ethanol solvent was used as a negative control. Stunted bacterial development is indicated by the formation of an inhibited zone around the wellbore. If the inhibited zone formed around the wellbore increases, the antibacterial activity produced by the sample will be even greater.

Table 4. Diameter of Inhibition Zones

Type of Bacteria	Inhibition Zone Diameter (mm)						
	Negative Control	K-soap		Fatty Acids		Methyl Esters	
		1 %	2 %	1 %	2 %	1 %	2 %
Escherichia coli	-	18.80	18.90	15.75	17.45	-	8.15
Staphylococcus aureus	-	14.65	16.90	13.25	16.35	-	8.75

Based on the results of the inhibited zone diameter (**Table 4**), it can be seen that antibacterial activity of castor oil derivatives against *Escherichia coli* bacteria is higher than that of *Staphylococcus aureus.* In addition it can be seen that potassium soap has the highest antibacterial activity compared to its fatty acids and methyl esters. This is due to differences in polarity of the three castor oil derivatives. The higher the polarity of a substance, the higher its antibacterial properties. Potassium soap which has the highest polarity has the highest antibacterial activity followed by fatty acids and the weakest methyl esters.

4. Conclusion

Castor oil can be transformed to derivatives, i.e K-soap (solid, melting point 168–175°C), FFAs (liquid, boiling point 210°C), and FAMEs (liquid, boiling point 170°C). The antibacterial activity of castor oil derivatives against *Escherichia coli* bacteria is higher than that of *Staphylococcus aureus*. The antibacterial activity of K-soap has a higher activity than free fatty acids and higher free fatty acids than fatty acid methyl esters.

The result of this research implies that potassium salt from castor oil can be used is the control of wound, skin infection, throat infection in the skin caused by *Staphylococcus aureus,* while the fatty acids from castor oil have the potential as an antibacterial agent.

References

[1] Xu Y, Hanna M A and Isom L 2008 *Open Agric. J.* **2** 54
[2] Axelsson L, Franzén M, Ostwald M, Berndes G, Lakshmi G and Ravindranath N H 2012 *Biofuels, Bioprod. Biorefining* **6** 246
[3] Donate P M 2014 *Chem. Biol. Technol. Agric.* **1** 4
[4] Joyce Omohu O 2017 *Eur. J. Biophys.* **5** 62
[5] Barbosa D D C, Serra T M, Meneghetti S M P and Meneghetti M R 2010 *Fuel* **89** 3791
[6] Al-Mamun M A, Akter Z, Uddin M J, Ferdaus K M K B, Hoque K M F, Ferdousi Z and Reza M A 2016 *BMC Complement. Altern. Med.* **16** 211
[7] Momoh O, Oladunyamoye M K and Adebolu T T 2012 *J. Pharmacogn. Phyther.* **4** 40
[8] Poonam K and Pratap S K 2012 *Int. Res. J. Pharm.* **3** 209
[9] Ayuba L, Agboire S, Gana A K, Ishaq M, Aliyu U, Affiniki G and Manjang J I 2017 *Adv. Food Sci. Eng.* **1** 95
[10] Kuppala R, Govindarajan M, Tambat R, Patel N, Nandanwar H, Bhutani K K and Kartha K P R 2016 *RSC Adv.* **6** 3700
[11] Desbois A P and Smith V J 2010 *Appl. Microbiol. Biotechnol.* **85** 1629
[12] Sutrisno, Retnosari R, Marfu'ah S and Fajaroh F 2019 *Key Eng. Mater.* **811** 40
[13] Sutrisno, Retnosari R and Marfu'ah S 2019 *IOP Conf. Ser. Earth Environ. Sci.* **299** 012004
[14] Kabara J J, Swieczkowski D M, Conley A J and Truant J P 1972 *Antimicrob. Agents Chemother.* **2** 23
[15] Kawahara T, Takita M, Masunaga A, Morita H, Tsukatani T, Nakazawa K, Go D and Akita S 2019 *Int. J. Mol. Sci.* **20** 312
[16] Zhao Y, Liu W, Zhu J and Pei X 2018 *Chem. Eng. Sci.* **184** 149
[17] Gocen T, Haman Bayarı S and Haluk Guven M 2017 *J. Mol. Struct.* **1150** 68

Experimental and Theoretical Study of Pinostrobin as Copper Corrosion Inhibitor at 1 M H_2SO_4 Medium

Saprizal Hadisaputra[1,*], Agus Abhi Purwoko[1], Aliefman Hakim[1], Rosita Wati[2], Dina Asnawati[2], Yuniar Ponco Prananto[3]

[1]Chemistry Education Division, Faculty of Treacher Training and Science Education, University of Mataram, Jalan Majapahit No 62, Mataram, 83125, Indonesia.
[2]Department of Chemistry, Faculty of Mathematics and Natural Sciences, University of Mataram, Jalan Majapahit No 62, Mataram, 83125, Indonesia.
[3]Department of Chemistry, Faculty of Mathematics and Natural Sciences, Brawijaya University, Malang 65145, East Java, Indonesia.

[*]Corresponding author: rizal@unram.ac.id

Abstract. The effect of variations in concentration and temperature on the efficiency of pinostrobin corrosion inhibition of copper in 1M H_2SO_4 was studied using an experimental and theoretical approach. Pinostrobin was isolated from the *Boesenbergia rotunda L* rhizome and was tested for corrosion inhibition of copper in acidic medium. Variations in the concentration of 300 ppm, 400 ppm, and 500 ppm and temperature (308 K, 318 K, and 328 K) were carried out in the weight loss method of corrosion inhibition test. Pinostrobin crystals were isolated with a yield of 57.65 %, a melting point of 98.5 °C and with a purity of 100 %. The optimum corrosion inhibition efficiency of pinostrobin is 65.71 % at a concentration of 500 ppm and a temperature of 328 K. The activation energy value of Ea < 80 $kJ.mol^{-1}$ indicates the adsorption of pinostrobin on copper surfaces is physical and spontaneously $\Delta G^{o}ads < -20$ $kJ.mol^{-1}$. The effect of substituent on the efficiency of corrosion inhibition from pinostrobin was studied using density functional theory. The addition of NH_2 increased the efficiency of corrosion inhibition to 73.07 %, whereas the addition of NO_2 substituents decreased the efficiency of inhibition to 60.97 %. Experimental and theoretical studies have a good correlation in explaining the efficiency of corrosion inhibition from pinostrobin.

1. Introduction

Corrosion is an electrochemical process that occurs between metals and corrosive environments such as sulfuric acid that will reduce the quality of the metal. The corrosion process results in economic and environmental losses [1]. The rate of corrosion can be inhibited by the addition of inhibitors. Of the various types of inhibitors, organic inhibitors are preferred because they are efficient, environmentally friendly and non-toxic. Organic corrosion inhibitors usually have heteroatom N, O, P, S and bonds which allow the inhibitor to coat the metal surface to inhibit corrosion [2-9].

The use of natural ingredients as corrosion inhibitors has been widely used [10-15]. One of them is pinostrobin which is isolated from the Boesenbergia pandurata Roxb rhizome [16]. Pinostrobin qualifies as a good corrosion inhibitor and has not been widely tested as a corrosion inhibitor [17]. Therefore, this study focuses on the application of pinostrobin as a corrosion inhibitor to copper metal in 1 M H_2SO_4 medium by weight loss method. In addition to experimental tests, a theoretical study was also carried out to examine the effect of substituents on the efficiency of corrosion inhibition of pinostrobin compounds. Theoretical studies have proven capable of bridging aspects that are difficult to explain experimentally [18-21].

2. Methodology

2.1. Experimental Section

2.1.1. Pinostrobin Isolation. Isolation of pinostrobin was carried based on modification of methods from the literatures [17]. The Boesenbergia rotunda L rhizome powder was socleted with n-hexane solvent. The extraction results are then evaporated using a rotary evaporator. The obtained viscous extract was allowed to form crystals and purified by recrystallization with methanol. Pinostronin crystals were further identified by the melting point test and gas chromatography-mass spectroscopy GCMS.

2.1.2. Weight Loss Test. The prepared copper plates were prepared, weighed and then immersed in 1 M H_2SO_4 medium and an inhibitor solution with concentrations of 0, 300, 400 and 500 ppm and at temperatures of 308 K, 318 K and 328 K. Soaking of the copper plate was then washed with distilled water and acetone then dried and weighed. Inhibition efficiency values are calculated based on the equation

$$\%IE = \frac{C_{Ro} - C_{Ri}}{C_{Ro}} x100\% \quad (1)$$

where %IE is the efficiency of inhibition, C_{Ro} is the rate of corrosion without inhibitors and C_{Ri} the rate of corrosion with inhibitors.

2.2. Theoretical Section

All geometry optimizations of pinostrobin compounds are carried out using Gaussian 09 [23]. Geometry optimization is not followed by re-optimization in the solution phase because it has little effect on the structure and energy, so that only a single point of the solution phase is carried out on the gas phase geometry [24-28]. Calculation of quantum chemical parameters based on Koopman's theorem which explains the relationship between ionization potential (I), electron affinity (A) and orbital energy (E_{HOMO} and E_{LUMO})[29-30]. Koopman's equation is as follows:

$$I = - E_{HOMO} \quad (2)$$

$$A = - E_{LUMO} \quad (3)$$

Electronegativity is obtained based on equation 4 [31]:

$$\chi = \frac{E_{HOMO} + E_{LUMO}}{2} \quad (4)$$

Furthermore, the inhibition efficiency is calculated based on the equation [32]:

$$I_{add}\% = \frac{I_{PI} - I_{\chi-PI}}{I_{PI}} X100\% \quad (5)$$

$$IE_{add}\% = I_{add}\% - IE_{PI}\% \quad (6)$$

$$I_{theor}\% = IE_{PI}\% + \%IE_{add}\% \quad (7)$$

IOP Conf. Series: Materials Science and Engineering **833** (2020) 012010 doi:10.1088/1757-899X/833/1/012010

where I_{PI} is the percentage of ionization potential, Ieadd% is the percentage of inhibition efficiency, IE_{PI}% is the percentage of inhibition efficiency of experimental results and IE_{theor}% is the theoretical efficiency of inhibition. The structure of pinostrobin is depicted in Figure 1.

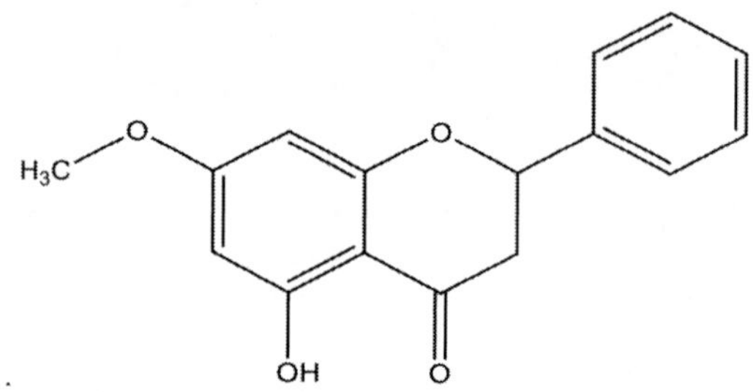

Figure 1. Structure of Pinostrobin (PN).

3. Results and Discussion

3.1. Experimental Study

Pinostrobin was successfully isolated from boesenbergia rotunda L rhizome with a yield of 57.65 %. The melting point test shows that the isolated pinostrobin melting point is 98.5 °C. Pinostrobin identification was continued using GC-MS and the pinostrobin chromatogram is depicted in Figure 2. It shows a peak with a percentage of 100 % which means that pinostrobin has high purity. Spectroscopic results also obtained m/e = 270 according to the relative molecular mass of pinostrobin [33].

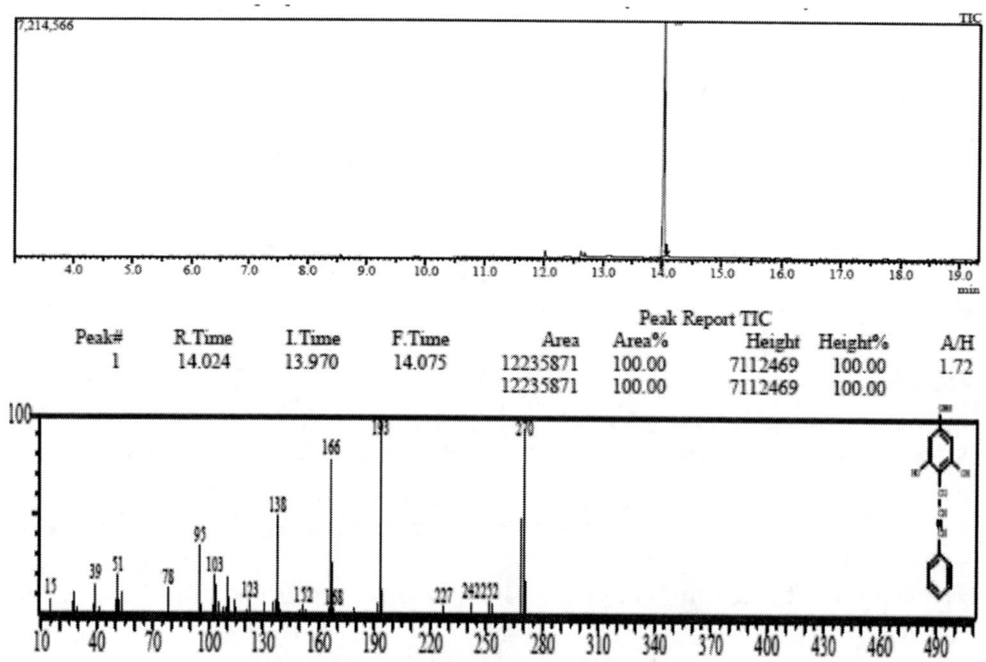

Figure 2. Pinostrobin chromatogram.

Weight loss test begins with the process of immersion of copper plates in a 1 M H_2SO_4 corrosive medium with pinostrobin concentrations of 0, 300, 400 and 500 ppm. Figure 3 shows the effectiveness of pinostrobin inhibition has increased at concentrations of 300 and 400 ppm. The amount of

IOP Conf. Series: Materials Science and Engineering **833** (2020) 012010 doi:10.1088/1757-899X/833/1/012010

pinostrobin that coats the copper surface increases so that the attack of corrosive solutions is blocked. The optimum inhibition efficiency of pinostrobin in copper corrosion is 65.35%.

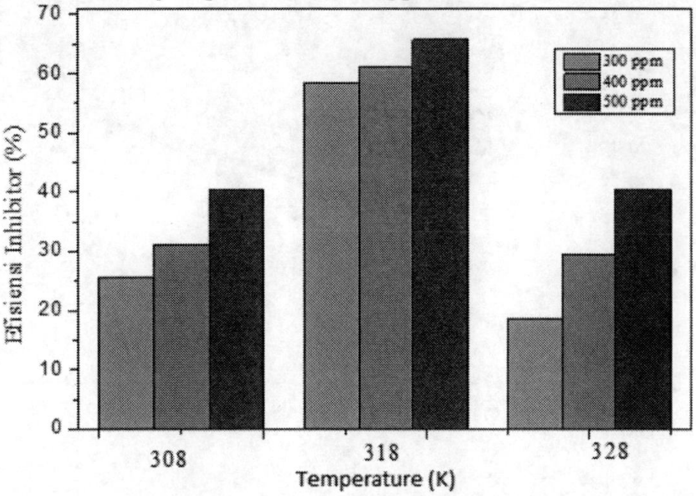

Figure 3. The concentration of inhibitors versus the efficiency of pinostrobin corrosion inhibition.

Activation energy E_a, change in enthalpy values ΔH and entropy ΔS and adsorption free energy ΔG°ads are used to determine the overall energy changes in physical and chemical processes between the initial and final states in the corrosion inhibition process. Positive ΔH and ΔS values indicate the reaction is endothermic but takes place spontaneously. This is reinforced by the negative value of ΔG°ads indicating the reaction takes place spontaneously shown in Table 1.

Table 1. Activation energy Ea, enthalpy of change ΔH and entropy ΔS values.

Concentration (ppm)	Ea ($kJ.mol^{-1}$)	ΔH ($kJ.mol^{-1}$)	ΔS ($J.mol^{-1}$)
0	-16.35	0.42	50.43
300	-4.89	7.78	71.21
400	-12.45	4.74	60.79
500	12.57	10.39	76.79

Tabel 2. Value of free energy adsorption.

Temperature (K)	K_{ads}	ΔG^0_{ads} ($kJ.mol^{-1}$)
308	0.3607	-7.70
318	1.2555	-11.20
328	0.2205	-6.80

Table 2 shows the process of pinostrobin adsorption on copper surfaces takes place spontaneously because the value of adsorption free energy increases with increasing temperature. At 318 K temperature, there was a decrease in adsorption free energy of -11.2 kJ / mol. A decrease in the value of ΔG°ads at 318 K indicates an equilibrium has occurred. A negative ΔG°ads value indicates that the adsorption process is a spontaneous process. The value of ΔG°ads <-20 kJ / mol indicates that adsorption occurs physically, while the value of ΔG°ads> -20 kJ / mol indicates the occurrence of chemisorption [34]. Table 3 shows that ΔG°ads is negative, which means that the adsorption of pinostrobin on the copper surface is physically adsorption.

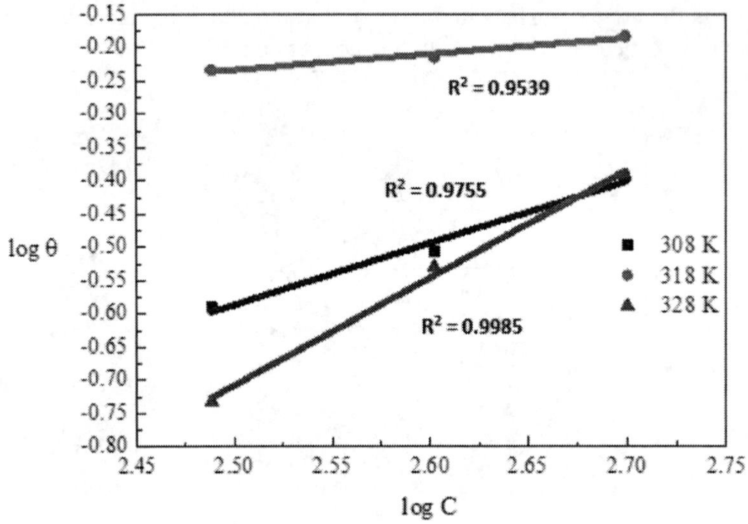

Figure 4. Freundlich isotherm adsorption

The mechanism of interaction between pinostrobin and the copper surface can be determined based on isotherm adsorption. Figure 4 shows the log θ vs. log C plot at different temperatures showing the adsorption of pinostrobin on the copper surface following the Freundlich adsorption. Adsorption isotherm Freundlich explained that on the surface of the metal a multilayer layer is formed from the inhibitor molecule and is heterogeneous, that is, each active group on the metal surface has a different ability to adsorb. This also corresponds to the negative value of ΔG^0ads which indicates the adsorption occurs physically.

3.2. Theoretical Study

Table 3 shows the validation of the theoretical calculation method by comparing the bonding distance and the binding angle based on theoretical and experimental. The average difference between the bonding distance and the binding angle of pinostrobin between experiment and theoretical is relatively small 0.021 Å and 1.753°, respectively. This shows a match between theoretical results and experiments. Based on these results, the density functional theory method at B3LYP/6-311++G(d, p) theory level meets the requirements for use in the system under study.

Table 3. Comparison of bond lengths (Å) and angles (°) between experiment [33] and theoretical

Bonds	Exp* (Å)	Theory (Å)	Angle	Exp* (°)	Theory (°)
O1-C2	1.441	1.444	C2-O1-C9	119	116.5
O2-C4	1.232	1.242	O1-C2-C3	112	110.0
C3-C4	1.530	1.518	C3-C2-C1'	114.6	113.0
C5-C6	1.405	1.389	O2-C4-C3	121	120.8
C6-C7	1.384	1.401	C3-C4-C10	118.1	115.7
C8-C9	1.405	1.390	O3-C5-C10	119.8	120.5
C1'-C2	1.353	1.511	C5-C6-C7	115.5	119.5
C2'-C3'	1.400	1.394	O4-C7-C8	125.4	123.3
C5'-C6'	1.390	1.395	C7-C8-C9	119.1	118.5
C3'-C4'	1.400	1.396	O1-C9-C10	121.2	121.6
C1'-C6'	1.406	1.399	C4-C10-C5	124.3	121.1
C9-C10	1.384	1.415	C5-C10-C9	116.8	118.1
C7-C8	1.401	1.405	C2-C1'-C6'	121.5	119.7

Table 4. Quantum chemical parameters the B3LYP method on the 6-311++ G(d, p) basis set.

Parameters	Compounds		
	PN	$PN\text{-}NH_2$	$PN\text{-}NO_2$
E_{HOMO} (eV)	-6.3856	-5.6705	-6.8461
E_{LUMO} (eV)	-1.4756	-1.5943	-2.1891
E*gab* (eV)	-4.9100	-4.7620	-4.6570
I (eV)	6.3856	5.6705	6.8461
χ (eV)	3.9306	3.6324	4.5176
ΔN(Ev)	0.1118	0.2079	-0.0080
IE_{theor} (%)	65.71	73.07	60.09

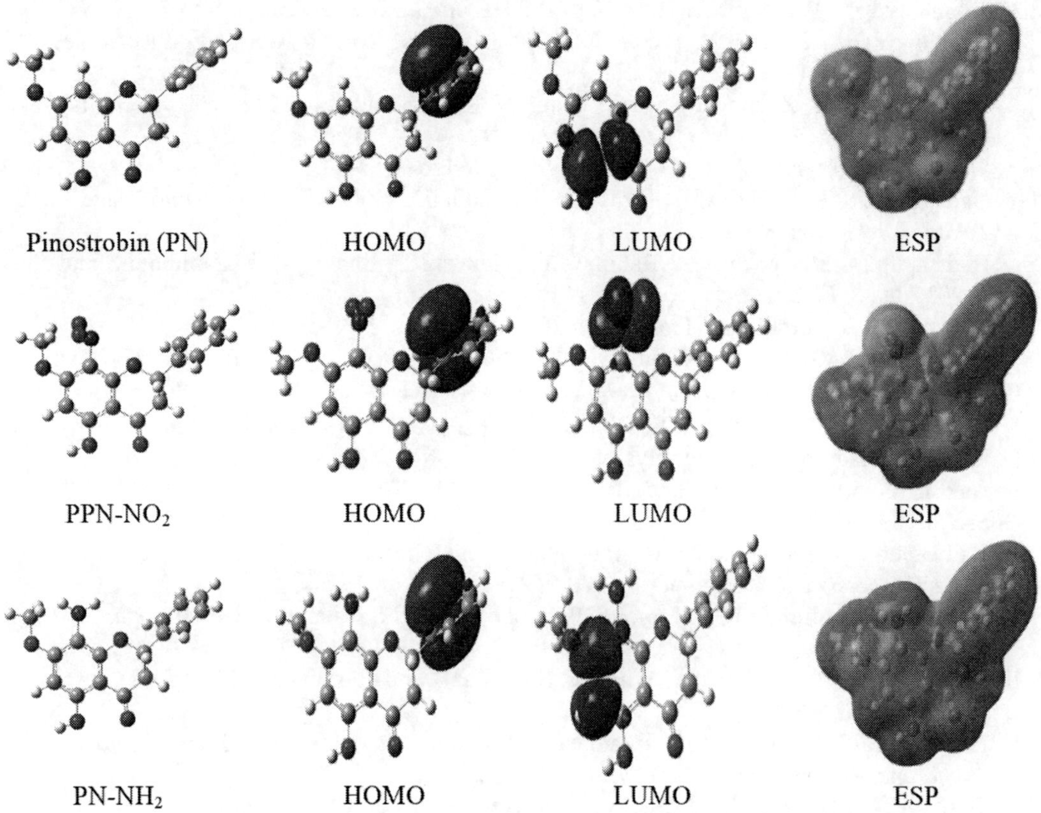

Figure 5. Visualization of pinostrobin HOMO, LUMO, and ESP and their derivatives.

4. Conclusion

The efficiency of pinostrobin corrosion inhibition in acid medium is influenced by concentration and temperature. Weight loss test on isolated pinostrobin produces a maximum inhibition efficiency of 65.71%. The result is less efficient considering the pinostrobin concentration applied is also high at 400 ppm. Pinostrobin adsorption on copper surfaces follows Freundlich isotherm adsorption and physical adsorption. Efforts to improve the efficiency of pinostrobin corrosion inhibition can be done by adding an electron donor group to pinostrobin. The addition of NH_2 for instant increase the efficiency of corrosion inhibition by 8 %. In contrast, the addition of electron withdrawal group NO_2 causes a decrease in efficiency of up to 5%.

Acknowledgments
This research was financially supported by Hibah Penelitian Dasar RISTEKDIKTI Indonesia 2019, and their support is gratefully acknowledged.

References

[1] Revie R W 2008 *Corrosion and corrosion control: an introduction to corrosion science and engineering* (John Wiley & Sons).
[2] Qiang Y, Zhang Z, Guo L, Xu S, Feng L, Obot I B and Chen S 2017 *J. Clean. Prod.* **152** 17-25.
[3] Mobin M and Rizvi M 2017 *Carbohydr. Polym.* **160** 172-183.
[4] Douadi T, Hamani H, Daoud D, Al-Noaimi M and Chafaa S 2017 *J. Taiwan Inst. Chem. Eng.* **71** 388-404.
[5] Bhawsar J, Jain P, Valladares-Cisneros M G, Cuevas-Arteaga C and Rani M 2018 *Int. J. Electrochem. Sci,* **13** 3200-3209.
[6] Albrakaty R H, Wazzan N A and Obot I B 2018 *Int. J. Electrochem. Sci,* **13** 3535-3554.
[7] Mendonça G L, Costa S N, Freire V N, Casciano P N, Correia A N and de Lima-Neto P 2017 *Corros. Sci.* **115** 41-55.
[8] Guo L, Obot I B, Zheng X, Shen X, Qiang Y, Kaya S and Kaya C 2017 *Applied Surf. Sci,* **406** 301-306.
[9] Shetty S K and Shetty A N 2017 *J. Mol. Liq.* **225** 426-438.
[10] Hadisaputra S, Purwoko AA, Ilhamsyah I, Hamdiani S, Suhendra D, Nuryono N and Bundjali B 2018 *Int. J.* Corros. *Scale* Inhib. **7** 633-647
[11] Hadisaputra S, Purwoko A A, Rahmawati, Asnawati D, Ilhamsyah I, Hamdiani S and Nuryono N 2019 *Int. J. Electrochem. Sci.,* **14** 11110 – 11121
[12] Marzorati S, Verotta L and Trasatti S P 2019 *Molecules, 24* 48.
[13] Noor E A, Al-Moubaraki A H and Al-Ghamdi A A 2019 Arab J Sci Eng., **44** 237-250.
[14] Raghavendra N 2019 *J. Bio-and Tribo-Corrosion,* **5** 54.
[15] Bouraoui M M, Chettouh S, Chouchane T and Khellaf N 2019 *J Bio-and Tribo-Corrosion,* **5** 28.
[16] Hakim A, Kadarohman A and Syah Y M 2015 *J. Chem. Edu,* **93** 193-196.
[17] Hakim A, Andayani Y and Rahayuan B D 2018 *Journal of Physics: Conference Series* **1095** 012039
[18] Hadisaputra S, Purwoko A A, Hamdiani S and Nuryono N 2019 *IOP Conference Series: Materials Science and Engineering* **509** 012129
[19] Hamdiani S, Rohimah I H, Nuryono, Purwoko AA and Savalas L R T 2019 *Asian. J. Chem,* **31** 555-558
[20] Hadisaputra S, Purwoko A A, Wajdi F, Sumarlan I and Hamdiani S 2019 *Int. J. Corros. Scale Inhib.,* 8, 3, 673–688
[21] Hadisaputra S, Purwoko A A, Rahmawati, Hamdiani S, Prananto Y P and Nuryono 2019 *IOP Conference Series: Materials Science and Engineering* **546** 032011
[22] Frisch M J, Trucks GW, Schlegel H B, Scuseria G, Robb M A, Kudin KN, Strain MC, Farkas O, Tomasi J, Barone V, Cossi M, Cammi R, Mennucci B, Pomelli C, Adamo C, Clifford S, Ochterski J, Petersson G A, Ayala P Y, Cui Q, Morokuma K, Malick D K, Rabuck A D, Raghavachari K, Foresman J B, Cioslowski J, Ortiz J V, Stefanov B B, Liu G, Liashenko A, Piskorz P, Komaromi I R, Gomperts R, Martin L, Fox D J, Keith T, Al-Laham M A, Peng C Y, Nanayakkara A, Gonzalez C, Challacombe M P, Gill MW, Johnson B, Chen W, Wong M W, Andres J L, Gonzalez C, Head-Gordon M, Replogle E S, Pople JA 2004 Gaussian 03; (Gaussian, Inc. Wallingford, CT, 6492).
[23] Hadisaputra S, Canaval L R, Pranowo H D and Armunanto R 2014 *Indones. J. Chem.* **14** 199-208.
[24] Saha S K, Hens A, Murmu N C and Banerjee P 2016 *J Mol Liq.* **215** 486-495.
[25] Hadisaputra S, Canaval L R, Pranowo H D, and Armunanto R 2014 *Monatsh Chem Chem Mon.* **145** 737-745.

[26] Purwoko A A, Setiawati V R and Hadisaputra S, 2019 *IOP Conference Series: Materials Science and Engineering* **509** 012130
[27] Hadisaputra S, Pranowo H D and Armunanto R 2012 *Indones. J. Chem.* **12** 207-216.
[28] Hadisaputra S, Hamdiani S, Kurniawan M A and Nuryono N 2017 *Indones. J. Chem.* **17** 431-438.
[29] Koopmans T 1934 *Physica*, **1**(1-6), 104-113.
[30] Foresman J B and Frisch A 1996 Exploring chemistry with electronic structure methods: a guide to using Gaussian.
[31] Pauling L 1960 *The Nature of the Chemical Bond* Ithaca, NY: Cornell university press, p. 3175-3187.
[32] Obayes H R, Alwan G H, Alobaidy A H, Al-Amiery M, Kadhum A A and Mohamad A B 2004 *Chem. Cent. J.* **8** 21.
[33] Yamovoi V I, Kul'magambetova E A, Kulyyasov A T, Turdybekov K M and Adekenov S M 2001 *Chemistry of natural compounds*, **37** 424-427.
[34] Lukovits I, Kalman E and Zucchi F 2001 *Corrosion*, **57** 3-8.

IOP Conf. Series: Materials Science and Engineering 833 (2020) 012011 doi:10.1088/1757-899X/833/1/012011

The Effect of CMC, Agar, and Konjac on the Characteristics of Durian Seed Starch Edible Film

Muhammad Husain Haekal[1], Lizda Johar Mawarani[1*]

[1] Institut Teknologi Sepuluh Nopember, Indonesia

*Corresponding authors: mhaekal96@gmail.com, lizda@ep.its.ac.id

Abstract. Edible film as synthetic plastic food wrap alternatives continuously developed. Durian seed starch is one of the materials used on edible film fabrication. This research studied on durian seed starch-based edible film with the addition of carboxymethyl cellulose (CMC), agar, and konjac with a mass fraction of 10%, 20%, 30%. The observed characteristics were swelling degree and tensile strength of the fabricated film. The edible film solution was made using ratio of flour and additives, water, and glycerol of 1:20:0.3 (w/w), stirred for 20 minutes at 70°C. The results showed that agar addition on edible film made the film could not be fabricated, konjac addition increases the swelling degree and the tensile strength, while CMC addition increases the swelling degree and decreases the tensile strength. Agar and CMC are not compatible to be the additive on the durian seed starch edible film. The most optimal results were with the addition of 30% of konjac with the tensile strength of 9.4 MPa and swelling degree of 1073.81%

1. Introduction

The research of edible film as synthetic plastic substitute on food packaging has been extensively studied. Starch has been widely used as edible film because it produces an odorless, tasteless, transparent film with good mechanical properties [1].

Durian is a fruit that is famous in Southeast Asia. In Indonesia, durian production continues to increase every year [2]. However, only its flesh that mainly used and the seed is thrown away. A study shows that durian seed has high starch content and predicted to have greater mechanical properties than mung bean based edible film, which its properties is comparable to HDPE, LDPE, PP, and PS [1][3].

The research of starch based edible film focussed on increasing its mechanical strength while decrease its susceptibility to water. Some study has been conducted to increase the performance of durian seed starch edible film. A study shows that water vapor transmissibility decrease with the addition of CMC on durian seed starch edible film but its tensile strength and elongation also decreasing [4]. Other studies by Ginting (2016), added chitosan as the additives on durian seed edible film and shows that its addition increase the tensile strength of the edible film but reduce its elongation [5]. Addition of agar and konjac has not been done on durian seed edible film. However, the addition of agar on cassava starch edible film increase its elongation [6]. In this study, the addition of CMC, agar, and konjac will be evaluated.

2. Method

2.1. Material

The material in this study is durian seed, aquadest, glycerol, CMC, agar, and konjac.

2.2. Durian seed starch preparation
Durian seed flour was made by drying the durian seeds by peeling the outside parts of the seeds first. Seeds that have been thinly sliced are then dried at a temperature of 45 °C for 24 hours. Dried seeds were weighed and mixed with water as much as 5 times the weight of seeds and then mashed using a blender for 5 minutes until they become porridge. The pulp is then filtered and the filtrate is taken. The filtrate is allowed to settle and washed with water, carried out until the water is washed clear. The slurry is dried in an oven at 45 °C for 24 hours, then mashed and sieved with a 200 mesh sieve.

2.3. Edible film preparation
Durian seed flour, water, and glycerol were mixed at a ratio of 1: 20: 0.3 (w/w) by stirring for 15 minutes at room temperature. The solution is heated at 70°C for 20 minutes until it thickens. The solution is poured into an acrylic mold with a depth of 3 mm with the size of the mold following the size of the standards of the tests to be performed.

2.4. Characterization of edible film
Edible film then characterized by swelling test and tensile test. The swelling test was carried out according to ASTM D570 and the tensile test was carried out according to ASTM D882.

3. Results and Discussion
Figure 1 shows the surface of each film formed. The addition of agar on durian seed starch makes the film could not be formed properly. The crack occurs because the film shrinks during the drying process and becomes brittle [7]. No defect found on the other samples.

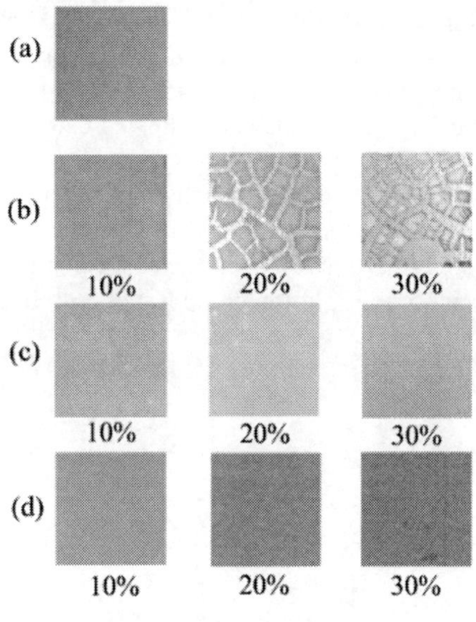

Figure 1. The surfaces of the samples after dried and peeled off (a) without addition, (b) agar addition, (c) konjac addition, (d) CMC addition.

3.1. Swelling test
The addition of CMC and konjac increases the swelling degree of the film. In Figure 6, it appears that the sample with the addition of CMC has a greater effect on increasing the degree of swelling, ranged from 700.98% to 1842.97%. the addition of konjac in the film increase the swelling ranged from 420.25% to 1073.81%. Compared to the film without any addition, the swelling degree of film with CMC addition should lessen the interaction of the film with water but it was increased instead [4]. From

the previous study, at high concentration, the CMC might congregate, hence it could not be fully dispersed in the solution thus creates a path for water to interact with the film and increase the film swelling degree [8].

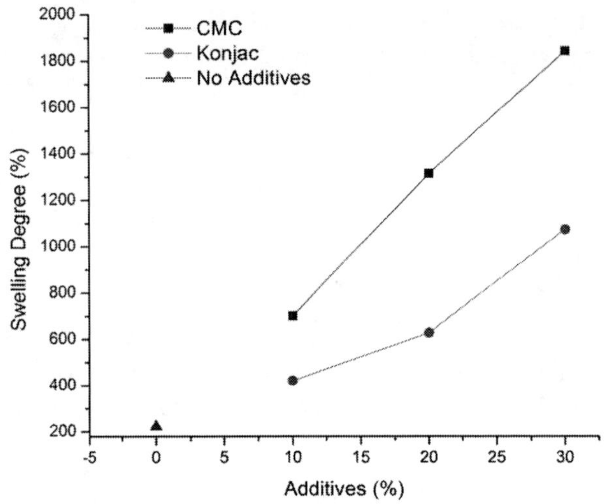

Figure 2. Swelling degree of durian seed starch film.

3.2. Tensile test

The tensile strength of the film without addition was greater than the one that has been reported by Pimpa (2012). This was because the difference between gelatinization which higher temperature would decrease the tensile strength. The tensile strength with konjac addition ranged from 9.4 MPa to 13 MPa, while CMC addition 5.8 MPa to 7.2 MPa. This test indicates that the highest tensile strength was with the addition of 30% konjac of 13 MPa.

The results of the CMC addition on the film decrease the tensile strength. There was declination in both strength and elongation in the addition of 30% CMC. On the lower addition, the result was similar to the previous study, which concludes that the decrease could possibly indicates that the CMC addition was already past its limits, therefore the solution could not distribute evenly and tends to congregate on itself [4].

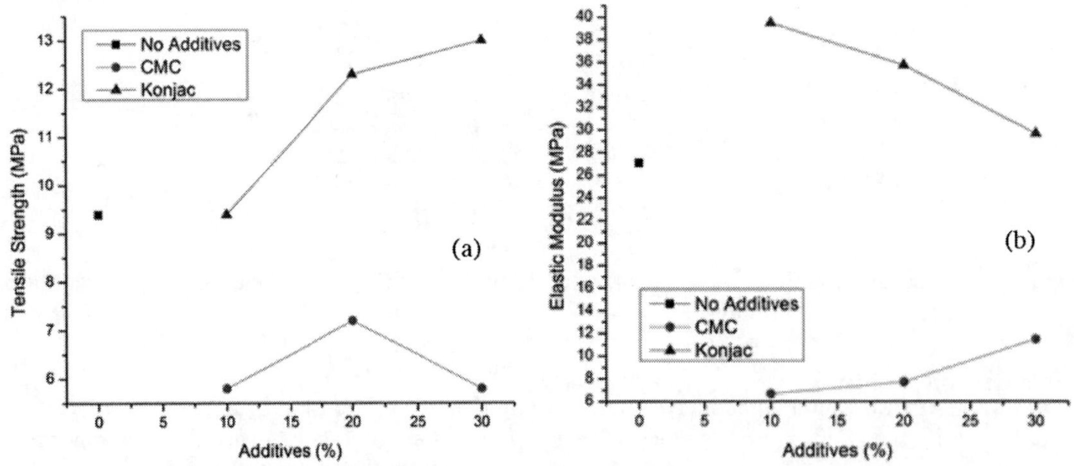

Figure 3. Tensile test results (a) tensile strength, (b) elastic modulus.

4. Conclusion

Based on the current study, the agar is not a compatible addition on durian seed edible film because it cracks the fabricated film. Furthermore, the addition of CMC worsens the film performance. Konjac addition not only shows an increase in its mechanical properties but also increase its swelling degree of the film. The best mechanical property of konjac addition is 30%.

References

[1] Rompothi O, Pradipasena P, Tananuwong K, Somwangthanaroj A and Janjarasskul T 2017 *Carbohydrate Polymers* **157** 748–56
[2] Dirjen Hortikultura Kementerian Pertanian 2015 *Statistik Produksi Hortikultura Tahun 2014* [in bahasa Indonesia]
[3] Tongdang T 2008 *Starch – Stärke* **60** 199–207
[4] Pimpa W, Pimpa C and Junsangsree P 2012 *Advanced Materials Research* **506** 311–4
[5] Y. A. Ginting, Muhammad Hendra S and Hasibuan, Rosdanelli and Siagian 2016 "Pengaruh Variasi Temperatur Pemanasan Larutan Pati Terhadap Sifat Kekuatan Tarik dan Pemanjangan pada Saat Putus Bioplastik Pati Biji Durian (Durio zibehinus)," [in bahasa Indonesia]
[6] The D P, Debeaufort F, Voilley A and Luu D 2009 *Journal of Food Engineering* **90** 548–58
[7] Obara S and McGinity J W 1995 *Int. J. Pharm.* **126** 1–10
[8] Ghanbarzadeh B, Almasi H, and Entezami A A 2010 *Innov. Food Sci. Emerg. Technol.* **11** 697–702

IOP Conf. Series: Materials Science and Engineering **833** (2020) 012012 doi:10.1088/1757-899X/833/1/012012

Antioxidant and Xanthine Oxidase Inhibitory Activities of Kecapi (*Sandoricum koetjape* (Burm.f) Merr.) Leaf Extract

Fajar Nur Hamzah[1], Subandi[1], Wawan Sujarwo[2], Abdi Wira Septama[3] and Tjandrawati Mozef[3]*

[1]Dept of Chemistry, Faculty of Mathematics and Natural Sciences, State University of Malang, Malang
[2]Bali Botanic Gardens, Indonesian Institute of Sciences, Tabanan
[3]Research Centre for Chemistry, Indonesian Institute of Sciences, Serpong

*Corresponding author: tjandrawati@lipi.go.id

Abstract. *Sandoricum koetjape* has been used as folk medicine for the treatment of several diseases. The study was undertaken to determine antioxidant property of *S. koetjape* extract. The xanthine oxidase inhibitory effect was also evaluated. The dried powder of *S. koetjape* leaves was extracted with methanol. The extract was then successively partitioned with different solvents: n-hexane, ethyl acetate, buthanol and water. Phytochemical, total phenolic contents (TPC) and total flavonoid compounds (TFC) were analysed. DPPH scavenging assay was conducted to evaluate antioxidant activity. Xanthine oxidase inhibitor were determined using spectrophotometer. The results showed that amongst of tested extract, methanol extract contained high amount of phenolic and flavonoid compounds. For antioxidant and XO inhibitor, ethyl acetate fraction revealed strong activity with IC_{50} values of 3.70 and 50.18 µg/mL, respectively. XO inhibition by ethyl acetate extract was found to be a competitive inhibition that mimics inhibition mechanism of allopuriol. LC-MS/MS chromatogram exhibited that the extract contains dehydrotumulosic acid, methyl ophiopogonanone A, quercetin-4′-glucoside, luteolin-7-β-D-glucopyranosid and neokurarinol. The result of this study indicated that the potency of *S. koetjape* extract as antioxidant and XO inhibitor.

1. Introduction

Antioxidant activity can prevent several chronic diseases such as cell damage by reactive oxygen species (ROS) from the environment and those that are induced by the activity of several enzymes in the body [1, 2]. Antioxidants also has a role in the inhibition of oxidoreductase enzymes groups such as xanthine oxidase (XO) to reduce uric acid levels in the body [3]. Xanthine oxidase catalyzes oxidation of hypoxanthine to xanthine becoming uric acid in purine base metabolism pathway [4]. Increased levels of uric acid in the blood lead to gout incidence in which uric acid forms crystals in the joints, bones, tissues and organs of the human body [5]. Recently, the prevalence of gout is around 1-4% of the total population. In western countries, cases of gout occur around 3-6% in men and 1-2% in women per total population. The number of gout sufferers' worldwide increases gradually due to diet, lack of exercise, obesity and metabolic disorders [6]. One of medications for gout, can be done by reducing uric acid synthesis in the body, using xanthine oxidase inhibitors (XOI). Allopurinol is an XOI that is used clinically but still has many side effects such as hypersensitivity, Stevens Johnson syndrome and is toxic to the kidneys [4]. Therefore, exploring the other drugs, especially herbal drug, is still important.

Kecapi or *Sandoricum koetjape*, a tropical plant belonging to family *Meliceae,* is abundantly found in Southeast Asian region. [7]. The seeds of Kecapi contain sandoricin and 6-hydroxidandoricin that can be used as pesticide [8]. The fruit of this plant exhibited anti-tumor activity [9]. Conic acid, koetjapic acid and indicic acid in the bark of this plant possessed anti-cancer activity [10]. It has been reported that the fruit extract has high antioxidant activity [2]. However, less information is available about xanthine oxidase inhibitor which induces gout. Therefore, the present study is conducted to determine antioxidant, XOI activities as well as total phenolic and flavonoid contents from this plant. Possible bioactive compounds are also identified using LC-MS/MS spectroscopy.

2. Experiment Methods

2.1. Materials and Instrumentation

Sandoricum koetjape leaves were collected from Bali Botanic Gardens. Methanol, chloroform and formic acid were technical grade, while acetonitrile was LC grade. Potassium iodide, iodine, mercury(II) chloride, glacial acetic acid, zinc powder, hydrochloric acid, sulfuric acid, Folin-Ciocalteu, sodium nitrite, aluminum chloride, sodium bicarbonate, sodium hydroxide, sodium dihydrogen phosphate, dimethyl sulfoxide) were obtained from Merck. DPPH (2,2-diphenyl-1-picrylhydrazyl), xanthine and xanthine oxidase were from Sigma Aldrich. Allopurinol was purchased from Indofarma. The apparatus used in this study were rotary evaporator (Buchi), UV-Vis Spectrophotometer (Agillent), ELISA Reader (Multiscan Go, Thermo Scientific), LC-MS/MS (Agıllent).

2.2. Procedure

2.2.1. Extraction. Samples were dried in the oven at 50°C, reduced to powder using grinder and weighed. The powder was then macerated using methanol with a solvent replacement every 24 hours. The filtrates were concentrated using a rotary evaporator at 50°C at rotation of 70-75 rpm. The concentrated extract was weighed and labelled as crude extract. The yield of crude extract were determined following equation 1, where W_1 is weight of crude extract after concentrated and W_2 is weight of dried leaf.

$$\%Yield = \frac{W_1}{W_2} \times 100\% \quad (1)$$

2.2.2. Phytochemical Test. The phytochemical screening was conducted following the Indonesian Ministry of Health protocols [12] to identify flavonoid, terpenoid and steroid, saponin and alkaloid. a) The qualitative test for flavonoids was done by taking a few milligrams of extract, added with 4 mL of ethanol, 2 mL of test solution was added by 0.5 g of Zn powder and 2 mL of 2 N HCl, then was incubated for a minute, then was added by concentrated HCl per drop to form an intensive red color which means positive flavonoids. b) Qualitative test for tannins in the extract was done by taking a few milligrams of extract, then was added with hot water and boiled for 5 minutes, then added by a few drops of $FeCl_3$ 1%. Positive reaction was shown in green-violet. c) Qualitative test for terpenoids and steroids in extracts was carried out by taking a few milligrams of crude extract then was dissolved in 2 mL chloroform in a test tube then added by 10 drops of anhydrous acetic acid and 3 drops of concentrated sulfuric acid. Positive reactions are shown in Green for steroid and in purplish red as triterpenoids. d) Qualitative test for saponins in the extract was done by taking a few milligrams of extract and added by hot water and shaken vigorously until intensive foam formed. The addition of a few drops of HCl 2 N into foam does not disappear, then indicates positive saponins. e) Qualitative test for the alkaloid (Buchard test) was done by taking about 500 mg of extract then was added with 1 mL of HCl 2 N and 9 mL of water and heated in a water bath for 2 minutes. The filtrate was taken as many as a few milliliters into the drip plate and added with the Burchard reagent (2 g iodine and 4 g KI dissolved in 100 mL distilled water) a few drops. A positive reaction is shown by the formation of brown-black precipitate. Qualitative Meyer's alkaloid test was performed by taking about 500 mg extract and was added 1 mL of HCl 2 N and 9 mL of water and then heated in a water bath for 2 minutes. The filtrate was taken as many as a few milliliters into the drip plate and was added a few drop of Meyer reagent (1.358 g $HgCl_2$ in 60 mL of distilled water and 5 g KI in 10 mL of distilled water). The positive reaction is indicated by the formation of white or yellow precipitate.

2.2.3. Partition of Crude Extract. One gram of crude extract of kecapi was successively partitioned using n-hexane (Hex), ethyl acetate (EtOAc), buthanol (BuOH) and water. Fractions were concentrated under reduced pressure at a temperature setting of 50 °C.

2.2.4. Determination of Total Phenolic Compounds (TPC). Total phenolic content was performed according to previous study with slight modification [13]. In brief, Extract stock solution was made with a concentration of 1000 µg/mL. The sample was piped 250 µL into a test tube then added 250 µL of Folin Ciocalteu reagent. A number of 3.5 ml of distilled water was added to the sample and was incubated for 8 minutes at room temperature. Then, the sample was added 0.75 mL of 20% Na_2CO_3 (20 grams Na_2CO_3/100 mL) and added to 5 mL with distilled water. The samples were incubated for 2 hours then absorbance was measured with a UV-Vis spectrophotometer at a wavelength of 765 nm. The concentrations of phenol in the samples were determined by the gallic acid calibration curve with a concentration of 0, 5, 10, 20, 30 µg/mL.

2.2.5. Determination of Total Flavonoid Contents (TFC). From the stock solution of 1000 µg/mL sample was taken 500 µL then added 2 mL of distilled water and 150 µL of 5% $NaNO_2$ (5 g $NaNO_2$ /100 mL). The samples were incubated for 5 minutes then added with 10% $AlCl_3$ (10 g $AlCl_3$ dissolved in glacial acetic acid up to 100 mL) of 150 µL. The samples were incubated for 6 minutes. A total of 2 mL of 1 M NaOH (2.5 g/100 mL) was added to the sample and then made final volume of 5 mL by water. Total flavonoid contents were measured using the UV-Vis spectrophotometer at a wavelength of 510 nm. The flavonoid concentration in the sample was determined by the calibration curve of the standard quercetin with concentrations of 0, 10, 30, 50, 100 µg/mL [1].

2.2.6. DPPH Scavenging. Crude extracts and the fractions of the kecapi were made a solution stock of 1000 ppm in methanol. Stock DPPH solution was made 4 mg/25 mL in methanol. All test substances are made in a standard series from 0.5 to 50 µg/mL. The screening results were compared with blank and quercetin standards with 1000 µg/mL stock with a series of concentrations of 1, 2, 3, 4 and 5 µg/mL. Percent inhibition was calculated using bellow formula (2):

$$\%DPPH\ inhibition = \frac{A\ blank - A\ sample}{A\ blank} \times 100\% \qquad (2)$$

IC_{50} values were determined by drawing a linear regression curve y = ax + b where y is the percent of inhibition and x for the concentration of samples [14].

2.2.7. Xanthine Oxidase Inhibitory Activity Test. Stock solutions of crude extract and fractions were made in 100% DMSO of 10,000 µg/mL. Stock sample solutions were tested by making crude extract concentration series (10, 20, 50, 100 µg/mL), n-hexane and water fractions series (20, 50, 80, 100 µg/mL), ethyl acetate and butanol fractions series (20, 50, 100, 200 µg/mL) with a final reaction volume of 300 µL in a 96-well microplate and a final DMSO concentration of maximum 1% in diluting phosphate buffer at pH 7.4. Then 10 µL 0.05 units/mL of XO enzyme was added to the mixture and pre-incubated at 25 °C for 15 minutes. After pre-incubation, 0.375 mM 100 µL of xanthine was added and incubated at 25 °C for 10 minutes. The reaction was then stopped by adding 100 µL of 1 N HCl. Samples were measured by Multiscan Go at a wavelength of 295 nm. Percent inhibition was determined by equation (3)

$$\% \, XO \, Inhibition = \frac{(A-B)-(C-D)}{(A-B)} \times 100 \, \% \quad (3)$$

Where A is the absorbance of without inhibitors, B is absorbance of the control of substrate, C is the absorbance of inhibitor tests, D is absorbance of inhibitors and substrates. Allopurinol solutions was made in a concentrations series of 0.5, 0.8, 2.0, 4.0 ppm and a final volume of 300 μL [15].

2.2.8. Determination of Enzyme Inhibition Kinetic. Variations in xanthine concentration were made with the same inhibition test procedure, measurements were made based on time variations during the incubation process. Absorbance values were read at specified intervals and expressed as t values to get the v values. The obtained absorbance values were used to make the Lineweaver-Burk line curve, with the y axis is 1/v and the x axis is 1/[Substrate] to determine the value of Km, Vmax and the type of inhibition.

2.2.9. Identification of Candidate Compunds by LC-MS/MS. About 5 mg of the test sample was weighed and dissolved into HPLC grade methanol at a concentration of 10000 μg/mL then injected into LC-MS/MS. The effluent used was 0.1% formic acid: water (eluent A) and acetontril: 0.1 formic acid (eluent B). The column used was C-18.

3. Results and Discussion

3.1. Result of Preparation and Phytochemical Screening

The stability of phenolics and flavonoids of leaves material can be affected by the drying process [16], thus the drying must be performed at temperatur of only 50°C. Beside of that, the refining has also been done to expand the surface area of the leaves material so that the surface contact of the material with the solvent will be greater and can facilitate the penetration of the solvent into the sample, then can attract more compounds [17]. The extraction temperature was carried out at room temperature in order to prevent destruction of secondary metabolites [18]. In this study methanol was selected due to its ability to afford more polyphenol components in plants, besides its boiling point is relatively lower than ethanol or water [19]. It has been known that several phytochemical compounds including flavonoid and polyphenol possessed antioxidant activity by the formation of ROS or reducing free radicals directly [20]. Based on phytochemical screening (Table 1), it indicated that the extract contains several secondary metabolites such as flavonoids, tannins, steroids, saponins, and alkaloids.

Table 1. The Result of phytochemical screening

Phytochemical Test					
Flavonoids	Tanins	Steroids	Saponins	Alkaloids	
				Buchard	Meyer
+++	++	+++	++	+	+

Notes: + (low contents), ++ (moderate contents), +++ (high contents)

3.2. TPC, TFC, Antioxidant Activities and XO Inhibitory Activities

The methanol extract was then successively partitioned using different solvents, starting from the most non-polar solvent to the most polar solvent. Partition was conducted to observe the distribution of compounds which have potential to be used as antioxidant and anti-gout. The process is also allowing to eliminate other compounds that are suspected to be antagonistically against compounds that have the potential as anti-gout. The result of antioxidant activity, TPC, TFC and XO inhibition of the crude extract and the fractions result, are shown in Table 2.

Table 2. Yield, TPC, TFC, DPPH scavenging and XO inhibitory activities of Kecapi leaf extract, and Quercetin or Allopurinol as Standard

Samples/ Fractions	**Yield (%)**	**TPC (%)**	**TFC (%)**	**IC_{50} DPPH (μg/mL)**	**IC_{50} XO (μg/mL)**
Crude extract	66.83	33.67	22.20	8.14	54.85
n-hexane fraction	62.60	11.97	43.09	31.08	74.03
Ethyl acetate fraction	5.67	35.72	75.13	3.70	50.18
n-Butanol fraction	24.34	9.15	18.79	12.62	138.05
Water fraction	3.16	6.22	22.75	28.65	142.11
Quercetin	-	-	-	3.20	-
Allopurinol	-	-	-	-	1.07

It can be seen that ethyl acetate fraction exhibit strong activity as antioxidant and XOI compared to others with IC_{50} values of 3.70 and 50.18 μg/mL. Moreover, this fraction also has high percentage in total phenolic and flavonoid contents (35.72 and 75.13%, respectively). Interestingly, this active fraction revealed more potent antioxidant effect compared to quercetin as positive control. Thus, ethyl acetate fraction was selected for further assay, e.g. XO inhibitory mechanism study.

3.3. Xanthine Oxidase Inhibitory Mechanism

The xanthine oxidase inhibition mechanism of ethyl acetate fraction was carried out using the Lineweaver-Burk equation curve. Based on the result, it showed dose-dependent manner. This indicates that the extract had a potency as XO inhibitor. It has been reported that some polyphenol compounds enable to inhibit XO [4]. According to the Lineweaver-Burk curve (Figure 1), the ethyl acetate fraction gave a competitive inhibition. This type of inhibition mimics inhibition mechanism of allopurinol.

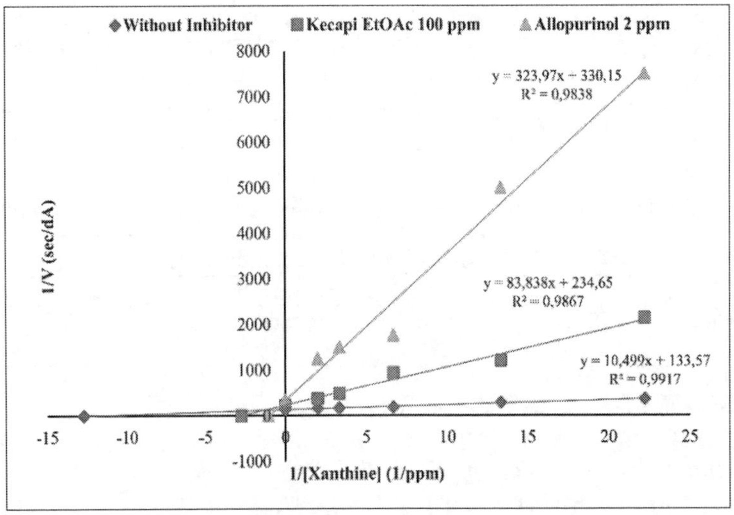

Figure 1. Lineweaver-Burk plot of XO inhibition

Table 3. Km and Vm of XO inhibitory activity of ethyl acetate fraction

Variables	**Vm (μg/mL/sec)**	**Km (μg/mL)**
Without Inhibitors	0.007487	0.078603
Allopurinol 2 μg/mL	0.003029	0.981281
Kecapi EtOAc 100 μg/mL	0.004262	1.526827

3.4. Identification of Candidate Active Compounds in The Crude Extract using LC-MS/MS

In order to know the profiling of secondary metabolites in the extract, identification was carried out using LC-MS/MS analysis. The principle of detection is separation of the components in the liquid chromatography column, while the detection was carried out on the abundant compounds in the crude extracts. The selected peaks are the highest peak, with the correction of the solvent blank, and then the m/z of selected peak was adjusted to the Agillent software database of LC-MS/MS. Analysis of crude extracts was carried out by considering the results of phytochemical tests as the first step in the exploration of a natural ingredient as the gout medecine.The results of detection of the predicted active compounds in crude extracts are shown in Figure 2.

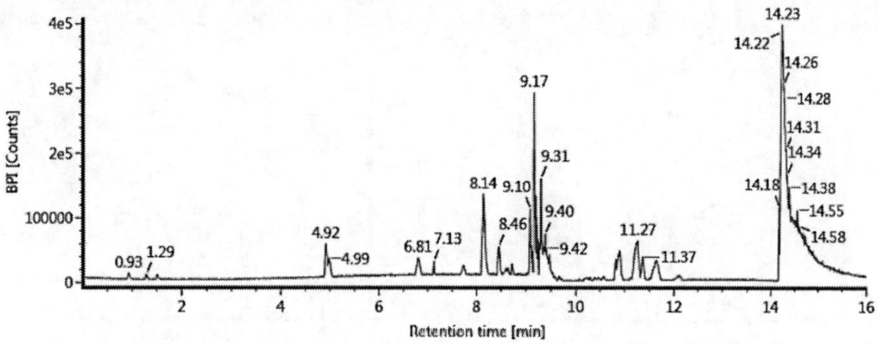

Figure 2. Chromatograms of methanol extract of *S. koetjape*

The results are presented in the Table 4 and Fig 3. Dehydrotumulosic acid is a triterpenoid group. This compound has been reported having activity as antihyperglycemia [21]. Phenolic compounds are not the only major components for XO inhibition and DPPH radical inhibiton activities, a previous study proves that terpenoids have XO inhibitory activity and antioxidant effects on DPPH free radicals [15]. The methyl ophiopogonanone is a homoisoflavone that has quite high antioxidant activity [22], quercetin and luteolin are flavonoid groups which have activity as competitive inhibitors of XO [23], while neocurarinol has been reported to modulate gamma-aminobutyric acid type A receptors [24].

Table 4. Compounds Identification based on LC-MS/MS data

Compounds	Retention Time (min)	*m/z*	Mass (Da)
Dehydrotumolosic acid	1.51	485.3565	484.35526
Methyl ophiopogonanone A	1.16	365.1024	342.11034
Quercetin-4′-glucoside	5.09	465.1001	464.09548
Luteolin-7-β-D-glucopyranoside	5.41	449.1047	448.10056
Neokurarinol	5.98	493.2244	470.23045

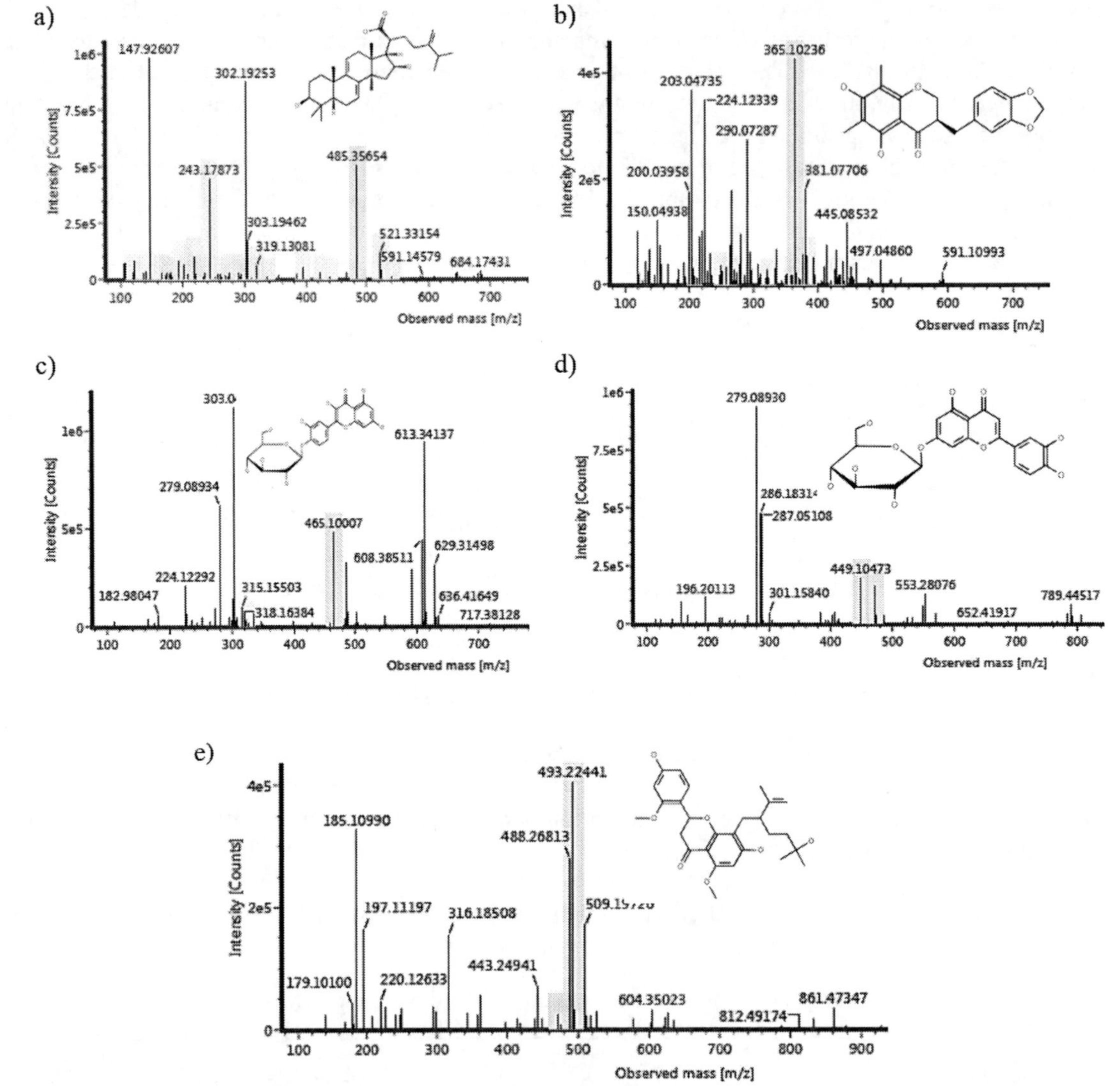

Figure 3. Mass Spectra and structures of candidate active compounds in the kecapi leaf crude extract. Dehydrotumulosic acid (a), Methyl ophiopogonanone A (b), Quercetin-4′-glucoside (c), Luteolin-7-β-D-glucopyranoside (d) and Neocurarinol (e)

4. Conclusion

It can be concluded that *S. koetjape* leaf extract has a potency as antioxidant and xanthine oxidase inhibitors. The extract contains high amount of phenolic and flavonoid compounds. The type of xanthine oxidase inhibition by ethyl acetate fraction is competitive inhibition, and based on spectroscopic analysis, predicted active compounds in the extract particularly are dehydrotumulosic acid, methyl ophiopogonanone A, quercetin-4′-glucoside, luteolin-7-β-D-glucopyranoside and neokurarinol. However, further study is still needed to elucidate others active components.

References

[1] Baba S A and Malik S A 2015 *J. Taibah Univ. Sci.* **9** 449–54
[2] Anantachoke N, Lomarat P, Praserttirachai W, Khammanit R and Mangmool S 2016 *Evidence-based Complement. Altern. Med.* **2016** 1-14

IOP Conf. Series: Materials Science and Engineering **833** (2020) 012012 doi:10.1088/1757-899X/833/1/012012

[3] Shukor N A A, Ablat A, Muhamad N A and Mohamad J 2018 *Int. J. Food Sci. Technol.* **53** 1476–85
[4] Abdullahi A, Hamzah R, Jigam A, Yahya A, Kabiru A, Muhammad H, Sakpe S, Adefolalu F, Isah M and Kolo M 2012 *J. Acute Dis.* **1** 126–9
[5] Saigal R and Agrawal A 2015 *J. Assoc. Physicians India* **63** 56–63
[6] Ragab G, Elshahaly M and Bardin T 2017 *J. Adv. Res.* **8** 495–511
[7] Lehra K S, Goyal S, Bajwa B S, Kaur R and Singh S 2012 *Indian Drugs* **49** 5–11
[8] Powell, R., Mikolajczak, K., and Zilkowski B 1991 *J. Nat. Prod.* **54** 241–6
[9] Sim K Y and Lee H T 1972 *Phytochemistry* **11** 3341–3
[10] Kaneda N, Pezzuto J M, Kinghorn D, Farnsworth N R, Santisuk T, Tuchinda P, Udchachon J and Reutrakul V 1992 *J. Nat. Prod.* **55** 654–9
[11] Ismail I S, Ito H, Hatano T, Taniguchi S and Yoshida T 2003 *Phytochemistry* **64** 1345–9
[12] Depkes RI 1995 *Farmakope Indonesia edisi IV* Jakarta Kementrian Kesehatan RI
[13] Singleton V L and Rossi J A J 1965 *Am. J. Enol. Vitic.***16** 147
[14] Yen G C and Chen H Y 1995 *J. Agric. Food Chem.***43** 27-32
[15] Mamat N, Jamal J A, Jantan I and Husain K 2014 *J. Sains Malaysiana* **43** 1827–33
[16] Luliana S, Purwanti N U and Manihuruk K N 2016 *J. Pharm. Sci. Res.* **3** 120–9
[17] Husni E, Suharti N and Atma A P T 2018 *J. Sains Farm. Klin.* **5** 12–6
[18] Sarker S D, Latif Z and Gray A 2005 *Natural Product Isolation 2nd edition* New Jersey Humana Press
[19] Mohammedi Z and Atık F 2011 *Int. J. Pharma Bio Sci.* **2** 609–15
[20] Shah, B.N., A.K S 2010 *Textbook of Pharmacognosy and Phytochemistry* New Delhi Elsevier
[21] Li T H, Hou C C, Chang C L T and Yang W C 2011 *Evidence-based Complement. Altern. Med.* **2011** 1-8
[22] Wang Y, Liu F, Liang Z, Peng L, Wang B, Yu J, Su Y and Ma C 2017 *Iran. J. Pharm. Res.* **16** 357–65
[23] Pauff J M and Hille R 2009 *J. Nat. Prod.* **72** 725–31
[24] Yang X, Baburin I, Plitzko I, Hering S and Hamburger M 2011 *J. Mol. Divers.* **15** 361–72

IOP Conf. Series: Materials Science and Engineering **833** (2020) 012013 doi:10.1088/1757-899X/833/1/012013

Antibacterial activity of *Achromobacter sp.* and *Bacillus sp.*, bacterial endophytes derived from Mangrove *Ceriops tagal* (Perr.) C.B.Robb

Yuli Haryani[1,*], Rahmiwati Hilma[2], Noviza Delfira2, Tetty Martalinda[3], Fifi Puspita[4], Amelia Friska[1], Dita Juwita[1], Analdi Farniga[1], Fri Ardi[5]

[1]Chemistry Department, Faculty of Mathematics and Natural Sciences, University of Riau
[2]Chemistry Department, Faculty of Health and Natural Sciences, Muhammadiyah University
[3]Biology Department, Faculty of Mathematics and Natural Sciences, University of Riau
[4]Agrotechnology Department, Faculty of Agriculture, University of Riau
[5] Faculty of Pharmacy, Andalas University

[*]Corresponding author: yuli.haryani@lecturer.unri.ac.id

Abstract. A total of 95 bacterial endophytes have been isolated from mangroves plant in Bengkalis coastal area, Riau Province, Indonesia. Mangrove plant samples were identified as *Bruguiera* sp. and *Ceriops tagal* (Perr.) C.B.Robb. Antagonism study showed that 4 out of 95 isolates were possessed antibacterial activity based on their ability to inhibit the growth of *Vibrio alginolyticus* and *Staphylococcus aureus;* while none of them was able to inhibit *Escherichia coli.* According to the results of 16s rRNA genes amplification, the closest species relative of the 4 active isolates was *Achromobacter insolitus* (isolate no. 34), *Bacillus siamensis* (isolate no. 39), and *Bacillus subtilis* (isolate no. 88 and 89). This is the first report on bacterial endophytes associated with Mangrove in Riau Province, Indonesia.

1. Introduction

Antibiotic resistance, the ability of specific bacteria to survive in the presence of an antibiotic resulting in drug inefficiency and persistent infections, has been studied for a long time. More than a hundred different types of antibiotics discovered have been used to treat numerous infectious diseases; however, drug resistance does exist. The world faces the crisis in antibiotic discovery as it is predicted by 2050, the increased number of resistant pathogens will cause 10 million deaths per year. Not only for humans, but the antibiotic resistance is also a big issue in the environment since the increase of its number has been accelerated in recent years caused by the discharge of antibiotics into the environment [1-3]. It is given that new antibiotics are needed to combat drug-resistant pathogens. Numerous studies on plants and their derivatives against hundreds of microbial strains have reported that they found remarkable antimicrobial potential [4-10].

Plants and their associated microorganisms have evolved together to adapt to a given environment. One species of the plant could be a host of thousands of microorganisms which are categorized as epiphytes and endophytes. Recently, endophytes - microorganisms that live in the inner tissues of plants without causing any external symptoms have been under increased investigation due to their complex relationship with the host. The presence of endophytes in the host plants provides several benefits, particularly growth promotion and protection from pathogens. It could be due to the metabolites of endophytes that have been reported to inhibit a number of microorganisms. Several studies reported the ability of endophytes to produce host-derived metabolites of importance such as alkaloids, terpenoids, phenolic, steroids, and tannins that serve as a potential candidate for antimicrobial, anticancer, and other biological activities [11-14].

Mangroves are an important ecosystem. They provide crucial ecosystem services, including fisheries, shoreline shields, and bioremediation of wastes [15]. Since they strive under extreme conditions and straddle the terrestrial and aquatic biomes, mangrove environments are a potential source to explore new microbiota with extensive applications in pharmaceutical science [16; 17]. Indonesia has 27% of the world's total mangrove forest, equivalent to 4.25 million ha [18], and 143 billion ha are in Riau Province. Considering the virtual lack of studies on endophytes derived from Mangrove in Riau Province, it reported here for the first time the bacterial endophytes from mangrove plant at Bengkalis coastal area.

2. Methodology

2.1. Sampling site and bacterial endophytes isolation

Fresh and healthy roots, stem barks, and leaves of mangrove plants were taken randomly from Sei Pakning (Siak Regency at 1°20'55"N, 102°9'33"E) and Tenggayun Beach (Bengkalis Municipality at 1°32'3"N, 101°54'58"E), and then were sent to Riau University Herbarium for identification. Some roots were brought to the laboratory for endophytic bacteria isolation in separate sterile polyethylene bags. Each root sample was washed in running tap water. The surface of each sample was sterilized by immersing each of them in 70% ethanol (2 minutes), soaked in 5.25% peroxide acid (2 minutes), rinsed in sterile distilled water (three times), and surface-dried with sterile filter paper. Aliquots of the last sterile distilled water used in the final rinse were inoculated onto the surface of Nutrient Agar (NA) plates supplemented with ketoconazole 0.3g/100ml for 18-24 h to confirm that the surface sterilization is successful. Each of those surface-sterilized samples was cut into fragments and each fragment was put onto the surface of ketoconazole-supplemented NA plates and then incubated at 37 °C for 24-48 h. Any bacterial growth around the incubated fragment samples was then taken and purified (streaked) onto other sterile NA plates, incubated overnight at 37 °C, all single colonies with different characters were taken and maintained as bacterial endophyte isolates.

2.2. Antagonism test

All endophytes isolated were screened for the antagonism in vitro against *Escherichia coli, Staphylococcus aureus,* and *Vibrio alginolyticus.* One colony of each bacterial endophyte was transferred (drawn as a circle of 6mm diameter) on the surface of Mueller Hinton Agar containing each pathogenic bacteria (OD 0.1) and incubated for 24-48 h. All treatments were performed in triplicate. Inhibition zones produced around each endophyte were measured and bacterial endophytes showing good antibacterial activity were then subjected to macroscopic, microscopic analysis, gram staining, and molecular species identification.

2.3. 16S rRNA Gene Amplification and Sequencing

DNA template of each active isolates was extracted using Presto™ Mini gDNA Bacteria kit. The amplification of 16S rRNA genes was carried out with two primers (24F/1541R) and performed in 50 μL reaction volumes of 5.0 μL 10x PCR Buffer, 2 μL of DNA, 2.5 μL of 2 mM dNTPs, 0.2 μL of 5 U/μL Taq Polymerase, 1 μL of each primers and 38.3 μL of sterile distilled water. The reaction was carried out in an amplification condition of 95 °C at 5 min for pre-denaturation followed by 35 cycles of denaturation (95 °C; 1 min), annealing (50 °C; 1 min) and extension (72 °C; 1 min and 30s). The final extension was at 72 °C (10 min). Gel electrophoresis was carried for the PCR products using 1% agarose with TBE buffer at 80 V for 25 min. PCR product was then sent for sequencing at Genetica Science Indonesia. Finally, the 16s rRNA gene sequence homology was then analyzed using the Mega BLAST program (https://blast.ncbi.nlm.nih.gov/Blast.cgi). Sequence homology was compared with that of other microorganism sequences of GenBank. The phylogenetic tree was constructed by the neighbor-joining method using the MEGA software version 6.06 based on 1000 bootstraps.

3. Results and Discussion

A total of 95 bacterial endophytes have been isolated from root samples of Mangroves which were identified as *Ceriops tagal* (Perr.) C.B.Rob and *Bruguiera* sp. All isolates were screened for their ability to inhibit the growth of three pathogens studied. Out of 95 isolates, we found 30 isolates (Table 1) gave inhibition zones. There were 24, 4, and 5 isolates that were able to inhibit the growth of *Vibrio alginolyticus, Escherichia coli* dan *Staphylococcus aureus*, respectively. All positive isolates were then used for antagonism test and the result showed that isolate number 34, 88, and 89 were active to inhibit *V. alginolyticus;* isolate number 34 and 39 were produced inhibition zones against *S. aureus*; while none of them was active against *E. coli* (Table 1; Figure 1).

Table 1. Bacterial endophytes, sources, and their antibacterial activities

No.	Isolate No.	Mangrove samples	Antibacterial screening			Antagonism result (diameter of inhibition zone; mm)		
			V. alginolyticus	*E. coli*	*S. aureus*	*V. alginolyticus*	*E. coli*	*S. aureus*
1	1	*Bruguiera* sp.	+	-	-	n.d	n.d	n.d
2	2	*Bruguiera* sp.	+	-	-	n.d	n.d	n.d
3	4	*Bruguiera* sp.	+	-	-	n.d	n.d	n.d
4	5	*Bruguiera* sp.	+	-	-	n.d	n.d	n.d
5	6	*Bruguiera* sp.	+	-	-	n.d	n.d	n.d
6	7	*Bruguiera* sp.	+	-	-	n.d	n.d	n.d
7	11	*Bruguiera* sp.	+	-	-	n.d	n.d	n.d
8	23	*Bruguiera* sp.	+	-	-	n.d	n.d	n.d
9	26	*Bruguiera* sp.	+	+	-	n.d	n.d	n.d
10	27	*Bruguiera* sp.	-	+	-	n.d	n.d	n.d
11	31	*Bruguiera* sp.	+	+	-	n.d	n.d	n.d
12	33	*Ceriops* tagal	+	-	-	n.d	n.d	n.d
13	34	*Ceriops* tagal	+	-	+	11.3±0.5	n.d	9.7±0.4
14	37	*Ceriops* tagal	+	-	-	n.d	n.d	n.d
15	39	*Ceriops* tagal	-	-	+	n.d	n.d	10±0.5
16	41	*Ceriops* tagal	-	-	+	n.d	n.d	n.d
17	52	*Ceriops* tagal	+	-	-	n.d	n.d	n.d
18	53	*Ceriops* tagal	+	-	-	n.d	n.d	n.d
19	62	*Bruguiera* sp.	+	-	-	n.d	n.d	n.d
20	65	*Bruguiera* sp.	+	-	-	n.d	n.d	n.d
21	68	*Bruguiera* sp.	+	-	-	n.d	n.d	n.d
22	69	*Bruguiera* sp.	+	-	-	n.d	n.d	n.d
23	71	*Bruguiera* sp.	-	+	-	n.d	n.d	n.d
24	72	*Ceriops* tagal	-	-	+	n.d	n.d	n.d
25	73	*Ceriops* tagal	+	-	-	n.d	n.d	n.d
26	74	*Ceriops* tagal	+	-	-	n.d	n.d	n.d
27	81	*Ceriops* tagal	+	-	-	n.d	n.d	n.d
28	88	*Ceriops* tagal	+	-	-	9.1±0.5	n.d	n.d
29	89	*Ceriops* tagal	+	-	-	8.9±0.1	n.d	n.d
30	92	*Ceriops* tagal	-	-	+	n.d	n.d	n.d

n.d: not-detected

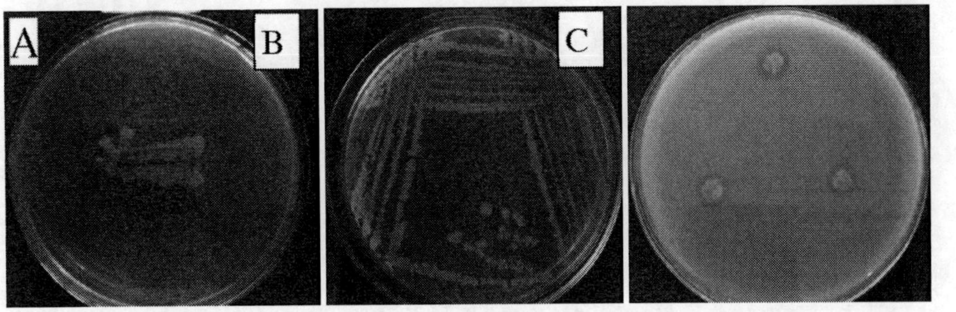

Figure 1. Overnight inoculated sterilized-surface sample (A), Single colonies (B), inhibition zones (C)

According to the antagonism data, this study selected four isolates i.e., isolate no.34, 38, 88, and 89 as active isolates to be identified further for morphology and molecular species identification. Isolate number 34 is only the one with coccus morphology, while others were bacilli. Interestingly, all of them were isolated from *Ceriops tagal* (Perr.) C.B.Rob; and isolate number 34 showed a potential source of broad-spectrum antibiotics since it could inhibit the growth of both gram-negative and gram-positive bacteria.

The electrophoresis product of amplificated DNA was found ~1500 bp. Species identification was carried out by comparing DNA sequences obtained to the DNA sequences in the NCBI GenBank Database (Table 2). According to the Max score and identity, the closest species of isolate no.34, 39, 88, and 89 are *Achromobacter insolitus, Bacillus siamensis, Bacillus subtilis,* and *Bacillus subtilis*, respectively. The nearest phylogenetic neighbour sequences for 16s rRNA gene sequences of the 4 isolated bacterial endophytes were shown in Figure 2.

Table 2. Closest species relative of the four active isolates based on sequence similarity using BLAST program in the GenBank database

Description of species	Max score	Query cover	E-value	Per. Ident
Isolate No. 34				
Achromobacter xylosoxidans strain ES-6	2382	100%	0.0	97.14%
Achromobacter insolitus strain Y2P1	2382	100%	0.0	97.14%
Achromobacter denitrificans strain USDA-ARS-USMARC-56712	2378	100%	0.0	97.07%
Isolate No. 39				
Bacillus amyloliquefaciens strain X030	2616	100%	0.0	100.00%
Bacillus velezensis strain WRN014	2616	100%	0.0	100.00%
Bacillus subtilis isolate K1	2616	100%	0.0	100.00%
Bacillus siamensis	2616	100%	0.0	100.00%
Isolate No. 88				
Bacillus velezensis strain SRCM100072	2616	100%	0.0	100.00%
Bacterium CulalenE33	2616	100%	0.0	100.00%
Bacillus amyloliquefaciens subsp. plantarum YAU B9601-Y2	2616	100%	0.0	100.00%
Bacillus subtilis strain ZJ06 16S	2616	100%	0.0	100.00%
Bacillus siamensis	2611	100%	0.0	99.93%
Isolate No. 89				
Bacillus siamensis strain SCSIO 05746	2531	100%	0.0	98.76%
Bacillus velezensis strain SRCM100072	2531	100%	0.0	98.76%
Bacterium CulalenE33	2531	100%	0.0	98.76%
Bacillus amyloliquefaciens subsp. plantarum YAU B9601-Y2	2531	100%	0.0	98.76%
Bacillus subtilis strain ZJ06	2531	100%	0.0	98.76%

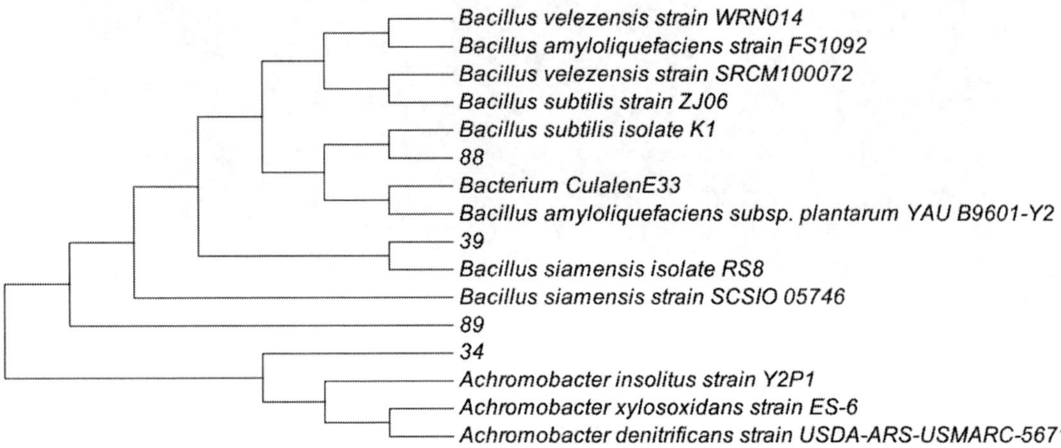

Figure 2. NJ tree – Kimura-2-parameter model – 1000x bootstrap

Isolate no. 34, which was able to inhibit both gram-positive and gram-negative bacteria studied, has the closest species relative with *Achromobacter insolitus.* The genus *Achromobacter*, of the family *Alcaligenaceae*, originally contained a single type species, *Achromobacter xylosoxidans.* Through times, several species were added to the family based on a polyphasic taxonomy study [19,20], including *Achromobacter xylosoxidans* and *Achromobacter denitrificans.* Further study is needed to extract and characterize any compounds produced by each isolated bacterial endophyte.

4. Conclusions

Bacterial endophytes of mangrove plants are alternative sources in finding some antibacterial compounds because 4 of 95 endophytes isolated could inhibit the growth one to two of pathogens studied. According to the results of 16s rRNA genes amplification, the closest species relative of the four active isolates was *Achromobacter insolitus* (isolate no. 34), *Bacillus siamensis* (isolate no. 39), and *Bacillus subtilis* (isolate no. 88 and 89).

Acknowledgment

This study was financially supported by a Fundamental Research Grant from the Indonesian Ministry of Higher Education through Riau University Research and Community Development Centre with contract no. 748/UN.19.5.1.3/PT.01.03/2019, and this support is highly appreciated.

References

[1] Niu G and Li W 2019 *Trends in Biochemical Sciences* **44**(11) 961-972.
[2] Trotter AJ, Aydin A, Strinden MJ, and O'Grady J 2019 *Current Opinion in Microbiology* **51** 39–45
[3] López-Jácome E, Franco-Cendejas R, Quezada H, Morales-Espinosa R, Castillo-Juárez I, González-Pedrajo B, García-Contreras R 2019 *Current Opinion in Pharmacology* **48** 48–56
[4] Pedrós-Garrido S, Clemente I, Calanche JB, Condón-Abanto S, Beltrán JA, Lyng JG, Whyte P 2020 *Food Control* **107** 106768.
[5] Abral H, Ariksa J, Mahardika M, Handayani D, Aminah I, Sandrawati N, and Ilyas RA 2020 *Food Hydrocolloids* **98** 105266
[6] Mohotti S, Rajendran S, Muhammad T, Strömstedt AA, Adhikari A, Burman R, Gunasekera S 2020 *Journal of Ethnopharmacology* **246** 112158.
[7] Fernandez V, Bengardino M, Jagus RJ and Aguero MV 2020 *Lwt* **117** 108622.
[8] Suroowan S, Jugreet BS, and Mahomoodally MF 2019 *South African Journal of Botany* **126** 282–308

IOP Conf. Series: Materials Science and Engineering **833** (2020) 012013 doi:10.1088/1757-899X/833/1/012013

[9] Ayaz M, Ullah F, Sadiq A, Ullah F, Ovais M, Ahmed J, and Devkota HP 2019 *Chemico-Biological Interactions* **308** 294–303

[10] Afzal I, Khan Z, Sikandar S, and Shahzad S 2019 *Microbiological Research* **221** 36–49.

[11] Singh DK, Sharma VK, Kumar J, Mishra A, Verma SK, Sieber TN, and Kharwar RN 2017 *Scientific Reports*, **7**(1), 1–14

[12] Gouda S, Das G, Sen SK, and Shin H 2016 *Frontiers in Microbiology* **7** 1–8.

[14] Deepika VB, Murali TS, and Satyamoorthy K 2016 *Microbiological Research* **182** 125–140.

[15] Extent MM and Approach A 2018 *Remote Sensing* **10**(1466) 1–20.

[16] Yao H, Sun X, He C, Maitra P, Li X, and Guo L 2019 *Microbiome* **7**(57) 1–15.

[17] Narendran R and Kathiresan K 2016 *Biocatalysis and Agricultural Biotechnology* **6** 189–194

[18] Arobaya A dan Wanma 2006 *Warta Konservasi Lahan Basah* **14 (4)** 4-5.

[19] Kuncharoen N, Muramatsu Y, Shibata C, Kamakura Y, Nakagawa Y and Tanasupawat S 2017 *International Journal of Systematic and Evolutionary Microbiology* **67**(1) 37–41.

[20] Abdel-Rahman HM, Salem AA Moustafa MMA and El-Garhy HAS 2017 *Acta Physiologiae Plantarum* **39**(2) 39–61.

Antifungal Activity of Red Dragon Peel (*Hylocereus polyrhizus*)

Rudi Hendra[1*], Lidya Masdeatresa[1], Muhammad Almurdani[1], Rizky Abdulah[2] and Yuli Haryani[1]

[1] Department of Chemistry, Faculty of Mathematics and Natural Sciences, University of Riau, Pekanbaru, Indonesia
[2] Faculty of Pharmacy, University of Padjadjaran, West Java, Indonesia

*Corresponding author: rudi.hendra@lecturer.unri.ac.id

Abstract. Red dragon fruit (*Hylocereus polyrhizus*) is a species of plant that provides natural pigment which is betalain. Betalain is one of the pigments that give a natural color to flowers and fruits. Moreover, of betalain in red dragon fruit peel, there is a bioactive compound such as polyphenol and flavonoid with antioxidant and antifungal activities. This research was objected to extract pigments and other secondary metabolites from red dragon fruit peel using ethanol, methanol, dichloromethane, and ethyl acetate solvents, continued with determined potential activity against *Candida albicans* by using Resazurin Microtiter Assay (REMA) method. Furthermore, total phenolics compounds (TPC) and total flavonoid compounds (TFC) were determined from all the extracts. The results of antifungal activity showed various inhibition percentage against the fungal at 500 ppm. The ethyl acetate showed the highest activity among the extracts with 74.27% inhibition followed by pigment extracts (60.88%). Furthermore, from test TPC and TFC determination, the ethyl acetate possessed the highest amount with the value of 34.38 mg gallic acid equivalent/ g fresh weight and 28.62 mg quercetin/ g fresh weight, respectively. Therefore, to find the active compounds, multiple analysis is needed.

1. Introduction

Plants, when extracted, can produce dyes and the extracts have been used for different purposes such as the textile or food industry [1]. One potential plant that can be used as a natural dye is dragon fruit (*Hylocereus cacti*) or by another name Pitaya. Wybraniec and coworkers reported that dragon fruit has a potential as a source to produce red pigment and it's called betalains [2]. Betalain is a water-soluble pigment that gives color to flowers and fruits. Betalain pigments are divided into two groups, namely betacyanin which produces purplish red and betaxanthin which produce yellow-orange colors [2, 3].

In the dragon fruit peel, it is also containing pigments which are usually only discarded as food waste and have not been used optimally. This is very unfortunate because the fruit of the dragon fruit itself has several advantages. According to Wu et al. (2006), the peel possessed polyphenol compounds with antioxidant activities [4]. Whereas according to Rahmawati (2016), the peel contains betacyanin, flavonoids, and phenol. Also, dragon fruit skin contains some vitamins, terpenoids, flavonoids, thiamine, niacin, pyridoxine, cobalamin, phenolic, carotene, and phytoalbumin which are thought to have antioxidant benefits and can also be potential for antimicrobial activities [5]. From previous studies, the fruit possessed antimicrobial activities against pathogenic microbial with different assay [6, 7]. Nowadays, there is growing interest in pharmaceutical industries to investigate plants as a source of active compounds with fewer side effects to human especially in combating infection diseases [8, 9].

Due to these facts, the exploration of new alternative medicines derived from plants is required. Flavonoids are one of phenolic group which has antimicrobial activity [10]. These compounds act to inhibit nucleic acid synthesis, cytoplasmic membrane function, and energy metabolism [11]. Therefore, this study was conducted to evaluate the antifungal activities of dragon fruit peel extracts against pathogenic fungal as well as to determine their total phenolics and flavonoids compounds.

2. Experimental

2.1. Extraction
The fresh peels were cleaned, cut into small pieces, blended and kept in a refrigerator. One kg of the peels was extracted by using maceration with 80 % (v/v) ethanol and HCl 0.1 N for 12 hours. The mixture was sonicated for 25 minutes and then filtered. The extract was concentrated and obtained pigment extract. The extract was partitioned with n-hexane, dichloromethane, and ethyl acetate to obtain n-hexane, dichloromethane, ethyl acetate, and pigment extracts. Furthermore, the residue was extracted by using maceration with methanol for 24 hours. The mixture was filtered and concentrated to obtained methanol extract. In the same manner, the methanol extract was partitioned to obtained n-hexane, dichloromethane, ethyl acetate extracts.

2.2. Antifungal Activity
The antifungal activities of the extracts were assayed against *C. albicans* by the microdilution method followed by Sarker *et al.* 2007 [12, 13].

2.3. Total Phenolics and Flavonoids Compounds
Total phenolic compounds were determined according to Hendra et al. [9] and the results were expressed as mg gallic acid equivalents/g fresh weight (DW). Total Flavonoid Assay The total flavonoid compounds in each extract was determined according to Hendra *et al.* [9] and expressed as mg quercetin equivalents/g fresh weight.

3. Results and Discussion
In this research, the pigment from the dragon fruit peel was extracted by using ethanol 80% and 0.1 N HCl and obtained red pigment. The ethanol was used because from previous publication showed that this aqueous organic solvent was optimal solvent to extract betalains. In this extraction, the acid was added due to increasing the stability as well as to avoid oxidation by polyphenoloxidase. [14] In this research, after the pigment extraction, the residue was air-dried and then applied to maceration with methanol and followed by liquid-liquid extraction to obtained n-hexane, dichloromethane, and ethyl acetate extracts.

The antifungal activity against Candida albicans by using microdilution method. In this assay, the resazurin was added as a microbial growth indicator. The addition of resazurin in the assay can calculate existence of visual antimicrobial activity (qualitative) which is through color change. A sample or extracts which have antimicrobial activity if the color is changed from pink to blue, whereas it does not have an activity or weak activity, the color stays in pink. This resulted from the oxyreductase enzyme in the microbe reduce blue resazurin into pinkish resofurin.[15] Furthermore, the antimicrobial activity can be determined by measuring the optical density to give the inhibition growth. The antimicrobial activity from various extracts of the species in 500 ppm was presented in Table 1.

All extracts exhibited moderate to low inhibition activity toward the fungal. The ethyl acetate showed the highest activity among the extracts followed by the pigment extract, but these activities are lower than the positive control (Table. 1). Based on our knowledge, there is no report regarding antifungal activity from the peel of these fruits. Furthermore, the table showed that the extracts exhibited various amounts of TPC and TFC. The ethyl acetate showed a high amount of TPC and TFC with a value of 34.38 mg gallic acid equivalent/ g fresh weight and 28.62 mg quercetin/ g fresh weight, respectively.

Table 1. The antimicrobial activity from various extracts of the species in 500 ppm.

Extracts	% Inhibition [a]	Total Phenolic compounds (TPC) [b]	Total Flavonoids compounds (TFC) [c]
Pigment	60.88	29.07	25.00
n-Hexane	44.95	16.29	14.27
DCM	47.23	27.54	25.95
Ethyl acetate	74.27	34.38	28.62
Ketoconazole	90.97	-	-

[a]. at concentration 500 ppm; [b] mg gallic acid eq./g fresh weight; [c] mg quercetin eq./ g fresh weight.

This study showed that the extract with high antifungal activity showed high TPC and TFC. This might be there is a correlation between phenolic and flavonoid compounds with antifungal activity. Previous reports showed that some medicinal plants contain phenolic and flavonoid compounds had antimicrobial activities [16-18]. Rodriguez Vaquero *et al.* (2007) investigated properties of some phenolic in different wine as antibacterial gallic, caffeic, vanillic acid, rutin, and quercetin) of different wine against pathogen bacteria and the result showed that they could inhibit the growth of *E. coli*. Further, these activities related to the concentration of phenolic compounds present in the wine [19]. The presence of flavonoid compounds in the ethyl acetate extracts suggests antifungal activities, for example, kaempferol and quercetin showed antibacterial activity against *Bacillus sp.* and other human pathogenic bacteria.[17, 20-22]. Therefore, to find the bioactive compounds from the extracts, an ongoing study is conducted by using chromatography and spectroscopy techniques.

4. Conclusions

The extracts responded differently toward antifungal and TPC and TFC. It can conclude that the pigment and the extracts from the peel can be classified as a moderate source of potent natural antifungal agents and it might be abundant of phenolic and flavonoids compounds. Therefore, further investigation is needed.

Acknowledgments
This study was supported by Ministry of Research, Technology and Higher Education in DRPM 2019 Contract no 742/UN.19.5.1.3/PT.01.03/2019 and this support is highly appreciated.

References

[1] Pujilestari T. 2016. *Majalah Ilmiah*. **32(2)**:93-106.
[2] Wybraniec S, Nowak-Wydra B, Mitka K, Kowalski P, Mizrahi Y. 2007. *Phytochemistry*. **68(2)**:251-9.
[3] Wybraniec S, Platzner I, Geresh S, Gottlieb HE, Haimberg M, Mogilnitzki M. 2001.Betacyanins fr. *Phytochemistry*. **58(8)**:1209-12.
[4] Wu L-c, Hsu H-W, Chen Y-C, Chiu C-C, Lin Y-I, Ho J-aA. 2006. *Food Chem*. **95(2)**:319-27.
[5] Shinta DY, Hartono A. 2018. *Sainstek: J. Sains & Tekn.*. **9(1)**:26-39.
[6] Nurmahani M, Osman A, Hamid AA, Ghazali FM, Dek MP. 2012. *Int. Food Res. J.*. **19(1)**:77.
[7] Ramero J, Waing K, Valentino M. 2017. *Int J Biol Pharm Allied Sci*. **6(6)**:1169-80.
[8] Cragg GM, Newman DJ, Snader KM. 1997. *J. Nat. prod*. **60(1)**:52-60.
[9] Hendra R, Ahmad S, Sukari A, Shukor MY, Oskoueian E. 2011. *Int. J. Mol. Sci*. **12(6)**:3422-31.
[10] Cowan MM. 1999. *Clinical microbiology reviews*. **12(4)**:564-82.
[11] Cushnie TT, Lamb AJ. 2005. *Int. J. Antimicrob. Agents*. **26(5)**:343-56.
[12] Elshikh M, Ahmed S, Funston S, Dunlop P, McGaw M, Marchant R. 2016. *Biotech lett*. **38(6)**:1015-9.
[13] Putri R, Hendra R, Teruna HY. 2019. *Pharmacol. and Clin. Pharm. Res.*. **10**:100.
[14] Azeredo H. 2009. *Int J Food Sci Tech*. **44(12)**:2365-76.

IOP Conf. Series: Materials Science and Engineering **833** (2020) 012014 doi:10.1088/1757-899X/833/1/012014

[15] Sarker SD, Nahar L, Kumarasamy Y. 2007. *Methods*. **42(4)**:321-4.
[16] Rahman MAA, Moon S-S. 2007. *Arch. Pharmacal Res*. **30(11)**:1374-9.
[17] Hendra R, Keller PA. 2017. *J. Nat. Prod*. **80(7)**:2141-5.
[18] Hendra R, Willis A, Keller PA. 2019. **33(14)**:1997-2003.
[19] Vaquero MR, Alberto MR, de Nadra MM. 2007. *Food control*. **18(2)**:93-101.
[20] Cushnie TPT, Lamb AJ. 2005. *Int J Antimicrob Agents*. **26(5)**:343-56.
[21] M Calderon-Montano J, Burgos-Morón E, Pérez-Guerrero C, López-Lázaro M. 2011. *Mini Rev Med Chem*. **11(4)**:298-344.
[22] Hendra R, Keller PA. 2016. *Aust. J. Chem*. **69(8)**:925-7.

The Reduction of Cadmium Concentration by *Oscillatoria* Microalgae in a Culture Media

Hentiana[1], Doni Setiawan[1], Risfidian Mohadi[2], Hermansyah[2], and Hilda Zulkifli[1,*]

[1]Biology Department, Faculty of Mathematics and Natural Sciences, Universitas Sriwijaya, Jalan Raya Palembang Prabumulih KM32, Inderalaya, South Sumatera, Indonesia 30662.
[2]Chemistry Department, Faculty of Mathematics and Natural Sciences, Universitas Sriwijaya, Jalan Raya Palembang Prabumulih KM32, Inderalaya, South Sumatera, Indonesia 30662.

*Corresponding author: hilda.zulkifli@yahoo.com

Abstract. Microalgae has functional group such as carboxylic (-COOH) and hydroxyl (-OH) which play an important role as adsorbent of heavy metal. Reduction of cadmium concentration by *Oscillatoria* microalgae in a culture medium has been carried out. This research aimed to study the ability of *Oscillatoria* microalgae as a biological adsorbent agent of cadmium heavy metal using BG-11 media. There were four treatments culture media with six repetitions consisting of P_0 = 100 mL medium + 10,000 individual *Oscillatoria*; P_1 = 1 mg/L Cd(II) + 100 mL medium + 10,000 individual *Oscillatoria*; P_2 = 3 mg/L Cd(II) + 100 mL medium + 10,000 individual *Oscillatoria*; and P_3 = 5 mg/L Cd(II) + 100 mL medium + 10,000 individual *Oscillatoria*. The result has shown the increasing of *Oscillatoria* in BG-11 culture media starting from day 4 to day 7 day was around 1208-1599 ind/L/day. P_3 treatment which consist of 1 mg/L show the highest reduction of cadmium absorption with percentage of 24.05%, compared to the others.

1. Introduction

Heavy metal is a metal that has not more than 5 g/cm^3 of density. Heavy metal could react to form coordination bonding with ligands, i.e. OH, -COO-, -OPO$_3$H-, -C=O, -SH, -S-S-, -NH$_2$ and =NH [12]. Based on toxicology explanation, non-essential heavy metal is toxic metal in which their existence in the human body does not have any function yet such as Hg, Cd, Pb, Sn, and Cr(VI) and As. These heavy metals bring a negative effect on human health [20]. The effects of heavy metal could directly impact the body if it accumulated on food chain although in a very small concentration. Even though it needs a long time to transferred, in the end it will be harmful.

Cadmium is one of the heavy metals with a wide range of dissemination in nature. This metal has 48 atomic numbers, 112.40 atomic mass with 321°C of melting point and 765°C of boiling point [13]. In the natural water, cadmium (Cd) form complex bonding with organic or inorganic ligands which are Cd^{2+}, $Cd(OH)^+$, $CdCl^+$, $CdSO_4$, $CdCO_3$, and Cd organic [17]. Cadmium has a low boiling point and easy to concentrate in the atmosphere. Water could be contaminated by sediment and mining waste that contain cadmium. Meanwhile, when contacted with fume, it will produce air pollution [9].

There were many papers about heavy metal waste treatment to resolve the environmental contamination and poisoning risk for living organisms. Adsorption process could remove the heavy metal from water. This technique is more valuable compared to other; it is more economical and does not produce any toxic substances. Generally, the adsorption method worked because of the interaction between the metallic ion and functional group that exist in the surface of adsorbent through complex formation reaction. Moreover, it always happens in the surface of solid which has functional groups of –OH, -NH, -SH and –COOH [3]. In reality, the process of heavy metal waste handling could be solved by using microalgae as the absorption agent of the metal.

Microalgae is a eukaryotic autotrophic protista and it differentiated by cell wall flagella, chloroplast and food reserves. Microalgae has green photosynthetic pigment (chlorophyll), brown (fucoxanthin), greenish blue (phycobilin), and red (phycoerythrin). Moreover, microalgae life supplies 50% of aquatic oxygen and the main constituents of plankton in the humidity surface [2]. The morphology structure of microalgae can be a single cell or colony and survives in the various aquatic habitats. Sometimes, microalgae have unicellular or multicellular structure. The difference between microalgae and plant is that algae cell does not have clear function [15]. In this research, *Oscillatoria* is the microalgae that utilizes for metal absorption.

Oscillatoria is an alga from Cyanophyceae class included in Cyanophyte division which has clorophyll combination of green and blue (fucoxanthin) usually discovered in freshwater. Generally, this genus can adapt and survive in every environmental condition because it has good metabolism in adjusting the amount of chlorophyll with their cell [5,18]. *Oscillatoria* has specific, bluish-green color and forms long, straight and soft filament. Fragmentation is the way of *Oscillatoria* to multiple. When one of the cells died in the long filament, this cell will seperated into 2 or more pieces. The process is known as hormogonium. Fragmentation also comes from cell wall separation with trachoma [11]. *Oscillatoria* develops in culture media to produce the abundance cell of it.

Media is the main material that utilized for isolation and culture which composed by the mixing of food substances or nutrients as the energy source for microorganism [21]. Sylvester [19] stated that the purpose of microalgae cultivation in media culture is to reproduce new individual. The materials were absorbed by cell that utilized for metabolism. In the bioenergy process, nutrient has a function as electron acceptor. The chemical energy from metabolism was spent on reproduction, spore formation, and biosynthesis.

The mechanism of heavy metal intake divided in two ways which are adsorption and absorption. The processes are ion exchange and ion bonding of heavy metal on the surface by the functional group. Generally, *Oscillatoria* cell wall made from cellulose that has functional group such as hydroxyl which can bind together with heavy metal [6]. Absorption was worked slower than adsorption through active transport. Heavy metal was bonded with protein and it collected in vacuoles of *Oscillatoria* [4].

2. Experimental Method

2.1. Treatment Test

This treatment was done in different place with six times of repetitions:
Treatment P_0 = 1000 mL media + 10,000 *Oscillatoria*
Treatment P_1 = 1 mg/L Cd(II) + 1000 mL media + 10,000 *Oscillatoria*
Treatment P_2 = 3 mg/L Cd(II) + 1000 mL media + 10,000 *Oscillatoria*
Treatment P_3 = 1 mg/L Cd(II) + 1000 mL media + 10,000 *Oscillatoria*

2.2. BG-11 Media as Culture

The synthesize of BG-11 media, as shown in Table 1, was made by adding 10 mL of stock solution 1 and 2, 1 mL of trace metal solution into Erlenmeyer continued by adding water until 1000 mL and finally homogenized. This solution mixture was then heated in an autoclave at 121°C with 15 lbs pressure for 15 minutes.

2.3. Glassware Sterilization

All glassware are washed, cleaned and then sterilized in an autoclave at 121°C with 15 lbs pressure for 15 minutes.

Table 1. The Composition of BG-11 Media.

Composition	Amount (g/mL)
Stock 1 (Per 500 ml):	
$NaNO_3$	75.0 g
Stock 2 (per 500 ml):	
K_2HPO_4	2.0 g
$MgSO_4.7H_2O$	3.75 g
$CaCl_2.2H_2O$	1.80 g
Citric acid	0.30 g
Ammonium ferric citrate green	0.30 g
EDTANa2	0.05 g
$Na2CO_3$	1.00 g
Trace metal solution (per 1.000 ml):	
H_3BO_3	2.86 g
$MnCl_2.4H_2O$	1.81 g
$ZnSO_4.7H_2O$	0.22 g
$Na_2MoO_4.2H_2O$	0.39 g
$CuSO_4.5H_2O$	0.08 g
$Co(NO_3)_2.6H_2O$	0.05 g
Medium	
Stock solutions 1	10 mL
Stock solution 2	10 mL
Trace metal solution	1.0 mL
Aquadest	1.000 mL

2.4. *Cadmium Stock Solution*

Cadmium stock solution was made based on dilution formula:

$$M_1 \times V_1 = M_2 \times V_2 \tag{1}$$

note:
M_1: initial concentration
M_2: calculated concentration
V_1: initial volume
V_2: calculated volume

2.5. *Oscillatoria Cultivation in BG-11 Media*

The pristine algae cultivation prepared from 6 bottles that filled up with BG-11 media until 995 mL which has been sterilized. 10,000 *Oscillatoria* germ put in each bottle and each of it connected with a generator and illuminated by neon light 36 watts with 3000 lux of light intensity for seven days continuously.

2.6. *Oscillatoria Inoculum Calculations*

The initial germ of *Oscillatoria* is 10.000 ind/L. The population of culture stock was counted by microscope with 10 times of magnification using Sedgewick Rafter Counting (SRC) formulae:

$$V_1 = N_2 \text{ x } V_2 / N_1 \tag{2}$$

note:
V_1= Calculated inoculum volume (mL)
V_2= Culture media volume (1.000 mL)
N_1= Stock abundance (unit/mL)
N_2= Calculated abundance cell (unit/mL)

2.7. *Oscillatoria Treatment Step*

Twenty-four bottles were initially prepared, 6 bottles for control contain only BG-11 media and another 18 bottles contain 995 BG-11 media mixed with prepared cadmium. Next, all bottles were sterilized in an autoclave at 121°C with 15 lbs pressure for 15 minutes. 10,000 of *Oscillatoria* were then placed in every sterilized bottle and the growing rate of *Oscillatoria* was counted every day for seven days.

2.8. *Oscillatoria Growing Rate Calculation*

The growing rate of *Oscillatoria* was counted using SRCC, while the specific algae growth rate is counted by the Vonshak formula:

$$\mu = (\ln N_t - \ln N_0)/t \quad (3)$$

note:

μ = Specific growing rate (days^{-1})
N_0 = Initial *Oscillatoria* cell solidity (ind/mL)
N_t = Calculated *Oscillatoria* cell solidity (ind/mL)
t = time interval N_0 to N_t (days)

The concentration of cadmium was analyzed by Atomic Absorption Spectroscopy (AAS) AA-7000 Shimadzu.

3. Result and Discussion

3.1. *The growing Rate of Oscillatoria*

Oscillatoria growing rate was carried out using *Sedgwick Rafter Counting Cell* (SRCC) under the microscope with 40 times magnification. The observation of *Oscillatoria* growing rate finished in day 7 inside BG-11 media. Figure 1 describes the growing rate of *Oscillatoria* at P_0 treatment during the research. From the beginning until the last days of observation, *Oscillatoria* growth was increased, from 136 Ind/L/day in day 1 to 1599 Ind/L/day in day seventh. These number of *Oscillatoria* growth is proof that BG-11 media is suitable for algae cell growth. Based on Figure 1, the optimum condition of *Oscillatoria* harvest time is the seventh day. Isnanstyo and Kurniastuti [10] explain that the best harvest time for microalgae is in the last exponential phase. Currently, in this phase, microalgae are in the optimum condition, it has high nutrient content of the cell. Between day 4th and 7th, the total abundance of *Oscillatoria* cell is significantly increases and it called as the exponential phase. The cells are actively multiply through the division in the exponential phase, it causes the rapid cell growth.

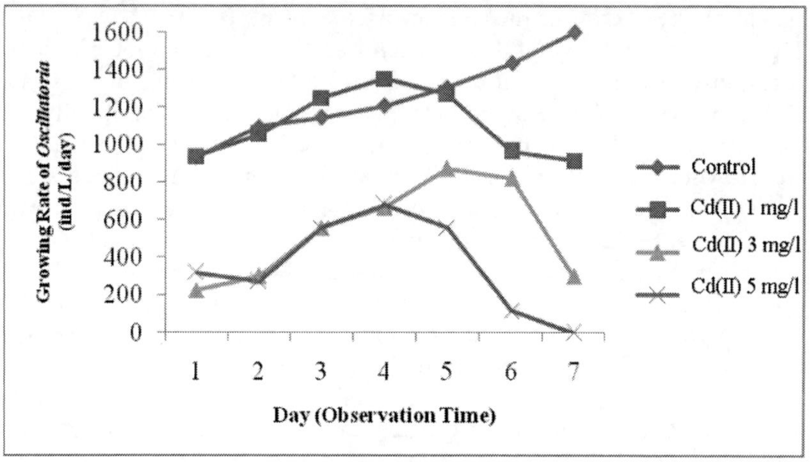

Figure 1. The growing rate of *Oscillatoria (*Ind/L/day) in BG-11 media.

The abundance of *Oscillatoria* was calculated each day for seven days. The growing rate was counted by adding heavy metal Cd(II) using variation: 0; 1; 3; 5 mg/L. The purpose of this growth data retrieval was to know the ability of *Oscillatoria* to absorb Cd(II) and the effect Cd(II) on *Oscillatoria*. The algae growth was measured by BG-11 media with six repetitions. At the beginning of culture, 1000 Ind/L/day of *Oscillatoria* was used and the average of algae growth that obtained in the first day reach 938 Ind/L/day with P_0 (control) treatment. P_0 (control) test in the second day was produced 1098 Ind/L/day and slightly increase in the third day. However, the increasing number was not significantly happened because *Oscillatoria* was in [7] the adaptation phase with the culture media. [16] It was concluded that in the adaptation phase by adding inoculum into the culture media in the population will not giving any changes. The population did metabolism without cell division, so the cell density was not increased quickly.

In the fourth day of *Oscillatoria* cell growth in BG-11 media which contain 1 mg/L of Cd(II) represent its optimum condition or exponential phase. In fact, *Oscillatoria* can tolerate that amount of Cd(II) and create 1345 Ind/L/day of a cell. Explain that, exponential phase duration depends on inoculum volume, growth speed, media and the surrounding which support the algae development. Exponential phase marked by rapid growth period, cell division with a constant rate, the equilibrium between nutrient absorption, and the increasing number of microalgae. Moreover, Cd(II) almost has the same characteristic with zinc (Zn) and this is the reason why Cd(II) can substitute Zn in enzymatic reaction through enzymatic structure changing and affect the activation process. Moreover, a small amount of Zn could stimulate the development of the plant. These abilities of Cd(II) could replace the function of Zn in the synthesis of carbonic anhydrase enzymes and generate hydrogen ion for cell division [1].

The growth of *Oscillatoria* between day fifth and sixth with 3 mg/L of Cd(II) was likely stable. This static condition caused by the lacking supply of nutrient for *Oscillatoria* that used for the division process. Compared to day seventh, the amount of *Oscillatoria* was rapidly decreased to 300 Ind/L/day. In a constant condition, the composition of microalgae was significantly changed because the limited amount of nitrate in culture media which multiplied the carbohydrate compound, compared to that of protein. This affect cell reproduction rate equal with the dead rate, which means cell addition and elimination was insignificantly different.

The growth of *Oscillatoria* algae with 5 mg/L was decreased dramatically until the day of observation. After a deep observation through the microscope, the cell was burst and died which was proved by the changing of color of the cell. At the beginning, the culture color was green and turned into light green to eventually pale green. The increasing number of Cd and Hg concentration will disturb the growth of algae cell. Furthetmore, the standar deviasion each treatment (control, Cd(II) 1, 3, and 5 mg/L) are lower than the average value. The avarage value are 1247; 1107; 535; and 358 Ind/L/day while the number of standar deviasion are 221; 180; 265; and 252 Ind/L/day. This data explain the small posibility of error in the conducted experiment.

Based on Figure 1, the conclusion of this research was the adaptation process happened in the first to third day. Some parameters which influence the cell growth were the size of inoculum and the condition of growth media. If a cell was placed in a lack of nutrients media, it will need a longer time of adaptation because the cell needs to produce an enzyme which is the same as the nutrient.

Oscillatoria growth depends on the given nutrients and environmental condition. The environmental conditions are temperature, aeration, light intensity, and pH. In this research were used 6 as constant pH for each test. Moreover, aeration also applied in the observation with water circulation to deliver the same amount of nutrient for each alga in the culture media. Microalgae also utilized the light for photosynthesis because the light intensity is important for the growth of algae [14]. The culture was finished at 25°C as the optimum temperature of *Oscillatoria* growth likewise every species has its optimum temperature. Hariati [8] stated that temperature is an important factor for the green blue algae spreading and behavior. Generally, 20°C-30 °C is the optimum temperature in most of green blue algae. BG-11 media as a place for growing the alga made from phosphor and potassium which came from KH_2PO_4 and K_2HPO_4. The function of KH_2PO_4 compound was as the source of phosphor for the synthesis of energy activation cell substances. Meanwhile, potassium role was to strengthen the algae cell, algae adding cell, nutrient absorption, and metabolism process.

3.2. The percentage of Cd(II) in the culture media

The observation of decreasing Cd(II) concentration in each volume 1 mg/L, 3 mg/L, dan 5 mg/L of culture media was analyzed by Atomic Absorption Spectrophotometer (AAS) and it is presented Table 2.

Based on Table 2, the highest decreasing concentration of Cd(II) was 1 mg/L with reduction percentage approximately 24.05%. The higher concentration of heavy metal was inversely proportional to the capability of *Oscillatoria* to absorb it. The dimension of concentration was declining from 1 ppm, 3 ppm, 5 ppm in a row because the absorption of the metal ion was decreasing. This data became proof when the concentration of heavy metal is high it will obstruct the growth of *Oscillatoria.*

The mechanism of Cd(II) absorption by *Oscillatoria* was divided into two processes which are adsorption and absorption. The processes are ion exchange and ion bonding of heavy metal in the surface by the functional group. Generally, *Oscillatoria* cell wall made from cellulose that has functional group such as hydroxyl that binds together with heavy metal. Cellulose has good potential as metal ion catcher because OH cluster interacts with adsorbate like Cd(II). Absorption happens through active transport. Heavy metal was accumulated in the cell and bind with protein i.e. metallothionein and phytochelatin in the vacuole.

Cadmium is a heavy metal that pollutes environmental if the amount exceeds the normal tolerance. Microalgae were utilized as bio-sorbent in bioremediation agent and there were so many works that observe about algae i.e. *Scenedesmus dimorphus* sp. Fauziah [4] reported that *Scenedesmus dimorphus* sp could be used as cadmium absorbent.

Table 2. The percentage of Cd(II) reduction in day seventh.

No	Concentration of Cadmium Cd(II) mg/L.			Reduction percentage (%)
	Initial	Final	Absorb	
1	1	0.7595	0.2405	24.05 %
2	3	2.8385	0.1615	5.38 %
3	5	4.8826	0.1174	2.34%

4. Conclusion

Oscillatoria has the ability in absorbing cadmium with absorption percentage of 24.05% for 1 mg/L of Cd(II). The growth rate in BG-11 media was relatively good in the day seven in which it can duplicate themselves into 1599 Ind/L/day especially in control media.

References

[1] Anderson R A 2005 Algal culturing technique London 245-250 *Elsevier Academic Press* London
[2] Barsanti L and Gualtieri P 2014 *Algae: Anatomy, Biochemistry, And Biotechnology Second Edition* (Jakarta: Widya Graha *LIPI*)
[3] Blais J F B, Dufresne, and G Mercier 2000 *Revuedes Sciences de l'Eau* **12(4)** 687-711
[4] Fauziah 2011 (Jakarta: Universitas Syarif Hidayatullah)
[5] Guiry M D and Guiry G M 2014 *Algaebase* World-wide Electronic Publication (National University of Ireland: Galway)
[6] Gupta R, P Ahuya, S Khan, R K Sakena, and H Mohapatara 2000 *Current science* **78(8)**
[7] Hadiyanto and Azim M 2012 *Mikroalga Sumber Pangan Dan Energi Masa Depan* (Semarang: Undip Press)
[8] Hariati R 2008 *Jurnal Biologi.* **10(1)** 19-22
[9] Herman D Z 2006 *Jurnal Geologi Indonesia* **(1)1** 31-36
[10] Isnanstyo, Alim, and Kurniastuty 1995 *Teknik Kultur Phytoplankton Zooplankton Pakan Alam Untuk Pembenihan Organisme Laut. (*Yogyakarta*)*
[11] Khoirul A A 2013*Cyanobacteri (Alga hijau-biru) (*Malang: Universitas Brawijaya)
[12] Nugroho A 2006 *Makar Teknologi* **10(2)**
[13] Palar H 2004 *Pencemaran dan toksikologi logam berat* (Jakarta: Rineka cipta)
[14] Regisata M, Litaay M, and Umar R M 2017 *Bioma: Jurnal Biologi Makassar* **2(1)**
[15] Ronaldo, Silalahi I H, and Wahyuni N 2013 J. K. K **2(3)** 148-152

IOP Conf. Series: Materials Science and Engineering **833** (2020) 012015 doi:10.1088/1757-899X/833/1/012015

[16] Sahabuddin, Amin M, and Jompa H 2013 Respon Pertumbuhan Populasi *Oscillatoria* sp. Pada Peningkatan Konsentrasi Karbodiosida Antripogenik (*Prosiding Forum Inovasi Teknologi Akuakultur*)
[17] Sanusi H S 2006 *Kimia Laut Proses Fisik Kimia dan Interaksinya dengan Lingkungan (*Bogor: Departemen Ilmu dan Teknologi Kelautan Fakultas Perikanan dan Ilmu Kelautan Institut Pertanian Bogor)
[18] Sari E W 2011 Isolasi Dan Identifikasi Mikroalga Cyanophyta Dari Tanah Persawahan Kampung Sampora, Cibionong, Bogor *Skripsi* (Jakarta: Fakultas Sains Dan Teknologi Universitas Islam Negeri Syarif Hidayatullah)
[19] Sylvester B D, D Nelvy, and Sudjiharno 2002 *Dalam* Seri Budidaya Laut No. 9.
[20] Vouk V 1986 *General Chemistryof Metals In: Freiberg L, Nordberg G F and Vouk V B (Eds) (*New York: Handbook on the Toxicology of Metals)
[21] Wijoseno T 2011 Uji Pengaruh Variasi Media Kultur Terhadap Tingkat Pertumbuhan dan Kandungan Protein, Lipid, Klorofil, dan Karotenoid pada Mikoalga *Skripsi* (Depok: Universitas Indonesi)

Phytochemistry and Antibacterial Activity Evaluation of Genitri (*Elaeocarpus ganitrus*)

Retno Indriatie[1,2], Siti Mudaliana[1], Febriyana Rizky Hapsari[1], Masruri MASRURI[2*]

[1]UPT Materia Medica, Jl. Lahore Batu, Indonesia
[2]Chemistry Department, Faculty of Mathematics and Natural Sciences, Brawijaya University. Jl. Veteran 65145 Malang, Indonesia

*Corresponding author: masruri@ub.ac.id

Abstract. The *Elaeocarpus ganitrus* has local name as genitri. It has been used traditionally as traditional medicine. This paper reported the phytochemistry and their antibacterial activity on *Staphylococcus aureus* and *Escherichia coli*. Several extracts have been afforded using high speed extraction technique using methanol, ethyl acetate and n-hexane as solvents. The extract is composed of alkaloid, tannin, and flavonoid. The negative result is afforded for saponin test. The MIC of all extracts in both bacteria are above 10 mg/ mL. Moreover, the quantitative analysis using spiking method using liquid chromatography found a quercetin and rutin in minor quantity of some sample extract.

1. Introduction

The plant of genitri (Javanesse) has a taxonomy name as *Elaeocarpus ganitrus* Roxb. Ex G. Don and a synonym as *Elaeoscarpus serratus* L. It is categorized as a high plant which commonly growth in forest. The distribution of this plant is getting wider, not only in West and Central Java but also the other places in Indonesia [1]. Traditionally, part of genitri tree, such as seed, bark, leave, and flowers was applied and used as source of herbs and a medicine for some diseases [2, 3]; such as antidepressant [2], antihypertensive [4], antifungal [5], anti-diabetic [6], and some other mental disorder diseases [3].

Recent report by Singh et al. [7] disclosed antianxiety activity of the flower extract prepared from genitri flower. Several solvents were used to extract, such as using petroleum ether, chloroform, and ethanol. The ethanol extract showed antianxiety with a safe dose until 5000 mg/kg of animal body weight. Interestingly, using 25 mg/kg of doses still indicate the antianxiety activity. Further separation and purification from the extract gave a 3,3',4',5,7-pentahydroxyflavone or known as quercetin composed the extract. This molecule belongs to the secondary metabolite class of flavonoid. Furthermore, this class of flavonoid was also able to be used for antibacterial agent. The mechanism relates to inactivation process of type III protein secretion substrate [8]. For example antibacterial activity of quercetin isolated from lotus leave [9], and also quercetin glycoside from narrow leaf vetch [10]. Quercetin also enhance the antibacterial activity by loading it in poly-lactic-co-glycolic acid nanoparticle both for in vitro and in vivo evaluation [11], and improve the antibacterial for silver nanoparticle [12]. The similar structure to quercetin glycoside reported previously [10] was known as rutin (Fig 1). Rutin also has been reported active as antibacterial agents [13, 14]. In this paper, will be reported the phytochemicals of genitri seed extract grown in Malang, East Java and evaluation of the presence of flavonoid quercetin and rutin (structure in Fig 1) in the extract. And also, evaluation their antibacterial activity in *Staphylococcus aureus* and *Escherichia coli*.

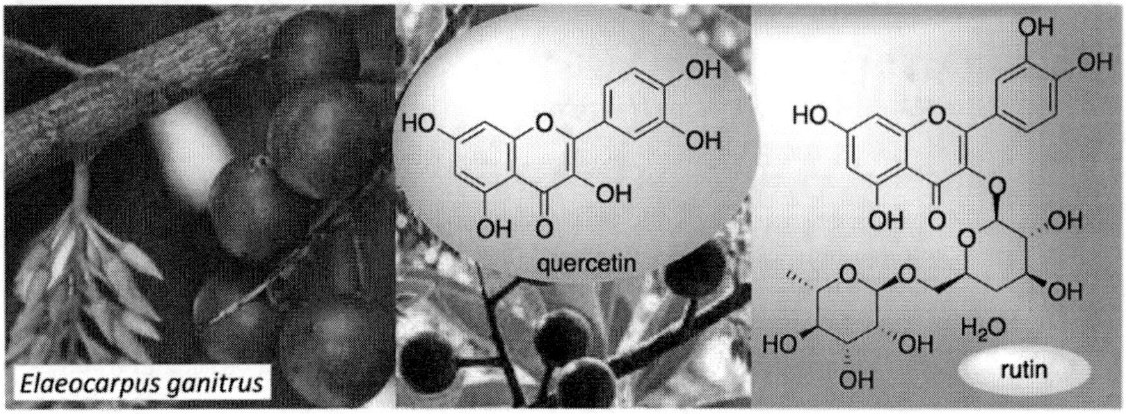

Figure 1. Genitri (*Elaeocarpus ganitrus*) seed and the molecular structure of quercetin and rutin.

2. Experimental section

2.1. Material and chemical

Sample of genitri seed is collected from Balai Materia Medica Plants. The taxonomy name is assessed by the same institution. The fresh seed is washed, grinded, and dried in oven before extraction process. Isolate sample of bacteria *Staphylococcus aureus* ATCC25923 and *Escherichia coli* was prepared from the collection by the same institution. Other chemical used for research include antibiotic gentamycine, Mueller Hilton agar (MHA), BHI broth media, methanol, methanol 70%, ethyl acetate, and n-hexane.

2.2. Instrument analysis

The ultra-high-performance liquid chromatography was operated for flavonoid analysis using UFLC (Shimadzu). Solvent used as mobile phase is methanol/water (63:37) isocratic with flow rate of 1 mL/min and column C18 (5 um) with dimension of 4.6x150 mm, and UV detector set at 254 and 260 nm. The standard of quercetin and rutin applied for analysis was 200 ppm.

2.3. Extraction procedure

The procedure for extraction was undergone using speed extractor apparatus (Buchi E-916, Switzerland). A 5.0 g of seed dried powder of genitri in sample holder was placed in speed extractor apparatus. Then, 50 mL of methanol was flowed under nitrogen at the constant pressure of 100 bar. The process was circulated for three times or 3-cycles. The solvent extract was further evaporated under rotary high vacuum evaporator to afford methanol extract of genitri seed. Separated procedure was conducted using different solvents, i.e. methanol 70%, ethyl acetate, and n-hexane. These processes provided methanol 70% extract, ethyl acetate extract, and n-hexane extract, respectively. For further analysis, similar concentration of extract (10 mg/mL) were prepared using 10 mg of extract dissolved in 1.0 mL of water.

2.4. *Phytochemical evaluation*

2.4.1. Flavonoid test. 1.0 mL of extract sample in a 10 mL of reaction tube was added with water. This mixture was heated for 5 min, and 1.0 mL of aliquot was taken in reaction tube. This mixture was added with 5.0 mg of magnesium powder, and a few drops of 37% hydrochloric acid solution. Positive result of flavonoid was indicated by the formation of orange, pink, or red solution.

2.4.2. Alkaloid test. 1.0 mL of extract sample in a 10 mL of reaction tube was added with water. This mixture was heated for 5 min and divided into three portions in different reaction tube. Each tube was added with 1.0 mL of different reagents, i.e. Meyer reagent, Dragendorf reagent, and Libermann-Burchart reagent. The positive alkaloid test was indicated by the presence of a white precipitate.

2.4.3. Saponin test. 1.0 mL of extract sample in a 10 mL of reaction tube was added with water. This mixture was heated for 5 min. 1.0 mL of filtrate was added with hot water and shaken rigorously and added with 1.0 mL of 37% hydrochloric acid solution. The positive saponin was indicated by permanent foam.

2.4.4. Tannin test. 1.0 mL of extract sample in a 10 mL of reaction tube was added with water. This mixture was heated for 5 min, and 1.0 mL of aliquot was taken in reaction tube and added with 1.0 mL of iron(III) chloride solution. The positive result for tannin was indicated with the presence of blue or black precipitate.

2.5. *Antibacterial activity evaluation*

The procedure to evaluate the antibacterial activity of genitri seed extract used a well diffusion technique. Two types of bacteria were used, i.e. *Staphylococcus aureus* and *Escherichia coli*. The bacteria isolate was grown in BHI broth media and incubated at 37 °C for 24 hours. Then, the bacteria culture in MHA media was spread with cotton swap aseptically. The well in media was made by using cock borer which was a 6.0 mm in diameter. Each well was dropped with 100 uL of the tested extract sample, 100 uL of distillated water as a negative control sample (NC), and gentamycin 10 mg/mL as a positive control. Then, each well was incubated for 24 hours at 37 °C, and a clear area around the well was measured using digital micrometer as an antibacterial activity.

The minimum inhibitory concentration (MIC) was undertaken using similar procedure above. But several different concentrations of extracts were evaluated, i.e. 10, 5.0, 2.5, 1.25, 0.62, and 0.31 mg/mL. Each sample was tested using a 100 uL of volume. The MIC value was determined based on the lowest extract concentration that was unable to inhibit the bacteria growth.

Table 1. Phytochemical analysis result of genitri extracts.

Extract sample genitri seed	Flavonoid test	Tannin test	Saponin test	Alkaloid test	
				Meyer	Dragendrof
Methanol 70% extract	+	+	+	+	+
Methanol extract	+	+	+	+	+
Ethyl acetate extract	-	-	-	-	-
Hexane extract	-	-	-	-	-

3. Result and discussion

Secondary metabolite class compounds are divided into phenolic, flavonoid, alkaloid, and terpenoid. Tannin and flavonoid are classified into phenolic. Meanwhile saponin is classified to terpenoid. Phenolic class is generally soluble in polar solvent such as metanol and water. But terpenoid, especially triterpenoid can be extracted using non polar solvent such n-hexane. The result for genitri

(*Elaeocarpus ganitrus*) seed phytochemical analysis is tabulated in Table 1. This analysis detects the secondary metabolite composed in the extract sample based on the color visualisation or precipitate resulted. The positive results indicate the presence of the secondary metabolites. The extracts of methanol and methanol 70%, indicate a positive result for flavonoid, tannin, saponin, and alkaloid test. However, the ethylacetate and n-hexane extract did not give a positive result. Previously, phenolic groups of flavonoid was also reported [3,7] from the ethanol extract of root and leaves part of genitri.

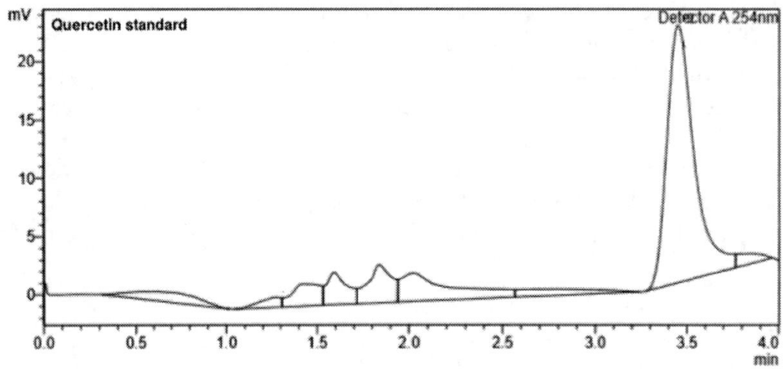

Figure 2. Chromatogram of quercetin standard.

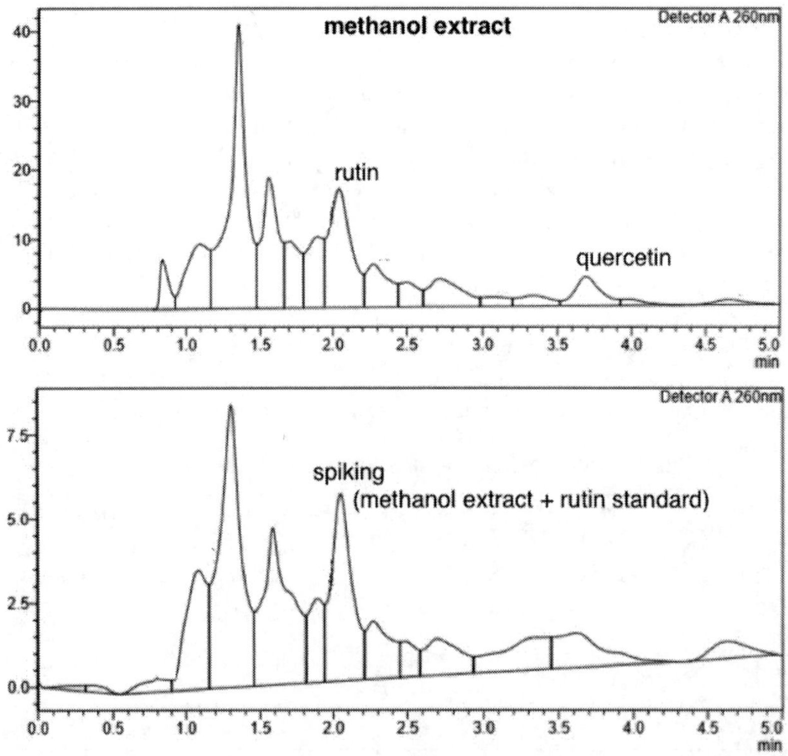

Figure 3. Chromatogram of methanol extract and spiking methanol extract with rutin.

Further flavonoid analysis was undertaken by using liquid chromatography. The representative chromatogram was depicted in Fig 2 and Fig 3. The quercetin and rutin was used as standard. Both are detected at retention time 3.45 and 2.02 minute, respectively (Fig. 3). Meanwhile, the composition of both flavonoid contained in each extract sample of genitri seed are summarized in Table 2. It was found that quercetin composed the methanol extract in the highest quantity (9.51%). Following by ethyl acetate (3.56%) and methanol 70% extract (2.24%). However, no quercetin was detected in the n-hexane extract. On the other hand, the highest rutin quantity was detected composed the methanol 70% extract (6.04%). Following by methanol extract (5.71%), and ethyl acetate extract (5.23%). And it was not found rutin in the n-hexane extract.

Tabel 2. The quantity of phenolic quercetin contained in the genitri extract.

Extract sample of genitri seed	Quercetin detected		Rutin detected	
	Retention time (min.)	Percentage (%)	Retention time (min.)	Percentage (%)
Methanol 70% extract	3.546	2.24%	2.026	6.04%
Methanol extract	3.477	9.51%	2.020	5.71%
Ethyl acetate extract	3.472	3.56%	2.011	5.23%
Hexane extract	-	0	-	0
Quercetin standard	3.453	95%	-	-
Rutin standard	-	-	2.021	97%

In some reports, the presence of quercetin, rutin, and other flavonoid compounds correlated to their bioactivity [15–17], such as growth inhibitor of bacteria [9, 10, 14]. The antibacterial activity of the genitri extract is tabulated in Table 3, meanwhile the minimum inhibitory concentration (MIC) is summarized in Table 4. Each sample tested has a concentration of 10 mg/mL, including the standard antibiotic of gentamycine.

Tabel 3. Antibacterial activity evaluation of the extract from genitri seed.

Sample extract of genitri seed	Growth inhibition (mm)					
	Escherichia coli			*Staphylococcus aureus*		
	Genitri	PC	NG	Genitri	PC	NC
Methanol extract 70%	0	32.50±4.95	0	0	26.50±4.95	0
Crude methanol extract	0	29.00±5.66	0	0	28.50±0.71	0
Ethyl acetate extract	0	27.50±0.71	0	0	31.50±2.12	0
Hexane extract	0	21.00±4.24	0	0	26.50±3.54	0

Note: PC is positive control contains gentamycine 10 mg/mL. NC is negative control contains solvent without sample.

Tabel 4. Minimum inhibitory concentration (MIC) of genitri extracts.

Sample extract of genitri seed	Minimum inhibitory concentration (mg/mL)	
	Escherichia coli	*Staphycoccus aureus*
Methanol extract 70%	>10	>10
Crude methanol extract	>10	>10
Ethyl acetate extract	>10	>10
Hexane extract	>10	>10

The genitri extract with 10 mg/mL concentration did not inhibit the bacterial growth (Table 3). Meanwhile the gentamycine give a strong activity in both bacteria tested. Moreover, the minimum inhibitory concentration (MIC) has value above 10 mg/mL for all extract (Table 4).

IOP Conf. Series: Materials Science and Engineering **833** (2020) 012016 doi:10.1088/1757-899X/833/1/012016

4. Conclusion

The extract of genitri seed is composed of flavonoid, tannin, saponin, and alkaloid. All these secondary metabolites can be isolated using methanol 70%, methanol, and ethyl acetate as solvent. The flavonoid contain in genitri seed is quercetin, and detected in low concentration. However, the antibacterial evaluation to *Staphylococcus aureus* and *Escherichia coli* give insignifficant activity.

References

[1]. Rohandi, A., & Gunawan, G. 2015. *Jurnal Ilmu Kehutanan*, **8** (1), 25–33.
[2]. Dadhich, A., Jasuja, N. D., Chandra, S., & Sharma, G. 2014. *Int. J. Pharm. Sci. Res.*, **5** (7), 2807.
[3]. Singh, B., Ishar, M. P. S., Sharma, A., Arora, R., & Arora, S. 2015. *Isr. J. Plant Sci.*, **62** (4), 265–276.
[4]. Sakat, S., Wankhede, S., Juvekar, A., Mali, V., & Bodhankar, S. 2009. *Int. J. PharmTech Res.*, **1** (3), 779–782.
[5]. Singh, B., Chopra, A., Ishar, M., Sharma, A., & Raj, T. 2010. *Indian J. Pharm. Sci.*, **72** (2), 261.
[6]. Hule, A. K., Shah, A. S., Gambhire, M. N., & Juvekar, A. R. 2011. *Indian Jo. Pharmacol.*, **43** (1), 56.
[7]. Singh, B., Sharma, V., Pal Singh Ishar, M., & Sharma, A. 2013. *Nat. Prod. J.*, **3** (3), 224–229.
[8]. Tsou, L. K., Lara-Tejero, M., RoseFigura, J., Zhang, Z. J., Wang, Y.-C., Yount, J. S., Lefebre, M., Dossa, P.D., Kato, J., Guan, F., & Lam, W., 2016. *J. Am. Chem. Soc.*, **138** (7), 2209–2218.
[9]. Li, M., & Xu, Z. 2008. *Arch. Pharmacal Res.*, **31** (5), 640–644.
[10]. Waage, S. K., & Hedin, P. A. 1985. *Phytochemistry.*, **24** (2), 243–245.
[11]. Sun, D., Li, N., Zhang, W., Yang, E., Mou, Z., Zhao, Z., Liu, H. & Wang, W. 2016. *J. Nanoparticle Res.*, **18** (1), 3.
[12]. Sun, D., Zhang, W., Mou, Z., Chen, Y., Guo, F., Yang, E., & Wang, W. 2017. *ACS Appl. Mater. Interfaces*, **9** (11), 10047–10060.
[13]. Arima, H., Ashida, H., & Danno, G. 2002. *Biosci. Biotechnol. Biochem.*, **66** (5), 1009–1014.
[14]. Farahani, A. S., & Taghavi, S. M. 2018. *Eur. J. Plant Pathol.*, **151** (2), 527–531.
[15]. Masruri, M., Lutfillah, M., Sumaryanto, A., & Retnowati, R. 2014. *J. Trop. Life Sci.*, **4** (3), 161–165.
[16]. Masruri, M., Baihaqi, M. A., Riyanto, S., & Srihardyastutie, A. 2017. *AIP Conf. Proc.,* **1823**, 020015.
[17]. Indriatie, R., Mudaliana, S., Masruri, M. 2019. *IOP Conf. Ser. Mater. Sci. Eng.,* **546**, 042013.

Thiocyanate Doping in Gel-Growth Cobalt Oxalate Crystals

Mohammad Misbah Khunur, Dini Tri Wahyuni, Gigih Wahyu Kurniawan, Yuniar Ponco Prananto*

Department of Chemistry, Faculty of Mathematics and Natural Sciences, Brawijaya University, Jl. Veteran 01 Malang, Indonesia 65145

*Corresponding author: prananto@ub.ac.id

Abstract. The presence of other anions, as impurities and/or doping agents, may alter crystallization and the properties of the crystal. The objective of this research is to examine the effect of thiocyanate doping in the synthesis of cobalt oxalate crystals grown in silica gel. The synthesis was conducted at room temperature for 12 weeks in a U-tube glass filled with silica gel. The gel was prepared from Na_2SiO_3 at pH 5 using dilute HNO_3 with a gelling time of 5 days. Supernatant solutions of $(NH_4)_2C_2O_4$ and $Co(NO_3)_2$/KSCN were added on each side of the tube. The $Co(NO_3)_2$ and KSCN were firstly mixed before adding onto the gel to increase the possibility of thiocyanate doping in the crystals. Various molar ratio of Co(II):$C_2O_4^{-2}$:SCN^- were used (1:1:0, 1:1:1, 1:1:2, 1:1:3, 1:1:4, and 1:1:5). FT-IR and SEM were used to analyse the crystal. The result shows that red block crystals were formed. FTIR analysis for all molar ratios reveals that they all give considerably identical spectra with an absence of thiocyanate peaks around 2200 cm^{-1}, suggesting that all molar ratios give an identical product of cobalt(II) oxalate hydrate. SEM analysis shows that the size of the crystals was significantly bigger than previously reported crystals. The Co(II) centre binds to the oxalate ligand, and not to the thiocyanate, due to chelating effect. However, inconclusive findings were observed regarding the influence of the molar ratio on the crystals yield. The use of a higher concentration of thiocyanate (fivefold molar ratio) did not affect the composition of the product, and the product remains crystallized as cobalt(II) oxalate hydrate.

1. Introduction

Cobalt oxalate crystal is commonly used as magnetic materials [1]. The crystal may be used as a precursor in the synthesis of multipurpose Co_3O_4 nanoparticles [2]. Due to its electronic properties, the cobalt(II) oxalate can be used as a molecule-based magnet or single-chain magnet. Oxalate ions as building blocks have versatile coordination mode, which allows two metal ions connected by the oxalate bridge. Electronic and magnetic properties of cobalt(II)-oxalate is predicted to be influenced by introducing π electron donor ligand, such as thiocyanate (SCN).

Due to the very low solubility of cobalt(II) oxalate in water, which is 2.68×10^{-9} [3], the compound is directly precipitated as amorphous powder. Thus, it is difficult to introduce thiocyanate into cobalt(II) oxalate complex by directly mixing the oxalate and Co(II) solutions. A 1D polymeric structure of $[Co(H_2O)_2(C_2O_4)]_n$ (Figure 1) was reported previously in which the oxalate acts as bridging ligands between two Co metal ions [4,5].

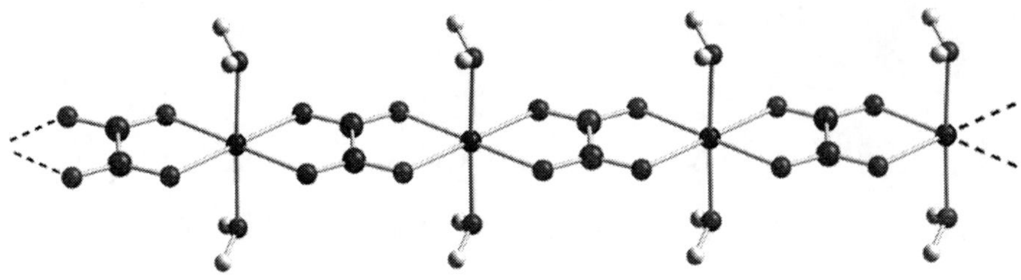

Figure 1. Polymeric structure of cobalt(II) oxalate hydrate (blue = cobalt(II), black = carbon, red = oxygen, white = hydrogen).

Silica gel as a growth medium of metal complexes possesses fascinating features in the synthesis of a low soluble compound due to its ability in controlling crystals nucleation by lowering the diffusion rate of ions [6]. The stability of Si–O–Si networks in the gel also provide space for crystal nuclei to grow into bigger and better single crystals [6]. Previously, we have studied the growth of lead(II) oxalate hydrate [7] and cobalt(II) oxalate crystals in silica gel [8]. The results show that although the crystallisation took in months, U-tube technique gave bigger and better crystal shape than the single-tube technique. Meanwhile, in the case of cobalt(II) oxalate hydrate by in the single-tube technique, direct precipitation or crystallisation of the product may be postponed by the use of different solvents, such as methanol, ethanol, or acetonitrile. In addition, both U-tube and single-tube techniques gave an identical product of cobalt(II) oxalate hydrate.

The work reported here is aiming to examine the effect of thiocyanate doping in the synthesis of cobalt(II) oxalate crystals grown in silica gel using U-tube technique. As frequently reported that the presence of other anions, as impurities, doping agents and/or metal salts, may influence the reaction equilibrium, and thus alter the crystallization and the properties of the crystals [9-11].

2. Materials and Methods

2.1. Materials and Measurements

All chemicals (Merck) were used as received without further purification, namely cobalt(II) nitrate hexahydrate, potassium thiocyanate, nitric acid, sodium oxalate, and sodium silicate. Crystals obtained from all molar ratio were analysed by Fourier Transform–Infrared Spectrophotometry (FTIR-8601PC) in Brawijaya University – Malang (using KBr pellet method in the range of 4000 – 400 cm^{-1}) and Scanning Electron Microscope (FEI Inspect S50) in State University of Malang - Indonesia.

2.2. Preparation of Gel and Supernatant Solutions

Silica gel was prepared from mixing Na_2SiO_3 0.5 M to 100 mL HNO_3 1 M until pH 5. A 30 mL of the solution was then transferred in a U-tube transparent glass (inner diameter = ± 2.5 cm) and stand at room temperature for 5 days until the gel is formed. Meanwhile, the supernatant solutions, solution A and B, were prepared in 1:1 ratio. Solution A was prepared by diluting 200 mg $Co(NO_3)_2{\cdot}6H_2O$ (0.687 mmol) in 20 mL of water, while solution B was prepared by diluting 92.1 mg $Na_2C_2O_4$ (0.687 mmol) in 20 mL of water. The molar ratio of Co^{2+}:$C_2O_4^{-2}$: SCN^- were varied (1:1:0; 1:1:1; 1:1:2; 1:1:3; 1:1:4; and 1:1:5). For ratios of 1:1:1 to 1:1:5, the solution A was also added with KSCN (66.8 mg; 133.6 mg; 200.4 mg; 267.2 mg; and 334.0 mg, respectively).

2.3. Crystal Growing

After the gel was ready, solution A and solution B were added separately on each side of the U-tube (Figure 2). Two sets of reactions were prepared for every molar ratio. After that, the tubes were closed with aluminium foil and the reactions were stood at room temperature for 12 weeks. Isolation of the

IOP Conf. Series: Materials Science and Engineering **833** (2020) 012017 doi:10.1088/1757-899X/833/1/012017

crystals was done by firstly remove both solutions, then hot water was added to dissolve the gel and finally filtered off. The crystals were dried in an oven (50-60 °C) for 3 hours and save it in a desiccator.

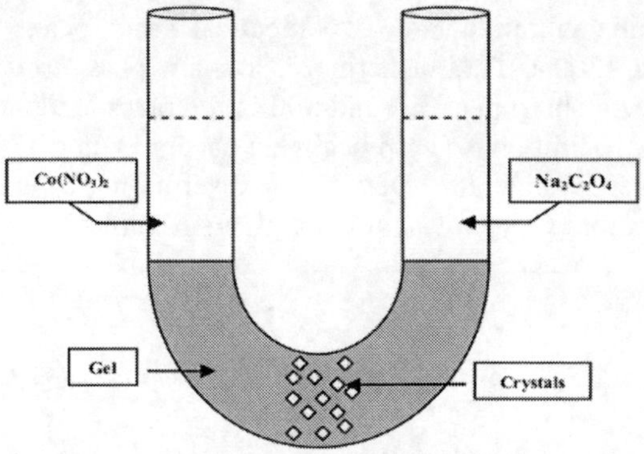

Figure 2. Illustration of gel-growth cobalt(II) oxalate crystals.

3. Result and Discussion

3.1. Crystal Growing

In week 6, several tiny single crystals started to form in the gel in every molar ratio, in which the area where crystal grew was approximately in between the two solution pathways or in the middle of the gel, as shown in Figure 3. The crystals grew bigger in the following weeks and red block single crystals were yielded by the end of week 12. Theoretically, the growth time can be shortened by increasing the concentration of both solutions.

The average yield of the obtained crystals is given in Table 1. Based on physical examination under the microscope, crystals obtained from all molar ratios have identical colour (red-orange) and morphology, indicating the uniformity of the compound in bulk product. The crystal yields were varied from 40 to 100 mg regardless of the molar ratio. At this stage, the inconclusive correlation between the thiocyanate concentrations and the crystal yields is noted. The isolation technique that was used to separate the crystal from the gel may influence the yield since the crystals were considerably tiny and fragile.

Figure 3. Crystal (dark spots) growing in silica gel, taken from the ratio of 1:1:5.

Table 1. The average yield of the obtained crystals for each molar ratio.

Ratio	1:1:0	1:1:1	1:1:2	1:1:3	1:1:4	1:1:5
Yield (mg)	90	50	40	70	100	90

3.2. Analyses of the crystal

Crystals obtained from all molar ratios were analysed by FT-IR and SEM. The IR spectra of the crystals from all molar ratios (Figure 4) shows that there is a sharp peak close to 1600 cm^{-1} which corresponds to carbonyl (C=O) stretching vibration, and two identical sharp peaks of the C–O symmetric and asymmetric stretch around 1300 cm^{-1}. Furthermore, a sharp peak around 800 cm^{-1} shows O–C–O asymmetric vibration, while a sharp peak around 500 cm^{-1} corresponds to Co–O stretch vibration, as also reported by Wang [2]. In addition, a sharp peak that occurs around 3300 cm^{-1} indicates that there is an O–H vibration of water as hydrate. Details of absorption peaks, wave numbers (cm^{-1}) and interpretations of IR spectra for every molar ratio are given in Table 2.

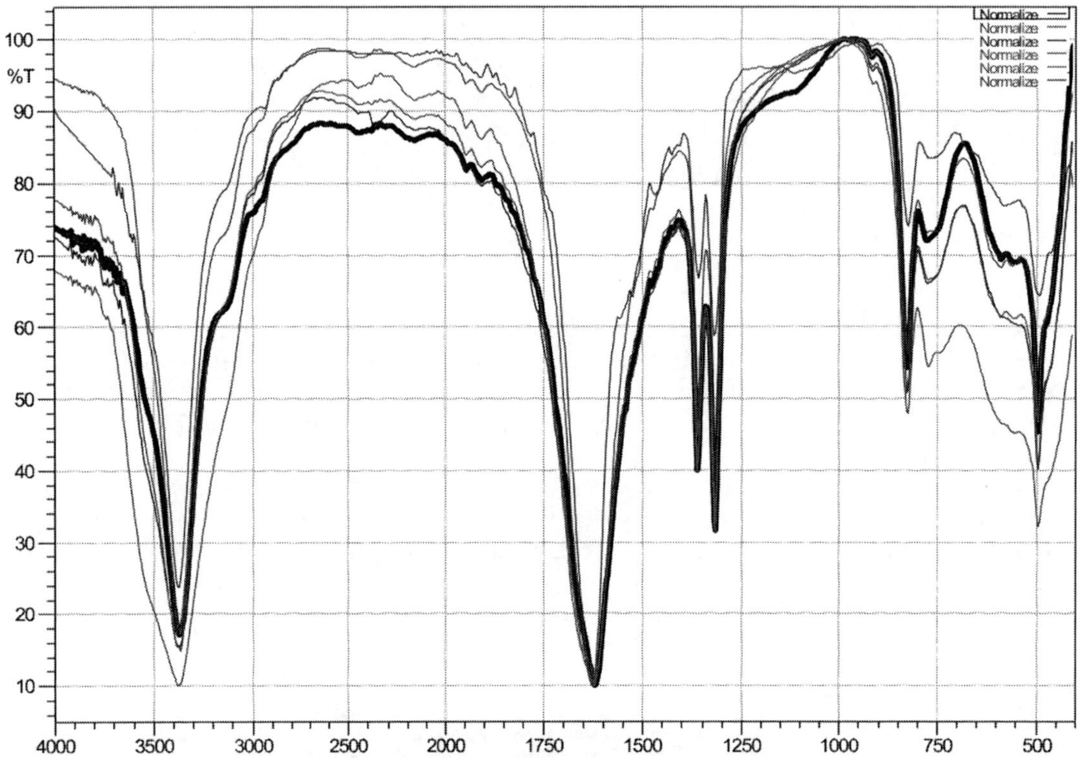

Figure 4. IR spectra of crystals obtained from all ratios (blue = 1:1:0; black = 1:1:1; violet = 1:1:2; red = 1:1:3; green = 1:1:4; brown= 1:1:5).

Table 2. IR spectra of the crystals obtained from all ratios.

Co(II) : $C_2O_4^{-2}$: SCN^-						Interpretation	References
1:1:0	**1:1:1**	**1:1:2**	**1:1:3**	**1:1:4**	**1:1:5**		
3369	3369	3373	3364	3364	3362	ν (O-H)	[2, 4, 5, 8]
1618	1620	1626	1618	1622	1624	ν (C=O)	[2, 4, 5, 8]
1362	1360	1358	1360	1360	1360	ν (C–O) asymmetric	[2, 4, 5, 8]
1317	1317	1317	1317	1317	1317	ν (C–O) symmetric	[2, 4, 5, 8]
825	825	824	825	825	825	ν (O–C–O) asymmetric	[2, 4, 5, 8]
494	494	490	494	494	494	ν (Co–O)	[2, 4, 5, 8]

Initially, the thiocyanate doping was intended to replace the water ligand to alter the magnetic properties of the cobalt(II) complex. Analogue to this is a 1D polymeric complex of oxalate-bridged M(II) (M = Co, Ni, Cu) with aminopyridine ligand [12]. However, crystals obtained from all ratios shows identical infrared spectra (Figure 4) and confirm the absence of specific bands that correspond to SCN vibration, such as a sharp peak of CN group around 2200 cm^{-1}. If sharp peak around 2200 cm^{-1} was observed, the presence of SCN in the crystal, either as coordinating ligand to cobalt(II) or as counter ions of the complex in the crystal lattice is confirmed. Therefore, it is concluded that crystals from all molar ratios were in fact a complex of cobalt(II) oxalate hydrate as reported previously by our group [7,8] and others [2,4,5,13]. Although the thiocyanate was firstly mixed with the cobalt(II) solution to increase the possibility of producing thiocyanate-cobalt(II) complex, the doping (until molar ratio of 1:1:5) was not affecting the final product and the oxalate ions remains having high preference to bind to the Co(II) centre than to the thiocyanate ions due to chelating effect. Higher thiocyanate concentration (up to ten-fold of the oxalate) might be needed to increase the possibility of introducing the ion into the cobalt(II) complex successfully.

Since all molar ratio gave identical product, the SEM analysis was conducted only to one crystal, which is from the ratio of 1:1:5. SEM image of the crystals is presented in Figure 5. The crystals have long block morphology and red-orange colour, in which the size is significantly bigger than that of identical crystals reported by Wang *et al.* [2], Bacsa *et al.* [4], and Bowen *et al.* [13]. The SEM image (mag = 400 ×) also gives more evidence that the use of silica gel as growth media, especially using the U-tube method, can lead to a bigger and better single crystal of insoluble salts or metal complexes.

Figure 5. SEM image of crystal at Co(II):oxalate:SCN ratio of 1:1:5.

For further research, the use of a higher concentration of thiocyanate or different anion should be investigated. Other variables, such as pH of gel, type of cobalt(II) salts and/or solvents also need to be considered. Other techniques for introducing thiocyanate ions into the complex, such as solvothermal, might also need to be explored.

4. Conclusion

Various molar ratio of Co^{2+}:$C_2O_4^{-2}$: SCN^- (1:1:0 to 1:1:5) gives identical red-orange block crystals of cobalt(II) oxalate hydrate. Although the thiocyanate ions were not successfully doped into the crystals or influence the final product, bigger crystals were obtained by using this U-tube gel technique. In the final product, the oxalate ions remain to have a high preference to bind to the Co(II) centre than to the thiocyanate ions due to chelating effect.

Acknowledgments

All author thanks Chemistry Department of Brawijaya University, Indonesia for the laboratory facilities and funding support. MMK and YPP are greatly acknowledge the PPIKID of Brawijaya University and the committee of IC2MS-2019 for publication support.

IOP Conf. Series: Materials Science and Engineering **833** (2020) 012017 doi:10.1088/1757-899X/833/1/012017

References

[1] Singh B P and Singh B. 2000. *Bull. Mater. Sci.* **23** (1) 11.
[2] Wang D, Wang Q, and Wang T. 2011. *Inorg. Chem.* **50** 6482.
[3] Lide D R. 2010. *CRC Handbook of Chemistry and Physics.* 90th ed. CRC Press Florida USA.
[4] Bacsa J, Eve D, and Dunbar K. R. 2005. *Acta Crystallogr. Sect. C.* **61** 58.
[5] García-Couceiro U, Castillo O, Luque A, Beobide G, and Roman, P 2004 *Inorg. Chim. Acta.* **357** 1 339
[6] Khunur M M, Wahyuni D T, and Prananto Y P. 2011. *Adv. Nat. Appl. Sci.* **5** 5 467.
[7] Henisch H. K. 1988. *Crystal in Gel and Liesegang Rings.* Cambridge University Press Australia.
[8] Prananto Y P, Khunur M M, Wahyuni D T, Shobirin R A, Nata Y R, and Riskah E 2013 *Bull. Chem. React. Eng. Catal.* **7** 3 199.
[9] Khunur M M and Prananto Y P. 2019. *IOP Conf. Series: Mater. Sci. Eng.* **546** 062012.
[10] Park H M, Hwang I H, Bae J M, Jo Y D, Kim C, Kim H-Y, Kim Y-M, and Kim S-J. 2012. *Bull. Korean Chem. Soc.* **33** 5 1517.
[11] Hu J, Liao C, Chen S, Jin P, Zhao J, and Zhao H. 2014. *Inorg. Chem. Commun.* **43** 126.
[12] Castillo O, Luque A, Roman P, Lloret F, and Julve M. 2001. *Inorg. Chem.* **40** 22 5526.
[13] Bowen P, Pujol O, Jongen N, Lemaitre J, Fink A, Stadleman P, and Hofmann H. 2010. *Nanoscale.* **2** 2470.

Isolation and Optical Properties of Natural Pigments from Purple Mangosteen Peels

Yehezkiel Steven Kurniawan[1], M. Riza Ghulam Fahmi[1], and Leny Yuliati[1,2,]*

[1]Ma Chung Research Center for Photosynthetic Pigments, Universitas Ma Chung, Villa Puncak Tidar N-01, Malang 65151, East Java, Indonesia
[2]Department of Chemistry, Faculty of Science and Technology, Universitas Ma Chung, Villa Puncak Tidar N-01, Malang 65151, East Java, Indonesia

*Corresponding author: leny.yuliati@machung.ac.id

Abstract. Purple mangosteen (*Garcinia mangostana*) has several biological applications such as anticancer, antitubercular and antioxidant agents. In this work, we isolated and studied the optical properties of the natural pigments from the purple mangosteen peels. To isolate the natural pigments, the mangosteen peels were macerated using distilled water, ethanol, or acetone for 24 h. The extracts were filtrated and characterized using spectrophotometers of Fourier transform infrared (FTIR), ultraviolet-visible (UV-Vis), and spectrofluorometer. The extracts gave the FTIR vibration peaks of O-H, C-H sp^3, C=O, C=C, and C-O functional groups, while absorption peaks at 210–374 nm were observed in the UV-Vis spectra of the extracts due to the presence of mangostins, anthocyanins, and phenolic acids. The three-dimensional fluorescence spectra showed that the excitation and emission peaks of the mangosteen peels extracted with ethanol were found at 444 and 498 nm, respectively, while that extracted with distilled water gave no significant fluorescence peaks. On the other hand, the mangosteen peels extracted with acetone gave the strongest emission intensity at 472 and 502 nm due to the most intense color intensity. This study provided useful information about the optical properties of natural pigments extracted from purple mangosteen peels through a simple isolation technique.

1. Introduction

Purple mangosteen (*Garcinia mangostana*) is a widely abundant tropical fruit in tropical countries. In general, it can be found in the Malay Archipelago especially in Indonesia. Purple mangosteen has been well recognized due to its excellent biological activities [1], therefore purple mangosteen is well-known as the queen of fruits [2]. Orozco *et al.* and Palawakong *et al.* reported that the biological activities of purple mangosteen are generated by xanthones, phenolic acids, anthocyanins, flavonoids, tannins, and carotenoids [3, 4]. Xanthones especially their prenylated and oxygenated compounds exhibit excellent antimicrobial activity against *Mycobacterium tuberculosis*, *Staphylococcus aureus*, *Candida albicans*, *Plasmodium falciparum*, *etc*. Because of that, mangosteen has been thoroughly investigated and employed as antitubercular and antimalarial agents [5, 6]. Furthermore, the anticancer, antioxidant, anti-inflammatory, anti-tumorigenic and anti-obesogenic activities of purple mangosteen have been also evaluated either in *in vitro* or *in vivo* studies [7–9].

In contrast to the excellent biological activities of the purple mangosteen flesh, the purple mangosteen peels were rarely utilized and thrown as the waste. Some researchers are putting effort to minimize the waste and utilize them for other potential applications [10–12]. Therefore, in the present work, the natural pigments from purple mangosteen peels were extracted through a maceration method and their optical properties were studied for a possible application as photonic and dye-sensitized materials. Distilled water, ethanol and acetone were employed as the media for natural pigments extraction process and the obtained extracts were characterized using spectrophotometers of Fourier transform infrared (FTIR), ultraviolet-visible (UV-Vis), and spectrofluorometer.

2. Experimental Section

Purple mangosteen fruits were bought from a supermarket in Malang, East Java. On the other hand, ethanol and acetone were obtained in EMPLURA grade from Merck. At first, as much as 1 kg of purple mangosteen fruit was cleaned with distilled water to remove the soil and other impurities. The peel of purple mangosteen was separated from its flesh, chopped and dried at 25 °C for 20 h. Then, 5 g of purple mangosteen peel was macerated with 50 mL of either distilled water or ethanol or acetone separately for 24 h. Afterwards the extract was characterized with a Fourier transform infrared (FTIR) spectrophotometer (JASCO FTIR-6800), an ultraviolet-visible (UV-Vis) spectrophotometer (JASCO V-760), and a spectrofluorometer (JASCO FP-8500).

3. Results and Discussion

In the present work, the natural pigments were extracted from purple mangosteen peels and their optical properties were investigated. Through a simple maceration method, three extracts of the purple mangosteen peels were successfully obtained and their photograph is shown in Figure 1. Figure 1 shows that the yellow color intensity was increased in the order of using distilled water, ethanol and acetone solvents. From the naked eye investigation, it could be predicted that the concentration of the natural pigments in acetone was higher than ethanol and was also higher than distilled water extract.

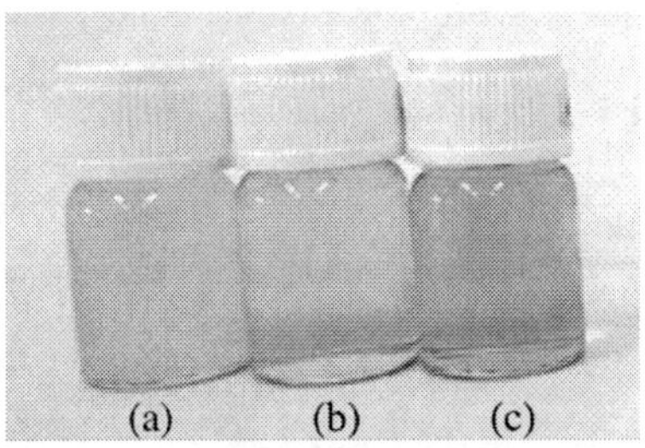

Figure 1. The photograph images of the purple mangosteen peel extract in (a) distilled water, (b) ethanol and (c) acetone.

To identify the functional groups of the extracted natural pigments, the FTIR spectra of the extract was measured by attenuated total reflectance (ATR) method. Before spectrum recording, the extract was treated using nitrogen gas flow to remove the used solvents. The FTIR spectra of the extracts are shown in Figure 2. The FTIR spectrum of the distilled water extract (Figure 2(a)) gave six main absorption peaks at 3319, 2972, 2872, 1735, 1375, and 1054 cm^{-1}. The presence of alkyl groups (R-CH_2-R') was indicated by the presence of C-H sp^3 stretching (2972 and 2872 cm^{-1}) and -CH_3 bending (1375 cm^{-1}) peaks, while the presence of carbonyl ester and hydroxyl groups was confirmed by the presence of C=O stretching (1735 cm^{-1}), C-O stretching (1054 cm^{-1}) and O-H stretching (3319 cm^{-1}), respectively.

The FTIR spectrum of the ethanolic extract (Figure 2(b)) gave eight main absorption peaks at 3440, 2938, 2828, 1686, 1419, 1363, 1207 and 1052 cm^{-1}. The presence of alkyl groups was shown by the presence of C-H sp^3 stretching (2938 and 2828 cm^{-1}), -CH_2- bending (1419 cm^{-1}), and -CH_3 bending (1363 cm^{-1}) peaks, while the presence of the carboxylic acid functional group was confirmed by the presence of O-H stretching (3440 cm^{-1}), C=O stretching (1686 cm^{-1}), and C-O ester (1027 cm^{-1}). On the other hand, the FTIR spectrum of the acetone extract (Figure 2(c)) gave five main absorption peaks at 2969, 1738, 1365, 1215 and 1052 cm^{-1}. The presence of alkyl groups was shown by the presence of C-H sp^3 stretching (2969 cm^{-1}), and $-CH_3$ bending (1365 cm^{-1}) peaks, while the presence of the ester functional group was confirmed by the presence of C=O stretching (1738 cm^{-1}), and either C-O ether (1215 cm^{-1}) or C-O ester (1052 cm^{-1}) functional groups.

From the FTIR spectra, both distilled water and ethanolic extracts showed that they contained polar natural pigments as the carboxylic acid and/or alcohol groups were detected. However, due to the low polarity of acetone as a non-polar solvent, the non-polar natural pigments were extracted as confirmed by its FTIR spectrum. This is in agreement with "like dissolve like" principle in which polar solvent gave higher possibility and suitability to extract polar natural pigments and vice versa [13].

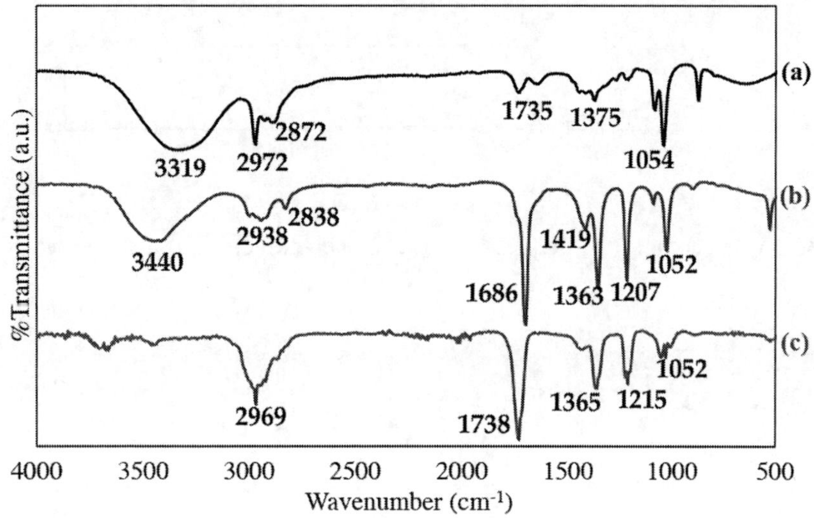

Figure 2. The FTIR spectra of the purple mangosteen peel extract in (a) distilled water, (b) ethanol and (c) acetone.

The optical properties of each extract were studied by UV-Vis and fluorescence measurements. The UV-Vis spectra of the extracts of purple mangosteen peels are shown in Figure 3 (a), (b) and (c) when using distilled water, ethanol and acetone as the media, respectively. The UV-Vis spectrum of the distilled water extract gave weak absorption peaks at 210, 283 and 374 nm (Figure 3 (a)), while the ethanolic extract gave a little bit stronger absorption peaks at 240, 256, 320 and 360 nm (Figure 3 (b)). Among the media, acetone extract gave the strongest absorption intensity. The absorption peaks were observed at 248, 258, 312, and 354 nm (Figure 3 (c)), in which agreed with the color appearance of the extract. The UV-Vis spectrum of the distilled water extract showed the absorption peaks at 210 and 374 nm, corresponding to the presence of the mangostin compounds (including α-mangostin, β-mangostin, γ-mangostin, gartanin and garcinone E), while the absorption peak at 283 nm corresponding to the presence of anthocyanin pigments (including cyanidin-3-sophoroside and cyanidin-3-glucoside) [14–16]. On the other hand, the characteristic absorption peak of cinnamic acid (256 nm), sinapinic acid (240 and 320 nm) and syringaldehyde (240 and 360 nm) were found at the ethanolic extract [17–19]. Similar to the ethanolic extract, the acetone extract also contained these pigments. However, their absorption peaks, *i.e.* cinnamic acid (258 nm), sinapinic acid (248 and 312 nm) and syringaldehyde (248 and 354 nm) were shifted due to the solvatochromic effect. The chemical structures of these compounds are shown in Figure 4, while further investigation is still required to confirm these natural pigments either by liquid chromatography-mass spectrometry (LC-MS) or nuclear magnetic resonance (NMR) analysis.

IOP Conf. Series: Materials Science and Engineering **833** (2020) 012018 doi:10.1088/1757-899X/833/1/012018

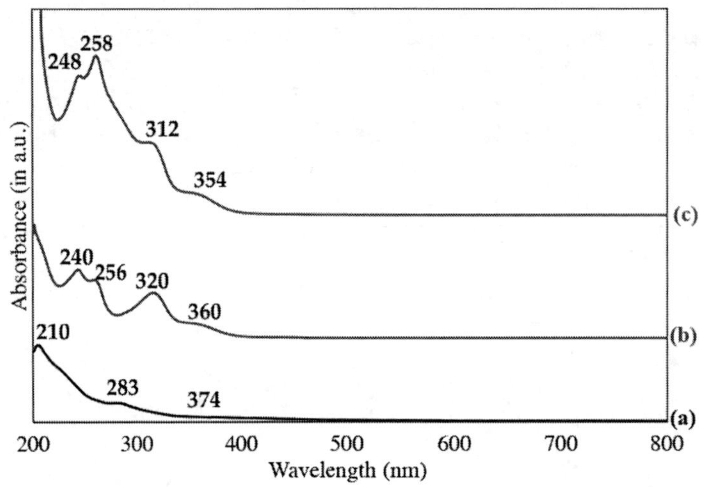

Figure 3. UV-Vis spectra of the purple mangosteen peel extract in (a) distilled water, (b) ethanol and (c) acetone.

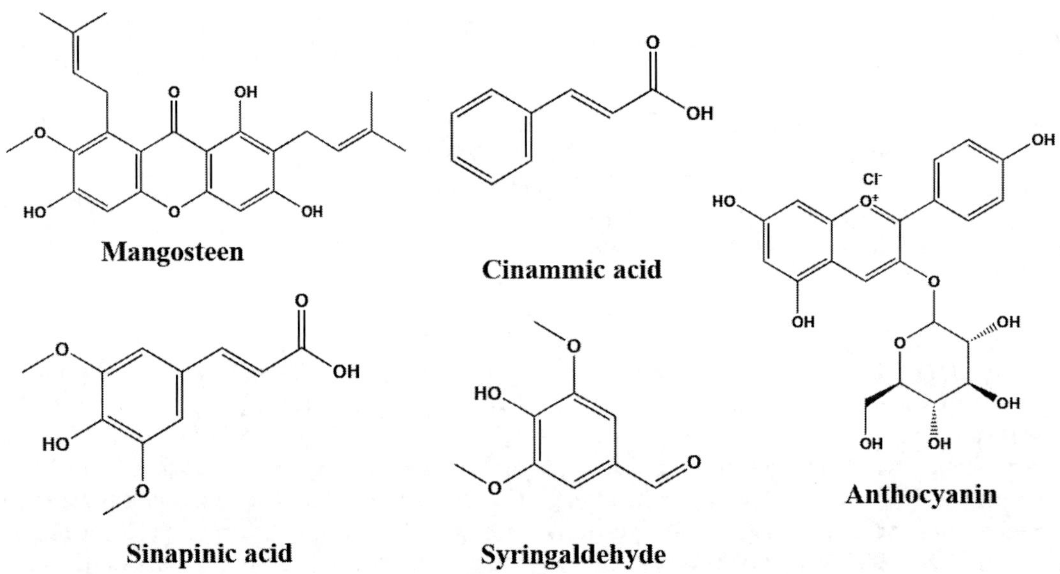

Figure 4. Chemical structures of the natural pigments found in the extracts of purple mangosteen peel.

The three-dimensional fluorescence spectra of the extracts are shown in Figure 5. The x-axis of the three-dimensional fluorescence spectrum shows the emission wavelength and the y-axis shows the excitation wavelength. Meanwhile, the z-axis shows the fluorescence intensity from 0 (black color) to 200 (red color) units. The three-dimensional fluorescence spectrum of the distilled water extract gave a very low fluorescence intensity, while ethanolic and acetone extract gave a stronger fluorescence spectrum as indicated by the color appearance.

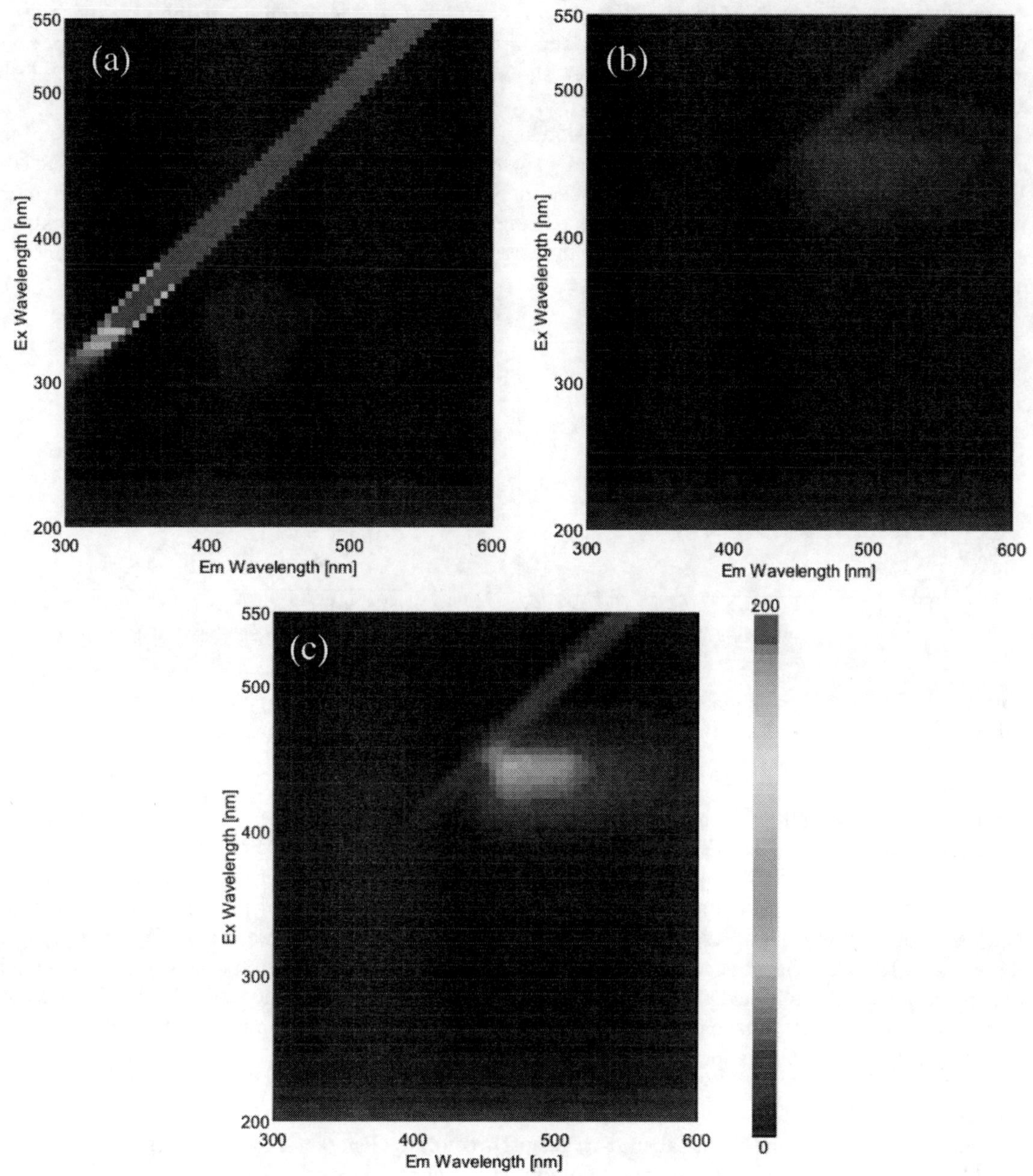

Figure 5. Three-dimensional fluorescence spectra of the purple mangosteen peel extract in (a) distilled water, (b) ethanol and (c) acetone.

In good agreement to the three-dimensional fluorescence spectra, the two-dimensional fluorescence spectrum of distilled water extract showed no significant absorption peak. The two-dimensional fluorescence spectra of ethanolic and acetone extracts are shown in Figure 6 (a) and (b), respectively. Corresponding to its three-dimensional spectrum, the ethanolic extract gave the main excitation and emission signals at 444 and 498 nm. Meanwhile, the acetone extract gave a single excitation signal at 452 nm and three emission signals at 472, 502 and 667 nm. According to their UV-Vis spectra shown in Figure 3 (b) and (c), the ethanolic and acetone extracts contained cinnamic acid, sinapinic acid and syringaldehyde compounds. However, these major compounds did not seem to be the emissive ones. It was found that the ethanolic extract gave a similar fluorescence spectrum to that of the anthocyanin compounds [20], while the acetone extract showed similar fluorescence spectra to those of the anthocyanin and chlorophyll compounds [21].

The intensity of either excitation or emission signals of acetone extract was higher than those of the ethanolic extract, which was in good agreement with the naked-eye observation as well as their UV-Vis spectrum. These experiments serve a useful preliminary data for the optical properties of the isolated natural pigments from purple mangosteen peels. Since both ethanolic and acetone extracts show a remarkable fluorescence intensity, they are promising to be used as the photonic material and/or dye-sensitizer agents in the near future.

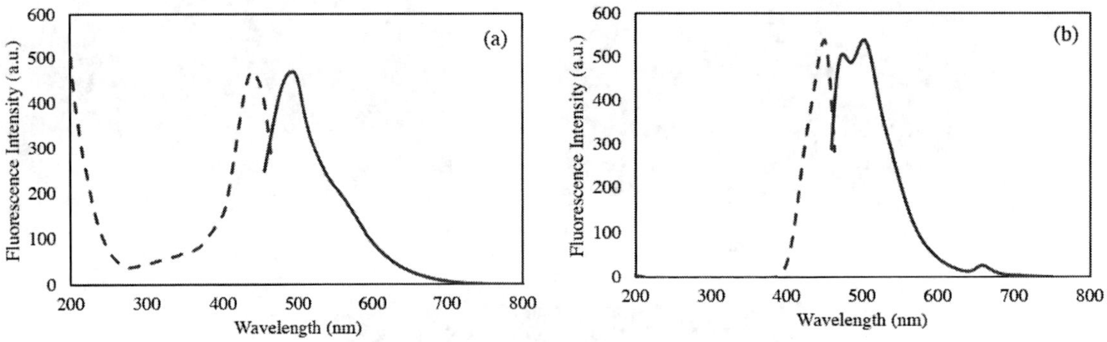

Figure 6. The two-dimensional fluorescence spectra of the purple mangosteen peel extract in (a) ethanol and (b) acetone. Dash line: excitation spectrum. Solid line: emission spectrum.

4. Conclusions

Isolation of the natural pigments from purple mangosteen peels has been successfully carried out through a maceration technique employing distilled water, ethanol, and acetone as the solvents. From the FTIR spectra, both distilled water and ethanolic extracts contained polar natural pigments, yielding a pale yellow color, while the acetone extract contained non-polar natural pigments, giving stronger yellow color intensity than those of the distilled water and ethanolic extracts. The UV-Vis spectra of the extracts revealed that the isolated natural pigments were dominated by mangosteen, anthocyanin and phenolic acid compounds. While the distilled water extract gave no remarkable fluorescence signal, the acetone extract gave the highest fluorescence intensity in which agreed well with the naked-eye observation and UV-Vis spectrum analysis.

Acknowledgement

The authors appreciate the financial support from Directorate General of Strengthening Research and Development, Ministry of Research, Technology and Higher Education of the Republic of Indonesia via the Higher Education Excellent Applied Research scheme (PTUPT 2019, No. 041/SP2H/LT/MULTI/L7/2019 and No. 005/MACHUNG/LPPM/SP2H-LIT-MULTI/III/2019).

References

[1] Chen G, Li Y, Wang W and Deng L 2018 *Expert Opin. Ther. Pat.* **28** 415
[2] Aizat WM, Ahmad-Hashim FH and Syed Jaafar SN 2019 *J. Adv. Res.* **20** 61
[3] Orozco FG and Failla ML 2013 *Nutrients* **5** 3163
[4] Palakawong C and Delaquis P 2018 *J. Food Process. Pres.* **42** e13744
[5] Fahmi MRG and Kurniawan YS 2019 *J. Exp. Clin. Microbiol.* **2** 29
[6] Failla ML and Gutiérrez-Orozco F 2017 Mangosteen Xanthones: Bioavailability and Bioactivities *Fruit and Vegetable Phytochemicals: Chemistry and Human Health (2nd*

edition) ed EM Yahia (Chichester, UK; Hoboken, NJ: John Wiley & Sons) chapter 9 pp 165–182

[7] Kondo M, Zhang L, Ji H, Kou Y and Ou B 2009 *J. Agr. Food Chem.* **57** 8788

[8] Akao Y, Nakagawa Y, Iinuma M and Nozawa Y 2008 *Int. J. Mol. Sci.* **9** 355

[9] Ibrahim MY, Hashim NM, Mariod AA, Mohan S, Abdulla MA, Abdelwahab SI and Arbab IA 2016 *Arabian J. Chem.* **9** 317

[10] Kurniawan YS, Adhiwibawa MAS, Setiyono E, Fahmi MRG and Lintang HO 2019 *Indones. J. Nat. Pigm.* **1** 49

[11] Jung HA, Su BN, Keller WJ, Mehta RG and Kinghorn AD 2006 *J. Agr. Food Chem.* **54** 2077

[12] Chaverri JP, Rodriguez NC, Ibarra MO and Rojas JMP 2008 *Food Chem. Technol.* **46** 3227

[13] Walker EB 2007 *J. Sep. Sci.* **30** 1229

[14] Maiaugree W, Lowpa S, Towannang M, Rutphonsan P, Tangtrakarn A, Pimanpang S, Maiaugree P, Ratchapolthavisin N, Sang-aroon W, Jarernboon W and Amornkitbamrung 2015 *Sci.Rep.* **5** 15230

[15] Chin YW, Kinghorn AD 2008 *Mini Rev. Org. Chem.* **5** 355

[16] Aizat WM, Hashim FHA, Jaafar SNS 2019 *J. Adv. Res.* **20** 61

[17] Zadernowski R, Czaplicki S and Naczk 2009 *Food Chem.* **112** 685

[18] Lamb CJ and Rubery PH 1975 *Anal. Biochem.* **68** 554

[19] Pardo I, Chanaga X, Vicente AI, Alcalde M and Camarero S 2013 *BMC Biotechnol.* **13** 90

[20] Singh V, Mishra AK 2016 *J. Mater. Chem. C* **4** 3131

[21] Lamb JJ, Rokke G, Marriott MFH 2018 *Photosynthetica* **56** 105

IOP Conf. Series: Materials Science and Engineering **833** (2020) 012019 doi:10.1088/1757-899X/833/1/012019

The Carbazole Compounds Endowed with Phosphonic Anchoring Group for Sensitizer in Dye-Sensitized Solar Cells

Y Kusumawati[1,*], B S Pamungkas[1]

[1]Chemistry Department, Faculty of Science, Institut Teknologi Sepuluh Nopember, ITS Campus Sukolilo, Keputih, Sukolilo, Surabaya, 60111

*Corresponding author: y_kusumawati@chem.its.ac.id

Abstract. The sensitizer dye contacts onto the TiO_2 surface through an anchoring group in a Dye-Sensitized Solar Cells (DSSC). The thermal stability of the anchoring group and TiO_2 binding contact is important to be considered in order to obtain the cell stability. Phosphonic acid is an anchoring group that has a good binding-contact thermal stability with TiO_2 surface. In this research we investigated through the computational study the electronic properties of carbazole derivative compounds endowed with the phosphonic anchoring-groups (CP). A series of 10 CP derivatives was studied which has a different functional donor group. The electronic properties were investigated using the TD-DFT method with the CAM-B3LYP basis set. The substitution of carboxyl group with the phosphonic one showed the absorption spectrum to the blue shift (5-70 nm) which gave a less beneficial effect to the DSSC performance. The addition of donor functional group aniline into the CP (aniline-CP) could shift back the absorption spectrum to the red-shift. The DSSC parameter also has been calculated including the electron injection driving force ΔG_{inject} and the open circuit votage (eV_{OC}). Overall the aniline-CP has the best performance among the observed compounds which showed the highest ΔG_{inject} (-0.84872 eV) and eV_{OC} (2.763 V).

1. Introduction

According to the average solar radiation that is received every year, Indonesia is one of country that is promising to conduct the solar cell device development [1]. Among the solar cells devices, Dye-Sensitized Solar Cell has attracted many researches due to its economic value. The cell comprises of two electrodes, dye-sensitized semiconductor as a photoanode and platinized transparent conductive oxide (TCO) as a counter electrode that are sandwiched between liquid electrolytes [2,3].

They are many key parameters that influence the cell performance. One of the key parameters is the electron injection from the Lowest Unoccupied Molecular Orbital (LUMO) to the conduction band semiconductor [4–6]. To guarantee the highest injection efficiency, the good binding between dye and semiconductor surface become the most important thing. The commonly dye that is used as sensitizer in DSSC have carboxyl group as anchoring site to attach the dye on the semiconductor surface. Aside the carboxyl group there are also the others anchoring site that has been studied, for instance, phosphonic, pyridine, tetracyanate, hydroxyquinoline, catechol or salicylic acid [7–11]. Among these, phosphonic group is promising because it is has tridentate binding site that will make the dye attach strongly to the semiconductor surface [7,8]. Nilsing et. al have calculated that the phosphonic functional group was attached stronger to the TiO_2 surface compare to the carboxylic group. The stronger interaction of anchoring site to the semiconductor will more facilitate the electron injection from the LUMO dye onto the conduction band semiconductor [9].

In this research we investigated the electronic properties of a series of carbazole dyes that has phosphonic group through a computational study. The carbazole dye is interesting to study because they have proven as a good candidate for sensitizer in DSSC application. The observed carbazole dyes have two anchoring site. The effect of carboxyl group substitution with the phosphonic to the electronic properties has been investigated. The addition of donating electron group to the carbazole compounds also has been simulated to improve their electronic properties. The cell parameters, including light harvesting efficiency, electron injection driving force and open circuit voltage also have been studied.

2. Computational Details

2.1. Molecule Model

The observed molecule were constructed from the carbazole dyes which has two anchoring site, namely CBC0 (Figure 1). The carboxylate functional group in the anchoring site position then were replaced by phosphonic functional group. The replacement one carboxylate group by phosphonic functional group created a compound namely CBP1 whereas the replacement two carboxylate functional groups with two phosphonic groups in the two positions of anchoring site created a compound namely CBP2. Then 10 donor functional groups were added into these carbazole compounds. The 10 donor functional groups are presented in Figure 2, whereas the resulted derivatives compounds are presented in Table 1.

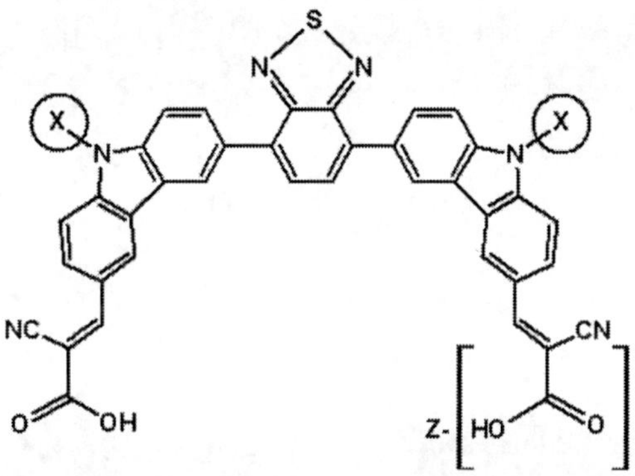

Figure 1. Basic structure of observed carbazole dye. The basic compound, CBC0, is the compound with X = H and the anchoring site = -COOH. The replacing one –COOH with the $–PO_3H_2$ result CBP1 and the replacing both of –COOH with the $–PO_3H_2$ result CBP2.

2.2. Computation Methods

The calculations in this research have been conducted using the Gaussian 16 code [12]. The optimization geometries using the CAM-B3LYP functional [13]. This functional is the best choice for such a huge molecule because contain a good long range-corrected [14,15]. After obtaining the optimized geometry, the TD-DFT calculation was performed to investigate the electronic properties [16]. The AVOGADRO [17] and GAUSSUM [18] were used to help the calculation result visualization including HOMO-LUMO localization and UV-VIS absorption spectrum.

3. Result and Discussions

3.1. The effect of phosphonic group substitution in the anchoring site

The spectrum UV-VIS of the CBC0, CBP1 and CBP2 which was obtained from the TDDFT calculation is shown in Table 1. Table 1 shows that the substitution of carboxyl groups with the phosphonic groups gives the wavelength maximum shift to the high energy. This condition is avoided

IOP Conf. Series: Materials Science and Engineering **833** (2020) 012019 doi:10.1088/1757-899X/833/1/012019

in the Dye-Sensitized Solar Cells (DSSCs). This result guides us to substitute the carboxyl groups with phosphonic ones only in one site.

Table 1. The derivatives compounds of CBP1 investigated in this research.

No	Compound Name	Replacing X
1	Methoxy-CBP1	$-OCH_3$
2	Amine-CBP1	$-NH_3$
3	Hexoxy-CBP1 ($-OC_6H_{13}$)	$-OC_6H_{13}$
4	Aminhexyl-CBP1	$-NHC_6H_{13}$
5	Aniline-CBP1	Aniline 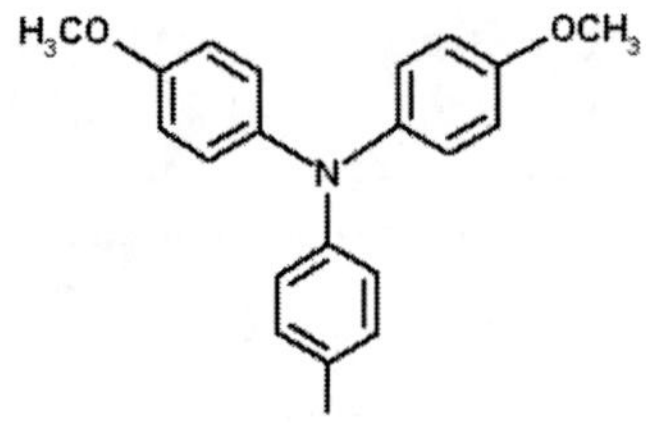
6	Phenonthiazine-CBP1	Phenothiazine 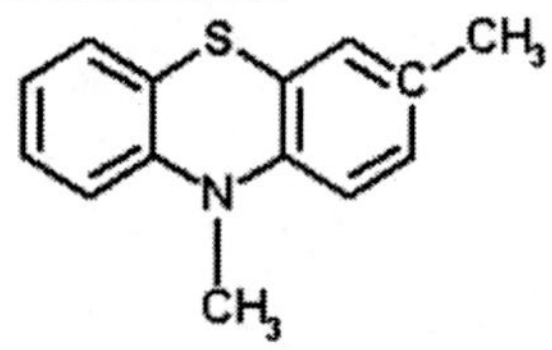
7	Fluorene-CBP1	Fluorene 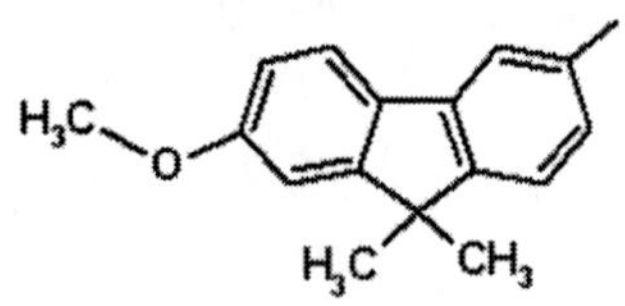
8	Silole-CBP1	Silole 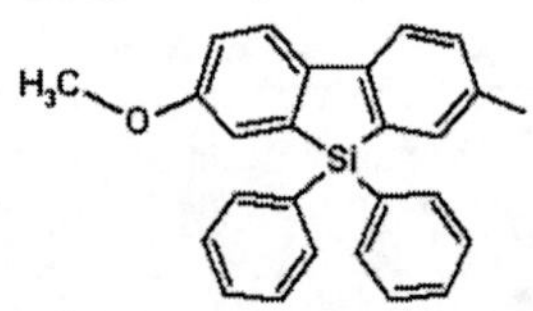
9	Pyrrole-CBP1	Pyrrole 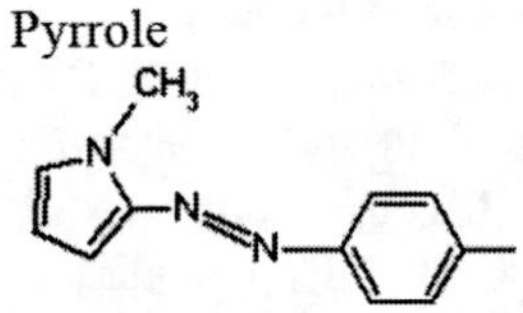
10	Methoxy-CBP1	Triazine 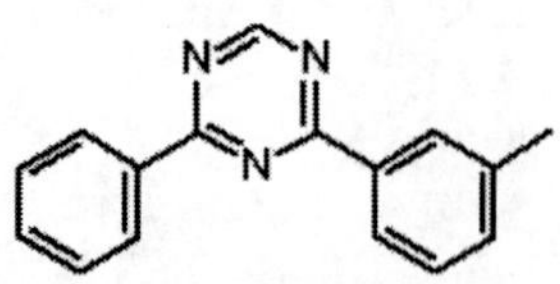

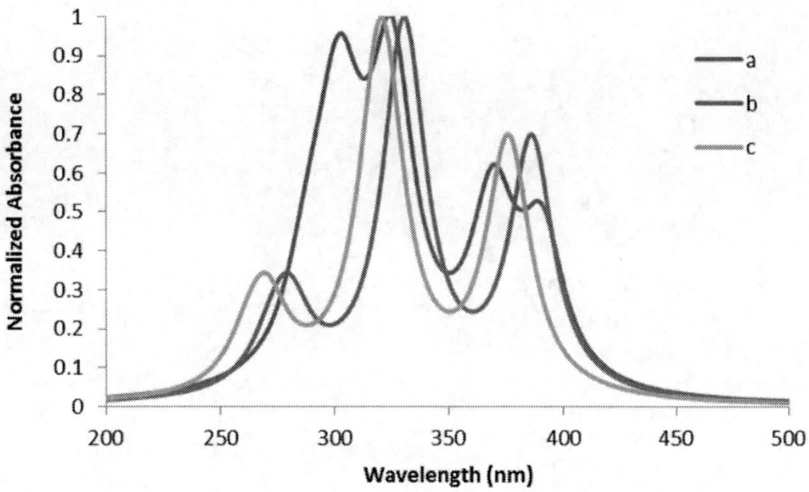

Figure 2. The calculated UV-VIS Spectrum of (a) CBC0, (b) CBP1 and (c) CBP2

3.2. The effect of donating electron groups to the electronic properties

The calculated spectrum UV-VIS for the CBP1 derivatives is shown in Figure 3. Figure 3 shows that most of the observed compounds has three peaks on their UV-VIS absorption spectrum at about 270 nm, 350 nm and 400 nm.

The electronic transition characteristic obtained from the calculation is presented in Table 2. The observed dyes show the electron transition from HOMO to LUMO except fluorine-CBP1, silole-CBP1 and pyrrole-CBP1. They show the electron transition from HOMO-2 to LUMO for fluorine-CBP1 and HOMO to LUMO+2 for silole-CBP1 and pyrrole-CBP1. Table 2 also informs that the maximum wavelength of anliline-CBP1 show a red shift compare to the CBP1 one.

To further understand the electronic transition character, the visualization of orbital molecules that involve in the electronic transition the HOMO-LUMO profile has been visualized. The visualization shows that all the substituted phosphonic group compounds have the LUMO that close to the carboxylic groups (for example the LUMO of CBP1 is presented in Figure 4.a), except the phenothiazine-CBP1 which has the LUMO near to the phosphonic group (Figure 4.b). This fact indicates that only phenotizine-CBP1 which will show a good electronic injection when the phosphonic group is attached to the semiconductor surface. Although that, the others still have a good candidate to be applied as sensitizer in DSSC because their LUMO position is located on the other side of the anchoring site, the carboxyl group. For the last case, the phosphonic group only has a role as the anchoring site but the electron injection will occur through the carboxyl side.

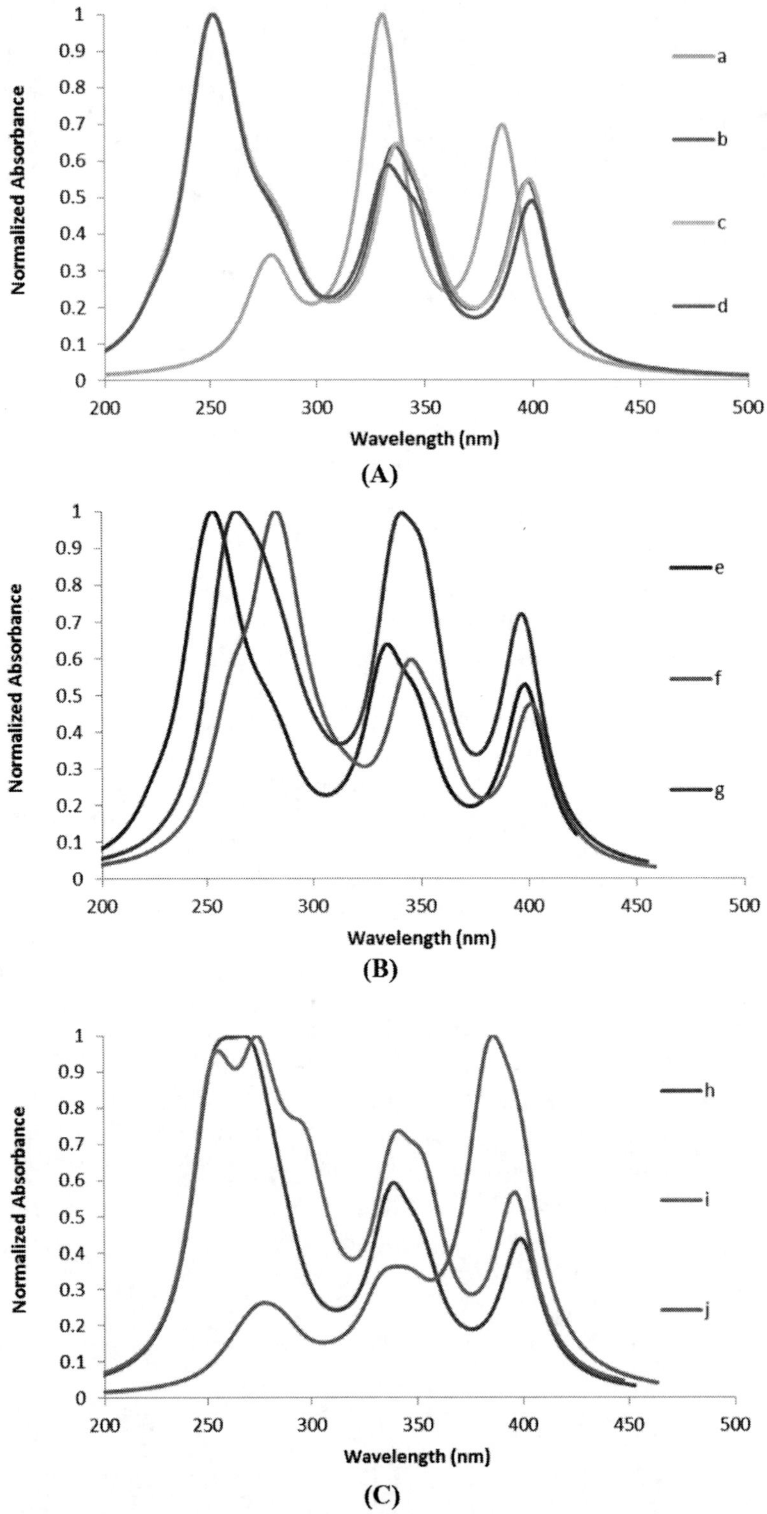

Figure 3. Calculated spectrum UV-VIS of the CBP1 derivatives, **A**: (a) methoxy-CBP1, (b) hexoxy-CBP1, (c) amine-CBP1, (d) hexyl-amine CBP1. **B**: (e) aniline-CBP1, (f) phenothiazine-CBP1, (g) fluorine-CBP1. **C**: (h) silole-CBP1, (i) pyrrole-CBP1, (j) triazine-CBP1.

Table 2. The electronic transition characteristic of the observed compound

No	Compound name	λ (nm)	*f*	Electronic transition character
1	CBC0	401,68	1,213	HOMO -> LUMO 72%
2	CBP1	401,50	1,1301	HOMO -> LUMO 73%
3	Methoxy-CBP1	397,61	1,1282	HOMO -> LUMO 71%
4	Amine-CBP1	399,44	1,0463	HOMO -> LUMO 68%
5	Hexosi-CBP1 (-OC_6H_{13})	398,33	1,1411	HOMO -> LUMO 72%
6	Aminhexyl-CBP1	397,83	1,0882	HOMO -> LUMO 75%
7	Aniline-CBP1	400,76	1,1171	H-2-> LUMO 61%
8	Phenonthiazine-CBP1	337,90	1,0331	HOMO -> LUMO+2 33%
9	Fluorene-CBP1	337,47	1,1635	HOMO -> LUMO +2 37%
10	Silole-CBP1	395,92	1,0692	HOMO -> LUMO 60%
11	Pyrrole-CBP1	397,52	1,5324	HOMO -> LUMO 41%
12	Triazine-CBP1	395,33	1,0764	HOMO -> LUMO 68 %

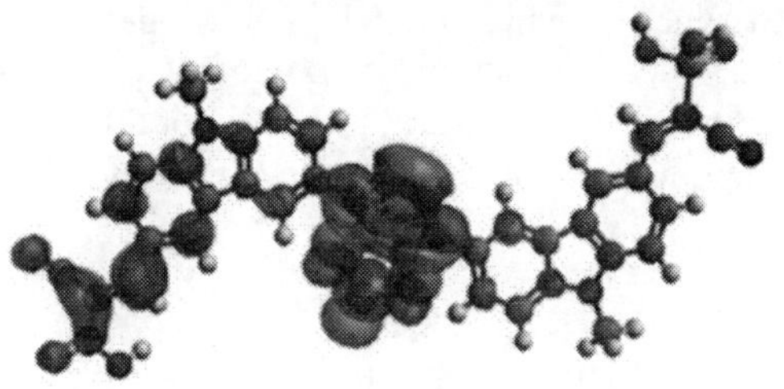

(a)

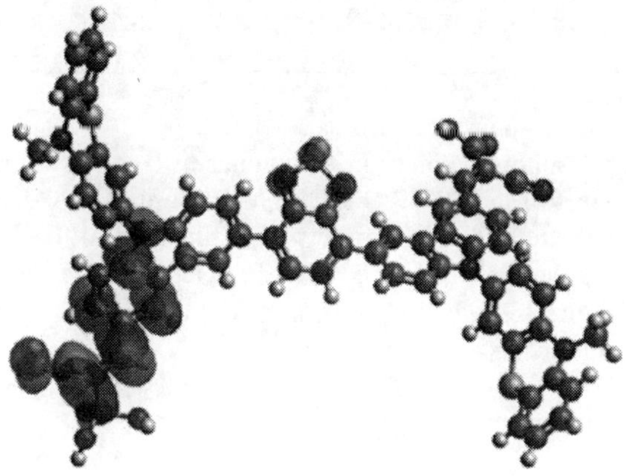

(b)

Figure 4. The LUMO profile of (a) CBP1 and (b) phenothiazine CBP-1.

3.3. The effect of donating electron groups to the DSSC parameter

The calculated DSSC parameter is presented in Table 3. Table 3 displays the DSSC parameters including the oscillator strength (*f*), electron injection driving force ($\Delta G_{injection}$) and open circuit voltage (V_{oc}). The oscillator strength influences the light harvesting efficiency (*LHE*). The molecules which have a higher oscillator strength will have a higher light harvesting efficiency [19]. Table 3 also informs that all the CBP1 derivatives show the $\Delta G_{injection}$ decrease except for aniline-CBP1 which show an almost the same value with the basic compound CBC0. We assume that the replacing of carboxyl moiety with the phosphonic one decreases the withdrawn of electron, but the present of three methoxy phenyl with the nitrogen atom in the centre of aniline moiety support the good electron donating that induce a $\Delta G_{injection}$ increasing. The other DSSC parameters of aniline-CBP1, oscillator strength and V_{OC}, also show a good value that makes this compound as a good candidate to be used as sensitizer in DSSC. The aniline also can be further studied as an electron donating moiety for the other basic compounds.

Table 3. The calculated DSSC parameters of the investigated compounds

Compound name	f	E_{ox}^{dye} (eV)	$E_{\lambda max}$ (eV)	E_{ox}^{dye*} (eV)	ΔG_{inject} (eV)	Voc (V)
CBC0	1,213	6,674	3,54602	3,12798	-0,87202	2,741
CBP1	1,1301	6,659	3,08802	3,57098	-0,42902	2,763
Methoxy-CBP1	1,1282	6,726	3,11822	3,60778	-0,39222	2,734
Amine-CBP1	1,0463	6,686	3,10392	3,58208	-0,41792	2,766
Hexoxy-CBP1 ($-OC_6H_{13}$)	1,1411	6,713	3,11262	3,60038	-0,39962	2,737
Aminhexyl-CBP1	1,0882	6,703	3,11652	3,58648	-0,41352	2,761
Aniline-CBP1	1,1171	6,245	3,09372	3,15128	-0,84872	2,763
Phenonthiazine-CBP1	1,0331	6,586	3,12662	3,45938	-0,54062	2,755
Fluorene-CBP1	1,1635	6,669	3,11012	3,55888	-0,44112	2,754
Silole-CBP1	1,0692	6,636	3,13152	3,50448	-0,49552	2,759
Pyrrole-CBP1	1,5324	6,616	3,11892	3,49708	-0,50292	2,743
Triazine-CBP1	1,0764	6,732	3,13622	3,59578	-0,40422	2,746

4. Conclusions

The phosphonic moiety in the anchoring site position of the investigated carbazole compound gave a blue shift of the UV-VIS spectrum maximum wavelength which is avoided for the DSSC application. The insertion of aniline moiety gave the red shift back of the UV-VIS spectrum maximum wavelength. This research also revealed that the insertion of aniline as an electron donating showed a good value of oscillator strength, $\Delta G_{injection}$ and V_{oc} that makes this compound as an interesting compound to be studied further.

Acknowledgments

KEMENRISTEKDIKTI (Ministry of Research, Technology and Higher Education of the Republic of Indonesia) is acknowledged for supporting the research funding.

References

[1] Rumbayan M, Abudureyimu A, Nagasaka K 2012. *Renewable and Sustainable Energy Reviews.* **16**(3):1437–49.

[2] Hagfeldt A, Boschloo G, Sun L, Kloo L, Pettersson H 2010. *Chemical Reviews* **110**(11):6595-66633.

[3] Sugathan V, John E, Sudhakar K 2015. *Renewable and Sustainable Energy Reviews.* **52**:54–64.

[4] Fang H, Wu Y, Kuhn DL, Zander Z, DeLacy BG, Rao Y 2018. *Chemical Physics.* **512**:93–7.

[5] Akın S, Açıkgöz S, Gülen M, Akyürek C, Sönmezoğlu S. *RSC Adv.* 2016. **6**(88):85125–34.

[6] Thavasi V, Renugopalakrishnan V, Jose R, Ramakrishna S 2009. *Materials Science and Engineering: R: Reports.* **63**(3):81–99.

[7] Abate A, Pérez-Tejada R, Wojciechowski K, Foster JM, Sadhanala A, Steiner U 2015. *Phys Chem Chem Phys.* **17**(28):18780–9.

[8] Baktash A, Khoshnevisan B, Sasani A, Mirabbaszadeh K 2016. *Organic Electronics.* **33**:207–12.

[9] Nisling M, Persson P, Ojamae L. 2005. *Chemical Physical Letters.* **415**: 375-380.

[10] Zhang L, Cole JM 2015. *ACS Appl Mater Interfaces.* **7**(6):3427–55.

[11] Ladomenou K, Kitsopoulos TN, Sharma GD, Coutsolelos AG 2014. *RSC Adv.* 2014 **4**(41):21379–404.

[12] M. J. Frisch, G. W. Trucks, H. B. Schlegel, G. E. Scuseria, M. A. Robb, J. R. Cheeseman, G.

Scalmani, V. Barone, B. Mennucci, G. A Petersson, H. Nakatsuji, M. Caricato, X. Li, H. P. Hratchian, A. F. Izmaylov, J. Bloino, G. Zheng, J. L. Sonnenberg, M. Hada, M. Ehara, K. Toyota, R. Fukuda, J. Hasegawa, M. Ishida, T. Nakajima, Y. Honda, O. Kitao, H. Nakai, T. Vreven, J. A, Jr. Montgomery, J. E. Peralta, F. Ogliaro, M. Bearpark, J. J. Heyd, E. Brothers, K. N. Kudin, V. N. Staroverov, R. Kobayashi, J. Normand, K. Raghavachari, A. Rendell, J. C. Burant, S.S. Iyengar, J. Tomasi, M. Cossi, N. Rega, M. J. Millam, M. Klene, J. E. Knox, J. B. Cross, V. Bakken, C. Adamo, J. Jaramillo, R. Gomperts, R. E. Stratmann, O. Yazyev, A. J. Austin, R. Cammi, C. Pomelli, J. W. Ochterski, R. L. Martin, K. Morokuma, V. G. Zakrzewski, G. A. Voth, P. Salvador, J.J. Dannenberg, S. Dapprich, A. D. Daniels, Ö. Farkas, J. B Foresman, J. V. Ortiz, J. Cioslowski, D.J. Fox, (2009) Gaussian 09, Revision D.01, Gaussian, Inc., Wallingford CT.

[13] T. Yanai, D. P. Tew, N. C. Handy 2004. *Chem. Phys. Lett.* **393**, 51–57

[14] S. Tortorella, M. Mastropaqua, M., A. Cardone, M Pastore, M., F. De Angelis 2016. *J. Phys. Condens. Matter.*, **28** 074005 (1-11)

[15] Kusumawati Y, Massin J, Olivier C, Toupance T, Ivansyah AL, Martoprawiro MA, Prijamboedi B, Radiman CL, Pauporte T 2017. *J PhotochemPhotobiolChem* **341**:69–77

[16] E. Runge, E. K. U. Gross, Density-Functional Theory for Time-Dependent Systems 1984, *Phys. Rev. Lett.* **52**, 997–1000

[17] Hanweel MD, Curtis DE, Lonie DC, Vandermeersch, T., Zurek, E., Hutchison, G. R 2012. *Cheminformatics* 4:17

[18] O'Boyle NM, Tenderholt AL, Langner KM 2008 *J. Comp. Chem* 29: 839-845

[19] Chaiamornnugool P., Tontapha S., Phatchana R., Ratchapolthavisin N., Kanokmedhakul S., Sang-aroon W., Amornkitbamrung V 2017. *Journal of Molecular Structure* **1127**: 145–155.

A Potency of Microcellulose from Pineapple Leave Isolated by Hydrolysis-Assisted Sonication

Luqi Khoiriyah Latif, Masruri Masruri*, Barlah Rumhayati

Chemistry Department, Faculty of Mathematics and Natural Science, Brawijaya University, Jl. Veteran 65145 Malang, Indonesia.

*Corresponding author: masruri@ub.ac.id

Abstract. Cellulose has been reported have many functions for separation and adsorption of heavy metals and organic materials waste or pollutants. This paper reported the preparation of micro size cellulose from the waste of pineapple leaves. The strategy involves delignification, bleaching, and hydrolysis process. The hydrolysis step plays an important role for cutting of the cellulose chain, and ultrasonication assisted hydrolytic procedure was applied. In addition, the isolated product was characterized by means of FTIR spectrometry to identify functional group, Scanning Electron Microscopy to inform morphology features, and Particle Size Analyzer to inform the size of cellulose particle. The yield of initial isolated cellulose was 62.07%. Then, through the ultrasonication process using 30% of sulphuric acid for 1 hour at room temperature resulted about 3.497% yield of microcellulose. The FTIR show O-H, C-O-specific band for cellulose in 3402 cm^{-1}, 1385 and 1200-1300 cm^{-1}, respectively. This finding lead for further application of this micro/nano-cellulose for heavy-metals-and dye pollutants adsorbent.

1. Introduction

Pineapple (*Ananas comosus* L. Merryl) easily grow at Indonesia, a tropic country. Increasing of pineapple fruit production in Indonesia 28.64% in 2017 from the previous year 2016 [2]. Then, pineapple leaf biomass waste is easily obtained, abundant, renewable, and high in cellulose content. Pineapple leaves basically consists of substances of cellulose, hemi-cellulose, lignin and pectin. Cellulose can be obtained from various biomass source. The diversity of cellulose biomass sources and the extraction process will affect the type of cellulose. Pineapple fiber consist of cellulose (78-83%), hemicellulose (19%) higher than lignin, around (5-15%), pectin 1%, waxes (2-3%) and ash (1%) [1,25-27].

Cellulose compose of a linier homopolysaccharide, D-glucopyranose monomers trough (1-4) glycosidic bond. Every monomer has three hydroxyl groups. Physical properties of cellulose controlled by this important functional group. Directing the crystalline packing not only controlled by hydrogen bonds but also reigning physical properties of cellulose [8]. Fundamental structural of cellulose is micro fibrils. The chains of poly-(1,4)-D-glucosyl residues aggregate to form a fibril which is a long thread-like bundle of molecules laterally stabilized by intermolecular hydrogen bonds [10-11].

Cellulose is a semicrystalline polymer. It differs from hemicellulose and lignin, amorphous polymer. The linkage between carbohydrate groups and lignin are ester-type bond and carboxyl. First, carboxyl group of uronic acid in hemicellulose connected by ester-type bond with lignin's hydroxyl. Alkali solution had great ability to destroy this linkage. Second, hydroxyl group of lignin and carbohydrate form ether-type bond. This bond cannot be damaged by alkali solution [30].

Diameter range of single cellulose microfibrils are 2-20 nanometers. Each microfibril can be considered as a string of cellulose crystals linked along the micro fibril axis by disordered amorphous domains [6,12]. Commonly microfibril term is used in high-level plants not only the length was decreased several microns, but also lose their structure. This condition depends on various diameters of microfibrils. Microfibril also used to refer nanofibrils and nanofibers. Aggregates of cellulose microfibrils will form bulky-microfiber cellulose (MFC) [13]. At the time when acid hydrolysis in progress, cellulose microfibrils lead lie athwart fission along the amorphous regions and the use of sonication yield in a rod-like material with a relatively low aspect ratio referred as cellulose whiskers [24]. Some term for cellulose whiskers puts in nano whiskers [15], nanorods [16], and rod-like cellulose crystals [17]. The individual cellulose crystals have powerful hydrogen bonding, further push re-aggregation [18] that guide to another cellulose structure called microcrystalline cellulose (MCC).

Various types of cellulose particles are obtained based on the source and method of the extraction. There is no standardization of nomenclature, so ambiguity is found in the literature [4,5]. Here we focus on extracting cellulose from pineapple leaves. The next research, we will compare between the result hydrolyzed pineapple leaves with sonication and without sonication. We predict that sonication will produce larger cellulose surface area than without sonication.

Many applications are carried out for large surface area of polymer, such as an adsorbent. For further research micro cellulose produced will be tested as heavy metal adsorbents in the water system.

2. Experiment

2.1. Materials
Pineapple leaves was garnered from pineapple seller at Malang, East Java, Indonesia, distilled water, filter paper. Reagents: 6% sodium hypochlorite; 30% sulphuric acid; 6% sodium hydroxide solution.

2.2. Procedures

2.2.1. Cellulose Isolation.
a) Delignification.
Pineapple leaves was dried at 50 °C up to 48 hours, then cut in small pieces, and next delignification process was done using 6% sodium hydroxide at ratio 1:10 (w/v) at 70 °C for 4 hours. The mixture was washed until gain neutral pH.
b) Bleaching process.
The neutral residue was bleached using 6% sodium hypochlorite for 1 hour at 70 °C, using 1:42 ratio. Then, it washed until achieve neutral pH, and dried at 50 ^{O}C.
c) Acid Hydrolysis.
The neutral residue was hydrolyzed using 30% sulphuric acid at room temperature for 1 hour at ratio 1:20, then washed until achieved neutral pH.

2.2.2. Preparation of micro-cellulose.
Preparation of micro-cellulose using Ultrasonication Cleaner DC 150 DC 150 H. In the ultrasonication stage, two variations were given. First variation using sulphuric solvent, while the other with water solvent. First variation using 30% sulphuric acid (v/v) at room temperature for 1 hour at ratio 1:20.

2.3. Identification

2.3.1. FTIR-Fourier Transform Infrared spectroscopy. The FTIR spectra produced by Fourier Transform Infrared Spectrometer Shimadzu 8400S. Thin pellets were prepared with KBr powder. Then blended with the sample using wavelength range from 500-4000 cm^{-1}.

IOP Conf. Series: Materials Science and Engineering **833** (2020) 012020 doi:10.1088/1757-899X/833/1/012020

2.3.2. Scanning Electron Microscopy (SEM). The surface morphology of the pineapple leaves was measured by SEM. Sample in dry condition. Voltage of operation was 20kV.

3. Results and discussion

3.1. Composition of Pineapple Leaves Fiber (PLF) at each chemical treatment.
During the delignification process there has been a change in the color of the solution. Sodium hydroxide solvent which was originally colorless turned blackish. This indicates that lignin and hemicellulose were extracted using sodium hydroxide. The elimination of lignin and hemicellulose was calculated by percent amount in the table.

Figure 1. Fresh PL

Figure 2. Dried PL

Figure 3. Dried Chip PL

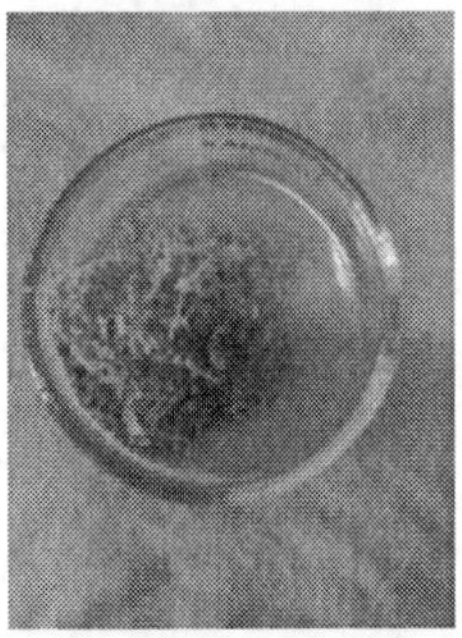

Figure 4. Powder PLF

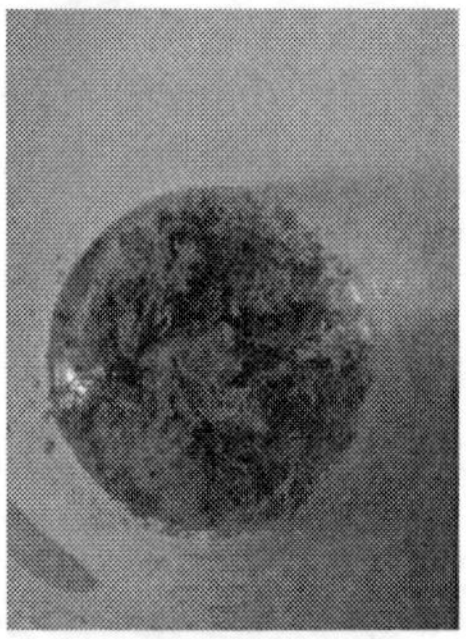

Figure 5. Delignification using 6% NaOH

Figure 6. Bleaching using 6% NaOCl

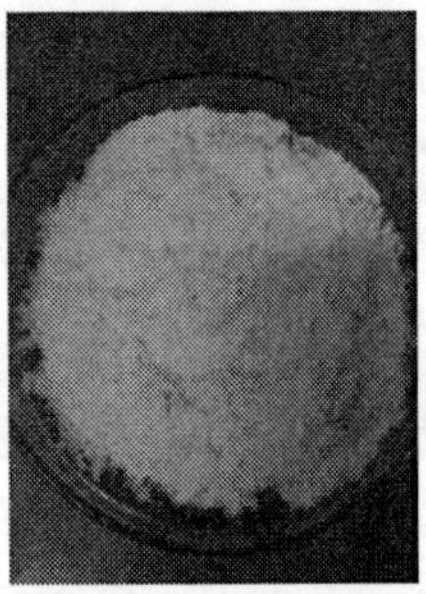

Figure 7. Hydrolysis without sonication

Figure 8. Hydrolysis-sonication

Figure 9. H_2O-sonication

NaOH solution is the best blooming solution in the pulping process. This is due to the suitability of Na^+ cations to infiltrate pores between fields of cellulose lattice. Infiltration of competing particles causes expansion followed by the entry of water molecules. No specific calculation was performed to distinguish between lignin and hemicellulose. The extracted amount is shown in table 1.

Table 1. Chemical Composition of Pineapple Leaves (PL)

Treatment	Cellulose (%)	Hemicellulose-Lignin (%)
Raw material*	81.27	15.77
Delignification	35.33	8.91
Bleaching	64.67	3.22
Acid hydrolysis	70.11	2.31
Sonication:		
H_2SO_4	72.17	1.77
H_2O	80.13	0.22

*Composition from literature [1, 25-27]

Hydroxyl anion can increase the pH of solution indicates that hemicellulose was dissolved. Increasing temperature can increase the rate of delignification (removal of lignin). However, using temperature above 160 ^{o}C can cause cellulose degradation.

3.2. Analysis Functional Group using Fourier Transform Infrared Spectroscopy (FTIR)

Dying sample will support the sample analysis. If the amount of the sample in water was high, the hydroxyl group of the water will be read.

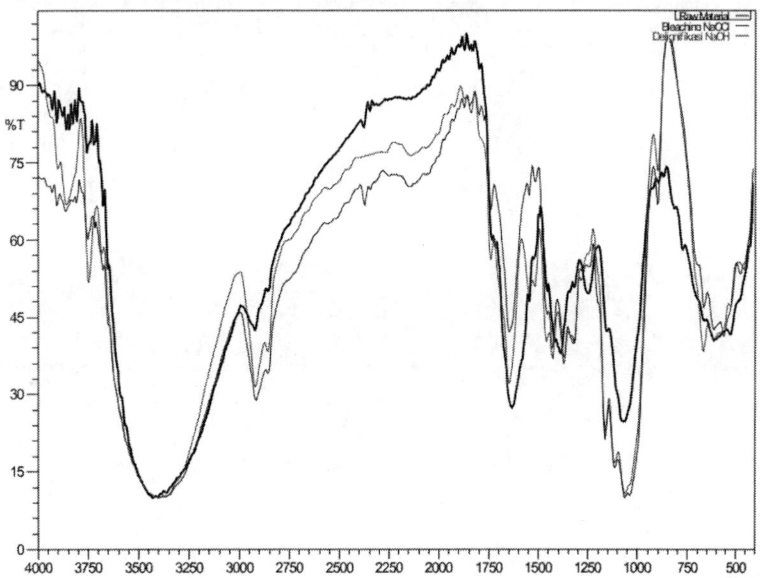

Figure 10. FTIR spectra of raw material (powder) pineapple leaves (black line), delignification using NaOH 6% (red line), bleaching using NaOCl 6% (blue line).

Based on the purpose of getting cellulose, the lignin and hemicellulose of the compounds must be removed. FTIR helps read the condition of the sample through the changes of functional groups composition from lignin and hemicellulose. Table 2 shown the specific uptake areas of lignin and hemicellulose.

Table 2. Functional Group of Pineapple Leaves (PL)

Steps	Stretching -OH (cm^{-1})	Vibration C-H (cm^{-1})	Absorbed water (cm^{-1})	Aromatic ring vibration of Lignin (cm^{-1})	C-C stretching (cm^{-1})
Bulk materials	3402.96	2920.79	1632.43	1248.62	1061.54
Delignification	3399.10	2918.86	1645.93	-	1059.61
Bleaching	3364.39	2916.93	1645.93	-	1059.61
Acid hydrolysis	3397.17	2911.15	1647.86	-	1061.54
Sonication:					
H_2SO_4	3360.53	2915.00	1638.21	-	1061.54
H_2O	3449.25	2922.72	1634.36	-	1063.47

Hydrolysis process did not initiate a new bond, which showed at figure 10. Based on the FTIR spectra, there are polysaccharide rings of cellulose, which showed as C-H and C-O vibration at 1382 cm^{-1}. The characteristic peak was found at 1200-1300 cm^{-1}, which is specific for aromatic ring vibration. This peak will disappeared after pulping and bleaching.

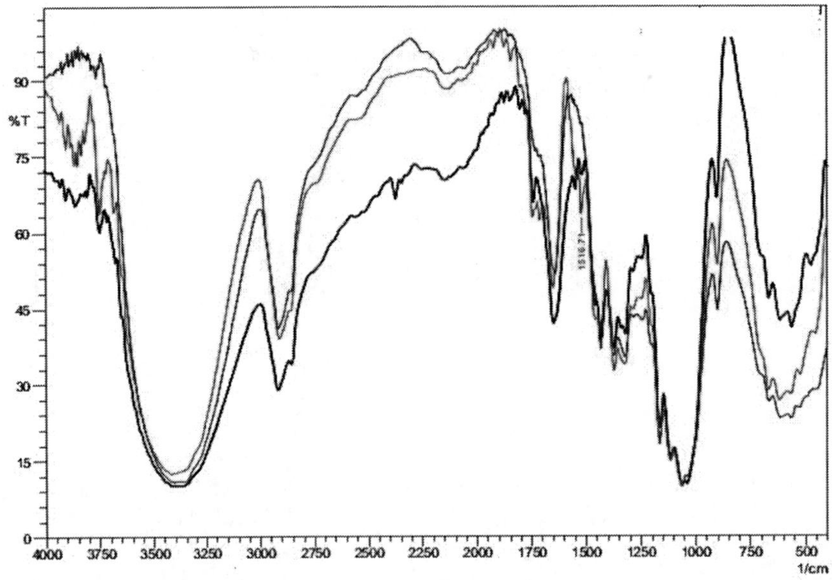

Figure 11. FTIR cellulose-bleaching (black line), hydrolysis-without sonication (red line), hydrolysis with sonication (blue line)

Based on figure 11, the band located at 1734.65 cm^{-1} is related to C=O stretching of acetyl and uronic ester groups of the hemicellulose or the ester linkage of carboxylic groups of ferulic and p-coumaric acid of lignin and/or hemicellulose [24,31]. Lignin and hemicellulose are still present in small quantities. Then, the peak of lignin at 1516,71 cm^{-1} disappeared after sonication process (blue line).

3.3. Scanning Electron Microscopy (SEM)

Figure 12 shown the dried macrofibril resembles structure like a bowl which consist of xylem and phloem tissues. The delignification process has damaged the xylem and phloem cell walls, and turned it become fibers. Lignin and hemicellulose were removed by dissolving it in 6% NaOH solution through delignification process (figure 13). Addition of 30% H_2SO_4 as hydrolysis agent cause the fibers start separated from one another indicating that the hydrogen bonds were broken (figure 14).

There are several diameters of cellulose obtained from the hydrolysis effect, such as 1.018 µm and 2.480 µm. Sonication using 30% H_2SO_4 make cellulose organize itself in a linier form (figure 15). The range of measured diameter are 1982 µm - 3715 µm. Sonication using water solvent causes irregular shape changes (figure 16). This is likely to be an interaction through hydrogen bonds between cellulose and water as solvent.

Figure 12. Dried macrofibril

Figure 13. Delignification using NaOH 6%

Figure 14. Hydrolysis using H_2SO_4 30% without sonication

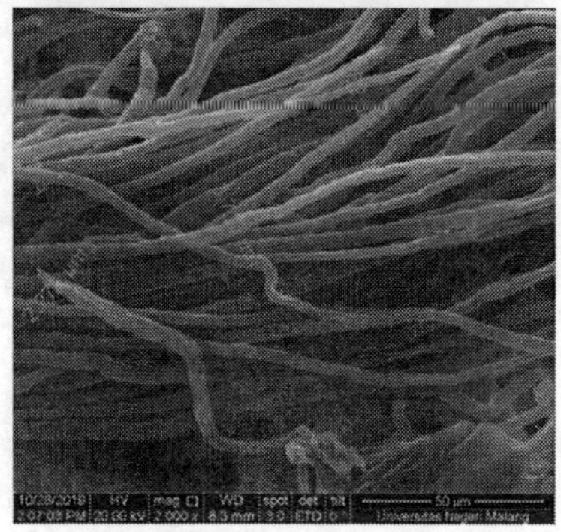

Figure 15. Hydrolysis using H_2SO_4 30% with sonication

Figure 16. Sonication using H_2O

4. Conclusions

The procedure used in this research was capable for producing pure cellulose. Hydrolysis is proven to be able to cut the macro-cellulose chain become micro-sized. The FTIR provides conclusive information that the elimination of lignin and hemicellulose were occurred. Hydrolysis technique using sonication help cellulose polymer eliminate amorphous region (which shown at SEM image). Other measurement still need to determine the size of the cellulose. Attainment of nano size is needed for further study using variations in temperature, acid type, and concentration. Increasing cellulose surface area therefore increases the chance for heavy metal adsorption.

Acknowledgment
Research material and funding assistance supported by supervisor.

References

[1] Brinchi L, Cotana F, Fortunati E and Kenny J M 2013 *Carbohydrate Polymer* **94** 154-169
[2] www.bps.go.id
[3] Moon R J, Martini A, Nairn J, Simonsen J and Youngblood J 2011 *Chem. Soc. Rev.* **40** 3941–3994
[4] Klemm D, Kramer F, Moritz M, Lindstrm T, Ankerfors M, Gray D, and Dorris A 2011 *Angew. Chem. Int. Ed.* **50** 5438 – 5466
[5] Siqueira G, Tapin-Lingua S, Bras J and da Silva Perez D, Dufresne A 2010 *Cellulose* **17** 1147–1158
[6] Siro I and Plackett D 2010 *Cellulose* **17** 459–494
[7] Chirayil C J, Mathew L and Thomas S 2014 *Adv. Mater. Sci.* **37** 20-28
[8] John, M J and Thomas S 2001 *Carbohydrate Polymers* **71** 343–364.
[9] Hu X Zhao M and Huang H 2010 *Water Environment Research* **82** 733-741
[10] Andresen M, Johansson L-S, Tanem B S and Stenius P 2006 *Cellulose* **13** 665 –677
[11] Stenstad P, Andresen M, Tanem B J and Stenius P 2008 *Cellulose* **15** 35-45
[12] Samir M A S A, Alloin F and Dufresne A 2005 *Biomacromolecules* **6** 612-626
[13] Svagan J A, Samir M A S A and Berglund L A 2007 *Biomacromolecules* **8** 2556-2563
[14] Mahardika M, Abral H, Kasim A, Arief S and Asrofi M 2018 *Fibers* **6** 28.
[15] Rodriguez N L G, Thielemans W and Dufresne A 2006 *Cellulose* **13** 261 –270
[16] Dujardin E, Blaseby M and Mann S 2003 *J. Mater. Chem.* **13** 696–699
[17] Iwamoto S, Kai W, Isogai A, and Iwata T 2009 *Biomacromolecules* **10** 2571–257
[18] Levis S R and Deasy B 2001 *International Journal of Pharmaceutics* **213**: 13 – 24
[19] Ngah W S W and Hanafiah M A K N 2008 *Bioresource Technology* **99**: 3935–3948
[20] Weng C-H and Wu Y-C. 2012 *Journal Environmental Engeenering* **138** 286-292.
[21] Shukla S R and Pai R S 2005 *Bioresources Technology* **96**, 1430-1438.
[22] Yamuna M and Kamaraj M 2016 *International Journal of ChemTech Research* **9** 544-550.
[23] Holan Z R and Volesky B 1995 *Appl. Biochem. Biotechnol.* **53** 133–142
[24] Alemdar A and Sain M 2008 *Biosource Technology* **99** 1664-1671
[25] Mukherjee P S and Satyanarayana K G 1986 *J. Mater. Sci.* **21** 51-56
[26] Sharma U 1981 *Carbohydr. Res.* **97** 323-329
[27] Saha S C, Das B K, Ray P K, Pandey S N and Goswami K, 1990 *Text. Res. J.* **60** 726-731
[28] Solo A A M, Masruri M, Rumhayati B 2018 *J. Pure App. Chem. Res.* **7** 239-246
[29] Firmansyah D, Rumhayati B, Masruri M 2017 *Int. J. of ChemTech Research* **10** 674-680
[30] Wong S and Shanks R, 2009 *Biodegradable Polymer Blends and Composite from Renewable Resources* (New Jersey:John Wiley ans Sons, Inc.)
[31] Sun X, Xu F, Sun R C, Fowler P and Baird M S 2005 *Carbohydrate Research* **340** 97-106

Effects of Growth Medium on Extracellular Secretion of Human Epidermal Growth Factor in *Escherichia coli* by Co-expression with *Bacillus cereus* Phospholipase C

F P U Latifah[1], A Indriyani[1], R D Pratiwi[2], Sriwidodo[3], IP Maksum[1,*]

[1]Department of Chemistry, Faculty of Mathematics and Natural Sciences, Universitas Padjadjaran, Indonesia
[2]Research Center of Biotechnology, Indonesian Institute of Science (LIPI), Bogor, Indonesia
[3]Department of Pharmacy and Pharmaceutical Technology, Faculty of Pharmacy, Universitas Padjadjaran, Indonesia

*Corresponding author: iman.permana@unpad.ac.id

Abstract. Human Epidermal Growth Factor (hEGF) is a small, mitotic growth polypeptide that promotes the proliferation of various cells and is widely applied in clinical practices, especially in therapeutic uses of wound healing. Since it has a lot of benefits, production of recombinant hEGF (rhEGF) in a large scale is needed. Some methods have been used in this protein production, one of them was the production of rhEGF using extracellular secretion in *Escherichia coli*. Previous research have been done using co-expression method with phospholipase C from *Bacillus cereus* to increase the amount of rhEGF. Phospholipase C *B. cereus* have been used in several protein expression and was proved that it could increase the secretion of recombinant protein through hydrolytic mechanism of cell membrane. In addition, growth condition is one of some major factors which can affect the yields of produced protein. Different compositions of bacterial growth medium often lead to different result. This paper studies how rich-nutrient Terrifict Broth (TB) medium and Luria Bertani (LB) medium produced different rhEGF results when it was co-expressed with phospholipase C *B. cereus*. rhEGF was characterized using SDS-PAGE and confirmed by western blot using anti-mouse EGF, and its concentration was measured using ELISA. rhEGF was successfully characterized after co-expression in TB medium and the concentration was 503.48 µg/mL. rhEGF was better produced in TB medium rather than in LB medium since TB medium has richer composition.

1. Introduction

Diabetes mellitus is a metabolic disorder disease characterized by chronic hyperglycemia due to defect in insulin secretion, insulin action, or both [1]. One of the most devastating and significant complications of diabetes and is associated with neuropathy and/or peripheral of the lower limb in patients suffering diabetes mellitus is diabetic foot ulcer (DM). The prevalence of diabetic foot ulcer in the diabetic population is around 4–10% [2]. Generally, diabetic foot ulcer is hard to be cured, therefore latest treatments have been done to cure this problem by using Human Epidermal Growth Factor protein to accelerate wound healing [3]. A previous study reported that recombinant hEGF can be considered as therapeutic agents in wound healing of diabetic foot ulcer and it was proven that the recovery of ulcer can be done in only about 8 weeks [4]. A number of recombinant proteins have been reported that they can be used as therapeutic agents. In addition, the market sales of recombinant pharmaceutical proteins has gradually increased. Increasing the needs of recombinant proteins as therapeutic agents suggests the needs of large amounts production of recombinant proteins with low production costs using recombinant DNA technology [5]. Extracellular expression in *Escherichia coli* host are the most commonly used methods [6]. Since it has limited space of the periplasmic membrane, this type of secretions in *E. coli* has several disadvantages due to the low secretion products [7]. In another case, extracellular expression in *E. coli* can be enhanced through hydrolytic membrane mechanism by co-expression with *Bacillus cereus* Phospholipase C, especially for protein with low molecular weight [8]. Since hEGF is only 6.2 kDa, which is one of the low molecular proteins in size, therefore in this study we enhanced the secretion of recombinant hEGF by co-expression methods with *B. cereus* Phospholipase C.

Like many other expression processes, the amount of produced protein depends on the survival and the ability of the host cell in producing the protein during cultivation. In addition, the survival of the cells is much affected by culture conditions. One of the crucial conditions affecting the amount of cell density and its survival is the composition of growth medium. Recombinant *E. coli* can be grown at high density in general media such as Luria Bertani broth (LB), the synthetic M9 minimal salt medium, Super broth (SB), and Terrific broth (TB) [9] in 37 °C, and pH between 6.4 a nd 7. 2[10]. The most common used media are TB and LB since they are economically effective, and their compositions are easy to be found. A study used those two types of media in expressing a fleunder growth hormone (fGH) and reported that the use of TB medium produced higher amount of fGH since TB has richer composition than LB [11]. Therefore, in this research we optimized the growth medium using LB and TB in producing higher amount of hEGF during co-expression process by *B. cereus* phospholipase C.

2. Materials and Methods

2.1. Bacterial strain, vector construction and materials

This research used *E. coli* BL21 (DE3) strain that was taken from our laboratory stock. Other reagents including medium compositions, antibiotic, inducer, and SDS-PAGE reagents were obtained from Sigma Aldrich (Singapore). Western blot reagents and analysis were supported by Research Center of Biotechnology, Indonesian Institute of Science (LIPI, Bogor, Indonesia). The construction of plasmid pETDuet-1 which harbor genes encoding *B. cereus* phospholipase C and hEGF was designed by our team and purchased from Genescript, Novagen (Madison, WI, USA).

The gene encoding *B. cereus* phospholipase C was gained from protein data bank (accession code: 1AH7) and it was designed to be inserted to multiple cloning site 1 (MCS1) in the plasmid. Besides, genes encoding PelB and hEGF were elaborated from our previous research [12] and were designed to be inserted to multiple cloning site 2 (MCS2). hEGF gene was fused with gene encoding PelB signal peptide in C terminal in order to help the migration of hEGF through Sec-dependent secretion mechanism.

2.2. Cultivation conditions

The plasmid described before was used to change chemically competent *E. coli* BL21(DE3) cells. These engineered *E. coli* strains were cultured in shake flasks to produce test proteins. A 5 mL of Luria Bertani medium (1% [w/v] tryptone; 0.5% [w/v] yeast extract; 1% [w/v] NaCl) was added with 100 µg/mL ampicillin. The medium was then inoculated with frozen glycerol stock from engineered *E. coli* strain (100 µL), and then incubated in a rotary shaker (200 rpm) at 37 °C for up to 8 h. Aliquots from this seed culture (5% [v/v]) were used to inoculate 50 mL of Luria Bertani and Terrific Broth medium (0.5% [w/v] glycerol; 1.2% [w/v] tryptone; 2.4% [w/v] yeast extract; 0.23% [w/v] KH_2PO_4; and 1.64% [w/v] K_2HPO_4) supplemented with 100 µg/mL ampicillin, which was then shaken in a rotary shaker (200 rpm) at 37 °C. Then isopropyl β-d-thiogalactopyranoside (IPTG) inducer was added to a final concentration of 0.1mM when the optical density of culture at 600 nm (OD600) increased to

1.5, and incubation was continued at 25 °C [8]. The each of culture samples were then collected every 20 h in cultivation medium, periplasm, and cytoplasm.

2.3. Protein characterization and analysis

The collected culture was then centrifuged in 10,000 rpm in 4 °C for 30 minutes and the supernatant was gained as medium fraction. Some of the pellets was dissolved with TSE buffer (5% [v/v] Tris-Cl pH 8.0, 1 M; 20% [w/v] sucrose; 10% [v/v] EDTA 0.5 M) to extract the periplasm and then incubated in the ice bath for 20 minutes. Periplasmic fraction was gained as the supernatant after its being centrifuged in 20.000 rpm, 4 °C, for 30 minutes. Some other pellets from the first centrifugation was dissolved in lysis buffer (1% [v/v] Tris-Cl pH 7.5, 1M; 0.2% [v/v] 0.5 M EDTA) and the suspension was then disrupted using ultrasonic disintegration to obtain proteins inside cytoplasm. Then it was centrifuged in 10,000 rpm in 4 °C for 30 minutes and the supernatant was collected. Protein characterization was performed by Tricine SDS-PAGE (gel concentration: 15%) following the newly developed protocol [13]. The sample was then confirmed by western blot using Ig anti-mouse EGF (as the primary antibody) and IgG anti-mouse EGF (as the secondary antibody) in a nitrocellulose membrane. The proteins in the membrane were then visualized by alkaline phosphatase substrate. The concentration of rhEGF in cultivation medium was then measured by ELISA which was purchased from Qiayee (Bio-Techne, USA).

3. Results and Discussions

The genes encoding PLC and PelB-hEGF were inserted to pETDuet-1 plasmid and this recombinant plasmid was used to transform *E. coli* BL21(DE3). The modified *E. coli* strain was then cultivated in shake flasks using two different mediums with different compositions. After 61 h of expression in those different mediums, hEGF was produced in both medium but with different concentration. In expression using LB medium, rhEGF was produced in low level. This data was supported by SDS-PAGE result which was not showing band around 6.2 kDa (below 10 kDa) in medium fraction. This result also shows that the localization process of rhEGF from cytoplasm to periplasm and culture medium was not successfully happened (Figure 1).

Figure 1 shows us that rhEGF was still localized in the periplasm even after 61 hours post induction. The probability of hEGF which might present in the medium fraction was low since there was no band produced in the medium fraction in 20, 40, and 61 hours post induction. This result also shows that PLC was not well characterized around 28 kDa both in medium and periplasm. Looking up to the cytoplasm, SDS-PAGE result shows the present of PLC and rhEGF band (Figure 2). This result tells us that both PLC and rhEGF were actually produced even since 20 h post induction but they were not migrated successfully to periplasm and culture medium.

Different result was obtained from expression in TB medium. Figure 3 shows that hEGF band below 10 kDa was produced in higher intensity than the band produced in figure 1. This means that the co-expression process in TB medium produced rhEGF in higher concentration than in LB medium. Besides, figure 3 tells us about the migration of rhEGF from priplasm to the culture medium. In 20 h after induction, rhEGF was present in periplasm but not in culture medium. After 40 h, it was present both in periplasm and culture medium, indicating the migration of the protein. Looking up to 61 h after induction, all of rhEGF in periplasm has been moved to the culture medium since there was no rhEGF band in periplasm. In conclusion, co-expression in TB medium produced rhEGF in high amount and provided a good migration of the protein to the culture medium.

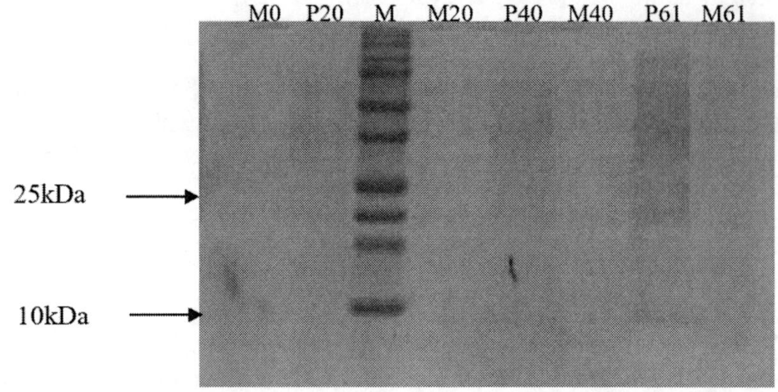

Figure 1. The SDS-PAGE analysis of the culture supernatant and periplasmic fraction in expression with LB medium. M0: medium fraction before induction, P20: periplasmic fraction 20 h after induction, M: protein marker, M20: medium fraction 20 h after induction, P40: periplasmic fraction 40 h after induction, M40: medium fraction 40 h after induction, P61: periplasmic fraction 61 h after induction. M6: medium fraction 61 h after induction.

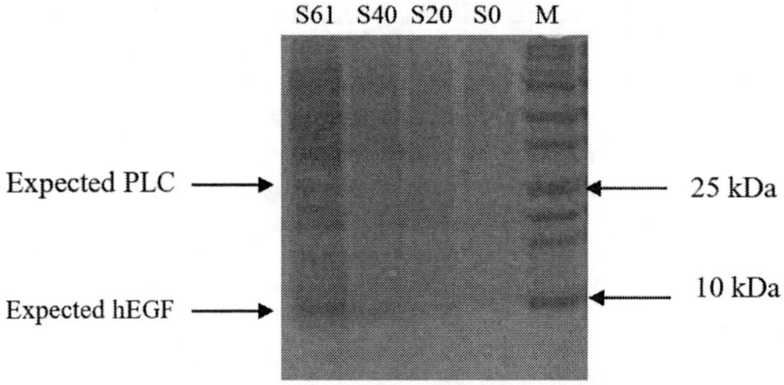

Figure 2. The SDS-PAGE analysis of cytoplasmic fraction in expression with LB medium. S0: cytoplasmic fraction before induction, S20: cytoplasmic fraction 20 h after induction, S40: cytoplasmic fraction 40 h after induction, S61: cytoplasmic fraction 20 h after induction.

The rhEGF production was better using TB medium than LB medium. This might be happened because of richer composition which exist in TB. TB medium contains higher concentration of tryptone and yeast extract (having roles as carbon source as amino acids and vitamins) than LB medium (TB: 1.2% tryptone, 2.4% yeast extract, while LB: 1% tryptone, 0.5% yeast extract). Besides, TB has phosphate buffering and glycerol as carbon source. Higher content of carbon sources can be used to support the growth of biomass yields up to five times higher than LB. Regarding to the results of volumetric recombinant protein, TB is superior when compared to LB as a result of higher cell density [14] and/or increased specific productivity per biomass. On the other hand, amino acids form a primary carbon source in LB medium which then be constituted by catabolizable amino acids, while the fermented sugar concentration is equivalent to 0.5% yeast extract which is estimated to be less than 100 μM. As a result, growth in the LB medium stops when the culture's optical density at 600nm (OD600) reached 7 when the amino acids catabolism has been depleted. Catabolism of amino acids lead to alkalinization of the medium up to pH 9 due to excessive ammonium excretion [14-15]. The optimal growth temperature for *E. coli* is 37 °C, and pH between 6.4 and 7.2 [10]. Thus, significant pH fluctuation of the medium will affect the survival of the cell and lead to cell death. As a result, cells cannot express the protein and the production of protein will be stopped, resulting the low yield of protein. In TB medium, the presence of phosphate buffering will attenuate pH drifts during cultivation, resulting the prevention of cell death caused by pH fluctuation.

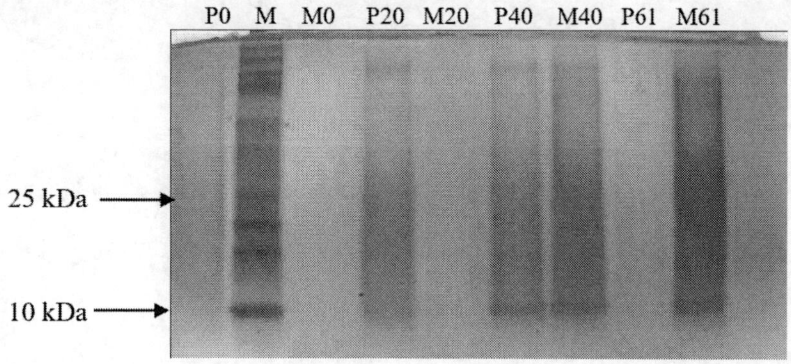

Figure 3. The SDS-PAGE analysis of the culture supernatant and periplasmic fraction in expression with TB medium. P0: periplasmic fraction before induction, M: protein marker, M0: medium fraction before induction, P20: periplasmic fraction 20 h after induction, M20: medium fraction 20 h after induction, P40: periplasmic fraction 40 h after induction, M40: medium fraction 40 h after induction, P61: periplasmic fraction 61 h after induction. M61: medium fraction 61 h after induction.

Another characterization was performed by western blot. This method was done to confirm the presence of rhEGF through binding activity with its antibody. Figure 4 shows the result of western blot and a band was shown up, confirming that the expressed protein was right. The concentration of rhEGF in TB medium culture media was measured using ELISA, with the concentration of 503.48 µg/mL.

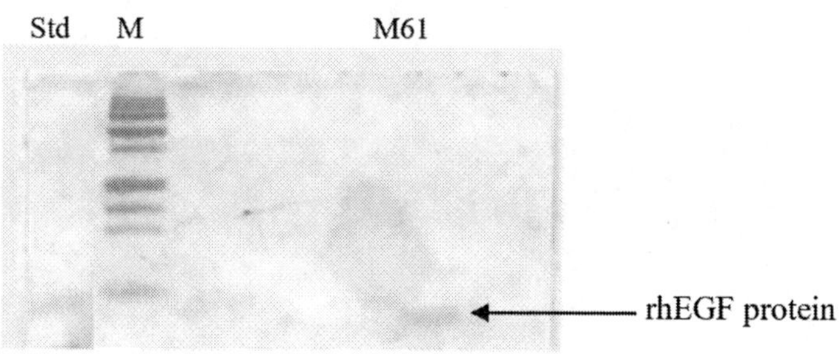

Figure 4. Western blot analysis result. Std: hEGF standard, M: protein marker, M61: culture medium 61 h after induction after co-expression in TB medium.

4. Conclusion

In this study showed the optimized culture medium for production of rhEGF protein. TB medium was found to be the best medium since it can support better growth and produce more recombinant protein than LB medium. This medium may also be suitable to produce other high-yield recombinant proteins. This result may also be a consideration in continuing the research by scaling up the volume fermentation so that we can produce rhEGF in high amount. Then, the large number of active recombinants rhEGF produced by this method can be utilized as a therapeutic candidates of diabetic foot ulcer.

Acknowledgements
The authors are thankful to KEMENRISTEKDIKTI for supporting this research through Master Thesis Grant (PTM) and also to Research Center of Biotechnology, Indonesian Institute of Science (LIPI) for facilitating us in doing our research.

References
[1] Kharroubi A T and Darwish H M 2015 *World J Diabetes* **6** 867.
[2] Alexiadou K and Doupis J 2012 *Diabetes Ther* **3** 1-15.
[3] Setyorogo S and Trisnawati S 2013 *J. Ilm. Kesehat.* **5** 6-11.
[4] Joon P H, Heun D J, and Yun WK 2006 *Ann. Plast. Surg.* **56** 394-98.
[5] Merlin M, Gecchele E, Capaldi S, Pezzotti M and Avesani L 2014 *Biomed Res. Int.* 2014 1-14.
[6] Maksum I P, Utama E, Sriwidodo, and Subroto T 2017 *Journal of Pharmaceutical Sciences and Research* **9** 1007-16.
[7] Ni Y and Chen R 2009 *Biotechnol Lett* **31** 1661–70.
[8] Su L, Jiang Q, Yu L, and Wu J 2017 *Microb. Cell Fac.* **16** 1–11.
[9] Madurawe R D, Chase T E, Tsao E I, and Bentley W E 2000 *Biotechnol. Prog.* **16** 571-76.
[10] Xu B, Jahic M, Blomsten G, and Enfors S O 1999 *Appl. Microbiol. Biotechnol.* **51** 564-71.
[11] Choi T J and Geletu T T 2018 *Journal of Genetic Engineering and Biotechnology* **16** 347-55.
[12] Sriwidodo, Riswanto N, Maksum I P, Rostinawati T, & Subroto T 2017 *International Journal of Research in Pharmaceutical Sciences* **8** 33–40.
[13] Haider S R, Reid H J, and Sharp B L 2012 *Methods Molecular Biology* **869** 151-160.
[14] Losen M, Frolich B, Pohl M, and Buchs J 2004 *Biotechnology Progress* **20** 1062-68.
[15] Sezonov G, Joseleau-Petit D, and D'Ari R 2007 *Journal of Bacteriology* **189** 8746–49

IOP Conf. Series: Materials Science and Engineering **833** (2020) 012022 doi:10.1088/1757-899X/833/1/012022

The Effect of Rice Bran on Triglyceride Levels and Histopatologic Aorta in Rat (*Rattus norvegicus*) of High Cholesterol Dietary Model

Chanif Mahdi[1,*], Putri Citrawati[2], Viski Fitri Hendrawan[2]

[1]Faculty of Mathematics and Natural Sciences University of Brawijaya, Indonesia
[2]Faculty of Veterinary Medicine, University of Brawijaya, Indonesia

*Corresponding author: chanifmahdi@gmail.com

Abstract. Cholesterol is a steroid compound found in animals and humans. Total cholesterol normal levels in rat is 10-54 mg/dL and triglyceride 26-145 mg/dL. Rice bran has a crude fiber content and antioxidants that can be used to resolve these conditions. This study aims to determine the effect of rice bran as therapy in the rats (*Rattus norvegicus*) high-cholesterol dietary model towards triglyceride levels and histopathologic of the aorta. In this research, rats (*Rattus norvegicus*) were divided into 5 groups with 4 repetitions. Group A as a negative control, group B was rat which fed with high cholesterol, group C was high-cholesterol diet model of rat with the 16% rice bran therapy/rat/day, group D was high-cholesterol diet model of rat with the 38% rice bran therapy/rat/day, and group E was high cholesterol diet model of rat with the 57% rice bran therapy/rat/day. Rice bran therapy performed for 21 days. Triglyceride levels were calculated by the GPO-PAP method and aortic histopathologic features were observed with HE staining (*Hematoxylin-Eosin*). The effect of rice bran on blood triglyceride levels was analyzed by one way ANOVA (Analysis of Variance) with α=5% and aortic histopathologic analysis was qualitatively descriptive. The results showed that rice bran therapy could significantly reduce triglyceride levels ($p < 0.01$) with a dose of 57%/rat/day to decrease triglyceride levels. Rice bran therapy also repaired aortic tissue histopathology of rat with high cholesterol diet model.

1. Introduction

Feeding high fat and cholesterol often occur in animals such as pet dogs and cats. One example is the granting of pet food in the form of meat and offal that can increase blood cholesterol levels in dogs and cats [1].

Cholesterol is a steroid compound that is found in animals and humans [2]. Normally, the levels of total cholesterol in the rat are 10-54 mg/dL [3]. Hypercholesterolemia is a situation where an increase in cholesterol levels in the blood is beyond normal level. The level of hypercholesterolemia on cats and dogs measured as much as 13% [4], whereas the dogs in Western countries stated as hypercholesterolemia when the level of their cholesterol measured 25% to 44% [5].

High cholesterol levels in the blood will turn into bile acids. The more bile acids will lead to the higher production of free radicals. This causes internal antioxidants are not being able to handle it [6]. High free radicals in the body will produce oxidative stress thus causing damage to the cells, tissues, or organs [7]. One of the damaged organs is the aorta. Free radicals also led to a high concentration of lipids so that the levels of triglycerides will increase. Normally, the levels of triglycerides in rats are 26-145 mg/dL [8].

The use of drugs in hypercholesterolemia animals is successfully controlling and lowering the cholesterol levels in the blood, but causing side effects in the long term [3]. Therefore, this research uses rice bran as a therapy for reducing cholesterol levels in the blood. Bran is the outer layer of rice that contains protein, fiber, fat, and vitamin E [9]. Rice brans contain some antioxidants, including tocopherol, tocotrienol, and oryzanol [10]. Types of rice bran used are white rice the rice bran. The crude fiber of rice bran in white (15.06 ± 2.09%) higher than the red rice bran (13.44 ± 0.40%) [9].

Rough fibers can accelerate the rate of feed-in bind bile acids absorption of fat-free so hampered and excretion of fats including cholesterol through the feces increased. Fibers will decrease bile acids in the bod, thus the body will establish a new bile acid from cholesterol in the blood, as a result, the cholesterol level in the blood will decrease [11]. Antioxidants in bran can reduce free radicals on the body [10], the levels of triglycerides in the body will decrease, and cells, tissues, or organs damaging of the aorta will be reduced. Based on those reasons, the purpose of this research is to investigate the effect of rice bran as a therapy in high cholesterol rats (*Rattus norvegicus*) diet model against the levels of triglycerides and to determine its effect and its effect on the level of cell damage in the aorta through the description of histopathology.

2. Material and Methods

2.1. *Tools and Materials*

The tools used in this study, among others, spectrophotometer, microscopes, glass object, hot plate, Buechner funnel, Erlenmeyer flask, compressors, porcelain bowls, scales, filter paper, oven, electric furnace.

Materials needed in the study were the acid cholate, pork, quail egg yolk, eosin, hematoxylen, distilled water, 1.5 N NaOH 0.3 N H_2SO_4, NaCl physiological, the white male rats (*Rattus norvegicus*), Wistar strain, 8 weeks with weight ± 100 g, BR1, feed and drinking water.

2.2. *Preparation of Experimental Animals*

The experimental animals used in the study were 20 male white rats (*Rattus norvegicus*) of Wistar strain aged 8 weeks with an average body weight of 100 g. They had been adapted for 7 days and feed with standard feed once a day and were drink by BR1 ad libitum.

2.3. *Analysis of The Proximate Content of Crude Bran Fiber*

Samples weighed ± 1 gram (A), then put in the Erlenmeyer flask. After that, 50 mL of H_2SO_4 0.3 N was added, mixed for 30 minutes, then 25 mL of 1.5 N NaOH was added to the mixture, and the mixture was boiled. The filter paper was placed on the Buchner funnel which has been known to weigh in (B). The solution was filtered by using a Buechner funnel and rinsed with 50 mL of hot water, and then solution was filtered.

Hydrochloric acid (HCl) 0.3 N by as much as 50 mL was placed into a Buechner funnel and left on for 1 minute with the compressor. Residue in a Buchner funnels was rinsed 5 timed with hot water, then 5 mL acetone was poured into the mixture, and incubated for 1 minute. Porcelain cup was heated for 1 hour in the 105 °C preheated oven and then cooled in a desiccator for 10-15 min, and weighed out (C). Filter paper containing the residue was placed in the porcelain bowls and then dried in an oven for 1.5 hours at 105 °C and cooled in the desiccator for CA for 30 minutes, and it was weighed out (D). Porcelain bowls were put into the 550 °C electric furnace for 2 hours. The electric furnace was turned off and left until the temperature is 0 °C. After that, the porcelain bowls were removed and placed in desiccator for 15 min and weighed out (E). The calculation formula is [12]:

$$\text{Crude Fiber} = \frac{\text{D}-\text{E}-\text{B}}{A} \times 100\% \tag{1}$$

2.4. Acclimatization of Experimental Animals

Rats in Group of B, C, D, and E were given high cholesterol feeding. Feeding a high cholesterol diet was conducted using a method that was reported by Larasathi and co-workers [13], as much as 2 grams of oil, 0.02 grams of acid cholate, 1 gram of quail egg yolk that has been heated at a temperature of 100 °C, and 2 mL of water. High cholesterol diet was given starting at 8th day for 35 days as much as 3.02 grams/2 mL.

2.5. Induction of Rice Bran Therapy

Bran therapy was given in Group C, D and E. The dose of each group was: the 16% rice bran therapy/rat/day for group C, 38% rice bran therapy/rat/day for group D, and 57% rice bran therapy/rat/day for group E.

2.6. Isolation of Blood Serum

Blood was taken at each group on the 22nd and 44th day. Blood sampling on the 22nd day was performed in sinus orbitalis, and blood sampling in 44th day were done by sacrificing them and collecting their blood from the heart. After the blood was collected, blood was centrifuged at 3000 rpm for 15 min, in order to obtain blood serum[3].

2.7. Measurement of Levels Of Triglycerides

Determination of the triglyceride levels using the method of GPO-PAP (Gliserolphosphat Oxidase-Aminoantipyrine Peroxidase). 500 µl of serum drawn using micropipette and inserted into the sample cup and the examination was being conducted. Absorbance reading was performed using Spectrophotometer Biosystem A15 at wavelength of 500 nm [8].

2.8. Preparation and observation of Aorta Histopathology

At the end of study, the aorta of the rats was taken after the blood was collected. After that, the aortic samples were washed using a 0.7% NaCl [3]. Steps in making preparations according to the Putra [14] and Rahmawati [15] i.e. the fixation, dehydration, clearing, impregnation, embedding and blocking, cutting and shaping object-glass, hematoxylin and eosin staining, and mounting. The observation was carried out using a Olympus microscope with magnification of 400×.

2.9. Data Analysis

The analysis of the data used in this research in the form of quantitative data in order to determine the effect of the grant of triglyceride levels against the bran is analyzed using one way ANOVA (Analysis of Variance). If there is a significant difference, then followed by the Tukey test with $\alpha = 5\%$. Analysis of histopathology data from cell damage of the aorta was descriptive qualitative.

3. Results and Discussion

3.1. Effects of Bran Rice to Triglyceride Levels of Rats (Rattus norvegicus) High Cholesterol Diet Models

The measurement method using triglyceride levels GPO-PAP (Gliserolphosphate Oxidase-Aminoantipyrine Peroxidase) aims to find out the effects of the rice bran in lowering levels of triglycerides in white rats (*Rattus norvegicus*) high cholesterol model. The results of level measurements of triglycerides that have been obtained are statistically analyzed using one way ANOVA test followed by Tukey test with $\alpha = 5\%$. Triglycerides levels test were carried out twice, pre-therapy and post-therapy analysis. Post-therapy data were analyzed statistically which can be seen in Table 1.

Table 1. Triglyceride levels in white rats (*Rattus norvegicus*) high cholesterol diet model

Treatment groups	Average of triglyceride levels (mg/dL)
Group A(negative control)	89.00 ± 10.39^{bc}
Group B (positive control)	186.00 ± 3.92^{d}
Group C (treatment 1)	100.75 ± 29.34^{c}
Group D (treatment 2)	60.75 ± 6.65^{ab}
Group E (treatment 3)	33.75 ± 2.63^{a}

*Notes: Different notations show real differences between treatments

The statistical analysis results from triglycerides levels post-therapy with one way ANOVA test shows that the treatment of the rice bran in white rats (*Rattus norvegicus*) was able for lowering levels of triglycerides ($p < 0.01$), and it is indicated by the presence of 4 types of notation on the Tukey test Advanced (Table 5.1). Triglyceride levels in Group A (89.00 ± 10.39^{bc} mg/dL) as negative control was highly significant differences with Group B and E, but no significant difference in groups C and D. While in Group B (186.00 ± 3.92^{d} mg/dL) as positive control, triglyceride levels differ markedly with those in groups A, C, D, and E. Group C (100.75 ± 29.34^{c} mg/dL) triglyceride levels had significant differences with those B, D, and E groups, but there is no significant difference with that in group A. Group D (60.75 ± 6.65^{ab} mg/dL) triglyceride levels resulted in differences with those in groups B and C, but differs markedly with those in Groups A and E. Group E triglyceride levels (33.75 ± 2.63^{a} mg/dL) had significant differences with those in groups A, B, and C, but there is no significant differences with Group D.

Normal triglyceride levels in the rat were 26-145 mg/dL [16]. Oil pork contains 38-43% of saturated fatty acids and cholesterol [17]. Quail egg yolk has a very high content of cholesterol that is 2,139.17 mg/100 gr [18]. Acid cholate accelerates the increasing of cholesterol in the blood [19]. An increase in cholesterol in the blood will increase free radicals in the body [1], thereby causing an increase in triglyceride levels in the blood [8].

Group C was high-cholesterol diet model of rat with the 16% rice bran therapy/rat/day, group D was high-cholesterol diet model of rat with the 38% rice bran therapy/rat/day, and group E was high cholesterol diet model of rat with the 57% rice bran therapy/rat/day has decreased levels of triglycerides. Calculation based on average test results triglyceride levels pre-and-post therapy-therapy on each of these treatment groups. Notes that group C decreased by 93 mg/dL, from 193.75 to 100.75 mg/dL. Group D decreased by 107.25 mg/dL, from 168 became 60.75 mg/dL. Group E decreased by 131.75 mg/dL, from 165.5 to 33.75 mg/dL. Group C, D, and E have triglycerides levels that classified as normal, but group E had the highest cholesterol reduction, therefore, the dose given to group E was the most effective dose in reducing triglyceride levels.

Rice bran therapy can lower triglyceride levels shown through changes in the levels of triglycerides from pre-to post-therapy therapy. Changes in the levels of triglycerides because bran contains rough fiber and antioxidants that can help lower triglyceride levels. Crude fibers in the bran that is used for the therapy contains 21.20% and strong antioxidants activity with the IC_{50} of 51.79 μg/mL. The IC_{50} value is a parameter used to determine the concentrations of the antioxidant compounds in inhibiting 50% of oxidation [20]. Very strong antioxidant activity when the values of $IC_{50}<$ μg/mL, 50 strong IC_{50} value when 50-100 μg/ml, while IC_{50} value when 100-150 μg/mL, and weak in IC_{50} values 150-200 μg/mL [21]. Antioxidant compounds found in rice bran is γ-oryzanol.

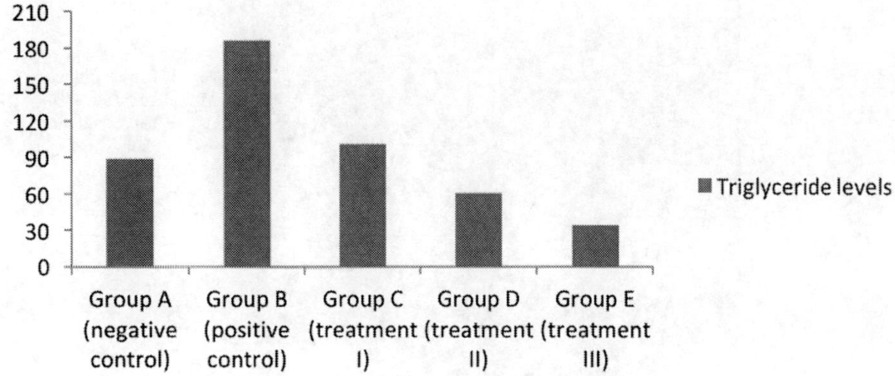

Figure 1. Graph of triglyceride levels

Crude fiber has the ability to bind bile acids and therefore cannot be absorbed and circulated back. Crude fibers that have binding bile acids will then enter into the colon to degrades and excreted. Fibers degradation in the intestinal microorganisms capable of producing propionic acid or other short-chain fat can inhibit the synthesis of fatty acids and cholesterol [22]. Therefore, this will decrease cholesterol and triglycerides will also cause decreased in number.

γ-oryzanol has the ability to lower triglycerides and cholesterol [23]. γ-oryzanol is nonpolar antioxidants that functions inhibit the concentration of lipids and preventing oxidative stress. The higher the antioxidant levels will cause a higher rate of inhibition of lipid concentration by free radicals [24].

3.2. Effects of Bran Rice to Aorta Histopathological Profiles of Rats (Rattus norvegicus) High Cholesterol Diet Models

Observations were performed using microscopes Olympus 400x zoom in. The change that occurs in the aorta of each group can be seen in the description of the histopathology following aortic (Figure 5.2).

An overview of histology aorta Group A (negative control) (Figure 5.2A) is profile of the aorta in the healthy rats. Tunica adventitia of the aorta in Group A shows no fat cells and no inflammation. In addition, in the tunica media shows smooth muscle cell proliferation. The profiles of histopathology aorta in Group B (positive control) (Figure 5.2B) indicates the presence of inflammatory cells and fat cells in tunica adventitia. Moreover, there are some foam cells formation in tunica media. In tunica intima looks the existence of smooth muscle cell proliferation. Fat cells formation are indicated by yellow arrows, inflammatory cells are indicated by a red arrow, foam cells are indicated by blue arrow, and the proliferation of smooth muscle cells is indicated by green arrow.

The description of the aorta histopathology in Group C (treatment 1) (Figure 5.2C) showed fat cells without the presence of inflammatory cells followed in tunica adventitia, indicated by a yellow arrow, on the media there is a foam cell tunica, indicated by blue arrow, and in the presence of visible tunica intima proliferation of smooth muscle cells, indicated by green arrow. The histopathological description of Group D (treatment 2) (Figure 5.2D) shows that there are some fat cells formation and there are no cells inflammation in tunica adventitia. This figure shows that there are no tunica foam cells, but it can be seen that there are tunica intima proliferation of aortic smooth muscle cells.

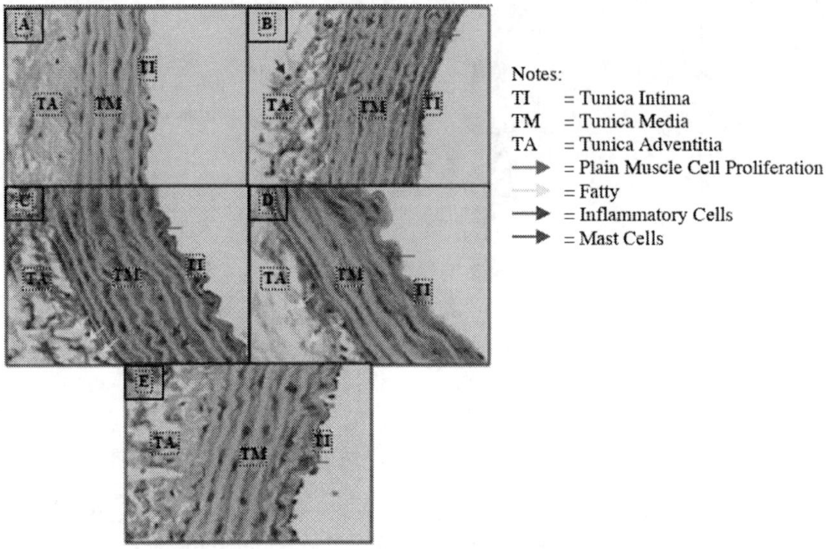

Figure 2. Aortic histology (A) Group A (negative control), (B) Group B (positive control), (C) Group A (Treatment of 1), (D) Group B (Treatment of 2), and (E) Group C (Treatment of 3).

The description of the aorta histopathology of Group E (treatment 3) (Figure 5.2E) indicates the absence of inflammatory cells in fatty and tunica adventitia; furthermore, there is no foam cells in tunica media. But the visible presence of the proliferation of smooth muscle cells on intima tunica. Presence of the smooth muscle cell proliferation likely due to still existing of macrophages in it or the presence of staining his cell so it looks less than perfect conduct proliferates. The fat cells in the cell will be carried by HDL to the liver for inflammatory cells metabolized and the numbers are dwindling because of the inflammation is reduced, while the foam cells will have emigrated to the lymphatic system so that the amount will be reduced. An overview of the histopathology cohort in group E approaching the normal aorta picture that shown by Group A as a negative control.

In tunica adventitia, there are fat cells formation and inflammations signs. Lipids accumulation is characterized by the presence of fat cells. Fat cells have large and flat-bottom vacuole line, and located in the peripheral [26]. Cell inflammation is characteristic of the existence of inflammation in the area. Lipids accumulations and inflammation in tunica adventitia aorta caused by the granting of a diet high in cholesterol. One of the materials used as a diet high in cholesterol is oil derived from pork lard. Fat pork contains saturated fatty acids. Saturated fatty acids that cause high cholesterol levels increase [27]. High cholesterol levels in the blood cause cholesterol plaque. The cholesterol plaque attached to the walls of the blood vessels causing the onset of inflammatory [28].

In tunica media, cell foam is formed. Foam cells are formed from macrophages that to phagocyte LDL (Low-Density Lipoprotein) oxidized [29]. The process of the formation of foam cells begins with endothelial function disturbance due to the existence of the increased formation of free radicals of oxygen that disable the nitric oxide. Fats in the blood vessels will undergo chemical changes due to the free radicals resulting in oxidized LDL. Oxidized LDL will be captured by macrophages via scavenger receptor macrophage continuously and will be transformed into foam cells [27].

In tunica intima the existence of smooth muscle cell proliferation was shown. The proliferation of smooth muscle cells is increasing the number of smooth muscle cells derived from cleavage of one cell into two cells. The proliferation occurs caused by macrophages which secrete growth factors [29]. Damage to endothelial dysfunction led to endothelial adhesion molecules, thus, stimulating that cause monocytes to endothelial bonding. Monocytes penetration to tunica intima through fissures and activated endothelial become macrophages. Macrophages release growth factors that will trigger the proliferation of smooth muscle cells [30].

High cholesterol in the blood will cause the production of bile acids increase. Increasing bile acids will increase the amount of free radicals production. Free radicals generated from the reduction of oxygen on the reaction of 7-α-hydroxylation from the process of the bile acids formation [20]. Free radicals may lead to oxidative stress formation. Oxidative stress resulting in cell damage [7]. Damage to these cells led to cuts in the endothelial cells. The wound on the endothelial cells resulting in inflammatory reaction and lead to the occurrence of vasodilator which increases the permeability of the endothelial cells against a variety of materials in the plasma. As a consequence, this will form cavities between cells, and inflammatory cells and fatty infiltration in tunica adventitia [3].

Group C, D, and E that received rice bran treatment with the dose of 16%/rat/day, 38%/rat/day, and 57%/rat/day, respectively, showed improvements in tunica adventitia and tunica media. It is indicated by a devastating drop in fat cells and inflammatory in tunica adventitia, as well as reduced cell foam in tunica media. Foam cells are in tunica media due to impaired endothelial function due to free radicals. Fat cells and inflammation are in tunica adventitia due to the presence of sores on endothelial cells caused by damage to the cell damage of cells due to increased oxidative stress in the body. Increased oxidative stress is caused by an imbalance between free radicals and antioxidants in the body, therefore, the body needs extra antioxidants from outside in order to balance the number of free radicals.

Rice bran contains rough fiber and antioxidants that are able to repair damage to endothelial cells. It repairs mechanism that is to speed up the process of disposal of bile acids out of the body and also adds antioxidant in the body to prevent the occurrence of oxidative stress. The content of antioxidants in the form of γ-oryzanol bran increases the metabolism of cholesterol [31]. γ-oryzanol can lower cholesterol levels and may inhibit the oxidation of cholesterol [32]. Decreased cholesterol levels will cause the synthesis of bile acids is declining. A decrease in the synthesis of bile acids will reduce the number of free radicals. The result of the decrease in free radicals is a decrease in oxidative stress. A decrease in oxidative stress will repair damaged cells so inflammatory, fatty, and foam cells decrease.

4. Conclusion

The current study concludes that rice bran is significantly decreased the levels of triglycerides and repair the damage of aorta cell in hypercholesterolemia rats (*Rattus norvegicus*) that were fed with high cholesterol diets, with the effective dose was 57% of rice bran/rat/day. However, this study requires further studies using other rice bran varieties in reducing triglyceride levels and repairing aortic cells due to a high cholesterol diet. Results of this study proves that rice bran fiber is a promising alternative treatment for hypercholesterolemia.

References

[1] Irawan, R. H. W., C. Mahdi, dan M. C. Padaga. 2014. *Pengaruh Pemberian Yogurt Susu Kambing sebagai Pencegahan untuk Hiperkolesterolemia melalui Pengamatan Malondehida (MDA) dan TNF-α pada Jantung Hewan Model Tikus (Rattus norvegicus).* Program Kedokteran Hewan. Universitas Brawijaya.

[2] Widyaningsih, W., A. Prabowo, dan Sumiasih. 2010. *Pengaruh Ekstrak Etanol Daging Bekicot (Achantina fulica) terhadap Kadar Kolesterol Total, HDL, dan LDL Serum Darah Tikus Jantan Galur Wistar.* Jurnal Sains dan Teknologi Farmasi, 15 (1): 1-10.

[3] Riesanti, D. G., M. C. Padaga, dan Herawati. 2014. *Kadar HDL, Kadar LDL, dan Gambaran Histopatologi Aorta pada Hewan Model Tikus ((Rattus norvegicus) Hiperkolesterolemia dengan Terapi Ekstrak Air Benalu Mangga (Dendrophthoe pentandra).* Program Kedokteran Hewan. Universitas Brawijaya.

[4] Maulana, B., M. C. Padaga, dan D. K. Wuragil. 2014. *Pengaruh Pemberian Susu Kambing untuk Pencegahan Hiperkolesterolemia: Studi Ekspresi Inducible Nitric Oxide Syntase (iNOS) dan Gambaran Histopatologi Duodenum pada Tikus (Rattus norvegicus) Model Hiperkolesterolemia.* Program Kedokteran Hewan. Universitas Brawijaya.

[5] Rufaida, F., Aulanni'am, S. Murwani. 2014. *Profil Kadar Kolesterol Total, Low Density Lipoprotein (LDL), dan Gambaran Histopatologi Aorta pada Tikus (Rattus norvegicus) Hiperkoleterolemia dengan Terapi Ekstrak Air Benalu Mangga (Dendropthoe Pentandra)*. Program Kedokteran Hewan. Universitas Brawijaya.

[6] Sugiarto, M., M. C. Padaga, dan D. K. Wuragil. 2014. *Efek Terapi Yogurt Susu terhadap Ekspresi Inducible Nitric Oxyde Synthase (iNOS) dan Kadar Malondialdehida (MDA) pada Aorta Hewan Model Tikus (Rattus norvegicus) Hiperkolesterolemia*. Program Kedokteran Hewan. Universitas Brawijaya.

[7] Susantiningsih, T 2015 *JuKe Unila* **5** 89-93.

[8] Kinenda, D. P. M. 2017. *Pengaruh Pemberian Ekstrak Kasar Buah Tomat (Solanum lycopersicum L.) terhadap Kadar Trigliserida dan Gambaran Histopatologi Hepar pada Model Hewan Hewan Coba Tikus Putih (Rattus norvegicus) Model Hiperkolesterolemia* [Skripsi]. Fakultas Kedokteran Hewan. Universitas Brawijaya.

[9] Iriyani, N. 2011. *Sereal dengan Substitusi Bekatul Tinggi Antioksidan*. Fakultas Kedokteran. Universitas Diponegoro.

[10] Indra, I. R. dan B. Panunggal. 2015 *Journal of Nutrition College* **4** 171-179.

[11] Teru, V., M. H. Natsir and E. Widodo 2017 *Jurnal Ilmu-Ilmu Peternakan*, **27** 76-82.

[12] Setiawan, A. 2010. *Kandungan Serat Kasar dan Protein Kasar Bekatul yang Difermentasi Acidothermus Cellulolyticus dan Aspergillus Terreus dari Cairan Rumen Sapi* [Skripsi]. Fakultas Kedokteran Hewan. Universitas Airlangga.

[13] Larasathi, A. S., C. Mahdi, dan M. C. Padaga. 2014. *Pengaruh Terapi Yogurt Susu Kambing terhadap Kadar Malondialdehida (MDA) dan Ekspresi Tumor Necrosis Factor Alpha (TNF-α) Organ Hati Hewan Model Tikus (Rattus norvegicus) Hiperkolesterolemia*. Program Kedokteran Hewan. Universitas Brawijaya.

[14] Putra, M. M. A. 2016. *Pengaruh Pemberian Minyak Jelantah terhadap Ketebalan Dinding Aorta Abdominal Tikus Putih (Rattus norvegicus) Jantan Galur Sprague dawley* [Skripsi]. Fakultas Kedokteran. Universitas Lampung.

[15] Rahmawati, Y. W. 2015. *Pengaruh Pemberian Ekstrak Metanol Daun Kayu Kuning (Arcangelisia flava (L.) Merr) terhadap Histopatologi Aorta Tikus Wistar Hiperlipidemia* [Skripsi]. Fakultas Farmasi. Universitas Jember.

[16] Gani, N., L. I. Momuat, and M. M. Pitoi 2013 *Jurnal MIPA UNSRAT Online*, **2** 44-49.

[17] Kusumastuty, I. 2014 *Indonesian Journal of Human Nutrition* **1** 50-56.

[18] Chahyanto, B. A., Rimbawan, S. A. Marliyati, and W. Winarsih 2016 *Jurnal Sain Veteriner* **34** 50-59.

[19] Paulina, A. J., E. Asni, dan M. Gaffar 2015 *Jom FK* **2** 1-15.

[20] Arauna, Y. 2013. *Studi Kadar Trigliserida dan Gambaran Histopatologi Hepar Hewan Model Tikus (Rattus norvegicus) Hiperkolesterolemia yang Diterapi dengan Ekstrak Air Benalu Mangga (Dendrophthoe petandra)* [Skripsi]. Program Kedokteran Hewan. Universitas Brawijaya.

[21] Sayekti, N. A. and N. Rustanti 2014 *Journal of Nutrition College* **3** 125-133.

[22] Lung, J. K. S. and D. P. Destiani. 2017. *Farmaka Suplemen* **15** 53-62.

[23] Agustina, E. 2017 Klorofil **1** 38-47.

[24] Tuarita, M. Z., N. F. Sadek, Sukarno, N. D. Yuliana, and S. Budijanto. 2017. *Pengembangan Bekatul sebagai Pangan Fungsional: Peluang, Hambatan, dan Tantangan*. Fakultas Teknologi Pertanian. Institut Pertanian Bogor.

[25] Mumpuni, P. D. 2013. *Analisis Kadar Tokoferol, γ-Oryzanol, dan β-Karoten serta Aktivitas Antioksidan Minyak Bekatul Kasar* [Skripsi]. Fakultas Kedokteran. Universitas Diponegoro Semarang.

[26] Permana, H. 2007. *Sel Adiposit sebagai Organ Endokrin*. Fakultas Kedokteran. Universitas Padjadjaran Bandung.

[27] Maramis, R., M. Kaseke, dan G. N. Tanudjadja 2014 *Jurnal e-Biomedik* **2** 431-435.

[28] Kartika, Y. D. A. 2017. *Pengaruh Pemberian Ekstrak Buah Tomat (Solanum lycopersicum L.) terhadap Kadar LDL dan Gambaran Histopatologi Aorta pada Model Hewan Coba Tikus Putih (Rattus norvegicus) Model Hiperkolesterolemia* [Skripsi]. Fakultas Kedokteran Hewan. Universitas Brawijaya.
[29] Welinsa, F., E. Asni, Z. Malik, and Ismawati. 2014. *Jom FK* **2** 1-11.
[30] Lintong, P. M., 2009. *Jurnal Biomedik* **1** 12-22
[31] Minatel, I. O., F. V. Francisqueti, C. R. Correa, and G. P. P. Lima. 2016. *Antioxidant Activity of γ-Oryzanol: A Complex Network of Interactions*. Vol. 17: 1107-1121.
[32] Nashriana, N., B. Wirjatmadi, and M. Adriani. 2015. *Jurnal Kedokteran Brawijaya* **28** 208-212.

Antibacterial Activity of Free Fatty Acids, Potassium Soap, and Fatty Acids Methyl Esters from VCO (Virgin Coconut Oil)

T P Mena[1], Sutrisno[1*], S Marfu'ah[1]

[1]Chemistry Department Faculty of Mathematics and Natural Sciences, Universitas Negeri Malang Jl. Semarang No. 5 Malang 65341

[*]Corresponding author: sutrisno.kimia@um.ac.id

Abstract. Virgin coconut oil is edible oil obtained from fresh coconut meat (*Cocos nucifera* L.) and is widely used for health such as for alternative healthy diets, anti-bacterial, anti-virus and anti-fungal. This is because VCO contains a lot of MCFA (Medium-Chain Fatty Acids) such as lauric acid (65.84%), myristic acid (13.16%) and caprylic acid (7.08%). The antibacterial ability of MCFAs is one of the most studied, but free fatty acids from the hydrolysis of VCO have not been done. The objective of the research is to derivatize VCO and determine the potential antibacterial activity of the product. The results of the research are as follows. (1) K-soap (solid, white, melting point 46-124 °C). (2) Free fatty acids (liquid, colorless, boiling point 266-296 °C, density 0.93 g.mL^{-1}, refractive index 1.44, viscosity 27.18 cP, acid value 184.54, saponification value 296.59 and ester value 112.05). (3) Fatty acid methyl ester (liquid, colorless, boiling point 260-262 °C, density 0.89 g.mL^{-1}, refractive index 1.43, viscosity 3.36 cP, acid value 0.33, saponification value 268.6 and ester value 268.27). VCO and its derivatives at concentrations of 2 and 1% are active as antibacterial agents against *Escherichia coli* and *Staphylococcus aureus*.

1. Introduction

After careful observed, the use of natural resources as raw materials for chemicals is generally used for food, textiles, building material, energy (fuel), and drugs or cosmectics. Natural resources are classified as biological natural resources (i.e. carbohydrates, lignocellulose, and triglycerides) and non-biological natural resources (i.e. minerals, petroleum, and natural gas). As one of the abundant renewable resources, triglycerides are used as raw materials in making soap [1], free fatty acids [2], foodstuffs, and biodiesel [3].

Triglycerides can be divided into two based on their sources, namely vegetable oil and fish oil. Chemical or enzymatic engineering of oils, especially edible vegetable oils can produce a variety of fine chemicals that are beneficial to human life. The benefits of making fine chemicals from edible vegetable oil can avoid products that are toxic and safe to use. One of the most abundant edible vegetable oils in Indonesia is VCO.

VCO is a processed natural oil obtained from a fresh and ripe coconut core. Recent research shows that VCO has anti-inflammatory activity, which protects the skin by increasing its endurance function [4]. VCO has medium-chain fatty acids and is rich in saturated fatty acids, such as lauric acid which has antibacterial activity [2][5][6]. Other studies report that VCO inhibits the growth of

Staphylococcus aureus by destructive mechanisms against bacterial cell walls and increases the ability of the cell's immune system [7].

Lauric acid (C_{12}) is a saturated fatty acid that most inhibits the growth of Gram-positive organisms [8]. Lauric acid at a concentration of 5% can inhibit the growth of bacteria *S. aureus*, *B. cereus*, *S. Typhimurium* and *E. Coli* [9]. The antibacterial properties of free fatty acids are used by many organisms to defend against parasitic or pathogenic bacteria, with the main target being the cell membrane, disrupting the electron transport chain, and oxidative phosphorylation [10].

In the context of antibacterial activity, for oil and fatty acids there are various results. For example tamarin oil (*Tamarindus indica*), free fatty acids are antibacterial against *E. coli* and *S. Aureus* [11] while the oil is inactive [12]. Coconut oil (*Cocos nucifera*), palm oil (*Elaeis guineensis*), babassu (*Attalea speciosa*), murumuru (*Astrocaryum murumuru*), tucuma (*Astrocaryum vulgare*), and Cuphea oil (*Cuphea ignea*) do not show effects on bacteria but their hydrolysis results are active against *C perfringens*, *E. cecorum*, *L. monocytogenes*, and *S. Aureus* [5]. Antibacterial agents can be applied in medicine, agriculture, food preservation, and the use of conventional antibiotics [10]. In addition to hydrolysis with acids, oils can be hydrolyzed with bases to obtain their salts which are useful as an antibacterial soap, such as potassium laurate.

Potassium laurate has high antibacterial activity against *S. mutans* and has great potential as an antibacterial agent [13]. Potassium laurate also shows antifungal activity against *P. pinophilum* which is a medium-chain fatty acid salt [14]. Lauric salts from magnesium, calcium, and manganese are effective in sterilizing the bacteria *S. aureus* and *P. acnes* while the cobalt, nickel and copper salts from lauric can sterilize the bacteria *S. aureus*, *P. acnes* and *S. Epidermidis* [15]. Sodium lauryl ether sulfate and sodium lauryl sulfate as a synthetic detergent can remove biofilms from *S. Aureus* [16].

Studies on VCO from coconut oil in Indonesia and its derivatives as antibacterial have never been reported. This study aims to investigate the antibacterial activity of VCO and its derivatives (K-soap, FA, and FAME) against Gram-negative (*E. coli*) and Gram-positive (*S. aureus*) bacteria and examine the relationship of polarity with their antibacterial activity.

2. Experimental Procedure

2.1. Chemicals, equipment, and instrumentations

2.1.1. Chemicals.

VCO is obtained from markets that are sold online at Tokopedia - Indonesia under the Sofia brand name. All reagents and solvents (i.e. potassium hydroxide, sodium chloride, hydrochloric acid, methanol, anhydrous magnesium sulfate, ethanol, n-hexane, chloroform, oxalic acid, and sodium carbonate) were obtained from Merck and the quality of the analysis.

2.1.2. Equipments and instrumentations.

A set of Iwaki reflux equipment, Pyrex titration tool, Thermo Scientific heater and stirrer, DA-20D Ulvac vacuum pump, Memmert oven, Kokusan H-103N Centrifuge, Capillary-Viscosimeter Schott Gerate, Refractometer Abbe, Iwaki Pycnometer, Hot plate HP-30S HP Shimadzu, Fischer Scientific Melting Point Apparatus, IR-prestige 21 Shimadzu and GCMS-QP2010S Shimadzu.

2.2. Experiment

2.2.1. Saponification reaction.

Amount 20g of VCO and 100mL of 3M potassium hydroxide solution was refluxed at 80°C for 3 h. The mixture is added with 100mL saturated sodium chloride solution and stirred until a precipitate is obtained. The formed product is filtered and washed with distilled water (2 x 100mL), then dried at 80°C.

2.2.2. Soap acidification.
The synthesized K-soap (10 g) of VCO was added with 40 mL of distilled water and added drop by drop of hydrochloric acid 1 M until the emulsion used completely. The mixture is separated, filtered, and purified with distilled water until the remaining is not acidic. After that, centrifuged at 3000 rpm for 10 minutes and the upper layer taken.

2.2.3. trans-Esterification reaction.
A total of 20 g of VCO, 9 mL of methanol, and 0.2 g of potassium hydroxide were refluxed at 60 °C for 4 hours, then cooled, and allowed to stand for 24 hours then obtained two layers. The two layers are separated, accommodated in different containers and tested for acidity with litmus paper. The top layer is washed with warm distilled water and then dried with anhydrous magnesium sulfate.

2.2.4. Characterization and identifications.
The characterization of VCO and its derivatives includes: (1) visual state and color, (2) boiling point is determined by boiling point equipment, (3) melting point by melting point apparatus, (4) density with a pycnometer, (5) viscosity with Capillary-Viscosimeter tubes, (6) refractive index with an Abbe refractometer, (7) saponification value, acid value, and esters value by the method in Indonesian National Standards [17], and (8) solubility test with n-hexane, chloroform, ethanol, methanol, and water. Identification of VCO and its derivatives includes: (1) IR spectra conducted by Shimadzu FT-IR, and (2) GC-MS conducted by GC-MS Shimadzu QP2010S device.

2.2.5. Antibacterial activity [18].
2.2.5.1 Preparation NA (Nutrient Agar) Media. 5 g of NA was dissolved in 250 mL of distilled water and heated until clear. Furthermore, this solution is heated to a temperature of 115°C for 15 minutes, and then stored at room temperature for 1 x 24 h.
2.2.5.2 Preparation NB (Nutrient Broth) Media. Bacto peptone (0.1 g) and beef extract (0.06 g) were dissolved in 20 mL distilled water and separated into 3 mL test tubes. The tes tubes were heated at 115°C for 15 minutes and stored at room temperature for 1 x 24 h.
2.2.5.3 Inoculation. The *S. aureus* and *E. coli* bacteria were transferred to NB media, dissolved, and then incubated for 1 x 24 h at 37°C. The media was standardized with 0.5 McFarland and applied to NA.
2.2.5.4 Planting samples and testing activities. Sample (20 μL) were input into holes in NA with a diameter of 0.6 mm and stored at 37°C for 1 x 24 h. Finally, samples observed and the inhibition zone diameter is calculated.

3. Result and Discussion

The data of physico-chemicals properties from the results of VCO experiments and their derivatives are presented in the **Figure 1** and **Table 1**. Based on visual observations, it appears that there is a change from VCO which was initially colorless and liquid to be solid and white. This shows that the lathering reaction was successful and produced a new product, fatty acid potassium salts (K-soap) from VCO. K-soap from VCO is synthesized to produce a liquid and colorless product which indicates that a new product has been formed. From the results of the acidification process, it can be seen that the synthesis of soap into fatty acids has been successful. Visual changes from the *trans*-esterification process do not show changes in state and color. Therefore, it is necessary to have other parameters such as boiling point, density, viscosity, and refractive index.

The *trans*-esterification product has been successfully characterized by a boiling point value, refractive index, viscosity, and density different from the initial product. The difference in the physical properties of a compound is influenced by intermolecular forces and molecular weights [19]. VCO has a higher molecular weight (calculated as $C_{42}H_{74}O_6$) than its methyl ester (calculated as $C_{13}H_{26}O_2$) so breaking bonds requires more energy. In the solubility test, the order of the nature of the polarity from low to high, namely VCO, VCO methyl ester, VCO fatty acid, and VCO K-soap.

Figure 1. Physical perfomance of VCO and its derivatives

Table 1. Characterization of physicochemical properties of VCO and its derivatives

Properties	VCO	K-soap	Fatty acid	Methyl ester
State	Liquid	Solid	Liquid	Liquid
Color	Colorless	White	Colorless	Colorless
Boiling point (°C)	274–336		266–296	260–262
Melting point (°C)		46–124		
Density (g.mL^{-1})	0.93		0.93	0.89
Viscosity (cP)	45.99		27.18	3.36
Refractive index	1.45		1.44	1.43
Saponification value, SV (mg KOH/g sample)	320.92		296.59	268.6
Acid value, AV (mg KOH/g sample)	0.93		184.54	0.33
Ester value (SV – AV)	319.99		112.05	268.27
Solubility in:				
n-hexane	soluble	insoluble	soluble	soluble
chloroform	soluble	insoluble	soluble	soluble
ethanol	soluble	emulsified	soluble	soluble
methanol	insoluble	emulsified	soluble	soluble
water	insoluble	emulsified	insoluble	insoluble

The IR spectrophotometric analysis of VCO and its derivatives, the results of the IR spectra are listed in **Figure 2, 3, 4**, and **Figure 5**. The characteristic bands in this IR spectra that distinguish functional groups from each substance are listed in **Table 2**. Based on the results of the interpretation of the IR spectra (**Table 3**) there are several typical absorption bands. VCO has an absorption area at wavenumbers 2852.72, 2922.16 and 2954.95 cm^{-1} with strong intensity which indicates the vibration strain of the C-H alkane bond and wavenumber 1745.58 cm^{-1} with strong and sharp intensity which indicates the presence of strain vibration C=O as ester. The synthesized potassium soap has a wavenumber of 1573.91 cm^{-1} with a strong and sharp intensity which indicates the vibrations of stretching asymmetric C=O bonds as carboxylic salts. The synthesized fatty acid has a wavenumber of 1697.36 cm^{-1} with strong and sharp intensity which shows the vibration stretching of the C=O bond as carboxylic acid. The synthesized methyl ester has a wavenumber of 1743.65 cm^{-1} with a strong and sharp intensity which indicates the vibration stretching of the C=O bond as an ester. VCO derivative was successfully synthesized based on the results of IR spectra interpretation.

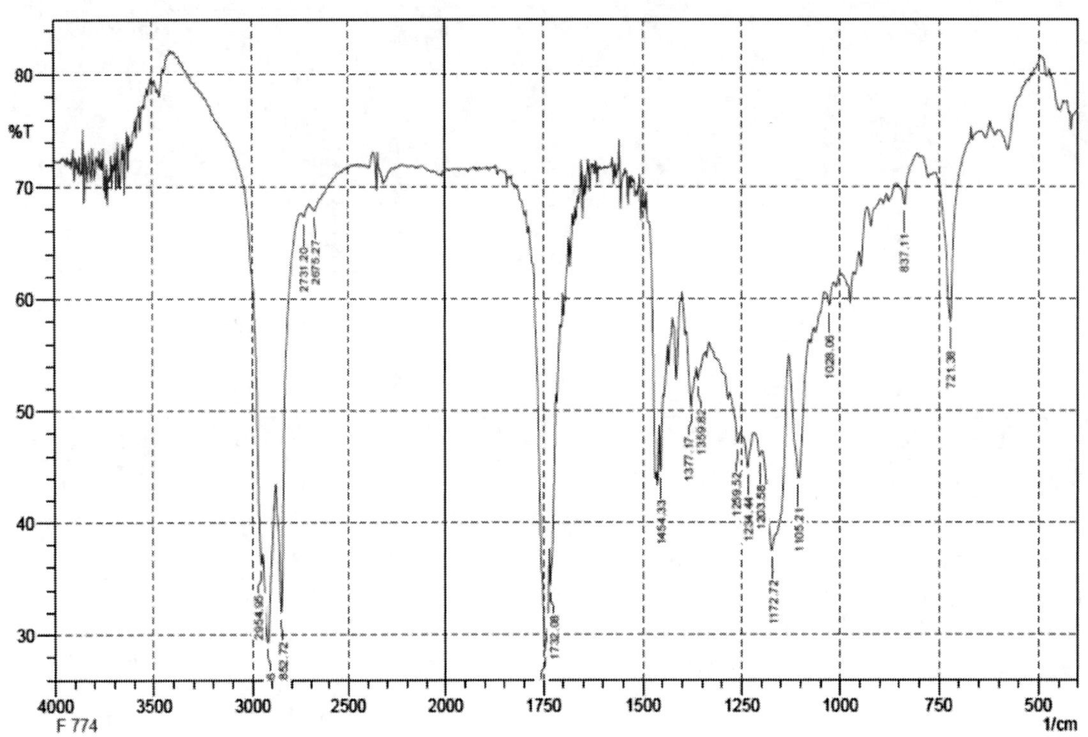

Figure 2. Infra-Red spectrum of VCO

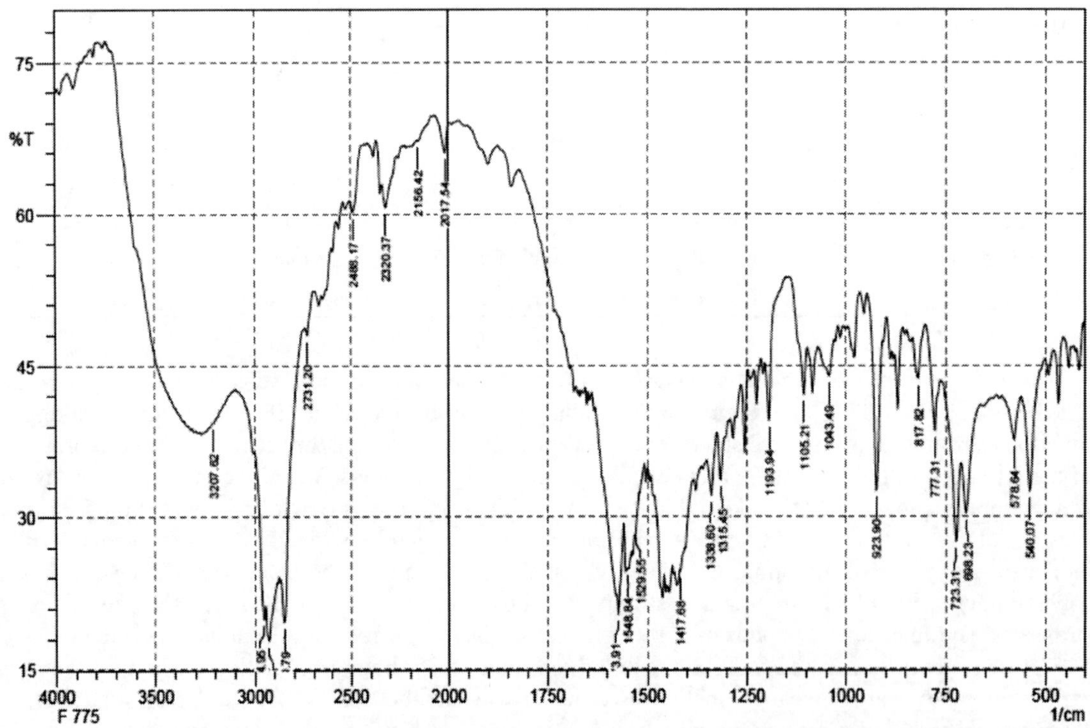

Figure 3. Infra-Red spectrum of fatty acids potassium salt (K-soaps) from VCO

IOP Conf. Series: Materials Science and Engineering **833** (2020) 012023 doi:10.1088/1757-899X/833/1/012023

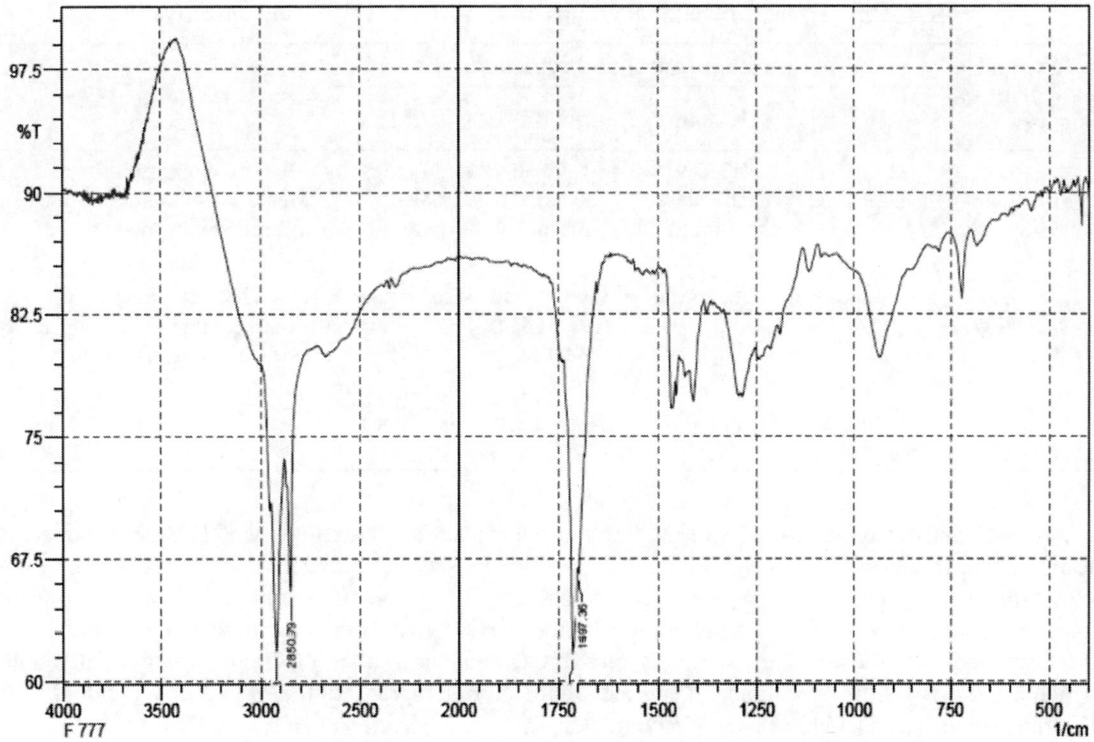

Figure 4. Infra-Red spectrum of free fatty acids (FFAs) from VCO

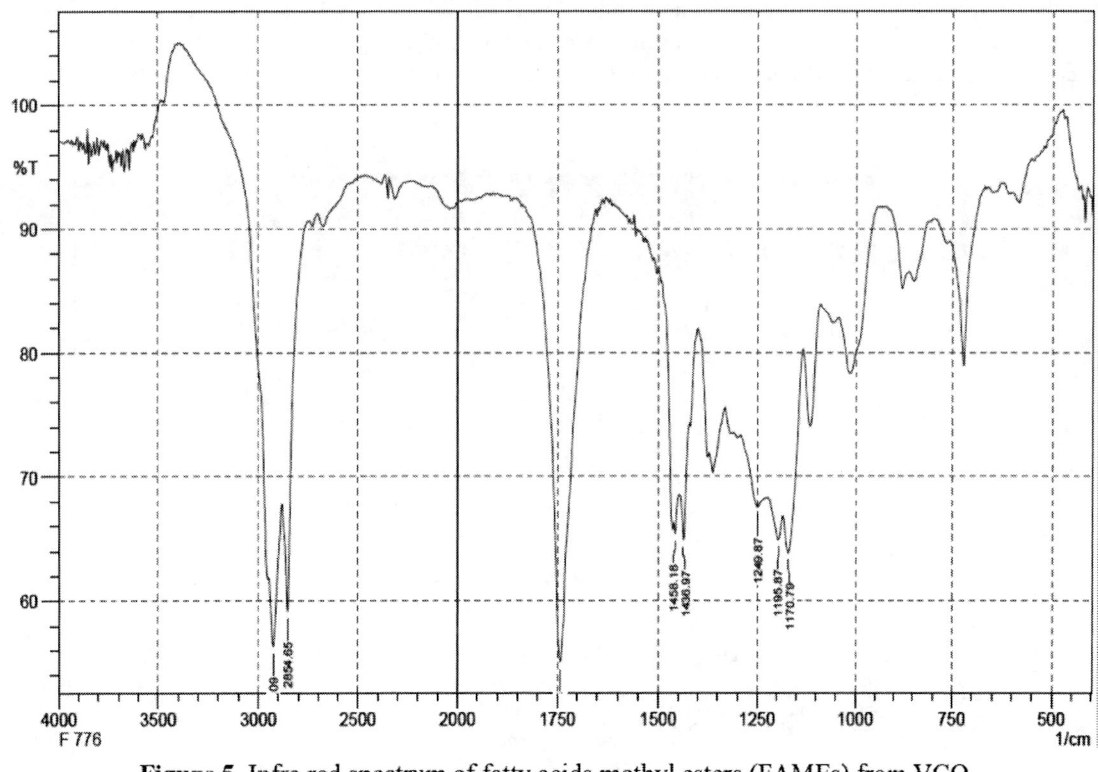

Figure 5. Infra red spectrum of fatty acids methyl esters (FAMEs) from VCO

Table 2. Typical bands functional groups on IR spectra of VCO and its derivatives

Typical absorption band	Wave number (cm^{-1}) and typical bands				Remarks
	VCO	K-soaps	FFAs	FAMEs	
C=O stretching	1732.08 (sharp, strong)	1573.91 (sharpless, strong)	1697.36 (sharp, strong)	1743.65 (sharp, strong)	Electron polarizability cause weakness of the C = O bond, shifts to the lower wave number
O-H stretching	not appear	not appear	3100 – 2600 (very broad, medium)	not appear	The very broad band characterizes O-H stretches with H bonds in a carboxylic acid
O-K stretching	3207.62 (sharp, strong)	not appear	Not appear	not appear	

GC-MS analysis of the FAMEs of VCO trans-esterification results with MeOH/KOH produced 9 components, (Figure 6). The peak with the largest area (peak 4, t_R 25,913 minutes), from the mass spectrum analysis indicates it is a lauric acid methhl ester. Therefore, it is proven that lauric acid is the most component in VCO. The overall composition of the fatty acids making up the VCO identified as the fatty acid methyl esters are listed in **Table 3**. GC data shows that VCO consists of 9 fatty acids which are shown by the appearance of 9 peaks on the chromatogram. In the MS results, it can be seen that the synthesized VCO have the main content, namely lauric acid.

Based on the results of the experiment, the synthesized fatty acids have a higher acid value than VCO which shows that the fatty acids have been successfully synthesized from VCO. However, the synthesized methyl esters have lower acid value than VCO caused by free fatty acids in VCO which have formed fatty acid methyl esters through esterification with a potassium hydroxide catalyst. VCO has the highest saponification value showing that the triglyceride content in VCO is the highest compared to the synthesized fatty acids and methyl esters. For ester value, VCO shows the highest number because VCO contains a fairly high ester.

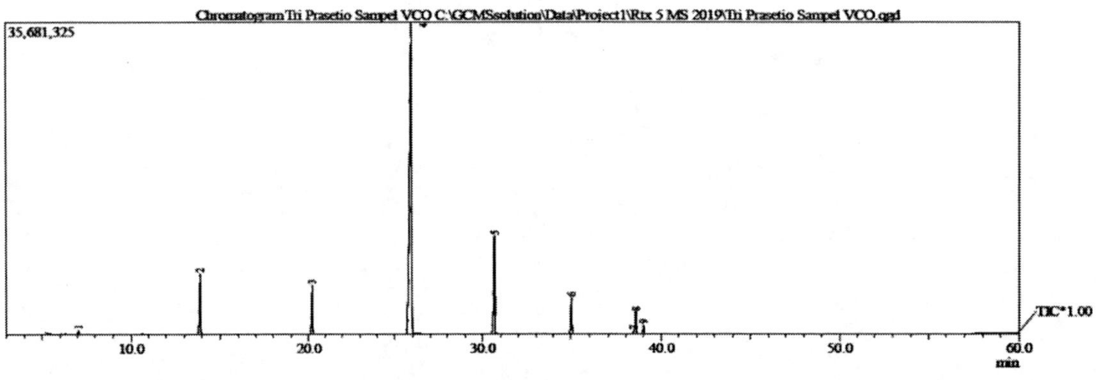

Peak Report TIC

Peak#	R.Time	I.Time	F.Time	Area	Area%	Height
1	7.052	6.983	7.133	1426108	0.35	438278
2	13.891	13.733	13.975	28599981	7.08	6858220
3	20.224	20.075	20.300	22352886	5.53	5553896
4	25.913	25.600	26.025	266131603	65.84	35536810
5	30.688	30.475	30.792	53186475	13.16	11122493
6	35.035	34.875	35.142	16815127	4.16	4069203
7	38.425	38.333	38.475	1500804	0.37	327983
8	38.565	38.475	38.692	10317149	2.55	2521805
9	38.988	38.883	39.075	3886654	0.96	977295
				404216787	100.00	67405983

Figure 6. GC-MS Chromatogram of fatty acids methyl ester from VCO

IOP Conf. Series: Materials Science and Engineering **833** (2020) 012023 doi:10.1088/1757-899X/833/1/012023

Table 3. Fatty acids content in VCO

Peak	Retention time (second)	Area% as relative content (%)	Fatty acid (that identified as FAMEs)	Fatty acids symbol
1	7.052	0.35	hexanoic acid (caproic acid)	C6:0
2	13.891	7.08	octanoic acid (caprylic acid)	C8:0
3	20.224	5.53	decanoic acid (capric acid)	C10:0
4	25.913	65.84	dodecanoic acid (lauric acid)	C12:0
5	30.688	13.16	tetradecanoic acid (myristic acid)	C14:0
6	35.035	4.16	octadecanoic acid (stearic acid)	C18:0
7	38.425	0.37	octadeca-9,12-dienoic acid (linoleic acid)	C18:2
8	38.565	2.55	hexadeca-9-enoic acid (palmitoleic acid)	C16:1
9	38.988	0.96	tetracosanoic acid (lignoceric acid)	C24:0

Analysis of mass spectrometry by EI-MS (electron impact mass spectrometry) for each peak of the nine peaks of GC results obtained by mass spectra as shown in **Figure 7** to **Figure 15.** The appropriate content of fatty acids from their fatty acid methyl esters as listed in **Table 3**, obtained from interpretation of each mass spectrum from nine peaks, and subsequently refered to WILEY229.LIB.

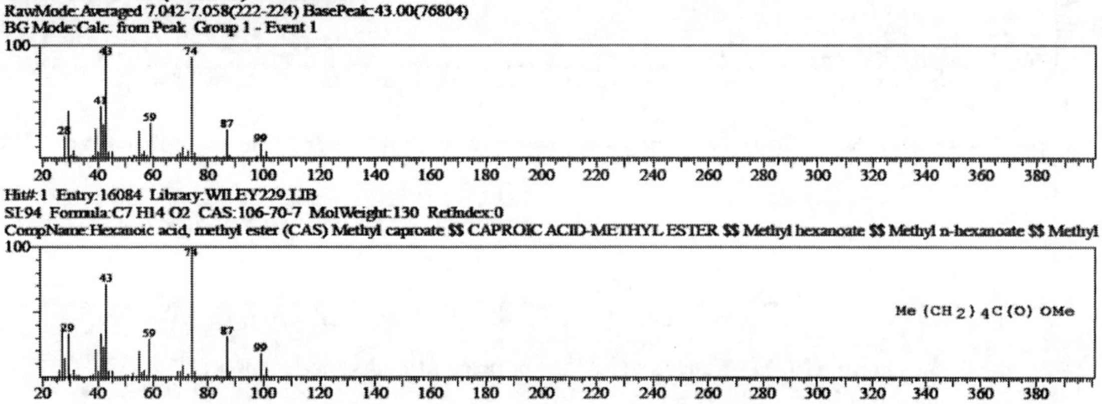

Figure 7. Mass spectra (EI-MS) of peak 1 (methyl hexanoate)

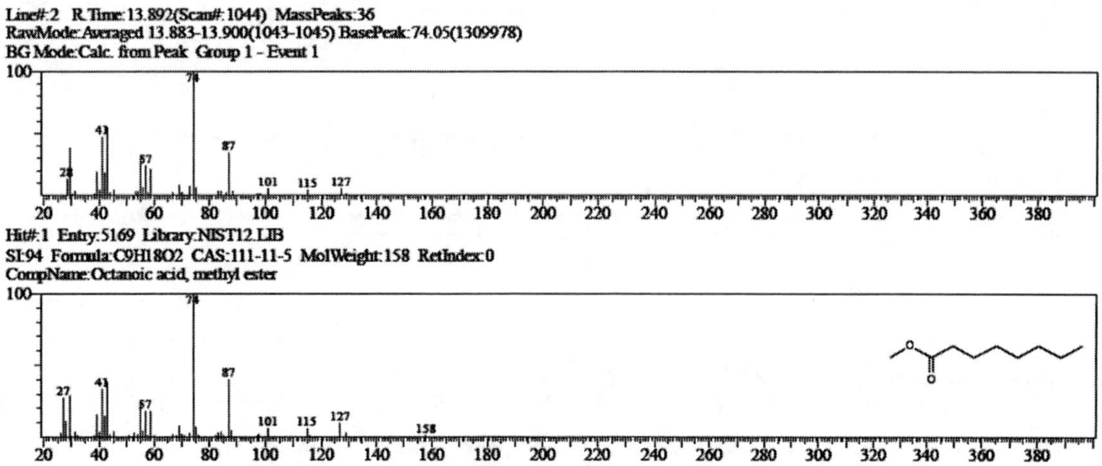

Figure 8. Mass spectra (EI-MS) of peak 2 (methyl octanoate)

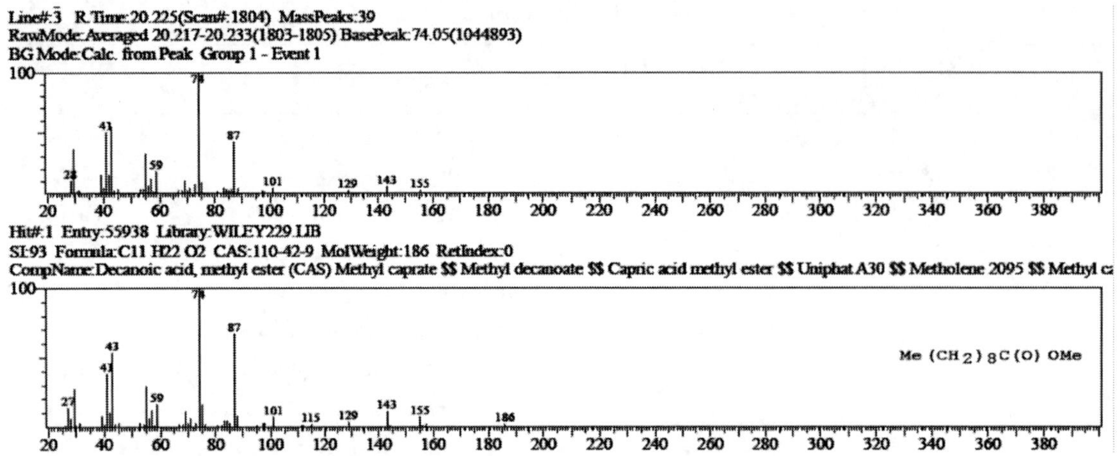

Figure 9. Mass spectra (EI-MS) of peak 3 (methyl decanoate)

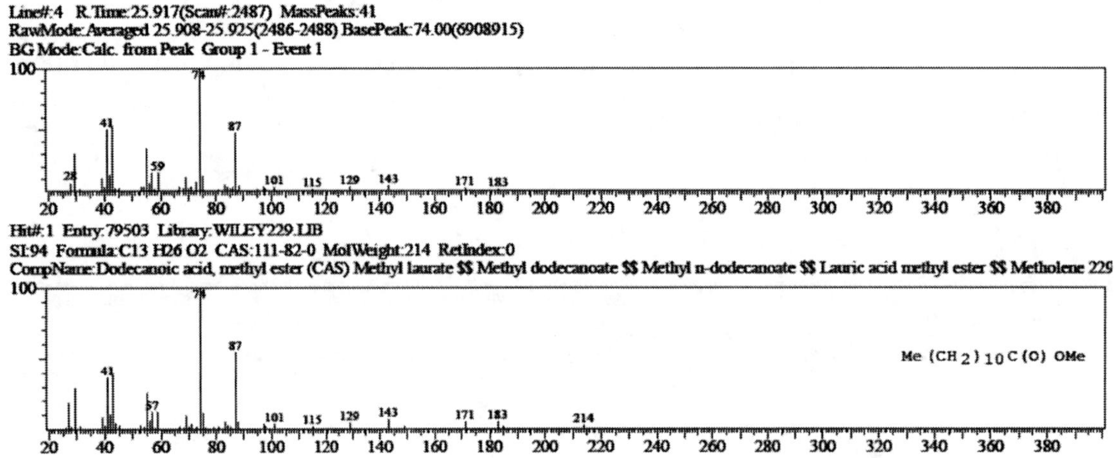

Figure 10. Mass spectra (EI-MS) of peak 4 (methyl dodecanoate)

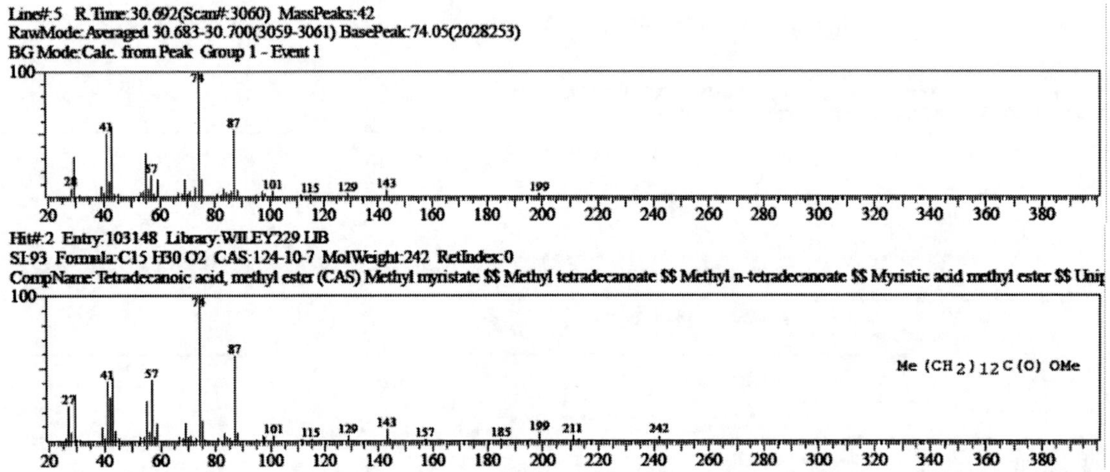

Figure 11. Mass spectra (EI-MS) of peak 5 (methyl tetradecanoate)

IOP Conf. Series: Materials Science and Engineering **833** (2020) 012023 doi:10.1088/1757-899X/833/1/012023

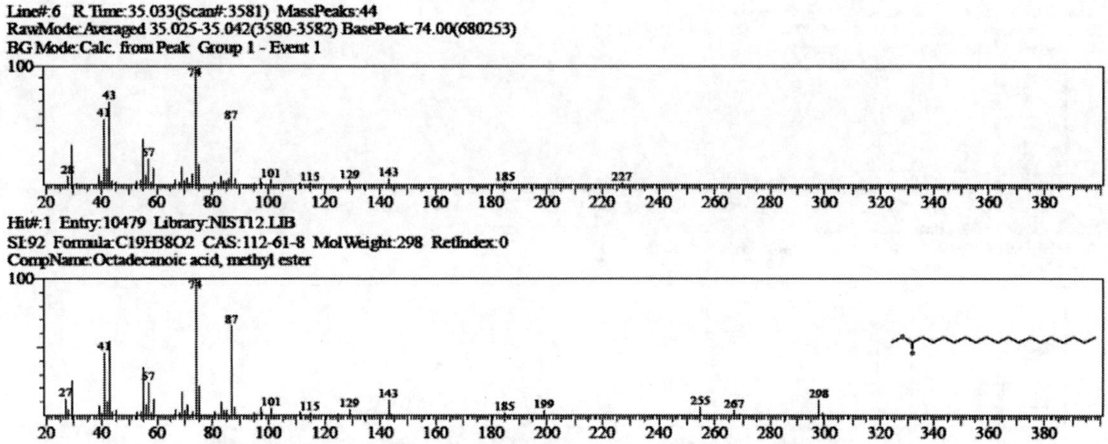

Figure 12. Mass spectra (EI-MS) of peak 6 (methyl octadecanoate)

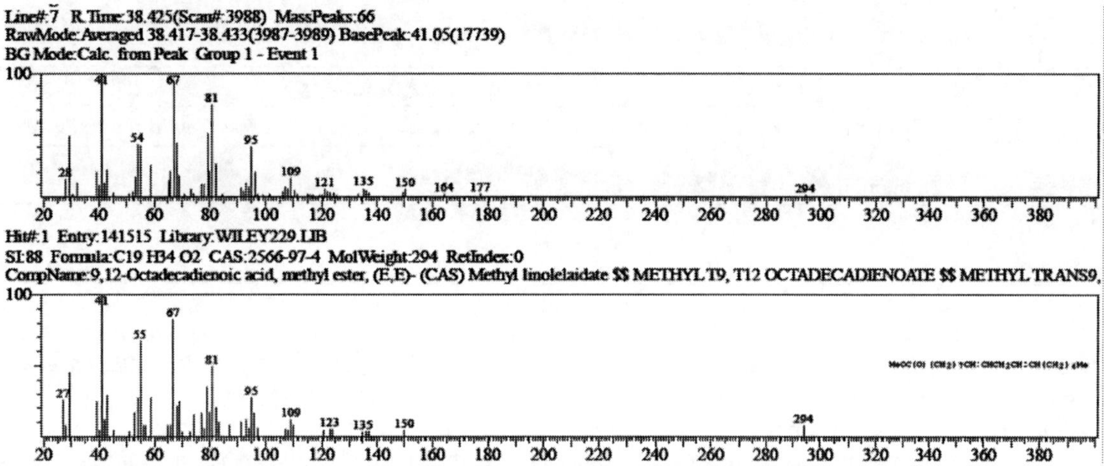

Figure 13. Mass spectra (EI-MS) of peak 7 (methyl 9,12-octadecadienoate)

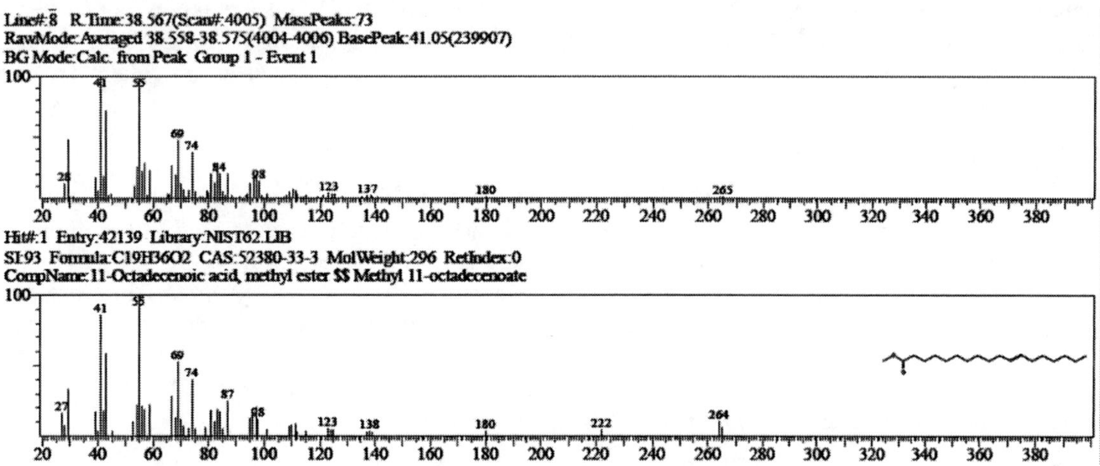

Figure 14. Mass spectra (EI-MS) of peak 8 (methyl 11-octadecenoate)

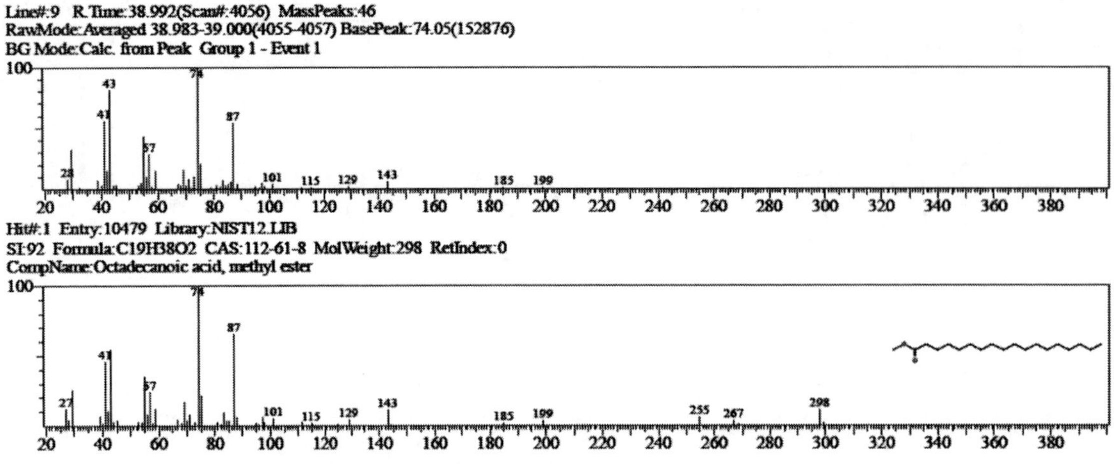

Figure 15. Mass spectra (EI-MS) of peak 9 (methyl octadecanoate)

Table 4. Test the VCO derivative antibacterial activity

Bacteria	Clear area diameter (mm)						
	Negative control (ethanol 60%)	K-soap		Free fatty acids		Fatty acid methyl ester	
		1%	2%	1%	2%	1%	2%
E. coli	-	8.85	10.9	11	12.25	-	7.75
S. aureus	-	10	11,5	10,4	14.2	-	8.9

Fatty acid potassium salt (K-soap), free fatty acids, and fatty acid methyl esters synthesized were tested for their antibacterial activity against *E. coli* bacteria as Gram-negative bacteria and *S. aureus* as Gram-positive bacteria with ethanol 60% as the solvent. The testing method used is the well diffusion method, shown by the formation of a clear zone around the wellbore. If the clear zone that forms around the wellbore increases showed the presence of antibacterial activity by the sample whose results are shown in **Table 4**. Based on the calculation of the clear zone diameter, it can be seen that fatty acids have the highest antibacterial activity compared to soap and methyl esters, the antibacterial activity of VCO derivatives more effective against Gram-positive bacteria (*S. aureus*). Medium-chain saturated fatty acids such as lauric acid have antibacterial activity[2] and are more active against Gram-positive bacteria than Gram-negative [10]. The fatty acids synthesized from VCO are more specific for Gram-positive bacteria as shown by the formation of a larger clear zone diameter than Gram-negative bacteria.

4. Conclusion

New products have been formed which are shown by the different values of the physicochemical properties and are supported by the IR spectra data of each product. The content of VCO fatty acids have been shown from the results of GC-MS. Of the three synthesized products (K-soap, free fatty acids, and fatty acid methyl esters from VCO) have antibacterial activity against Gram-negative bacteria (*E. coli*) and Gram-positive bacteria (*S. aureus*). This work proposes an increase in the study of antibacterial activity against VCO derivatives by testing in other bacteria such as *S. epidermidis*. VCO derivative products can potentially be a good fine chemical because they are produced from easily found substances that are not toxic.

References

[1] Danha G, Muzenda E and Maotsela T 2019 *Procedia Manuf.* **35** 541

[2] Nguyen V T A, Le T D, Phan H N and Tran L B 2017 *Journal of Lipids.* **2017**. 7170162
[3] Rodriguez G and Beckman E J 2020 *Fluid Phase Equilib.* **503** 112303
[4] Varma S R, Sivaprakasam T O, Arumugam I, Dilip N, Raghuraman M, Pavan K B, Rafiq M and Paramesh R 2019 *J. Tradit. Complement. Med.* **9** 5
[5] Hovorková P, Laloučková K and Skřivanová E 2018 *Czech J. Anim. Sci.* **3** 119
[6] Nguyen T A V., Le T D, Phan H N and Tran L B 2018 *Scientifica.* **2018** 9120942
[7] Cahya D, Tri C, Isrina S and Salasia O 2019 *Heliyon* **5** e02612
[8] Kabara J J, Swieczkowski D M, Conley A J and Truant J P 1972 *Antimicrob. Agents Chemother.* **2** 23
[9] Odel F, Siswanta D and Nurwening E 2016 *Procedia Chem.* **18** 132
[10] Desbois A P and Smith V J 2010 *Appl Microbiol Biotechnol.* **85** 1629
[11] Sutrisno, Retnosari R and Marfu'ah S 2019 *IOP Conf. Ser.: Earth Environ. Sci.* **299** 012004
[12] Sutrisno, Retnosari R, Marfu'ah S and Fajaroh F 2019 *Key Engineering Materials* vol 811 40
[13] Masuda M, Era M and Kawahara T 2015 *Biocontrol Science.* **20** 209
[14] Ra M E, Akai S S, Anaka A T and Awahara T K 2015 *Japan J. Food Eng.* **16** 99
[15] Yamamoto Y, Morikawa T, Kawai T and Nonomura Y 2017 *ACS Omega* **2** 113
[16] Kawahara T, Takita M, Masunaga A and Morita H 2019 *Int. J. Mol. Sci.* **20** 312
[17] Badan Standardisasi Nasional 1988 *Standar Nasional Indonesia* (SNI 01-3555-1998)
[18] Jorgensen J H and Ferraro M J 2009 *Medical Microbiology.* **49** 1749
[19] Yang Y, Zhao Y, Xu M, Wu N, Yao Y, Du H, Liu H, Tu Y, Zhao Y, Xu M, Wu N, Yao Y, Du H, Liu H and Tu Y 2018 *Food Hydrocolloids.* **10** 016

Removal of Cr(VI) Using Biochitin from White Shrimp (*Litopenaeus vannamei*) Shell Modified by Dithizone

Lutfiyatul Mukhlisah[1], Barlah Rumhayati[1*] and Sasangka Prasetyawan[1]

[1] Departement of Chemistry, Faculty of Science, Brawijaya University, Indonesia

*Corresponding author: rumhayati_barlah@ub.ac.id

Abstract. Biochitin is a chitin produced from the sequential fermentation of white shrimp shells using *Lactobacillus plantarum* followed by *Bacillus thuringiensis*. Biochitin was then modified by adding 0.015 g dithizone/ g adsorbent in toluene followed by reflux for 4 hours at 70°C. The purpose of this study was to determine the optimum conditions for removal of Cr(VI) using the dithizone modified biochitin (DiBch). This research was carried out using 25 mL Cr(VI) 50 mg/L mixed with DiBch adsorbents in various masses (0.1; 0.2; 0.3; 0.4 and 0.5 gram). The mixture was stirred for various time (0, 2, 4, 6 and 8 hours). Furthermore, the adsorption of electroplating waste with known concentrations was also carried out (0.92; 9.38; 92.39; 986.42 and 13215.94 mg/L). The results of this study indicate that Cr(VI) can be removed optimally using 0.4 g DiBch with stirring time of 2 hours. The removal efficiency of Cr(VI) from electroplating waste reached 83.45% ± 0.78.

1. Introduction

Chromium is one of the metals that has a higher concentration in waste from electroplating, paint, coloring, the paper industry, and others. This metal is very toxic and exposure in high concentrations can cause epigastric pain, nausea, vomiting, severe diarrhea and bleeding. The maximum level of chromium contaminants for drinking water is 0.05 mg/L [1]. Thus, to reduce the concentration of heavy metal ions to environmentally acceptable levels, effective and efficient separation methods must be developed. Of all the various water treatment techniques, adsorption is generally more desirable to remove heavy metal ions because it has high efficiency, using a simple method, low cost, and can be used widely. Adsorption can be done on heavy metals using various types of adsorbents such as chitin extracted from shrimp shells.

Chitin can be obatined from the white shrimp (Litopenaeus vannamei) shell obtained from the waste of fish processing plants. Shrimp shell waste consists of three main components namely protein (25% -44%), calcium carbonate (45% -50%), and chitin (15% -20%) [2]. Therefore the waste needs to be handled more seriously, especially because the waste contains chemical compounds that can be used for the basic ingredients of adsorbents.. Shrimp shell can be used as an adsorbent because it has an active group that is the acetamide group ($NHCOCH_3$) and has a reactive amine group and stable chelation [3]. In this study, the process of chitin extraction was carried out using fermentation method because this method uses only a few chemicals that are harmful to the environment and shrimp shell waste can be utilized more efficiently at a low cost. Chitin obtained can be modified with dithizone.

Previous research by Mudasir showed that immobilization of dithizone in chitin resulted in a more selective and higher adsorption capacity for Cd (II) ions compared to unmodified chitin[4]. Dithizon is one of the ligands that can be used to modify the surface of chitin, because it contains many N, -NH and -SH donor atoms which are sensitive to metal ions. The use of dithizone in making adsorbents has aimed to increase the selectivity and adsorption capacity of chitin for removal of heavy metal. Hence, this study aimed to determine the optimum condition for the removal of Cr(VI) using biochitin modified by dithizone (DiBch).

2. Experiment

2.1. Chemicals and Instrumentation
Equipment used in this study included: Laminar Air Flow (LAF), cool box, oven, blender, 80 mesh sieves, magnetic stirrers, glass apparatus, shaker, cuvette, pH meter, UV-Vis 1601 Shimadzu spectrophotometer, Fourier Transform Infrared (FTIR) Spectrophotometer - 8400S Shimadzu brand, Scanning Electron Microscopy (SEM), Particle Size Distribution 1090 Dry Rinse.

Materials and chemicals used were: white shrimp (*L.vannamei*) shell, *L. plantarum* in MRSA (deMann Rogosa Sharpe Agar) modified SHES (Shrimp head extract solution), 2% glucose, 10% glucose, 1% salt, *B. Thuringiensis,* SA (sialic acid) on NA (Nutrient Agar) modified SHES, NaOH 4% (Merck 106498.1000), distilled water, 1 M HCl, 4% NaOCl (Merck 105614), dithizone, toluene (Merck 108325), ethanol (Merck 100983), diphenylcarbazide crystal (Merck 103091), acetone (Merck 100012), $K_2Cr_2O_7$ crystal (Merck 104865), 10% H_2SO_4 (Merck 109073), electroplating wastewater samples.

2.2. Biochitin Extraction using Fermentation Method
Biochitin was extracted from the white shrimp shell using sequential fermentation method which was published elsewhere [5]. In principle, the culture of 15 hours *L. plantarum* was used in SHSA-modified MRSA with 2% glucose. About 500 mL starter was then added with 500 g of shrimp shell flakes (1: 1) [2]. Next 10% glucose and 1% salt were added to the Erlenmeyer. Then 30 hours fermentation was carried out at 37 °C. Chitin was then washed with water and dried in the oven overnight with a temperature of 80 °C. A total of 3% (w/v) dried chitin was then added to 10% (v/v) of the 24-hours culture of *B. thuringiensis* SA on the SHES-modified NA medium with 10% glucose. The mixture was incubated at 37 °C with agitation speed at 200 rpm for 81 hours. Chitin was then washed with water and dried in the oven overnight at 80 °C.

2.3. Modification of Biochitin with Dithizone
To prepare the modified biochitin, the following procedure was applied: 2.0 g of biochitin was added to 40 mL of toluene and mixed with 0.015 grams of dithizone in a 500 mL flask. The mixture was refluxed and stirred for 4 hours at 70 ° C. The product was filtered and washed sequentially with toluene, ethanol, and water several times until the filtrate did not show the characteristics of dithizone color. The immobilized Dithizone biochitin (DiBch) was then dried in an oven at 60 ° C for 12 hours and filtered through an 80-mesh filter.

2.4. Adsorption Study of Cr(VI)
Adsorption of Cr(VI) solution was carried out using batch adsorption methods. Cr(VI) solution with a concentration of 50 mg / L was added to DiBch 0.0; 0.1; 0.2; 0.3; 0.4 and 0.5 g . The mixture was stirred for 0, 2, 4, 6 and 8 hours at 100 rpm. Afterwards, the mixture was filtered using filter paper and the concentration of Cr(VI) in the filtrate was analyzed by a spectrophotometer visible at a wavelength of 540 nm. The number of metal ions adsorbed can be calculated by the following equation,

$$Q = \frac{[(C_0 - C_A)V]}{m}. \quad (1)$$

where Q is the amount of Cr(VI) adsorbed onto unit amount of the adsorbents (mg/g), Co and C_A are the initial and final concentrations of Cr(VI) (mg/L), respectively, V is the volume of the aqueous phase (mL), and m is the weight of DiBch adsorbent.

Furthermore, the percentage of adsorption can be calculated by the following formula,

$$\%\,Adsorption = \frac{m_0 - m_t}{m_0} \times 100\% \quad (2)$$

where m_0 and m_t are the initial and the final mass of Cr(VI) in the solution, respectively.

3. Result and Discussion

3.1. The Equilibrium Time for Cr(VI) Adsorption using Dithizone Modified Biochitin

The equilibrium time of adsorption is the time where the rate of adsorption equals to the desorption rate. At this condition, the surface of adsorbent is filled by the adsorbate. It can be seen from figure 1, the adsorption of 50 mg/L Cr(VI) with 0.2 g of DiBch adsorbent at the initial hours experienced a high increase because at the beginning of adsorption the entire pore surface remained empty and the Cr(VI) ions would be attached to the surface of the adsorbent so that the rate of adsorption will take place quickly. The rate of adsorption was increased slowly up to 2 hours. As shown in Figure 1, the equilibrium time for Cr(VI) adsorption by DiBch was 2 hours with the Cr(VI) adsorbed was 3.71 mg/g.

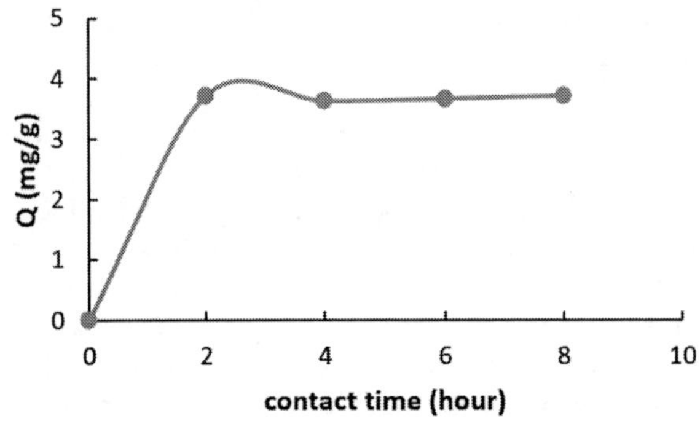

Figure 1. The influence of adsorption time to the amount of Cr(VI) adsorbed (mg/g) on the surface of DiBch adsorbent.

After 2 hours, the rate of adsorption was slightly decreased in equilibrium with the desorption rate. During the adsorption, the Cr(VI) ions were either attached to the active sides available on the surface of the adsorbent or filled the pore of adsorbent.

3.2. The Influence of DiBch Mass

The influence of the DiBch mass to the adsorption of Cr(VI) was carried out by using 25 mL of Cr(VI) solution of 50 mg/L mixed with the DiBch adsorbent at the various mass of 0.1; 0.2; 0.3; 0.4 and 0.5 g. Adsorption was conducted by stirring the mixture for 2 hours. Figure 2 shows that increasing the mass of adsorbent affected the efficiency of Cr(VI) adsorption. The number of active sites on the surface of adsorbent will increase with the increasing of the adsorbent mass. The percentage of Cr(VI) adsorbed reached its optimum value at 62.83% when using 0.4 grams of DiBch as adsorbent. Further increase of the adsorbent mass decreased the efficiency of Cr(VI) adsorbed due to the saturation of the active site of the adsorbent which forms lumps in the adsorbent thereby reducing the surface area of the adsorbent [8].

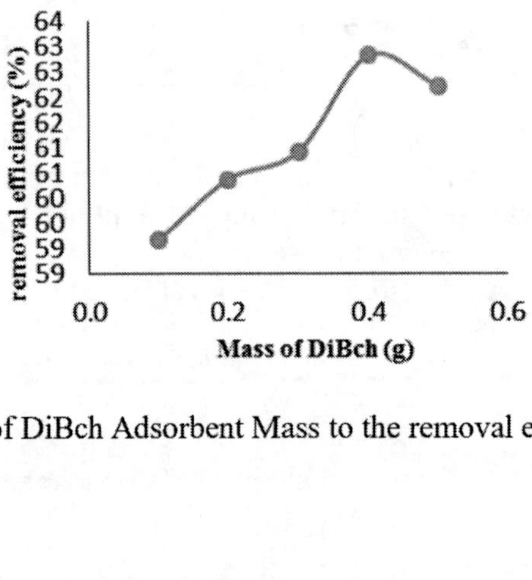

Figure 2. The Influence of DiBch Adsorbent Mass to the removal efficiency of Cr(VI).

3.3. The Influence of Cr(VI) Initial Concentrations

Adsorption on Cr(VI) in electroplating waste was carried out by means of 0.4 g of DiBch adsorbent mixed with 25 mL of electroplating waste at various concentrations of 0.92; 9.38; 92.39; 986.42 and 13215.94 mg/L. Figure 3 shows that the optimum adsorption occurred at a concentration of 986.42 mg/L with the percentage of Cr(VI) removal was 83.45%. With increasing Cr(VI) concentration, the adsorption efficiency decreases, because the ability of DiBch to adsorb Cr(VI) is already optimum. In other words, the maximum capacity of DiBch adsorbents has been reached at a concentration of 986.42 mg/L.

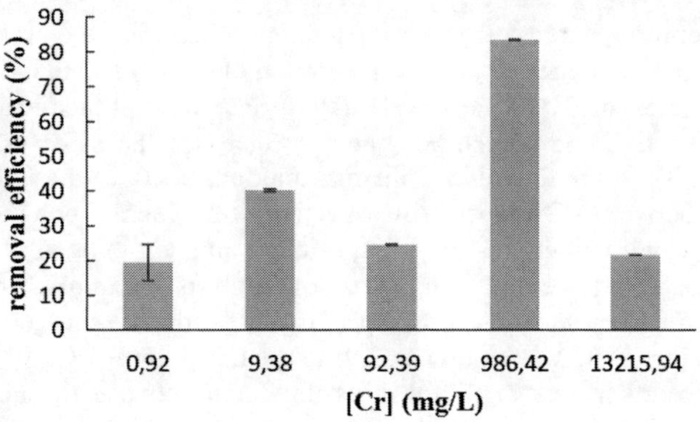

Figure 3. The influence of initial concentration of Cr(VI) in the electroplating waste to the Cr(VI) removal efficiency.

The decrease in adsorption efficiency is caused by the unproportioned number of DiBch active sites for interaction with the Cr(VI) at high concentration [9]. This is consistent with Langmuir's theory which says that the active sites on the surface of the adsorbent has an amount proportional to the surface area of the adsorbent, so that if the active site on the surface of the cell wall of the adsorbent is saturated with adsorbates, then the addition of adsorbates concentration can no longer increase the adsorption capacity of the adsorbent. Therefore, the adsorption efficiency of Cr(VI) was decreased at concentrations more than 986.42 mg/L. It means the optimum removal efficiency of Cr(VI) could be reached by means of initial concentration of CrVI) was 986.42 mg/L.

Adsorption mechanism of Cr(VI) on the DiBch can be studied from the adsorption isotherms. We use the Langmuir and Freundlich adsorption isotherms to figure out the mechanism of Cr(VI) adsorption with the linear fit showed in Figure 4. For the Langmuir line graph, the linearization model was obtained from a plot between 1/Ce and 1/Qe. It can be seen that the graph obtained is linear with equal y = 67.193x – 1.8916, with a correlation value (R^2) of 0.993. The Freundlich Isotherm graph was obtained from the linearization line from the plot between log Ce and log Qe which shows a linear equation y = -2.055x + 2.556 with a value of R^2 = 0.8374.

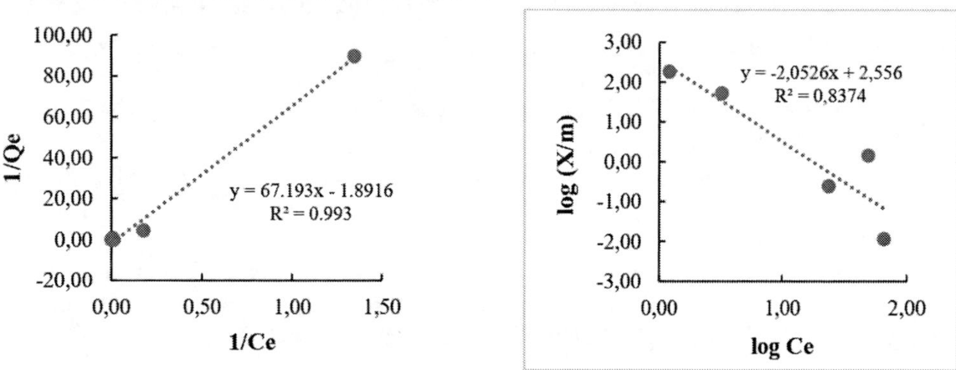

Figure 4. Graph of Langmuir Isotherm (left) and Freundlich (right) of Cr(VI).

Table 1. Parameters of Langmuir and Freundlich Adsorption Isotherm.

Langmuir Isotherm	Freundlich Isotherm
Qmax (mg/g) = 0.53	Kf = 359.75
b (L/mg) = 0.39	n = 0.49
R^2 = 0.993	R^2 = 0.8374

It can be seen from the Figure 4 that the Langmuir isotherm has a correlation value (R^2) which is close to 1. Hence, it can be suggested that the adsorption of Cr(VI) on the DiBch follows the Langmuir isotherm. The adsorption of Cr(VI) on the DiBch surface occurs chemically in one layer with the assumption that the active site on the adsorbent surface has the same energy and the adsorption occurred reversible [10]. From this model, the maximum capacity of the DiBch adsorbent for Cr(VI) was 0.53 mg/g. The surface of DiBch consists of active sites which responsible to interact with Cr(VI) oxyanions, including hydroxyl groups of chitin, primary amine groups of chitin and dithizone, and –SH groups of dithizone. At lower pH than the pzc of chitin (6.7) and pKa of dithizone (pKa 1 = 4.5), the active groups undergo protonation resulting in partially positive charge of the sites. At this study, the pH of electroplating waste was as low as 3.0, so the oxyanions of Cr(VI) could be attached to the surface of DiBch adsorbent due to the electrostatic force between the negative charge of Cr(VI) oxyanions and the positive charge of adsorbent sites.

4. Conclusion

The results showed that the dithizone modified biochitin (DiBch) can be used to adsorb Cr(VI) up to 986.42 mg/L by using 0.4-gram adsorbent for 2 hours with the percentage of Cr(VI) removal was 83.45% ± 0.78. The Cr(VI) adsorption mechanism using DiBch follows the Langmuir isotherm equation model which shows the adsorption capacity of DiBcH was 0.53 mg/g.

References

[1] J. Kotaś and Z. Stasicka 2000 *Environmental Pollution* **107** pp 263–283
[2] M. S. Rao, J. Muñoz, dan W. F. Stevens 2000 *Applied Microbiology and Biotechnology* **54** pp 808–813
[3] R. Muzzarelli 1969 *Talanta* **16** pp 1571–1577
[4] G. Raharjo, I. Tahir, dan E. T. Wahyuni 2008 *Journal of Physical Science* **19** pp 16
[5] A. Rosmawati, B. Rumhayati, dan A. Srihardyastutie 2019 *IOP Conf. Ser.: Mater. Sci. Eng.*, **546** pp 022019
[6] A. Tabak, N. Baltas, B. Afsin, M. Emirik, B. Caglar, dan E. Eren 2010 *J. Chem. Technol. Biotechnol* **85** pp 1199–1207
[7] H. Panda, N. Tiadi, M. Mohanty, dan C. R. Mohanty 2017 *South African Journal of Chemical Engineering* **23** pp 132–138
[8] Y. R. Wijayanto 2010 *thesis* pp 7
[9] W. Hua, Y. Lou, W. Xu, Z. Cheng, X. Gong, dan J. Huang 2016 *Appl Microbiol Biotechnol*, **100** pp 879-891
[10] Shriver, D.F. and Atkins, P.W 1999 *Inorganic Chemistry 3rd Edition* (Oxford: University Press)

Acknowledgements

Authors would like to thank to the Republic of Indonesia Ministry of Research, Technology, and Higher Education for the research funding under the research scheme of Penelitian Tesis Magister, with the contract number of 055/SP2H/LT/DRPM/2019, 11 March 2019.

Emulsifier and antimicrobial activity against *Propionibacterium acnes* and *Staphylococcus epidermidis* of oxidized fatty acid esters from hydrolyzed castor oil

A Nabilah[1], S Handayani[1], S Setiasih[1], D U C Rahayu[1] and S Hudiyono[1,2]

[1]Department of Chemistry, Faculty of Mathematics and Natural Sciences, Universitas Indonesia, Depok 16424, Indonesia

[2]Corresponding author: sumi.hudiyono@sci.ui.ac.id

Abstract. This study aimed to synthesize esters from oxidized fatty acids produced by castor oil hydrolysis as emulsifiers and antimicrobial compounds. Castor oil was hydrolyzed using KOH and the fatty acids were then oxidized using $KMnO_4$. The success of oxidation proven by determined the iodine number. Esterification was conducted with varied alcohols, namely methanol, ethanol, isopropanol, and 1-butanol using $ZnCl_2$ as catalyst and mole ratio was 1:2. The conversion percentage of esterification was determined using titrimetric method and products were characterized using FTIR. From the hydrolysis of castor oil, 84% of fatty acids were produced. Decreasing iodine number from 43.38 mg/g to 13.11 mg/g and increasing intensity of the -OH group absorption in the FTIR spectrum showed the success of fatty acids oxidation. Emulsifier test showed all products have emulsifier ability and emulsions were stable up to 24 hours with a water-in-oil (w/o) emulsion type. The best ability as an emulsifier demonstrated by methyl ester. Antimicrobial assay against *Propionibacterium acnes* and *Staphylococcus epidermidis* showed all ester products could inhibit the growth of both bacteria. Largest inhibition zone obtained from isopropyl ester for *P. acnes* by 16 mm and butyl ester for *S. epidermidis* by 17 mm.

1. Introduction

Castor oil is a vegetable oil extracted from the seeds of the castor plant (*Ricinus communis L.*) and has been used since ancient times as an important base material in the oleochemical industry [1]. The world market demand for castor oil and its products continue to increase due to several advantages, among them are the renewable sources, not competing with food, biodegradable, low production costs, and environmentally friendly [2].

In Indonesia, the production of *Ricinus communis L.* in 2015 reached 1.5 thousand tons [3]. However, it's processing into a product with a high economic value has not been maximized, because there is no refined castor oil processing industry and modified castor oil processing industry [4].

Castor oil is more unique than other vegetable oils because of its fatty acid content. The highest content of fatty acids in castor oil is ricinoleic acid, about 87-90%. Ricinoleic acid is an unsaturated fatty acid consisted of 18 atom carbon with one double bond at carbon 9 and one hydroxyl group at carbon 12. The hydroxyl group makes a ricinoleic acid difference than other fatty acids because it has three reaction sites, namely the hydroxyl group, the double bond, and the carboxylic group. So ricinoleic acids can undergo many more reactions than other fatty acids [1].

IOP Conf. Series: Materials Science and Engineering **833** (2020) 012025 doi:10.1088/1757-899X/833/1/012025

In the cosmetic industry, ricinoleic acid and its derivatives were widely used because of their properties as an emulsifier and antimicrobial agent.

A study was conducted in 2018 to identify the emulsifying and antimicrobial activity against *Propionibacterium acnes* and *Staphylococcus epidermidis* of oxidized fatty acids ethyl ester obtained from hydrolysis of castor oil. The results of the study concluded that oxidation of fatty acids before esterification with ethanol generated a better emulsifying and antimicrobial activity of the fatty acids ethyl ester [5]. However, it was not yet known whether the length of alkyl carbon affects its ability as an emulsifier and its antimicrobial activity. The hypothesis is the longer the carbon alcohol chain used in esterification can produce better emulsification and antimicrobial activity.

In this research, variations of alcohol were used in the esterification of oxidized fatty acids of hydrolyzed castor oil to determine their effect on emulsifier and antimicrobial activity against *P. acnes* and *S. epidermidis*. Alcohols used were methanol, ethanol, 2-propanol, and 1-butanol and the esterification was performed using $ZnCl_2$ as a catalyst.

2. Materials and methods

2.1. Materials
The materials used were castor oil obtained from Heansa Kimia, potassium hydroxide, ethanol 96%, hydrochloric acid, potassium permanganate, sodium hydroxide, chloroform, Wijs solution, potassium iodide, aquadest, sodium thiosulfate, methanol p.a, ethanol p.a, 2-propanol, 1-butanol, anhydrous zinc chloride, n-hexane, phenolphthalein, starch, eosin, clindamycin, DMSO, beef extract, peptone, nutrient agar, *Propionibacterium acnes* and *Staphylococcus epidermidis* cultures obtained from Biochemistry Laboratory, Department of Chemistry, Universitas Indonesia.

2.2. Methods

2.2.1. Castor oil hydrolysis. 300 g castor oil and 192.6 mL of 5 M KOH solution in 96% alcohol were mixed until homogenous then heated in an oil bath at 70°C for 1 hour accompanied by stirring. The mixture was then cooled at room temperature and then 250 mL of 5 M HCl was added to the mixture. Then the mixture was transferred to a separating funnel and allowed to stand for 24 hours to form two phases. The fatty acids from hydrolysis in the upper phase were then filtered [6].

2.2.2. Fatty acid oxidation. 120 mL of hydrolyzed fatty acids were cooled to 25-27 ° C then added by the alkaline solution of $KMnO_4$ (60 mL of 0.5 M $KMnO_4$ with 360 mL of 2 M NaOH, pH 12) slowly accompanied by stirring for 30 minutes. The mixture was protected from temperature changes and direct light exposure. The mixture was then left for 24 hours until split into two phases. The organic phase contained MnO_2 precipitate was separated and filtered [7].

2.2.3. Determination of iodine number. Iodine number was determined using a standard Wijs method [8].

2.2.4. Esterification and determination of conversion percentage. Oxidized fatty acid and alcohol (mole ratio of 1:2) were placed in a two neck flask equipped with a condenser reflux and a thermometer and then added by 0.3% $ZnCl_2$. The mixture was stirred until homogeneous and the reflux system was run for 6 hours at 60 °C. Ester product extraction was carried out using methanol and n-hexane. Methanol was removed by heating and oxidized fatty acids ester was obtained [9]. The titrimetric method was used to determine the conversion percentage of esterification. Unreacted fatty acid was moved into a 10 mL volumetric flask and the volume was adjusted using n-hexane. Thereupon 1 mL aliquot of the solution was titrated with 0.1 N NaOH.

2.2.5. FTIR analysis. Identification using FTIR was carried out for castor oil, fatty acids from hydrolysis, oxidized fatty acids, and ester products.

2.2.6. Emulsifier ability test and emulsion type determination. 0.1 g of sample was put into a mixture of oil and water in the proportions according to table 1 and then shaken to formed an emulsion. The emulsion formed was observed in its stability after 24 hours.

Table 1. Variations of Water-Oil Composition.

Variation 1						Variation 2					
Tube Number	1	2	3	4	5	Tube Number	1	2	3	4	5
Water (mL)	2	2	2	2	2	Water (drops)	2	4	6	8	10
Oil (drops)	2	4	6	8	10	Oil (mL)	2	2	2	2	2

The type of emulsions was determined by observing the mixture of emulsion and 1 drop of eosin dye under a microscope.

2.2.7. Antimicrobial assay using the disk diffusion method. 200 μL suspensions of bacteria (*P. acnes or S. epidermidis*) with a cell density of 1 x 10^8 cells / mL were placed aseptically into a sterile petri dish. Then 20 mL of nutrient agar (± 45-50 °C) was poured in and was shaken until homogeneous, then allowed to harden. Sterile disc paper was placed on top of media and 4 μL of the sample was dropped onto disc paper. The dish containing each of the test bacteria then incubated at 37 °C for 24 hours. The transparent area surrounding the disc paper was measured. DMSO, n-hexane, and methanol were used as negative control and 0.5% clindamycin antibiotic was used as a positive control.

3. Results and discussion

3.1. Hydrolysis

Hydrolysis of castor oil was conducted in an alkaline solution to generate more yield of fatty acids because basic hydrolysis is irreversible, whereas acidic hydrolysis is reversible [10]. In basic hydrolysis, -OH as a nucleophile attacks carbon from the carbonyl group followed by the release of the leaving group. Deprotonation causes the formation of carboxylic anions and forms fatty acid salts (soap). This process occurs gradually from triglycerides to diglycerides by breaking the ester bond on carbon number 1, then becoming monoglyceride by breaking the ester bond on carbon number 3 and finally being released three fatty acids from breaking the ester bond number 2 and forming glycerol [11].

In this study, KOH was used as nucleophile and ethanol acts as a reaction medium. The addition of KOH was done slowly because of its exothermic characteristics. The reaction was carried out at temperature 70 °C based on a study [6] that it was the optimum temperature of the castor oil hydrolysis reaction. Free fatty acids obtained from the resulting soap by neutralization that was carried out using excessive HCl. The hydrogen cation of HCl will be captured by the alkanoic anion to form free fatty acids. The addition of HCl was done slowly accompanied by stirring without heating to prevent dehydration in OH groups adjacent to the double bond [12].

From the hydrolysis of 900 grams of oil, 723.64 grams of fatty acids were obtained with a yield of 84%. Fatty acids from the hydrolysis of castor oil are dark yellow and viscous.

3.2. Oxidation of fatty acids

Oxidation of fatty acids aims to formed two hydroxyl groups (diols) from carbon 9 double bonds which can increase the polarity of ricinoleic acid. The oxidizing agent used in this study was diluted alkaline $KMnO_4$. To prevent the fatty acids undergo further oxidation, the reaction took place at pH 12 and temperature 25°C due to the very strong oxidizing properties of $KMnO_4$ [10]. At the end of the reaction, a mixture of fatty acids and black MnO_2 precipitate was obtained and then separated by filtration.

The success of the oxidation reaction was determined by the iodine number test. A decrease in iodine number indicated that the number of double bonds was decreased and thus the reaction was successful. In this study, the iodine value of oxidation products decreased from 43.38 mg/g to 13.11 mg/g. Ricinoleic acid was oxidized to produce 9,10,12-trihydroxy stearic acid.

3.3. Esterification and Conversion Percentage
Esterification of oxidized fatty acids was carried out chemically using variations of alcohols namely methanol, ethanol, isopropanol, and 1-butanol and $ZnCl_2$ as a catalyst. The ratio between oxidized fatty acids to alcohol used in this study was 1: 2 based on the previous study that this ratio gave the most yield [5].

A Lewis acid, $ZnCl_2$, was used as a catalyst replacing the common catalyst used in general esterification, sulfuric acid. This replacement aims to avoid the dehydration of OH groups obtained from previous oxidation by strong acids which can cause the breakdown of fatty acid chains. In the process, the carboxylic acid was bound to the acidic site of the catalyst through its carbonyl carbon and formed an activated complex. Then ester was produced by nucleophilic attack from alcohol together with water release as a by-product, and $ZnCl_2$ is recovered at the end of the reaction [13]. After the reaction, two solution phases were formed, the upper was the ester and the lower was the water phase.

The conversion percentage was determined using equation (1):

$$\%\ \text{Conversion} = \frac{\textit{reacted fatty acid (mmol)}}{\textit{total fatty acid (mmol)}} \times 100\% \qquad (1)$$

Mmol of unreacted fatty acid was obtained from equation (2):

$$\text{Mmol} = \frac{V \times N \times Vufa \times DF}{\textit{Volume of titrated fatty acid (mL)}} \qquad (2)$$

Where:
V: Volume of NaOH (mL)
N: Normality of NaOH
V ufa: Volume of unreacted fatty acid (mL)
DF: Dilution Factor
V tfa: Volume of titrated fatty acid

The mmol of reacted fatty acid can be obtained from the subtraction of total fatty acid to unreacted fatty acid.

Based on table 2 which shows the conversion percentage of esterification, the highest conversion percentage was obtained from esterification between oxidized fatty acid and ethanol.

Table 2. Conversion percentage of esterification.

Alcohol used in Esterification	Conversion Percentage (%)
Methanol	91
Ethanol	98
Isopropanol	92
1-Butanol	75

3.4. Identification using FTIR
FTIR spectrum of castor oil and its fatty acids are showed in figure 1. The absorption band for -OH groups showed at the wavenumber 3600-3200 cm^{-1} and the C = O carbonyl absorption band appears at the wavenumber 1725-1700 cm^{-1} for the fatty acids spectrum, shifted from 1750-1717 cm^{-1} in castor oil spectrum. Based on the FTIR spectrum of oxidized fatty acids in figure 2, the presence of -OH absorption was seen sharper and wider than its fatty acids spectrum, indicated that the double bond of the fatty acids has been oxidized to diols [10].

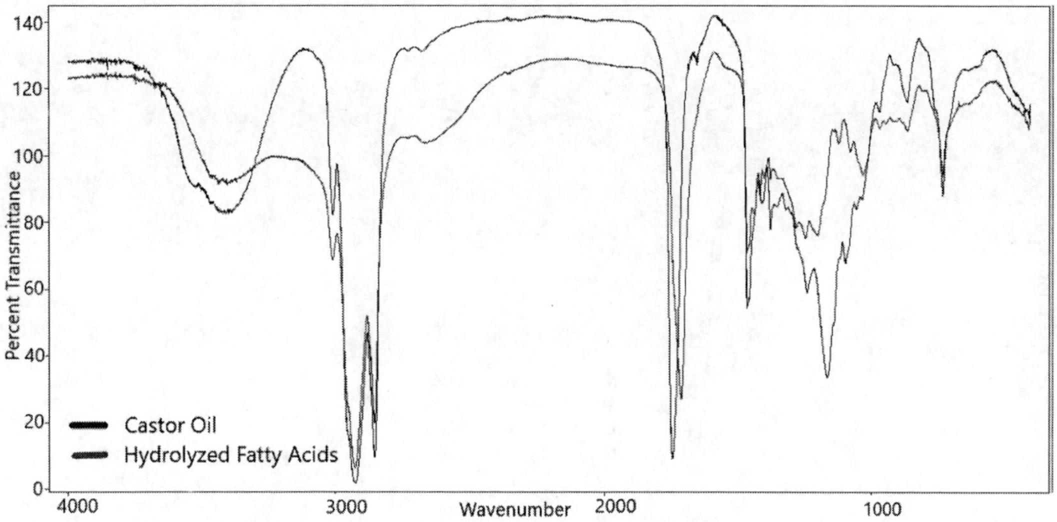

Figure 1. FTIR spectra of castor oil and its fatty acids.

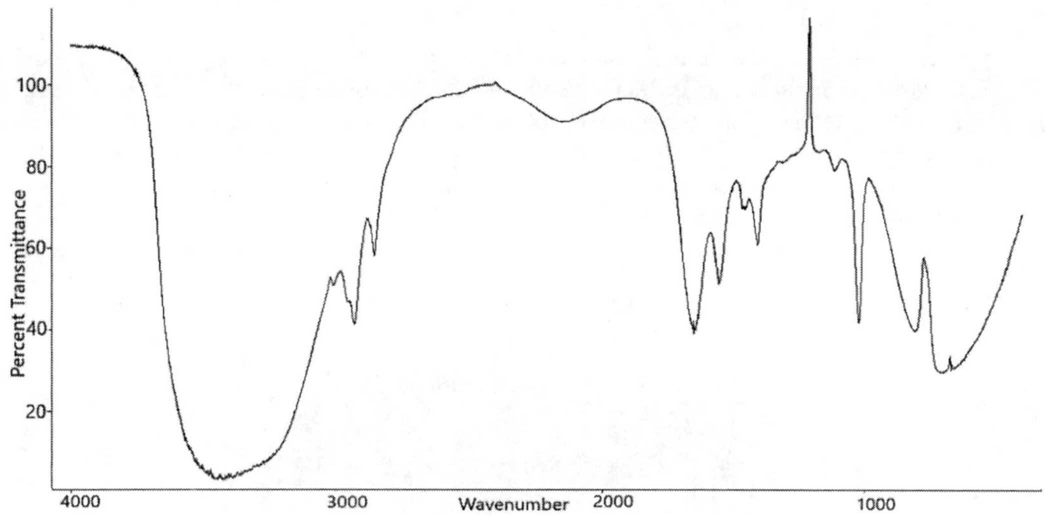

Figure 2. FTIR spectra of oxidized fatty acids.

Based on figure 3 below, the spectrum of all oxidized fatty acid esters showed the presence of O-H absorption bands at wavenumbers 3600–3200 cm^{-1}, and C = O ester at wavenumber 1750-1717 cm^{-1} shifted to a larger wavenumber than the oxidized form. The absorption for C-O-C at wavenumber 1300-1000 cm^{-1} also shows the typical ester absorption [10].

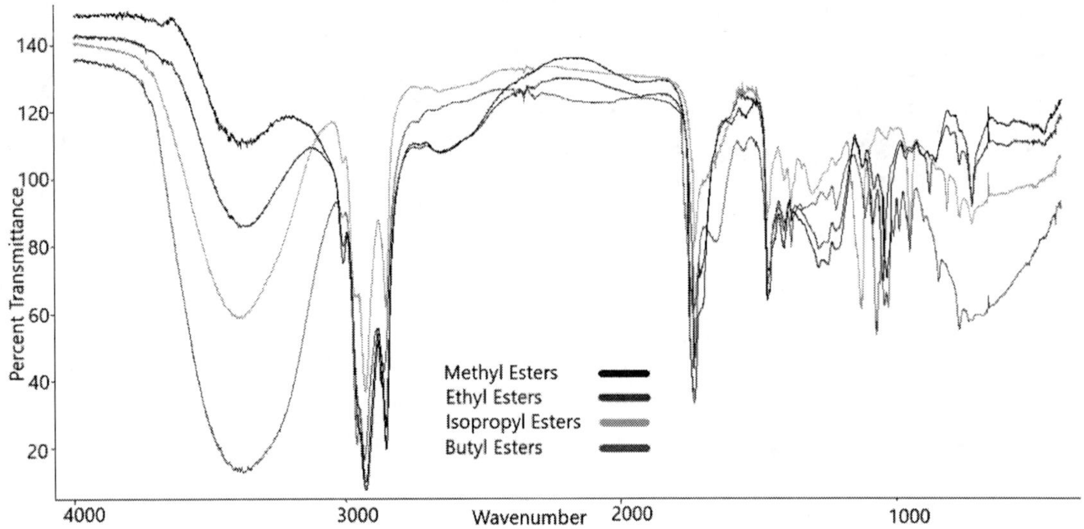

Figure 3. FTIR spectra of oxidized ricinoleic acid esters.

3.5. Simple Qualitative Test as Emulsifier

In this study, a simple emulsifier test was conducted using two types of water-oil mixtures which were mixtures with water's volume greater than the oil's volume and mixtures with oil's volume greater than the water's volume. The samples tested were fatty acids from castor oil hydrolysis, oxidized fatty acids, and ester products. A mixture of oil and water was used as control. After 24 hours, the emulsion with fatty acids as emulsifier was seen separated into the oil phase and the water phase. Meanwhile, the emulsion formed with the addition of the oxidized fatty acid showed a more stable emulsion. These results showed that the oxidation of fatty acids can generate a better emulsifying ability of the fatty acids.

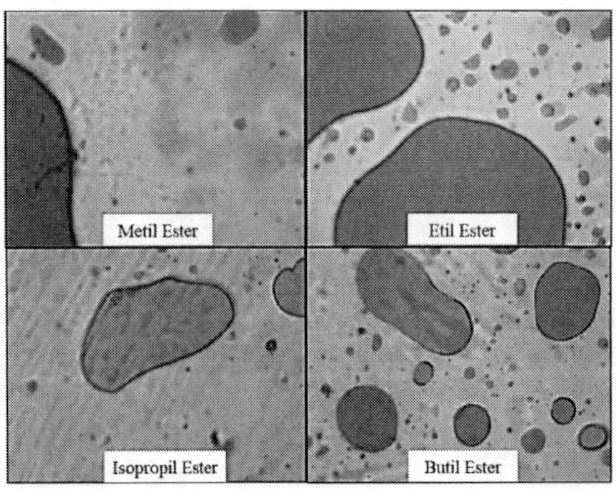

Figure 4. Results of determination of emulsion type.

Emulsions formed with the addition of oxidized fatty acids esters as its emulsifiers generally maintained its emulsified form. Methyl ester as emulsifier formed the most stable emulsion compared to other esters as an emulsifier, while isopropyl ester emulsion showed the most unstable emulsion The results of observations of the type of emulsion produced can be seen in figure 4. Red droplets showed eosin dye dissolved in the water phase, surrounded by the clear color that was the oil phase. Thus it can be concluded that the type of emulsion produced from oxidized fatty acids esters was water-in-oil (w/o) emulsion.

3.6. Antimicrobial Assay

Antimicrobial assay against gram-positive bacteria, *Propionibacterium acnes* and *Staphylococcus epidermidis,* was carried out using disc diffusion method. The bacterial growth inhibition by antimicrobial substances was indicated by the diameter of the clear area around the disc paper. The strength of the antimicrobial activity of the sample was determined based on the classification of the effectiveness of antimicrobial compounds in table 3 [14]. The inhibition zone diameter data of the samples tested in this study shows in table 4.

Table 3. Classification of effectivity of antimicrobial compounds.

Inhibitory zone diameter	The response of growth barriers
>20 mm	Strong
16-20 mm	Medium
10-15 mm	Weak
<10 mm	Not effective

Table 4. Results of antimicrobial assay.

Sample	*Propionibacterium acnes*		*Staphylococcus epidermidis*	
	Inhibitory zone (mm)	Effectiveness	Inhibitory zone (mm)	Effectiveness
Castor Oil	-	Not effective	-	Not effective
Fatty Acids	10	Weak	8	Not effective
Oxidized Fatty Acids	21	Strong	22	Strong
Methyl ester	13	Weak	14	Weak
Ethyl ester	12	Weak	13	Weak
Isopropyl ester	16	Medium	16	Medium
Butyl ester	12	Weak	17	Medium
Clindamycin 0,5%	10	Weak	11	Weak
DMSO	-	Not effective	-	Not effective
Methanol	-	Not effective	-	Not effective
Ethanol	12	Weak	12	Weak
Isopropanol	7	Not effective	8	Not effective
1-butanol	10	Weak	10	Weak
n-hexane	-	Not effective	-	Not effective

From the data above, it can be seen that oxidized fatty acids have the strongest antibacterial activity. The antimicrobial activity of oxidized fatty acids was higher than fatty acids from the hydrolysis of castor oil. This indicated that the addition of two hydroxyl groups increases the polarity of fatty acids and their ability to interact with bacterial cell membranes. Meanwhile based on the classification of the effectivity of antimicrobial compounds in table 3, among the esters produced the strongest antibacterial activity was shown by oxidized isopropyl ester with medium strength against *P. acnes* and oxidized butyl ester with medium strength against *S. epidermidis*. Other ester products were classified as weak antimicrobial agents against both bacteria. These results showed that the ester products could be a potential antimicrobial compound, although there was not any trend seen in the chain length of alcohols used in esterification towards antimicrobial properties. Generally, the mechanism of how the ester class of antimicrobial agents in bacterial membrane lysis is similar to how emulsifier works, which relies on hydrophobic and hydrophilic balance. But from these results, we could imply that besides the chain length from the hydrophilic group and the bacteria tested, several things such as the position of the hydrophobic part and the structure of the alkyl chain used maybe contributes to the antimicrobial activity of the substances. More study is needed to clarify this.

4. Conclusion

Esterification of oxidized fatty acids from hydrolyzed castor oil with various alcohols has been accomplished and proved by the presence of absorption of C = O ester groups at 1750-1735 cm^{-1} and C-O-C at 1300-1000 cm^{-1} in FTIR spectrum. All esterification products could act as emulsifiers with emulsion type water-in-oil emulsion (w/o). The emulsions formed were stable up to 24 hours and the best activity was shown by methyl esters. All esters also have antimicrobial property against both bacteria, *P. acnes* and *S. epidermidis* and the largest inhibitory zone obtained for isopropyl ester against *P. acnes* by 16 mm and butyl ester against *S. epidermidis* by 17 mm.

Acknowledgments

We appreciate Universitas Indonesia for financially supporting this research by Hibah Publikasi Internasional Terindeks 9 (PIT 9), Universitas Indonesia No. NKB.003/UN2.R3.1/HKP.05.00/2019.

References

[1] Alwaseem H, Donahue C J , Marincean S 2014 *J. Chem. Ed.* **91** (4) 575-578
[2] Mubofu E B 2016 *Sust. Chem. Processes* **4** (1) 11
[3] BPS (Statistics Indonesia) 2018 *Produksi Perkebunan Rakyat Menurut Jenis Tanaman (ribu ton) 2000-2018* available at https://www.bps.go.id/dynamictable/2018/06/21/1476/produksi-perkebunan-rakyat-menurut-jenis-tanaman-ribu-ton-2000-2018-.html
[4] Alamsyah R, Isyanti M and Yuniarti 2007 *J.o. Agro-Based Industry* **24** 21–31
[5] Khairani A 2018 *Sintesis Enzimatik Senyawa Etil Ester Asam Lemak Hasil Hidrolisis Minyak Jarak dan Produk Oksidasinya Menggunakan Lipase Candida rugosa E.C. 3.1.1.3 serta Uji Aktivitasnya sebagai Emulsifier dan Antimikroba* (Depok: Faculty of Mathematics and Natural Sciences Universitas Indonesia)
[6] Ramadhan IF 2010 *Pembuatan asam risinoleat melalui hidrolisis minyak jarak dalam larutan basa* (Depok: Faculty of Engineering Universitas Indonesia)
[7] Rifqi A, Siadi K and Sudarmin 2014 *Indo J. Chem. Sci.* **3**, 198-202
[8] Ketaren S 1989 *Pengantar Teknologi Minyak dan Lemak Pangan* (Depok: UI Press)
[9] Mostafa N A, Maher A and Abdelmoez W 2013 *Advances in Bioscience and Biotechnology* **4** 900-907
[10] Smith J G 2011 *Organic Chemistry 3rd ed* (New York: Mc Graw Hill)
[11] Asakuma Y, Kawanami O, Maeda K., Kuramochi H and Fukui K. 2011 *Int. J. o.Thermodynamics* **14** (4) 193–196
[12] Ajiwinanto M 2015 *Hidrolisis minyak jarak kepyar (Ricinus communis) dengan katalis larutan NaOH dan dolomit teraktivasi* (Bogor: Institut Pertanian Bogor)
[13] Sandro L B, Miguel J D, Gabriela R H, Stanlei I K, Adriano C M B and Christielle C 2006 *Applied Catalysis A: General* **313** (2), 146-150
[14] Greenwood 1995 *Antibiotic susceptibility (sensitivity) test, antimicrobial and chemotherapy* (New York: Mc Graw Hill Company)

Electron transfer study between hydrogen molecules and graphene surface

Nasri, RR D J N Subagyono, R Gunawan*

Inorganic and Physical Chemistry Laboratory, Chemistry Department, Faculty of Mathematics and Natural Sciences, Mulawarman University

*Corresponding email: gunawan@fmipa.unmul.ac.id

Abstract. Electron transfer studies on the interaction of graphene surface and hydrogen (H_2) molecules have been carried out. The calculations were based on the Density Functional Theory (DFT) with a basis set of Generalized-Gradient-Corrected PBE method approach (Perdew-Burke-Ernzerhof). The results showed that the optimum distance (lowest potential energy) between hydrogen molecules on the surface of the graphene was 3.2 Å, and the partial density of state (PDOS) calculation showed that the hydrogen-graphene system band gap was 2.51 eV. The electron transfer shown by the isosurface value indicates that the electron transfer from the *p* sub orbitals from the surface of the graphene to the hydrogen molecular orbitals was +0.3603 e/Å.

1. Introduction

Experiments related to hydrogen compounds as energy resources in the modern era, have attracted great attention from researchers, especially on their use as energy and their storage in a material[1-6]. The nature of hydrogen gas that is reactive and flammable, causing difficulties in the storage process.

Research on the role and synthesis of graphite as a hydrogen storage medium was initiated by Chambers et al. [7]. Then, Kim and Park [8] observed that the micropore size of the porous carbons had an important role in the hydrogen adsorption at 77 K. In 2002, Browning et al. [9], carried out a synthesis of carbon nanofiber which have a high potential as a hydrogen storage medium because it was able to adsorb hydrogen by more than 4% by weight, but not yet close to the target of 6.5% by weight (the ratio of the weight of hydrogen stored in the media system, with a volume density of 62 kg H_2/m^3).

Modeling H_2 storage in a carbon-based material, specifically graphene and nanotubes, remains a challenge both in laboratory research and modeling. Computational studies on the investigations of hydrogen storage with a graphene system are not many, so the interaction of hydrogen molecules on graphene system is worth to be studied. Therefore, this research is focused on computational studies of H_2 gas interaction with graphene material and the transfer electron during the interactions.

2. Methods

2.1. Computational Details

Determination of electron structure and interaction energy was calculated through the UNIX-based PWScf Quantum Espresso Schrodinger program [10]. Visualization of the program uses the UNIX-based XcryDen program[11]. In the study of interactions between hydrogen molecules on the surface of graphene, three positions of hydrogen molecules with the surface of graphene are chosen, namely:

IOP Conf. Series: Materials Science and Engineering **833** (2020) 012026 doi:10.1088/1757-899X/833/1/012026

top (one hydrogen atom of the hydrogen molecule was above the carbon atom), bridge (one hydrogen atom was between two carbon atoms), and hollow (one hydrogen atom was at the top of the six carbon atom resonance ring). Calculations for different orientation of hydrogen molecules, eg parallel hydrogen position or helicopter position above the graphene surface, were not carried out.

2.2. Electronic Structure Calculations

In assessing the nature of solid system interactions, a change-correlation potential energy function was performed, which was based on the Density Functional Theory (DFT) with a basis set of Generalized-Gradient-Corrected PBE method approach (Perdew-Burke-Ernzerhof) [12] This approach was chosen because it is more accurate for the atomic/molecular interaction system on a graph solid surface.

3. Result and Discussions

Calculation of the potential energy surface from the interaction of hydrogen molecules and the surface of the graphene was conducted by choosing the same distance between all points on the graphene surface o and the hydrogen molecule. The potential energy surface curve (Figure 1) showed that hydrogen molecules in the top position produced low interaction energy (marked in blue), whereas hydrogen molecules in the bridge and hollow positions produced medium and high interaction energies (marked in white and red, respectively). This shows that the hydrogen molecule has the highest stability in the top position compared to the bridge and hollow positions. The difference in the interaction energy of each position can be expressed as follows $E_{hollow} > E_{bridge} > E_{top}$.

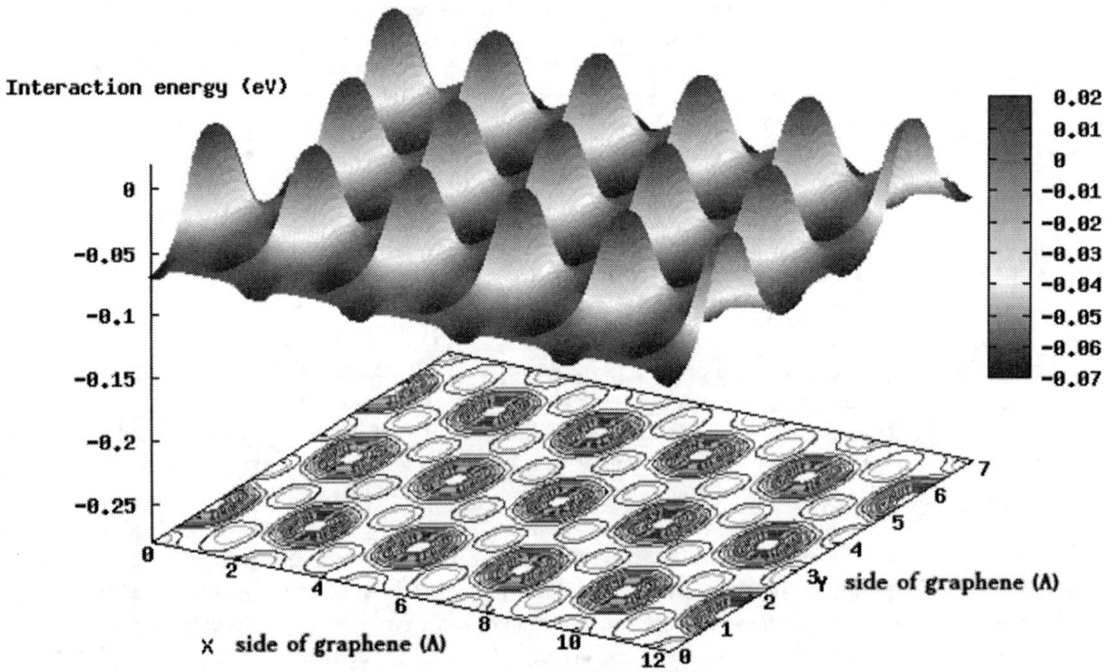

Figure 1. The contour and the potential energy surface of interaction between H_2 molecules and graphene as a function of distance between H_2-graphene.

Calculation of the optimum distance between the hydrogen molecule and the graphene surface was done after calculating the value of the potential energy surface above, and the calculation was performed on the hydrogen molecule at the top position. Interactions between hydrogen molecules on the surface of graphene produced the optimum distance (lowest energy) (3.2 Å), with measurements ranging from 2.8 Å to 3.6 Å (Figure 2).

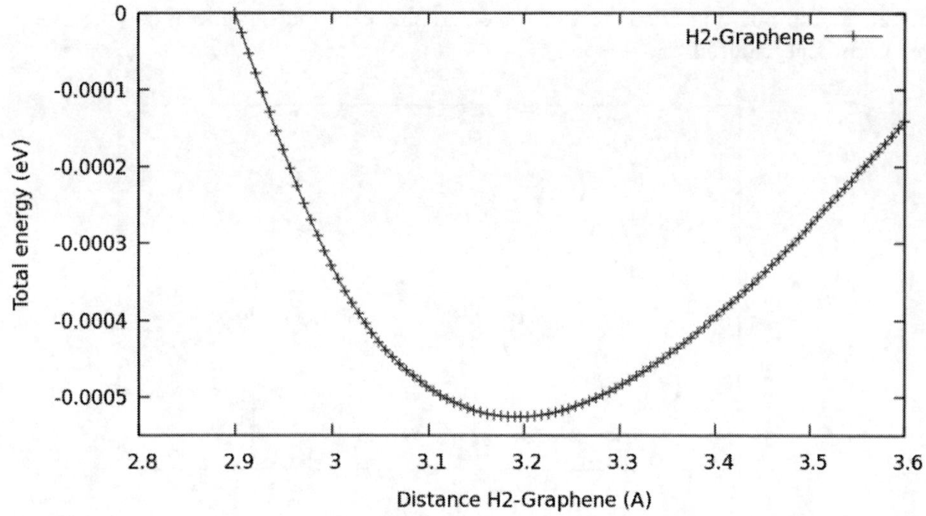

Figure 2. The potential energy of interaction of H_2 molecules and graphene as a function of distance between H_2-graphene.

Calculation of electron density between several hydrogen molecules on the surface of graphene was performed by calculating pDOS (partial Density of State). which was a band gap calculation of the interaction between hydrogen molecules and graphene surface. This calculation showed a fairly strong interaction between s orbitals from hydrogen molecules and p orbitals from the surface of graphene. There was a uniform distribution of s orbitals of the hydrogen molecule, both in the valence band and in the conduction band, which showed the attraction of electrons from the hydrogen molecule by the graphene surface. In the valence band there was a weak interaction between the s orbitals of the hydrogen molecule and the s orbitals of each carbon atom in the energy region of -9.5 eV. This was indicated by the green color of the s orbital of the non-dominant carbon atom (Figure 3). Meanwhile, there was a strong contribution from the interaction of p atoms of carbon atoms with s molecules of hydrogen molecules. This was observed both in the valence band (from -10 eV to - 1.8 eV) and the conduction band (from -1.8 eV to 3 eV). Whereas the Fermi energy, which was the band gap energy, was measured at -1.8 eV. So that, the interaction between the hydrogen molecule and the graphene surface produced a band gap of 2.51 eV.

To support the PDOS calculation data above, an isosurface calculation was performed between the hydrogen molecule and the graphene surface. By scanning the surface of graphene perpendicularly, the electron transfer occurred can be observed (Figure 4). The interaction between hydrogen molecules on the surface of graphene produced an electron transfer energy of 0.3506 e / Å from the hydrogen molecule's s orbitals. Electron transfer from hydrogen molecular s orbitals occurred because of the attractive force of atoms on the carbon's *p* orbitals from the surface of graphene. However, the overlap between the hydrogen molecular s orbitals and the carbon atom p orbitals did not form sp^3 hybridization. This only resulted in a change in the character of carbon atoms from sp^2 to sp^3 of each graphene surface carbon atom that interacted with a hydrogen molecule. The red color in Figure 4 shows strong electron transfers with values of 0.3506 e / Å to 0.1804 e / Å. The electron transfer value was not strong enough to form covalent bonds with sp^3 hybridization at the top position of the hydrogen molecule above the graphene surface. The blue color in Figure 4 shows that the covalent bonds in hydrogen molecules had non-polar characters (electronic transfer value of 0.0030 e / Å) while the light blue color indicates that covalent bonds in hydrogen molecules changed from non-polar to polar character (electron transfer value of 0.0231 e / Å). This character change indicates a strong interaction between the hydrogen molecule at the top position with the surface of graphene so that graphene has the potential to be a hydrogen adsorbent material.

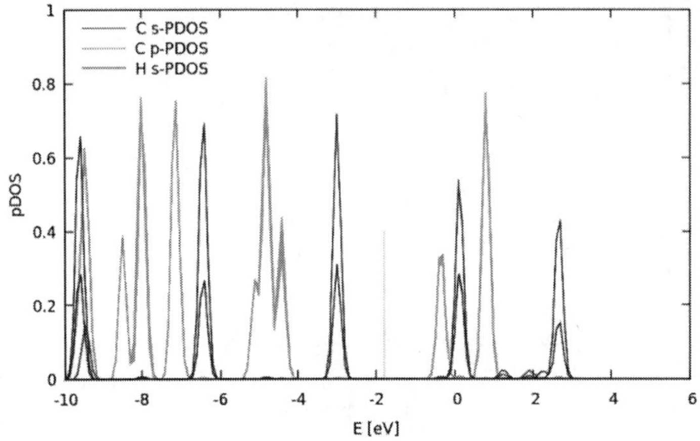

Figure 3. Partial density of state (PDOS) of the interactions between hydrogen molecules and graphene surface.

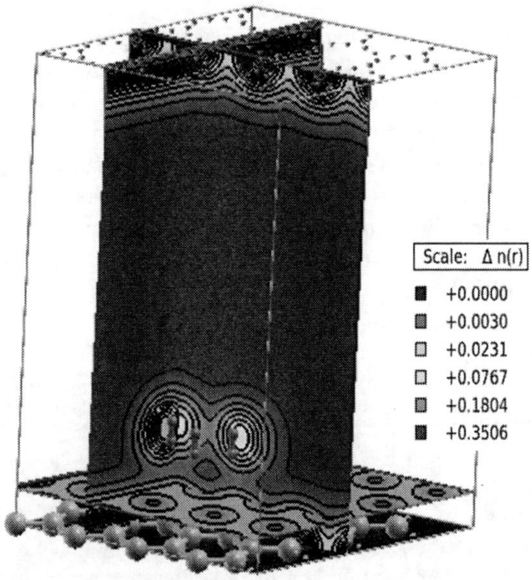

Figure 4. Electron transfer energy of the interactions between H_2 molecules and graphene.

4. Conclusions

Interactions between hydrogen molecules on the surface of graphene produced electron transfer energy of 0.3506 e/Å from graphene p orbitals to hydrogen molecular orbitals. This interaction was considered as chemical adsorption because the interaction distance was 3.2 Å and there was a change in character from sp^2 to sp^3 carbon atoms from graphene.

Acknowledgments

The author would like to thank the Ministry of Research, Technology and Higher Education of the Republic of Indonesia through the Magister Grant (Contract number:069/SP2H/LT/DRPM/2019) for financial support in the project.

References

[1] Kunowsky M, Marco-Lozar J P and Linares-Solano Á 2014 *Adv. Sci. Technol.* **93** 102.
[2] Dündar-Tekkaya E and Yürüm Y 2016 *Int. J. Hydrogen Energ* **41** 9789.
[3] Abe J O, Popoola A P I, Ajenifuja E and Popoola O M 2019 *Int. J. Hydrogen Energ* **44** 15072.
[4] Alfadlil B R, Knowles G P, Parsa M R, N Subagyono R D J, Daniel and Chaffee A L 2019 *J. Phys.: Conf. Series* **1277** 012024.
[5] Kaur M and Pal K 2019 *J. Energy Storage* **23** 234.
[6] Zhang M, Lv H, Kang H, Zhou W and Zhang C 2019 *Int. J. Hydrogen Energ* **44** 25777.
[7] Chambers A, Park C, Baker R T K and Rodriguez N M 1998 *The J. Phys. Chem. B* **102** 4253.
[8] Kim B-J and Park S-J 2007 *Jo. Coll. Inter. Sci.* **311** 619.
[9] Browning D J, Gerrard M L, Lakeman J B, Mellor I M, Mortimer R J and Turpin M C 2002 *Nano Lett.* **2** 201.
[10] Giannozzi P, Baroni S, Bonini N, Calandra M, Car R, Cavazzoni C, Ceresoli D, Chiarotti G L, Cococcioni M, Dabo I, Dal Corso A, de Gironcoli S, Fabris S, Fratesi G, Gebauer R, Gerstmann U, Gougoussis C, Kokalj A, Lazzeri M, Martin-Samos L, Marzari N, Mauri F, Mazzarello R, Paolini S, Pasquarello A, Paulatto L, Sbraccia C, Scandolo S, Sclauzero G, Seitsonen A P, Smogunov A, Umari P and Wentzcovitch R M 2009 *J. Phys.: Condensed Matter* **21** 395502.
[11] Kokalj A 1999 *J. Mol. Grap. Model.* **17** 176, 215.
[12] Perdew J P, Burke K and Ernzerhof M 1996 *Physic. Rev. Lett.* **77** 3865.

Acetylacetone as A Potential Chemosensor for Rapid Detection of Cu(II) in Aqueous Media

Antonius Agung Nugroho[1], Yehezkiel Steven Kurniawan[1], and Leny Yuliati[1,2*]

[1]Ma Chung Research Center for Photosynthetic Pigments, Universitas Ma Chung, Villa Puncak Tidar N-01, Malang 65151, East Java, Indonesia
[2]Department of Chemistry, Faculty of Science and Technology, Universitas Ma Chung, Villa Puncak Tidar N-01, Malang 65151, East Java, Indonesia

*Corresponding author: leny.yuliati@machung.ac.id

Abstract. Monitoring of Cu(II) ions is one of the most critical issues in wastewater treatment. Since the application of diketone-derived compounds as the Cu(II) chemosensor has not been addressed yet, in this present work, acetylacetone (penta-2,4-dione) was employed as the chemosensor to detect Cu(II) ions in the aqueous media. Ultraviolet (UV)-visible spectra showed that the blue colored-Cu(II) solution gave absorption at 811 nm, while the acetylacetone only gave absorption at UV region of 272 nm. Interestingly, the blue Cu(II) solution immediately turned to green solution after the addition of Cu(II) solution into the ligand solution, giving a new peak at 748 nm due to formation of Cu(II)-acetylacetone complex. This large blue shifting suggested the potential application of the acetylacetone as a colorimetric sensor of Cu(II), which gives benefit in the rapid detection. The limit of detection (LoD) and limit of quantification (LoQ) were found to be 0.15 and 0.51 mM, respectively. Furthermore, the presence of other metal ions, such as Na(I), K(I), Mg(II), Mn(II), Ni(II), Al(III), Au(III) and La(III), could be neglected since they gave low extents of interferences up to less than 10%. These results demonstrated that the acetylacetone is a potential chemosensor for Cu(II) detection in the aquatic environment.

1. Introduction

Heavy metals pollution is one of the most serious problems in the aquatic environment. This kind of pollution came from many sources, such as industries, energies, and so on [1]. Copper as one of the most abundant heavy metal ions is pivotal in our daily life because it is massively used for electronic appliances. In the environment, copper is mostly found in cationic form as Cu(II) ions, and they contaminate the aquatic environment [2-4]. An excessive amount of Cu(II) intake could lead to several health problems, such as anemia [5], irritation [6], and liver dysfunction [7]. Because of these negative effects, Cu (II) ion monitoring is a critical issue in wastewater treatment.

Many pieces of research have been done to monitor the Cu(II) concentrations either in the laboratory or environmental samples using a chemosensor agent due to the simple procedure. Phapale *et al.* synthesized a chemosensor for Cu(II) and Fe(III) recognition using pyrene based molecules [8], while Patil *et al.* employed phthalimide based compound for Cu(II) detection using spectrophotometer [9]. The drawbacks of these researches were the complicated and time-consuming synthesis process. Furthermore, the pyrene compounds are highly toxic to our environment, and thus other simple and non-toxic chemosensor agents should be evaluated.

On the other hand, diketone-based molecules are simple and environmentally friendly chemical, which are able to act as chelating agents for certain metal ions. Among them, 1,3-diketone compounds are potential to be a Cu(II) ions chemosensor because they form a stable complex with Cu(II) ions [10] and they have been thoroughly investigated for Cu(II) extraction through a liquid-liquid extraction process [11]. Even though the complexation of Cu(II) and 1,3-diketone derivatives have been extensively studied, the usage of diketone derivatives as chemosensor agent for Cu(II) detection and quantification in the aqueous media has not been addressed yet. Therefore, in this work, a simple 1,3-diketone derivative, acetylacetone (penta-2,4-dione), was applied as a chemosensor agent for Cu(II) detection and quantification in the aqueous media. At first, the ultraviolet-visible (UV-vis) spectra of the ligand before and after addition of Cu(II) ions at various concentrations were recorded to determine the limit of detection (LoD) as well as the limit of quantification (LoQ). Afterwards, the interference effect due to the presence of the other metal ions was also studied.

2. Experiment

2.1. Materials

All of the used chemicals, such as acetylacetone (Merck, China), acetonitrile (Merck, Germany), $CuCl_2$ (Merck, UK), NaCl (Merck, Denmark), KCl (Merck, USA), $MgCl_2$ (Merck, Germany), $MnCl_2$ (Merck, Germany), $NiCl_2$ (Merck, Germany), $AlCl_3$ (Merck, Germany), $AuCl_3$ (Merck, Germany) and $LaCl_3$ (Merck, UK) salts were purchased in pro analytical grade and used without further purification. The chemical structure of acetylacetone (penta-2,4-dione) ligand is shown in Figure 1.

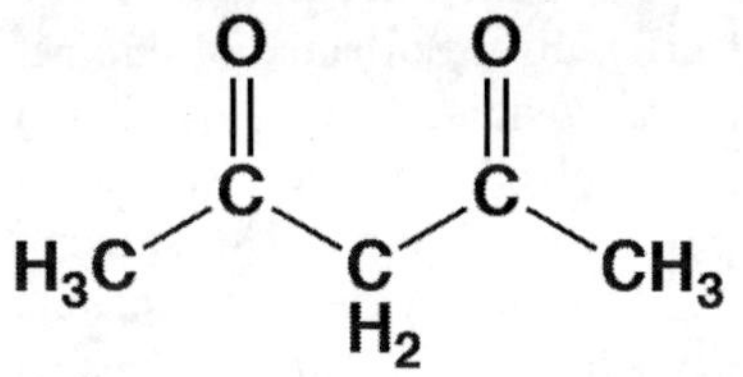

Figure 1. Chemical structure of acetylacetone ligand.

2.2. Evaluation of Cu(II) detection using acetylacetone as the chemosensor agent

A 50 mM solution of acetylacetone was prepared by dissolving 0.51 mL of acetylacetone (MW: 100.13 g mol^{-1}) in 100 mL of H_2O:CH_3CN mixture in 1:1 volume ratio (v/v). Then 1 mL of the ligand solution was then mixed with 1 mL of 10 mM $CuCl_2$ solution in H_2O:CH_3CN 1:1 v/v media. The absorbance of the solution was measured using a UV-visible spectrophotometer (UV-Vis, JASCO V-760) at 200–900 nm.

2.3. LoD and LoQ measurements of Cu(II) detection with acetylacetone

The LoD and LoQ values of Cu(II) sensing with acetylacetone were determined using a calibration curve. The ligand solution was mixed with $CuCl_2$ solution at various concentrations (1, 2, 3, 4, 5, and 6 mM) in 1:1 volume ratio. The absorbance of each solution was measured with UV-Vis spectrophotometer. The calibration curve was obtained by plotting the absorbance of the solution at 748 nm versus the Cu(II) concentrations. The LoD and LoQ values were calculated using equations (1) and (2), whereas σ is the standard errors of the blank solution while a is the slope of the calibration curve, respectively.

$$Limit\ of\ Detection\ (LoD) = \frac{3\,\sigma}{a} \quad (1)$$

$$Limit\ of\ Quantification\ (LoQ) = \frac{10\,\sigma}{a} \quad (2)$$

2.4. Interference study of Cu(II) sensing in the presence of other metal ions

The mixture of binary metal ions, *i.e.* Cu(II) and each metal ion of (Na(I), K(I), Mg(II), Mn(II), Ni(II), Al(III), Au(III), or La(III)) was prepared by mixing 300 μL of 10 mM Cu(II) solution, 300 μL of 10 mM other metal ion solution, and 400 μL mixture of H_2O:CH_3CN 1:1 v/v to obtain a bimetal solution at 0.3 mM concentration for each metal ions. Then the 1 mL of the ligand solution was added into the binary metal solution. The absorbance of the mixture was measured using the UV-vis spectrophotometer. The percentage of interference was calculated by equation (3), where $[Cu(II)]_{initial}$ and $[Cu(II)]_{interference}$ showed the concentration of detected Cu(II) in the absence and presence of other metal ions.

$$Extent\ of\ Interference\ (\%) = \frac{[Cu(II)]_{initial} - [Cu(II)]_{interference}}{[Cu(II)]_{initial}} \times 100 \qquad (3)$$

3. Results and Discussion

In the present work, acetylacetone was employed as the chemosensor agent for Cu(II) detection in H_2O:CH_3CN 1:1 v/v media. The Cu(II) ion solution has a blue color, however, interestingly, after the addition of Cu(II) solution into the ligand solution, the color of the mixture was immediately changed to green color. The photographic images of the ligand and Cu(II) solutions and their mixture are shown in Figure 2. The color changes would indicate the difference in their UV-Vis absorption spectrum. Figure 3 shows the UV and visible spectra of the acetylacetone ligand before and after the addition of Cu(II). As shown in Figure 3(a), the acetylacetone has an absorbance peak in the UV region at 272 nm. This absorbance peak shall correspond to $n \rightarrow \pi^*$ electronic transition of carbonyl group [12]. After the addition of Cu(II), the absorbance of that peak was quenched and shifted to 282 nm. Furthermore, a new peak on the visible region at 748 nm was observed and this peak was obviously different from either the acetylacetone ligand (no absorption peak at visible region) or the Cu(II) solution (an absorption peak at 811 nm). The quenching and shifting phenomena may correspond to the stable complexation between acetylacetone and Cu(II) ion forming a Cu-acetylacetone complex at room temperature. The spectral changes were observed within a second, demonstrating that rapid sensing of Cu(II) ions was achieved.

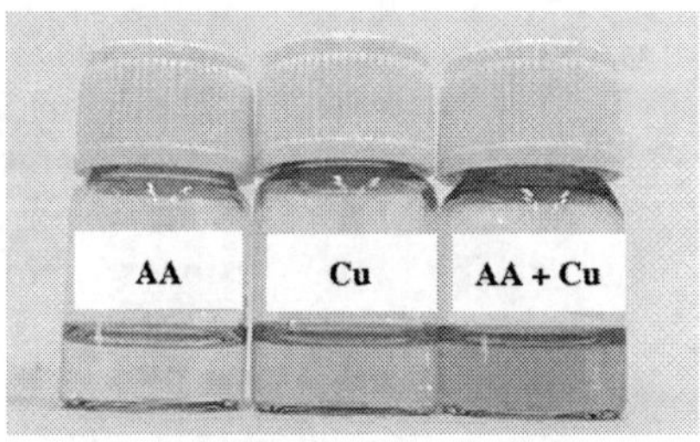

Figure 2. Photographic images of the acetylacetone solution (AA), Cu(II) solution (Cu), and their mixture (AA + Cu) solution.

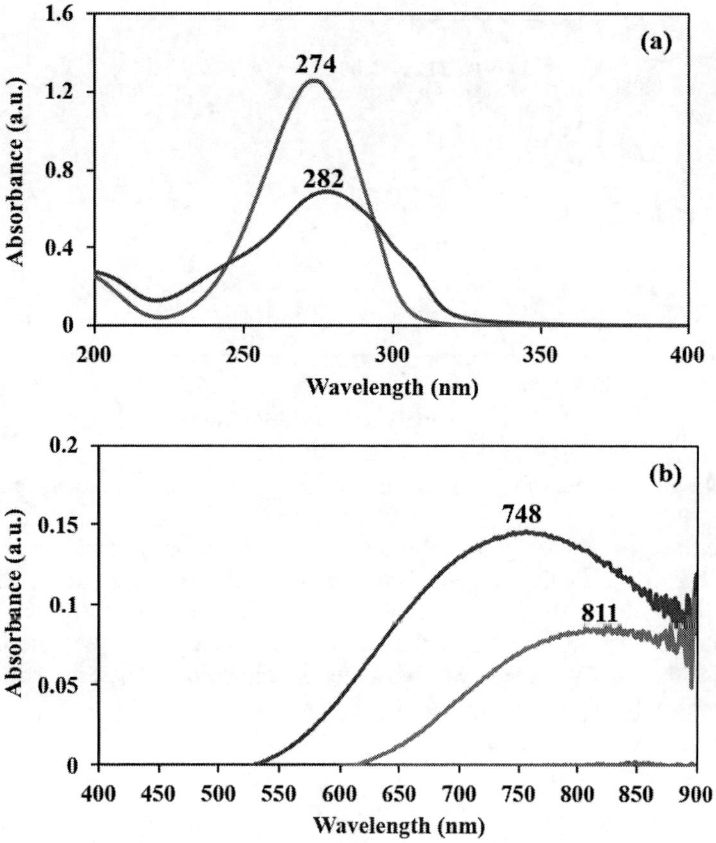

Figure 3. (a) UV spectra of acetylacetone before (blue line) and after (red line) the addition of Cu(II) solution. (b) Visible absorption peak of acetylacetone solution (blue line), Cu(II) solution (grey line) and their mixture (red line).

The new absorption peak at the visible region of 748 nm gave us information about the successful detection of Cu(II) by the acetylacetone via the formation of the Cu(II)-acetylacetone complex. In order to investigate the possibility to quantify the Cu(II) concentrations by monitoring the intensity of this new absorption peak, the absorbance of the solution at 748 nm was plotted against the Cu(II) concentrations, which were varied to 0, 0.5, 1.0, 1.5, 2.0, 2.5, and 3.0 mM. The plot of this calibration curve is shown in Figure 4. It was clear that a strong linear correlation between the concentration of Cu(II) and absorbance of the mixture at 748 nm could be observed with R^2 value of 0.9959.

In order to evaluate the performance of the acetylacetone, the LoD and LoQ values for Cu(II) sensing application were determined using equations (1) and (2) aforementioned above. The LoD and LoQ of acetylacetone were found to be 0.15 and 0.51 mM, respectively. These values are good enough since they are smaller than 1 mM, suggesting that the acetylacetone could be potential for detection of wastewater treatment containing Cu(II) with such concentration range limit. Further study was then also carried out to investigate the effect of other metal ions on the Cu(II) detection and quantification. The Na(I) and K(I) ions were selected as the representative of monovalent cations, the Mg(II), Mn(II), and Ni(II) ions were selected as the representative of divalent cations while the Al(III), Au(III) and La(III) ions were selected as the representative of trivalent cations.

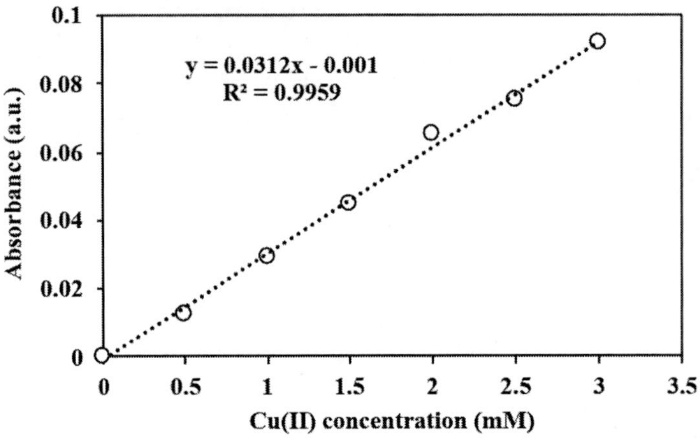

Figure 4. Calibration curve of Cu(II) sensing using acetylacetone as the chemosensor agent.

The photograph images of the ligand solution after the addition of each metal ion are shown in Figure 5. It gave clear evidence that the color changes were observed only for Cu(II) ion and Au(III) ions. While the yellow color of the ligand after addition of Au(III) ions was caused by the initial color of Au(III) ions solution, the green color of the Cu(II)-acetylacetone complex did not come from the color of Cu(II) since the Cu(II) solution was originally blue in color. This result strongly indicated that the acetylacetone was only highly selective for Cu(II) ions.

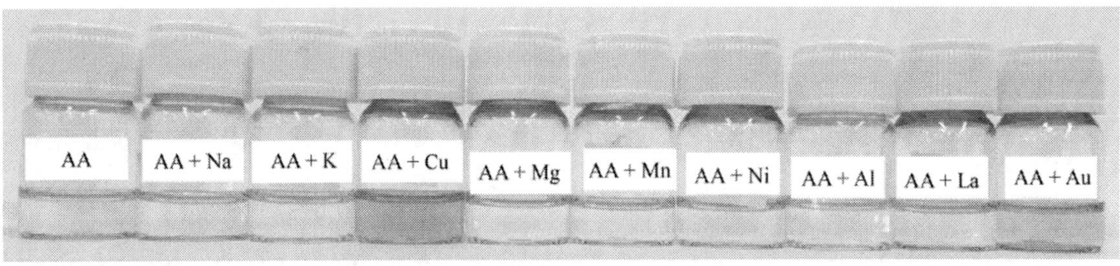

Figure 5. The photographic images of the acetylacetone solution (AA) mixture with the various metal ions. From left to right: ligand only and ligand after addition of Na(I), K(I), Cu(II), Mg(II), Mn(II), Ni(II), Al(III), La(III) and Au(III) ions, respectively.

The interference of the other metal ions, *i.e.* Na(I), K(I), Mg(II), Mn(II), Ni(II), Al(III), Au(III) and La(III) on the Cu(II) quantification was studied by measuring the absorbance of Cu(II)-acetylacetone complex at 748 nm in the absence and presence of other metal ions. Figure 6 shows the absorbance of the Cu(II)-acetylacetone complex in the absence and presence of the other metal ions in the aqueous media. It was found that the absorbance of Cu(II)-acetylacetone complex was not much affected by the presence other metal ions, as indicated by similar absorbance values. The data labels shown in the Figure 6 represent the extent of interference of other metal ions in the of Cu(II) quantification. The extent of interference for monovalent cations, *i.e.* Na(I) and K(I) ions were 0.69 and 4.97%, respectively. The extent of interference for divalent cations, *i.e.* Mg(II), Mn(II), and Ni(II) ions were 3.79, 9.82, and 6.46%, respectively, while the extent of interference for trivalent cations, *i.e.* Al(III), Au(III) and La(III) ions were 9.00, 4.01, and 4.66%, respectively. Even though there was no clear correlation on the effect of the cations to the extent of interference, it is worthy to note that the presence of the other metal ions in general gave a low extent of interference (less than 10%).

Therefore, this study revealed that the acetylacetone could act as a potential Cu(II) chemosensor agent in the aqueous media with a rapid detection and quantification process.

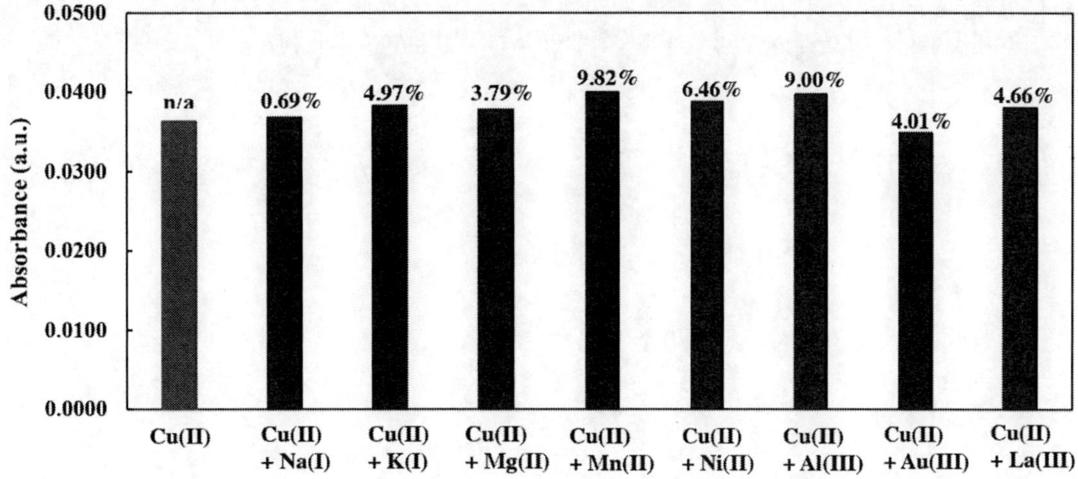

Figure 6. The absorbance at 748 nm of the bimetal solution (Cu(II)+M(n+)) using acetylacetone as the chemosensor agent. The data labels show the interference extent percentages of other metal ions on the Cu(II) quantification.

4. Conclusions

The acetylacetone was found as a potential chemosensor agent for Cu(II) detection and quantification in H_2O:CH_3CN 1:1 v/v media. After the addition of Cu(II) ions solution into the ligand solution, the color of the mixture was immediately changed from blue to green color within a second, demonstrating that a rapid sensing of Cu(II) ions was successfully achieved. It was also found that the LoD and LoQ values for Cu(II) sensing using acetylacetone were 0.15 and 0.51 mM, respectively, which are acceptable values for wastewater containing Cu(II). Furthermore, the presence of the other metal ions gave the extent of interference below 10%, confirming that a selective Cu(II) detection and quantification had been established. These findings showed that acetylacetone is a potential Cu(II) chemosensor agent with a rapid detection and quantification process in the aqueous media.

Acknowledgements

We acknowledged Dr.Eng. Hendrik Oktendy Lintang, S.Si., M.Eng. for the fruitful discussion in the preliminary work of this study. This research is supported by LPPM Universitas Ma Chung with Mandiri Research Grant No. 045/MACHUNG/LPPM-LIT-MANDIRI/IV/2019.

References

[1] Zwolak A, Sarzynska M, Szpyrka E and Stawarczyk, K 2019 *Water Air Soil Pollut.* **230** 164
[2] Aksu A 2015 J. Environ. Health Sci. Eng. **13** 79
[3] Li J, Wang J, Hu H, Ma Y, Zang L and He J 2016 *J. Environ. Sci.* **39** 155
[4] Harvey P J, Handley H K and Taylor M P, 2016 *Environ. Res.* **151** 275
[5] Sharma S, Toppo A, Harbhajanaka B R A and Jyotsna P L 2010 *Indian J. Hematol. Blood Transfus.* **26** 101
[6] Hosytnek J J and Maibach H I 2004 *Toxicol. Mech. Methods* **14** 205
[7] Ibrahim N L 2013 *Anat. Physiol.* **4** 1
[8] Phapale D, Gaikwat A and Das D 2017 *Spectrochim. Acta* A. **178** 160

[9] Patil P, Sehlangia S, Patil A, Pradeep C, Sahoo S K and Patil U 2019 *Spectrochim. Acta* A. **220** 117
[10] Almeida J C, Paixao D A, Marzano I M, Ellena J, Pivatto M, Lopes N P, Ferreira A M D C, Pereira-Maia E C, Guilardi S and Guerra M 2015 *Polyhedron* **89** 1
[11] Radzyminska-Lenarcik E and Witt K 2017 *Separ. Sci. Technol.* **53** 1223
[12] Min-Chul Y, Choi Y S, and Kim S K 1999 *Chem. Phys. Lett.* **300** 207

IOP Conf. Series: Materials Science and Engineering **833** (2020) 012028 doi:10.1088/1757-899X/833/1/012028

Synthesis of $ZnO/rGO/TiO_2$ Composite and Its Photocatalytic Activity for Rhodamine B Degradation

Haniffudin Nurdiansah[1*], Rena Eka Firlyana[1], Diah Susanti[1], Hariyati Purwaningsih[1]

[1]Department of Materials Engineering, Faculty of Industrial Technology, Institut Teknologi Sepuluh Nopember, Jl. Raya ITS, Keputih, Sukolilo, Surabaya 60111, Indonesia.

*Corresponding author: haniffudin@mat-eng.its.ac.id

Abstract. Nowadays the growth of the textile industry in Indonesia is increasing. An example of dangerous textile waste is Rhodamine-B. Therefore, a method to treat Rhodamine B waste is needed. One method that can be used is photocatalysis. ZnO and TiO_2 semiconductors are known as good photocatalyst material because they have a small band gap energy. In addition, reduced graphene oxide (rGO) has a potential due to its good conductivity and large surface area. This study aims to synthesize $ZnO/rGO/TiO_2$ composite. Analysis of the composite shows peaks correspond to diffraction pattern of ZnO, TiO_2, and rGO. Based on scanning electron microscope analysis, the surface morphology of the composite was in the form of sheets (rGO), agglomerates (TiO_2), and hexagonal nanorod (ZnO). Further confirmation by EDX analysis shows elemental compositions of Zn, Ti, O, C (18.44, 7.44, 46.01, 28.11, respectively). FTIR analysis resulting in bands of C-O at 998 cm^{-1}, Ti-O-Ti at 506 cm^{-1}, and Zn-O at 497 cm^{-1}. Photocatalysis test using the composite gave the optimum value of Rhodamine-B degradation of 94.72 % under UV light irradiation for 5 hours.

1. Introduction

Today, the increasing number of textile industry in Indonesia indirectly increases the production of textile dye waste. The existence of this waste can cause environmental damage. This is because textile dyes are non-biodegradable organic substances that are easily accumulated in soils and waters [1]. One of the dyes that is usually used in the textile industry is Rhodamine B. This dye is a dangerous chemical that is not allowed to be mixed either in food, drinks, or cosmetics. If this molecule enters the human body, it can cause irritation of the skin, eyes, and liver cancer.

Various methods have been done to overcome this waste, for example catalytic ozonation [2], osmosis process [3], coagulation process [4], and etc. However, those several methods only transferred dye pollutants from one phase to another and unable to solve the problem [5]. A promising method to solve this problem is photocatalysis using semiconductor materials. With this method, dyes are broken down into compounds that are simpler and safer for the environment.

In this study, ZnO and TiO_2 semiconductors were used. ZnO was chosen because it has a small band gap of ~ 3.3 eV, chemically stable, and biodegradable [6]. TiO_2 was chosen because it has a high and stable photocatalyst activity and has the ability to absorb high energy with an energy gap value of 3.2 eV [7]. To increase the photocatalytic efficiency, graphene material can be added. This is because graphene is an extraordinary material with 2D structure and covalent bonds which make graphene has good properties including electron mobility of 200000 $cm^2\ v^{-1}\ s^{-1}$, surface area of 2630 $m^2\ g^{-1}$, and thermal conductivity of 5000 $W\ m^{-1}\ K^{-1}$ [8]. Thus, ZnO and TiO_2 semiconductors combined with reduced graphene oxide (rGO) to improve its photocatalytic properties for Rhodamine B degradation is investigated and discussed in this paper.

2. Experimental Details

2.1. Materials

All the chemicals used were of analytical grade without further purification. Graphite powder (>99 purity), H_2SO_4 98%, Natrium Nitrate ($NaNO_3$), Kalium Permanganate ($KMnO_4$), Hydrogen Peroxide (H_2O_2), Zinc Powder, Titanium (III) Chloride ($TiCl_3$), Barium Chloride ($BaCl_2$), Natrium Chloride (NaCl), Ammonia (NH_4OH) 10% and Rhodamine B were obtained from SAP Chemicals.

2.1.1. Synthesis of Graphite Oxide.

The method used to oxidize graphite to graphite oxide in this study is the modification of the Hummer's Method. Synthesis begins with stirring 2 grams of graphite with 80 ml of 98 % H_2SO_4 for 2 hours in an ice bath at temperature of around 5 °C. After that, 4 grams of $NaNO_3$ and 8 grams of $KMnO_4$ were added gradually for one hour. The addition process turns the solution to purplish green. Then the solution was stirred for one hour~~,~~ resulting in homogeneous solution. The solution ~~is~~ was removed from the ice bath and stirred at 35 °C for 20 hours. In this process, the solution turn to dark brown and thickens. After that, the addition of 200 ml of distilled water ~~is~~ was carried out in stages. After 1 hour of stirring, 30 % H_2O_2 was added to the solution. The addition of H_2O_2 aims to stop the oxidation process and cause the color of the solution to turn yellow. The solution ~~is~~ was left in a stirring condition for 30 minutes, then the solution ~~is~~ was centrifuged to separate the precipitate. After that, 5 % HCl ~~is~~ was added to remove the remaining metal ions. Then it periodically washed until neutral pH ~~is~~ was obtained. To confirm the presence of sulfate ions, 1 M $BaCl_2$ titration was performed. If clean white precipitate appeared, it indicated that there were still sulfate ions left. During the washing process, the color of the solution would turn darker due to peeling of graphite to graphite oxide. The neutral solution was then dried at 120 °C for 12 hours to obtain graphite oxide sheet.

2.1.2. Synthesis of ZnO/rGO/TiO_2 composite.

Prior to composite synthesis, TiO_2 was firstly prepared using $TiCl_3$ precursor. The process was started by stirring 4.5 ml of $TiCl_3$ and 9 ml of NaCl for 30 minutes at 45 °C. Then, the solution was heated at 200 °C for 5 hours to produce a powder product. Meanwhile, the previously formed graphite oxide through the Hummer method with variations of 15 %wt was stirred for 1 hour so that the graphite oxide sheet was disintegrated and turned into smaller flakes. Then, it was cultivated for 2.5 hours so that the small debris dissolved in distilled water and graphene oxide compound was formed. After that, the stirring process was carried out simultaneously with the addition of 1 gram of Zn powder. After 1 hour stirring, 1 gram of powder product from the heating $TiCl_3$ solution was added into solution. Next, 8 ml of 10 % NH_4OH was added until precipitate was formed. The solution was then washed with distilled water and calcined at 700 °C for 4 hours to get the ZnO/rGO/TiO_2 composite.

2.1.3. Characterization and Testing

2.1.3.1. Characterization of ZnO/rGO/TiO_2 composite

To assess the characteristics of ZnO/rGO/TiO_2 composite, X-Ray Diffraction (XRD) (Philips Analytical) test was performed at 2θ angle of 10°– 90° with λ Cu-Kα 1.54060 Å to determine the crystal structure. Scanning Electron Microscope (SEM) and Energy Dispersive Spectroscopy (EDS) Inspec S50 was used to observe the morphology and composition. Fourier Transform Infrared Spectroscopy (FTIR) (Scientific Nicolet iS10) test was done to determine the functional groups of the composite at wavenumber range 3500-500 cm^{-1}.

2.1.3.2. Photocatalysis testing of $ZnO/rGO/TiO_2$ composite for Rhodamine B degradation

Firstly, Rhodamine B powder was weighed in 0.01 gram then dissolved into 100 mL of distilled water to obtain a solution of Rhodamine B with a concentration of 100 mgL^{-1}, then pipetted 2; 4; 6; 8; and 10 mL into a 100 mL measuring flask and diluted with distilled water to obtain a solution with a concentration of 2; 4; 6; 8; 10 mgL^{-1}. Then the maximum wavelength of rhodamine B and sample of 2; 4; 6; 8; 10 mgL^{-1} is determined by using a UV-Vis spectroscopy. Next, 0.2 gram of ZnO, TiO_2, ZnO/TiO_2 (50:50 %wt), rGO/ZnO (15:85 %wt), rGO/TiO_2 (15:85 %wt), and $ZnO/rGO/TiO_2$ (42.5:15:42.5 %wt) were dissolved into Rhodamine B solution. The irradiation time was variate 1-5 hours with the help of UV Light 20 Watt, and the sample was taken every 1 hour. Next, Ultra-Violet Visible Spectroscopy (UV Vis) test was conducted to determine the optimum wavelength of all the sample and absorbance after degradation process. Then using equation 1 to calculate the % of degradation

$$(\%\ D) = \frac{C_0 - C_t}{C_0} \times 100\% \qquad (1)$$

Where % D = % Degradation, C_0 = Initial concentration before irradiation, C_t= Concentration after t hour irradiation.

3. Results and Discussion

3.1. XRD Result

XRD analysis was used to determine the phases contained in the material. The XRD results of graphite, graphite oxide and rGO are shown in Figure 1. A sharp peak (002) and high intensity at 2θ = 26.43° with *d*-spacing of 3.37 Å correspond to a graphite peak with a hexagonal crystal structure (JCPDS 03-065-6212). The graphite crystallinity is very good, as indicated by the high intensity that shows better crystallinity.

Next, the graphite was oxidized to form graphite oxide which has a different peak (001) compared to that of graphite at 2θ of 12.45° with lower intensity but bigger *d*-spacing of 7.10 Å.

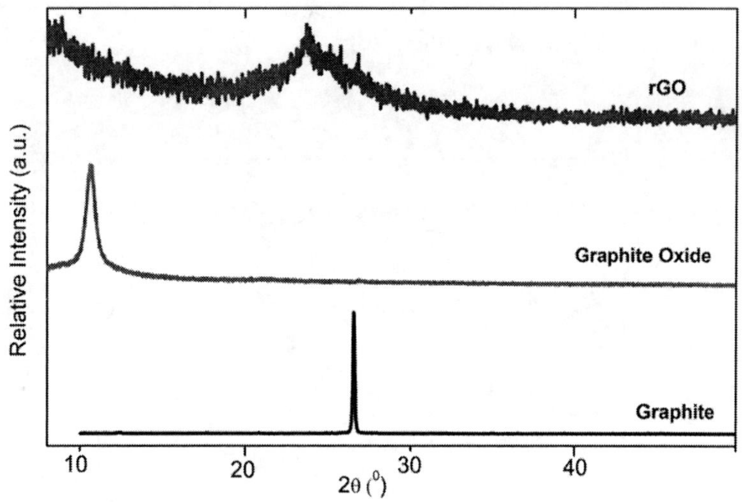

Figure 1. XRD pattern of graphite, graphite oxide, and rGO.

After the reduction process has been done, graphite oxide was converted to reduce graphene oxide (rGO). The rGO diffraction peak observed at peak (002) at 2θ of 24.04 ° with *d*-spacing of 3.38 Å, and lower intensity than graphite and graphite oxide, as well as broadening peaks. It indicates that the

IOP Conf. Series: Materials Science and Engineering **833** (2020) 012028 doi:10.1088/1757-899X/833/1/012028

structure is amorphous. The XRD results for graphite, graphite oxide and rGO are in accordance to the research conducted by Mishra [9].

The results of XRD for TiO_2, ZnO, and the composite are shown in Figure 2. TiO_2 samples formed diffraction peaks (110) at 2θ of 27. 056°, (101) at 2θ of 35.988°, and (200) at 2θ of 38.636°, which shows that the dominant structure of TiO_2 formed is *rutile* (JCPDS 21-1276) [10].

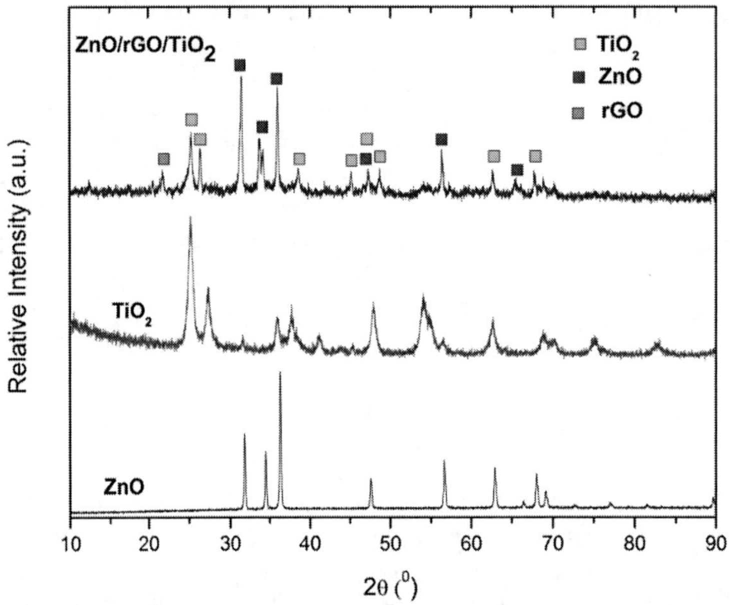

Figure 2. XRD diffraction pattern of ZnO, TiO_2 and ZnO/rGO/TiO_2 composite.

In the ZnO samples, the highest peaks were formed at 31.75°, 34.44°, 36.25° that correspond to (100), (002) and (101), respectively. Based on these data, the formed ZnO structure is hexagonal wurtzite (JCPDS 79-2205) [11]. In addition, there is a new peak in rGO at 2θ of 23.9° [12] indicating that the composite has been successfully synthesized.

3.2. FTIR Result

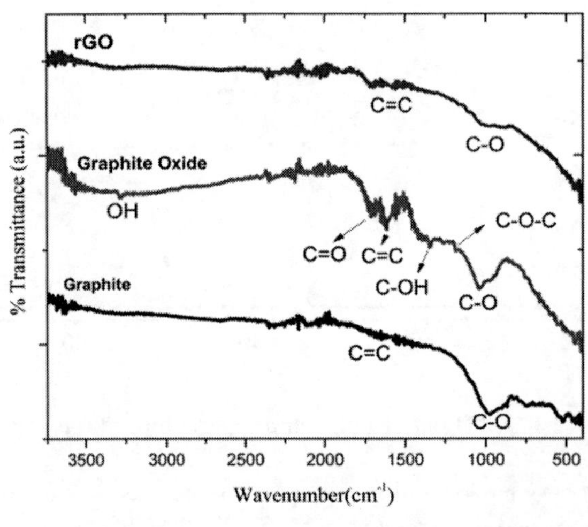

Figure 3. FTIR spectra of graphite, graphite oxide, and rGO.

Figure 3 shows the results of FTIR analysis for graphite, graphite oxide, and rGO. Graphite has functional groups that display bands at 1577 cm^{-1} (C=C), and at 1003 cm^{-1} (C–O). Graphite oxidation produces graphite oxide structure, in which the new bonds with O and H are formed, such as C=O at 1830 cm^{-1}, C–OH at 1233 cm^{-1}, C–O–C at 1300 cm^{-1}, and OH at 3293 cm^{-1}. After that, the reduction of graphite oxide will eliminate the O and H bonding groups, so that the remaining functional groups are C=C and C–O. It indicates that the reduction occurs imperfectly. This result is in accordance with previous research conducted by Yan, *et al* [13].

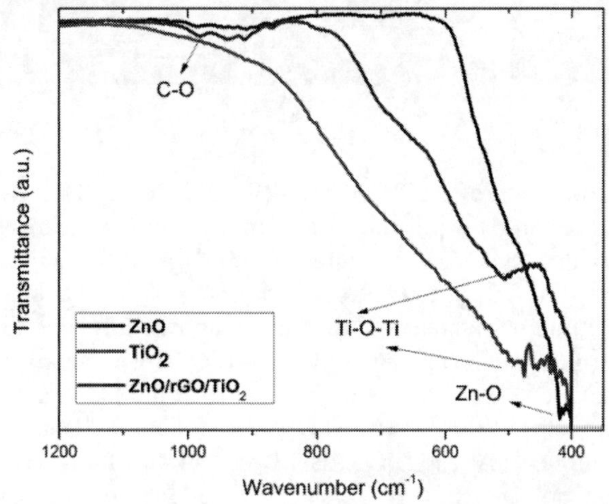

Figure 4. FTIR spectra of ZnO, TiO_2 and composite ZnO/ rGO/ TiO_2.

The IR spectra of ZnO/rGO/TiO_2 composite are shown in Figure 4. There are C–O peak at 998 cm^{-1} and C=C peak at 1689 cm^{-1} indicating the formation of rGO.Moreover, the composite also has Ti–O–Ti peak at 506 cm^{-1} [14], and Zn–O peak at 497 cm^{-1} [15]. Based on the IR spectra of the composite, it can be said that the composite has been formed .

3.3. SEM and EDX Result

Figure 5 shows the SEM test results for graphite (A), Graphite Oxide (B) and rGO (C). Graphite has a thick morphology in the form of flakes. When graphite is oxidized, the flakes will turn to transparent sheets, which resembling the graphite oxide. Furthermore, the reduced graphite oxide will form a thin sheet structure, This mechanism is known as chemical exfoliation, which is the reduction in the number of layers of graphite, resulting in a thin structure in the form of a monolayer, or few layers of graphene in the form of rGO.

Figure 5. SEM images of graphite (A), graphite oxide (B), and rGO (C) mag 20000x.

Figure 6. SEM images of ZnO (A), TiO_2 (B) and ZnO/ rGO/ TiO_2 composite (C) mag 10000x.

Figure 6 shows the SEM result of ZnO (A), TiO_2 (B) and ZnO/rGO/TiO_2 composite (C). ZnO has morphology of hexagonal nanorods, while TiO_2 has a morphology of nearly spherical agglomerate. In the composite, it appears that there are structures in the form of hexagonal nanorods which is the characteristics of ZnO, agglomerated structure which shows TiO_2, and thin and transparent sheet which shows rGO. Thus, it can be summarized that the composite has been formed.

The EDX analysis results are shown in Figure 7. At ZnO, a comparison value of %At Zn:O is 62.18:37.82. The high value of %At of Zn is probably due to the fact that a small amount of Zn was left in the composite. The ratio value of %At Ti:O is 39.15:60.85, close to the atomic ratio of TiO_2. For the composite, the ratio of %At Zn:Ti:O:C is 18.44:7.44:46.01:28.11. The %wt of C is 12.81%, from the addition of 15% graphene oxide. It can be concluded that the compound has been formed.

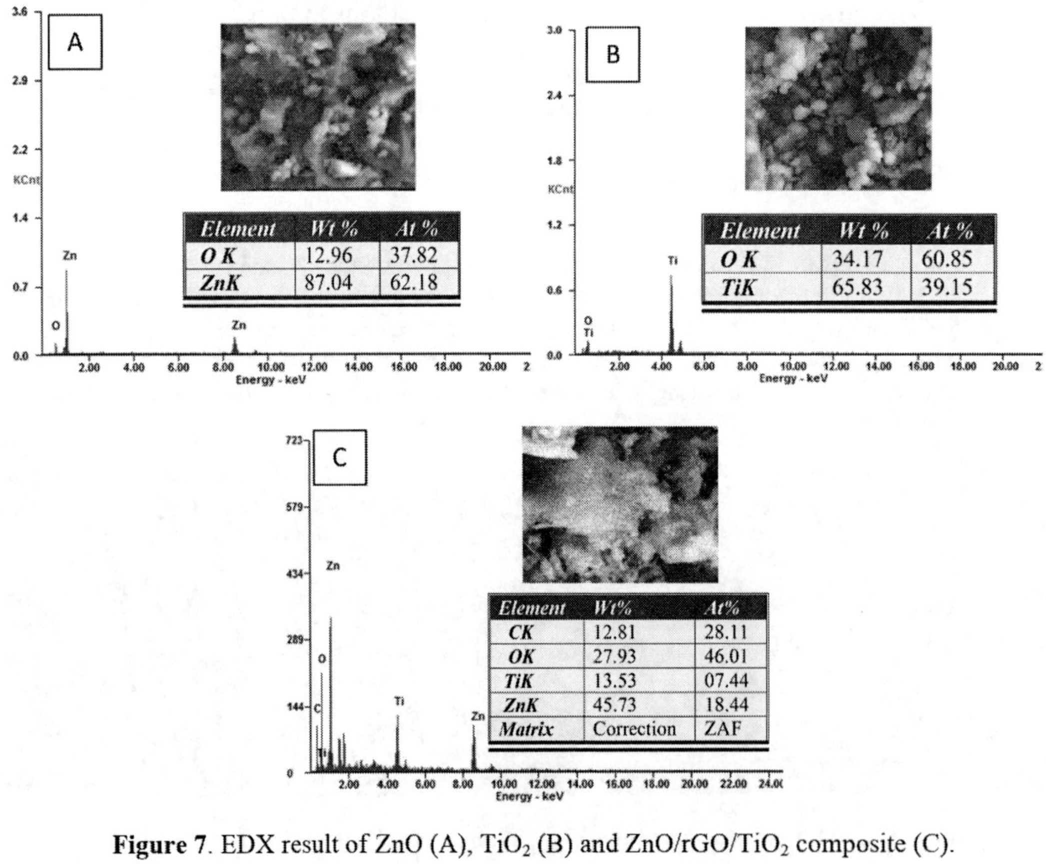

Element	Wt %	At %
O K	12.96	37.82
ZnK	87.04	62.18

Element	Wt %	At %
O K	34.17	60.85
TiK	65.83	39.15

Element	Wt%	At%
CK	12.81	28.11
OK	27.93	46.01
TiK	13.53	07.44
ZnK	45.73	18.44
Matrix	Correction	ZAF

Figure 7. EDX result of ZnO (A), TiO_2 (B) and ZnO/rGO/TiO_2 composite (C).

3.4. UV VIS and Photocatalysis Result

From Figure 8 (A) it is observed that the highest absorbance of UV Light after 5 hours irradiation was shown by ZnO. This indicates that there are still many Rhodamine B that have not been degraded.

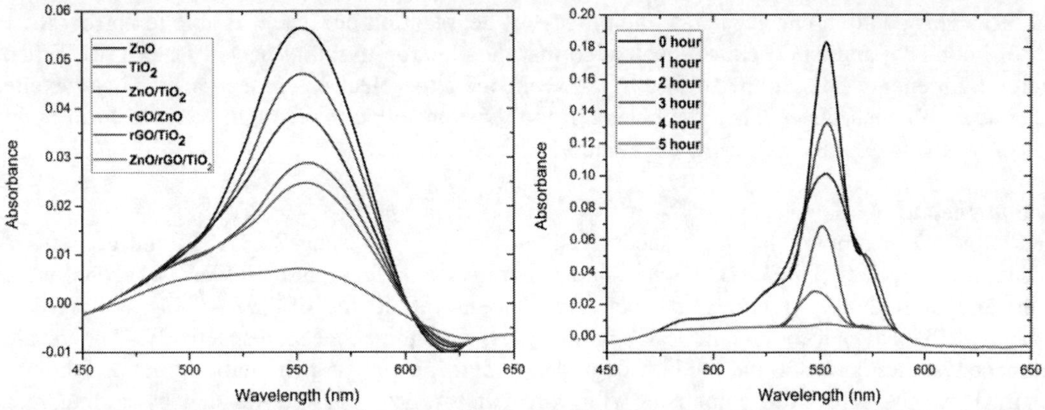

Figure 8. (A) Absorption profile of ZnO, TiO_2, ZnO/TiO_2 (50:50 %wt), rGO/ZnO (15:85 %wt), rGO/TiO_2 (15:85 %wt), and ZnO/ rGO/ TiO_2 (42.5:15:42.5 %wt) after 5 hours irradiation; (B) absorption profile of ZnO/rGO/TiO_2 in various irradiation time of 1-5 hours

The lowest absorbance value was shown by ZnO/rGO/TiO_2 composite, which indicates that only small amount of Rhodamine-B remains. Figure 8 (B) shows the absorbance value of ZnO/rGO/TiO_2 composite in various irradiation time. As the irradiation time increases, the absorbance value decreases, indicating that more Rhodamine-B concentration was degraded.

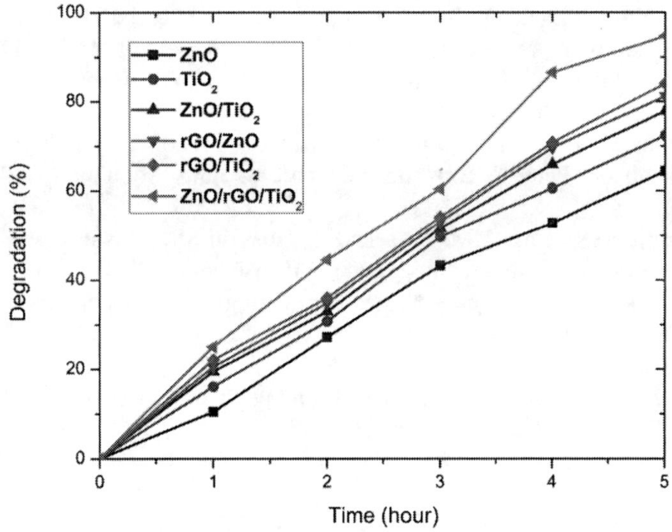

Figure 9. Percentage of degradation profile of Rhodamine-B by ZnO, TiO_2, ZnO/TiO_2 (50:50 %wt), rGO/ZnO (15:85 %wt), rGO/TiO_2 (15:85 %wt), and ZnO/rGO/TiO_2 (42.5:15:42.5 %wt) after 5-hours irradiation.

The degradation percentage results in Figure 9. show that the lowest degradation value after 5-hours irradiation was 64.42 %, which is correspond to ZnO. In TiO_2, the degradation value was 72.41%. When ZnO and TiO_2 were combined, the degradation value was increased to 77.87%. The interesting point is, when rGO was added to ZnO or TiO_2, it increases the degradation value, to 81.18% (rGO/ZnO) and to 84.11% (rGO/TiO_2). When the three catalysts were combined as ZnO/rGO/TiO_2, the degradation percentage reaches the highest value of 94.73%.

The addition of rGO, which has a wide active surface area, increases the formation of super anions and OH^-. This addition can accelerate the process of degradation because it is able to narrow the band gap of both TiO_2 and ZnO semiconductors so that the electron mobility occurs faster [16]. When the photon light energy falls on the surface of the semiconductor, electrons in the valence band excited to the conduction band faster. Then the holes left in the valence band react with water to form hydroxyl radical, which is responsible for the degradation of Rhodamine B.

4. Conclusion

ZnO/rGO/TiO_2 composite has been successfully synthesized by using Zn, $TiCl_3$, and graphite oxide powder as precursors. The XRD analysis shows that the ZnO, TiO_2, and rGO has hexagonal wurtzite, rutile and amorphous structures, respectively. The morphologies of ZnO, TiO_2, and rGO were hexagonal-shaped nanorods, spherical agglomerates, and thin sheets, respectively. The results of photocatalysts test showed that the best degradation percentage using irradiation time of 5 hours was obtained by ZnO/rGO/TiO_2 composite with value of 94.73%. This shows that by adding reduced graphene oxide, the active surface area of the catalyst will be much greater because reduced graphene oxide has a large surface area. It also increases the formation of super anions and OH^-, which can accelerate the process of degradation.

References

[1] Shivatharsiny Rasalingam, Rui Peng, and Ranjit T. Koodali, " Journal of Nanomaterials, vol. 2014, Article ID 617405, 42 pages
[2] Hu E, Wu X, Shang S, Tao XM, Jiang SX, Gan L.. J Clean Prod 2016;112:4710e8.
[3] Zhao P, Gao B, Xu S, Kong J, Ma D, Shon HK, et al. Chem Eng J 2015;264:32e8.
[4] Huang X, Gao B, Yue Q, Zhang Y, Sun S. Sep Purif Technol 2015;154:108e14.
[5]Nuengmatcha P, Chanthai S, Mahachai R, Oh WC. Dyes Pigments 134:487–497.
[6]Yangyang Zhang, Manoj K. Ram, Elias K. Stefanakos, and D. Yogi Goswami, Journal of Nanomaterials, vol. 2012, Article ID 624520, 22 pages, 2012.
[7]Dette, Christian & Perez-Osorio, Miguel & Kley, Christopher & Punke, Paul & Patrick, Christopher & Jacobson, Peter & Giustino, Feliciano & Jung, Soon Jung & Kern, Klaus. (2014). Nano letters. 14.
[8]Bhuyan MSA, Uddin MN, Islam MM, Bipasha FA, Hossain SS. Int Nano Lett 6:65–83
[9]Mishra, Ashish & Sundara, Ramaprabhu. (2011). AIP Advances. 1.
[10]Thangavel, Karthik & Rathinamoorthy, R. & Murugan, Ramachandran. (2012). Journal of Industrial Textiles. 42. 99-117..
[11]Johra FT, Jung WG. Appl Catal A 2015;491:52e7.
[12]Chen, You-Hu & Niu, Yufang & Tian, Tian & Zhang, Juan & Wang, Yuanfei & Li, Yunpeng & Qin, Lu-Chang. Chemical Physics Letters. 677. 143-147.
[13]Yan, Jincan & Zeng, Lydia & Ren, Tianhui & Van Der Heide, Emile. Biosurface and Biotribology. 1.
[14]Bobrova, A.M., Zhigun, I.G., Bragina, M.I. et al. J Appl Spectrosc (1968) 8: 59.
[15]Sowri Babu, K., Ramachandra Reddy, A., Sujatha, C. et al. J Adv Ceram (2013) 2: 260.
[16]Vallejo William, Rueda Angie, Díaz-Uribe Carlos, Grande Carlos and Quintana Patricia R. Soc. open sci.

Oxalic Acid from Corn Stalk for Photocatalytic Reduction of Cr(VI)

Ana Nurjanah[1], Barlah Rumhayati[2*], Adam Wiryawan[2]

[1]Chemistry Postgraduate program, Chemistry Department, Faculty of Sciences, Brawijaya University Indonesia
[2]Chemistry Department, Faculty of Sciences, Brawijaya University Indonesia 65415

[*]Corresponding author: rumhayati_barlah@ub.ac.id

Abstract. Cr(VI) is toxic species of chromium in the aquatic environment. The reduced species, Cr (III), has lower toxicity than Cr(VI). The change of Cr(VI) species to Cr(III) can be conducted by photocatalytic reduction using oxalic acid as an oxidant with the aid of Fe(III) as a catalyst. Oxalic acid can be isolated from the biomass waste of corn stalk. This research aimed to isolate oxalic acid from corn stalk biomass as an oxidator for Cr(VI) photocatalytic reduction. The oxalic acid was isolated from corn stalks powder using NaOH, precipitated with the addition of $CaCl_2$, acidified with H_2SO_4, and then crystalized to form the oxalic acid crystals. The photocatalytic reduction was carried out in a closed reactor equipped with two of 30-watt UV lights. About 30 mg of oxalic acid was used to reduce 300 mg/L Cr(VI) with the presence of 4 ml Fe(III) 100 mg/L as a catalyst. The process was conducted for 120 minutes while stirring. The reduction efficiency was calculated from the differentiation of the initial Cr(VI) concentration to the final ones. Cr(VI) concentrations were measured as Cr(VI)-diphenylcarbazide complex compound at pH 2 that could be detected at 540 nm. The results showed that the isolated oxalic acid has hydroxyl and carbonyl groups at 3441 cm^{-1} and 1627 cm^{-1} respectively identical with the oxalic acid standard. The crystals have melting point at $103^{0}C$. It could be used effectively to reduced Cr(VI) up to 45.9 %.

1. Introduction

The concentration and toxicity of dissolved chromium and its mobilization in the environment depend on its oxidation state [1]. The toxicity level of Cr(VI) is about 100 times higher than the toxicity level of Cr(III). Hence, the reduction of Cr(VI) becomes one of the ways to reduce its toxicity. In the aqueous system, Cr(VI) is dissolved as divalent oxyanions of chromate (CrO_4^{2-}) and dichromate ($Cr_2O_7^{2-}$) while Cr(III) is deposited or absorbed by organic and inorganic compounds. Cr(VI) can cause many health problems. Direct contact with skin and eye causes irritation in the digestive and respiratory tract, characterized by a burning sensation, sneezing and coughing. Based on the Indonesian regulation (PP No. 82 of 2001), the maximum concentration of total chromium in drinking water and fishery water is 0.05 mg/L. Removal of Cr(VI) has been devoted through such process including ion exchange, adsorption using activated carbon, and reduction with the help of bacteria. However, such methods have disadvantages of the need for very high energy and/or chemicals. This weakness makes the photocatalytic method more prospective and superior for its application to reduce Cr(VI) compared to another methods [2].

The photocatalytic method can be applied to reduce Cr(VI) by the addition of Fe(III) catalysts in a homogeneous state. The catalysts have high activity and selectivity. Based on [3], the Fe catalyst can reduce Cr(VI). Another study prove that using a catalyst Fe(III) with the addition of oxalic acid and citric acid can reduce Cr(VI) ions [2]. Fe(III) forms a complex compound with organic substances such as oxalic acid. This compound reduces Cr(VI) to Cr(III) while the organic substances are reduced to carboxyl radicals and start the new cycle. The oxalic acid is a derivative of carboxylic acid, which contains 2 carboxyl groups located at the ends of a straight carbon chain that has a molecular formula namely $C_2H_2O_4$.

Oxalic acid can be isolated from corn stalk biomass. Corn stalk contains cellulose in large enough quantities. Cellulose is a long chain carbon compound which can be cracked into simpler carbon compounds using strong alkali [4]. One of the carbon compounds produced in the cracking process is oxalic acid. According to previous studies, oxalic acid obtained from rice husk as much as 15% [5]. Therefore, the aims of this present research were to isolate oxalic acid from corn stalk biomass, determine its characteristics, and investigate the efficiency of photocatalytic reduction of Cr(VI) using the isolated oxalic acid.

2. Research Methods

2.1. Materials and Apparatus

The materials used in this study included corn stalk that obtained from Banyuwangi, sodium hydroxide (Emsure-Merck), calcium chloride (Merck), sulfuric acid (AR), nitric acid (Merck), diphenylcarbazide (Merck), acetone (AR), potassium dichromate (Merck).

The apparatus used in this study were a set of glassware, reactors, ovens, thermometers, UV-Vis (Genesys 10S VV-Vis Spectrophotometer). The photocatalytic reduction was conducted using a reactor (Figure 1) that has a size of 110 x 35 x 50 cm. The reactor box (1) was made from wood and was equipped with a magnetic stirrer (2), sample holder (3), beaker glass (4), and UV lights of 30 watts (5). The lights and magnetic stirrer were connected with electric power using a current connection (6).

2.2. Isolation of Oxalic Acid from Corn Stalk

Corn stalk was cleaned from dirt, dried, and then grinded. About 15 g corn stalk powder was put in a two neck flask and then added with 250 mL NaOH 2 M. The mixture was then heated at 90^{0}C while stirring using magnetic stirrer for 3 hours. After heating was completed, the mixture was then cooled and filtered. The resulting filtrate was diluted to 400 mL [4].

As much as 25 mL of the filtrate was added with 25 mL of $CaCl_2$ 1M to form calcium oxalate deposits. The calcium oxalate was dissolved with 100 mL of H_2SO_4 2M. Afterwards, the solution was heated to 70°C for 1 hour. After being cooled for 24 hours in the refrigerator, oxalic acid deposits were formed in the form of white needle crystals. The oxalic acid crystals were then separated from the liquid and dried at 70 ^{0}C for 30 minutes. [4]. Afterwards, the functional groups and melting point of the oxalic acid crystals were characterized using FTIR spectroscopy and melting point apparatus. The yield of oxalic acid was calculated using equation (1).

$$\text{yield} = \frac{\text{mass of oxalic acid}}{\text{mass of corn stalk}} \times 100\% \quad (1)$$

2.3. Photocatalytic reduction of Cr(VI)

As much as 30 mg of oxalic acid was mixed with 25 mL of Cr(VI) solution of 300 mg/L, 2 mL of Fe(III) 100 mg/L in a 50 ml flask. The mixture was then irradiated with 60 watts UV lights in the reactor for 120 minutes. The amount of Cr(VI) reduced can be determined from the difference between the initial and final concentration of Cr(VI). Cr(VI) concentrations can be determined spectrophotometrically through the formation of complex compounds with diphenyl carbazide. The complex compounds formed were measured at a wavelength of 540 nm.

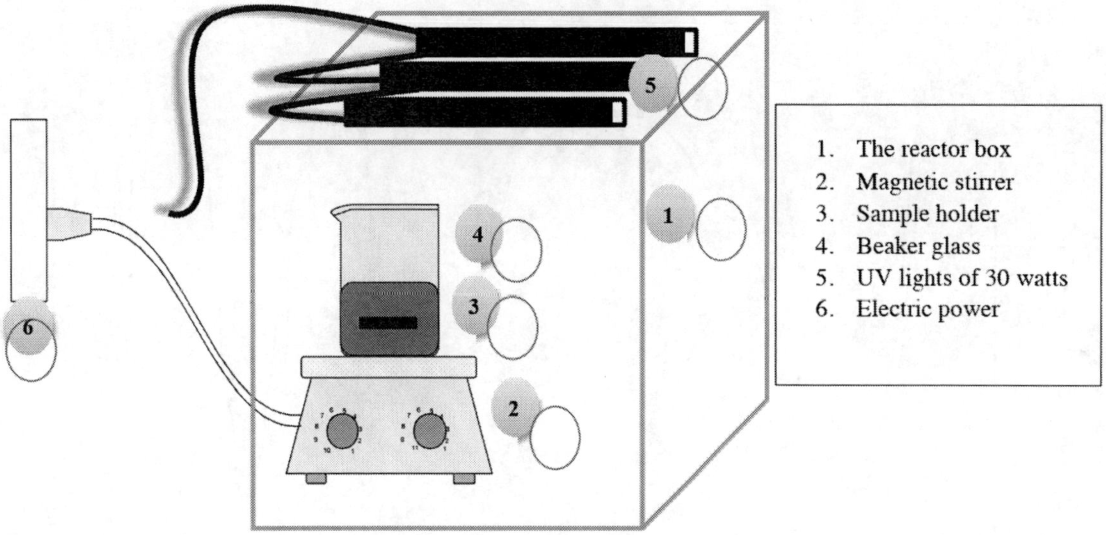

Figure 1. Photocatalytic Reactor.

3. Result And Discussion

3.1. Characteristic of Isolated Oxalic Acid

Oxalic acid could be isolated from cellulose of biomass such as corn stalk. Oxalic acid produced from the cellulose of Banyuwangi corn stalk was about 30% (calculated using Eq.1). Characterization of the isolated oxalic acid was carried out for its functional groups and melting point. The functional groups of the oxalic acid crystal were characterized using FTIR spectrophotometry with the spectra as shown in Figure 2. It can be seen from Figure 2 that the spectra of isolated oxalic acid from corn stalk was identical with the spectra of oxalic acid standard although there is a slightly shifted in the wave numbers. The standard of oxalic acid spectra has strong and sharp absorption of hydroxyl group at 3441.53 cm^{-1}, the carbonyl group at 1694.15 cm^{-1} , and the vibrational strain of the C-O group at wave number of 1258.26 cm^{-1}. Meanwhile, the isolated oxalic acid from corn stalk contained hydroxyl group vibrations at 3411.84 cm^{-1}, carbonyl group vibration at 1627.81 cm^{-1}, and strong vibration of C-O group at 1124.42 cm^{-1}. Based on the similarity of vibrations wave of hydroxyl and carbonyl groups between standard oxalic acid and the isolated oxalic acid from corn stalk powder, it can be concluded that the resulting compound is oxalic acid.

The analysis of the melting point test for isolated oxalic acid has been done using the Melting Point Apparatus. The melting point of oxalic acid was observed at 100.3^{0}C. The melting point of oxalic acid standard is between 106 and 107 ^{0}C. The difference in melting point is probably due to the results of crystallization which are not yet pure or there are still many impurities, this is supported by the results of the FTIR analysis as shown in Figure 2.

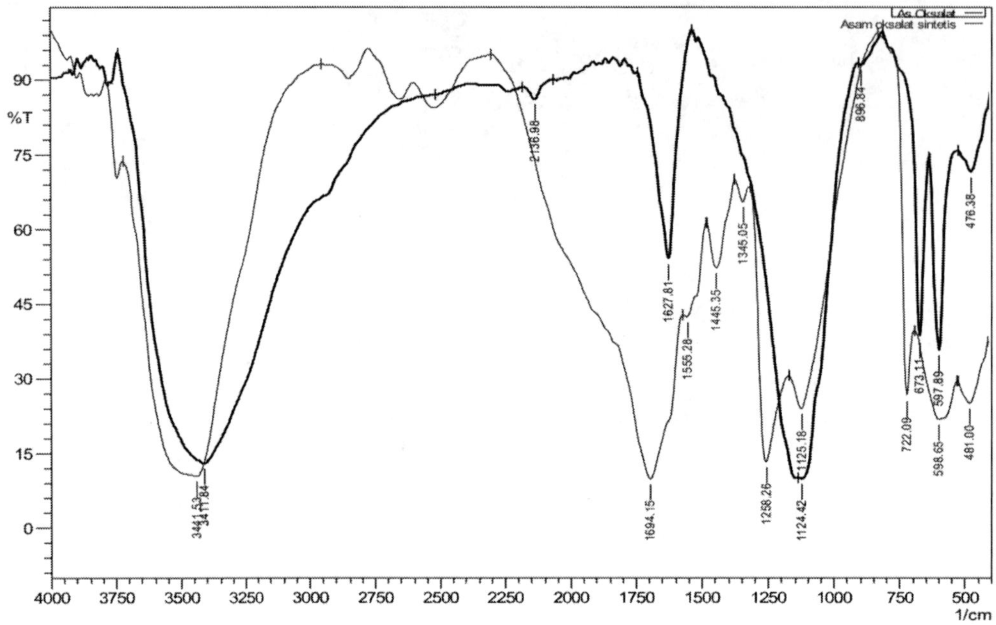

Figure 2. FTIR spectrum of oxalic acid standard (blue) compares to the isolated oxalic acid (black).

3.2. Photocatalytic Reduction of Cr(VI) with the isolated Oxalic Acid

The amount of oxalic acid as the oxidant in the photocatalytic reduction of Cr(VI) affects the efficiency of reduction. In this study, the dose of oxidant was varied to reduce Cr(VI) at the initial concentration of 300 mg/L.

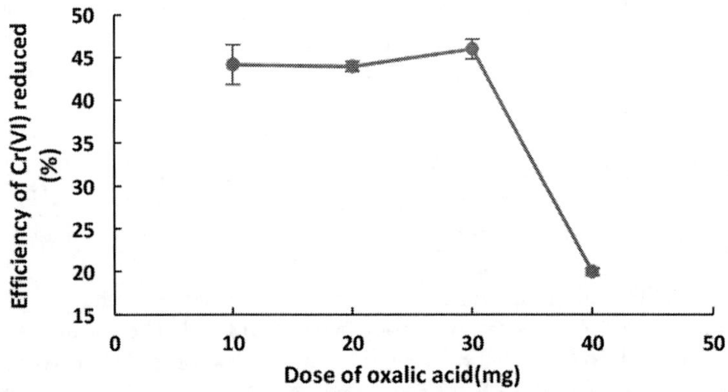

Figure 3. The effect of oxidant dose to the efficiency of Cr(VI) photoreduction. (Photoreduction was conducted using 25 ml Cr(VI) 300 mg/l mixed with 2 ml Fe(III) 100 mg/l as catalyst and stirred for 120 minutes under UV lights of 60 W).

A shown in Figure 3, the oxidant dose did not affect the efficiency of Cr(VI) reduction significantly at a low mass of 10-30 mg. The optimum dose of oxalic acid to reduce 300 mg/L Cr(VI) was 30 mg with the percentage of Cr(VI) reduction was up to 45.97%. Increasing the mass of the oxidant has decreased the efficiency of Cr(VI) reduction. Oxalic acid as much as 0.04 g only reduced Cr(VI) up to 19.95%. This is because the more dose of oxalic acid added, the solution becomes more turbid, so the intensity of the light used for illuminating the Fe decreases. It causes a decrease in reduced Cr(VI) ion [7]. The process of photocatalytic reduction of Cr(VI) ions using Fe(III) as a catalyst can be assumed to follow the scheme as shown in Figure 4. At first, oxalate forms a Fe(III)-oxalate complex compound. Irradiation of this compound results in Fe (II) and carbonyl radicals. Both compounds reduce Cr(VI) to Cr (III). The secondary photochemical process of carbonyl radical produces hydroxyl radicals, oxygen radicals, and peroxide radicals. These radicals are strong oxidants that contribute to the reduction of Cr(VI) [8].

Possible reactions of Cr(VI) reduction is [2]:

$$2HCrO_4^- + 3C_2O_4^{2-} + 14H^+ \xrightarrow{h\nu, Fe(II,III)} 2Cr^{3+} + 6CO_2 + 8H_2O \quad (2)$$

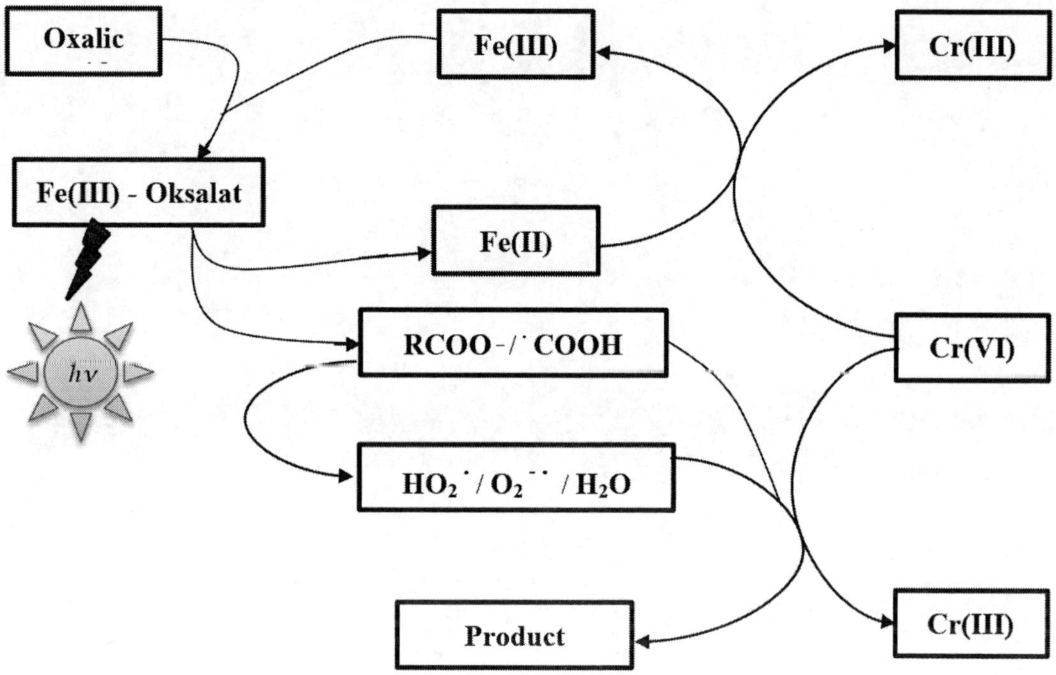

Figure 4. The mechanism of Cr(VI) photocatalytic reduction with oxalic acid as oxidant and Fe(III) as catalist[8].

4. Conclusion

Oxalic acid could be isolated from the corn stalk with the yield production was about 30%. The isolated oxalic acid has similar IR spectra as the oxalic acid standards with the functional groups available were hydroxyl and carbonyl groups. The lower melting point of isolated oxalic acid than the standard could be due to the presence of impurities. The isolated oxalic acid could be used as an oxidant for the photocatalytic reduction of Cr(VI). By using 30 mg the isolated oxalic acid, Cr(VI) at a concentration of 300 mg/L could be reduced for up to 45.9%.

References

[1] Juan J T, Maria A G, Marta I L 2004 *Environ. Sci. Technol* **38** 5 p. 1589–1594
[2] Stephan J H, Hans U L, Bruce R J 1997 *Environ. Sci. Technol* **31** 1 p. 160–170
[3] Robert J and George R H 1992 *Environ. Sci. Technol.* **26** 2 p. 307–312
[4] Afriandi, Fajri A, Idral A 2015 *Jom FTEKNIK* **3** 1 p. 1–7
[5] Elda M and Effan K 2008 *Journ. of Chem Eng. Sriwijaya University* **20** 2 p. 55–63
[6] Pamilia C, Oktarisky, Rangga W 2008 *Journ. of Chem Eng. Sriwijaya University* **15** 4 p. 36-43
[7] Raudhotul I, Sri W, Rosyid R 2019 *Jurnal Crystal : Publikasi Penelitian Kimia dan Terapannya* **1** 01 p. 55–66

IOP Conf. Series: Materials Science and Engineering **833** (2020) 012029 doi:10.1088/1757-899X/833/1/012029

[8] Xianghea F, Shimin D, Lixian Z 2012 *Korean Chem. Soc* **33** 11 p. 3691–3695

Acknowledgments

The author thanks the Faculty of Science, Brawijaya University, for Research Funding below "Hibah Guru Besar dan Doktor Tahun 2019." Scheme based on SK Dekan FMIPA No. 1597 Tahun 2019.

Mechanical Properties of Alginate Based Biopolymers as Wound Dressing Material

Ane Nurjanah[1*], Muhammad Bachri Amran[1], Rusnadi[1]

[1]Department of Chemistry, Faculty of Mathematics and Natural Science, Institut Teknologi Bandung.

*Corresponding author: 12001ane@gmail.com

Abstract. The utilization of biopolymers has been widely accepted in many areas due to its eco-friendly. Especially in medical care, biopolymers are receiving great attention and are considered as a potential for wound healing. Wound dressing material is one of medical needs and the demand continue to raise. One of the most widely used biopolymers for wound dressing is alginate. To improve alginate properties as wound dressing material, various alginate modifications with nanoparticles or synthetic polymers have been developed. In this paper alginate biopolymer will be modified with ZnO nanoparticles and synthetic polymer, such as poly (ethylene glycol) dimethacrylate (PEGDMA) as one of the candidates of wound dressing material. Alginate can go through swelling and because of its crystallinity is low so that it affects its mechanical properties. In order to increase the compatibility, ZnO needs to be modified with PEGDMA which is crosslink with alginate to strengthen the bonding network of materials. Characterization of alginate-based wound dressing material was performed using FTIR, XRD and SEM. In addition, optimization tests of mechanical and antibacterial resistance properties were performed to meet detailed studies of the effect of ZnO nanoparticles and PEGDMA interpenetrated into alginate to the properties of the wound dressing material.

1. Introduction

Alginate is a linear biopolymer consisting of two monomeric units, namely D-manuronic acid and L-guluronic acid. Alginate is found in all types of brown algae (Phaeophyta) which is one of the main components making up the cell wall. Alginate has a characteristic as wound dressing material because it is non-toxic, biocompatible and biodegradable. In addition, some wound dressing materials use alginate as raw material because its increase the growth of new cell tissue and reduce inflammation, thereby accelerating wound healing. However, alginate has relatively low mechanical resistance and antibacterial properties [1]. Thus, further modifications are needed, for example with nanoparticles or synthetic polymers. ZnO nanoparticle (zinc oxide) is one of the materials that have been widely synthesized into nanoparticles [3,8,9]. This material has potency to be applied in the biomedical field because of its good antibacterial properties.

In this paper, the characterization of ZnO nanoparticles was carried out and modified [3,8] with alginates and synthetic polymers, poly (ethylene glycol) dimethacrylate (PEGDMA) as a candidate for wound dressing material. Alginate can experience swelling and low crystallinity so it affects the mechanical properties. Interpenetration of ZnO nanoparticles into the alginate matrix is able to limit the volume of the alginate polymer chain with the surface of ZnO to improve its mechanical and antibacterial properties. To increase the compatibility, ZnO needs to be modified by synthetic polymers that can cross-link with alginates so that they are able to strengthen the bonding network of

IPN material [3]. The characterization of alginate-based wound dressing material was carried out using FTIR and SEM-EDS. Its mechanical resistance properties were characterized and initial studies on its antibacterial properties was carried out.

2. Material and Methods

2.1. Material

The equipment used in this study are measuring flask, Erlenmeyer flask, measuring cup, funnel, drop pipette, analytical balance, chemical beaker, drip plate, reflux tool set, mechanical stirrer, desiccator, incubator, X-ray diffractometer, Fourier Transform Infrared Spectrophotometer (FTIR), Scanning Electron Microscope (SEM), Transmission Electron Microscope (TEM), UV-Vis Spectrophotometer and Instron IX series testing machines. The materials used are sodium alginate, aquades, zinc nitrate hexahydrate ($Zn(NO_3)_2.6H_2O$), sodium hydroxide, hexamethylene diisocyanate (HDI), ethylene glycol dimethacrylate (EGDMA), dibutyltin dilaurate (DBTDL), nitrogen gas, and Favigraph.

2.2. Methods

Sodium alginate as much as 0,75 grams was dissolved in 100 mL of water, then added 0,25 grams of ZnO and PEGDMA synthesis results with a ratio of 1: 1 to form a gel. Alginate-ZnO-PEGDMA material that has been formed into a membrane is then characterized by FTIR and SEM and tested for its mechanical strength using a Favigraph.

3. Results

3.1. Synthesis and characterization of Alginate-ZnO-PEGDMA

PEGDMA has been used in medical applications, but as a synthetic polymer in wound dressing material, it has never been done. Wang (2016) used synthetic poly (ethylene glycol) methacrylate (PEGMA) synthetic polymers modified with agarose and ZnO to produce wound dressing material [8]. Alginate, ZnO and EGDMA concentrations used in the synthesis of this IPN material were 1 mmol, 4 mmol and 20 mmol. AZP material was obtained from synthesis in the form of white powder as much as 4.9959 grams. The following image of AZP material synthesized is shown in Figure 1:

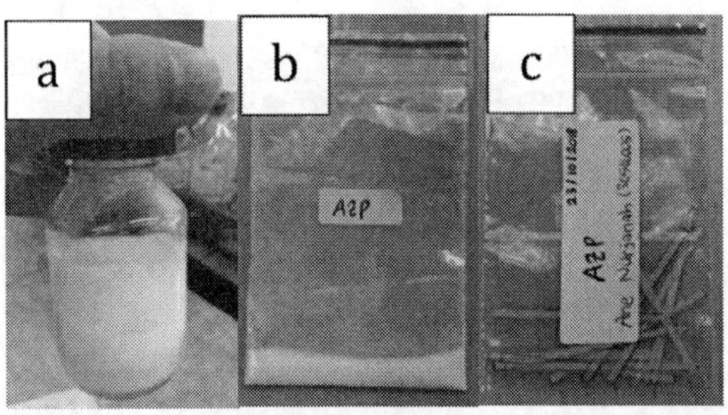

Figure 1. Synthesis of Alginate-ZnO-PEGDMA (AZP) IPN material: (a) gel form, (b) powder form, and (c) membrane form.

Material characterization was carried out using FTIR and SEM-EDS. The mechanical strength and antibacterial properties were characterized to be studied as wound dressing material. The following is a spectrum of IR Alginate-ZnO (AZ) and Alginate-ZnO-PEGDMA (AZP) synthesis results:

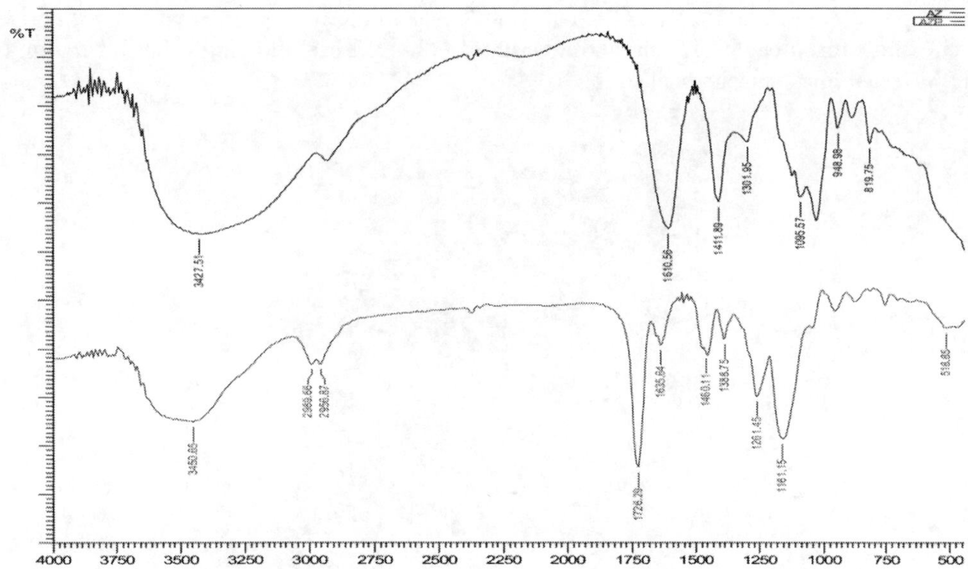

Figure 2. Alginate-ZnO (AZ) and IR spectra.

3.2. Synthesis of Alginate-ZnO-PEGDMA (AZP)

Based on the Figure 3, the morphology of AZP material was successfully observed at 10000x and 40000x magnification. ZnO sample particle distribution seems to have non-uniform shape. This is because the whole sample can be said experiencing agglomeration between its constituent components.

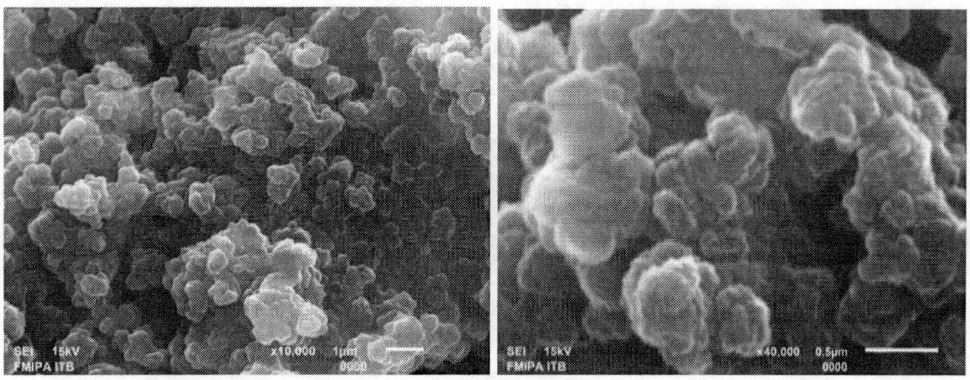

Figure 3. Morphology of Alginate-ZnO PEGDMA (AZP) at magnifications of 10000x and 40000x.

SEM imaging results generally describe the synthesis of AZP which has a spherical shape "like a flower" with a gap between the particles looks quite clear. The flower-like form of AZP morphology may be displayed by ZnO nanoparticles, as in a study reported by Mun Lam (2017) regarding the synthesis of Ag-ZnO composite material whose morphology is flower-shaped with several petals growing radially from the center and well dispersed.

3.3. Synthesis and characterization of Alginate-ZnO-PEGDMA

Analysis of mechanical resistance properties was studied to determine the performance of the material. The instrument used in the analysis of the nature of this resistance was used Favigraph with a distance of 2.5 mm and a tensile speed of 10 mm/minute. The samples tested were 2, namely Alginate-ZnO

(AZ) and Alginate-ZnO-PEGDMA (AZP). The sample was in the form of gel that has been dried into a thinly cut film with a length of 3 cm and a width of 0.2 cm. The following is an illustration of the sample to be tested for its mechanical strength.

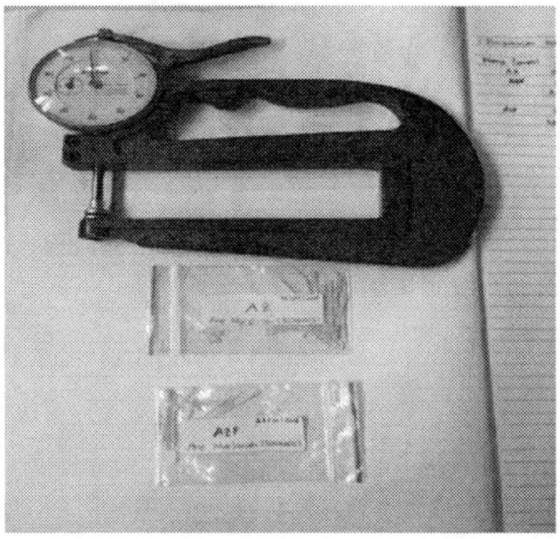

Figure 4. AZ (Alginate-ZnO) and AZP (Alginate-ZnO-PEGDMA) samples to be tested for mechanical strength.

AZP material has the tensile strength value that is closest to the average value of the mechanical resistance properties of AZP material. We plot the linear lines, x (E yield) and y (F yield) at the points of 0.96 and 300.58. Then, the value of tensile strength and elasticity were obtained and has been converted to 26.27 MPa.

Table 1. Mechanical strength of AZP material

Number of Sample	Tensile Strength (MPa)	Elasticity (MPa)
1	20.47	2477.19
2	20.60	2726.81
3	22.97	1839.88
4	26.63	2785.78
5	26.84	2215.83
6	27.65	1900.54
7	28.53	1890.13
8	29.79	2873.97
9	32.99	1687.53

The converted elasticity values can be obtained by making a linear line plot so that the E yield and F yield can be determined. AZP material with sample is the closest to its elasticity value with an average AZP material elasticity value of 2266.40 MPa. The elasticity value was obtained by dividing the value of the tensile strength by the value of E yield. AZP material (E yield, F yield) linear line plot with sample code 5 is at 1.21 and 302.92. After the tensile strength value is converted into MPa units, then the AZP material elasticity value is obtained with sample code 5 of 2215.83 MPa.

4. Discussion

4.1 Synthesis and characterization of Alginate-ZnO-PEGDMA

Based on Table 2, the characterization with ftir for alginate-ZnO (AZ) and alginate-ZnO-PEGDMA (AZP) shows strong and wide absorption for oh groups which are seen at wave number 3427.51 cm-1 and when there is interaction with pegdma a wave number shifted to 3450.65 cm^{-1}. The presence of significant pegdma can be seen in coo- uptake at wave number 1726.29 cm^{-1}, while its interaction with az material which is indicated by shifting some of its main peaks can be described as follows, the c = c alkene peak of the az material experiences a shift in the wave number and decreases its intensity at the wave number 1635.64 cm^{-1} which indicates that PEGDMA has interacted with az. The presence of pegdma influences the absorption of oh groups in AZ material as a result of interactions between az molecules with PEGDMA at wave numbers of 1460.11 cm^{-1} and 1386.75 cm^{-1}. The interaction of pegdma with AZ material is more dominant in shifting the wave numbers of important groups of AZ material such as the coc ether uptake from wave number 1095.57 cm^{-1} to 1261.45 cm^{-1}, then the wavelength for ZnO absorption shifts to 1161.15 cm^{-1} [3].

Table 2. Identification of functional groups in the IR spectrum of Alginate-ZnO-PEGDMA (AZP)

Number of Waves (cm-1)	Identification Functional Group
3450.65	O-H
2956.87-2989.66	CH alkane
1726.29	C-O-O-
1635.64	C=C alkene
1386.75; 1460.11	O-H
1261.45	C-O-C ether
1161.15	ZnO

In general, the azp particle size distribution produced varies with a range of 105-171 nm. Vasquez et al (2016) state that particles having a diameter <1.000 nm can be accepted as nano-sized carriers that can be used in the pharmaceutical industry. The process of material synthesis using natural materials is greatly influenced by the concentration used [3,8]. The greater the concentration of alginate used, the greater the average AZP particle size produced. because the more the concentration of alginate, the more functional groups available to reduce metal ions, which then produce a large number of particles formed so that the distance between particles is smaller, thus allowing interactions between particles to form particles with a larger size.

4.2 Mechanical properties of Alginate-ZnO-PEGDMA material

Table 1 shows the tensile strength and elasticity data of AZP material. The tensile strength of AZP material is increasing with a relatively irregular increase between one sample code sheet and the next sample code. The increase in tensile strength values lasted until the code snippet of sample 9, meaning that the AZP material samples used in testing the mechanical properties of this sheet were 9 sheets. Unlike the tensile strength values, AZP material elasticity data display random values on each sheet. There are even several sheets of samples that are significantly different. The difference in tensile strength values between the sample sheets can also be caused by the inhomogeneity that occurs in AZP material dispersion [8].

5. Conclusions

ZnO nanoparticles have been successfully synthesized and characterized. Their particle size was determined so that they can be modified with alginate and poly (ethylene glycol) dimethacrylate (PEGDMA) and can be used as wound dressing material. Alginate-ZnO-PEGDMA (AZP) material has been successfully synthesized and characterized. Its mechanical resistance and antibacterial properties have been characterized. The study of mechanical properties was confirmed through tensile strength and elasticity of 26.28 MPa and 2,266.41 MPa, respectively.

Acknowledgments
The research was conducted at the Analytical Chemistry Laboratory of the Bandung Institute of Technology, which was funded by the Education Fund Management Institute (LPDP) of the Ministry of Finance of the Republic of Indonesia. Testing the nature of mechanical endurance was carried out at the Central Textile Office of the Ministry of Industry.

References
[1] Chiang C. Y and Chu C. C 2015 *J. Carbohydrate Polymers.* **119** 18-25
[2] Da Cunha C. B, Klumpers D. D, Li W.A, Koshy S. T, Weaver J. C, Chaudhuri O, Granja P. L and Mooney D. J 2014 *J. Biomaterial.* **35** 8927-8936
[3] Khalid A, Khan R, Ui-Islam M, Khan T and Wahid F 2017 *J. Carbohydrate Polymers.* **164** 214-221.
[4] Liu Y and Kim H. I 2012 *J. Carbohydrate Polymers.* **89** 111–116
[5] Lu G, Ling K, Zhao P, Xu Z, Deng C, Zheng H, Huang J and Chen J. 2010 *Wound Repair Regen.* **18** 70–79.
[6] Stickler D. J, Lear J. C, Morris N. S, Macleod S. M, Downer A, Cadd D. H and Feast W. J. 2006 *J. Appl. Microbiol.* **100** 1028–1033.
[7] Vicentini D. S, Jr Smania A and Laranjeira M. C. M 2010 *Mater. Sci. Eng.* B **30** 503–508.
[8] Wang J, H. Hu Z, Yang J and Wei J. Li. 2016 *J. Material Science and Engineering.* C **61** 376-386.
[9] Wang Z, Kale G. M, Yuan Q and Ghadiri M 2012 *Rsc Advances.* **2** 9993-9997.

Antifungal Activity from Co-Culture of a Local Fungus of Tropical Peat Swamp Soil, *Penicillium* sp. LBKURCC34 with Gram-Negative and Gram-Positive Bacteria

Yuana Nurulita*, Yuharmen, Andy Dahliati, Yum Eryanti, Supridianto, Khairulinas, Yuli Haryani, & Titania Tjandrawati Nugroho

Chemistry Department, Faculty of Mathematics and Natural Sciences, University of Riau, Pekanbaru 28293

*Corresponding author: ynurulita@lecturer.unri.ac.id

Abstract. Microorganism is important producer of novel bioactive natural products, particularly in the field of drug discovery. Co-culture methods is one of powerful emerging tools for enhancing the chemical diversity of microorganisms. This research used the local culture collection of *Penicillium* sp. LBKURCC34, the fungus isolated from peat soil of primary forest at Giam Siak Kecil Bukit Batu (GSKBB) - Biosphere Reserve in Riau Province, to produce secondary metabolites secreted to their growth media that was cultivated by two different Gram of bacterial pathogen, *Escherichia coli* and *Staphylococcus aureus*. The 14 days fermentation was carried out and the media was extracted with ethyl acetate. The ethyl acetate crude extract was evaporated, then the concentrate dissolved in methanol. Antifungal, *Candida albicans* test was performed by the disc diffusion and the resazurin-based turbidimetric method. This study found that the crude extract of the co-culture with *S. aureus* could inhibit *C. albicans* growth, while that extract of the co-culture with *E.coli* could not do. The value of minimum inhibitory concentration (MIC) of the potential extract was less than the positive control, Ketoconazole. It only has potency as bacteriostatic extract.

1. Introduction

Recently, the antifungal resistance is an worldwide problem due to the increase in microbial antibiotic resistance in clinical settings and the limited findings of new antibiotic compounds. Resistance to antifungals is a major health problem with an antifungal resistant rate of dispersion not supported by the discovery of new antifungal compounds [1]. Pathogenic fungi are stronger and existing antifungal compounds no longer have any effect. the resazurin-based turbidimetric method.

Natural products are the main source for developing of drugs and other chemical agents. One of the resourceful method to find out new native compounds especially from microorganisms, is by mixing several microorganisms in one culture production media (called co-culture method). This method can increase the productivity of known compounds [2] as well as discover new compounds. Furthermore, on the side of biotechnology, co-culture fermentation could increase the yield as well as improve the control of product qualities and the prospect of applying cheaper substrates [3]. Several compounds have been known as an active compound that produced by co-culture fermentation such as emericellamides A and B [4], mycotoxin [5], and 2,5-diketopiperazines [6].

In this study, co-culture of potential local isolate of peat swamp soil, *Penicillium* sp. LBKURCC34 was reported with modification of fermentation media using co-culture with two different Gram of bacterial pathogen, *Staphylococcus aureus* (represent as Gram-positive bacterium) and *Escherichia coli* (represent as Gram-negative bacterium). The activity of antifungal, *Candida albicans* was assessed using diffusion method (disc) and the potential extract was measured to find out its minimum inhibitory concentration (MIC) and minimum fungicidal concentration (MFC).

2. Materials and methods

2.1. Antibiotics production of fungal strain stimulated by pathogen bacteria

Antibiotics production was performed using co-culture of fresh isolates of *Penicillium sp.* LBKURCC34 and bacterial pathogen of *S. aureus* and *E. coli* as elicitors. Generally, the method was based on Lee et al., [7] with some modification. Firstly, fresh spores of *Penicilliun sp.* LBKURCC34 were inoculated into flasks with an initial spore concentration of 7 x 10^{12} spores/ml (OD 0.34). The flasks contained Potato Dextrose (PD) medium (50 ml). This initial inoculum was incubated at 30°C for 7 days on a rotary shaker (125-150 rpm). At day 7, these two flasks of 50 ml PD initial inoculum and biomass that had grown were transferred to new flasks containing 1 liter of production media. The composition of 1 liter of production media (pH 5.6-5.8) included glucose 20 g, yeast extract 5 g, peptone 5 g, $MgSO_4$ 0.5 g, and KH_2PO_4 1 g. Incubation was continued with shaking for 5 minutes everyday at 30°C for 14 days incubation. Pathogen bacterial elicitors were added on Day 3 which was prepared by 1 loop of pathogen colonies of *S. aureus* and *E. coli* had been inoculated into tubes with 5 ml of nutrient broth (NB), then incubated for 18-24 hours until they reached OD 1.0.

2.2. Extraction

After 14 days of fermentation, the mycelium was separated from the medium (Inoculum A for *S. aureus* addition and Inoculum B for *E. coli* addition) by filtration using a Whatman® GF/C filter (CAT No. 1822-055). The filtrates were extracted with ethyl acetate (media filtrates – ethyl acetate 2:1). The ethyl acetate extracts were dried with evaporation at room temperature in a rotary vacuum evaporator. The residue obtained from every 1 l fermentation media was dissolved in 1 ml methanol (MeOH). It resulted in Extract A (co-culture with *S. aureus*) and Extract B (co-culture with *E. coli*)

2.3. Antifungal activity screening against Candida albicans, using the disc diffusion and the resazurin-based turbidometric methods.

2.3.1. The disc diffusion. Ethyl acetate extracts of co-culture: Extract A and Extract B were dissolved in MeOH (Merck 1.06009.1000, pa.) based on the concentration calculation. That extracts were applied to a sterile 6-mm antibiotic blank disc and left to dry at room temperature (30°C), before transferring to agar plate that have been spread with *Candida albicans*. The cultures were incubated at 37°C for 1-3 days, and the inhibition zones were measured (mm, started the edge of disc). The assays were performed in triplicates. As positive controls, ten µl of a 3 µg/µl Amoxicillin (Sigma A8523-5G) in MeOH were used. As negative controls 10 µl of the solvent MeOH were used.

2.3.2. The resazurin-based turbidometric method: This method was based on Teh el al. [8]. Resazurin solution was prepared by dissolving 0.0675 g in 10 ml sterile distilled water, then mixed by vortex mixer to ensure homogeneity. The solution was sterilised using a syringe filter 0.2 µm and kept in a brown bottle. In a 96-well flat-bottom microtiter plate, for pathogen fungal culture, the assay composed triplicate of Extract A and B, control positive amoxicillin, control positive tetracycline, and control negative dimethyl sulfoxide (DMSO). In this project, each extract was used 12 concentration of 4000; 2000; 1000; 500; 250; 125; 62.5; 31.25; 15.62; 7.81; 3.91; and 1.95 µg/ml. Each well contained 140 µl muller hinton broth (MHB), sample 20 µl, pathogen bacteria 20 µl (in MHB after 1 day incubation with OD 0.1). The solution was homoganized. The next day, resazurin solution 20 µl was added, then the plate was incubated for 48-36 hours. The lowest concentration prior to colour change was considered as the minimum inhibitory concentration (MIC). The solution 100 µl on MIC well was spread on the

IOP Conf. Series: Materials Science and Engineering **833** (2020) 012031 doi:10.1088/1757-899X/833/1/012031

nutrient agar plate, then incubated for 2-5 days. Minimum fungicidal concentration (MFC) was considered to the concentration that there was no fungal growth on the plate.

3. Result and discussion

The development of the fermentation method employed more than one microbial strain to constitute synthetic co-culture in order to improve bioproduction has been applied in many biotechnology approaches and widely studied. It has been reported in several projects that the method has been successful to enhance metabolite productivity. Nonaka et al [2] combined *Penicillium pinophilum* FKI-5653 with *Trichoderma harzianum* FKI-5655. While Onaka et al., [9] mixed *Streptomyces* species with mycolic-acid containing bacteria, Schroeckh et al. [10] revealed that biosynthesis of archetypal polyketides in *Aspergillus nidulans* could be stimulated by 58 soil-dwelling actinomycetes. Furthermore the capacity of *Penicillium* sp. LBKURCC34 (local isolate from peat swamp soil of Riau), to produce active compounds among other isolates has also been proven [11]. To increase the productivity and the potency to find new compounds from this isolate, this study has tried to modify the fermentation method using elicitor biotic, bacterium of Gram-negative and Gram-positive. After 14 days of incubation in production media, bacteria biotic elicitor changed the appearance of fermentation media. Media *Penicillium* sp. LBKURCC34 in single cultivation showed clear yellowish media (Picture is not showed). Compared with that media, mixed of bacteria and *Penicillium* sp. LBKURCC34, resulted in cloudy yellowish media (Figure 1). Precipitation of bacteria could be found at the bottom of the flask. After 14 days of incubation, the media was extracted by ethyl acetate after filtration from mycelium.

Figure 1. Production media of co-culture *Penicillium* sp. LBKURCC34 with pathogen bacteria. (a) is co-culture with *Staphylococcus aureus* and (b) is co-culture with *Escherichia coli.*

Dilution and diffusion methods are easy. They provide a high level of accuracy and reproducible result for antimicrobial tests [12]. Dilution method, fungal cells irreversibly reduced the blue dye resazurin to a pink resofurin by oxidoreductase [13]. In the investigation of the activity against *Candida albicans*, Extract A that is formed from co-culture of *Penicillium* sp. LBKURCC34 with *S. aureus* inhibited stronger than Extract B (co-culture of *Penicillium* sp. LBKURCC34 with *E. coli*) (Table 1). At the same concentration (38 µg/disc), this value was slightly lower than the inhibition value of positive control, amoxicillin, because amoxicillin actually has a better activity as antibacterial, not antifungal. The obtained results of the antifungal activity using diffusion and dilution method (MIC value, Table 2) confirmed that there was an activity on Extract A. Table 2 demonstrates the MIC of the extracts and the positive controls against *C. albicans*. It was observed higher MIC values of Extract A compared with Extract B and the positive control, amoxicillin and ketoconazole. Although the mechanism is not clear yet, it seems that Extract A perhaps could damage the cell walls of fungi [14].

Table 1. Antifungal activity of the extracts toward *Candida albicans*.

Name of extract	Inhibition (mm)	
	Extract A (Co-culture with *S. aureus*)	Extract A (Co-culture with *S. aureus*)
Extract concentration 1 (57 µg/disc)	8.50 ± 0.43	1.13 ± 0.02
Extract concentration 2 (38 µg/disc)	7.80 ± 0.53	1.01 ± 0.02
Extract concentration 3 (19 µg/disc)	0.20 ± 0.10	0.00 ± 0.00
Positive control, amoxicillin (38 µg/disc)	8.10 ± 0.00	
Positive control, ketoconazole (38 µg/disc)	25.90 ± 0.36	
Negative control (methanol)	0.00 ± 0.00	

Table 2. The values of MIC and MFC of the extract A towards *Candida albicans*.

Name of extract	MIC (µg/ml)	MFC (µg/ml)
Extract A	2000	>4000
Positive control, amoxicillin	>4000	>4000
Positive control, ketoconazole	4000	>4000

However, further investigation is needed to explore the potency of this extract as well as the method to confirm the promising of microbial isolate to contribute on finding new lead compounds as a new antifungal. A deeper study regarding this part is needed to investigate such as isolation and characterization of the active compounds and regulation of fungal – bacterial secondary metabolism in terms of the biosynthetic gene cluster (BCG) using genome mining method. Further work will be continuing to assess the potency of local fungus to produce candidate antifungal compounds.

4. Conclusion

The present investigation has shown that the co-culture fermentation of the fungi *Penicillium* sp. LBKURCC34 and *S. aureus* has the potency to produce bioactive as antibiotics, while the same method using *E. coli* as an elicitor biotic did not showed better potency. The research needs further investigation to identify the structure of active compounds using isolation and characterization of the active compounds.

Acknowledgment

The Ministry of Research, Technology and Higher Education of Indonesia has been supported this project by providing research funding through the Research Grants Program of Fiscal Year 2019 (Penelitian Dasar) under contract No. 775/UN.19.5.1.3/PT.01.03/2019 for Yuana Nurulita.

References

[1] Lin J, Nishino K, Roberts M C, Tolmasky M, Aminov R, and Zhang L 2015 *Front. Microbiol.* **59(1)** 5
[2] Nonaka K, Abe T, Iwatsuki M, Mori M, Yamamoto T, Shiomi K, Omura S, and Masuma R 2011 *J. Antibiot.* **64** 769
[3] Bader J, Mast-Gerlach E, Popovic M K, Bajpai R, and Stahl U 2010 *J. Appl. Microbiol.* **109** 371
[4] Oh D, Kauffman C A, Jensen P R, and Fenical W 2007 *J. Nat. Prod.* **70** 525
[5] Sweeney M J and Dobson A D W 1998 *Int. J. Food Microbiol.* **43** 141
[6] Yu L, Ding W, Wang Q, Ma Z, Xu X, Zhao X and Chen Z 2017 *Tetrahedron Lett.* **73** 907
[7] Lee H J, Lee J H, Hwang B Y, Kim H S and Lee J J 2001 *Arch. Pharm. Res.* **24(5)** 397
[8] Teh C H, Nazni W A, Nurulhusna A H, Norazah A and Lee H L 2017 *BMC Microbiol.* **17** 36
[9] Onaka H, Mori Y, Igarashi Y, and Furumai T 2011 *Appl. Environ. Microbiol.* **77(2)** 400

IOP Conf. Series: Materials Science and Engineering **833** (2020) 012031 doi:10.1088/1757-899X/833/1/012031

[10] Schroeckh V, Scherlach K, Nutzmann H-W, Shelest E, Schmidt-Heck W, Schuemann J, Martin K, Hertweck C, and Brakhage A A 2009 *PNAS* **106(34)** 14558
[11] Nurulita Y, Yuharmen, and Dahliaty A 2017 *Laporan penelitian LPPM Universitas Riau*
[12] Yoshida S, Kasuga S, Hayashi N, Ushiroguchi T, Matsuura H, and Nakagawa S 1987 *Appl. Environ. Microbio.* **53** 615
[13] Elshikh M, Ahmed S, Funston S, Dunlop P, McGaw M, Marchant R, and Banat I M 2016 *Biotechnol Lett,* **38** 1015
[14] Amsler K, Santoro C, Feleno B, Bush K, and Flamm R, 2010 *J. Clin. Microbiol* **48** 3353

Essential Oil Extraction of *Cananga odorata* Flowers using Hydrodistillation and Steam-Water Distillation Processes

Ika Oktavianawati

Dept. Chemistry, Fac. Math & Science, University of Jember

*Corresponding author: ika.fmipa@unej.ac.id

Abstract. This paper discussed about the effect of distillation processes in the extraction of cananga essential oil from *Cananga odorata*.In this research, cananga oils were extracted by steam-water distillation and hydrodistillation processes. The equipment used in this current research is a distillation home scale set-up plant which further will be applied on cananga local farmer in Sukorambi region, Jember district. Therefore, preliminary optimization process for this distillation equipment was required. The distillation batch (retort) contained 2 kg of cananga flowers and 20 L of water. Based on research result, the distillation equipment could run hydrodistillation process for 3 hours, while in steam-water distillation set-up could only run for 2 hours. The volume of water in distillation batch was a limiting factor in both processes. In general, the yield of essential oil was increase along the longer of distillation time that have been applied. However, the result showed that essential oil from hydrodistillation for 1 hour had the best quality compared to other distillation conditions, in terms of its volatile compound contents. Oxygenated hydrocarbons content in cananga oil from 1 hour hydrodistillation process (56.5%) was highest compared to 2 and 3 hour hydrodistillation (35.48% and 22.36%, respectively), and even to essential oil from steam-water distillation process (43.25%). Generally, extra grade cananga oil was markedly by the high content of esters, and linalool, but low in β-caryophyllene content. These characteristics could be found in cananga oil obatined from hydrodistillation process for 1 hour which contained major volatile compounds, i.e. geranyl acetate (16.58%), benzyl benzoate (13.38%), β-caryophyllene (16.35%), germacrene D (15.3%), and farnesol (9.54%). However, linalool was only found in cananga oil from steam-water distillation. Therefore the current distillation condition, hydrodistillation for 1 hour, would be applied in distillation of cananga oil at Jember local community.

1. Introduction

Cananga is a kind of tree native to parts of Asia, including Indonesia. This plant is known for its two famous variaties, Java cananga (*Cananga odoratum forma macrophylla*) and ylang-ylang (*Cananga odoratum forma genuina*), besides another varian dwarf ylang-ylang, locally known as kenanga perdu (*Cananga odoratum forma fructicosa*).This cananga tree can be found in relatively huge amount at Karangpring village, Jember district. Previously, Karangpring village is wellknown for producing natural resource, rose. In fact, cananga flower is also to be another feasible local product that can be explored from this village.

Cananga oil is derived from flower part of cananga tree. This oil was produced from cananga flowers in a yield of 1.5% - 2.5% [1,2]. Cananga oil has some benefits such as aromatherapy, soul relaxation, lowering blood pressure tension, pharmaceutical, perfumery, cosmetics, anti aging, relieving acne, hair conditioner, flavoring agent, fixative and soap additive [2,3,4,5]. Some research have analyzed chemical compounds of cananga essential oil correlated to its bioactivities[6,7,8]. Oxygenated hydrocarbons are targetted to be extracted from plant tissues in order to provide higher bioactivities[9,10]. Extraction methods to obtain essential oil influence quantity and quality of the oil. Favorite conventional method to extract essential oil is distillation, which are in the form of hydrodistillation, steam-water distillation, and steam distillation. In addition, some research also have focused on essential oil fractionation to offer valuable and qualified essential oil in a series of fractionated groups toward time of distillation or purification by partition on non polar solvents [3,11-15].

Businesses of essential production in Indonesia are still require an encouragement due to some constrains on its marketing section, including low quality of the oil products, domestic marketing, and unstable prices [16].In 2003, price of cananga oil was around Rp 500.000,-/kg [1], but in a range of 2005-2014, it varied as much as €60 to €120/kg on the average of all cananga oil grades in Comoros [17]. In fact, US import of ylang ylang and cananga oil from Indonesia was about 1.5 tonnes (in 2005), and increased up to 3.2 tonnes (in 2007), but decreased significantly in the following years into 1.0 tonnes (in 2012). ITC also mentioned that premium ylang ylang oil imported from Madagascar to US showed a stable high price ranged from US$96 into US$225/kg in 2006 – 2012, but was dropped in the following year into US$149/kg [18]. Nowadays, cananga oil price is up to 1.2 million rupiahs per litre from local distillerin Indonesia [19] and can be marketed as much as US$ 175.00/kg by Ultra International B.V [20] in Netherland. This price is quite satisfying and stable compared to prices of other essential oil products.

Distillers of cananga oil consider two parameter on deciding quality of cananga oil, ester number and saponifikasi value. Distillation process is normally carried out in fractions to differentiate its qualities and usages. First oil obtained from distillation is called as extra grade which is comprised of 40% of total oil produced and used as based oil in high class perfumes such as Aqua di Gio (Giorgio Armani), Poison (Christian Dior) and Champ Elysee (Guerlin)[2]. The following fractions, first grade, second grade, and third grade, are cananga oils with lower quality in proposed usage as cosmetic based, soap, and detergen. Some research apply steam-water distillation on cananga oil and fractionate them in some uptake based on certain time interval, i.e in the first two hour of distillation, four hours from the first oil uptake, six hours from the second oil uptake, and 8 hours from the third oil uptake[11,12].It is correlated to previous report by Buccellato [3] which mentioned that ylang flowers resulted in ylang-ylang oil yields in a series of oil sampling (intake) time are 0.33% (1 hour distillation: extra grade), 0.33% (3 hours after first oil intake: first grade), 0.38% (5 hours after second oil intake: third grade); 1.41% (9 hours after third intake: fourth grade). This ylang-ylang extra grade is rich in linalool, which is responsible for citrus and floral aromas, methyl benzoate is responsible for the marker odor of ylang-ylang flower, benzyl acetate and geranyl acetate which contribute to fruity floral odor.

Extra grade cananga oil is majority consisted of esters and ethers also aromatics, while the lower grade contains mainly of sesquiterpenes. Ylang-ylang oil contains higher level of β-linalool, benzyl acetate and other esters [21], but lower level of β-caryophyllene compared to cananga oil. Third grade of ylang-ylang oil would be similar to the quality of cananga oil.In detail, Ekundayo [13] has compared the quality of ylang-ylang oil [22] and cananga oil [3]. Lawrence [22]identified linalool (1.0-19.03%), β-caryophyllene (6.8-10.6%), germacrene-D (10.23%), and esters of farnesyl acetate (1.88%), 3-methylbut-2-enyl acetate (0.13-3.2%), geranyl acetate (3.5-6.7%), linalyl acetate (~ 0.2%), cinnamyl acetate (1.01%), methyl benzoate (1.0-8.7%), methyl salicylate (0.1-1.34%), benzyl acetate (2.8-25.1%), benzyl benzoate (2.2-16.4%), benzyl salicylate (1.9-5.2%) to be present in ylang-ylang essential oil. In the other hand, Buccellato [3] emphasized that cananga oil has lower quality than ylang-ylang oil since it only contains linalool (1.7%), geranyl acetate (1.8%), benzyl benzoate (2.9%), benzyl salicylate (0.1%), and a bulky β-caryophyllene (37.0%). Rachmawati et al [23] also mentioned some major chemical compounds in Java cananga oil, including β-caryophyllene (19.39%), germacrene-D (13.36%), linalool (11.28%) and α-humulene (9.46%).

2. Methods

2.1. Materials

Cananga flowers were obtained from Karangpring village, Jember district and collected on October 2018. Cananga plant was determined in Balai Konservasi Tumbuhan Kebun Raya Purwodadi, Lembaga Ilmu Pengetahuan Indonesia (Indonesian Institute of Sciences) as *Cananga odorata* (Lmk) Hook.f. & Thoms. Samples for distillation were the whole parts of flowers and a mixture of young and mature cananga flowers.

2.2. Procedures

Distillation equipment was set up according to Figure 1. Main kettle provided dual function extraction methods only by modifying steam filter inside kettle. Kettle could be a chamber of hydrodistillation when steam filter was taken out from kettle, and be a chamber of steam-water distillation when steam filter was put in the kettle.Samples were directly contacted with water inside kettle during heating in hydrodistillation process, while in steam-water distillation, samples were steamed in the same kettle and contacted with hot water vapours only.

(a)

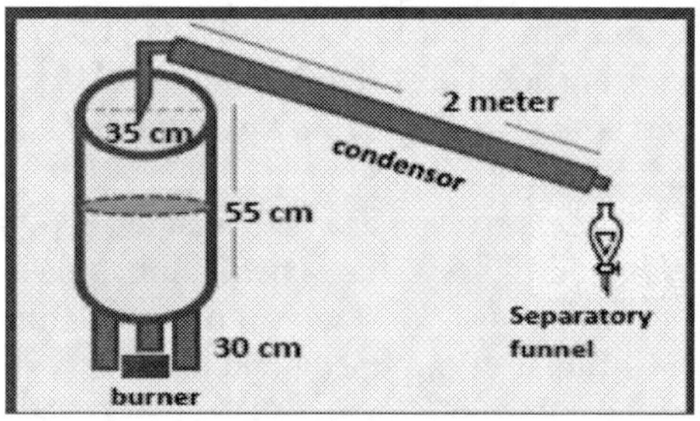

(b)

Figure 1. (**a**) Real picture of distillation reactor; (**b**) Schematic representation of distillation reactor for extraction of cananga oil

Each distillation process employed 2 kg of cananga flowers with the amount of water as much as 20 L. Steam-water distillation required maximum of 2 hour distillation to avoid an empty kettle since water was run out after 2 hours. Hydrodistillation method was run for longer time of distillation up to 3 hours since water still presented in kettle at the end of distillation.

Cananga essential oil was separated from aqueous phase in distillate using separatory funnel. Cananga oil was presented in upper layer (organic phase), while lower layer contained aqueous phase which commonly called as cananga water. Cananga oil was then added for a few of magnesium sulfate anhydrate to absorb water residues left in the oil. Further, this cananga oil was subjected on chemical analysis using GCMS (Gas Chromatography Mass Spectrometry).

3. Results and Discussion

Hydrodistillation of cananga flowers resulted in cananga essential oils with percent yield of 0.33% (1 hour), 0.46% (2 hours), and 0.67% (3 hours), while steam-water distillation for 2 hour process has produced 0.60% of cananga oil. Previous research by Ferdiansyah et al [24] showed that eight hour steam distillation of cananga flower results in 1.95% of cananga oil, while Rachmawati [23] also obtain 1.066% of cananga oil from similar process to Ferdiansyah et al [24].

Table 1. Compositions of cananga oils in various distillation methods.

No	Compound Names	Type	Composition (%)			
			SWD 2 h	HD 1 h	HD 2 h	HD 3 h
1	p-methylanisole	LOC	0.62	5.07	0.17	0.21
2	methyl benzoate	LOC	-	0.33	-	-
3	β-linalool	OM/LOC	8.89	0.17	-	-
4	nonanal	LOC	-	0.11	-	-
5	benzyl acetate	LOC	0.1	0.11	-	-
6	geraniol	OM/LOC	2.03	3.65	0.13	0.11
7	Z-citral	OM/LOC	0.1	0.53	-	-
8	bicyclogermacrene	S	0.21	0.14	0.22	0.25
9	δ-elemene	S	0.1	-	0.14	0.16
10	eugenol	OM/LOC	1.77	0.56	0.11	-
11	geranyl acetate	OM/LOC	7.79	16.58	7.1	4.26
12	β-elemene	S	0.59	0.44	0.66	0.71
13	β-caryophyllene	S	20.31	16.15	25.1	28.46
14	β-gurjunene	S	0.5	0.36	0.5	0.54
15	alloaromadendrene	S	0.31	-	-	-
16	8-methylene-dispiro[2.1.2.4]undecane	HC	-	0.28	-	-
17	1-(2-methylene-3-butenyl)-1-(1-methylenepropyl) cyclopropane	HC	-	-	0.22	0.24
18	α copaene	S	-	-	0.43	0.43
19	α-amorphene	S	0.44	0.32	-	-
20	α-humulene	S	6.5	5.42	8.13	8.96
21	δ-cadinene	S	0.14	0.14	0.2	0.15
22	germacrene D	S	22.57	15.3	22.79	25.76
23	farnesene	S	3.78	3.16	4.59	5.36
24	acetyl eugenol	OM/HOC	0.19	-	-	-
25	γ-cadinene	S	-	-	0.19	0.2
26	aromadendrene VI	S	1.46	1.78	1.43	1.4
27	1S-cis-calamenene	S	0.2	-	-	-
28	(-)-isoledene	S	-	0.21	0.21	0.19
29	α-cubebene	S	-	-	0.09	0.09
30	elemol	OS/HOC	0.31	0.34	0.27	0.18
31	cis-3-hexenylbenzoate	HOC	0.23	0.24	0.16	0.11
32	germacrene D-4-ol	OS/HOC	0.69	0.75	0.24	0.18
33	patchulane	S	0.15	0.15	0.14	0.15
34	spathulenol	OS/HOC	0.33	0.49	0.32	0.28
35	α-cedrol	OS/HOC	0.18	0.19	0.34	0.36
36	δ-cadinol	OS/HOC	0.63	0.71	0.91	0.9
37	α-cadinol	OS/HOC	0.67	0.79	1.08	1.06
38	α-copaene-11-ol	OS/HOC	-	-	0.11	-
39	farnesol	OS/HOC	6.93	9.54	8.47	6.42
40	farnesal	OS/HOC	-	0.1	0.11	0.11
41	benzyl benzoate	HOC	9.2	13.38	12.47	9.93
42	10-(acetylmethyl)-(+)-3-carene	OM/HOC	-	0.11	0.16	0.15
43	Farnesyl acetate	OS/HOC	0.51	0.38	0.53	0.59
44	2-(acetylmethyl-(+)-3-carene	OM/HOC	-	0.15	0.2	0.18
45	benzyl salicylate	HOC	1.32	1.57	1.68	1.47
46	geranyl isobutyrate	OM/HOC	0.26	0.29	0.42	0.46

SWD 2 h: steam-water distillation for 2 hours; HD 1 h: hydrodistillation for 1 hour; HD 2 h: hydrodistillation for 2 hours; HD 3 h: hydrodistillation for 3 hours.

Quality of those cananga oils were analyzed based on their chemical compound compositions using GCMS [14,15,21]. Based on previous research by Buccellato [3] and Ekundayo [13], it was mentioned that high quality of cananga oil relies on the content of esters and its marker compounds, such as methyl benzoate and benzyl acetate, two dominant esters in cananga oil, and also β-linalool. Table 1 listed the number of chemical compounds together with their percentages presented in cananga oils. Some esters presented in these cananga oils were methyl benzoate, benzyl acetate, benzyl salicylate, geranyl format, benzyl benzoate, and geranyl acetate (as major esters).

Table 2. List of the compound types which determine the quality of in cananga oils.

Components	Composition (%)			
	SWD 2 h	HD 1 h	HD 2 h	HD 3 h
Non-oxygenated hydrocarbons	57.26	43.85	65.04	73.05
Oxygenated hydrocarbons	42.75	56.14	34.98	26.96
Esters	19.60	32.88	22.36	16.82
β-linalool	8.89	0.17	-	-

SWD 2 h: steam-water distillation for 2 hours; HD 1 h: hydrodistillation for 1 hour; HD 2 h: hydrodistillation for 2 hours; HD 3 h: hydrodistillation for 3 hours.

Therefore, it can be summarized in Table 2 that one hour and two hour hydrodistillation contained 32.88% and 22.36% of esters, respectively, while in two hour steam-water distillation contained only 19.60% of esters. Interestingly, β-linalool could be obtained in significant amount only from steam-water distillation method as much as 8.89%, while from one hour hydrodistillation obtained only for 0.17%. No β-linalool can be found in cananga oil resulted from longer time of hydrodistillation processes.

It also can be seen from Table 2 that oxygenated hydrocarbons were much higher in hydrodistillation for an hour process (56.14%), and were reduced significantly in a longer time of distillation processes (two hours and three hours). Steam-water distillation for 2 hours brought about 42.75% of oxygenated hydrocarbons which much higher than hydrodistillation product in similar time of distillation process. Oxygenated hydrocarbons content was important to know since it was responsible for bioactivity of cananga oil.On the other hand, non-oxygenated hydrocarbon contents of cananga oils increased during the longer time of hydrodistillation method. β-caryophyllene, germacrene-D, α-humulene and farnesene were four non-oxygenated hydrocarbons that increased in time to time during hydrodistillation process.

Table 3. Classes of essential oil compounds in cananga oils.

Components	Composition (%)			
	SWD 2 h	HD 1 h	HD 2 h	HD 3 h
Non-oxygenated monoterpenes (M)	-	-	-	-
Oxygenated monoterpenes (OM)	21.03	22.04	8.12	5.16
Non-oxygenated sesquiterpenes (S)	57.26	43.57	64.82	72.81
Oxygenated sesquiterpenes (OS)	10.25	13.29	12.38	10.08
Light oxygenated compounds (LOC)	21.30	27.11	7.51	4.58
Heavy oxygenated compounds (HOC)	21.45	29.03	27.47	22.38
Hydrocarbons non-terpenoid and aromatic (HC)	-	0.28	0.22	0.24

SWD 2 h: steam-water distillation for 2 hours; HD 1 h: hydrodistillation for 1 hour; HD 2 h: hydrodistillation for 2 hours; HD 3 h: hydrodistillation for 3 hours.

Table 3 shows the detail of groups on essential oil compounds based on data on Table 1. Oxygenated monoterpenes (OM) is defined as monoterpenes containing oxygen atom in their structures which then could be in a form of alcohol or ester monoterpenes. This term also applied for oxygenated sesquiterpenes (OS) group. Light oxygenated compounds (LOC) is defined as oxygenated compounds with low molecular weight and boiling point properties, including some oxygenated monoterpenes and oxygenated aromatics. Heavy oxygenated compounds (HOC) consisted of oxygenated sesquiterpenes, other oxygenated aromatics and monoterpenes with higher molecular weight and boiling points. Interestingly, no monoterpenes could be found in all cananga oils. Data on Table 3 also supported information on the previous paragraphs whether the highest quality of cananga oil was obtained from hydrodistillation process for an hour. The presence of 22.04% of oxygenated monoterpenes, 13.29% of oxygenated sesquiterpenes, and 27.11% of light oxygenated compounds were leading to a judgement for quality characteristic of cananga oil.

According to those information, it can be suggested that one hour hydrodistillation of cananga flowers could gain first grade quality of cananga oil, compared to the following hours of hydrodistillation oil products, in term of the highest content of oxygenated hydrocarbons and esters. However, steam-water distillation method was quite satisfying on producing cananga oil with a huge amount of β-linalool, and higher amount of oxygenated compounds and esters compared to hydrodistillation method in the same time of distillation, 2 hours running.

4. Conclusion

Yield of cananga oils increased with the longer time of hydrodistillation processes was took place. However, it had found that the best quality of cananga oil was obtained in one hour hydrodistillation process since it was counted for the highest content of esters, such as geranyl acetate and benzyl benzoate. The non-oxygenated hydrocarbon contents in cananga oil increased considerably during the longer time of hydrodistillation, which majority were consisted of β-caryophyllene, gemacrene-D, α-humulene dan farnesense. Nevertheless, β-linalool was available in significant amount up to 8.89% only from steam-water distillation processes.

Acknowledgement

I would like to acknowledge University of Jember for providing community engagement programme grant in scheme of Program Pengabdian Kemitraan 2018 between LP2M University of Jember and cananga farmer community on Karangpring village, Jember district. Big thanks also deliver for all members of Atsiri Squads at Organic Chemistry Laboratory, University of Jember.

References

[1] Julianto T S 2016 *Minyak Atsiri Bunga Indonesia* (Yogyakarta: Deepublish) 101-114
[2] Trubus 2009 *Trubus info kit: Minyak Atsiri* Vol **07** Juni (Depok: PT. Trubus) 100
[3] Buccellato F 1982 *Perfum. Flavor.* **7** 9-13
[4] Salvatore-Battaglia 2019 *Essential oil monograph: Ylang ylang* 1-6, www.perfectpotion.com.au, http://www.salvatorebattaglia.com.au/wp-content/uploads/2019/01/A4_Monograph_Ylang-ylang_ONLINE-1.pdf
[5] Tan L T H, Lee L H, Yin W F, Chan, C K, Kadir H A, Chan K G, Goh B H 2015 *Evid.-Based Complementary Med* 2015 1-30
[6] Elmhalli F, Palsson K, Orberg J, Grandi G 2018 *Exp. Appl. Acarol.* **76** 209-220
[7] Maulidya R, Aisyah Y, Haryani S 2016 *Jurnal Teknologi dan Industri Pertanian Indonesia* **8** 2 53-60
[8] Pujiarti R, Widowati T B, Kasmidjo, Sunarta S 2015 *Jurnal Ilmu Kehutanan* **9** 1 3-11
[9] Stashenko E E, Torres W, Morales J R M 1995 *J. High Resolut. Chrom.* **18** 101-104
[10] Giang P M, Son P T 2016 *Am. J. Essent.* **4** 4 09-11
[11] Balai Penelitian Tanaman Rempah dan Obat 1998 *Penanganan dan penyulingan Bunga Ylang-Ylang* Unpublished paper, Report (Bogor: Kerjasama Perum Perhutani dengan Balai Penelitian Tanaman Rempah dan Obat)
[12] Mallavarapu G R, Gurundutt K N, Syamasundar K V 2016 *Essential Oils in Food Preservation, Flavor and Safety, Part II* (Elsevier Inc) Chapter 99. Ylang ylang (*Cananga odorata*) oils
[13] Ekundayo O 1989 *J. Essent. Oil Res.* **1** 5 223-245

[14] Gaydou E M, Randriamiharisoa R, Bianchini J P 1986 *J. Agr. Food Chem.* **34** 481
[15] Gaydou E M, Randriamiharisoa R, Bianchini J P, Llinas J R 1988 *J. Agr. Food Chem.* **36** 574-579
[16] Alighiri D, Eden W T, Supardi K I, Masturi, Purwinarko A 2016 *IOP Conference Series: JPCS* **824** (012021) 1-5
[17] ITC 2014 Ylang ylang oil – a review of production from Comoros http://www.intracen.org/uploadedFiles/intracenorg/Content/Exporters/Market_Data_and_Information/Market_information/Market_Insider/Essential_Oils/Country%20study_Comoros%20ylang%20ylang%20oil.pdf
[18] ITC 2014 Ylang ylang oil – the essential oil of the flowers of *Cananga odorata* http://www.intracen.org/uploadedFiles/intracenorg/Content/Exporters/Market_Data_and_Information/Market_information/Market_Insider/Essential_Oils/YLANG%20YLANG%20OIL%20%20The%20essential%20oil%20of%20the%20flowers%20of%20Cananga%20odorata.pdf
[19] Personal Communication with Distiller of essential oil August 2019
[20] Ultra Internasional B V 2019 *Product Market Report* (Netherlands) http://ultranl.com/market/market-report-autumn_2019/#scrollspy-883
[21] Stashenko E E, Prada N Q, Martinez J R 1996 *J. High Resolut. Chrom.* **19** 353-358
[22] Lawrence B M 1986 *Perfum. Flavor.* **11** 5 111-125
[23] Rachmawati R C, Retnowati R, Juswono U P 2013 *Kimia Student Journal* **1** 2 276-282
[24] Ferdiansyah P P, Zulfikar, Mahfud 2010 *Analisis pengaruh arah aliran steam dan massa bunga kenanga untuk mendapatkan minyak kenanga yang memiliki kualitas dan rendemen optimum dengan menggunakan metode distilasi uap (steam distillation)* Skripsi: unpublished paper (Surabaya: Fakultas Teknologi Industri, Institut Teknologi Sepuluh November)

Consistency of Spectra and Antibacterial Activity of The Extract Mixture of *Curcuma longa*, *Zingiber officinale*, and *Syzigium aromaticum*

Dwika Putri Pangesti, Masruri MASRURI*

Chemistry Department, Faculty of Mathematics and Natural Sciences, Brawijaya University. Jl. Veteran 65145 Malang, Indonesia

*Corresponding email: masruri@ub.ac.id

Abstract. Finding new antibiotic based on local resources is always a challenging effort, since Indonesia almost totally import the antibiotics. Then, antibiotic based jamu is an alternative. Jamu is a traditional herb prepared from local plants and some rhizomes, such as *Curcuma longa*, *Zingiber officinale*, and *Syzigium aromaticum*. These plants have been reported for their antibacterial capability. This paper is focused on the evaluation of antibacterial of these mixtures and their spectral consistency. The method was initiated by extraction of each plants/rhizome using ethanol and water. Each extract was mixed with equal ratio, and further evaluation for growth inhibiting activity in *Staphylococcus aureus* and *Escherichia coli*. Meanwhile, the spectral analysis was determined using FTIR and UV-Vis spectrophotometry. The result showed that the mixture of ethanol extract gives a slightly better activity than that using water extract. Furthermore, the UV-Vis spectra of the mixture from ethanol extracts indicates different band absorption in 439 and 417 nm, but no absorption observed in water extract in this range. In addition, the ethanol mixture extract also gives new band (FTIR spectra) in between 1010-1045 cm^{-1}, that these are not observed in water extract. In short, it can be summarized that ethanol extraction process give better extraction procedure and provide better antibacterial activity.

1. Introduction

The traditional medicine derived from local herbs, named as jamu, has paid many attentions, especially for their easily prepared and biological activity [1]. Some jamu recipe was derived from rhizome such as *Curcuma longa, Zingiber officinale* and other species from this genus. Their function is for healthy drink, diet [2] and antibacterial [3]. The other important herb was *Syzigium aromaticum*. It was reported having activity to kill pathogenic bacteria [4] and anticancer [5].

The bioactivity of the herbs always correlates to their chemical composition. For example, the species of *Curcuma longa* correlate to the curcuminoid molecule [6] contain in the herb. Most of these class molecules are phenolic aromatic compounds with conjugation bond. Meanwhile, the *Zingiber officinale* contains beta-sesquiphellandrene, bisabolene and farnesene [7]. These molecules are classified into terpenoid group with mostly contain aliphatic hydrocarbon and few double bonds in their structure. On the other hand, *Syzigium aromaticum* mostly contain phenolic compound such as eugenol and beta-caryophylene [8]. All the compound groups are categorized as the non-hydrophilic compound and was not easily dissolved in water solvent. However, traditionally, jamu is prepared using water as the main solvent.

This paper reports the consistency of the mixture extract derived from *Curcuma longa, Zingiber officinale,* and *Syzigium aromaticum* toward their bioactivity as antibacterial. The extract was prepared using two different solvents i.e. ethanol and water. Meanwhile, their consistency of chemical extracted by both solvents were compared from the FTIR and UV-Vis spectra evaluation to the reference. Recently, the FTIR and UV-VIS methods has been applied to detect the adulteration of traditional plant for herbs and also for chemometric analysis of the main secondary metabolite [9–11]. The specific region, known as the fingerprint area in FTIR spectra [9,10] can be used to detect the consistency the herb sample to the reference. For example the common herb for antibacterial contain a phenolic groups [12]. This has double bond of aromatic and hydroxyl group band absorption. Meanwhile the UV-Vis technique, characterized the common band absorption due to electronic transition from the specific functional groups such as carbon-carbon double bond, double bond in aromatic ring [13], conjugated and non-conjugated double bond, and heterocyclic functional group [11] related their bioactivity [6,7,11,13,14].

2. Experiment section

2.1. Material and chemical

The sample of herbs used for the research was bought from local market in Malang, East Java, i.e. rhizome of *Curcuma longa, Zingiber officinale,* and a dried flower of *Syzigium aromaticum*. Meanwhile the chemicals used for the research was ethanol (Merck) and distilled water.

2.2. Instrument for analysis

Some instruments applied for analysis include UV-Vis spectrophotometer (Shimadzu UV-1600 series). Analysis using double beam mode and water was applied as solvent or as a mentioned differently. The FTIR spectrophotometer (Shimadzu 8400S) was operated using potassium bromide pellet.

2.3. Extraction procedure

The extraction procedure using maceration techniques with water and/or ethanol as the solvent. A dried-powder mixture of herbs sample, i.e. *Curcuma longa, Zingiber officinale* and *Syzigium aromaticum* were mixed in 250 mL of Erlenmeyer flask. The quantity of each sample was 4.0 g (dry basis). Then, the mixture was added with a 100 mL of water and further stirred until homogeneous mixture was formed and rested at room temperature for 48 h. Then, the mixture was filtered off, and the filtrate was prepared directly as sample for further analysis (FTIR and UV-VIS spectrophotometry) and antibacterial evaluation. Similar separated procedure was also conducted but using ethanol as solvent for extraction.

2.4. Antibacterial activity evaluation

The procedure for testing antibacterial activity using disc diffusion technique. It was undertaken to measure the capability of the sample to inbihit the growth of *Escherichia coli* and *Staphylococcus aureus* [15]. The sample used was a mixture extract prepared from ethanol and water extraction process. The variation concentration used for each extract was 0.0; 3.13; 6.25; 12.5; 25.0; 50.0; and 100%, respectively. These sample were loaded into paper disc by immersion. Then, the mixture extracts were put aseptically into the media contain bacteria. It was further inoculated at 37 °C for 24 h. Then, the clean area surround the paper disc was measured as the quantity correlate to the activity of sample inhibit the bacteria growth.

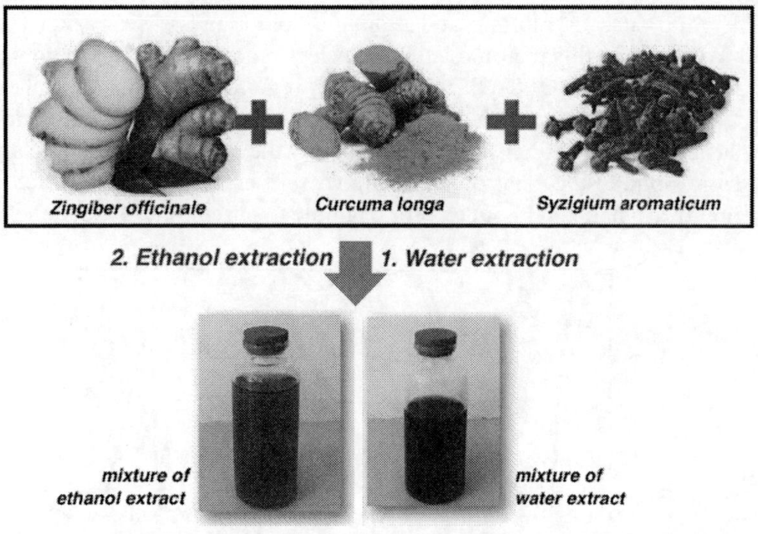

Figure 1. Schematic process for preparation of the mixture of herbs.

3. Result and discussion

The extraction process of the herbs mixture i.e. *Curcuma longa, Zingiber officinale,* and *Syzigium aromaticum* using water as a solvent provide a crude extract with deep-brown in colour. Meanwhile the ethanol extraction give light brown solution (Figure 1). The deep-brown of water extract has strongly smell of eugenol or olive oil. Meanwhile, the ethanol extract has a light smell of olive oil and a spice smell of ginger.

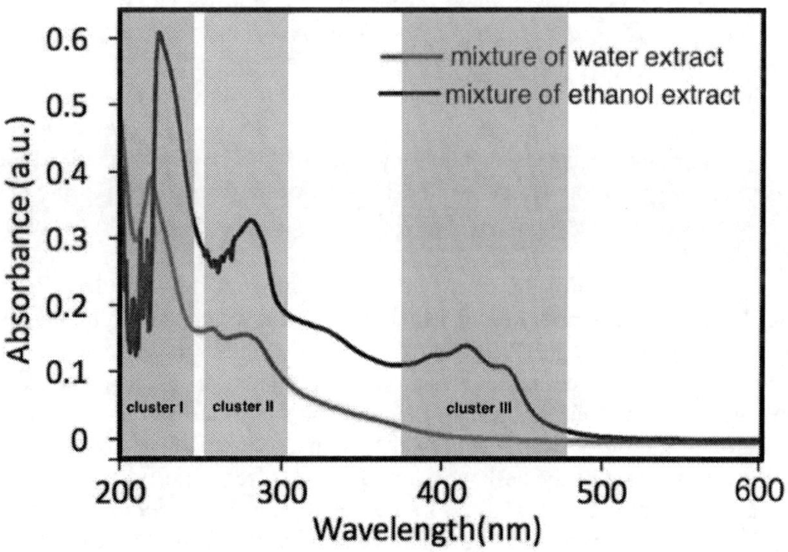

Figure 2. UV-Vis spectra of the extract of water and ethanol from mixture of *Curcuma longa, Zingiber officinale,* and *Syzigium aromaticum.*

The UV-Vis spectra of the both extract have a pattern band in between 200 and 600 nm (Figure 2). Both extracts also has similar band pattern in two cluster, i.e. in cluster I and cluster II. Cluster I has band absorption in 200-240 nm and this band absorption relate to the electronic transition of the double bond electrone in group such as olefinic and aromatic compounds. Meanwhile cluster II has band absorption in 250-310 nm, and this region relate to the electronic transition from non-bonding electrone in carbonyl (C=O) and carboxyl (O=C-OH) group. Both region is predicted correlate to the extracted compound and classified as phenolic, such as eugenol, gingerol, tannin, and tannic acid from the sample. However, in the cluster III (380-480 nm), it was found only the ethanol extract give an absorption. This region generally correspond to the electronic transition of several compound, which contain extention of conjugation from the aromatic compounds. For example in curcuminoid molecules.

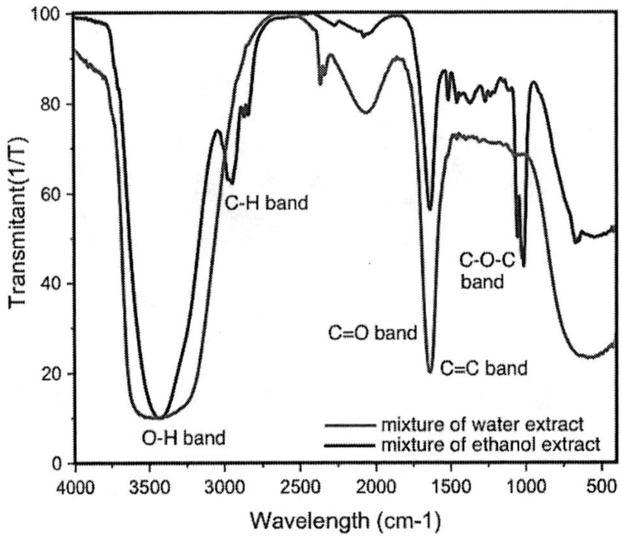

Figure 3. The FTIR spectra of the extract of water and ethanol from mixture of *Curcuma longa, Zingiber officinale,* and *Syzigium aromaticum.*

Moreover, it was detected the different band absorption in FTIR spectra (Figure 3). The ethanol extract give a two absorption band in between 1000-1080 cm^{-1}. These bands correlate to C-O-C bending vibration, that are commonly observed in methoxy-group (-OCH_3) attached in aromatic compound. However, it was not recorded in the water extract. The rest of the bands in FTIR are mostly similar for both extract. The hydroxyl group is detected at 3470 cm^{-1}, the double bond vibration is detected in about 1650-1620 cm^{-1} [16].

The consistency of correlation of the extracted compound from the mixture of of *Curcuma longa, Zingiber officinale,* and *Syzigium aromaticum* using ethanol and water as solvent, are also relate to their bioactivity to inhibit the growth of *Escherichia coli* and *Staphyloccus aureus* bacteria. The summary of antibacterial activity of both extract in *Escherichia coli* is tabulated in Table 1. Overall, ethanol extract of the herbs mixture has better inhibition than that showed for water extract. The highest activity for *Escherichia coli* growth inhibition was using 100% of ethanol extract, meanwhile the using the same concentration, the water extract of the herbs mixture does not affect significantly.

On the other hand, the antibacterial activity for both extract in *Staphylococcus aureus* is tabulated in Table 2. The similar activity is reported using ethanol extract of the mixture of *Curcuma longa, Zingiber officinale,* and *Syzigium aromaticum*. The highest activity is provided from ethanol extract at concentration 100% with inhibition value 11.27±0.01 mm. However, using similar concentration of water extract only provide activity 6.31±0.02 mm. The representative photo during antibacterial evaluation in both bacteria using both extract mixture is depicted in Figure 4.

Table 1. Antibacterial activity of the mixture of ethanol and water extract in *Escherichia coli.*

No.	Mixture of ethanol extract (%)	Zone inhibition (mm) in *E. coli*				X±SD
		I	II	III	IV	
1	100	7.88	7.91	7.93	7.89	7.90±0.20
2	50	7.23	7.19	7.23	7.22	7.22±0.02
3	25	6.74	6.75	6.74	6.73	6.74±0.01
4	12.5	6.22	6.18	6.19	6.19	6.19±0.02
5	6.25	6.07	6.06	6.05	6.05	6.06±0.01
6	3.13	6.01	6.02	6.01	6.01	6.01±0.01
7	0	6.00	6.00	6.00	6.00	6.00±0.00
No.	**Mixture of water extract (%)**	**Zone inhibition (mm) in *E. coli***				**X±SD**
		I	II	III	IV	
1	100	6.54	6.49	6.53	6.53	6.52±0.02
2	50	6.21	6.23	6.19	6.19	6.21±0.02
3	25	6.09	6.08	6.11	6.08	6.09±0.01
4	12.5	6.00	6.01	6.00	6.01	6.01±0.01
5	6.25	6.00	6.00	6.00	6.00	6.00±0.00
6	3.13	6.00	6.00	6.00	6.00	6.00±0.00
7	0	6.00	6.00	6.00	6.00	6.00±0.00

Note: X is the average number, and SD is the standard deviation.

Table 2. Antibacterial activity of mixture of ethanol and water extract in *Staphylococcus aureus.*

No.	Mixture of ethanol extract (%)	Zone inhibition (mm) in *S. aureus*				X±SD
		I	II	III	IV	
1	100	11.29	11.26	11.27	11.27	11.27±0.01
2	50	7.97	7.97	7.95	7.97	7.97±0.01
3	25	6.34	6.35	6.34	6.34	6.34±0.01
4	12.50	6.12	6.15	6.14	6.13	6.14±0.01
5	6.25	6.00	6.00	6.00	6.00	6.00±0.00
6	3.13	6.00	6.00	6.00	6.00	6.00±0.00
7	0	6.00	6.00	6.00	6.00	6.00±0.00
No.	**Mixture of water extract (%)**	**Zone inhibition (mm) in *S. aureus***				**X±SD**
		I	II	III	IV	
1	100	6.29	6.33	6.29	6.32	6.31±0.02
2	50	6.15	6.16	6.16	6.16	6.16±0.01
3	25	6.03	6.05	6.05	6.03	6.04±0.01
4	12.50	6.00	6.00	6.00	6.00	6.00±0.00
5	6.25	6.00	6.00	6.00	6.00	6.00±0.00
6	3.13	6.00	6.00	6.00	6.00	6.00±0.00
7	0	6.00	6.00	6.00	6.00	6.00±0.00

Note: X is the average number, and SD is the standard deviation.

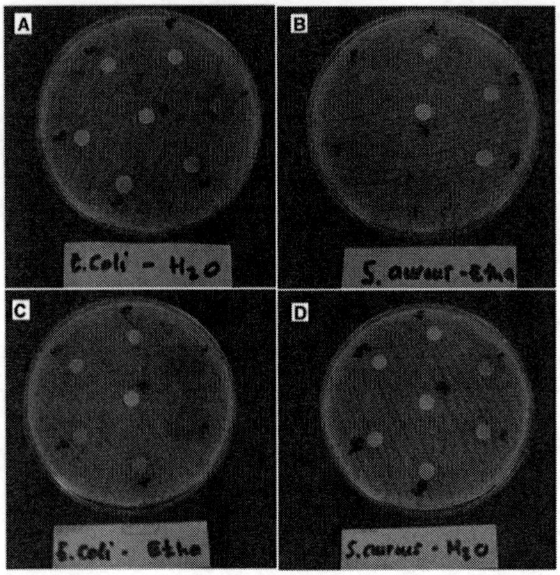

Figure 4. Antibacterial evaluation of extract to *Escherichia coli* and *Staphylococcus aureus*. (A-D) is mixture of water extract and (B-C) is mixture of ethanol extract.

4. Conclusion

In short, the extraction process of the mixture *Curcuma longa*, *Zingiber officinale*, and *Syzigium aromaticum* using water and ethanol solvent provide a different mixture of chemical composition extracted. The water extract mostly provides a deep brown of phenolic-eugenolic smell, meanwhile ethanol extract provides a light brown and spice ginger or curcuminoid smell. The antibacterial activity also in consistence to the chemical extracted from both solvents. This finding suggests to further explored for possible different bioactivity evaluation.

References

[1] Dhami, N., Mishra, A. D. *J. Herb. Med.* **2015**, *5* (2), 118–127.
[2] Eigner, D., Scholz, D. *J. Ethnopharmacol.* **1999**, *67* (1), 1–6.
[3] Marini, E., Di Giulio, M., Magi, G., Di Lodovico, S., Cimarelli, M. E., Brenciani, A., Nostro, A., Cellini, L., Facinelli, B. *Phytother. Res.* **2018**, *32* (3), 488–495.
[4] Oluwasina, O. O., Ezenwosu, I. V., Ogidi, C. O., Oyetayo, V. O. *AMB Express* **2019**, *9* (1), 20.
[5] Chua, L. K., Lim, C. L., Ling, A. P. K., Chye, S. M., Koh, R. Y. *Plant Foods Hum. Nutr.* **2019**, *74* (1), 18–27.
[6] Singh, G., Kapoor, I., Singh, P., De Heluani, C. S., De Lampasona, M. P., Catalan, C. A. *Food Chem. Toxicol.* **2010**, *48* (4), 1026–1031.
[7] El-Ghorab, A. H., Nauman, M., Anjum, F. M., Hussain, S., Nadeem, M. *J. Agric. Food Chem.* **2010**, *58* (14), 8231–8237.
[8] Massodi, M. L. E., Mime, L. C., Fogang, H. P. D., Djikeng, F. T., Karuna, M. S. L., Womeni, H. M. *J. Food Stab.* **2018**, *1* (1), 1–13.
[9] Sim, C. O., Hamdan, M. R., Ismail, Z., Ahmad, M. N. *J Anal. Chim. Acta* **2004**, *1*, 14.
(10) Rohman, A., Nugroho, A., Lukitaningsih, E., Sudjadi. *Appl. Spectrosc. Rev.* **2014**, *49* (8), 603–613.
[11] Kucharska-Ambrożej, K., Karpinska, J. *Microchem. J.* **2019**, 104278.
[12] Rohman, A., Sudjadi, D., Ramadhani, D., Nugroho, A. *Res. J. Med. Plants* **2015**, *9* (4), 179–186.
[13] Lima, L. A., Siani, A. C., Brito, F. A., Sampaio, A. L. F., Henriques, M. das G. M. O., Riehl, C. A. da S. *Quím. Nova* **2007**, *30* (4), 860–864.

[14] Sahu, R., Saxena, J. *Indian J. Adv. Chem. Sci.* **2014**, *2* (4), 300–302.
[15] Indriatie, R., Mudaliana, S., MASRURI, M. *IOP Conf. Ser. Mater. Sci. Eng.*, **2019**, Vol. 546, p 042013.
[16] Masruri, M., Pangestin, D. N., Ulfa, S. M., Riyanto, S., Srihardyastutie, A., Rahman, M. F. *IOP Conf. Ser. Mater. Sci. Eng.*, **2018**, Vol. 299, p 012072.

Edible Coating Development of Durian Seeds Starch and Glucomannan with The Addition of Essential Oil As An Antimicrobial to Increase Shelf Life of Tomato and Cauliflower

Andhika Suryo Prabowo[1], Lizda Johar Mawarani[1*]

[1]Institut Teknologi Sepuluh Nopember, Indonesia

*Corresponding author: andhikaojzan@gmail.com, lizda@ep.its.ac.id

Abstract. Edible coating can increase shelf life of fruit and vegetable that coated by minimizing water loss. This study aims to determine the effect of the addition of essential oil to edible coating has been made from durian seed starch and glucomannan. The essential oil that is eucalyptus oil and virgin coconut oil (VCO) added with variations up to 2 wt% as antimicrobial to bacteria and fungi on tomatoes and cauliflowers. The results showed that the smallest degree of swelling is obtained on the sample with 2 wt% VCO amount 457.52%. All samples not found fungal growth and bacterial contamination in oil less samples amounted to 3.1 x 10^3 CFU/mL, therefore safe to consume according to BPOM no.16 of 2016. Organoleptic results showed that all samples could be accepted by all panellists except samples of VCO 2%, EO 1% and EO 2% were somewhat disliked. Edible coating application on tomatoes and cauliflower can increase their shelf life until 10 days.

1. Introduction

Vegetable and fruit contain nutrients that support the health of the body. According to data released by the Badan Pusat Statistik (BPS) in January 2017, three out of four Indonesians consume fruit and almost all Indonesians (97.29%) have consumed vegetables [1]. The island of Java is the largest vegetable and fruit producing region in Indonesia and the distribution process of vegetable and fruit from Java to other islands requires various transportation modes such as pickup trucks, refrigerated trucks, and ships with refrigerated containers [2]. However, the transportation modes have a long travel time so that it can affect to the quality of vegetable and fruit that are brought.

Edible coating can extend the shelf life and improve product quality [3]. Starch can be used to form edible coating [4] but it hydrophilic so that bacteria can attack the food products and damage them. To prevent this, an antimicrobial that edible is essential oils [5][6]. Essential oils are secondary metabolites from plants obtained in the flowers, leaves, seeds, bark, fruits, and roots or rhizomes [7]. Based on previous research, the addition of 0.7% cinnamon essential oil, lemongrass 0.7%, lemon 3%, and betel 0.1% were done for each sample of Manalagi fresh-cut apples to inhibit bacterial growth and maintain shelf life [8].

Starch as the raw material for edible coatings can be obtained from durian seeds. The total consumption of durian fruit will certainly produce a lot of durian seed waste. The content of starch in the seeds of durian is high enough so that they can be processed into flour which has a high economic value and becomes a potential product diversification of wheat flour [9]. In this study, an edible coating was made with the addition of 30% glucomannan and essential oils namely eucalyptus and virgin coconut oil. The edible coating applied to tomatoes and cauliflower.

2. Method

2.1. Material
The material used in this study was durian seed starch, glucomannan, glycerol, aquades, virgin coconut oil (VCO) and eucalyptus oil.

2.2. Edible coating preparation and application
Durian seed starch, glucomannan, glycerol, and aquades were mixed at a ratio of 1: 0.3: 0.3: 20(w/w) for 15 minutes at room temperature. The solution is heated at 70°C for 20 minutes until it thickened then added a variations of VCO and eucalyptus oil amount 0%, 0.1%, 0.5%, 1% and 2%. The method of applying the edible coating was dip coating: tomatoes and cauliflower dipped in edible coating solution for 5 minutes then lifted and air dried for one day.

2.3. Characterization of edible film
The swelling test carried out according to ASTM D-570-098. To find out the size of the edible coating shrinkage itself, an observation was carried out using mica as a substitute for the fruit skin that was coated. Mica size 3 x 3 cm with 3 samples of each variation. Samples are weighed every day to find out the amount of depreciation that occurs over 10 days. The degree of swelling can be found by the following equation:

$$W(\%) = \frac{m_1 - m_0}{m_0} \text{x}100\% \tag{1}$$

where W (%) is the degree of swelling, m_1 is the final mass of the sample (g) and m_0 is the initial mass of the sample (g).

Shrinkage testing was done by observing tomatoes and cauliflower (3 samples of each variation) with edible coating application and the tomatoes and cauliflower without coating as a control variable for 10 days with daily monitoring started one day after the sample coated. The data obtained in the form of weight from the sample. From the data obtained, then calculate the value of the weight loss using the following equation

$$\text{Weight loss } (\%) = \frac{m_0 - m_1}{m_0} \text{x}100\% \tag{2}$$

Microbiological testing to determine bacterial and fungal contamination in edible coatings is still in accordance with the standard or not. Organoleptic testing to determine the response of panelists' acceptance of fruits and vegetables that are coated based on color, texture, aroma, taste and general acceptance.

3. Results and discussion
Edible coating solution from durian seeds with a variety of brownish-colored essential oils for variation of 0% and VCO and whitish color for eucalyptus oil variations. Before being applied to tomatoes, an edible coating solution is printed on the film to be tested for the degree of swelling and shrinkage.

3.1. Degree of Swelling and Shrinkage of Edible Coatings
Edible coating samples with the addition of eucalyptus oil cannot be measured the degree of distortion because the sample is destroyed after being immersed in water. Degree of swelling of the samples with VCO addition (Figure 1) showed that the addition of VCO reduced the degree of swelling.

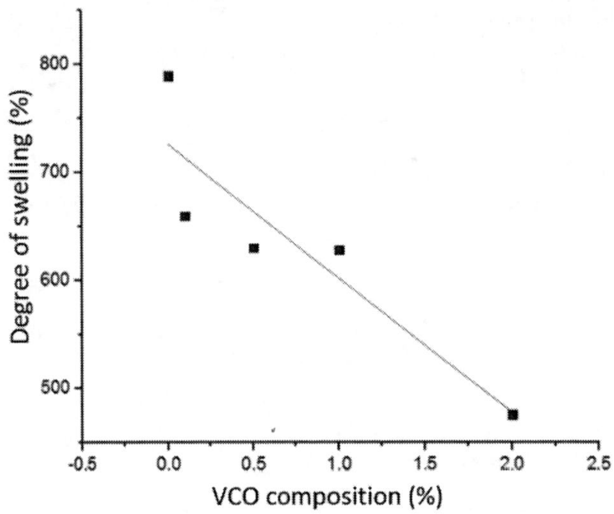

Figure 1. Effect of VCO composition on the degree of swelling

Results of the shrinkage test seen in Figure 2 for edible coating shrinkage with the addition of VCO and eucalyptus oil. Based on Figure 2 edible coating has decreased from the first day to the third day because the coating still wet and not yet dry completely. Meanwhile, after the third day the weight loss of coatings not too large and tend to be stable or no longer shrink.

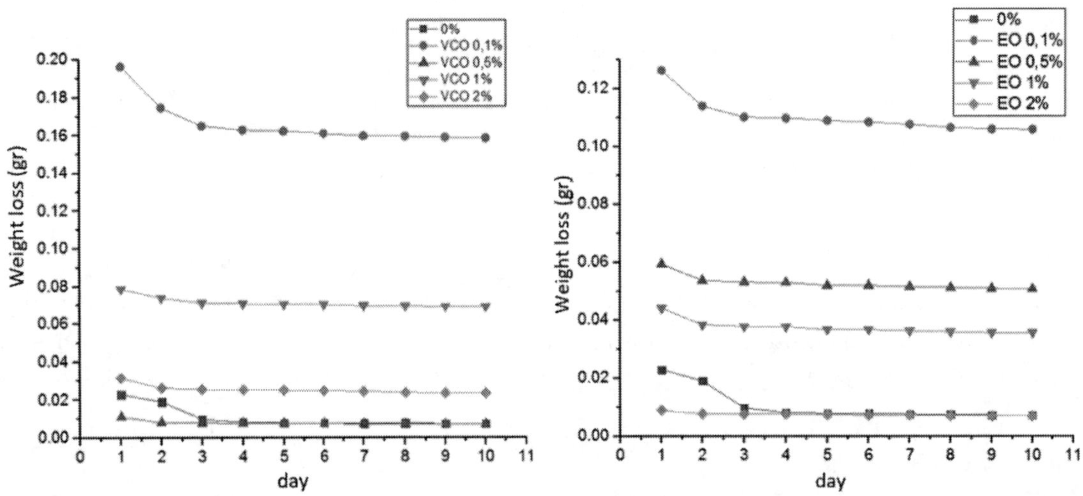

Figure 2. Weight loss of edible coatings on mica

3.2. Effect of Edible Coating on Tomatoes and Cauliflower Shelf Life

The weight loss observed in tomatoes and cauliflower coated edible coating with the addition of VCO and eucalyptus oil for ten days are showed in Figure 3 and Figure 4. Figure 3 showed that the lowest weight loss was in tomato sample with the addition of 1% VCO and sample with the addition of 2% eucalyptus oil, while the largest in tomato without coating. From Figure 4 seems that the lowest weight loss in the cauliflower sample with the addition of 2% VCO and sample with the addition of 0.5% eucalyptus oil, while the largest in the cauliflower sample without the addition of coating.

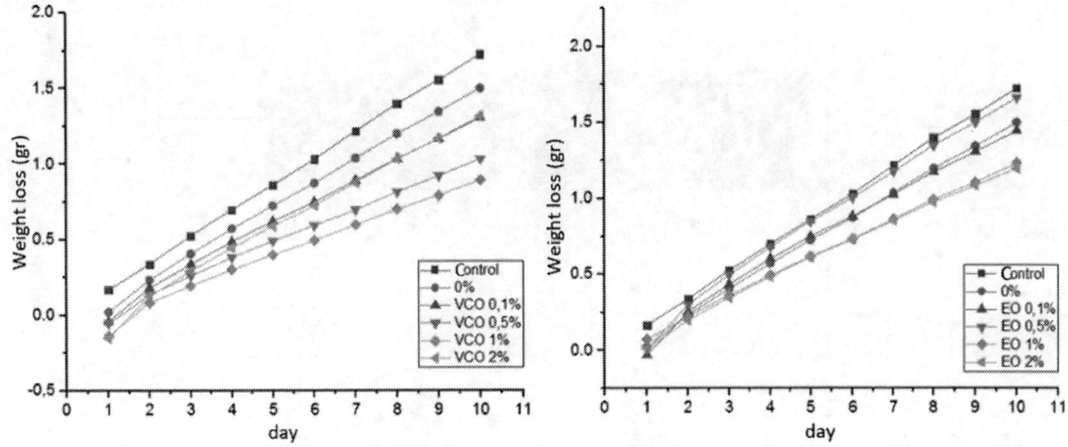

Figure 3. Weight loss of tomatoes with edible coating

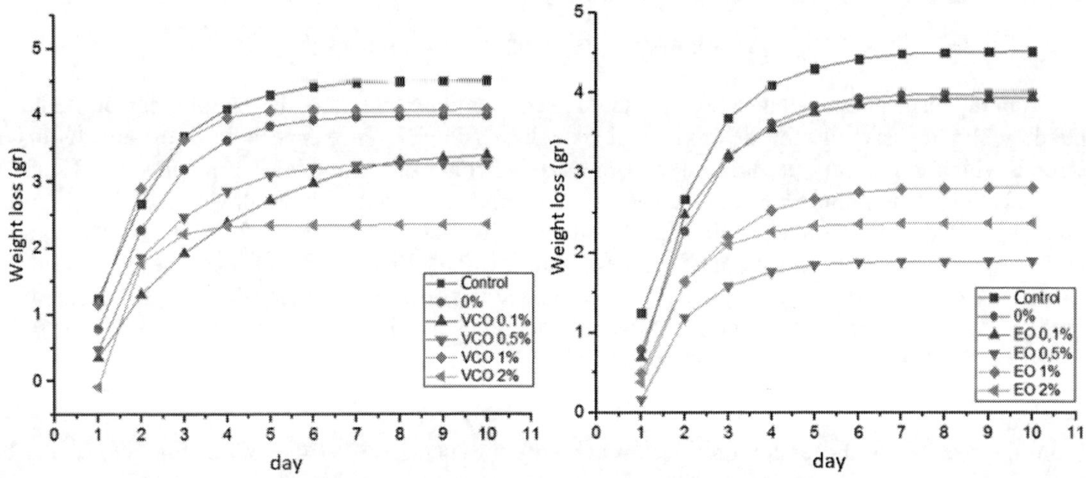

Figure 4. Weight loss of cauliflower with edible coating

3.3. Microbiology and Organoleptic Tests

Microbiological testing has been carried out by the TPC (total plate count) method to count the number of bacterial and fungal colonies in the samples. The test sample for bacterial TPC was selected 0% sample assuming the sample with the addition of oil has a smaller number of colonies than the 0% sample. While for fungal TPC, 3 samples were selected: 0%, VCO 2% and EO 2%.

Results of the test obtained that the number of colonies formed in the TPC of bacteria is 3.1×10^3 meet to the regulation of the Head of the Indonesian Drug and Food Control Agency No. 16 of 2016 concerning "Microbiological Criteria in Processed Food".It means the 0% edible coating samples are safe for consumption because the allowable bacterial colony limit is a minimum of 10^3 CFU/mL and a maximum of 10^4 CFU/mL. While for fungal TPC, the results are as shown that during the incubation period of 7 days there was no growth of fungal colonies.

Organoleptic testing is only done on tomatoes by coating the tomatoes with an edible coating of each variation. For each variation used 1 tomato, after dipped tomatoes sliced in half and stored for 1 day so that the edible coating layer can be formed perfectly. After the edible coating is formed, the tomatoes can be sliced back into 13 parts and will be tested by 13 panelists who have been selected to test in terms of color, texture, flavor and taste of each variation.

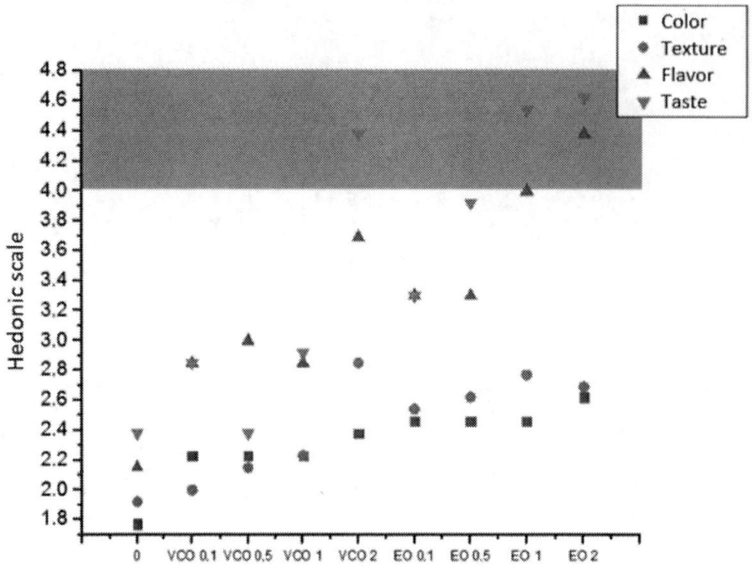

Figure 5. Organoleptic test results

From the results obtained it can be said that the color and texture of all the samples can be said to be liked by all panelists only for the aroma and taste there is a difference especially in the eucalyptus oil sample which is not very preferred due to the strong aroma of eucalyptus oil making it difficult for panelists to feel the sample because the taste of the sample is mixed with the flavor. Based on Figure 5 that on VCO 2% the average panelists stated rather dislike the taste while for EO 1% and 2% both the aroma and taste of the average panelist stated rather dislike. With this result it can be concluded that based on the color and texture of all the samples passed the organoleptic test while for the aroma of the sample 1% and 2% EO did not pass the organolpetic test while in terms of taste the sample VCO 2%, 1% EO and 2% did not pass the organoleptic test.

4. Conclusion

From the results and discussion on the current study, the increasing VCO composition decreasing the degree of swelling edible coating. The results of bacterial contamination in samples without oil amounted to 3.1×10^3 CFU/mL while fungal contamination in all samples found no fungus growth so it safe to consume according to BPOM No. 16 of 2016. Addition of VCO and eucalyptus oil are proven to be able to prevent microbial growth, especially in fungi. Organoleptic results showed that all samples were acceptable to all panelists except for the 2% VCO sample, 1% EO and 2% EO rather unpopular. Application of edible coatings on tomatoes and cauliflower can reduce weight loss and increase shelf life.

References

[1] Statistik, Paparan BPS Konsumsi Buah dan Sayur 2017 Dirjen Hortikultura Kementerian Pertanian, Statistik Produksi Hortikultura Tahun 2014-2015. [in bahasa Indonesia]
[2] Broto, W. (2009). *Teknologi Penanganan Pascapanen Buah untuk Pasar.* Jakarta: Badan Penelitian dan Pengembangan Pertanian.
[3] Aloui H and Khwaldia K 2016 *Comprehensive Reviews in Food Science and Food Safety* **15** 6 1080–1103.
[4] C Pagella, G Spigno, D M De Faveri 2002 *Food and Bioproducts Processing* **80** 3 193-198
[5] Garcia L C, Pereira L M, de Luca Sarantópoulos C I G and Hubinger M D 2011 *Packaging Technology and Science* **25** 7 413–425.
[6] Swamy M K, Akhtar M S and Sinniah U R 2016 *Evidence-Based Complementary and Alternative*

IOP Conf. Series: Materials Science and Engineering **833** (2020) 012034 doi:10.1088/1757-899X/833/1/012034

Medicine 1–21.

[7] M. Butnariu, I Sarac 2018 *J. Biotechnology and Biomedical Science* **1** 4 35-43

[8] Hikmatyar E P, Utama N A and Setiawan C K 2017 *Kajian Berbagai Minyak Atsiri dalam Edible Coating Berbasis CMC Sebagai Antibakteri Fresh-Cut Apel Manalagi* (Malus sylvestris Mill). Yogyakarta: Universitas Muhammadiyah Yogyakarta.

[9] A H Mulyati *et al* 2018 *J. Phys.: Conf. Ser.* **1095** 012001

IOP Conf. Series: Materials Science and Engineering **833** (2020) 012035 doi:10.1088/1757-899X/833/1/012035

Carbon dioxide occupancies inside ice XVII structure from grand-canonical Monte Carlo simulation

Irwansyah Putra Pradana[1], Diah Mardiana[1], Lukman Hakim[1, *]

[1] Department of Chemistry, Faculty of Science, Brawijaya University, Jl. Veteran, Malang, 65145, Indonesia.

*Corresponding author: lukman.chemist@ub.ac.id

Abstract. Clathrate hydrate is a promising material that can be used to trap the carbon dioxide (CO_2) as a mean for the greenhouse-gas emission control. A compromise between experimental findings and density functional theory calculations was recently made to determine the occupancy of the newly reported CO_2 clathrate hydrate of ice XVII structure. In this work, a hybrid isobaric Grand-Canonical Monte Carlo simulation is performed as a direct approach to determine CO_2 occupancy inside ice XVII structure. The simulation results show that the CO_2-to-water ratio starts at about 1:3.55 under lower pressure and ends at about 1:4 under high pressure. The potential energy of CO_2-water interaction as a function of CO_2 molecule displacement inside the voids shows a cage-like character, and the orientation of CO_2 molecules inside the spiral void is shown to be well-ordered. The simulation results support the experimental observation and provide molecular insight into the structure of CO_2 molecules inside the ice XVII structure.

1. Introduction

Investigation of new methods and materials for controlling CO_2 emission to the atmosphere has been widely carried out to reduce the impact of greenhouse gas on global climate change [1–3]. Carbon dioxide clathrate hydrates are non-stoichiometric compounds of ice-like crystalline water whose voids are occupied by CO_2 molecules [4]. The ice-like structure is termed as the host, while the CO_2 is termed as the guest. This material is viewed as a potential candidate to trap the excess CO_2 for long-time storage.

The CO_2 clathrate hydrate has been known to exist as a cubic clathrate-I (CS-I) structure at low-pressure and low-temperature regions [5]. Upon compression, the clathrate then decomposes to ice VI and dry ice at high pressure. However, Hirai and co-workers reported the emergence of a new phase of CO_2 clathrate hydrate under pressure region between the stable CS-I and ice VI-CO_2(s) [6]. Further investigation shows that the structure of this new CO_2 clathrate hydrate is the same with the new hydrogen clathrate-hydrate of the C_0 structure reported by Efimchenko [7]. The extraction of guest molecules from this clathrate-hydrate under vacuum produced a metastable ice structure termed as ice XVII [8,9].

A refined experimental work using neutron and synchrotron X-ray diffraction shows that the water lattice in ice XVII consists of five-membered rings of hydrogen-bonded water molecules. The water molecules form hexagonal helices with a diameter of 0.610 nm, measured from the center of the oxygen atoms, creating a spiral tunnel where guest molecules such as CO_2 and hydrogen can reside. The experimental results, however, encountered by uncertainty in the number of carbon dioxide that occupies the voids of ice XVII due to the mobility of guest molecules, and it was settled to be 1:3.55 with the results of density functional theory calculation performed at CO_2-to-water ratio of 1:4 [10]. This occupancy value describes the capacity of CO_2 trapping capacity that can be carried out inside the ice XVII structure, thus an intriguing issue from the industrial perspective, especially since the stable pressure region for this clathrate-hydrate structure is practically feasible at the order of several kbar [11].

In this work, the occupancy of CO_2 as a function pressure is directly determined using a hybrid type Monte Carlo (MC) simulation in grand-canonical and isobaric ensembles. The method has been reported to successfully reproduce the occupancy of hydrogen, neon, and carbon dioxide in the clathrate hydrate system [12–15]. The simulations also provide information on the CO_2 configuration inside ice XVII, as well as the potential energy profile of CO_2 molecule inside the cavity of ice XVII.

2. Research Methods

2.1. Potential Models and Structure of Ice XVII

All potential energies are defined as the sum of intermolecular pair-interaction. The non-bonded interaction is expressed as the summation of Lennard-Jones (LJ) 12-6 term and the electrostatic term. A rigid body model for CO_2 according to the work of Zhigang and co-workers [16] is employed and is referred as Z3D model onward. The interaction between water molecules is described by TIP4P/ice potential [17]. This potential model is derived from the TIP4P model that is often used to study water and ice, and the TIP4P/ice is tailored to provide better accuracy in reproducing the phase diagram of ice. Detailed parameters for Z3D and TIP4P/ice models are shown in Table 1. The CO_2-water interaction is calculated according to the Lorentz-Berthelot mixing rule. A switching function, starting from separation distance of 0.6655 nm, is applied to the pair potential energy and gives a smooth truncation at 0.8655 nm [18].

The proton-disordered ice XVII structure, containing 1296 water molecules, is generated according to the Bernal-Fowler rule using the GenIce package [19]. The simulation cell of CO_2 hydrate of ice XVII is considered a rectangular prism with a dimension of 3.7381 nm × 3.2906 nm × 3.5787 nm. The periodic boundary condition for all simulations are implemented in all directions.

2.2. Simulation Methods

2.2.1. MD Simulations and the Chemical Potential of CO_2 Fluid

The grand canonical ensemble corresponds to the equilibrium state between CO_2 inside the voids of ice XVII structure with CO_2 in the fluid phase outside, marked by the equality of CO_2 chemical potential in both phases. The chemical potential of ideal-gas is not appropriate to describe the CO_2 fluid under high pressure and low temperature; thus, an isobaric-isothermal molecular dynamics (MD) simulations are carried out using GROMACS 2018.7 package [20] to obtain the isotherm pressure-density curves of CO_2 fluid. The equation motion is settled using the algorithm of velocity-verlet at 2 fs time step. The temperature is maintained using the Nosé-Hoover thermostat, while the pressure is maintained using the Parrinello-Rahman barostat. The sampling is done over the 1.0 ns time evolution.

For a single-component and single-phase system, the Helmholtz energy per molecule, a is given by the sum of ideal and non-ideal parts as

$$a = k_B T\left[\ln\left\{\left(\frac{h^2}{2\pi m k_B T}\right)^{3/2}\right\} - 1 + \ln\left(\frac{h^2}{4\pi^2 I k_B T}\right)\right] + \int_0^{\rho} \frac{P - \rho k_B T}{\rho^2}\,\mathrm{d}\rho. \tag{1}$$

where m and I are the mass and moment of inertia of CO_2, respectively. The chemical potential of CO_2, μ_g, is then given as

$$\mu_g = a + \frac{P}{\rho} \tag{2}$$

where density ρ at a given pressure, P is obtained from the NPT MD simulation results.

Table 1. Parameters for CO_2 and water potential models used in this research. In TIP4P/ice, M refers to the negative charge located on a bisector of two OH bonds.

Model	Site	σ / 0.1 nm	ε / kJ mol^{-1}	q / e
CO_2/Z3D	C	2.7918	0.2398	+0.5888
	O	3.0000	0.6871	−0.2944
TIP4P/ice	H	-	-	+0.58970
	O	3.1668	0.8822	-
	M	-	-	−1.17940

2.2.2. Isobaric Grand-Canonical Monte Carlo Simulation

The occupancy of CO_2 molecules inside the cavity of ice XVII structure is directly investigated using a hybrid-type of Monte Carlo simulation in grand-canonical isobaric-isothermal ensemble (GC/NPT MC). The attempts to insert or eliminate CO_2 are carried out with the same probability under constant chemical potential. A trial insertion of the CO_2 is performed on random position and orientation represented by polar (θ) and azimuthal (ϕ) angles. The insertion of CO_2 is conducted with the probability

$$\max\left\{\frac{\exp[\beta(\mu'_g - w)]\, V \sin\theta (2\pi\beta m_g/h^2)^{3/2}}{N_g + 1}, 1\right\}, \tag{3}$$

where w is the interaction energy between CO_2 and the surrounding water molecules, μ'_g is the chemical potential of CO_2 that does not include the contribution from the free-rotational motion. A trial elimination of CO_2 is conducted by choosing an existing CO_2 randomly, and the trial is accepted with the probability

$$\max\left\{\frac{\exp[\beta(-\mu'_g + w)] N_g}{V \sin\theta (2\pi\beta m_g/h^2)^{3/2}}, 1\right\}. \tag{4}$$

In this Monte Carlo simulation, one MC step consists of a trial CO_2 molecule insertion or elimination with an equal probability, followed by five attempts to move a guest molecule or a water molecule that is randomly chosen, and a trial volume change based on the pressure. Each simulation is performed for at least 10^8 MC steps.

3. Results and Discussion

The isotherm pressure-density curve of CO_2 modeled here at 273.15 K in the range of 10 to 2000 MPa is depicted in Figure 1. At low pressure, the chemical potential of CO_2 fluid is shown to be less than its ideal gas counter-part, emphasizing the intermolecular attractive-force that tends to lead to CO_2 condensation. As the pressure increases beyond 100 MPa, the chemical potential of CO_2 deviates rapidly from the ideal gas curve due to the strong intermolecular repulsion. The curves shape provides a realistic representation of CO_2. In GCMC simulation, the chemical potential of CO_2 at a certain pressure is fixed according to this isotherm.

IOP Conf. Series: Materials Science and Engineering **833** (2020) 012035 doi:10.1088/1757-899X/833/1/012035

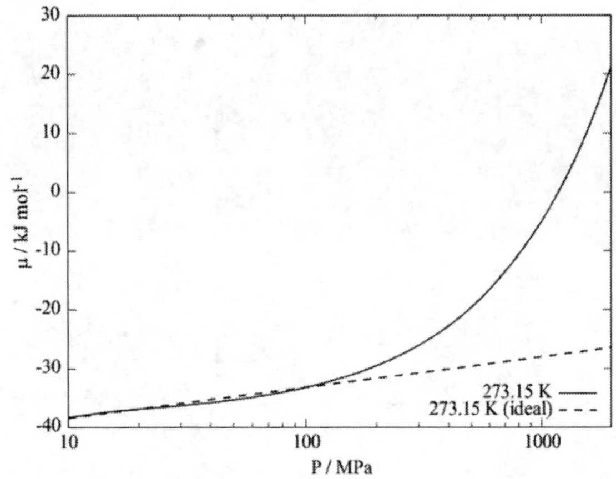

Figure 1. Pressure dependence of chemical potential of CO_2 fluid. The ideal gas curve (dashed-line) is provided here as a reference.

The typical depiction of clathrate hydrate where guest molecules are encaged inside the host structure is somewhat less relevant in the case of CO_2 hydrate of ice XVII structure because of the cavity shape in ice XVII that resembles a spiral tunnel. Figure 2 shows the potential energy curve arising from the interaction of a single CO_2 molecule with the surrounding water molecules inside the cavities of ice XVII. The potential energy is calculated as a function of CO_2 displacement from its equilibrium position, obtained by quenching the system at 0 K, and averaged over all CO_2 molecules. The displacement is classified according to each axis direction. As shown in Figure 2, as the CO_2 molecule moves 1 Å away from its equilibrium position, at any axis direction, the rapid increase of potential energy resembles a cage-like profile whose radius is between 1.0 - 1.5 Å. Figure 2(c) also suggests that the movement along *z*-axis is also limited with similar cage-like property. The description may be confusing because the voids in ice XVII have a spiral-tunnel shape that stretches along the *z*-axis. However, those potential energy curves are built under the absence of thermal motion of water molecules; thus, it should not be interpreted that CO_2 cannot move across the tunnel. Nevertheless, the information provided by Figure 2 is useful to understand the size of the cavities of ice XVII, whose size may change due to the thermal motion of water molecules. Figure 2 also shows that the orientation of CO_2 molecule does not significantly alter the profile of cage-like cavity. Note that the shift of potential energy curve due to orientational change in Figure 2 is due to the way that the displacement is always measured from the center carbon. At certain orientation, e.g. at $\theta = \pi/2$, the oxygen atom becomes very close to the tunnel "wall" along *x*-axis; thus, the displacement on *x*-axis direction becomes more limited.

The Isobaric Grand-Canonical Monte Carlo simulations of CO_2 inside ice XVII at 273.15 K are carried out in the pressure range between 0.05 and 2.0 GPa. The occupancy of CO_2 is expressed as the ratio of the number of CO_2-to-water molecules, N_w/N_g and is shown in Figure 3. The occupancy of CO_2 inside ice XVII is found near 1:3.55 at low pressure, which is similar to the experimental finding. However, uncommon to clathrate hydrate, Figure 3 shows a slight increase in the N_w/N_g curve with the increase of pressure, which means that the number of CO_2 inside ice XVII decreases during the course of compression. The possible cause is that the shrinking of system size due to compression increases the free energy of occupancy to a degree that is not favorable even with the increase of chemical potential. The simulation also suggests that the CO_2 hydrate of ice XVII is mechanically stable at a range of 0.05 - 1.6 GPa. Further compression beyond 1.6 GPa, although it increases the CO_2 occupancy due to the increase in chemical potential of CO_2 fluid, leads to a distortion in the ice XVII structure. This observation is consistent with the stable region of CO_2 clathrate hydrate of ice XVII structure in the phase diagram that was reported by Michl [11]. It is reasonable to argue that further distortion at higher pressure may lead to a phase separation that develops into the formation of ice VI and dry ice.

IOP Conf. Series: Materials Science and Engineering **833** (2020) 012035 doi:10.1088/1757-899X/833/1/012035

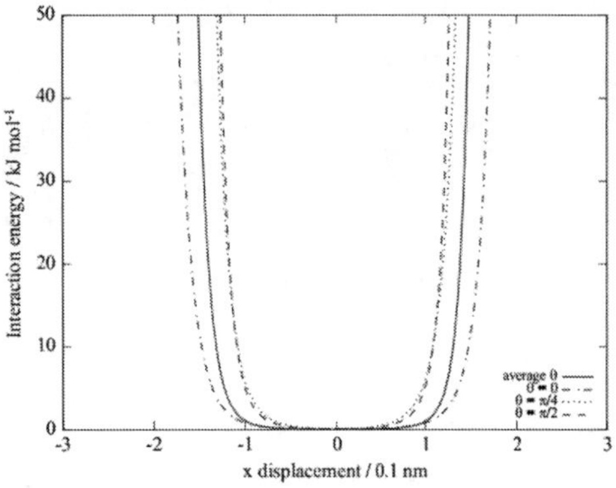

(a)

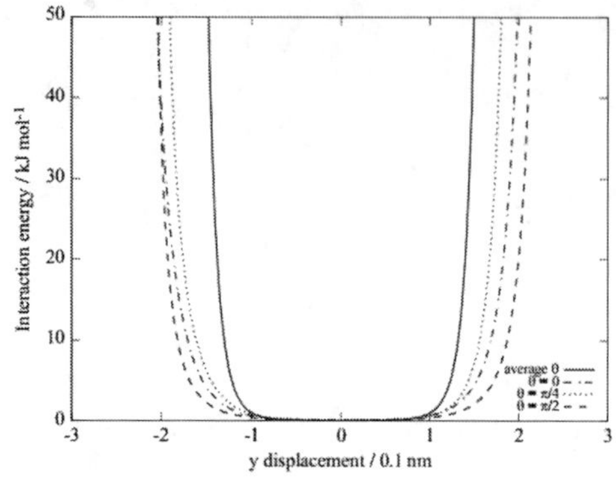

(b)

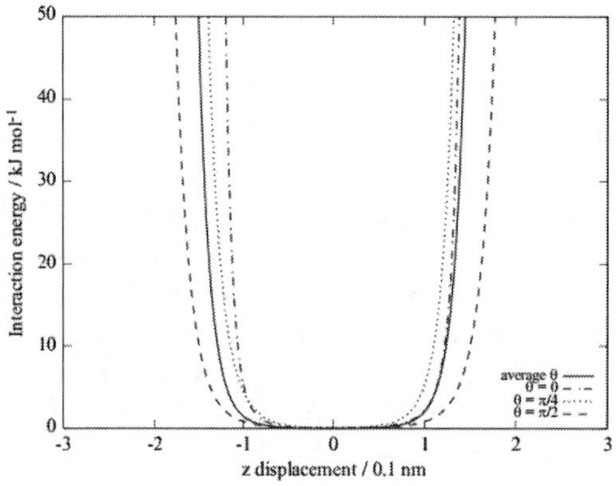

(c)

Figure 2. The potential energy curve of interaction between a CO_2 molecule with the surrounding water molecules in ice XVII. The potential energy is shown as a function of CO_2 displacement from its equilibrium position along the (a) *x*-direction, (b) *y*-direction, and (c) *z*-direction.

An insight to the CO_2 configurations inside the ice XVII cavities is obtained by analyzing the angular distribution CO_2 molecules at its maximum occupancy. The orientation of a linear molecule such as CO_2 can be simply expressed in terms of two angles in spherical coordinate: θ and ϕ. The θ angle describes the orientation of CO_2 against the z-axis, and the ϕ angle describes the orientation of CO_2 on the *xy*-plane. Figure 4 shows that CO_2 molecules tend to arrange themselves in configurations with $\theta = \pi/4$ and $3\pi/4$. These angles indicate a zig-zag orientation of CO_2 inside the tunnel. The distribution also suggests that CO_2 molecules tend not to align themselves closely parallel to *z*-direction ($\theta = 0$, or $\theta = \pi$). Furthermore, Figure 2 suggests that the cavity size is too small for stacking two or more CO_2 in the circular plane of the tunnel; thus, the only plausible configuration is to have CO_2 molecules lining across the tunnel in the *z*-direction with zig-zag orientation. In addition, the orientation of CO_2 in *xy*-plane described by ϕ angle is less defined as the orientation angle is broadly distributed, except for a peak between $\pi/2$ and $3\pi/4$.

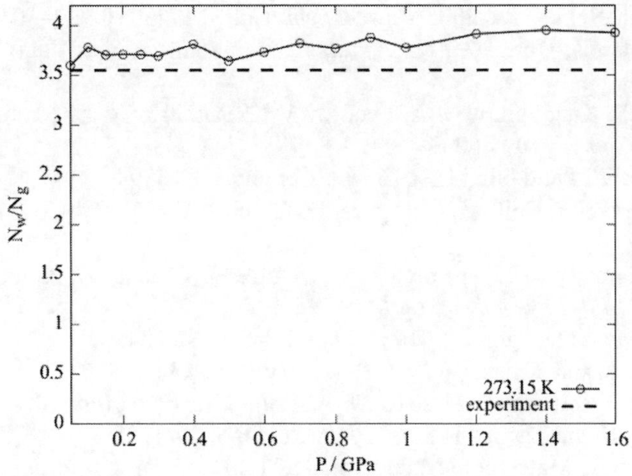

Figure 3. Pressure dependence of CO_2-to-water molar ratio in CO_2 hydrate of ice XVII structure, obtained from isobaric grand-canonical Monte Carlo simulation.

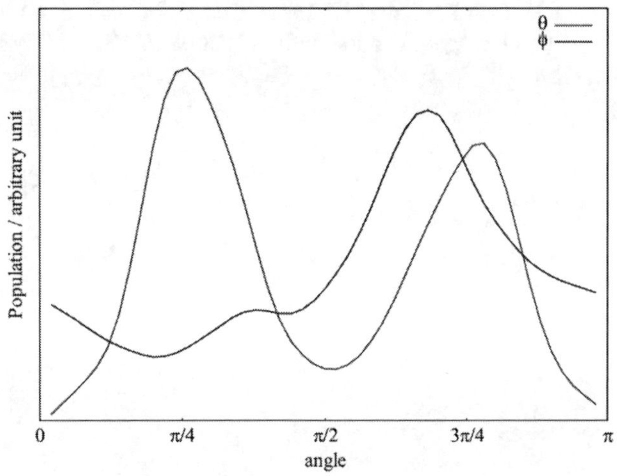

Figure 4. Distribution of orientational angles of CO_2 molecules inside ice XVII structure.

4. Conclusion

The isobaric grand-canonical MC simulation is useful as a direct approach to determine the occupancy of CO_2 hydrate and provide an insight into the CO_2 configurations. The occupancy of CO_2 inside ice XVII structure from simulations results is found to be aligned with the experimental findings, but the number of CO_2 molecules inside ice XVII is shown to be unusually decreasing during the course of compression. The simulations further show that CO_2 hydrate of ice XVII structure is mechanically stable at the low-pressure region up to 1.6 GPa, which is in line with the hydrate stable region in the phase diagram.

References

[1] Sanz-Pérez E S, Murdock C R, Didas S A and Jones C W 2016 *Chem. Rev.* **116** 11840–76
[2] Ekwurzel B, Boneham J, Dalton M W, Heede R, Mera R J, Allen M R and Frumhoff P C 2017 *Clim. Change* **144** 579–90
[3] Pradana I P and Hakim L 2019 *IOP Conf. Ser. Earth Environ. Sci.* **239** 1–7

[4] Sloan E D and Koh C 2007 *Clathrate Hydrates of Natural Gases, Third Edition* vol 20074156
[5] Loveday J S and Nelmes R J 2008 *Phys. Chem. Chem. Phys.* **10** 937–50
[6] Hirai H, Komatsu K, Honda M, Kawamura T, Yamamoto Y and Yagi T 2010 *J. Chem. Phys.* **133** 12–5
[7] Efimchenko V S, Kuzovnikov M A, Fedotov V K, Sakharov M K, Simonov S V. and Tkacz M 2011 *J. Alloys Compd.* **509** 860–3
[8] del Rosso L, Celli M and Ulivi L 2016 *Nat. Commun.* **7** 13394
[9] del Rosso L, Grazzi F, Celli M, Colognesi D, Garcia-Sakai V and Ulivi L 2016 *J. Phys. Chem. C* **120** 26955–9
[10] Amos D M, Donnelly M E, Teeratchanan P, Bull C L, Falenty A, Kuhs W F, Hermann A and Loveday J S 2017 *J. Phys. Chem. Lett.* **8** 4295–9
[11] Michl J, Sega M and Dellago C 2019 *J. Chem. Phys.* **151** 104502
[12] Hakim L, Koga K and Tanaka H 2010 *Phys. Rev. Lett.* **104** 115701
[13] Hakim L, Koga K and Tanaka H 2010 *Phys. A Stat. Mech. its Appl.* **389** 1834–8
[14] Hakim L, Koga K and Tanaka H 2010 *Phys. Rev. B* **82** 144105
[15] Matsuo M, Takii Y, Matsumoto M and Tanaka H 2012 *J. Phys. Soc. Japan* **81** 1–6
[16] Zhang Z and Duan Z 2005 *J. Chem. Phys.* **122** 1–15
[17] Abascal J L F, Sanz E, Fernández R G and Vega C 2005 *J. Chem. Phys.* **122** 1–9
[18] Ohmine I, Tanaka H and Wolynes P G 1988 *J. Chem. Phys.* **89** 5852–60
[19] Matsumoto M, Yagasaki T and Tanaka H 2017 *J. Comput. Chem.* 61–4
[20] Hess B, Kutzner C, Van Der Spoel D and Lindahl E 2008 *J. Chem. Theory Comput.* **4** 435–47

Acknowledgments
The research work is supported by research grants from DPRM Indonesian Ministry of Research, Technology, and Higher Education.

Preliminary Study on The Development of Preconcentration Method of Cu(II), Co(II), Ni(II), and Cr(III) Ions in Water Samples Using Nanomagnetite Coated by Carboxymethyl *kappa*-Carrageenan (CMKC)

Nuridhia Nisa Purnama[1], Irma Kartika Kusumaningrum[1, *], Anugrah Ricky Wijaya[1], Yudhi Utomo[1], Munzil Arief[1], Rohima Nostia[1], Lutfiyah Findiyani[1]

[1]Chemistry Department, Faculty of Mathematics and Natural Sciences, State University of Malang, Jl. Semarang No. 5 Malang

[*]Corresponding author: irma.kartika.fmipa@um.ac.id

Abstract. Flame Atomic Adsorption Spectrophotometry (FAAS) is one of the instruments that is often used to determine metal ion concentration in solution. FAAS uses simple operational procedure for metal ions concentration determination and produces accurate metal ion concentrations. Unfortunately, the accuracy of metal ion determination test decreases if the concentration of metal ions is too low, whereas the threshold for metal ions in drinking water is very low. Preconcentration is a method for increasing the concentration of metal ions in sample. An effective preconcentration method needs to be studied to improve the accuracy in the determination of metal ions concentration in water at low concentration using FAAS. This research aims to determine the effectiveness of metal ions preconcentration in water samples using nanomagnetite coated by CMKC. The research was conducted in four stages, 1) synthesis of nanomagnetite coated by CMKC; 2) preconcentration of samples containing Cu(II), Co(II), Ni(II), and Cr(III); 3) measurement of metal ion concentration with FAAS; 4) effectivity of preconcentration process and preconcentration optimum conditions. The result showed that nanomagnetite coated by CMKC can be used to preconcentration of metal ions Cu(II), Co(II), Ni(II), and Cr(III) in water samples by utilizing desorption-adsorption mechanism. Desorption was done using 0.1 N HNO_3 as a desorption solvent without variation in concentration and desorption time preconcentration optimum conditions for each metal ion were as follows, adsorption condition of Ni(II) was pH 7 and 60 minutes; Cr(III) was pH 7 and 40 minutes; Co(II) was pH 7 and 60 minutes; Cu(II) pH 6 and 40 minutes. The effectiveness of preconcentration at optimum conditions for each metal ions are: %Cr(III) > %Cu(II) > % Co(II) > %Ni(II), successively.

1. Introduction

Flame atomic absorption spectroscopy (FAAS) was very useful to determine a large number of elements, especially at trace levels. FAAS has widely used to analyze variety of sample matrices including water, soils, and biota. It is a reputable technique that is inexpensive and delivers accurate results even in a complex matrix [1]. In addition, this instrument is often accepted as a suitable instrumental technique for the measurement of metal ion because of its speed and ease of operation [2]. Many researchers use FAAS to determine the concentration of metal ion because of its low cost, easy usage, and it is available in most of laboratories [3]. The test results of metal ions concentration determination have a high accuracy. Unfortunately, the accuracy of metal ion determination test decreases if the concentration of metal ions is at trace level. The widespread distribution of metal ion is a source of contamination problems, resulting in the low concentration. As a result, a treatment must be done when dealing with a very low-level sample. As in drinking water, the lower the concentration, the higher the risk of contamination.

The levels of heavy metal ions in drinking water are generally at μg/L level. There are many other method for precise and accurate analysis of heavy metal ions in drinking water, such as Neutron Activation Analysis (NAA), Inductively Coupled Plasma-Mass Spectrometry (ICP-MS), Inductively Coupled Plasma-Atomic Emission Spectrometry (ICP-AES), X-ray Fluorescence Spectrometry (XFS), and Graphite Furnace Atomic Absorption Spectrometry (GFAAS). The accuracy of those instruments are much better than FAAS, but those are only available in a few laboratories in Indonesia. It caused the sample testing costly and it would be processed in a long time, due to the lack of laboratories that provide those instruments. Beside that, laboratories have those instruments mostly, so it would be better that we optimize the use of FAAS to determine levels of heavy metal ions in drinking water. The use of FAAS is more ideal for routine use in laboratories to produce an accurate and precise measurements. However, the sensitivity of FAAS is not high enough to determine metals at trace levels. Hence, its sensitivity should be firstly increased using either *simple* modified systems or some preconcentration procedures [4].

Preconcentration is a method for increasing the concentration of metal ions in water due to its-low concentration. There are variety of procedures for preconcentration of metal ions, such as solid phase extraction (SPE), liquid-liquid extraction (LLE), coprecipitation, and cloud point extraction (CPE) which have been developed [5]. Among them, SPE is one of the most reliable and simple preconcentration methods for the enrichment of trace metal ions in water. SPE method is based on the contact of a homogeneous liquid sample through a flask containing a selective absorber holds metal ions [6].

The absorber that used in this research is nanomagnetite coated by carboxymethyl *kappa* carrageenan. Magnetite (Fe_3O_4) is one of many magnetic materials that widely investigated because of their unique properties, inexpensive, and easy to synthesize for MRI, sensor, and adsorbent. Nanomagnetite has a high surface area that produces numerous active sites. However, nanoparticle Fe_3O_4 preparation is a problematic since Fe_3O_4 could agglomerate and leads to the decrease in the active sites [7]. Therefore, coating them with organic surfactants (polymer) such as carboxymethyl *kappa*-carrageenan could prevent the agglomeration [8].

The effectiveness of preconcentration method needs to be studied to improve the accuracy of the measurement using FAAS. This research aims to determine the effectiveness of preconcentration of metal ions in water samples using CMKC coated nanomagnetite. This research was conducted in four stages, 1) synthesis and characterisation of nanomagnetite coated by CMKC; 2) preconcentration of samples containing Cu(II), Co(II), Ni(II), and Cr(III) ions using CMKC coated nanomagnetite; 3) measurement of metal ion concentration with FAAS; 4) determination of the effectivity of preconcentration process and optimum preconcentration condition.

2. Experimental Procedure

2.1. Materials, Equipments and Intrumentations

2.1.1. Material. *Kappa*-carrageenan was obtained from CV Makmur Sejati, Malang, Indonesia. Sodium hydroxide, ethanol 96%, isopropyl alcohol, acetic acid (glacial, 96%), sodium chloride, nitric acid were bought from Merck. Iron(III) chlrodide hexadyrate, iron(II) sulphate hexahydrate, and trichloroacetic acid were bought from Sigma Aldrich. Distilled water, demineralized water, and powder of metal ions (e.g. Cu(II), Co(II), Ni(II), and Cr(III)) were provided by Chemistry Department of Universitas Negeri Malang.

Equipment and Instrumentation. In this study was used analytical balance (Ohaus PA224), laboratory glassware, hotplate with magnetic stirer (IKA C-MAG HS 7), oven, a set of vacuum pump and buchner funnel. Instruments used to characterization nanomagnetite by CMKC. Fourier Transform Infrared spectrophotometer (Shimadzu Irrestige 21) was conducted between 400 - 4000 cm^{-1}. Scanning Electron Microscopy (FEI Inspect-S50) used to determine morphology and particle size. Magnetic properties of nanomagnetite were determined by Vibrating Sample Magnetometer (Oxford 1.2H). Crystallinity of nanomagnetite was determined by X-Ray Diffraction (Pananalytical E'xpert Pro).

2.2. Synthesis and Characterisation of Nanomagnetite Coated by CMKC

Nanomagnetite was synthesized by firstly mixing $FeCl_2$ 0.5M (10 mL) and $FeSO_4$ 0.25 (10 mL) which then precipitated by slow addition of 40 mL of NaOH 2M. Then, the mixture was stirred for 30 minutes, filtered off, neutralized with distilled water until pH 7 was reached and dried in the oven for 5 hours at 50 °C [9]. Study about optimization of synthesis of CMKC was done [8], the optimum condition of synthesis of CMKC from 1 g of *kappa*-carragenan need 7.2 g of sodium hydroxide for alkalization step and 1.5 g of trichloroacetic acid for etherification step due to their mol ratio. To coating nanomagnetite with CMKC, first step is preparing CMKC suspension by dissolved 1 g of CMKC in 500 mL of water and stirred for 3 hours. 1 g of nanomagnetite that dissolved in 100 mL of water was added into CMKC suspension and stirred for 2 hours at room temperature and filtered. The residue was neutralized by distilled water until *pH* 7 was reached and dried in the oven at 50 °C for 5 hours [10]. Fourier transform infrared spectroscopy (FTIR) operated from 400 cm^{-1} to 4000 cm^{-1} at room temperature used KBr pellet method. SEM machine working at EHT 25kV of accelerating voltage was used to determine morphology and particle size of nanomagnetite, whereas, accelerating voltage for nanomagnetite coated by CMKC was 20kV. XRD pattern of nanomagnetite crystal was obtained at 2θ range of 10-90° by a Philips, X'pert, Cu Kα radiation (λ= 0.154 nm) source operated at 35 mA and 40 kV. Sampling technique to select nanomagnetite and nanomagnetite coated by CMKC done by magnetic bar. Selected sample was measured using VSM at room temperature.

2.3. Preconcentration of Samples Containing Cu(II), Co(II), Ni(II), and Cr(III)

Standard solution of each metal ion has different concentration due to lack of metal powder stock in laboratory. Metal ions weighed and made in to solution with concentration 5 ppm for Cu(II), 20 ppm for Cr(III), 30 ppm for Ni(II), and 50 ppm for Co(II). Preconcentration procedure of this research was done by adding 0.1 g of nanomagnetite coated by CMKC into 50 mL of sample solution that contain each metals ion. There are three stages for preconcentration procedure, first preconcentration was conducted at room temperature by varying pH of metal ion solution (e.g. pH 3, 5, 7) to determine the optimum condition of preconcentration. Standard solution of 0.1 M NaOH and 0.1 M HCl were used for pH adjustment. Once the optimum pH was determined, this pH was used for the second preconcentration at different adsorption time to determine the effect of time at metal ions adsorption by put them on the shaker for 20, 40 and 60 minutes at room temperature. The third preconcentration aimed to determine effectivity of preconcentration procedure at the optimum pH and contact time. After nanomagnetite added into the sample solution, nanomagnetite was filtered and the filtrate (filtrate 1) was kept for the further analysis. Nanomagnetite washed with 15 mL demineralization water, filtered, and the filtrate was thrown. Then, for the desorption, nanomagnetite eluted with 20 mL of 1 M HNO_3 and the filtrate (filtrate 2) kept for further analysis [11]. These procedures conducted only once as preliminary study for further research to develop preconcentration method of metal ions in drinking water.

2.4. Measurement of Metal Ion Concentration With FAAS

Initial concentration measurement of standard solution before adsorption was needed to determine the real concentration of metal ion in the solution. Filtrate 1 is meant to be a filtrate after nano magnetite-CMKC adsorb metal ion, measurement of filtrate 1 would determine the concentration of metal ion that still remain in solution. Meanwhile, filtrate 2 is meant to determine metal ion concentration after metal ion un-attached from nanomagnetite-CMKC in 20 mL of 0,1N HNO_3. Standard solution, filtrate 1, and filtrate 2 was measured by using Flame Atomic Absorption Spectroscopy (Shimadzu, AA7000). This procedure conducted only once without any repetition, so there would be no statistical data in this study.

2.5. Determination of The Effectivity of Preconcentration Process and Optimal Preconcentration Condition

The effectivity of preconcentration process was determined by analysing the concentration that was obtained from the third preconcentration procedure, which conducted at the optimum pH and contact time of each metal ion. The adsorption capacity of nanomagnetite-CMKC was calculated using Equation (1), the percentage of metal ion removal was calculated using Equation(2), the desorption capacity was calculated using Equation(3), and concentration increasing was calculated using Equation (4):

$$Q_{adsorption}\left(\frac{mg}{g}\right) = \frac{(C_{initial} - C_{remain})}{m_1} . V_1 \tag{1}$$

$$\%removal = \frac{(C_{initial} - C_{remain})}{C_{initial}} . 100 \tag{2}$$

$$Q_{desorption}\left(\frac{mg}{g}\right) = \frac{C_{desorption}}{m_2} . V_2 \tag{3}$$

$$Concentration\ Increasing = \frac{C_{desorption}}{C_{initial}} \tag{4}$$

where V_1 is the volume of metal ion solution (L), m_1 is the weight of nanomagnetite-CMKC used to adsorb (g), V_2 is the volume of HNO_3 as releasing media of metal ion (L), and m_2 is the weight of nanomagnetite-CMKC after adsorption. The adsorption capacity is the amount of metal ion that attached to per gram of nanomagnetite-CMKC. The desorption capacity is the ampunt of metal ion desorbed or un-attached from per gram of nanomagnetite-CMKC.

3. Result and Discussion

3.1. Synthesis and Characterisation of Nanomagnetite Coated by CMKC

Nanomagnetite coated by CMKC is one of surface modifications to prevent nanomagnetite from agglomeration. Agglomeration could decrease the active site in nanomagnetite that could be binding with metal ions. Coating them with carboxymethyl *kappa*-carrageenan is not only to prevent agglomeration, but also to increase their ability to bind to the metal ions, because CMKC has negative charge to charge-balance the metal ions. Nanomagnetite was synthesized by co-precipitation method and dried it in the oven for several hours at 50°C. Meanwhile, CMKC was synthesized in two steps, alkalization and etherification steps. FTIR analysis was done to nanomagnetite coated by CMKC, nanomagnetite, and CMKC itself. FTIR analysis aimed to make sure that carboxyl groups in *kappa*-carrageenan was converted into hydroxyl groups and formed carboxymethyl *kappa*-carrageenan. The carboxyl groups formation can be seen through the comparison between FTIR analysis of *kappa*-carrageenan as shown in Figure 1 and carboxymethyl *kappa*-carrageenan as shown in Figure 2.

Figure 1 showed the hydroxyl groups (O-H) absorption peak appeared at 3205.69 cm^{-1} in *kappa*-carrageenan with high intensity. The O-H absorption peak was shifted at 3197.98 cm^{-1} in CMKC spectra, while the intensity of hydroxyl groups in CMKC was decreased and converted into carboxyl groups as shown from a new absorption peaks appeared at 1774 cm^{-1} and 1417 cm^{-1} in Figure 2. These two peaks indicate that C=O groups was formed, in which these peaks were not observed in the *kappa*-carrageenan spectra. Hence, the carboxymethyl *kappa*-carrageenan was successfully synthesized.

IOP Conf. Series: Materials Science and Engineering **833** (2020) 012036 doi:10.1088/1757-899X/833/1/012036

FTIR analysis was also done to determine the functional groups contained in magnetite (Fe_3O_4) as shown in Figure 3. According to specific peaks appeared in Figure 3, it is indicated that there was a bonding between iron (Fe) and oxygen (O), observed at 441 cm^{-1} and 586 cm^{-1} which is a signature peak of Fe-O. In similar to Fajaroh (2011) research, a signature peaks of Fe-O at 580 cm^{-1} and 442 cm^{-1} was resulted [12].

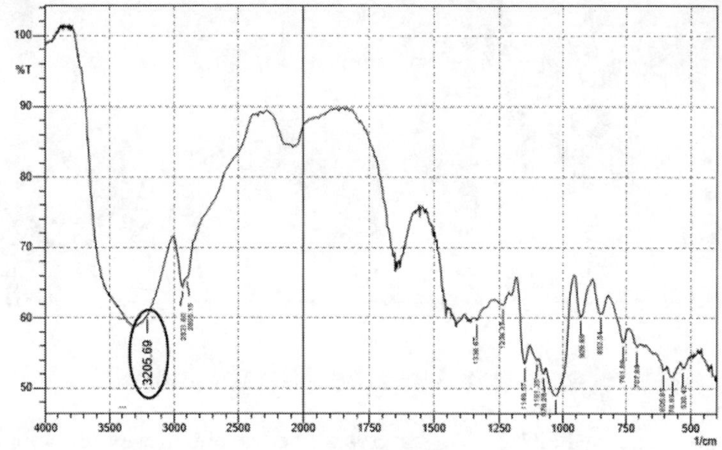

Figure 1. FTIR Analysis of *kappa*-carrageenan.

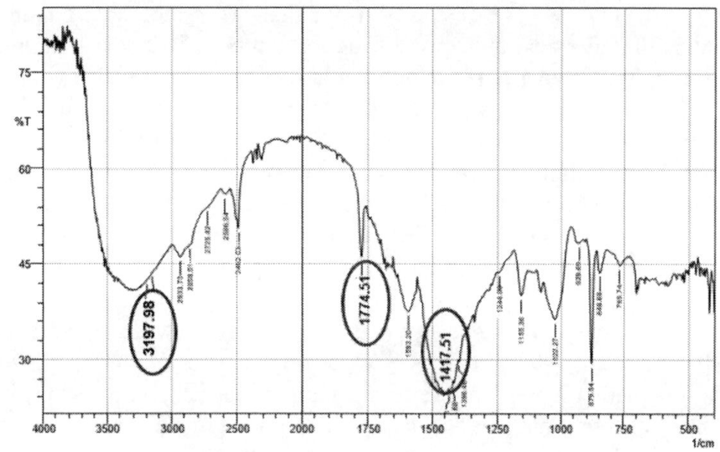

Figure 2. FTIR Analysis of CMKC.

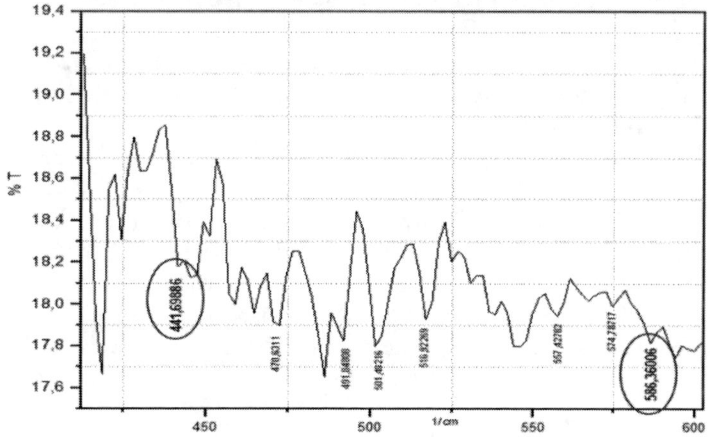

Figure 3. FTIR Analysis of Nanomagnetite.

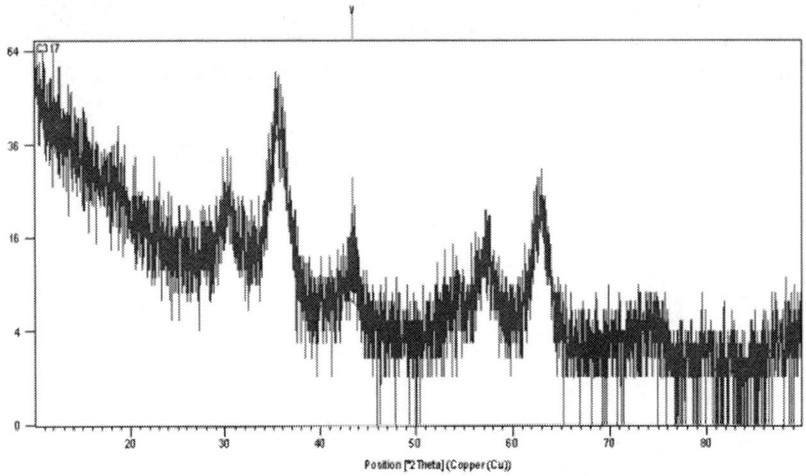

Figure 4. XRD Analysis of Nanomagnetite.

Structure of nanomagnetite Fe_3O_4 has to be crystalline, to obtain a pure nanomagnetite [13] To determine the crystallinity of nanomagnetite, powder X-ray diffraction (XRD) analysis has been conducted and shown in Figure 4. The results showed that the synthesized nanomagnetite has peak intensity and 2θ angles that identical to X-ray diffraction pattern of standard nanomagnetite Fe_3O_4 of JCPDS (Joint Committee for Powder Diffraction Standard - Number 19-0629) which were: 8.2°, 30.1°, 35.4°, 37.1°, 53.4°, 56.9°, 62.5°, 65.7°, 70.9°, and 73.9°. High intensity that resulted from θ to 2θ angle because of impurities existed. Nanomagnetite made from mixture of iron(II) sulfate and iron(III) chloride that precipited by sodium hydroxide. The existance of sodium(I) in nanomagnetite could possibly made diffraction pattern in below 2θ has high intensity. Amplitudes would be scattered by an electron from the donation of Sodium(I). The standards were used as comparison to confirm the product was truly the synthesized nanomagnetite. The data showed that the 2θ angles were appeared at 30°, 35°, 42°, 57°, 63°, and 74°, these peaks almost identical to the result reported by Fajaroh (2011) in which the nanomagnetite has 2θ angles from 30.5° to 63.1°, which correspond to crystalline nanomagnetite [12].

Iron has a strong magnetic moment due to four unpaired electrons in its 3d orbitals, when crystals were formed from iron, different magnetic states could happen [14]. Magnetization value is important to make sure that nano magnetite has magnetic ability to attract metal ions. Figure 5 showed VSM analysis result of nanomagnetite Fe_3O_4 at 25 °C. The saturation of nanomagnetite Fe_3O_4 magnetization was 59 emu/g and for nanomagnetite coated by CMKC was 23.5 emu/g. An excellent magnetic response was necessary for preconcentration metal ions in water. From the previous research [15], the saturation magnetization is about 22 emu/g, indicating an enough magnetic response to attract metal ions in preconcentration procedures.

The SEM analysis of nanomagnetite Fe_3O_4 is represented in Figure 6. As can be seen, the particles have relatively uniform structure. The nanomagnetite was agglomerate of a cluster, it was the reason that nanomagnetite has to be coated with CMKC. The diameter mean is about 31 nm, meanwhile for nanomagnetic coated by CMKC has diameter mean of about 51 nm. The diameter of nanomagnetic coated by CMKC was bigger and was not agglomerated into a cluster. Instead, it is smoother and more separated than the uncoated nanomagnetite. Both type of nanomagnetite are still categorized in nano size. Magnetite would categorized as nanomagnetite if the diameter was under 100 nm [16].

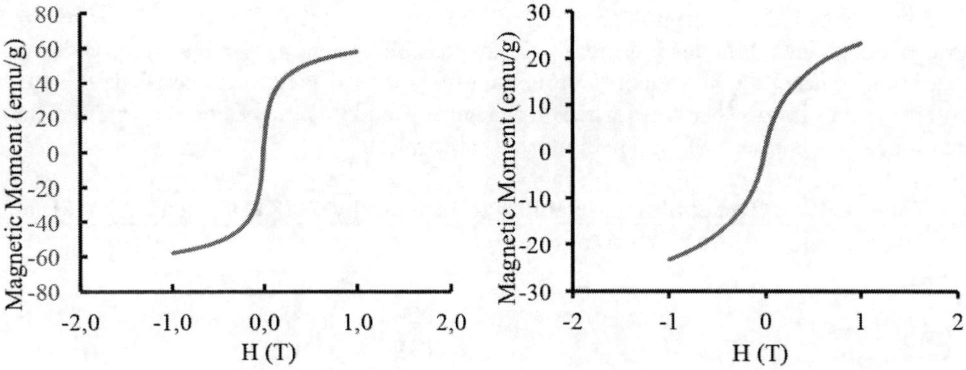

(a) nanomagnetite. (b) nanomagnetite coated by CMKC.

Figure 5. VSM Analysis of Coated and Uncoated Fe_3O_4 Nanomagnetite.

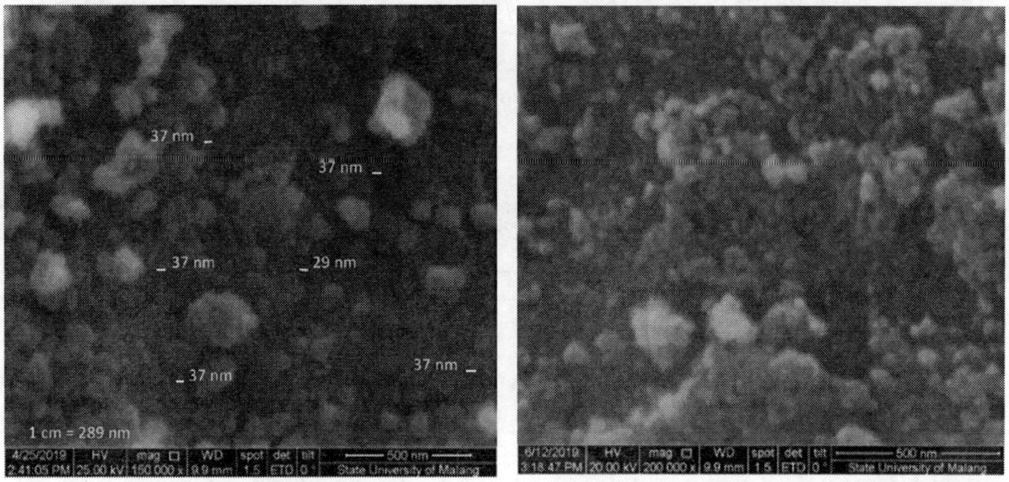

(a) nanomagnetite **(b) nanomagnetite coated by CMKC**

Figure 6. SEM Analysis of Nanomagnetite.

3.2. Preconcentration of Samples Containing Cu(II), Co(II), Ni(II), and Cr(III)

pH of aqueous solution is an important parameter that controlled in metal ion adsorption processes as reported by several authors in the literature. The effect of solution pH (e.g. pH 3, 5, 7) on the adsorption of Cu(II), Co(II), Ni(II), and Cr(III) ions from aqueous solutions use nanomagnetite coated by CMKC as shown in Table 1. Experiments were carried out at 298 K with a contacting time of 40 minutes. The results demonstrate that adsorption of metal ions dependent on the pH. The adsorption capacity increased by increasing the pH. The most optimum pH of this experiment is 7 and used for the next preconcentration procedure. Real concentration of each metal ion slightly different with the theoretical one, due to lack of accuracy and precision of analytical balance. The sensitivity of FAAS instrument nd the material it self also influenced the different of concentration. Standard solution made in very low concentration as a simulation for preliminary study to develop preconcentration method of metal ions in drinking water. As we knew that metal ions in drinking water at trace levels.

Contact time was the second important parameter to determine the effectivity of preconcentration procedures. pH 7 was the most optimum condition to be used in the second preconcentration. Contact time were varied: 20, 40, and 60 minutes. The results showed that each metal has their own optimum contact time. According to Table 2, 40 minutes is the optimum contact time for Cu(II) and Cr(III), meanwhile 60 minutes is the optimum contact time for Co(II) and Ni(II). When the adsorption has reached equilibrium, the adsorption power of nanomagnetite-CMKC would be decreases, so the ability to adsorb metal ions will also decrease. It is possible when nanomagnetite-CMKC reaches the saturated state caused the adsorbent to erode and the attached metal ions could be separate from the active adsorbent group. Therefore, when the metal ion and adsorbent are contacted at the optimum contact time, the adsorbent will gradually lose its absorption.

Table 1. The Effect of pH on Metal Ion Adsorptions by CMKC-Coated-Nanomagnetite.

Metal	**Cu(II) (ppm)**			**Co(II) (ppm)**		
pH	**3**	**5**	**7**	**3**	**5**	**7**
$C_{initial}$	5.53	5.53	5.53	43.31	43.31	43.31
C_{remain}	1.31	0.23	0.0019	15.75	10.04	2.19
$C_{desorption}$	9.03	11.93	18.57	18.15	27.13	39.17
Metal	**Ni(II) (ppm)**			**Cr(III) (ppm)**		
pH	**3**	**5**	**7**	**3**	**5**	**7**
$C_{initial}$	27.1	27.10	27.1	18.76	18.76	18.76
C_{remain}	4.32	2.14	0.51	17.55	2.75	0.04
$C_{desorption}$	2.95	4.83	6.39	5.86	7.40	12.53

$^{a}C_{initial}$ = concentration of initial sample before adsorption
$^{b}C_{remains}$ = concentration of filtrate 1
$^{c}C_{desorption}$ = concentration of filtrate 2.

Table 2. The Effect of Contact Time on Metal Ion Adsorptions by CMKC-Coated-Nanomagnetite.

Metal	**Cu(II) (ppm)**			**Co(II) (ppm)**		
Contact Time (minutes)	**20**	**40**	**60**	**20**	**40**	**60**
$C_{initial}$	5.53	**5.53**	5.53	43.31	43.31	**43.31**
C_{remain}	0.05	**0.0019**	0.04	3.95	2.19	**0.32**
$C_{desorption}$	5.37	**18.57**	8.74	51.77	39.14	**104.56**
Metal	**Ni(II) (ppm)**			**Cr(III) (ppm)**		
Contact Time (minutes)	**20**	**40**	**60**	**20**	**40**	**60**
$C_{initial}$	27.1	27.1	**27.1**	18.755	**18.76**	18.755
C_{remain}	0.0482	0.50685	**0.00837**	0.1667	**0.039**	0.3421
$C_{desorption}$	-	-	**7.0017**	-	**12.53**	-

$^{a}C_{initial}$ = concentration of initial sample before adsorption
$^{b}C_{remains}$ = concentration of filtrate 1
$^{c}C_{desorption}$ = concentration of filtrate 2.

The effectivity determination of preconcentration procedures was carried out by using the optimum condition, which is at pH 7 and 40 minutes for Cu(II) and Cr(III), while Ni(II) and Co(II) at pH 7 and 60 minutes. Table 3 presented data of adsorption capacity and percentage of metal ion removal, meanwhile the data for desorption capacity and concentration increasing are presented in Table 4. According to the data, the adsorption capacity of nanomagnetite-CMKC around 2.76 – 21.50 mg/g. Adsorption capacity is defined to the amount of metal ion that attached to per gram of nanomagnetite-CMKC. Percentage of metal ion removal showed that nanomagnetite-CMKC has high result which is above 99% for all of metal ions.

Based on Table 4, desorption capacity is defined the amount of metal ion that desorbed or un-attached from nanomagnetite-CMKC. Metal ions released by addition of 20 mL of HNO_3. HNO_3 was chosen as desorption solution because it produces a lot of H^+ ion that could deprotonated carboxyl groups of the adsorbent that bind with metal ions. In contrast with adsorption capacity, the result of desorption capacity was slightly different, in which the highest desorption capacity was shown by Cr(III) ion (12.50 mg/g) and the lowest percentage was shown by Cu(II) ion (3.71 mg/g).

The different between initial concentration and desorption concentration was used to determine the increasing of sample concentration as presented in Table 7. The highest value was shown by increasing concentration was Cr(III) ion, because Cr(III) ion has the biggest atomic radius that made them easier to be un-attached from the nanomagnetite than other metal ions.

Table.3 The effectivity of metal ion adsorption at the optimum condition for each metal ions.

Metal Ion	pH 7 and 40 minutes			
	$C_{initial}$ (ppm)	C_{remain} (ppm)	Adsorption Capacity (mg/g)	% Metal Ion Removal
Cu(II)	5.53	0.0019	2.76	99.97
Cr(III)	18.755	0.0391	9.36	99.80
Metal Ion	pH 7 and 60 minutes			
	$C_{initial}$ (ppm)	C_{remain} (ppm)	Adsorption Capacity (mg/g)	% Metal Ion Removal
Ni(II)	27.1	0.00837	13.55	99.97
Co(II)	43.31	0.32	21.50	99.26

Table 4. The effectivity of metal ion desorption at the optimum condition for each metal ions

Metal Ion	pH 7 and 40 minutes			
	$C_{initial}$ (ppm)	$C_{desorption}$ (ppm)	Desorption Capacity (mg/g)	Increasing Concentration
Cu(II)	5.53	18.57	3.71	3.36
Cr(III)	13.9	62.5	12.50	4.50
Metal Ion	pH 7 and 60 minutes			
	$C_{initial}$ (ppm)	$C_{desorption}$ (ppm)	Desorption Capacity (mg/g)	Increasing Concentration
Ni(II)	27.1	35.2	7.04	1.30
Co(II)	43.31	104.56	10.46	2.41

4. Conclusion

Nanomagnetite Fe_3O_4 were synthesized by co-precipitation method and coated with carboxymethyl *kappa*-carrageenan. These materials used for preconcentration method in a water artificially contaminated by metal ions, such as Cu(II), Co(II), Ni(II), and Cr(III). Nanomagnetite has mean particle size of 31 nm and 51 nm for nanomagnetite coated by CMKC. Saturation magnetization of nanomagnetite was 56 emu/g meanwhile for nanomagnetite coated by CMKC was 23,5 emu/g. The most optimum condition to adsorb metal ions was pH 7 in 40 minutes for Cu(II) and Cr(III) ions, while pH 7 in 60 minutes for Ni(II) and Co(II). The highest adsorption capacity of nanomagnetite-CMKC is Co(II) ion (21.50 mg/g) and the lowest is Cu(II) ion (2.76 mg/g), meanwhile for the highest desorption capacity is Cr(III) ion (12.50 mg/g) and the lowest is Cu(II) ion (3.71 mg/g). This preconcentration procedure effectively remove metal ions at trace level and the preliminary study on this research could be developed in further research.

Acknowledgement

We thank to LP2M for granting our research by PNBP program of Universitas Negeri Malang. We also acknowledge colleagues who have been involved in the laboratory work.

References

[1] Janusa M A and Beck J N 2002 *Appl. Spectrosc. Rev.* **37** 137–86
[2] Shakerian F, Haji Shabani A M, Dadfarnia S and Shabani M 2013 *J. Chem.* **2013**

[3] Ipeaiyeda A R and Ayoade A R 2017 *Appl. Water Sci.* **7** 4449–59
[4] Arslan Y, Kabak B, Trak D and Kendüzler E 2018 *Anal. Sci.* **34** 1143–7
[5] Naeemullah, Kazi T G, Shah F, Afridi H I, Khan S, Arian S S and Brahman K D 2012 *J. Anal. Methods Chem.* **1**
[6] Moghimi A and Yari M 2019 1–18
[7] Drahansky M, Paridah M ., Moradbak A, Mohamed A ., Owolabi F abdulwahab taiwo, Asniza M and Abdul Khalid S H . 2016 *Adv. Surf. Eng. Res.* **i** 13
[8] Kartika Kusumaningrum I, Marfuah S, Fadilah M N, Nazriati, Ricky Wijaya A and Kurniawan F 2019 *IOP Conf. Ser. Earth Environ. Sci.* **276** 012021
[9] Dorniani D, Hussein M Z Bin, Kura A U, Fakurazi S, Shaari A H and Ahmad Z 2012 *Int. J. Nanomedicine* **7** 5745–56
[10] Mahdavinia G R, Etemadi H and Soleymani F 2015 *Carbohydr. Polym.* **128** 112–21
[11] Giraldo L, Erto A and Moreno-Piraján J C 2013 *Adsorption* **19** 465–74
[12] Fajaroh F, Setyawan H, Widiyastuti W and Winardi S 2012 *Adv. Powder Technol.* **23** 328–33
[13] Fajaroh F, Setyawan H, Sutrisno, Nazriati and Wonorahardjo S 2014 *AIP Conf. Proc.* **1586** 179–82
[14] Daoush W M 2017 *J. Nanomedicine Res.* **5** 3–8
[15] Pourghazi K, Diva M A and Karimi S 2017 **2** 223–9
[16] Pham X N, Nguyen T P, Pham T N, Tran T T N and Tran T V T 2016 *Adv. Nat. Sci. Nanosci. Nanotechnol.* **7**

Toxicity and Radical Scavenger Properties of Various Extracts of Sponge *Clionidae sp.* Kangean Islands

Moh. Farid Rahman[1*], Masruri MASRURI[1], Alyaa Farrah Dibha[1]

[1] Chemistry Department, Faculty of Science, Brawijaya University, Jl. Veteran Malang, 65145, Indonesia

[*]Corresponding author: mfaridrhmn@gmail.com

Abstract. Sponge *Clionidae sp.* is a multicellular animal that has active secondary metabolites that are used for survival. Sponge has cytotoxic, anticancer and antitumor properties and can be used in the pharmaceutical field, but the toxicity and antioxidant levels must be done to determine the uses level. Toxicity levels of sponge *Clionidae sp.* against *Artemia salina L.* and the IC_{50} value of the antioxidant was tested against DPPH radicals. Extraction of sponge *Clionidae sp.* was carried out with sonication using methanol, n-butanol, ethyl acetate and n-hexane as solvents. Then, the crude extract was tested for toxicity levels by the BSLT method and the antioxidant test by the DPPH method. Sponge extracts with ethyl acetate has the highest LC_{50} at 62.50 ppm and n-butanol has the highest IC_{50} values of 77.34 ppm. Identification of extract sponge with ethyl acetate was carried out using a UV-Vis spectrophotometer, resulting a maximum wavelength at 407.5 nm, ethyl acetate sponge extract showed the presence of O-H stretch groups 3327 cm^{-1}, C-H stretch 2922 cm^{-1}, C=C aromatic 1462 cm^{-1}, and C=O 1711 cm^{-1} by FT-IR. The dominant compound in ethyl acetate extract that identified in LC-HRMS are chlorhexidine, 1,3,7-octanetriol, tobramycin, stigmatellin Y and maraniol.

1. Introduction

As many as 71% of the earth's surface is covered by sea water that makes sea have a variety of natural resources. Marine organisms are very diverse both biological and animal including algae, corals, sponges and other intervertebrate animals [1]. Indonesia is a country that rich in natural products, one of that is a sea sponge. Sugumaran, et al [2] stated that marine organisms, especially sponges are the best source for antimicrobial and antitumor substances. However, until now the utilization of marine products has not been maximized especially in the pharmaceutical field [3]. Marine metabolites and other natural products work in competition with their environment. Therefore, many marine organisms have evolved into efficient metabolites and exhibit high toxicity or biological activity even at very low concentrations [2]. Bioactive metabolic substances are often found in certain concentrations produced by marine sponges [4].

Sponge can be the most promising organism for drug development because it has the ability to synthesize a variety of organic products and can be a major source of biologically active marine products [5]. The last few decades, sea sponge has been considered as a material to find bioactive chemicals related to the presence of primary and secondary metabolites in it. Sea sponges produce a high concentrations of cytotoxic substances by secrete a toxic mucus that can kill other marine species as a defense against themselves [6]. Compounds from marine sources such as sponges have bioactive properties such as antitumor, anticancer, antiproliferation, cytotoxic and antibiotics. These compounds have been isolated from marine sources having chemical compounds such as phenols, alkaloids, terpenoids, polyester, and other secondary metabolites found in sea sponges, seaweed and dinoflagellates [7].

Figure 1. Sponge Clionidae *sp.*

Sponge from the *Clionidae* family is a type of digging organism that can dig calcareous substrates such as rocks, coral reefs or oyster shells. Previous research of sponge *Clionidae sp.* reported by Palermo, et al. in 1998 that the sponge *Clionidae sp.* from the genus *Cliona* was isolated and was identified had an aromatic hexacyclic alkaloid, Storniamides A-D. In addition, it was also reported that the isolate *Cliona chilensis* also contained a tripeptide alkaloid namely Celenamide E. [8].

Brine Shrimp Lethality Test (BSLT) is a method for screening the active compounds from natural materials. BSLT method are used to determine the LC_{50} value and detect the toxicity level of compounds [9]. The antioxidant activity was tested using the DPPH radical solutions 2,2-Diphenyl-1-pikrilhidrazil, in the presence of antioxidant compounds that able to donate the H atom or the electron, the absorbance of the solution will be decrease and form a stable molecule. The antioxidant test using DPPH method is used to determine the IC_{50} value. IC_{50} value is the concentration of the extract to scavenging 50 % of radical[10]

Antioxidant test on sponge extract needs to be done to determine the antioxidant activity of the dominant compound containing phenolic or heteroaromatic compounds on the dominant component of the extract [11]. A drug must be among the lowest drug concentrations in plasma to get the desired drug working. These compounds can be used as drugs if the using does not exceed the LC_{50} value of the toxicity value. Therefore, a toxicity test is necessary to be evaluated [12].

Based on the background above, this study was conducted with the aim of identifying secondary metabolite compounds contained in the sponge extract of the family *Clionidae sp.* and conducting toxicity tests and their antioxidant tests.

2. Experimental Section

2.1. Chemicals and Instrumentation

The materials used in this study were sponge *Clionidae sp.* from Kangean Island, ethyl acetate (Smart Lab), n-hexane (Full time), n-butanol (Smart Lab), ethanol 96% (Merck) and methanol (Full time), Dimethyl Sulfoxide (DMSO) solution (Sigma Aldrich), sea water and *Artemia salina L.* and 2,2-Diphenyl-1-picrylhydrazyl (Sigma-Aldrich). The instrument used in this research were sonicator, rotary evaporator (IKA RV 10 digital), UV Vis Spectrophotometer (Shimadzu 1600), FTIR spectrophotometer (Shimadzu 8400S, extract was analyzed using KBr pellet), LC HRMS (Thermo Fisher Scientific).

2.2. Sponge Collection

Sponges were collected from Kangean Island, Madura, Indonesia. Sponges were stored in a freezer and cleaned with ethanol 96 %. After that sponge cutted into small pieces and dried in the room temperature (not exposed to direct sunlight) for 2-3 days.

2.3. Sponge Extraction

The sponges were grounded to a fine powder. The powdered sponge was extracted with different solvents, n-hexane, ethyl acetate, n-butanol and methanol with sonicator for 1 h. The extracts were filtered through filter paper and concentrated with rotary evaporator. The crude extract was stored at 5 ^{o}C and used in toxicity and antioxidant assay.

2.4. Toxicity Assay with BSLT (Brine Shrimp Lethality Test) Method
Subramani, et al. in 2013 [13], was used to study the toxicity test of sponge with different solvents using *Artemia salina L.* method. The toxicity properties known by the mortality number of *Artemia salina L.*, and the LC_{50} (Lethal Concentration 50) can be determined.

2.4.1. The Hatching of Brine Shrimp Larvae. Brine shrimp were hatched in 150 mL of seawater under aeration condition for 48 h. After 48 h the eggs hatched and shrimp larvae were ready to be used for toxicity testing of sponge extract.

2.4.2. Preparation of Sample Solution. The crude extracts of methanol, n-butanol, ethyl acetate were made a concentration of 200 ppm as a stock solution by adding 1% Dimethyl Sulfoxide (DMSO) to the total solution. Then, a dilution was carried out of 5 ppm; 10 ppm; 25 ppm; 50 ppm; 75 ppm; 100 ppm; 125 ppm; 150 ppm by using sea water.

2.4.3. Toxicity Test of Sponge Extract with Shrimp Larvae. Twenty larvae shrimp were putted into vial bottle and the 5 mL of test solution was added with respective concentrations; 5 ppm; 10 ppm; 25 ppm; 50 ppm; 75 ppm; 100 ppm; 125 ppm; 150 ppm, then observed after 24 h and observed the number of dead larvae. 1% DMSO and sea water were added as blank control and compared with the test solution with a concentration of 5 ppm; 10 ppm; 25 ppm; 50 ppm; 75 ppm; 100 ppm; 125 ppm; 150 ppm. LC_{50} by making a curve between % mortality and concentration. % mortality was calculated by the equation:

% mortality = (Number of initial larvae-Number of dead larvae)/ Number of initial larvae x 100%

2.5. Antioxidant Test of Sponge Extract with DPPH Radical Methods

2.5.1. Preparation of Solution. Crude sponge extract was made a concentration of 250 ppm as a stock solution. Then, the solution is carried out to a concentration of 12.5 ppm; 37.5 ppm; 62.5 ppm; 87.5 ppm; 125 ppm; 150 ppm; 187.5 ppm using methanol.

2.5.2. Antioxidant Test of Sponge Extract. The antioxidant test with the DPPH method was adapted from Hanani, et al. in 2005 [3], it was carried out by adding 4 mL test solution with a concentration of 0 ppm; 12.5 ppm; 37.5 ppm; 62.5 ppm; 87.5 ppm; 125 ppm; 150 ppm; 187.5 ppm; 250 ppm into a vial bottle and 1 mL DPPH 0.1 mM. The test control blank was used by adding 4 mL methanol and 1 mL DPPH 0.1 mM. Control blanks and test solutions were added with DPPH 0.1 mM and incubated at 37 °C for 30 minutes. Measurements were taken at 515 nm. IC_{50} values were calculated by making a curve between % inhibition and concentration. % inhibition was calculated by the equation [14]:

$$\%\ inhibition = \frac{(Absorbance\ blank - absorbance\ sample)}{absorbance\ blank} \times 100\ \% \quad (1)$$

2.6. Characterization of Sponge Extract

2.6.1. Characterization of Sponge Extract Using UV-Vis Spectrophotometer. The method for characterizing sponge extract products with UV-Vis spectrophotometer was adapted from Pore, et al. in 2016 [15] crude sponge extract was dissolved in methanol and then measurements were taken at wavelengths of 200-800 nm. The maximum wavelength is obtained from the maximum absorbance value. The instrumentation that used was UV-Vis Spectrophotometer Shimadzu 1600.

2.6.2. Characterization of Sponge Extract Using FTIR Spectrophotometer. Sponge extract was mixed with KBr and made the pellets and inserted into the sample holder. Then, infrared light will pass through the sample so that the molecules in the sample will vibrate. The system on the computer will analyze the data in the form of spectrum in the form of %T and wavenumbers. The data was used to predict the functional groups which are constituent compounds of sponge extract.

2.6.3. Characterization of Sponge Extract Using LC-HRMS. Samples that were ready to be filtered with milipore and injected with syringes were 5 μL with a flow rate of 0.7 mL/min.

3. Result and Discussion

3.1. Crude Sponge Extract

From the extraction of sponges with various solvents, the results are described in the Table 1.

Table 1. Sponge extracts yields in various solvent.

No	Solvent	Initial Mass (g)	Extract Mass (g)	% Yield
1	Methanol	24	1.89	7.875
2	n-Butanol	24	0.40	1.67
3	Ethyl Acetate	24	0.32	1.33
4	n-Hexane	24	0.11	0.45

The polar extract which is methanol extract has the highest yield (7.875 %) and n-butanol extract (1.67 %), this means that many polar compounds are distributed in to the polar solvent. Meanwhile, ethyl acetate and n-hexane extract have 1.33 % and 0.45 % of yield, respectively; which means the distribution into non polar solvent.

3.2. Toxicity Test Value of Sponge Extract with Artemia salina L.

Based on the results of toxicity test, the linear equation is obtained to determine the LC_{50} value of the toxicity test. The LC_{50} value are described in the Table 2.

Table 2. LC_{50} value of Toxicity Test.

No	Sponge Extract	LC_{50} (ppm)
1	n-Butanol	101.77
2	Methanol	95.72
3	n-Hexane	75.61
4	Ethyl Acetate	62.50

The LC_{50} values in the toxicity test with BSLT method for n-hexane, ethyl acetate, butanol and methanol sponge extracts. Each sponge extract obtained LC_{50} values with concentrations <1000 ppm. In this study, it was known that each extract had toxic properties with an LC_{50} value <1000 ppm. Ethyl acetate sponge extract has the lowest concentration LC_{50} value among other sponge extracts of 62.50 ppm. The toxicity category can be seen from Table 3.

Table 3. Toxicity Category [9].

Categories	LC_{50} values
Supertoxic	$\leq$ 5 mg/kg
Very Toxic	5-50 mg/kg
Toxic	50-500 mg/kg
Toxic Medium	0.5-6 g/kg
Mild Toxic	5-15 g/kg
Practically non-toxic	>15 g/kg

From Table 3, ethyl acetate, n-hexane, methanol and n-butanol extracts have a toxic category.

3.3. Antioxidant Test Value of Sponge Extract with DPPH Radicals

Figure 2 was describing free radical DPPH scavenging activity of various extract sponge *Clionidae sp.*

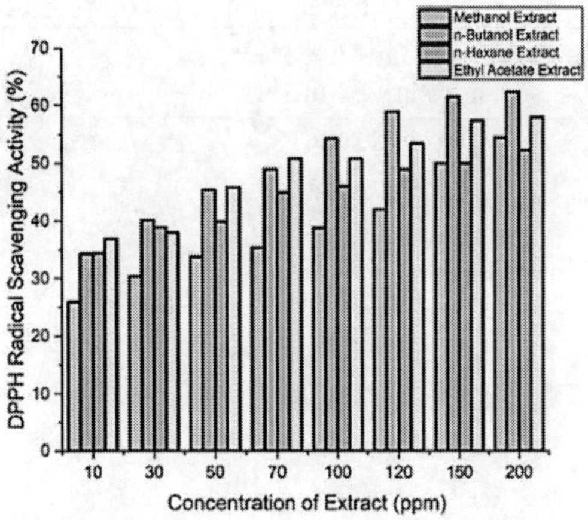

Figure 2. DPPH Radical Scavenging of Various Extract Sponge in Difference Concentration.

Table 4. The IC_{50} Value of Antioxidant Test.

No	Sponge Extract	IC_{50} (ppm)
1	Methanol	172.27
2	n-Hexane	161.17
3	Ethyl Acetate	98.96
4	n-Butanol	77.34

The IC_{50} value was described at Table 4. IC_{50} was defined as the concentration of the extract sponge solution scavenging 50% of radical. n-Butanol extract has the highest IC_{50} value 77.34 ppm which means has the higher ability of free radical scavenging using the free radical DPPH method.

3.4. Characterization Result

Table 5. Characterization result with UV – Vis and FTIR Spectrophotometer.

No	Sponge Extract	Maximum Wavelength (nm)	Wavenumber (cm^{-1})
1	Ethyl Acetate	407.5	3327 (Broad-Intense; O-H), 2922 (Sharp-Intense; C-H sp^3). 1711 (Sharp-Intense; C=O)
2	n-Hexane	405	3414 (Broad-Intense; O-H), 2924 (Sharp-Intense; C-H sp^3). 1740 (Sharp-Intense; C=O)
3	Metanol	405	3375 (Broad-Intense; O-H), 2918 (Sharp-Intense; C-H sp^3), 1638 (Sharp-Intense; C=O)
4	n-Butanol	401.5	3377 (Broad-Intense; O-H), 2922 (Sharp-Intense; C-H sp^3). 1705 (Sharp-Intense; C=O)

IOP Conf. Series: Materials Science and Engineering **833** (2020) 012037 doi:10.1088/1757-899X/833/1/012037

Table 6. The Dominant Compound of Ethyl Acetate Extract.

RT (min)	Chemical Formula and Area	Compound Name dan Relative Mass	Chemical Structure
26.392	$C_{22}H_{30}Cl_2N_{10}$ (6739874.85)	Chlorhexidine (505.4 g/mol)	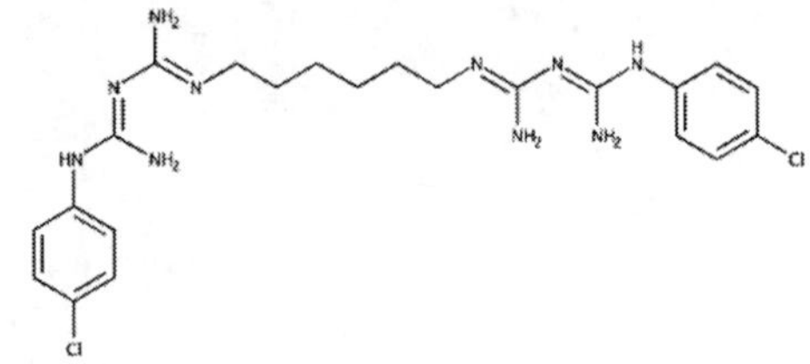
0.898	$C_8H_{18}O_3$ (4278293.48)	1,3,7-Octanetriol (162.23 g/mol)	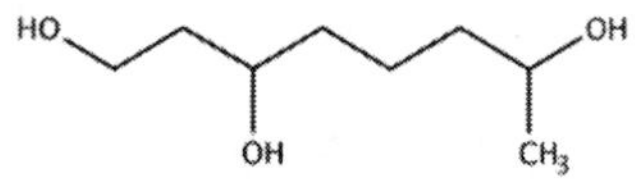
21.533	$C_{18}H_{37}N_5O_9$ (5894042.49)	Tobramycin (467.5 g/mol)	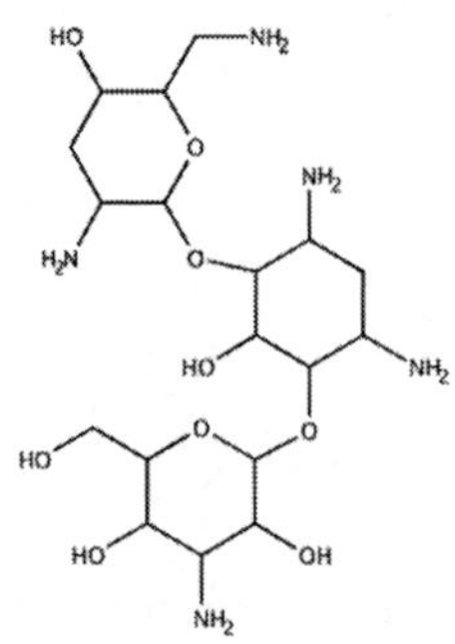
19.215	$C_{29}H_{40}O_6$ (3759166.65)	Stigmatellin Y (484.28 g/mol)	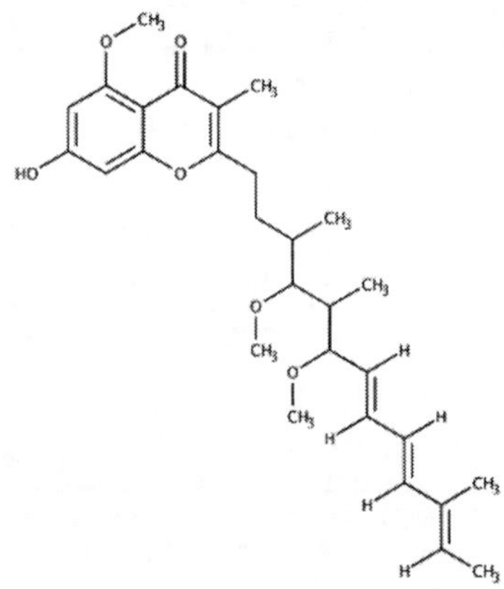
12.99	$C_{12}H_{12}O_3$ (2619117.25)	Maraniol (204.07 g/mol)	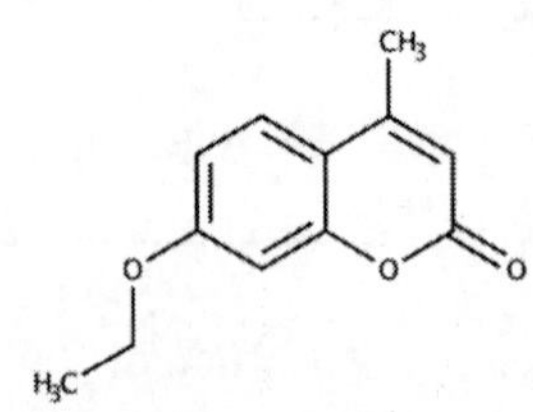

Chlorhexidine is a biguanide compound used as an antiseptic agent with antibacterial activity. Chlorhexidine reacts with the negative charge of microbial cell surface, thereby destroying the integrity of the cell membrane [16]. 1,3,7-Octanetriol belongs to the organic compounds known as a fatty alcohol. 1,3,7-Octanetriol can be found at plant and animals [17].

Maraniol and stigmatellin Y are compounds that have chromone groups. Chromone group is a natural compound consisting of 4H-benzopyran-4-one. Chromone group can be found in several species of higher organism. Chromone group has antioxidant activity by reacting directly with free radicals and through the chelation process of Fe (II) and Cu (II). Chromone compounds or chromone compound derivatives are known with their anticancer properties. The treatment of cancer, especially breast cancer with chromone compounds, makes modulation of drug transportation more efficient to a higher level of chemotherapy [17].

Tobramycin is identified as part of a complex antibiotic. Tobramycin can inhibit protein synthesis by binding with 30S ribosomes from ribosomes of prokaryotic cells. Tobramycin has antibacterial properties against *Staphylococcus pneumoniae* and tobramycin are active against many aerobic and facultative gram-negative bacteria including *Enterobacteriaceae*, *Pseudomonas aeruginosa* and *Acinetobacter sp.*[18,19].

Conclusions

Sponge extract with ethyl acetate solvent has the highest LC_{50} value 62.50 ppm and the other sponge extracts obtained toxic properties with LC_{50} values concentrations <1000 ppm. Meanwhile, sponge extract with n-butanol solvent has the highest IC_{50} value 77.34 ppm which means has the higher ability of free radical scavenging. Furthermore, the dominant compound of ethyl acetate extracts are chlorhexidine, 1,3,7-octanetriol, tobramycin, stigmatellin Y and maraniol.

References

[1] G. Ercolano *et al* 2019 *Mar. Drugs* **17** 1 31
[2] M. Sugumaran *et al* 2010 *Mar. Drugs* **8** 12 2906–2935
[3] E. Hanani *et al* 2005 *Maj. Ilmu Kefarmasian* **2** 3 127–133
[4] G. Mohan *et al* 2016 *Biotechnol. Rep.* **11** 44–52
[5] G. Beedessee *et al* 2012 *Environ. Toxicol. Pharmacol.* **34** 2 397–408
[6] S. Perdicaris *et al* 2013 *Nat. Prod. Chem. Res.* **01** 03
[7] G. M. Suarez-Jimenez *et al* 2012 *Mar. Drugs* **10** 12 963–986
[8] J. A. Palermo *et al* 1998 *J. Nat. Prod.* **61** 4 488–490
[9] M. S. Lestari *et al* 2015 *J. of Chem and Pharm Research* **7** 4 866-872
[10] H. Seradj *et al* 2012 *Iranian Journal of Pharmaceutical Sci.* **8** 4 249-255
[11] K. Chairman *et al* 2012 *Asian Pac. J. Trop. Dis.* **2** 3 234–238
[12] Nuryati 2017 *Farmakologi* (Jakarta: Kementerian Kesehatan Republik Indonesia) chapter 1 pp 7
[13] R. Subramani *et al* 2013 *Asian Pac. J. Trop. Biomed.* **3** 4 291–296
[14] N. W. S. Agustini *et al* 2017 *KnE Life Sci.* **3** 4 147
[15] T. S. Pore *et al* 2016 *Res. J. Life Sci. Bioinforma. Pharm. Chem. Sci.***1** 6
[16] Chlorhexidine *National Center for Biotechnology Information. PubChem Compound Database* U.S. National Library of Medicine pubchem.ncbi.nlm.nih.gov/compound/9552079.
[17] (3R,7R)-1,3,7-Octanetriol *National Center for Biotechnology Information. PubChem Compound Database* U.S. National Library of *Medicine* pubchem.ncbi.nlm.nih.gov/compound/85279741
[18] N. F. L. Machado *et al* 2010 *Curr. Bioact. Compd.* **6** 2 76–89
[19] J. AL Buensalido 2014 *J. Pharmacovigil.* **02** 02

Evaluation of The Stability and In Vitro Anti-inflammatory Activity of Partially Purified Bromelain Nanoemulsion

Fransiskus Randy[1], Siswati Setiasih[1], Mahdi Jufri[2], Sri Handayani[1], Sumi Hudiyono[1*]

[1]Department of Chemistry, Faculty of Mathematics and Natural Sciences, Universitas Indonesia, Depok, Indonesia
[2] Faculty of Pharmacy, Universitas Indonesia, Depok, Indonesia

*Corresponding author: sumi.hudiyono@sci.ui.ac.id

Abstract. Bromelain is an enzyme belonging to cysteine protease. It is known for topical use as debridement for inflammation treatment. Nanoemulsion of bromelain (NEB) aims to solve this problem and also improve its stability. NEB was made by high emulsification method using homogenizer and sonicator. Nanoemulsion physicochemical stabilities were evaluated. The bromelain as an active compound was isolated from the core of pineapple (*Ananas comosus* [L.] Merr) which was purified by ammonium sulphate fractionation and then dialysis followed. Nanoemulsion formula 2 was found to be the most suitable formula for encapsulating bromelain with good physicochemical stability after 30 days of storage. This formula showed clear visual appearance with globule diameter of 37.42 nm with polydispersity index of 0.40. The result showed that the fraction of bromelain obtained from each purification step showed an increase in specific activity. The specific activities of fractionation using 20-50% ammonium sulphate and dialysis were 152.05 and 266.58 U/mg, respectively. The purest bromelain fraction was encapsulated into nanoemulsion formula 2. The bromelain fraction's anti-inflammatory activity was tested based on HRBC membrane stabilization method. The result showed that bromelain dialysis fraction shows the ability as anti-inflammatory agent with stability percentage of 79.76%.

1. Introduction

Pineapple is one of Indonesia's most leading fruit with a production of 1.73 tons in 2015 [1]. Indonesia exports canned pineapple to various countries and a lot of waste is made such as pineapple peel, cores, stem, and crown. The pineapple wastes were only use for cattle feed even though it contains bromelain. Bromelain is a proteolytic enzyme with sulfhydryl group as an active site. Bromelain is used as analgesic, anti-tumors and anti-inflammatory agent [2]. Inflammation is a vascular tissue's biologic response against the endanger impulse such as pathogen, irritant, and cell damage to initiate the healing process [3]. One of the methods of bromelain's application as anti-inflammatory agent is via transdermal route. Transdermal is an administration route where active compounds are delivered across the skin to give systemic effect.

The main problem of transdermal route is its penetration rate through the outer skin layer [4]. Nowadays, nanoemulsion is one of the most promising way for transdermal drug delivery. Nanoemulsion can increase the bioavailability due to its big surface area and small free energy. This advantage can prevent the emulsion from creaming, flocculation, coalescence, and sedimentation process [5]. Transdermal drug delivery system can prevent first pass metabolism, estimate activity duration, minimize side effect, increase pharmacologic and physiologic response, and increase patient comfort.

Bromelain can be loaded in nanoemulsion by spontaneous emulsification [6]. The small droplet size of emulsion can increase bromelain penetration levels. The nanoemulsion particle size is determined by formula of substance mixture. Accordingly, this study focuses on the formulation, evaluation physicochemical properties, stability, and anti-inflammatory activity of bromelain-loaded nanoemulsions.

2. Methods

2.1. Materials
The bromelain used in this research was isolated from pineapple cores (*Ananas comosus* [L.] Merr), which purchased from Induk traditional market, Jakarta. All chemical substances and reagents were purchased from Merck.

2.2. Bromelain isolation, ammonium sulphate precipitation, and dialysis
The process of isolation of ammonium sulfate, and dialysis were conducted by the previous method with slight modification [7].

2.3. Determination of bromelain specific activity
The specific activity was determined by dividing the total proteolytic activity with total protein contents. The total protein contents were determined by Lowry method and the proteolytic activities were determined by Kunitz method with some modifications [8].

2.4. Preparation of bromelain-loaded nanoemulsion
The bromelain-loaded nanoemulsion was produced by using spontaneous emulsification method. Water phase such as tween 80, span 60, and aquadest were homogenized at 3000 rpm using homogenizer (Omni-Multimix Inc., Malaysia). Oil phase consist of oleic acid and vitamin E acetate were dissolved in crude palm oil (CPO) and virgin coconut oil (VCO). The oil phase was added into the water phase and then the mixtures were homogenized. Methylparaben and propylparaben were dissolved in propylene glycol and then added into the emulsions. The compositions of emulsions were using the formulation design in Table 1 and these were homogenized at 4000 rpm for 20 min. To make nanoemulsions, the emulsions were sonicated at 24kHz, 27 °C for 60 min by using ultrasound Hielscher UP400s. After physicochemical properties determination and stability test, the most stable nanoemulsion were used as a carrier of bromelain. The bromelain was added to the nanoemulsions and then stirred using magnetic stirrer. The bromelain-loaded nanoemulsions were sonicated for 30 min until the color is clear.

2.5. Evaluation of Nanoemulsion Physicochemical Properties and Stability
Organoleptic evaluation (consists of color, odor, clarity), pH, and cycling test of stability was carried out at 40 °C, 27 °C, 4 °C of nanoemulsion. The organoleptic evaluation was weekly observed for 4 weeks (Agoes, 2012). Density of nanoemulsions was determined by using pycnometer at room temperature. The viscosity of nanoemulsions was determined by using the Brookfield viscometer and spindle 1 with speed of 0.5, 1, 2, 2.5, 4, 10, 20 rpm to get rheology data [9]. The nanoemulsions type was conducted by dilution method, where the nanoemulsions were added into water (1:10 w/w) and oil (1:10 w/w). If the nanoemulsion soluble in water, it will become oil in water type (O/W). While, if the nanoemulsion soluble in oil, it will become water in oil type (W/O). In addition, the particle size of nanoemulsions was carried out by using Zetasizer Nano S (Malvern) at the second week after the nanoemulsion making.

2.6. *HRBC In Vitro Test of Anti-inflammatory Activity*

Human red blood cells (HRBC) were obtained from PMI Depok which already contains anticoagulants. Determination of anti-inflammatory activity was done by Oyedapo method [10]. Some substances were mixed by following the Table 2 and the absorbance was measured using spectrophotometer. The HRBC membrane stability was calculated by following equation,

$$\%\ \text{HRBC membrane stability} = \left(\frac{A1-A2}{A1}\right) \times 100\%, \qquad (1)$$

where, A1 = absorbance of aspirin (positive control) and A2 = absorbance of sample.

Table 1 Formulation design of nanoemulsion.

No.	Substance	Nanoemulsion formula (%b/b)		
		1	2	3
1.	Tween 80	25	35	30
2.	Span 60	25	-	25
3.	Oleic acid	-	5	-
4.	*Crude Palm Oil*	5	5	5
5.	VCO	5	5	5
6.	Propylene glycol	5	5	5
7.	Methylparaben	0.3	0.3	0.3
8.	Propylparaben	0.6	0.6	0.6
9.	Vitamin E acetate	0.01	0.01	0.01
10.	Aquadest	54.09	44.09	49.09

Table 2. Composition of HRBC anti-inflammatory test mixture.

Blank	Positive control	Sample
2 mL of hyposaline	2 mL of hyposaline	2 mL of hyposaline
1 mL of phosphate buffer 0.15 M pH 7.4	1 mL of phosphate buffer 0.15 M pH 7.4	1 mL of phosphate buffer 0.15 M pH 7.4
0.5 mL of HRBC suspense (10%v/v)	0.5 mL of HRBC suspense (10%v/v)	0.5 mL of HRBC suspense (10%v/v)
-	1 mL of aspirine	1 mL of bromelain
	incubated for 30 min at 37 °C	
	centrifuged at 3000 rpm for 30 min	
	absorbance measurement at λ560 nm	

3. Result and Discussion

3.1. *Bromelain isolation, ammonium sulphate precipitation, and dialysis*

Bromelain specific activities are shown at Table 2. In line with purification steps, the specific bromelain activities were increased. The highest specific activity of ammonium sulfate salt fraction is in the second fraction which the purification level increases about 2.30 times compared to crude enzyme. The second fraction was dialyzed using membrane semipermeable to omit the salt content in bromelain. The principle of dialysis is diffusion, which the salt and small protein non-bromelain will penetrate throughout the membrane. The result has similar trend with previous study [6, 7], which the specific activity was increased from 63.65 to 266.58 U/mg with the purification levels 4.04 compared to crude enzyme. The dialyzed F2 is used as active compound in nanoemulsion preparation.

3.2. *Evaluation of nanoemulsions physicochemical and stability*

The result of organoleptic evaluation and density determination of nanoemulsions are reported at the Table 4. The density between the three nanoemulsions is found to be quite similar. These three nanoemulsions are odorless and different in color appearance. The phase separation only occurs in the first formula at the third week. Therefore, the first formula is rejected.

Table 3. Bromelain specific activities from each purification step.

Fraction	Volume (mL)	Total		Specific activity (U/mg)	Purification levels (times)
		Protein content (mg)	Proteolytic activity (U)		
Crude enzyme	410	3355.84	50.84	66.01	1
F1 (0-20%)	4	36.01	0.25	144.41	2.19
F2 (20-50%)	6	120.72	0.80	152.05	2.30
F3 (50-80%)	2	16.10	0.25	63.65	0.96
Dialyzed F2	8	120.76	0.45	266.58	4.04

Table 4. Organoleptic and density observation of nanoemulsions.

Formula	Density	Week	Color	Phase separation	Odor
		0	White	No	Odorless
		1	White	No	Odorless
1	1.04 g/mL	2	White	No	Odorless
		3	White	Yes	Odorless
		4	White	Yes	Odorless
		0	FG medium white	No	Odorless
		1	FG medium white	No	Odorless
2	1.03 g/mL	2	FG medium white	No	Odorless
		3	FG medium white	No	Odorless
		4	FG medium white	No	Odorless
		0	FG NCP transparent white	No	Odorless
		1	FG NCP transparent white	No	Odorless
3	1.05 g/mL	2	FG NCP transparent white	No	Odorless
		3	FG NCP transparent white	No	Odorless
		4	FG NCP transparent white	No	Odorless

The determination of nanoemulsion type was conducted by dilution method in water phase and oil phase (Figure 1). The nanoemulsion is perfectly soluble in water as dispersion medium. This concluded that the nanoemulsion type is water in oil (W/O) and suitable with the purpose of the skin use.

Figure 1. Observation of nanoemulsion dilution in; a) water, b) oil.

Based on the data from Zetasizer Nano S, formula 2 has Z-average 37.42 nm with polydispersity index (PdI) 0.40 (Figure 2). This result fits the previous research which the droplet size for nanoemulsion is about 10-200 nm [11]. The formula 1 has droplets size similar with formula 3 which has big pdI about 0.51 with Z-average 141.1 nm.

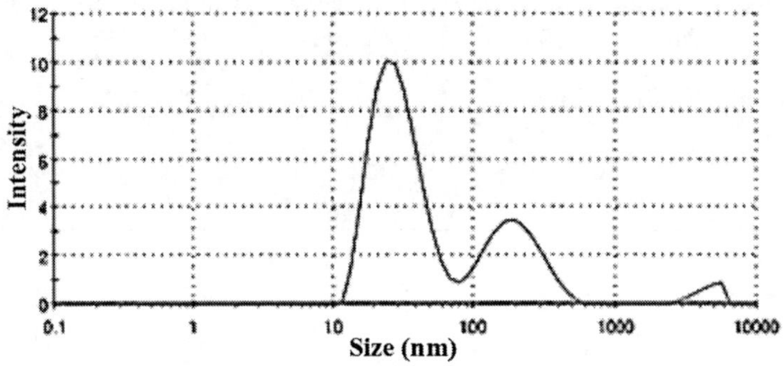

Figure 2. Droplet size of formula 2 nanoemulsions.

pH evaluation results can be seen in the Figure 3. The desired pH of nanoemulsions is about the pH value of skin, 4.5-7.0. The most probable fit for that requirement is the second formula, which pH value is in the middle of the range. The viscosity value of formula 1, 2, and 3 are 600, 200, 800 cps, respectively. It is caused by the different amount of surfactants that used. All of the nanoemulsions have similar flow behavior type that is plastic fluid (Figure 4). Based on the Figure 4, it can be seen that the rheology of nanoemulsion is plastic. In this flow property, the cream could not flow until it sheared stress over the certain yield value. The plastic flow type shows an increasing of shear rate which can cause rearrangement of tween 80 polymer chains into straight chain. This condition can lead the system resistance weaken and the viscosity decrease. Figure 5 shows that the nanoemulsions are stable in stored conditions at 40 °C, 27 °C, and 4 °C.

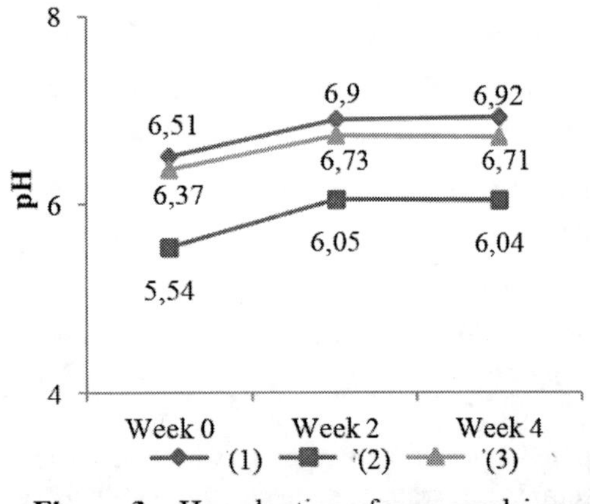

Figure 3. pH evaluation of nanoemulsions.

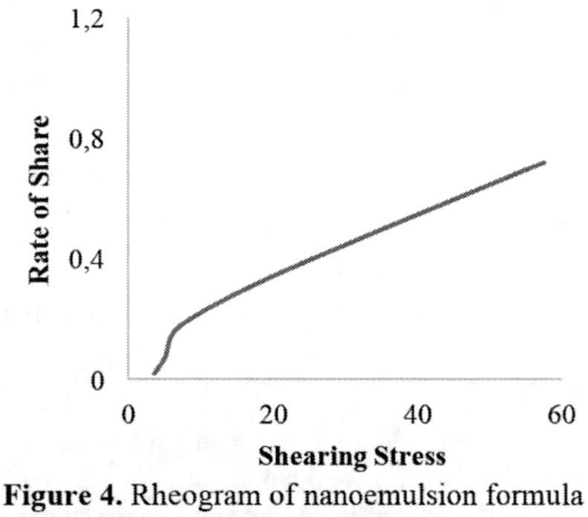

Figure 4. Rheogram of nanoemulsion formula 2.

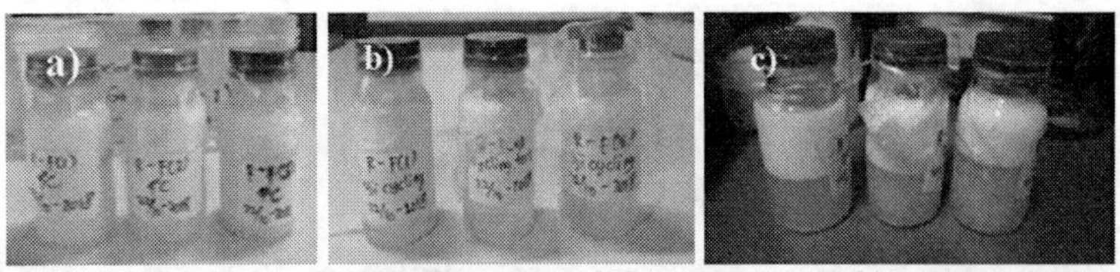

Figure 5. Physical stability of nanoemulsion at a) 4 °C, b) 27 °C, c) 40 °C.

3.3. HRBC In Vitro Test of Anti-inflammatory Activity

The absorbance in the solution sample indicated a HRBC lysis process in which the inflammation was represented. The stabilization of HRBC is monitored by absorbance value of hemoglobin that produced from HRBC lysis process [10]. The lower absorbance value showed a more stable HRBC.

Based on the data result of anti-inflammatory activities at the Figure 6, bromelain-loaded nanoemulsion has the highest stability HRBC membrane (79.76%).

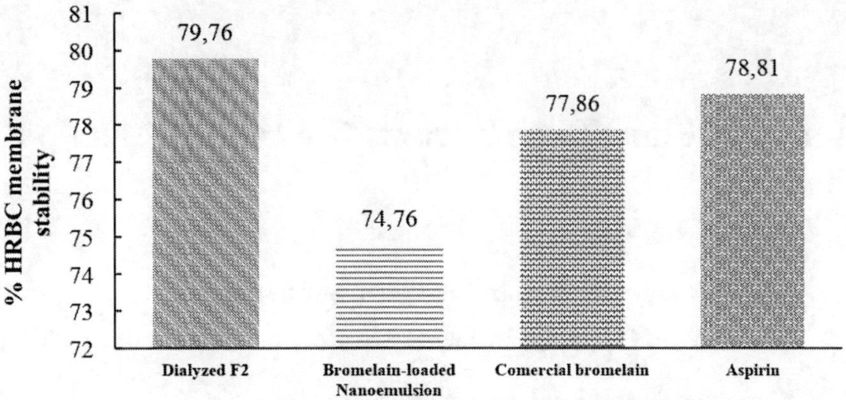

Figure 6. In vitro anti-inflammatory test of bromelain nanoemulsion by using HRBC method

4. Conclusion

The specific activity of bromelain can be increased by using ammonium sulfate precipitation and dialysis. The dialyzed F2 bromelain has the highest specific activity. Based on nanoemulsions physicochemical properties and stability test, formulation 2 is the most suitable nanoemulsion to load dialyzed F2 bromelain. The bromelain nanoemulsion has high anti-inflammatory activity.

Acknowledgements
This research was financially supported by the grant of International Publication Indexed 9 (PIT 9) program of the Univeritas Indonesia 2019.

References

[1] Respati E. 2016 *Outlook nenas* (Jakarta: Pusat Data dan Sistem Informasi Pertanian Sekretariat Jenderal Kementerian Pertanian 2016)
[2] Manzoor Z, Nawaz A, Mukhtar H, Haq I 2016 *Braz. arch. biol. Technol.* **59** 1–16
[3] Chen L, Deng H, Cui H, Fang J, Zuo Z, Deng J, Li Y, Wang X, Zhao L 2018 Oncotarget **9**(6) 7204-7128
[4] Jagannath, SS, Manohar SD, Bhanudas SR 2013 *World J. Pharm. Pharm. Sci.* **3**(2) 1068-80
[5] Gupta PK, Pandit JK, Kumar A, Swaroop P, Gupta S 2010 *T. Ph. Res.* **3**(1) 117-138
[6] Rachmawati H, Sulastri E, Iwo MI, Safitri D, Rahma A. J 2016 *Nano Res.* **40** 156-166
[7] Setiasih S, Prabowo HA, Budianto E, Hudiyono S 2018 *J. App. Pharm. Sci.* **8**(10) 17-24
[8] Musfiroh FF, Ilyas NM, Setiasih S, Handayani S, Hudiyono S 2018 IOP Conf. Ser.: Mat. Sci. Eng. **299** 1-4
[9] Kumar KK, Sasikanth K, Sabareesh M, Dorababu N 2011 *Asian J. Pharm. Clin. Res.* **4** 93-98
[10] Oyedapo OO, Akinpelu BA, Akinwunmi KF, Adeyinka MO, Sipeolu FO 2010 *Int. J. Plant Physiol. Biochem.* **2**(4) 46-51
[11] Devarajan V, Ravichandran V 2011 *Int. J. Compr. Pharm.* **4**(01) 1-6

IOP Conf. Series: Materials Science and Engineering **833** (2020) 012039 doi:10.1088/1757-899X/833/1/012039

The Potency of Plant Dye Extracts for Halal Detection on Consumed Animal Fats

Rurini Retnowati*, Hermin Sulistyarti, Suratmo

Department of Chemistry, Faculty of Science, Brawijaya University, Indonesia

*Corresponding author: rretnowati@ub.ac.id

Abstract. One point of the halal concepts is the food does not contain any lard or any pigs derivatives. UV-vis spectrophotometry using natural dyes, among other methods, can be developed for animal fat analysis. One factor that influence the analysis is the solubility of animal fats and natural dyes in organic solvents. This research was conducted to study the differences of UV-vis spectrum profile in animal fats (chicken, beef, lamb, pork) using various natural dyes (namely *Curcuma longa Linn.*, *Curcuma zanthorriza*, *Curcuma heyneana*, *Zingiber montanum*, *Uncaria*, *Caesalpinia sappan*, and *Areca catechu*) in different type of solvents. The research consists of four main steps: 1) screening of plant dyes in several solvents with different polarity, 2) screening of animal fat in several solvents with different polarity, 3) screening of animal fats mixed with plant dyes in selected solvents, and 4) analysis of animal fats mixed with plant dyes in selected solvents by UV-Vis spectrophotometry. The results showed that plant dyes from *C. longa* Linn., *C. zanthorrhiza*, *A. catechu*, and *Uncaria*, well as all animal fats used in this work, were soluble in ethyl-acetate and isopropanol. UV-vis spectrum profile of the animal fats with turmeric dyes in ethyl-acetate for chicken, beef, pork, and lamb fats, shows peaks at λ of 484, 792, 488.5, and 741 nm with absorbance of 3.913, 0.816, 3.524, and 0.175 respectively. Meanwhile in isopropanol, the profile for chicken, beef, pork, and lamb fats have λ of 751, 787, 712, 499 nm with absorbance of 0.007, 1.012, 0.479, and 3.913, respectively. Therefore, the UV-vis spectrum profile of animal fat with *C. longa* Linn dyes in ethyl-acetate can distinguish lard from the beef and lamb fats. In isopropanol, the lard cannot be distinguished from the chicken and beef fats, but it can differentiate lard from the lamb fats. The *C. zanthorrhiza* dyes in ethyl-acetate cannot distinguish lard from the chicken, beef, and lamb fats, but *C. zanthorrhiza* dyes can distinguish lard from the beef fat using isopropanol as the solvent. For *A. catechu* and *Uncaria* dyes in ethyl-acetate and isopropanol, they all cannot distinguish lard from the chicken, beef and lamb fats.

1. Introduction

Exploration of animal fat analysis by UV and FTIR has been developed mainly because of the efficiency and simplicity of the procedure. However, these methods have limitations in which all UV spectrum profiles of animal fats are similar, as is the analysis with FTIR. Those analyses cannot identify the type and composition of each fatty acid component of a sample [1]. Fat consists of several kinds of fatty acids, both types and composition, including saturated fatty acids and unsaturated fatty acids. One type of fatty acid is distinguished by the number of carbon atoms (the length of the short chain of carbon atoms, the number of double bonds, structural configuration). The differences in the structure of these fatty acids cause differences in solubility in organic solvents with different polarity.

Fatty acid analysis can be done qualitatively by adding a fat-soluble diazonium dye, and the mixture of these compounds gives an orange to deep red color. Color intensity that occurs is proportional to the levels of unsaturated fatty acids in fatty acids. The color reaction is caused by the coupling reaction (merging) of diazonium with compounds in the samples that have double bonds. The azo-couple compound that produced is a compound which has an extension of π electron delocalization system and absorbs UV radiation in the visible area [2].

Preliminary research has been carried out for qualitative analysis of fatty acids in various vegetable fats (coconut, palm, soybean, corn, olive oils) using the diazonium reagent, producing an orange to orange-red colour. The increase in the intensity of the orange colour obtained is proportional to the levels of unsaturated fatty acids (oleic, linoleic, linolenic acids) in these vegetable fats, each of which are coconut, palm, soybean, corn and olive fat [3].

This analysis method was developed to identify animal fats that have different compositions in solvents with different polarity. To differentiate the absorption intensity, an analysis is performed by UV-vis spectrophotometry. However, this method has limitations in distinguishing color density with a short range [4]. In addition, the use of diazonium compounds as coloring reagents is carcinogenic, hence their use must be limited and reduced [5].

Plant dyes can be used as natural dyes, an alternative for reducing the use of harmful synthetic dyes. Yellow fat-soluble dyes include curcumin and lutein, orange (zeaxanthin, α and β-carotene) and red (lycopene and fucoxanthin), while green ones are chlorophyll a and b and pheophytin a and b. The solubility of fat-soluble dyes in organic solvents corresponds to their polarity, and the difference in fat polarity is proportional to the levels of unsaturated fatty acids, the higher the levels of unsaturated fatty acids, the higher the polarity, thereby increasing the solubility of the dye in the polar solvent, which is also proportional to the increase in intensity. It is noticed that plant dyes have limitations in its stability. It is influenced by pH, oxidizing agents, light and storage time. In addition, the intensity range of the resulting mixture is relatively less sharp.

Preliminary results of qualitative analysis of fatty acids in vegetable oils, with different levels of saturated fatty acids (coconut and palm oils) using yellow, orange and red dye reagents from plant extracts have been carried out. The results showed a difference in color intensity that was proportional to the levels of unsaturated fatty acids [3]. These results will be applied to the fat of some meat sold in the market, which are known to have different levels of saturated fatty acids. In other words, it is expected to be used for the detection of halal animal meat.

This study aims to develop a new analytical technique for the detection of halal consumption of fatty meat with plant dye reagents using UV-vis spectrophotometry in several stages. Generally speaking, pork fat has relatively higher levels of unsaturated fatty acids than other consumed meat fats. The plant dyes produce sharp intensities and will be used for reagents in the qualitative analysis of consumed animal fats (chicken, beef, goat, pork). Fat extract was obtained from the extraction by the Folch method, using solvents with different polarity. Differences in color intensity from UV-vis spectrophotometry were analyzed. The differences in the profile of UV-vis spectrum (wavelength, intensity) of each selected animal fat in solvents with a certain polarity were used to distinguishing animal fat in a mixed meat fat [6]. In addition, digital image analysis is also performed, in which the color intensity profile of consumed animal fats, RGB values, and Gray Value digital imaging are discussed.

2. Experimental Details

2.1 Materials and Instrumentations

All chemicals are in analytical grade, namely methanol, ethanol, isopropanol, ethyl-acetate, *n*-hexane, and chloroform. A technical grade of isopropanol was also used as comparison. Natural dyes from several plants (*Curcuma longa Linn* – code A1, *Curcuma zanthorriza* – code A2, *Curcuma heyneana* – code A3, *Zingiber montanum* – code A4, *Uncaria* – code A5, *Caesalpinia sappan* – code A6, *Areca catechu* – code A7) and animal fats (chicken, beef, pork, lamb) were obtained from a local grocery store in Malang, Indonesia.

UV-vis spectrophotometer (Shimadzu 1601), 13 MP digital camera, and *Image J 1.50i* free software (Wayne Rasband, National Institution Institutions of Heath, USA) were used in this work.

2.2 Screening of natural plant dyes in several solvents with different polarity

The screening was done by dissolving the dyes in nonpolar (chloroform, *n*-hexane), semi-polar (ethyl-acetate) and polar (methanol, ethanol, isopropanol) solvents. A 2 g of each natural dyes powder (from A1 to A7) was added into Erlenmeyer flask and added with 20 mL of the solvent (chloroform, *n*-hexane, ethyl-acetate, methanol, ethanol, isopropanol). The solution was then homogenized using a vortex mixer and allowed to stand at room temperature for 1 hour. The solubility of each dye in each solvent was observed. The absorbance of each dye was then measured by UV-vis spectrophotometry to determine its maximum wavelength (λ_{max}). Next, the color stability of the dyes in several solvents was checked by measuring the absorbance at λ_{max} of each dye solution after one week, two weeks, and three weeks.

2.3 Screening of animal dyes with plant dyes in selected solvents

The screening was done by dissolving the animal fat in nonpolar (chloroform, *n*-hexane), semi-polar (ethyl-acetate) and polar (methanol, ethanol, isopropanol) solvents. A 2 mL of each animal fat (chicken, beef, pork, and lamb) was added into reaction tube and added with 4 mL of the solvent (ethyl-acetate and isopropanol) the solution was then homogenized using a vortex mixer and rested for 5 minutes. The solubility of each animal fat in ethyl-acetate and isopropanol solvent was observed. The absorbance of each animal fat was then measured by UV-vis spectrophotometry to determine its maximum wavelength (λ_{max}).

2.4 Digital image profile of halal detection of plant dyes extracts in various animal fats

Each solution of a mixture of animal fat and each natural dye extract which was dissolved in ethyl-acetate and isopropanol is captured using a digital camera with 13 Megapixel resolution. The results of digital images are stored in computer data and saved in .JPEG (Joint Photographic Experts Group) file format. The digital images were analyzed using the free Image J 1.50i software (Wayne Rasband, National Institutional Institutions of Heath, USA). The color parameters obtained from this software are the average value of RGB (Red Green Blue). The RGB value is obtained through plugins> analyze> RGB Measure (Ferreira and Rasband, 2011). The percentage of each value of R, G, and B is calculated using the formula: % R = ((mean R / (mean R + mean G + mean B)) x 100%, percentages G and% B are obtained in the same above. Digital RGB values used to determine the difference in color intensity of a mixture of animal fat solutions and natural dyes extracts based on the range of RGB values.

3. Results and Discussion

3.1 The solubility screening of natural dyes in non-polar, semi-polar and polar solvents

The solubility of natural dyes was carried out using non-polar, semi-polar and polar solvents. The results showed that the non-polar chloroform dissolves more natural dyes compared to that of *n*-hexane. Solubility test using semi-polar solvents shows that all-natural dyes have good solubility in ethyl-acetate. As for polar solvents, all-natural dyes have good solubility in methanol and ethanol, as well as in isopropanol. However, the A5 natural dye was slightly soluble in isopropanol.

Solubility of natural dyes in various solvents can be identified from the intensity of the color produced. Thus, from the screening results (Table 1), it is found that natural dyes are soluble in several solvents, namely methanol, ethanol, isopropanol and ethyl acetate.

Table 1. Solubility of natural dyes in several solvents.

Plant Dyes	*n*-Hexane	Chloroform	Ethyl-acetate	Methanol	Ethanol	Isopropanol (technical)	Isopropanol (analytical)
A1	Slightly soluble	Soluble	Soluble	Soluble	Soluble	Soluble	Soluble
A2	Slightly soluble	Soluble	Soluble	Soluble	Soluble	Soluble	Soluble
A3	Slightly soluble	Soluble	Soluble	Soluble	Soluble	Soluble	Slightly soluble
A4	Insoluble	Soluble	Soluble	Soluble	Soluble	Soluble	Slightly soluble
A5	Insoluble	Slightly soluble	Soluble	Soluble	Soluble	Slightly soluble	Slightly soluble
A6	Slightly soluble	Slightly soluble	Soluble	Soluble	Soluble	Soluble	Slightly soluble
A7	Slightly soluble	Soluble	Soluble	Soluble	Soluble	Soluble	Soluble

Note: A1 = C. longa Linn., A2 = C. zanthorriza, A3 = C. heyneana, A4 = Z. montanum, A5 = Uncaria, A6 = C. sappan, A7 = A. catechu.

Analysis of natural dye extracts using UV-vis spectrophotometer produces the maximum wavelength (λ_{max}) and absorbance data of each extract of natural dyes using *n*-hexane and chloroform as the solvent, as presented in Table 2. The natural dyes extract in *n*-hexane has λ_{max} range of 580-670 nm with the highest λ_{max} was detected in *C. zanthorriza* (745 nm with absorbance of 0.011). The natural dyes extract in chloroform has λ_{max} range of 665-670 nm. In addition, λ_{max} of natural dyes extracts in chloroform has a close difference of λ_{max} among each other.

Table 2. Absorbance and λ_{max} results of extracts by UV-vis.

No	Solvent	Sample	λmax (nm)	Abs.	No	Solvent	Sample	λmax (nm)	Abs.
1	*n*-Hexane (3)	A1.3	389	1.698	4	Methanol (1)	A1.1	748.4	0.018
		A2.3	745	0.011			A2.1	746.8	0.019
		A3.3	581.6	0.009			A3.1	722	0.038
		A4.3	324.8	3.104			A4.1	725.6	0.012
		A5.3	669	0.020			A5.1	663.8	0.109
		A6.3	593.8	0.004			A6.1	747	0.015
		A7.3	653.4	0.020			A7.1	739	0.565
2	Chloroform (7)	A1.7	670.2	0.079	5	Ethanol (2)	A1.2	748.6	0.008
		A2.7	666.0	0.113			A2.2	745	0.011
		A3.7	669.0	0.048			A3.2	570	0.091
		A4.7	667.8	0.068			A4.2	727.0	0.008
		A5.7	667.0	0.119			A5.2	665.2	0.085
		A6.7	670.2	0.042			A6.2	569.4	0.182
		A7.7	665.0	0.304			A7.2	569.8	0.918
3	Ethyl-acetate (6)	A1.6	465.2	3.436	6	Isopropanol (technical grade) (4)	A1.4	747.2	0.008
		A2.6	473.8	0.012			A2.4	746.2	0.008
		A3.6	570	0.091			A3.4	741.4	0.012
		A4.6	451.6	0.018			A4.4	744.2	0.019
		A5.6	665.2	0.085			A5.4	665.6	0.125
		A6.6	569.4	0.182			A6.4	547.5	0.286
		A7.6	569.8	0.918			A7.4	741	0.083

Stability test of natural dyes extract in various solvents was carried out to determine the maximum wavelength (λ_{max}) and absorbance values of each natural dye extract in weeks 1, 2 and 3. Table 3 and

IOP Conf. Series: Materials Science and Engineering **833** (2020) 012039 doi:10.1088/1757-899X/833/1/012039

Figure 1 showed that the natural dyes were stable in *n*-hexane, chloroform, ethyl-acetate, methanol, ethanol and isopropanol was stable until the 2nd week.

Table 3. Stability test result of natural dyes extracts in several solvents.

Solvent	Sample Code	Week of observation					
		1		2		3	
		λmax (nm)	Abs	λmax (nm)	Abs	λmax (nm)	Abs
n-Hexane	A1.3	389	1.698	389.4	0.895	406	2.255
	A2.3	745	0.011	388	1.553	491	0.365
	A3.3	581.6	0.009	286.4	2.698	681.5	0.025
	A4.3	324.8	3.104	315	3.740	315.6	3.559
	A5.3	669	0.020	222	2.507	408	0.234
	A6.3	593.8	0.004	229.8	3.025	404	0.104
	A7.3	653.4	0.020	350.2	2.619	548	1.673
Chloroform	A1.7	670.2	0.079	669.5	0.110	508	1.704
	A2.7	666.0	0.113	744.5	0.046	487.5	1.285
	A3.7	669.0	0.048	669	0.059	463	0.158
	A4.7	667.8	0.068	773	0.038	470.5	0.609
	A5.7	667.0	0.119	667.5	0.210	478	0.485
	A6.7	670.2	0.042	795	0.055	477.5	0.729
	A7.7	665.0	0.304	666.5	0.446	549	1.814
Ethyl-acetate	A1.6	465.2	3.436	464.5	3.436	470	3.374
	A2.6	473.8	0.012	474.5	3.612	471	3.436
	A3.6	570	0.091	406.5	2.960	410	2.872
	A4.6	451.6	0.018	406	3.175	406	3.135
	A5.6	665.2	0.085	664	0.725	665.5	0.308
	A6.6	569.4	0.182	426	2.069	426	2.094
	A7.6	569.8	0.918	396	3.763	529	1.860
Methanol	A1.1	748.4	0.018	507.5	3.913	506	3.957
	A2.1	746.8	0.019	455	3.311	457	3.175
	A3.1	722	0.038	491.5	3.612	457.5	3.135
	A4.1	725.6	0.012	453	3.311	663.5	0.124
	A5.1	663.8	0.109	664	0.148	491	3.436
	A6.1	747	0.015	484	3.612	489	3.436
	A7.1	739	0.565	496	4.000	499	3.913
Ethanol	A1.2	748.6	0.008	507	3.913	508	3.763
	A2.2	745	0.011	486.5	3.524	484	3.524
	A3.2	570	0.091	380.6	2.659	350.8	2.972
	A4.2	727.0	0.008	436.5	3.315	447	3.135
	A5.2	665.2	0.085	665.5	0.043	665.5	0.028
	A6.2	569.4	0.182	467.5	3.612	471.5	3.524
	A7.2	569.8	0.918	443.5	3.215	444	3.101
Isopropanol	A1.4	747.2	0.008	481	3.612	493	3.612
	A2.4	746.2	0.008	478	3.436	469.5	3.436
	A3.4	741.4	0.012	752	0.001	404.5	2.422
	A4.4	744.2	0.019	429	2.960	440.5	3.010
	A5.4	665.6	0.125	666.5	0.062	666	0.058
	A6.4	547.5	0.286	547.5	0.453	547	0.2898
	A7.4	741	0.083	420	1.542	419.5	1.527

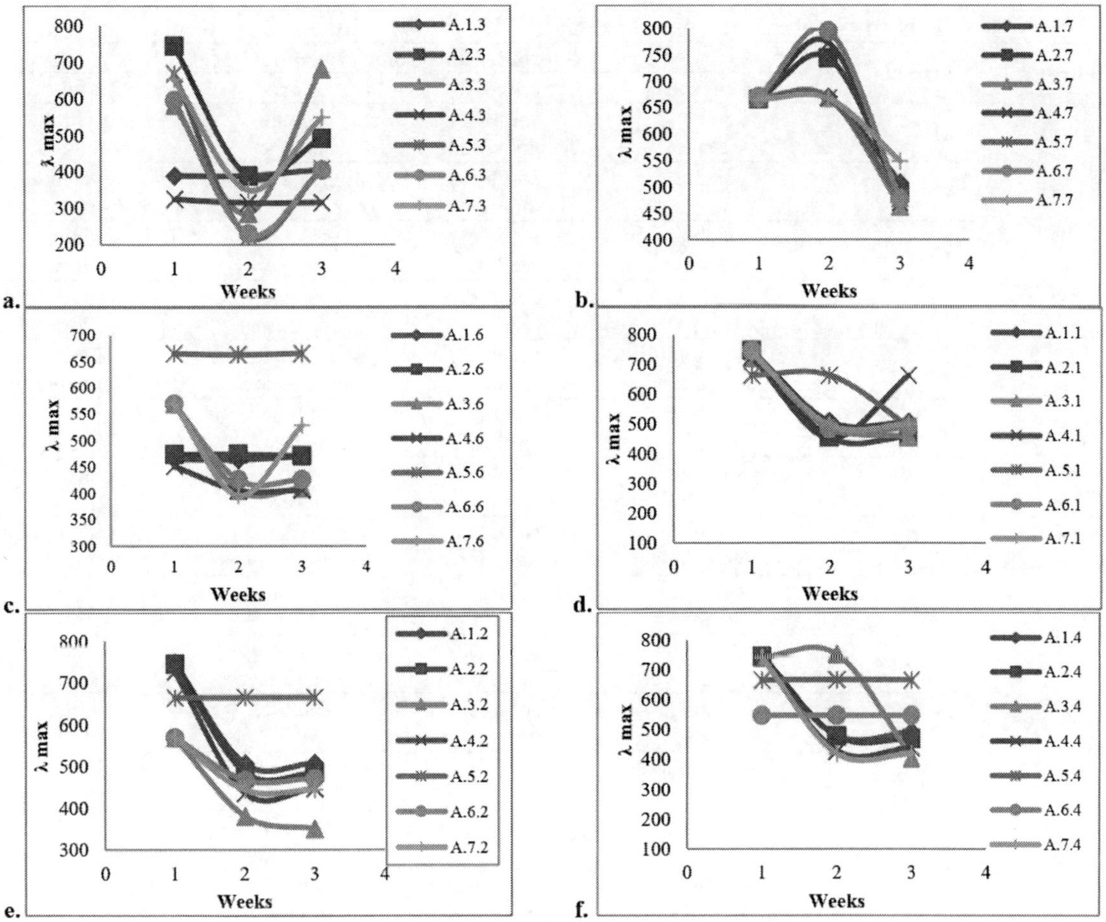

Figure 1. Stability test of natural dyes in several solvents **a**. *n*-hexane; **b**. chloroform; **c**. ethyl-acetate; **d**. methanol; **e**. ethanol; **f.** isopropanol.

The solubility screening test of natural dye extracts were also conducted in various solvents (non-polar, semi-polar and polar). Based on the result (Table 4), it can be concluded that ethyl-acetate and isopropanol are suitable solvents to be used in the solubility of animal fats – natural dye extracts mixture due to its good solubility.

Table 4 shows that the *C. longa* Linn., *C. heyneana, Uncaria,* and *A. catechu* dye extracts have the best solubility in animal fat using ethyl-acetate, as observed from the fully dissolving of each natural dye extract and color changes in response to animal fats. Meanwhile, in isopropanol, the *C. longa* Linn, *C. heyneana*, *C. sappan* and *A. catechu* dye extracts are slightly soluble in animal fat and form 2 layers with a color change in the lower layer of each animal fat.

3.2 Analysis of animal fats with plant dyes in selected solvents by UV-Vis spectrophotometry

The ethyl-acetate and isopropanol solvents which has good solubility in natural dye and fat extracts mixture were further used in the next stage. The extract of natural dyes – animal fat mixture in ethyl-acetate has λ_{max} range of 792-367 nm. Meanwhile in isopropanol, the λ_{max} was obtained in a range of 798-414 nm. The lowest λ_{max} was detected for *A. catechu* dyes which was 367.8 nm with an absorbance of 2.647. These results are presented in Table 5.

Table 4. Solubility test of animal fat in different dye extracts and solvents.

Solvent	Extracts	Solubility and Color of Animal Fat			
		Chicken	**Beef**	**Pork**	**Lamb**
Ethyl-acetate	A1	soluble, orange	soluble, feculent orange	soluble, orange	soluble, orange
	A2	soluble, orange	soluble, orange	slightly soluble, orange (2 layers)	slightly soluble, orange (2 layers)
	A3	soluble, yellow	soluble, yellow	soluble, yellow	soluble, yellow
	A4	soluble, yellow	soluble, yellow	soluble, yellow	soluble, yellow
	A5	slightly soluble, pale yellow	soluble, yellow	slightly soluble, pale yellow	soluble, feculent yellow
	A6	soluble, yellow	soluble, yellow	soluble, yellow	soluble, yellow
	A7	soluble, yellow	soluble, yellow	soluble, yellow	soluble, yellow
Isopropanol	A1	soluble, orange	soluble, yellow	slightly soluble, orange (2 layers)	soluble, orange
	A2	soluble, yellow	soluble, yellow	soluble, yellow	soluble, yellow
	A3	slightly soluble, orange (2 layers)	soluble, light orange	slightly soluble, orange (2 layers)	soluble, orange
	A4	soluble, yellow	soluble, yellow	soluble, yellow	soluble, yellow
	A5	slightly soluble, pale yellow	soluble, yellow	slightly soluble, pale yellow	soluble, yellow
	A6	soluble, yellow	slightly soluble, orange (2 layers)	slightly soluble, orange (2 layers)	slightly soluble, orange (2 layers)
	A7	slightly soluble, yellow (2 layers)	slightly soluble, orange (2 layers)	soluble, orange	slightly soluble, orange (2 layers)

Note: A1 = C. longa Linn., A2 = C. zanthorriza, A3 = C. heyneana, A4 = Z. montanum, A5 = Uncaria, A6 = C. sappan, A7 = A. catechu.

Table 5. UV-Vis results of animal fat mixed with natural dyes extracts in selected solvents

Solvent	Dyes	Animal Fat							
		Chicken		**Beef**		**Pork**		**Lamb**	
		λ_{max} (nm)	Abs.	λ_{max} (nm)	Abs.	λ_{max} (nm)	Abs.	λ_{max} (nm)	Abs.
Ethyl-acetate	A1	484	3.913	792	0.816	488.5	3.524	741	0.1755
	A2	742.5	0.056	487.5	4.000	782.5	1.489	780.5	2.613
	A3	484	3.763	445	3.436	436.5	3.135	454	3.436
	A4	411.5	3.311	789	0.067	440	3.135	789.5	0.384
	A5	667.5	0.109	666.5	0.309	662.5	0.085	665.5	0.734
	A6	639	0.026	753.5	0.071	745.5	0.046	464.5	3.612
	A7	376.8	3.101	387.2	3.311	367.8	2.647	394	3.763
Isopropanol	A1	751	0.007	787	1.012	712	0.479	499	3.913
	A2	754	0.048	471	4.000	752.5	0.063	771.5	0.042
	A3	797	0.796	460.5	3.913	760	0.270	762	0.170
	A4	783.5	0.692	749	0.024	783.5	1.459	750	0.310
	A5	794	0.602	766.5	2.237	777	0.471	661	1.086
	A6	793.5	0.815	476	4.000	784.5	0.663	791	0.138
	A7	794	1.043	460.5	3.913	788	0.276	774	0.319

Note: A1 = C. longa Linn., A2 = C. zanthorriza, A3 = C. heyneana, A4 = Z. montanum, A5 = Uncaria, A6 = C. sappan, A7 = A. catechu.

3.3 Digital image profile of halal detection of plant dye extracts in various animal fats

Digital image profiles were used to quantify and differentiate the % RGB values (red, green, blue) of animal fat mixed with plant dye extracts in ethyl-acetate and isopropanol solvents. Thus, it is expected to be able to determine the profile differences of digital image of animal fat mixed with plant dye extracts. The results of the % RGB values are presented in Table 6.

Table 6. RGB Percentage of animal fat in natural dyes extracts.

Solvent	Extract	RGB	RGB Mean Values of Animal Fat			
			Chicken	Beef	Pork	Lamb
Ethyl-acetate	A1	Red	144.958	170.304	166.824	166.419
		Green	80.881	107.123	102.184	98.486
		Blue	31.273	31.053	36.838	41.636
		% Red	56.38	55.21	54.55	54.29
		% Green	31.46	34.73	33.41	32.13
		% Blue	12.16	10.07	12.04	13.58
	A3	Red	146.510	148.386	140.660	140.794
		Green	90.595	82.817	75.025	77.078
		Blue	5.448	4.671	5.656	4.860
		% Red	60.40	62.91	63.55	63.21
		% Green	37.35	35.11	33.90	34.61
		% Blue	2.25	1.98	2.56	2.18
Isopropanol	A1	Red	171.853	184.926	149.552	162.023
		Green	81.853	122.917	52.456	72.023
		Blue	20.043	20.480	4.945	19.371
		% Red	62.78	56.32	72.26	63.94
		% Green	29.90	37.44	25.35	28.42
		% Blue	7.32	6.24	2.39	7.64
	A3	Red	170.304	176.042	143.217	157.533
		Green	81.199	94.542	44.049	71.616
		Blue	15.083	21.458	5.236	17.074
		% Red	63.88	60.28	74.40	63.98
		% Green	30.46	32.37	22.88	29.09
		% Blue	5.66	7.35	2.72	6.93

Table 6 showed that the %RGB mean values of each animal fat – plant dye extract in ethyl-acetate and isopropanol are distinguishable. The percentage of the RGB value showed the intensity of the color produced after the addition of extracts of natural dyes on each animal fat. However, not all plant extracts have the potential to be halal detectors for animal fat due to the difference of each extract dye. From the result, *C. longa* Linn and *C. heyneana* extracts have the potential as halal detectors for consumed animal fats in isopropanol solvent. However, consumed animal fats were less soluble in isopropanol. Lard (fat of pork) has the highest %RGB value as shown by the digital image of *C. longa* Linn and *C. heyneana* extracts in isopropanol. Meanwhile, the highest %RGB was shown only on *C. heyneana* in ethyl-acetate.

4. Conclusion

This research result showed that the extract of plant dyes from A1 (*C. longa* Linn) and A3 (*C. heyneana*) in ethyl-acetate and isopropanol solvents, by combination of UV-vis and digital image spectrophotometer methods, can be developed to distinguish halal fats (chicken, beef, lamb) from a non-halal fats (lard or fat of pork).

Acknowledgment
The author would like to thank the Professor and Doctor Research Grant Program 2019 (Brawijaya University) for the research funding and the organizing committee of the 2nd IC2MS for the assistance.

References

[1] Ardilla, D., Taufik, M., Tarigan, D. ., Thamrin, M., Razali, M., & Siregar, H. 2018 Analisis Lemak Babi Pada Produk Pangan Olahan Menggunakan Spektroskopi UV-Vis. *Jurnal Teknologi Pangan Dan Hasil Pertanian*, **1**(2), 111–116.

[2] McMurry, John 1999 *Organic Chemistry* (Fifth Edit). America: Thomson Learning Brooks/Cole.

[3] Rurini, R., Effendi, M. H., and Kurniawan, F 2019. *Pengaruh Pelarut pada Identifikasi Profil Lemak Hewani Menggunakan Spektrofotometri UV-Vis-Citra Digital Untuk Deteksi Produk Halal.*

[4] Prabawati, S. and Fajriati, I. 2019 *Analisis Lemak Sapi Dan Lemak Babi Menggunakan Gas Chromatography (Gc) Dan Fourier Transform Infra Red Spectroscophy Second Derivative (Ftir-2d) Untuk Autentifikasi Halal. 8.*

[5] Othman, R., Zaifuddin, F. A. M., and Hassan, N. M. 2015 *Potential Sources For Lipid Soluble Food Colorants From Selected Malaysian Traditional Vegetables*. **19**(7).

[6] Berhe, D. T., Eskildsen, C. E., Lametsch, R., Hviid, M. S., van den Berg, F., & Engelsen, S. 2016 Prediction of total fatty acid parameters and individual fatty acids in pork backfat using Raman spectroscopy and chemometrics: Understanding the cage of covariance between highly correlated fat parameters. *Meat Science* **111** 18–26.

IOP Conf. Series: Materials Science and Engineering **833** (2020) 012040 doi:10.1088/1757-899X/833/1/012040

The Preventive Effect of Ethanolic Extract of Rome Beauty Apple Peel (*Malus sylvestrys Mill*) towards Protease Activity and Jejunal Histopathology of Rat (*Rattus norvegicus*) Exposed to Lead Acetate

Anna Roosdiana[1*], Analis Wisnu Wardhana[2], Tarsisius Handaru Cahyo Putro[2]

[1]Department of Chemistry, Faculty of Mathematics and Natural Sciences, University of Brawijaya, Malang, Indonesia
[2]Faculty of Veterinary Medicine, University of Brawijaya, Malang, Indonesia

*Corresponding author: aroos@ub.ac.id

Abstract. The exposure of lead to body can be through inhalation, digestion and adsorption. Lead intoxication can lead to jejuna damages and increase protease activity due to reactive oxygen species (ROS) production. Apple peel extract contains flavonoids that can scavenge ROS. This research aims to investigate the effect of ethanolic extract of Rome Beauty apple peel to protease activity and jejuna histological feature of rat exposed to lead acetate. This research used 20 rats, that divided into 5 treatment groups, which were negative control group (healthy rat), positive control group (exposed to lead acetate 10 mg/rat/day for 14 days), and preventive therapy group (administrated the apple peels extract with doses of 28 mg/200 g BW, 56 mg/ 200g BW, 112mg /200g BW for 21 days and also lead acetate with 10mg/rat/day for 14 days on day 15th until 28th). Jejunal histological features was stained using HE stain which then observed using a microscope. Protease activity was measured by using spectrophotometric method. The data of protease activity was statistically analyzed using ANOVA, followed by Tukey test (α = 5%). The results showed that the administration of ethanolic extract of Rome Beauty apple peels at best dose of 112 mg/200 g BW can reduce protease activity and improve jejuna histopathology. The conclusion of this study was that apple peels extract of Rome Beauty can be used to prevent the increase of protease activity and the damage of rat jejuna exposed to lead acetate.

1. Introduction

Lead is a heavy metal used for manufacturing batteries, paints, greases and oil for car engine, farm machinery and pipes. Lead is important toxic metal because of the danger to both human and animal health. Animals exposed to lead poisoning mainly via oral routes, the pastures and water contaminated by lead which is accidentally consumed and can cause intoxication. Lead poisoning seldom occur by airways due to the prolonged inhalation of lead fumes and absorption of lead through the skin [1]. Lead toxicity in adult cow is at dose of 400-600mg/kg (single dose), however, it defends of the lead compounds. Chronic toxicity occurs on cow, whether fed grass containing Pb 390 mg/kg daily. Cow and poultry kept near landfill are the most at risk of lead poisoning because they are inquisitive and commonly taste the new finds. In addition, in cow meat was found Pb levels of 0.05-1.05ppm [2], while in chicken meat was 0.56-4.83ppm [3].

When commercial animals contaminated by lead are consumed by human, that may cause health problem. Lead enter the body via digestive tract that will be absorbed by intestine. Symptom of lead poisoning in gastrointestinal system include anorexia, abdominal pain, vomiting and constipation [4]. The accumulation of absorbed toxic material can disrupt the adsorption system in small intestine especially jejuna. The presence of lead can form free radical and increase ROS level in body without the equal increase of endogenous antioxidant leading to the occurrence of oxidative stress, that consequently, can damage cell and organs. Cell damage can trigger macrophage activity and inflammatory cell to produce pro-inflammatory cytokines such as TNF-α, then it activates neutrophils to produce protease [5].

Current therapy of lead poisoning can be conducted by using chelating agents which can bind lead and facilitate its release. The chelating agent, meso-2,3–dimercaptosuccinic acid (DMSA) is used orally, while ethylene diamine tetra acetate (EDTA)is used intravenously and intramuscularly. Those chelating agents can reduce blood lead levels. However, there is no information of the changes of morphological organ [6]. In order to overcome lead poisoning in animals, the effort may use natural product such as apple peel instead of apple flesh. Apple peel contains flavonoid, ascorbic acid, phenolic compound, glutathione and L-cysteine [7]. It has been previously reported that ethanolic extract of apple peel at dose of 28mg/kg BW can reduce MDA levels and fat degeneration in rat hepatocytes induced by streptozotocin [8].

In this work using rat model, the rats were orally induced by lead acetate and received preventive therapy with ethanolic extract of Rome beauty apple peel. We reported the activity of jejuna protease activity and jejuna histopathology changes exposed to different dose of ethanolic extract of Rome beauty apple.

2. Materials and Methods

Eight to ten weeks old, male Wistar rats weighing about 180-200g were allowed to acclimatize for 7 days prior to initiation of the study. The animals were housed in completely controlled environment (room temperature 25 ± 2 °C and 12 h light and dark cycle) with free access to water and food. The Research ethics committee (animal care and use committee) of Brawijaya University has approved the study protocol (No. 1087 –KEP-UB).

2.1 Chemicals and Instrumentations

The materials used in this study were fruit peel of Rome beauty Apple which obtained from local market and determined by PT. Materia Medica, Batu City, East Java. Other materials used were protease reagents (PBS-Tween solution, PMSF, Trichloro acetic acid (TCA), buffer phosphate pH 7,0, absolute ethanol, bovine serum albumin (BSA), L-tyrosine, paraffin, 10%paraformaldehyde (PFA), xylol. paraffin block, PBS –azida solution pH 7.4, hematoxylin and eosin.

2.2 Experimental Animals and Design

Twenty male Wistar rats were divided into five groups. Each group consist of 4 rats. Group I was the control group (healthy rats). Group II was the positive control (Pb poisoning). Group III, IV and V rats were designated as experimental group and received ethanolic fruit peel extract of apple orally in a dose of 28, 56 and 112/200gBW/day, respectively by gavage for 21 days. Group II, III, IV and V received lead acetate at a dose of 10mg/200gBW/day for 14 days on day 15th until 28th. At the end of experiments, all rats were sacrificed, and jejunum organ were isolated for determining protease activity and observing histopathology feature.

2.3 Isolation of protease and measurement of protease activity

The isolated jejunum organ was weighted for 0.5 g and cut into small pieces by using surgical scissors. Then it was added with 1mL of PBS-tween solution: PMSF (9:1), crushed in cold mortar and transferred to sterile micro tube that used addition of 2 mL of PBS-tween solution: PMSF (9:1). The mixture was put into a vortex vibrator for 10 min, sonicated for 10 min, and centrifuged for 10 minute 6000 rpm.

Supernatant was separated and precipitated by absolute ethanol at volume ratio of 1:1 and allowed for 24 hours. The mixture was centrifugated at 10000 rpm for 15 minutes. Finally, the precipitate was protease which was dried until free ethanol and dissolved in 0.02M Tris-HCl solution pH 6.5. The protease activity was measured using spectrophotometer based to Walter Method with equation:

$$\text{protease activity} = \frac{\text{tyrosin}}{\text{Mr tyrosin}} \times \frac{V}{P \times q} \times fp \qquad (1)$$

where V = total volume, P= volume protease, q= time reaction (minutes) and f_p =dissolution factor.

2.4 Jejuna histopathological Observation

Histopathology observation of rat jejunum were carried out by using a light microscope Olympus BX51 with magnifications of 400x. The observations were especially in mucosal layer with vile, the presence of hyperplasia goblet cells, epithelial erosion, infiltration of macrophage and neutrophils.

2.5 Statistical Analysis

Data of protease activity were expressed as the mean ± standard deviation of the mean and comparison between different treatments was carried out using one-way ANOVA through software SPSS 22.0 followed by Tukey test. Significance was accepted at $p<0.05$.

3. Results and Discussion

Protease has been considered to act primarily as degradative enzyme in extracellular space, however their biological actions in tissues and cells suggest important roles as a part of body hormonal communication system during inflammation and immune response. The lead acetate induction to rats can damage jejunum and increase the jejuna protease activity. The protease activity can be measured by spectrophotometry. The effect of preventive therapy of ethanolic extract of apple peel to jejuna protease activity of rats induced by lead acetate were displayed in Table 1.

Table 1. Profile of jejuna protease activity in control rats, Pb poisoning rats, and Pb poisoning rats with preventive therapy with ethanolic extract of Rome beauty apple peel.

Group	Protease activity (μmol/mL min)
1. Negative control	4.08 ± 0.08^{a}
II. Pb poisoning	5.99 ± 0.37^{c}
III. Therapy 28mg/kgbw	4.84 ± 0.38^{b}
IV. Therapy 56 mg/kgbw	4.45 ± 0.17^{ab}
V. Therapy 112 mg/kgbw	$4.25 \pm 0{,}16^{a}$

*different letters (a-c) show significant statistical different effects in each group (p<0.05).

As shown at Table 1, the negative control rats show jejuna protease activity of 4.08±0.08 μmol/mL min, which is significantly different compared to positive control rats. Protease is normally an intracellular enzyme which has a function as signal transduction and cell regulation to activate polypeptide hormone, development factor and body defense. In development factor, protease has a function as collagen assembly from procollagen, cell proliferation, proteolytic control and plays a role in apoptosis. Moreover, protease can be used as an indicator of severity of inflammation [9].

The increase of protease activity of rat jejunum induced by lead acetate was caused by lead inhibition of δ ALAD activity that can increase the accumulation of ALA. The increase of ALA resulted on the formation of free radicals, peroxide radical, superoxide radicals and hydroxyl radicals [10]. Free radicals cause cell damage, thus triggering the activity of macrophages in tissue to secrete pro inflammatory cytokines, namely tumor necrosis factor (TNF-α). The TNF- α subsequently activate neutrophils to produce protease.

Therapy with ethanolic extract of apple peel may reduce the elevated jejuna protease activity. Protease activity decreased with the increasing doses of ethanolic extract of apple peel. Statistical test results showed that there were significantly differences (p>0.05) between protease activity in rat jejunum induced with lead acetate and rat jejunum with therapy. It suggests that flavonoid in ethanolic extract enable to act as an antioxidant, especially as hydroxyl radical scavenger and suppress the increase of protease activity. The role of flavonoid as antioxidants directly is by donating hydrogen electron to free radicals and increasing the expression of endogenous antioxidant genes [11]. In addition, ethanolic extract of apple peel contains vitamin C that can act as antioxidant. The suppression of oxidative stress can decline the activation of NF-KB, consequently, TNF-α cannot be produced. The decline of TNF-α will decrease the activity of neutrophils as first immunity defense that will suppress the protease activity [12].

The damage of rat jejuna exposed to lead acetate can be observed by the changes of histopathological jejuna instead of jejuna protease activity. Free radicals are the result of normal cell metabolism; however, lead poisoning may interfere the balance between free radicals and endogenous antioxidants leads to situation of oxidative stress. The oxidative stress can cause damage to jejunum tissue and jejunum dysfunction. The histopathology image of jejunum was observed to determine both the level of damage and organ repair. Comparison of histopathological jejunum image from normal rats, lead poisoning rat and therapeutic rats were displayed in Figure1. In negative control rats, there are neat arrangement of columnar epithelial cell and goblet cells in tunica mucosa and the infiltration of inflammatory cells do not exist. Normally, jejunum consists of four layers, tunica mucosa, tunica sub mucosa, tunica muscularis and tunica serous [5].

The positive control rats (K+) showed a different image of histopathogical jejunum mucosa compare to negative control as shown by the damage of villous structure, hyperplasia goblet cell, epithelial cell erosion and rupture of villi and infiltration of macrophages and neutrophils. Similar finding by Sharma and Barber, that lead induction can change histopathology in small intestinal mucosa such as damage to villous structure, epithelial cell erosion and infiltration of inflammatory cells. Villi are the most important part of the small intestine with function of absorption in digestive tract. Intact villi will facilitate the transportation of nutrient to the body. Damage to villous structure occurred due to erosion of epithelial cell that will lose a portion of mucosa [13]. Goblet cells spread between absorptive cells that functions of producing mucus to protect mucosal epithelium and facilitate the movement of food ingested which is not absorbed. The synthesized mucus consists of a mixture of water, glycoprotein, glycolipids, electrolytes, enzymes, salts and glandular secretions. Lead induction will cause intestinal response to toxic compounds by multiplying goblet cells to produce a lot of mucus as jejunum defense. Therefore, some hyperplasia goblet cell existed in positive control rats

Improvement in histopathologic image of rat jejuna induced by lead acetate can be seen in therapeutic groups. Improvement is evident from the decline of erosion of epithelial cells, hyperplasia goblet cells and infiltration of inflammatory cell (macrophages and neutrophils). In therapeutic groups treated with a dose of 112mg/kg bw show an improvement in the histopathological features of rat jejunum which is almost the same as negative control group where the villous structure is neatly arranged, hyperplasia goblet cell and inflammatory cell are not found. It suggests that flavonoid and vitamin C in ethanolic extract of apple peel can neutralize free radicals and prevent the occurrence of protein, DNA and lipid peroxidation, thus consequently, can prevent the damage of rat jejuna.

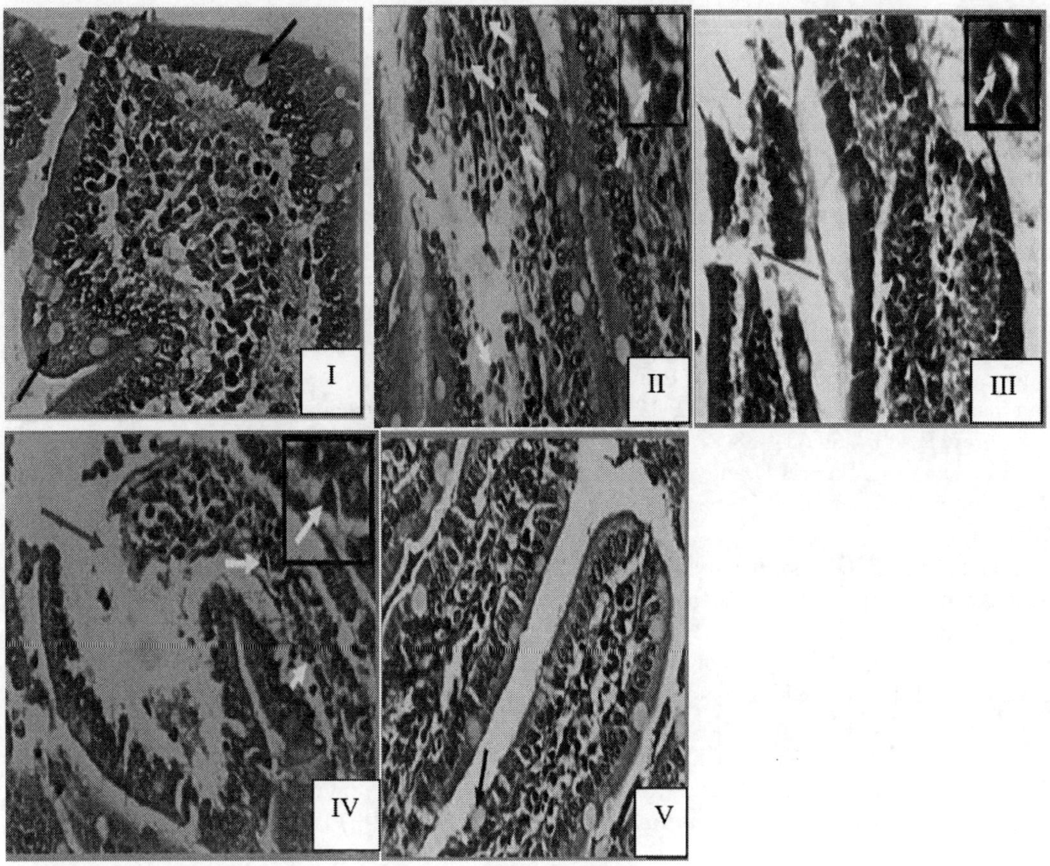

Figure 1. Histopathology features of normal rat jejuna (I), Pb poisoning rats (II),Pb poisoning rats and treated with ethanolic extract of apple peel (III, IV, V) at magnifications of 400x. goblet cells ( ), infiltration of inflammatory cells ( ), villous erosion ( ), villous rupture ( ).

4. Conclusion

It can be concluded that ethanolic extract of apple peel has bioactive compounds as preventive agents to lead poisoning. A dose of 112 mg/200g BW/day is the effective dose to prevent damage of rat jejuna induced by lead acetate based on the decrease of protease activity and improvement of jejuna histopathological feature in the form of reduced goblet cell hyperplasia, infiltration of inflammatory cells and epithelial cell damage in villi compared to positive group. The results indicate that ethanolic extract of apple peel is potential to be used to prevent lead poisoning. Further study is necessary to package the extract. Instead of giving solution of apple peel extract by force feeding, it needs for giving therapy in simple way such as giving the extract in different form, like capsule, tablet, or feed mixture. Therefore, the stability of extract in the package should be also examined.

References

[1] Traverso S D, Loretti A P, Donini M A and Driemeier D 2004 *Arq. Bras. Med. Vet. Zootec.* **56** 418-421.
[2] Janardani N MK, Berata I K, and Kardena I M 2018 *Indonesia Medicus Veterinus.* **7** 42-50 In Indonesian.
[3] Priyono O 2013 Study of Pb levels and distributions in chicken meat, liver, and Kidney, Faculty of Veterinary medicine, Bogor Agricultural University. In Indonesian.

IOP Conf. Series: Materials Science and Engineering **833** (2020) 012040 doi:10.1088/1757-899X/833/1/012040

[4] Lubis B, Rosdiana N, Nafianti S, Rasyianti O and Panjaitan F M 2013 Relationship of lead poisoning and Iron Deficiency in Children. CDK 200, **40**(1) 17-21 In Indonesian.

[5] Huang C, Šali A, Stevens RL, Regulation and function of mast cell proteases in inflammation. J Clin Immunol. 1998; **18** 169–183.

[6] Mason LH, Harp JP and Han D Y 2014 Pb neurotoxicity: Neuropsychological effects of lead toxicity. *BioMed Research International* 2014: 1-8.

[7] Lata, B., and Tomala K. 2007 Apple peel as a contributor to whole fruit quantity of potentially healthful bioactive compound. cultivar and year implication. *J. Agric. Foods Chem.* **55**(26) 10795-10802.

[8] Andriani S R D 2016 The effect of ethanolic peel extract of apple Manalagi (Mallus Sylvestris Mill) towards MDA levels and hepatic histopathology feature of diabetic Type 1 rat. Faculty of Veterinary Medicine, University of Brawijaya. In Indonesian.

[9] Chapman H A, Riese R J and Shi G P 1997 Emerging roles for cysteine protease in human biology. *Annu. Rev. Physiol.* **59** 63 – 88.

[10] Gurer-Orhan H, SabirH U and OzgunezH 2004 Correlation between clinical indicators of lead poisoning and oxidative stress parameters in control and Lead exposed workers toxicology. **195** 147-154.

[11] Purwaningsih I, Sapriani R dan Indrawati R 2018 Antioxidant Activity of methanolic extract of Daun Kesum (*Polygonum minus* Huds.) by DPPH method. *Jurnal Laboratorium Khatulistiwa*. **2**(2) 161-165. In Indonesian.

[12] Chattopadhyay I, Bandyopadhyay U, Biswas K, Maity P and Banerjee R K 2006 Indomethacin inactivates gastric peroxides to induce reactive-oxygen- medicated gastric mucosal injury and curcumin protects it by preventing peroxidase inactivation and scavenging reactive oxygen. *Free Radical Biology Med*. **40** 1397-1408

[13] Sharma R and Barber I2014 Lead toxicity and postnatal development of gastrointestinal tract. *Universal Journal of Environmental Resarch and Technology*. **4**(3) 121-133.

Sulphate Removal from Wastewater in Constructed Wetland Ecotechnology Using Pumice Amended in the Sand Media

Philiphi de Rozari[1], Maria Agusta Dua Monang[1], Denik Sri Krisnayanti[2] Bibiana Dho Tawa[1]

[1]Department of Chemistry Faculty of Science and Engineering Nusa Cendana University, Kupang, Jalan Adisucipto Penfui Kupang, NTT
[2]Department of Civil Engineering Faculty of Science and Engineering Nusa Cendana University, Kupang, Adisucipto Penfui Kupang, NTT

Corresponding author: phderozari@yahoo.com or p_derozari@staf.undana.ac.id

Abstract. Sulphate is a part of sulphur compounds which potentially inhibit plant growth and microbial activities in receiving surface water. Thus, it is important to remove concentration of sulphate from wastewater to acceptable concentration before the water is released into aquatic system. Constructed wetlands (CWs) are good option of wastewater treatments due to their low-cost and eco-technology. This research aimed to investigate the removal of sulphate from domestic wastewater using sand media amendment with pumice in CWs. In this experiment, six variations of CWs consisting sand and pumice planted with lemongrass (*Cymbopogon citratus*) were established. Domestic wastewater was loaded into CWs with loading rate 2 L/day. Sulphate concentrations were determined in inflows and outflows. The results showed that all treatment significantly remove sulphate concentrations from the inflow. The highest removal efficiency was in the media with 100% of pumice.

1. Introduction

As a capital city of East Nusa Tenggara Province (ENTP), Kupang is experiencing problems with the decline in water quality mainly caused by urbanization and population growth. Several rivers in the city of Kupang such as Dendeng and Liliba Rivers have been contaminated by domestic wastewater, which affects the aesthetics of the river such as the colour, smell and taste of water. One of the pollutants caused aesthetic problems in surface water is sulphate. This compound is commonly found in wastewater and natural water body either in surface water or ground water. However, excess of sulphate, in some circumstances, could lead to detrimental effect to human health particularly diarrhoea [1]. In limited oxygen condition, sulphate could produce hydrogen sulphide (H_2S) and organic sulphur that result in environmental problems such as aesthetic issues in water body, water pipe corrosion and plant phytotoxicity [2]. Thus, it is crucial to remove sulphate from domestic wastewater until recommendable concentration before this compound is discharged to water body.

Constructed wetland (CW) is an alternative technology that potentially uses for wastewater treatment. Recently, CWs have been widely used for removing different types of polluted waters because of their low operational and maintain cost, effective pollutant removal and ecotechnology [3-5]. CWs are categorised as free water surface (FWS) and subsurface flow (SSF) regimes [4]. Meanwhile, SSFs are further classified into horizontal flow (HF) and vertical flow (VF). The SSFs are widely used for wastewater treatment technology because the treatment processes occur effectively during contact with the media and around the roots and rhizomes of the plants [6]. In addition, SSFs have more capability to reduce odour problems, effectiveness of land area uses, and higher capacity to adsorb pollutant [4].

In SSFs, plants and media play a significant role in removing pollutants. Plants contribute in removing pollutants via production of roots and rhizome for suitable place to microorganisms, filtration processes, plant uptake and absorption processes [4, 7]. The effectiveness of plants to remove pollutants depends on plant species and plant density [8]. Media, in other hand, plays numerous processes in removing pollutants, particularly through controlling water infiltration, filtering particulate maters, providing surface areas for nutrients which are source of energy for microorganisms and encouraging the growth of microorganisms and plants [9, 10].

Removal of pollutants in SSF system occurs through various interactions among media, microorganisms, plants, the atmosphere and the wastewater itself [4]. The pollutant is removed through physical, chemical and biological mechanisms. These mechanisms take place simultaneously. The sulphate removal in SSF system could be via several mechanisms such as precipitation with heavy metals such as Zn and Fe, microbial oxidation and plant uptake [11-13]. Several authors have investigated the use of media and plants for sulphate removal. Chen et al. [14] have designed five SSF CW gravel media microcosms, with unplanted control, litter added microcosm, planted (*Thypa latifolia*) microcosms, and planted plus litter added microcosms to investigate the effect of plants and litters in removal sulphate. The results showed that the presence of *Thypa latifolia* gave minor contribution in removing sulphate either in the media with additional carbon sources or without additional carbon source. San Miguel-Espinosa et. al. [15] used rice husk as a control media and the same media planted with *Thypa angustifolia* and *Puccinellia frigida* to examine the effect of rice husk and plants in removing sulphate. The results showed that control media was more significantly effective in removing sulphate compared to those planted media. The sulphur balance analysis conducted by Hou et al. [16] using constructed wetlands planted with *Thypa latifolia* revealed that 43 – 71 % of sulphate removal was found in the gravel and only 8.84 – 2.34 % was uptaken by the plant. It can be inferred that the presence of *Thypa latifolia* and *Puccinellia frigida* in CWs had limited effect on sulphate removal. Regarding the application of CWs in ENTP, it is interesting to explore local media and endemic plants that effectively reduce pollutants from domestic wastewater because of the use of local media in CWs could reduce the investment cost significantly. Thus, it is important to investigate the potential local media and endemic plant in removing sulphate. Lemongrass (*Cymbopogon citratus*) is classified as perennial grasses and could store higher nutrients from wastewater [17]. The lemongrass also effectively removed suspended solids [18] and has socio-economic purposes particularly for traditional medicine [19]. Since the lemongrass is abundantly available in ENTP, it is crucial to study the effectiveness of CWs planted with lemongrass in removing in removing pollutant. This research aims to investigate the use of pumice amended in sand media planted with lemongrass (*Cymbopogon citratus*) in CW system for sulphate removal from domestic wastewater.

2. Material and Methods

2.1 Chemicals

All chemical were analytical grade chemicals. $BaCl_2$ used in this study was obtained from Merck. Solution containing sulphate 100 mg/L was prepared by adding 1.479 g of Na_2SO_4 (Merck) to a 1000 mL measuring flask and filling with deionised water to the mark. Buffer solution was prepared by adding 30 g of $MgCl_2.6H_2O$ (Merck), 5 g of $CH_3COONa.3H_2O$ (Merck), 1 g of KNO_3 and 20 mL of CH_3COOH (99%) (Merck) to a 1 L measuring flask containing 500 ml deionised water. The solution was then filled with deionised water to the mark.

IOP Conf. Series: Materials Science and Engineering **833** (2020) 012041 doi:10.1088/1757-899X/833/1/012041

2.2 Experimental design and operation

The experiment was conducted at Department of Chemistry, Nusa Cendana University starting from February – June 2019. Six vertical flow (VF) mesocosms containing media planted with lemongrass (*Cymbopogon citratus*) and barren media with combination of pumice amended in the sand were constructed (Table 1). The VF mesocosms were created from polyvinyl pipe column with inner diameter and height of 10.2 cm and 55 cm, respectively. Figure 1 shows the schematic design of VF mesocosm. The media introduced in VF mesocosms consisted of 5 cm layer of gravel (diameter 10 – 20 mm) placed at the bottom of each VF mesocosms, followed by 40 cm of layer media in the middle and 5 cm of gravel at the top. The bottom of VF mesocosms was fitted with the hose and connected to bucket for collecting the outflow. Domestic wastewater was stored in 50 L containers and fed six VF mesocosms. The VF mesocosms were loaded daily with domestic wastewater with hydraulic loading rate of 2 L/day. The media used in this research were pure sand, pure pumice and combination among them with and without planted with lemongrass (Table 1).

Table 1. Media composition in VF mesocosms.

No.	Media treatment	% Media		Plant
		Sand	Pumice	
1	CW1	100%	0%	-
2	CW2	90%	10%	-
3	CW3	50%	50%	-
4	CW4	0%	100%	-
5	CW5	90%	10%	+
6	CW6	50%	50%	+

+ = planted with lemongrass (*Cymbopogon citratus*)

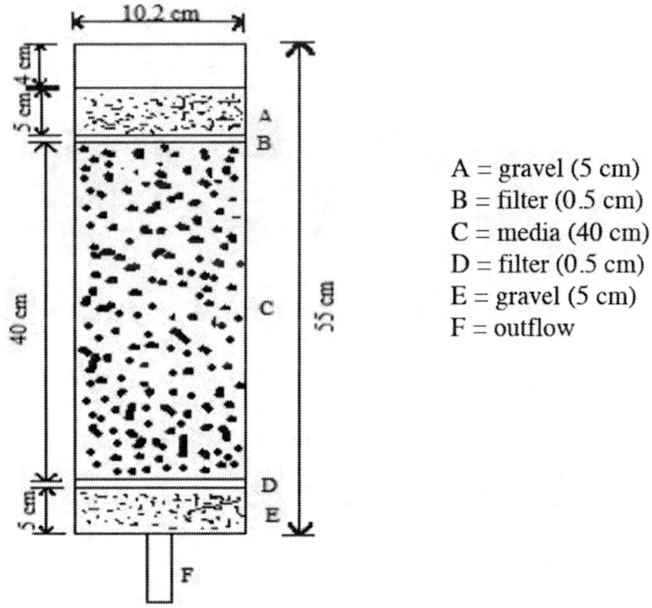

Figure 1. Schematic of VF mesocosm.

2.3 Media characterisation

The sand was sampled from Takari in Kupang Disctrict and the pumice was obtained from Tuamese Village, Biboki Anleu Subdistrict, The District of North Centre Timor (Timor Tengah Utara).

Characterization of sand and pumice were carried out before these materials are applied as media in CW systems. The parameters characterized were included cation exchange capacity, organic matter, pH, FTIR and XRF. CEC was determined using silver thiourea methods [20]. The analysis of pH is based on Methods of Analysis for Soils of Arid and Semi-Arid Regions [21]. Soil organic matter was analysed using gravimetric methods according to ASTM D 2974.

2.4. Water sample and collection
The inflows and outflows were collected every week for the first five weeks starting from February to April 2019. The collection of the samples was then carried out every two weeks from April to June 2019. Sulphate concentration was measured in each sampling event. Sulphate samples were analysed using turbidimetry based on the national Indonesian standard methods SNI 06-6989.20-2004, meanwhile pH was measured using pH meter based on the standard methods for the examination of water and wastewater [19].

2.5. Data analysis
SPSS 21 software was applied to analyse the data obtained from the experiment. Sulphate concentrations from inflows and outflows were determined by the mean and standard deviations. Then, the performance of various types of media either planted or unplanted with lemongrass in removing sulphate was calculated using the percentage removal efficiency (equation 1), where C_{in} and C_{ef} were inflow and outflow concentrations, respectively.

$$(\%R) = \frac{C_{in} - C_{ef}}{C_{in}} x100\% \qquad (1)$$

The significant difference of the media treatment was statistically analysed using One-way ANOVA. Tukey HSD post-hoc tests were then conducted to deeply address the significant difference of each treatment. In all cases, a significant level of $\alpha = 0.05$ was used.

3. Result and discussion
In this study, sulphate in domestic wastewater was treated using VF mesocosms containing six different types of media with combination of sand and pumice (Table 1). In this study, the media was firstly characterized and the inflow and outflow concentrations of sulphate in the six different types of media were determined.

3.1. Media Characterisation
As shown in Table 2, pumice had a higher organic matter than sand while water content of sand and pumice determined by gravimetric methods was almost the same which was 1.6 and 1.1 %, respectively. The pH of sand and pumice were 8.05 and 8.17, respectively. This indicated that these media slightly in the base environment. The results of CEC measurement revealed that pumice had the higher CEC compared to sand. The CEC in the media is influenced by soil pH, organic matter, and type of clay. In this research, the higher CEC in the pumice compared to sand media could be influenced by the presence of organic matter. The results of mineral content using XRF (Figure 2) showed that the main elements of pumice were SiO_2, Fe_2O_3, CaO, Al_2O_3, and K_2O with percentage of 49.0, 16.2, 11.9, 11.0, and 8.4%, respectively. For sand media, the main elements were CaO (41.9%), SiO_2 (38.0%), Fe_2O_3 (14.9%), and TiO_2 (1.5%).

Table 2. Characterization of sand and pumice.

Parameters	Media	
	Sand	Pumice
Organic matter (%)	3.1	10.9
H_2O (%)	1.6	1.1
pH	8.05	8.17
CEC (cmol(+)/kg)	47.76	50.88

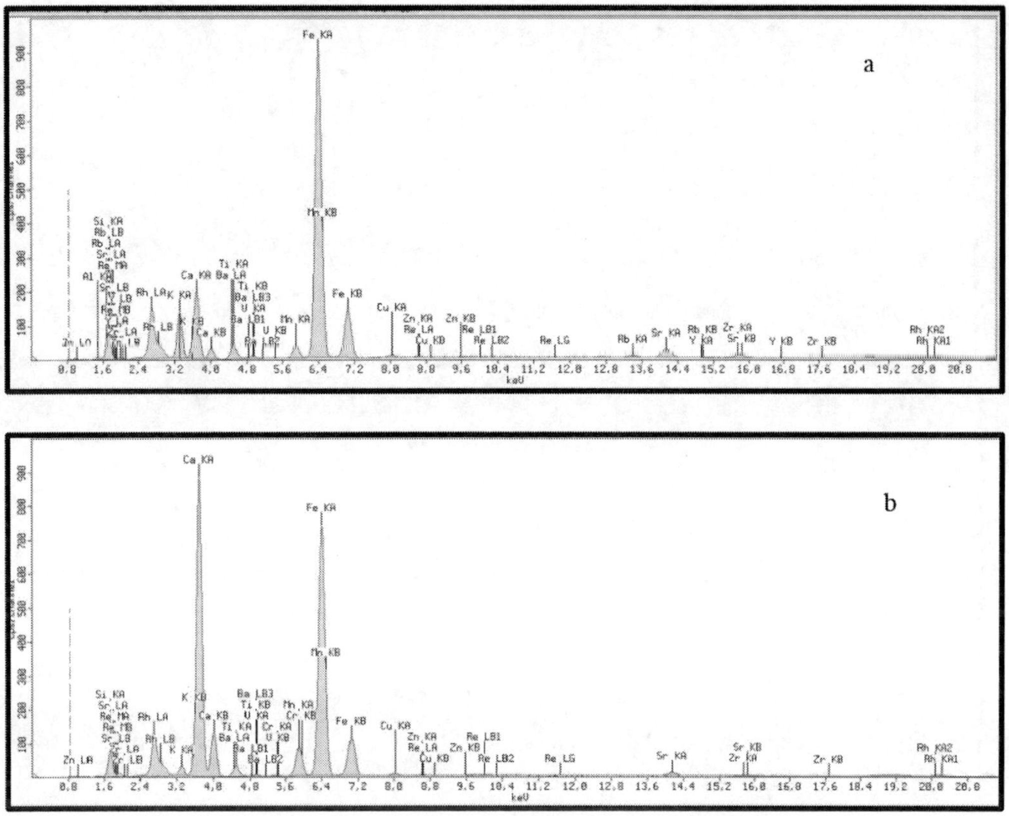

Figure 2. XRF spectra of (a) pumice and (b) sand.

The results of FTIR spectra for pumice and sand are shown in Figure 3. As shown in Figure 3a, FTIR spectra for sand media sample contained several peaks. The band at 1035 cm^{-1} and 1166 cm^{-1} were assigned to Si-O-Si stretching. The presence of Si-O was supported by the peaks at 518 cm^{-1} and 462 cm^{-1}. This indicated that siloxane and silicon compounds could be available in sand media sample. This spectrum also showed the presence of H_2O represented by O-H stretching at 3400 cm^{-1}. Several peaks were appeared in the pumice spectra (Figure 3b). The peak at 1031.95 cm^{-1} was attributed to Si-O-Si stretching. The O-H stretching was identified at 3458.48 cm^{-1} indicating that the pumice contained H_2O. The bands at 459 cm^{-1} and 435 cm^{-1} were assigned to Si-O-Al and Si-O-Si asymmetry bend, respectively. It could be concluded that the pumice contained siloxane and silicon compounds [23].

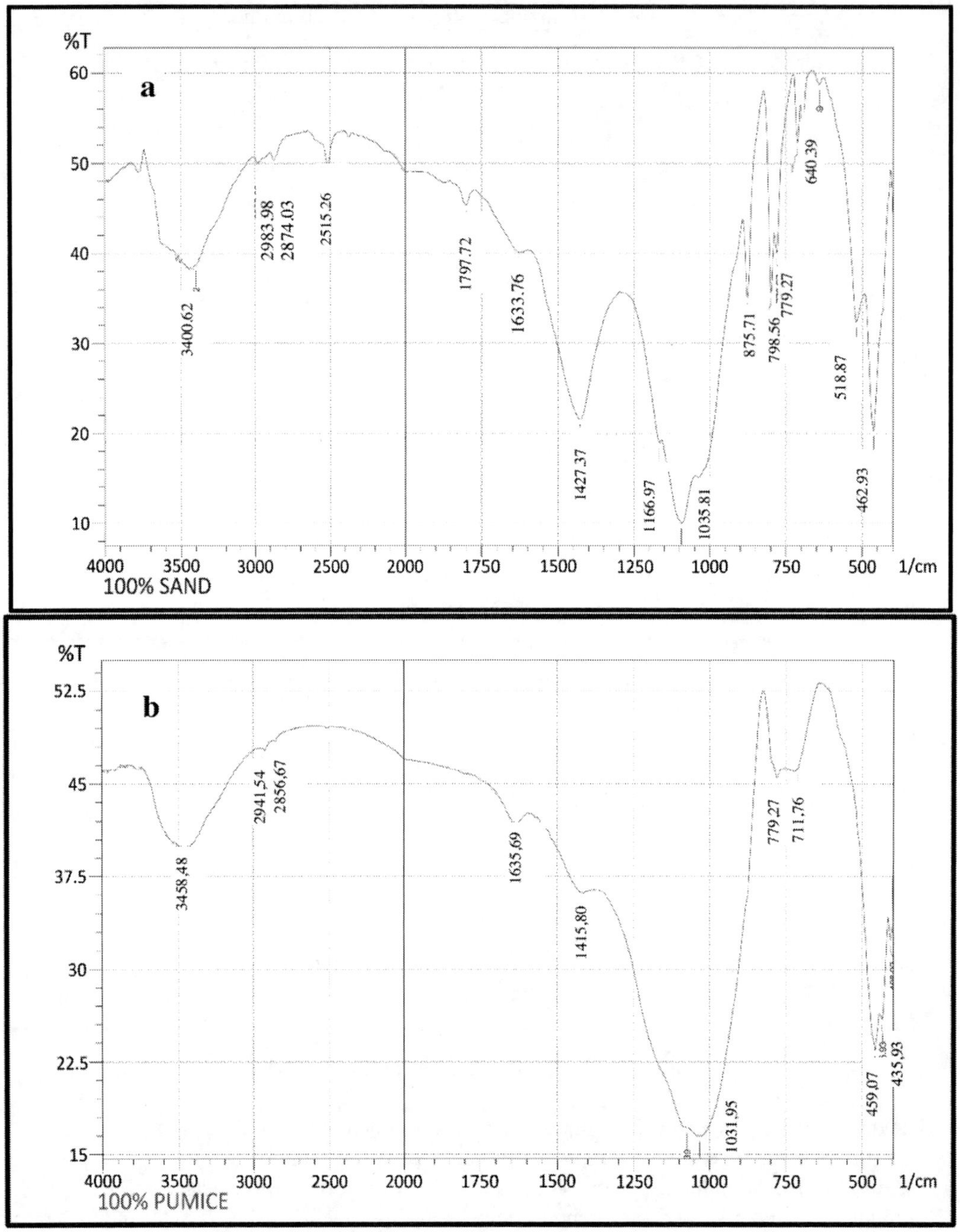

Figure 3. FTIR Spectra of (a) pumice and (b) sand.

3.2. Sulphate removal

Figure 4 shows the inflows and outflows of sulphate concentrations in the VF mesocosms with six media treatments loaded with domestic wastewater. Inflow concentrations were fluctuated ranging from 17 mg/L to 31 mg/L meanwhile sulphate concentrations in the outflow ranged from 1.5 mg – 14.7 mg/L (Figure 4). The highest outflow concentration of sulphate was observed in media containing 50% sand and pumice (CW3) while the lowest outflow concentration was observed in the media with 100% pumice (CW4). This indicated that the highest removal efficiency was obtained in the 100% pumice media (CW4), while the media containing 50% sand and pumice (CW3) was the poorest performance. One-way ANOVA analysis showed that significant differences were observed between inflow and outflow concentrations indicating that the six various VF mesocosms had a better performance in reducing pollutants. However, there were no significant differences in sulphate concentrations among six VF mesocosms. This suggested that either addition of pumice in sand media or the presence of plants in VF mesocosms did not significantly influence the performance of VF mesocosms in removing sulphate. This result was in line with the result obtained by [2] which revealed that the presence of plants only had a minor effect for sulphate removal.

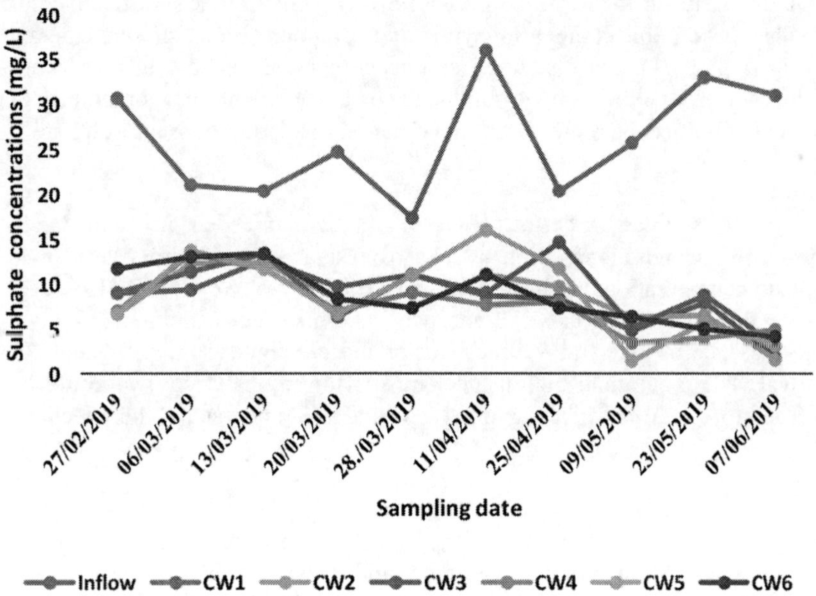

Figure 4. Sulphate concentrations (mg/L) in different types VF mesocosms loaded with domestic wastewater (the values shown in the figure are in average).

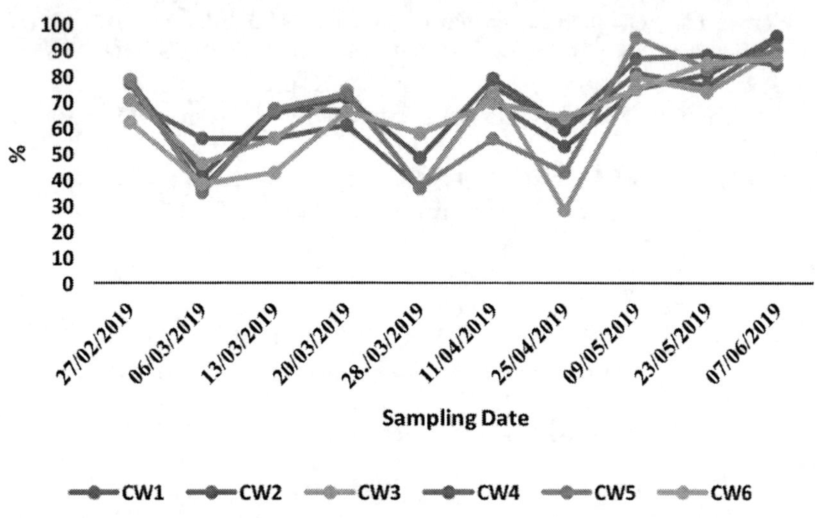

Figure 5. Percentage of sulphate removal (%) from domestic wastewater.

It was revealed that the removal efficiency of sulphate in six VF mesocosms were in the range of 62 – 69 % (Figure 5). Several possible mechanisms for sulphate removal in CW systems included plant uptake, hydrogen sulphide emission due to the presence of bacterial sulphate reduction, precipitation and adsorption of sulphur compound in media due to the presence of calcium and iron [14, 24, 25]. Since VF mesocosms had aerobic condition, emission of hydrogen sulphide might not significantly contribute in removing sulphate. The main possible mechanisms occurring in sulphate removing could be the adsorption and precipitation which occurred simultaneously because all the media was in the higher pH value (approximately around 8) and contained high proportion of Fe and Ca. The precipitation mechanisms, particularly due to the availability of Fe, might not provide significant contribution because of higher pH value. In addition, competition with other compounds could affected the removal efficiency of sulphate. Since all the media contained silicon compound and the inflows were obtained from domestic wastewater, competition among sulphate and other elements and compound cannot be avoided. This condition could reduce the removal efficiency of all media in removing sulphate.

4. Conclusion

This study revealed successful performance of pumice amended in sand media planted with *Cymbopogon citratus* in removing sulphate ions from aqueous solution. All six media significantly removed sulphate concentrations with the removal efficiency of 62 – 69 %. However, no significant differences were observed among these VF mesocosms. This suggested that addition of pumice in sand media and the presence of plants in VF mesocosms did not significantly affected VF mesocosms for sulphate removal. The main contribution mechanisms for sulphate removal could be the adsorption mechanism. Therefore, further investigation is needed to understand the mechanism of sulphate removal.

Acknowledge

This research is financially supported by Directorate of Research and Community Service of the Directorate General of Strengthening Research and Development at the Ministry of Research, Technology and Higher Education (Grant No 47/UN15.19/PL/2019). The authors would like to acknowledge Ministry of Research, Technology and Higher Education for funding this research project.

References

[1] López P, Pérez-Rodríguez I, Estrany F, Devesa R *2017 J Water Supply: Res T.* **66**(8) 598-605.
[2] Chen Y, Wen Y, Zhou Q, Huang J, Vymazal J, Kuschk P 2016 *Water Res.* **102** 572-81.
[3] Abou-Elela SI, Golinelli G, El-Tabl AS, Hellal MS *Water Sci Technol.* **69**(1) 38-47.
[4] Kadlec RH, Wallace S 2008 *Treatment Wetlands*. CRC press
[5] Garcia J, Rousseau DP, Morato J, Lesage E, Matamoros V, Bayona JM 2010 *Crit Rev Env Sci Tec.* **40**(7) 561-661.
[6] Shutes RBE 2001 *Environ Int.* **26** (5-6) 441-7.
[7] Sandoval L, Zamora-Castro SA, Vidal-Álvarez M, Marín-Muñiz JL 2019 *Applied Sciences.* **9**(4) 685.
[8] Saeed T, Sun G. A 2012 *J. Environ Manage.* **112** 429-48.
[9] Gorgoglione A, Torretta V 2018 *Sustainability.* **10**(11) 3910.
[10] Valipour A, Ahn Y-H 2016 *Environ Sci Pollut Res.* **23**(1) 180-97.
[11] Wu S, Chen Z, Braeckevelt M, Seeger EM, Dong R, Kästner M 2012 *Water Res.* **46**(6) 1923-32.
[12] Wu S, Jeschke C, Dong R, Paschke H, Kuschk P, Knöller K 2011 *Water Res.* **45**(20) 6688-98.
[13] Faulwetter JL, Gagnon V, Sundberg C, Chazarenc F, Burr MD, Brisson J 2009 *Ecol Eng.* **35**(6) 987-1004.
[14] Chen Y, Wen Y, Zhou J, Tang Z, Li L, Zhou Q 2014 *Water Res.* **59** 1-10.
[15] San Miguel-Espinosa C, Rámila C, Leiva E, Lizama-Allende K 2019 *Bioresources.* **14**(1) 363-75.
[16] Hou J, Guo W, Wen Y, editors. 2018 *IOP Conference Series: Earth and Environmental Science.*

IOP Conf. Series: Materials Science and Engineering **833** (2020) 012041 doi:10.1088/1757-899X/833/1/012041

[17] Rao E. P. 2013.:*Plant Sciences Reviews 2012*. 105IOP Publishing.
[18] Wanyama J, Herremans K, Maetens W, Isabirye M, Kahimba F, Kimaro D, Poesen J, Deckers J 2012 *Soil Use and Manage.* **28** 409-418.
[19] Ekpenyong C E, Akpan E, Nyoh, A. 2015 *Chin J of Nat Medicines.* **13** 321-337.
[20] Pleysier J, Juo A 1980 *Soil Sci.* 1980;129(4):205-11.
[21] Bashour II, Sayegh AH 2007 *Methods of analysis for soils of arid and semi-arid regions.* Rome: FAO; 2007.
[22] APHA 2005 *Standard methods for the examination of water and wastewater. 21 ed.* Washington DC, USA:
[23] Anbalagan G, Prabakaran A, Gunasekaran S 2010 *J. Appl Spectrosc.* **77**(1) 86-94.
[24] Baldwin DS, Mitchell A 2012 *Water res.* **46**(4) 965-74.
[25] Rickard D, Morse JW 2005 *Mar chem.* **97**(3-4) 141-97.

Particle Size Effect on the Water Vapour – Activated Tamarind Seeds (*Tamarindus indica L.*) Toward the Adsorbent Physical Properties

Mimi Salmawati[1], Prabasti Kusumoning Gati[1], Zubaidah Ningsih[1], Diah Mardiana[1*]

[1]Dept. of Chemistry, Faculty of Science, Brawijaya University, Malang, Indonesia.

*Corresponding author: mdiah@ub.ac.id

Abstract. The increasing problem of dyestuff liquid waste in the textile industry occurs due to the low adsorption of textiles toward dyes. Therefore, various efforts have been made to enhance the fabric's absorption. One method that has been done is by adding biomass-derived dye-binding additives which act as an adsorbent. One of potential biomasses is tamarind seeds. The purpose of this study is to determine the effect of particle size of tamarind seeds, which is physically activated, on the adsorbent character. The variation of adsorbent particle size applied were -80 + 100 mesh (177-149 μm), -100 + 120 mesh (149- 125 μm), -120 + 150 mesh (125-99.3 μm) and less than 150 mesh (<99.3 μm). Tamarind seed was activated using water vapour at high pressure prior to the application. The adsorbent characters were studied based on particle size distribution, powder porosity, average relative molecular mass, and hydrophilicity. In addition, tamarind seed chemical structure was analysed using Fourier Transform Infra-Red (FTIR) spectrophotometry, while adsorbent potency was studied in the form of an adsorbent thin film. The results showed that based on the analysis of particle size distribution, adsorbents with the best activation results was obtained by particle size of 125-149 μm. The highest porosity was achieved by particle that smaller than 99.3 μm and the highest average relative molecular mass, based on the Mark–Houwink approach, was obtained by particle size of 125-149 μm. The highest hydrophilicity, which was determined based on the contact angle, was obtained by particle sizes of 99.3-125 μm.

1. Introduction

Waste dyes in the textile industry are one of the main sources of wastewater pollution. 5-10% of the dyestuffs used in the textile dyeing process are unutilized as waste. This is due to the low adsorption of dye substances in the textile substrate. One of efforts made to increase the fabric's absorption on dyes is to use an adsorbent as a dye-binding additive that can be obtained from biomass [1,2].

Tamarind seeds are one of the biomasses which is lately have been widely developed as an adsorbent. Some application of tamarind seeds are as adsorbents, among others, in metal adsorption, reduce turbidity on the surface of the water and adsorb methylene blue from the dye waste [3–6].

The tamarind seeds consist of 20-30% seed bark, commonly called testa and 70-75% kernel or endosperm [7,8]. Kernel of tamarind seeds contains a lot of carbohydrates, in the form of polysaccharides ranging from 65-72% [9]. The structure of carbohydrates in tamarind seeds has a hydroxyl group as the main functional group and acts as a reaction point for the formation of polymers[10]. The presence of hydroxyl groups in the carbohydrate structure of tamarind seeds gives a role in the interaction with hydroxyl groups in fabric fibers through the formation of hydrogen bonds.

The adsorbent character greatly influences the efficiency of the adsorption process. The adsorbent character can be modified through activation, both physically and chemically. Physical activation using water vapor at high pressure can increase the specific surface area of tamarind seed kernel powder by 18.42% and increase its porosity. The purpose of this study is to determine the influence of the particle size of the adsorbent powder from the tamarind seed kernel on the physical properties. The adsorbent characteristics were studied based on the distribution of particle size, porosity, average molecular mass based on viscosity and hydrophilicity. As support, chemical structure analysis was carried out by FTIR spectrophotometry.

2. Experimental

2.1. Apparatus and materials
The materials used in this research are tamarind seeds (*Tamarindus indica* L.) obtained from PT. Bintang Jaya, Surabaya, East Java, Indonesia.

The instruments used in this research were particle size analyzer (PSA-Cilas Particle Size Analyzer 1090 D), contact angle analysis using goniometer (contact angle measurement system) and identification of functional groups using FTIR spectrophotometer (Shimadzu FTIR 8400S).

2.2. Tamarind seeds preparation and activation
The tamarind seed kernels were crushed using a grinder so that they are in the form of powder. The powders were sieved using filters of various size, i.e. 80 mesh (177 μm), 100 mesh (149 μm), 120 mesh (125 μm) and 150 mesh (99.3 μm) to obtain variations in the size of tamarind seed kernel powders. A total of 20 g of powdered kernel tamarind seeds of various sizes were then placed in containers made of filter paper and their masses were measured. Furthermore, the container which has been filled with acid seed kernel was placed above the filter, so it does not touch the water at the bottom of the autoclave. After closing the autoclave, it was then heated at a fixed temperature. When the water vapor starts to come out, the vapor outlet was closed, and the heating was continued for 30 minutes. After completion (no more steam comes out), the autoclave lid was opened. The resulting activation powder was then removed and placed in a desiccator. After 2 to 3 hours, the powder and the container were weighed. This activation method was carried out for powders of all different sizes.

2.3. Characterization of adsorbent
The activated adsorbent powders were characterized for particle size distribution, adsorbent porosity, average relative molecular mass, hydrophilicity, and functional group analysis. Particle size distribution was analysed using particle size analyser (PSA). The porosity of the adsorbent powders is estimated from the ratio of bulk density to dimensional density. The average relative molecular mass of the adsorbent is determined using the Ostwald viscometer. Hydrophilicity was determined based on the contact angle of the liquid on the surface of the solids observed using a goniometer device connected to the camera. The solid used comes from a thin film formed from an acid seed gel. Goniometer results in the form of images that were further processed using contact angle measurement software. Furthermore, the identification of functional groups was done by Fourier Transform Infra Red (FTIR) spectrophotometry method at 400-4.000 cm^{-1}.

3. Results and Discussion

3.1. Particle size distribution
The particle size of the activated tamarind seed kernel powders was analyzed using a particle size analyzer (PSA), and the results of the analysis, as a particle volume distribution curve, are presented in

Figure 1. Distribution patterns for curves a and b shows non-symmetric distributions, whereas curves c and d tend to form symmetric distributions. The results of the curve interpretation are listed in Table 1.

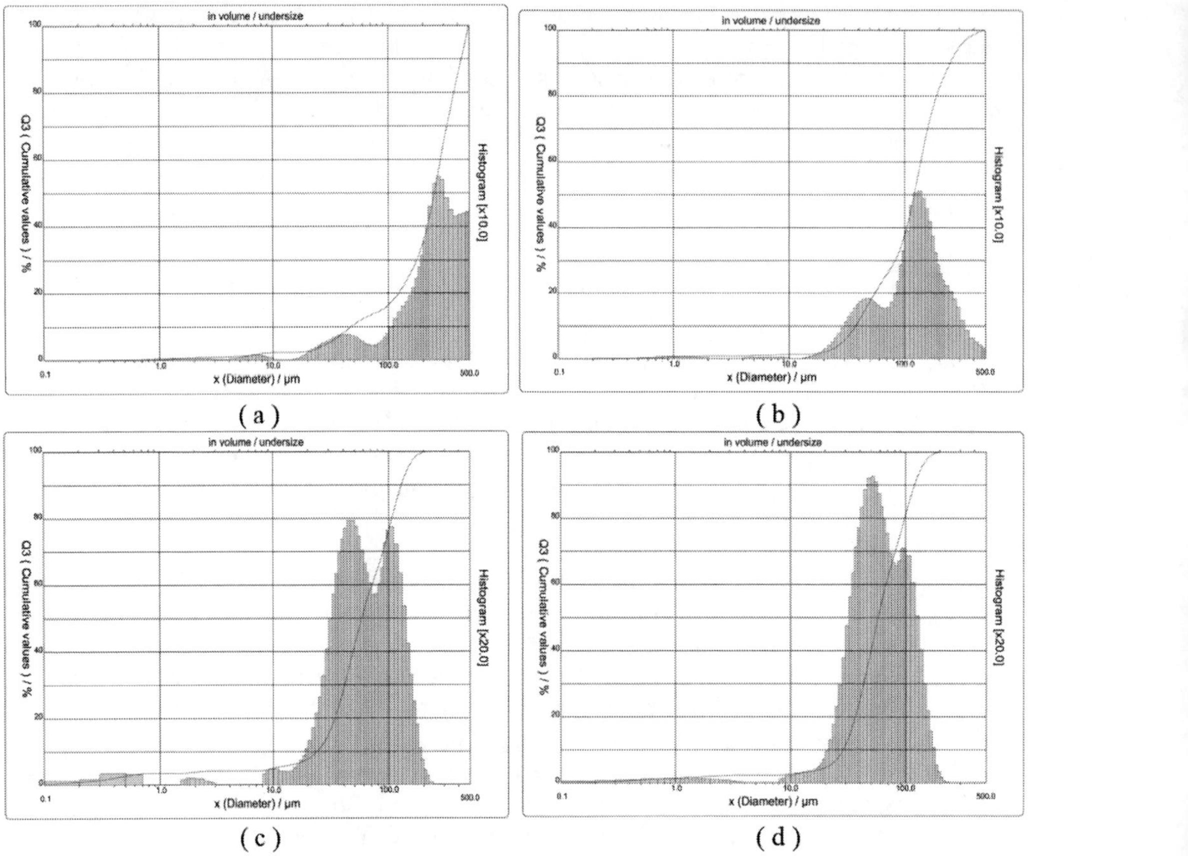

Figure 1. Particle size distribution graph (a = 177-149 µm; b = 149-125 µm; c = 125-99.3 µm; d = < 99.3 µm).

Table 1. Percentage distribution of particle size.

Particle size (µm)	Diameter percentage 10 (%)	50 (%)	90 (%)
177-149	47.02 µm	249.1 µm	430.65 µm
149-125	37.46 µm	120.99 µm	249.23 µm
125-99.3	23.97 µm	57.81 µm	129.05 µm
< 99.3	28.57 µm	57.35 µm	119.2 µm

An important parameter of particle size is the 50% distribution, which gives information about the mean value of the particle size. Activation causes an increase in particle size in the size of 149-177 µm tamarind seeds kernel powder. This is likely due to the phenomenon of particle agglomeration. The smaller the particle size, the symptoms of agglomeration tend to decrease. The two smallest particle sizes which are smaller than 99.3 µm and 99.3-125 µm have low resolution. Therefore, the two-particle sizes are still not well separated. The best activation results are for sizes 125-149 µm because they provide median, mean and mode values that are almost the same as the particle size before activation.

IOP Conf. Series: Materials Science and Engineering **833** (2020) 012042 doi:10.1088/1757-899X/833/1/012042

3.2. Porosity
The porosity in this experiment is approached from the value of bulk density. Density is determined by picnometry in solvents with different polarities so that the powder does not dissolve. Figure 2 shows the results of the measurement of the bulk density.

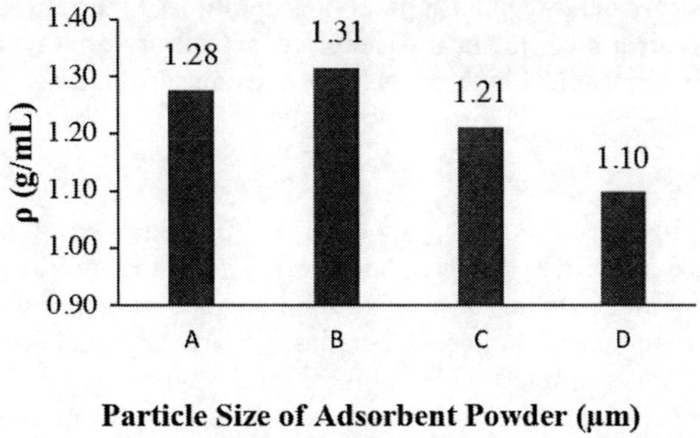

Figure 2. Density of adsorbent powder each particle size (A = 177-149 μm; B = 149-125 μm; C = 125-99.3 μm, D = <99.3 μm).

Bulk density, in this research, is determined based on the average density of powders for a certain volume in a specific medium. The medium used is toluene. Bulk density includes the density of solids also accommodates the presence of empty space in the form of pores. Therefore, bulk density can be used as a reference to estimate porosity. In Figure 2, the particle size of 125-149 μm has the highest density so it has the lowest porosity. In Figure 1(b), the particle size distribution appears to widen, so that the low porosity is not caused by particle size. Most of the natural polymers such as polysaccharide have relatively inflexible chains. Therefore, the size is no longer the dominant factor, because the polymers chain is no longer in the form of a flexible random coil. The transition in configuration, namely chain entanglement has occurred. Therefore, the polymer will be denser, empty space is reduced and the density is higher [11].

3.3. Average relative molecular mass
In Figure 3, the highest intrinsic viscosity was achieved for particles with a size of 125-149 μm.

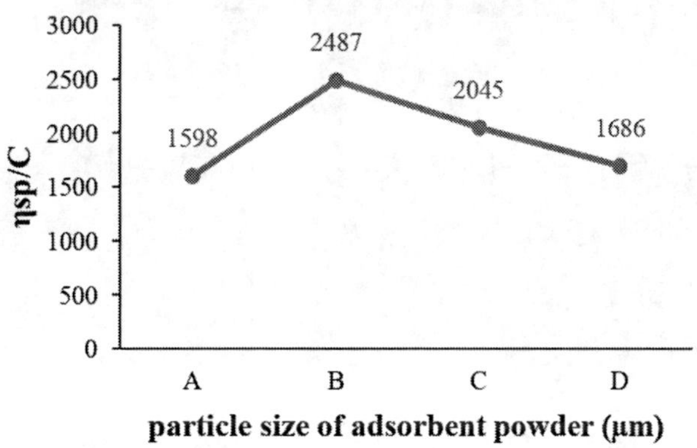

Figure 3. Gel viscosity of each particle size of the adsorbent powder (A = 249.1-120.99 μm; B = 120.99-57.81 μm; C = 57.81-57.35 μm D = <57.35 μm).

Polysaccharides with high hydroxyl groups in the water medium absorbs water and forms a gel. High density causes higher friction between polymer molecules so that the flow rate was also slower. This causes the disperse system to become more viscous and provides the highest intrinsic viscosity. Although the polysaccharide molecule shape is not the only influential factor but based on the Mark–Houwink approach, the particle size has the highest average molecular mass. Tamarind polysaccharide has been known not to form a gel, except in appropriate conditions. One of the conditions is by removing galactose residues from tamarind xyloglucan. Therefore, activation using water vapor at high pressure and temperature is likely to be able to remove galactose residues from the polysaccharides of tamarind kernel seeds [12].

3.4. Hydrophilicity

The hydrophilicity of a surface shows the ability of polar fluids, such as water, to spread on the surface of a solid. This ability is expressed as the spreading coefficient. If the coefficient is small, the liquid will form droplets at a certain angle. On this basis, hydrophilicity is determined based on the analysis of the contact angle. The observation results of contact angles for various particle sizes are shown in Figure 4.

Figure 4 is a form of water droplets when dropped on the surface of a thin film of tamarind kernel powder. The droplet shape is then processed using contact angel software (contact angle measurement software) to obtain the contact angle value in Table 2.

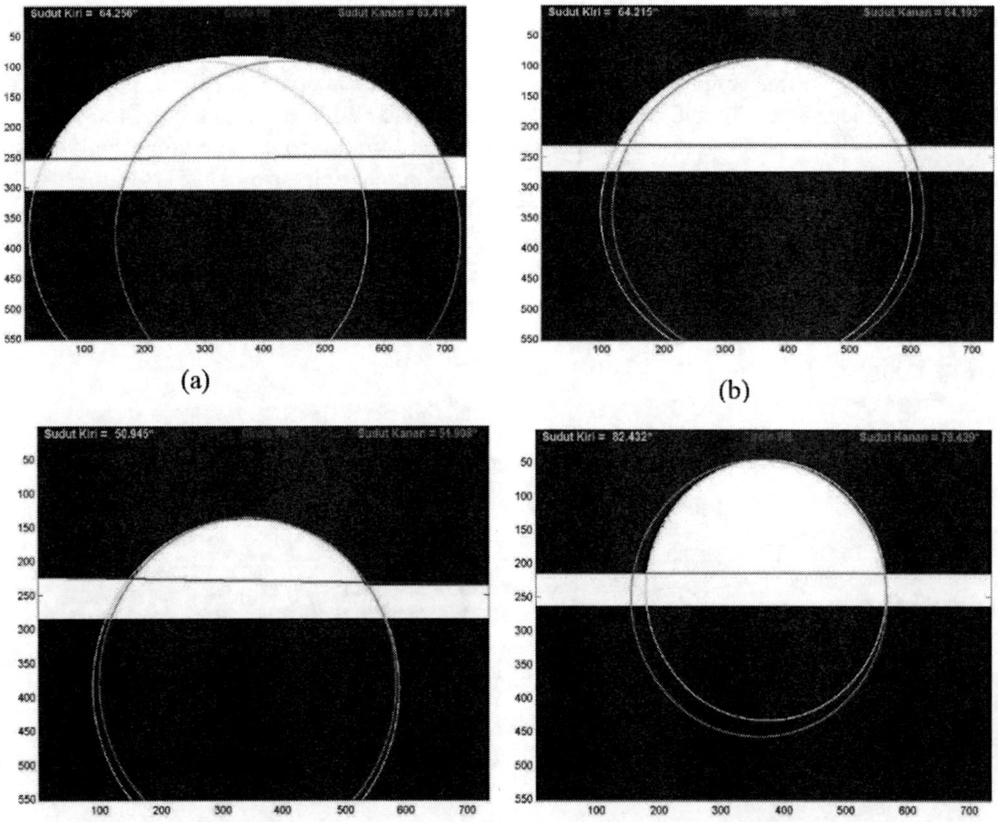

Figure 4. Liquid droplets on thin films that have been processed using contact angle measurement software (A = 249.1-120.99 µm; B = 120.99-57.81 µm; C = 57.81-57.35 µm D = <57.35 µm).

Overall, all thin films are hydrophilic because they have a contact angle that smaller than 90°. For larger particles, such as 125-149 µm and 149-177 µm, they both have the same hydrophilicity. The most hydrophilic thin film was produced from particle sizes of 99.3-125 μm. Whereas if the particle size is smaller than 99.3 μm, thin films tend to have higher hydrophobicity.

Table 2. The contact angle of the film for each particle size.

Particle size (μm)	Contact angle	
	left angel	right angel
249.1-120.99	64.2561	63.41442
120.99-57.81	64.2152	64.1928
57.81-57.35	50.9454	51.90771
<57.35	82.4322	79.42909

The spreading coefficient of water on the thin film surface is a balance between cohesion force and adhesion force. The cohesion force is originated from the attraction between water molecules while the adhesion force is due to attraction of water with the thin film surface. Thin films are hydrophilic and have a high affinity for water because acid kernels are polysaccharides that have polar functional groups, namely hydroxyl. However, the rate of spread is determined by the surface tension of the water and the surface tension of the thin film and the interface tension of both. The higher the difference in surface tension of thin films compared to water, the affinity for water increases. The surface tension of the film is affected by the film structure and surface morphology. It is suspected that this factor resulting in particle size of 99.3-125 μm being more hydrophilic. Whereas when the size is smaller than 99.3 μm, the adhesion force decreases due to lower molecular mass. Thus, the forming films have lower surface tension. As a result, the affinity for water is reduced [13].

3.5. Functional groups analysis

Figure 5 is the spectrum of the adsorbents using the FTIR spectrophotometer. The patterns on the three spectrums are identical. Complete spectrum interpretation is presented in Table 3. The main difference occurs in the stretching vibrations for hydroxyl groups in the region of 3300-3400 cm^{-1}.

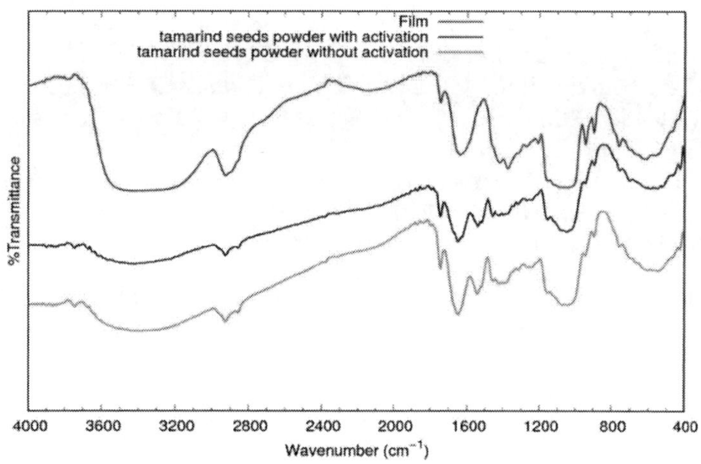

Figure 5. Functional groups analysis using FTIR spectrophotometer on tamarind seed powder and film.

The three spectra exhibit the character of a wide hydroxyl functional group, indicating a hydrogen-bond hydroxyl group. This is also supported by the bending vibration of hydroxyl functional group. Hydrogen bonds occur both inter and intramolecularly. The intensity of the hydroxyl group uptake for powder is lower than that of the film. The difference in intensity indicates the effect of activation which causes an increase in the number of hydroxyl groups. It is assumed that the diffusion of water vapor into the cavity will reduce the hydrogen bonds, hence hydroxyl groups are relatively free.

Table 3. Identification functional groups of adsorbent powder and film.

Functional groups	Wavenumber (cm^{-1})		
	Kernel powder without activation	Kernel powder activated	Film
O-H stretching	3405	3380	3283
C=O stretching	1744	1744	1744
C=C stretching	1648	1646	1642
O-H bending	1387	1395	1374
C-O-C	1044	1048	1044

4. Conclusion

Based on particle size analysis, the adsorbent powder with the best activation results is in the adsorbent powder particle size 125-149 µm. The highest porosity was obtained from adsorbent made of particle size smaller than 99.3 µm and the largest density was shown by the adsorbent made of particle size of 125-149 µm. The highest hydrophilicity was produced on thin films from adsorbents prepared from particle size of 99.3-125 µm.

References

[1] Abo-Shosha M H, Ibrahim N A, Allam E and El-Zairy E 2008 P *Carbohydrate Polymers* **74** 241–249
[2] Yagub M T, Sen T K and Ang H M 2012 *Water, Air, & Soil Pollution* **223** 5267–5282
[3] Mopoung S, Moonsri P, Palas W and Khumpai S 2015 *The Scientific World Journal* **2015** 1–9
[4] Sa'id S, Mohammed K, Adie D B and Okuofu C A 2016 *Bayero Journal of Pure and Applied Sciences* **9** 236
[5] Pardede A S D, Siburian A M and Pandia S 2014 *Medical Sciences* 5
[6] Jamion N A and Hashim I N 2018 *Journal of Fundamental and Applied Sciences* **9** 102
[7] Chacón-Fernández M G, Hernández-Medel M R, Bernal-González M, Durán-Domínguez-de-Bazúa M C and Solís-Fuentes J A 2019 *Grasas y Aceites* **70** 333
[8] Chaudhary H and Singh V 2018 *Fibers and Polymers* **19** 2514–2523
[9] Khounvilay K and Sittikijyothin W 2012 *Food Hydrocolloids* **26** 334–338
[10] del Real A, Wallander D, Maciel A, Cedillo G and Loza H 2015 *Carbohydrate Polymers* **117** 11–18
[11] S.F. Sun 1994 *Chemistry of Macromolecules* (Canada: John Wiley & Sons)
[12] Yoko Nitta and Katsuyoshi Nishinari 2005 *J. Biol. Macromol* **5** 47–52
[13] Chen N, Maeda N, Tirrell M and Israelachvili J 2005 *Macromolecules* **38** 3491–3503

The Effect of K_2O Concentration in K_2O/Al_2O_3 Catalyst on Methyl Ester (Biodiesel) Synthesis from CPO Off Grade with Ultrasonic Wave

Aman Santoso[1], Anugrah Ricky Wijaya[1], Chandra Fetty Purwaningtyas[1], Dedek Sukarianingsih[1], Rini Retnosari[1], Sumari Sumari [1]

[1]Department of Chemistry, Faculty of Mathematics and Natural Sciences, State University of Malang, Jl. Semarang No. 5 Malang, 65145, Indonesia

*Corresponding author: aman.santoso.fmipa@um.ac.id

Abstract: Fatty acid methyl esters can be made through transesterification of vegetable oils with homogeneous or heterogeneous catalysts. Ultrasonic waves and the type of catalyst affect the transesterification reaction. The purpose of this study was to determine the effect of ultrasonic waves on the transesterification of CPO with the K_2O/Al_2O_3 catalyst. Research stages: CPO off grade characterization, activation and characterization of natural zeolite, preparation and characterization of K_2O/Al_2O_3 catalysts, and oil esterification. Transesterification of CPO with K_2O/Al_2O_3 catalyst (3% w/w) at various levels of K_2O assisted by ultrasonic waves. The characterization of transesterification results includes density, viscosity, refractive index, and acid number. The use of ultrasonic waves reduces the transesterification reaction time from 120 to 35 minutes. The optimum conditions of methyl ester synthesis were achieved in the K_2O/Al_2O_3 catalyst with K_2O levels of 35% w/w, with a yield of 88.06%. Characteristics of the synthesized methyl ester include density, viscosity, refractive index, and acid number respectively are 0.891 g/mol, 7.86 cSt, 1.454 and 0.465 which are in accordance with SNI biodiesel.

1. Introduction

Energy demand always increase every year. The need for energy is still largely fulfilled from the earth's oil. Fuel from petroleum is a non-renewable energy source, so to fulfill the needs of fuel oil, energy sources alternative that can be renewed are needed [1]. Biodiesel is one of energy source alternative that is recommended as a substitute for petroleum. Biodiesel is a transesterification of long chain carboxylic acids which can be synthesized from vegetable oils or animal oils through esterification or transesterification processes. Transesterification synthesis is carried out by reacting triglycerides and alcohols assisted by acidic or basic catalysts in the process of transesterification and reacting carboxylic acids with methanol in the esterification process [2]. Triglycerides and carboxylic acids can be found in several types of vegetable oils and animal oils. Vegetable oil which is quite abundant in Indonesia is palm oil. Palm oil is divided into groups according to quality. Low quality palm oil has high levels of free fatty acids and does not meet the standards as raw material for production so that it ends up being waste. High levels of free fatty acids in low quality palm oil have the potential as a raw material in the production of biodiesel and make it possible to use this biodiesel feedstock in this research [3].

In general, biodiesel is synthesized through esterification and transesterification reactions [4]. Two things that are not considered in the transesterification process even though it affects the results of the transesterification synthesis are the stirring process and the catalyst used to accelerate the reaction rate in the synthesis process. The transesterification process usually used a homogeneous catalyst such as a KOH or NaOH solution [5]. The use of homogeneous catalysts has a disadvantage, which is very difficult to separate from the product because it has the same phase. The process of separating the catalyst from the product requires a lot of water in its purification so that the waste water from biodiesel purification can be corrosive and pollute the environment [6] Homogeneous catalyst with carboxylic acids will form alkaline soap and water which can reduce the quality of transesterification produced. Alkaline soap and water formed in the transesterification process can be minimized with heterogeneous catalysts in this study. Heterogeneous catalysts have several advantages including being easily separated, minimizing the formation of water and alkaline soap, resistant to high temperatures, and not corrosive [7].

The heterogeneous catalyst that has been developed at the moment is the heterogeneous catalyst K_2O/Al_2O_3. Al_2O_3 , which is a solid compound that can increase the strength of the base [8] so it can be functioned as a good catalyst support in the transesterification reaction. This study includes the impregnation of KOH into Al_2O_3 then calcined at high temperature into a heterogeneous catalyst K_2O/Al_2O_3 and can produce 96.8% transesterification yield [9].

The oil, catalyst and alcohol in the transesterification reaction can not mixed well, so it needs the energy to help the mixing process. Biodiesel synthesis is generally carried out with mechanical stirrers. The stirring is less optimal and requires a long time. In this study the stirring process was assisted by ultrasonic waves thereby reducing reaction time [10][11].

Ultrasonic waves have a high intensity or frequency, causing small bubbles on the liquid media. When the maximal pressure of bubble approaches, the bubble experiences a split called cavitation [12]. Cavitation produced by ultrasonic waves can help the reaction between triglycerides with alcohol in the formation of alkyl esters. The yield of transesterification results using ultrasonic waves is higher than using conventional stirring. Based on this background, the propose of this research is transesterification of low quality palm oil with K_2O/Al_2O_3 catalysts and ultrasonic waves [9].

2. Method

This research is a type of laboratory experimental research. This research went through several stages, such as characterization of low quality palm oil, activation and characterization of natural zeolite, preparation and characterization of K_2O/Al_2O_3 catalysts. Rafination low quality palm oil, determination of free fatty acid levels of refined palm oil resulting from K_2O/Al_2O_3 refined palm oil rafination, transesterification of esterified palm oil with K_2O/Al_2O_3 catalyst with the help of ultrasonic waves, and characterization of transesterification results include density, viscosity, refractive index, and acid number.

2.1. Tools and Material

The tools used in this research are beaker, burette, static glass, clamps, fillers, measuring flasks, analytical balance with 0.01 gram accuracy, glass bottles, spray bottles, dropper pipettes, volume pipettes, Erlenmeyer, porcelain plates, crucible, separating funnel, three-neck flask, 100 ^{0}C thermometer, glass funnel, watch glass, mortar and pestle, stir bar, spatula, universal indicator paper, filter paper, warp plastic, warp paper, label paper, aluminum foil, a set of refluxes, a set of ultrasonic wave instruments, a set of Buchner tools, centrifuges, hot plate and magnetic stirrer, furnaces, ovens, 100 mesh size sieves, Abbe refractors, Ostwald Viscometers, SEM-EDX instruments, XRF instruments, XRD instruments and GC-MS instruments.

The materials used in this study are natural mordenite zeolite obtained from south Malang, low quality palm oil (CPO off grade) from Lodoyo Blitar, methanol, ethanol, 37% HCl, $MgSO_4$ anhydrous, indicator of phenolphthalein, KOH, Al_2O_3 powder, and distilled water.

2.2. Characterization of low-quality palm oil

Low quality palm oil obtained from PT. Sawit Arum Madani in Sutojayan Blitar Indonesia. Characterization of low-quality palm oil includes density, refractive index, viscosity, and free fatty acid levels.

2.3. Synthesis and characterization of K_2O/Al_2O_3 catalyst in various percent of K_2O catalyst

K_2O/Al_2O_3 catalyst is made by wet impregnation method. Potassium hydroxide solids were dissolved with distilled water and then Al_2O_3 solids was added with various ratio of KOH : Al_2O_3 as follows 15:85, 20:80, 25:75, 30:70. Each mixture of KOH and Al_2O_3 was stirred for 3 hours then roasted for 24 hours at 110 °C, then calcined at 500 °C for 3 hours [4], [13]. Preparation of Al_2O_3 catalysts was characterized by XRD and SEM-EDX to determine the morphology of the catalyst topography, and whether K_2O was formed.

2.4. Rafination Low Quality Palm Oil

The rafination of low quality palm oil in this study aims to fade the color, remove the impurities and reduce the odors. Rafination is carried out by heating the low quality palm oil at 175 °C then added with natural zeolite which has been activated as much as 2% (w/w) and stirring using a magnetic stirrer for 1 hour. The oil which has been finished with activated natural zeolite is separated by centrifuge for 15 minutes at 3000 rpm.

2.5. Esterification Palm Oil from Rafination result

The esterification process in this study was to convert free fatty acids into transesterification. Fifty g of refined low quality palm oil is put into a triple neck flask that has been strung together with reflux and magnetic stirrer. The oil is heated to 40 °C, then methanol is added in a mole ratio of oil: mole of methanol (1:12), added H_2SO_4 1% w/w and stirred for 3 hours at 60 °C. The mixture formed is transferred into a separating funnel and allowed to stand for 4 hours to form 2 layers [7][14].

2.6. Transesterification of Palm Oil with K_2O/Al_2O_3 catalyst

The transesterification process in this study aims to convert triglycerides into transesterification. Twenty g of low quality palm oil produced by esterification was put into a three neck flask and heated at 60 °C, methanol was added with a mole ratio of oil: mole of methanol (1:12) and added with 3% w / w of K_2O/Al_2O_3 catalyst. The K_2O content in the catalyst varies from 15%, 20%, 25% and 30% w/w. The variations of the mixture included ultrasonic wave reactors with various stirring time ratios of 25 minutes, 35 minutes and 50 minutes.

3. Result and Discussion

3.1. Rafination Palm Oil with Activated Natural Zeolite

Rafination is carried out using activated natural zeolites. The characterization of palm oil before being transformed and after finalization was carried out in this study to see the differences that occur in oil. Low quality palm oil used in this study has two integrated phases, namely the solid phase and the liquid phase. The refining process in this study was carried out by heating up to 175 ° C to liquefy the solid quality low-quality palm oil to be more homogeneous. Warming up to 175 ° C with stirring is able to remove the water content found in the oil sample. Furthermore, natural zeolite which has been activated as much as 2% by weight of the refined oil is added. Activated zeolite is able to absorb impurities contained in palm oil and absorb natural dyes present in palm oil so that the color changes occur before and after refining. Color changes in low quality palm oil can be seen in Figure 1.

Table 1. Characterization Results of Low-Quality Palm Oil.

Parameter	**Value before**	**After rafinations**
Density (g/mL)	0.918	0.919
Viscosity (cSt)	25.90	25.90
Refractive index	1.463	1.463

(a) **(b)**

Figure 1. Palm oil (a) before rafination and (b) after rafination.

3.2. Esterification Palm Oil from Rafination Result

Transesterification of low quality palm oil contains high free fatty acid content for example 10.92%, must be carried out in two stages. started with esterification. The purpose of the esterification reaction is to convert free fatty acids into alkyl esters, so the free fatty acids do not become soap during the transesterification reaction process. The reaction of the accumulation of free fatty acids into unwanted soaps in the transesterification reaction process is as follows

$$RCOOR' + H_2O \rightleftharpoons RCOOH + R'OH$$

The transesterification process can be carried out if the free fatty acids found in low quality palm oil samples are less than 2%. The esterification process in this study was carried out by using 1% w/w H_2SO_4 catalyst and methanol. Methanol is used because it has the shortest chain so that it has a higher reactivity than other types of alcohol. High methanol reactivity makes it easy for acid catalysts to polarize methanol. The general esterification reaction is as follows [15]

$$RCOOH + CH_3OH \overset{H^+}{\rightleftharpoons} RCOOCH_3 + H_2O$$

The reaction that occurs is a reversible reaction that requires the addition of one of the reactants in excess, so the reaction goes to the product. This study uses a 1 mole ratio of oil mole: alcohol. The results of the esterification were tested for free fatty acid levels. The transesterification reaction process is then carried out if the result is less than 2%. The levels of free fatty acids resulting from esterification can be seen in Table 2.

Table 2. Esterification of Free Fatty Acid (FFA) Levels.

Number	**sample (g)**	**FFA (%)**
1	1.049	1.646
2	1.044	1.182
Average		1.414

In this study, the rafination results of the esterification reaction can convert the free fatty acids contained in low-quality palm oil from 10.92% to 1.41%. The mechanism of the esterification reaction converts fatty acids into transesterification are shown in Figure 2.

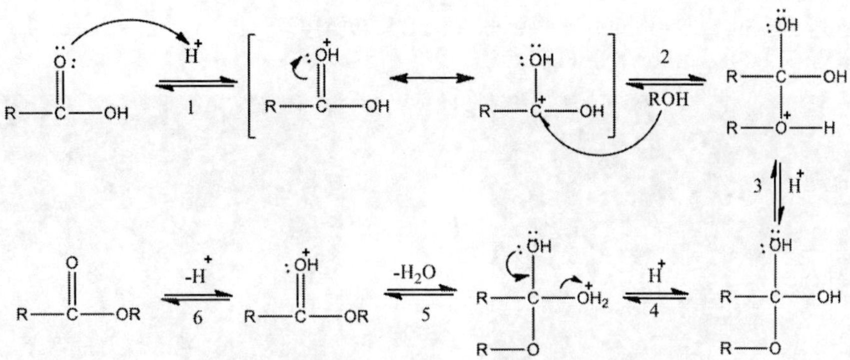

Figure 2. The mechanism of esterification reaction.

3.3. Synthesis of K_2O/Al_2O_3 Heterogeneous Catalysts at various levels of K_2O

Heterogeneous catalyst used in this study was KOH impregnated into Al_2O_3. Calcination is carried out [16] until KOH decomposes to K_2O by heating at 500 °C for 3 hours which serves to decompose KOH into K_2O in the pore Al_2O_3 with the following reaction.

$$2KOH \longrightarrow K_2O + H_2O$$

The catalyst preparation results in various K_2O/Al_2O_3 mass ratios are shown in Figure 3. In Figure 4 the XRD of 15% K_2O and 20% K_2O catalysts showed the typical peaks with K_2O compounds at 2θ = 31, 39, 51, 55, 62. The XRD results are in accordance with the XRD results of the K_2O compound from the K_2O/Al_2O_3 catalyst. The results of XRD K_2O/Al_2O_3 characterization with 15% K_2O and 20% K_2O levels can be seen in Figures 4 and 5.

Figure 3. K_2O/Al_2O_3 catalyst with fraction K_2O (1) 15%, (2) 20%, (3) 25% and (4) 30%.

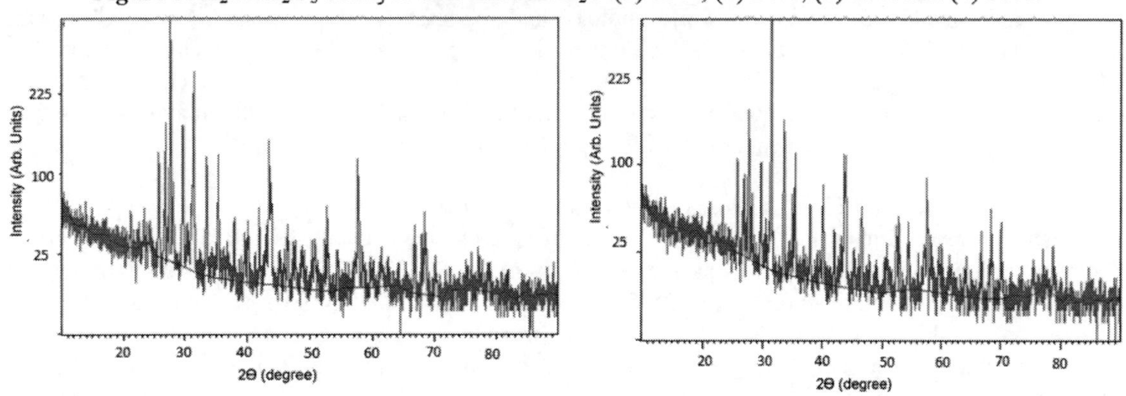

(a) **(b)**

Figure 4. XRD K_2O/Al_2O_3 catalyst with **(a)** 15% K_2O and **(b)** 20% K_2O.

The morphological and topological structure of K_2O/Al_2O_3 with a bar scale of 100 μm with a magnification of 1.000× is shown in Figure 5.

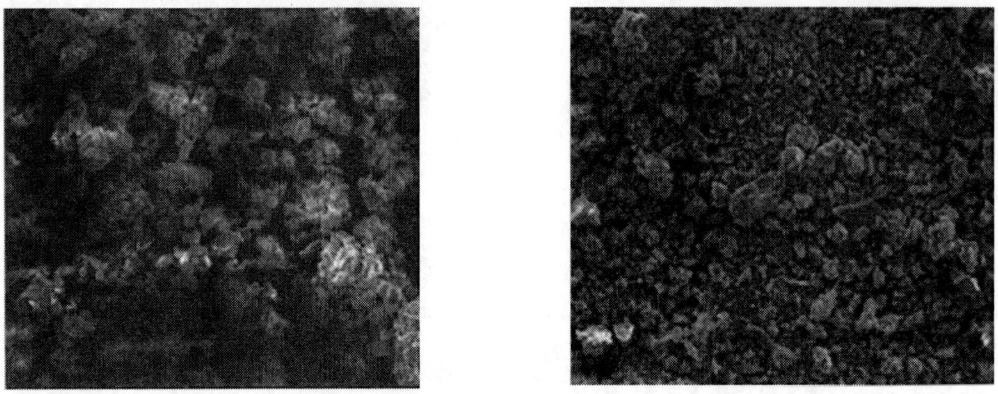

(a) **(b)**

Figure 5. The structure of XRD K_2O/Al_2O_3 catalyst with **(a)** 15% K_2O and **(b)** 20% K_2O.

Comparison of morphological and topological structure sizes with a 20 μm bar scale with magnification of 5.000 × is shown in Figure 6.

(a) **(b)**

Figure 6. Catalyst particle size with **(a)** 15% K_2O fraction and **(b)** 20% K_2O fraction.

Figure 5 and Figure 6 show the morphology and topography of the K_2O/Al_2O_3 catalyst with variations in K_2O levels of 15% and 20%. Figure 5a and Figure 6a are K_2O/Al_2O_3 catalyst with 15% K_2O content and Figure 5b dan Figure 6b are K_2O/Al_2O_3 catalyst with 20% K_2O content. As shown in Figure 6, the catalyst size at 15% K_2O fraction is between 3.129-21.460 μm with an average of 8.824 μm and at a fraction of 20 percent 2.434-14.901 μm with an average of 6.88 μm.

3.4. Transesterification Using K_2O/Al_2O_3 Catalysts with Ultrasonic Waves

This study was conducted to determine the effect of heterogeneous catalyst K_2O/Al_2O_3 and the effect of time variations on yield using ultrasonic waves. In this study, 20%, 25%, 30% of K_2O variations of K_2O/Al_2O_3 catalyst and reaction time of 25 minutes. 35 minutes. and 50 minutes were used. Ultrasonic waves have a high intensity so that when liquid medium propagates, it will produce alternating pressure which causes small bubbles in the liquid media. At maximum bubble pressure, bubbles undergo a split called cavitation [17]. Cavitation generated by ultrasonic waves can help the reaction process between triglycerides and alcohol in the transesterification reaction. The series of processing tools for the transesterification process using ultrasonic waves is shown in Figure 7.

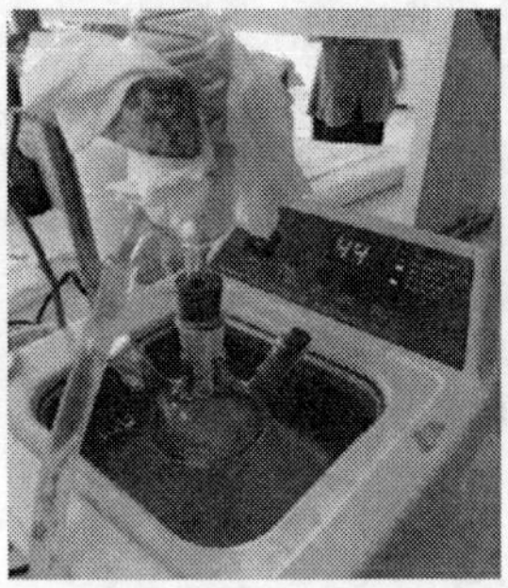

Figure 7. Series of Transesterification Process Tools Using Ultrasonic Waves.

The yield of transesterification results from low quality palm oil using heterogeneous catalyst K_2O/Al_2O_3 in various mass ratio assisted with ultrasonic waves at time variations can be seen in Figure 8.

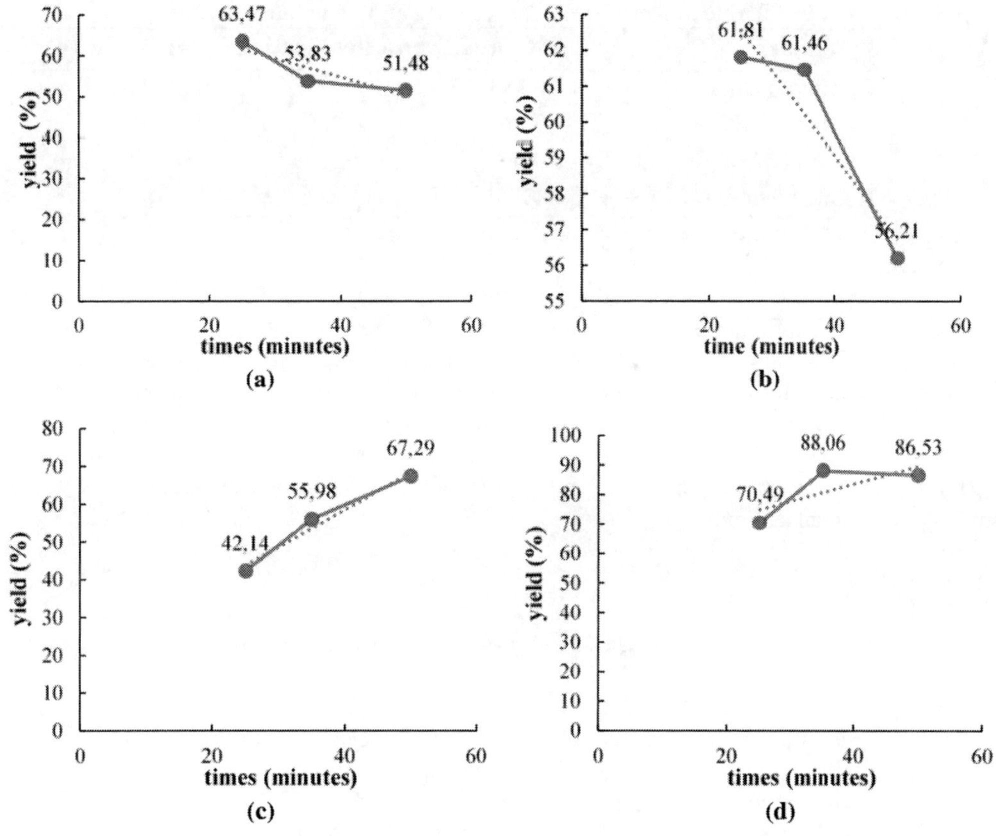

Figure 8. Result of transesterification results with K_2O/Al_2O_3 catalyst with (a) 15% content, (b) 20% content, (c) 25% content, and (c) 30% content of K_2O to Reaction time variations.

Figure 8a and Figure 8b show with the increasing time, the percentage of reaction results tends to decrease. Transesterification with 15% and 20% K_2O catalysts was suspected to undergo a saponification reaction, thereby reducing the reaction results. Figure 8c and Figure 8d show the results of transesterification using heterogeneous catalyst K_2O/Al_2O_3 with the help of ultrasonic waves at various time variations have increased. The increasing of yield indicates the longer time needed for the transesterification process at 25% K_2O levels on heterogeneous K_2O/Al_2O_3 catalysts. The smaller the saponification reaction increases the transesterification result. Fig. 12 shows the yield of transesterification using heterogeneous catalyst K_2O/Al_2O_3 with the help of ultrasonic waves at various time variations that have increased and then decreased. The three-time variations result experienced the highest yield compared to other catalyst variations. Increasing and decreasing results indicate the longer time required for the transesterification process at 30% K_2O levels on heterogeneous catalyst K_2O/Al_2O_3 experiences low lathering reactions so that the resulting transesterification increases compared to other catalyst variations.

3.5. Characterization of Transesterification Results

The characterization results of low-quality palm oil transesterification with the K_2O/Al_2O_3 catalyst with ultrasonic waves as in Table 3. As shown in Table 3, the results of low-quality palm oil transesterification with K_2O/Al_2O_3 catalyst with ultrasonic waves have characteristics that are close to SNI Biodiesel 2015. This is in accordance with Santoso 2019 that the results of the synthesis of methyl ester with K_2O/Al_2O_3 catalyst products are in accordance with SNI biodiesel.

Table 3. Results of Transesterification Characterization.

Parameter	Results Transesterification	SNI Biodiesel
Density (g/mL)	0.891	0.85-0.89
Viscosity (cSt)	7.86	2.3-6.0
Free Fatty acid (mg KOH/g)	0.465	Maks 0.5
Refractive index	1.454	1.3-1.4

4. Conclusions

Synthesis of K_2O/Al_2O_3 can be carried out by the wet impregnation method of KOH and Al_2O_3 then calcined at a high temperature of 300 °C for 3 hours. The XRD results an angle of 10-90 ° at $2\theta = 31$, 39, 51, 55 and 62 which are the typical peaks of the K_2O compound. The yield of transesterification results is influenced by the K_2O content in the catalyst. The optimum conditions of the reaction are achieved at 30 % K_2O levels in the K_2O/Al_2O_3 catalyst with 35 minutes of reaction time. The yield of K_2O/Al_2O_3 obtained 88.06 % (w/w). The characterization result of transesterification has a density of 0.891 g/mol, a viscosity of 7.86 cSt, a refractive index of 1.454 and an acid number of 0.465 and showed the potential as biodiesel.

References

[1] R. M. Moawia et al., 2019 *Advances in Chemical Engineering and Science* **9** 281
[2] A. Santoso, Sumari, D. Sukarianingsih, and R. M. Sari 2018 IOP *J. Phys. Conf. Ser.* **1093** 012047.
[3] F. Anguebes-Franseschi *et al.* 2016 *Open J. Ecol.* **06** 13
[4] O. J. Alamu, M. A. Waheed, and S. O. Jekayinfa 2007 *Energy Sustain. Dev.* **11** 77.
[5] B. K. Abdalla and F. O. A. Oshaik 2013 *Agric. Sci.* **04** 85
[6] Knothe, Gerhard 2005 *Fuel Process. Technol.* **86** 1059
[7] T. W. Chung and Y. J. Chen 2015 *J. Eng. Technol* **03** 105
[8] K. Y. Chew, W. L. Tan, N. H. H. Abu Bakar, and M. Abu Bakar 2017 *Int. J. Energy Environ. Eng.* **8** 47
[9] A. Santoso, Sumari, U. Urfa Zakiyya, and A. Tiara Nur 2019 *IOP Conf. Ser. Mater. Sci. Eng.* **515** 012042.
[10] Chin-Chiuan Lin , Ming-Chien Hsiao, Peir-Horng Liao 2012 *Journal of Sustainable Bioenergy Systems* **2** 117
[11] L. F. Chuah *et al.* 2016 *J. Clean. Prod.* **136**, 14
[12] S. Sumari, F. Fajaroh, I. Bagus Suryadharma, A. Santoso, and A. Budianto 2019 *IOP Conf. Ser. Mater. Sci. Eng.* **515** 012075

IOP Conf. Series: Materials Science and Engineering **833** (2020) 012043 doi:10.1088/1757-899X/833/1/012043

[13] S. Ashraf, M. Hussain, M. W. Mumtaz, and M. Shuaib 2017 *Int. J. Altern. Fuels Energy* **1** 14
[14] E. Lotero, Y. Liu, D. E. Lopez, K. Suwannakarn, D. A. Bruce, and J. G. Goodwin 2005 *Ind. Eng. Chem. Res.* **44** 5353
[15] R. Arora 2016 *Chem. Biochem. Eng. Q.* **29** 617
[16] Sumari, F. Fajaroh, A. Santoso, and R. K. Wardani 2018 IOP J. *Phys. Conf. Ser.* **1093** 012036,
[17] A. Santoso, C. F. Aghnia Hanindita, Sumari, and I. Budi Rachman 2019 IOP *Conf. Ser. Mater. Sci. Eng.* **515** 012082

IOP Conf. Series: Materials Science and Engineering **833** (2020) 012044 doi:10.1088/1757-899X/833/1/012044

Preparation of Curcumin Nanoemulsion in Soybean Oil – Tween 80 System by Wet Ball Milling Method

Shobbu Ibabas Sholihat[1], Ellya Indahyanti[1], Maria Lucia A.D Lestari[2], and Zubaidah Ningsih[1*]

[1] Dept. of Chemistry, Brawijaya University, Malang
[2] Dept. of Pharmaceutics, Airlangga University, Malang

*Corresponding author's email: zubaidah@ub.ac.id

Abstract. Curcumin nanoemulsion which has a particle size of 20-200 nm, is one of the curcumin drug delivery system that can increase curcumin solubility and bioavailability. The previous method of curcumin nanoemulsion formulation is modified thin-film hydration followed by sonication that has a long process and using organic solvent. This study aims to make curcumin nanoemulsion by wet ball milling method in soybean oil – Tween 80 system. Nanoemulsion is expected to have smallest particle size, high curcumin loading capacity and good stability during storage. Preparation of curcumin nanoemulsion is done by wet milling method which consists of curcumin as drug, soybean oil as curcumin solvent, tween 80 as stabilizer, water as dispersed medium and milling beads as milling media on the vial with various milling time. Particle size and polydispersity index are monitored using dynamic light scattering while nanoemulsion morphology is observed using digital imaging microscope. Results show that optimum milling times is 24 h and the highest curcumin added which produces less than 500 nm particles is 300 mg which is stable during 60 days of storage. Particle size of the nanoemulsion is ranging from 127 – 338 nm. In addition, wet ball milling methods is relatively simple and easy to apply.

1. Introduction

Indonesia is one of the countries that has various plant as a potential drug. One of the plant is turmeric [1]. Based on Badan Pusat Statistik [2], production of turmeric in Indonesia is 128.682 ton each year with the harvested area of 64.967.310 m^2. This makes turmeric becomes one of the most abundant plant in Indonesia. Turmeric contains curcumin, a compound which provides effect on health [3]. Curcumin is mainly polyphenolic compound that is obtained from turmeric rhizome's [4]. Curcumin potentially developed as an antidiabetic, anti-inflammation and antioxidant agent [5]. Nevertheless, curcumin has low solubility and bioavailability in water and unstable in storage which make curcumin application as drug is limited. Therefore, curcumin is designed as nanoemulsion drug delivery system [6].

Nanoemulsion is a colloidal system, oil dispersed in water, with an average diameter between 20-200 nm. Nanoemulsion can increase the solubility and bioavailability of a drug because of its relatively small size which enhances the distribution of drugs in the body. In the formation of nanoemulsion, oil and water are required [7]. The oil used in this research is soybean oil which has capacity to solubilize 2.4 mg of curcumin per mL [8]. To stabilize nanoemulsion, a stabilizer (surfactant) is required to prevent aggregation of nanoemulsion [7]. The surfactant used in this research is Tween 80, which is a type of non-ionic surfactant and generally recognized as safe surfactant which is non-toxic and non-irritating chemicals [10].

Various studies have been carried out in the preparation of curcumin nanoemulsions. Anuchapreeda et al. [8], proposes research on the preparation of lipid nanoemulsions combined with curcumin for cancer therapy. The method used is the modified thin film hydration followed by sonication. In that study, nanoemulsion was produced with an average particle diameter of 47-55 nm. Preparation of nanoemulsion with modified thin film hydration has a long procedure and uses harmful organic solvents. Therefore, preparation of nanoemulsion in this study uses wet ball milling method, a simple and straight forward method [11]. Wet ball milling method is the process of making particles smaller than the size of the original material by continuously milling the particles using milling beads. The longer the milling time used, the smaller the particle size. Nevertheless, at a certain time the particle size will enlarge due to friction that occurs between particles which indicates the stability of nanoemulsion [12]. Therefore, the optimum milling time and stability of nanoemulsion are very important to be determined in the study.

Based on the description and previous research above, this research is conducted in preparation of curcumin nanoemulsion in the soybean oil - Tween 80 system by wet ball milling method. This research is aimed to obtain curcumin nanoemulsion with small particle size and stable during the storage time.

2. Experimental

2.1. Apparatus and materials

Apparatus that were used is 50 mL beakers, 100-1000 μL micropipettes (Vitlab), tips, magnetic stirrers, glass stirrers, stirrer plates, milling beads, aluminum foil, vial bottles, eppendorf tubes, Vortexer Cleaver (Scientific Ltd.), Delsa™ Nano C Particle Analyzer (Beckman Coulter), Digital Imaging Microscope (Olympus BX51).

Materials that was used is curcumin and Tween 80 were purchased from Sigma-Aldrich, soybean oil, purified water, methanol, ethanol 96%.

2.2. Preparation of Curcumin Nanoemulsion

Pre-emulsion solution was made by mixing 1 mg curcumin, 62.5 mg Tween 80, 200 μL soybean oil, and 5 mL purified water in beaker glass. Following this step, 2.5 mL milling beads and 1 magnetic stirrer were added to the vial bottle. Pre-emulsion solution then added to the vial bottle. The solution was milled above the stirring plate with various milling time i.e 1, 4, 8 and 24 h. During every determined milling time, 200-300 μL curcumin nanoemulsion was taken from vial bottle and saved in the Eppendorf tube. Next, the Eppendorf tubes were saved at 4°C and at room temperature 25°C and further analyzed to measure the particle size and polydispersity index. The same procedure was carried out in making variations of curcumin loading with a mass of 100, 200, 300, 400, and 500 mg. In the milling process, optimum milling time was applied.

2.3. Measurement of Particle Size and Polydispersity Index

Particle size and polydispersity index were determined using Delsa™ Nano C Particle Analyzer Beckman Coulter instrument, using the principle of dynamic light scattering. Measurements was made using distilled water at 25°C, refractive index 1.3328, viscosity 0.8878, dielectric constant 78.3 and in disposable cells. At each measurement, 50 μL of curcumin nanoemulsion solution is dissolved in 2450 μL of purified water. Previously, an optimization method had been performed to obtain a comparison of the curcumin nanoemulsion composition with the solvent used.

Nanoemulsion stability test as a function of time was carried out by comparing the particle size and polydispersity index between nanoemulsion at day 1 and nanoemulsion stored until day 60. The effect of storage method at room temperature and at 4°C were also determined by comparing the same parameters as previously mentioned.

2.4. Nanoemulsion Morphology
The morphology of curcumin nanoemulsion was recorded using Digital Imaging Microscope Olympus BX51. The magnification used was 400 times.

3. Results and Discussion
Preparation of curcumin nanoemulsion in this research is carried out using wet ball milling method by mixing all drug nanoemulsion forming materials consist of oil, water, curcumin, and milling media into a vessel. Milling process is carried out continuously at various milling time i.e 1, 4, 8, and 24 h. The sample then being analyzed for its nanoemulsion size, polydispersity index, morphology, and stability.

Nanoemulsion size and polydispersity index are measured using Delsa^{TM} Nano C Particle Analyzer with a dynamic light scattering principle. Both nanoemulsion size and polydispersity index are dependent on milling time variation. Nanoemulsion size indicates the average diameter of nanoemulsion in the system which decreases while milling time increases (Figure 1). Our results show that particle size is linearly reduced as the milling time increases. At 1 h, the particle size is 533 nm and reduced to 127 nm at 24 h milling time. Therefore at 24 h milling time, curcumin nanoemulsion has the smallest size than the shorter milling time. This indicates that the rate of nanoemulsion size reduction obeys first-order kinetic theory.

Polydispersity index indicates the dispersity and homogeneity of nanoemulsion size which is ranging from 0 to 1. A homogenous and well-dispersed colloidal system has polydispersity index lower than or approaching 0,1. Figure 2 shows polydispersity index in which at 1, 4, and 8 h milling time the value of polydispersity index is almost identical (0,30). At 24 h milling time, polydispersity index produces smaller value than the previous results which is 0,15. Therefore 24 h milling time provides the best colloidal system of curcumin nanoemulsion.

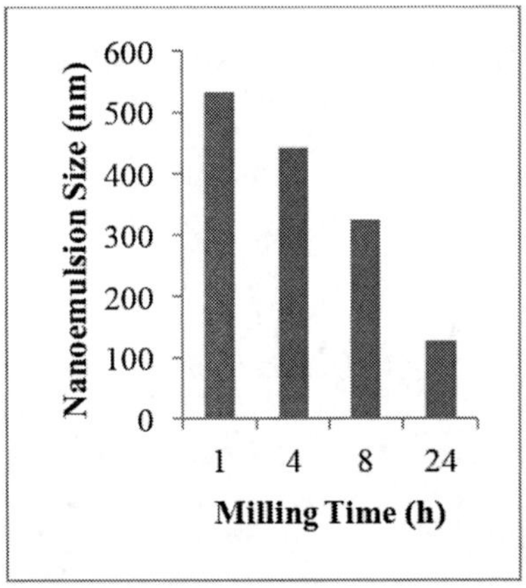

Figure 1. Effect of milling time to nanoemulsion size.

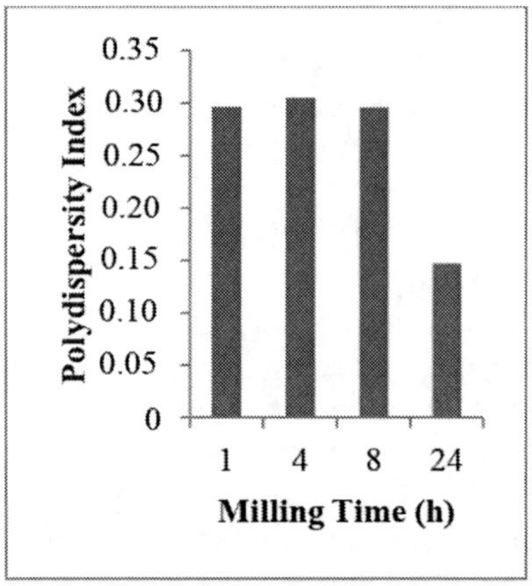

Figure 2. Effect of milling time to polydispersity index.

To further confirm the results, nanoemulsion morphology is analyzed using Digital Imaging Microscope. The images obtained support nanoparticle size and polydispersity index results where it can be seen that the particle size is reduced. Figure 3 (a-c) indicates that the particle size decreasing slightly at 1,4 and 8 h milling time while 24 h milling time image (Figure 3 d) signifies the smallest particle size.

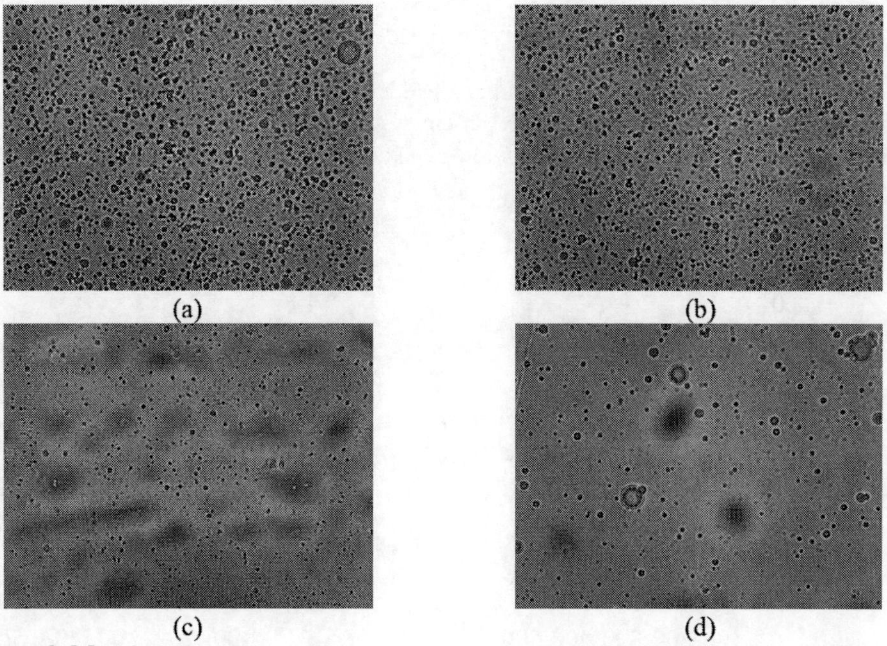

Figure 3. Morphology of curcumin nanoemulsion at various milling time. (a) 1 h, (b) 4 h (c) 8 h (d) 24 h.

Table 1. Stability of curcumin nanoemulsion particle size toward milling time.

No	Milling Time (h)	Nanoemulsion Size (nm)		
		Day 1	Day 60, 4°C	Day 60, 25°C
1	1	532.7	985.6	637.7
2	4	443.2	398.5	491.6
3	8	325	1096.3	273.8
4	24	126.7	120	120

Table 2. Stability of polydispersity index toward milling time.

No	Milling Time (h)	Polydispersity Index		
		Day 1	Day 60, 4°C	Day 60, 25°C
1	1	0.297	0.363	0.32
2	4	0.305	0.319	0.326
3	8	0.296	0.429	0.299
4	24	0.147	0.101	0.151

To monitor the stability of 1 mg curcumin nanoemulsion, particle size and polydispersity index are observed as temperature (4°C (fridge temperature) and 25°C (room temperature) and storage time function (60 days). Both temperature and storage time affect the change of nanoemulsion size and polydispersity index because it causes coalescence at the nanoemulsion system. At 24 h milling time, nanoemulsion size and polydispersity index are relatively identical between day 1 and day 60 observation (Table 1 and Table 2). Meanwhile, at 1, 4, and 8 h milling time both nanoemulsion size and polydispersity index are fluctuated randomly. This implies that larger nanoemulsion size and polydispersity index at day 1 influence the stability of nanoemulsion. It is possible that the repulsion between particles is minimum which facilitates coalescence during storage. On the other hand, small particle size and polydispersity index produce higher repulsion effect which hinders the coalescence between particle. The results signify that curcumin nanoemulsion is stable at 24 h milling time during 60 days storage both at 4°C and 25°C.

To further improve the amount of curcumin embedded in soybean oil droplets while maintaining the size of the particles, more curcumin is added in the system. Figure 4 and 5 show the particle size and polydispersity index of curcumin nanoemulsion when various curcumin amount (100, 200, 300, 400 and 500 mg) are incorporated in the system.

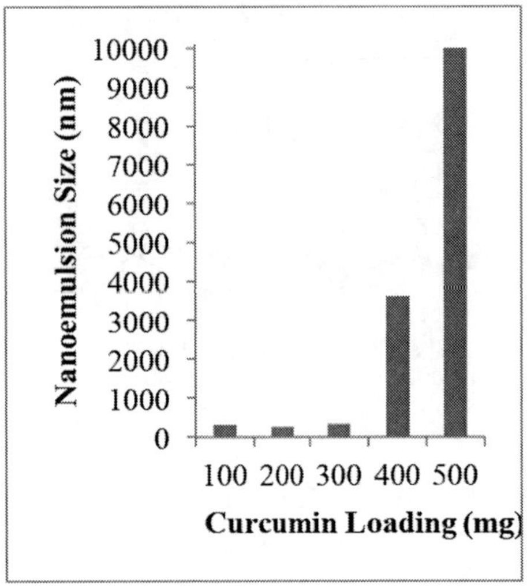

Figure 4. Effect of curcumin loading to nanoemulsion size at optimum milling time.

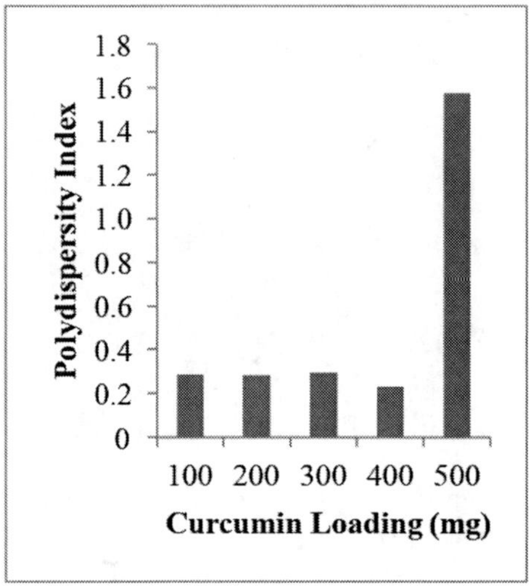

Figure 5. Effect of curcumin loading to polydispersity index at optimum milling time.

The addition of 100, 200 and 300 mg produces particle size ranging from 255 – 338 nm which is still classified as particle size in the nanoemulsion system. Meanwhile, the addition of 400 and 500 mg of curcumin dramatically change the particle size to more than 3000 nm which is classified as particle size in the conventional emulsion. Polydispersity index at 100, 200, 300 and 400 mg are relatively the same at around 0,2 and dramatically increases at 500 mg. Thus, the maximum curcumin that can be added while maintaining nanoemulsion size is 300 mg.

4. Conclusion

Curcumin that was designed as nanoemulsion using wet ball milling method can improve the solubility of curcumin in water. Optimum milling time in the preparation of curcumin nanoemulsion is 24 h. On the optimum milling time, curcumin nanoemulsion is stable at 4°C and 25°C during 60 days storage time. The maximum curcumin that can be added to form nanoemulsion is 300 mg.

References

[1] Hastati S, Hadju V, Alam G and Nusratuddin 2015 *Int. J. Sci. & Technol. Res.* **4** 4
[2] Badan Pusat Statistik n.d. https://www.bps.go.id/publication/2018/10/05/fb684e53549e5fa3fc174c8d/statistik-tanaman-biofarmaka-indonesia-2017.html (accessed March 23, 2019).
[3] Kocaadam B and Şanlier N 2017 *Critical Reviews in Food Science and Nutrition* **57** 2889–2895
[4] Niranjan A and Prakash D 2008 *J Food Sci Technol* **45** 9
[5] Bansal SS, Goel M, Aqil F, Vadhanam MV and Gupta RC 2011 *Cancer Prevention Research* **4** 1158–1171
[6] Rajsekhar P, Bharani RA, Angel KJ, Ramachandran M and Rajsekhar SPV 2015 *Research Journal of Pharmaceutical, Biological and Chemical Sciences* **7**.

IOP Conf. Series: Materials Science and Engineering **833** (2020) 012044 doi:10.1088/1757-899X/833/1/012044

[7] Pathak K, Pattnaik S and Swain K 2018 *Application of Nanoemulsions in Drug Delivery. Nanoemulsions* Elsevier 415–433

[8] Anuchapreeda S, Fukumori Y, Okonogi S and Ichikawa H 2012 *Journal of Nanotechnology* **2012** 1–11

[9] Dave N and Joshi T 2017 *Research India Publication* **13** 10.

[10] Hvattum E, Yip WL, Grace D and Dyrstad K 2012 *Journal of Pharmaceutical and Biomedical Analysis* **62** 7–16

[11] Bensebaa F 2012 *Nanoparticle Technologies: From Lab to Market* Academic Press

[12] Malamatari M, Taylor KMG, Malamataris S, Douroumis D and Kachrimanis K 2018 *Drug Discovery Today* **23** 534–547

IOP Conf. Series: Materials Science and Engineering **833** (2020) 012045 doi:10.1088/1757-899X/833/1/012045

Characterization of Slow Pyrolysis Products of *Macaranga motleyana*: Effect of Sample Size

RR Dirgarini Julia Nurlianti Subagyono[1*], Rika Puspitasari[1], Ari Susandy Sanjaya[2], Wiwin Suwinarti[3]

[1] Chemistry Department, Faculty of Mathematics and Natural Sciences, Mulawarman University
[2] Chemical Engineering Department, Faculty of Engineering, Mulawarman University
[3] Forestry Department, Faculty of Forestry, Mulawarman University

*Corresponding email: dirgarini@fmipa.unmul.ac.id

Abstract. The characterization of biocrude oil produced by slow pyrolysis of fast growing wood *Macaranga motleyana* has been conducted. Slow pyrolysis was carried out in a fixed bed pyrolysis reactor at 500 °C for 1 hour with a variety of wood sample size. The decrease of a sample size from 20 to 40 mesh resulted in an increased in the percent yields of liquid product, which were 28.5 wt.% and 36.1 wt.%, respectively. Characterization of the biocrude oil showed that the pH of the biocrude oil was 2, the density of biocrude oil was 1.132 g/mL, while the viscosity value of the biocrude oil was 53.6 cSt. Analyses of the biocrude oil using a Gas Chromatography-Mass Spectrometry (GC-MS) revealed that the biocrude oil derived from *Macaranga motleyana* wood contained phenol and its derivatives, such as eugenol and compounds resulted from thermal degradation of cellulose or hemicellulose. The use of sample with a smaller sample size resulted in biocrude oil containing more chemical compounds.

1. Introduction

Exploration and exploitation of fossil fuels on a large scale in recent decades is not balanced with the formation process in a short time. This prompted some researchers to look for alternative energy substitutes for fossil fuel energy sources. One alternative source of substitute for fossil energy can come from the bark, logs, grass or other plants. *Macaranga* spp wood is one of the endemic plants of East Kalimantan that has the potential as an alternative energy source. *Macaranga* is a pioneer plant that grows in tropical forests which are classified as fast-growing plants and are easily found in large numbers [1]. *Macaranga motleyana* contains lignin, cellulose and hemicellulose which are potential to be fuel energy sources [2].

One method for converting lignocellulosic material into an energy source is pyrolysis. Pyrolysis is a thermochemical conversion process at temperatures from 650 to 800 K and at low pressures that occur in the absence of air and reducing gases to convert material (e.g biomass) to liquid (bio-oil), solids (charcoal), and gas [3-5]. Biocrude oil or liquid smoke is a blackish-colored liquid resulting from the pyrolysis of biomass such as wood, bark and other biomass [6-10]. Pyrolysis yield is affected by temperature, time and particle size. It was found that particle size can affect the value of the activation energy of mixed wood sawdust pyrolysis, the yield of phenolic compounds and linear hydrocarbons, and the production of CO and CO_2 gases [11]. In the slow pyrolysis process, biomass is heated at low heating rates and long vapor residence time [12]. This method produces not only liquid products but also char that can be used for several applications. Previously, slow pyrolysis of *Macaranga gigantea* produced a quite substantial amount of liquid products [6]. Thus, slow pyrolysis of the other species of *Macaranga* (e.g. *Macaranga motleyana*) is worth studying. *Macaranga motleyana* (and also other species) is also worth exploiting because this plant is considered a weed. The use of *Macaranga motleyana* as a fuel producing material will increase the added value of this plant. In this report, the physical and chemical characteristics of slow pyrolysis products of *Macaranga motleyana* with several sample sizes are analyzed and studied.

2. Materials and methods

2.1. Sample preparations

Macaranga motleyana samples were obtained from the Education Forest of Mulawarman University in Taman Hutan Raya Bukit Soeharto, Loa Janan, Kutai Kartanegara, Indonesia. The bark of the wood sample was peeled. The wood sample was then chopped and dried in the sun for ± 7 days until the water content in the sample was reduced. After that, the sample was mashed using a hammer mill to 20 and 40 mesh.

2.2. Sample analysis

The ash content and moisture content of *Macaranga motleyana* were calculated using the ASTM methods. The wood lignin content (Klason lignin) was determined by TAPPI T 222 om-88 method [13] and determination of α–cellulose content were performed using TAPPI T 9 m-54 method [14].

2.3. Pyrolysis experiments

Slow pyrolysis of *Macaranga motleyana* wood was carried out using a laboratory-scale batch reactor as described in previous studies [6]. About 250 g of wood samples (20 and 40 mesh size) were charged into the sample container in the pyrolysis reactor. Slow pyrolysis runs were conducted at 500 °C for 1 hour. The time needed to reach the pyrolysis temperature from room temperature was about 15 minutes. The resulting liquid product consisted of two phases. The top layer was the aqueous phase and the blackish-brown bottom phase was the organic phase (bio-oil). The two phases were separated by a separating funnel and then weighed. The yields of organic phase, aqueous phase, and solid residue were calculated as described in Sinaga et al. [6]. Characterization of the organic phase was performed using a Gas Chromatography-Mass Spectrometry (GCMS QP 2010S SHIMADZU). The acidity (pH), viscosity, and density of the organic phase were determined.

3. Results and discussions

Ash content, moisture content, lignin, cellulose and hemicellulose contents of *Macaranga motleyana* are shown in Table 1. Cellulose and hemicellulose contents in *Macaranga motleyana* samples are considerably high, so the wood species has the potential to produce large quantities of pyrolysis products because cellulose and hemicellulose can produce volatile compounds that are easy to condense [7]. The content of lignin in *Macaranga motleyana* wood can determine the temperature and time of pyrolysis because lignin has a very complex structure. Wood with a high lignin content will tend to produce aromatic compounds (e.g. phenol and its derivatives) and charcoal [8].

Table 1. Moisture content, ash content and chemical content of *Macaranga motleyana.*

Wood species	Moisture content (%)	Ash content (%)	Hemicellulose (%)	α-cellulose (%)	Lignin (%)
Macaranga motleyana	11.05+0.06	0.76±0.04	3.4 ± 1.6	68.8 ± 1.6	37.6 ± 1.1

Table 2. Yields (%) of organic phase (bio-crude oil), aqueous phase, and char from slow pyrolysis of *Macaranga motleyana.*

Sample size	Organic phase (%)	Aqueous phase (%)	Char (%)
20 mesh	2.5±0.2	26.2±0.7	13.3±0.7
40 mesh	5.4±0.5	30.7±1.2	11.6±0.7
	Liquid yield (%)		Char (%)
18 mesh*	20.88		30.00
60 mesh*	3.25		22.63

*Slow pyrolysis of *Imperata Cylindrica* at 500 °C [15]

Table 3. List of identified compounds in the chromatograms of the organic phase obtained from slow pyrolysis of *Macaranga motleyana.*

20 mesh					40 mesh				
No.	Compounds	RT (min)	%area		No.	Compounds	RT (min)	%area	
1	Phenol	12.0	6.7	*	1	Butyrolactone	9.2	9.3	**
2	3-methyl-1,2-cyclopentanedione	13.6	1.4	**	2	Phenol	12.0	5.1	*
3	2-methyl phenol	14.7	3.6	*	3	2-methyl phenol	14.7	2.3	*
4	Mequinol	15.7	8.9	*	4	Mequinol	15.7	4.2	*
5	2,6-dimethyl phenol	16.5	1.1	*	5	2,6-dimethyl phenol	16.5	2.3	*
6	2,3-dimethyl phenol	17.9	8.1	*	6	2,3-dimethyl phenol	17.9	3.2	*
7	2-methoxy-4-methyl phenol	19.2	8.5	*	7	2-methoxy-4-methyl phenol	19.2	5.8	*
8	4-ethyl-2-methoxy phenol	21.8	6.6	*	8	2-ethyl-4-methyl phenol	20.9	1.5	*
9	4-(1-methyl propyl) phenol	24.1	3.9	*	9	4-ethyl-2-methoxy phenol	21.8	4.4	*
10	2-methoxy-4-propyl phenol	24.4	3.2	*	10	2,6-dimethoxy phenol	24.0	3.1	*
11	Hexadecanoic acid	24.5	2.2	**	11	Hexadecanoic acid	24.5	8.5	**
12	Eugenol	25.6	1.4	*	12	Eugenol	25.6	4.1	*
13	4-hidroxy-3-methoxy benzoic acid	26.7	5.5	**	13	4-hidroxy-3-methoxy benzoic acid	26.7	2.2	**
14	Dodecane	27.9	1.8	**	14	1,2,3-trimethoxy-5-methyl benzene	28.7	1.5	
15	1,2,3-trimethoxy-5-methyl benzene	28.7	1.7		15	Dodecanoic acid	29.8	2.3	**
16	Dodecanoic acid	29.8	1.9	**	16	Isopropyl myristate	35.7	4.7	**
17	2-heptadecanone	37.4	1.2	**	17	2-heptadecanone	37.4	1.7	**
18	Methyl ester hexadecanoic acid	37.9	3.5	**	18	Methyl ester hexadecanoic acid	37.9	5.4	**
19	9-octadecanoic acid	41.4	1.9	**	19	9-octadecanoic acid	41.4	3.8	**
20	Methyl ester-9-octadecanoic acid	41.5	1.1	**	20	Methyl ester-11-octadecanoic acid	41.5	2.2	**

*lignin derived compound, ** cellulose/hemicellulose derived compound

Pyrolysis of *Macaranga motelyana* produced gas, liquid products and char. Wood pyrolysis with a particle size of 40 mesh produced higher yields of organic phase and aqueous phase and a smaller char yield than that with a particle size of 20 mesh (Table 2). This is presumably due to pyrolysis of wood with smaller particle size (40 mesh) (thus, greater surface area) will cause a faster heat transfer process, so that the thermal decomposition of lignin, cellulose and hemicellulose is more optimal. It was found that if lower particle size feedstock is used, slow pyrolysis may lead to the increase in gas production (lower yields of liquid and solid) due to cracking of heavy molecules (tar) [15](Table 2). Thus, it is important to note that the selection of sample size must consider the desired pyrolysis product. The organic phase (biocrude oil) produced has dark, aromatic, smoke-like characteristics. Biocrude oil produced has a pH of 2, a density of 1.132 g / mL and a viscosity of 53.6 cSt. The high acidity of the organic phase was probably because of the presence of acetic acid and other acids resulting from thermal degradation of lignin, cellulose and hemicellulose (Table 3). The density value depends on the compounds in the sample, if the sample contains large molecular weight compounds, then the density value obtained is higher. The viscosity of *Macaranga motleyana* biocrude oil is different from the viscosity of biocrude oil from other biomass samples [2]. This is because the viscosity value is affected by the type of sample used, pyrolysis temperature, water content in oil and the pyrolysis reactor used [16].

Chromatograms of *Macaranga motleyana* wood bi crude oil are presented in Figure 1. In the chromatogram of 20 mesh biocrude oil there were 65 compounds observed, whereas in that of 40 mesh biocrude oil there were 82 compounds were observed. This shows that the sample size affects the composition of compounds produced in pyrolysis of the wood sample. The compounds with the highest 20 percent area identified in the *Macaranga motleyana* biocrude oil were shown in Table 3. These compounds were from decomposition and degradation of lignin, cellulose, and hemicellulose. The formation of products derived from hemicellulose and cellulose, e.g. hexadecanoic acid, ketones, occurred at around 200-260 °C and 240-350 °C [2]. Phenol and its derivatives and aromatic hydrocarbons were thermal degradation products of lignin, which occurred around 400 °C[17].

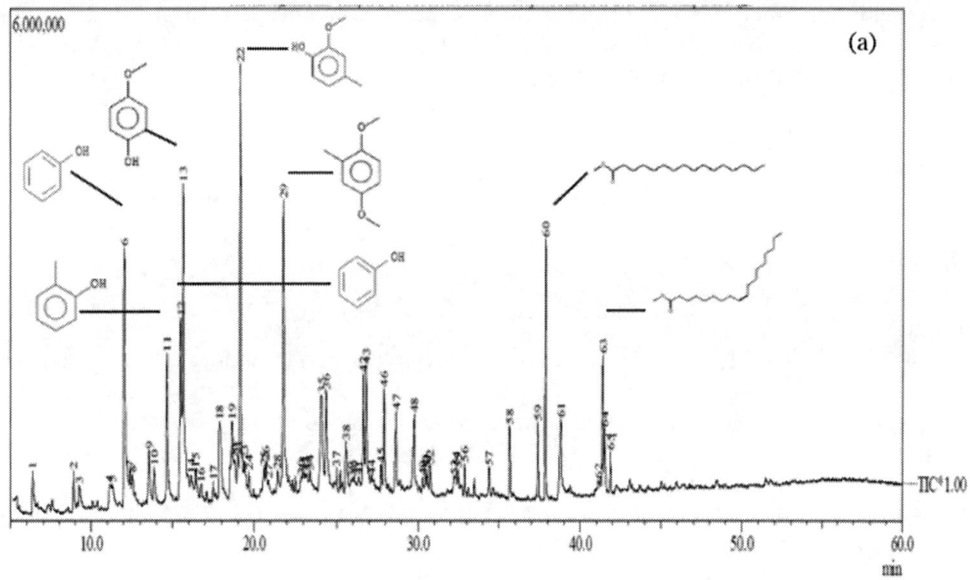

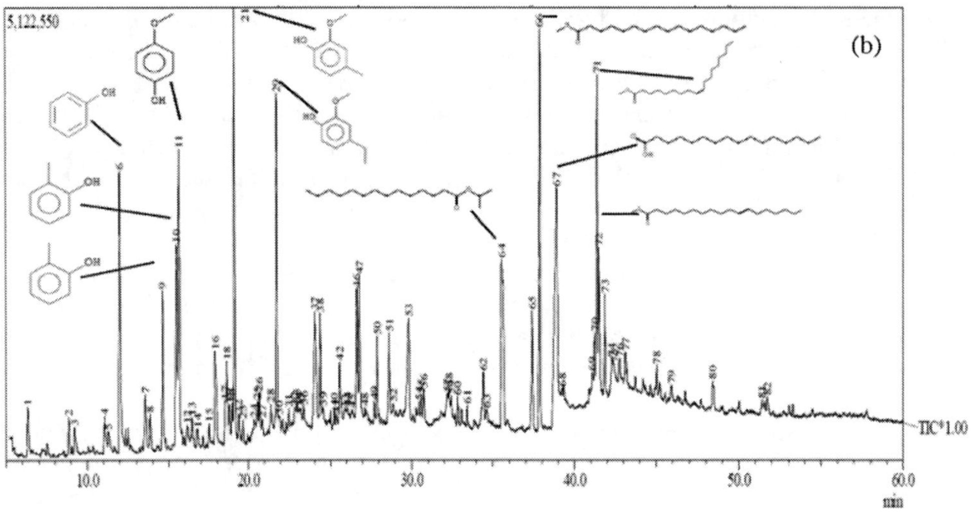

Figure 1. Chromatograms of the organic phase obtained from slow pyrolysis of *Macaranga motleyana* (a. 20 mesh; b. 40 mesh)

4. Conclusions

The wood particle size affected the yields of the liquid products and the solid residue and also the chemical composition of the liquid products. Pyrolysis of 40 mesh *Macaranga* wood produced higher pyrolysis yields and biocrude oil containing more chemical components than that of 20 mesh sample. The compounds in *Macaranga motleyana* wood biocrude oil were the result of thermal degradation and decomposition of cellulose, hemicellulose and lignin.

Acknowledgment

We would like to thank the Ministry of Research, Technology and Higher Education of the Republic of Indonesia for financial support in the project (the Fundamental Research Scheme, contract number:069/SP2H/LT/DRPM/2019).

References

[1] Amirta R, Nafitri S I, Wulandari R, Yuliansyah, Suwinarti W, Candra K P and Watanabe T 2016 Comparative characterization of Macaranga species collected from secondary forests in East Kalimantan for biorefinery of unutilized fast growing wood *Biodiversitas* **17** 116-23

[2] Mohan D, Pittman C U and Steele P H 2006 Pyrolysis of wood/biomass for bio-oil: a critical review *Energ Fuel* **20** 848-89

[3] Alonso D M, Bond J Q and Dumesic J A 2010 Catalytic conversion of biomass to biofuels *Green Chem* **12** 1493-513

[4] Bridgwater A V 1994 Catalysis in thermal biomass conversion *Appl Catal Gen* **116** 5-47

[5] Demirbas A 2001 Biomass resource facilities and biomass conversion processing for fuels and chemicals *Energ Convers Manag* **42** 1357-78

[6] Sinaga R J, Wijayanti A, Asih R D K, Subagyono R D J N and Sanjaya A S 2019 Slow pyrolysis of fast growing wood Macaranga gigantea *Journal of Physics: Conference Series* **1277** 012001

[7] Subagyono D J N, Marshall M, Fei Y, Jackson W R and Chaffee A L 2015 Thermo-chemical reactions of algae, grape marc and wood chips using a semi-continuous/flow-through system *Fuel* **158** 927-36

[8] Subagyono D J N, Marshall M, Jackson W R, Fei Y and Chaffee A L 2016 Thermochemical Reactions of Blue Gum and Fossil Wood with CO/H_2O: Some Mechanistic Comments *Energ*

IOP Conf. Series: Materials Science and Engineering **833** (2020) 012045 doi:10.1088/1757-899X/833/1/012045

Fuel **30** 1039-49
[9] Di Blasi C, Signorelli G, Di Russo C and Rea G 1999 Product distribution from pyrolysis of wood and agricultural residues *Ind Eng Chem Res* **38** 2216-24
[10] Liu C, Wang H, Karim A M, Sun J and Wang Y 2014 Catalytic fast pyrolysis of lignocellulosic biomass *Chemical Society Reviews* **43** 7594-623
[11] Suriapparao D V and Vinu R 2018 Effects of Biomass Particle Size on Slow Pyrolysis Kinetics and Fast Pyrolysis Product Distribution *Waste and Biomass Valorization* **9** 465-77
[12] Jahirul M I, Rasul M G, Chowdhury A A and Ashwath N 2012 Biofuels Production through Biomass Pyrolysis —A Technological Review *energies* **5** 4952-5001
[13] TAPPI 2006 Acid-insoluble lignin in wood and pulp: Reaffirmation of T 222 om-02. In: *Chemical Properties,* (New York: Technical Association of the Pulp and Paper Industry)
[14] TAPPI 1999 Alpha-, beta- and gamma-cellulose in pulp: T 203 cm-99. In: *Alpha-, beta- and gamma-cellulose in pulp,* (New York: Technical Association of the Pulp and Paper Industry)
[15] Khasri A, Mohd Jamir M R, Sulaiman M H and Tuan Abdullah T A 2012 Slow Pyrolysis Of Imperata Cylindrica In a Fixed Bed Reactor *International Journal of Biological, Ecological and Environmental Sciences (IJBEES)* **1** 176-80
[16] Demirbas A 2007 The influence of temperature on the yields of compounds existing in bio-oils obtained from biomass samples via pyrolysis *Fuel Processing Technology* **88** 591-7
[17] Brebu M, Cazacu G and Chirila O 2011 Pyrolysis of lignin-a potential method for obtaining chemicals and/or fuels *Cellulose Chemistry and Technology* **45** 43-50

IOP Conf. Series: Materials Science and Engineering **833** (2020) 012046 doi:10.1088/1757-899X/833/1/012046

Analysis of Fe in Coral Reefs for Monitoring Environmental Areas of Prigi Coast Waters Using the Tessier-Microwave Method

Cahyanti Wulan Suci[1], Anugrah Ricky Wijaya[1,*]

[1]Department of Chemistry, Faculty of Mathematics and Natural Sciences, Malang State University, Malang, Indonesia
Jl. Semarang No.5, Malang City, East Java, 65115

*Corresponding author: anugrah.ricky.fmipa@um.ac.id

Abstract. This study revealed the analysis of Fe content in the coral reef of Prigi Coast to monitor its waters as an environmental biomonitoring agent. Fe can accumulate into the coral structure through food, the mucous tissue, the mesenterium filament, and calcium element substitution that built-in the each of Fe coral fractions. Fe can be leached in the coral reef fractions using the Tessier-microwave method. Tessier microwave method divided of Fe in the ion exchange (fraction 1), carbonate bound (fraction 2), reduction (fraction 3), oxidation (fraction 4) and residual (fraction 5), respectively. The Tessier-microwave method results detected of Fe in the fraction 3 only, indicated that all of the Fe in the coral reefs was bound to the reduction fraction. The Fe content in the fraction 3 at 6 selected locations of Prigi Coast was 127.10 ppm, 39.12 ppm, 56.64 ppm, 78.10 ppm, 11.46 ppm, and 47.62 ppm, respectively. According to several calculations of CF, PLI and I_{geo} indexes, it indicated the contaminated of Fe in the coral reefs, especially at reduction fraction.

1. Introduction

Coral bleaching caused by heavy metals is a major threat to coral reefs. Heavy metal concentrations at a certain level will cause bleaching in corals[1]. Heavy metals accumulated in the coral aragonites structure are associated with ion-exchange reactions, where carbonate ions in aragonites are replaced by heavy metal ions[2], as in the following equation:

$$CaCO_3(S) + aX^{-}(aq) \rightleftharpoons CaX(s) + CO_3^{a-}(aq)$$

Heavy metals accumulate into the coral reef through four mechanisms. First, through food, coral reefs eat zooplankton, where zooplankton is affected, and heavy metals accumulate in it. Second, through the mesenterial filaments that capture foods, not only zooplankton but also particulates which are subsequently and periodically digested by coral reefs[3]. Third, through mesentrium filaments that take heavy metals bound from the sediments directly. Fourth, with the substitution of the calcium element by certain metals, organic material fragments containing heavy metals that enter the pores of the coral structure[4]. The coral reef and sediment structure can assimilate the metal over a period of hundreds of years so that it can be used for monitoring purposes[5–8].

Considering of monitor natural and anthropogenic sources of Fe, meteorites and terrestrial highly contain Fe as a heavy metal, which is 90.85%[9]. The Fe concentration in basaltic rocks was 86.5 ppm, in granite rocks with high calcium of 29.6 and in granite rocks with low calcium of 14.2 ppm[9]. This shows that Fe exists in nature in large quantities. Fe contained in coral reefs can either come from natural sources or anthropogenic factors [10]. Examples of anthropogenic factors that increase the Fe content in coral reefs include port and tourism activities, industrial development around the sea and mining areas. There is so many marine natural products are applied by researcher to reduce heavy metals and pollution including Fe and material organic in the human body[5,10,11].

Some researcher used the BCR and Tessier-microwave methods to analyze the heavy metals bound in each fraction [6,12–14]. The Tessier-microwave method is divided into 5 fractions with shorter stages, namely the ion exchange fraction, the carbonate bound fraction, the fraction reduction, the oxidation fraction and the residual fraction. Research using this method was conducted to determine the location of Fe bound in coral reefs and its status for environmental monitoring detailed.

2. Method

The coral reef analyzed is the Acropora reef in the Prigi Coast waters. Each coral reef sample was taken from 6 different locations (Figure 1). We used those locations as a preliminary mapping of Fe to explore more detailed Fe minerals.

The Tessier-microwave procedures have been adapted and modified by Wijaya et al. (2019)[12]. The coral reef sample was dried in an oven at 60°C until a constant weight achieved. Then it was smoothed out and is weighed as much as 1 gram and treated with the Tessier-microwave method. In the first fraction, 1 gram of dry coral was added to 4 mL $MgCl_2$ and was heated in the microwave at 50% power irradiation (399 W) for 3 minutes. The mixed coral and chemicals were then centrifuged and filtrated as the solution in the fraction 1. In fraction 2, CH_3COONa has added about 4 mL and heated in the microwave at 50% power irradiation (399 W) for 3 minutes. After microwaved, the centrifuged and filtrated samples were taken as a solution in fraction 2. Then, 10 mL $NH_2OH.HCl$ was added and heated in the microwave at 50% power irradiation (399 W) for 3 minutes and repeated them at two times. The sample was then centrifuged to obtain filtrate for the solution in the fraction 3. There is no residue in the fraction 3, Fe in the fraction 4 and fraction 5 were almost completely not detected. The solutions in the fraction 1-3 were analyzed of Fe using the AAS instrument.

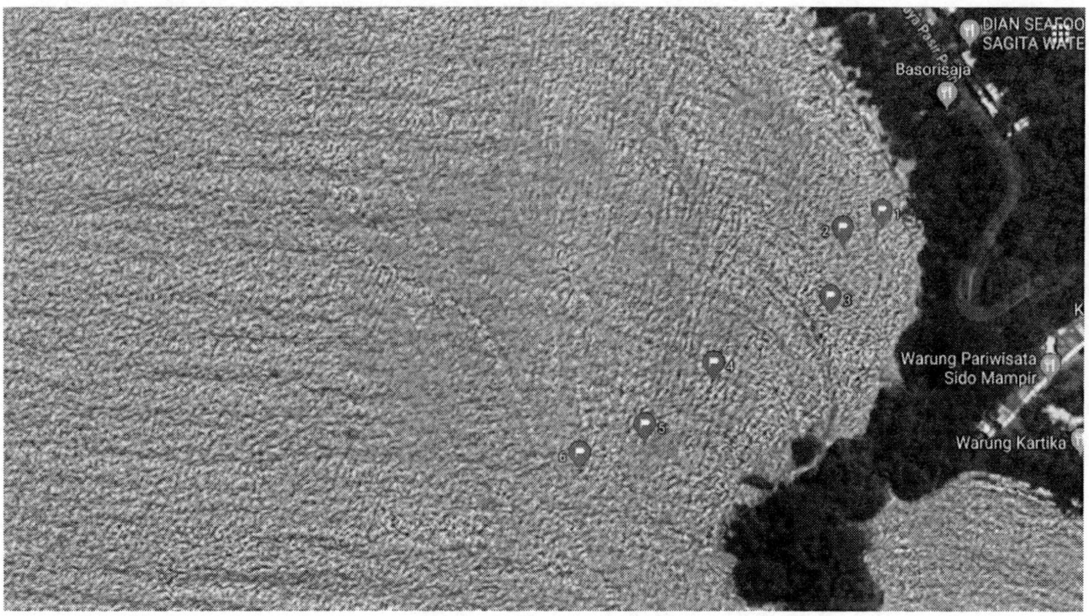

Figure 1. Six sampling locations in Prigi Coast Waters area.

3. Results and Discussion

The Tessier-microwave method was applied to analyze the Fe from the fraction 1 to fraction 5 by using specific solvents. It started from fraction 1 to fraction 3, all of their residues were completely dissolved by attacking solvents. But, Fe in fraction 4 and 5 could not continue analyzed. The applied of Tessier-microwave method in the coral reef is different from their applied in sediment. Fraction 4 and and 5 contained humic acid and silica-alumina. It suggested the investigated coral reef did not content of humic acid and silica-alumina

The reaction of each fraction shows different chemical strengths in the level fraction. In the ion-exchange fraction is known as a mobile fraction and has the lowest strength and easily dissolved in the waters (fraction 1). The carbonate bound fraction has a greater strength in binding Fe than ion-exchange fraction (fraction 2). The third fraction or reduction fraction has a greater strength in binding the metal where a stronger acid was added.

Figure 2 shows the Fe concentrations in fraction 1, 2, and 3. The contents of Fe dominantly were recorded at fraction 3. The ranged concentrations of Fe in the fraction 3 was recorded from 22.46 to 127.1 ppm. All of Fe concentrations in Fraction 1 and 2 were low.

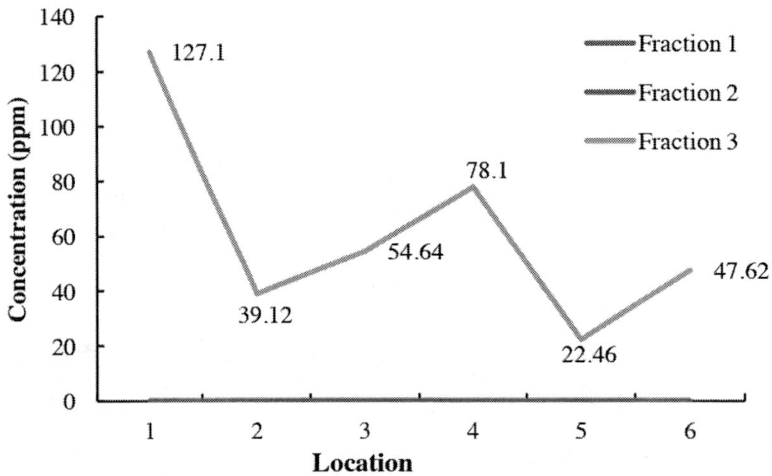

Figure 2. Fe concentrations in coral reefs at each location.

This indicates that a few Fe content in coral reefs was bound to the ion exchange fraction through ion exchange reactions with Mg^{2+}. In the fraction 2, it indicates there was also a few Fe binding by CH_3COO^- in the coral reef. In fraction 3 (reduction fraction), the various concentrations of Fe were dominantly obtained from all sampling locations. The Fe detection signifies that all reaction that occurs in fraction 3 undergoing the reduction of Fe^{3+} to Fe^{2+} due to the attacking of $NH_2OH.HCl$ in coral reef samples. The fraction 3 as the final fraction in coral reefs indicates that there is neither Fe bound to organic material (humic acid) as fraction 4 nor bound to silica-alumina or fraction 5. There is no Fe bound to fraction 4 and 5 because the coral reef has been totally leached of Fe in the fraction 3. It suggests there is no Fe effect contributed from natural sources associated with coral reefs. The geochemical fraction of silica and alumina was found in the natural sources. The chemical reactions of Fe in the reducible fraction in the coral reefs may be described as follows:

1. $\text{Coral-}MnO_2\text{-}Fe_{(s)} + 2\ NH_2OH.HCl_{(aq)} \rightarrow Mn^{2+}{}_{(aq)} + Cl_{2(g)} + Pb^{2+}{}_{(aq)} + H_2O_{(l)} + 2NH_{3(g)}$

2. $\text{Coral-}Fe_2O_3\text{-}Fe_{(s)} + 2NH_2OH.HCl_{(aq)} \rightarrow 2Fe^{2+}{}_{(aq)} + Cl_{2(g)} + Pb^{2+}{}_{(aq)} + 2H_2O_{(l)} + 2NH_{3(g)}$

3. $\text{Coral-}Fe(OH)_3\text{-crystal-}Fe_{(s)} + 2NH_2OH.HCl_{(aq)} \rightarrow 2Fe^{2+}{}_{(aq)} + Cl_{2(g)} + Pb^{2+}{}_{(aq)} + H_2O_{(l)} + 2NH_{3(g)}$

IOP Conf. Series: Materials Science and Engineering **833** (2020) 012046 doi:10.1088/1757-899X/833/1/012046

Diversing and fluctuating Fe concentrations in coral reefs can be caused by a variety of factors, including varied sampling locations. The coral reefs where near the coast are more exposed to anthropogenic activities compared to those are far from the coast or human activities. Furthermore, this can be due to the various periods of Fe accumulation in coral reefs. Consequently, if the coral reefs used for the analysis are in a highly contaminated area, the reefs will grow earlier and receive Fe exposure than coral reefs. The subsequently high Fe content in the coral reef was recorded at 1, 4, 3, 6, 2 and 5 locations respectively.

Table 1. The CF[a], PLI[b], and Igeo[c] indexes of Fe.

Location	CF	PLI	Igeo
1	8.95	127.10	2.58
2	2.75	39.12	0.88
3	3.84	54.64	1.36
4	5.50	78.10	1.87
5	1.58	22.46	0.07
6	3.35	47.62	1.16

[a]Non polutant (CF<1), pollutant (CF>1)
[b]uncontaminated (0<PLI≤1), uncontaminated to moderately contaminated (1<PLI≤2), moderately contaminated (2<PLI<3), moderately to highly contaminated (3<PLI≤4), highly contaminated (4<PLI≤5), or very highly contaminated (PLI>5).
[c]uncontaminated (Igeo≤0), uncontaminated to moderately contaminated (0<Igeo≤1), moderately contaminated (1<Igeo≤2), moderately to heavily contaminated (2<Igeo≤3), heavily contaminated (3<Igeo≤4), heavily to extremely contaminated (4<Igeo≤5), or extremely contaminated (Igeo>5)

CF calculation for Fe refers to the Earth's Crust database[9]. CF calculation results of Fe in the coral reef obtained from all sampling locations have a value greater than 1 which indicates that the entire sampling location is in "contaminated Fe" status. In addition, the CF value at each location was proportional to the PLI and Igeo values. Pollution area was indicated by CF status > 1 and PLI was categorized with highly contaminated area status. While the status at Igeo varies, locations 2 and 5 are categorized into the uncontaminated status until moderate contamination, locations 3, 4 and 6 are in moderate contamination area and location 1 is in moderate status until contaminated categorized.

The new findings of these researches indicate that the contaminated of Fe in the coral reefs in Prigi Coast is caused by human activities or anthropogenic factors that occur around its waters. The future research is needed to use Pb Isotope linked by Fe to search the originated anthropogenic input. In addition, the Prigi coast waters where closes to the residence area, a national port and is highly visited by tourists were suspected as a source of anthropogenic activities to distribute of Fe in the Prigi coast water.

4. Conclusion

The Fe was bound from fraction 1 to fraction 3 in the Prigi coral reef leached by the Tessier-microwave method. Only Fe in the fraction 3 was dominantly detected by AAS suggesting Fe was associated in the reduction fraction in the coral reef. The indexes PLI and CF calculation at the fraction 3 showed that the Prigi Coast waters were in contaminated status. Almost in the sampled location, the contaminated coral reefs by Fe indicated from anthropogenic inputs. The future research is needed to search for detailed anthropogenic input to seek the contaminated Fe using Pb isotopes associated with Fe in the coral reefs.

References

[1] Bielmyer K, Grosell M, Bhagooli R, Baker C, Langdon C, Gillette P and Capo R 2010 *Aquat. Toxicol.* **97** 125–33
[2] Ichikuni M 1979 *Chem. Geol.* **27** 207–14
[3] Madkour A 2013 *Arab. J. Geosci.* **6** 1739–52
[4] Dodge E and Gilbert R 1984 *Mar. Biol.* **82** 9–13
[5] Esslemont G 2000 *Mar. Chem.* **69** 69–74
[6] Wijaya A, Ohde S, Shinjo R, Ganmanee M and Cohen D 2019 *Arab. J. Chem.* **12** 780–99
[7] Sazzad H, Wijaya A, Tanaka K and Ohde S 2010 *Afr. J. Biotechnol.* **9** 5373–82
[8] Wijaya A, Semedi B, Lusiana R, Armid A and Muntholib M 2019 *J. Braz. Chem. Soc.* **30(5)** 915–929
[9] Turekian K and Wedepohl H 1961 *Geol Soc Am Bull* **72** 175–92
[10] Ariadi R, Putri W, Wijaya A, Siswanta D, Mudasir M and Santosa J 2017 *Orient. J. Chem.* **33** 2913–9
[11] Zaaboub N, Oueslati W, Amine M, Abdeljaouad S, Javier F and Lopez A 2014 *Chem. Speciat. Bioavailab.* **26** 1–12
[12] Wijaya A, Arifani R, Kusumaningrum I, Budiasih E, Daupor H, Hossain S and Sanjaya E 2019 *IOP Conf. Ser. Mater. Sci. Eng.* **515** 012015
[13] Wijaya A, Farida I, Wonorahardjo S, Utomo Y, Daupor H, Hossain S and Kunisue T 2018 *J. Phys. Conf. Ser.* **1093** 012002
[14] Wijaya A, Farida I, Sakbaniah A, Rahmawati A, Budiasih E, Daupor H, Hossain S and Kunisue T 2018 *J. Phys. Conf. Ser.* **1093** 012013

Acknowledgements

This research is funded by the 2019 Masters Thesis Grand-research, Indonesian Ministry of Research, Technology and Higher Education.

Development of Indirect Spectrophotometric Method for Mercury Determination Based on The Formation of Iron(III)-Thiocyanate Complex

Hermin Sulistyarti[1,2*], Rurini Retnowati[1,2], Erwin Sulistyo[1,3], Eka Ratri Wulandari[1,4], Hikmanita Lisan Nashukha[1]

[1]Research Group of Low Cost and Automated Method and Instrumentation, Department of Chemistry, Faculty of Science, Brawijaya University, Malang, 65145, Indonesia
[2]Department of Chemistry, Faculty of Science, Brawijaya University, Malang, 65145, Indonesia
[3]Mechanical Engineering, Faculty of Engineering, Brawijaya University
[4]Vocational Program, Brawijaya University

*Corresponding author: hermin@ub.ac.id

Abstract. A new spectrophotometric method for mercury determination has been developed based on the complexation of mercury(II) in ascertained excess of thiocyanate. The remaining thiocyanate is reacted with iron(III) to form a red iron(III)-thiocyanate complex which gave maximum absorbance at 460 nm. The concentration of mercury(II) is inversely proportional to the decrease absorbance of the complex. The method was optimized to the concentrations of thiocyanate and iron(III). Selectivity of the method has also studied based on percent recovery of mercury(II) in the presence of common metal ions in the tailing waters of gold mine. Under the optimum conditions, the method showed linear correlation to mercury(II) concentration from 1-30 mg L^{-1} with detection limit of 0.58 mg L^{-1}. The method was not interfered in the presence of all metal ions studied up to 50 mg L^{-1}, except of silver which was tolerated at 10 mg L^{-1}. The developed method has been satisfactorily applied for mercury(II) determination in synthetic and tailing waters of gold mine samples.

1. Introduction

Mercury pollution has become a global concern as its toxicity can cause severe health problems. The exposure of volatile metal mercury can induce neurological and behavior disorder, harm the kidney and thyroid, pneumonitis, which in excessive exposure led to death [1]. The wide spread of mercury contamination has been observed in several places in Indonesia which mainly generated from small-scale gold mining. Therefore, monitoring mercury in the small scale gold mining environment is important to prevent the health hazard and global environmental pollution.

Determination of mercury usually uses cold-vapour atomic absorption [2-4]. This method has high accuracy and precision in determining small amount of mercury (up to ng/L) but this method need a specific expertise in analysis and operated using expensive instrumentation. Because of those factors, spectrophotometric method for mercury determination is popularly selected and mostly developed in developing countries [5]. Spectrophotometry employing complexing agent for mercury determination has long reported and well known. The earliest method was reported employing thiocyanate to form

Published under licence by IOP Publishing Ltd

colorless mercuric thiocyanate complex in aqueous or 1-butanol solutions detected using ultra violet spectrophotometer at 281 nm [6]. Other spectrophotometry methods based on complexation of mercury were reported involving complexing agents of dithizone, iodide, iodide-rhodamine B, thiocyanate-rodhamine, diacetyl monoxime isonicotinoyl hydrazone (DMIH), diphenylthiocarbazone, and anthrone phenylhydrazone (APH) [5-10]. Determinations of mercury using dithizone, iodide-rhodamine B, and thiocyanate-rhodamine B were done by involving extraction of the complex with organic solvent for detection [9-10]. Whereas the use of DMIH and diphenylthiocarbazone required synthesis and condensation procedures prior to detection [5,12]. Thus, an alternative spectrophotometric method with no requirement for extraction, synthesis, and condensation procedure is required in order to develop a simple and fast analysis. Efforts for developing environmentally friendly methods for mercury determination have been reported based on the formation of tetraiodomercurate(II) and blue starch-iodine [13], which gave linearity measurement from 1-9 mgL^{-1} mercury(II).

This paper reports a new spectrophotometry for mercury(II) determination based on the fact that mercury(II) ion forms colorless mercuric thiocyanate complex in the presence of excess thiocyanate [6-7], in which in the presence of iron(III) ion, the remaining thiocyanate can form a red complex of iron(III)-thiocyanate, $Fe(SCN)]^{2+}$ at pH below 2 [2]. The principle of determination of mercury(II) in this experiment is based on the decrease of $[Fe(SCN)]^{2+}$ absorbance which proportional to the concentration of mercury(II). The chemical parameters such as SCN^- and Fe(III) concentration are optimized to increase the sensitivity of measurement. The linearity, sensitivity and validation test were also done in this experiment to assure the accuracy and precision of this new method.

2. Method

2.1. Apparatus and Reagents

All spectrophotometric measurements were carried out using single beam spectrophotometer (Spectronic-20). De-ionized water was used for preparation of all solutions. Stock solutions of mercury(II), thiocyanate, iron(III) were prepared by dissolving the appropriate amount of $HgCl_2$ (Merck), KSCN (Merck), and $Fe(NO_3)_3$ (Merck) respectively in de-ionized water. HNO_3 (Merck) was used to attain acidic condition. All chemicals and solvents used were of analytical reagent grade and working solutions were prepared by appropriate dilutions of the stock solution.

2.2. Optimization of Thiocyanate (SCN^-) Concentration

Thiocyanate was used as the complexing agent for mercury(II) and iron(III) ions. Thiocyanate concentration was optimized by mixing 0.2 mL of 100 mgL^{-1} Hg(II) solution with 1mL of 1 M HNO_3 solution in 10 mL volumetric flask. Then, the solution was added with various concentration of thiocyanate solution in the range concentration of 4 to 22 mgL^{-1}, mixed thoroughly and add 0.58 mL of 1000 mgL^{-1} Fe^{3+} solution and add de-ionized water up to 10 mL. The final solution was measured with spectronic-20 at 460 nm.

2.3. Optimization of Iron(III) Concentration

Iron(II) ion was used to complex the free (remaining) thiocyanate and formed Iron(III)-thiocyanate complex used for detection. Optimization of iron(II) concentration was conducted similarly to procedure 2.2 using optimum concentration of thiocyanate; however, the concentration of iron(III) was varied from 9-58 mgL^{-1}.

2.4. Linearity of Mercury(II) Measurement

In order to determine the linearity of mercury(II) measurement, a series concentration of mercury(II) from 1-30 mgL^{-1} were determined under optimum concentration of thiocyanate and iron(III) ions under acidic condition (pH 1) and detected spectrophotometrically at 460 nm. The absorbance of iron(III)-thiocyanate complex from all solutions with different concentration of mercury(II) were plotted vs the concentration of mercury(II), and examined the correlation coefficient towards linear relation.

2.5. Selectivity Test

Selectivity test was done by adding the foreign ions (Co^{2+}, Cu^{2+}, Pb^{2+}, and Ag^{+}) with various concentrations of 0, 1, 5, 10 and 50 mgL^{-1} into mercury solution. Then, the absorbance of these solutions was measured by spectrophotometric detection at 460 nm and the obtained absorbance was compared to the absorbance of mercury(II) solution with no foreign ion.

2.6. Validation Test

The method was validated by applying to synthetic samples and natural sample taken from gold mine tailing soil. The tailing soil was destructed using concentrated nitric acid and diluted 100 times to suit the linear calibration. All samples were treated with the same procedure, then the results were compared to the results obtained from CV-AAS analysis.

3. Results and Discussion

Determination of mercury (II) is based on the decrease absorbance of $[Fe(SCN)]^{2+}$ complex in the presence of mercury(II). When mercury(II) is reacted with excessive thiocyanate ion (SCN^-) in acidic condition (pH < 2), some of SCN^- ions react to Hg(II) ions forming colorless mercuric thiocyanate complex which may exist as $Hg(SCN)^+$, $Hg(SCN)_2$, $Hg(SCN)_3^-$, and $Hg(SCS)_4^{2-}$. Under high concentration of thiocyanate ions, it is considered that the $[Hg(SCN)_4]^{2-}$ complex is the predominating ion in aqueous solutions [6-7]. The remaining SCN^- in the presence of iron(III) ion forms a red complex of iron(III) thiocyanate, $[Fe(SCN)]^{2+}$[2,6]. The principal reaction of mercury determination in the proposed method is shown in equation (1) and (2). The absorbance of $[Fe(SCN)]^{2+}$ complex was measured using visible spectrophotometer at 460 nm. The higher concentration of Hg(II) in solution the more SCN^- ions bound to Hg(II) ions; and thus the remaining free SCN^- is smaller. The smaller amount of SCN^- results smaller product of the red complex of $[Fe(SCN)]^{2+}$. The decrease of absorbance of $[Fe(SCN)^{2+}$ is proportional to the concentration of mercury(II).

$$Hg\,(II) + 4SCN^-_{(excessive)} \rightarrow [Hg(SCN)_4]^{2-}_{(colorless)} \quad (1)$$
$$SCN^-_{(excess)} + Fe^{3+} \rightarrow [Fe(SCN)]^{2+}_{(red)} \quad (2)$$

Preliminary study showed that various concentration of mercury(II) under this principle of determination produced red colors with different intensity. The higher concentration of mercury(II) showed lower intensity of complex color as shown in Fig. 1.

Figure 1. Color intensity of $Fe(SCN)^{2+}$ complex in the presence of 0-30 mgL^{-1} Hg(II)

3.1. Optimization of Thiocyanate (SCN^-)

The optimization of thiocyanate concentration was aimed to obtain sufficient amount of SCN^- to complex with Hg(II) and Fe(III) ions. The concentration of SCN^- was optimized in the range from 4 to 22 mgL^{-1} and the results are shown in Fig. 2. Based on Fig. 2, higher concentration of SCN^- produce higher amount of $[Fe(SCN)]^{2+}$ complex shown by increasing of absorbance. The absorbance increased up to 12 mgL^{-1} and relatively stable for further addition (22 mgL^{-1}). In this experiment, 15 mgL^{-1} thiocyanate was selected as optimum condition and used for further measurements.

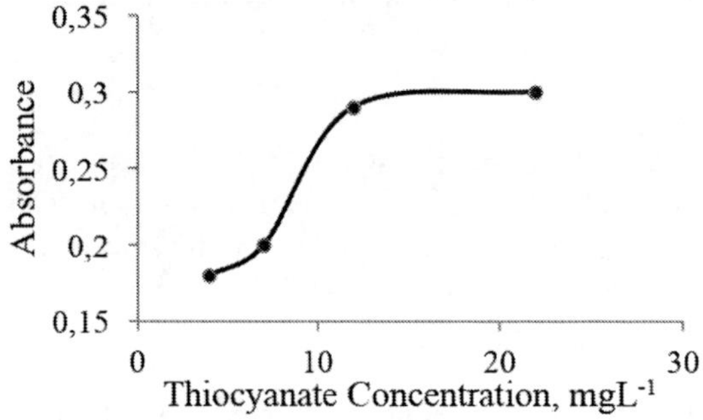

Figure 2: The effect of thiocyanate concentration on absorbance of $[Fe(SCN)]^{2+}$.

3.2. Optimisation of Iron(III) Concentration

The optimization of iron(III) concentration was carried out in order to obtain sufficient concentration for complexing the free thiocyanate. The influence of iron(III) concentration to complex with remaining SCN^- to form $[Fe(SCN)]^{2+}$ red complex is illustrated in Fig. 3, in which the higher concentration of Fe^{3+} produced higher intensity of red complex $[Fe(SCN)]^{2+}$ and subsequently resulted higher absorbance. In this research, iron(III) concentration of 40 mgL^{-1} which gave stable and acceptable intensity was selected for reproducible measurements and sensitivity reasons.

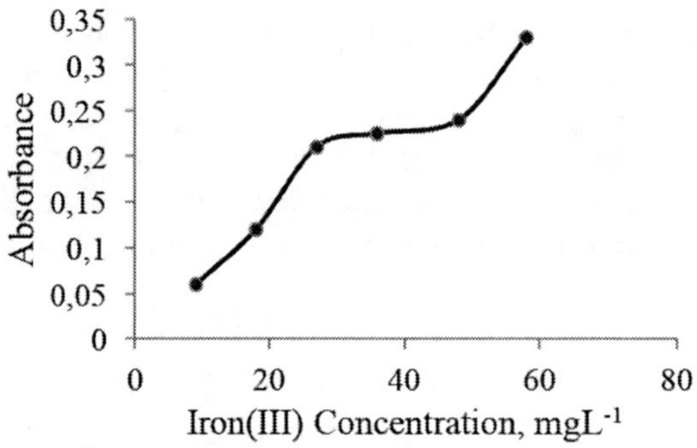

Figure 3: The effect of iron(III) concentration on $[Fe(SCN)]^{2+}$ absorbance.

3.3. Linearity of mercury(II) measurement

Under the optimum conditions outlined above, using concentrations of 15 mgL^{-1} thiocyanate and 40 mgL^{-1} iron(III)), the method was found to give two linear calibrations for mercury(II) concentration (Fig. 4) with first linear range from 1–10 mgL^{-1} ($y = -0.013x + 0.393$) with coefficient determination, $R^2 = 0.998$, and second linear range from 10-30 mgL^{-1} ($y = -0.013x + 0.393$) with R^2 of 0.992. The detection limit of this method determined from the mean of the blank with three times standard deviation was 0.96 mgL^{-1} ($n = 10$).

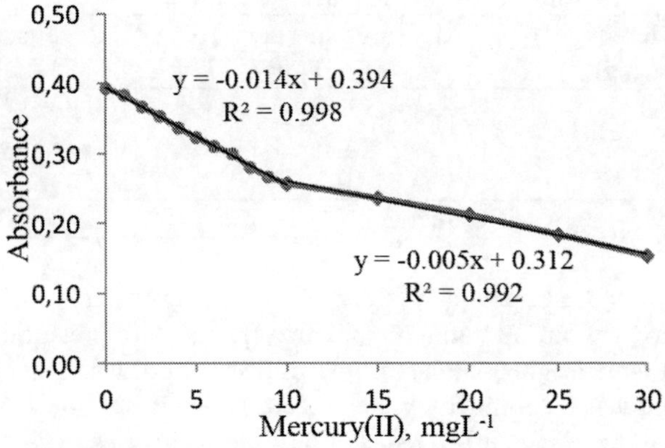

Figure 4: Linearity of measurements of mercury(II).

3.4. Selectivity of measurement

The selectivity of the proposed was examined by monitoring the influence of foreign ions commonly present in tailing gold mine, such as Co^{2+}, Cu^{2+}, Pb^{2+}, and Ag^{+} on the measurement of mercury(II). The effect of these ions were determined by measuring the absorbance of mercury(II) solution in the absence and in the presence of those foreign ions as shown in Fig. 5. As depicted in Fig. 5, the presence of Co^{2+}, Cu^{2+}, and Pb^{2+} ions up to 50 mgL^{-1} did not significantly interfere the measurement of mercury(II). However, the method only tolerated the presence of Ag^{+} up to 10 mgL^{-1}. This is because silver(I) ion reacted with thiocyanate forming silver thiocyanate, Ag(SCN), precipitation with low constant solubility product ($Ksp_{AgSCN} = 1.0\times10^{-12}$) [5].

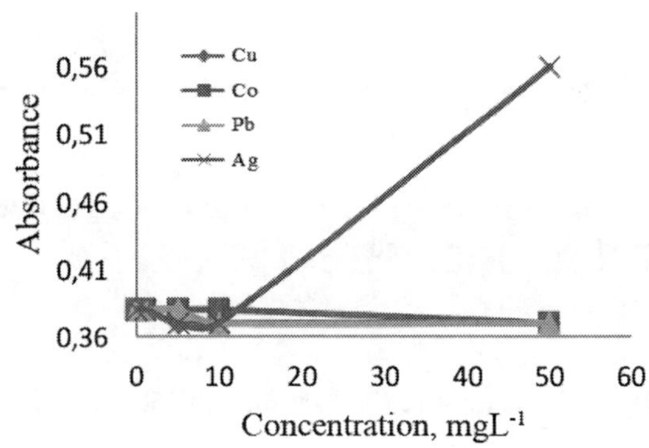

Figure 5: Selectivity of measurement of mercury(II).

3.5. Method Validation

In order to validate the proposed method, it was applied to determine concentration of mercury(II) in synthetic samples and natural samples of soil tailing waste obtained from small scale gold mining using standard addition method. In this method, three different concentrations of mercury(II) were added separately to three synthetic and tailing samples, and then the recoveries of mercury(II) in the sample after addition of the mercury(II) standards were determined using the proposed method. The method showed very good recoveries for synthetic samples with average percent recoveries of 102.48% as depicted in Table 1.

Table 1. Validity test using synthetic samples.

Mercury(II) (mgL^{-1})	Measured mercury(II) (mgL^{-1})	Recovery (%)
1	1.00	100.00±1.79
5	5.27	105.41±1.80
10	10.20	102.03±2.30
Average recovery		102.48±1.37

Validity test for the proposed method was applied to tailing water from small scale gold mining with 100 times dilution (without addition of mercury(II) standard) resulting concentration of 3.85 mgL^{-1} (Table 2); thus it gave original concentration of 385 mgL^{-1}. Using standard addition technique the validity of the method was determined by the percent recoveries of the sample after the addition of three different concentration of mercury(II) standards. The recoveries of the three standard additions are displayed in Table 2 depicting that the proposed method showed excellent recoveries with average percentage of 105.26.

Table 2. Recoveries of mercury(II) from tailing water by standard addition technique.

Sample*	Added Mercury(II) (mgL^{-1})	Measured mercury(II) (mgL^{-1})	Recovery (%)
Sample	0	3.85 mgL^{-1}	-
Sample	1	5.05 mgL^{-1}	105.13±0.18
Sample	2	6.24 mgL^{-1}	108.00±1.88
Sample	3	6.95 mgL^{-1}	102.65±1.95
		Average recovery	105.26±1.87

*Tailing water with 100 dilution factor.

4. Conclusion

The spectrophotometric method outlined above allows simple and sensitive method for determination of mercury(II) concentration. Based on the validation test, this method also showed very good accuracy and precision shown by the high recoveries of close to 100 % with low standard deviations. This result suggests that the method mentioned above is prospective to be used for monitoring of mercury(II) concentration in water and waste water samples.

Acknowledgement

The authors are grateful to University of Brawijaya Malang for facilitating research and Ministry of Research, Technology and Higher Education of the Republic of Indonesia for financial support.

References

[1] Krisnayanti B D, Anderson C W N, Utomo W H, Feng X, Handayanto E, Mudarisna N, Ikram H and Khususiah 2012 *Journal of Environmental Monitoring* **14** 2598

[2] Greenberg A E Lenore S C and Andrew D R 1992 *Standard Methods for the Examination of Water and Waste Water* 17th ed. (Washington: American Public Health Association)

[3] Hageman P L 2007 *Determination of Mercury in Aqueous and Geologic Materials by Continuous Flow-Cold Vapor-Atomic Fluorescence Spectrometry (CVAFS)*, Book V, Chapter 2 D (Virginia: US Geological Survey)

[4] Leal L, Elsholz O, Forteza R and Cerdà V 2006 *Analytica Chimica Acta* **573-574** 399–405

[5] Gadikota C R, Devanna N and Chandrasekhar K 2011 *International Journal of Chemistry* **3** 227-232

[6] Markle G E and Boltz D F 1954 *Analytical Chemistry* **26** 447–9

[7] Kolthoff I M and Sandell E B 1952 *Textbook of Quantitative Inorganic Analysis* 3rd ed. (New York: Macmillan Co) p 548

IOP Conf. Series: Materials Science and Engineering **833** (2020) 012047 doi:10.1088/1757-899X/833/1/012047

[8] Muralidhara B 1998 *Turk. J. Chem.* **22** 215-219
[9] Hernandez-Córdoba M López-Garcia I and Sánchez-Pedreño C 1984 *Microchim. Acta* **84** 467-475
[10] Loo AYW Lay YP Kutty MG Timpe O Behrens M and Hamid SBA 2012 *Sains Malays.* **41** 2 213-218.
[11] Veerana V, Prasad ARG, and Rao VS 2011 *An. Univ. Bucuresti. Chimie* **20** 1 57-64
[12] Ahmed M and Alam MS 2003 *Spectroscopy* **17** 45–52
[13] Sulistyarti H, Sulistyo E, Sutrisno, and Rismiarti Z 2019 *Alchemy* **15** 1 149-164

IOP Conf. Series: Materials Science and Engineering **833** (2020) 012048 doi:10.1088/1757-899X/833/1/012048

Indirect Spectrophotometry Method Based on the Formation of Chromium(VI)-Diphenylcarbazide Complex for Determination of Hydroquinone in Cosmetics

Hermin Sulistyarti[1,2*], Puspita Mufidah Sari [1,2], Syamaidzar[1,2], Rurini Retnowati[1,2], Herman Tolle[3], Adam Wiryawan[2]

[1]Research Group of Low Cost and Automated Method and Instrumentation, Departement of Chemistry, Faculty of Science, Brawijaya University, Malang, 65145, Indonesia

[2]Departemen of Chemistry, Faculty of Science, Brawijaya University, Malang, 65145, Indonesia

[3]Information Technology, Faculty of Computer Science, Brawijaya University, Malang, 65145, Indonesia

*Corresponding author: hermin@ub.ac.id

Abstract. This research is focused on the development of a new indirect spectrophotometry for the determination of hydroquinone in cosmetics based on the capability of hydroquinone as reducing agent for chromium(VI). The principle of the determination is based on the decrease of the color intensity of the red-violet complex of chromium(VI)-Diphenylcarbazide (DPC) due to the reduction of the available chromium(VI) by hydroquinone to chromium(III). Thus, only the remaining chromium(VI) reacted with DPC to form a red-violet complex detected by spectrophotometry at 542 nm. All chemical parameters were optimized with respect to sensitivity and selectivity of measurements. Under the obtained optimum conditions (reaction time of 5 minutes, chromium(VI) of 1 mgL^{-1}, sulfuric acid of 0.05 M, and DPC of 0.0125%), the method gave linear calibration for hydroquinone concentration from 0.2-2.5 mgL^{-1} (R^2 = 0.996). The proposed method also provided good sensitivity shown by low values of LoD and LoQ (0.07 and 0.1 mgL^{-1}) and satisfactory selectivity towards the common interfering ions present in cosmetics. Satisfactory results with recoveries of close to 100 % were observed when the method was applied to facial whitening and black spot removal cosmetics. This result suggests that the developed method is able to be used for monitoring of hydroquinone in skin lightening cosmetics.

1. Introduction

Whitening skin cosmetics is one of popular facial cosmetics in Asia and Africa as skin lightening agent, especially for woman with dark skin or hyperpigmentation. This type of cosmetics commonly contains active ingredients of hydroquinone, kojic acid, arbutin, glutathione, recorcinol, mercury, retinol, niacin amide, and ascorbic acid. Hydroquinone has been used for decades in creams, gels and lotions for the treatment of the hyperpigmentary disorders of the skin. Hydroquinone works by inhibiting tyrosinase for production of skin pigment (melanin); thus, skin remains fair and bright. However, research showed that hydroquinone can cause serious health complications and for prolong use it can permanently damage the skin and potentially cause cancer [1-3]. Therefore, European Union and NADFC of Indonesia banned the utilization of hydroquinone in cosmetics. In Indonesia hydroquinone is only allowed to be used in hair (0.3 %) and nail dyes (0.02 %), while in facial whitening cream or facial removal cream has been banned since 2008 [4].

Several analytical methods for the determination of hydroquinone in cosmetics are described, including spectrophotometry [5-9], flow injection analysis (FIA) with spectrophotometric detection [10-13], and chromatography techniques [14-16]. The present study is aimed to describe a new indirect spectrophotometry method for hydroquinone in cosmetics based on the capability of hydroquinone to reduce chromium(VI) to chromium(III), which in the presence of 1,5-diphenylcarbazide (DPC), the remaining chromium(VI) formed an intense red-violet chromium(VI)-DPC complex. The intensity of the color was reduced proportionally to the available hydroquinone. Thus, the concentration of hydroquinone can be measured spectrophotometrically based on the decrease of absorbance of the red-violet chromium(VI)-DPC complex at 542 nm. The use of chromium(VI) was considered as it showed a fast and sensitive oxidizing agent for hydroquinone [6], thus only a small concentration was used and expected to be safe for environment. In order to achieve the optimum performance of the method, the common parameters which affect redox reaction of hydroquinone and chromium(VI) as well as the stability of complex formation of chromium(VI)-DPC were optimized with respect to sensitivity. The interfering ions frequently exist with hydroquinone in cosmetics were also studied for selectivity of the method. The method was also validated by applying to determine hydroquinone in facial whitening and black spot removal creams.

2. Tools and Materials

Spectrophotometric measurements were carried out using UV-Vis 10S Genesys Spectrophotometer. Materials used in this research were hydroquinone (Sigma-Aldrich), potassium dichromate (Merck), diphenylcarbazide ($C_{13}H_{14}N_4O$, Merck), acetone (AR), sulfuric acid (Merck), ascorbic acid ($C_6H_8O_6$, Merck), lead nitrate (Merck), resorcinol (Merck), and silver nitrate (Merck). All of the chemicals applied for the research were pro-analytical grade without any purification process. De-ionized water was used for preparation of all solutions.

3. Method

3.1. Optimization of maximum wavelength

Optimization of maximum wavelength of chromium(VI)-DPC complex was conducted by mixing chromium(VI) in acidic condition with DPC solution to form a red-violet complex; then, the solution was scanned using UV-Vis 10S Genesys Spectrophotometer at wavelengths in the range of 400 to 800 nm.

3.2. Optimization of stability of complex.

The stability of complex chromium(VI)-DPC was optimized by monitoring the absorbance of the complex at 542 nm in the range time of 2-40 minutes. This optimization was done by mixing 0.4 mL of 10 mgL^{-1} hydroquinone solution with 1 mL of 10 mgL^{-1} chromium(VI) solution, 0.5 mL of 1 M H_2SO_4, and 0.2 mL of 0.25 % DPC, followed by dilution with de-ionized water to obtain 10 mL solution. The absorbance of chromium(VI)-DPC complex was then read at 542 nm using UV-Vis 10S Genesys Spectrophotometer from 2 to 40 minutes. The optimum time to obtain the stable complex was determined by the maximum absorbance, and this optimum condition was used for the next experiments.

3.3. Optimization of H_2SO_4 concentration.

The optimum concentration of acid (H_2SO_4) was done similarly to procedure 3.2, under the optimum time; but the concentration of H_2SO_4 was varied from 0.02 to 0.3 M. The optimum acid concentration was determined by the maximum absorbance, and this optimum condition was used for further experiments.

3.4. Optimization of chromium(VI) concentration.

The optimum concentration of chromium(VI) was conducted similarly to procedure 3.3, but under optimum acid concentration and the concentration of chromium(VI) was varied from 0.2 to 3 mgL^{-1}. The optimum chromium(VI) concentration was determined by the minimum absorbance which showed adequate remaining chromium(VI) for detection, and this optimum chromium(VI) concentration was used for the subsequent experiments.

3.5. Optimization of DPC concentration.

The optimum concentration of DPC was carried out using similar procedure to 3.4, but under optimum chromium(VI), and the concentration of DPC was varied from 0.005 to 0.0625 %. Chromium(VI) concentration which gave maximum absorbance was selected as optimum DPC concentration for further experiments.

3.6. Linearity of Measurement

The linear range of measurement was conducted by measuring absorbance of hydroquinone sample from 0.2-3 mgL^{-1} under the all optimum conditions outlined in procedures 3.1 to 3.5. The absorbance of complex chromium(VI)-DPC from all hydroquinone concentrations were plotted versus the concentrations of hydroquinone and examined the coefficient determination towards linear correlation.

3.7. Selectivity of Method

In order to obtain selectivity of the method, the effect of interfering ions commonly present in cosmetics, such as ascorbic acid, resorcinol, Pb^{2+} and Ag^{+} on the absorbance of chromium (VI)-DPC were studied by adding each of those interfering substances to a certain hydroquinone concentration. The selectivity was defined by the percent error of measurement of the mixture compared to hydroquinone sample without interfering ions.

3.8. Method Validation

The proposed method was validated using standard addition procedure, followed by calculation of percent recoveries of samples after the addition of standard hydroquinone. The validity was defined with percent recovery of close to 100 %.

4. Result and Discussion

The reaction of determination of the proposed method was based on two main reactions: (1) the capability of hydroquinone to reduce chromium(VI) to chromium(III); (2) the formation of chromium(VI)-DPC complex. The redox reaction between hydroquinone and chromium(VI) under acidic condition is shown in equation (1), where hydroquinone can be oxidized to benzoquinone and chromium(VI) was reduced to chromium(III). The high potential reduction standard of chromium(VI) to chromium(III), $E^0 = +1.33$ V, is expected to give complete oxidation of hydroquinone to benzoquinone, supported by high positive potential reaction of $E^0 = +0.631$ V.

$$3C_6H_4O_2H_2 + Cr_2O_7^{2-} + 8H^+ \rightarrow 3C_6H_4O_2 + 2Cr^{3+} + 7H_2O \qquad E^0 = +0.631\ V \qquad (1)$$

In this method, chromium(VI) was added quantitatively in a certain excess to ensure reduction all hydroquinone. After the oxidation of hydroquinone has taken placed, the remaining chromium(VI) was reacted with DPC to form a red-violet chromium(VI)-DPC complex as depicted in equation (2).

$$Cr_2O_7^{2-} + 3\ DPC \rightarrow Cr(VI)\text{-}DPC_3 \text{ (red-violet complex)} \qquad (2)$$

4.1. Optimization of wavelength for measurement

Wavelength for measurement was optimized in order to select a wavelength which gave maximum absorbance with maximum sensitivity for measurement. The product of reaction, chromium(VI)-DPC complex, was scanned in the range of visible wavelength from 400 to 800 nm using UV-Vis 10S

Genesys Spectrophotometer and the obtained absorption spectra is shown in Figure 1. Based on Fig. 1, chromium(VI)-DPC complex showed maximum absorbance at wavelength of 542 nm, which correlate to the complement color of red-violet [17] as obtained by the previous work [18]. Therefore, 542 nm was selected as the maximum wavelength used for the experiments.

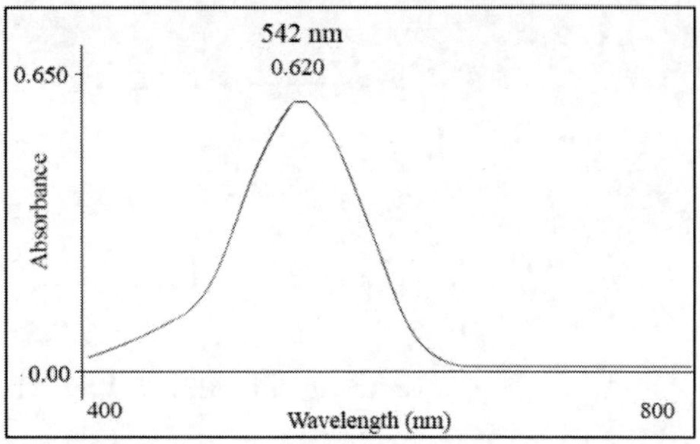

Figure 1. The maximum wavelength of chromium(VI)-DPC complex.

4.2. Optimization of Stability of Complex.
Stability of complex was optimized in order to obtain the reaction time to achieve the equilibrium which gave maximum product of reaction of chromium (VI)-DPC complex shown by its maximum color intensity and maximum absorbance [19]. Figure 2 shows the relatively constant absorbance of chromium(VI)-DPC complex monitored from 2 to 40 minutes of reaction. Therefore, 2 minute reaction with absorbance value of 0.697 was selected for further measurements.

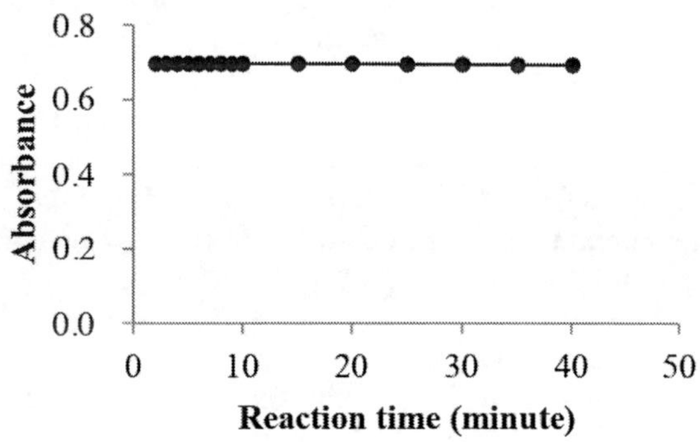

Figure 2. The optimization of stability of complex.

4.3. Optimization of H_2SO_4 Concentration.
Optimization of acid (H_2SO_4) concentration was conducted in order to achieve maximum oxidation of hydroquinone by chromium(VI), which thus reduces the presence chromium(VI); therefore, the optimum of acid concentration was shown by the minimum absorbance of chromium(VI)-DPC complex. As shown in Equation 1, chromium(VI) requires acidic condition for oxidizing hydroquinone. Besides, the stability of chromium(VI)-DPC complex is affected by the acidity of solution. According to APHA (American Public Health Association), chromium(VI)-DPC complex formed in interval acidity of pH 1.6 to 2.2 [20]. Furthermore,the form of chromium(VI) can be various from H_2CrO_4, $HCrO_4^-$, CrO_4^{2} to $Cr_2O_7^{2-}$ depends on the acidity of the solution. Thus, optimization of H_2SO_4 concentration was also aimed to ensure the form of chromium(VI) in the solution as dichromate

($Cr_2O_7^{2-}$) [21]. Figure 3 performs the effect of sulfuric acid concentration from 0.05 to 0.30 M to the absorbance of chromium(VI)-DPC complex, which shows relatively constant absorbance of chromium(VI)-DPC complex. Sulfuric acid concentration of 0.05 M, which correlates to pH suggested by APHA standard was chosen as optimal and used for further optimization.

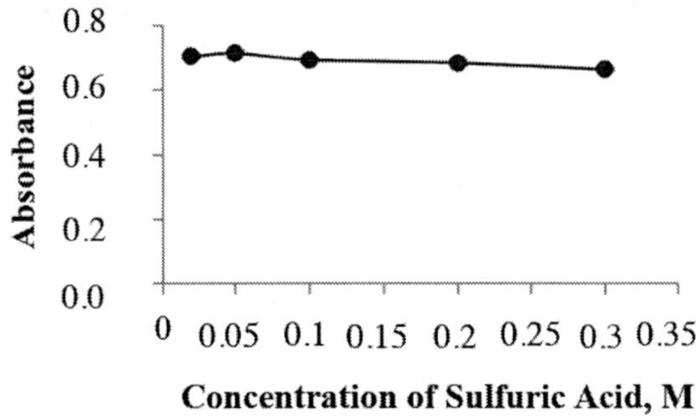

Figure 3. Optimization of H_2SO_4 concentration.

4.4. Optimization of chromium(VI) concentration.

Optimization of the concentration of chromium(VI) was carried out to determine the concentration of chromium(VI) required for oxidizing the available hydroquinone in the sample with adequate excess to form complex with DPC for detection. The higher concentration of chromium(VI) gives the more remaining chromium(VI) and increases the color intensity shown by the high absorbance of chromium(VI)-DPC complex. The optimization of chromium(VI) concentration was monitored based on the absorbance of chromium(VI)-DPC complex, and the result is depicted in Figure 4. As expected, the higher concentration of dichromate produced higher absorbance of the complex. In this method, the adequacy of chromium(VI) for oxidizing hydroquinone is indicated by the presence of the excess of Cr(VI) observed as red-violet complex of Cr(VI)-DPC. Thus, the optimum chromium(VI) was determined by the minimum Cr(VI) which gave acceptable absorbance of Cr(VI)-DPC. In this study, 1 mgL^{-1} chromium (VI) provided sufficient amounts to oxidize hydroquinone with minimum excess of chromium(VI) sufficient for detection. Therefore, 1 mgL^{-1} chromium(VI) was selected for further experiments. The use of a minimum excess of chromium(VI) was also considered to fulfill the spirit of green chemistry.

4.5. Optimization of DPC concentration.

Optimization of DPC is required to determine the suitable concentration of DPC to react with the remaining chromium(VI) forming maximum chromium(VI)-DPC complex. It was reported that the formation of chromium (VI)-DPC complex can be achieved under mole ratio of 1:8 for Cr(VI):DPC [13]. From the experimental results (Figure 5) showed that the absorbance of complex increased by increasing DPC concentration from 0.005 to 0.0125 %, but it slightly decreased for higher concentration of 0.025% and relatively constant up to 0.0625 %. The concentration of 0.0125 % DPC which have mole ratio of more than 1 to 8 and gave the highest absorbance of chromium (VI)-DPC complex was chosen as the optimum concentration of DPC and used for the subsequent experiment.

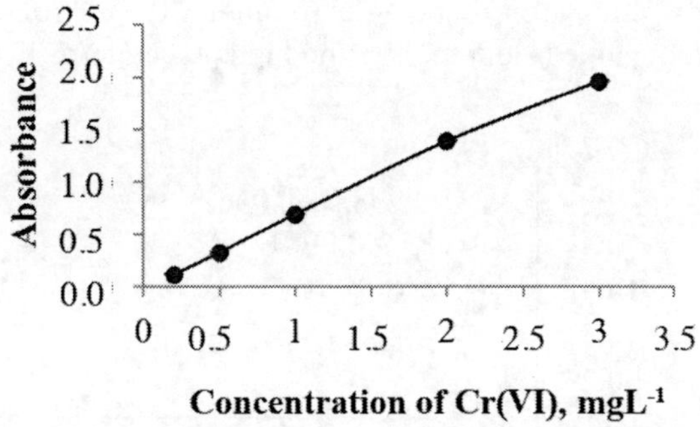

Figure 4. The optimization of Cr(VI) concentration.

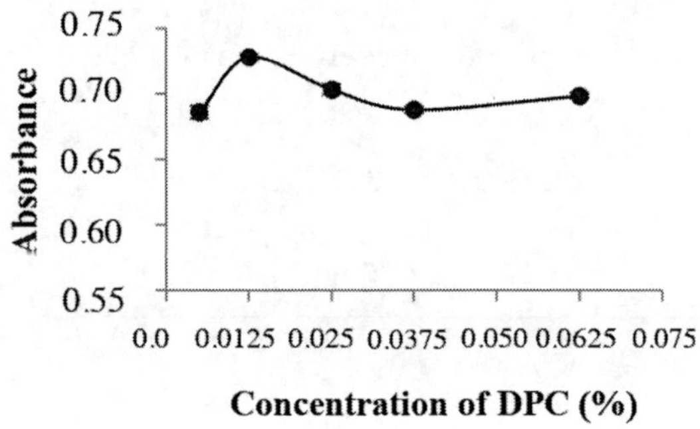

Figure 5. The optimization of DPC concentration.

4.6. Linearity of Measurement

The linearity of measurement was conducted by measuring various concentration of standard hydroquinone from 0.1 to 3 mgL^{-1} under the obtained optimum parameters, i.e. 1 mgL^{-1} dichromate as chromium(VI), 0.05 M H_2SO_4, 0.0125% DPC, and spectrophotometric detection at 542 nm. Then, the delta (Δ) absorbance (absorbance difference) between blank solution and sample (hydroquinone) solution was plotted versus hydroquinone concentration. Figure 6 shows that the method performed linear correlation (R^2, of 0.9961) between the concentration, x, and the delta absorbance, y, with linear equation of y=0.2293x+0.0362 for concentrations of hydroquinone from 0.2 to 2.5 mgL^{-1}.

4.7. Selectivity of Method

The selectivity of method was tested by monitoring the effect of the interfering substances to the absorbance of the standard hydroquinone, and the difference of absorbance was performed as percent error. This procedure was conducted by adding substances which commonly present in whitening cosmetics, such as silver and lead ions, resorcinol, and vitamin C (ascorbic acid) to standard hydroquinone solutions. Under the addition of 0-5 mgL^{-1} of each substance, the method was not interfered by silver and lead ions, as well as resorcinol, supported by percent error of less than 8 %. Based on the t-test with a confidence level of 95 % and $\alpha = 0.05$, the recoveries of hydroquinone in the presence of the three substances did not show significant different shown by the smaller value of t_{count} compared to t_{table}. The presence of ascorbic acid in hydroquinone sample up to 1 mgL^{-1} can be tolerated by the proposed method; however, higher concentrations of ascorbic acid (2.5-5 mgL^{-1}) resulted in high percent error of much more than 10 % Table 1). The interfering effect of ascorbic acid to the measurement of hydroquinone in consequence of its nature as strong reducing agent which capable to behave as hydroquinone in reducing chromium(VI) to chromium(III); thus gave results to higher recoveries of hydroquinone (> 0.5 mgL^{-1}) and high percent errors.

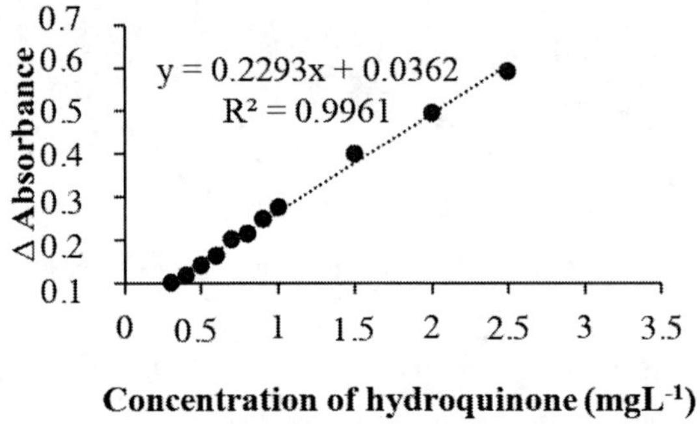

Figure 6. Linearity of the method

Table 1. The selectivity of method

[HQ] (mgL^{-1})	Interfering ions	[Interfering ions] (mgL^{-1})	Measurable concentration (mgL^{-1})	% Error
		0.00	0.50 ± 0.03	0.00
		0.50	0.51 ± 0.05	2.00
	Ag^{+}	1.00	0.48 ± 0.13	4.00
		2.50	0.47 ± 0.04	6.00
		5.00	0.47 ± 0.07	6.00
		0.00	0.50 ± 0.04	0.00
		0.50	0.49 ± 0.04	2.00
	Pb^{2+}	1.00	0.49 ± 0.01	2.00
		2.50	0.49 ± 0.03	2.00
		5.00	0.48 ± 0.06	4.00
0.50		0.00	0.50 ± 0.03	0.00
		0.50	0.51 ± 0.02	2.00
	Resorcinol	1.00	0.48 ± 0.04	4.00
		2.50	0.47 ± 0.03	6.00
		5.00	0.46 ± 0.03	8.00
		0.00	0.50 ± 0.04	0.00
		0.50	0.51 ± 0.05	2.00
	Ascorbic acid	1.00	0.54 ± 0.01	8.00
		2.50	0.78 ± 0.03	56.00
		5.00	1.05 ± 0.01	110.00

*An average of 3 repetitions

4.8. Validation Method

The proposed method was validated by applying to determine hydroquinone concentration in four types of cosmetics (A, B, C, and D) consisted of facial whitening creams and black spot removal creams, followed by standard addition procedure. In this standard addition procedure, three different concentrations of hydroquinone were added separately to each of the four extract creams and the hydroquinone concentrations were determined using the proposed method. The recoveries of the hydroquinone are summarized in Table 2 showing that the proposed method provided good accuracy shown by the high % recoveries of close to 100 % with good precisions (typical SD ≤ 0.1 and RSD < 10 %). The limits of detection and quantification (LoD and LoQ) were determined by measuring blank solutions, and the LoD and LoQ values were derived from the addition of the signal of blank with 3 times of the standard deviation for LoD and with 10 times of the standard deviation for LoQ, resulting respective values of 0.07 mgL^{-1} and 0.10 mgL^{-1}. These low values of LoD and LoQ showing that the proposed method provided adequate sensitivity required for the determination of hydroquinone in whitening cosmetics. Table 2 also showed that the hydroquinone in all cosmetics used in this study were measurable under the developed method. Therefore, the proposed method can be considered as an alternative method for hydroquinone determination in cosmetics.

Table 2. The validity of method

Cosmetic samples	HQ addition concentration (mgL^{-1})	Measurable concentration±SD (mgL^{-1})	% Recovery
	0.00	0.53 ± 0.05	-
A	0.30	0.85 ± 0.04	103.77
	0.50	1.06 ± 0.03	105.66
	0.00	0.50 ± 0.06	-
B	0.30	0.80 ± 0.03	100.00
	0.50	1.00 ± 0.02	100.00
	0.00	0.40 ± 0.10	-
C	0.20	0.60 ± 0.04	100.00
	0.40	0.79 ± 0.08	97.50
	0.00	0.23 ± 0.05	-
D	0.20	0.44 ± 0.04	104.30
	0.40	0.63 ± 0.06	100.00

*An average of 3 repetitions

5. Conclusion

Indirect spectrophotometry method for the determination of hydroquinone in whitening cosmetics based on the decrease absorbance of the chromium(VI)–DPC complex has been successfully developed. This method has high selectivity towards Ag^{+}, Pb^{2+}, and resorcinol, except of ascorbic acid and has been applied to cosmetics sample with satisfactory results (recoveries of close to 100 %). This result suggests that the method mentioned above is able to be used for monitoring of hydroquinone in skin lightening cosmetics.

References

[1] Couteau C and Coiffard L 2016 *Cosmetics* 3 **27** 1-16
[2] Dadzie OE and Petit A 2009 *J. Eur Acad. Dermatol. Venereol.* **23** (7) 741-50
[3] Kamakshi, R., 2012, *J. Cosmet.Sci.*, 63, 43-54.
[4] National Agency of Drug and Food Control (NADFC), 2011, *Peraturan No.03.1.23.668.11.07331*, Metode Analisis Kosmetik, BPOM RI, Jakarta.
[5] Al-Mosulli BB Al-Jomaili AS and Al-Janabi HS 2015 *IJS* **56** 265-74
[6] Moldovan Z Popa DE David IG Buleandra M and Badea IA 2017 *Journal of Spectroscopy* **2017** 1-9
[7] Gavcia PL Santoro MIRM Singh AK and Kedor-Hachmann ERM 2007 *J. Pharmac. Sci.* **43** 397-404
[8] Odomasu RO and Ekwe TO 2010 *Afr. J. Pharm. Pharmacol.* **4** 5 231-34
[9] Ibrahim S Damayanti S and Riani Y 2004 *Acta Pharmaceutica Indonesia* **29** 1 28-33
[10] Ruedo ME Sarabia LA Herrero A and Ortiz MC 2003 *Anal. Chim. Acta* **479** 173-84

[11] Fahmi MI Sulistyarti H Mulyasuryani A and Wiryawan A 2019 *J. Pure App. Chem. Res.* **8** 1 53-61
[12] Trenggamayunelgi FS Sulistyarti H and Retnowati R 2019 IOP *Conf. Ser. Mater. Sci. Eng.* **546** 032031 1-10
[13] Alshibani MMO Sulistyarti H and Sabarudin A 2019, *J. Pure App. Chem. Res.* **8** 3 208-216
[14] Siddique S Parveen Z Ali Z and Zaheer M 2012 JCDSA **2** 224-28
[15] Desiderio C Osscini L and Fanali S 2000 *J. Chromatography A* **887** 489-96
[16] Soekmin J and Kim Y 2005 *Bull. Korean Chem. Soc.* **26** 5 819-22
[17] Mullertz A Perrie Y and Rades T 2016 *Analytical Techniques in Pharmaceutical Sciences*, Springer, New York.
[18] Suryati L Sulistyarti H and Atikah 2015 *J. Pure App. Chem. Res.* **4** 1 34-41
[19] Huheey JE Keiter RL 1993 *Inorganic Chemistry Principles of Structure and Reactivity* **4th** Edition Harper Collins College Publisher New York
[20] Greenberg AE Lenore SC and Andrew DR 1992 *Standard Methods for the Examination of Water and Waste Water 3500-Cr* American Public Health Association Washington
[21] Balan C Irina V and Doina B 2013 *Chromium (VI) Removal from Aqueous Solution by Purolite Base Anion-Exchange Resins with Gel Structure* Chemical Industry & Chemical Engineering Quarterly **19** 4 615-28
[22] Wandoyo VW and Mudasir R 2006 *Indo. J. Chem.* **6** 2 150-54

Acknowledgments
The authors are grateful to University of Brawijaya Malang Indonesia for financial support through "*Hibah Guru Besar dan Doktor 2019*" research grants and research facilities.

Potency of Modified Cassava Flour as Binder and Thickener in Formulation of Instant Infant Porridge using Fortificant of Natural Folic Acid

Agustine Susilowati*, Yati Maryati and Aspiyanto

Research Center for Chemistry, Indonesian Institute of Sciences, Kawasan PUSPIPTEK, Serpong, Tangerang Selatan – 15314

*Corresponding author : agustine_1408@yahoo.co.id

Abstract. A dry mixing process in preparation instant and pour infant porridges fortified by natural folic acid of A and B, and added by modified cassava (mocav) flour as thickener and binder had been conducted. Fortificant of natural folic acid of A and B are a mixture of soy bean tempeh, nixtamalized yellow corn and fermented broccoli, and a mixture of mung bean tempeh, nixtamalized yellow corn, and fermented broccoli. This experiment aims to optimize by adding mocav flour and the best type of natural folic acid in preparing instant and pour infant porridges for complementary breastfeeding (CBF) and its characteristic on the whole composition, particularly folic acid, characteristic of folic acid monomer, volatile compounds, particle size, and particle size distribution on instant and pour infant porridges. The experiment were done on mocav flour with concentrations of 0, 6, 12, 18, 24, and 30% (w/w of base formula of infant porridge) with fortificant of natural folic acid of A and B, and pouring to produce instant and pour infant porridges A and B. The result of experiment showed that based on folic acid, process optimization was achieved on fortificant A at mocav concentration 18% (w/w, based formula of porridge) and increase folic acid 532.26% (5.30-folds), dissolved protein 226.62% (2.27-folds), total sugars 120.44% (1.20-fold), however drop total solids 4.61% and reducing sugars 14.94% in instant infant porridge compared to each components concentration in base formula of infant porridge. In this condition, volatile compounds of instant infant porridge and pour infant porridge dominated by acetic acid and its derivatives (44.21%), and furan and its derivatives (29.812%).

1. Introduction

Modified cassava (mocav) flour is a product of cassava flour processed by fermentation using the principles of modification of cassava cells. The growth of microbial enzymes during fermentation cause characteristic change of flour, such as increasing viscosity, gelation ability, power rehydration, and dissolves easily. The whole mocav characteristics are possibility affected by types and Lactic Acid Bacteria (LAB) culture concentration, and fermentation time [1]. In general, mocav is used as substitution material of wheat flour for food products, such as bakery, noddle, and cookie to get non-gluten food products. Today, the essential role of mocav as food additives for thickener, binder, and emulsifier are not explored yet. Bill of the Republic of Indonesia Number 18 Year 2012 concerning Food and Government Regulation Number 28/2004 on Food Safety, Quality and Nutrition Chapter II, Part Two concerning Food Additives differentiates between thickener or emulsifier or stabilizer agent as modified starch through acid/basic/oxidation/bleached modification and enzymatic modification, which can act as monostarch phosphate or hydroxy propyl distarch phosphate, however, it is not yet as specific type to modify it enzymatically [2]. Mocav is possibly modified starch type in enzymatic because cassava is prepared by fermentation by LAB.

A series of formulation process is performed in preparing infant porridge as complementary breastfeeding (CBF) by adding fortificant of natural folic acid from mixture of soy beans (*Glycine soya*) tempeh and mung beans (*Phaseolus radiatus L.*) fermented by *Rhizopus oligosporus* C1, fermented broccoli (*Brassica oleraceae* L.) by kombucha culture, and yellow dent corn (*Zea mays identata*)by nixtamalized by $Ca(OH)_2$ solution [3]. Adding mocav to this formulation is not only conducted as a thickener in order to produce safer infant porridge products, but also mocav has function as a carrier of folic acid. Mocav effects on the existence of the folic acid in products as instant porridge after pour porridge and has an important essential role in composition and inorganic quality.

Folic acid (also known as folate, vitamin M, vitamin B9 or folacin, pteroyl-L-glutamic acid, pteroyl-L-glutamate, pteroylmonoglutamic acid, C19H19N7O6) [3] having the similar basic molecular structure consisting of 3-parts structure, pteridine ring, p-aminobenzoic acid (p-ABA), and glutamic acid molecule [4], sensitive on oxidative degradation, light, mechanical treatment and high temperature [5]. Adding folic acid is generally performed through fortification of folic acid as a result of synthesis by mixing into instant porridge flour, using natural folic acid as a result of a process aiming to explore the potency of natural resource (legumes, vegetables, cereals), and outcoming as nutritional sources [6]. Folic acid is an essential component used to the growth and development of infants in the periconceptual period and during pregnancy. Folic acid has essential role as precursor in synthesis of nucleotide to remethylation homocysteine on the growth in cell division that is important for producing red blood cell [7;8] and prevent its emerging Neural tube defects (NTDs), such as *spina bifida* and anencephaly on infants. Application of fortificant of natural folic acid in preparing instant infant porridge for CBF has become a priority in its essential role as a functional food to reach smartness level. Infant porridge is food with semi solid shape with a composition according to staple food being proposed to adapt a digestive system of baby with various categories, such as instant porridge powder, ready-to-eat porridge with a result of the pour, and ready-to-cook porridge [9]. Infant porridge powder is diluted with equivalent to folic acid content according to RDA, 80 µg/day or 30 – 50 µg per serve [10]. Formulation of instant infant porridge powder and ready-to-consume it through fortificant of natural folic acid by adding mocav at various concentrations is enabled to affect on composition, particularly folic acid, particle size, and domination of folic acid monomer.

The experiment activity aims to find out effect of the best concentration of mocav as thickener and carrier of folic acid at different type of fortificant in formulating instant infant porridge powder and ready-to-consume pour porridge on the whole compositions, particularly the best folic acid and it's characteristic on the present of the folic acid monomer, type of volatile compounds, particle size, and particle size distribution.

2. Materials and Methods

2.1. Materials and Equipments

Main materials utilised for this study were [3] distilled water, soy beans, mung beans, broccoli and spinach (local), yellow dent corn (*Zea mays* var. indentata) and white dent corn (local, corn plantation center, South Tangerang), sucrose (local), inoculum of *Rhizopus oligosporus* strain C1 and Kombucha culture (Research Center for Chemistry – LIPI), wheat flour (local, Bogasari), corn starch (Honig, Nederland), powder milk (local), butter (local), baking powder (local), salt and vanilla (local), and perforated plastic (local). Whereas, chemicals utilised in this study were $Ca(OH)_2$ solution (E.Merck), standard folic acid (Aldrich), hydrochloric acid (E.Merck), sodium nitrite (E.Merck), and methanol (E.Merck). All the chemicals with reagent grade used in this study were procured through local market [1]. Important equipment utilised in this study were homogenizer (Ultra-Turrax, Ika Labortechnik, T50, Jane & Kunkel, Germany), balance (Fujitsu, Japan), autoclave (CHENG YI, LS – 50 L, China), , blender (local, National), flow chamber (local), incubator (local), cabinet dryer (local), and sieves of 80 and 100 mesh (Retsch, Germany). Important instruments for analysis were Liquid Chromatography-tandem

Mass Spectrometry (LC-MS) (Mariner Biospectrometry) with LC (Hitachi L 6200), Particle Size Analyzer (PSA) (Horiba Nano Partica-Backman) [2], GC-MS (Shimadzu, Japan). and UV-vis Spectrophotometer (Model RF-550, Shimadzu, Japan),

2.2. Experimental Design

This experiment activity was carried out by adding mocav with concentration 0, 6, 12, 18, 24 and 30% (w/w, based formula of infant porridge) on porridge formula consisting of base formula and fortificant A and B, in which fortificant A are mixture of soy bean tempeh, nixtamalized yellow dent corn and fermented broccoli, and fortificant B is mung beans tempeh, nixtamalized yellow corn and fermented broccoli to get instant infant porridge A and instant infant porridge B. In this similar time, for each treatment is performed at instant or pour porridge powder in warm-hot water (± 80 – 90 °C) on a 50 g : 160 mL ratio as being done for pour porridge C and D. Analyses were performed on initial fortificant materials and instant/pour infant porridge on total sugars (Phenol Sulphate method), total solids (Gravimetric method)[11], dissolved protein (spectrophotometry) [12], folic acids (LC-MS) [13] and reducing sugars (Somogyi-Nelson method) [14], . Identification on folic acid was conducted by LC-MS (Mariner Biospectrometry) tandem with LC (Hitachi L 6200) [15], particles size distribution (Particle Size Analyser PSA) [16], and volatile compounds using GC-MS (Shimadzu, Japan) [17]. Experimental data processing were evaluated using descriptive way from a yield of average in triplicate analyses from the duplicate process.

2.3. Procedure

2.3.1. Fermentation process of soybeans, mung beans tempeh, and broccoli

A number of soybeans or mung beans are sorted, washed, boiled for 30-45 minutes, cooled, and soaked at pH 5 overnight. The following day, processed bean were peeled, washed, inoculated with *Rhizopus oligosporus* tempe inoculum strain-C1 at a concentration of 0.2% (w / w) uniformly, packed in perforated plastic, and incubated at room temperature (28 - 30 ° C) for 24 - 36 hours. Meanwhile, fermentation of broccoli is conducted by blanching broccoli at 80 °C for 5 minutes, and pulverized at the ratio of broccoli and water of 1:4 to produce broccoli suspension as a substrate for get folic acid. Broccoli suspension was then poured by appropriate broccoli inoculum with types of broccoli of 15 % (v/w broccoli suspension) and sucrose 10 % (w/w, broccoli suspension), and stored in a closed container with cheesecloth in dark room at room temperature for 3 days. All activities were conducted aseptically. Biomass produced was fermented broccoli suspension [18].

2.3.2. Nixtamalization process

A number of horse dent- yellow corn was washed and soaked in water with a ratio of corn and water 1: 4. Then soaked yellow corn, then added Ca (OH) 2) 20% (w/w dissolved corn protein), cooked at 90 ° C for 60 minutes , cooled, and rinsed to remove excess cooking water and lime, left, grounded and filtered through 80 mesh to produce nixtamalized yellow corn powder [18]

2.3.3. Mixing, homogenizing, drying, size reducing and sieving fortificant

The preparation process of mixing tempe paste and fermented broccoli was done in a ratio of 1: 1 [18], then homogenized (8000 rpm, 30 minutes) to produce a mixture of soybean tempeh and broccoli fermentation. Nixtamalized corn was added into mixture of tempeh paste and fermented broccoli on a 1:1 ratio and homogenized (8000 rpm, 30 minutes) to have resulted in fortificant paste, in treatment was subsequently dried by means of cabinet dryer (50 °C, 24 hours), size reduction, and sieved via 60 mesh so that it is yielded desired powder as fortificant powder of A and B.

2.3.4. Preparing instant/pour infant porridge using mocav as binder and thickener of folic acid

Instant porridge for baby food was prepared according to Recommended Dietary Allowances (RDA) on folic acid in accordance with SNI [10]. Folic acid intakes according to RDA or range of 400 µg per day or 100 – 150 µg per serving of porridge. On this base formula of porridge was added fortificant of folic acid A and B equivalent to folic acid 1,000 μg/mL (w/w of based formula of porridge). The base formula of infant porridge consisted of rice flour, skim milk powder, wheat flour, sugars of 70.12, 7.08, 5.26, 17.54 % (w/w of 100 g of base formula), and mocav in various concentrations as shown in experimental design. Instanization process was subsequently carried via by mixing rice flour and flour in the same ratio (1 part) by adding water (7 parts) at gelatination temperature (± 90 °C) in order to form paste, drying at 50 °C for 24 hours, siege, sieving via 60 mesh, adding skim milk and sugars. Pouring is carried by stirring 50 g instant porridge in 160 mL hot water (80 – 90 °C).

2.3.5. Identification and analysis of folic acid using LC-MS
Samples instant infant porridge contain fortificant A and B (known as instant infant porridge A and B), and pour infant porridge contain fortificant C and D (known as pour infant porridge C and D). Oligomer was analyzed by a LC-MS using Mariner Biospectrometry. LC system was integrated with Q-TOP MS through Electrospray Ionization (ESI) system, in which scan mode was carried out in a mass spectra ranging m/z 100 – 1200 at 140 °C. LC (Hitachi L 6200) was carried out on a 15 mm x 2 mm C18 (RP 18) Supelco column and particles size of 5 μm. Solvent types were a mixture of 80 parts of methanol and 20 parts of water with flow rate 0.1 mL/minute and injection volume of 5 uL [15].

3. Results and Discussions

3.1. Characteristics of materials
Instant and pour infant porridges use fortificant from mixtures of soy tempeh or mung beans tempeh, fermented broccoli, and nixtamalized yellow dent corn as a source of folic acid. Fortificant is related to the composition of each material and a series of processes being reached. **Table 1** showed Composition in preparation of fortificant of natural folic acid.

Table 1. Composition of materials in preparation fortificant of natural folic acid.

Kind of material	Components				
	Folic acid (μg/mL)	Total solids (%)	Dissolved protein (mg/mL)	Reducing sugars (mg/mL)	Total sugars (mg/mL)
Mung bean tempeh	766.56	53.49	1.80	22.60	286.36
Soy bean tempeh	568.52	44.02	1.15	23.60	160.60
Fermented broccoli	57.98	8.09	0.98	28.28	316.21
Nixtamalized yellow corn	466.81	54.46	0.30	12.40	316.14

Preparation of instant infant porridge is performed by using fortificant of soy or mung beans tempeh, fermented broccoli, and nixtamalized yellow dent corn as source of natural folic acid. From composition of each material, the highest concentration of folic acid (766.56 μg/mL) from mung beans was higher than that soy beans tempeh (568.52 μg/mL), fermented broccoli (57.98 μg/mL) and nixtamalized yellow corn (466.81 μg/mL). This matter is related with presence of dissolved protein in mung beans tempeh (1.80 mg/mL) being higher compared with soy beans tempeh (1.15 mg/mL), fermented broccoli (0.98 mg/mL) and nixtamalized yellow dent corn (0.3 mg/mL). Difference in dissolved protein reflect activities of yeast protease enzyme of R. ologosporus C1 degrading protein to smaller and simpler amino acids, such as glutamic acid with high solubility in water [19].

Glutamic acid is part of the whole structure of the folic acid molecule besides para-aminobenzoate (p-ABA) and pteridine heterocycle [20]. This fermentation increases dissolved protein and folic acid concentrations in soy and mung bean tempeh compared with before fermentation, dissolved protein and folic acid of broccoli [21]. As a source of carbohydrate, nixtamalized yellow dent corn generates the highest concentration of total sugars (316.14 mg/mL) compared with other materials. On this material has been done a nixtamalization process of corn in order to raise its folic acid [22]. Nixtamalized yellow dent corn produces the highest total solids concentration (54.46%), in which other materials have also essential role on the whole composition and texture of fortificant. On reducing sugars, fermentation process of broccoli by Kombucha culture generates the highest concentration of reducing sugars (18.28 mg/mL), being higher compared with soy beans or mung beans tempeh and nixtamalized yellow dent corn. This fermentation uses sucrose as a source of carbon to produce folic acid through synthesis *de novo* [23] by Kombucha culture converting component of vegetable, such as polyphenol. Remain reducing sugars produced from this synthesis is a soluble component in biomass, that is the source of energy and affects on the taste of application products.

3.2. *Effect of formulation on fortificant characteristic*

Process of material mixing, homogenizing, drying, size reduction and sieving on mixtures of soy beans or mung beans tempeh (1 part), fermented broccoli (1 part), and nixtamalized yellow corn (1 part) produce fortificant with different composition, as tabulated in **Table 2.**

Table 2. Composition of materials in the formulation of instant or pour porridge using mocav as binder and thickener of folic acid.

Kind of material	Component				
	Folic acid (μg/mL)	Total solid (%)	Dissolved protein (mg/mL)	Reducing sugar (mg/mL)	Total sugar (mg/mL)
Fortificant A*	101.89	92.91	5.80	28.00	78.28
Fortificant B*	130.54	93.07	6.00	19.14	89.65
Base formula of infant porridge**	109.09	96.07	1.39	27.18	159.76
Mocav flour	22.26	96.13	1.17	5.50	162.90

Legend : *fortificant powder A : mixture of soy beans tempeh, fermented broccoli and nixtamalized yellow corn; B : mixture of mung bean tempeh, fermented broccoli and nixtamalized yellow corn.

Fortificants A and B are suspension of smooth powder with color gradation from pale brown to dark brown. The whole appearances of fortificant showed effect of processes related, particularly drying process producing reaction of browning due to melanoidine brown pigment emerged by result of reaction between sugars and protein [19]. Composition of fortificant B indicates folic acid (130 μg/mL), total solids (93.07%), total sugars (89.65 mg/mL), and dissolved protein (6.0 mg/mL) being higher, however reducing sugars (19.14 mg/mL) being lower compared with fortificant A. This matter is caused by difference in type of material and interaction amongst each process. Formulating and drying decrease folic acid, dissolved protein, and total sugars as consequence of denaturation, high temperature, however they increase total solids caused by evaporation of water mass.

Mocav flour functioning as food additive for carrying folic acid and thickening porridge have lower concentration of folic acid (22.26 μg/mL) compared with base formula of porridge (109.09 μg/mL). This matter is caused by flour formula which is combination of rice flour, wheat flour, skim milk powder and sugars so that it contributes possibility sufficient significant on the whole concentrations of folic acid. Base formula of porridge has also the highest concentration of total sugars (159.76 mg/mL) compared with other components showing main source of porridge energy, meanwhile reducing sugars indicates sugars having ability to reduce sugars to fructose and glucose contributing on taste of porridge. Figures 1a to 1h showed subsequently soy beans tempeh, mung beans tempeh, nixtamalized yellow corn, fermented broccoli, fortificant A, fortificant B, base formula of porridge flour and mocav flour, fortificant A, fortificant B, base formula of instant porridge, and mocav flour.

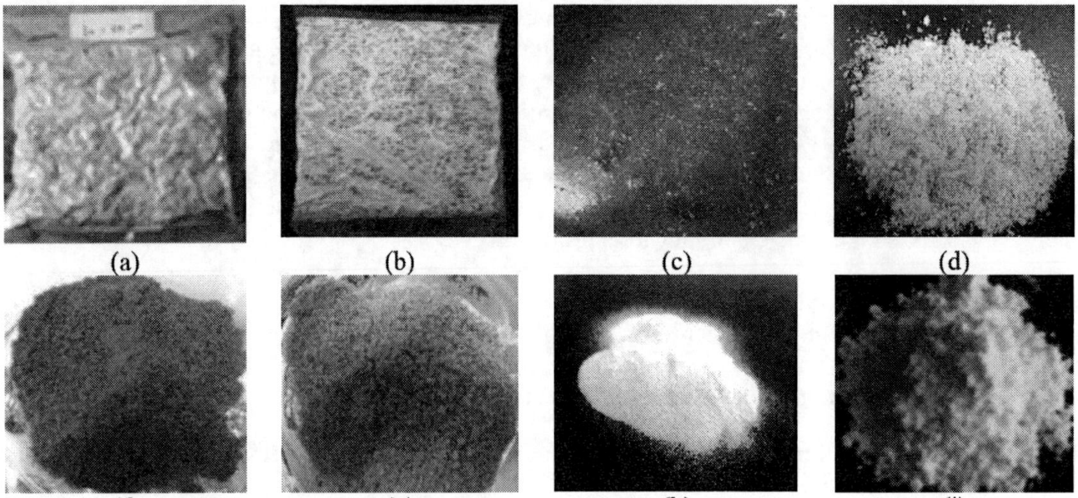

(a) (b) (c) (d)
(f) (g) (h) (i)

Figure 1. (a) Soy beans tempeh, (b) mung beans tempeh, (c) fermented broccoli, (d) nixtamalized yellow corn, (e) fortificant A, (f) fortificant B, (g) base formula of instant porridge, and (h) mocav flour.

3.3. Effect of adding mocav flour on the composition of instant and pour infant porridge

3.3.1. Folic acid (μg/mL)

Adding mocav flour on the formula of instant infant porridge by using fortificant A and B produce instant infant porridge, as shown in **Figure 2a.** Mocav flour concentration becoming more and more high increases folic acid in instant infant porridge to the best binding tolerance, as shown in **Figure 2b.**

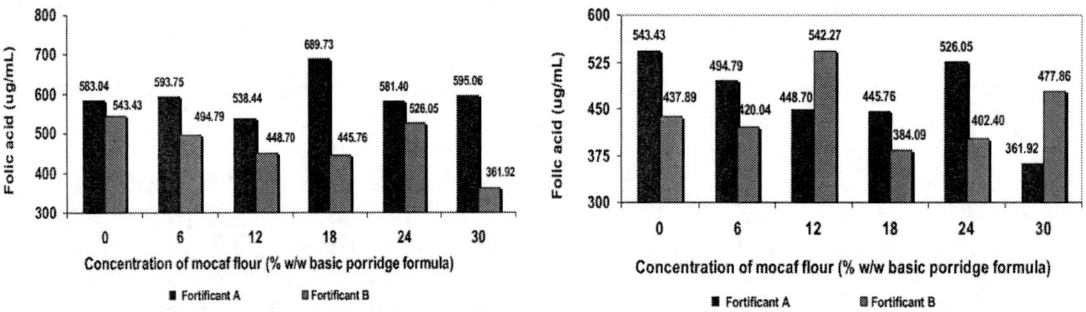

(a) (b)

Figure 2. Relationship between mocav concentration and type of fortificant on the recovery of folic acid in (a) instant infant porridge and (b) pour infant porridge.

Optimization is achieved on instant porridge with fortificant A at mocav concentration of 18% and gives folic acid 689.73 μg/mL. On this optimum concentration of mocav, formulation system is able to increase folic acid 532.26% (5.3-folds) compared with folic acid in the base formula of instant infant porridge (109.09 μg/mL). Increasing folic acid is possibly caused by mocav flour, which is able to absorb folic acid through the dry mixing process (6000 rpm for 5 – 10 minutes). On higher concentration of mocav is enabled to be occurred denaturation of folic acid by dry mixing system. Reaction rate of pouring on folic acid of pour porridge is demonstrated in Figure 3b, in which optimization of folic acid (542.27 μg/mL) is reached at fortificant B with mocav concentration of 12% being higher compared with folic acid of porridge (526.05 μg/mL) in fortificant A at the optimum concentration of mocav (24%). Increasing folic acid content in pour porridge is not only caused by the factor of fortificant type, but also by its occurence of gelatinization causing swelling flour granular so that it binds more much

IOP Conf. Series: Materials Science and Engineering **833** (2020) 012049 doi:10.1088/1757-899X/833/1/012049

folic acid. Dropping folic acid is also caused by a lower concentration of material (50 g instant porridge/160 mL) or instant porridge and hot water ratio of 1 and 3. In the optimum condition of instant porridge (fortificant A with mocav concentration of 18%), pouring process at 50 g instant powder/160 mL hot water decreases folic acid of pour porridge 35.37%.

3.3.2. Dissolved protein (mg/mL)

The term dissolved protein refers to the presence of folic acid by 3-part structure containing *p*-aminobenzoic acid (*p*-ABA) molecule linked at one end to a pteridine ring and at the other end to one glutamic acid molecule [20]. Adding mocav flour with concentration becoming more and more increase seems to increase the dissolved protein of instant porridge. Optimization is reached on instant porridge with fortificant A (4.54 mg/mL) at mocav concentration of 18% being higher compared with optimization by fortificant B at mocav concentration of 30% (4.16 mg/mL), as shown in Figure 3a. The difference in dissolved protein is not only caused by interaction among fortificant, base formula of porridge and mocav, but also by unstable protein property because of high solubility in water and treatment of dry mixing so that it is the occurrence of lysis which is not detected by Lowry method [12]. On the optimum mocav concentration (18%), formulation system using fortificant A is able to increase dissolved protein of instant porridge 226.62% (2.26-folds) compared to dissolved protein in base formula of instant infant prior to process of mixing (1.39 mg/mL).

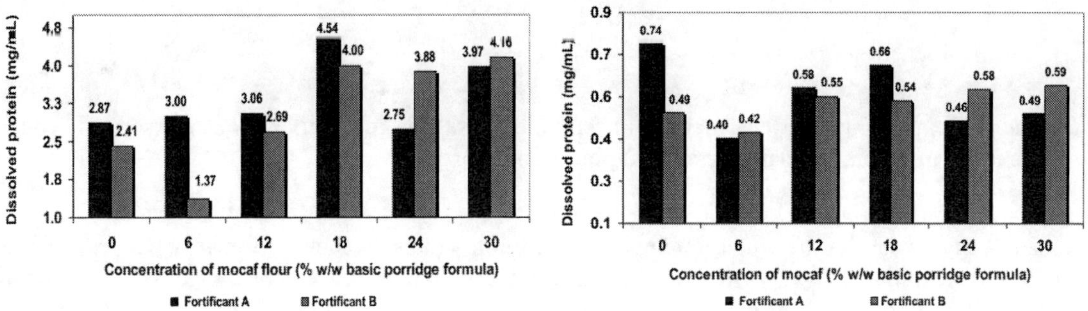

(a) (b)

Figure 3. Relationship between mocav concentration and type of fortificant on recovery of dissolved protein in (a) instant infant porridge and (b) pour infant porridge.

Trend of reaction rate on binding dissolved protein seems on pouring process of instant porridge, as displayed in Figure 3b. This matter is indicated by optimizing mocav concentration of 18% in pour porridge with fortificant A (0.66 mg/mL) being higher compared with fortificant B at mocav concentration of 30% (0.59 mg/mL). Increase of dissolved protein is not only caused by the factor of fortificant type, but also by dissolving amino acids, whereas decline of dissolved protein is caused by denaturation in using hot water (± 90 °C). In the optimum dissolved protein of instant porridge (fortificant A with mocav concentration of 18%), pouring process at 50 g instant porridge/160 mL hot water drops dissolved protein of pour porridge 587% (5.9-folds) from 4.54 mg/mL to 0.66 mg/mL.

3.3.3. Reducing sugar (mg/mL)

A reducing sugar is any sugar that is capable of acting as a reducing agent because it has a free aldehyde group or a free ketone group. All monosaccharides are reducing sugars, such as glucose, fructose and galactose. They act as a source of energy, which eases to be absorbed and affect directly on taste of infant porridge. Adding mocav flour (30%) to instant porridge with fortificant B shows optimization of reducing sugars being higher compared with instant infant porridge with fortificant A at mocav concentration of 12% (45.42 mg/mL), as demonstrated in Figure 4a. On this optimum condition, formulation system is able to increase reducing sugars of instant porridge of 87.16% compared with reducing sugars in base formula of instant infant porridge before mixing process (27.18 mg/mL).

Increase of reducing sugars is caused by contribution of mocav, meanwhile decrease of reducing sugars is possibility caused by dry mixing system or interaction amongst other components to form compounds being undetected according to Somogyi Nelson method [13]. On pouring process of instant porridge, increase of mocav concentration in pour porridge shows similar trend, in which reducing sugars with fortificant B shows the highest mocav concentration 30% (41.60 mg/mL), whereas pour porridge with fortificant A reaches optimum condition at mocav concentration 18% (27.82 mg/mL), as shown in Figure 4b. On this optimum condition, reducing sugars in instant porridge (fortificant B, mocav concentration 30%) followed by pouring process at 50 g instant porridge/160 mL hot water decreases reducing sugars of pour porridge 18.25% from 50.89 mg/mL to 41.60 mg/mL.

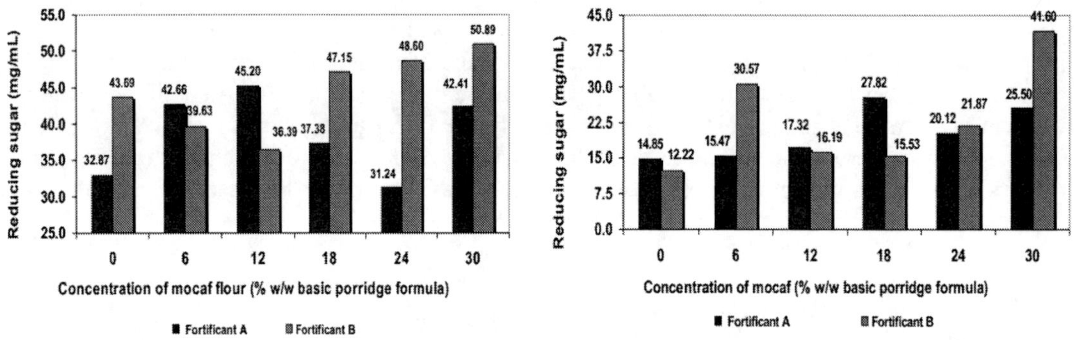

(a) (b)

Figure 4. Relationship between mocav concentration and type of fortificant on recovery of reducing sugars in (a) instant infant porridge and (b) pour infant porridge.

3.3.4. Total Sugars (mg/mL)

Total sugars acts as the whole references for carbohydrates as source of energy. Increase of total sugars concentration in instant infant porridge fluctuates, however increases on both types of fortificants. Instant infant porridge with fortificant B achieves optimization condition (420.29 mg/mL) at mocav concentration 30% being higher compared with optimization with fortificant A at mocav concentration 24% (418.32 mg/mL), as showed in **Figure 5a.**

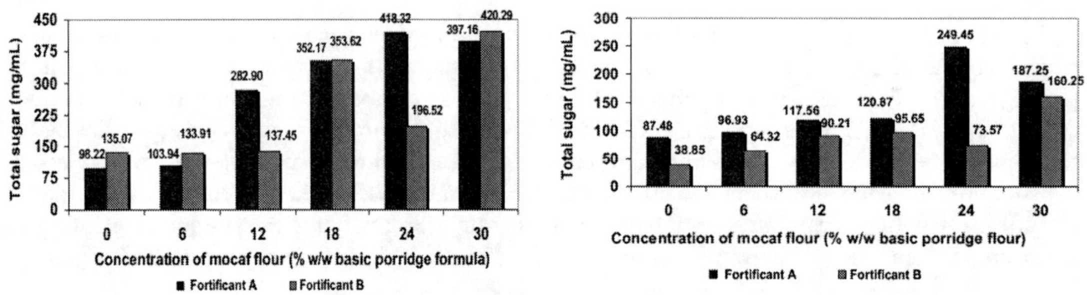

(a) (b)

Figure 5. Relationship between mocav concentration and type of fortificant on recovery of total sugars in (a) instant infant porridge and (b) pour infant porridge.

In this condition, formulation system is able to increase total sugars (420.29 mg/mL) of instant porridge of 163.53% (1.63-folds) compared with total sugars of base formula of instant porridge before mixing process (159.76 mg/mL). Difference in optimization condition of total sugars at both types of fortificant are possibility caused by difference in composition of formulation, particularly on fortificant, in which total sugars in fortificant B being higher (89.65 mg/mL) compared with fortificant A (78.28 mg/mL). It had been known that base formula of infant porridge consist of rice flour, wheat flour, skim

IOP Conf. Series: Materials Science and Engineering **833** (2020) 012049 doi:10.1088/1757-899X/833/1/012049

milk powder, and sugars, or in other words almost all compositions are carbohydrate (> 94.74% of the whole formulas). On pouring process shows optimization of total sugars (249.45 mg/mL) is reached by pour porridge with fortificant A at mocav concentration 24% being higher compared with fortifikan B (160.25 mg/mL) at mocav concentration 30%, as indicated in Figure 5b. This matter indicates that gelatinization process of fortificant A is more efficient because it needs mocav flour with lower concentration (24%) in order to get a higher concentration of total sugars compared with gelatinization using fortificant B (30%). In this optimum condition, total sugars of instant porridge (fortificant B, mocav concentration 30%) and pouring process at 50 g instant porrige/160 mL hot water drops total sugars of pour porridge 40.84% from 420.29 mg/mL to 249.45 mg/mL.

3.3.5. Total solids (%)

Total solids is a measurement that includes the whole components both dissolved, suspended and settleable solids in water. It acts as a reference on nutrition, texture, organoleptic quality, and life time of food products. Adding mocav flour with concentration becoming more and more increase produces an optimum concentration of total solids of instant porridge with fortificant A (91.64%) at mocav concentration of 18% being higher compared with fortificant B at mocav concentration of 12% (91.42%), as displayed in **Figure 6a**.

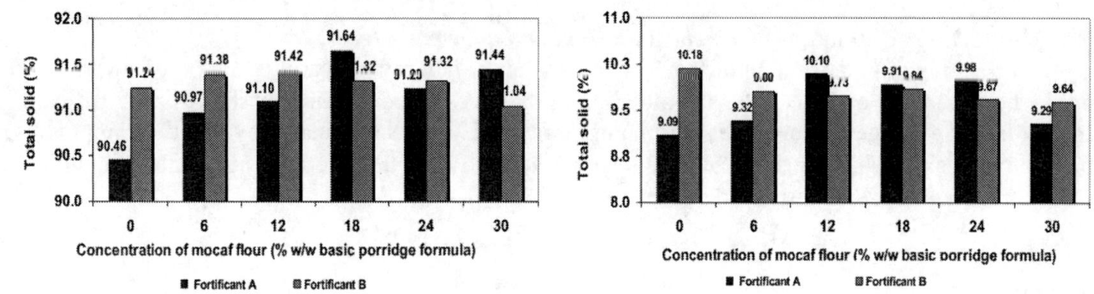

(a) (b)

Figure 6. Relationship between mocav concentration and type of fortificant on recovery of total solids in (a) instant infant porridge and (b) pour infant porridge.

In this condition, the formulation system declines total solids 4.61% compared with total solids in base formula (96.07%). Declining total solids is possibility caused by interaction amongst initial fortificant, base formula, and mocav. Fortificant A has lower concentration of total solids (92.91%) than that fortificant B (93.07%), however by contributing base formula (96.07%) and mocav (18% w/w) cause higher concentration of total solids of instant porridge compared with instant porridge with fortificant B. Formulation result shows that total solids in fortificant B, base formula, and mocav at concentration 12% are 93.07%, 96.07%, and 96.13%, respectively. This matter shows that dry mixing system needs mocav with higher concentration, despite lower concentration of total solids in fortificant to get instant porridge with the optimum concentration of total solids. Similar trend of reaction rate seems to process pouring, in which optimization condition (10.10%) is reached at pour porridge with fortificant A at mocav concentration 12% being higher compared with optimization condition at fortificant B (9.84%) with mocav concentration 18%, as shown in Figure 6b. This matter showed that using mocav as thickener for fortificant A is more efficient than that fortificant B. On the optimum condition, total solids of instant porridge (fortificant A, mocav concentration 18%) and pouring process at 50 g instant porridge/160 mL boiled water drops total solids of pour porridge from 91.64% to 9.91% (80.82%).

3.4. Optimum process condition on formulation of instant and pour infant porridges

From assessment on formulation process in preparation of instant infant porridge and pour porridges by adding mocav flour as carrier of folic acid and thickener, the best formulation based recovery of folic

IOP Conf. Series: Materials Science and Engineering **833** (2020) 012049 doi:10.1088/1757-899X/833/1/012049

acid is reached by using fortificant A (mixtures of soy beans tempeh, fermented broccoli, and nixtamalized yellow corn at mocav flour concentration 18%, w/w base formula of porridge). This formula is generated instant and pour infant porridge from similar formula with composition of folic acid 689.73 and 445.76 μg/mL, dissolved protein 4.54 and 0.66 mg/mL, total solids 91.64 and 9.91%, total sugars 352.17 and 120.87 mg/mL, and reducing sugars 37.38 and 27.82 mg/mL, respectively. Figures 7a and 7b displays instant infant porridge and pour porridge using fortificant A and adding mocav flour 18% (w/w base formula of proridge).

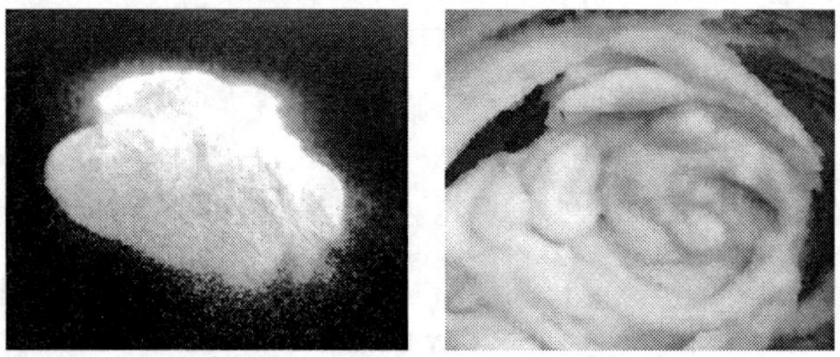

(a) (b)

Figure 7. (a) Instant infant porridge and (b) pour porridge from the optimum formulation process.

3.5. Identification of volatile compounds in instant and pour porridge

Identification of volatile compounds is conducted on instant infant porridge and pour porridge based on the best recovery of folic acid at formulation of mocav with concentration 18% using fortificant A. Instant infant porridge demonstrates chromatogram with 19 peaks dominated by volatile compounds at peak 3 and peak 4, which are also found at peaks 7, 9, and 10 as ethyl/methyl ester acetic acid with total concentration 44.21%, as displayed in **Figure 8a**.

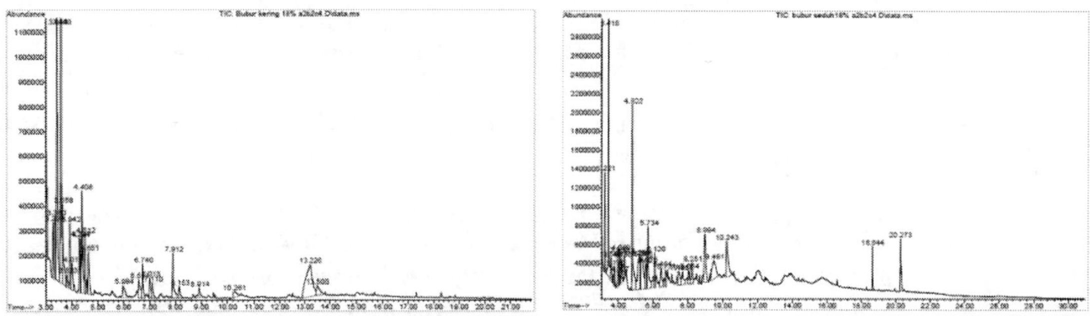

(a) (b)

Figure 8. Chromatogram of volatile compounds at (a) instant infant porridge and (b) pour infant porridge at optimum condition .

These compounds are ester, which is possibility get from mocav flour. It had been known that mocav flour is produced from fermentation process of Cassava flour by *Lactobacillus* sp., in which this process is enabled to yield acetic acid, lactic acid, and their derivative compounds (ethyl/methyl ester). Other volatile compounds are furan compounds and their derivatives, such as furfural, 5-hydroxymethylfurfural, -D-Glucopyranose, 4H-Pyran-4-one with total 8.75%. These compounds are yielded from the whole flours in formulation (rice flour, wheat flour, mocav) and on fortificant of folic acid are calculated as carbohydrate, their fatty acid derivatives (propanoic acid, 2-propanone, propanoic acid, propylamine with total 11.81%), butanoic acid and their derivatives (2,3-Butanediol) with total 6.412%, and other methyl ester (mercaptoacetone, 3-methoxy-2,2-dimethyloxirane, thylhydroxylamine) with total 6.553%.

Pour infant porridge displays chromatogram with 28 peaks, as showed in Figure 8b dominated by volatile compounds at peak 1, 2, 12, and 17 as ethyl/methyl ester acetic acid, such as acetic acid, fluoroacetic acid, hydrazinecarboxylic acid with concentration total 17.811%, furan compound and their derivatives furfural, 2-furanmethanol, 2(5H)-furanone, 2H-pyran, 4H-pyran-4-one, tetrahydro-4H-pyran, 5-hydroxymethylfurfural with total 29.812%, alcohol derivatives (1-butanol, ethanol, ethanol, propylene glycol, 1,2,15-pentadecanetriol, 1,3-butanediol) with total 16.127%, fatty acids (n-hexadecanoic acid, 9,12-octadecadienoic acid, 9,12-octadecadienoic acid) with total 5.84%, and other volatile compounds, such as methyl ester (3-amino-2-oxazolidinone, L-1-butyne, 2-cyclopenten-1-one, 1,3dioxane, 1,3-diaminoguanidine, formic acid, 2-ethoxyethyl acetate, 1-propanamine) with total 19.96%. It had been seemed that dry mixing process is dominated by acetic acid and their derivatives, and only a less part of sugar (as furan and their derivatives), fatty acid (propanoic acid and butanoic acid), and other methyl ester, meanwhile pouring process is dominated by sugars (as furan and its derivative), the presence of alcohol compounds, less acetic acid and fatty acids. The whole dominant volatile compounds from both type of infant porridges are indicated in Table 3.

Table 3. Dominant volatile compounds from instant infant porridge and (b) pour infant porridge at optimum condition (fortificant A and mocaf flour 0.3% w/w base formula of porridge).

Kind of porridge	RT*	Kind of volatile compound	(%)**	Kind of porridge	RT*	Kind of volatile compound	(%)**
	3.29	2-Propanone	2.87		3.22	Propanoic acid	2.09
	3.38	Silanediol,	2.72		3.41	L Acetic acid,	10.60
	3.44	Acetic acid	16.40		3.59	LFluoroacetic acid	2.34
	3.60	Acetic acid,	25.48		3.79	L 1-Butanol,	0.59
	3.65	Propanoic acid	5.23		3.983	Methyl glyoxal	2.05
Instant infant porridge	3.82	Propylamine	0.63	Pour infant porridge	4.02	Butyric acid hydrazide	0.47
	3.94	2,3-Butanediol	2.57		4.08	Propanoic acid,	11.86
	4.01	2,3-Butanediol	1.87		4.18	L 1-Butyne	2.10
	4.29	Thylhydroxylamin	1.86		4.34	LFurfural	1.78
	4.34	Acetic acid	2.33		4.82	Furanmethanol	11.85
	4.40	Furfural	4.86		5.29	Acetic acid	3.19
	4.51	2-dimethy loxirane	2.82		5.39	Propylene Glycol	6.58
	4.65	2-Propanone,	3.08		5.60	2(5H)-Furanone	2.36
	5.98	Mercaptoacetone	1.87		5.67	Glyceraldehyde	1.40
	7.10	Butanoic acid	1.01		5.73	2-Cyclopenten	5.05
	8.91	4H-Pyran-4-one	0.99		6.12	L Acetic acid	1.93
	10.25	Hydroxy metylfurfural	1.14		6.48	Pentadecanetriol	2.64
	13.22	Butanoic acid,	11.52		6.73	6L 2H-Pyran	0.997
	13.49	D-Glucopyranose,	1.76		7.48	Methylpropyl ester	2.537
					7.71	L Formic acid	1.980
					8.06	1,3-Butanediol	1.451
					8.25	1-Propanamine	2.934
					8.99	4H-Pyran-4-one	7.072
					9.49	,4-Furandiol,	6.324
					10.24	Hydroxymethylfurfural	6.939
					18.64	n-Hexadecanoic acid	2.061
					20.27	Octadecadienoic acid	3.779

Legend: *Retention time; **concentration.

3.6. Identification of folic acid monomer in instant and pour infant porridges

Identification on standard folic acid is found at 1 peak (T 3.3) with retention time of 0 – 10 minutes with relative intensity 100%, in which at mass spectra m/z 439 – 448 from T 3.3 displays domination of monomer with molecular weight (MW) of 442.37 Dalton (Da.) and relative intensity 100%, as demonstrated in Figures 9a and 9b. It had been known MW of folic acid is 441 Da. By means of LC-MS instrument, it had been given that a compound has possibility of M^+, M^+ Na^+, $2M^{++}$ or $2M^+$, Na^+ stated as difference in MW. Operation condition of LC-MS is injection volume 5 μL, flow rate 0.2 mL/min. with eluent mixture of methanol and water at a 80 : 20 ratio using column of C-8 (15 mm x 2 mm) [14]. Identification on folic acid monomer is performed on samples of instant infant porridge and pour porridge from the best treatment based on folic acid using treatment with fortificant A at mocav flour concentration 18%. Figures 9c and 9e demonstrate chromatogram of instant infant porridge and pour infant porridge dominated by peak T 1.5 with retention time range 0 – 10 minutes and relative intensity 100%. Figure 9d displays mass spectra from T 1.5 giving 7 folic acid monomer dominated by monomer with MW 442.4175 Da. and m/z range 441.95 – 443.11, and relative intensity 100%. Figure 9f shows mass spectra m/z from T 1.5 yielding 7 folic acid monomer dominated by 2 monomer with MW 442.4774 and 442.8006 Da. and m/z range 441.94 – 442.94, and relative intensity 100%.

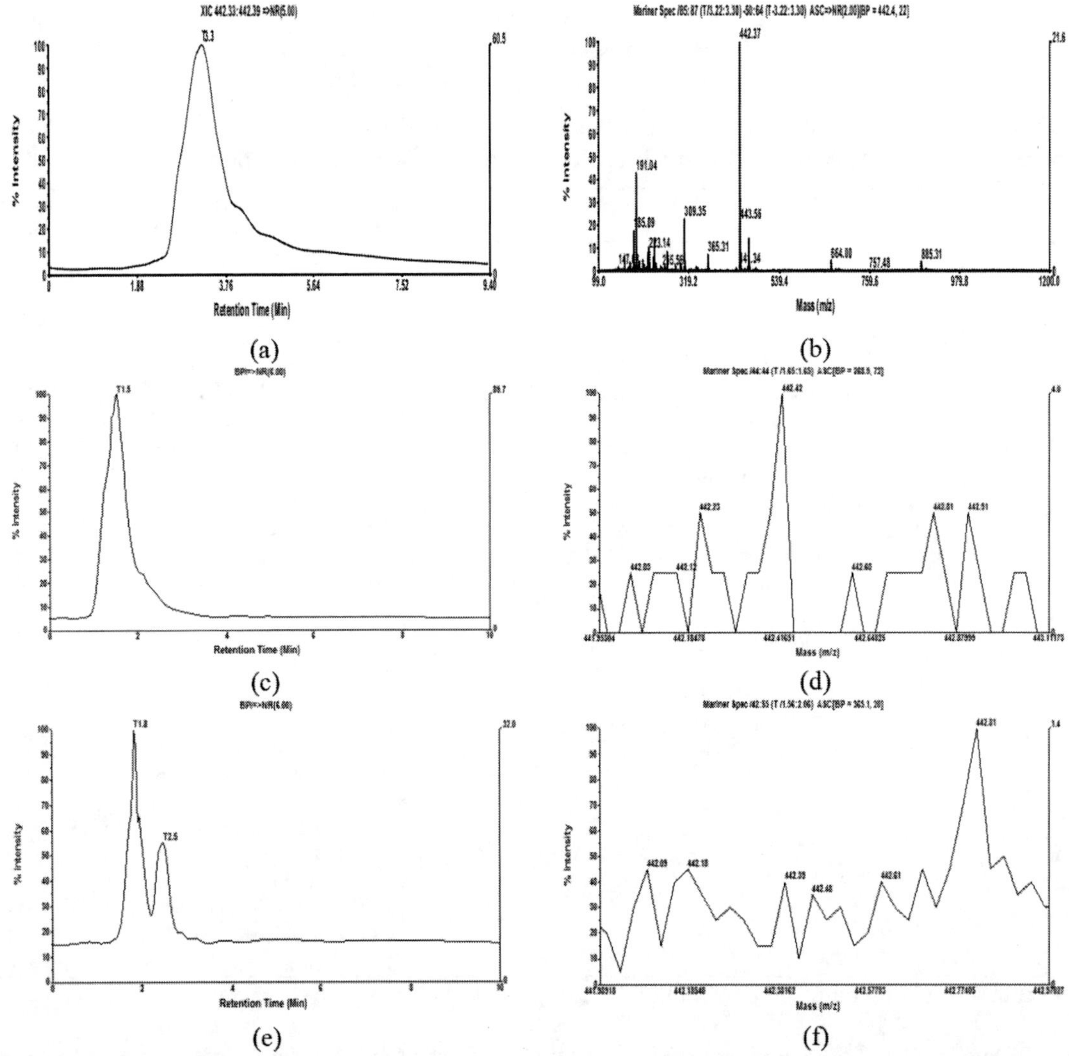

Figure 9. (a) Chromatogram and (b) mass spectra of standard folic acid, (c) chromatogram and (d) mass spectra of instant infant porridge, (e) chromatogram and (f) mass spectra of pour infant porridge at optimum condition .

3.7. Particle size distribution

Particle size distribution from instant infant porridge shows particles with diameter size (Ø) in the range of 5.122 – 10.097 μm (< 10.00 μm) with quantity 3% or average Ø 10.27 μm with quantity 10% per volume of material, Ø ranged from 11.565 – 890.0 μm (< 100.00 μm) with quantity 50% and Ø ranged from 11019.515 – 5000 μm (< 5000 μm) with quantity 90%. In other words, particles of instant infant

IOP Conf. Series: Materials Science and Engineering **833** (2020) 012049 doi:10.1088/1757-899X/833/1/012049

porridge are found 100% per volume of material with Ø ranged from 5.122 – 5000 μm or in the whole diameter sizes (Ø) are dominated by particles with Ø 124.70222, as shown in Figure 10a. Particle size distribution in pour infant porridge displays particles with Ø in the range of 4.472 – 10.097 μm (< 10.00 μm) with quantity 4.6% or average Ø 9.38 μm with quantity 10% per volume of material, Ø in the 11.565 – 1019.515 μm or average Ø 28.54 μm with quantity small than 100.00 μm with quantity 50%, and Ø ranged from 2009.68 to 5000 μm (< 5000 μm) or average Ø 295.45 μm with quantity 90% or the whole Ø are dominated by Ø of 126.63 μm, as demonstrated in Figure 10b.

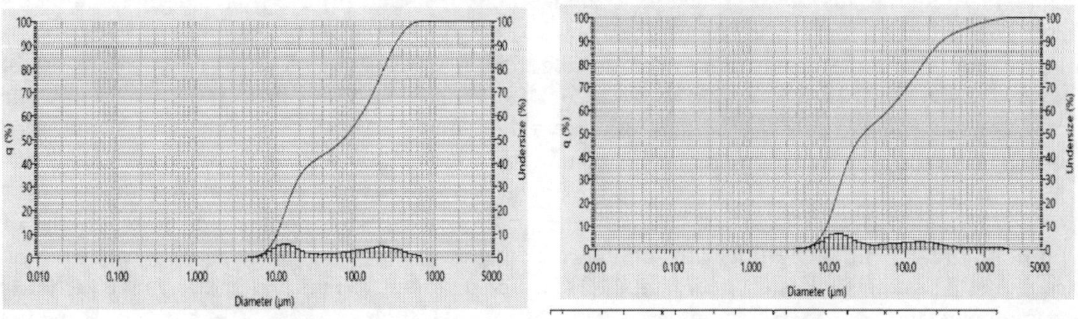

(a) (b)

Figure 10. Distribution of particles size from (a) instant infant porridge and (b) pour infant porridge in optimum condition (fortificant A and mocav flour 18% w/w basic porridge formula).

Table 4 shows particle size and particle size distribution in both instant and pour infant porridges from the optimum formulation condition. Difference in particle size and particle size distribution from both kinds of porridges are related with texture, in which instant infant porridge is dry granule due to result of dry mixing process with average particle size of 126.63 μm. Meanwhile, pour infant porridge is suspension like pasta produced by pouring instant infant porridge by boiled water causing average particle size in pour porridge (126.63 μm) being larger compared with average particle size in instant porridge (124.70 μm) because of its occurrence of gelatinization process. The particle size distribution of instant infant porridge indicates larger particle size compared with each quantity (q) of pour instant porridge, as shown in **Table 4.**

Tabel 4. Particle size and particle size distribution in both instant and pour infant porridges at the optimum formulation condition by adding mocav as carrier of folic acid and thickener.

Kinds of porridge*	Diameter of particle size on cumulative (%)	
	q (%)	Undersize (μm)
Instant infant porridge	10.00	10.27
	50.00	71.80
	90.00	323.82
	Σ	124.70
Pour infant porridge	10.00	9.38
	50.00	28.54
	90.00	295.45
	Σ	126.63

IOP Conf. Series: Materials Science and Engineering **833** (2020) 012049 doi:10.1088/1757-899X/833/1/012049

4. Conclusion

Mocav flour has potential use as food additive functioning as carrier of folic acid and thickener in preparation of instant infant porridge for complementary breast feeding (CBF) using fortificant of natural folic acid. Mocav flour concentration becoming more and more increase will increase composition of instant infant porridge fluctuately to optimization condition achieved. Based folic acid, optimization condition is achieved for fortificant A at mocav concentration 18% (w/w basic porridge formula) yielding instant infant porridge and pour infant porridge with composition of folic acid 689.73 and 445.76 µg/mL, dissolved protein 4.54 and 0.66 mg/mL, total solids 91.64 and 9.91%, total sugars 352.17 and 120.87 mg/mL, and reducing sugars 37.38 and 27.82 mg/mL, respectively. In this optimum mocav concentration, formulation system is able to increase folic acid 532.26% (5.3-folds), dissolved protein 226.62% (2.27-folds), total sugars 120.44% (1.2-folds), however decrease total solids 4.61% and reducing sugars 14.94% in instant infant porridge compared with each components concentration in base formula of infant porridge. In this condition, instant infant porridge gives dominant folic acid monomer with MW 442.4175, whereas pour infant porridge gives 2 folic acid monomers with MW 442.4774 and 442.8006 Da., and relative intensity 100%. Volatile compounds from instant infant porridge and pour infant porridge are dominated by acetic acid compound and its derivative (Thioacetic acid, methyl ester, ethyl ester) with total 44.21% and sugar compounds and its derivative (as Furfural, 2-Furanmethanol, 2 (5H)- Furanone, 2H- Pyran, 4H -Pyran-4-one, Tetrahydro-4H-pyran, 5-Hydroxymethylfurfural) with total 29.812%. Particle distribution in instant infant porridge showing 10, 50 and 90% per volume per weight have particle size of 10.27, 71.80 and 323.82 µm with average 124.70 µm, while particle distribution in pour infant porridge displaying 10, 50 dan 90% per volume per weight have particle size of 9.38, 28.54 and 295.45 with average 126.63 µm.

References

[1] Rahayu, E.S. 2010. Current Research in Indonesia. Skripsi Sarjana. UGM.Yogyakarta.

[2] Departemen Kesehatan Republik Indonesia. 1992., Edisi II, Jilid II. Peraturan Menteri Kesehatan Republik nIndonesia Nomor :722/Menkes/Per/IX/88 Tentang Bahan Tambahan Makanan.

[3] Agustine Susilowati, Aspiyanto Aspiyanto and Yati Maryati. 2018. Proceeding of the 3rd International Seminar on Chemistry (ISOC 2018), AIP Conf. Proceeding. 2049, 030006-1–030006-11; https://doi.org/10.1063/1.5082507 Published by AIP Publishing. 978-0-7354-1775-5/30.00.

[4] De Crécy-Lagard, V., El Yacoubi, B., de la Garza, R. D., Noiriel, A., and Hanson, A. D., 2007, BMC Genomics, 8 245.

[5] Delchier, N., Ringling, C., Maingonnat, J. F., Rychlik, M., Renard, C. M. G. C. 2014., 157: pp.439 – 47. Accessed at 25 April 2017.

[6] Agustine Susilowati, Yati Maryati, Puspa D Lotulung dan Aspiyanto. 2018., Jurnal Applikasi Teknologi Pangan, 7(2) PISSN 2089-7693, e ISSN 24605921.

[7] Baley, L.B. 2009. Folate in Health and Disease. Boca Roton, CRC Press.

[8] McKenzie, S.B. 1999. Textbook of Hematology, Edisi 2. Baltimore, Williams adn Wilkins. (1999) : 179-199.

[9] U.S Food and Drug Administration (FDA). 2018., FDA waebsite https://www. acessdata. fda. gov/ scrips/ fdcc/index.cfm.set. indirectadditives & id=METHYLOLEATE. Accessed 13 May 2018.

[10] Anonymus. 2012a., National Institutes of Health. http://ods.od.nih.gov/factsheets/folate.asp, Diakses 8 December 2014.

[11] AOAC. 2016. Official Methods of Analysis of the Association of Analytical Chemistry, A. O. A. C, Inc., Washington D.C.

[12] Lowry, O. H., Rosebrough, N. J., Farr, A. L., Randall, R. J. 1951. J. Biol. Chem. 193 (1), 265 – 276. PMID 14907713

[13] Hatanaka, C and Kobara, Y. 1980. Agric. Biol. Chem. 44(12), 2943-2949, 1980.

[14] Ruengsitagoon, W. & Hattanat, N. 2012, The 4th Annual Northeast Pharmacy Research Conference, Thailand, 2012

[15] Eichhorn, P. and Knepper, T. P., 2001, Journal of Mass Spectrum, June: 36 (6), 677 – 684. DOI: 10.1002/jms.170.

[16] Backman. 2018. Backman Coulter Life Sciences, © Backman Coulter Inc.,USA.

[17] Hussain, S.Z and Maqbol, K. 2014. GC-MS: INT J CURR SCI, 13, P 116-126.

[18] Agustine Susilowati, Zalinar Udin, Puspa Dewi Lotulung, Yati Maryati, Aspiyanto Aspiyanto, Andreas Andreas, Sri Fahmiati. 2018., Proceedings of the 4th International Symposium on Applied Chemistry 2018 AIP Conf. Proc. 2024, 020008-1–020008-11; https://doi.org/10.1063/1.5064294, Published by AIP Publishing. 978-0-7354-1744-1/$30.00.

[19] Belitz & Grosch. 2002.. 2nd edition Springer-Verlag Berlin Heidelberg. ISBN 3-540-15043-9, pp 379-393.

[20] Quilivan, E.P., S. Roje, G. Basset, Y. Shahar-Hill, J.F. Gregory, and A.D. Hanson. 2003.. J. Bio. Chem. 278:20731-20737.

[21] Agustine Susilowati, Aspiyanto, Yati M, Hakiki M and Puspa Dewi L. 2017. Proceeding of the 3 rd International Symposium on Applied Chemistry 2017 (ISAC 2017), Volume 1904, Jakarta 23-25 October 2017, AIP Publishing, Melville, New York, 2017. ISBN 978-0-7354-1471-6 &ISSN 0094-243X. https://doi.org/10.1063/1.5011882.

[22] Susilowati A, Lotulung P D, Maryati Y and Aspiyanto 2018 Modification process in nixtamalization of folic acid-rich dent corn (Zea mays identata) and its identification as smart food fortificant MATEC Web of Conferences 154 01017

[23] Rossi M., Amaretti A and Raimondi S, 2011, Nutrients 3 (1) 118 – 134.

IOP Conf. Series: Materials Science and Engineering **833** (2020) 012050 doi:10.1088/1757-899X/833/1/012050

Anti-inflammatory Activity of Neuropathic Pain Reducing Herbal Medicine Based on Edema Inhibition of CARR-induced Sprague Dawley Paws

Dewi Tristantini*, Aisyah Hanifah and Raiska Bani Pramadhanya

Department of Chemical Engineering, Faculty of Engineering, Universitas Indonesia, Depok 16424, Indonesia

Corresponding author: detris@che.ui.ac.id

Abstract. Neuropathic pain is nerve pain whose treatment still has side effects when used for the long term. Regulation and modulation of the immune system is needed to reduce the inflammatory reaction that triggers of the pain. Increasing neural tension may induced with nerve inflammation which producing compounds responsible for signalling pain, fever and heat. Ginger (*Zingiber officinale* Roscoe), cloves (*Syzygium aromaticum* L.) and nutmeg (*Myristica fragrans* Houtt) have been known to contain phenolic compounds that are efficacious as anti-inflammatory agents. This study aims to prove the anti-inflammatory activity of neuropathic pain reducing herbs consisting of a combination of the three ingredients. Result from herbs formulation and reflux extraction then analysed for anti-inflammatory activity using in vivo & Winter method with Sprague-Dawley white male rats (Rattus norvegicus) and quantitative phenol content. Anti-inflammatory analysis done by 6 groups of rats, which is normal control, negative control, positive control, dosage I (given herbal extract 1.125 mL/200 g BW), dosage II (given herbal extract 2.25 mL/200 g BW) and dosage III (given herbal extract 4,5 mL/200 g BW). Anti-inflammatory of 10 g herbs extract in 250 mL solvent then observed through the inhibition of oedema formation on rats' paw along the time. Dosage III group has the higher inhibition percentage as in 75.76%, which almost equivalent as normal control group (82.28%).

1. Introduction

Neuropathic pain was affected 7–10 % of the general population. Neuropathic pain usually divided according to the cause of nerve injury. This illness mostly caused either by nerve system structure located in the vertebral area or genetic Neuropathic pain is a condition in which part of the nerve system became tense mostly caused by over sensitive stimulants which emitted pain sensation. Neural tension pain sufferers generally feel electrocuted and stabbing pain sensation causing depression, anxiety, until sleep disturbances [1] [2].

Stimulants such as infection, pressure, and viruses are damaging the nerve which can stimulate inflammatory response by immune cells [3]. Neuropathic pain is usually caused as an effect of other diseases such as cancer, diabetes mellitus, spinal degenerative, and diseases caused by viral infections [4]. The inflammatory response involves several pathways such as prostaglandin and interleukin synthesis, protein receptor adhesive response, and platelet-activation factors. The mentioned stimulants will activate the phospholipids hydrolysis process of membranes into arachidonic acid. The arachidonic acid then will be synthesized by COX enzymes into prostaglandins PGE_2, PGH_2. The presence of these nociceptor stimulation can cause excruciating pain [5][6].

Until lately, efforts to deal with neuropathic pain that are commonly carried out are the administration of antidepressant drugs such as amitriptyline in the first line and anticonvulsant drugs such as gabapentin and carbamazepine as a second line [7]. Both types of pharmaceutical drugs can control several types of seizures also treat nerve pain but have some side effects, especially for the long-term use. Surgery can also give permanent medication. However, surgery requires a large amount of risk and has the possibility that the pain will reappear in a certain time.

Empirically, stated herbal medicine cured neuropathic pain which consisted of ginger (*Zingiber officinale* Roscoe), cloves (*Syzygium aromaticum* L.), and nutmeg (*Myristica fragrans* Houtt) were practically used. These herbal ingredients have known to contain beneficial phenolic compounds from previous scientific journals. Phenolic compounds are plants secondary metabolism which are beneficial for human. The phenolic compounds in Ginger, Nutmeg, and clove, specifically quercetin can conduct an anti-inflammatory response by inhibiting COX enzymes, TNF-α and IL-1β which will reduce the inflammation [8].

Therefore, this study had the aim to investigate the anti-inflammatory activities of neuropathic pain reducing herbal medicine consist of multi herbs aqueous extract from nutmeg, clove, and ginger. The selection of plants considered in empirical traditional uses.

2. Materials and Method

2.1. Herbal Medicine Extraction

Herbal samples for the whole research such red ginger (*Zingiber officinale var rubrum*) rhizomes procured from Depok Indonesia traditional market, dried clove buds (*Syzygium aromaticum L.*) and nutmeg seed (*Myristica fragrans Houtt*) were obtained from local vendors in Maluku Indonesia. All herbs materials were fully washed and dried at room temperature, and then Clove and Nutmeg continues dried in an oven at a temperature of 50 °C for 15 minutes. The ginger rhizome already cleaned was peeled and sliced, while the clove and nutmegs crushed to a coarse powder using a ceramic mortar and pestle. The prepared 10 grams multi herbs (Clove, Nutmeg and Ginger) were extracted using a reflux apparatus with 250 mL distilled water at 80 °C for 90 minutes. Extracts were filtered using a strainer followed by five times used Axiva qualitative filter paper Grade-1. Liquid extract of the herbs was stored in an airtight glass bottles and stored at temperature 5 °C.

2.2. Experimental Animals

36 *Sprague-Dawley* white male rats (±200 g) were acclimatized for 7 days at QLab, Universitas Pancasila, Jakarta, Indonesia with 12 h light and dark cycle. All animals were provided with pellet diet and water *ad libitum*.

2.3. Animal Treatment

Animals were divided into six groups, each consisting six males. Group I was the normal control group, administered with water orally without any treatment; Group II was positive control group which orally administered with Diclofenac sodium 2%; Group III was negative control group which orally administered with Na CMC 0.5%; Group IV, V and VI were orally administered herbal extract at 1.125, 2.25, 4.5 mL/200 g b.wt.

2.4. Edema Weight Determination

After treated, the animals were injected by carrageenan (CARR) 1% as inflammation induction on the paw except for Group I. The animal's paws were weighed with plastometer 1 hour before the injection as initial normal weight. The animals were then weighed in 1-hour interval for 5 hours observation. Injection period were observed as hour 0, and the volume differences were gradually taken.

IOP Conf. Series: Materials Science and Engineering **833** (2020) 012050 doi:10.1088/1757-899X/833/1/012050

2.5. Edema Inhibition Determination

The edema weights that were recorded within the observation period were counted for the edema weight reduction within time which represent the inflammation inhibition. The paws volume differences were calculated to the initial paws weight, and the area under curve were then taken to indicates the edema inhibition percentage of which the herbal medicine took responsibility.

3. Results and Discussion

The edema weight throughout the 5-hour observation period of CARR-induced rats' paw can be used as an indicator of anti-inflammatory activity because of the weight differences overtime indicates the inflammatory inhibition. If the CARR-induced edema weight reduced over time, it means that the inflammation was inhibited by the oral medicine treatment. The greater the weight reduction, the better the anti-inflammatory activity of medicine took place.

Herbal medicine showed significant edema weight reduction over time. Shown in Figure 1 and Table 1 where they can reduce the edema back into initial normal weight.

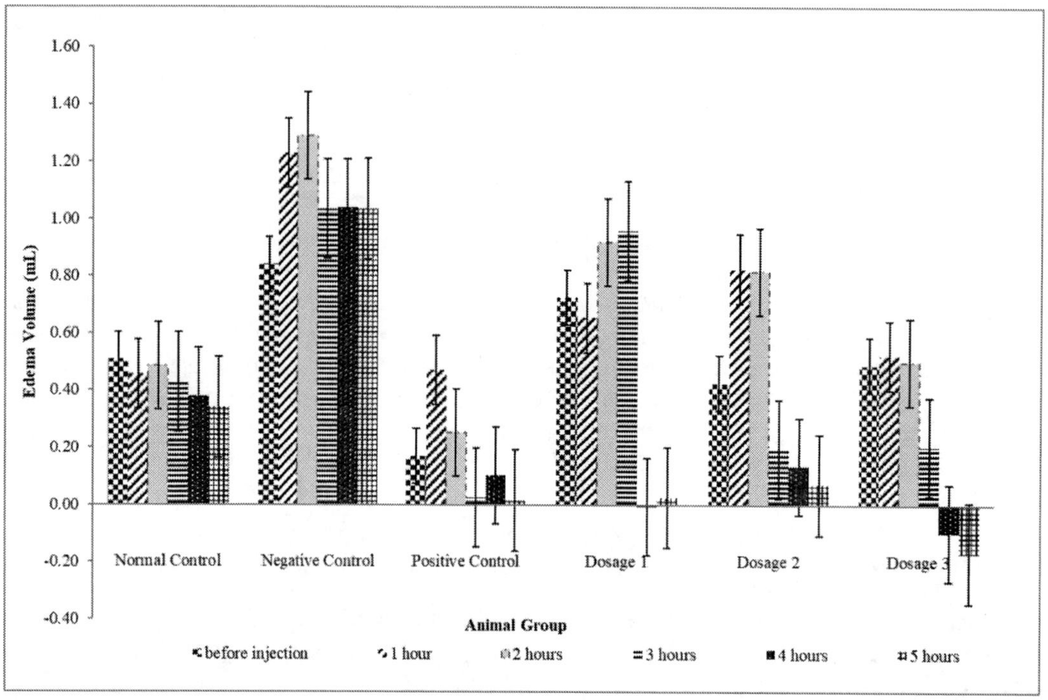

Figure 1. Effect of 5 hours CARR-induced rats' paws after administration of herbal medicine on edema weight.

Table 1. Edema volume of animal group within the observation time.

Animal Group	Edema Volume (mL)					
	Time (hour)					
	0	1	2	3	4	5
Normal	0.51	0.46	0.48	0.43	0.38	0.34
Negative	0.84	1.23	1.29	1.03	1.04	1.03
Positive	0.17	0.47	0.25	0.03	0.10	0.01
Dosage I	0.73	0.65	0.92	0.96	-0.01	0.03
Dosage II	0.43	0.83	0.81	0.19	0.13	0.07
Dosage III	0.49	0.52	0.50	0.20	-0.10	-0.17

The test results showed a decrease on edema weight in the group that was given the diclofenac sodium or herbal medicine dosage I, II and III. Decreased edema weight indicates inhibition of inflammation by the medication. This is occurred because of the medicine that has anti-inflammatory activity inhibits COX enzyme to synthesize phospholipids into pain-stimulating prostaglandin, therefore the inflammation response stopped, indicating with decreased edema weight. Phenolic acid in ginger (*Zingiber officinale* Roscoe), dried clove buds (*Syzygium aromaticum* L.) and nutmeg (*Myristica fragrans* Houtt) such as gingerol, shogaol, myristicin, eugenol, are known to clinically have the ability of anti-inflammatory response. Gingerol, shogaol, and other structurally related substances in ginger inhibit prostaglandin and leukotriene biosynthesis through suppression of 5-lipoxygenase or prostaglandin synthetase. Additionally, they can also inhibit synthesis of pro-inflammatory cytokines such as IL-1, TNF-α, and IL-8 [9]. The herbal medicine worked in synergic within its constituent ingredient to put anti-inflammatory activity to inhibit the prostaglandin synthesis pathway directly.

Based on test results, herbal medicine-administered rats showed anti-inflammatory activity more gradually compared to positive group in which gave immediate response, but as effective shown by the edema weight reduction over time. This happened because the herbal medicine needs longer time for it to work effectively after orally administered than pharmaceutical medicine. Dosage I, Dosage II and Dosage III were generally worked effectively 3 hours after CARR 1% induction, while the positive control group (Diclofenac sodium 2%) were able to suppress the inflammation immediately after the CARR induction.

The area under curves (Figure 2) for each animal group were calculated to determine the inflammation inhibition percentage (Table 2). The greater the area under curve, the least of inflammation inhibition took place. All of the area under curve were then deducted to the negative control group, to determine the effectiveness of medicines inflammation inhibition.

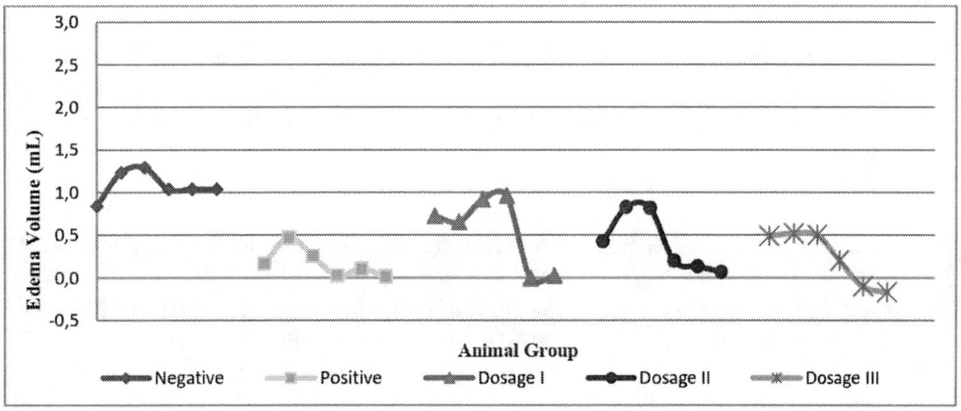

Figure 2. Area under curve of edema weight after inflammation injection.

Table 2. Edema inhibition of animal group within the observation time.

Animal Group	Edema Inhibition (%)							
	Rats Number						Avg	Rank
	1	2	3	4	5	6		
Normal	-	-	-	-	-	-	-	-
Negative	-	-	-	-	-	-	-	-
Positive	89.62	87.74	70.75	71.70	75.16	98.43	82.23	1
Dosage I	56.92	35.85	32.39	50.94	51.89	43.71	45.28	4
Dosage II	53.46	66.67	66.67	44.34	61.32	56.92	58.23	3
Dosage III	60.38	67.14	79.87	82.70	71.70	92.77	75.76	2

According to Table 2, dosage III (4.5 mL/200 g b.wt. of herbal medicine) showed highest anti-inflammatory activity (75.76%) compared to other herbal medicine dosages. Anti-inflammatory activity of dosage III gave parity result towards positive control group (Diclofenac sodium 2%) which showed 82.23% of inflammation inhibition. The highest anti-inflammatory activity was from positive control group (82.23%), followed by dosage III (75.76%), dosage II (58.23%) and dosage I as the lowest (45.28%).

Active compounds were identified in Neuropathic Pain Reducer extract by using UPLC-QTOF-MS show very complicated chemical system containing many types of compounds that has pharmacology effect analgesics, anti-inflammatory, anti-depressants and anti-convulsions that used as treatments for neuropathic pain. [1]. This shows that phenolic compounds contained in the constituent herbal ingredients can stimulate the release of anti-inflammatory response that will reduce pain. That means, the herbal medicine can be an alternative medication to reduce neural tension pain caused by inflammation as well as pharmaceutical drugs.

4. Conclusion

Neural tension reducing herbal medicine proven to have anti-inflammatory activity based on the CARR-induced rats paws edema inhibition ability with 45.28% inhibition on dosage I (1.125 mL/200 g), 58.23% on dosage II (2.25 mL/200 g) and 75.76% on dosage III (4.5 mL/200 g) as the highest. Dosage III (4.5 mL/200 g) gave parity results with positive control group (Diclofenac sodium 2%) by 13.52% difference. Although herbal medicine needs longer time to work effectively, but it was proven to be able to substitute pharmaceutical anti-inflammatory drugs as neural tension reducing herbal medicine.

References

[1] Hanifah A and Tristantini D 2019 *AIP Conference Proceedings* **2193** 030010
[2] Colloca L, Ludman T, Bouhassira D, Baron R, Dickenson A H, Yarnitsky D, Yarnitsky D, Freeman R, Truini A, Attal N, Finnerup, Eccleston C, Kalso E, Bennett D L, Dworkin R H and Raja S N 2017 *Nat Rev Dis Primers* **3** 17002
[3] Mani S 2017 *International Journal of Phytopharmacy Research* **7** 1 25
[4] Schmader K E 2002 *Clin J Pain* **18** 6 350-354
[5] Ricciotti E and FitzGerald G A 2011 *Arterioscler Thrombosis and Vascular Biology* **31** 5 986–1000
[6] Dubin A E and Patapoutian A 2010 *The Journal of clinical investigation* **120** 119 3760-3772
[7] O'Connor, A.B., and R.H. Dworkin 2009 *The American journal of medicine.* **122** 10 S22-S32.
[8] Yahfoufi N, Alsadi N, Jambi M, and Matar C 2018 *Nutrients* **10** 11 1618
[9] Mashhadi N S 2013 *International journal of preventive medicine*. **4** 1 S36-S42

Acknowledgments

The authors would like to thank Direktorat Jenderal Penguatan Riset dan Pengembangan, Kementerian Riset, Teknologi dan Pendidikan Tinggi Republik Indonesia for providing necessary financial support in carrying out this study by Penelitian Tesis Magister grant 2019 based on Decree Number 6/E/KPT/2019 with grant contract number NKB-1898/UN2.R3.1/HKP.05.00/2019.

Solvent Effect at Ibuprofen Adsorption Using Zinc Oxide Plate Rod-Like from Gelatine

Maria Ulfa*[1], Muh Ari Purnama Ali[2]

Dept. of Chemistry Education, Faculty of Teacher Training and Education Sebelas Maret University, Jl. Ir. Sutami 36A, Surakarta 57126, Central Java Indonesia

*Corresponding author: ulfa.maria2015@gmail.com

Abstract. The ubiquitous occurrence of several pharmaceuticals in sewage effluents has led to the considerable deterioration of some life forms and the quality of the receiving water bodies. Ibuprofen is a nonsteroid anti-inflammatory drug, represents a diverse class of drugs andthe most commonly used analgesics for the management of pain and/or inflammation associated with rheumatoid arthritis and osteoarthritis, muscle stiffness and pain, dental pain, migraine, and headache. The present work investigated the effect of different solvents on ibuprofen dissolution during the adsorption of Ibuprofen on to zinc oxide plate rod-like from its hexane, methanol, ethanol solutions, and water as co-solvents. The zinc oxide plate rod-like was synthesized using block copolymer and gelatine as a template and zinc sulphate as zinc precursor was then characterized by X-ray diffraction, scanning electron microscopy, and FTIR to analyse the structure and morphology. The impact of various solvents on the percentage removal (%) of Ibuprofen was determined by batch adsorption experiments. The data obtained were subjected to isotherm and kinetic analysis to describe the distribution of ibuprofen between the liquid and solid phases in the batch studies. The results obtained best fitted the Langmuir isotherm model with an adsorption capacity close to 110mg/g at room temperature with an initial concentration of 100 ppm. To sum up, for large-scale removal, ibuprofen treatment for aquatic, zinc oxide plate rod-like maybe an adsorbent in future.

1. Introduction

At this time, there has been a lot of environmental pollution detected, especially in waters, for example in gutters, surfaces, and groundwater. Non-steroidal anti-inflammatory drugs (NSAIDs), such as ibuprofen, is pollutants that are harmful to the environment. Medication in drinking water can cause severe damage [1]. The presence of drugs in drinking water has adverse effects on human health and the ecosystem. Now, ongoing research to find more advanced ways to eliminate these contaminants is exploring submerged membrane bioreactors, coagulation and flocculation, activated sludge treatment, wet land construction, photocatalytic oxidation, catalytic ozonation, adsorption, advanced oxidation process, nanofiltration, biological processes, reverse osmosis, ultra-filtration, and etc [2]. However, this method still cannot be used effectively and economically to remove drug pollutants from water.

To get rid of these pollutants, effective treatment is needed. From this method, adsorption will be used because this method is the most promising. This process has been applied to eliminate organic compounds in drinking water treatment[3]. Adsorption method has several advantages including; Adsorption can handle various levels of pollutants, is efficient, easy to operate, is not affected by toxins, and is suitable for sustainable processes; The adsorbent can be regenerated and used repeatedly and has a low initial cost[4]. Heterogeneous photocatalysis with activated semiconductors, such as zinc oxide (ZnO), is very possible for the process of decomposition of hazardous substances (pollutants) [5]. By using templates, such as zinc oxide, it is hoped that the adsorption process of ibuprofencan run optimally.

For solubility of ibuprofen, three variations of solvent will be used, namely n-hexane, methanol, and ethanol. From its nature, each one is different. N-hexane is the lightest solvent in lifting the oil contained in grains and is volatile, making it easier to reflux. This solvent has a boiling point between 65 and 70 °C and a dielectric constant of 1.89D [6]. Ethanol often used as a solvent in the laboratory because it has relatively high solubility, inert and does not react with other components. Ethanol has a low boiling point, making it easier to separate oil from its solvent in the distillation process. It has a boiling point of 79°C, a density of 0.789g/mL, and a dielectric constant of 24.30D [7]. While methanol is the most widely used solvent in the process of isolating the organic compounds of natural material, it has a boiling point of 65°C, a density of 0.791g/mL, and a dielectric constant of 33.60D [8]. When viewed from polarity, methanol is the most polar solvent compared with the other two solvents, so that the potential to obtain the best results from methanol solvents cannot be denied [9].

In this study, the formation of zinc oxide plate rod-like was investigated using Pluronic F127 with gelatin as aco-template through the hydrothermal method. Pluronic F127 is a soft organic template with a relatively high dissolution rate and low mechanical strength [10]. For the solubility of ibuprofen in the absorption process, three variations of solvents n-hexane, methanol, and ethanol will be used using the ratio between water and solvent, which is 100%water; 1:9; 5:5; 9:1; and 100% solvent. To know the effect of the variation and concentration of the solvent on the absorption of ibuprofen, the zinc oxide material was characterized using the X-ray diffraction (XRD), Scanning Electron Microscopy (SEM), and FTIR methods. For the large-scale removal of ibuprofen treatment from the aquatic body, zinc oxide plate rod-like may be an adsorbent in future.

2. Materials and Method

2.1. Synthesis of Zinc Oxide Plate Rod-Like

EtOH was dropped into the F127 solution using a burette (1 drop of EtOH per 20 seconds) at room temperature. after all, ethanol has been dropped, the mixture is stirred for 4 hours 150 rpm using closed conditions (crepe plastic). During dropping, no air enters by making a plastic crepe hole and covering the mouth with a mask. Gelatine is added then stirred for 60 minutes.

$ZnSO_4$ / Zn $(NO_3)_2$ drops in to the F127 with a burette (1 drop per 10 minutes) at room temperature. After everything was dropped, the mixture was stirred for 20 hours 150 rpm closed condition (crepe plastic) during the dripping process, no air enters by making holes in the crepe plastic and covering the mouth with the mixture put into the hydrothermal reactor then oven at 100 ^{0}C for 24 hours then cool, the white solid was filtered and washed with distilled water so that the pH = 7.The solid was dried at a temperature of 100 ^{0}C for 24 hours, and calcined at 550 ^{0}C for 12 hours, then the solid was stored in a clear bottle.

2.2. Characterization of Zinc Oxide Plate Rod-Like

The resulting sample will then be characterized using a Shimadzu 2100 brand Fourier Transform Infrared (FTIR) spectrophotometer resolution 0.5cm^{-1}, Scanning Electron Microscopy (SEM) ZEIS EVO MA 10, and X-Ray diffraction (XRD). This characterization aims to determine the size and morphology of the particles. Then the sample was applied in Ibuprofen adsorption.

2.3. Adsorption of ibuprofen onZinc Oxide Plate Rod-Like

To find out the adsorption on a nanomaterial, ibuprofen solution was prepared in a solution of Water: Hexane in a ratio of 10: 0; 9: 1; 5: 5; 1: 9; 0:10 using 100 ppm concentration. The concentration of ibuprofen was monitored with a UV-Vis spectrophotometer at a wavelength of 291.5 nm. Calibration was done by preparing a standard solution of ibuprofen with a concentration of 5 to 100 ppm. Sixty mg of nano-ZnO was added to the ibuprofen solution and sterilized at room temperature for 60 minutes. A total of 3 mL of ibuprofen solution was taken at 5-minute intervals and then absorbance was measured using a Hitachi Japan UV-Vis U-2000 spectrophotometer. Then to determine the effect of the use of the solvent, the absorbent above process was carried out again using two other solvents, namely by replacing hexane to methanol and ethanol with the same concentration variations. The amount of ibuprofen absorbed by nano-ZnO is calculated using the equation,

$$qe = (Co - Ce)\left(\frac{v}{w}\right). \tag{1}$$

Co is the initial concentration of ibuprofen, *Ce* is the equilibrium concentration of ibuprofen, v is the volume of the solution tested for its absorption, and w is the amount of zinc oxide plate rod-like used for adsorption.

3. Results and Discussion

3.1. Characterization of Zinc Oxide Plate Rod-Like

Using blocks copolymer and gelatin in the co-templating method, produced Zink oxide plate rod-like, and its morphology can be seen using SEM. Figure 1a shows the morphology of Zink oxide plate rod-like produced using block copolymer and gelatin at 2500x magnification. The zinc oxide plate rod-like material consisted of non-uniform structures, and the percentage of zinc obtained from EDX analysisis 76.6%. Following the introduction of gelatin at 1:0.05 of the block copolymer to gelatin weight ratio in the synthesis mixtures, the SEM analysis in Figure 1b shows zinc oxide plate rod-like at 10000x magnification. The SEM analysis in Figure 1b shows zinc oxide appeared in plate shapes with a diameter in the range of 1.2 ± 0.8μm. The transformation between non-uniform structures to zinc oxide microspheres following the introduction of 0.05g of gelatin indicates the significant role of gelatin in directing the formation of zinc oxide structure. The SEM analysis proved that the presence of gelatin in the reaction mixture plays significant roles in directing the formation of microspherical zinc oxide and is also responsible for controlling the diameter of the zinc oxide plate rod-like.

(a) (b)

Figure 1. (a) Characterization results using SEM at 2500 × magnification, (b) Characterization results using SEM at 10000 × magnification

The XRD pattern on zinc oxide plate rod-like synthesized shown in Figure 2, the diffraction peaks of zinc oxide plate rod-like appeared at 2θ = 27° that corresponded to the diffraction of (001) plane. The (001) plane is significantly shifted to a higher diffraction angle when gelatin was used as co-template, suggesting lattice contraction that occurs due to the transformation from non-uniform zinc oxide to the microspherical structure. When plate rod-like of zinc oxides were produced at a low concentration of gelatin, high order geometric zinc oxide was formed based on a relatively small full width at a half-

IOP Conf. Series: Materials Science and Engineering **833** (2020) 012051 doi:10.1088/1757-899X/833/1/012051

maximum intensity of 001 plane. Changes in the concentration of gelatin also helped change the level of graphitization of zinc oxide as indicated by the decrease in intensity and FWHM from the peak, which shows the effect of gelatin in controlling the physical and chemical structure of zinc oxide plate rod-like.

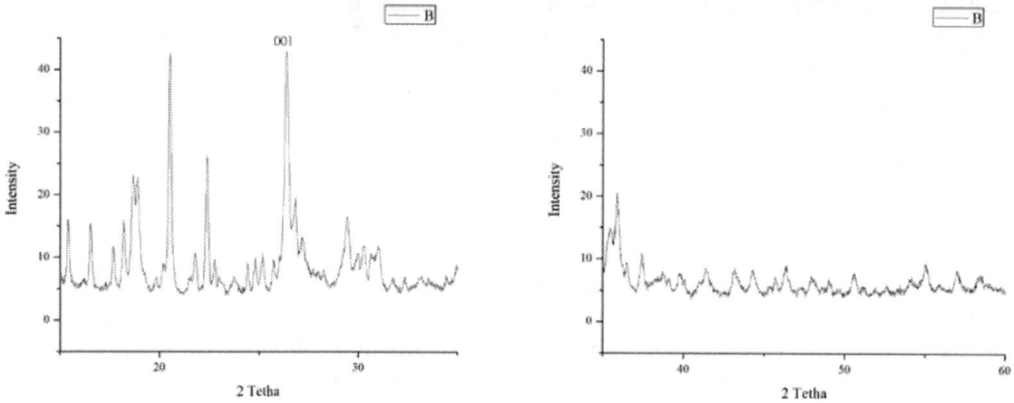

Figure 2. Characterization results of Zinc Oxide Plate Rod-Like with XRD at 2 Tetha, 10-35° (left) and 35-60° (right).

The synthesized plate rod-like of zinc oxide were also characterized using infrared analysis. At Figure 3, the broad peak centred at 3429.58 cm^{-1}, which is corresponded to the stretching vibration of the hydroxyl group, were observed in all nanomaterials. Analysis of plate rod-like of zinc oxide synthesized using gelatin as co-template shows that the intensity of the hydroxyl functional group is significantly increased with the concentration of gelatin. The presence of hydroxyl on the surface reflects the number of terminals bonded to an oxygen atom, due to the formation of surface defects. The increase of surface hydroxyl species in nano plate rod-like of zinc oxide when increasing the amount of gelatin indicates the contribution of gelatin on the surface defects and also enhanced the functionality of zinc oxide. The residual functional groups from gelatin decomposition occurred on the surface of the nanoparticle, which generally occurred on materials produced via hydrothermal reaction.

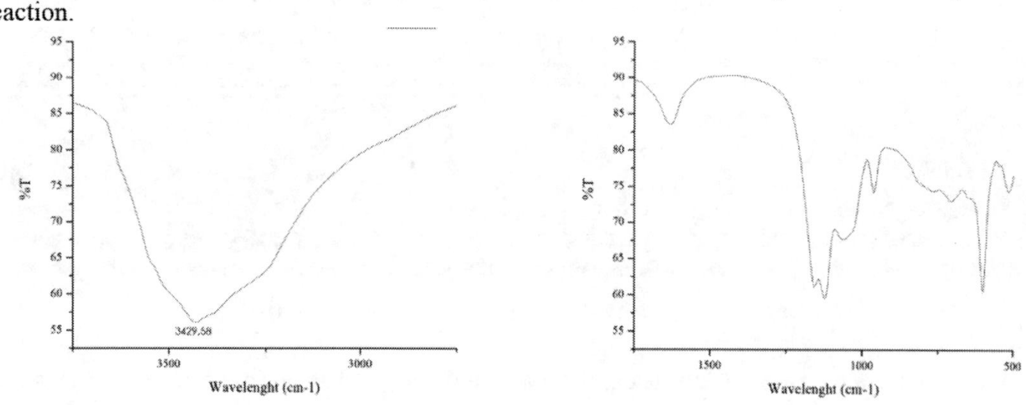

Figure 3. Characterization results of Zinc Oxide Plate Rod-Like with FTIR at wavenumber (left) 3600-2500 cm-^{1} and (right) 1200-500 cm^{-1}

3.2. Adsorption studies of ibuprofen on Zinc Oxide Plate Rod-Like

The potential use of plate rod- like of zinc oxide as an adsorbent for drug removal was investigated using ibuprofen as probe molecule with three solvent variations: methanol, ethanol, and n-hexane.

Figure 4 shows the absorption profile of ibuprofen on zinc oxide plate rod-like obtained from the variation of solvent (water:ethanol). High absorption of ibuprofen was observed on zinc oxide plate rod-like produced from water 100%. Figure 5 shows the adsorption profile of ibuprofen on zinc oxide plate rod-like obtained from the variation of solvent (water:methanol). High adsorption of ibuprofen was observed on zinc oxide plate rod-like also produced 100% from water. Figure 6 shows the absorption profile of ibuprofen on zinc oxide plate rod-like obtained from the variation of solvent (water:hexane). High absorption of ibuprofen was observed on zinc oxide plate rod-like produced from ratio water: n-hexane at 5:5. The absorption reached equilibrium at 330 min. Although infrared analysis revealed the high intensity of functional groups on the z zinc oxide plate rod-like when increasing the concentration of gelatin, there is no significant enhancement on the absorption of ibuprofen. Meanwhile, the adsorption of ibuprofen is significantly influenced by the surface area and crystallite size of the zinc oxide plate rod-like. Kinetic analysis of ibuprofen absorption was carried out using linear regressionanalysis. Linear regressionanalysis is a quick and simple method applicable to abroad array of kinetic absorption systems. The best fit model is determined based on the r^2 and the obtained equilibrium value is compared with the experimental value in which the pseudo-second-order kinetic model is well fitted with the absorption of ibuprofen on plate rod-like of zinc oxide. Variations of solvents in the adsorption of Ibuprofen with zinc oxide plate rod-like uses several variations of solvents (methanol, ethanol, and n-hexane), which are shown in Figure 7.

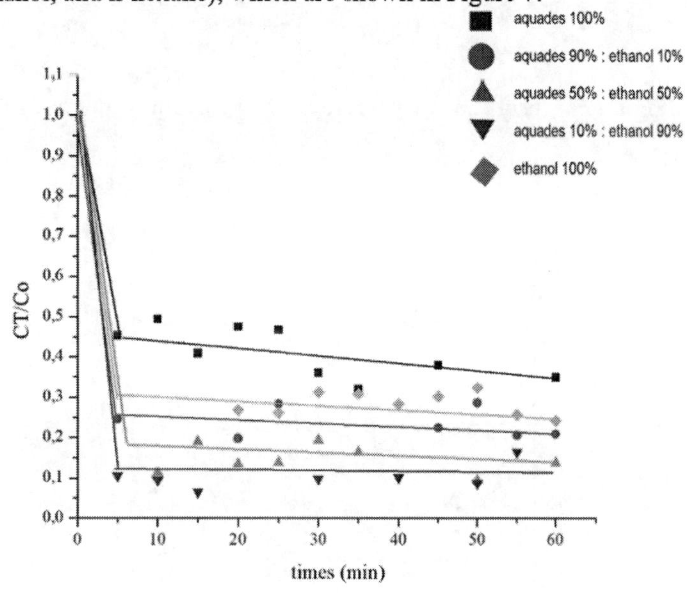

Figure 4. Absorption of ibuprofen using solvent: water and ethanol

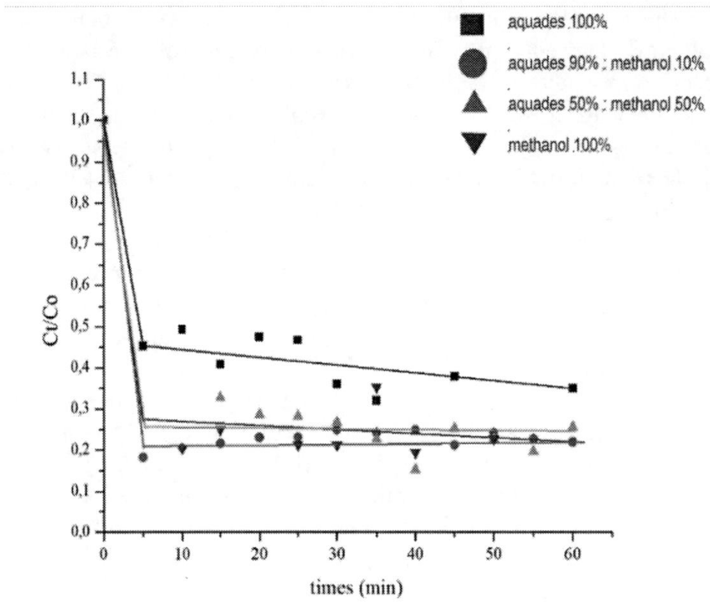

Figure 5. Absorption of ibuprofen using solvent: water and methanol

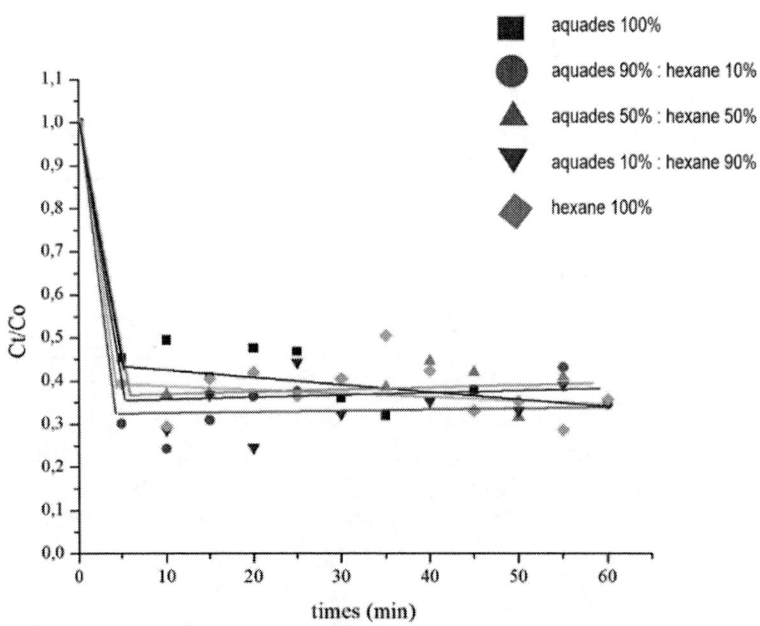

Figure 6. Absorption of ibuprofen using solvent: water and hexane

Figure 7 takes an example of the variation of solvents at ratio water:ethanol, methanol, and n-hexane of 1:9. Absorption high of ibuprofen was observed on zinc oxide plate rod-like resulting from water and hexane 9:1 ratio.

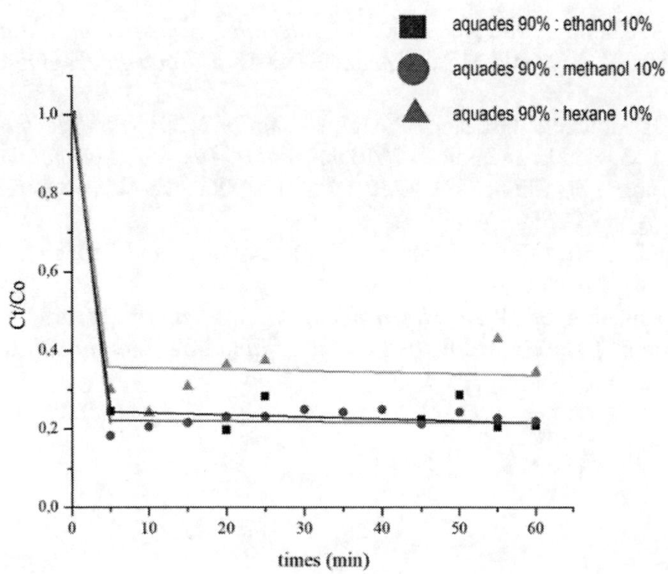

Figure 7. Ibuprofen absorption using solvent:methanol, ethanol and hexane ratio with water, 1: 9

In the curve, it can be seen that the methanol solvent is better than the two other solvents because it has the lowest Co/Ct, which means it has good absorption. Another reason is that methanol is also the most polar solvent compared to the other two, where the polariton methanol value is 33.60D, ethanol is 24.30D, and n-hexane is 1.89D. The character of the ibuprofen molecule as an organic molecule tends to dissolve easily in very polar solvents, such as methanol, because the interaction between polar-polar groups helps homogenization during absorption. The more polar the solvent, the faster it will dissolve the compound because ibuprofen can dissolve very easily in a solvent such as methanol. Not only facilitate dissolving, but methanol with short-chain clusters-OH will reduce the ibuprofen space barrier when interacting with the pore Zn plate to produce the highest absorption performance compared to the other two solvents.

4. Conclusion

Zinc oxide plate rod-like has been successfully synthesized using the hydrothermal method. The use of stem plates, such as zinc oxide, as an adsorbent in the absorption of ibuprofen is very dependent on the surface area and shape of ZnO. Effective absorption of ibuprofen is done using a variety of water:methanol and reaching equilibrium at 330 minutes. Methanol is selected as a better solvent compared to two other solvents, n-hexane and ethanol, because it has the lowest Co/Ct, which means it has good absorption. Another reason is that methanol is also the most polar solvent compared to the other two; methanol is more polar (33.60D) compared to ethanol (24.30D) and n-hexane (1.89D) because of the more polar the solvent, the faster it will dissolve compounds because ibuprofen can dissolve very easily in solvents such as methanol.

Acknowledgement

Authors acknowledge the Research Collaboration Indonesian Program (PPPKI) of SebelasMaret University2019 under contract Number 1362/UN27.21/PN/2019 for Maria Ulfa.

References

[1] Li Z A, Gomez L, Sellaoui J, Bedia A, Bonilla C, and Belver 2019 *J.Chemical Engineering* **371** 868-875
[2] Huang T C, Yin H E, Chiu W Y, and Lee C F 2014 *J. Thin Solid Films.* **552** 98-104

IOP Conf. Series: Materials Science and Engineering **833** (2020) 012051 doi:10.1088/1757-899X/833/1/012051

[3] Jia Z, Yue L, Zheng Y, and Xu Z 2008 *J. Materials Chemistry and Physics.* **107(1)** 137–141
[4] Ejaz M, Arfat Y A, Mulla M, and Ahmed J 2018 *J. Food Packaging and Shelf Life.* **15** 113–121
[5] Rauf M A, Hisaindee S, Graham J P, and Al-ZamLy A 2015 *J. Molecular Liquids* **211** 332–337
[6] Wang L, Liu G, Zou L, and Xue D 2010 *Journal of Alloys and Compounds* **493** 471–475
[7] Xu Y M, Tang Y P, Chung T S, Weber M, and Maletzko C 2018 *Separation and Purification Technology* **193** 165–174
[8] Yousefinejad S, Eftekhari R, Honarasa F, Zamanian Z, and Sedaghati F 2017 *J. of Molecular Liquids* **241** 861–869
[9] Sankar M, and Bhyrappa P 2019*J. Chemical Physics Letters* **132** 145-151
[10] Ramachandran T, and Hamed F 2018 *J. Physics and Chemistry of Solids* **121** 93–101

IOP Conf. Series: Materials Science and Engineering **833** (2020) 012052 doi:10.1088/1757-899X/833/1/012052

Computational study of Cu^{2+}, Fe^{2+}, Fe^{3+}, Mn^{2+} and Mn^{3+} binding sites identification on HSA 4K2C

Syahputra Wibowo*[1], Sutiman B Sumitro[1], Sri Widyarti[1]

[1]Biology Dept., Faculty of Mathematics and Natural Sciences, University of Brawijaya, Malang, Indonesia

*Corresponding authors: wibowo@student.ub.ac.id

Abstract. This research aims to computationally characterize HSA 4K2C protein and describe as well as its ability to bind transition metal ions. Data mining is performed to obtain HSA 4K2C from PDB and transition metal ions such as Cu^{2+} (ID: 27099), Fe^{2+} (ID: 27284), Mn^{2+} (ID: 27854), Mn^{3+} (ID: 105130) and Fe^{3+} (ID: 29936) from PubChem. The analysis consists of ProtParam, Motif Search, CFSSP, DLP-SVM, and docking. Docking used PyRx Autodock Vina. Analysis of receptor-ligand interactions used DS 2016. The results of the ProtParam analysis provide some information on HAS 4K2C, which is it has 585 amino acids with an isoelectric point (pI) of 5.67, an index of protein instability of 38.85, then total amino acids (aa) residues of negatively charged (Asp + Glu) are 98 while the positively charged ones (Arg + Lys) are 83. Motive Search shows that there are three HSA motifs namely motif 1 (aa. 551-575), motif 2 (aa. 353-377), and motif 3 (aa. 61-185). CFSP shows α-helix structure is the dominant structure compared to β-sheet, turn and coil in 4K2C. DLP-SVM shows two domain linkers where DL-1 (aa. 410-451) and DL-2 (aa. 96-122). Docking shows the ability of HSA 4K2C in binding metal ions such as Cu^{2+}, Fe^{2+}, Mn^{2+}, Mn^{3+} and Fe^{3+}.

1. Introduction

Scavenger can be found both endogenous and exogenous [1]. An example of endogenous scavenger is albumin (human serum albumin) which has the largest antioxidant capacity in human serum working well by sequestering transition metal ions or directly by binding and carrying scavenger [2]. Serum albumin is a very abundant protein in human blood [3]. Albumin has a molecular weight of 66.5 kDa which is a single chain with 585 amino acids. Albumin consists of three homologous structures with 67% of helix, each domain (I, II, and III) containing two subdomains i.e. A and B [4]. The ability of albumin as a scavenger is the ability of metal ions binding, which is known before that heme has pro-oxidant properties and albumin is an effective protein in the binding of heme. When it binds to albumin, the pro-oxidant is decreased which means albumin has the function as an antioxidant [5].

Ligands binding such as metal ions on a protein is capable of causing structural changes. The albumin structure can be used as a diagnosis of disease. In diabetes mellitus, it has been found that albumin in the form of HNA (human nonmercaptalbumin) increased while HMA (human mercaptalbumin) decreased. Conversion of the form of albumin affects the ability of albumin as scavenger [6]. Another example is cancer where there are differences in the structure of albumin in patients with cancer, compare with healthy people caused by the bonding of albumin that be the reason for allosteric modification in albumin [7]. The ability of albumin as a scavenger against various metal ions can be done not just with in vivo analysis but also in silico. Research using domain analysis, motive, and metal ion docking in HSA 4K2C has never been done so that it needs to be done in order to give an overview of the ability of albumin as a scavenger with a focus on metal ions sequestering.

2. Methods

2.1. Data Mining

Human Serum Albumin obtained from PDB database (Protein Data Bank) with ID 4K2C as HSA Ligand Free. Next datamining of metal ions obtained from PubChem such as Cu^{2+} ion (ID: 27099), Fe^{2+} ion (ID: 27284), Mn^{2+} ion (ID: 27854), Mn^{3+} ion (ID: 105130), and Fe^{3+} ion (ID: 29936) with molecular weight 63.546 gr/mol, 55.84 gr/mol, 54.938 gr/mol, 54.938 g/mol, 55.845 gr/mol, respectively. Data from PDB database is a PDB file that is a 3D conformation and a sequence file (fasta). As for data from Pubchem in the form of 3D conformer in SDF format. Data with formats (. pdb) and (. sdf) are used in docking and RAMPAGE analysis while (.fasta) format is used for ProtParam analysis, MOTIF Search, DLP-SVM, and CFSSP.

2.2. ProtParam and Motif Search

ProtParam is used in conducting computation of physicochemical properties analyzed from the sequence of proteins that we input. Sequence in Fasta or ID format and SwissProt/TrEMBL accession number. ProtParam is one of the tools integrated with SwissProt (web.expasy.org). The parameters are input in the form of molecular weight, theoretical pI, amino acid composition, atomic composition, and instability index [8]. The Motif Search uses PROSITE as part of its system integration. The MotifSearch is one of the tools provided by GenomeNet (genome.jp).

PROSITE is a method for determining the function of a protein sequence. The techniques used include alignment to get the pattern, motif, or fingerprint of the input proteins. PROSITE database consists of two ASCII files where the first one is (PROSITE. DAT) which is a file containing all the pattern and matrix information. The second file is (PROSITE. DOC) which contains the text information of each profile and pattern [9].

2.3. CFSSP and DLP-SVM

CFSPP (biogem.org/tool/chou-fasman) stands for (Chou & Fasman Secondary Structure Prediction Server) which is a dedicated Server to predict the secondary structure of the protein amino acid sequence. This server can predict α-helix, β-sheet, turn and coil. The algorithm used is divided into three, where any part consisting of six or more amino acid residues (Pa) ≥ 1.03 and (Pa > Pb) is predicted as helical. Then if the part of the amino acid sequence is five or more residue or (Pb) ≥ 1.05 with (Pb > Pa), it will be predicted as β-sheet. As for the turn is measured from the content of proline or glycine because both amino acids are commonly found in the turn area [10].

DLP-SVM is a server for domains linker predictor which is a specific amino acids sequence that become a barrier as well as interconnection between domains, so the desired interaction does not occur in those proteins. This server uses SVM or support vector machine which is a method for machine-learning that consists of three types namely SVM-long, SVM-Short and SVM-Joint which combines the predicted results of linkers [11].

2.4. Docking

Docking has a function in studying receptor and ligands interactions through the identification of a protein-compatible active site, then obtaining the best geometry of the complex between the ligands and its receptors and conducting calculation of energy interactions with ligands. As expected, the results of the docking analysis can help in designing effective ligands. The component that affects the success of a docking program is the scoring function and search algorithm [12]. The scoring component has a function in estimating the binding affinity between macromolecules and ligands. While the use of algorithms serves to determine the conformation of receptors interacting with ligands [13]. Docking was done with AutoDock Vina where Vina Search Space is used the same for all ligand with Center X (9.3232), Y (-23.360), and Z (5.6878) and Dimensions X (85.7204), Y (109.1736) and Z (79.3440).

3. Results and Discussion

3.1. Characterization of HSA 4K2C

Albumin as one of the proteins synthesized by hepatocytes as well as the continuous circulation has a total amount 60% to 65% of the overall plasma protein. Total amino acids were identified by ProtParam analysis. The results of the ProtParam analysis provided some information on the HSA 4K2C, it has an amount of 585 amino acids with an isoelectric point (pI) 5.67. The isoelectric point is the pH of a molecule when the electrical charge is neutral or uncharged. Further information is obtained by the negative amino acid residue (Asp + Glu) is 98 while the positively charged (Arg + Lys) is 83. Besides, the atomic composition is shown in Table 1. Further data obtained that 4K2C has a protein instability index of 38.85 which according to Walker (2005) if the protein has an index of instability under 40, then the protein is stable while above 40 means unstable. Stability of HSA 4K2C have been proved. Normal albumin before damage such as glycation is a stable protein, it is evidenced by molecular dynamics simulation by YASARA of human serum albumin in the period 15.000 ps [14]. Albumin until the end of the simulation still shows the initial conformation and does not occur unfolding. But when molecular dynamic simulation of glycated albumin (HSA 4K2C + glucose), there are a lot of conformational change happened including unfolding event between sidechain A and B of this glycated albumin model that represent the condition of albumin in diabetes mellitus.

Table 1. *Human Serum Albumin's* Atom Compositions (ProtParam).

Atom	Total
Carbon (C)	2936
Hydrogen (H)	4624
Nitrogen (N)	786
Oxygen (O)	889
Sulfur (S)	41
Formula	$\mathbf{C_{2936}H_{4624}N_{786}O_{889}S_{41}}$

High solubility of this protein in liquid media is caused by the existence of these amino acid residues. In terms of structure, it is known that human albumin is a protein monomer, whereby sub domains A and B have six and four α-helix connected with loops that are flexible [15]. The characterization of HSA 4K2C such as the domain linker, pattern of motives and secondary structure in this study showed that albumin is a protein complex with amino acids integrated in order to complete the function. Motif and domain linker are a characteristic of protein so it can be distinguished from other proteins in the database. By knowing the character as well as the position of amino acids from a protein then in the research involving the process of docking, researchers can choose which proteins from databases should be used for next analysis.

The results of Motif Search indicate that there are three patterns of protein HSA (Human Serum Albumin) (Fig. 1A). Albumin consists of two side chains in which the Fig. 1B, side chain A is green, and the side chain B is light blue. The first motif at 551-575 amino acid residual arrangement which colored by magenta (FAAFVEKCCKADDKETCFAEEGKKL) while the second motif is in the yellow residue 353-377 (YETTLEKCCAAADPHECYAKVFDEF). The last motif colored in red and the amino acid residues at 161-185 (YKAAFTECCQAADKAACLLPKLDEL). These motives are small sequences of protein fragments with highly conserved properties that have been used for predicting secondary and tertiary structure of a protein. These structure motifs also have application in drug design where it will provide ligand binding sites data or the active sites of protein.

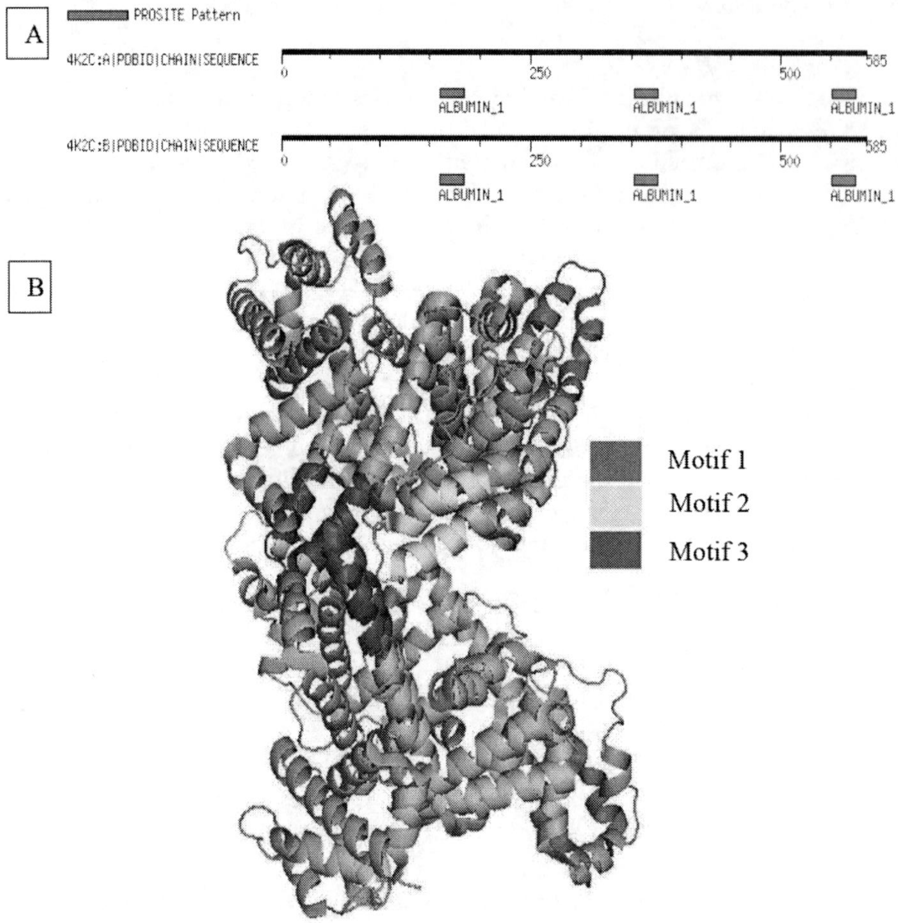

Figure 1. (A) Motif Search Result (B) Motif Search *Human Serum Albumin* Visualization with PyMOL.

Figure 2. Secondary Structure of HSA 4K2C.

The linker domain analysis using DLP-SVM. The domain linker is a loop area that separates two structural domains. The Linker is divided into two types i.e. short linker (SVM-Short) which contains less than equal to 9 amino acid residues. While long linker (SVM-Long) is more than 9 residues of amino acids. The combination is called SVM-All on this server, but it is used as a linker representation of SVM-Joint where the presentation of accuracy is much higher compared to SVM-Long and SVM-Short. Figure 3 shows the linker probability of the SVM-Long analysis shown in blue, with the SVM-Short (yellow) and SVM-All colored in red.

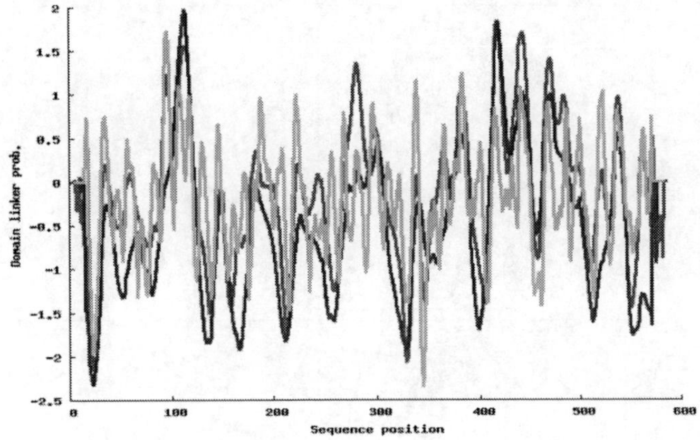

Figure 3. *SVM linkers*, Long (Blue), Short (Yellow), All (Red).

Figure 3 on SVM-All has a peak value of 1,835 in the 411-450 and 1,558 regions of the amino acid 90-122. SVM-Long has a 1,966 peak value in the 93-122 region. Then SVM-Short with a peak value of 1,720 in amino acid residue 92-100. Further SVM-Joint will be taken as final data for visualization using PyMOL. Figure 4 is a visualization of the SVM-Joint which is the highest predictor of the linker domain found in the HSA 4K2C. The results of the SVM Joint analysis consist of two linker domains that are the first DL-1 with 410-451 amino acid residues (RYTKKVPQVSTPTLVEVSRNLGKVGSKCCKHPEAKRMPCAED) then the DL-2 is composed of amino acid residue number 96-122 (PERNECFLQHKDDNPNLPRLVRPEVDV).

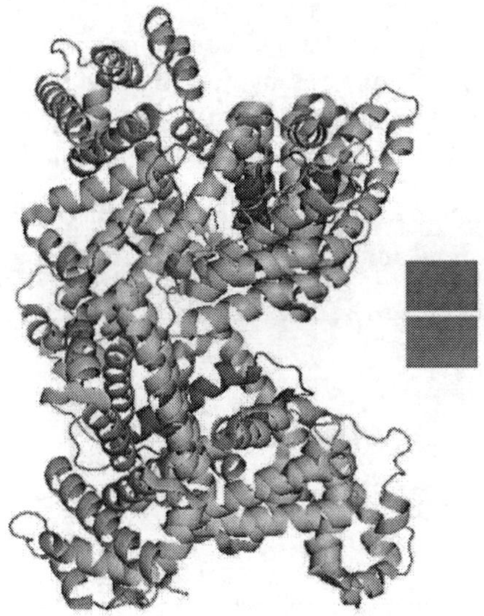

Figure 4. HSA 4K2C *Domain Linker* Visualization (SVM-Joint).

3.2. *Docking*

Docking analysis is to know the ability of albumin in metal ions sequestering. Before performing the preparation, the first thing to do is Ramachandran Plot (RAMPAGE) analysis (Fig. 5). Ramachandran Plot shows the quality of the proteins to be used based on the location of certain amino acid residues in four quadrants that are representations of proteins to be analyzed.

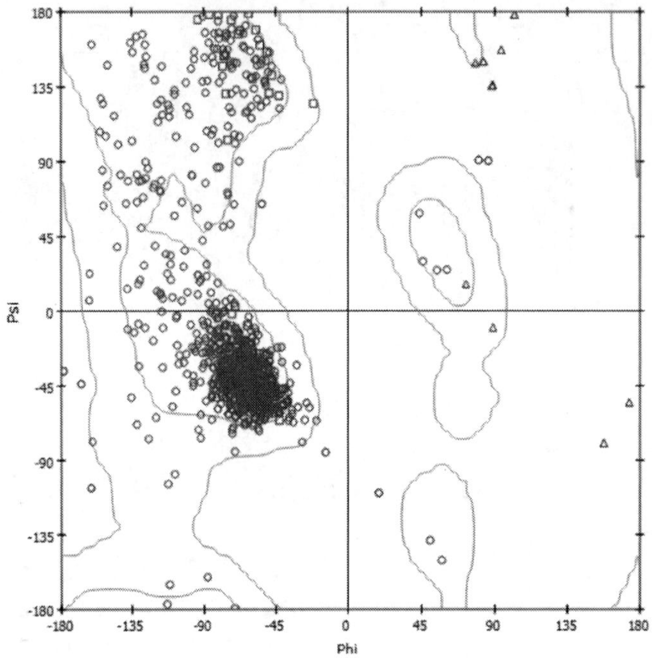

Figure 5. Ramachandran Plot of HSA 4K2C. RAMPAGE evaluation of residues show that Ser5(A), Glu60(A), Thr79(A), Gly85(A), Pro113(A), Ala364(A), Ser5(B), Arg81(B), Gly85(B), Ala362(B), Ala364(B), Glu400(B), Val469(B), Lys541(B) in outlier region or less than 1.2%.

Data from the Ramachandran Plot analysis indicates that HSA 4K2C can be used for docking because the amount of residue that is in place should be as much as 1061 or 91.9% followed by 6.8% of residues in allowed region. Then it is followed by docking using PyRx. Before the docking, ligand preparation is done by minimizing energy using Open Babel integrated with PyRx software. Open Babel also used for the conversion of ligand format from (. sdf) to (. pdbqt).

Table 2. Binding Energy of Ligands and HSA Protein.

Metal Ions	**Binding Energy (kcal/mol)**
Cu^{2+}	-1.2
Fe^{2+}	-1.2
Mn^{2+}	-1.2
Mn^{3+}	-1.2
Fe^{3+}	-1.2

As for protein did not need preparation because HSA 4K2C has no innate ligand nor water in the file. The docking results show albumin can bind a wide range of metal ions both copper ions, iron ions (II) and (III), as well as manganese ions (II) and (III). Binding Energy (Table. 2) of all ligands with HSA 4K2C shows the same value. However, the binding location of each ligand on the HSA 4K2C amino acid residue is different. Analysis of binding site is done using Discovery Studio (Table. 3) where the iron ions (II) bind to Gln29 (A), Tyr30 (A), and Gln32 (A). Meanwhile, manganese ion (II) binds to CYS34 (A), Arg144 (A), Leu31 (A), Gln32 (A), and His39 (A).

Table 3. Tabulation of binding site.

Metal Ions	Binding Site
Fe^{2+}	Gln29 (A), Tyr30 (A), Gln32 (A)
Fe^{3+}	Cys34 (B), Arg144 (B), Leu31(B), Gln32 (B)
Mn^{2+}	Cys34 (A), Arg144 (A), Leu31 (A), Gln32 (A), His39 (A)
Mn^{3+}	Gln29 (A), Tyr30 (A), Gln32 (A)
Cu^{2+}	Gln29 (A), Tyr30 (A), Gln32 (A)

Copper ions (II) bind to Gln29 (A), Tyr30 (A), and Gln32 (A). Manganese ions (III) interact with amino acid residues Gln29 (A), Tyr30 (A), and Gln32 (A). The last ion is Iron (III) where it binds to Cys34 (B), Arg144 (B), Leu31 (B), and Gln32 (B). In Figure 6 the protein-side chain of human serum albumin is divided into two namely side A (green) and side chain B (light blue). The type of bond between the Cu^{2+} ion and the amino acid residue in HSA is hydrogen bond acceptor, while the Fe^{3+} and Mn^{2+} bond type is divided into two namely hydrogen bond acceptor and metal donor. Then the type of bond between Mn^{3+} ions and Fe^{2+} is hydrogen bond acceptor.

The transition metal ions used in this study include iron ions (II) and (III), copper ion (II), manganese ions (II) and (III). Metal ions are often associated with the cause of various diseases in the human body, one of which is atherosclerosis [16]. Some examples of hazards from the accumulation of transition metal ions such as iron ions on human lesions and can contribute to the progression of an illness. Oxidative damage caused by multiple transition metal ions such as iron and copper ions capable of damaging the extracellular matrix components, the oxidation that occurs can even activate enzyme which capable to degrading matrix. One example of this enzyme is metalloproteinase [17].

From the results of this study HSA 4K2C can bind with a wide range of transition metal ions. The 3D visualization of the metal ion docking with HSA 4K2C can be seen in Figure 6. The binding energy of the docking result shows the same number of all simulated results i.e.-1.2 kcal/mol. The binding power of metal ions with proteins cannot only be determined by the binding energy. The interaction of metal and ligands conducted on the model, in this case, docking can weaken the true condition of the binding of metal ions in nature. The effect that occurs is called a shielding effect [18]. The binding energy of the transition metal ion docking results is also a character of the albumin protein. Binding sites of both human and bovine albumin to the metal ions consisted of four locations i.e. N-terminal site, Cys34 residue, side chains A and B [19].

The results of the NMR site A of Human Serum Albumin study constitute the MBS or multi-metal binding site [20]. It can explain an amino acid residue can bind to some metals above. As well as manganese ions (II) and iron (III) that interact with residue Cys34 both in the side chain A and B. Metal ions binding on Cys amino acid residues, can be caused due to cysteine have negative charge so that interactions are easier to occur compared to other metal ions that have a dipole charge interaction [21]. Metal ions binding on the amino acid residue of Gln and His is also due to the hydrophilic and polar properties of these two amino acid residues. The residue capability of certain amino acids as a metal donor as well as the acceptor is determined by the affinity of each amino acid. The amino acids Cys, Asp and Glu can be as donors and acceptors [21]. Site A as a multiple binding site indicated by amino acid residues that interact with metal ions such as Cu^{2+}, Fe^{2+}, and Mn^{3+}.

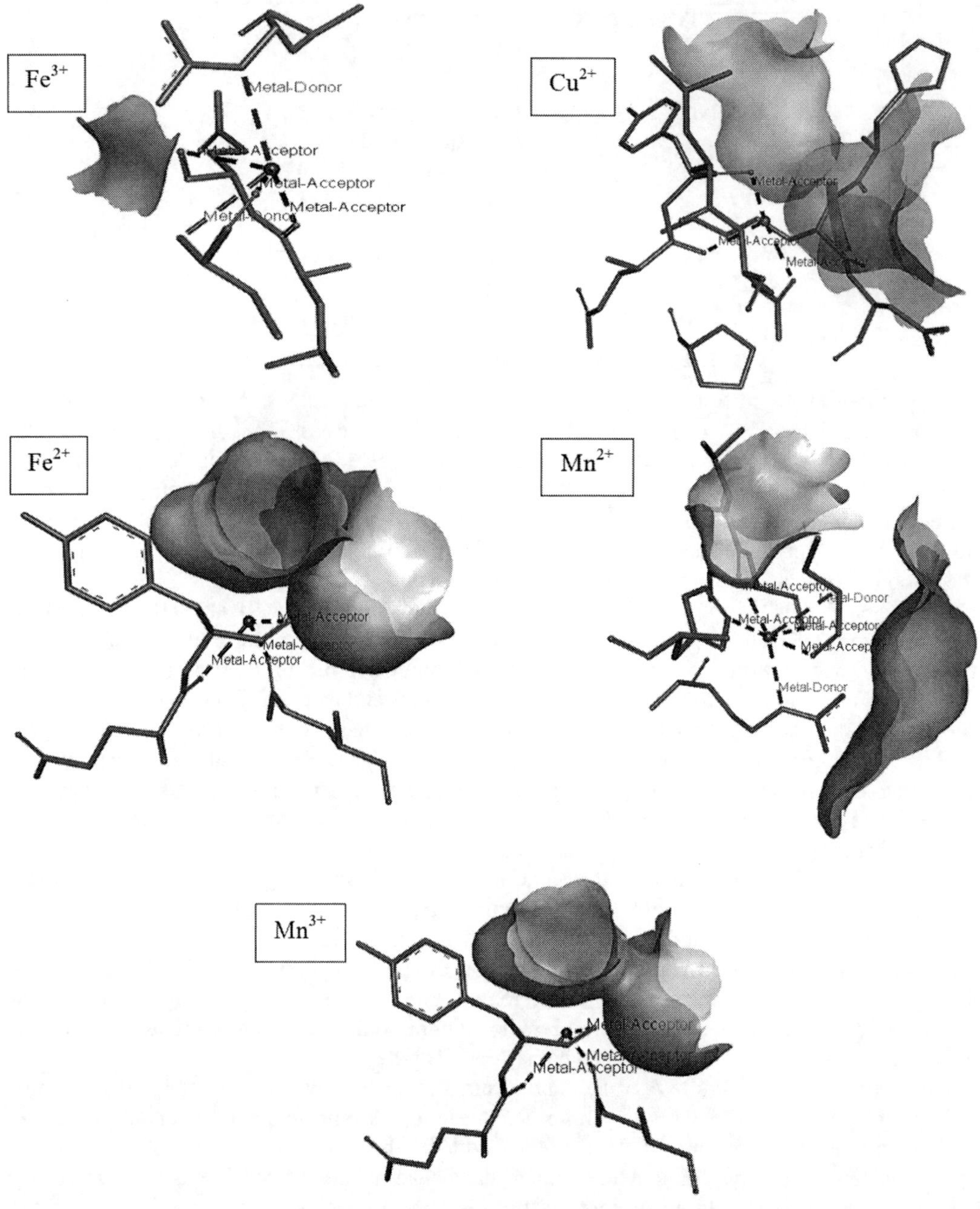

Figure 6. 3D Visualization of HSA 4K2C docking with metal ions.

The ability of albumin in free radical trapping is due to the presence of cysteine (Cys34) residue wherein as much as 70-80% of the residue in an adult contains the sulfhydryl group. The existence of such residue that makes albumin capable of binding to radical hydroxyl [22]. Cys 34 on a subdomain of IA is located in the slit of a protein surface and not related with disulfide bridge, but its sulfhydryl group is able to bind Hg^{2+}, Ag^{2+}, Au^{+} and Pt^{2+} [15]. The scavenger activity owned by albumin is able to protect the ligands from oxidation events, an example is the binding of unsaturated fatty acids [23]. Cys34 in albumin is capable to reduce thiol in human blood plasma [24]. In addition to the Cys amino acids there is also the Met in human albumin with the function of Cys residue as a free radical scavenger while the Met functions in a metal chelator [25]. It was found that 80% of the group – SH which is a strong scavenger of ROS in albumin comes from Cys-34 [26]. The presence of cysteine and methionine residues is essential in the antioxidant ability of human serum albumin. Research shows that when a mutation is performed on Cys34 the antioxidant capabilities of albumin fall by 30% and if free cysteine residues are placed in another position on the HSA 4K2C, the antioxidant capability still remain. Site 2 in HSA 4K2C is the target of oxidative stress, but lost or modified cysteine residue results in more easily albumin to be degraded [24].

4. Conclusion

The characteristic albumin as a stable protein also has a function as an endogenous scavenger that is closely related to sequestering metal ions. Albumin has MBS (Multi-Metal Binding Site) represented from the docking results of several metal ions that bind to the same amino acid residue in the side chain A. The ability of amino acid residue in albumin as free radical trapping like Cys34 has donor and acceptor metal ions binding type. The binding energy of transitional metal ions to HSA 4K2C has the same amount of energy as -1.2 kcal/mol, allegedly shielding effect plays a role in determining binding energy value.

References

[1] Murray R K, Granner D K and Rodwell V W 2006 *Harper's illustrated biochemistry 27th Edition* (Asia: Mc Graw Hills Companie)
[2] Fasano M, Curry S, Terreno E, Gilliano M, Fanali G, Narciso P, Notari S and Ascenzi P 2005 *IUBMB Life* **57** 787-96
[3] Xu X, Zhang L, Shen D, Wu H and Liu Q 200 *J Fluoresc* **18** 193-201
[4] Stewart A J, Blindauer C A, Berezenko S, Sleep D and Sadler P J 2003 *PNAS* **100** 3701-06.
[5] Quinlan G J, Martin G S and Evans T W 2005 *Hepatology* **41** 1211-19
[6] Suzuki E, Yasuda K, Takeda N, Sakata S, Era S, Kuwata K, Sogami M and Miura K 1992 *Diabetes Research and Clinical Practice* **18** 153-58
[7] Kazmierczak S C, Gurachevsky A, Matthes G and Muravsky V 2006 *Clinical Chemistry* **52** 2129-34
[8] Walker, J M 2005 *The proteomics Protocols Handbook* (USA: Humana Press)
[9] Sigrist C J A, Cerutti L, Hulo N, Gattiker A, Falquet L, Pagni M, Bairoch A and Bucher P 2002 *Brief Bioinform* **3** 265-74
[10] Kumar T A 2013 *Wide Spectrum* **1** 15-19
[11] Ebina T, Toh H and Kuroda Y 2008 *Peptide Science* **92** 1-8
[12] Mukesh B and Rakesh K 2011 *International Journal of Research in Ayurveda & Pharmacy* **6** 1746-51
[13] Serina J J C 2013 *Master Dissertation* Universidade de Madeira, Portugal
[14] Wibowo S, Widyarti S, Sabarudin A, Soeatmadji D W and Sumitro S B 2019 *Asian Journal of Phamaceutical and Clinical Research* **12** 276-82
[15] Otagiri M and Chuang V T G 2016 *Albumin in Medicine: Pathological and Clinical Applications* (Singapore: Springer)
[16] Horwitz L D and Rosenthal E A 1999*Vasc Med.* **4** 93–99
[17] Stadler N, Lindner R A and Davies M J 2004 *Arterioscler Thromb Vasc Biol* **24** 949-54
[18] Chen D, Li Y, Guo W, Li Y, Savidge T, Li X and Fan X 2019 *Physical Chemistry Chemical Physics* **21** 205-16
[19] Bal W, Sokolowska M, Kurowska E and Faller P 2013 *Biochim Biophys Acta* **12** 5444-55
[20] Blindauer C A, Lu J, Stewart A J, Sadler P J and Pinheiro T J T 2008 *Biochem Soc Trans* **36** 1317–21
[21] Dudev T and Lim C 2014 *Chem Rev* **114** 538-56
[22] Gutteridge J M 1986 *Biochim Biophys Acta* **869** 119-27
[23] Roche M, Rondeau P, Singh N R, Tarnus E and Bourdon E 2008 *FEBS Letters* **582** 1783–87
[24] Anraku M, Chuang V T G, Maruyama T and Otagiri M 2013 *Biochim Biophys Acta* **1830** 5465-72
[25] Bourdom E, Loreau N, Lagrostl and Blache D 2005 *Free Radic Res* **39** 15-20

IOP Conf. Series: Materials Science and Engineering **833** (2020) 012052 doi:10.1088/1757-899X/833/1/012052

[26] Nagumo K, Tanaka M, Chuang V T G, Setoyama H, Watanabe H, Yamada N, Kubota K, Tanaka M, Matsushita K, Yoshida A, Jinnouchi H, Anraku M, Kadowaki D, Ishima Y, Sasaki Y, Otagiri M and Maruyama T 2014 *PLOS-ONE* **9** e85216

In Silico Analysis of Saponin Isolates from Mesocarp of Cucumber (*Cucumic sativus* L.) and Purple Eggplant (*Solanum melongena* L.) as Pancreatic Lipase Inhibitor

Mely Wijaya[1] and Subandi*[1]

[1]Department of Chemistry Universitas Negeri Malang, Jl. Semarang 5, Malang Indonesia

Corresponding author: subandi.fmipa@um.ac.id

Abstract. Currently orlistat has been widely used as anti-obesity drug, because of its activity as a pancreatic lipase inhibitor. Two saponin isolates, from cucumber mesocarp and purple eggplant, also proved to be active as pancreatic lipase inhibitors in vitro. Based on spectrophotometric analysis, the two saponin isolates are thought to be Silphioside F and Cesdiurins I-III. The purpose of this study is to confirm the ability of the two compounds as pancreatic lipase inhibitor through in silico analysis, relative to orlistat. This study uses Python Molecular Viewer (PyMol), Python Prescription (PyRx) 0.8, and Discovery Studio software. As a ligand, 3D structure of Silphioside F and Cesdiurins I-III have been used. The orlistat as a comparative ligand molecule have also been used. 3D structure of porsine pancreatic lipase has been used as receptor molecule. The result of the analysis shown that the binding site of pancreatic lipase is relatively same as orlistat of Cesdiurins I-III molecule, but different for the Silphioside F molecule. The data indicated that in inhibiting pancreatic lipase, the two isolate compounds used different mechanism. However, against pancreatic lipase, both molecules have greater binding affinities each, compared to orlistat, which were -9.7 kcal/mol for Silhphioside F and -9.5 kcal/mol for Cesdiurins I-III, and orlistat only -7.6 kcal/mol. The latest data were in line with the in vitro analysis, that both isolates have greater inhibition power than orlistat.

1. Introduction

Obesity has become a global problem [1]. It is a risk factor for the emergence of degenerative diseases such as type 2 diabetes mellitus (T2DM), cardiovascular disease, cancer, and premature death [2,3,4]. As an anti-obesity drug, orlistat is a type of tetrahydrolipstatin, which is isolated from Actinomycetes (*Streptomyces toxytricini*), and works as a pancreatic lipase inhibitor [5]. The pancreatic lipase is an enzyme secreted by pancreas into the intestine and hydrolyzes 50-70% triglycerides into fatty acids [6]. Orlistat is a competitive inhibitor of pancreatic lipase. Its activity is determined by the presence of β-lactone rings in its carbon chain. The inhibitory activity of orlistat disappears when the lactone β ring is opened [7]. Orlistat consumption can cause several side effects such as hyperuricemia, diarrhea, nausea, myositis, stomach irritation, oily spots, flatulence, flatus, fecal incontinence, and dry skin [8]. Thus, many studies have used herbal sources to replace orlistat, as a pancreatic lipase inhibitor.

Previous studies have revealed the presence of saponins that active as pancreatic lipase inhibtor in methanol extracts of cucumber mesocarp [9] and purple eggplants [10]. The UV-Vis and FT-IR spectroscopic analysis suggested that the saponin isolated from cucumber mesocarp has a similar structure to saponin from *Acanthopanax senticosus* fruit, namely Silphioside F [11], while the saponin isolate from purple eggplant mesocarp has a similar structure to Cesdiurins I-III [12]. The purpose of

IOP Conf. Series: Materials Science and Engineering **833** (2020) 012053 doi:10.1088/1757-899X/833/1/012053

this study is to confirm the ability of the two saponin compounds as pancreatic lipase inhibitors relative to orlistat, through in silico analysis using molecular docking.

In molecular docking, the interaction between ligands and receptors will be studied by identifying the active site that matches the receptor, getting the best geometry of the ligand-receptor complex, and calculating the interaction energy in different ligands to get a more effective ligand [13]. By using molecular docking, the binding site and binding affinity of the two saponin isolate to the pancreatic lipase enzyme, relative to the standard ligand, orlistat, can be predicted.

2. Experiment Methods

This study used ASUSPRO P5430UA laptop with Intel® Core™ i7 6500U Processor, Windows 10 Pro-ASUS, Windows 10 Pro, memory DDR4 2133 MHz SDRAM, OnBoard Memory 4 GB , 1xDIMM socket for expansion up to 12 GB SDRAM, Python Molecular Viewer Program (PyMol), Python Prescription (PyRx) 0.8, and Discovery Studio.

The 3D structure of the receptor protein used was adjusted to the type of lipase used in in-vitro analysis [9,10], namely Pancreas Triacylglicerol Lipase P00591 (LIPP_PIG, with code 1ETH) downloaded from the UniProt Bank Protein Data database (https://www.uniprot.org/) in the PDB file format. The structure used is the A/C.chain

The 2D structure of orlistat and Silphioside F was obtained from PubChem (https://pubchem.ncbi.nlm.nih.gov), while the 2D structure of Cesdiurins I-III was obtained from Fouad et al .2008 [12]. The 3D structure of orlistat, Silphioside F and Cesdiurins I-III were obtained from the PubChem database in Sybil Data File format (sdf).

3. Result and Discussion

3.1. Sterilization of 3D Struture of The Enzyme

The 3D structure of Pancreatic lipase enzyme must be sterilized to eliminate the water content. All the ligands contained in the enzyme. This process is carried out using PyMol software. The results can be seen in Figure 1.

a b

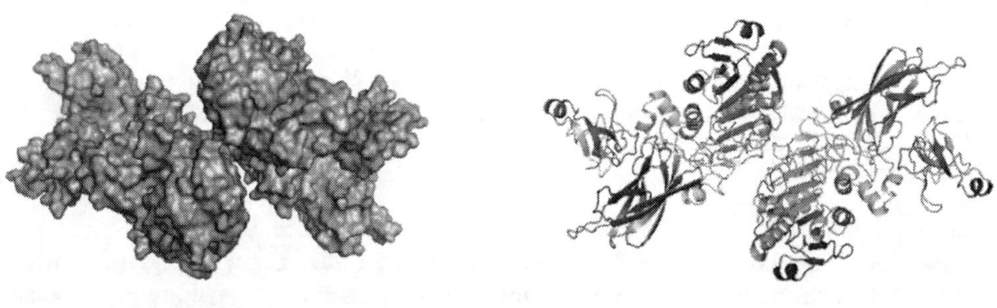

Figure 1. Sterilization results of the 3D structure of the porcine pancreatic lipase enzyme (Uniprot 2019) (surface form (a) and worm form (b))

3.2. Ligand Preparation

The 2D structures of orlistat, Siphiosite F was made in PubChem and Cestiurins I-III [13] (Figure 2), then converted to 3D form using the PubChem program (Figure 3). The structure is saved in the form of a sdf file, and read by the PyRx program. Then, it is used for the docking process.

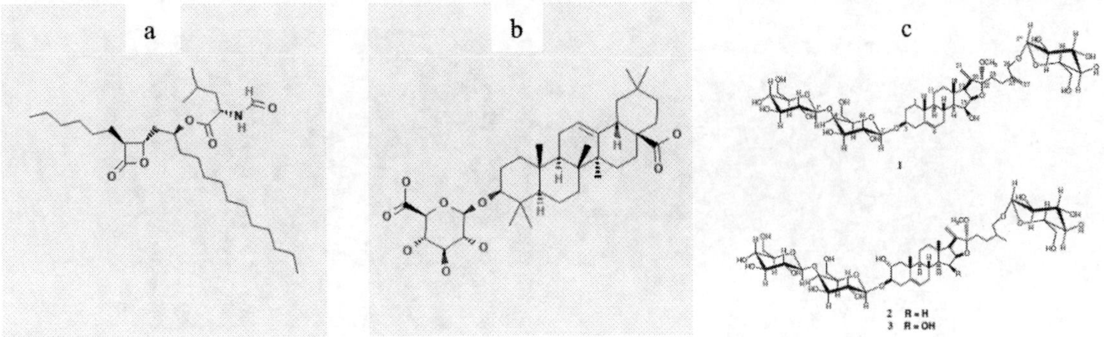

Figure 2. 2D Structure of Orlistat (a), Silphioside F (b) and Cesdiurins I-III (c).

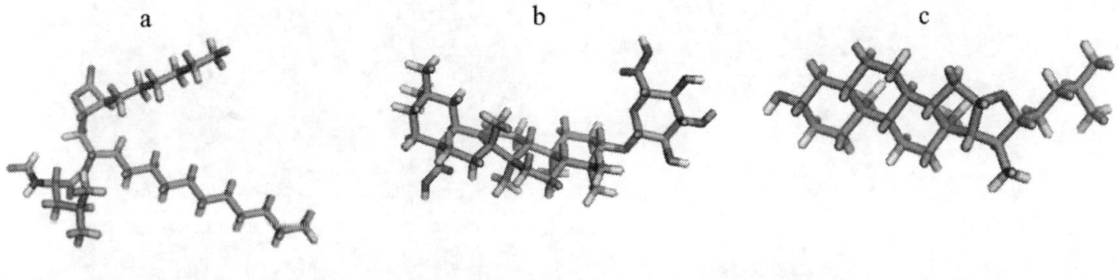

Figure 3. 3D Structure of (a) Orlistat, (b) Silphioside F and (c) Cesdiurins I-III visualized by PyMol software.

3.3. Binding Affinity Between Ligands and Pancreatic Lipase Enzyme Using Molecular Docking

Molecular docking process has been used to identify the activity of ligand compounds that has a potency as inhibitor against pancreatic lipase. Molecular docking analysis was performed using the Autodock Vina program in PyRx 0.8 software. This process produced the binding affinity (kcal/mol) value of a ligand to its receptor (enzyme). The higher absolute binding affinnity value was predicted to has higher inhibitory activity against the enzyme. The result of this molecular docking analysis has shown that Silphioside F and Cesdiurins I-III, which have a higher binding affinity to pancreatic lipase than orlistat has (Table 1). This data suggested that inhibitory activity of both saponin against pancreatic lipase is higher than orlistat.

Table 1. Binding affinity of ligands to receptor.

Ligand	Binding Affinity (kcal/mol)
Orlistat	-7.6
Silphioside F	-9.7
Cesdiurins I-III	-9.5

3.4. Binding Position of The Ligan on Pancreatic Lipase

The binding position of the ligand compound to the pancreatic lipase receptor was visualized using PyMol program, and the results were shown in Figure 4.

According to Figure 4, for the Pancreatic Lipase, the binding site of Silphioside F is different from the binding site of orlistat. The binding site of Cesdiurins I-III is relatively the same as orlistat. It has been known that the orlistat inhibitory mechanism is competitive [5]. This data suggests that the Silphioside F inhibition mechanism is non-competitive, while the inhibitory mechanism of Cesdiurins I-III is competitive.

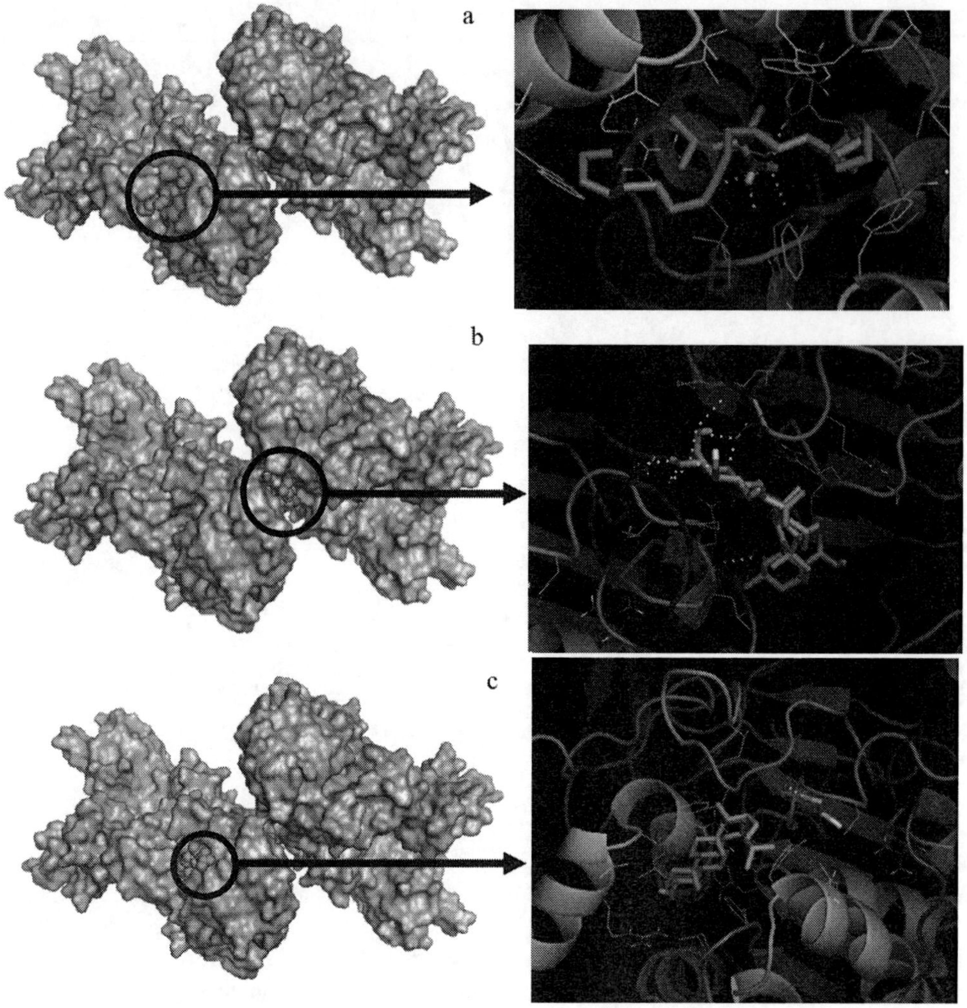

Figure 4. Binding Position of (a) Orlistat, (b) Silphioside F and (c) Cesdiurins I-III against pancreatic lipase enzymes in pigs visualized with PyMol software.

3.5. The Interaction of Ligand Compounds to Amino Acid Residues on The Binding Site of The Enzyme

This visualization used the Discovery Studio program to find out the bonding distances, bond types, and amino acid residues involved in the interaction between the ligand and target protein. The results of the analysis were presented in Figure 5 and Table 2.

The number and type of amino acids involved in the binding to each ligand which is the binding position of Silpioside F, is different from orlistat. Whereas, Cesdiurins I-III binding position is relatively same with the residues of ALA261, HIS264, VAL261, and PRO261 (Table 2 and Figure 5).

The overall result of this analysis was in line to the previous data which shows that inhibitory activity of the both predicted saponin isolates were higher than orlistat, at the same mass [9,10]. In addition, the isolate from other sources that also contain Silphioside F has activity as pancreatic lipase inhibitor [11]. All the data suggested that the isolates which contain one or both of the predicted saponin have a potency as anti-obesity drug.

IOP Conf. Series: Materials Science and Engineering **833** (2020) 012053 doi:10.1088/1757-899X/833/1/012053

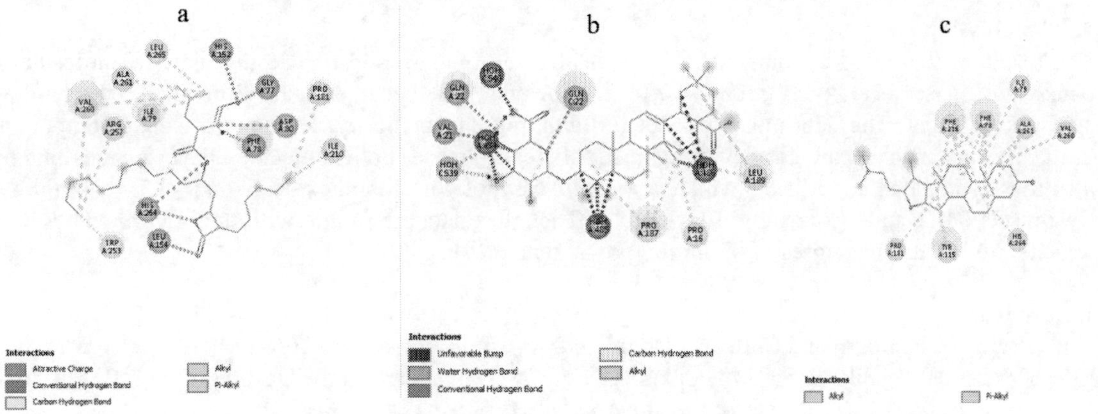

Figure 5. Types of bonding and amino acid residues involved in binding with pancreatic lipase (Orlistat (a), Silphioside F (b) and Cesdiurins I-III (c)).

Table 2. Types of numbers and bonding distances between Amino acid residues in pancreatic lipases involved and the intermolecular interactions with Orlistat, Silphioside F, and Cesdiurins I-III.

Ligand	Types and Residues of Amino Acids Involved	Bonding Distance (Å)	Interaction Type
Orlistat	VAL A:260	4.05	Alkyl, Pi-Alkyl, Carbon Hydrogen Bond
		4.37	Alkyl, Pi-Alkyl, Carbon Hydrogen Bond
	HIS A:152	2.58	Conventional hydrogen bond
	PHE A:78	2.01	Conventional hydrogen bond, Carbon Hydrogen Bond
	PRO A:181	4.41	Alkyl, Pi-Alkyl
	HIS A:264	5.03	Attractive Charge
		3.89	Attractive Charge
		3.01	Carbon Hydrogen Bond
	LEU A:154	2.86	Conventional hydrogen bond
Silphioside F	VAL A:21	2.01	Conventional hydrogen bond
	PRO A:187	4.87	Alkyl
		5.21	Alkyl
		5.05	Alkyl
	PRO A:16	4.72	Alkyl
	LEU A:189	3.83	Alkyl
		5.20	Alkyl
		4.65	Alkyl
Cesdiurins I-III	PRO A:181	5.31	Alkyl, Pi-Alkyl
	HIS A:264	4.74	Alkyl, Pi-Alkyl
	VAL A:260	4.71	Alkyl, Pi-Alkyl
	ALA A:261	3.92	Alkyl, Pi
		5.19	Alkyl, Pi
	ILE A:79	4.03	Alkyl, Pi
	PHE A:78	4.05	Alkyl, Pi
		5.39	Alkyl, Pi
		5.37	Alkyl, Pi
		4.49	Alkyl, Pi
	PHE A:216	4.64	Alkyl, Pi
		4.32	Alkyl, Pi
		3.75	Alkyl, Pi
		5.01	Alkyl, Pi

IOP Conf. Series: Materials Science and Engineering **833** (2020) 012053 doi:10.1088/1757-899X/833/1/012053

4. Conclussion

The result of this in silico analysis has shown that the binding site for Cesdiurins I-III molecule to pancreatic lipase is relatively same as orlistat. However, for the Silphioside F molecule, the binding site was different. The data indicated that in inhibiting pancreatic lipase, the two isolate compounds used different mechanism. However, against pancreatic lipase, both molecules have greater binding affinities, compared to orlistat, which are -9.7 kcal/mol for Silhphioside F and -9.5 kcal/mol for Cesdiurins I-III), and -7.6 kcal/mol for Orsilat. The latest data are in line with the in vitro analysis, in which both isolates have greater inhibition power than orlistat.

References

[1] Barry M Pompkin and Collen M Doak 1998 *Nutrition Reviews* **56** 106–114.

[2] Vasanti S. Malik, Barry M. Popkin, George A. Bray, Jean-Pierre Després, Frank B. Hu 2010 *Contemporary Reviews in Cardiovascular Medicine* **121** 1356-1364.

[3] Hu FB. 2008. Obesity Epidemiology. *New York, NY*: Oxford University Press.

[4] Darusman L.K., Pradono D.I., and Susanti A.I 2011 *Jurnal Natur Indonesia* **13** 142-154.

[5] Filippatos T. D, Derdemezis C. S., Gazi I. F, Na-kou E. S, Mikhalidis D. P, and Elisaf M. S. 2008 *Drug Safety* **31** 53-65.

[6] Delphin D.V, Haripriya R, Subi S, Jothi D and P. Thirumalai Vasan 2014 *World Journal Of Pharmacy And Pharmaceutical Sciences* **3** 1041-1048.

[7] Birari RB and Bhutani KK 2007 *Drug Discov Today* **12** 879 [PMID: 17933690].

[8] Yamamoto M, Shimura S, Itoh Y, Ohsaka T, Ega-wa M, and Inoue S 2000 *International Journal of Obesity and Related Metabolic Disorders* **24** 758-764.

[9] Subandi, Wijaya M, Sudarmo Tatas P B and Suarsini E 2018 *AIP Conference Proceedings* **2021**, 070016.

[10] Subandi, Zakiyaturrodliyah L, and Sudarmo Tatas P B 2019 *IOP Conference Series: Materials Science and Engineering* **509** 012139

[11] F. Li, W. Li, H. Fu, Q. Zhang and K. Koike. 2007. *Chem. Pharm. Bull.* **55** 1087-1089.

[12] Fouad M.A., Mohamed K.M., Kamel M.S., Matsunami K. 2008. Cesdiurins I-III, steroidal saponins from *Cestrum diurnum* L. *Journal of Natural Medicines* PubMed

[13] Mukesh B. Rakesh K 2011 *International Journal of Research in Ayurveda & Pharmacy*. **2** 1746-1751.

Effect of NaCl Addition and The Incubation Time on Gallic Acid Concentration in Cabbage Fermentation using *Lactobacillus plantarum* and The Potential as Antioxidant

Viky Arina Zuhria[1], Arie Srihardyastutie[1*], Anna Safitri[1,2], Sasangka Prasetyawan[1]

[1]Chemistry Department, Brawijaya University, Malang, 65151, Indonesia
[2]Research Centre for Smart Molecules of Natural Genetic Resources, Brawijaya University

*Corresponding author: arie_s@ub.ac.id

Abstract. Cabbage is one type of vegetable that is often consumed. Cabbage contains gallic acid and sulforaphane compounds which potentially used as an antioxidant. Gallic acid and sulforaphane are produced from the hydrolysis reaction, which occurs by fermentation. Therefore, this research was focused on the effect of NaCl addition and the incubation time on the gallic acid concentration during cabbage fermentation using *L. plantarum*. The addition of NaCl used were 0 %, 0.5 %, 1.0 %, 1.5 %, and 2.0 % (w/v). While the incubation time used were 2, 3, 4, 5, and 6 days. The fermentation conditions used were 5 % (v/v) inoculum volume and pH 6. Gallic acid in fermented cabbage (biomass and filtrate) were determined using Folin-Denis method. The antioxidant analysis was determined using DPPH method. The optimum conditions were obtained at the addition of 1.0 % NaCl for 4 days fermentation. During these conditions, the gallic acid produced in biomass and filtrate were equal to 4.726 mg/100 g FW and 147.857 mg/100 g FW, respectively. Sulforaphane and gallic acid have potential as antioxidants with IC_{50} valuesof 95.113 mg/L and 21.648 mg/L, respectively.

1. Introduction

Cabbage (*Brassica oleracea var. capitata*) contains tannin and glucoraphanin. Tannin could be degraded into glucose and gallic acid by tannase activity from bacteria, such as lactic acid bacteria [1], while glucoraphanin degraded into sulforaphane in the presence of myrosinase enzyme [2,3]. Gallic acid and sulforaphane reported as antioxidants [4,5].

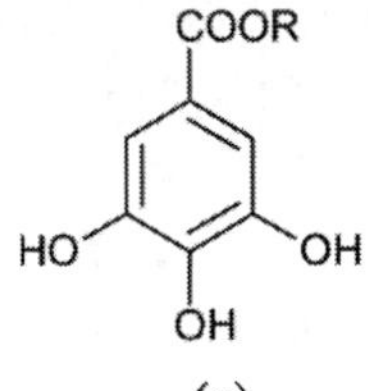

(a)

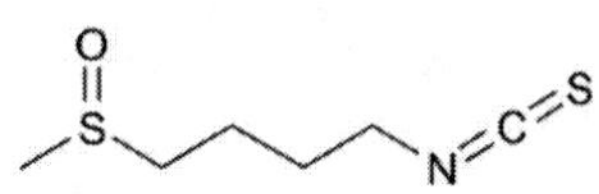

(b)

Figure 1. Structure of gallic acid (a) and sulforaphane (b)

Gallic acid can be an antioxidant compound because it has a hydroxyl group that can donate hydrogen atom to neutralize free radicals (see Figure 1 (a)) [6]. Sulforaphane can also be an antioxidant compound because oxygen atom in the methylene sulfoxide group can donate electrons to neutralize free radicals (see Figure 1 (b)) [7]. The myrosinase enzyme which functions to help hydrolysis glucosinolate can only be active due to cell wall damage. One way to prevent cell wall damage is the fermentation process [8]. Fermentation is a process of transformation, modification, or food processing, that cause changes in nutritional value, shelf life, and food safety [9]. Besides being able to hydrolysis glucoraphanin into sulforaphane, the fermentation process can also degrade antinutrients component in cabbage, namely tannin.

The fermentation of cabbage can be done using several types of lactic acid bacteria, such as: *Lactobacillus plantarum, Lactobacillus sakei, Leuconostoc mesenteroides, Pediococcus pentosaceus, Lactococcus lactis N8*, and *Lactobacillus lactis LAC67*. Based on these studies it was found that *Lactobacillus plantarum, Lactobacillus sakei*, and *Leuconostoc mesenteroides* have high effectiveness in the process of cabbage fermentation compared to other lactic acid bacteria [10]. Fermentation using *Lactobacillus plantarum* on media containing tannic acid 0.35 % (w/ v) at a temperature of 30 °C, pH 6, and agitation speed of 125 rpm can induce the production of tanase enzymes at 9.29 U/mL enzyme activity [11]. The optimum conditions of fermentation using *Lactobacillus plantarum* bacteria were: 5% (v/v) inoculum volume, pH 6, at 25 °C for 96 hours, can reduce tannin levels from 290.877 mg/g to 75.938 mg/g [12].

Besides several factors, such as inoculum volume, pH, and temperature, the fermented product also influenced by another factors, namely NaCl addition and incubation time [13,14]. Recent studies report the use of NaCl effect on microbial composition [15]. Excessive concentration of NaCl can damage the bacterial cell wall which then affect the ability of bacteria to secrete tannase enzymes which has the function to degrade tannins into reduced gallic acid [11,16]. Therefore, in this research, cabbage fermentation was carried out using *Lactobacillus plantarum* bacteria to hydrolysis one type of glucosinolate compound namely glucoraphanin to sulforaphane and degrade tannin to gallic acid which have potential as an antioxidant. The effect of NaCl addition and incubation time to fermentation product will also be studied.

2. Materials and Methods

2.1. Chemicals and Instrumentation

Ethanol 90 %, Na_2CO_3, dichloromethane, Folin-Denis reagent were obtained from Merck, and 2,2-diphenyl-1-picrylhydrazyl (DPPH) reagent was obtained from Sigma-Aldrich. The medium for *L. plantarum* cultivation was MRS media (de Mann Rogosa Sharpe) (OXOID). UV-Vis Spectrophotometry was used to determine gallic acid level and antioxidant potential.

2.2. Microorganism Preparation

Lactobacillus plantarum was supplied by Microbiology Laboratory, Brawijaya University. The bacteria were grown in MRSA and incubated for 48 h at 30 °C. The bacteria from MRSA was transferred into MRS broth and incubated for 26 h at 30 °C.

2.3. Fermentation

Fresh cabbage was prepared by slice off into strips and washing with water until clean. To study the effect of NaCl addition, each 100 g cabbage which had been prepared into five different fermentation vessels. Each vessel added with NaCl at various concentration of 0 %, 0.5 %, 1.0 %, 1.5 %, and 2.0 %. To study the effect of incubation time, five vessels of fermentation that have been filled with cabbage and NaCl in optimum condition were fermented with time variations from 2 until 6 days. Fermentation was carried out at room temperature, using an inoculum volume of 5 % (v/v), pH 6 in anaerobic conditions

2.4. Gallic Acid Determination

2.5 g of biomass from cabbage fermentation was pureed and extracted using 20 mL of hot water. The extract was stirred and heated on a hotplate for 30 minutes. The extract was filtered and the filtrate was added with hot water until the volume reached 25 mL. After that, 0.2 mL of extract was added 0.6 mL of distilled water and 0.2 mL of Folin-Denis reagent. After 5 minutes, 1 mL of sodium carbonate solution (8 %, w/v) and 1 mL of distilled water was added to the mixture. The solution was mixed and allowed to stand up to 30 minutes in a dark room. The solution was measured at a wavelength of 765 nm. Meanwhile, the filtrate from fermented cabbage was centrifuge for 10 minutes. The supernatant was used for gallic acid determination in the filtrate from fermentation using the same procedure with gallic acid determination in biomass. Gallic acid levels were determined by gallic acid standart curve (5 – 400 mg/L) [17].

2.5. *Antioxidant Test*
To test the antioxidant potential of gallic acid, the supernatant from the filtrate was diluted with various concentrations of 10 – 50 mg/ L. 6 mL of sample was added 4 mL of DPPH solution (20 μg/ mL). The solution was allowed to stand for 30 minutes in a dark room and measured at a wavelength of 515 nm. Meanwhile, to extract the sulforaphane, 2 g of the biomass was mashed using mortar and pestle. Then, the sample was extracted using dichloromethane (3 x 10 mL) solvent, and Na_2SO_4 solution was added. After that, the solution was evaporated using a rotary evaporator at 30 °C for 10 minutes. The extract was diluted with various concentrations of 20 – 100 mg/ L. The antioxidant potential determination in biomass was done using the same procedure as the antioxidant potential determination of gallic acid in filtrate.
Inhibition of DPPH was calculated using following formula [18]:

$$\% \text{ inhibition} = (A_0 - A/ A_0)\times 100 \quad (1)$$

notes: A = sample + DPPH absorbance; A_0= sample absorbance
The results of calculations were plotted into a curve and used to determine the value of IC_{50}.

3. Result and Discussion

3.1. *Optimization of NaCl Addition*
The fermentation process of cabbage can cause cell wall damage. The addition of NaCl which used in this study using various concentration between 0 – 2.0 %.

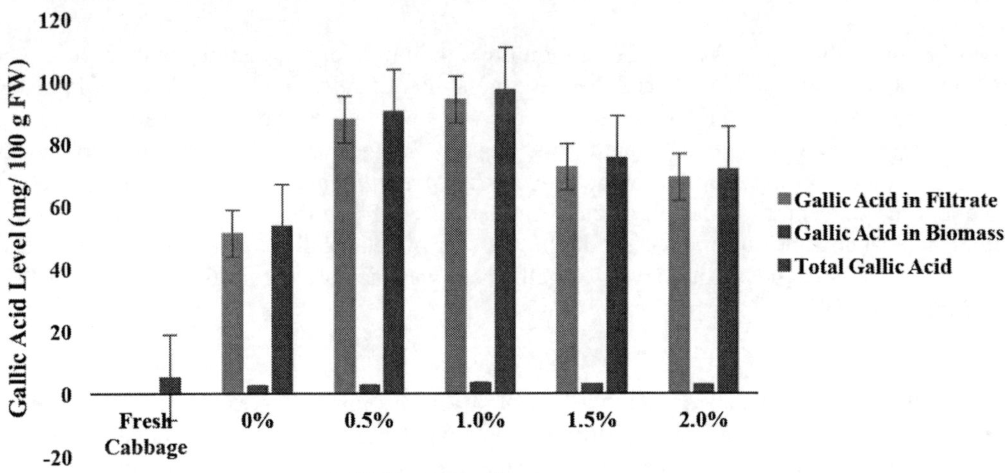

Figure 2. The effect of various NaCl concentration on gallic acid level during cabbage fermentation

IOP Conf. Series: Materials Science and Engineering **833** (2020) 012054 doi:10.1088/1757-899X/833/1/012054

Based on Figure 2, fermentation process can increase the levels of gallic acid in the filtrate, whereas the levels of gallic acid in biomass was decreased (significant value: 0.000 < 0.05). This is because gallic acid is easily soluble in water [19], so there are more gallic acid levels in the filtrate than biomass. The amount of NaCl addition also affects the levels of gallic acid production during the fermentation process. The more amount of NaCl added until 1% (w/v), the more production of gallic acid, both in the filtrate and biomass. However, at the addition of NaCl concentration of 1.5 %, gallic acid level was decreased, both in the filtrate and biomass. So, the optimum concentration of salt addition is 1.0%. At this condition, the total levels of gallic acid was increased from 4.983 mg/100 g FW into 97.236 mg/100 g FW.

3.2. Optimization of Incubation Time

Incubation time can affect fermented products. Therefore, various incubation time ranging from 2 – 6 days were studied for increase the resulting gallic acid.

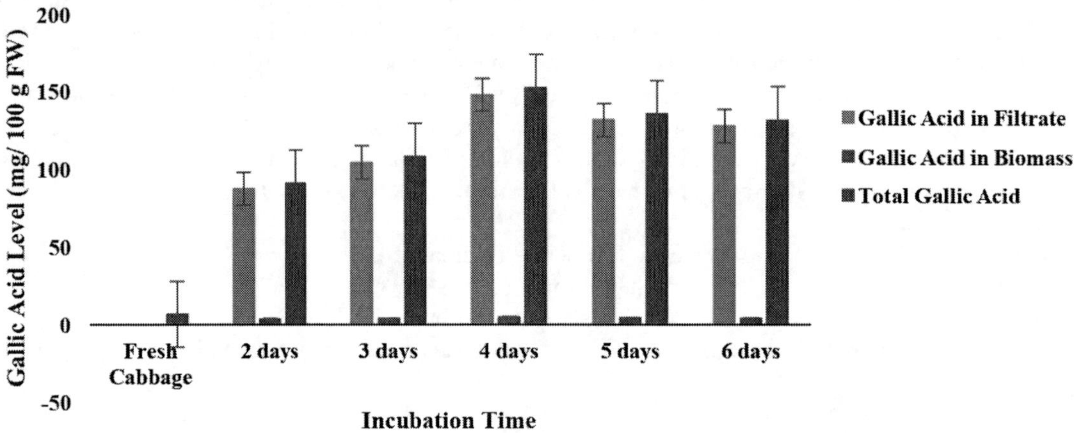

Figure 3. The effect of various incubation time on gallic acid level during cabbage fermentation

Figure 3 shows that the longer the incubation time, the levels of total gallic acid will also increase (significant value: 0.001 < 0.05). This is because the ability of *Lactobacillus plantarum* in producing tannase enzymes to degrade tannin will also increase. However, after 4 days, the resulting of total gallic acid was decreased. Based on the standard growth curve, *Lactobacillus plantarum* bacteria undergo a phase of death after 4 days, so that the activity of *Lactobacillus plantarum* in secreting the tannase enzyme and the concentration of gallic acid were decreased. From this result, it can be seen that the optimum incubation time to produce the highest levels of total gallic acid during the fermentation process is 4 days. For 4 days, total levels of gallic acid was increased from 6.462 mg/100 g FW into 152.583 mg/100 g FW.

3.3. The Antioxidant Potential

The main characteristic of antioxidants is their ability to capture the presence of free radicals and forming more stable free radicals [20]. To test the potential of compounds as antioxidants, it can be done using the DPPH method. Compounds that have potential as antioxidants can neutralize radicals from DPPH, where there is a change in the color of the solution from purple to colorless.

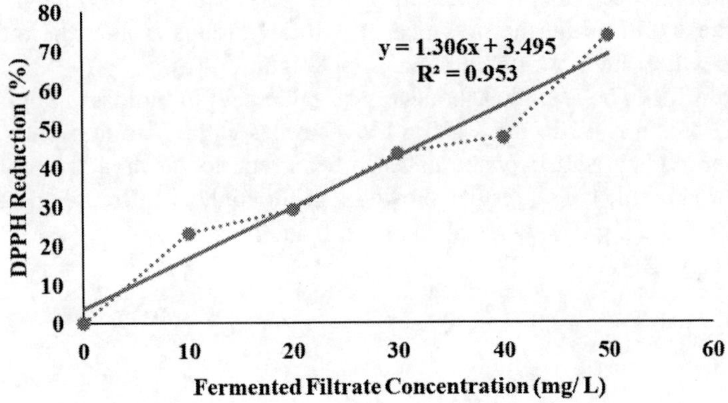

(a)

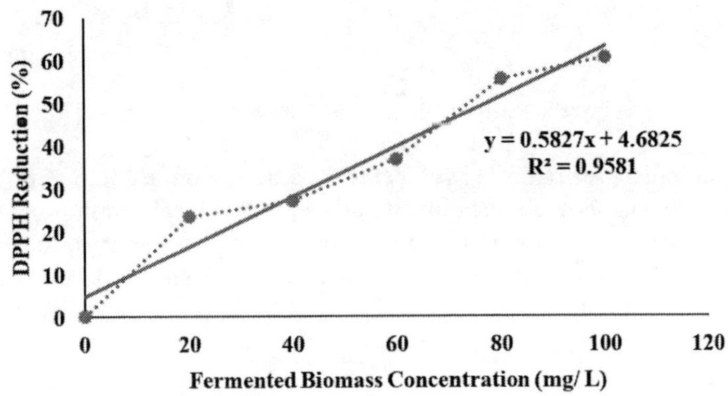

(b)

Figure 4. The antioxidant potential of filtrate (a) and biomass (b) from fermented cabbage

From Figure 4, it can be known that the IC_{50} of filtrate and biomass values are 35.595 mg/L and 77.777 mg/L, respectively. The IC_{50} values indicate how much concentration of compounds needed to inhibit 50% of free radicals. The higher the concentration of the sample, the higher the amount of DPPH reduction, which was marked by the decreasing of absorbance value.

Based on the results of determining gallic acid levels (figures 2 and 3), it can be seen that in the fermented filtrate at optimum conditions has high gallic acid levels, which is 147.857 mg/ 100 g FW. Thus, the possibility of active compounds in the filtrate that have the potential as antioxidants is gallic acid. In addition, there is also a study which states that gallic acid can act as an antioxidant [6].

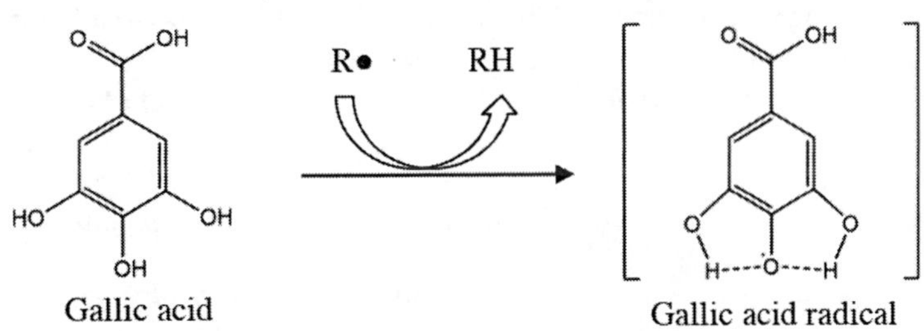

Figure 5. Gallic acid reaction as antioxidant [6]

Gallic acid can neutralize free radicals by donating hydrogen atom from the OH group. Radicals formed in gallic acid can be stabilized by the presence of two OH groups around the radicals, so gallic acid radicals are formed which are more stable radicals (see figure 5) [6].

In figures 2 and 3 it can be seen that the levels of gallic acid in biomass at optimum conditions are less than in filtrate, which is 4.726 mg / 100 g FW. Besides gallic acid in biomass which is an active compound as an antioxidant, sulforaphane has also been reported as an active compound that has the potential as an antioxidant [7]. The results showed that the highest sulforaphane levels were found in cabbage, which was 540 µg/ g FW, whereas in broccoli and brussels sprout were 220 µg/ g FW and 120 µg/ g FW, respectively [21].

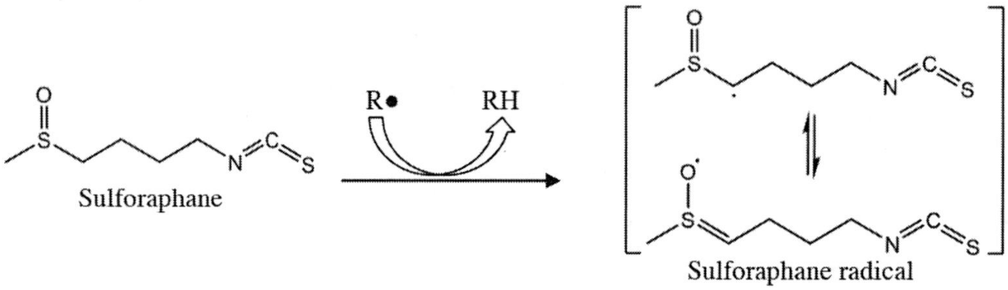

Figure 6. Sulforaphane reaction as antioxidant [7]

Sulforaphane can neutralize free radicals by donating hydrogen atom from the CH_2-SO-CH_2- group, and finally form a new more stable free radical due to an electron delocalisation process (see figure 6) [7]. It can be seen that filtrate and biomass from fermented cabbage contain gallic acid and sulforaphane active compound, where both of these compounds can react with source of free radicals in the body with the mechanism as shown in figure 5 and 6 [6,7]. So, that they have the potential as antioxidant [4,5].

4. Conclusion

This study reveals that the NaCl addition and incubation time can affect gallic acid level in cabbage by simultaneous fermentation using *Lactobacillus plantarum*. The results showed that the most favourable conditions to increase total gallic acid level by *L. plantarum* fermentation were: NaCl 1 g and 4 d incubation time, as for total gallic acid increase from 6.462 mg/100 g FW into 152.583 mg/100 g FW The antioxidant potential of filtrate and biomass were shown by their IC_{50} values of 35.595 mg/L and 77.777 mg/L, respectively.

References

[1] Ren B Mingbo W Qin W 2013 *Mol. Biol.* **425** 2737-2751

[2] Benzekri R Lamjed B Adele P Medji S Imen B Majdi H Ferid L 2015 *Industrial Crops and Products* **80** 228-234

[3] Juengel E, Holger H H E, Axel H, Jochen R, Felix K H, Chun, Roman A B 2018 R *Cancer Letters* **435** 121-126

[4] Kong L Bo L Chenghong Z Bing W Hongfei W Xioxia S Yang Y Xiang R Liangwei Y Hui K, Haiying M 2016 *Neurochemistry International* **100** p 52-61

[5] Gemede H F and Negussie R 2014 *International Journal of Nutrition and Food Sciences* **3** 4 284-289

[6] Bharti B Neha S Rita K 2015 *Royal Society of Chemistry.* **5** 27540-27557

[7] Yuan H Yao S You Y Xiao G You Q 2010 *Chinese Journal of Chemical Engineering* 2 **18** 312-321

[8] Vig A P Geetanjali R Tarunpreet S T Saroj A 2009 *LWT - Food Science and Technology* **42** 1561–1572

[9] Malaterre A S Fabienne R Patrick P 2017 *Food Research International* **104** 86-99

[10] Tolonen M S Rajaniemi J M Pihlava T Johansson P E J Saris E L, Ryhanen 2004 *Food Microbiology* **21** 167-179
[11] Natarajan K and Aravindan R 2009 *E-Journal of Chemistry* **6** 4 979-984
[12] Fahmia A R Arie S Sasangka P Anna S 2019 *Material Science and Engineering* **546**
[13] Ciska E and Dorothy R P 2004 *J. Agric. Food Chem.* **52** 7938-7943
[14] Johanningsmeier S, Roger F M, Henry P F, Roger L T 2007 *Food Microbiology and Safety* **72** 5 166-72
[15] Zinno P Barbara G Giuditta P Gianni P Chiara D 2017 *Food Microbiology* **63** 239-247
[16] Jimenez Natalia Maria E T Jose M M Blanca R Rosario M 2014 *Applied and Environmental Microbiology* 80 **10** 2991-2997
[17] Chandra S Shabana K Bharathi A Hemant L Min H Y Mahmoud A E and Ikhlas A K 2014 *Research Article.* **9**
[18] Demirbas A Bruce A W Ismail O 2016 *Material Letters* **179** 20-23
[19] Daneshfar A Hassan S G Nasrolah H 2008 *J. Chem. Eng. Data* **53** 776-778
[20] Prakash D Upadhyay G Pushpangadan P Gupta C 2011 *J Complement Integr Med* **8**
[21] Mohamed A F Amira A A M 2010 *Journal of Advanced Research* **1** 65-70

Purification of Keratinase from *Bacillus sp.* MD24 using Ammonium Sulfate Fractionation

U F Zuhriyah, E Susanti, and Suharti*

Department of Chemistry, Faculty of Mathematics and Natural Sciences, Universitas Negeri Malang, Jl. Semarang No. 5 Malang 65145, Indonesia

*Corresponding author: suharti.fmipa@um.ac.id

Abstract. A keratin degrading bacterium, *Bacillus sp* MD24, was isolated from soil. The crude keratinase produced by the bacterium has been reported to dehair goat skin. However, the dehairing process took 72 hours. In order to shorten the dehairing time it is necessary to increase the keratinase concentration. This could be done by optimizing keratinase production either finding the best fermentation media or optimizing fermentation condition. Another way to increase the concentration could be done by partially purifying the enzyme. Keratinase from *Bacillus sp* MD24 had been produced under submerged fermentation, however, it produced a relatively low amount of enzyme. Although an effort to increase enzyme production had been reported by solid state fermentation, the enzyme concentration was not enough for industrial purposes. This work aimed to increase enzyme concentration by partial purification through enzyme precipitation using ammonium sulphate. The research was conducted in three stages: (1) regeneration the bacterium, (2) production of keratinase, and (3) purification of keratinase with ammonium sulfate fractionation. Keratinase activity was measured by Anson method and protein concentration was measured by Lowry method. Enzyme purity was clarified using a combination of specific activity, purity level, and SDS-PAGE analysis. Based on the analysis result, ammonium sulphate did not act as a good precipitation agent for the keratinase. Two major bands were suggested as keratinase with an estimated molecular weight of 25 and 66 kDa as monomer and dimer form, respectively.

1. Introduction

Industrial process in which using chemicals has been affecting the environmental quality and health of people around the world. Enzymatic processes offer to replace these noxious process for a better life. In the leather industry, various enzymes specifically from microbial proteolytic are the most essentially used. Dehairing is an important step during leather processing. Hair is composed primarily of strong fibrous proteins named as keratin which contains a large amount of cysteine residues. The cysteine residues form a strong covalent disulphide bond. Enzymatically this strong chemical bond can be disrupted by a class of protease named as keratinase.

Microbial keratinases had been reported to be promising enzymes for application in dehairing process. Keratinolytic enzymes or called keratinase (E.C 3.4.21 / E.C 3.4.24 / E.C 3.4.99) is one of the serine- or/and metallo-hydrolase groups that disrupt the peptide and disulfide bond in the keratin. Various extracellular keratinase showed capability to degrade keratin, such as keratinase produced by *Bacillus amyloliquefaciens* S13 [1], *B. amyloliquefaciens* TCCC11319 [2], *Bacillus cereus* [3], *Bacillus cereus* PCM 2849 [4], *Bacillus sp* JM7 [5], *Bacillus sp* [6], *Bacillus licheniformis* [7], *Bacillus licheniformis* N5 [8], *Bacillus subtilis* LFB-FIOCRUZ 1266 [9].

Keratinases have a very high diversity in term of biochemical and biophysical properties among microorganisms isolated from different environmental. In addition, the needs of the global markets are increasingly becoming the main reason for the leather industry to produce leather quickly, the use of minimum materials, and good quality products. Therefore, the production of keratinase from new bacteria with better dehairing capabilities is a considerable prospect for industrial progress. A new *Bacillus sp.* was isolated from decomposed chicken feather containing soil, *Bacillus sp.* MD24 produces keratinase and the enzyme showed activity in dehairing of goat skin [10]. Based on the result of Scanning Electron Microscope Analysis, enzymatically and chemically dehaired goat skin were showed similar or improved surface of the skin, which makes the keratinase from *Bacillus sp.* MD24 as potential candidate for application specifically in industrial process such as leather industry to avoid problems due to the use of chemicals. However, the dehairing process took 72 hours and was reported that the enzyme concentration was still relatively low (0.3 U/mL). In order to shorten the dehairing time, it is necessary to increase the keratinase concentration.

The optimization of keratinase producing either finding the best fermentation media or optimization of fermentation condition had been attempted. The keratinase had been produced under *Submerge Fermentation* (SmF) with variations in additional carbon sources, pH, and temperature. The fermentation result showed that the activity of the enzyme was still low (6.09 U/mL). The *Solid State Fermentation* (SSF) method also had been applied, but the enzyme concentration produced by this method was not enough for industrial purposes [10][11]. Another alternative to increase the keratinase concentration is preconcentration of the enzyme by enzyme purification methods. The enzyme purification is a method for obtaining pure enzyme without the presence of contaminants such as residual metabolites from microbes, residual isolation media and other mineral salts. The loss of contaminants that will increase the concentration of the enzyme so that the enzyme activity increases. Contaminants are likely to interfere with the enzyme stability, so that reduce or eliminate its activity. This paper reports our work on keratinase purification from *Bacillus sp.* MD24 using precipitation method by partial ammonium sulfate fractionation.

2. Materials and methods

2.1. Keratinase production

Bacillus sp. MD24 was isolated from decomposed chicken feather containing soil [10]. The fermentation method was undertaken according to [11] procedures with modification. The bacteria was grown under solid state fermentation media at water concentration 500% with chicken feather as sole carbon and nitrogen sources, and maintained at 37 °C for three days. The crude extract keratinase was filtered and recovered in the culture filtrate by centrifugation at 10.000 rpm for 10 minutes at ± 4 °C.

2.2. keratinase Assay

Keratinase activity was assayed by the modified protease colorimetric method using tyrosine as the standard. One unit (U) of protease activity was defined as the amount of enzyme required to liberate 1 μmol of tyrosine at the defined assay conditions.

2.3. Protein Assay

Protein concentration was measured by the modified method of Lowry et al. [12] using bovine serum albumin (BSA) as the standard.

2.4. Keratinase purification

All operation was performed at ± 4 °C. The crude extract of keratinase was precipitated by the gradual addition of solid ammonium sulfate with gentle stirring in the following saturation ranges 0-25%, 25-50%, 50-75%, and 75-100%. The precipitate was collected by centrifugation at 10.000 rpm for 10 minutes. The precipitate was then dissolved in a minimal volume of 50 mM Tris-HCl $CaCl_2$ 2 mM buffer (pH 8.0) and desalted by dialysis against the same buffer. Keratinase activity and protein concentration from each fraction were assayed.

2.5. Electrophoretic methods
Sodium dodecyl sulphate – polyacrylamide gel electrophoresis 12% was carried out according to the Laemmli method [13] and stained with Coomassie brilliant blue.

3. Result and discussion

3.1. Purification of extracellular keratinase
A keratinase crude extract was purified from the fermentation filtrate of *Bacillus sp.* MD24. The crude extract was purified by precipitation using $(NH_4)_2SO_4$ fractionation (0-25%, 25-50%, 50-75%, and 75-100%). The activities and yield during the purification processes are summarized in Table 1. All fractions showed almost no specific activity and purity difference compared to the crude extract. Total activity of all fractions is 735.20 U, it is about 14.36% of total activity in the crude extract. Furthermore, in the remaining supernatant, it still contained keratinase activity of 4200 U or about 84.17% of total activity in the crude extract. It means ammonium sulphate did not act as a good precipitation agent for keratinase from *Bacillus sp* MD24. However, ammonium sulphate might act as a good precipitation agent of impurities protein which lead to increasing purity of keratinase.

Table 1. Summary of the purification of a keratinase from *Bacillus sp.* MD24.

Fraction	Total Enzyme Activity (U)	Total Protein (mg)	Specific Activity (U/mg of Protein)	Purity Level (fold)	Yield (%)
Crude Extract	5111.74	1493.80	3.42	1.00	100.00
0-25 %	45.63	12.87	3.54	1.03	0.89
25-50 %	184.44	53.53	3.44	1.00	3.61
50-75 %	290.69	75.66	3.84	1.12	5.69
75-100 %	213.46	59.40	3.59	1.05	4.17

The profile of protein content in crude extract and all fractions was analyzed by SDS-PAGE is shown in Figure 1. The profile showed strong bands were estimated molecular weight of 25 and 66 kDa. This is almost similar to the keratinase produced from *Bacillus amyloliquefaciens* strain S13, it has appeared two protein bands at 28 kDa and 47 kDa. It was reported that the two protein bands are monomers [1]. Compared to keratinase from the literature, the first protein band of keratinase from *Bacillus sp.* MD24 has an almost similar molecular weight, but the second protein band is different, which was higher than of other keratinases secreted by *Bacillus amyloliquefaciens* strain S13 (28 kDa and 47 kDa) [1], *Bacillus amyloliquefaciens* strain TCCC11319 (28 kDa) [2], *Bacillus subtilis* strain SCK6 (30.95 kDa) [14]. However, it is possible that two protein bands of keratinase from *Bacillus sp.* MD24 are monomeric and dimeric form respectively. In addition, other literature keratinase from *Bacillus subtilis* MTCC (9102) reported that has a molecular weight range between 64-69 kDa [15], Thus, the two bands suggested as keratinase. However, further procedure must be done such as chromatography or organic solvent precipitation, especially on the remaining fraction. SDS-PAGE also must be done to see the protein profile of the remaining fraction.

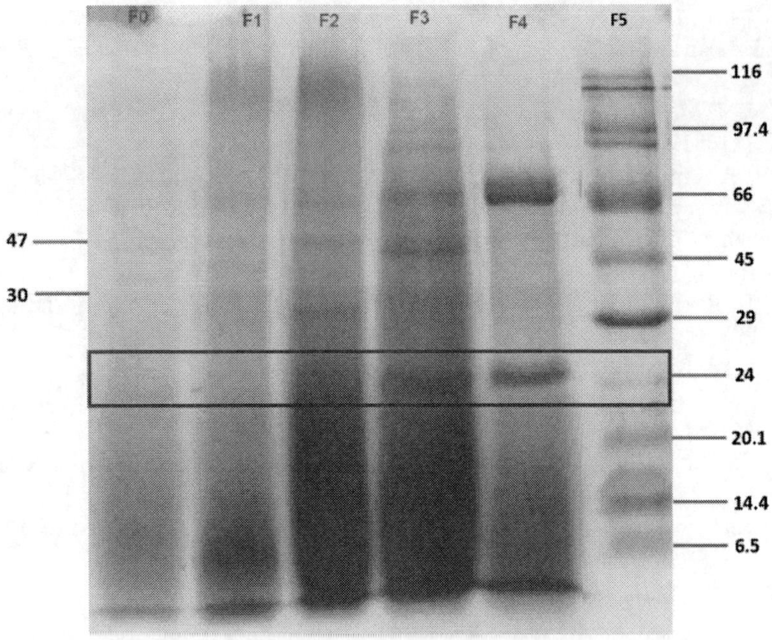

Figure 1. SDS-PAGE (12%) of purified keratinase from *Bacillus sp.* MD24. Lane F0: Crude Extract. Lane F1: Purified keratinase at ammonium sulphate between 0-25% saturation. Lane F2: Purified keratinase at ammonium sulphate between 25-50% saturation. Lane F3: Purified keratinase at ammonium sulphate between 50-75% saturation. Lane F4: Purified keratinase at ammonium sulphate between 75-100% saturation; and Lane F5: The standard molecular weight marker proteins (unstained broad range protein marker bioner).

4. Conclusion

A new keratinase produced by *Bacillus sp.* strain MD24, newly isolated from soil, was partially purified. Purification of the keratinase using ammonium sulfate fractionation methods. Based on the result, ammonium sulfate did not act as a good precipitation agent for the keratinase. Enzyme and protein assay were showed almost no specific activity and purity difference compared to the crude extract. The protein profile showed strong bands estimated molecular weight of 25 and 66 kDa.

Acknowledgments

PNBP funding has supported this research with the contract number of 20.3.116/UN32.14.1/LT/2019 which addressed to Dr. Suharti and Team, Department of Chemistry, Faculty of Mathematics and Natural Sciences, Universitas Negeri Malang.

References

[1] Hamiche S, Mechri S, Khelouia L, El Hattab M, Badis A and Jaouadi B 2019 *International Journal of Biological Macromolecules* **122** 758-769.
[2] Cao S, Li D, Ma X, Xin Q, Song J, Lu F and Li Y 2019 *World Journal of Microbiology and Biotechnology* **35**(122) 1–10.
[3] Arokiyaraj S, Varghese R, Ahmed B A, Duraipandiyan V and Al-Dhabi N A 2019 *Saudi Journal of Biological Sciences* **26** 378–381.
[4] Ciurko D P, Łaba W, Piegza M, Juszczyk P, Choińska-Pulit A and Sobolczyk-Bednarek J 2019 *Polish Journal of Chemical Technology* **21**(3) 53–59.
[5] Jin M, Chen C, He X and Zeng R 2019 *Acta Oceanologica Sinica* **38**(2) 87–95.
[6] Dada M T and Wakil S M 2019 *Biotechnology Journal International* **23**(1) 1–12.
[7] El-komy H M, Al-Dosary S K, El-Naghy M A, Abdelhamid M A A and Immam M M 2019

Journal of Advanced Biomedical and Pharmaceutical Sciences **2** 41-46.

[8] Hegazy W K and Abdel-Salam M S 2019 *Egyptian Journal of Genetics and Cytology* **48** 91-105.

[9] de Paiva D P, de Oliveira S S, Mazotto A M, Vermelho A B and de Oliveira S S 2019 *3 Biotech*, **9**(2) 1-12.

[10] Suharti S, Riesmi M T, Hidayati A, Zuhriyah U F, Wonorahardjo S and Susanti E 2018 *Pertanika Journal of Tropical Agricultural Science* **41**(3) 1449–1461.

[11] Nurkhasanah U and Suharti 2019 *IOP Conference Series: Earth and Environmental Science* **276** 012016.

[12] Lowry O H, Rosebrough N J, Farr A L and Randall R J 1951 *Journal of Biological Chemistry* **193** 265-275

[13] Chamberlin M, McGrath J and Waskell L 1970 *Nature* **228** 227–231.

[14] Tian J, Xu Z, Long X, Tian Y and Shi B 2019 *International Journal of Biological Macromolecules* **135** 119–126.

[15] Balaji S, Kumar M S, Karthikeyan R, Kumar R, Kirubanandan S, Sridhar R and Sehgal 2008 *World Journal of Microbiology and Biotechnology* **24**(11) 2741–2745.

IOP Conf. Series: Materials Science and Engineering **833** (2020) 012056 doi:10.1088/1757-899X/833/1/012056

Synthesis of 5-(4-bromobutyl)-2,3-dimethyl-1,4-benzoquinone and the Solubility Test in *n*-Octanol/Water Partition System

Nadiyah Zuhroh[1] and Siti Mariyah Ulfa[1*]

[1]Chemistry Department, Faculty of Science, Brawijaya University, Jl. Veteran, Malang, 65145, Indonesia

*Corresponding author's: ulfa.ms@ub.ac.id

Abstract. The modification structure of 1,4-benzoquinone is fascinating in the frame of drug design. The solubility properties of a new drug are important, especially for oral drug administration. The 2-isopropyl-5-methyl-1,4-benzoquinone (known as thymoquinone) which is the major constituent in *Nigella sativa* seed extract is known as an active compound with poor lipophilicity. Here, we reported the synthesis of 5-(4-bromobutyl)-2,3-dimethyl-1,4-benzoquinone which analog with thymoquinone having higher lipophilicity. The synthesis proposed by oxidation reaction of 2,3-dimethyl-1,4-hydroquinone **(1)** with H_2SO_4 followed by bromoalkylation reaction using bromopentanoic acid in the presence of $AgNO_3$ and $(NH_4)_2S_2O_8$. The oxidation product 2,3-dimethyl-1,4-benzoquinone **(2)** and alkylated-quinone 5-(4-bromobutyl)-2,3-dimethyl-1,4-benzoquinone **(3)** were obtained in 52.64% and 5.85%, respectively. The FTIR analysis of **3** showed the additional C-Br stretching at 562 cm^{-1}. The solubility test in *n*-octanol/water system using HPLC gave the log P value of 5-(4-bromobutyl)-2,3-dimethyl-1,4-benzoquinone **(3)** is 2.99, higher than thymoquinone. By this result, it is suggested that the modification of 1,4-benzoquinone by bromo alkylating agent increased the solubility of the compound.

1. Introduction

The development of herbal medicines in Indonesia is currently increasing since Indonesia has large biodiversity for the medicinal plant. One of the herbal medicines is black cumin or *Nigella sativa*. *Nigella sative* was found in Mediterranean countries and Southeast Asia, including Indonesia. *Nigella sativa* seed oil was reported to contain with 2-isopropyl-5-methyl-1,4-benzoquinone, trivially known as thymoquinone (18.4-24%). Thymoquinone is an active ingredient that was reported to have hepatoprotective through antioxidant and anti-inflammatory properties [1]–[3].

A new drug design should be considered the bioavailability issue, involved the balance of lipophilicity and hydrophobicity. Drugs should be able to penetrate the membrane cell and then give the biological effects. In previous research, oral administration of thymoquinone in rabbits showed a rapid elimination and slow absorption rate in the body. It also reported that the partition coefficient (log P) of thymoquinone is 2.55 [4]. According to Lipinsky *et al.*, drug candidates should have log *P* less than 5 [5]. This is the main reason for drug development in the first step.

The bioavailability of a drug candidate can be improved by modifying the chemical structure. Antonenko and the groups (2008), reported the modification of quinone derivatives by the addition of an alkyl group. The results of the modification showed increasing lipophilicity and antioxidant activity in the mitochondrial tissue [6]. In accordance with this report, Ulfa *et al.* published the bromoalkylation or alkyl phosphonium substitution into 1,4-benzoquinone. It is reported that the synthesized compound has an increase in its lipophilicity. The *in-silico* determination suggested the activity of 1,4-benzoquinone derivatives, such as, 3-(10-bromodecyl)-5-isopropyl-2-methyl-1,4-benzoquinone, 5-(7-bromoheptyl)-2,3-dimethyl-1,4-benzoquinone, and 5-(10-bromodecyl)-2,3-dimethyl-1,4-benzoquinone have the ability to work as anticancer drug candidate [7]–[10].

Based on this background, the synthesis of 5-(4-bromobutyl)-2,3-dimethyl-1,4-benzoquinone (**3**) was carried out. The structure of **3** is the mimic of thymoquinone with the longer alkyl group in order to increase the lipophilicity. The solubility of **3** in *n*-octanol/water system is determined and compared with thymoquinone.

2. Material and method

2.1. Material and instrumentation

2,3-Dimethyl-1,4-hydroquinone (**1**), bromopentanoic acid, $(NH_4)_2S_2O_8$ by Sigma Aldrich Singapore. $KBrO_3$, Na_2SO_4, $AgNO_3$ anhydrous by Merck, Singapore. A solution of H_2SO_4 2.5 M was prepared by diluted 98% H_2SO_4 from Smartlab, Indonesia. The solvents chloroform, n-hexane, diethyl ether, ethanol, acetonitrile, and n-octanol by Merck.

The instruments used are Ohaus analytical balance, IKA RV 10 digital rotary evaporator, Orbital Shake Brand Edmund Buhler SM 25, FTIR spectrophotometer Shimadzu 8400S, spectrophotometer UV-Vis double beam Shimadzu 1600, ^{1}H-NMR spectrometer Varian 600 MHz, and LC-20AD Prominence HPLC spectrophotometer from Shimadzu.

2.2. Method

2.2.1. Synthesis of 2,3-dimethyl-1,4-benzoquinone (2). The synthesis of 2,3-dimethyl-1,4-benzoquinone was carried out according to the previously published method [7], by reacting 9 mmol of 2,3-dimethyl-1,4-hydroquinone with 3 mmol of $KBrO_3$. 9 mL H_2O and 0.45 mL H_2SO_4 2.5 M was added while stirring at temperatures 40-50°C for 60 minutes. After completion, the reaction mixture was extracted using diethyl ether. The organic layer is separated and dried using anhydrous Na_2SO_4 then vacuum evaporated. The crude product was purified by column chromatography using SiO_2 column with eluent n-hexane: $CHCl_3$ (7:3 (v/v)). The product was analyzed using TLC, UV-Vis spectrophotometer, FTIR spectrophotometer, and ^{1}H-NMR.

2.2.2. Synthesis of 5-(4-bromobutyl)-2,3-dimethyl-1,4-benzoquinone (3). The synthesis of **3** is carried out by firstly reacting (2.1 mmol) of bromopentanoic acid and $AgNO_3$ (1 mmol) in 7 ml of the mixture of acetonitrile and water (2:1 v/v), then stirred until 90°C. After that, added with the solution of 2,3-dimethyl-1,4-benzoquinone (**2**) (2 mmol) and $(NH_4)_2S_2O_8$ (2 mmol) in 3 mL H_2O added dropwise to the mixture while stirring followed by heating at 80-90°C for 3 hours. The reaction mixture then extracted with diethyl ether, dried the organic layer using Na_2SO_4 anhydrous. The solvent was evaporated with a rotary evaporator. The product was purified by SiO_2 column chromatography using n-hexane: $CHCl_3$ (7:3, v/v). The product was analyzed using TLC, UV-Vis spectrophotometer, FTIR spectrophotometer.

2.2.3. Solubility test in n-octanol/water system. The solubility test is perform followed Andres methods by some modification [11]. Prior to the solubility test, n-octanol and water system (phosphate buffer pH 7.4) were prepared by saturation. Phosphate buffer solution pH 7.4 made by mixing 67 mL of 0.1 M $NaH_2PO_4.H_2O$ mixed with 330 mL $Na_2HPO_4.7H_2O$ 0.1 M, adjusted to pH of solution 7.4. The water system was saturated with n-octanol. The saturation is carried by taking a 20 mL water phase and added dropwise with n-octanol, shaking in separating funnel, and lay aside until complete separation then separated the water phase from an n-octanol layer. The n-octanol was also saturated with water with a similar procedure. The saturated n-octanol and water phase are then put in a separating funnel, shaken and then allowed to stand for at least 24 hours.

The synthesized compound **3** (10 mM) and thymoquinone are dissolved in saturated *n*-octanol, then the saturated aqueous phase is added with the volume ratio 3:7 (*n*-octanol: water). This solution is shaken with an orbital shaker for 1 hour at room temperature (200 rpm). The *n*-octanol phase and water phase are separated. The concentration in each solution is measured using HPLC methods.

2.2.4. Measurement of solubility test using HPLC. The HPLC analysis was carried out by injecting 2 μL of the sample into the HPLC instrument with a flow rate of 0.5 mL/min. The detector is set 254 nm for thymoquinone and 259 nm for compound **3**. The column temperature is set at 37°C using isocratic phase acetonitrile:water (70:30 (v/v)) and 0.5% of acetic acid. The log P was calculated by equation (1), where $C_{octanol}$ and C_{water} represent, the total drug concentration in octanol and water phase, respectively. Thus, total drug concentration is derived by peak areas of compound **3** in octanol (A_o) and in water (A_w) and the volumes of octanol (V_o) and in water (V_w) partition [11].

$$\text{Log P} = \log \frac{C\ octanol}{C\ water} = \log \frac{Ao/Vo}{Aw/Vw} \quad (1)$$

3. Result and discussion

3.1. *Synthesis of 2,3-dimethyl-1,4-benzoquinone (2)*

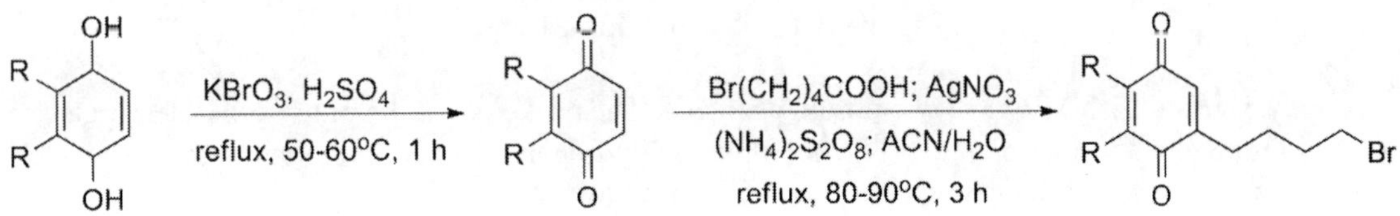

Scheme 1. The synthesis of 1,4-benzoquinone derivatives.

Table 1. The physical properties of the synthesized 1,4-benzoquinone derivatives

Compound	Rf[a]	Wavelength (nm)[b]	Yield (%)
2,3-dimethyl-1,4-hydroquinone (**1**)	0.01	204	-
2,3-dimethyl-1,4-benzoquinone (**2**)	0.18	245	52.64
5-(4-bromobutyl)-2,3-dimethyl-1,4-benzoquinone (**3**)	0.23	259	5.85

a) TLC analysis, mobile phase n-hexane: chloroform (7:3)
b) UV-Vis analysis, ethanol solvent

The synthesis of 2,3-dimethyl-1,4-benzoquinone (**2**) was conducted by oxidation 2,3-dimethyl-1,4-hydroquinone (**1**) using H_2SO_4 and $KBrO_3$. Compound **2** was obtained as yellow solid in 52.64% by comparing the experimentally obtain product with the theoretically calculated product. The thin layer chromatography analysis showed a greater *Rf* value of **2** compared to starting material **1**. It is suggested that compound **2** has low polarity compared to **1**. Analysis using UV-Vis spectrophotometer gave a longer wavelength for compound **2**. The existence of electronic excitation from n→π* and π→π* should influence the bathochromic shift for compound **2**. The analysis using the FTIR spectrophotometer (**Figure 1**) showed the formation of carbonyl groups (C=O) at 1657 cm^{-1}. The disappearance of the hydroxyl group at 3262-3225 cm^{-1} showed that the oxidation of hydroquinone **1** into benzoquinone **2** is successful [7].

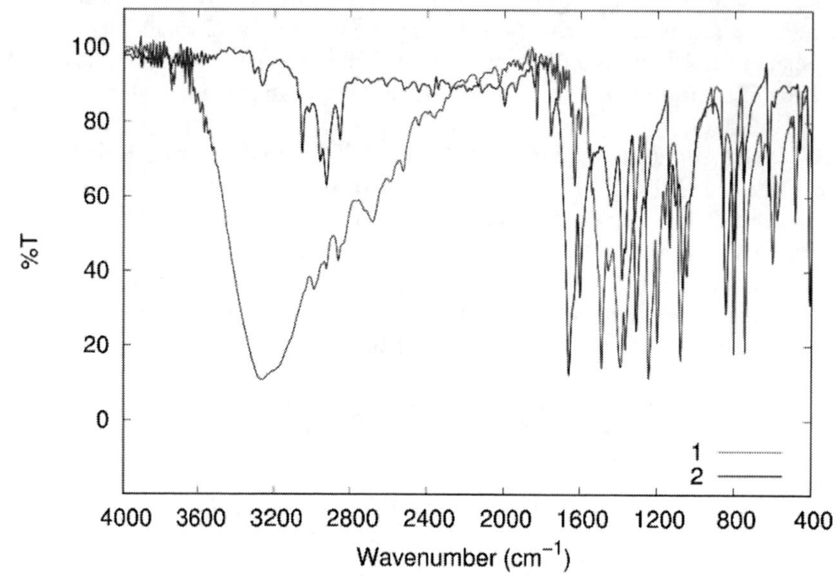

Figure 1. FTIR spectrum of **1** and **2**.

Analysis using ^{1}H-NMR gave two proton character. The peak at δ = 2.04 ppm (s, 6H) designated for methyl group and δ = 6.72 ppm (s, 2H) should be the hydrogen attached in the quinone ring. The ^{1}H-NMR spectrum of the oxidation product **2** is shown in **Figure 2**.

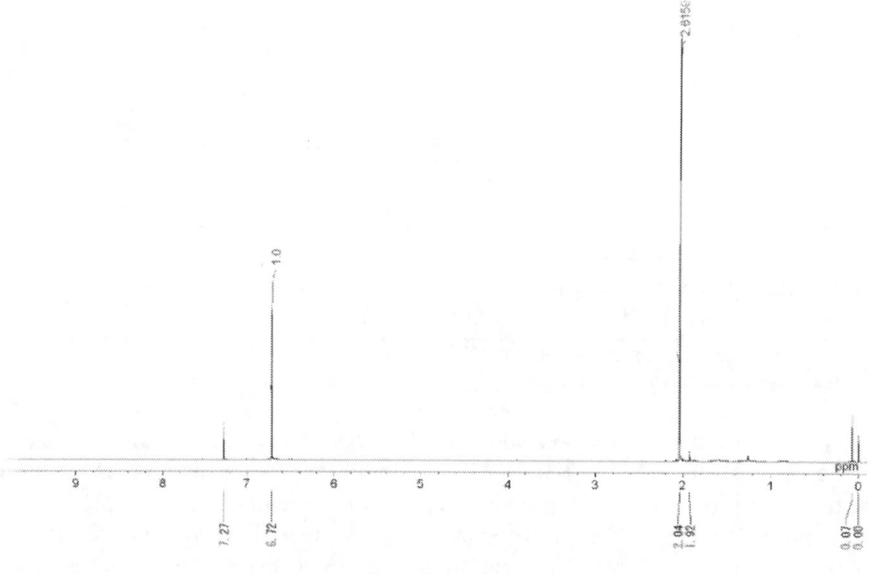

Figure 2. ^{1}H NMR of **2** in $CDCl_3$ (400 MHz).

3.2. Synthesis of 5-(4-bromobutyl)-2,3-dimethyl-1,4-benzoquinone (3)

The 5-(4-bromobutyl)-2,3-dimethyl-1,4-benzoquinone **(3)** was synthesized from previously prepared 2,3-dimethyl-1,4-benzoquinone **(2)** by bromoalkylation reaction using bromopentanoic acid. The obtained compound **3** is a yellow liquid in a 5.85% yield. The product was analyzed using TLC, UV-Vis, and FTIR spectrophotometer. TLC analysis showed that compound **3** has a greater *Rf* value than **2**, showed the product **3** has more non-polar properties. The UV-Vis analysis of **3** gave a greater

IOP Conf. Series: Materials Science and Engineering **833** (2020) 012056 doi:10.1088/1757-899X/833/1/012056

wavelength compared with **2**. The addition of alkyl in the beta position of the quinone ring is equal to 12 nm [11]. The additional wavenumber at 562 cm^{-1} showed the absorption of the halogen functional group (C-Br) (**Figure 3**). Based on the overall analysis, it can be suggested that the substitution of bromoalkyl is successful [7].

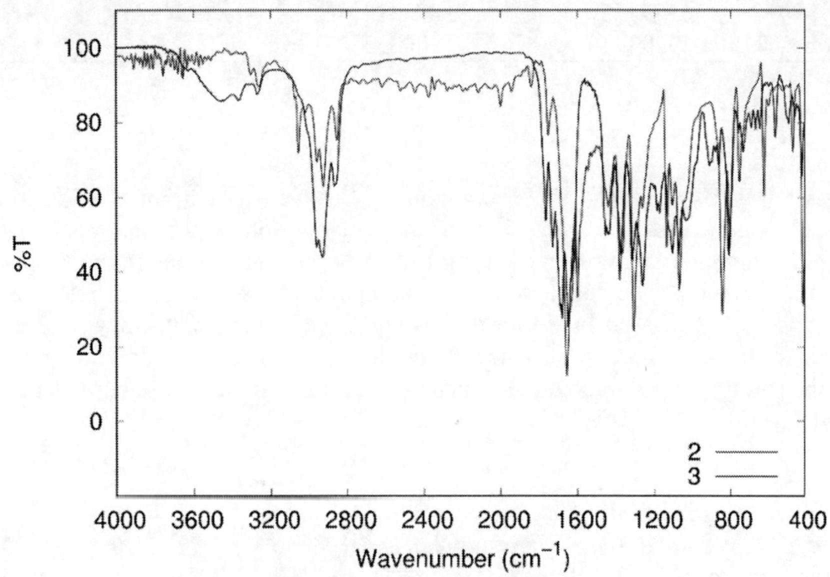

Figure 3. FTIR spectrum of **2** and **3**.

3.3. Solubility test in n-octanol/water partition system

The solubility test was carried out using the HPLC method. The HPLC profile of compound **3** in *n*-octanol and in water showed in **Figure 4**. The area in *n*-octanol and water was noted in **Table 2**.

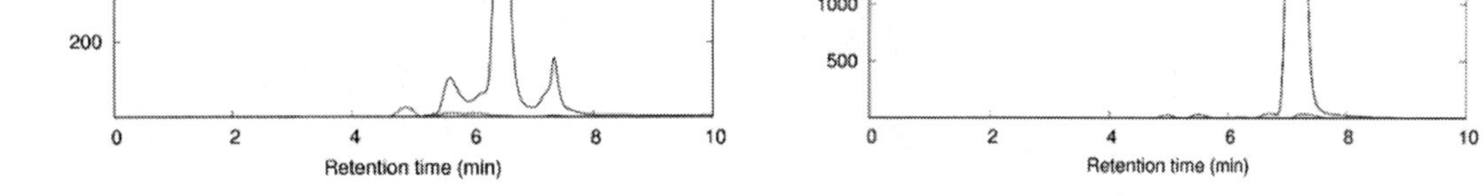

Figure 4. HPLC profile of compound **3** (left) and thymoquinone (right) in *n*-octanol/water phase.

The HPLC profile showed that there was a partition of compound **3** from *the n*-octanol system (blue) into the water system (red). Compared to the partition between compound **3** and thymoquinone (TQ), it showed that the partition of **3** is larger than TQ. Quantitative analysis using log P calculation gave the log P value of compound **3** is 2.99 slightly higher than TQ that is 2.21. According to computational calculation using Hyperchem software, it showed a similar trend, that is, log P of compound **3** is higher compared with TQ. This calculation prediction supports the *in vitro* result, suggested that the addition of bromo alkylating agent into 1,4-benzoquinone skeleton supported the increasing of lipophilicity [7].

IOP Conf. Series: Materials Science and Engineering **833** (2020) 012056 doi:10.1088/1757-899X/833/1/012056

Table 2. Solubility test analysis.

Compound	The area in octanol (A_o)[a]	The area in water (A_w)[a]	log P	
			In vitro[a]	Prediction[b]
3	36350375	87531	2.99	3.60
Thymoquinone (TQ)	63970392	920323	2.21	2.80

a) HPLC analysis
b) Predicted using Hyperchem software

4. Conclusion

The synthesis of 2,3-dimethyl-1,4-benzoquinone (**2**) and 5-(4-bromobutyl)-2,3-dimethyl-1,4-benzoquinone (**3**) was successful to carried out using oxidation followed by bromoalkylation reaction. The synthesized compound then analyzed using the HPLC method to determine the solubility in n-octanol and water system. The log P value for compound **3** is 2.99. This result is higher than thymoquinone, that is, 2.21. Based on this result, it is suggested that compound **3** which has an additional bromo-aliphatic chain compared to thymoquinone have an increasing lipophilicity which might contribute to the ability of the compound **3** to penetrate in the lipid bilayer, then it can be promoted to be a drug candidate.

References

[1] C C Woo *et al* 2012 *Biochem. Pharmacology* **83** 4 443–451
[2] M. Bule *et al* 2020 *Food Research International* **127** 108736
[3] M. Rashidmayvan *et al* 2019 *J Diabetes Metab. Disord.* **18** 2 453–459
[4] K. M. Alkharfy *et al* 2015 *Eur. J. Drug Metab. Pharmacokinet* **40** 3 319–323
[5] C. A. Lipinski *et al* 2012 *Advanced drug delivery reviews* **64** 4-17
[6] Y. N. Antonenko *et al* 2008 *Biochemistry (Moscow)* **73** 12 1273–1287
[7] S. M. Ulfa *et al* 2017 *AIP Conference Proceedings* 020102.
[8] F. Dwisari *et al* 2019 *IOP Conf. Ser.: Mater. Sci. Eng* **546** 062005
[9] A. Furoida *et al* 2019 *IOP Conf. Ser.: Mater. Sci. Eng* **546** 062008
[10] N. E. Setyatama *et al* 2018 *Proceedings of the 1st Inter. Conf. in One Health (ICOH 2017)*
[11] A. Andrés *et al* 2015 *European Journal of Pharmaceutical Sci.* **76** 181–191

Synthesis and Molecular Docking Study of Pyrazine-2-carboxylic acid Derivatives

Muhammad Zulqurnain[1], M. Riza Ghulam Fahmi[2], Arif Fadlan[1], Mardi Santoso[1,]*

[1]Departemen of Chemistry, Faculty of Science, Institut Teknologi Sepuluh Nopember, Kampus ITS Sukolilo Surabaya 60111, Indonesia
[2]Ma Chung Research Center for Photosynthetic Pigments, Universitas Ma Chung, Villa Puncak Tidar N-01, Malang 65151, Indonesia

*Corresponding author: tsv09@chem.its.ac.id

Abstract. The pyrazine-2-carboxylic acid derivatives **1a-c** with aromatic, cyclic, and aliphatic side chain were successfully synthesized. The structures of derivatives were confirmed by spectroscopic methods (FTIR, NMR, HRMS). The molecular docking was performed to determine the possible binding interaction between **1a-c** with *Mycobacterium tuberculosis* InhA protein. The derivative **1c** showed the lowest rerank score (-86.4047 kcal mol^{-1}) and it might correspond to the lowest experimental MIC value.

1. Introduction

Tuberculosis (TB), an old infectious disease caused by organisms of *Mycobacterium tuberculosis*, contaminates one-third of population and leads the cause of death worldwide [1]. About 10 million new TB were found all over the world in 2018 and 8% of these cases were experienced by Indonesian. TB cases in Indonesia have decreased from 2000 to 2018, but there has been an increase in new TB cases and relapses [2]. The TB cases have been complicated by the emersion of multi- and extensively-drug resistance (MDR, XDR) of the first-line TB medication including isoniazid, rifampicin, ethambutol, and pyrazinamide which usually consume in six months' duration [3], [4]. The development of potential and promising lead compounds with better activity for TB treatment then is needed.

The first-line of TB medication especially pyrazinamide gained special attention for the development of lead compounds for preventing drug resistance and shorten the treatment duration [5]. Pyrazinamide derivatives with substitution at C2 and C3 position in pyrazine unit have been synthesized and showed promising activity [6]. Further installation of different groups such as ethylphenyl, cycloheptyl, and octyl on the pyrazine carboxamide structure by utilization of thionyl chloride for amidation gave diverse derivatives with higher activity against *M. tuberculosis* [7]–[10]. Due to the safety of thionyl chloride in organic synthesis, several researchers then reported new systems including cerium catalyst in water, *N*-heterocyclic carbene (NHC) in toluene, and Fe-mont catalyst for amidation [11]–[14]. In order to find lead compounds for TB treatment and in continuation of our interest in anti-TB agents [15], the present study reported synthesis of pyrazine-2-carboxylic acid derivatives **1a-c** using Yamaguchi esterification and study their interaction with *M. tuberculosis* InhA protein by molecular docking.

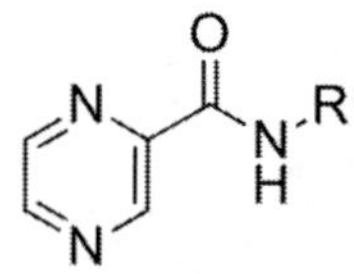

(1a) R = 4-$C_2H_5C_6H_4$
(1b) R = Cycloheptyl
(1c) R = Octyl

Figure 1. Structure of pyrazine-2-carboxylic acid derivatives **1a-c**.

2. Experimental Section

2.1. General information
Pyrazine-2-carboxylic acid, 2,4,6-trichlorobenzoyl chloride, 4-dimethyaminopyridine, 4-ethylaniline, cycloheptylamine, *n*-octylamine, triethylamine, and tetrahydrofuran were purchased from Sigma-Aldrich, Tokyo Chemical Industry, and Merck. Melting point analysis was performed on Fisher-John melting point apparatus. FT-IR Shimadzu 8400S was used for FTIR spectra measurement. NMR spectra were recorded in deuterated chloroform using TMS as standard. The mass spectroscopy analysis was achieved from a Waters Q-TOF Xevo instrument.

*2.2. General method for synthesis **1a-c***
1a-c was obtained by applying esterification method [15] with pyrazine-2-carboxylicacid (1 eq), triethylamine (1 eq), 2,4,6-trichlorobenzoyl chloride (1 eq), 4-dimethylaminopyridine (1 eq). The mixture was stirred at rt for 20 minutes, and after that 4-dimethylamino-pyridine (1 eq) and aniline or amine (0.25 eq) were added. The mixture was cooled after refluxed for 1 h, filtered, and extracted with dichloromethane. The organic part was washed with 5% HCl, 5% NaOH, 5% Na_2CO_3, water, and dried. **1b** as yellow solid (86%) was obtained after evaporation and **1a**, **c** were yielded as white solid (50% and 6%) after further purification of the crude with chromatotron.

2.3. Computational details
The interaction of pyrazine-2-carboxylic acid derivatives **1a-c** with *M. tuberculosis* InhA protein (PDB ID: 4DRE) using molecular docking were studied. Geometry of ligands were optimized with semi-Empirical (AM1) on Hyperchem 8.0.7. and protein and ligand reference (NAI) were prepared using Molegro Virtual Docking (MVD) 5.0. Docking of pyrazine-2-carboxylic acid **1a-c** was performed on MVD using MolDOck SE Algorithm to search the preferred docking pose (Number of LGA runs: 50). The ligands were docked on the same site as the standard ligand (NAI) and the docking results were visualized using MVD.

3. Result and Discussion

3.1. Chemistry
A critical issue in tuberculosis medication is the long duration of treatment (6 months) with five different types of drugs. Pyrazinamide derivatives have attracted attention and been developed as lead compounds to overcome these problems. Many pyrazinamide derivatives have been synthesized and showed promising activity against *M. tuberculosis*. In the present study, compounds **1a-c** were prepared by using Yamaguchi esterification to avoid the use of thionyl chloride.

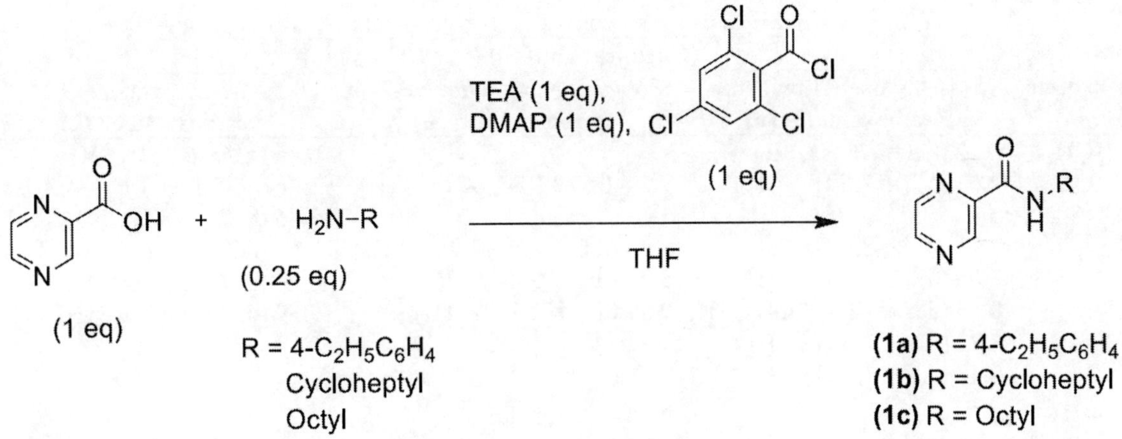

Scheme 1. Synthesis of **1a-c**.

1a-c were obtained by Yamaguchi esterification according to previous paper [15] (Scheme 1). In the Yamaguchi esterification, a mixed anhydride is produced from the reaction of carboxylic acid and Yamaguchi reagent (2,4,6-trichlorobenzoyl chloride). Further reaction of the anhydride with an alcohol in the presence of stoichiometric amount of DMAP generate the esters. In the previous paper we reported the synthesis of 2-phenylethyl esters from reaction of various carboxylic acids with 2,4,6-trichlorobenzoyl chloride to form aliphatic aromatic anhydrides which upon nucleophilic acyl substitution with 2-phenylethanol furnished the desired esters [15]. Here in this paper we utilized this method to generate pyrazine 2-carboxylic acid derivatives **1a-c** from a reaction of pyrazine-2-carboxylic acid with 2,4,6-trichlorobenzoyl chloride followed by reaction of the resulting anhydride with 4-dimethylamino-pyridine (1 eq) and aniline or amine. The pyrazine 2-carboxylic acid derivative **1b** was yielded after evaporation of the reaction mixture whereas the derivatives **1a,c** were produced after purification of the crude. The pyrazine-2-carboxylic acid derivatives **(1a-c)** was detected as $[M+H]^+$ species in ESI high resolution mass spectrometry (**1a** calcd 228.1137, found 228.1135; **1b** calcd 220.1450, found 220.1461; **1c** calcd 236.1763, found 236.1762). The IR spectra showed bands at 1659-1678 cm^{-1} and 3310-3358 cm^{-1} indicated amide carbonyl, NH-secondary groups, and the absence of hydroxyl group (Figure 2). The ^1H NMR spectra recorded in $CDCl_3$ confirmed the structure of **1a-c** which showed singlet signals for NH proton at 7.75-9.62 ppm and the carbonyl carbon appeared at 160.6-162.9 ppm in ^{13}C NMR spectra. The NMR data in details are presented in Table 1-2.

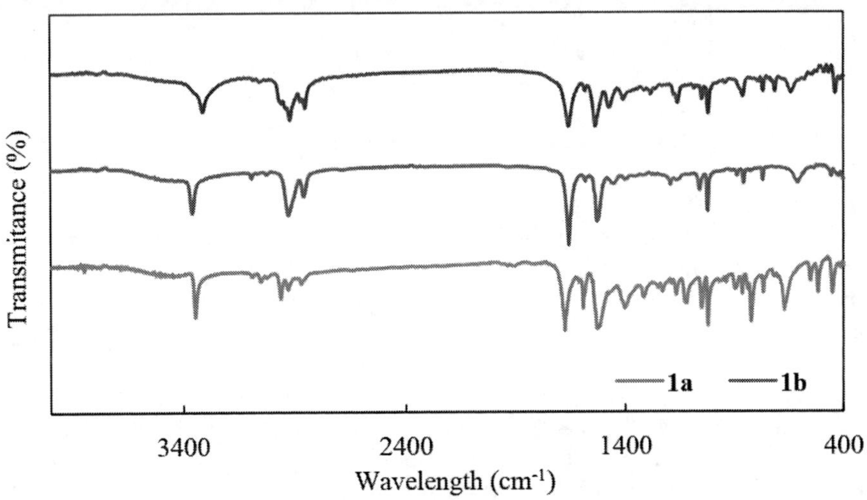

Figure 2. FTIR of pyrazine-2-carboxylic acid derivatives **1a-c**.

Table 1. ^{1}H-NMR data of pyrazine-2-carboxylic acid derivatives **(1a-c)**.

Proton	Chemical Shift (δ) (in $CDCl_3$, ppm)		
	N-(4-ethylphenyl)pyrazine-2-carboxamide **(1a)**	*N*-cycloheptylpyrazine-2-carboxamide **(1b)**	*N*-octylpyrazine-2-carboxamide **(1c)**
-CH$_3$-	1.22 (t, *J* = 7.6 Hz, 3H)	-	0.85 (t, *J* = 6.8 Hz, 3H)
-CH$_2$-	2.63 (q, *J* = 7.6 Hz, 2H)	1.54-1.69,1.96-2.02 (m,12H)	1.24-1.40, 1.58-1.65 (m,12H) 3.46 (q, *J* = 6.8 Hz, 2H)
ArH	7.21 (d, *J* = 8.4 Hz, 2H) 7.65 (d, *J* = 8.4 Hz, 2H) 8.56 (dd, *J* = 2.4, 1.4 Hz, 1H) 8.77 (d, *J* = 2.4 Hz, 1H) 9.49 (d, *J* = 1.4 Hz, 1H)	8.48 (dd, *J* = 2.4, 1.3 Hz, 1H) 8.71 (d, *J* = 2.4 Hz, 1H) 9.37 (d, *J* = 1.3 Hz, 1H)	8.50 (dd, *J* = 2.4, 1.4 Hz, 1H) 8.72 (d, *J* = 2.4 Hz, 1H) 9.39 (d, *J* = 1.4 Hz, 1H)
-CH-	-	4.10-4.15 (m, 1H)	-
NH	δ 9.62 (bs, 1H)	7.75 (bs, 1H)	7.80 (s, 1H)

Table 2. ^{13}C-NMR data of pyrazine-2-carboxylic acid derivatives **(1a-c)**.

Carbon	Chemical Shift (δ) (in $CDCl_3$, ppm)		
	N-(4-ethylphenyl)pyrazine-2-carboxamide **(1a)**	*N*-cycloheptylpyrazine-2-carboxamide **(1b)**	*N*-octylpyrazine-2-carboxamide **(1c)**
-CH$_3$-	15.8		14.2
-CH$_2$-	28.5	24.1 28.1 35.1	22.7 27.0 29.3 29.3 29.6 31.9 39.6
-CH-	-	50.5	-
ArCH	120.0 128.6 142.5 144.7 147.5	142.5 144.5 147.1	142.6 144.5 147.2
ArC	134.9 141.1 144.6	144.9	144.7
C=O	160.6	161.6	162.9

3.2. Molecular Docking Study

Molecular docking is an important method in drug discovery area due to its ability to model the interaction between small molecules with proteins at the atomic level. This method can be used to study the characteristics of small molecules at target binding sites as well as to explain the fundamental biochemical processes [16]. In this paper, the computational calculation was done to study the possible interaction of **1a-c** with *M. tuberculosis* InhA protein (PDB ID: 4DRE). Good binding interaction of **1a-c** with amino acid residues via hydrogen bonding and π-π interaction on the active site of protein was observed. Compound **1c** showed best activity with lowest rerank score (-86.4047 kcal/mol) as can be seen in Table 3. The derivative **1c** has three correlations with protein through hydrogen bonding (Glycine14, Gly14; Threonine39, Thr39; and Phenylalanine41, Phe41) and five connections via π-π interaction (Phe41; Leucine63, Leu63; Gly14; Thr39; and Isoleucin95, Ile95). The number of

IOP Conf. Series: Materials Science and Engineering **833** (2020) 012057 doi:10.1088/1757-899X/833/1/012057

interactions of **1c** is greater than with **1a-b** both through hydrogen bonding and π-π interaction. Compound **1c** could interact with the amino acid Thr39 of the protein through hydrogen bonding and π-π. This mode also showed by the amino acids Gly14 and Phe41. This interactions may be responsible for the lower rerank score of **1c** which might correlate with the MIC value in the experimental study.

Table 3. Docking results of **1a-c** with *M. tuberculosis* InhA protein (PDB ID: 4DRE).

Compounds	Rerank Score (kcal mol^{-1})	Interaction
1a	-79.3431	H-Bond: Val65 π-π: Gly96, Ile16, and Val65
1b	-73.5657	H-Bond: Gly96 π-π: Val65, Phe41, Gly96, and Ile95
1c	-86.4047	H-Bond: Gly14, Thr39, and Phe41 π-π: Phe41, Leu63, Gly14, Thr39, and Ile95

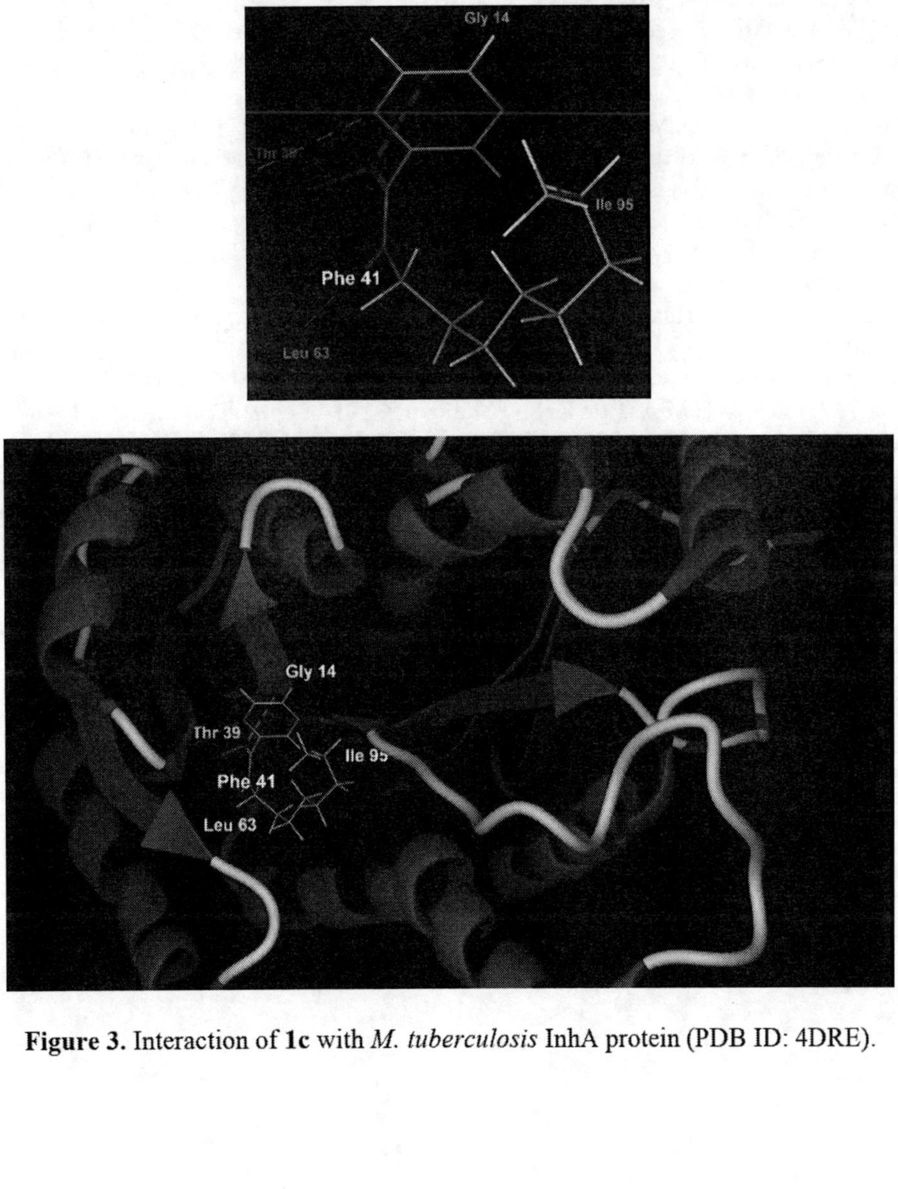

Figure 3. Interaction of **1c** with *M. tuberculosis* InhA protein (PDB ID: 4DRE).

4. Conclusion

Pyrazine-2-carboxylic acid derivatives **1a-c** were successfully synthesized by Yamaguchi method. Molecular docking study on *M. Tuberculosis* InhA protein (PDB ID: 4DRE) indicated that **1c** have the lowest rerank score (-86.4047 kcal mol^{-1}) and it might correspond with experimental MIC value.

Acknowledgments

Institut Teknologi Sepuluh Nopember and The Ministry of Research, Technology, and Higher Education are acknowledged for financial support.

References

[1] Smith I 2003 *Clin. Microbiol. Rev.* **16** 463–496
[2] Wourld Health Organization 2019 *Global Tuberculosis Report* (Geneva: WHO) p 206
[3] Abdel-Aziz M and Abdel-Rahman H M 2010 *Eur. J. Med. Chem.* **45** 3384–3388
[4] Wu S, Lan L, Jiang J, Ding X, Ho C-M, Lou Y, and Fan G 2019 *J. Pharm. Biomed. Anal.* **168** 44–54
[5] Zitko J, Jampílek J, Dobrovolný L, Svobodová M, Kuneš J, and Doležal M 2012 *Bioorganic Med. Chem. Lett.* **22** 1598–1601
[6] Doležal M, Zitko J, Kešetovičová D, Kuneš J, and Svobodová M 2009 *Molecules* **14** 4180–4189
[7] Zitko J, Servusová B, Paterová P, Mandíková J, Kubíček V, Kučera R, Hrabcová V, Kuneš J, Soukup O, and Doležal M 2013 *Molecules* **18** 14807–14825
[8] Onajole O K, Pieroni M, Tipparaju, S K, Lun S, Stec J, Chen G, Gunosewoyo H, Guo H, Ammerman N C, Bishai W R, and Kozikowski A P 2013 *J. Med. Chem.* **56** 4093–4103
[9] Semelkova L, Konecna K, Paterova P, Kubicek V, Kunes J, Novakova L, Marek J, Naesens L, Pesko M, Kralova K, Dolezal M, and Zitko J 2015 *Molecules* **20** 8687–8711
[10] Bouz G, Semelková L, Janďourek O, Konečná K, Paterová P, Navrátilová L, Kubíček V, Kuneš J, Doležal M, and Zitko J 2019 *Molecules* **24** 1212
[11] Zhou S, Yang S, and Huang G 2017 *J. Enzyme Inhib. Med. Chem.* **32** 1183–1186
[12] Tamura M, Tonomura T, Shimizu K, and Satsuma A 2012 *Appl. Catal. A Gen.* **417-418** 6–12
[13] Naredla R R, Dash B P, and Klumpp D A 2013 *Org. Lett.* **15** 4806–4809
[14] Ali A M, Siddiki H S M A, Kon K, and Shimizu K 2014 *Tetrahedron Lett.* **55** 1316–1319
[15] Maharani N and Santoso M 2015 *Adv. Environ. Biol.* **9** 214–218
[16] Meng X-Y Zhang H-X Mezei M and Cui M 2012 *Curr. Comput. Aided-Drug Des.* **7** 146–157

Electrochemical Behavior of Nickel Laterite Ores Dissolution in Sulphuric Acid Solutions

Faizinal Abidin[1,2], Abdul Hatta Gunawan Wibowo[1], Arya An Ambari[1], Sri Harjanto[1,*]

[1]Department of Metallurgical and Materials Engineering, Faculty of Engineering, Universitas Indonesia, Kampus UI Depok 16424, Indonesia
[2]Center for Mineral Resources Development Technology, Agency for The Assessment and Application of Technology, 820 Building, Earth System Technology (Geostech), Puspiptek Area, South Tangerang 15314, Banten – Indonesia

*Corresponding Author E-mail: sri.harjanto@ui.ac.id

Abstract. Hydrometallurgical process of laterite nickel ore is carried out by leaching techniques at high temperatures. At atmospheric temperatures, the acquisition of nickel is low. Therefore, this study aims to characterize the electrochemical laterite nickel ore using a solution of sulfuric acid and hydrochloric acid at atmospheric temperature. Sulfuric and chloride acid solutions were used with solution concentrations of 1 M, 2 M, 4 M and 6 M. The study was carried out by sample and solution preparation, characterization by SEM-EDAX and Petrography method, and electrochemical characterization using the OCP, EIS method and LSV. The characterization results showed that the highest dissolution rate in both solutions was obtained at a concentration of 6 M solution. Increased concentrations of sulfuric acid reduced the value of OCP testing, while an increase in the concentration of hydrochloric acid increased the value of OCP. The EIS test results showed that the lowest R2 or Rct value in sulfuric acid solution was obtained at 2M concentration of 269.59 Ω while in hydrochloric acid solution it was obtained at a concentration of 6M of 523.07 Ω. The results of the second LP acid solution showed the formation of a passive layer at each concentration. The highest dissolution rate occurred at a concentration of 6M in each solution where the corrosion rate of sulfuric acid was 1.00 mm / year, while in hydrochloric acid it was 9.55 mm / year. Increasing the concentration of the two solutions led to the breakdown of the passive layer on the surface which can be seen from the Nyquist curve, the value of R2 is lower, the value of Q1 is higher and the value of N is lower, but in the sulfuric acid solution the formation of passive layers returns to increasing concentration starting from 4M. The three electrochemical tests show an increase in concentration to the optimum concentration increasing the dissolution rate and breaking the passive layer on the surface.

1. Introduction

The purpose of the dissolution process is separate valuable metals from impurities [1]. The dissolution process will produce two products, namely the first product which is not soluble or precipitated in the form of impurities that will be cleaned and discarded or processed by products and the second product is a solution containing valuable metals [1]. Solution selectivity is an important factor. The selectivity of the solvent to the ore is influenced by the concentration of the solution, temperature, and contact time [3].

IOP Conf. Series: Materials Science and Engineering **833** (2020) 012058 doi:10.1088/1757-899X/833/1/012058

Hydrochloric acid is widely used as a leach reactant for laterite minerals because it is considered easier than other reactants. In addition, there are several advantages, namely that it is more optimal to dissolve nickel metal compared to other solutions, has higher leachability for complex ores, regeneration of leaching agents, removal of impurities such as iron, cobalt and magnesium [4]. In a previous study by Ahmet Goveli in 2016, to dissolve nickel metal in laterite minerals was carried out with a number of variations including variations in solution concentration, dissolution time, and dissolution temperature. In laterite leaching, the most optimal solution concentration is HCl 4 M with a nickel yield that can be extracted at 97.69% of the initial content. The most optimal time variable for this leaching is 4 hours with a nickel yield that can be extracted at 24.60% of the initial content. Whereas the most optimal temperature variable for this leaching is 100°C with a nickel yield that can be extracted at 54.82% of the initial content [4]. Another study conducted by Jinhui Li et al. in 2018, dissolve valuable metals from laterite nickel ore with hydrochloric acid solution with varying concentrations of hydrochloric acid, leaching time and leaching temperature. The most optimal variation of hydrochloric acid concentration is 4 M HCl with a nickel leaching rate of 80.1%. For the most optimal leach temperature variation is at 90°C with a leaching rate reaching 88%. Whereas the most optimal time variation is 90 minutes or 1.5 hours with leaching rate reaching 86%. From these results that the variable concentration of solution, temperature, and time of leaching is directly proportional to the leaching rate of laterite nickel ore [5].

The most previous research topics on electrochemical studies were only carried out on sulfide ores such as pyrite (FeS) and chalcopyrite ($CuFeS_2$), studies have never been carried out on laterite nickel ores [6] [7] [8]. This research was conducted to study the behavior of laterite nickel ore when leaching with leachate solution with different concentrations. In this study, sulfuric acid and hydrochloric acid solutions were used as leachate solutions with different concentrations

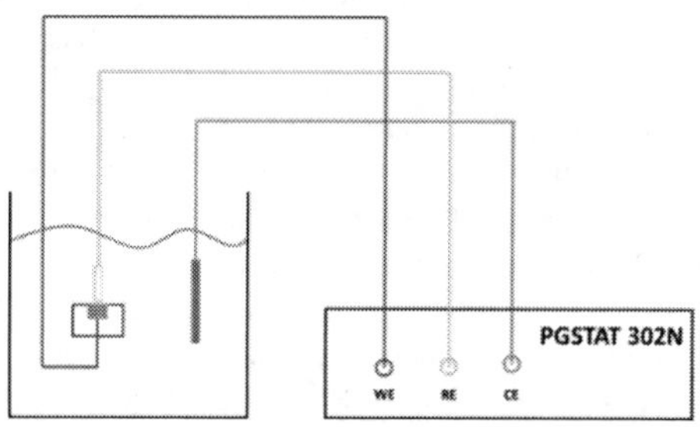

Figure 1. Schematic testing of electrochemical characterization

2. Materials and Methods

Laterite nickel ore were obtained from the island of Sulawesi. Scanning Electron Microscopy (Inspect F50) and Energy Dispersive X-Ray Analysis (AMETEK) as well as petrographic tests are carried out to characterize the chemical composition of the samples and types of laterite nickel ore, petrographic test was performed by Nikon Eclipse Ci-POL Polarization Microscope equipped DS-Ri2 Nikon Camera and integrated with NIS-Elements software is used to identification and measure content of mineralogy. The ore sample tested for SEM-EDAX is the same sample for the petrographic test, the results of the two tests to determine and assure the type of nickel ores used in this study.

The solution used for testing is sulfuric acid solution (H_2SO_4) concentration of 96% (Merck, EMSURE®). The solution is diluted using distilled water to reduce its concentration. The solution used to dissolve the sample was divided into concentrations 1 M, 2 M, 4 M and 6 M. Laterite nickel ore was then prepared into a working electrode for further testing. Polishing machine (METASERV, Universal Polisher) is used to prepare the surface of the sample before electrochemical studies are conducted. Silver chloride electrode (Ag / AgCl) is used as a reference electrode and platinum metal is used as an auxiliary electrode. Copper wire is used as a connecting electrochemical cell. Samples of laterite nickel ore were prepared for the electrochemical characterization process. Large samples are cut into small sizes. The samples that have been cut are each mounted using a resin that is added with a hardener in a ratio of 15: 1. The sample surface was prepared by polishing with # 500, # 1000, and # 1500 and the surface exposed to the solution was only 1 cm^2 in size. Potentiostat-galvanostat (Metrohm Autolab, PGSTAT302N) integrated with Metrohm NOVA Autolab software was used to perform the electrochemical measurement, open circuit potential (OCP) was always measured until the electrode potential stabilized to a chance in the electrode potential of no more than 0.1 $mv.min^{-1}$. After steady OCP was observed electrochemical impedance spectroscopy (EIS) experiments were performed at OCP using an AC perturbation of ± 5 mV in the frquency range 10 kHz – 10 mHz. The electrochemical characterization test scheme is shown in Figure 1.

3. Result and Discussion

Based on the petrographic test results, laterite nickel ore samples used are peridotite rocks with harzburgite types undergoing serpentinization with typical "mesh" and "sea and island" textures. Peridotite rocks based on laterite nickel ore profiles are found in the bedrock and bottom saprolite layers.

Table 1. Mineralogy of Laterite Nickel Ores

	Mineralogy	Percentage (%)
Primary	Olivine	40
	Enstatite	20
	Picotite	3
Secondary	Serpentine	30
	Goethite	3
	Iddingsite	3
	Smectite	1

The main minerals contained in the sample are olivine ($(Mg^{2+}, Fe^{2+})_2SiO_4$), pyroxene variation of enstatite ($Mg_2Si_2O_6$) with augite thin lamella ($(Ca, Na)(Mg, Fe, Al)(Si, Al)_2O_6$), and picotite ($(Fe, Mg)(Al, Cr)_2O_4$), while the secondary minerals contained in the sample are serpentine – antigorite variations ($Mg_3(Si_2O_5)(OH)_4$), iddingsite ($MgO{\cdot}Fe_2O_3{\cdot}3SiO_2{\cdot}4H_2O$), goethite ($Fe^{3+}O(OH)_4$), smectite, talc ($Mg_3Si_4O_{10}(OH)_2$), and magnetite ($Fe_3O_4$). The secondary mineral that predominates is serpentine – antigorite variation that is spread on the rock matrix with a sea texture and is present with either goethite or magnetite. No samples were found for the primary ore mineral nickel. The results of mineral identification in laterite nickel ore samples can be seen in Table 1.

Table 2. Elements detected by EDAX scanning.

Element	Wt%	At%
C	4.92	9.10
O	28.69	41.03
Mg	19.56	18.71
Al	1.64	1.35
Si	24.54	20.19
Ca	6.36	3.53
Ti	3.17	1.49
Fe	9.69	4.01
Ni	1.43	0.59

The SEM-EDAX test results show the presence of nickel content in laterite nickel ore. The nickel contained in laterite nickel ore is concentrated in certain parts. The microstructure and elements contained are shown in Figure 2. The chemical composition of the sample is shown in Table 2. The electrochemical study of the dissolution of laterite nickel ore produces graph data in the form of Nyquist plot, open circuit potential, and polarization curves for each different solution concentration used in the test.

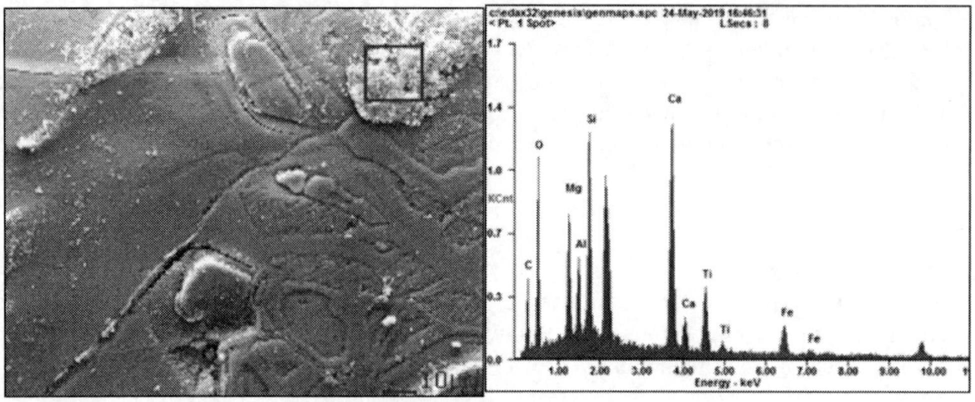

Figure 2. Surface microstructure of lateritic nickel ore from SEM observations from 10,000 magnifications (left) and concentrated elements at the EDAX scan point (right).

Figure 3 shows a graph of OCP with respect to time of testing at different concentrations. At a concentration of 1 molar, the curve fluctuates up to 50 seconds. Then the curve is sloping with a tendency to rise until 140 seconds. At 0-15 seconds, there is a potential increase from 0.124 to 0.129 V vs Ag/AgCl. At a concentration of 2M, the curve decreased without a change in slope up to 300 seconds from a potential of 1.328 - 0.1 V vs Ag/AgCl. Stability is achieved at a potential of 0.1 V vs Ag/AgCl at 300 seconds. At 4M concentration, the curve increases regularly until 300 seconds from a potential of 0.075 to 0.08 V vs Ag/AgCl. At a concentration of 6M, the curve has decreased until 300 seconds from a potential of 0.094 to 0.083 V vs Ag/AgCl. Stability is achieved at 300 seconds at a potential of 0.083 V vs Ag/AgCl. The effect of the concentration of the solution on OCP from laterite nickel ore is shown in Figure 4. OCP decreases when adding concentrations from 1M to 4M. After that, the OCP did not experience a significant change from 4M to 6M concentrations.

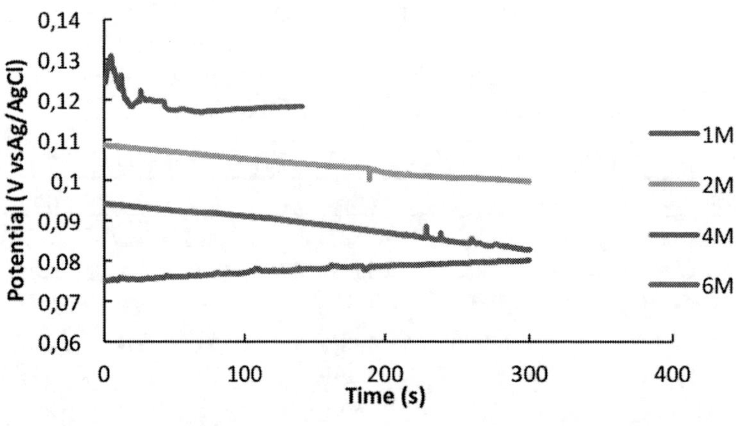

Figure 3. OCP Graph of laterite nickel ore at room temperature.

Testing with the LSV method produces the linear polarization curve shown in Figure 5. Electrochemical polarization parameter data from the LSV graph is shown in Table 3. The corrosion rate of the dissolution of the sample from each solution concentration is also shown in Figure 6. The EIS test results have been fitted using the equivalent electrical circuit shown in Figure 7. From the results of the fittings, the EIS test parameter data is shown in Table 4. Resistance of R1 solution is related to testing the high frequency range of 10,000 Hz. The charge transfer resistance R2 at the film layer is related to testing the intermediate frequency range 100-1 Hz with the constant phase element CPE1/Q1 as the charge capacitance at the film layer.

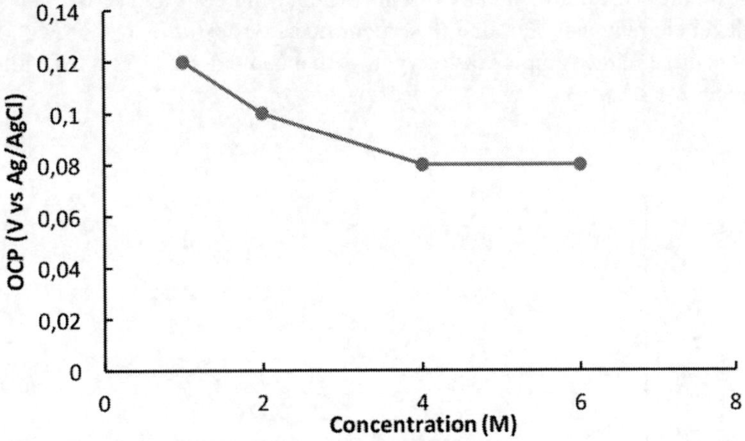

Figure 4. Effect of sulphuric acid concentration on laterite ore OCP

The nickel that is deposited in lateritic nickel comes from olivine and orthopyroxene in ultramafic rocks which are the source of nickel, then it is sentinized and undergoes lateritization due to interacting with sea water or during alteration. Olivine undergoes hydration and turns into amorphous silica, serpentine, and limonite. Nickel accumulates in serpentine through cation exchange, especially with Mg^{2+} ions, when the olivine reaction becomes smectite, goethite, serpentine, and talc. The nickel accumulation process is further characterized by the formation of nickel-enriched filo-silicate minerals such as kerolite (Ni-talc), neupoite (Ni-serpentine), and pimelite (Ni-smectite) [9]. Nickel concentration can also be associated with goethite through the adsorption process [10].

Table 3. Electrochemical polarization parameters of laterite nickel ore dissolution in H_2SO_4 solution at room temperature.

Concentration (M)	jcorr (μA/cm^2)	Ecorr (mV)	Corrosion Rate (mm/year)
1	9,74	47,67	0,113
2	12,03	103,04	0,140
4	10,37	29,78	0,120
6	86,24	-52,34	1,002

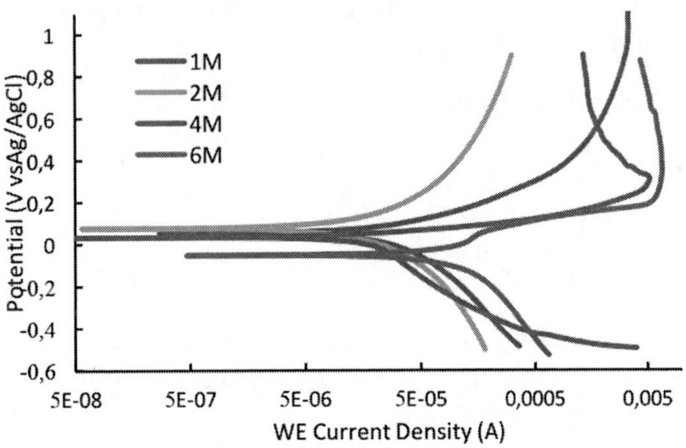

Figure 5. Linear polarization curves of LSV test results at sulphuric acid solutions

From the EDAX test results, it can be concluded that laterite nickel ore samples contain elements of oxygen, magnesium, silica, titanium, iron, and nickel. Previously it was found that the minerals making up of laterite nickel ore were dominated by olivine primary minerals and serpentine secondary minerals. The detected nickel element can be found in serpentine as a substitute ion for Mg^{2+}. The content of the element most contained in the sample was oxygen with a content of 28.69 wt% while the nickel content contained as much as 1.43 wt%.

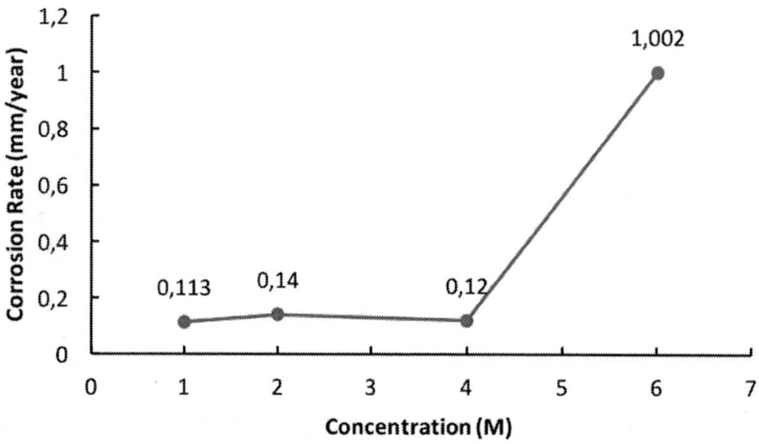

Figure 6. Graph of the relationship between corrosion rate and solution concentration.

From the OCP test data, fluctuations can be expected from the formation and separation of corrosion products formed on the metal surface. However, the corrosion products that are formed need to be characterized again [6]. A steady increase in potential indicates the formation of a corrosion product where a passive layer might have formed on the surface of the sample [7] [8]. A gentle curve shows a stable corrosion-free potential where corrosion occurs throughout the sample surface. When viewed in the Ni-O-H pourbaix diagram, the OCP value is in the nickel dissolution area, which is above -0.288 vs Ag/AgCl. When compared with the pourbaix diagram, the OCP value of each solution concentration is in the dissolving region. OCP value is potential when the sample begins to react or undergo dissolution. Increasing the concentration results in a decrease in the potential OCP of the sample, indicating that laterite nickel ore is dissolved faster with increasing concentration.

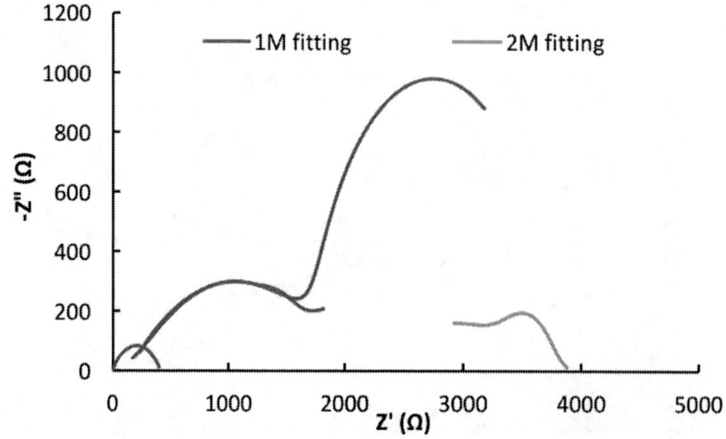

Figure 7. Nyquist Plot of Laterite Nickel Ores at Sulphuric Acid

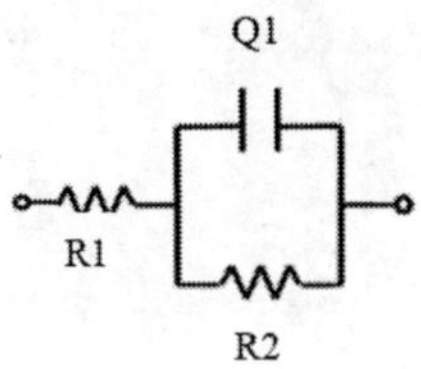

Table 4. Element values of EIS test results.

Element	Parameter	Solution			
		1M	2M	4M	6M
R1	R (Ohm)	144,01	2122,3	10,1	0,175
Q1	Y_0 (Mho/Ohm^{-1})	9,11x10^{-5}	3,49x10^{-5}	1,48x10^{-3}	4,44x10^{-4}
	N	0,406	0,853	0,131	0,508
R2	R (Ohm)	1780,09	269,59	1,10x10^{12}	403,49

Passivation areas on the polarization graph are shown by a potential increase but current that tends to remain or decrease. The highest anodic reaction is shown at the highest current point occurring in the active region or linear curve. However, at the dissolution of 2M concentration, pseudo-passive behavior is shown which is shown on a linear curve with a steep slope in the potential range of 0.2 V to 0.92 V. This behavior can be associated with partial passive layer loss due to certain ions [10]. In the Ni-OH pourbaix diagram, the dissolution of Ni does not occur when the potential is below -0.738 V vs Ag/AgCl and the hydrogen evolution reaction occurs at a potential of -0.288 V vs Ag/AgCl at pH below 8. Of the four concentrations, it can be said that the curve the reduction in the graph shows the hydrogen evolution reaction. The evolution of hydrogen causes a lack of protons on the surface of the sample, causing inhibition at concentrations of 1 M, 2 M and 6 M. While at 4 M concentration, inhibition does not occur that can be seen from the reduction curve that does not experience significant potential changes at the lowest potential up to potential of -0.5 V vs Ag/AgCl [9]. Corrosion rate did not change significantly when the concentration of the solution increased to 4 M. The rate increased rapidly when the concentration of the solution was increased from 4 M to 6 M. This shows that corrosive dissolution occurred in the 6 M solution. Dissolution of nickel ore at room temperature did not increase significantly when the concentration increases, but the increase in dissolution is more influenced by an increase in temperature [11].

The formation of a semicircular curve in the nyquist diagram is related to the charge transfer resistance where the magnitude of the charge transfer resistance is related to the diameter of the formed curve [12]. At the dissolution of 1 M and 4 M concentrations, two curves are formed at high frequency and at low frequency. These two curves can show the existence of double layers and layers of corrosion products on the surface [13]. The formation of a semicircular curve at high frequency can also indicate the formation of the main dissolving product of laterite nickel ore on the surface due to the immersion of the ore at the beginning of dissolution [10].

From the N value in the table it can be explained that the dissolution of the sample at a solution concentration of 1 M and 6 M has an N value close to the warburg impedance (N = 0.5) where the surface layer of the film pivots. In a solution with a concentration of 4 M, the value of the parameter N approaches the value of 0. A decrease in the value of N indicates a more severe attack on the exposed surface. The decrease in the value of Q1 capacitance with increasing concentration also indicates the formation of a film layer. Formation of the film layer is the cause of decreased capacitance [10].

4. Conclusions

The samples tested were serpentinized harzburgite peridotite rocks. This type of rock has the main mineral content of olivine ((Mg^{2+}, Fe^{2+})$_2$$SiO_4$) and pyroxene variation enstatite ($Mg_2Si_2O_6$). From SEM and EDAX results also confirmed the main minerals contained in the sample where the dominant elements detected were O, Si, and Mg. The results of all electrochemical characterization showed that the surface of laterite nickel ore formed a passive layer that experienced destruction when the concentration of the solution rise to optimum concentration which are 4 M for sulphuric acid, then an increase in the rate of dissolution exceeding optimum concentration caused the re-formation of the previously destroyed passive layer. Ore samples used in dissolving with different concentrations are recommended using the same sample. This is because the use of different samples can affect the test where each sample can have a different surface texture.

References

[1] Crundwell F K, Moats M, Robinson T, Ramachandran V, and Davenport W 2011 Extractive metallurgy of nickel, cobalt and platinum-group materials (Amsterdam: Elsevier)

[2] Crundwell F K, Moats M S, Ramachandran V, Robinson T G and Davenport W G 2011 Overview of the Hydrometallurgical Processing of Laterite Ores *Extractive Metallurgy of Nickel, Cobalt and Platinum Group Metals* 117–22

[3] Ray H S, Ghosh A 1991 Principles of Extractive Metallurgy 2nd edition (New Delhi: New Age International)

[4] Goveli, A. 2016 Nickel Extraction From Gordes Laterites by Hydrochloric Acid Leaching (Middle East Technical University)

[5] Li J, Li D, Xu Z, Liao C, Liu Y and Zhong B 2018 *Journal of Cleaner Production* **179** 24–30

[6] Olvera O G, Rebolledo M, Mora N, Dixon D G and Asselin E 2018 *Hydrometallurgy* **179** 110–7

[7] Almeida T D C, Garcia E M, Silva H W A D, Matencio T and Lins V D F C 2016 *International Journal of Mineral Processing* **149** 25–33

[8] Rafsanjani-Abbasi A and Davoodi A 2016 Electrochemical Characterization of Natural Chalcopyrite Dissolution in Sulfuric Acid Solution in Presence of Peroxydisulfate *Electrochimica Acta* **212** 921–8

[9] Robb L J 2005 *Introduction to ore-forming processes* (Hoboken, NJ: Wiley-Blackwell)

[10] Elias M 2002 Nickel Laterite Deposits-Geological Overview, Resources and Exploitation. Centre for Ore Deposit Research, University of Tasmania, Special Publication **4** 205-220

[11] Mohammadreza F, Mohammad N and Ziaeddin S S 2014 *International Journal of Mining Science and Technology* **245** 43–8

[12] Bhattarai J, Mikhaylov V, Neupane D, Nepal B, Demchenko A, Stine K 2017 Carbohydrate–Protein Interactions Studied Using Electrochemical Impedance Spectroscopy. In book: Electrochemical Impedance Spectroscopy: Methods, Analysis and Research, Publisher: Nova Science Publishers, Editors: Jennie Brock

[13] Hiroyoshi N, Kuroiwa S, Miki H, Tsunekawa M and Hirajima T 2004 *Hydrometallurgy* **74** 103–116

Acknowledgement

Authors would like to thank the Directorate of Research and Public Engagement, Universitas Indonesia for funding this research under PITTA Project 2019 under contract No. NKB-0788/UN2.R3.1/HKP.05.00/2019.

Voltammetric Determination of Paracetamol using Polyvinyl Alcohol (PVA)-Fe_3O_4 Modified Glassy Carbon Electrode

Robi'atul Andawiyah[1], Ani Mulyasuryani[1*], Hermin Sulistyarti[1]

[1]Department of Chemistry, Faculty of Mathematics and Natural Science, Brawijaya University

*Corresponding e-mail: mulyasuryani@ub.ac.id

Abstract. Modifications of glassy carbon electrodes (GCE) as working electrodes have been carried out by polyvinyl alcohol (PVA)-Fe_3O_4 membranes. The PVA-Fe_3O_4 membrane was superimposed on the surface of a glassy carbon electrode. This study influences the percentage of paracetamol in the membrane, the effect of modulation time, the effect of modulation amplitude, the effect of scan rate, the effect of pH and supporting electrolytes. The performance evaluation of the modified electrode was based on a voltammogram of cyclic voltammetry (CV) and the analysis was carried out with a differential pulse voltammetry (DPV) method. The effect of pH was studied in the range of 2-7 by Britton – Robinson buffer. The results showed that the percentage of paracetamol in the membrane affected the diffusion of current. The best performance of GCE modified was produced by a membrane with paracetamol percentage of 3% (w/w). The optimum operational condition of the instrument was at modulation amplitude of 10 mV, the modulation time of 0.1 s and the scan rate of 12.5 mV /s. pH affects the shift of the anodic peak of paracetamol in a more negative direction. The results of the analysis of paracetamol using DPV method showed that more sensitivity was achieved when phosphate buffer as a supporting electrolyte was used. The sensitivity of the GCE modified by the PVA-Fe_3O_4 membrane was 0.2922 μA/μM in the linear concentration range of 0-100 μM and limit of detection (LoD) of 8 μM.

1. Introduction

Paracetamol is one of the headaches and fever medications that is circulating in the community [1-3]. The spectrophotometry method is a fast, simple and inexpensive method of determining paracetamol, but this method has several limitations [4]. The presence of spectral disturbances caused by other components in the drug, the maximal wavelength of paracetamol 244 nm, can be disturbed by propyphenazone,266 nm, and caffeine,273 nm [5]. Therefore, the method of determining paracetamol in drugs requires a more selective method that is more sensitive.

Paracetamol is an electroactive compound that can be detected using electroanalytical methods such as voltammetry. Voltammetry method has high sensitivity, precision, inexpensive cost, low limit of detection and wider linear range [6]. Determination of paracetamol using voltammetry has been widely used in the blood samples, water, medicine and mixed compounds. The results showed a recovery range of 97-102% and highly reproducible response [7-10]. The oxidation reaction of paracetamol is shown in Scheme 1.

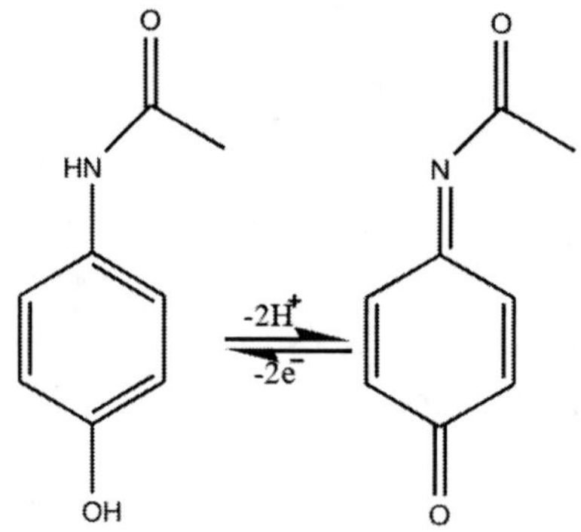

Scheme 1. The oxidation reaction of paracetamol to be N-acetyl-p–quinonimine in acidic condition [11].

Modification of carbon electrodes using molecular imprinted polymer (MIP) technique can increase the sensitivity and selectivity of paracetamol determination. The presence of specific molecules in the membrane allowing only specific species of sizes and shapes diffused to the surface of the electrode. In this study, polyvinyl alcohol (PVA) is used as base polymer, paracetamol as template molecule, glutaraldehyde as crosslinker and citric acid as catalyst.

Polyvinyl alcohol (PVA) is one of the basic polymers in the manufacture of selective membrane. PVA is biocompatible and nontoxic material. It also has mechanical and pH resistance and formidable hydrophilicity. It has been widely used in the field of biosensors and sensors, such as determination of 2,4-DNT vapor and paracetamol [12-13].

Magnetite (Fe_3O_4) is a superparamagnetic metal oxide that is biocompatible and low toxicity. It has high surface reaction activity, high catalytic efficiency and high conductivity so it can increase electron transfer [14]. Modification of carbon electrodes with magnetite has been studied for the determination of glucose, dopamine, and ascorbic acid [14-15]. He (2017) modified glassy carbon electrode with magnetite and graphene to increase the sensitivity of sulphanilamide determination.

Herein, an electrochemical sensor for paracetamol in MIP PVA-Fe_3O_4modified glassy carbon electrode will be studied. Some conditions for this study are the effect of the differential of paracetamol concentration as template in membrane, the effect of pH and supporting electrolyte.

2. Materials and Methods

2.1. Materials and Equipment

The material used in this experiment were paracetamol (Sigma Aldrich), polyvinyl alcohol, glutaraldehyde 50% (Sigma Aldrich), citric acid, magnetite (Fe_3O_4), sodium hydroxide (Merck), sodium dihydrogen phosphate, phosphate acid (Merck), acetic acid (Merck), boric acid, ethanol (Merck), hydrobath aqua demineralization, potassium chloride, and chloride acid (Merck).

The equipment in this study were galvanostat potentiostat (Autolab PGSTAT204), glassy carbon electrode as working electrode, platinum (Pt) wire as counter electrode, and silver chloride Ag/AgCl as reference electrode.

2.2. The Preparation of Membrane

The membrane was made in different percentage of paracetamol in membrane. The compositions of MIP PVA were 5% PVA 0.9 mL, 4% glutaraldehyde 0.1 mL, 5% citric acid 0.1 mL and 1% paracetamol 54 μL , 110 μL and 167 μL for percentage of paracetamol 1; 2; 3 % (w/w) in membrane, respectively. Meanwhile, the compositions of MIP PVA-Fe_3O_4 were same for other composition except 1% paracetamol 55 μL, 113 μL, 171 μL for percentage of paracetamol 1; 2; 3% (w/w) in membrane, respectively, and addition of 23 μL Fe_3O_4 0.5%. All compositions were mixed and stirred at 50 °C for 30 minutes. The removal template using cyclic voltammetry technique as shown in 2.3.

2.3. Modification of GCE
The GCE (length = 6.5 cm; d = 2.0 mm). The GCE was rinsed with ethanol and dried at 50°C for 20 min. Next, the GCE was immersed three times in mixture that left in hot condition (50°C). The first and the second immersion were dried at 50°C for 10 min then the last immersion was dried at 50°C for 2 h. The paracetamol was removed to form a template by cyclic voltammetry for 5 cycles.

2.4. Cyclic Voltammetry (CV) and Different Pulse Voltammetry (DPV)
The study of paracetamol percentage in membrane used paracetamol 0.05 mM solution in pH 4 Britton Brinson buffer. The electrochemical behaviors of paracetamol in modified electrode were studied by cyclic voltammetry (CV) and analyzed quantitatively by different pulse voltammetry (DPV). The potential range of CV was -1.200 - -1.200 V and the scan rate was 0.05 V/s. Meanwhile, the potential range of DPV was 0.2 – 1.2 V with amplitude of 0.1 V, modulation time of 0.12 s and scan rate of 0.025 V/s. In the next step, the DPV method was used to measure pH and supporting electrolyte effects. The settings were potential range of 0.2-1.2 V, amplitude of 0.1 V, modulation time of 0.1 s and scan rate of 0.0125 V/s.

3. Results

3.1. The effect of paracetamol concentration in membrane
The percentage difference of paracetamol in the membranes produce different current response. Measurements were carried out by the DPV method and analyzed based on the height of anodic peak currents.

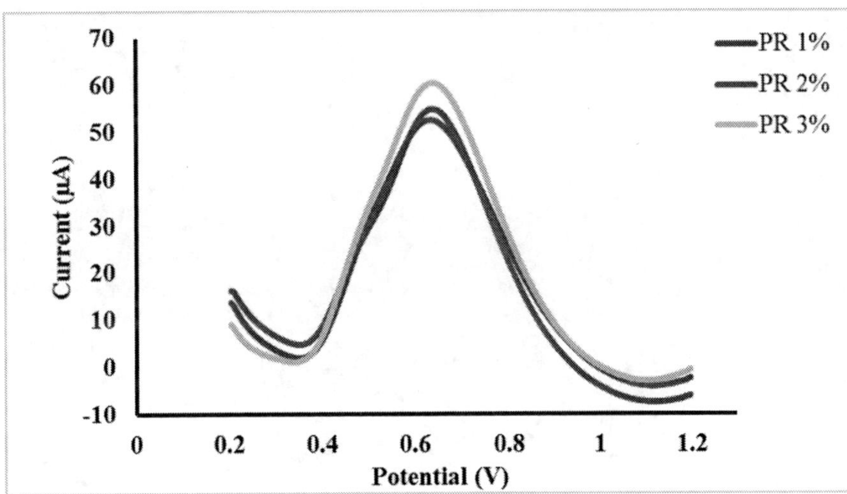

Figure 1. Voltammogram DPV of the effect of percentage of paracetamol (PR)% (w/w) in membrane with paracetamol 1; 2; and 3% (w/w).

The Figure 1 showed the relationship of percentage of paracetamol in the membrane with the anodic peak current. The high percentage of paracetamol generates higher current. This is due to the increase of the number of templates in the membrane according to the percentage of paracetamol. However, the higher percentage of paracetamol in membrane may decrease the selectivity of the sensor because the greater number of templates resulted in the diffusion of other components or

IOP Conf. Series: Materials Science and Engineering **833** (2020) 012059 doi:10.1088/1757-899X/833/1/012059

interferents on the surface electrode if their potential is sufficient. For this case, the optimum of percentage of paracetamol in the membrane was 3% (w/w).

Table 1. Anodic peak current of paracetamol using working electrode; MP1; MP2; and MP3*.

Membrane	Paracetamol % (w/w)	E_{pa} (V)	I_{pa} (μA)
MP1	1%	0.62805	52
MP2	2%	0.63309	57
MP3	3%	0.63309	60

*MP1, MP2, MP3 are membrane with percentage of paracetamol 1; 2; and 3% (w/w), respectively.

The anodic peak potential of paracetamol in different membranes are shown in Table 2. The E_{pa} shifts slightly to positive and there is no significant difference between MP1, MP2 and MP3. Thus, the percentage difference of paracetamol doesn't give the specific change for electrochemical oxidation of paracetamol.

3.2. Optimization of Different Pulse Voltammetry Parameters

The parameters of DPV have to be optimized for the analytical application. There are three parameters of modulation amplitudes, modulation times and scan rates.

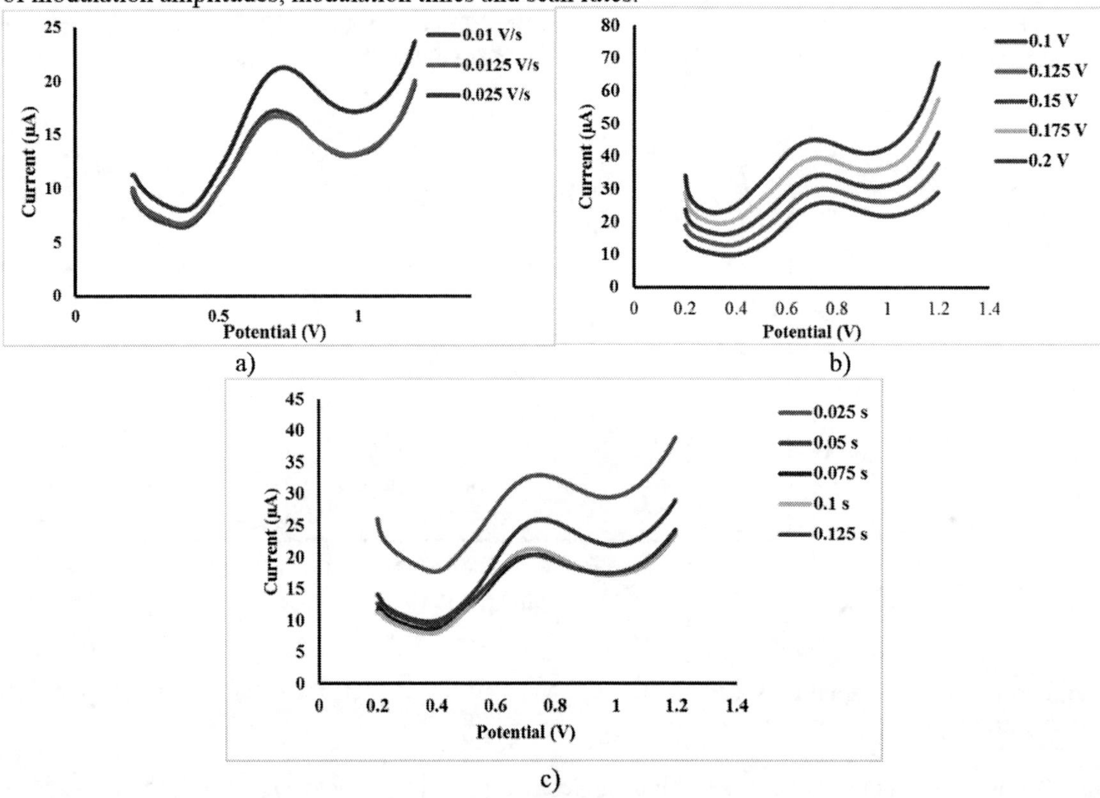

Figure 2. Voltammogram of optimization of DPV a) modulation amplitude (0.1 -0.2 V); b) modulation times (0.025 – 0.125 s) and scan rate (0.01 – 0.025 V/s)

The height of anodic current peak from the different modulation amplitude is shown in Figure 2(a). The peak current increases with the increasing modulation amplitude and the height of peak current.

Thus, voltammogram of DPV 0.1 V is chosen as the optimum of modulation amplitude. The modulation time is the duration of the potential pulse. Figure 2(b) indicates that there is no different high peak current between 0.075; 0.1; and 0.125 s. The highest anodic peak current is obtained at 0.025 s. The modulation time at 0.1 s is used for the next application. The scan rate influences the height of anodic peak current because it can increase the rate of controlling time. The observation of electrochemical oxidation of paracetamol can't be shown at the higher scan rate more than 0.025 V/s. For the next application, 0.0125 V/s is chosen for measurement.

3.3. The effect of pH and supporting electrolytes

The electrochemical oxidation of paracetamol in different pH by cyclic voltammetry (CV) and differential pulse voltammetry (DPV) can be viewed in Figure 3. The anodic peak potential shifts from 0.6 V to 0.3 V with increasing pH value. It indicates that the voltammetric determination of paracetamol is pH dependent. The CV voltammogram showed one anodic peak potential indicating the irreversible reaction [17]. Meanwhile, the DPV voltammogram showed that the highest anodic peak current occurred at pH 2, so it is chosen as optimum pH for measurement.

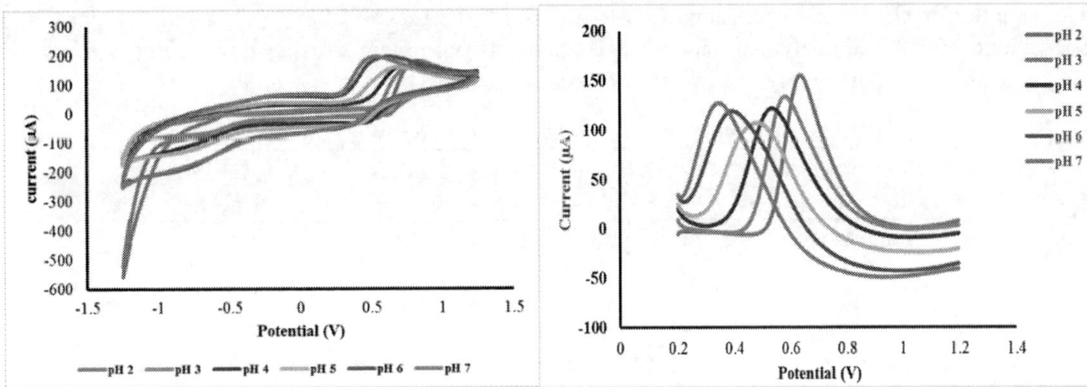

Figure 3. CV (a) and DPV (b) voltammogram of pH measurement of paracetamol 1 mM in Britton Robinson buffer 0.04 M pH 2-7.

The electrolyte solutions are used to reduce obstructions from solution and added conductivity. It also used for controlling potential during measurement to reduce the effect of electron migration. Based on the data in Table 2, the highest peak current is carried out with Britton Robinson buffer 0.04 M. However, the supporting electrolyte phosphate buffer has the lowest background current, thus, it is chosen as optimum supporting electrolyte.

Table 2. Peak potential and current of paracetamol 1 mM in different supporting electrolytes

Supporting electrolyte	E_{pa}	I_{pa} (μA)
BR 0.04 M	0.67337	174.76
Phosphate buffer 0.04 M	0.64819	172.35
HCl 0.1 M	0.653229	168.93

3.4. The Performance of GCE-PVA-Fe_3O_4

To understand the relationship between concentration and peak current, DPV technique was used to determinate paracetamol in concentration range from 0 – 100 μM. The determination is under the optimized experimental condition using unmodified carbon electrode (GCE), MIP PVA and MIP PVA-Fe_3O_4modified glassy carbon electrode (as shown in Figure 4).

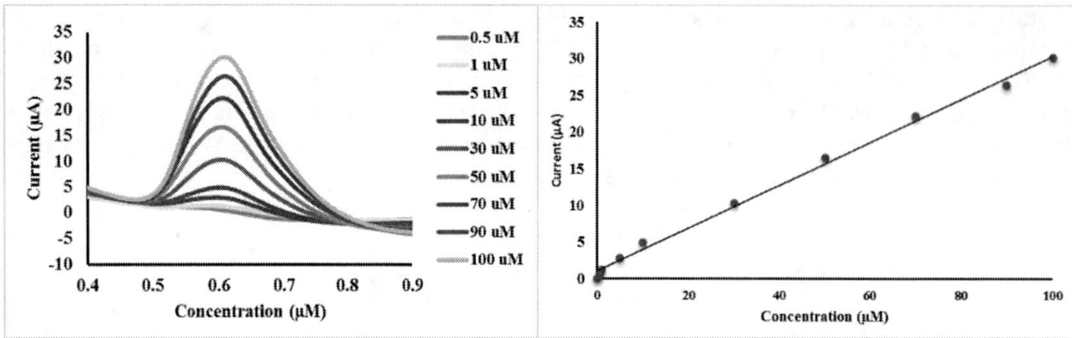

Figure 4. a) DPV voltammogram and b) standard curve of Paracetamol 0 – 100 µM in phosphate buffer 0.04 M pH 2 measured by GCE-PVA-Fe_3O_4.

The anodic peak current values versus concentration of paracetamol plots is shown in Figure 4(b). The equation of the linear regression plots is Ip (µA) = 0.2922x + 1.0827 (µM) with the correlation coefficient 0.9958.The analytical parameters of paracetamol obtained using three difference working electrodes produced difference performance (as shown Table 3).

Table 3. Analytical performance of difference working electrode

Parameters	GCE	GCE-PVA	GCE-PVA-Fe_3O_4
Sensitivity/ (µA/µM)	0.4871	0.2695	0.2922
Correlation coefficient	0.9799	0.9906	0.9958
LoD/ (µM)	17	12	8
Linear range/ (µM)	0 – 100 µM	0 – 100 µM	0– 100 µM

The modified glassy carbon electrode with PVA and PVA-Fe_3O_4 can increase the performance of determination compared to unmodified glassy carbon electrode. The highest correlation coefficient is MIP PVA-Fe_3O_4 with the limit of detection of 8 µM.

4. Conclusion

The simple carbon electrode which increase the performance of paracetamol determination using voltammetry technique was established. The best concentration of paracetamol in membrane was 3% (w/w). The best DPV parameters were obtained at modulation amplitude 0.1 V, modulation time 0.1 s and scan rate 0.0125 V/s. The pH value can influence the peak current and potential of paracetamol. The highest peak current occurred at pH 2. The supporting electrolyte of phosphate buffer was chosen for the next measurement. The modified glassy carbon electrode with PVA-Fe_3O_4 has the highest correlation coefficient with LoD 8 µM.

References

[1] A. Mao, H. Li, D. Jin, L. Yu, and X. Hu 2015 *Talanta.* **144** 252–7
[2] C. M. Kuskur, K. Swamy BE, and Jayadevappa H 2015 *J. Anal. Bioanal. Tech.* **6** 1-6
[3] G. Yang, L. Wang, J. Jia, and D. Zhou 2012 *J. Solid state Electrochem* **16** 2967-77
[4] H. Montaseri, and P. B.C. Forbes 2018 *Trends in Analaytical Chemistry* **108** 122-134
[5] K. Delvadiya, R. Kimbahune, P. Kabra, Sunil K, P. Patel 2011 Int J Pharm Pharm Sci **3** 170-174
[6] A. Ejaz and S. Jeon 2017 *Electrochimica Acta* **245** 742–751

[7] S. Berto 2018 *Electrochimica Acta* **284** 279–286
[8] Tanuja SB, Kumara SBE and Vasantakumar PK 2018 Ind. Chem. **4**(124) 1-10
[9] M. Akbari, H. Shayani-Jam, M. R. Yaftian, and M. Parinejad 2018 *J. Electroanal. Chem.* **827** 160–6
[10] L. Ozcan and Y. Sahin 2007 Sensors Actuators B Chem. **127**(2) 362–369
[11] P. Alagarsamy, R. Settu, S.M. Chen, T.W. Chen, I.S. Hong, and M. M. Rao 2018 *Int. J. Electrochem. Sci.***13** 7930-8
[12] M. F. Koudehi, and S. M. Pourmortazavi 2018 *electroanalysis* **30** 1-10
[13] Nandhita A., Manokaran J., and Balasubramanian N 2014 *Res. J. Chem. Environ.* **18** 54-61
[14] A. L. Kavitha, H. G. Prabu, S. A. Babu, and S.K. Suja 2013 *Journal of Nanoscience and Nanotechnology* **13** 98-104
[15] C. Fernandez, Z. Heger, R. Kizek, T. Ramakrishnappa, A. Borun, and N. H. Faisal 2015 *Int. J. Electrochem. Sci.* **10** 7440-52
[16] B. He, and S. Yan 2017 *Int.J. Electrochem. Sci.* **12**, 3001-11
[17] C. Engin, S. Yilmaz, G. Saglikoglu, S. Yagmur, and M. Sadikoglu 2015 *Int. J. Electrochem. Sci.* **10** 1916–25

IOP Conf. Series: Materials Science and Engineering **833** (2020) 012060 doi:10.1088/1757-899X/833/1/012060

The Effect of Aging and Crystallization Time on the Synthesis and Characteristics of Zeolite-Y from Malang-Quartzite Silica

Sukrawati Arni[1], Sumari Sumari[1]*, Aman Santoso[1], Tasya Tamara[1]

Department of Chemistry, Faculty of Mathematics and Science, Universitas Negeri Malang, Jl. Semarang No. 5 Malang 65145, Indonesia

Corresponding author: sumari.fmipa@um.ac.id

Abstract. The purpose of this study is to determine the effect of aging and crystallization time on the synthesis and characteristics of zeolite-Y. The raw material of silica used was obtained from the silica extraction of Malang quartzite by leaching and sol-gel methods. The extracted silica was analyzed using XRF and XRD. Synthesis of zeolite-Y was carried out using hydrothermal method by mixing NaOH, $Al(OH)_3$, SiO_2, and H_2O according to the molar ratio of the elements in zeolite-Y. The aging process was carried out with time variations of 12, 24, 48 hours and the crystallization process were carried out with time variations of 6, 9, 12 hours at a 150 °C. The synthesized zeolite-Y was analyzed using XRF, XRD, FTIR, and SEM. The results showed that quartzite silica could be extracted by leaching and sol-gel methods with a yield of 99.1 wt%. The optimum condition of synthesis zeolite-Y was obtained at the aging time of 12 hours and the crystallization time of 12 hours.

1. Introduction

In general, researchers obtain silica from the extraction of other natural materials such as perlite stones [1], diatomite [2], shale rock [3], and bentonite [4]. However, to the best of our knowledge, research on the use of quartzite as a precursor of zeolite catalysts has not been reported yet. In this work, Malang quartzite stone will be used as the source of silica as the precursor of the zeolite-Y catalyst.

Silica or silicon dioxide (SiO_2) is an abundant compound in nature. Generally, the source of silica that has been widely used in industry is quartz sand or silica sand. Other sources of silica can also be found in rice husk ash which contains 86.9% of silica [5] or Lapindo mud (Indonesia) with contains 53.03% of silica [6]. The other compounds in silica sand are Al_2O_3, Fe_2O_3, CaO, TiO_2, V_2O_5, Cr_2O_3, MnO, CuO, SrO, and ZrO_2. XRF results in the preliminary test of this work show that silica content in Malang quartzite is 75.7%. The result also shows that silica is the most abundant element in Malang quartzite compared to other compounds. Based on the XRF results, Malang quartzite rocks have the potential as the source of silica for further development.

Zeolite is an alumina-silica crystal with a three-dimensional porous structure [2]. Zeolites are divided into natural and synthetic zeolites. Natural zeolite has been widely used; thus the number is decreasing. Generally, natural zeolites are used for fertilizer and water purification, while synthetic zeolites are widely used in the industry. However, in Indonesia, synthetic zeolites have not been widely produced, and most zeolites in Indonesia are generally imported. Synthetic zeolites are widely used for chemical processes in the chemical industry such as catalysts [7], ion exchangers [8], [9], and adsorbents in waste treatment [5]. High-purity, high-quality synthetic zeolites are needed in the industry. The raw material for making zeolites is silica and aluminium. One of the potential sources of silica is quartzite. Zeolites can be synthesized with different crystal structures, fixed pore sizes, Si/Al ratio frameworks and acid centres that can be adjusted to gain the targeted catalytic properties.

Previous study showed that zeolite-Y was successfully synthesized using *Pluronic* P123 block copolymer as a template reported [10] using diatomite from Bolivian as silica base material [2], and using perlite glass as basic ingredient of silica [1]. In this research, silica extracted from quartzite was used as a base material. Quartzite silica stone is extracted by leaching and sol-gel methods. In this study, the synthesis of zeolite-Y using silica extracted from Malang quartzite stone is discussed. The purpose of this study was to determine the effect of aging and crystallization time on the characteristics of the synthesized zeolite-Y.

2. Method

2.1. Materials

SiO_2 was obtained from quartzite stones (Malang, East Java). HCl (analytical grade) and NaOH (analytical grade) were purchased from Merck. $Al(OH)_3$ was purchased from Sigma-Aldrich. All chemicals were used without further purification.

2.2. Synthesis of Zeolite-Y

Silica were obtained from the extraction of Malang quartzite stone by mixing 22.5 grams of its powder with 30 grams of NaOH in a hotplate until the mixture become completely melted and well mixed to get sodium silicate. The sodium silicate then was dissolved in 150 mL DI water before being filtered. The resulting filtrate then were titrate with HCl p.a before being filtered again to get the residual precipitate. The as-obtained residual precipitate was dried at 100° C for 12 hours to obtain silica. The impurity of as-obtained amorphous silica was observed to be as high as 99.1%. Synthesis of zeolite-Y was carried out using a procedure that has been reported [11] with a $NaOH:Al(OH)_3:SiO_2$: H_2O molar ratio of 58:58:134:384. The synthesis was carried out by dissolving 2 grams of SiO_2 in 40 mL NaOH 2 N, stirred for 30 minutes until it fully dissolved to obtain sodium silicate. Sodium aluminate was prepared by dissolving 2 grams of $Al(OH)_3$ in 60 mL of 7 M NaOH followed by refluxing the mixture until it completely dissolved. Sodium silicate and sodium aluminate were then mixed together for 30 minutes. The mixture then stands for aging for 12, 24, and 48 hours at RT before transferred into a hydrothermal Teflon bottle equipped with a stainless-steel autoclave for the crystallization process. The crystallization was performed at 150 °C for 12, 9, and 6 hours. After the aging and crystallization completed, the mixture was filtered off, and the residual precipitate was washed with DI-water. The precipitate was then dried for 24 hours at 80 °C to form zeolite-Y.

2.3. Characterisation of Zeolite-Y

In order to obtain the as-synthesized zeolite-Y phase information, the samples were characterized with X-ray Diffraction (XRD, PANalytical Brand, type X'pert-Pro) using Cu-Kα under 40 kV and 35 mA, the 2θ range was set between 10° to 90°. The microstructures and morphologies of the samples were observed using scanning electron microscopy (SEM, FEI Brand, type inspect S50) at 20 kV with magnification 10000x, and 50000x. X-ray Fluorescence were recorded on a MiniPal (PANalytical type MiniPal 4) to observed the Si and Al atoms and its atomic weight ratio. FTIR analysis were carried out on an IR spectrophotometer (Shimadzu Brand, type IR-prestige 21), with scanning range between 4000-400 cm^{-1}, using KBr pellet method.

3. Results and Discussion

3.1. Synthesis and Characterization of Zeolite-Y

The silica was reacted with sodium alumina for the synthesis of zeolite-Y. The XRD results showed that zeolite-Y has been formed, as proven by comparing the diffraction pattern of synthesized zeolite-Y with that of a standard zeolite-Y [12]. The diffraction pattern for the synthesized zeolite-Y is presented in Figure 1.

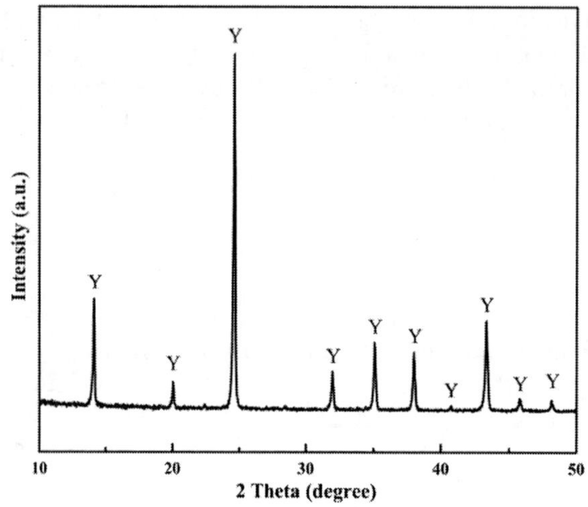

Figure 1. Diffractogram of as-synthesized zeolite-Y.

The synthesis of zeolite-Y was carried out at 150 °C with a variation of aging time and crystallization time. Crystallization and aging time are the factors that influenced the formation of zeolite-Y. Comparison of 2θ and *d*-spacing between zeolite-Y in various aging time and crystallization time are shown in Table 1.

Table 1. Value of 2θ and d-spacing of zeolite-Y at various aging and crystallization time.

Zeolite-Y standard		Z-12-12 (ta = 12; tc = 12)		Z-24-12 (ta = 24; tc = 12)		Z-48-12 (ta = 48; tc = 12)		Z-48-9 (ta = 48; tc = 9)		Z-48-6 (ta = 48; tc= 6)	
2θ (°)	*d-spacing* (Å)	2θ (°)	*d-spacing* (Å)	2θ (°)	*d-spacing* (Å)	2θ (°)	*d-spacing* (Å)	2θ (°)	*d-spacing* (Å)	2θ (°)	*d-spacing* (Å)
14.61	6.064	14.15	6.259	14.29	6.199	14.25	6.215	14.33	6.182	14.22	6.228
20.71	4.288	20.06	4.427	20.15	4.408	20.09	4.420	20.24	4.387	20.04	4.431
24.06	3.699	24.62	3.616	24.70	3.604	24.65	3.611	24.53	3.630	24.57	3.623
31.95	2.801	31.95	2.802	31.99	2.798	31.93	2.802	31.78	2.816	31.82	2.813
35.29	2.543	35.09	2.558	35.11	2.556	35.05	2.560	35.28	2.544	34.91	2.570
37.82	2.379	38.00	2.368	38.01	2.367	37.95	2.371	38.20	2.356	37.80	2.380
41.28	2.187	40.73	2.215	-	-	-	-	40.91	2.206	-	-
43.51	2.080	43.32	2.089	43.31	2.089	43.25	2.092	43.51	2.080	43.05	2.101
44.84	2.021	45.80	1.981	45.78	1.982	45.69	1.986	-	-	-	-
48.03	1.894	48.18	1.889	48.13	1.891	48.06	1.893	48.40	1.881	47.86	1.901

ta = aging time (hours); tc = crystallization time (hours)

From Table 1, there is no peak observed at 2θ of 41.28° for the aging times of 24 hours (tc = 12), 48 hours (tc = 12), 48 hours (tc = 6). Meanwhile, in the aging time of 12 hours a peak at 2θ of 40.73° was observed or shifted 0.54° from the standard angle. In the tc of 9-hour, there is a peak at 2θ of 40.90° or shifted 0.38° from the standard. Moreover, at 2θ of 44.84°, there was no peak observed in of 9 hours and 6 hours, whereas in the 12 hours crystallization time there was a peak at 2θ of 45.68°, which shifted 0.84° from the standard. This result indicated that the number of peaks displayed by the 2 hours aging time are greater than that of the 24 or 48 hours. There are also more peaks exhibited by the 12- and 9-hours crystallization times compared to that of 6 hours crystallization time.

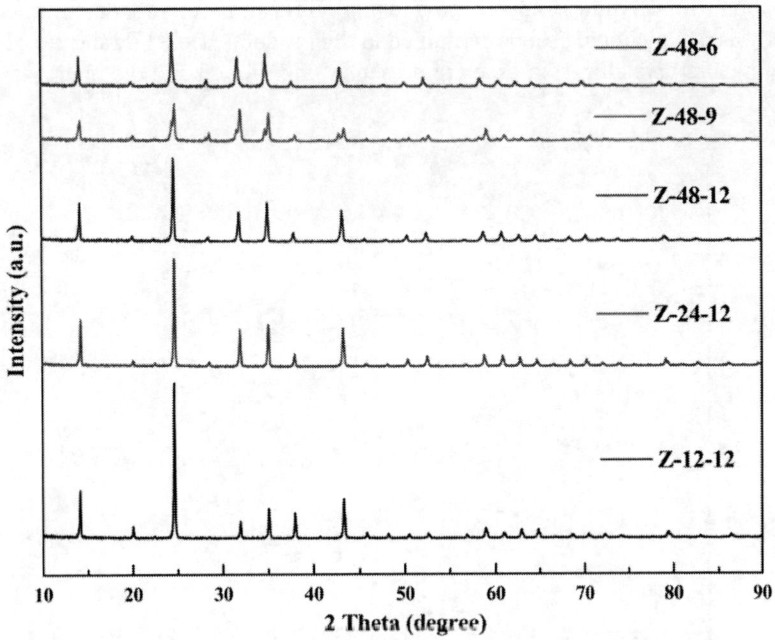

Figure 2. Zeolite-Y diffractograms at various aging and crystallization times.

Figure 2 shows that the peaks in zeolite-Y with a variation of aging time of 12 hours and crystallization time of 12 hours has the highest intensity compared to others. It means that although zeolites-Y successfully made at all 5 different conditions, the optimum results was obtained when the synthesis of zeolite-Y was done at 12 hours aging time and 12 hours crystallization time.

XRF analysis conducted on zeolite-Y samples aims to determine the relative concentration levels of the elements in the synthesized zeolite. Thus, the Si/Al molar ratio of the zeolite can be identified. The Si/Al molar ratio of zeolite-Y was in range of 2.045 to 2.087. Complete data of the Si/Al molar ratio is presented in Table 2.

Table 2. The Mole Ratio of Si/Al of Zeolite-Y.

Zeolite-Y	**Time**		**Element composition (%)**		**Moles**		**Molar ratio of Si/Al**
	Aging (hours)	Crystallization (hours)	Si	Al	Si	Al	
Z-12-12	12	12	66.9	31.1	2.389	1.152	2.073
Z-24-12	24	12	65.5	30.9	2.339	1.144	2.045
Z-48-12	48	12	66.8	31.1	2.385	1.152	2.071
Z-48-9	48	9	65.0	30.3	2.321	1.122	2.068
Z-48-6	48	6	67.1	31.0	2.396	1.148	2.087

The reported molar ratio of Si/Al of zeolite-Y was in the range of 1.5 to 3 [13]. The synthesized zeolite-Y in this work has the largest Si/Al molar ratio of 2.087 (48 hours aging time and 6 hours crystallization time) and the smallest Si/Al molar ratio of 2.045 (24 hours aging and 12 hours crystallization time). Based on this XRF data, which is in the expected range of Si/Al, it can be stated that zeolite-Y was formed in all five variations of aging time and crystallization time.

Characterization of the synthesized zeolite-Y using FTIR aims for support the successful synthesis through the analysis of functional groups contained in the product. The FTIR spectra of zeolite-Y in five variations of aging and crystallization time are presented in Figure 3 and the interpretation is presented in Table 3.

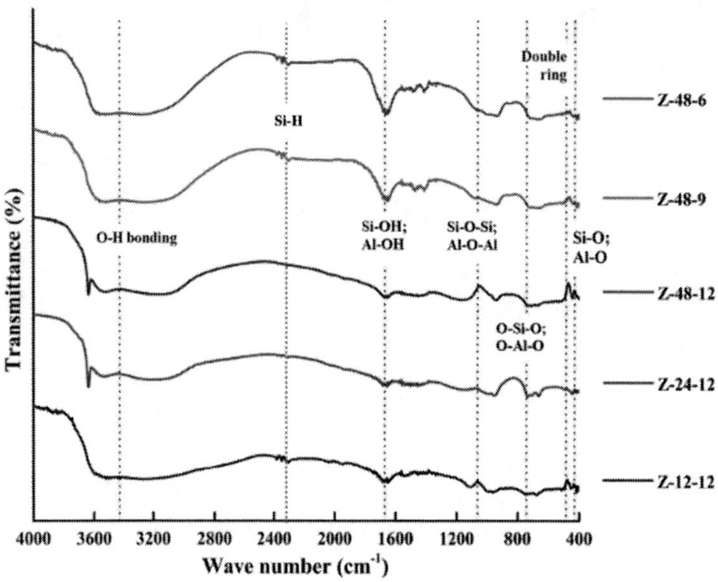

Figure 3. IR spectra of zeolite-Y at various aging and crystallization time.

Table 3. The Interpretation of FTIR spectra of the synthesized zeolite-Y at various aging and crystallization time.

Wave numbers (cm^{-1})	Ref. (cm^{-1})	Wave numbers (cm^{-1})					Interpretation
		Z-12-12 (ta = 12; tc = 12)	Z-24-12 (ta = 24; tc = 12)	Z-48-12 (ta = 48; tc = 12)	Z-48-9 (ta = 48; tc = 9)	Z-48-6 (ta = 48; tc = 6)	
500 - 418	450	418.55	447.49	447.49	418.55	-	Buckling vibration T-O (Si-O or Al-O)
650 - 500	574	501.49	609.51	561.29	567.07	549.61	Double ring
820 - 650	722	744.52	736.81	740.67	719.45	719.45	Symmetric stretching vibrations O-Si-O and O-Al-O
1250 - 950	1025	1103.86	1141.43	1149.57	1053.13	-	Asymmetric stretching vibrations Si-O-Si/Al-O-Al
1655 -1610	1646	1614.42	1693.5	1645.28	1631.78	1614.42	Buckling vibration Si-OH/Al-OH
2400 -2300	2375	2380.16	2378.23	2347.37	2380.16	2380.16	Buckling vibration Si-H
3200 -3600	3459	3271.27	8,53468.01	3444.87	3444.87	3442.94	Bonding O-H

ta = aging time (hours); tc = crystallization time (hours)

The IR spectra of synthesized zeolite-Y shows some wave numbers of functional group indicating the formation of zeolite-Y. For instance, the IR spectra of zeolite-Y with aging time for 12 hours and crystallization time of 12 hours shows a peak at wave number of 1103.86 cm^{-1} that corresponds to asymmetric stretching vibrations of Si-O-Si/Al-O-Al, a peak at wave number of 744.52 cm^{-1} that

IOP Conf. Series: Materials Science and Engineering **833** (2020) 012060 doi:10.1088/1757-899X/833/1/012060

corresponds to symmetrical stretching vibrations of O-Si-O/O-Al-O, and a peak at wave number of 418.55 cm^{-1} that corresponds to Si -O-Si or Al-O-Al vibration. However, the peaks spectra of zeolite-Y produced at aging time of 12 hours and crystallization time of 12 hours was higher. This data indicated that the obtained zeolite-Y under these conditions was more successful. SEM analysis for the synthesized zeolite-Y were presented in Figure 4, Figure 5, and Figure 6.

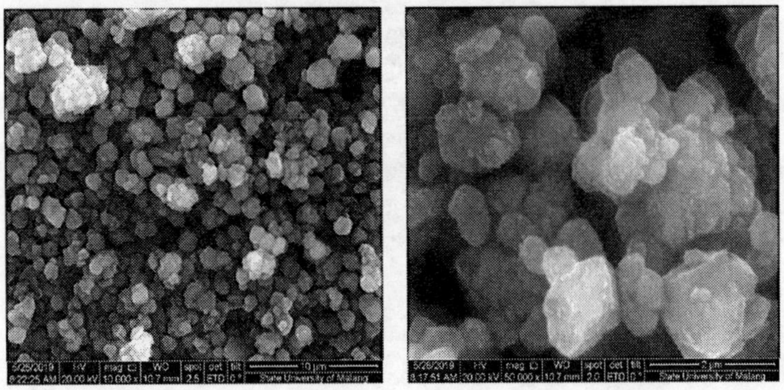

(a) (b)

Figure 4. SEM zeolite-Y with 12 hours aging time and 12 hours crystallization time with (a) 10.000x and (b) 50.000x magnification.

(a) (b)

Figure 5. SEM zeolite-Y with 24 hours aging time and 12 hours crystallization time with (a) magnification of 10.000x and (b) 50.000x magnifications.

(a) (b)

Figure 6. SEM zeolite-Y with 48 hours aging time and 12 hours crystallization time with (a) magnification of 10.000x and (b) 50.000x magnifications.

Based on the SEM data in Figure 4, Figure 5, and Figure 6, It was known that the diameter of the crystal unit is 1.2 micrometres with octahedral shape. It was in line with the characteristics of zeolite-Y reported by previous researchers [1], [14] where the crystalline form of zeolite-Y is octahedral.

4. Conclusions

Zeolite-Y has been successfully synthesized using a hydrothermal method in variations of aging time between 12 to 48 hours and crystallization time of 6 to 12 hours. The optimal synthesis conditions of zeolite-Y were 12 hours aging time and 12 hours crystallization time with mole ratio Si/Al of 2.073. The FTIR spectra shows asymmetric stretching vibrations of Si-O-Si/Al-O-Al at wave number of 1103.86 cm^{-1}. Symmetrical stretching vibrations of O-Si-O/O-Al-O appeared at wave number of 744.52 cm^{-1} and the bonding vibrations of Si-O-Si/Al-O-Al was appeared at wave number of 418.55 cm^{-1}. The SEM data shows that the produced zeolite-Y has an octahedral crystalline structure which also supports that the synthesis of zeolite-Y is successful.

References

[1] Christidis G. E. and Papantoni H. 2008 *Open Mineral. J.* **2** 1 1–5
[2] Garcia G., Cardenas E., Cabrera S., Hedlund J., and J. Mouzon 2016 *Microporous Mesoporous Mater.* **219** 29–37
[3] Doyle A. M., Alismaeel Z. T., Albayati T. M., and Abbas A. S. 2017 *Fuel* **199**
[4]. Soetaredjo F. E., Ayucitra A., Ismadji S., and Maukar A. L. 2011 *Appl. Clay Sci.* **53** 2 341–6
[5] Trivana L., Sugiarti S., and Rohaeti E. 2015 *J. Sains &Teknologi Lingkung.* **7** 2 66–75
[6] Fadli A. F., Tjahjanto R. T., and Darjito 2013 *Kim. Student J.* **1** 2 182–7
[7] Sumari S., Fajaroh F., Suryadharma I. Bagus, Santoso A., and Budianto A. 2019 *IOP Conf. Ser. Mater. Sci. Eng.* **515** 1
[8] Santoso A., Sumari, Joharmawan R., and Hutami L. B. 2019 *IOP Conf. Ser. Mater. Sci. Eng.* **509** 1
[9] Sumari S., Fajaroh F., Yahmin, Sholihah N., Santoso A., and Budianto A. 2019 *IOP Conf. Ser. Mater. Sci. Eng.* **515** 1
[10] Zhao J., Wang G., Qin L., Li H., Chen Y., and Liu B. 2016 *Catal. Commun.* **3** 98–102
[11] Belaabed R., Elabed S., Addaou A., Laajab A., Rodríguez M. A., and Lahsini A. 2016 *Bol. la Soc. Esp. Ceram. y Vidr.* **55** 4 152–8
[12] Treacy M. M. and Higgins J. B. 2001 *Collection of Simulated XRD Powder Patterns for Zeolites* (Elsevier) p 184
[13] Le T., Wang Q., Pan B., Ravindra A. V., Ju S., and Peng J. 2019 *Microporous Mesoporous Mater.* **284** 476–485
[14] Bortolatto L. B. *et al.* 2017 *Microporous Mesoporous Mater.* **248** 214–221

Photocatalytic Performance of Bismuth Oxide Prepared by Citric Acid-Fueled Solution Combustion on Decolorisation of Organic Dye Molecules

Yayuk Astuti*, Dhina Anggraeni, Adi Darmawan

Chemistry Department, Faculty of Science and Mathematics, Diponegoro University, 50275, Semarang, (Central Java) Indonesia

e-mail: yayuk.astuti@live.undip.ac.id

Abstract. Bismuth oxide can be applied as photocatalyst. The purpose of this study was to synthesize bismuth oxide using the solution combustion method, identify the characteristics of bismuth oxide using FTIR and XRD, and determine the photocatalytic activity of the obtained bismuth oxide to decolour Remazol Black B and Rhodamine B dye molecules. The results showed that bismuth oxide was formed with yellow colour after calcination convinced with the presence of Bi-O and Bi-O-Bi functional groups and having 95.38% similarity using Match software. The photocatalytic activity test shows that the obtained bismuth oxide enabled to decolour RBB, RhB dyes well.

1. Introduction

Bismuth Oxide is a material that has many advantages and can be applied in various fields including solid oxide fuel cells, gas sensors, high temperature super conductor materials, functional ceramics and photocatalyst materials [1]. One of methods for bismuth oxide synthesis is solution combustion [2]. This method has various advantages such as fast and simple process, as well as easy to do [3].

The solution combustion method is a synthesis method whose main components consist of oxidants and fuels, with processes that involve high temperatures [3]. Fuels commonly used in the solution combustion method are citric acid [2], urea [4], glycine [5], and hydrazine [6]. While, the oxidant used in the method solution combustion is metal nitrate because the nitrate group acts as an oxidizing agent and has great solubility and allows for the homogenization of larger product yields [4].

Y. Astuti et al [7] synthesized bismuth oxide using solution combustion method from bismuth subnitrate (Bi5O(OH)9(NO3)4) and citric acid as oxidant and fuel, respectively. The results showed that bismuth oxide enable to decolor methyl orange by 38% for 12 hours with a UV lamp. In addition, by this method, synthesis of bismuth oxide with bismuth nitrate pentahydrate as oxidant and citric acid as fuel was also undertaken [8]. Specifically, the effect of a variation of moles ratio of fuel-oxidant on the photocatalytic activity of obtained products for the decolorization of methyl orange dyes was investigated. The results showed that the moles ratio 1 presented a good photocatalytic activity in methyl orange decolorization. Therefore, in this paper the performance of bismuth oxide synthesized using solution combustion method with moles ratio 1 for decolorization of other dye molecules, namely, rhemazol black B (RBB) and rhodamine B (RhB) will be presented. Both dyes are usually applied for Batik coloring. It should be borne in mind that almost 40% dyes applied for Batik coloring are disposed into environment [9]. In a huge amount of these dyes makes the environment unfriendly for human. Therefore, it must be overcome, for example by using photocatalyst bismuth oxide. The investigation is expected to contribute on science about performance of bismuth oxide in decolorizing other dyes besides methyl orange.

2. Experimental Methods

2.1. Materials
The materials used were bismuth nitrate pentahydrate from Sigma-Aldrich USA, distilled water, nitric acid from Merck Germany, citric acid Merck Germany, Remazol Black B and Rhodamine B (RhB) from Sigma-Aldrich USA.

2.2. Research Procedure

2.2.1. Synthesis of Bismuth Oxide. Bismuth oxide synthesis was undertaken using the solution combustion method. As much as 2.42 g bismuth nitrate pentahydrate was dissolved in 10 mL nitric acid 65% then added 1.26 g of citric acid. The mixture was stirred until dissolved and then heated at temperature of 300 °C for 8 hours. Obtained powder (Figure 1a) was allowed to cool then calcined for 4 hours at 700 °C. This procedure is based on Y Astuti report [8].

2.2.2. Product Characterization. The obtained products were then characterized using FTIR and XRD. FTIR analysis was carried out on the products before and after calcination to identify the presence of Bi-O or Bi-O-Bi functional groups. The obtained product after calcination was also characterized using XRD to convince that bismuth oxide was successfully formed.

2.2.3. Photocatalytic Activity. Photocatalytic activity of product after calcination was then determined by applying it to decolour RBB and RhB dye molecules. As much as 0.1 g of synthesized bismuth oxide was added into 50 mL RBB 25 ppm and subsequently was irradiated using UV-A (3.94-3.1 eV) at time variation of 1, 2, 3, 4, and 5 hours. This procedure was applied for RhB (5 ppm) solution as well. After irradiation, the solution was centrifuge to separate bismuth oxide powder and dye solution. The filtrate was then analyzed using spectrophotometer to determine the absorbance change of the solution after photocatalysis. This absorbance was further used to calculate the percentage of decolorization by applying formula:

$$decolourization\ percentage = \left(\frac{C_0 - C_t}{C_0}\right) \times 100\ \% \qquad (1)$$

with C_0 = initial concentration of dye molecule (ppm), C_t = final concentration of dye molecule (ppm) at t time.
In addition, to convince that prepared bismuth oxide enabled decolourising RBB and RhB, these dyes were also irradiated using UV-A for 5 hours without the presence of bismuth oxide. Photocatalysis of both dyes using commercial bismuth oxide was also undertaken at a time variation of 0.5 to 3 hours with 0.5 hour interval.

3. Result and Discussion

3.1. Synthesis and structural analysis of the obtained products
The result showed that the yellow powder in the obtained product after heating for 8 hours, Figure 1a, is very little indicating bismuth oxide was formed. It is dominated by black and dark green powder indicating that there is still carbon in the sample derived from citric acid. After calcination for 4 hours at a temperature of 700oC, the mixed color powder turned to yellow powder as shown in Figure 1b. It is indicated that bismuth oxide was formed.

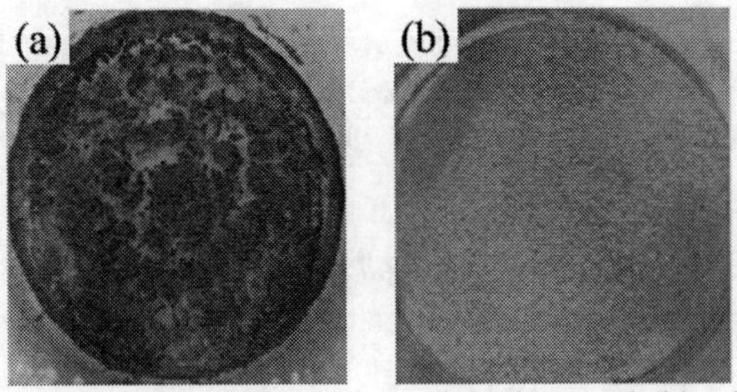

Figure 1. The obtained products (a) after 8 hour heating and (b) after 4 hour calcination.

The obtained products before and after calcination were analyzed using FTIR and the results can be seen in Figure 2. Product before calcination showed there is a peak at 850 cm-1 wavenumber attributed to Bi-O-Bi functional group [10] and a peak at the wavenumber of 1387 cm-1 indicating the existence of the Bi-O vibration mode [10]. These peaks convinced that bismuth oxide was formed in the product before calcination even though the product was physically not only yellow but mixed color as presented in Figure 1a. The mixed color product indicating there are still many impurities derived from bismuth nitrate pentahydrate and citric acid as starting material shown by the appearance of many peaks besides Bi-O-Bi and Bi-O. The peaks at wave numbers 1080 and 1132 cm-1 on sample before calcination are assigned as vibration modes of CO stretching [11] coming from citric acid. Moreover, peaks at wavenumbers 1525 and 1587 cm-1 were also present attributed to vibrations of OH functional group [11]. The stretching OH vibration was also observed at wavenumber of 3455 cm^{-1}. In addition, peak at 1300 cm-1 is also assigned as vibrations of N-O from NO3 [12]. Due to the impurities presence, the product was then calcined at 700 °C. FTIR spectra among the products and starting materials can be seen in Figure 2.

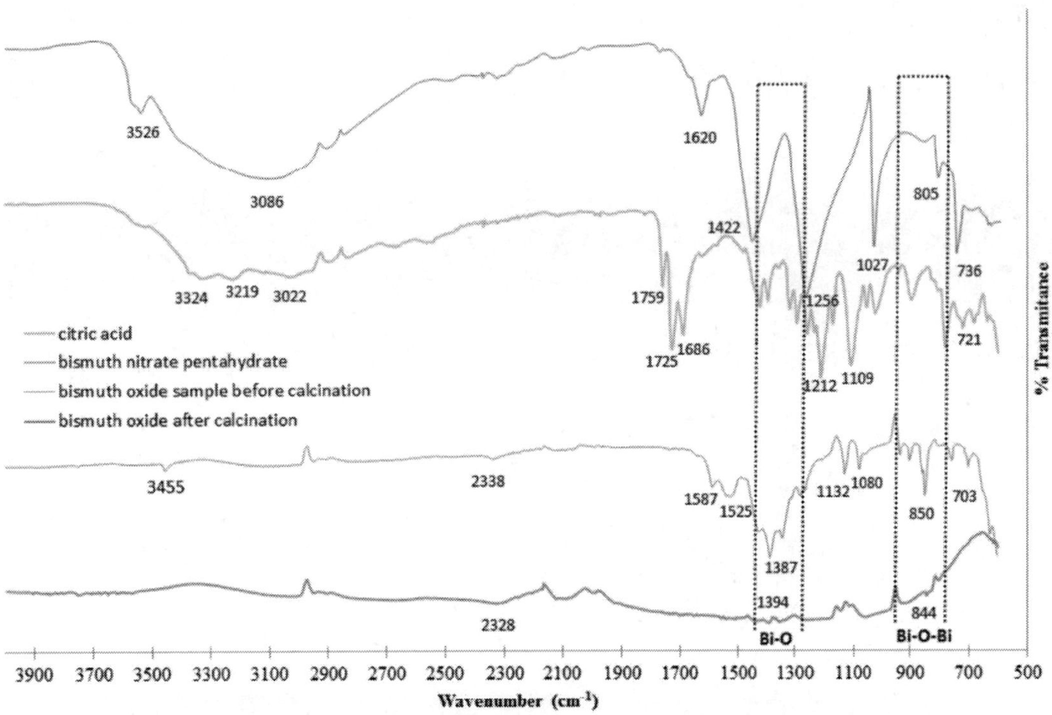

Figure 2. FTIR spectra of the obtained products before calcination, citric acid, bismuth nitrate pentahydrate and the obtained products after calcination.

FTIR spectrum of sample after calcination showed that there is a peak at wavenumber of 844 and 1394 cm-1 indicating the formation of Bi-O-Bi bond and Bi-O respectively [10]. These results are in

IOP Conf. Series: Materials Science and Engineering **833** (2020) 012061 doi:10.1088/1757-899X/833/1/012061

accordance with FTIR spectra of pure bismuth oxide [8, 13]. In the meantime, other peaks were not observed in this FTIR spectrum. Therefore, these peaks confirmed that the functional groups of the impurity largely disappeared, and the product obtained after calcination be purer than the product before calcination.

Fu et al [12] reported that in the wave number of 1300~ there was not only a Bi-O group but there was also a NO3- group as stated previously. The FTIR spectrum of sample before calcination has a peak with sharper intensity than that of sample after calcination. The intensity of this peak in both samples can be calculated through the ratio of absorbance approach at the wavenumber of around 1400-1300 cm-1 with a control wavenumber of about 2300 cm-1. The comparison of peak intensity of both samples is presented in Table 1.

Table 1: Comparison of absorbance in wave numbers of 1300 cm-1 in samples before and after calcination and control group

Treatment	Wavenumber absorption 1300 cm^{-1}: Control group
Before Calcination	4.18
After Calcination	1.25

Based on Table 1, it can be seen that the ratio of absorbance decreases after calcination. The decrease is assumed to occur because the NO3- group observed at wavenumber of 1300 cm-1 was lost as a result of the calcination process.

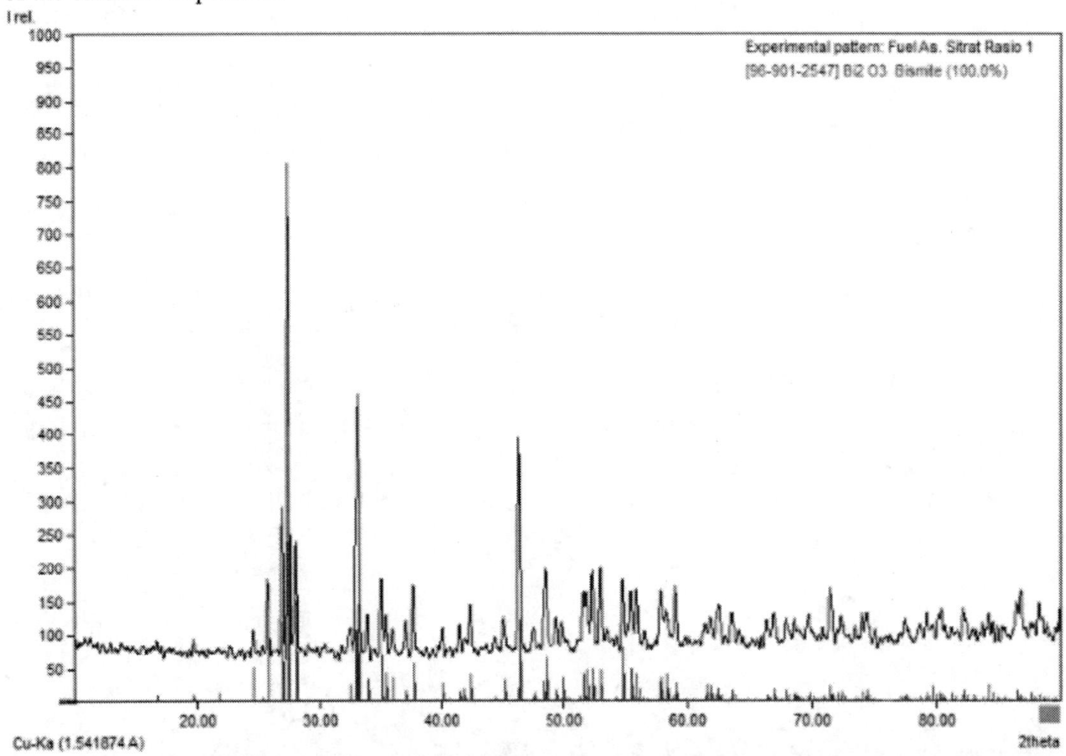

Figure 3. Diffractogram of bismuth oxide analysis results match.

The crystal structure of product after calcination was analyzed using XRD. The diffractogram was then analyzed using match software to determine the content of the product. As seen in Figure 3, the product has 95.38% similarity with reference Malmros indicating the presence of bismuth oxide

IOP Conf. Series: Materials Science and Engineering **833** (2020) 012061 doi:10.1088/1757-899X/833/1/012061

(Bi2O3) [14]. This result confirmed that bismuth oxide was formed after calcination process and the impurity was almost 4.62%.

3.2. Photocatalytic Activity

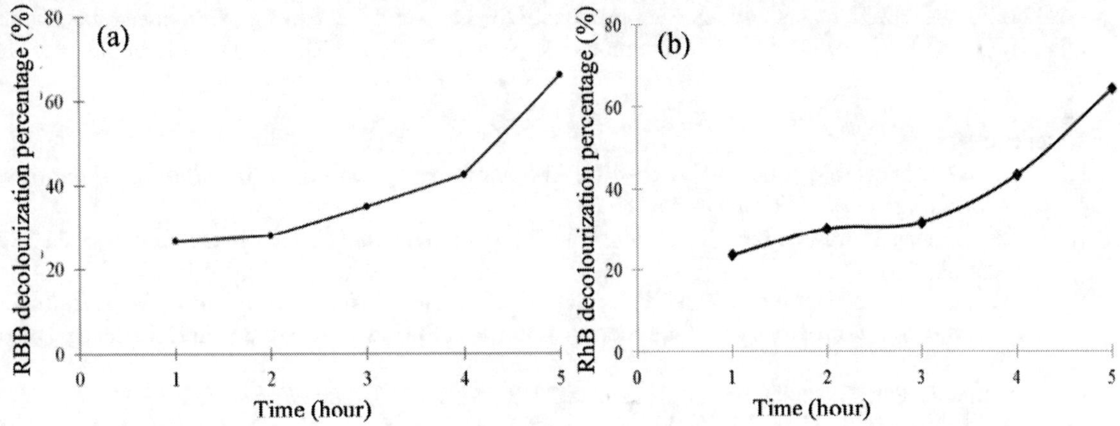

Figure 4. Photocatalytic activity of prepared bismuth oxide on decolorization of (a) RBB, and (b) RhB.

The photocatalytic activity of bismuth oxide was then investigated by applying in decolorization of RBB and RhB dye molecules. Figure 4 showed the decolorization percentage of dye molecules by bismuth oxide. It shows that after 5 hours irradiation, decolorization of RBB was around 66 % while RhB was 64 %. The difference of decolorization percentage may be due to the differences in dye molecules stability. Meanwhile, decolorization of RBB and RhB by commercial bismuth oxide as presented in Figure 5 showed that after 3 hours irradiation, decolorization percentages of RBB and RhB were 66 and 79%, respectively. This result indicated that resulting products content impurity.

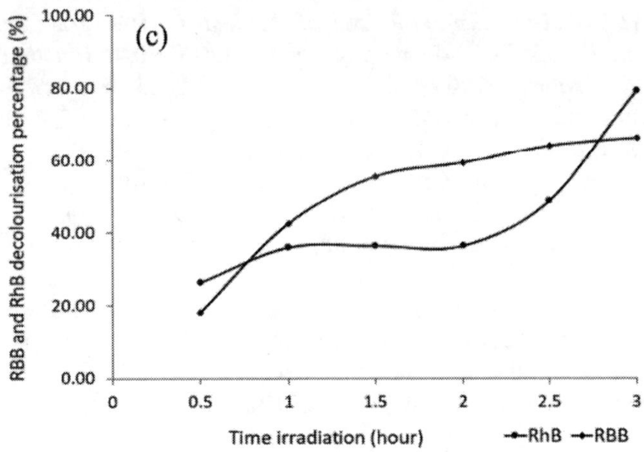

Figure 5. Photocatalytic activity of commercial bismuth oxide on decolorization of RBB and RhB.

Furthermore, in order to clarify that prepared bismuth oxide enabled to decolorized RBB and RhB, these dyes were irradiated using UV-A without presence of bismuth oxide. The result showed that after irradiation for 5 hours, only 1.4% and 9% of RBB and RhB were decolorized.

IOP Conf. Series: Materials Science and Engineering **833** (2020) 012061 doi:10.1088/1757-899X/833/1/012061

Decolorization of dyes by bismuth oxide occurs when the light source illuminates the surface of the semiconductor bismuth oxide so that resulting in electrons (e−) to be excited from the valence band to the conduction band, and forms a hole (h+) in the valence band. Holes in the valence band react with hydroxyl anions (OH−) to form hydroxyl radicals (HO•). The electrons in the conduction band react with O2 from the atmosphere and form superoxide radical anions (O2•-) which then react with H+ form HO2•. After that, HO2• reacts with electrons to produce HO2−which further reacts with H+ to form H2O2 oxidants. H2O2 reacts with O2•− to produce •OH, OH− and O2. Strong oxidizing agents such as OH−, HO •, and H2O2 oxidize organic dye molecules such as RBB and RhB into simpler molecules [15-17].

References

[1] Jalalah M, Faisal M, Bouzid H, Park J-G, Al-Sayari S and Ismail A A 2015 *Journal of Industrial and Engineering Chemistry* **30** 183–9

[2] La J, Huang Y, Luo G, Lai J, Liu C and Chu G 2013 *Particulate Science and Technology* **31** 287–90

[3] Alves, A., C.P. Bergmann, and F.A. Berutti 2013 *Novel Synthesis and Characterization of Nanostructured Materials*, Engineering Materials., Springer-Verlag Berlin Heidelberg: Brazil. p. 85.

[4] Jadhav L, Patil S, Jamale A and Chavan A 2013 *Materials Science Forum* **757** 85–98

[5] Shamsudin, Ili & Helwani, Zuchra & Abdullah, A.Z. & Wiheeb, A.D. & Othman, M.R. 2013 *Malaysian Journal of Analytical Sciences* **17** 171-175.

[6] Patil K C 1993 *Bulletin of Materials Science* **16** 533–41

[7] Astuti Y, Fauziyah A, Nurhayati S, Wulansari A, Andianingrum R, Hakim A and Bhaduri G 2016 *IOP Conference Series: Materials Science and Engineering* **107** 012006

[8] Astuti Y, Fauziyah A, Widiyandari H and Widodo D S 2019 *Rasayan Journal of Chemistry* **12** 2210–7

[9] Sastrawidana I, W. Lay B, Miftah Fauzi A, and Santosa D 2012 *Jurnal Ilmu Lingkungan (Journal Of Environmental Science),* **3** [online].

[10] Bandyopadhyay S and Dutta A 2017 *Journal of Physics and Chemistry of Solids* **102** 12–20

[11] Petit T and Puskar L 2018 *Diamond and Related Materials* **89** 52–66

[12] Fu H, Pan C, Yao W and Zhu Y 2005 Visible-Light-Induced Degradation of Rhodamine B by Nanosized Bi2WO6 *The Journal of Physical Chemistry B* **109** 22432–9

[13] Astuti Y, Elesta P, Widodo D, Widiyandari H, and Balgis R 2020 *Bulletin of Chemical Reaction Engineering & Catalysis* **15** 104-111

[14] Malmros G, Fernholt L, Ballhausen C J, Ragnarsson U, Rasmussen S E, Sunde E and Sørensen N A 1970 *Acta Chemica Scandinavica* **24** 384–96

[15] Sood S, Umar A, Mehta S K and Kansal S K 2015 *Ceramics International* **41** 3355–64

[16] Liu X, Deng H, Yao W, Jiang Q and Shen J 2015 *Journal of Alloys and Compounds* **651** 135–42

[17] Prasetya N B A, Haris A and Gunawan G 2012 *Molekul* **7** 143

Acknowledgement

The authors would like to thank the Ministry of Research, Technology and Higher Education, Republic of Indonesia for financial support through scheme of "Penelitian Dasar" in 2018 (No. 101-71 / UN7.P4.3 / PP / 2018) and 2019 (No. 101-71 / UN7.P4.3 / PP / 2019).

IOP Conf. Series: Materials Science and Engineering **833** (2020) 012062 doi:10.1088/1757-899X/833/1/012062

Coordination Polymers Containing a Glycine-Derived Trimellitic Acid Imide

Michael J. Beards, David R. Turner*

School of Chemistry, Faculty of Science, Monash University, Australia

*Coresponding author: david.turner@monash.edu

Abstract. A trimellitic imide ligand ($GlyTMI^{2-}$), derived from glycine, has been used in the formation of three coordination polymers as a means towards understanding the role of the aromatic core of the ligand in determining structure. The polymeric structures *poly*-$\{(N(CH_3)_2H_2)[Zn_2(GlyTMI)_2(OAc)(DMF)_{0.5}(CH_3OH)_{0.5}]\cdot CH_3OH\}$ (**1**), *poly*-$[Cd_2(GlyTMI)_2(DMF)_3]$ (**2**), and *poly*-$\{[Cd_2(GlyTMI)_2(4,4'\text{-bipy})_2]\cdot 2H_2O\cdot DMF\}$ (**3**) have different dimensionalities, attributed to the change in size of the metal ion (Zn *vs*. Cd).

1. Introduction

Coordination polymers, infinite arrays of metal ion/clusters and organic ligands, have been explored since the concept of their deliberate geometric design was demonstrated by Robson in the late 1980s [1]. The predictions made in this seminal paper have largely proven accurate with coordination polymers, and their porous sub-set metal organic frameworks (MOFs), being exploited for numerous applications including guest storage, separations and conductive properties [2].

Whilst there is much research directed at the applications of coordination polymers, there is also a significant body of work dedicated to understanding the self-assembling construction of these materials, known as crystal engineering [3]. Control of the final topology can be somewhat designed by the synthetic choices that are made such as the metal ions used (with preferential coordination geometries) and the connectivity and directionality of the organic ligand.

In recent years, our group has explored the use of diimide-based ligands for the construction of both coordination polymers and discrete supramolecular species [4,5]. Typically with rigid cores, these ligands have shown a great diversity in the structures that form yet with repeatable motifs evident in most cases. The use of amino acids to form the imide groups gives a synthetic versatility to the approach and, specifically with chiral amino acids, demonstrates a steric influence that impacts the substructures that are formed [6]. Work using naphthalenediimides has shown a propensity for the overall structures to be dominated by π-interactions, and our intention in this current work was to examine whether smaller aromatic ligands would be subject to this same driving force.

Herein we report the synthesis and application of a small, low symmetry monoimide ligand, $GlyTMI^{2-}$, derived from benzene-1,2,4-tricarboxylic acid (trimellitic acid) anhydride and contrast its structural behavior to that of the larger diimide analogues. $H_2GlyTMI$ has previously been reported in the synthesis of discrete Mn^{II}, Ni^{II} and Zn^{II} coordination complexes $[M(HGlyTMI)(OH_2)_4]$ and a series of hydrogen-bonded salts [7,8]. Other related derivatives of this ligand have also been reported in coordination complexes/polymer such as the N-benzoate [9], N-pyridyl [10], N-aminobenzene [11], and N-picoyl derivatives [12].

2. Experimental Details

2.1. Synthesis

All reagents and solvents were purchased from standard commercial suppliers and used without purification. Basic zinc acetate was synthesized according to a literature procedure. NMR spectra were collected using a Bruker Avance 400 operating at 400 MHz. Mass spectrometry was performed using a Micromass Platform Electrospray system using DMSO as the mobile phase. IR spectra were collected using an Agilent Cary 630 diamond ATR spectrometer.

2.1.1. H_2GlyTMI. Trimellitic anhydride (1.50 g, 7.81 mmol) and glycine (0.59 g, 7.81 mmol) were mixed in 10 mL of DMF and stirred at 120 °C overnight. The resulting yellow mixture was then placed in 100 mL of 5% HCl solution, with a white precipitate forming over the course of an hour. All precipitates were collected by filtration and dried under vacuum. Yield 0.83 g, 43%. m.p. 263-265 °C. Found C, 52.92; H, 2.77; N, 5.71 %, $C_{11}H_7NO_6$ requires C, 53.02; H, 2.83; N, 5.62 %. δH (400 MHz, d_6-DMSO): 4.35 (s, 2H), 8.06 (d, J=7.8 Hz, 1H), 8.28 (s, 1H), 8.40 (d, J=7.8 Hz, 1H). FT-IR, ν_{max} /cm^{-1}: 2860w, 1693s, 1405m, 1282m, 1249m, 1100m, 962m, 738m, 723m, 693m. m/z (ES^-) 248.0 ($[M-H]^-$, 50%).

2.1.2. Poly-{($N(CH_3)_2H_2$)[Zn_2(GlyTMI)$_2$(OAc)(DMF)$_{0.5}$(CH_3OH)$_{0.5}$]·CH_3OH} (1). H_2GlyTMI (10 mg, 40.1 μmol) and [$Zn_4(OAc)_6O$] (9 mg, 13.4 μmol) were added to a 2:1 mixture of DMF:methanol (3 mL) and sonicated to dissolve. The solution was heated at 70 °C for 48 hours in a heat block to yield a colourless crystalline material. Isolated yield 11.1 mg, 59%. FT-IR, ν_{max}/cm^{-1}: 3094w, 1713m, 1624s, 1374s, 1107m, 973m, 751m, 721s. Bulk purity confirmed by PXRD (Figure 1).

2.1.3. Poly-[Cd_2(GlyTMI)$_2$(DMF)$_3$] (2): H_2GlyTMI (10 mg, 40.1 μmol) and $Cd(NO_3)_2{\cdot}4H_2O$ (19 mg, 80.2 μmol) were added to a 2:1 mixture of DMF:methanol (3 mL) and sonicated to dissolve. The solution was heated at 85 °C for 2 weeks in a heat block to yield a colourless crystalline material. Isolated yield 2.9 mg, 29%. FTIR, ν_{max} /cm^{-1} 3400w, 1694s, 1566m, 1378s, 963m, 756s, 721s. Bulk purity confirmed by PXRD (Figure 1).

2.1.4. Poly-{[Cd_2(GlyTMI)$_2$(4,4'-bipy)$_2$]·$2H_2O$·DMF} (3): H_2GlyTMI (10 mg, 40.1 μmol), $Cd(NO_3)_2{\cdot}4H_2O$ (19 mg, 80.2 μmol) and 4,4'-bipyridine (6 mg, 40.1 μmol) were added to a 2:1:1 mixture of DMF:methanol:water (3 mL) and sonicated to dissolve. The solution was heated at 85 °C for 48 hours in a heat block to yield a colourless crystalline material. Isolated yield 15.7 mg, 54%. FTIR, ν_{max}/cm^{-1}: 3469w, 1703s, 1599s, 1580s, 1560s, 1377s, 1317m, 973m, 805s, 757m, 816s. Bulk purity confirmed by PXRD (Figure 1).

2.2. X-Ray Crystallography

Crystallographic data for compounds **1** and **3** was collected using the MX1 beamline at the Australian Synchrotron operating at 17.4 KeV (λ = 0.7108 Å).[13] Data collection was controlled using the BluIce software [14], and initial data processing and reduction was conducted using the XDS software suite [15]. Data for compound **2** was collected using an Oxford Gemini diffractometer using graphite-monochromated Cu-Kα radiation (λ = 1.54084 Å), with data processing conducted using CryAliPro [16]. All structures were solved by Direct methods using SHELXS-2014 and refined against F^2 using SHELXL-2018 [17]. X-Seed was used as a graphical interface [18]. All non-hydrogen atoms were refined using an aniostropic model. All hydrogen atoms are refined using a riding model. Data are deposited with the Cambridge Crystallographic Data Centre (www.ccdc.cam.ac.uk). Crystallographic details and refinement parameters are given in Table 1.

Compound **1** contains a mixed MeOH/DMF ligand position (modelled with fixed 50:50 occupancies and no OH hydrogen atom included on MeOH) and a disordered acetate ligand (modelled with refined occupancies 65:35).

Compound **3** contains a disordered, non-coordinated DMF molecule (modelled with refined occupancies 63:37) and an apparent disordered water molecule (modelled with fixed 50:50 occupancies and no hydrogen atoms included in its model).

Table 1. Crystallographic details and refinement parameters for all compounds.

Compound	**1**	**2**	**3**
Formula	$C_{29}H_{30.5}N_{3.5}O_{16}Zn_2$	$C_{31}H_{31}Cd_2N_5O_{15}$	$C_{45}H_{37}Cd_2N_7O_{15}$
FW	814.81	938.41	1140.61
Crystal system	Monoclinic	Triclinic	Triclinic
Space group	$P2_1/c$	P-1	P-1
a / Å	7.8620(16)	9.3117(14)	8.3660(17)
b / Å	17.021(3)	11.3770(15)	11.644(2)
c / Å	24.724(5)	17.587(2)	24.395(5)
α / °	90	74.549(11)	99.73(3)
β / °	95.62(3)	83.317(11)	92.73(3)
γ / °	90	82.085(11)	110.76(3)
V / Å^3	3292.7(12)	1772.5(4)	2175.2(9)
Z	4	2	2
μ / mm^{-1}	1.538	10.294	1.060
F(000)	1668	936	1144
2θ range collected	2.91 to 55.888	5.232 to 133.628	3.482 to 55.898
Index ranges	$-10 \leq h \leq 10$	$-8 \leq h \leq 11$	$-11 \leq h \leq 11$
	$-22 \leq k \leq 22$	$-13 \leq k \leq 13$	$-15 \leq k \leq 15$
	$-32 \leq l \leq 32$	$-20 \leq l \leq 20$	$-32 \leq l \leq 32$
Refs. collected	53158	15005	72971
Independent refs.	7794	6229	10360
R_{int}	0.0525	0.1012	0.0305
GooF	1.068	1.062	1.048
Final R indexes [I>=2σ (I)]	$R_1 = 0.0497$	$R_1 = 0.0545$	$R_1 = 0.0462$
	$wR_2 = 0.1376$	$wR_2 = 0.1361$	$wR_2 = 0.1067$
Final R indexes [all data]	$R_1 = 0.0582$	$R_1 = 0.0747$	$R_1 = 0.0572$
	$wR_2 = 0.1432$	$wR_2 = 0.1507$	$wR_2 = 0.1133$

3. Results and Discussion

3.1. Synthesis

The diacid H_2GlyTMI was synthesized by reaction of glycine with trimellitic anhydride in DMF (Scheme 1). Precipitation from the reaction mixture using HCl afforded the product in modest yield. Three coordination polymers were synthesised using H_2GlyTMI, namely poly-{$(N(CH_3)_2H_2)$[Zn_2(GlyTMI)$_2$(OAc)(DMF)$_{0.5}$(CH_3OH)$_{0.5}$]·CH_3OH} (**1**), poly-[Cd_2(GlyTMI)$_2$(DMF)$_3$] (**2**), and poly-{[Cd_2(GlyTMI)$_2$(4,4'-bipy)$_2$]·2H_2O·DMF} (**3**). All compounds were synthesized in DMF:MeOH or DMF:MeOH:H_2O at elevated temperatures and produced crystalline materials suitable for single crystal X-ray diffraction. The synthesis of **3**, which incorporates 4,4'-bipyridine, was deliberately conducted after observing the 2D structure of **2** and the potential to bridge the sheets with a charge-neutral co-ligand.

3.2. Structural Descriptions

3.2.1. Poly-{$(N(CH_3)_2H_2)$[Zn_2(GlyTMI)$_2$(OAc)(DMF)$_{0.5}$(CH_3OH)$_{0.5}$]·CH_3OH} (1). The structure of **1** was solved and refined in the monoclinic space group $P2_1/c$, with the asymmetric unit containing one formula unit. The compound contains a one-dimensional anionic coordination, with the charged

IOP Conf. Series: Materials Science and Engineering **833** (2020) 012062 doi:10.1088/1757-899X/833/1/012062

balanced by dimethylammonium cations (a product of the hydrolysis of DMF). The coordination polymer contains a dinuclear zinc node with the metal ions bridged by one acetate ligand and two carboxylates from the imide ends of two GlyTMI ligands (Figure 2). The benzoate ends of the two unique GlyTMI ligands coordinate in a monodentate manner to the metal centres (one to each metal). One of the metal ions adopts a distorted tetrahedral geometry, with four carboxylates coordinated to it, whilst the other has a trigonal bipyramidal geometry and has an additional coordinated solvent molecule. There is disorder associated with the solvent position, which is modelled as 50:50 DMF:MeOH with no restraints applied. Pairs of GlyTMI ligands bridge the bimetallic nodes, in a cyclic motif reminiscent of our earlier work using NDI ligands.[4,6] The two ligands are not parallel, although a closest C···C distance of 3.57 Å indicates that there is π-stacking between the two phthalimide ring systems. There are also π-π interactions between adjacent 1D chains with a shortest C···C distance of 3.44 Å. The dimethylammonium counter cation and methanol molecule are involved in a hydrogen bonding chain that bridges across a bimetallic node between a carboxylate and an imide oxygen (Figure 2). It is interesting to note that the GlyTMI ligand, whilst much shorter and with a smaller aromatic profile than the related NDI ligands, still forms a macrocyclic motif between nodes, and that π-interactions appear to remain a dominant influence on the crystal structure.

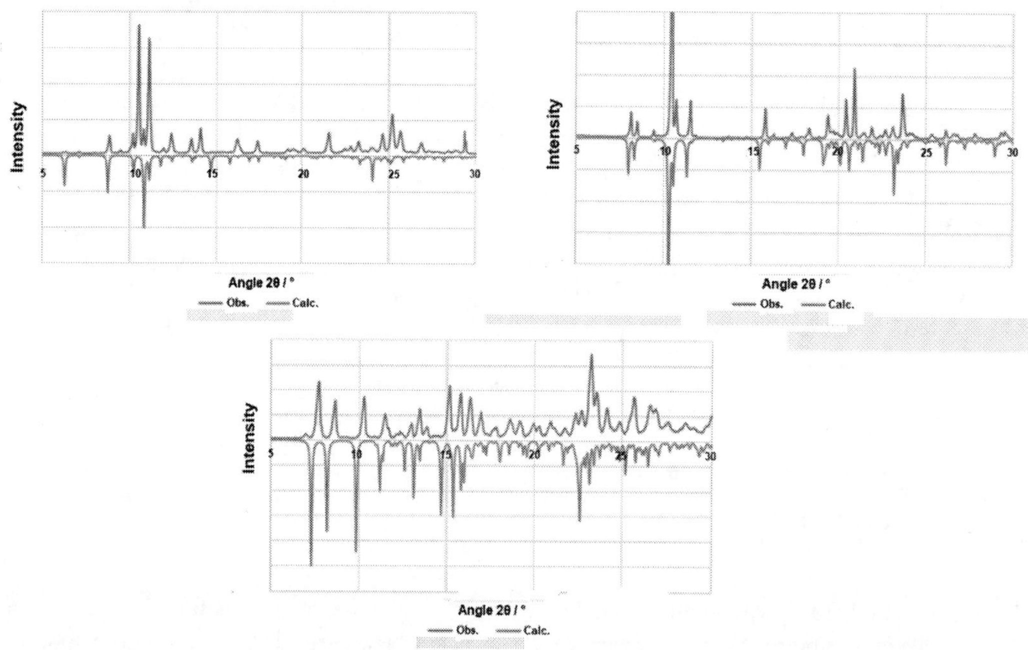

Figure 1. PXRD comparisons of experimental data at room temperature *vs.* calculated patterns from low temperature single crystal data; top left, compound **1**; top right, compound **2**; bottom, compound **3**.

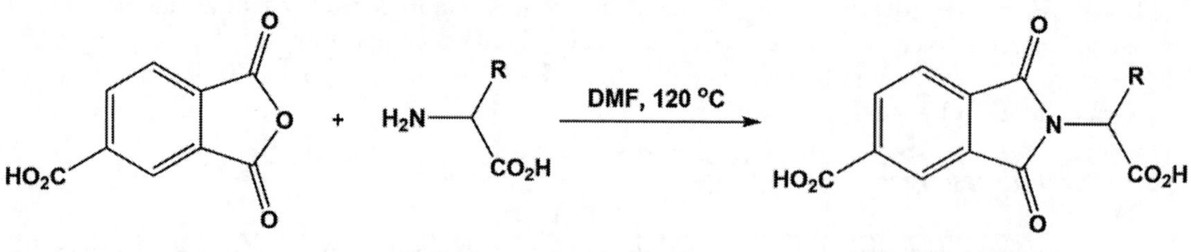

H2GlyTMI

Scheme 1. Synthesis of the dicarboxylic acid H_2GlyTMI.

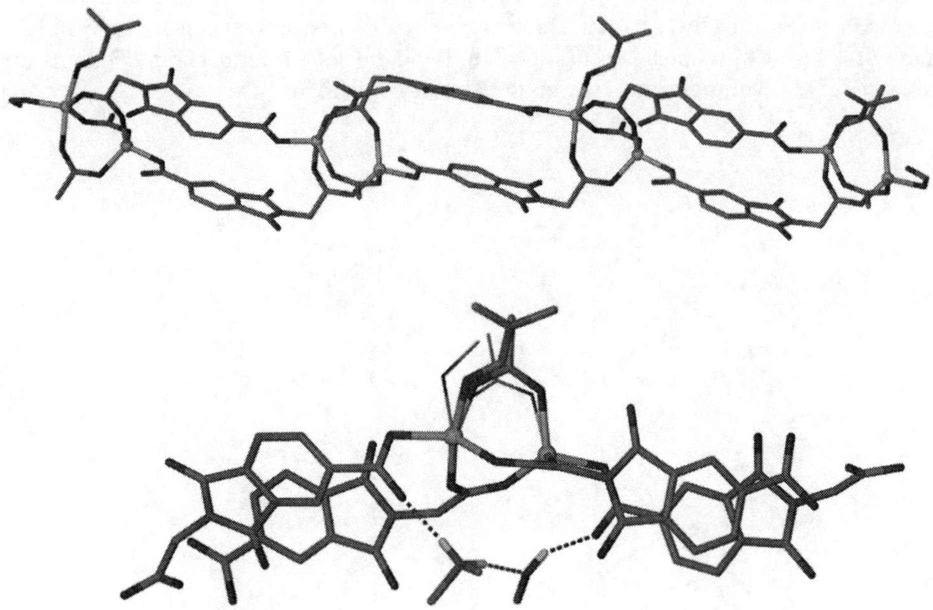

Figure 2. The anionic 1D coordination polymer in the structure of poly-{$(N(CH_3)_2H_2)[Zn_2(GlyTMI)_2(OAc)(DMF)_{0.5}(CH_3OH)_{0.5}]\cdot CH_3OH$} **1** (top) and the hydrogen bonding motif involving the dimethylammonium cation and methanol molecule (bottom) and the disorder acetate and solvent positions. Hydrogen atoms are omitted for clarity.

3.2.2. Poly-[$Cd_2(GlyTMI)_2(DMF)_3$] (2). The structure of **2** was solved and refined in the triclinic space group *P*-1, with the asymmetric unit containing one formula unit. The two crystallographically unique metals both adopt a distorted octahedral coordination geometry and form a bimetallic node (Figure 3). The metal ions in the node are bridged by one benzoate group and one glycine-derived carboxylate (in a chelating-bridging coordination mode). The nodes are bridged in one direction by a carboxylate to give a 1D metal-carboxylate chain propagating parallel to the crystallographic *a* axis. The chains are connected by the GlyTMI ligands, and overall the coordination polymer is a 2D sheet. The GlyTMI ligands are arranged such that they are parallel and aligned with the imides oriented in the same direction within a given sheet (inversion between adjacent sheets means that the orientation of ligands between sheets is antiparallel). The adjacent GlyTMI ligands have closest C···C contacts of *ca.* 3.4 Å and 3.6 Å between the crystallographically unique pairs, indicating that the formation of the sheet is likely driven by π-π interactions. The ligated DMF molecules point above and below the plane of the sheet.

3.2.3. Poly-{[$Cd_2(GlyTMI)_2(4,4'\text{-bipy})_2$]$\cdot 2H_2O\cdot DMF$} (3). The structure of **3** was solved and refined in the triclinic space group *P*-1, with the asymmetric unit containing one formula unit. The 3D coordination polymer contains two types of bimetallic nodes, both of which contain coordination from only one end of the GlyTMI ligands in addition to coordination by the 4,4'-bipyridine ligands (Figure 4). These nodes have near identical coordination environments and connect to six other bimetallic nodes (two of their 'own kind' through the bipyridyine ligands and four others through the GlyTMI ligands). The metal-bipyridine chains run in diagonally opposing directions, with π-π ineractions within these 'double-pillars' and the GlyTMI ligands bridge between these chains, arranged in pairs with π-π interactions, rather than in an infinite stacking arrangement as found in the structure of **2**. There is only one crystallographically unique pair of GlyTMI ligands; these are non-parallel and have a closest C···C contact of *ca.* 3.6 Å. There is one well-ordered water molecule that forms hydrogen bonding interactions between a carboxylate and an imide oxygen atom. There is also a disorder DMF position that occupies the space between adjacent GlyTMI pairs, potentially with some interaction to the nearly parallel ligands, and a disordered water position was located. It can be seen from packing diagrams that there are no accessible solvent voids/channels within the 3D network (Figure 4).

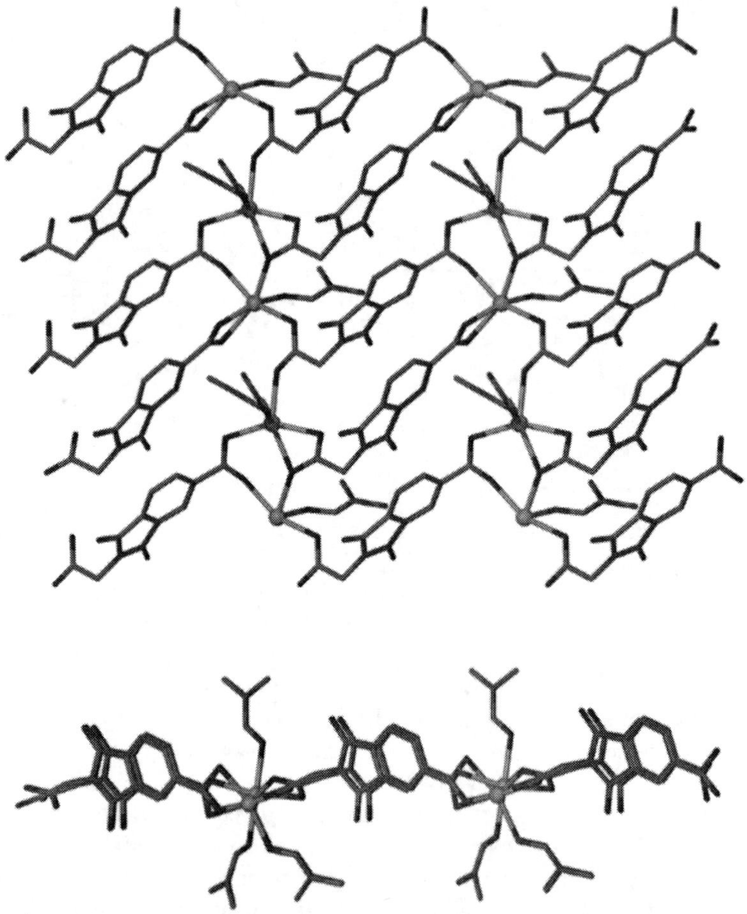

Figure 3. Top-down and side-one views of the 2D coordination polymer in the structure of poly-$[Cd_2(GlyTMI)_2(DMF)_3]$ **2**. Hydrogen atoms are omitted for clarity.

3.3. Discussion

The three coordination polymers that have been isolated containing the $GlyTMI^{2-}$ ligand differ in their dimensionality; compound **1** is 1D, compound **2** is 2D, and compound **3** is 3D. This difference can in some way be attributed to their constituent parts. Compound **1** contains zinc(II) which has a smaller ionic radius than cadmium(II) and is satisfied by 4- and 5-coordinate geometries, thereby limiting the dimensionality when carboxylate ligands are employed. Additionally, the serendipitous incorporation of acetate prevents propagation in some directions as it is not able to bridge in the same manner as GlyTMI. Compound 2 contains the larger cadmium(II) ion and is able to be 4-connecting through the GlyTMI ligands and form a 2D sheet. The coordination polymer appears to be 'templated' to some extent by the infinite arrays of face-to-face π-interactions between these ligands. It is clear that our attempts to lessen the influence of such interactions by using a smaller ligand are ineffective. Given the solvent ligands protruding from the top and bottom of the 2D sheets in compound **2**, we decided to

IOP Conf. Series: Materials Science and Engineering **833** (2020) 012062 doi:10.1088/1757-899X/833/1/012062

employ 4,4'-bipyridine as a co-ligand, to hopefully displace these DMF ligands and bridge the 2D networks to form a porous material. This strategy met with only partial success in compound **3**, with no coordinated solvent and the 4,4'-bipyridine ligands bridging between bimetallic nodes; however, the 2D sheet of **2** did not persist and a close-packed 3D coordination polymer resulted.

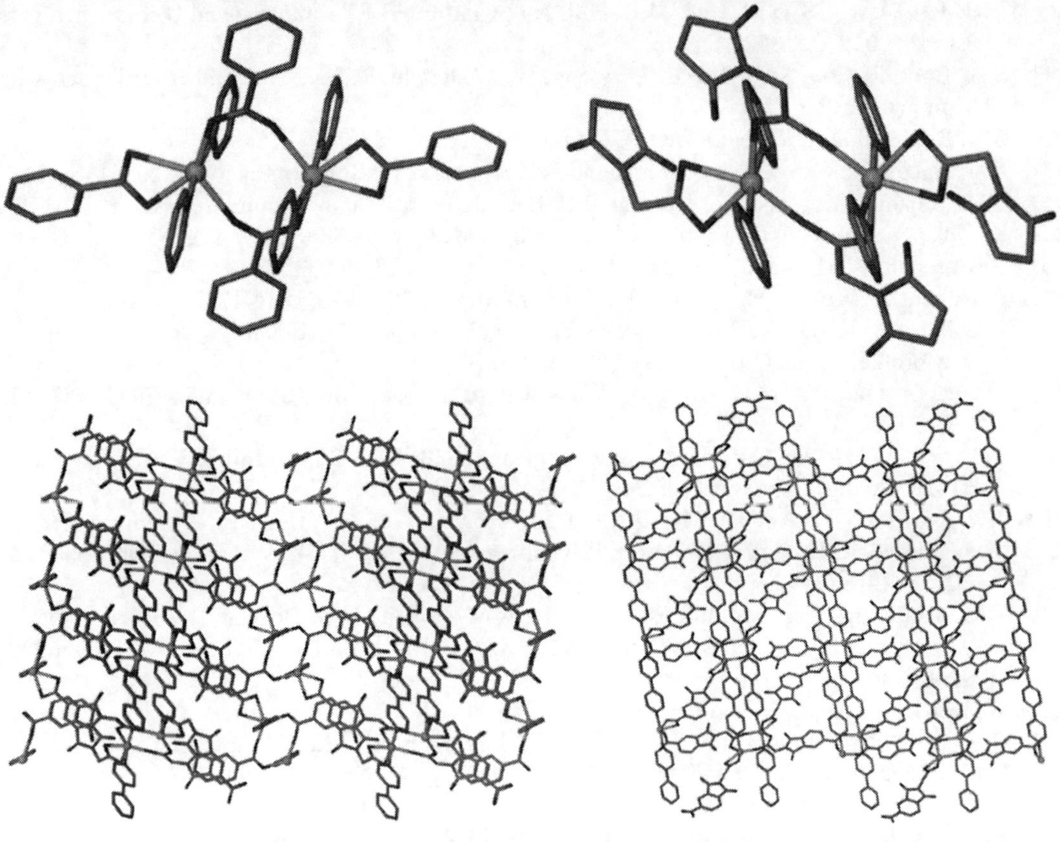

Figure 4. The two unique bimetallic nodes in the structure of poly-{[$Cd_2(GlyTMI)_2(4,4'\text{-}bipy)_2$]·$2H_2O$·DMF} **3** (top) and two views of the 3D coordination polymer along the crystallographic *b* and *a* axes (bottom left and right, respectively). Hydrogen atoms and solvent molecules are omitted for clarity.

4. Conclusion

Coordination polymers formed using the $GlyTMI^{2-}$ ligand show the promise of this class of ligands in the formation of metal-organic materials. It is evident that π-interactions are a dominant force in controlling the structures that are obtained, although the influence of metal ion and co-ligands appears to have a significant influence on the dimensionality of the networks that form.

References

[1] B.F. Hoskins, R. Robson, J. Am. Chem. Soc., 1989, 111, 5962-5964.
[2] H. Furukawa, K.E. Cordova, M. O'Keeffe, O.M. Yaghi, Science, 2013, 341, 1230444.
[3] R.-B. Lin, S. Xiang, B. Li, Y. Cui, G. Qian, W. Zhou, B. Chen, Coord. Chem. Rev., 2019, 384, 21-36.
[4] G. R. Desiraju, J. Am. Chem. Soc., 2013, 135, 9952–9967.
[5] S.A. Boer and D.R. Turner, CrystEngComm, 2017, 19, 2402-2412.
[6] S.A. Boer and D.R. Turner, Cryst. Growth Des., 2016, 16, 6294-6303.
[7] S.A. Boer, Y. Nolvachai, C. Kulsing, L.J. McCormick, P.J. Marriott and D.R. Turner, Chem. Eur.

J., 2014, 20, 11308-11312.
[8] S.A. Boer, C.S. Hawes and D.R. Turner, Chem. Commun., 2014, 50, 1125-1127.
[9] S.A. Boer and D.R. Turner, Chem. Asian J., 2019, 14, 2853-2860.
[10] S.A. Boer, K.F. White, B. Slater, A.J. Emerson, G.P. Knowles, W.A. Donald, A.W. Thornton, B.P. Ladewig, T.D.M. Bell, M.R. Hill, A.L. Chaffee, B.F. Abrahams and D.R. Turner, Chem. Eur. J., 2019, 25, 8489-8493.
[11] S.A. Boer, R. Cox, M.J. Beards, H. Wang, W.A. Donald, T.D.M. Bell and D.R. Turner, Chem. Commun., 2019, 55, 663-666.
[12] S.A. Boer and D.R. Turner, Chem. Commun., 2015, 51, 17375-17378.
[13] N. Kyratzis, W. Cao, E.I. Izgorodina and D.R. Turner. CrystEngComm, 2018, 20, 4115-4126.
[14] N. Barooah, A. Karmakar, R.J. Sarma, J.B. Baruah, Inorg. Chem. Commun., 2006, 9, 1251-1254.
[15] K. Deka, N. Barooah, R.J. Sarma, J.B. Baruah, J. Mol. Struc., 2007, 827, 44-49.
[16] N. Barooah, W.M. Singh, J.B. Baruah, J. Mol. Struc., 2008, 875, 329-338.
[17] P. Nugent, L. Wojtas, M.J. Zarorotko, J. Chem. Cryst., 2001, 41, 1834-1838.
[18] J.A. Perman, K. Dubois, F. Nouar, S. Zoccali, L. Wojtas, M. Eddaoudi, R.W. Larsen, M. J. Zaworotko, Cryst. Growth Des., 2009, 9, 5021-5023.
[19] F. Liu, L. Duan, Y. Li, E. Wang, X. Wang, C. Hu, L. Xu, Inorg. Chim. Acta, 2004, 357, 1335-1359.
[20] J.-Y. Wu, P.-T. Yuan, C.-C. Hsiao, CrystEngComm, 2014, 16, 3128-3140. J.-Y. Wu, C.-C. Hsiao, M.-H. Chiang, Cryst. Growth Des., 2014, 14, 4321-4328.
[21] N.P. Cowieson, D. Aragao, M. Clift, D.J. Ericsson, C. Gee, S.J. Harrop, N. Mudie, S. Panjikar, J.R. Price, A. Riboldi-Tunnicliffe, R. Williamson, T. Caradoc-Davies, J. Synchrotron Radiat., 2015, 22, 187–190.
[22] T.M. McPhillips, S.E. McPhillips, H.J. Chiu, A.M. Cohen, A.M. Deacon, P.J. Ellis, E. Garman, A. Gonzalex, K. Sauter, R.P. Phizackerley, S.M. Soltis, P. Kuhn, J. Synchrotron Radiat., 2002, 9, 401–406.
[23] W. Kabsch, J. Appl. Crystallogr., 1993, 26, 795–800.
[24] CrysAlisPRO, Oxford Diffraction /Agilent Technologies UK Ltd, Yarnton, England.
[25] G.M. Sheldrick, Acta Crystallogr., Sect. A: Found. Crystallogr., 2015, 71, 3–8.
[26] G.M. Sheldrick, Acta Crystallogr., Sect. C: Struct. Chem., 2015, 71, 3–8.
[27] L.J. Barbour, J. Supramol. Chem., 2001, 1, 189-191.

Acknowledgement
Part of this work was conducted using the MX1 beamline at the Australian Synchrotron, part of ANSTO. The authors acknowledge use of facilities within the Monash X-ray Platform.

Preparation of *Cyperus diffusus* Vahl cellulose–polyanniline composite for dye removal

Ricky Jenihansen Burhan[1], Aswin Falahudin[1], Salprima Yudha S.[1,*], Irfan Gustian[1]

[1]Department of Chemistry, Faculty of Mathematics and Natural Sciences, Universitas Bengkulu, Jalan W.R, Supratman, Kandang Limun, Kota Bengkulu 38371A, Indonesia

[*]Corresponding author's e-mail: salprima@unib.ac.id; sp.yudha.s@gmail.com

Abstract. *C. diffusus* Vahl cellulose was used as an alternative in the preparation of cellulose-polyaniline composites. This involved the use of sodium hydroxide, hydrogen peroxide, and sulfuric acid as reagents for the isolation of *C. diffusus* Vahl cellulose. The isolated material was analysed using Fourier Transforms Infrared (FTIR) and the results showed the it has similar peaks pattern with the original cellulose as observed with 3333 cm^{-1} for OH stretching, 2889 cm^{-1} for C-H vibration, and several other indicators. The powder produced was used as a reaction partner for *in-situ* synthesis of *C. diffusus* Vahl cellulose-Polyaniline (*C. diffusus* Vahl cellulose-PANI) in dimethyl sulfoxide (DMSO) using ammonium persulfate $(NH_4)_2S_2O_8)$ as the oxidator. The solid material product was effective in metanil yellow adsorption from the corresponding solution up to 99%.

1. Introduction

The use of polymer is familiar to mankind from the past up to the present moment. In recent times, several developments have been made and one of the useful polymers is polyaniline. Despite its strength, this polymer has a weakness caused by its rigid-rod macromolecular chains [1]. Therefore, several studies have been conducted to increase its processability by forming composites with other polymers such as the popular cellulose [2]. For instance, cellulose fiber was impregnated with polyaniline under *in-situ* reaction by a common procedure using ammonium persulfate as polymerization reagent. The composites produced were able to adsorb meloxicam effectively compare to independent cellulose fiber and polyaniline [3]. Moreover, water hyacinth was also used as a natural source of cellulose and later modified with polyaniline using a blending method. The composite formed was further applied as a sensor for humidity measurement in the room. These findings provide further information on the usefulness of cellulose-PANI as a polymer composite [4]. Furthermore, even though it is possible to use bacterial cellulose purely because it was produced by specific bacteria, it was also modified using PANI under some varied conditions and the results showed the characteristic of the blended materials was affected by the attributes of the matrix and the conditions of the reaction [5].

Surfactant such as cetyltetraammonium bromide (CTAB) was also discovered to have effects on the morphology and other characteristics of cellulose-PANI composites. The condition of the reaction provides a soft template to introduce reactants with lamellar structure into the connection between surfactant and anion of APS [6]. Moreover, some composites were synthesized via *in-situ* chemical oxidation polymerization through the use of certain aniline precursor derivatives such as polyaniline (PANI), poly (N-methyl aniline) (PNMANI), and poly (N-ethyl aniline) (PNEANI)] combined with cellulose. The capacity of the composites as adsorbent of AR4 and DR23 dyes was tested and the best result was found with PANI/cellulose [7]. In addition, the synthetic method of *in-situ* polymerization of aniline to coat the bacterial cellulose (BC) using interfacial polymerization was also reported, for example, on isolated bacterial cellulose in water and organic phase [8]. Other results showed BC/PANI nanocomposite membranes with both conductive and electrode resistant sides were easily obtained. The concentration of the oxidant also provided a significant effect on their microstructures [9]. Further developments showed *Imperata cylindrica* L. cellulose was isolated and used in a cellulose-PANI composite under some variations of HCl concentration. In this case, $K_2Cr_2O_4$ was used as an oxidator instead of the common ammonium persulfate and the desired product was obtained with good properties while the electrical conductivity was observed to be depending on the HCl concentration. Finally, the best condition was found at 3 M HCl as indicated by its production of the highest electrical conductivity [10].

These reported efforts toward the production of cellulose-polyaniline composites with high adsorbent capacity showed there is a need for further research, especially on the sustainability of the materials. This research was, therefore, conducted to provide new information on the utilization of local material, *C. diffusus* Vahl, as a natural cellulose resource and the composite produced was applied as an adsorbent on metanil yellow dyes in synthetic waste.

2. Material and Methods

2.1. Materials
Aniline as a precursor for synthesis of polyaniline was purchased from Merck (98% purity) and used as purchased. HCl, NaOH, H_2SO_4, DMSO, H_2O_2, $(NH_4)_2S_2O_8$ (ammonium persulfate, APS), acetone, and metanil yellow were used as received. Demineralized water was used for preparing some solutions and for cleansing the solid materials prior to further steps. *C. diffuses* Vahl was collected from coastal area in Bengkulu city.

2.2 Procedure
The cellulose was extracted from *C. diffusus* Vahl by modifying a previously applied procedure [10-12]. The pieces of dried sample were ground to powder and placed in a round bottom flask after which 1 M NaOH was added at a ratio of 10 g/120 mL. The mixture was heated at 80 °C for 1.5 h and the solid material produced was cleansed using demineralized water. The cleansed solid materials in 50 mL H_2O_2 5% heated at 80°C for 20 minutes, cooled to room temperature followed by filtration. Furthermore, the solid material residue was added to 60 mL H_2SO_4 (2 N), stirred at room temperature for 30 minutes, then the solid material obtained was rinsed using demineralized water and later dried using an oven at 50 °C for 24 h for further analysis.

The *in-situ* synthesis of *C. diffusus* Vahl cellulose-polyaniline also followed the modifications of a standard procedure [3]. Typically, 0.5 g as-prepared cellulose was added to 25 mL dimethyl sulfoxide (DMSO) and stored at room temperature for 1 h followed by the addition of 0.5 g aniline and stirring for another 1 h. Moreover, 80 mL of 1 M hydrochloric acid was added to the mixture after which 4.86 g of APS was slowly added for 30 minutes [13]. The mixture was further stirred for 1 h and stored at room temperature for 24 h. It was later rinsed with 100 mL of 0.2 M HCl and 100 mL acetone. The solid material was dried at 60°C for 24 h and stored in a desiccator for further analysis and application. In order to identify the functional group of cellulose and *C. diffusus* Vahl cellulose-PANI, a Fourier transform infra-red (FTIR, Bruker) analysis was performed. The surface areas of the materials were measured using Brunauer, Emmett and Teller (BET) method (Quantachrome® ASiQwin™ analyzer).

The application of the material obtained for adsorption started with the preparation of the calibration curve by measuring the absorbance of standard solutions of 2, 4, 6, 8, and 10 mg/L metanil yellow using UV-Vis spectrophotometer (Cary 60 UV-Vis - Agilent). The adsorption ability of the polymers was evaluated by adding 10 mL of 10 ppm metanil yellow to 10 mg of the cellulose-PANI composite. After 90 minutes, the aliquot's absorbance was measured using a UV-Vis spectrophotometer. Furthermore, a control experiment was conducted using cellulose as an adsorbent following a similar procedure. All the experiments were repeated three times.

3. Results and Discussion

Figure 1 shows the natural cellulose was isolated from the *C. diffusus* Vahl which are grown in the coastal area and with limited reports showing their utilization for polymer development. The extraction processes involved the use of some chemicals such as sodium hydroxide, hydrogen peroxide, and sulfuric acid to remove lignin, bleaching, and hemicellulose hydrolysis, respectively as shown in Figures 1a-c [14]. The processes were conducted easily in every step and the material obtained was in yellow-white powder form. Moreover, *in-situ* polymerization of aniline was conducted in the presence of *C. diffusus* Vah cellulose and a desired black sticky composite material was produced as shown in Figure 1d.

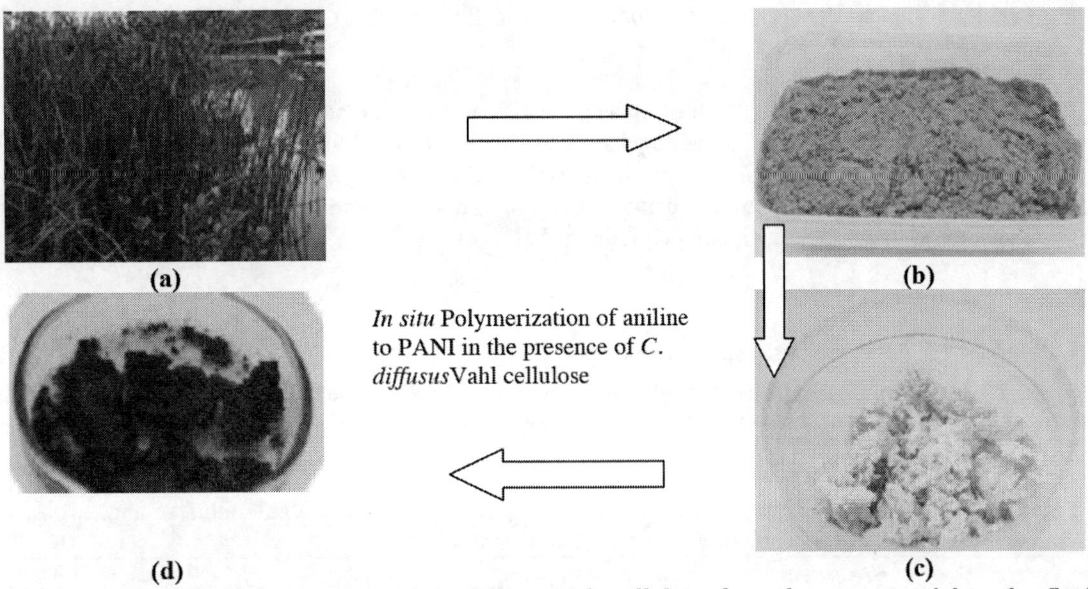

Figure 1. Flow chart of extraction of *C. diffusus* Vah cellulose from the raw material to the final composite product: (a) *C. diffusus* Vahl growth location, (b) *C. diffusus* Vahl powder, (c) final product of *C. diffusus* Vahl cellulose, and (d) *C. diffusus* Vahl cellulose-PANI composite.

Further investigation was conducted to identify the functional group changes of the cellulose after it was treated with aniline under *in-situ* polymerization and the results are shown in Figure 2. Some specific peaks are shown in Figure 2(a) with 3333 cm^{-1} found corresponding to O-H group, 2889 cm^{-1} to C-H stretching vibration while the sharp peak at 1026 cm^{-1} was corresponding to –C-O- bonding in the cellulose structure. These peaks are almost similar to the patterns of the reported cellulose [15]. However, Figure 2(b) shows a peak at 3226 cm^{-1} corresponding to N-H of polyaniline as well as 1666 cm^{-1} and 1436 cm^{-1} assigned as an aromatic chain for the double bond (C=C). In addition, the peak at 796 cm^{-1} was the area showing the C-H bonding of the cellulose and aniline aromatic rings.

Further experiment involved the application of extracted *C. diffusus* Vahl cellulose and *C. diffusus* Vahl cellulose-PANI composite as an adsorbent for metanil yellow solution. This is important because metanil yellow is widely used in many industries such as textiles and cosmetics. Unfortunately, they produce wastes that have an environmental impact due to their carcinogenic nature and inability to degrade naturally [16].

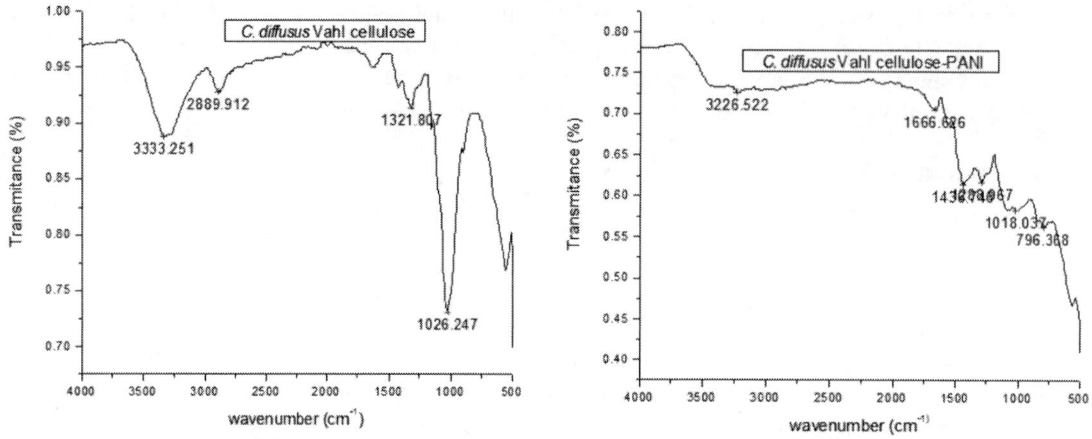

(a) **(b)**

Figure 2. FTIR spectra of (a) *C. diffusus*Vahl cellulose, (b) *C. diffusus*Vahl cellulose-PANI Composite.

This metanil yellow standard solution has a specific absorbance at 437 nm and a linear regression was obtained from the aforementioned peaks. This was further used to easily determine the amount of metanil yellow absorbed by *C. diffusus* Vahl cellulose and *C. diffusus* Vahl cellulose-PANI. The adsorption efficiency was calculated using equation (1) and the amount of metanil yellow adsorbed (mg/g adsorbent) was calculated using equation (2).

$$\textit{Adsorption efficiency} = \left(\frac{C_o - C_t}{C_o}\right) \times 100\% \quad (1)$$

$$\text{Adsorption capacity} = \frac{(C_o - C_t) \times V}{W} \quad (2)$$

Where C_0 is the initial concentration of metanil yellow and C_t is the final concentration after it has been treated with the adsorbent, V (L) is the volume of the solution, W (g) is the mass of absorbent used. The results are shown in Table 1.

Table 1. Adsorption efficiency and adsorption capacity of *C. diffusus* Vahl cellulose and *C. diffusus* Vahl cellulose-PANI composite toward metanil yellow.

Repetition	Adsorption efficiency of *C. diffusus*Vahl cellulose (%)	Adsorption capacity of *C. diffusus*Vahl cellulose (mg/g)	Adsorption efficiency of *C. diffusus*Vahl cellulose-PANI (%)	Adsorption capacity of *C. diffusus*Vahl cellulose-PANI (mg/g)
1	0.548	0.055	97.648	9.765
2	3.078	0.308	99.103	9.910
3	2.720	0.272	99.461	9.946
Average	2.115 ± 1.369	0.212 ± 0.137	98.737 ± 0.960	9.874 ± 0.096

Table 1 clearly shows the presence of PANI increased the adsorption efficiency of cellulose from 2.115% to an extremely high adsorption capability > 99%. The adsorption capacity of the *C. diffusus* Vahl cellulose-PANI composite (9.874 mg/g) was also higher than the corresponding *C. diffusus*Vahl cellulose in the absence of PANI (0.212 mg/g). This was further confirmed by visually differentiating between the results of both materials as shown in Figure 3. Although BET analysis show that the *C. diffusus* Vahl cellulose has higher surface area (40.108 m^2/g) than the *C. diffusus* Vahl cellulose-PANI composite (25.918 m^2/g), the *C. diffusus* Vahl cellulose-PANI composite has higher adsorption efficiency due to the higher porous matrix than the *C. diffusus* Vahl cellulose. The other possibility is the presence of specific chemical interaction between metanil yellow with *C. diffusus* Vahl cellulose-PANI could increase the adsorption efficiency. This research showed it is possible to use *C. diffusus* Vahl cellulose-PANI composite as an alternative adsorbent for metanil yellow and other dyes.

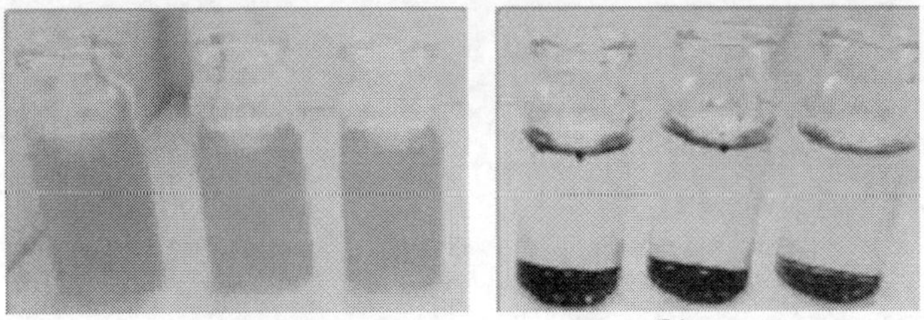

(a) **(b)**

Figure 3. Visual differentiation of metanil yellow treated with (a) *C. diffusus*Vahl cellulose, (b) *C. diffusus*Vahl cellulose-PANI.

4. Conclusion

C. diffusus Vahl from the Bengkulu coastal area has the potential to be used as an alternative cellulose source for manufacturing natural-based composites. It was found that *C. diffuses* Vahl cellulose-PANI to have an average of >99 % efficiency which is higher than the cellulose precursor

Acknowledgments

The authors would like to thank the Direktorat Riset dan Pengabdian Masyarakat (DRPM), Direktorat Jenderal Penguatan Riset dan Pengembangan, Kementerian Riset, Teknologi, dan Pendidikan Tinggi, Republik Indonesia, who give us funding under the scheme of Penelitian Tim Magister Fiscal Year 2019.

References

[1] Samuelson L A, Anagnostopoulos A, Alva K S, Kumar J and Tripathy S K 1998 *Macromolecules* 31 4376–4378.
[2] Huang J G, Ichinose I and Kunitake T 2005 *ChemCommun* 13 1717–1719.
[3] Dutra F V A, Pires B C,Nascimento T A, Mano V and Borges K B **2017** *RSC Adv.*, 7, 12639–12649.
[4] Putri N P, Kusumawati D H, Widiyanti N and Munasir 2018 Synthesis of polyaniline/cellulose composite as humidity sensor *Journal of Physics: Conference Series* **997** 012009
[5] Alonso E, Faria M, Mohammadkazemi F, Resnik M, Ferreira A and Cordeiro N 2018 *Carbohydrate Polym.* **183** 254 – 262.
[6] Abdi M M, Tahir P M, Liyana R and Javahershenas R 2018 *Molecules* **23** 2470
[7] Abbasian M, Niroomand P and Jaymand M 2017 *Journal of Applied Polymer Science* **134** 45352
[8] Lee H. –J, Chung T. -J, Kwon H. -J, Kim H. -J, and Tze W T Y 2012 *Cellulose* **19** 1251 – 1258.
[9] Lin Z, Guan Z and Huang Z 2013 *Ind. Eng. Chem. Res.* **52** 2869 – 2874.
[10] Yuningsih L M, Mulyadi D and Aripandi I 2017 *American J. Materials Sci.* **7(3)** 59 – 63.
[11] Hamisan A F, Abd-Aziz S, Kamaruddin K, Shah U K, Shahab N and Hassan M A 2009 *J. Agricultural Res.* **4** 250 – 256.
[12] Ahmad Z, Roziaizan N N, Rahman R, Mohamad A F and Ismail W I N W 2016 *MATEC Web of Conferences* **47** 05013
[13] Liu X, Zhou W, Qian X, Shen J and An X 2013 *Carbohydrate Polym.* **92** 659 – 661.
[14] He W, Tian J, Li J, Jin H and Li Y 2016 *BioResources* ***11*(4)** 8535 – 8547.
[15] Ciolacu D, Ciolacu F and Popa V I 2011 *Cellulose Chem. Technol.* **45** 13 – 21.
[16] Gupta S, Sundarrajan M and Rao K V K 2003 *Teratog. Carcinog. Mutagen.* **23,** S1, 301 – 312.

IOP Conf. Series: Materials Science and Engineering **833** (2020) 012064 doi:10.1088/1757-899X/833/1/012064

Synthesis of Copolymer of Chitosan with Acrylamide as an Adsorbent for Heavy Metal Waste Treatment

D Desnelli*, E Eliza, A Mara, and A Rachmat

Chemistry Department, Faculty of Mathematics and Natural Sciences, Sriwijaya University, Indralaya, 30662, Indonesia

*Corresponding author: desnelli@unsri.ac.id

Abstract. The aim of this research is to chemically modify chitosan to improve and enhance the ability as an adsorbent. Chitosan is a natural polymer produced from chitin deacetylation. The best performance of chitosan ~~is~~ as an adsorbent is in acidic solutions. Chemical modification was carried out by chitosan copolymerization with acrylamide. Copolymer synthesis was done by using microwave oven with the weight ratio of chitosan and acrylamide of 1: 4. The characterization of chitosan copolymer with acrylamide (chitosan-g-acrylamide copolymer) was conducted by FTIR spectroscopy and Scanning Electron Microscope (SEM). The analysis of the functional group by FTIR spectroscopy shows that the copolymer has been successfully synthesized. The surface analysis with SEM shows that the morphology of the copolymer surface is more homogeneous compared to that of chitosan.

1. Introduction

Chitosan is a biopolymer produced by the chitin deacetylation process. Chitin is a compound widely found in shrimp shells [1]. Chitosan is polycationic, so that it can be applied in various fields such as metal adsorbents, dyes, and cosmetics. Chitosan is biodegradable, biocompatible and non-toxic compound. Chitosan has biological and chemical properties. Chitosan has antimicrobial activity so it is widely used in biomedicine. Chitosan can interact very well with inorganic solutes (metal ions) and dyes, due to the presence of reactive groups such as -OH and $-NH_2$ [2].

However, chitosan has disadvantage of only dissolved in acidic solutions, insoluble in water and unstable in acidic solutions, so that their use is limited mainly as an adsorbent [3]. Therefore, researchers have done a chemical modification of chitosan [4,5]. Chemical modification of chitosan has been done to improve and enhance the ability of chitosan as an adsorbent, such as adding cross-linked material such as glyoxal, formaldehyde, and ethylene glycol [6]. Cross-linked material can stabilize chitosan in acidic solutions and improve its mechanical properties. Another method that has also been carried out by researchers to improve the ability of chitosan as an adsorbent is by copolymerization, namely grafting between chitosan with some vinyl monomers such as acrylic acid, methyl acrylate, acrylonitrile, vinyl chloride, acrylamide and others [7].

Copolymerization of the graft into chitosan is one of the most effective method to improve the performance of chitosan. Graft copolymers are used to improve the physical and chemical properties of natural and synthetic polymers for agricultural, biomedicine and other fields. There have been many studies on the graft copolymerization including copolymerization of chitosan with acrylonitrile, acrylamide, acrylic acid using cerium ammonium nitrate as an initiator, methyl acrylate using the initiator of potassium periodatcuprate(III) [3]. Copolymerization of chitosan with acrylamide has also been carried out by previous researchers, among others using the initiator of ammonium persulfate [9], and copolymerization of chitosan with acrylamide using gamma rays for the adsorption of Cu and Ni metals [8].

In this research, copolymerization of chitosan graft has been carried out into acrylamide, where copolymerization is conducted using the ammonium persulfate initiator. The copolymerization process is done by using microwave radiation heating using a microwave oven which is commonly used at home. Polymer synthesis using microwave has the advantage of short reaction time compared to the conventional heating. The results of chitosan graft copolymerization with acrylamide will later be applied for treatment of heavy metal waste, such as Pb metal, which is not included in this paper.

2. Materials and Methods

2.1. Materials
The materials used in this research are chitosan with 87.5% deacetylation degree obtained from the local market, acrylamide (Merck, Germany), ammonium persulfate (AR, China), 2% acetic acid, technical methanol obtained from local market in Indonesia and distilled water. All chemicals are used directly without further purification.

2.2. Synthesis of copolymers
The study was conducted with a weight ratio between chitosan and acrylamide of 1: 4. A 0.5-gram chitosan was placed into 2% acetic acid solution and stirred with magnetic stirrer until dissolved for 20 minutes in a Teflon container. Then, 2 grams of acrylamide were dissolved in 10 ml of distilled water, then poured into a Teflon container containing chitosan and stirred until the mixture was homogeneous. After that, 5% ammonium persulfate was added from the total weight of the sample and the mixture was stirred until homogeneous. Then, the Teflon container was placed into microwave oven by adjusting the microwave power that is 30%, 50%, and 70% by adjusting the reaction time every 2 minutes until the gel was formed. The time of gel formation was noted. After the reaction was complete, the formed gel was poured into a container containing excessive methanol to precipitate the product while stirring. The precipitate that has been obtained was washed with methanol for several times to remove the formed homopolymer. The precipitate was filtered, placed into a petri dish and then dried in a vacuum oven at 40 °C to a constant weight. Copolymers were then characterized using FTIR and SEM, while the percentage of grafting can be determined by equations (1) and (2), where W_0, W_1 and W_2 are initial mass of chitosan, mass of copolymer, and mass of acrylamide, respectively [11]:

$$\%G = (W_1 - W_0) / W_0 \times 100\% \quad (1)$$
$$\%E = (W_1 - W_0) / W_2 \times 100\% \quad (2)$$

2.3. Characterization
2.3.1. Fourier transformation infrared (FTIR) spectroscopy. FTIR analysis was performed on samples of chitosan, acrylamide and chitosan copolymers with acrylamide in the form of KBr pellets and measured at wavelengths of 400 - 4000 cm using FTIR SHIMADZU-8400.

2.3.2. Scanning electron microscopy (SEM). Scanning Electron Microscope (SEM) analysis was also performed on samples namely chitosan, acrylamide and chitosan copolymers with acrylamide using SEM JEOL JSM 6510 LA. Samples were measured in a powder form.

3. Result and Discussion

3.1. Synthesis of copolymer
Synthesis of chitosan copolymers with acrylamide was carried out using a microwave oven commonly used at home. Chitosan used in this work was originated from shrimp shells with deacetylation degree

of 87.5% that obtained from the local market. Copolymers were synthesized with a ratio of chitosan and acrylamide of 1:4. The initiator used for the synthesis of chitosan and acrylamide copolymers was ammonium persulfate. Copolymer synthesis was carried out with variations of power namely 30%, 50%, and 70%. The reaction is stopped until the solution is thickened or form a gel. This indicates that the copolymer has been formed. The initially white solution turns into a thick yellow solution. To stop the reaction, the gel obtained is poured into excessive methanol to form a precipitate. The precipitate formed were dried in a vacuum oven at a temperature of 40 °C until constant weight is obtained.

The time needed to form copolymers on various microwave power was varied. The average time required for the formation of the copolymer in microwave with 30%, 50% and 70% power, respectively were 24, 12, and 8 minutes. The greater the microwave power used, the faster the formation of copolymers. Due to greater power, the radicals were formed faster, so that of the copolymer. With the greater power, the reaction rate will be greater so that the time needed for the formation of copolymers is faster. Energy transfer is faster with the increasing of microwave power. In addition, the higher power microwave radiation indicates a higher temperature. Observations of the sample during the reaction show that with the increasing of microwave power, the time taken for the sample to boil is also faster. The time needed to boil each of the microwave with power of 30%, 50% and 70%, are 8, 4, and 2 minutes, respectively.

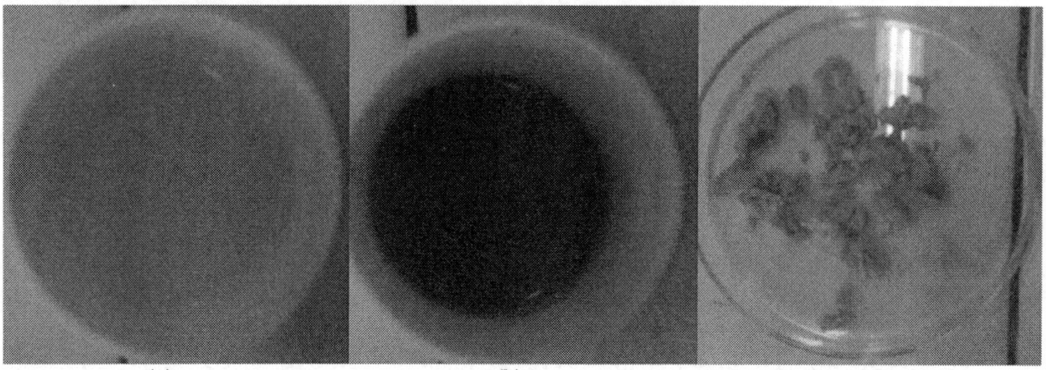

(a) (b) (c)

Figure 1. Sample (a) before the copolymerization reaction, (b) after the copolymerization reaction, and (c) after precipitated with methanol.

Synthesis of chitosan copolymer with acrylamide is a graft copolymerization in which acrylamide is grafted into the chitosan. The number of acrylamide monomers that can be grafted into chitosan can be determined by calculating the results of grafting (%G) and grafting efficiency (%E) based on equations (1) and (2). The results of grafting (%G) and grafting efficiency (%E) synthesis of chitosan and acrylamide copolymers at various microwave powers are presented in Table 1. Table 1 showed that %G and %E were decreased with the increasing of microwave power used. Due to the greater radiation power, the greater homopolymer reaction occurred thus %G and %E were decrease. A decrease in %G with the increasing of microwave power in the synthesis of chitosan-grafting-polyacrylamide was also reported by previous researchers, namely chitosan-polyacrylamide copolymers, which is carried out in various power of 60, 70, 80, 90 and 100%. The results showed a decrease in %G with the increasing of microwave power of 80% and 100%, due to more homopolymerization reactions [10].

Table 1. Effect of variation of microwave power on %G and %E.

Power of Microwave	%G	%E
30%	399.18	99.79
50%	367.06	91.76
70%	309.00	72.25

Chitosan copolymerization with acrylamide is free radical polymerization using the ammonium persulfate initiator. The mechanism of chitosan copolymerization with acrylamide which has been proposed by Machalova *et al* (2006) is presented in Figure 2. Based on the polymerization of free radicals in the initial stages of the reaction, the initiator undergoes initiation to form radicals. Radicals are formed due to heat generated in a microwave oven. Furthermore, a propagation reaction occurs where radicals from the initiator attack the chitosan functional groups and free the H atoms contained in the NH_2 and OH groups contained in chitosan. Thus, chitosan will form radicals. The chitosan radical then attacks the acrylamide monomer to forms another radical. In the final stage, the termination stage, there is a bond formed between the radicals from the polymer that has been grafted.

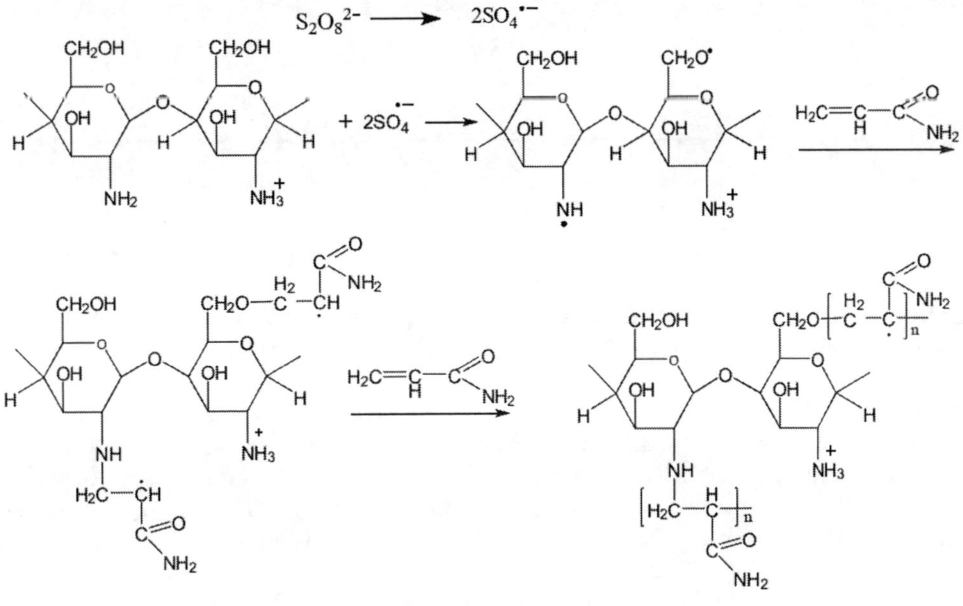

Figure 2. Mechanism of copolymerization of chitosan with acrylamide [14].

3.2. Characterization of copolymers

3.2.1. FTIR Spectroscopy Analysis. FTIR spectroscopic analysis was carried out on chitosan, acrylamide and chitosan-acrylamide copolymers and the spectra are shown in Figure 3. FTIR spectra for copolymers were taken from the results of copolymer synthesis at 70% microwave power. Figure 3 (a) shows the FTIR spectra of the chitosan. The absorption peak with high intensity at 3427.51 cm^{-1} is the stretching vibration from OH and NH_2 groups. The absorption at 2920.23 cm^{-1} and 2881.65 cm^{-1} shows the stretching vibration of C-H. The absorption peak at 1660.71 cm^{-1} is the C=O vibration of carboxylate group of chitosan, while the absorption peak at 1596.99 cm^{-1} is the absorption of N-H. Vibration stretching of C-N appears at the absorption peak of 1427.32 cm^{-1}.

Figure 3 (b) is an FTIR spectra of acrylamide. The emergence of absorption peaks at 3352.26 cm^{-1} and 3178.69 cm^{-1} is the peak of NH_2 vibration stretching from acrylamide. The absorption peak at 1674.21 cm^{-1} is the C = O absorption band of amide (I). The C-N absorption band of acrylamide appeared at 1429.25 cm^{-1}. While the absorption peak of N-H bending vibration of acrylamide appeared on 1610.56 cm^{-1}. The presence of the C = C double bond of acrylamide appeared on the absorption peak around 900-990 cm^{-1} and 3057 cm^{-1}.

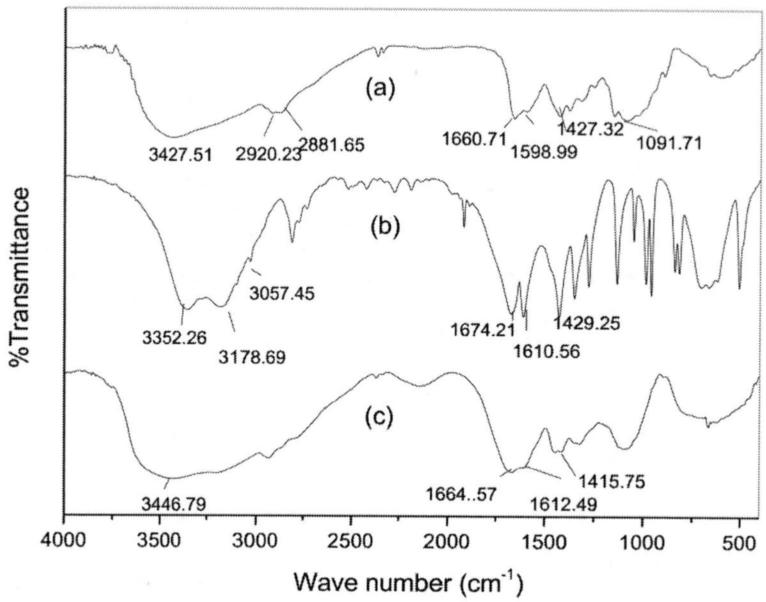

Figure 3. Spectra FTIR of (a) chitosan, (b) acrylamide, (c) copolymer of chitosan-acrylamide.

FTIR spectra of chitosan-acrylamide copolymers can be seen in Figure 3 (c). From the spectrum of the copolymer in Figure 3 (c), it is observed that the double bond C=C does not appear at wavelengths around 900-990 cm^{-1} and 3075-3095 cm^{-1}. This shows that the copolymerization reaction has taken place. Moreover, the spectra show the formation of copolymers is the shift in the absorption of C=O amide from 1674.21 cm^{-1} to 1664.57 cm^{-1}. There was a broad peak at wave number of 3446.79 cm^{-1} due to OH absorption. The broad absorption peak is also caused by an overlap between the N-H vibrations of chitosan and NH_2 from acrylamide. In addition, an indication of chitosan-acrylamide copolymers has been formed is the peak of C-N absorption at wave number of 1415.75 cm`1. Based on previous research showing that the peak of the adsorption of chitosan-acrylamide copolymer appeared in 3434.71 cm^{-1} which is OH stretching from polyacrylamide and chitosan. While the adsorption band at 1411.69 cm^{-1} is C-N stretching vibration from the chitosan-acrylamide copolymer [12]. Based on the FTIR spectra, the chitosan and acrylamide copolymers have been successfully synthesized using microwave oven.

The FTIR spectra of the copolymer that has been synthesized with microwave power variations of 30, 50, and 70% can be seen in Figure 4. From the spectra, it can be seen that the absorption peaks of the copolymers produced in the microwave power variation have identical absorption peaks. FTIR spectra in Figure 4 shows that the copolymer has been successfully synthesized in various microwave power.

3.2.2. SEM Analysis. Scanning Electron Microscope (SEM) analysis is a surface analysis that can shows the surface morphology of a material. The results of SEM analysis can be seen in Figure 5. SEM analysis was carried out for chitosan, acrylamide, and chitosan-acrylamide copolymers. The SEM analysis results (Figure 5) show the surface differences of chitosan, acrylamide and chitosan-acrylamide copolymers. Based on Figure 5 (c), chitosan-acrylamide copolymers with porous surfaces were observed. The formation of these pores is due to the grafting between chitosan and acrylamide.

This porous surface indicates that chitosan-acrylamide copolymers can be used as adsorbents. Based on Figure 5, it can be seen that the surface of the copolymer is more homogeneous compared to that of chitosan.

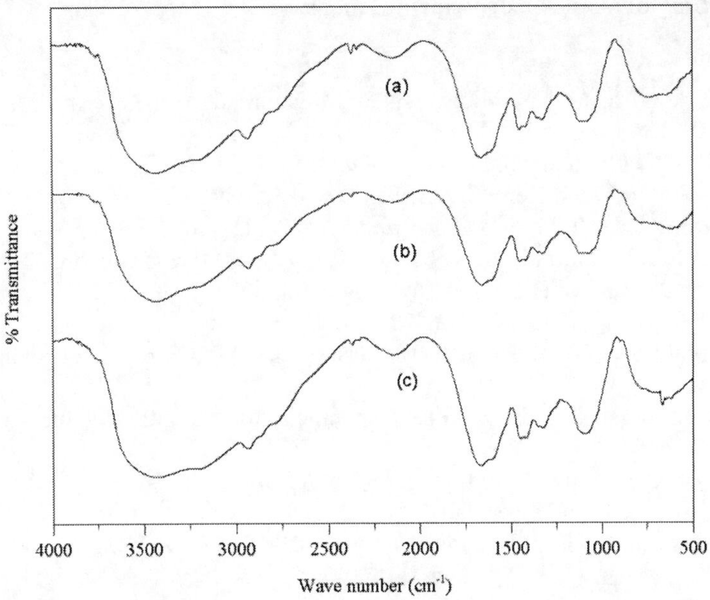

Figure 4. FTIR Spectra of chitosan-acrylamide copolymers synthesized by various microwave power of (a) 30%, (b) 50% and (c) 70%.

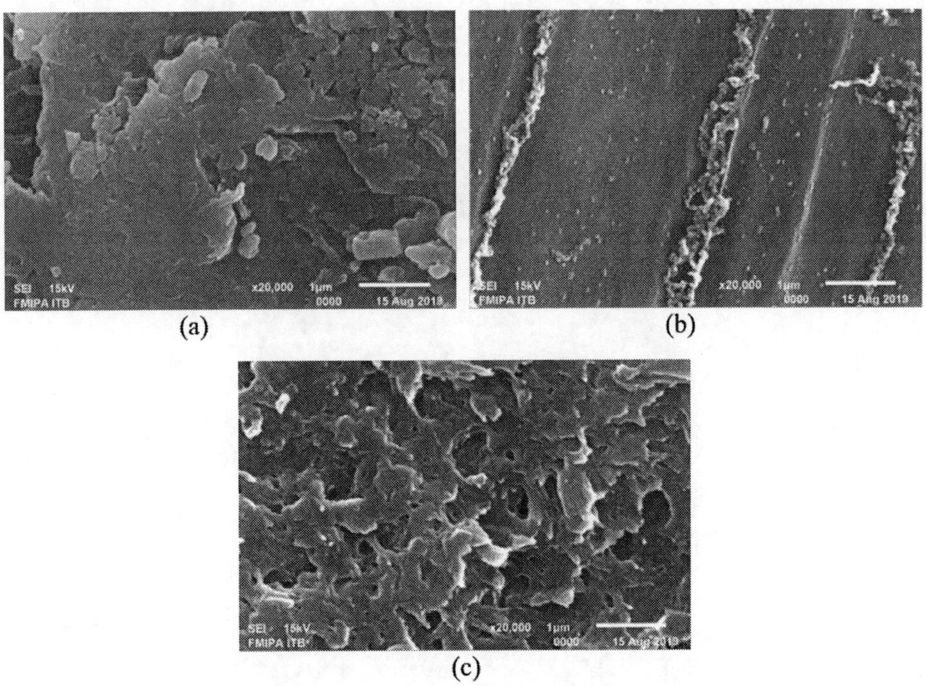

Figure 5. SEM (a) chitosan (b) acrylamide (c) copolymer of chitosan-acrylamide

4. Conclusion

The chitosan copolymerization with acrylamide using ammonium persulfate initiator was successfully synthesized in a microwave oven. The decrease in %G and %E as the microwave power increases was caused by the occurring of more homopolymer reactions.

References

[1] Abdulkarim A, Isa, M T, Abdulsalam S, Muhammad A J and Ameh A O *Bioresourc. Technol.* **99** 1359
[2] Wan Ngah W S and Fatinathan S 2008 *Chem. Eng. J.* **143** (1-3) 62
[3] Liu Y, Liu Z, Zhang Y and Deng K *J. Appl. Polym. Sci.* **89** 2283
[4] Feng H and Dong C M 2006 *Biomacromolecules* **7**(11) 3069
[5] Jayakumar R., Prabhaharan M., Reis R L and Mano J F 2005 *Carbohydr.* **62** 142
[6] Chiou M S, Ho P Y and Li H Y 2004 *Dyes Pigm.* **60** 69
[7] Peter M G 1995 *J. Macromol. Sci.* **32** 629
[8] Soleh A.S, Ibrahim A G, Abdelhai F, Elsharma E M, Metwally E and Siyam T 2017 *Radiat. Phys. Chem.* **134** 33
[9] Mochalova A, Zaborshchikova N, Zaitsev S, Semchikov Y, Zaitseva I and Pavlov G *Polym. Sci Ser A+* **48**(9) 918
[10] Singh V,Tiwari A, Tripthi D N and Sanghi R 2006 *Polym.* **47** 254
[11] Boonstra JD, Bekker A 1975 Ger Pat: 2 44 012
[12] Ali Z A, Venkatesan J, Kim S W and Sudha P N 2011 *Int. J. Environ. Sci.* **5**(1) 820

Acknowledgments

This research was funded by Sriwijaya University for Sains Teknologi dan Seni Sriwijaya University Research Grant No. 0016 / UN9 / SKL.LP2M.PT / 2019.

IOP Conf. Series: Materials Science and Engineering **833** (2020) 012065 doi:10.1088/1757-899X/833/1/012065

The Effect of Zr Addition on Microstructures and Hardness Properties of Zn-Zr Alloys for Biodegradable Orthopaedic Implant Applications

Moh Waqyan Ghani Fahmi, Achmad Fauzi Trinanda, Rizki Yuni Pratiwi, Sotya Astutiningtyas, and Ahmad Zakiyuddin·

Metallurgy and Materials Engineering, Faculty of Engineering, University of Indonesia, Depok, Indonesia

·Corresponding author: ahmadzakiyuddin@ui.ac.id

Abstract. The development of Mg and Fe based biomaterials in the past decade has been extensively studied as biodegradable material for medical applications. The development of this material is limited in terms of its suitability for clinical applications. Zn-based alloys began to be an alternative to be studied as a substitute for Mg and Fe based biomaterials. Zn-based alloys have a moderate degradation rate but have low mechanical properties, so other elements need to be added to improve their mechanical properties. In this study, the added element is zirconium (Zr) with a composition variation of 0.5%, 1%, and 2%. The alloying method used was casting with a temperature of 550°C. The results of the microstructure analysis, the addition of Zr to Zn alloys will form precipitates in the side of the grain boundaries and more addition of Zr composition, the smaller grain size formed. The grain size from pure Zn until the addition of 2% Zr in the sequence are 266.40 μm, 20.16 μm, 16.70 μm, and 15.85 μm. The XRD analysis, from the addition of Zr, will form the Zn phase and the intermetallic phase $Zn_{22}Zr$. The hardness value obtained from pure Zn until 2% Zr in the sequence are 35.162 HV, 41.988 HV, 42.324 HV, 57.112 HV.

1. Introduction

Biomaterial is a material that can interact with biological systems and substitute damaged bone tissue. The use of biomaterials as implant material in Indonesia is increasing along with the increasing cases of fracture that occur in Indonesia [1]. Biomaterials based on the type of material are divided into four, there are metals, polymers, ceramics and composites. Biomaterials that are often used in the medical world are materials with metal materials, such as stainless steel, titanium alloys, and cobalt alloys. The density and modulus of elasticity of non-biodegradable material are higher than bone which can make bones bear heavier loads so that their growth can be inhibited. In addition, these materials also require secondary surgery to prevent the side effects of not being able to decay and toxic in the body [2]. Therefore, we need a biodegradable material where the metal ions can metabolize in the body and not become toxic in the body, particularly in orthopedic implant applications. The kind of biodegradable materials divided into three, there are Mg-based alloy, Fe-based alloy, and Zn-based alloy [3].

The biodegradable material that has been widely researched is Mg-based alloy material. Mg-based alloy materials have the advantage of having mechanical properties similar to bone in the body in terms of yield strength, tensile strength, and elongation. However, this Mg-based alloy material also has the disadvantage of high corrosion levels. Corrosion can result in the formation of hydrogen gas which if

IOP Conf. Series: Materials Science and Engineering **833** (2020) 012065 doi:10.1088/1757-899X/833/1/012065

absorbed quickly can cause the effect of gas bubbles in vivo [4,5]. Fe-based biodegradable material was proposed in 2001 for non-rusty stents. This material also has mechanical properties that correspond to bones. But this material has a limitation that is the level of corrosion is too small which requires a long time to disappear. This can cause chronic inflammation, where the inflammatory precipitate can inhibit the completion of the healing process[6].

The development of Mg and Fe-based biomaterials in the past decade has been extensively studied as biodegradable material for medical applications. However, the extensive development of this material system experiences limitations or stagnation in terms of its suitability for clinical applications[7]. In the last three years, Zn-based biomaterials have begun to be proposed to be studied as alternative biodegradable materials besides Mg and Fe. Zinc is one of the essential elements in the body where there are more than 300 enzymes and proteins in the body. When zinc ions are degraded from the implant, they can still integrate into the body's metabolic activities without producing toxic effects [8]. The advantage of other Zn-based alloy biomaterials is the moderate rate of degradation because it has a passive layer of moderate stability and Zn-based alloys are easier to do casting because of their low melting point and low chemical reactivity [9]. However, pure Zn biomaterials still have shortcomings namely the strength and elongation of this material is still low (σUTS ~ 30MPa and ε <0.25%), so it is not enough for medical applications [10].

One way to improve the mechanical properties of pure Zn biomaterial is by adding other alloying elements to Zn-based materials, such as Mg, Al, Cu, Ca, Sr, and Zr. Zirconium (Zr) as an alloying element for Zn-based material has not been widely studied by people. Zr is one element with potential biocompatible and bio neutral elements. Zirconium is also one of the elements that has low systemic toxicity [11]. Based on the Zn-Zr phase diagram, the addition of a small amount of Zr in Zn can form the Zr-rich intermetallic phase ($Zn_{22}Zr$). The addition of 0.05% wt Zr in pure Zr can increase the strength (UTS) of the alloy to 157 MPa with 22% elongation. The amount of Zr released from Mg-Zn-Zr alloys results in a corrosion rate of 50 µg / day which is still acceptable to the human body [12]. Therefore, with the potential of Zr as a biocompatible element, this study will examine the effect of adding Zr in Zn alloys on the microstructure and hardness properties of the alloy material.

2. Experimental Procedures

2.1. Materials Preparation

The initial stages of the sample making process is doing alloy and mold designs. Materials that use in the Zn-Zr alloy preparation are commercially pure Zinc (99.95 wt%) and Zirconium (powder 99.9wt%) addition of 0.5%, 1%, and 2wt% that are melting in the graphite crucible followed by centrifugal casting at 550°C. After that, the Zn-Zr alloy in the form of the melt was poured into a graphite mold with rod-shaped product and cooled at room temperature. Then, homogenized the sample at around 380°C for 8-10 hours. After the sample has been homogenized, then the sample is cutting with a thickness of approximately 5 mm for the sample preparation process. The stages of sample preparation are mounting, grinding, polishing, and etching. The chemical substances that use in sample preparation are resin and hardener for mounting, alumina solution for polishing, and 2 mL of HNO_3 and 98 mL of distilled H_2O for etching.

2.2. Microstructure Analysis

The sample for microstructure analysis was ground using sandpaper from 240-2000 grit. After that, the polishing process used a polish machine with alumina solution. Then, the etching process was done using 2 mL of HNO_3 and 98 mL of distilled H_2O. The microstructure of Zn-Zr alloys was observed using an Optical Microscope Carl Zeiss Primotech and Scanning Electron Microscope FEI Inspect 50 to determine the shape of the grain, grain size, and intermetallic distribution in the matrix and grain boundaries. Then the X-Ray Diffraction test was done to determine the compound and phases in the sample.

IOP Conf. Series: Materials Science and Engineering **833** (2020) 012065 doi:10.1088/1757-899X/833/1/012065

2.3. Hardness Analysis
The hardness test was done using BUEHLER micro hardness tester with the applied load of 100 gf. The initial step taken in testing the hardness of Vickers was sample preparation firstly. After sample preparation was finished, then the sample was placed in the area to be tested. After that, the load was set to be used, in this study using a load of 100 gf. Then, the point was found to be carried out as a loading. Once deemed appropriate, then the start button was pressed to start the loading process. Waiting time for the loading process lasts for approximately 10-15 seconds. After loading was complete, then the diagonals formed as a result of loading were measured. After that, the value of the Vickers hardness was calculated.

3. Result and Discussion

3.1. Microstructures
The results of testing the microstructure of the Zn-Zr alloy can be seen in Figure 1

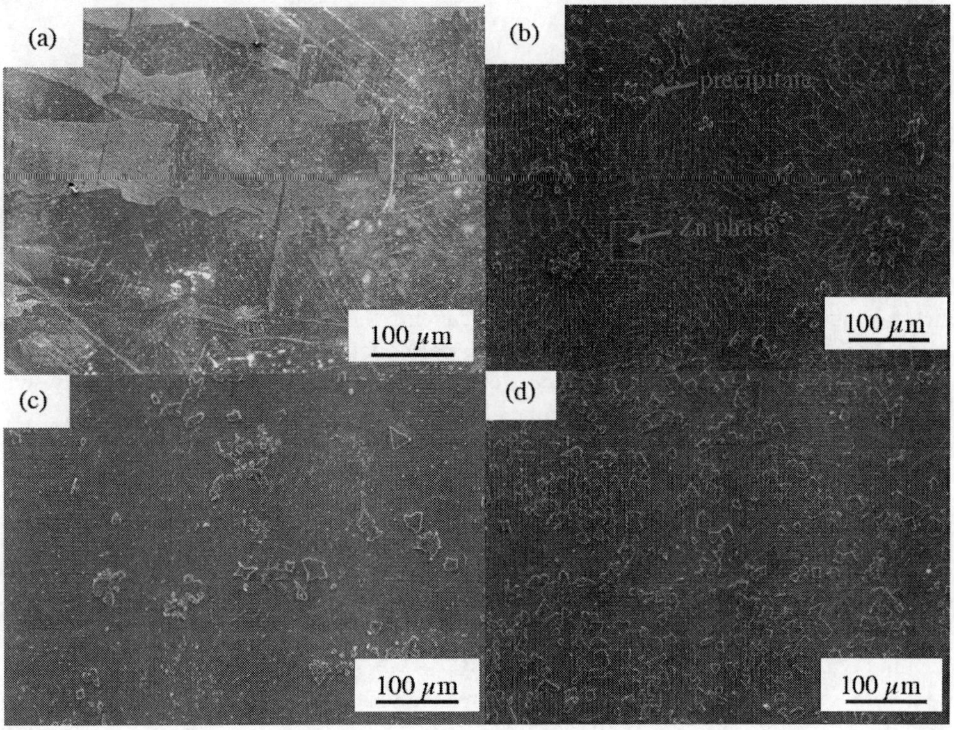

Figure 1. Microstructures of (a) pure Zn, (b) Zn-0.5Zr, (c) Zn-1Zr, (d) Zn-2Zr with SEM analysis.

In figure 1 (a) it can be seen that the grain size of the pure Zn microstructure test results is quite large and tends to be rough. In figure 1 (b), after adding 0.5% zirconium, a precipitate formation at the periphery is close to the grain boundary. The precipitates were formed due to the casting temperature (550°C) and during the homogenization process (380 °C), zirconium is not dissolved in the Zn matrix. Based on the Zn-Zr binary phase diagram, the precipitate is identified as the $Zn_{22}Zr$ phase [2].

In figure 1 (b) precipitates that are formed still tend to be small and the amount are still less. These precipitates are globular or oval-shaped and the distribution of the precipitates was not evenly distributed and still collects only in some parts of the grain boundary. In figure 1 (c) with the addition of zirconium 1%, the precipitates formed tend to be larger and more numerous. The shape of the precipitate looks sharp and the distribution of the precipitate looks more evenly distributed. In figure 1 (d) with the addition of 2% Zr, it appears that more precipitates are formed. The shape of the precipitate of this microstructure is that it looks more diverse, there are those in the form of sharp, round, and elongated. The distribution of precipitates is more evenly distributed. It can be concluded that the more zirconium content added in zinc alloys, the more precipitates are formed where the size larger and the distribution of precipitates are more evenly distributed at the edges of grain boundaries.

3.2. Grain Size.

After getting the result of the microstructure of the zirconium content addition variable, then the grain size was calculated. The grain size calculation was done using the Planimetric (Jeffries) method with the help of Fiji ImageJ software application and in accordance with ASTM E112 standards. The results of the grain size calculation can be seen in table 1.

Table 1. Grain size of Zn-Zr alloys.

Sample	Grain Size (μm)
Pure Zn	266.40
Zn-0.5Zr	20.16
Zn-1Zr	16.70
Zn-2Zr	15.85

From Table, 1 it can be seen that the more zirconium additions to the Zn-Zr alloys will make the grain size smaller. In other words, the addition of zirconium content in Zn-Zr alloys will reduce the grain size of the alloy. The most significant reduction in grain size occurs from pure Zn to Zn-0.5Zr alloys. The reduction in grain size is caused by the presence of precipitates in the alloy. The presence of precipitates in Zn-Zr alloys will inhibit the process of grain growth, so that nucleation will occur which makes the grain size smaller and more numerous. Reducing the grain size will increase the grain limit per unit volume and will reduce the path free from continuous slips. Small and fine grain size will strengthen the material by blocking the dislocation movement around the grain boundary, where the dislocation movement requires high stress to open or produce a new dislocation on the next item. Alloy with a small and fine grain size will make the strength and hardness of the alloy be increased compared to large and coarse grains. That is because alloys with small and fine grain size have a larger amount of surface in the total grain layer area which will inhibit the movement of dislocations.

3.3. XRD Result

XRD testing conducted in this study aims to determine the phases formed from each alloy. This XRD analysis was carried out using the help of Origin 2018 software. The results of XRD Zn-Zr alloy testing can be seen in Figure 2.

Figure 2 shows the comparison of the XRD test results for the addition of Zr levels in Zn-Zr alloys. The composition of Zr added to Zn-Zr alloy is 0.5% Zr in Zn-0.5Zr alloy, 1% Zr in Zn-1Zr alloy in Zn-1Zr alloy, and 2% Zr in Zn-2Zr alloy. In figure 2 it can be seen that the Zn-0.5Zr alloys have two phases that appear, namely the α Zn phase and the $Zn_{22}Zr$ intermetallic phase. This is consistent with the literature on the Zn-Zr phase diagram, where zirconium at 550°C is not dissolved into the Zn matrix, so at this temperature, the intermetallic phase $Zn_{22}Zr$ is formed. The peak height of the Zn phase is higher compared to the $Zn_{22}Zr$ phase due to the higher Zn concentration compared to Zr, so the Zn phase will be more dominantly formed. The peak of diffraction 2θ in the highest Zn phase is 43.28°. Some of the other 2θ diffraction peaks that occur in the Zn phase are 36.32°, 39.04°, 43.28°, 54.94°, 70.16° and 82.20°.

In the Zn-1Zr and Zn-2Zr alloys, the phases formed are the same as those in the Zn-0.5Zr alloys namely Zn phase and $Zn_{22}Zr$ phase. This is due to the addition of 1% Zr and 2% Zr when casting and homogenizing, the Zr element does not dissolve in the Zn matrix. The difference between Zn-0.5Zr and Zn-1Zr and Zn-2Zr is the height of the peak that occurs. In the Zn-2Zr alloy, the peak of the Zn phase is seen to be lower than the peak of the Zn phase in the Zn-0.5Zr alloy and the Zn-1Zr alloy. The phase intensity of Zn-2Zr alloys at 43.28° is 21438 a.u, whereas in Zn-0.5Zr and Zn-1Zr alloys are 36267 a.u and 36962 a.u. This happens because of the increasing composition of Zr in the Zn alloy. The intensity of the peak phase Zn will decrease with increasing composition combined.

While the peak height of the $Zn_{22}Zr$ phase in the Zn-2Zr alloy looks higher compared to that of the Zn-0.5Zr alloy and Zn-1Zr alloy. This happens because the levels of zirconium in Zn-2Zr alloys are higher than the levels of zirconium in Zn-0.5Zr and Zn-1Zr. Some of the 2θ phase diffraction peaks of $Zn_{22}Zr$ that occur in Zn-2Zr alloys are 37.73°, 40.44°, 42.52°, 44.38°, 45.94°, 76.42°.

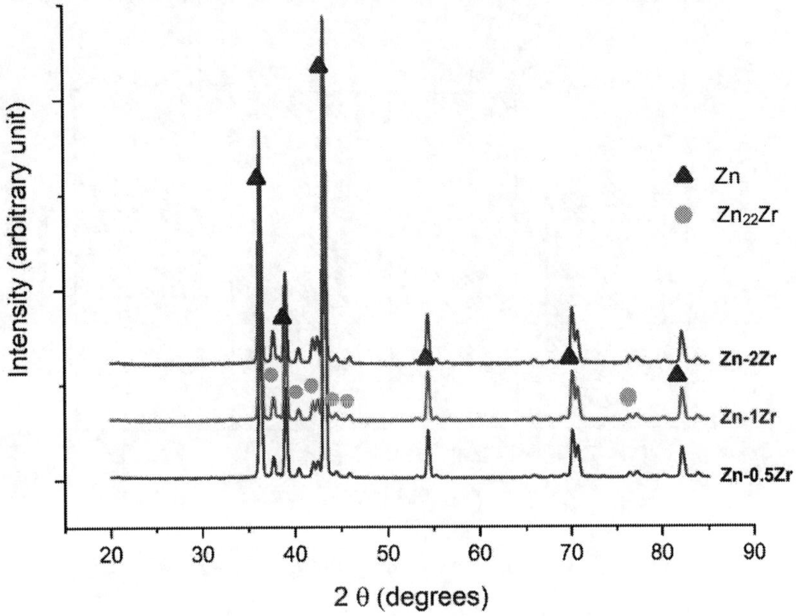

Figure 2. XRD result of Zn-Zr alloys.

3.4. Hardness Properties

Hardness testing was done to determine the comparison of the hardness value of each alloy and the comparison of the value of pure bone hardness. The method used for hardness testing in this study is using the Vickers microhardness method. The results of Vickers microhardness testing can be seen in figure 3.

In figure 3, it can be seen that the more zirconium content added in the Zn alloy, the harder the existing value of the alloy will increase. This is because the addition of the Zr element to the Zn alloy will result in smaller and finer grain sizes, the smaller and finer grain size can increase the hardness. The smaller and finer grain size occurs because zirconium does not dissolve in the Zn alloy during the homogenization process. The insoluble zirconium turns into precipitates which will inhibit the growth of grain size so that the grain will undergo nucleation that its size becomes smaller and more numerous.

Meanwhile the addition of other elements to the Zn-based alloy, the value of the resulting hardness tends to be higher than with the addition of Zr. As in the example of research conducted by HF Li, and friends, about the effect of adding Mg, Ca and Sr on Zn-based alloys, the hardness value of the alloys is 78.26 HV for Zn-1Mg alloys, 73HV for Zn-alloys 1Ca, and 61.88 HV for Zn-1Sr alloys [13].

From the results of the study of the addition of other elements to the Zn-based alloy, the value of Zn-Zr alloy hardness is below the value of the addition of other elements [14]. Whereas based on the ASM Metals Handbook, Vol 02 Properties and Selection of Nonferrous Alloys and Special-Purpose Materials, the hardness value of Zn-Zr alloys is between the average hardness values of Zn-based alloys which is between 42-163 HV. From these data, it can be concluded that the value of Zn-Zr alloy hardness is not too high, possibly suitable for use in biodegradable orthopedic implant applications. This is because if the alloy hardness value is too high it can damage the bone surface due to rubbing against the implant itself.

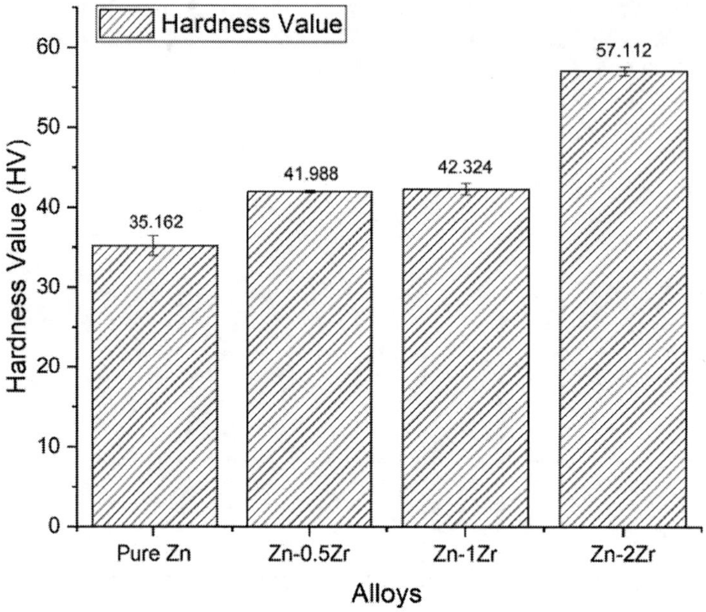

Figure 3. The hardness value graphic of Zn-Zr alloys.

4. Conclusion

Based on the results of research and analysis of data that has been done, it can be concluded that: adding zirconium content to zinc alloys with 0.5%, 1%, and 2% Zr content, can form precipitates in microstructures. Precipitates that are formed can inhibit the process of grain growth which makes grain size smaller and finer. The smaller and finer grain size makes the strength and hardness values of the alloy larger. The more zirconium content is added to the zinc alloy, the smaller the grain size will be produced. The grain size of pure Zn was 266.40 μm, Zn-0.5Zr was 20.16 μm, Zn-1Zr was 16.70, Zn-2Zr was 15.85 μm. The phases produced in the Zn-Zr alloys from the XRD test are the Zn phase and the $Zn_{22}Zr$ phase. The more zirconium content added to the Zn alloy, the greater the resulting of the hardness value. The value of pure Zn hardness produced was 35.162 HV, Zn-0.5Zr was 41.988 HV, Zn-1Zr was 42.324 HV, and Zn-2Zr was 57.112 HV. Zn-Zr alloy material can be a candidate alloy material suitable for use in biomedical applications because it has a hardness value that is relatively close to the hardness value of the bones (40.38HV)[15], besides that based on the literature Zn alloy material also has a moderate degradation rate[14].

Acknowledgements

The authors would like to thank the Direktorat Riset dan Pengabdian Masyarakat Universitas Indonesia (DRPM- UI) for the financial support under the grant of Hibah PITTA-B UI with contract number: 2535/UN2.R3.1/HKP.05.00/2018.

References

[1] B. Kemenkes RI, *Riset Kesehatan Dasar* 2013 (Jakarta: Badan Penelitian dan Pengembangan Kesehatan Kementerian Kesehatan RI) pp 101-106.

IOP Conf. Series: Materials Science and Engineering **833** (2020) 012065 doi:10.1088/1757-899X/833/1/012065

[2] M. J. Olszta *et al.* 2007 "Bone Structure and Formation : A New Perspective" *Met. Sci. Eng.* **58** pp. 77–116.

[3] D. Vojtech, J. Kubasek, J. Capek, and I. Pospisilova 2014 "Magnesium, Zinc and Iron Alloys for Medical Applications in Biodegradable Implants" *Metal.*

[4] S. Agarwal, J. Curtin, B. Duffy, and S. Jaiswal 2016 "Biodegradable Magnesium Alloys for Orthopaedic Applications : A Review on Corrosion, Biocompatibility and Surface Modifications," *Mater. Sci. Eng. C* **68** pp. 948–963.

[5] X. Gu and Y. Zheng 2010 "A Review on Magnesium Alloys as Biodegradable Materials," *Front. Mater. Sci.* **4** pp. 111–115.

[6] M. I. Rahim, S. Ullah, and P. P. Mueller 2018 "Advances and Challenges of Biodegradable Implant Materials with a Focus on Magnesium-Alloys and Bacterial Infections" *Metals (Basel).* **8** p. 532.

[7] N. Zhao, D. Zhu, N. C. Agricultural, N. C. Agricultural, and U. States 2016 "Endothelial Cellular Responses to Biodegradable Metal Zinc" *ACS Biomater. Sci. Eng.* **1** pp. 1174–1182.

[8] X. Liu, J. Sun, K. Qiu, Y. Yang, Z. Pu, and L. Li 2016 "Effects of Alloying Elements (Ca and Sr) on Microstructure, Mechanical Property and In Vitro Corrosion Behavior of Biodegradable Zn-1.5Mg Alloy" *J. Alloys Compd.* **664** pp. 444–452.

[9] J. Niu *et al.* 2016 "Research on A Zn-Cu Alloy as A Biodegradable Material for Potential Vascular Stents Application" *Mater. Sci. Eng. C* **69** pp. 407–413.

[10] J. Kubásek, J. Šerák, P. Novák, and D. Vojte 2011 "Mechanical and Corrosion Properties of Newly Developed Biodegradable Zn-based Alloys for Bone Fixation" *Acta Biomater.* **7** pp. 3515–3522.

[11] D. Hong *et al.* 2013 "In Vitro Degradation and Cytotoxicity Response of Mg-4%Zn-0.5%Zr (ZK40) Alloy as A Potential Biodegradable Material" *Acta Biomater.* **9** pp. 8534–8547.

[12] W. Maria, W. Bednarczyk, J. Kawa, and P. Ba 2018 "Effect of Zirconium Microaddition on The Microstructure and Mechanical Properties of Zn-Zr Alloys" *Mater. Charact.* **142** pp. 187–194.

[13] H. F. Li *et al.* 2015 "Development of Biodegradable Zn-1X Binary Alloys with Nutrient Alloying Elements Mg , Ca and Sr" *Sci. Rep.* **5** p. 10719.

[14] G. K. Levy, J. Goldman, and E. Aghion 2017 "The Prospects of Zinc as a Structural Material for Biodegradable Implants — A Review Paper" *MDPI J. - Met.* **7** p. 402.

[15] K. Y. S. Lee, K. M. C. Chin, S. Ramesh, C. Y. Tan, W. D. Teng, and I. Sopyan 2012 "Characterization of Forsterite Bioceramics" *Adv. Mater. Res.* **576** pp. 195–198.

IOP Conf. Series: Materials Science and Engineering **833** (2020) 012066 doi:10.1088/1757-899X/833/1/012066

The Effect of Acid Type in Mechanochemical Treatment of Rice Husk Ash for Concrete Application

J F Fatriansyah[1,*], S A Khairunnisa[1], D Dhaneswara[1], M Z Rahmatullah[1], B A Ramadhan[1]

[1]Metallurgical and Materials Engineering, Faculty of Engineering, Universitas Indonesia, Depok, Jawa Barat 16424

*Corresponding author: jakafajar@ui.ac.id

Abstract. Concrete was successfully produced using rice husk ash (RHA) as a partial cement substitute. RHA was mechanochemically treated using ball mill as mechanical force and acids (HCl and CH_3COOH) as chemical force. Mechanochemical treatment decreases particle size as demonstrated by particle size analyser test and reduces crystallinity of RHA as demonstrated by X-ray diffraction analyses. These characteristics make RHA more suitable as partial substitution of cement as well as pozzolanic material. The compressive strength of concrete increases 11-26 % by addition of RHA as cement substitution. The significant increase of compressive strength was obtained from RHA mechanochemically treated with HCl which has compressive strength of 14.8 MPa in comparison of standard concrete which has compressive strength of 11.72 MPa.

1. Introduction

Indonesia is an agricultural country which in 2017 produced paddy around 81 million tonnes [1]. Around 20 % of the total paddy production went as agricultural wastes. One of the agricultural wastes is rice husk. The rice husk contains 22% of silicon oxide (silica), 35 % carbon, 5% hydrogen, 36 % oxygen and the rests are nitrogen and sulphur [2]. Approximately for every ton of rice husk burning, around 0.19 ton of rice husk ash is produced [3]. This rice husk ash has greyish colour to white colour after complete burning. The burning process increases its silica content up to 90 % [3]. In addition, burning process below controlled temperature of 800 °C produces rice husk ash in its amorphous form [4]. Silica can be used as raw material for mesoporous silica, silicon carbide, silicon tetrachloride, pure silicon and zeolites even for hydrogen adsorption as shown in simulation [5–10].

Mechanical method using ball mills can be used to reduce the size of particles for example carbon [11–13] and amorphous silica grain/particle size [14]. Higher surface area of rice husk ash has high reactivity as active mineral mixing material for example as concrete mixing material [15]. The addition of rice husk ash in concrete increases compressive strength and impermeability of ultra-high-performance concrete (UHPC) due to silicas fine porous structure [16, 17]. Partial substitution of rice husk ash in Portland cement increases its working ability and mechanical durability on the economical point of view in construction industry [18–20]. In addition, substitution of some fraction in cement with rice husk ash can increase strength of concrete [4, 21- 24]. In this paper, we studied the effect of mechanochemical (mechanical plus chemical) treatment using ball mill as mechanical force and acids as chemical force in rice husk ash on mechanical properties of concrete produced partially using rice husk as cement substitution. We suggested that the mechanochemical treatment on rice husk ash (RHA) can enhance mechanical properties of concrete.

2. Experimental Methods

2.1 Sample

Rice husk was obtained from paddy mills industry in Leuwiliyang area, West Jawa - Indonesia. HCl and CH_3COOH was purchased from Mallinckrodt Pharmaceuticals and Emsure, respectively, and they were used without any further purification.

2.2 Preparations

Rice husk ash was dried under the sun for 8 hours to reduce water content. Dried rice husk was then burnt in a furnace for 5 hours at 700 °C resulting in a white powder. The white powder of RHA then mixed with 1M HCl with 10% w/w and ball milled with vertical planetary ball mill. In addition, the other sample of RHA powder was mixed with CH_3COOH with 10% w/w and ball milled with vertical planetary ball mill. The speed of ball mill was set to 20 rpm within 15 minutes. In total, there were 3 samples, RHA without ball mill and acid treatment (X), RHA milled with CH_3COOH (Y), and RHA with HCl (Z). The labelling of the samples is shown in Table 1.

Table 1. RHA samples labelling.

Name	Treatment
X	RHA without ball mill and acid
Y	RHA milled with ball mill and CH_3COOH addition
Z	RHA milled with ball mill and HCl addition

Concrete was made by mixing cement, water, fine aggregates and coarse aggregates. Three RHA samples were added in modified concretes with the concentration of 10% of cement mass before RHA added. The detail of composition is shown in Table 2.

Table 2. Concrete compositions.

Concrete types	RHA (g)	Cement (g)	Fine aggregates (g)	Coarse aggregates (g)	Water (g)
Concrete A	0	354.550	693.63	1007.34	195
Concrete X	35.455	319.095	693.63	1007.34	195
Concrete Y	35.455	319.095	693.63	1007.34	195
Concrete Z	35.455	319.095	693.63	1007.34	195

The concrete was made with the composition as detailed in Table 2. All of materials were mixed and water was added slowly in order to bind all of materials. Mixed material was then poured in concrete mould and stand at room temperature for 1 day, followed by preserve it in water for 28 days.

2.3 Characterization

RHA composition was determined using X-Ray Fluorescence (XRF) (Panalitycal Epsilon). The size of RHA particles were measured using Particle Size Analyzer (PSA) (Nanoplus Particulate Systems). Concrete compressive strength was measured in civil engineering laboratory of Universitas Indonesia.

3. Results and Discussion

First, RHA compositions were obtained using XRF. The results were presented in Table 3. Table 3 shows that silica content in RHA is 88.13 %. This result shows that burning temperature of 700 °C is suitable to reduce impurity in RHA. High silica content is important to improve concrete strength as well as a pozzolan material.

Table 3. RHA compositions measured using XRF.

Compound	Concentration (%)
SiO_2	88.13
K_2O	4.14
Fe_2O_3	2.32
P_2O_5	1.69
MgO	1.55
CaO	1.70
Cl	0.37
Ag_2O	0.12
Total	100

The effect of mechanochemical treatment on particle size is shown in Table 4. The particle sizes were determined by PSA for all RHA samples. It is demonstrated that the addition of acids along mechanical treatment can reduce the size of RHA particles significantly. The decrease of particle size is plummeted for the addition of strong acid HCl in which the mean size of particles was reduced from 42.00 μm (concrete X) to 27.36 μm (concrete Z).

Table 4. PSA results of RHA samples.

Concrete	Particle mean size (μm)
X	42.00
Y	32.85
Z	27.36

XRD results are shown in Figure 1 for all samples. The effect of mechanochemical treatment on particle crystallinity is shown in Table 5. All of RHA samples show almost identical diffraction pattern with one broadening peak with the centre around 22 degrees. This shows the characteristics of amorphous silica. However, the diffraction peak shows that all of the samples are not completely amorphous. Thus, the degree of crystallinity index needs to be quantified which can be calculated based on Equation 1 as follows

$$CI\ (\%) = \frac{Ic - Iam}{Ic} \text{ x } 100, \tag{1}$$

where CI (%) is the crystallinity index, I_c is the maximum diffraction intensity and I_{am} is the amorphous index of crystallinity intensity. The crystallinity index of all samples calculated from Eq. (1) is presented in Table 5. Table 5 shows that the mechanochemical treatment reduces the crystallinity index of RHA. According to Akasaki *et al.* [22], more amorphous material has pozzolanic activity than less amorphous material. High pozzolanic activity is desired for the cement substitution application. Thus, it is expected that the mechanochemically treated samples are more suitable as cement substitution in comparison with untreated RHA.

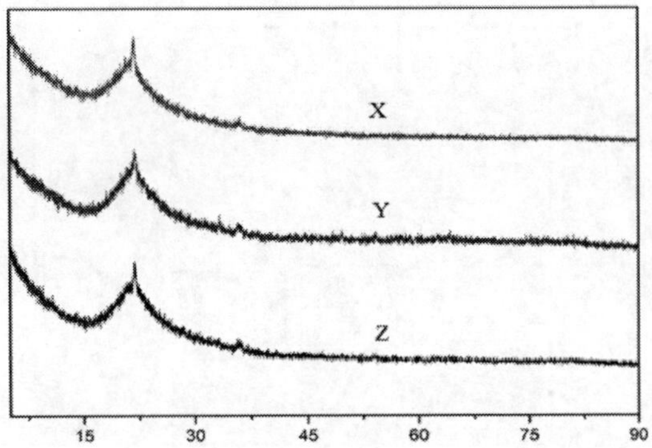

Figure 1. XRD diffraction pattern for RHA samples.

Table 5. Index of crystallinity of RHA samples.

Sample	Crystallinity index
X	53.08
Y	52.90
Z	52.82

The mechanical strength specifically measured in this work is compressive strength. The results are shown in Table 6 and Figure 2. Figure 2 shows that the substitution of RHA as pozzolanic material for cement in concrete yields higher compressive strength than that of unsubstituted concrete. Sample Z, which uses HCl, shows maximum compressive strength. It is suggested that smaller particle size and smaller crystallinity index resulting in higher compressive strength. Smaller particle size produces higher pozzolanic activity due to ability of smaller particles to fill empty space in concrete. Thus, it strengthens surrounding bonding as RHA transform into calcium silicate as a good binder after water addition. In addition, high crystallinity index in RHA shows low pozzolanic component inside because crystalline silica is not pozzolanic thus calcium silicate cannot be formed. As a result, low crystallinity index produces more calcium silicate as binder which in turn increase compressive strength of concrete. However, the fact that HCl treatment produces better result than that of CH_3COOH treatment is not yet understood. It is proposed that HCl produces smaller particle size due to stronger acidity nature in comparison with CH_3COOH thus increase the pozzolanic activity of RHA. The compressive strength of concrete increases 11-26 % by addition of RHA as cement substitution.

Table 6. Compressive Strength of Concrete.

Concrete	Compressive strength (Mpa)
Standard concrete	11.72 ± 0.42
X	13.03 ± 0.37
Y	12.92 ± 0.38
Z	14.8 ± 0.32

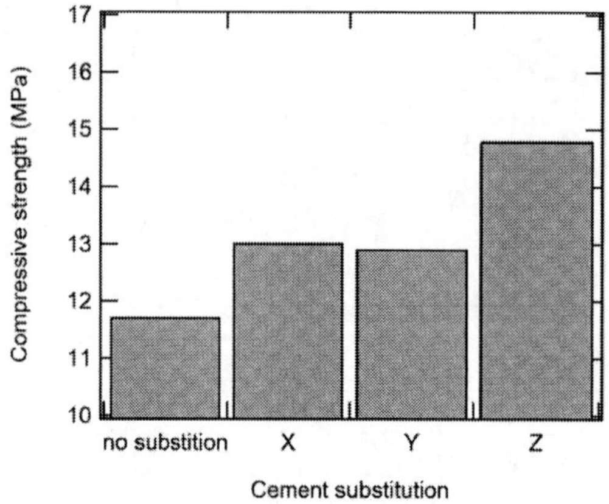

Figure 2. Compressive strength of concrete with the RHA substitution.

4. Conclusion

The mechanochemically treated RHA with ball mill and acids has been synthesized. The RHA was used as a cement substitution of concrete. Mechanochemically treated RHA shows smaller particle size and crystallinity index which is potential as pozzolanic material. The compressive strength of concrete increases 11-26 % by addition of RHA as cement substitution. The significant increase of compressive strength was obtained from RHA that mechanochemically treated with HCl with a compressive strength of 14.8 MPa (26 % higher than that of the standard concrete).

References

[1] Central Bureau of Statistics 2013 Keadaan Ketenagakerjaan (Jakarta: Central Bureau of Statistics)
[2] Nornikman H and Soh P J 2008 *Progress in Electromagnetics Research* **104** pp 145-166
[3] Sandhu R K and Siddique R 2017 *Construction and Building Materials* **153** pp 751–764
[4] Habeeb G A and Mahmud H Bin 2010 *Materials Research* ***13*(2)** pp 185–190
[5] Sun L and Gong K 2001 *Ind. Eng. Chem. Res* **40** pp 5861 – 5877
[6] Dhaneswara D, Fatriansyah JF, Putranto DA, Utami SA, Delayori F. 2018 *IOP Conference Series: Materials Science and Engineering* **Vol. 285** No. 1, p. 012032
[7] Fatriansyah J F, Dhaneswara D, Abdurrahman M H, Kuskendrianto F R, Yusuf M B. 2019 *IOP Conference Series: Materials Science and Engineering* 2019 **Vol. 478** No. 1, p. 012034
[8] Dhaneswara D, Siti Agustina A A A, Putranto D A, Delayori F and Fatriansyah J F. 2018 *IOP Conf. Series: Journal of Physics: Conf. Series* **1011** 012017
[9] Dhaneswara D, Fatriansyah J F, Wardana A K, Haqoh A N, Khairunnisa S A. 2019 *IOP Conf. Series: Materials Science and Engineering* **547** 012032
[10] Dhaneswara D, Fatriansyah J F, Yusuf M B, Abdurrahman M H, Kuskendrianto F R. *IOP Conf. Series: Materials Science and Engineering* **547** 012038
[11] Fatriansyah JF, Matari T, Harjanto S 2018 *Materials Science Forum* **Vol. 929**, p 50-55
[12] Fatriansyah JF, Dhaneswara D, Khairunnisa S A, Kusumawardhani D. 2019 *IOP Conf. Series: Materials Science and Engineering* **547** 012034
[13] Harjanto S, Fatriansyah J F, Noviana L N, Yunior SW 2018 *International Journal of Technology* **9(5)** pp 993-1005.
[14] Do J L and Friščić T 2017 *ACS Central Science* ***3*(1)** pp 13–19
[15] Wang W, Meng Y and Wang D 2017 *Kemija u Industriji* ***66*(3–4)** pp 157–164
[16] Huang H, Gao X, Wang, H and Ye H 2017 *Construction and Building Materials* **149** pp 621–628

[17] Zareei S A, Ameri F, Dorostkar F and Ahmadi M 2017 *Case Studies in Construction Materials* **7 pp** 73–81
[18] Yalçin N and Sevinç V 2001 *Ceramics International* **27(2)** pp 219–224
[19] Chopra D, Siddique R, Kunal 2015 *Biosyst. Eng.* **130** pp 72–80
[20] Chatveera B, Lertwattanaruk P 2011 *J. Environ. Manag.* **92** pp 59–66.
[21] Gosalvitr C, Srinophakun P, Kanchanapiya P and Danwittayakul S 2015 *Proceedings of the International Conference on Biological, Environment and Food Engineering* **(BEFE-2015)** pp 15-16.
[22] Akasaki J, Tashima M, Da Silva C A, Da Silva E, Barbosa M and Paya J 2005 *Congreso Nacional de Materiales Compuestos* 5
[23] Dhaneswara D, Fatriansyah J F, Kusumawardhani D P, Khairunnisa SA 2019 *IOP Conference Series: Materials Science and Engineering 578 (1), 012063*
[24] Fatriansyah J F, Dhaneswara D, Khairunnisa SA, Kusumawardhani D P, 2019 *IOP Conference Series: Materials Science and Engineering 547 (1), 012034*

IOP Conf. Series: Materials Science and Engineering **833** (2020) 012067 doi:10.1088/1757-899X/833/1/012067

A Green Method to Prepare Composite of Graphene Oxide-Manganese Oxide using a Modified Hummer's Method

Marselus Jeques Gros, Masruri MASRURI*, Rachmat Triandi Tjahjanto

Department of Chemistry, Faculty of Mathematics and Natural Sciences, Brawijaya University, Malang 65145, Indonesia.

*Corresponding author email: masruri@ub.ac.id

Abstract. A composite graphene oxide-manganese oxide (GO-MnO_x) was facilely fabricated in one-step preparation by modified Hummer's method. The Hummer's method for graphene oxide preparation produced manganese residue as by product, therefore, we modified the Hummers method to convert the manganese residue into value-added materials with aqueous sodium hydroxide (NaOH). The number of Mn on composite and surface chemistry was investigated. GO-MnO_x was characterized by atomic absorption spectroscopy (AAS), UV-Visible spectrometer, fourier transform infrared (FT-IR), X-ray diffraction (XRD), and transmission electron microscopy (TEM). Quantitative measurements showed that GO-MnO_x 1 and 2 have the number of Mn of 1.4231 and 2.0958 $\mu g/g^{-1}$, respectively. FTIR showed the presence of functional groups such as hydroxyl, epoxy, carboxyl, carbonyl and manganese oxide. XRD confirmed the interlayer spacing of GO-MnO_x synthesized by our techniques is lower than GO-MnO_x produced by other processes. TEM showed that GO-MnO_x is single layer yet and Mn species was not homogeneously distributed on composite.

1. Introduction

Manganese oxide (MnO_x), a transition metal oxide, has attracted the attention of researcher due to less toxicity, large natural abundance of manganese element, cheaper, environmentally-friendly and high redox potential [1,2]. Up to now, MnO_x has been applied in many fields such as catalyst, supercapacitor, adsorbent [3–5], etc. However, the bulk of MnO_x has low thermal and chemical stability, difficulties in separation from aquatic solution, and low surface area [6] which seriously limit their application. Therefore, MnO_x was added on supporting materials as well as zeolite [7], silica nanoparticles [8], and carbon-based materials (e.g. graphite oxide) [9] to increase the physicochemical properties such as preventing the aggregation and increasing the stability and catalytic activity [10]. Graphene oxide (GO) is a more promising material for preparation GO-MnO_x composite than carbon black and carbon nanotube (CNT). The presence of functional groups (hydroxyl, carboxyl, epoxy groups) on GO provides hydrophilic character and chemical reactivity which cause GO easily to bind with other composites [11,12]. Moreover, GO has superior physical and chemical properties such as high specific surface area, good mechanic and thermal properties, eco-friendly and biodegradable [13–15]. On the other hand, the segregation and agglomeration of GO caused by Van der Waals interactions can be prevented by addition of manganese oxide on GO [16]. Other studies have also proven that GO-MnO_x composite has high conductivity and the surface area can be easily separated from aquatic solution which is beneficial for the oxidation of organic molecules [6,17,18].

GO is synthesized from natural graphite or other carbon-based materials with various methods such as Hummers method [19], epitaxial growth [20] and chemical vapor deposition (CVD) [21]. However, the epitaxial growth and CVD needed high temperature reaction condition up to 1000°C and low yield [22]. Hummers method has also limitations in practical use such as the hazardous reagents are involved in the reactions (e.g., sodium nitrate or potassium chlorate) [23]. For this reasons, Hummers method with modification has been investigated and could be a promising method for the preparation of GO-MnO_x composite due to its simple, safe, and efficient reaction process, high oxidation level of graphite and less expensive [24]. Previous researches have reported the synthesis of GO-MnO_x with a modification of Hummers method. Nevertheless, a few of process required additional amount of potassium permanganate ($KMnO_4$) after the sonication of graphite oxide [10,25–27]. Recently, Suzuki and Nishina reported one-step preparation of graphite oxide-manganese oxide composite with a modification of Hummers method which used potassium hydroxide (KOH) to precipitate the residual Mn on graphite oxide without an excessively addition of $KMnO_4$ [9]. This method converted residual Mn which is a byproduct into a value material. Nevertheless, graphite oxide was not exfoliated with sonication process affecting the properties of surface chemistry.

In this work, one-step synthesis of GO-MnO_x was investigated using the modification of Hummers method with a ratio of $KMnO_4$:H_2SO_4 as oxidant and the sodium hydroxide (NaOH) to precipitate the residual Mn. The route of synthesis GO-MnOx can be divided into three stages. Firstly, natural graphite was oxidized by $KMnO_4$ and strong acid was added to improve the solubility of inorganic carbon in solution. Basically, inorganic carbon is inert at room temperature [28]. Secondly, the mixture was diluted with water solution and then graphite oxide was centrifuged and added NaOH to precipitate Mn salt on GO. Thirdly, the graphite oxide was exfoliated by sonicator to produce *mono/bilayer* graphene oxide. The composite was then characterized by atomic absorption spectroscopy (AAS), UV-Visible (UV-Vis) spectrophotometer, fourier transform infrared (FT-IR), X-ray diffraction (XRD), and transmission electron microscopy (TEM).

2. Experimental Section

2.1. Materials and Instruments

In this work, natural graphite powder was purchased from CV. Pratama. The high purity chemicals include 95% H_2SO_4 from Smart Lab Indonesia, KMnO4 and NaOH from Merck. The instrumentation used to analyse are the UV-Visible spectrophotometer (UV-1600 series), atomic absorption spectroscopy (Thermo Scientific, ICE 300 series), fourier transform infrared (Shimadzu, 8400S), X-ray diffraction (PANalytical XPERT-3), and transmission electron microscopy (JEOL JEM-1400 and HITACHI H-9500).

2.2. Synthesis of GO-MnO_x Composite

Natural graphite was used as the starting material to prepare GO-MnO_x by the modified Hummers method [19]. Three gram of graphite powder was added to a 95% (50 mL) H_2SO_4 and stirred for 15 min. $KMnO_4$ was slowly added to the mixture in an ice bath (<10°C) and stirred for 2 h at 35°C. The resulting mixture was added with DI water (150 mL) dropwise, while the temperature was maintained at 90°C. Under vigorous stirring the mixture was cooled down so that the temperature did not exceed 50°C. After 30 min, the suspension was added with DI water (20 ml) and centrifuged at 4000 rpm for 20 minutes. The supernatant was removed, and the suspension was added with DI water (150 mL). A solution of NaOH (3 M, 30 mL) was added slowly into the suspension in an ice bath at <5°C, and then the suspension was stirred for 15 min at room temperature. The suspension was filtered and washed with a sufficient amount of DI water, sequentially. The product was dried at an oven (<40°C) and fully dispersed in the DI water (300 mL). The resulting mixture was sonicated in a water bath sonicator operating at 100 W for 30 min. Then, the filtering and drying (<40°C) were carried out to collect the composite.

2.3. Designation of the prepared samples

GO-MnO_x 1 and GO-MnO_x 2 were synthesized by taking the same amount of sulfuric acid and potassium permanganate, as well as the concentration of NaOH. However, the sequence of centrifugation step between GO-MnO_x 1 and GO-MnO_x 2 was different. On the preparation of GO-MnO_x 2, the mixture firstly was cooled down in an ice bath (<5°C) after oxidized with $KMnO_4$ and diluted with water, and added slowly with NaOH, and then the final mixture was centrifuged.

Table 1. Synthesis parameters (the ratio of mass $KMnO_4$ and volume H_2SO_4 as well as the concentration of NaOH) and designations of the prepared samples.

Sample	Graphite (g)	$KMnO_4$ (g)	95% H_2SO_4 (mL)	30 mL NaOH (M)
GO-MnO_x 1	3	6	50	3
GO-MnO_x 2	3	6	50	3

3. Results & Discussion

3.1. Synthesis of GO-MnO_x Composite

The modification of surface and lateral dimension is profitable strategy to improve the physicochemical properties of GO-MnO_x materials. In this research, we prepared GO-MnO_x composite via Hummers method [19] with modification, as ilustrated in Figure 1 and showed in Figure 2. By permanganate and strong acid, graphite was oxidized and several graphitic structures are broken up into smaller fragments [29,30]. The specific oxidizing agent species attacking graphene layers is not studied systematically. Several researches reported that the green-colored dimanganese hepatoxide (Mn_2O_7) was oxidizing agent which can be isolated from $KMnO_4/H_2SO_4$ solution [12]. Manganese salts and the highly reactive dimanganese hepatoxide can be reduced with the centrifugation process [31]. This step also was carried out to acquire GO with high purity. In contrast, our process intended to convert the residual Mn into value-added materials. Here, our method investigated the effect of the addition of NaOH before and after the centrifugation step toward the number of Mn on GO-MnO_x composite. As seen in Table 2, GO-MnO_x 1 has lower concentration of Mn than GO-MnO_x 2, although on same condition (see experimental section). In GO-MnO_x 1, the residual Mn without oxidation has been reduced with supernatant caused by centrifugation process. Therefore, the number of Mn precipitated by NaOH was diminished. The number of Mn on composite can be seen in Table 2.

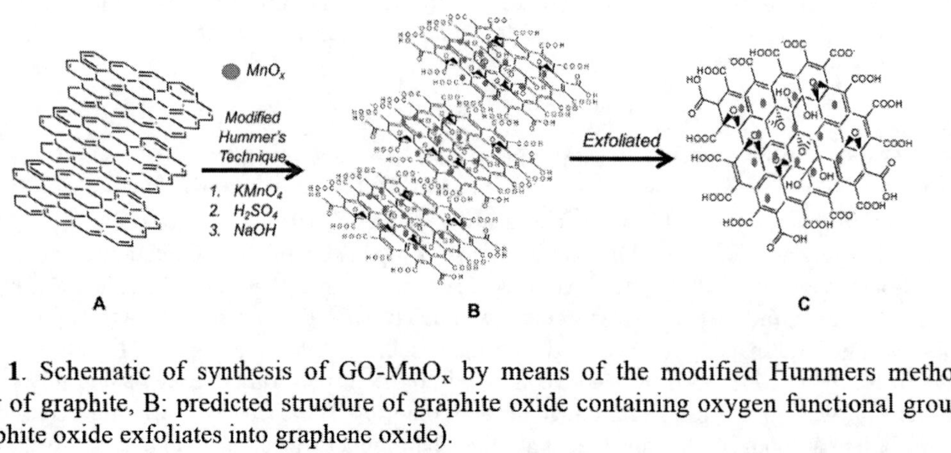

Figure 1. Schematic of synthesis of GO-MnO_x by means of the modified Hummers method (A: structur of graphite, B: predicted structure of graphite oxide containing oxygen functional groups, C: the graphite oxide exfoliates into graphene oxide).

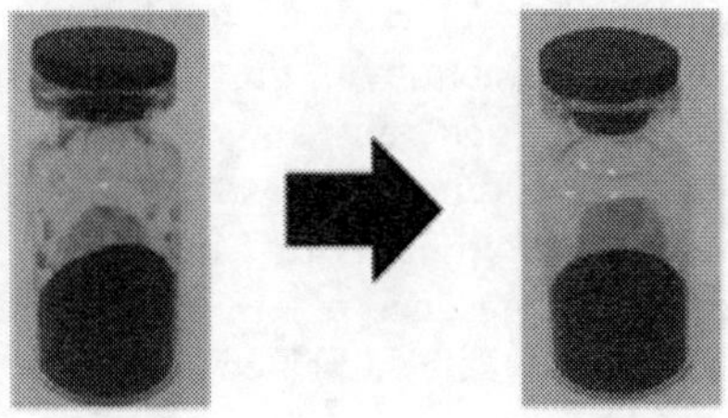

Figure 2. The image of graphite as starting material (A) and GO-MnOx as the product (B) in this research. The two materials have no difference significantly.

Table 2. The concentration of Mn on GO-MnO_x analyzed by AAS.

Sample Code	Mass of Sample (g)	Concentration of Mn (μg/g)
GO-MnO_x 1	0.50001	1.4231
GO-MnO_x 2	0.50003	2.0958

The reduction of lateral size of graphite oxide sheets to graphene oxide (several layers) could be attained via sonication process because the surface chemical properties of GO sheets changed can be negligible [32]. During sonication, the exfoliation process is occurred repeatedly [33]. As seen in XRD pattern (**Figure 5**), the low intensity of GO-MnO_x showed that graphite was exfoliated to form GO-MnO_x with silghtly layers. The formation and propagation of cracks which are catalysed by the presence of sp^3 defects in the basal of graphene lattice during sonication induced the physical breakdown of GO sheets [34]. As demonstrated in the following sections, the effect of sonication resulted in the acquisition of GO-MnO_x which differed by their surface chemical properties and crystal structure (as seen in the next section).

3.2. Characterization of GO-MnOx composite

The FTIR spectra of graphite and GO-MnO_x are given in Figure 3. IR spectra of graphite is similar to IR spectra of graphite in the reference [35]. The IR spectrum of GO-MnO_x showed characteristic peaks at ~1050, ~1200, ~1380, ~1700, and ~3400 cm^{-1}, ascribed to C–O–C (alkoxy), O=C–O (epoxy), C–OH (carboxyl), C=O (carbonyl), O–H (hydroxyl) stretching vibrations, respectively [36,37]. The peak at ~1570 cm^{-1} corresponded to C=C skeletal vibrations from unoxidized graphitic domains. Moreover, % transmittance of FTIR spectra shows the number of functional group vibrations. The presence of peak at ~500-700 cm^{-1} might be interpreted as Mn–O–C vibration [26,38]. These peaks (GO-MnO_x, 1 and 2) were similar to each other. However, the intensity of peaks was different and even disappeared in some cases. The low intensity of the peaks around ~1050, ~1380, ~3400, and ~500-700 cm^{-1} suggest that GO-MnO_x 1 and 2 might have a slightly low abundance of alkoxy, carboxyl, hydroxyl and manganese oxide, respectively. This means that composite has minor differences in changes in the chemical environment although have been added with oxidized agent or base concentration.

UV-Vis spectra for graphite and GO-MnO_x in the range 200–800 nm is investigated in water solution, as seen in Figure 4. The maximum absorbance of GO-MnO_x at λ = 230 nm is assigned to $\pi\rightarrow\pi^*$ transitions in conjugated systems [39,40]. A shoulder at λ = 290 nm often attributed to the $n\rightarrow\pi^*$ transition of carbonyl group on GO-MnO_x disappears [28], which might be due to a less numbers of conjugated systems with carbonyl. In contrast, the maximum absorbance for graphite was not detected probably due to the absence of the oxygen groups.

In the XRD pattern of graphite (Figure 5) it was clear that graphite showed a peak at about $2\theta = 26°$ with interlayer distance of d = 0.34 nm which was similar to graphite in the reference [41]. The XRD analysis of GO-MnO_x (1 and 2) showed the shifting of interlayer distance (d = 0.37 nm) at about 2θ =

24.5° which is still slightly bigger than interlayer distance of graphite. Comparing with the literature, this peaks were corresponded to the reduction of graphene oxide peaks with the fewer oxygen containing groups, whilst the peak of GO generally appeared at 2θ = 10-15° with interlayer distance about 0.8 nm [42]. This suggests that the functional oxygen groups on GO-MnO_x not fully reduced into graphene. Moreover, due to the low amounts of MnO_x in the composite, there are no signals ascribed to MnO_x species. However, the presence of MnO_x can be confirmed by FTIR. Furthermore, the broader peak of XRD with lower intensity indicated the low crystallinity (amorphous phase) of GO-MnO_x compared with graphite [43].

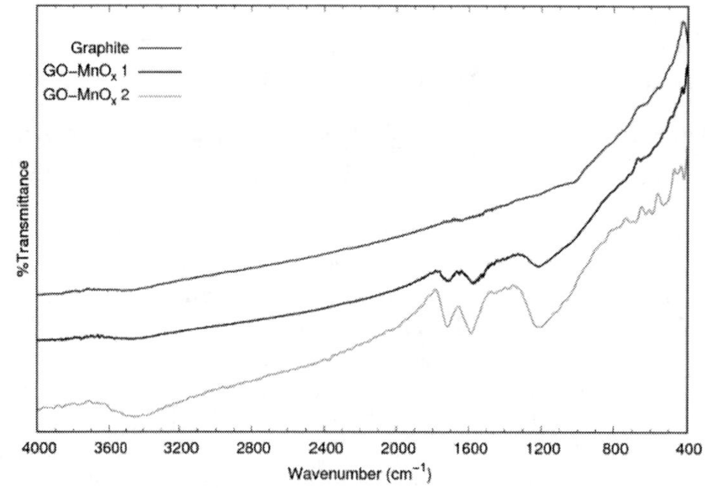

Figure 3. FTIR spectra of graphite and GO-MnO_x (1 and 2).

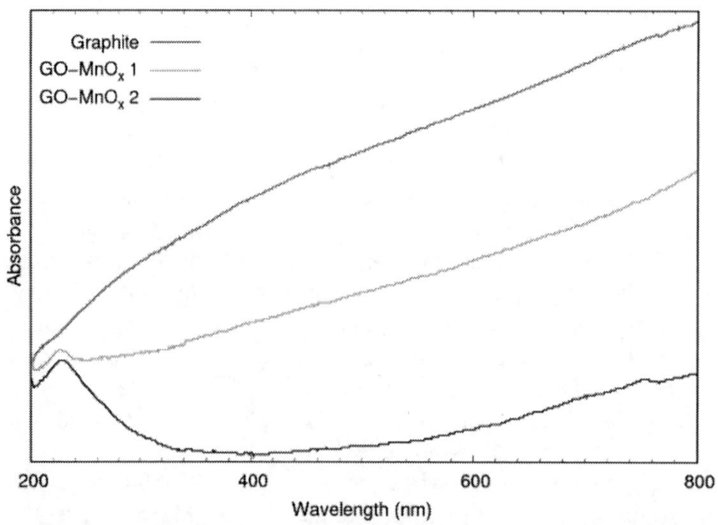

Figure 4. UV-Vis spectra of graphite and GO-MnO_x (1 and 2).

The morphology GO-MnO_x was characterized by TEM and presented in Figure 6. TEM images showed that GO-MnO_x has been exfoliated into a thin layer but not to a single layer. The images of composite showed to be crumpled, overlapped, and folded which clearly appeared in the edges might due to aggregation, as displayed in Figure 6. Moreover, due to little Mn eluted into graphene oxide, the TEM images were not confirmed the MnO_x crystal growth on the GO surface even on higher magnified of TEM image, in agreement with the obtained XRD pattern (Figure 5).

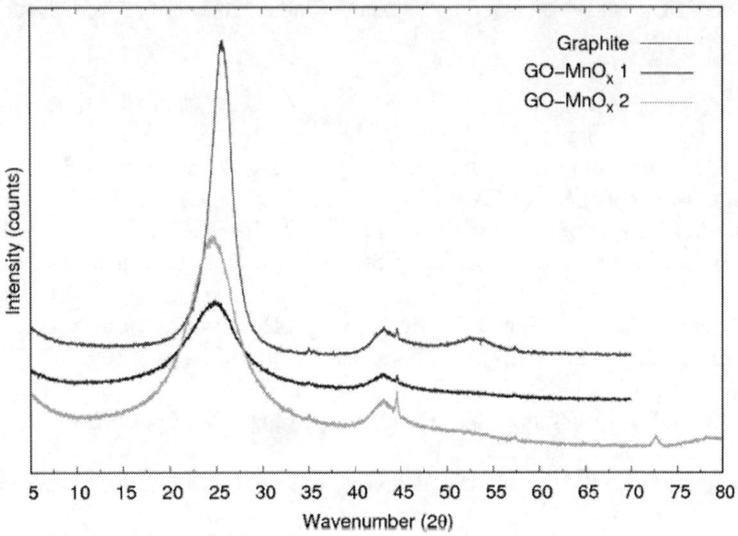

Figure 5. XRD patterns of graphite and GO-MnO_x.

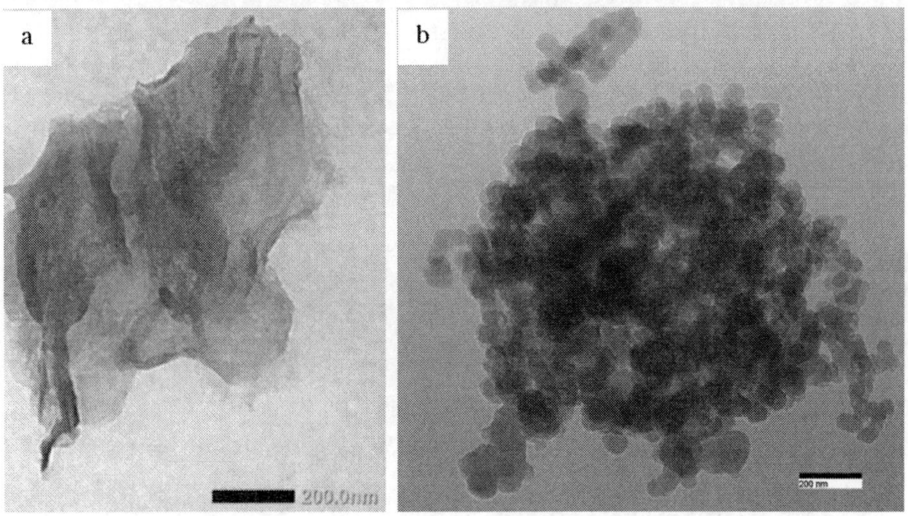

Figure 6. TEM images with magnification 20000x of GO-MnO_x 2 (a) GO with crystals suspected of manganese oxide (b), respectively.

4. Conclusions

In summary, the composites were successfully synthesized by modified of Hummer's technique. This is supported by AAS data which shows the slight number of Mn on GO-MnO_x. Moreover, XRD analysis shows the lower intensity of GO-MnO_x comparing with graphite which indicated that the several layers of graphite have been exfoliated into slight layers by sonication process.

Acknoledgements
We would like to thank the chemistry laboratory and LSIH of Brawijaya University regarding the FTIR, UV-Vis, and XRD analysis. The AAS and TEM analysis were obtained at the chemistry laboratory of State University of Malang, Gajah Mada University, and Bandung Institute of Technology, respectively.

References

[1] Saroyan H, Ntagiou D, Samanidou V and Deliyanni E 2019 *Chemosphere* **225** 524–34
[2] Xin Zhang Y, Long Guo X, Huang M, Dong Hao X, Yuan Y and Hua C 2015 *Journal of Physics and Chemistry of Solids* **83** 40–6
[3] Abdel Salam M 2015 *Chemical Engineering Journal* **270** 50–7
[4] Gu J, Fan X, Liu X, Li S, Wang Z, Tang S and Yuan D 2017 *Chemical Engineering Journal* **324** 35–43
[5] Sihaib Z, Puleo F, Garcia-Vargas J M, Retailleau L, Descorme C, Liotta L F, Valverde J L, Gil S and Giroir-Fendler A 2017 *Applied Catalysis B: Environmental* **209** 689–700
[6] Saroyan H, Kyzas G and Deliyanni E 2019 *Processes* **7** 40
[7] Wang L, Wang G, Zhang J, Bian C, Meng X and Xiao F-S 2017 *Nature Communications* **8** 15240
[8] Afzal S, Quan X, Chen S, Wang J and Muhammad D 2016 *Journal of Hazardous Materials* **318** 308–18
[9] Suzuki H and Nishina Y 2017 *Bulletin of the Chemical Society of Japan* **90** 74–8
[10] Park G, Bartolome L, Lee K G, Lee S J, Kim D H and Park T J 2012 *Nanoscale* **4** 3879
[11] Adán-Más A and Wei D 2013 *Nanomaterials* **3** 325–56
[12] Dreyer D R, Park S, Bielawski C W and Ruoff R S 2010 *Chem. Soc. Rev.* **39** 228–40
[13] An X, He X, Chen L and Zhang Y 2016 *Journal of Materials Chemistry B* **4** 6125–33
[14] Baishya P and Maji T K 2018 *International Journal of Biological Macromolecules* **115** 970–7
[15] Mao Y, Wen S, Chen Y, Zhang F, Panine P, Chan T W, Zhang L, Liang Y and Liu L 2013 *Scientific Reports* **3**
[16] Karthick S, Muralidharan S, Lee H-S, Kwon S-J and Saraswathy V 2019 *Cement and Concrete Composites*
[17] He Y, Chen W, Li X, Zhang Z, Fu J, Zhao C and Xie E 2013 *ACS Nano* **7** 174–82
[18] Hu M, Hui K S and Hui K N 2014 *Chemical Engineering Journal* **254** 237–44
[19] Hummers W S and Offeman R E 1958 *Journal of the American Chemical Society* **80** 1339–1339
[20] Lim J Y, Mubarak N M, Abdullah E C, Nizamuddin S, Khalid M and Inamuddin 2018 *Journal of Industrial and Engineering Chemistry* **66** 29–44
[21] Yamada T, Kim J, Ishihara M and Hasegawa M 2013 *Journal of Physics D: Applied Physics* **46** 063001
[22] Emiru T F and Ayele D W 2017 *Egyptian Journal of Basic and Applied Sciences* **4** 74–9
[23] Dimiev A M and Tour J M 2014 *ACS Nano* **8** 3060–8
[24] Marcano D C, Kosynkin D V, Berlin J M, Sinitskii A, Sun Z, Slesarev A, Alemany L B, Lu W and Tour J M 2010 *ACS Nano* **4** 4806–14
[25] Naghdi S, Sajjadi M, Nasrollahzadeh M, Rhee K Y, Sajadi S M and Jaleh B 2018 *Journal of the Taiwan Institute of Chemical Engineers* **86** 158–73
[26] Yang A, Zhu Y and Huang C P 2018 *Scientific Reports* **8**
[27] Zou J-P, Liu H-L, Luo J, Xing Q-J, Du H-M, Jiang X-H, Luo X-B, Luo S-L and Suib S L 2016 *ACS Applied Materials & Interfaces* **8** 18140–9
[28] Pendolino F and Armata N 2017 *Graphene Oxide in Environmental Remediation Process* (Cham: Springer International Publishing) pp 5–21
[29] Li Z, Zhang W, Luo Y, Yang J and Hou J G 2009 *J. Am. Chem. Soc.* **131** 6320–1
[30] Zhang L, Liang J, Huang Y, Ma Y, Wang Y and Chen Y 2009 *Carbon* **47** 3365–8

[31] Rodrigues A F, Newman L, Lozano N, Mukherjee S P, Fadeel B, Bussy C and Kostarelos K 2018 *2D Mater.* **5** 035020
[32] Liu S, Hu M, Zeng T H, Wu R, Jiang R, Wei J, Wang L, Kong J and Chen Y 2012 *Langmuir* **28** 12364–72
[33] Ravindra B K and Senthilkumar E 2017 *Research Journal of Chemistry and Environment* **21** 13
[34] Gonçalves G, Vila M, Bdikin I, de Andrés A, Emami N, Ferreira R A S, Carlos L D, Grácio J and Marques P A A P 2015 *Sci Rep* **4** 6735
[35] Choucair M, Thordarson P and Stride J A 2009 *Nature Nanotechnology* **4** 30–3
[36] Fathy M, Moghny T A, Abdou M M, El-Bellihi A-H A-A and Awadallah A E 2015 9
[37] Seredych M and Bandosz T J 2012 Manganese oxide and graphite oxide/MnO2 *Microporous and Mesoporous Materials* **150** 55–63
[38] Liang S, Teng F, Bulgan G, Zong R and Zhu Y 2008 *J. Phys. Chem. C* **112** 5307–15
[39] Cao H, Wu X, Yin G and Warner J H 2012 *Inorg. Chem.* **51** 2954–60
[40] Paredes J I, Villar-Rodil S, Solís-Fernández P, Martínez-Alonso A and Tascón J M D 2009 *Langmuir* **25** 5957–68
[41] Xu P, Yang Y, Qi D, Barber S D, Schoelz J K, Ackerman M L, Bellaiche L and Thibado P M 2012 *Phys. Rev. B* **86** 085428
[42] Ramesh P, Bhagyalakshmi S and Sampath S 2004 *Journal of Colloid and Interface Science* **274** 95–102
[43] Viculis L M 2003 *Science* **299** 1361–1361

IOP Conf. Series: Materials Science and Engineering **833** (2020) 012068 doi:10.1088/1757-899X/833/1/012068

Synthesis and Characterization of Chitosan Membrane as a Matrix in Drug Delivery of Curcumin

Budi Hastuti*and Alfiah Assya'adha

[1]Department of Chemistry Education, Faculty of Teacher Training and Education, Universitas Sebelas Maret

*Corresponding author: budihastuti@staff.uns.ac.id

Abstract. This study aims to find out how to synthesize the chitosan membrane contained curcumin and its characteristics. The characterization of chitosan membrane was determined use FTIR for functional group analysis, XRD for crystallinity and SEM to show the surface of morphology and cross-section on the chitosan membrane. Curcumin encapsulated into the chitosan membrane was performed by using UV-Vis spectrophotometry at λ 426 nm. The results of this study indicate that the chitosan membrane synthesized was obtained homogeneous clear yellow membrane sheets. The characterization of FTIR instrument obtained that chitosan membrane have peaks at wavenumber of 3425.58 cm^{-1} that show OH groups and 1627.92 cm^{-1} which show the presence of amine groups. While XRD characterization show chitosan membrane was semi-crystalline which was shown by a high peak intensity with angles 2θ 4.1604 and 2θ 3.8711. According to the SEM photo shows a smooth and even surface morphology. The optimum time for encapsulation of curcumin into the chitosan membrane is 60 minutes with % EE of 14%.

1. Introduction

Indonesia is a country that has rich in spices. So in ancient times when the Dutch colonized Indonesia aimed at mastering various types of spices. Spices are biological resources that are used by humans for various purposes in daily life. The benefits of spices, among others, as a food flavour enhancer, food flavouring and also as an herbal treatment. Research conducted by Hakim (2015) in Kopen, Banyuwangi, there are 27 types of spices used by the people there. Types of spices used for food flavoring: galangal, ginger, turmeric, chili, and lemongrass. The spices besides food seasonings are also used for herbal medicines [1].

Herbal medicine is a medicine made from natural ingredients using traditional processing without a mixture of chemicals. Usually herbal medicine is formed of powder, capsules, liquid (herbal medicine). One type of spice that is often used as medicine is turmeric. Turmeric (*Curcuma domestica VAL.*) is a plant that contains mainly curcuminoids and essential oils. Based on the results of research at the Research Institute for Medicinal Plants and Spices (Balittro) found that the content of curcumin in the turmeric rhizome averaged 19.92% [2]. Curcuminoids contain curcumin compounds and their offspring. The content of curcumin in turmeric is very useful for medicines.

Curcumin is an ingredient found in curcuminoids in the turmeric rhizome. The nature of curcumin as an anticancer, antitumor, antioxidant, anti-inflammatory, antimutagen and others [3]. In the field of pharmacology, curcumin is often used as a slow-release drug that is mixed in a particular matrix. Drug release is a process of release of the drug from the used ingredients. In the drug release process, the drug delivery system should ideally deliver on a controlled drug or controlled level for the desired duration. This is done to ensure the level of safety and patient needs [4]. Compounds that can be used as drugs must have certain characteristics such as absorption, distribution, metabolism, excretion, and toxicity (ADMET) [5]. However, curcumin also has the disadvantage that the level of solubility is low in the water so that it will limit its activity pharmacologically [6]. Therefore, needed a curcumin delivery matrix. This curcumin delivery matrix utilizes natural polymers.

The application of natural polysaccharides polymers is increase during recent years because of their unique biological and physical properties as carriers of drugs that can be used effectively [7]. Chitosan is a natural polymer which consist of polysaccharides that have an abundance second after cellulose and have a structure similar to cellulose. Some of polymers cellulose derivates are widely used in controlled release preparations as matrix in drug delivery systems [8]. Chitosan is a polysaccharide containing d-glucosamine and N-acetyl-D-glucosamine. Chitosan produced from the deacetylation of chitin which can form gels, films, microparticles, nanoparticles, and beads. Because of this ability, it can be used for matrix or drug release membrane [9].

Chitosan has unique properties that are biodegradable, biocompatible, and non-toxic [10]. Chitosan has polycationic properties by having amino and hydroxyl groups. The presence of chitosan active groups, namely amino and hydroxyl groups can bind toward molecules such as proteins and drugs. Chitosan can be used as a membrane material for slow-release preparations with superior properties.

Based on the background above, this research is intended to develop natural polymers as slow-release matrix drugs by making chitosan membranes. This membrane utilizes natural polymers that have unique properties that can interact with curcumin. Curcumin is encapsulated in chitosan membrane which is used as a drug delivery agent and used as a slow release which is suitable for overcoming the therapeutic problems needed.

2. Materials and Instrumentation

2.1. Materials
Chitosan (technical) powder, curcumin pro-analyst, the solutions used were of the type of pro-analyst namely glacial acetic acid 2% and 70% ethanol solution, and distilled water.

2.2. Instrumentation
The FTIR was used to see the functional groups of the membrane, SEM was used to determine the surface morphology of the membrane, to see the crystallinity use XRD, and UV-vis spectrophotometry to determine the concentration contained in the chitosan membrane.

3. Experiment Method

3.1. Preparation of chitosan membrane
0.5-gram chitosan powder dissolved in 25 ml of 2% acetic acid solution, stirred by using a magnetic stirrer for ± 1 hour. After the mixture is homogeneous then printed with printed media made from polypropylene and dried on the temperature of 50°C for ± 24 hours.

3.2. Determination of the maximum wavelength of curcumin solution and the calibration curve of curcumin solution
100 ppm of curcumin solution in 50 ml of 70% ethanol is varied to 10 ppm, 20 ppm, 30 ppm, 40 ppm, 50 ppm, 60 ppm, 70 ppm, 80 ppm, and 90 ppm. Finding the maximum wavelength of a curcumin solution by testing a 100-ppm curcumin solution at a wavelength of 300-600 nm. The results obtained wavelength of curcumin is 426 nm. After that, determine the absorbance of curcumin solution from 10 ppm to 100 ppm by using the λ maximum.

3.3. The loading of curcumin in the chitosan membrane

50 mg chitosan membrane put into a 25 ml solution of 100 ppm curcumin solution. Then, the curcumin solution stired using a magnetic stirrer and was taken 1 ml of the solution every 5, 10, 15, 30, 45, 60, and 120 minutes. Next, the concentration of curcumin solution determined using a UV-vis spectrophotometry at 426 nm.

3.4. Characterization of chitosan membrane

3.4.1. Functional group analysis with Fourier Transform Infrared (FTIR) spectrophotometer. The functional group of the membran was tested with FTIR. Samples analyzed in the form of chitosan membrane powder to characterize of functional groups in the chitosan membrane.

3.4.2. Structure Analysis of Surface by using SEM. Characterization SEM was carried out at the PPGL Research and Testing Laboratory. The analyzed sample is a 2×2 cm membrane size. the results obtained in the form of images from the front surface.

3.4.3. Analysis of crystallinity using XRD. The XRD testing was carried out at the UII Research and Testing Laboratory. Samples analyzed in the form of chitosan membrane powder that has been crushed first.

4. Result

This study has been synthesized chitosan membrane. Chitosan used in the form of brownish-yellow powder and a bit rough. While the chitosan membrane produced in the form of a clear yellow layer is presented in Figure 1.

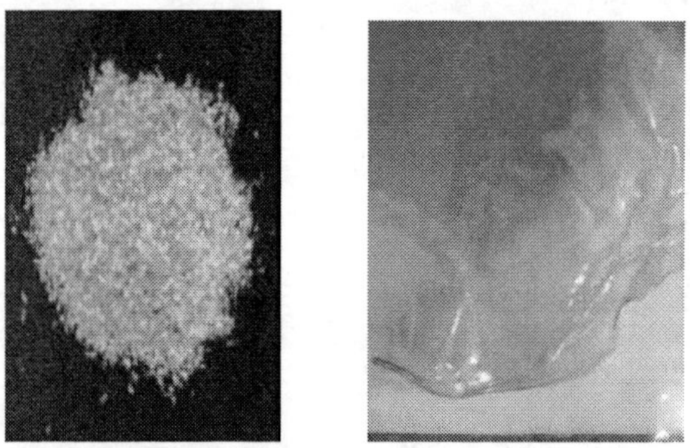

(a) **(b)**

Figure 1. (a) chitosan powder and (b) chitosan membrane.

Chitosan membranes characterized using the FTIR (Fourier Transform Infrared) instrument to determine the functional groups contained in the chitosan membrane. Based on the FTIR spectra of the chitosan membrane in Figure 2 showed that it has a wide absorption at 3425.58 cm^{-1} that indicates the vibration stretching -OH group. The band at 1627.92 cm^{-1} indicates the presence of a -COOH group and the band on 1519.91 cm^{-1} indicate of $-NH_3$ group.

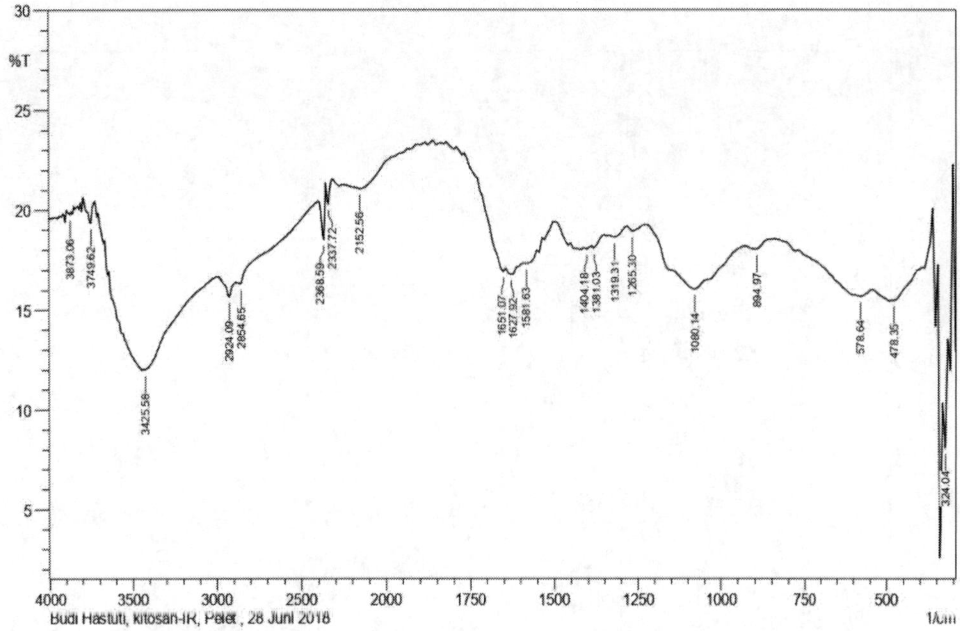

Figure 2. Results of FTIR spectroscopy of the chitosan membrane.

The results of chitosan membrane characterized by using SEM image to show the surface morphology of the membrane. In figure 3 it is showed the photo SEM of the chitosan membrane with a magnification of 500x. It can be seen that the chitosan surface has a smooth and flat surface.

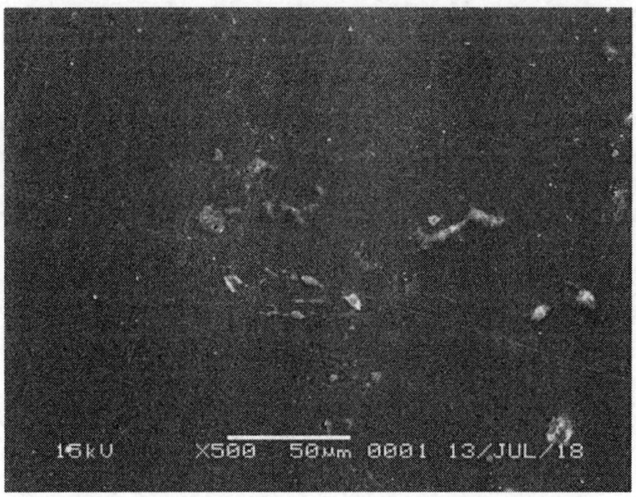

Figure 3. The SEM image surface of chitosan membrane with a magnification of 500x

Chitosan membrane that has been made is also analyzed in terms of crystallinity using XRD. Based on the XRD results shown in Figure 4, it provides information that the chitosan membrane has semi-crystalline properties. The characteristic diffractogram image of chitosan membrane has angles of 2θ on 15° and 21°.

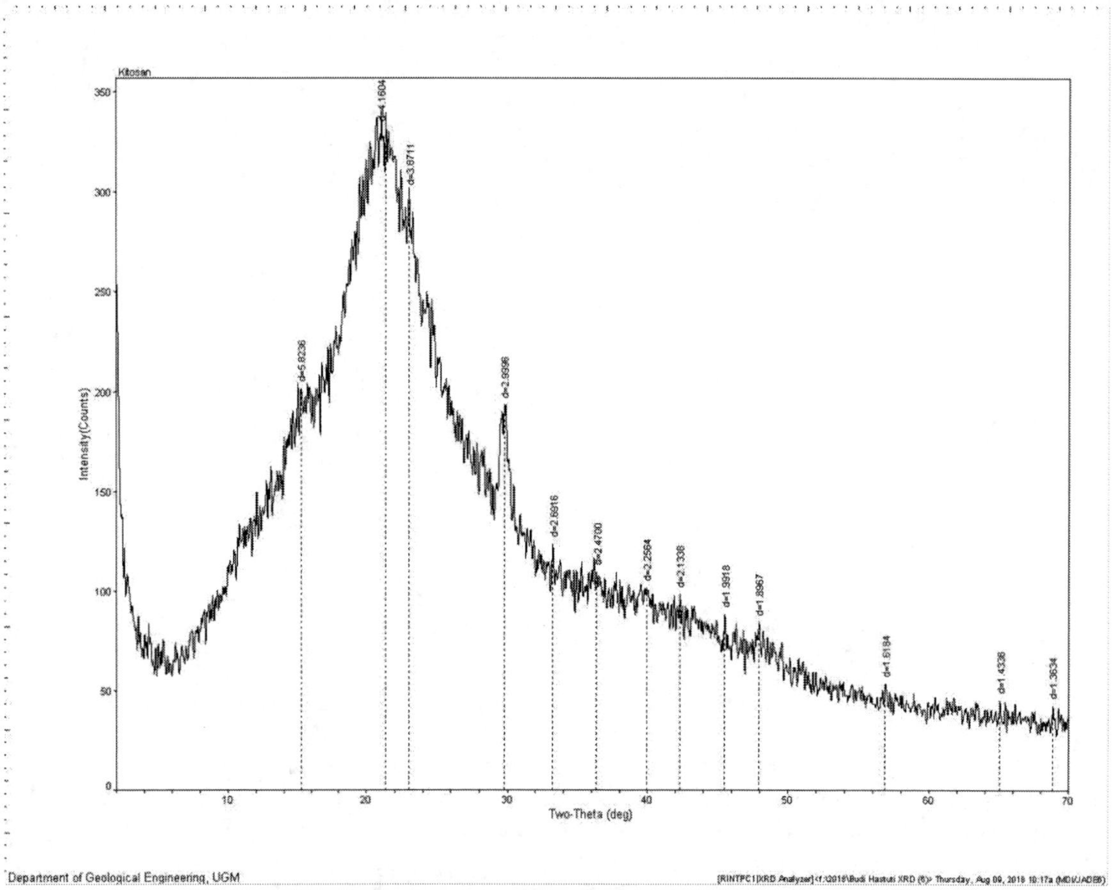

Figure 4. XRD results in chitosan membrane

Drug release is influenced by several factors, one of the most important factors being the loading time of the drug into the drug delivery matrix. The loading time determines the optimal time for the drug compound to be held in the pore of the drug delivery matrix so that the optimal drug concentration can be found in the drug delivery matrix. Loading is done by immersing 50 mg of chitosan membrane in 25 ml of 100 ppm curcumin solution. The immersion time are 5,10,15,30,45,60 and 120 minutes. Based on research conducted by Dash et al (2010) this loading has the purpose of trapping curcumin into the chitosan membrane as a conduit and stored until the release process is carried out [11]. Calculation of the percentage of Curcumin Efficiency of Encapsulation contained in the chitosan membrane using the following formula:

$$\%EE = \frac{[Curcumin]total] - [Curcumin\ loaded]}{[curcumin\ total]} \text{ x } 100\ \%. \quad (1)$$

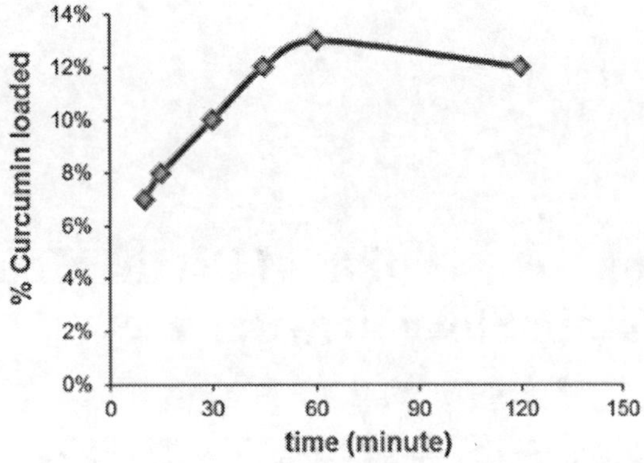

Figure 5. Graph of loading efficiency of curcumin over time

In Figure 5 above the graph shows% EE during the chitosan membrane loading of curcumin with various time variations. The optimum% EE yield of chitosan membrane loading against 100 ppm curcumin is 14% at 60 minutes. The efficiency pattern of the encapsulation of curcumin load over time shows that curcumin was initially inserted into the membrane increasing gradually. After reaching the optimal loading time, at 60 minutes with% EE is 14%, the loading of curcumin decreases, this shows that the loading of curcumin by the chitosan membrane has been saturated.

5. Conclusion

The results of this study indicate that the chitosan membrane synthesis obtained homogeneous clear yellow membrane sheets. Characterization of chitosan FTIR was obtained characteristic peaks on wave number 3425.58 cm^{-1} that shows OH groups and 1519.91 cm^{-1} which shows the presence of amine groups. While the XRD characterization of chitosan shows semi-crystalline which is shown by diffractogram image that has a high peak intensity with angles 2θ on 15° and 21°. According to the SEM image, it shows that chitosan has a smooth and even surface morphology. The optimum time for encapsulation of curcumin into the chitosan membrane is 60 minutes, with % EE of 14%.

Acknowledgment

The authors express gratitude to the Indonesian Ministry of Research and Technology Directorate of Higher Education for fundamental research grant and Universitas Sebelas Maret for the facility

References

[1] Hakim, Luchman, Jati Batoro, Kurniasih Sukenti 2015 *J-PAL.* **6** 133-142
[2] Rukmana, Rahmat 1997 *Kunyit* (Jogjakarta: Kanisius)
[3] Chattopadhyay I, Biswas K, Bandyopadhyay and Banwrjee RK 2004 *Current Science* **87** 44-53
[4] Singvhi G and Singh M 2011 *J. Pharm. Res.* **11** 77-84
[5] Akhtar F, Rizvi M M A, and Kar S K 2012 *Biotech.* **30** 310-320
[6] Sahu A, Bora U, Kasoju N and Goswami P 2008 *Acta Biomat.* **4** 1752-1761
[7] Saheb M, Fereydouni N, Nemati S, Barreto G, Johnston T P, Sahebkar A 2019 *Journal of Cellular Physiology* **50** 1-16
[8] Herdini, Darusman L K, Sugita P 2010 *Makara, Sains,* **14** 57-62
[9] Safdar R, Omar A A , Arunagiri A, Regupathi I, and Thanabalan M 2019 *Journal of Drug Delivery Science and Technology* **49** 642–659
[10] Kaban J, Bangun H , Dawolo A K , Daniel 2006 *Jurnal Sains Kimia* **10** 10–16
[11] Dash S, Murthy P, N Nath L, Chodury P 2015 *ActaPolo.Pharm. Res.***67** 217-223

Synthesis and Characterization of Pectin Membrane as a Matrix for Curcumin Sustained-Release

Budi Hastuti* **and Maulida Kurniawati**

[1]Department of Chemistry Education, Faculty of Teacher Training and Education. Universitas Sebelas Maret

*Corresponding email: Budihastuti@staff.uns.ac.id

Abstract. This study aims to synthesize pectin membranes, characterization of pectin membranes use FTIR, XRD, SEM, and to see the loading of curcumin in pectin membranes as a drug delivery system by contact time variations. Pectin membrane synthesize by dissolved 0.5 gram of pectin powder in 25 ml of 2% acetic acid solution then stirred for 1 hour. The solution is molded in membrane mold then oven for 24 hours at 50°C. Pectin membranes were characterized by FTIR to determine functional groups, XRD test to determine crystallinity, and SEM test to determine the surface morphology. The loading of curcumin in pectin membrane are carried out by contacting the pectin membrane into 100 ppm curcumin at various times of 5 to 120 minutes. Determination of curcumin concentration contained in pectin membrane use UV-Vis spectrophotometer at λ 426 nm. The results showed that the produced of pectin membrane was thin, pellucid, and homogeneous sheets. The results of FTIR characterization showed OH group at wave number 3448.72 cm^{-1} and carbonyl group at wave number 1604.77 cm^{-1}. XRD characterization result showed that the pectin membrane was amorf. SEM characterization results showed smooth and flat surface morphology. The optimum contact time in loading curcumin proses in pectin membrane was 60 minutes with a loading percentage of 10%.

1. Introduction

Curcumin is a compound extract from Curcuma Longa. It is often found in Indonesia commonly called turmeric which is used for herbs and herbal medicines. Curcumin contains bioactive compounds that are potentially used as drugs. Many studies have suggested various benefits of curcumin, including anti-cancer drugs, anti-inflammatory, antioxidant, antibacterial, antineoplastic, antiproliferative, etc. [1]. However, curcumin has several disadvantages including poor solubility and absorption, rapid metabolism, low bioavailability, etc. [2]. Therefore, it needs a matrix that can load curcumin and then release it according to the absorption target in order to function optimally.

Sustained-release designed to release a drug in the body slowly over an extended period of time [3]. Moreover, sustained-release is designed to release a therapeutic dose proper initial drug (loading dose) followed by drug release constant and slower. The speed of drug release is designed in such a way that the amount of medicine lost from the body due to the elimination is constantly replaced. The advantage is drug spread evenly into the blood without repeated dosing.

Pectin is a complex polysaccharide that has a main component of esterified D-galacturonic acid. Pectin is the main constituent of the cell wall of higher plants [4]. Pectin is not soluble in ethanol or other organic solvents, but it is soluble in water. The use of pectin as a matrix drug sustained release is based on its ability to form a gel. Pectin gel formation is caused by hydrogen bonds between hydroxyl groups around molecules and between carboxyl groups of pectin molecules [5]. Otherwise, pectin is a biopolymer which still exists in the digestive system and then degraded by the pectinolytic enzymes in the colon. So that pectin is suitable as a drug delivery matrix that targets the colon [6]. This research aims to synthesize pectin membranes, characterization of pectin membranes use FTIR, XRD, SEM, and to see the loading of curcumin in pectin membranes as a drug delivery system by contact time variations.

2. Materials and method

2.1. Materials
Pectin powder, Acetic Acid 2% (Mallincord), Ethanol 70% (Mallincord), Curcumin (Merck), and Distilled water.

2.2. Method

2.2.1. Synthesis of Pectin Membrane. 0.5-gram pectin powder dissolve in 25 ml acetic acid 2% then stirred for an hour till homogeneous. Gel pectin printed in polypropylene container then oven at 50°C for 24 hours to form pectin membrane. After that, pectin membrane characterized using FTIR, XRD, and SEM.

2.2.2. The Loading of Curcumin in Pectin Membrane. 50 mg of pectin membrane was immersed in a 25 ml curcumin solution 100 ppm then stirred. The percentage of loading was observed in time intervals of 5, 10, 15, 30, 45, 60, and 120 minutes. The concentration of curcumin loaded on pectin membrane was measured by UV Vis Spectrophotometer at λ 426 nm.

3. Results and discussion
Pectin membranes are prepared by dissolving pectin powder in 2% acetic acid solution without heating, then printed in a polypropylene container and dried with an oven. Drying in the oven aims to evaporate the solvent, acetic acid, so the remains is pectin in the forming of water free membrane. The pectin membrane produced is clear, flexible and not easily torn as shown in Figure 1a.

The procedures of sustained materials or drug sustained release is focused on the plate (film) model. The loading of curcumin into the pectin membrane by submersion 100 ppm curcumin solution in 70% ethanol into the pectin membrane for 2 hours. The dissolution of curcumin in 70% ethanol has the reason that curcumin can dissolve completely in 70% ethanol. Percentage of loading was observed in time intervals of 5, 10, 15, 30, 45, 60, and 120 minutes. The loading has the purpose of trapping curcumin into the pore of the delivered material and store it until the releasing process is carried out. In this study curcumin contained in the pectin membrane was yellow, thin, flexible and not easily torn that was shown in Figure 1b.

3.1. FTIR Characterization
FTIR characterization was used to know the functional group contained in the pectin membrane. The results of FTIR characterization showed OH group at wave number 3448.72 cm^{-1} and carbonyl group at wave number 1604.77 cm^{-1}. Pectin spectrum has typical absorption of a strong band between 1630-1600 cm^{-1} which is a stretching band of carboxylic ions. The wave number 1300 cm^{-1} showed COO- stretching group. The result of pectin membrane showed in Figure 2.

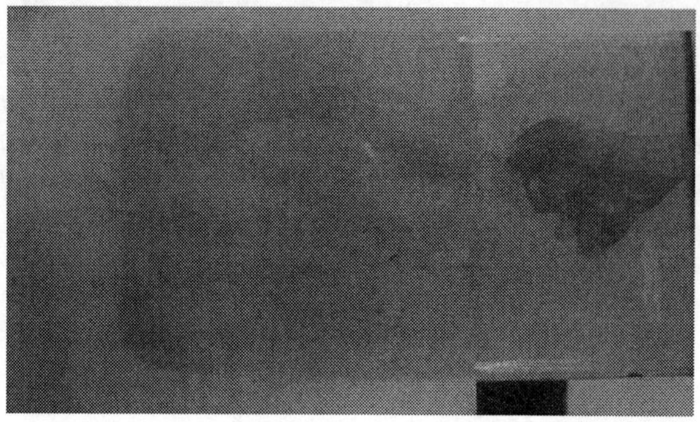

a b

Figure 1. a) pectin membrane, b) Pectin membran loaded curcumin

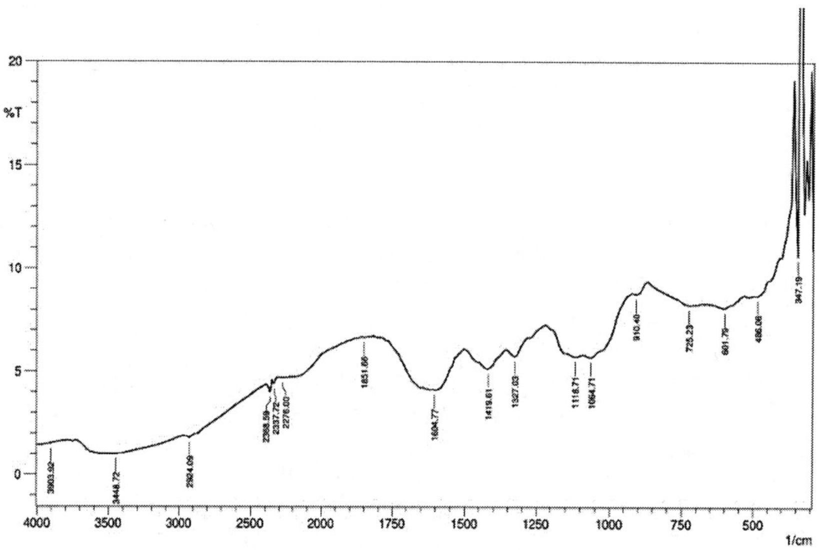

Figure 2. FTIR Characterization of Pectin Membrane

3.2. XRD Characterization

Pectin membrane was also analyzed in terms of crystallinity using XRD characterization. The thermogram showed that pectin membrane is semicrystalline that indicated by high peak intensity with angles 2θ at 22°. It caused that pectin membrane can load the curcumin.

3.3. SEM Characterization

Pectin membrane was analyzed by SEM characterizations in order to see the morphology of surface area which is showed in Figure 4. The morphology of pectin powder showed rough structure that distributed in the particle while membrane showed microporous smooth membrane. So pectin membrane has stable structure than pectin powder.

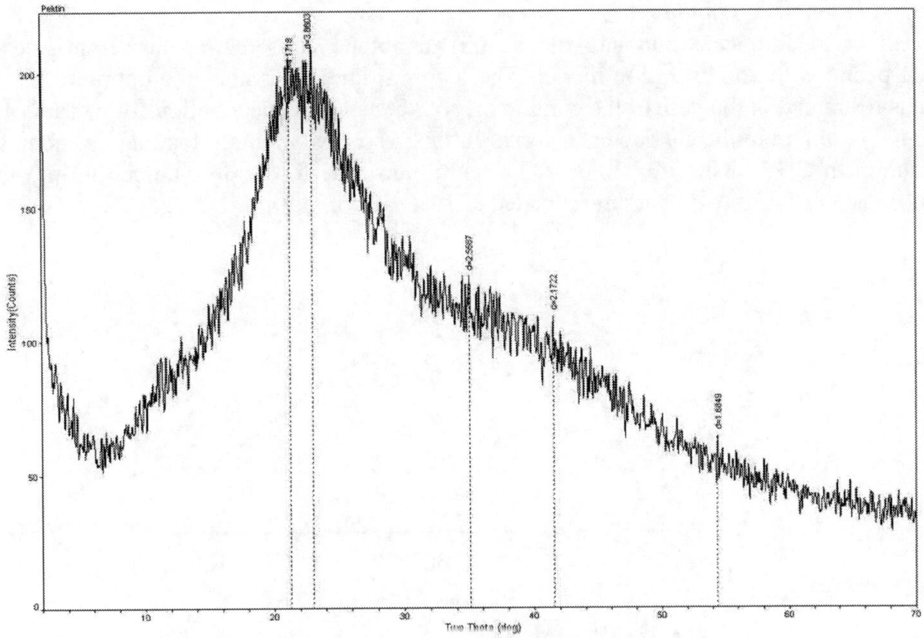

Figure 3. XRD Characterization of Pectin Membrane

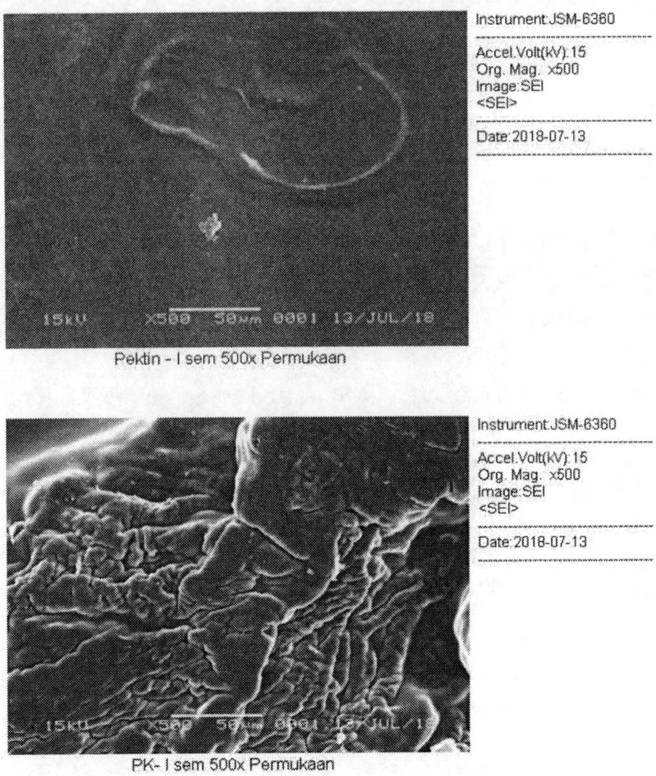

Figure 4. SEM Characterization of pectin membrane and pectin powder

IOP Conf. Series: Materials Science and Engineering **833** (2020) 012069 doi:10.1088/1757-899X/833/1/012069

3.4. Encapsulation of Curcumin

The process of loading curcumin into the pectin membrane aims to trap curcumin into the pore membrane pectin with an adsorption model. The loading time determines the optimum time of drug compounds restrained in the pore of the conductor, so that certain concentrations or masses of the drug can be found optimum in the conductor material. In this research curcumin loading by submersion 100 ppm curcumin in 70% ethanol for 2 hours. The optimum time for loading curcumin into the pectin membrane was 60 minutes with curcumin loaded at 10% shown in Figure 5.

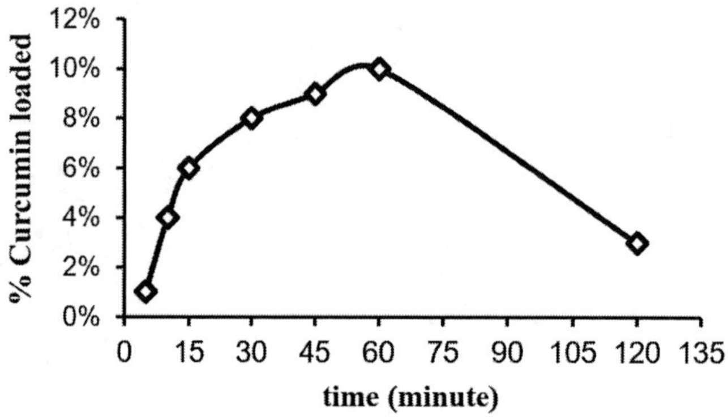

Figure 5. Graph of Loading Curcumin into Pectin Membrane

Calculation of the percentage efficiency of encapsulation using the formula:

$$\%EE = \frac{[Curcumin]total]-[Curcumin\ loaded]}{[curcumin\ total]} \text{ x } 100\ \%. \quad (1)$$

The optimum time for loading of curcumin into the pectin membrane is 60 minutes with the percentage efficiency of encapsulation is 10%. The efficiency pattern of the curcumin loading over time showed that curcumin was initially loaded gradually increased into the membrane. The loading of curcumin decreased after 60 minutes with a percentage efficiency of encapsulation is 10%. That proves the loading of curcumin by pectin membrane has been saturated.

4. Conclusion

The Pectin membrane has been successfully synthesized based on the result of FTIR, XRD, and SEM characterizations. The results of FTIR characterization showed that pectin membrane has an active functional group. There are carbonyl groups at wave number 1604.77 cm^{-1} and hydroxyl groups at wave number 3448.72 cm^{-1}. Based on XRD characterization pectin membrane was semi-crystalline and based on SEM characterization, the pectin membrane showed smooth and flat surface morphology. Curcumin can be loaded into pectin membrane with an optimum time of 60 minutes. The percentage efficiency of encapsulation is 10%.

Acknowledgment

The authors express gratitude to the Indonesian Ministry of Research and Technology Directorate of Higher Education for fundamental research grant and Universitas Sebelas Maret for the facility.

References

[1] Prasong S, Thidarat C, Naruemon K, Maneerat C and Duangporn W 2019 *J.Heliyon.* **5** e02222
[2] Mousumi K, Pritam S, Noyel G, Sharmistha C, Prasenjit M, Joydeep D and Parames C 2019 *J. Advanced Research.* **18** 162-172

IOP Conf. Series: Materials Science and Engineering **833** (2020) 012069 doi:10.1088/1757-899X/833/1/012069

[3] Susana R, Joyce T and Pedraza C 2018 *J.Nutrition & Intermediary Metabolism.* **14** 29-41
[4] Purwaningsih E 2016 *J.Kedokteran Yarsi.* **24** 203-211
[5] Hashem A, Mahnaz T, Mohammadifarc A and Hamishehkar H 2017 *J.Biological Macromolecules.* **97** 16–22
[6] Lena N and Havazelet B 2017 *J.Biological Macromolecules.* ***101*** 852–861

IOP Conf. Series: Materials Science and Engineering **833** (2020) 012070 doi:10.1088/1757-899X/833/1/012070

Synthesis of ZnO-Ag Nanocomposites trough Ultrasonication-Microwave Combination Method with Clove Leaf Oil

A Hisbiyah *[1], K Nisyak [1], Y A Prasetya[1], E D Iftitah[2] and A Srihardyastutie[2]

[1] Medical Laboratory Technology Department, STIKES Rumah Sakit Anwar Medika, Sidoarjo Indonesia

[2] Chemistry Department, Brawijaya University, Malang Indonesia

Corresponding author: yuihisbi@gmail.com

Abstract. ZnO and Ag nanoparticle have known their antibacterial activity especially their use in medical materials. In this study, ZnO-Ag Nanocomposite was synthesized by ultrasonication-microwave combination method with variation of reaction time using clove oil as their bioreductor. ZnO-Ag was prepared from ZnO Acetate as a source of ZnO and AgNO3 as a source of Ag. The crystallinity structure, average particle size, morphology, and composition of ZnO-Ag was characterized by X-Ray Diffraction, Scanning Electron Microscope, and Energy Dispersive X-ray Spectroscopy. X-ray diffraction pattern indicates that reaction time of 30 minutes have optimal synthesis results. The nanocomposite obtained consists of 43.2 % Ag nanoparticle, 17.5 % hexagonal Zincite, 14.6% Zinc Oxide, 14.5% wulfingite (deuterated), and 10.2% Zn(OH)2 with average particle size of 28.29 nm according to Scherer's equation. The result of scanning electron microscope showed that ZnO has an fiber shape and Ag has a round shape.

1. Introduction

Nanoparticle technology is becoming a very interesting technology to be developed. Nanoparticles are particles of 1-100 nm in size [1]. This material has broad uses because of its small particle size, one of which is the Ag and ZnO nanoparticles (AgNPs and ZnONPs). AgNPs has broad applications in the field of biomedicine, such as antibacterial [2], anticancer [3], antifungal [4], and antiviral [5]. However, the use of AgNPs with high concentrations can cause side effects because of its high toxicity [6]. Therefore, AgNPs can be combined with materials that have high biocompatibility with low toxicity on cell. ZnONPs is a metal oxide nanoparticle which has low toxicity and high biocompatibility [7], low cost, easily synthesized, and has almost the same uses as AgNPs in the biomedical field [8]–[10]. ZnO can improve some Ag capabilities, for example, increased antibacterial activity [11], photocatalytic activity so that several methods of making ZnO-Ag nanocomposites are beginning to be studied by researchers.

Research on the ZnO-Ag nanocomposite synthesis method that is environmentally friendly and economical continues to be developed until the maximum synthesis results are obtained. ZnO-Ag can be synthesized by natural materials as its bioreductor [12]. Plant extracts that have hidroxyl group (-OH), carboxyl group (-COOH), double bond, or other electron rich functional group can give a role as bioreductor [13]. Clove leaf oil is one of the essential oils from clove tree (*Syzigium aromaticum*) which is abundant in Indonesia [14]. The main content of clove leaf oil is eugenol at 60-68% [14],

[15]. Clove leaf oil has high antioxidant activity[16]. Therefore, clove oil is used as a bioreductor in the synthesis of ZnO-Ag nanocomposites in this research. So far, the method used for ZnO-Ag synthesis is conventional heating, but this method consumes time and energy. So this research uses the ultrasonication-microwave combination method for the synthesis of ZnO-Ag nanocomposites. The microwave method is a nanoparticle synthesis method that utilizes microwaves to heat the material so that reactions occur. The advantages of this method are energy efficiency, high reaction rates, rapid volumetric heating, high reaction rates, size and shape control by tuning reaction parameters, and energy efficiency [17][18]. Meanwhile ultrasonication method use ultrasonic radiation to disperse the molecule [18]. This paper was reporting the optimal reaction time in ZnO-Ag synthesis by the microwave method using clove leaf oil. The ZnO-Ag nanocomposite was characterized by X-Ray Diffractometer (XRD), Scanning Electron Microscope and Energy Dispersive X-ray Spectroscopy (SEM-EDX), while clove leaf oil was analyzed for phytochemical cost using Gas Chromatography-Mass Spectroscopy (GC-MS). XRD data analysis using Origin Pro and Match software.

2. Experimental Section

2.1 Material and Instrumentation

The chemicals used in this study were Silver Nitrate (Merck), Zinc acetate dihydrate (Merck), ethanol (Merck), clove leaf oil (CV. Nusaroma Depok), demineralized aqua (Bratachem), ammonia (Merck). The equipment used are glasses equipment, microwave reactor (Electrolux), Oven (DGG 9053A), Analytical Balance (Ohaus px224/E), ultrasonicator. The composition of clove leaf oil was analyzed by Gas Chromatography-Mass Spectroscopy (GC-MS Shimadzu QP2010). Morfology and atomic composition was observed by Scanning Electron Microscopy-Energy Dispersive X-ray Spectroscopy (SEM-EDX HITACHI FLEXSEM 1000). Molecular phase was observed using X-Ray Diffractometer spectroscopy (XRD Panalytical X'Pert Pro). XRD pattern was analyzed by Match and Origin pro software.

2.2 Synthesis ZnO-Ag with clove oil

ZnO-Ag nanocomposite was synthesized according to previous report with modification [19][20]. A 15 mL of clove leaf oil was added to 50 mL of ethanol in a 250 mL of erlenmeyer flask. The mixture was ultrasonicated for 30 minute until homogen, and was further added with 100 mL of Zinc acetate dihydrate 3 x 10^{-3} M slowly. Furthermore, It was added with ammonia until pH 9 and was further ultrasonicated until homogen. Then, it was reacted in microwave reactor at power level of 300 W, temperature of 30°C for 30 minutes. Furthermore, silver nitrate was added to the mixture drop by drop while ultrasonication is taking place for 30 minutes. Then, it was reacted in microwave reactor with five variations of reaction time for 10, 15, 20, 25, and 30 minute. The mixture was filtered and washed with ethanol until a colorless filtrat was afforded. ZnO-Ag nanocomposite was obtained and further dried at 130°C.

2.3 Characterization of reaction product

ZnO-Ag characterization was carried out to see the composition, morphology, particle size, and the degree of crystallinity. Phase identification, particle size, and degree of crystallinity were analyzed from XRD data. The size of nanocomposite particles is determined through the Scherrer equation. The phase composition is analyzed using the Match! Version 3.8.2, the degree of crystallinity is determined through equation (2), the XRD diffraction area is determined using the Origin pro software.

$$d_{diffr} = (0{,}9\lambda) / (\beta\cos\theta) \quad (1)$$

$$\%\ \text{Crystallinity} = (\text{Area of crystalline} / \text{Total area}) \times 100\% \quad (2)$$

Furthermore, ZnO-Ag was analyzed for morphology and elemental composition using SEM-EDX.

IOP Conf. Series: Materials Science and Engineering **833** (2020) 012070 doi:10.1088/1757-899X/833/1/012070

3. Result and Discussion

3.1 Synthesis ZnO-Ag with clove oil using ultrasonication-microwave method

ZnO-Ag nanocomposite synthesis reaction is an exothermic reaction between the oxidizer with the bioreductor, ZnO acetate and AgNO3, in this study ZnO acetate and $AgNO_3$ act as oxidizing agents. ZnO and Ag synthesis process is carried out in one container or called one pot synthesis. Phytochemicals such as polysaccharides, polyphenolic compounds, vitamins, amino acids, alkaloids, terpenoids could help reducing metal ions or metal oxides to 0 valence metal NPs[21]. Clove leaf oil could worked as a bioreductor in the synthesis of ZnO-Ag nanocomposites because clove leaf oil contains several compounds that can work as Zn^{2+} ion complexing agents in solution and reduce ZnO Acetate and $AgNO_3$ to ZnO and Ag [22][23]. The clove leaf oil content was analyzed using GC-MS and the results of the analysis are shown in Figure 1. Based on the results of GC-MS, clove leaf oil consists of 71.41% eugenol; 25.96% trans-Caryophyllene; 2.63% alpha-Humulent.

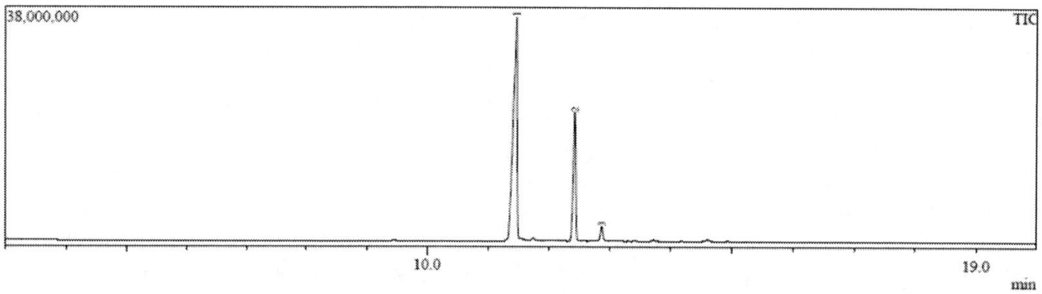

Figure 1. Total Ionic Chromatogram (TIC) of clove leaf oil.

3.2 Phase identification and Cristallinity

ZnO-Ag nanocomposites can be synthesized from ZnO Acetate and AgNO3 using a microwave-ultrasonication combination method with clove leaf oil as a bioreductor. In this research, the synthesis process was carried out with five variations of reaction time to determine the optimal reaction time. The nanocomposite was characterized by XRD to obtain a diffractogram in Figure 2. Figure 2 showed that there is a difference in peak intensity values and there is no significant 2 theta shifts in each diffractogram.

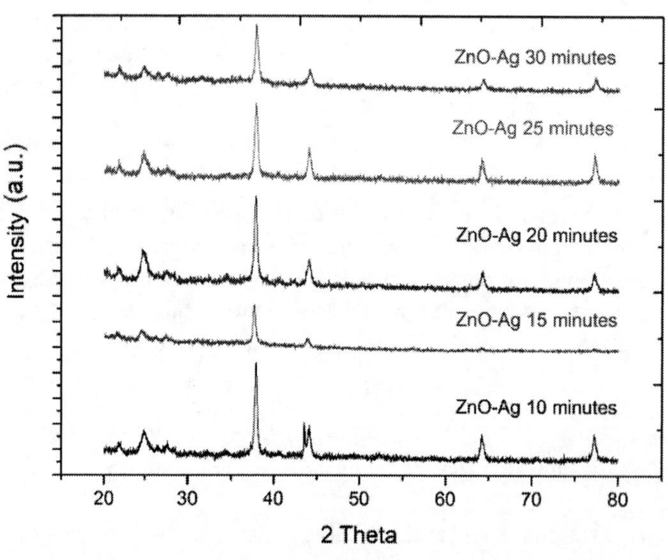

Figure 2. Diffractogram ZnO-Ag nanocomposites.

Nanocomposite phase identification was done using Match! software from the diffractogram to obtain the results that shown in Figure 3. Match! software showed that there are differences in the composition of nanocomposite elements at each reaction time variation. Globally the nanocomposite consists of five phases that is Hexagonal Zinc Oxide, Hexagonal Zinc Oxide (Zincite), Orthorhombic $Zn(OH)_2$ (Wuelfingite), trigonal (hexagonal axes) Zn $(OH)_2$, and cubic Ag. Based on the results of the phase identification, nanocomposites with the most AgNP composition were obtained at the reaction time for 20 minutes, which was 67%. While nanocomposite with the most ZnO-NP composition was obtained during the reaction time for 10 minutes, which was 44.7%. The side product of this synthesis was $Zn(OH)_2$ which is most produced during the reaction time for 15 minutes, it was consist of 23.7% trigonal (hexagonal axes) $Zn(OH)_2$ and 51.4% orthorhombic $Zn(OH)_2$ (Wuelfingite). $Zn(OH)_2$ was intermediate product of ZnO synthesis reaction. Based on these results, the composition of $Zn(OH)_2$ was inversely proportional to the composition of ZnO, so there was a possibility of an equilibrium reaction between ZnO and $Zn(OH)_2$ during the synthesis process. Hexagonal ZnO and Zincite was the best phase for antimicrobial application [24]

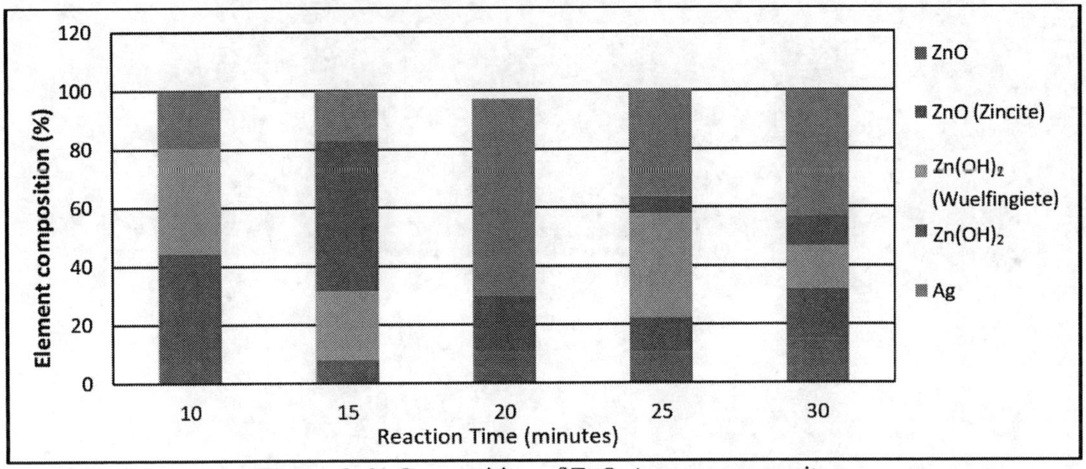

Figure 3. % Composition of ZnO-Ag nanocomposite.

The average particle size was calculated using the Scherrer equation and % crystallinity was calculated using the origin pro and Ms. Excel software. The results of the calculations are shown in Table 1. Nanocomposites with the smallest average particle size were obtained at a reaction time of 20 minutes, which was 19.66 nm. Whereas nanocomposite with the largest average particle size was obtained at a reaction time of 10 minutes, which was 36.44 nm, so the five nanocomposites meet the nanoparticle size criteria, which is 1-100 nm. The five nanocomposites have a low degree of crystallinity. The highest degree of crystallinity as much as 28% was obtained at the reaction time of 20 minutes, while the lowest degree of crystallinity of 15% was obtained at the reaction time of 15 minutes.

Table 1. Average particle size of ZnO-Ag nanocomposite.

No	Time (minutes)	Average particle size (nm)	Cristallinity (%)
1	10	36.44	23
2	15	22.64	15
3	20	19.66	28
4	25	32.64	27
5	30	28.29	20

3.3 Morphology of ZnO-Ag

Morphological studies were carried out to determine the shape of nanocomposites and the position of Ag and ZnO in nanocomposites. Morphological studies were carried out through SEM-EDX analysis on ZnOAg-NP10, ZnOAg-NP20, ZnOAg-NP30. The selection of these three nanocomposites was based on the % ZnO and Ag compositions that were successfully synthesized. Figure 4 showed the morphology of the three nanocomposites consisting of round and fiber shapes. The EDX was used to determine the location of the elements Zn, Ag, and O in the nanocomposite. The EDX results showed that Ag (round shape) can be deposited into ZnO (fiber shape). Based on Figure 4, Ag-NP in ZnOAg-NP20 is more evenly distributed on ZnO-NP with smaller particle sizes than ZnOAg-NP10 and ZnOAg-NP30. The AgNP particle was aggregated in ZnOAg-NP30 so that the average particle size is greater than the other two nanocomposites.

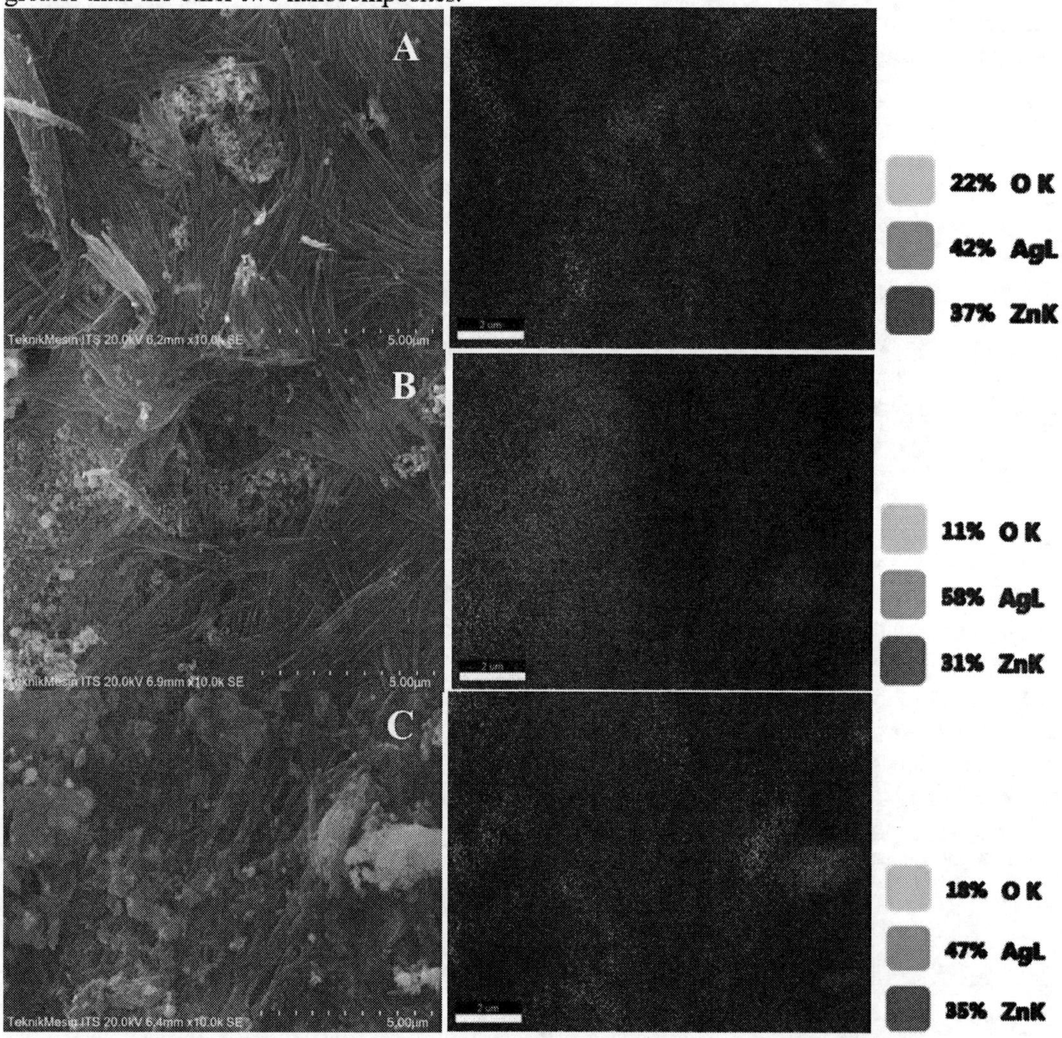

Figure 4. SEM of: (A) ZnOAg-NP10; (B) ZnOAg-NP20;(C) ZnOAg-NP30.

4. Conclusions

Ag-ZnO nanocomposite was synthesized by ultrasonication-microwave method using clove leaf oil at different reaction time 10, 15, 20, 25, 30 minutes. Characterization of ZnO-Ag nanocomposite by XRD and SEM-EDX revealed that the best Ag-ZnO nanocomposite was obtained from 20 minute reaction time. The ZnO-Ag nanocomposite consists of 67.0 % Ag nanoparticle, 10.0 % hexagonal Zinc Oxide, 19.0% Zinc $(OH)_2$, with average particle size of 19.66 nm. The five nanocomposites have a low degree of crystallinity. Therefore, for further research, nanocomposites need to be calcined.

References

[1] Khan, I., Saeed, K., Khan, I. 2017 *Arab. J. Chem.* **12** 908-931.
[2] Qais, F. A., Shafiq, A., Khan, H. M., Husain, F. M., Khan, R. A., Alenazi, B., Alsalme, A., Ahmad, I. 2019 *Bioinorganic Chemistry and application.* **5** 1-11.
[3] Al-sheddi, E. S., Farshori, N. N., Al-oqail, M. M., Al-massarani, S. M., Saquib, Q., Wahab, R., Musarrat, J., Al-khedhairy, A. A., Siddiqui, M. A. 2018 *Bioinorganic Chemistry and application.* 1-12.
[4] Paiva, K., Fonseca, G., Canteri, P., Souza, D., Oliveira, A. G., Durán, N., Nakazato, G., Cristina, M., Sérgio, R., Almeida, D., Aparecido, L. 2019 *Int. J. Food Microbiol.* **291** 79–86.
[5] Pangestika, R., Ernawati, R. 2017 *The Veterinary Medicine International Conference.* 536–548.
[6] Asharani, P. V, Low, G., Mun, K., Hande, M. P., Valiyaveettil, S. 2009 *ACS Nano.* **3** 279–290.
[7] Nohynek, G. J., Antignac, E., Re, T., Toutain, H. 2010 *Toxicol. Appl. Pharmacol.* **2** 239–259.
[8] Sirelkhatim, A., Mahmud, S., Seeni, A. 2015 *Nano-Micro Lett.***7** 219–242.
[9] Bisht, G., Rayamajhi, S. 2016 *Nanobiomedicine.* **3** 1-11.
[10] Linhari, R. C., Pietro, R., Chiavacci, L. A. 2019 *Colloids Surfaces B Biointerfaces.* **177** 440-447
[11] Ismail, M. M., Ibraheem, L. Q., Marzoog, T. R., Ahmed, D. S., Humadi, M. D. 2015 *Chemistry and Material Research.* **7** 57–65.
[12] Zare, M., Namratha, K., Byrappa, K. 2018 Int. J. Sci. Res. Sci. Technol. **4** 1636-1640.
[13] Hasan, S. 2015 *Research Journal of Recent Sciences.* **4** 7–10.
[14] Kapahang, A. 2017 *Journal of Medicinal Plants Studies.* **5** 27–32.
[15] Barroso, Â. G., Deans, S. G., Dorman, H. J. D., Figueiredo, A. C. 2000 *Flavour Fragr. J.* 12–16.
[16] Pérez-rosés, R., Risco, E., Vila, R., Peñalver, P., Canigueral, S., Pérez-rosés, R. 2016 *J. Agric. Food Chem.* **64** 4716-4724.
[17] Grewal, A. S. 2014 *Int. Res J Pharm. App Sci.* **3** 278-285.
[18] Taylor, P., Pholnak, C., Sirisathitkul, C., Danworaphong, S., Harding, D. J. 2014 *Ferroelectrics.* 37–41.
[19] Azizi, S., Mohamad, R., Shahri, 2017 *Molecules.* **22** 1–13.
[20] Pinjari, D. V, Pandit, A. B., Mhaske, S. T. 2016 *Indian j. Chem. Technol.* **23** 221–226.
[21] Agarwal, H., Kumar, S. V., Rajeshkumar, S. 2017 *Resour. Technol.* **3**, 406–413.
[22] Rajakannu, S., Shankar, S., Perumal, S., Subramanian, S. 2015 *Int. J. Curr. Microbiol. App. Sci.* **4**, 944–952.
[23] Zare, M., Namratha, K., Alghamdi, S., Hussein, Y., Mohammad, E., Hezam, A., Zare, M., Drmosh, Q. A., Byrappa, K. 2019 *Scientific Reports.* **9** 1–15.
[24] S. Fakhari, M. Jamzad, and H. Kabiri Fard 2019 *Green Chem. Lett. Rev.***12** 19–24.

Acknowledgements

The author would like to thank the Ministry of Research, Technology and Higher Education who has provided research grant 2019 through *Penelitian Kerjasama Perguruan Tinggi* with contract number 113/SP2H/LT/DRPM/2019; 141/SP2H/LT/ MONO/L7/2019; 006/STIKESRSAM/SP-Research/III/2019. The author also thanks Dr. Elvina Dhiaul Iftitah, M.Sc. and Dr. Arie Srihardyastutie, M.Kes. from Universitas Brawijaya who provide a guidance and direction in this research.

IOP Conf. Series: Materials Science and Engineering **833** (2020) 012071 doi:10.1088/1757-899X/833/1/012071

Blending Polymer Cellulose Acetate/Polysulfone For Resistance of Protein Fouling

Dwi Indarti*, Dana Iswara Putra, Bambang Piluharto, D.Setyawan

Dept. Chemistry, Faculty of Mathematics and Natural Sciences, University of Jember. Jl. Kalimantan 37, Jember 68121

*Corresponding author: indartidwi.fmipa@unej.ac.id

Abstract. Blending polymer CA/PSf was prepared using N.N-dimethylacetamide (DMAc) as solvent and PEG400 as an additive in membrane fabrication. The phase inversion method using immersion precipitation process was carried out to make the membrane CA/PSf. Membrane CA/PSf characterized by measuring the contact angle of the membrane to determine the membrane hydrophilicity. Performance of membrane is measured using water flux and rejection of dextran 70 kDa and Bovine Serum Albumin (BSA) 67 kDa. The contact angle of the membrane CA/PSf which amounted to 50° until 63°, so it can be categorized as hydrophilic membranes, because the value of the contact angle is lower than 90°. The resulting water flux membranes CA/PSf from 3.8708 $L.m^{-2}.h^{-1}$ to 10.462 $L.m^{-2}.h^{-1}$. Dextran rejection generated which is equal from 40.24% to 89.53%, while the BSA rejection of membrane from 36.21% to 81.68%. Increasing the composition of PSf produces contact angle and water flux increased, but the rejection of dextran and BSA decreased. Resilience fouling membranes CA/PSf is measured by Relative water Flux Reduction (RFR) and Fouling Resistance. RFR membranes CA/PSf which amounted from 9.445% to 32.75%, while the membrane fouling resistant CA/PSf ranged from 0.6725 to 0.9055. RFR is higher when the amount of PSf increased and indicating that their membranes are prone to fouling, it also can be evidenced by the decreased value of resistance fouling when the composition of PSf increase.

1. Introduction

Cellulose acetate membrane (CA) is a hydrophilic membrane has fouling resistant but has low mechanical properties. Polysulfone membrane (PSf) is a hydrophobic membrane that has good mechanical properties but not resistant to fouling [1]. Polymer molecular weight affects the mechanical strength of the membrane. The CA in this study has a molecular weight of 30 kDa, while PSf has a molecular weight of 35 kDa, so the mechanical strength of the PSf is stronger than that of CA. The low mechanical properties of CA membranes and fouling problems on PSf membranes can be improved by blending the two polymers in the manufacture of membranes. Performed variations of CA/PSf polymer variations 95: 5, 90:10, 85:15, 80:20 and 75:25 [2]. The CA/PSf polymer blend membrane has salt rejection of 7.54% to 20.75% and 3.98% for CA membrane salt rejection without blending with PSf. CA/PSf blending polymer can increase membrane rejection value compared to membrane without blending process. The CA/PSf membrane in the [2] study, has a salt flux of 7.69 $L.m^{-2}.h^{-1}$ to 9.93 $L.m^{-2}.h^{-1}$. Salt flux increases when the CA polymer composition decreases, because the membrane pore size increases when the CA composition decreases, so that the salt flux will increases.

CA/PSf polymer blend membrane added polyethylene glycol (PEG) additive which functions as a pore forming agent on the membrane. The [3] study stated that CA / NMP membranes without PEG produced 29.5 $L.m^{-2}.h^{-1}$ water flux and Human Serum Albumin (HSA) rejection of 87%. The addition of PEG400 was 5% and 10% resulting in water flux of 164.4 $L.m^{-2}.h^{-1}$ and 252 $L.m^{-2}.h^{-1}$ and BSA rejection of 70% and 45%. The CA/PSf blending polymer in this study aims to produce membranes with good mechanical properties and resistant to membrane fouling.

Fouling is the deposition of a dissolved particle on the membrane surface or membrane pores caused by the interaction of dissolved particles with membranes. Membrane fouling phenomenon is generally very easy to occur on hydrophobic membranes. [4] stated that PSf / PEG /DMAc membranes had a fouling resistance value of Bovine Serum Albumin (BSA) of 0.825. The fouling resistance value is <1 which indicates fouling on the membrane. Membrane fouling is also influenced by the hydrophilicity properties of membranes. Hydrophilicity of membranes can be seen from the value of membrane contact angle. Membranes with low hydrophilicity will be susceptible to membrane fouling.

The method for improving fouling was done by [5], namely by blending polymers in the manufacture of membranes. Blending polymer PES / CA / PVP which has Fouling Degree (FD) and Flux Attenuation (FA) values of 21% and 30%, while FD and FA values on PES membranes without blending with CA are 38 % and 49% [5]. High FD and FA values can indicate that the membrane is susceptible to fouling. Based on the research, blending polymer CA / PES on membranes is more resistant to fouling than PES membranes without the blending process with CA.

2. Methods

2.1. Materials

Polysulfone p.a. 35 kDa, cellulose acetate (39.9 wt%, MW 30 kDa), N,N-dimethyl acetamide (DMAc) = 0.98 g / mL, $NaH_2PO_4.2H_2O$ (pa), BSA 67 kDa, and dextran 70 kDa were puchase from sigma aldrich. Phenol p.a, sulfuric acid, Comasie Brilliant Blue (CBB), ethanol 95%, 85% phosphoric acid were purchase from merck, poly(ethylene glycol) 400 (PEG400) was obtain from Brataco Chemika, aluminum foil, filter paper, electric tape and distilled water.

Membranes preparation using phase inversion techniques. Content of polymers and solvents used is 25% and 75%, respectively. The polymer solution was prepared according to Table 1. The CA/PSf polymer mixture was dissolved in a solvent until homogeneous, then PEG400 was added as much as 5% and stirred the mixture until homogeneous.

Table 1. CA/PSf membrane research design.

Composition	w_{CA} (gram)	w_{PSf} (gram)	V_{DMAc} (mL)	V_{PEG} (mL)
95/5	2.85	0.15	9.18	0.53
90/10	2.7	0.3	9.18	0.53
85/15	2.55	0.45	9.18	0.53
80/20	2.4	0.6	9.18	0.53
75/25	2.25	0.75	9.18	0.53

Casting solution (dope solution) is left until no air bubbles in the membrane. The dope solution is poured into a glass plate membrane mold which has been given tape on the edge of the glass plate to adjust the thickness of the membrane. Next, a glass plate that immersed into a coagulation tub containing distilled water, so that a membrane will form.

2.2. Hydrophilicity and morphology of membranes

The hydrophilicity test of CA/PSf polymer blend membrane with various variations can be determined by the contact angle of membrane. A meter contact angle with the help of image J. software was used

IOP Conf. Series: Materials Science and Engineering **833** (2020) 012071 doi:10.1088/1757-899X/833/1/012071

to measure the membrane contact angle. Membrane morphology was observed from cross-section SEM (Scanning Electron Microscopy).

2.3. *Flux Test*

The membrane was cut in a circle according to the diameter of the ultrafiltration dead-end device. Membrane compaction was carried out before water flux testing. Compaction was carried out by inserting approximately 50 mL of distilled water into the filtration cell and tightly closed then given a pressure of 2 bars. Distilled water that pass through the membrane every mL are noted the flow time needed to obtain a constant flux. Volume flux is determined by the equation:

$$J_v = \frac{V}{A\,x\,t} \quad (1)$$

2.4. *Rejection Coefficient*

The effectiveness of the membrane can be known from the membrane rejection coefficient. Membrane rejection was measured using 70 kDa dextran solution and 67 kDa BSA at 2 bar pressure. The dextran concentrations in the retentate and permeate sections were obtained through visible spectrophotometer analysis using H_2SO_4 reagents and 5% phenol, while the BSA concentrations were analyzed using the Bradford method. Rejection was determined using the equation (2) which C_p is permeate concentration, and C_f is feed concentration):

$$R = \left(\frac{1 - C_p}{C_f}\right) x\ 100\% \quad (2)$$

2.5. *Fouling Test*

Fouling test of ultrafiltration was carried out by measuring water flux before and after filtration using BSA solution. A foulant solution and 50 mL BSA solution were put into an ultrafiltration tool set and then BSA was filtrated for one hour. Membrane that has been filtered with BSA soaked and rinsed using distilled water and then wiped the surface of the membrane using a tissue, then measured water flux in the same way using three times repetition. Membrane fouling was determined by measuring the value of water flux before and after filtration using BSA to calculate the RFR (Relative water Flux Reduction) and fouling resistance with the equation (3) and (4) [7]. J_{am} is flux before adsorption and J_{ads} is flux after adsorption.

$$RFR = \frac{J_{am} - J_{ads}}{J_{am}}\ x\ 100\% \quad (3)$$

$$Fouling\ Resistance = \frac{J_{ads}}{J_{am}} \quad (4)$$

3. Result and Discussion

The membrane in this study was made by mixing polymer cellulose acetate (CA) with a polysulfone (PSf) called a blending polymer. CA/PSf polymer blend membrane was made using a phase inversion technique, where the polymer which is soluble in its solvent will experience a change from liquid phase to solid through controlled conditions.

(a) (b) (c)

Figure 1. Preparation of polymer blend CA/PSf; (a) casting solution (*dope)*; (b) immersion precipitation; (c) membrane.

Figure 1 shows a series of processes for making CA/PSf membranes starting from the process of making dope solution (Figure 1 a). CA has physical properties in the form of white powder, whereas PSf has physical properties in the form of yellowish crystals. The resulting dope solution is then printed on a glass plate followed by an immersion precipitation process in a coagulation bath (figure 1b). Polymers must dissolve in the solvent so that the liquid-liquid de-mixing process can occur. De-mixing is the initial compaction process in membrane formation due to the exchange of solvents and non-solvents on the membrane in a coagulation bath resulting in a white membrane matrix (Figure 1c) [6].

Table 2. Physical properties of membrane CA/PSf.

CA/PSf Variation	Picture	Color	Thickness (mm)
95/5		White	0.051± 0.000577
90/10	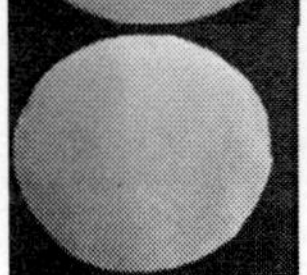	White	0.060± 0.000577
85/15	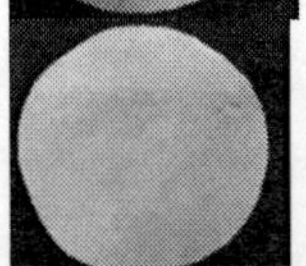	White	0.065 ± 0.00116
80/20	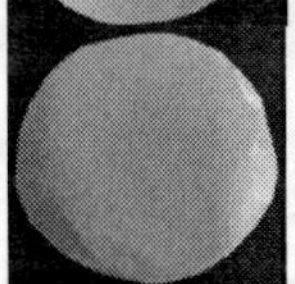	White	0.070± 0.000577
75/25	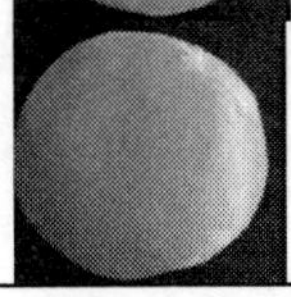	White	0.078 ± 0.00116

Table 2 shows that the CA/PSf membrane has physical properties in the form of white solids but has different thicknesses. The difference in membrane thickness in various CA/PSf compositions in a polymer blend solution can be explained by the mechanism of membrane formation. The membrane in the coagulation bath undergoes a liquid-liquid de-mixing process that is influenced by the composition of the CA/PSf in a polymer blend solution, so that when the solution is put in the coagulation bath, it causes phase separation. In this case, the process of exchanging solvents with non-solvents namely DMAc and water (non-solvent). The speed of solvent diffusion to non-solvents affects the formation of

IOP Conf. Series: Materials Science and Engineering **833** (2020) 012071 doi:10.1088/1757-899X/833/1/012071

the top layer of the membrane. The the amount increasing of PSf produces a layer of top layer on a thicker membrane. The thicker top layer causes the membrane thickness to increase.

3.1. Effect of CA/PSf composition on the contact angle (θ) and morphology of the membrane

Contact angle is an angle formed between the surface of the membrane and water droplets on the membrane. Contact angle measurements can be used to determine the hydrophilicity of the membrane. The contact angle value <90° indicates that the membrane is hydrophilic, while the contact angle> 90° indicates that the membrane is hydrophobic [8]. The effect of the composition of the CA and PSf polymers on the membrane contact angle can be seen in table 3.

Table 3. CA/PSf membrane contact angles.

CA/PSf Variation	Picture	Contact Angle
95/5	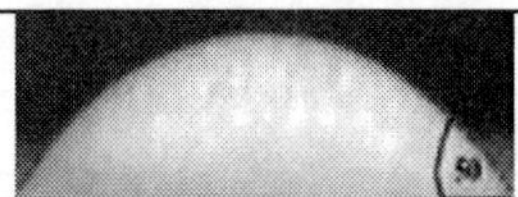	50°
90/10		52°
85/15	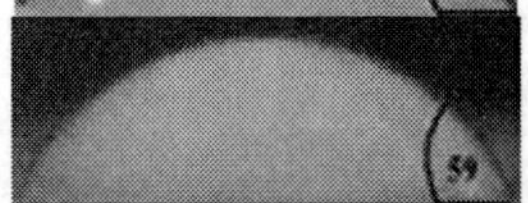	59°
80/20		61°
75/25	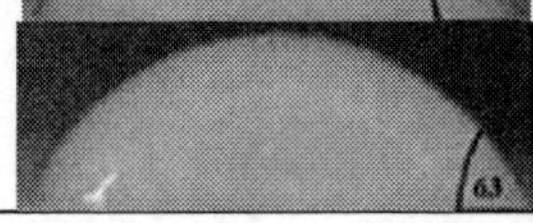	63°

Table 3 shows membranes with different CA and PSf compositions will produce different membrane contact angle values. As shown in Table 3, the more the number of PSf in the composition of CA/PSf, the greater the value of contact angle membrane. The greater the contact angle of the membrane is due to the increasing number of hydrophobic polymers (PSf). The contact angle produced on the membrane shows the effect of wettability which is the result of interaction between the membrane solids and water. The more hydrophilic the membrane, the smaller the contact angle. Increased PSf causes interactions between the membrane solids with small water, so that the wettability effect on the membrane will be small and produce a high contact angle.

The membrane contact angle can be used as a reference in determining membrane hydrophilicity. Table 3 shows membranes with variations in the composition of CA/PSf have different hydrophilicity properties. The membrane hydrophilicity is inversely proportional to the membrane contact angle. A small membrane contact angle value indicates that the membrane has high hydrophilicity properties, whereas membranes with large contact angles indicate that membrane hydrophilicity properties are low. The more the amount of PSf, the lower the hydrophilicity of the membrane, because it has a high membrane contact angle value. That is caused by the increasing number of hydrophobic polymers (PSf), so that it has a high contact angle value.

The amount increasing of PSf in addition to increasing the contact angle or hydrophilicity can also increases the membrane pore. The morphology of the membrane cross section shows the same pore shape of the three variations, namely finger like (Figure 2). Besides the larger pore size of the membrane and the shape of the finger-like pore, an increase amount of PSf causes a decrease in the uniformity of the supporting layer.

IOP Conf. Series: Materials Science and Engineering **833** (2020) 012071 doi:10.1088/1757-899X/833/1/012071

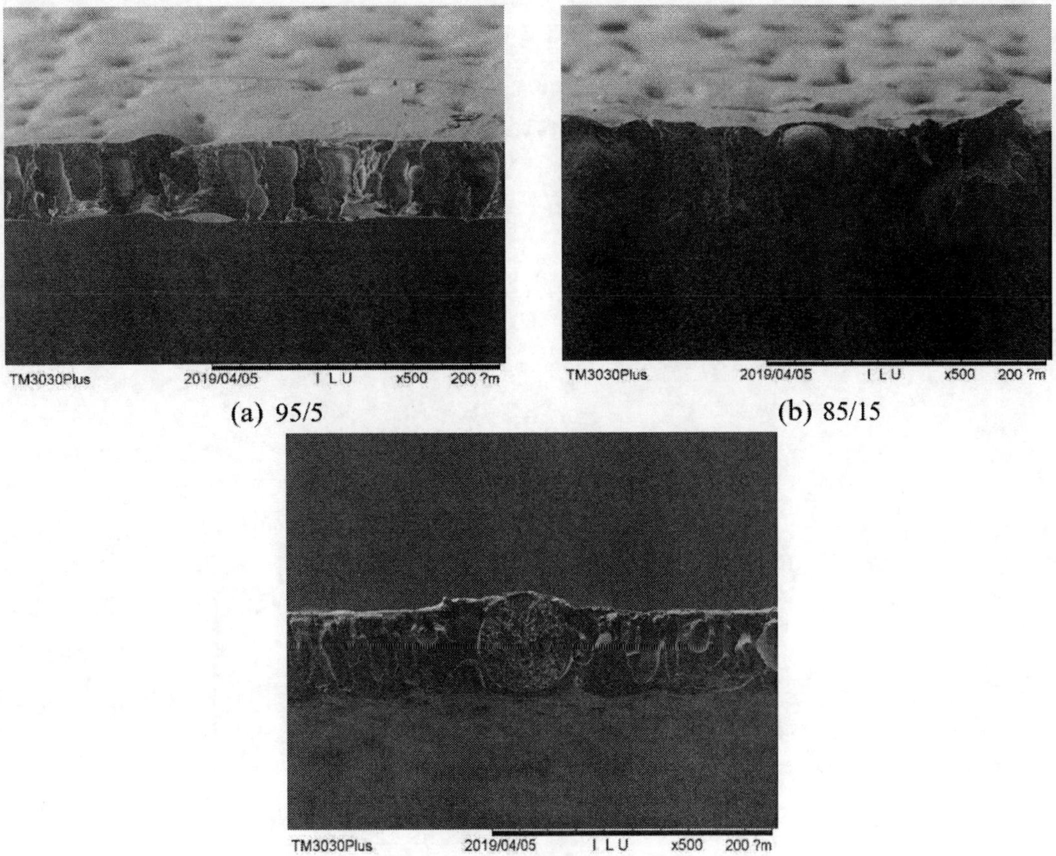

(a) 95/5 (b) 85/15

(c) 75/25

Figure 2. Morphology of membrane blend CA/PSf.

3.2. Effect of CA/PSf on water flux

Flux is the volume of feed solution that passes through the membrane per unit area of membrane and per unit time. Membranes with high flux values indicate that the membrane has good performance, conversely if the membrane has a low flux, the membrane has a poor performance. The determination of membrane flux is carried out after the membrane has compacted. At the time of compacting, the membrane flux decreases until a certain time the flux becomes constant. The membrane that has been compacted will produce membrane pores in the geometry and size that remain during the filtration process at a certain pressure. Membrane compaction was done using distilled water at 2 bar pressure.

Figure 3 shows that the more polysulfone, the shorter the compacting time. According to research by [2], the greater the amount of cellulose acetate in the membrane in the composition of CA/PSf , the smaller pores will be produced. The small pore size in the membrane makes it harder for water molecules to pass through the membrane. Water molecules that are difficult to pass through the membrane cause the rearrangement of the pore in the membrane to be longer, thus increasing the compacting time. However, when the amount of Polysulfone increases, the pore size of the membrane increases, so that water molecules will easily pass through the membrane which will reduce membrane compacting time.

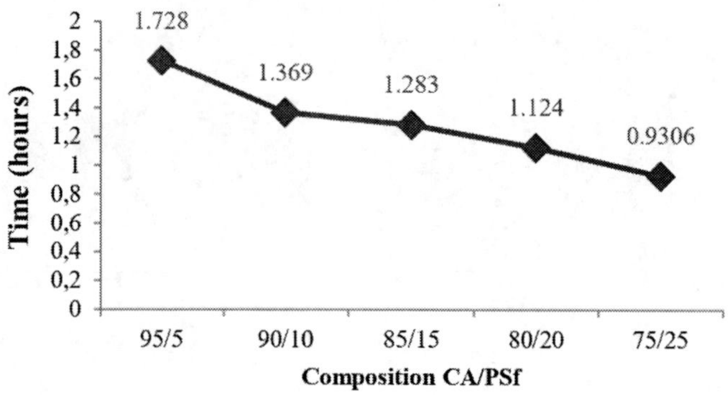

Figure 3. Membrane compacting time.

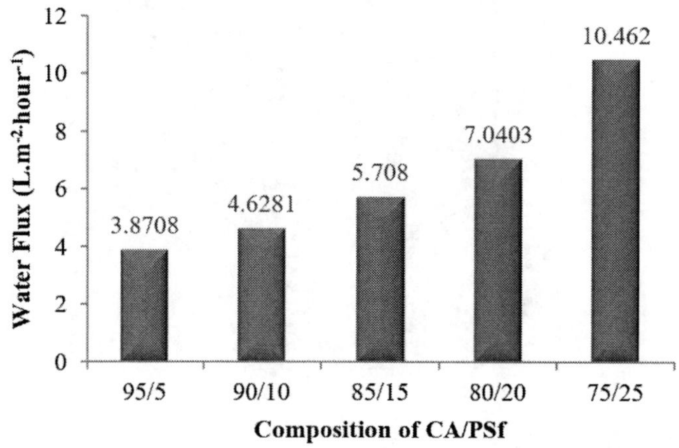

Figure 4. Effect of CA/PSf on Water Flux.

Figure 4 shows that the water flux increases with the amount increasing of Polysulfone in CA/PSf. This shows that at the same time, variations in the composition of CA/PSf affect the water flux. Water flux is affected by pore size and membrane porosity where flux will increase with the increasing of the pore size or porosity of the membrane. High water flux indicates that the greater the amount of Polysulfone, the larger the pore size of the membrane.

The pore size of the membrane will affect the flux. The increasing number of polysulfone in CA/PSf polymer blending is likely to result in larger pore sizes marked by the high-water flux produced due to the CA/PSf polymer blending. Each polymer has different solubility parameters in DMAc. The difference in solubility parameter ($\Delta\delta$) CA with DMAc is 2.14 while PSf is 0.63, so that DMAc will further dissolve PSf because it has a small $\Delta\delta$. The different solubility parameters cause the polymer dissolution process is less than perfect. CA and PSf will experience interactions during the dissolution process with DMAc. The interaction between CA and PSf produces a large aggregate. The larger polymer aggregate produces a larger membrane pore. The more the amount of PSf in the composition of CA/PSf, the more PSf can interact with CA, resulting in the formation of large polymer aggregates that are also more numerous, causing the membrane pores produced to be larger.

3.3. Effect of CA/PSf on Membrane Rejection

Membrane rejection (R) is one of the parameters to express the selectivity of a membrane. Rejection is a measurement of membrane selectivity for passing or holding a molecule [6]. R values range from

100% (when the solute in the feed solution can be held completely) and 0% (when the solute can pass completely through the membrane). Determination of rejection was done using 70 kDa dextran and 67 kDa BSA concentrating 1000 ppm at 2 bar pressure. Rejection was influenced by permeate concentration and the concentration of feed solution. The concentration of permeate and feed solution in this study were measured using a visible spectrophotometer.

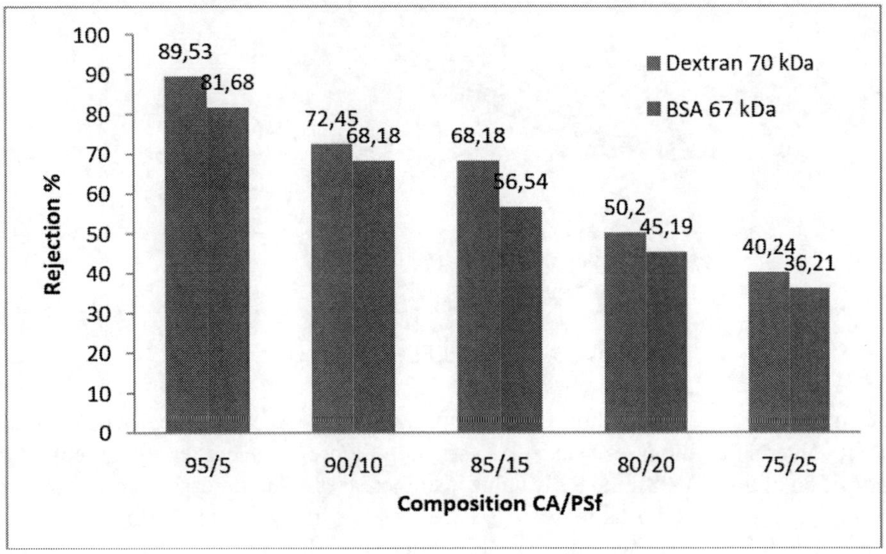

Figure 5. CA/PSf membrane rejection.

Figure 5 shows the increasing amount of Polysulfone in the composition of CA/PSf produces low rejection. Rejection of the polymer blend CA/PSf membrane is higher when the amount of Polysulfone is smaller. High rejection caused by a small membrane of pore size. Smaller pore size can hold more solutes (dextran and BSA), so particles of the solute are difficult to pass through the membrane which causes high membrane rejection. The greater amount of Polysulfone can produce a larger pore size. Larger membrane pore sizes can easily pass dissolved particles in the bait. The greater amount of solute that passes through the membrane results in lower membrane rejection.

Membrane rejection and water flux have an inverse relationship, where the larger rejection membranes, the smaller water fluxes. Rejection and flux are strongly influenced by the pore size of the membrane. Membranes with large pore sizes have small rejection and high flux values, whereas membranes with small pore sizes have high rejection and low flux [9]. Large membrane pore size will easily pass the feed particles to pass through the membrane, so that membrane rejection becomes small and membrane permeability will increases.

3.4. Resistant Fouling Membrane Polymer Blend CA/PSf

The fouling resistance of the CA/PSf polymer blend membrane can be measured based on the fouling and RFR resistance values of the membrane. Fouling resistance and RFR (relative water flux reduction) are indicators for evaluating membrane performance. RFR is the amount of reduction of water flux in the membrane that has undergone fouling with BSA. RFR can also be used to evaluate fouling resistance of membranes. A large RFR value indicates that a decrease in the value of water flux is getting bigger. A membrane with a small RFR value indicates that the membrane has good fouling resistance, whereas a membrane with a high RFR value indicates a lower membrane fouling resistance. A membrane with a fouling resistance value equal to or greater than 1 indicates that the membrane is resistant to fouling [7].

IOP Conf. Series: Materials Science and Engineering **833** (2020) 012071 doi:10.1088/1757-899X/833/1/012071

This research tested fouling resistance on membranes that had been carried out with protein filtration namely BSA 67 kDa.

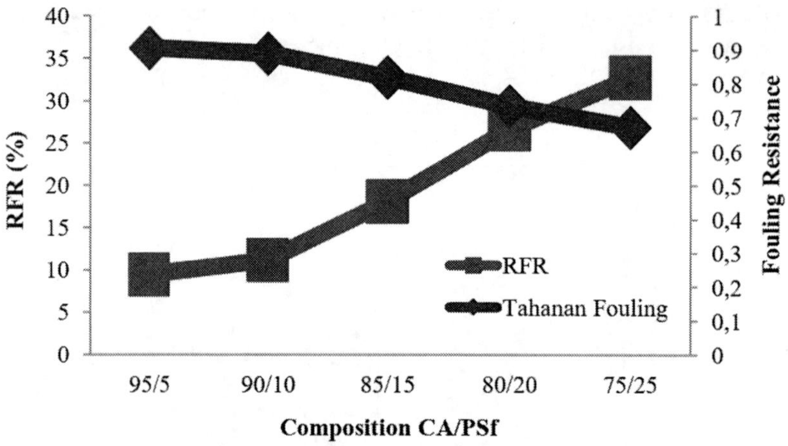

ss

Figure 6. RFR and Fouling Resistance.

Figure 6 shows the effect of variations in the composition of CA/PSf on membrane fouling resistance based on RFR values and fouling resistance. As shown in Figure 5, the higher the amount of Polysulfone in the composition of the CA/PSf, the RFR value also increases. The increased RFR value indicates that the membrane has experienced a large decrease in flux after being filtered with BSA. Membrane that has been filtered with BSA is carried out on soaking process in distilled water, so that the protein fouling on the membrane surface can be minimized. However, after measuring the water flux, the flux membrane has decreased. A decrease in flux indicates that the membrane fouled inside the pores of the membrane. Fouling that occurs in the CA/PSf polymer blend membrane happened because the membrane contains Polysulfone material, which is hydrophobic, so that the possibility for fouling of the membrane can occur. Polysulfone is a hydrophobic material, the material can adsorb protein strongly, so that during the filtration process there will be a large decrease in flux. Hydrophobic interactions between proteins and membrane surfaces can cause irreversible fouling, which can reduce the efficiency of the filtration process [8]. The decrease in water flux is caused by fouling of the membrane, so that the membrane pores are covered by feed particles, which are proteins that cause water difficult to pass through the membrane, so the flux decreases. As the amount of Polysulfone in the composition of CA/PSf increases, the RFR value also increases. An increase in RFR value indicates a membrane with a large amount of Polysulfone is more susceptible to fouling.

The fouling resistance of the membrane can also be seen from the value of the fouling resistance. Membranes with high fouling resistance values indicate that the membrane is more resistant to fouling and vice versa. As the amount of Polysulfone increases, the fouling resistance value decreases. This shows that membranes with more polysulfone are more prone to fouling, which can be measured by the high RFR value and low resistance value of membrane fouling. The membrane fouling resistance is also influenced by the membrane hydrophilicity which is determined based on the membrane contact angle values. Membranes with high hydrophilicity properties will be more resistant to fouling compared to membranes that have low hydrophilicity properties. The amount of polysulfone produces a high membrane contact angle, so that the membrane's hydrophilicity decreases causing susceptibility to be fouling as evidenced by the low resistance value of membrane fouling.

4. Conclusion

The results of the study showed the contact angle of the membrane increases with increasing number of PSf in the composition of CA/PSf , so that the hydrophilicity of the membrane is lower. The value of water flux and membrane rejection is influenced by the composition of CA/PSf in the membrane. The amount of PSf increases resulting in a greater value of water flux and rejection (dextran 70 kDa and BSA 67 kDa) getting smaller and vice versa. Increasing the amount of PSf in the membrane results in high RFR and low fouling resistance. making it vulnerable to fouling and vice versa. Membrane that is resistant to fouling has a low RFR value and high fouling resistance.

Reference

[1] Cheryan M 1998 *Ultrafiltration and Microfiltration Handbook* 2nd edition (United States America: CRC Press)
[2] Munirah N, Mazira M A, Ramlah T M and Sharifah A 2014 *Trans Tech Publications* **594-595** 281- 5.
[3] Saljhoughi E. Amirilargini M and Mohammadi T 2010. *Desalination*. **262** 72-8.
[4] Karlina. 2014. *Karakterisasi Fouling Protein Pada Membran Polisulfon Tersulfonasi (SPSf)*. (Jember: Universitas Jember)
[5] Sun Z and Chen F 2016. *Int. J. Biological Macromoecules*. **13** 21-7.
[6] Mulder M 1996 *Basic Prinsiple of Membran Technology*. 2nd edition (Dordrecht: Kluwer Academic Publisher)
[7] Susanto H, Balakhirshnan M and Ulbricht M 2006. *J. Membrane Sci* **288** 157-67.
[8] Yuan Y and Lee T R 2015 *Surface Science Technology* (USA: University of Houston)
[9] Velu S, Muruganandam L and Arthanareeswaran G 2015 *Brazlian J. Chem. Eng.* **32** 179-89.

The Structural and Optical Band Gap Energy Evaluation of TiO_2-Fe_2O_3 Composite

Rendy Muhamad Iqbal[1*], Dyah Ayu Pramoda Wardani[1], Luqman Hakim[2], Akhmad Damsyik[1], Riaa Safitri[1], Hamzah Fansuri[3]

[1]Department of Chemistry, Faculty of Mathematics and Natural Science, Universitas Palangka Raya, Kampus UPR Tunjung Nyaho, Palangka Raya 73111, Indonesia
[2]Department of Physics, Faculty of Mathematics and Natural Science, Universitas Palangka Raya, Kampus UPR Tunjung Nyaho, Palangka Raya 73111, Indonesia
[3]Department of Chemistry, Faculty of Science, Institut Teknologi Sepuluh Nopember, Kampus ITS Sukolilo, Surabaya 60111, Indonesia

Email: iqbal.rm@mipa.upr.ac.id

Abstract. Solar cells, one of which is Dye Sensitizer Solar Cell (DSSC), are promising alternative energy in the 2000s. Compared to many other energy sources, solar cell does not emit waste (clean energy). An important component that generates electricity in DSSC is photoanode. TiO_2, the most widely used materials as photoanodes in the current era, is widely used photoanode although the energy gap of the TiO_2 band is quite large (3.43 eV). This study aims to reduce the energy gap of TiO_2 band by forming composites with Fe_2O_3. Various compositions of TiO_2 and Fe_2O_3 fine powders mixtures were dispersed in methanol and stirred until homogeneous in order to synthesize the photoanode. The methanol was then evaporated to obtain a dried mixture of solids, followed by calcination to form composites. The resulted composites were characterized by XRD and Diffuse Reflectance UV-Vis Spectrophotometer. XRD analysis results showed that the composites are composed of TiO_2 and Fe_2O_3 mixtures. Diffuse Reflectance UV-Vis analysis results showed that there was a change in bandgap energy due to the mixing of TiO_2 with Fe_2O_3. The lowest band gap energy is 2.67 eV at 1: 4 ratios. The smallest and largest refractive index values calculated by the Ravindra relation are 2.0725 and 2.4286, respectively. The refractive index value is the reversely proportional of the band gap energy value, the greater the band gap energy, the smaller the refractive index value.

1. Introduction

Energy requirement is increasing every year so that in the future petroleum is no longer able to fulfil energy necessary. In particular, the demand for electricity produced by the Diesel Power Plant will be threatened if its oil will be depleted in the future. Thus, the development of alternative energy sources became the main choice by researchers in the 21st century.

Alternative energy sources can be create using potential natural resourced in Indonesia, one of which is the promising solar cells (dye sensitizer solar cell). If term of sustainability, a technology that utilizes photons from the sun does not have restrictions such as petroleum that can run out in the near future [1]. DSSC technology is very promising to be developed in areas that have high continuous sunlight intensity. Palangka Raya is one of the cities in Kalimantan and has a position close to the equator. Although it does not exactly on the equator line, the intensity of sunlight in Palangka Raya is high enough so that it has great potential in the development of DSSC as an alternative energy to produce electricity.

One of the devices in DSSC is a semiconductor-based material which called as photoanode. The material interact with photons to produce electron transfer from the valence band to the conduction band [2], as well as the formation of current electric due to holes and electron movements in the semiconductor structure. The most well-known and widely used semiconductor material in DSSC is the expensive difficult to synthesize SiGe. Some other materials have been reported to replace SiGe such as ZrO_2 [3], Cu-doped TiO_2 [4], and TiO_2 [5] with the band gap energy above 3 eV which is very active in UV light. However, the materials require more energy to produce the electron movement from the valence band to the conduction band. One way to reduce band gap energy is through the doping process. However, the process requires more energy which uses heating at a certain temperature to form a new material. On the other hand, combinations with materials that have lower band gap energy through the physical mixing process to form a composite have not been widely reported.

According to Apte *et al.* [6], Fe_2O_3 has a band gap energy of 2.2 eV. Hence, it has the potential to be combined with TiO_2 to reduce the gap energy of the pit through the process of physical mixing with alcohol dispersants and followed by the calcination process to form a composite. The making of the composite offers a new alternative to the use of visible light-based photoanodes, this is because visible light has a wider wavelength range than the UV rays (Vis: 400 - 800 nm). This is strongly supported by the location of the city of Palangka Raya which has continuous very good sunlight intensity throughout the year because it is closer to the equator. Based on the above explanation, this research will examine the structure of TiO_2-Fe_2O_3 and the effect of the TiO_2/Fe_2O_3 ratio on the gap energy of the pit.

2. Methodology

2.1. Materials

The materials that were used in this works were TiO_2 (Merck, 99%), Fe_2O_3 (Merck, 99%), and methanol (Aldrich, 99.6%).

2.2. Synthesis and Characterization of TiO_2-Fe_2O_3 Composite

TiO_2 and Fe_2O_3 were mixed in a beaker with mass ratio of 1: 1, 1: 2, 1: 3, and 1: 4. Each solid mixture with different of mass ratio was then dispersed into 50 mL methanol and stirred for 2-3 hours until homogeneous mixture was obtained. The homogeneous mixture was allowed to settle at room temperature and then decanted to separate the dispersant from the solid TiO_2-Fe_2O_3 precipitate. The precipitate was dried in an oven at 90 °C for 1 hour to vaporize the methanol which was still trapped on the surface of TiO_2-Fe_2O_3 composite particles before being calcined at 500 °C for 1 hour.

TiO_2-Fe_2O_3 composites were characterized by Panalytical X-Ray Diffraction (XRD) using a Cu-Kα light source ((λ = 1.5406 Å) which is operated at 30 mA and 40 kV. The step size of measurement was set at 0.02°. The TiO_2-Fe_2O_3 composites were also characterized by SEM-EDX to observe the morphology and elemental mapping. UV-Vis Diffuse Reflectance spectrophotometer Shimadzu-2450 was used to determine the maximum wavelength and band gap energy of the TiO_2-Fe_2O_3 composite. The band gap energy was evaluated using Equation 1, where A is a constant, h is the photon energy, and E_g is the optical band gap energy. E_g can be observed from the extrapolation of Tauc plot when the curve $(\alpha h\upsilon)^2$ as y axis = 0 [4].

$$(\alpha h\upsilon) = A\,(h\upsilon - E_g)^{1/2} \quad (1)$$

After E_g was successfully determined by the Tauc plot, the E_g value was used to determine refractive index (n) using Ravindra relation as shown in Equation 2 [7].

$$n = 4.084 - 6.02E_g \quad (2)$$

3. Result and Discussion

3.1. Structural Study

X-Ray Diffraction pattern of TiO_2-Fe_2O_3 is shown in Figure 1. According to JCPDS Card No 21-1272 for TiO_2 and JCPDS Card No 39-1346 for Fe_2O_3, the identity peaks of TiO_2 appeared at 2θ angles of around 25°, 32°, 47°, 53°, and 63° while the peaks for Fe_2O_3 are observed at 2θ angles of around 33°, 35°, 40°, 49°, and 62°. The diffraction pattern of the composite has identical profile with the diffraction pattern from both JCPDS standards which indicates that the TiO_2 and Fe_2O_3 are both present and merged to form a new composite material.

The XRD analysis of composite in several mass ratios of TiO_2:Fe_2O_3 resulting in different intensity in their diffraction pattern. As shown in Figure 1, the intensity of TiO_2 peak decreases as the amount of Fe_2O_3 in composite increases.

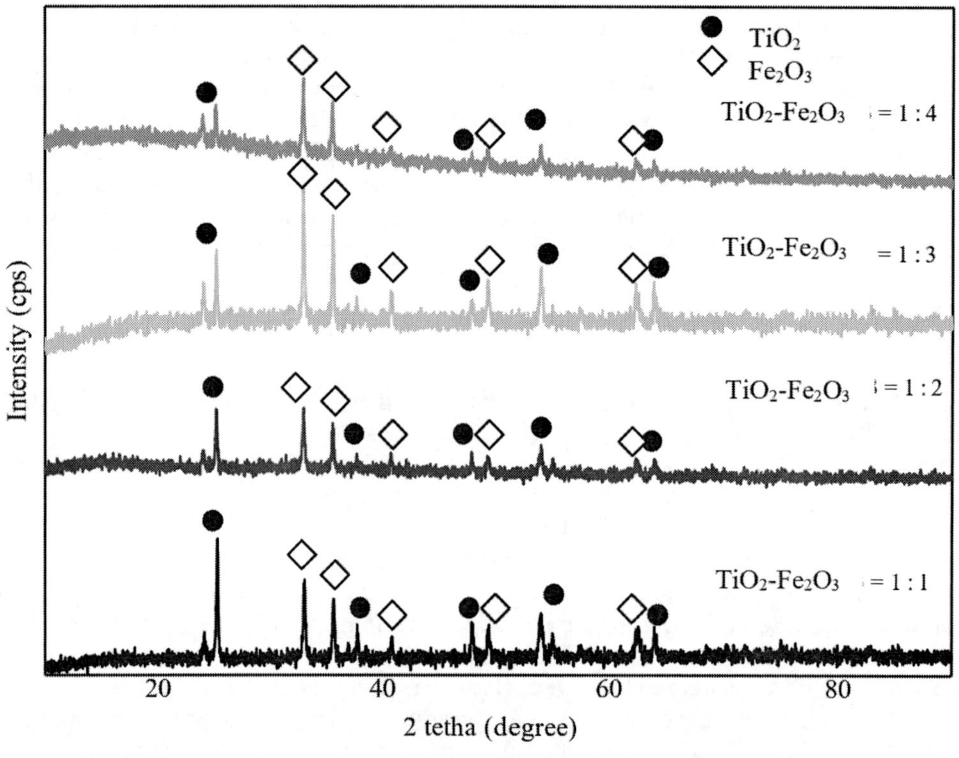

Figure 1. XRD pattern of TiO_2-Fe_2O_3 composite.

Figure 1 also shows that heating the mixture at 500°C did not change both crystal structures. The diffractogram exhibits that each oxide has their signature peaks which represented by the diffraction pattern and it can be estimated that the composite is likely having new properties compared to the starting materials [8].

Microstructural characterization was also carried out with SEM-EDX, the results are shown in Figure 2. The figure shows that TiO_2 and Fe_2O_3 materials in a ratio of 1:2 are uniformly mixed as indicated by the spreading of the same shape of all particles. In addition, elemental mapping shows the distribution of elements. Oxygen is distributed to all parts with brighter color intensity, and Fe also spread with greater intensity compared to Ti. This is due to the different composition of the material forming the

TiO_2-Fe_2O_3 composite. The intensity of Fe is higher than Ti which correspond to Fe_2O_3 present in the composite than TiO_2.

Figure 2. Microstructure and Elemental Mapping of TiO_2-Fe_2O_3

3.2. Optical properties evaluation

Optical properties is important for photoanode material which will be used in Dye Sensitizer Solar Cell (DSSc) system. The band gap energy can be determined by Tauc plot from absorption spectra, in which the band gap energy plays an important role on electron movement from the valence band to the conduction band [9]. For the material with low band gap energy, it needs electromagnetic radiation with longer wavelength [10]. In contrast, material with high band gap energy will require electromagnetic radiation with a shorter wavelength to impuls the movement of electrons. It is expressed with Equation 3, in which E is energy, *h* is planck constant, *c* is the velocity of light, and λ is wavelength.

$$E = hc.\lambda^{-1} \tag{3}$$

IOP Conf. Series: Materials Science and Engineering **833** (2020) 012072 doi:10.1088/1757-899X/833/1/012072

Based on Equation 3, energy has a value that is inversely related to the wavelength. Thus, the band gap energy is one of the parameter which interesting to study. Tauc plot of TiO_2-Fe_2O_3 with different ratio in Figure 3 shows that the composition directly influence the band gap value. Higher Fe_2O_3 content reduce the band gap of composite and also shows that the formation of TiO_2-Fe_2O_3 composites produce new band gap value. According to Riyani *et al.* [11], when TiO_2 with gap energy = 3.43 eV was combined with Fe_2O_3 (band gap energy = 2.2 eV), the band gap of the composite will decreased. At TiO_2 and Fe_2O_3 ratio of 1:1, the band gap energy is 3.24 which was lower than the pure TiO_2. The gap energy is also becoming closer to Fe_2O_3 band gap at the highest composition of Fe_2O_3 *i.e.* 2.67 eV.

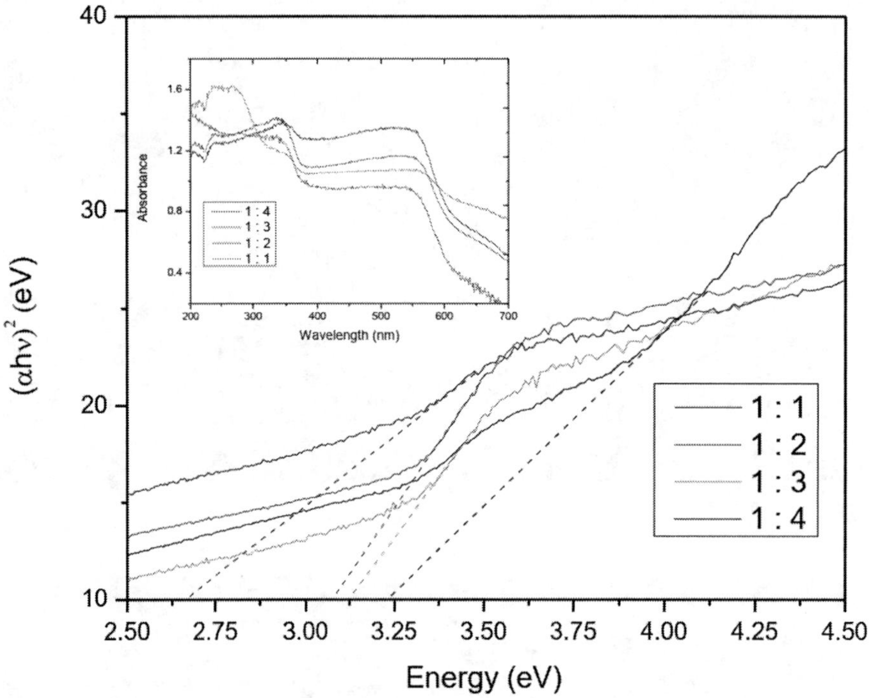

Figure 3. Optical absorption spectra and Tauc plot of TiO_2-Fe_2O_3 with different ratio.

The band gap energy influenced optical absorption properties in solid materials. The optical absorption spectra of the composite are shown in Figure 3. The peak around 200-300 nm only appears for composite with TiO_2 and Fe_2O_3 ratio of 1:1. This is due to TiO_2 to still dominate the optical properties for its composition. The UV light has a high energy and electron in the valence band of TiO_2 which make the electrons easy to move [12]. The band gap energy of composite with all ratio is shown in Table 1.

Table 1. The Band Gap Energy of TiO_2-Fe_2O_3

Material	Band Gap Energy (eV)	Reference
TiO_2-Fe_2O_3 (1:1)	3.24	This work
TiO_2-Fe_2O_3 (1:2)	3.08	This work
TiO_2-Fe_2O_3 (1:3)	3.12	This work
TiO_2-Fe_2O_3 (1:4)	2.67	This work
TiO_2	3.43	[11]
Fe_2O_3	2.2	[6]

The optical absorption of composite also appears on visible area. TiO_2: Fe_2O_3 with ratio of 1: 4 has a high absorbance due to the high content of Fe_2O_3. It will give more advantages to the composite if it is used as a photoanode which active under visible light.

The composition of composite also plays an important role to determine optical and structural properties. The different composition of composite obtained many results from some parameters such as the refractive index. Refractive index describes how fast the light travels through the material [13]. It can be determined by Ravindra relation which uses band gap energy value as one of the parameters. Figure 4 shows the relation of composition with refractive index and band gap energy.

Based on calculation by using Ravindra equation, as shown in Figure 4, the highest refractive index is 2.4286 and the lowest one is 2.0752. The refractive index decreases as the Fe_2O_3 content in composite increased. Inversely, the band gap energy increases as the Fe_2O_3 content in the composite is getting higher.

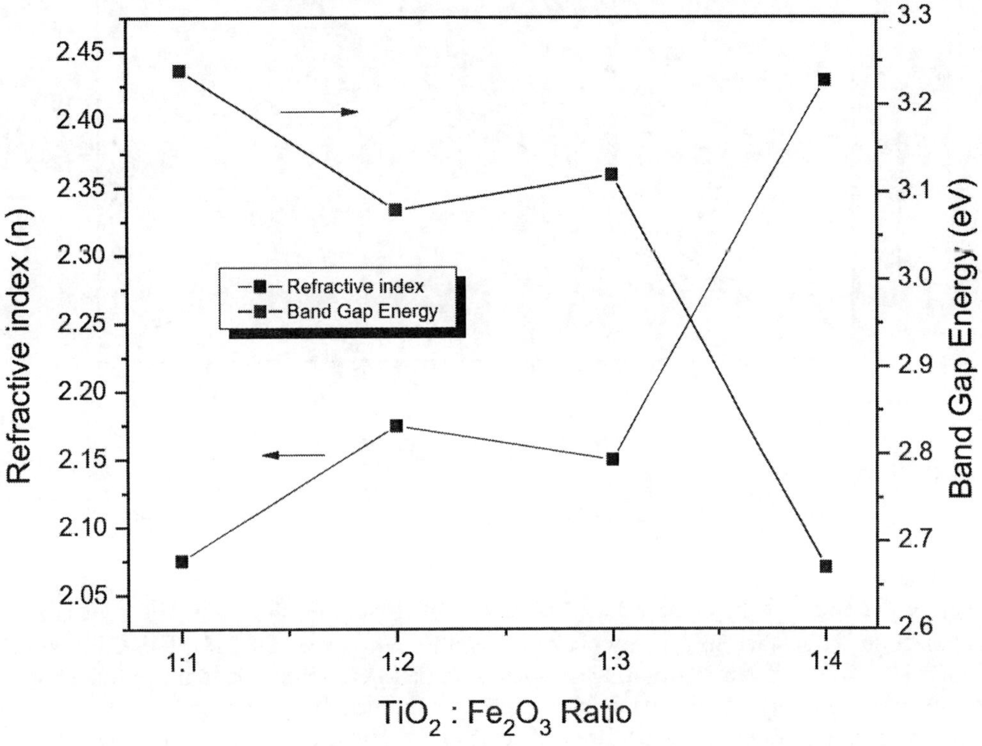

Figure 4. The effect of composite content on band gap energy and refractive index

Another property which important to study is dielectric constant which has a direct relation to the band gap energy. The dielectric constant was classified as the high frequency of dielectric constant which denoted as ε_∞ and static dielectric constant was denoted as ε_o. The high-frequency dielectric constant and static dielectric constant were calculated by Equations 4 and 5, respectively.

$$\varepsilon_\infty = n^2 \quad (4)$$

$$\varepsilon_o = - 33.26876 + 78.61805E_g - 45.70795E_g^2 + 8.32449\, E_g^3 \quad (5)$$

Figure 5 shows the result of a high frequency dielectric constant (HFDC) calculation and static dielectric constant (SDC) which plotted into a graph. As shown in Figure 5, the highest value of HFDC is 5.8981 while the lowest one is 4.306. On the other hands, the highest value of SDC is 24.764 and the lowest one is 9.243. The HFDC was reversely proportional to band gap energy, which means that when the band gap energy increases, the HFDC decreases. Inversely, the SDC is suitable with the increasing of band gap energy of TiO_2-Fe_2O_3.

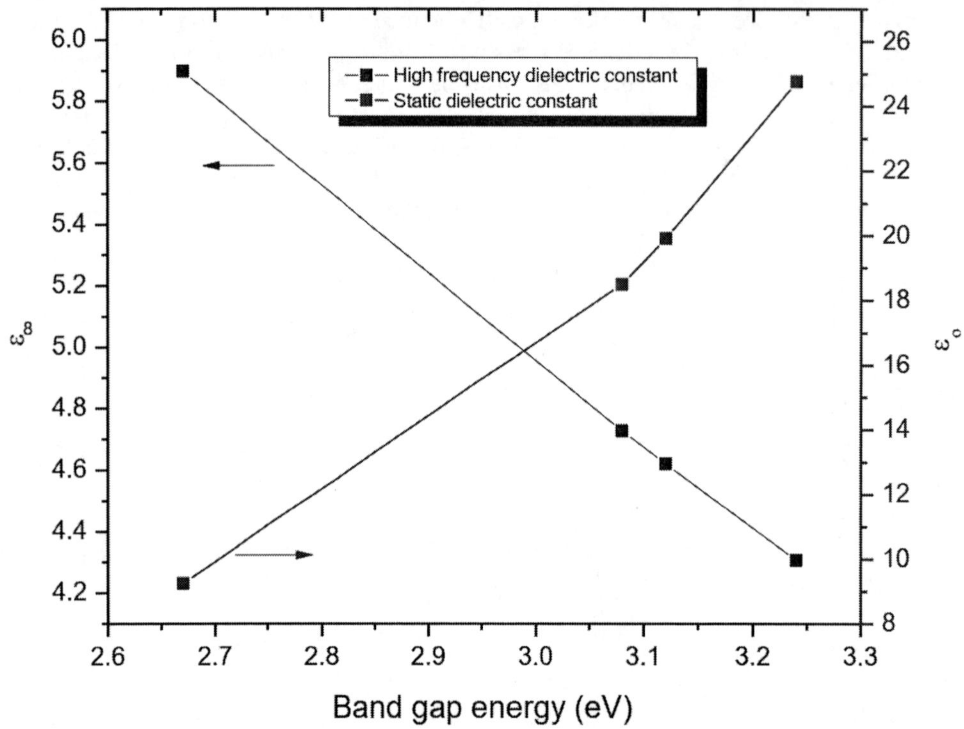

Figure 5. The relation of band gap energy and dielectric constant.

4. Conclusion

The composite of TiO_2-Fe_2O_3 was successfully synthesized in this work. The result of XRD characterization shows that the identity peak is fit with the peaks of TiO_2 (JCPDS Card No 21-1272) and Fe_2O_3 (JCPDS Card No 39-1346). The micrograph and elemental mapping results show that the composite has regularity on the particle's shape and uniformity of element in all areas of SEM micrograph. Analysis results using Diffuse Reflectance UV-Vis analysis shows that there was a change in band gap energy. The lowest band gap energy is 2.67 eV for TiO_2:Fe_2O_3 of 1:4. The refractive index, calculated by Ravindra equation, shows that the lowest value is 2.0752 and the highest value is 2.4286. The refractive index was reversely proportional to the band gap energy, which means that when the band gap energy increases, the refractive index decreases.

Acknowledgments

The authors thank to LPPM Universitas Palangka Raya for financial support under "Penelitian Dosen Pemula" grant with contract no 473/UN.24.13/PL/2019. We also thank to Nikita Turnip, Poppy Enzelica, and Jonedi for their help in this work.

References

[1] K. Nazeeruddin, E. Baranoff, and M. Gra, 2011. *Sol. Energy*, 85 1172–1178
[2] Y. Jiang *et al.* 2016 *J. Power Sources*, 327 465–473
[3] A. M. H. Flores *et al.* 2018 *J. Mater. Sci. Mater. Electron.*,1-8

[4] D. Dahlan, S. Khatijah, A. Usra, A. Bajili, and A. Ali, 2017 *Phys. E Low-dimensional Syst. Nanostructures*, 91 185–189
[5] G. Tio and B. Dye 2015 *JoP*, 1 5–11
[6] B. . K. S.K., Apte, S.D. Naik, R.S. Sonawane 2007 *J. Am. Ceram. Soc.*, 414, 2006–2008
[7] J. Singh, V. Verma, R. Kumar, and R. Kumar, 2019 *Results Phys* 3, 102-106
[8] G. H. Li, L. Yang, Y. X. Jin, and L. D. Zhang 2000 *Thin Film solid* 368 63–167
[9] J. Vidal *et al.*, 2012 *Appl. Phys. Lett.*, 100 1–6
[10] I. Cihan, S. Akin, H. Akyildiz, and S. Sonmezoglu, 2018 *Sol. Energy* 69 196–205
[11] D. W. Riyani, Kapti, Setyaningsih, Tien, Dwiasi, 2012 *Prosiding Seminar Nasional* 978–979.
[12] K. Balachandran, R. Venckatesh, R. Sivaraj, K. V Hemalatha, and R. Mariappan, 2015 *Mater. Sci. Semicond. Process* 35 59–65
[13] M. Ahmed and I. Dincer, 2018 *Int. J. Hydrogen Energy* 1-34

Characterization of alginate-chitosan membrane as potential edible film

Nurul Ismillayli[1], Surya Hadi[1], Ni Komang Tri Dharmayani[1], Rochmad Kris Sanjaya[2], Dhony Hermanto[1,*]

[1]Departement of Chemistry, Faculty of Mathematics and Natural Sciences, University of Mataram, Indonesia, 83125
[2]Departement of Pharmacy, Faculty of Health Sciences, University of Kediri, Kediri-East Java, Indonesia, 64115

*Corresponding author: dhony.hermanto@unram.ac.id

Abstract. Alginate-chitosan based biopolymer for possible application as edible film coating has been studied. Alginate hydrosol and chitosan hydrosol with mass ratio of 1:1 were mixed to form a thin membrane and then dried. The obtained alginate-chitosan membrane was confirmed using FTIR spectrophotometers. Characterization of the membrane, which includes thickness, tensile strength, water vapor sorption, resistance to pH change and antimicrobials properties, were conducted. It was showed that the interaction of alginate and chitosan in the membrane occurred through the electrostatic interaction of the carboxylic group of alginate and ammonium groups of chitosan. At the same thickness, the alginate-chitosan membrane tensile strength was higher and more resistant to pH changes than both native alginate and chitosan membranes. Furthermore, the alginate-chitosan membrane has good antibacterial potential against gram-positive bacteria (Staphylococcus aureus) and gram-negative bacteria (Escherichia coli). It is expected that the alginate-chitosan membrane has the potential application for safe and efficient fruit coating.

1. Introduction

Recently, there has been a shift in the use of petrochemical based synthetic food packaging to biodegradable packaging due to their serious ecological problems [1]. The most promising and versatile biodegradable polymers, chitosan, has been used as food coatings in terms of their ability to form strong, elastic, flexible and difficult to tear edible films [2]. Chitosan is a good barrier for gas and water vapor because of its matrix structure and antimicrobial properties that inhibit the growth of food-destroying microbes [3, 4]. Chitosan coating inhibits maturation by reducing the production of ethylene and carbon dioxide on fresh lychee, star fruit, and tomatoes [5]. Alginate membranes are also used as wrappers for fish, fruit [6], and meat to preserve them [7]. However, both of them have their own weaknesses, chitosan polymers undergo leaching under acidic conditions and alginate polymers undergo swelling when interacting with water. This weakness can be overcome through the formation of alginate-chitosan membranes.

Alginate-chitosan membranes are formed from polycationic chitosan and polyanionic alginate through electrostatic interactions [8]. With its dense matrix structure, the membrane inhibits the decay process by blocking oxygen inhalation. The formation of ethylene in fruit can be stimulated by mechanical damage and infection. By coating the alginate-chitosan membrane, the fruit is protected from physical impact because the dense matrix structure increases the tensile strength of the membrane. The use of alginate-chitosan membrane has fulfilled food safety and price worthiness because the membrane only consists of alginate and chitosan without other toxic materials, in accordance with the safe edible film criteria [9]. The simple preparation technique makes it easy for large scale application both for home and large industries in creating smart bioplastic and environmentally friendly food packaging. In this study, alginate-chitosan film will be synthesized, and the physical, chemical, and antibacterial properties of formed film will be examined.

2. Materials and methods

2.1. Materials

Chitosan (deacetylation > 90%, Mol. wt. 300 KDa-400 KDa), sodium alginate (Mol. wt. 12Kda-40 KDa), *S. aureus* (ATCC 25923) and *E. coli* (ATCC 25922) were obtained from Sigma-Aldrich (USA). Hydrochloride acid (37%), glacial acetic acid (98%), sodium hydroxide were obtained from Merck (Germany).

2.2. Method

2.2.1 Film preparation. The alginate-chitosan film (1:1 in mass) was prepared based on the procedure described by Hermanto et al. [10]. Briefly, hydrosol of chitosan at 1% (w/v) were prepared in 20 mL glacial acetic acid under stirring at 400 rpm using a magnetic stirrer (IKA Combimag REO Drehzahl Magnetic Stirrer Electronic, Germany), while alginate was dissolved in deionized water until hydrosol at concentration of 4% (w/v) was formed. Then, both alginate and chitosan hydrosols were left out overnight to remove air bubbles. The chitosan hydrosol (100 mL) was added to the alginate solution and added 8 mL of 32% HCl (100 mL) at 25 °C, homogenized using a homogenizer (IKA T18 Ultra Turrax, Germany) for 90 seconds. Following, NaOH 10% (w/v) was added till the pH of the suspension was elevated to 5.28 and homogenized again for 90 seconds. The mixture was then transferred in the glass plate and dried at room temperature for 72 h.

2.2.2 Membrane characterization regarding physical, chemical and antibacterial properties. The film thickness was determined with a digital micrometer (MDC-25S, Mitutoyo) at 5 random points. The film tensile strength and elongation at break were determined with Instron Universal test Machine (Model 4210, Canton MA). Film resistance to pH change was evaluated by comparing both tensile strength and elongation film immersing at different pH solution. Functional groups of films were characterized using the KBr pellet method at a compression pressure of 2500 Ib/m^2 on an FTIR spectrophotometer (FTIR 1600 Perkin Elmer Co, Japan). Water vapor sorption was determined using gravimetry method, according to Bueno et al. [11]. The experiment was performed in triplicate by measuring the weight gradual increase 1 cm x 1 cm film (in 95% relative humidity at 32 °C). Vapor sorption capacity (V_c) was calculated based on the following Equation 1 below,

$$V_c = \frac{(W_t - W_0)}{W_0} \times 100 \quad (1)$$

W_0 is the initial weight of the film and W_t is the weight at t time of the film.

The disk diffusion method was used to determine the antibacterial activity of the alginate-chitosan film against both *E. coli* and *S. aureus* as the one implemented by Kaya et al. [12]. As a reference standard 0.5 McFarland was used to adjust bacterial susceptibility. Suspension of bacterial culture was wiped onto a plate inside agar of Müeller-Hinton. On agar medium that has been inoculated with bacteria, paper discs were placed, then loaded with 20 mL hydrosol film. It was incubated for 24 hours at 37 °C. As positive control, it was used 10 mg gentamicin per disc and deionized water as a negative control. Inhibition zone diameter on a clear area of the paper disc was measured. All experiments were done in triplicate.

3. Result and discussion

3.1. Formation of the alginate-chitosan film

Finding out molecular interactions in the film between chitosan and alginate can be done by FTIR spectrophotometry. FTIR spectra interpretation of chitosan, alginate and alginate-chitosan films are described in Table 1.

Table 1. FTIR spectra interpretation of films.

Chitosan spectra (cm^{-1})	Alginate spectra (cm^{-1})	Alginate-chitosan spectra (cm^{-1})	Vibration
3459	3459	3459	N-H and O-H vibrations
2924	2930	2924	symmetric C-H vibration
2852	2864	2870	asymmetric C-H vibration
1589	1628	-	Primary amine N-H bending
-	-	1578	$-COO^-$ group vibrations
-	-	1550	presence of NH_3^+ ions
1426	1420	-	CH_2 bending
-	-	1398	Electronic interaction of alginate-chitosan
1375	-	-	CH_3 symmetric deformation
1325	-	-	C-N vibration
-	1305	-	O-C-H vibration
1260	-	-	hydroxyl groups
1157	-	-	Amino groups
1081 and 1028	-	1081	C-O vibration
-	1028	1028	C-O-C vibration
-	950	-	uronic acid residues
897	-	897	CH bending
-	891	-	manuronic acid residues

There is absorption in the wave number 3459 cm^{-1} which indicates the presence of the -OH group of alginates and $-NH_2$ chitosan, 2924 and 2870 cm^{-1} are the absorption of CH (sp^3), 1578 cm^{-1} is the vibration of the $-COO^-$ group. The loss of amine group absorption in 1157 cm^{-1} indicates that the chitosan amine group has been protonated and interacts with the carboxylate group of alginate [13]. This is strengthened by the appearance of peaks in the area of 1550 cm^{-1} which shows the presence of protonated amine ($-NH_3^+$ ion) [14]. Peak intensity at wave number of 1398 cm^{-1} also shows electrostatic interactions on alginate-chitosan film [15], with alginate and chitosan mass ratio of 1: 1. The presence of the absorption band mentioned above shows the existence of ionic bonds of protonated amine and the carboxylate groups.

Chitosan consists of amine groups which have intrinsic pKa value of 6.5, while alginates consist of carboxyl groups that undergo ionization at pH of the solution above its pKa of 4.7 [10, 16]. Therefore, at pH 5.28 chitosan presence at protonated form while alginate is in carboxylate form which is negatively charged. When both hydrosols are mixed in a 1: 1 ratio, electrostatic interaction of protonated amine and carboxylate has occurred [17].

3.2. Physical, chemical, and antibacterial properties

The film mechanical properties *i.e.* tensile strength, elongation and Modulus Young are important parameters for understanding material behavior during functional application. The Young Modulus describes the stiffness of material. The greater the Young modulus values of the material, the higher the level of stiffness. As part of mechanical strength, tensile strength is defined as the maximum tension achieved when the test of tension was carried out [12]. Meanwhile, elongation at break describes the elasticity of the film. In this study, for alginate-chitosan film, Young Modulus and tensile strength value was observed higher but its elongation was lower than both chitosan and alginate films, as described in Figure 1. It is indicated that alginate-chitosan film stiffer than its native single films. It is possible that the alginate-chitosan film has a tighter structure since protonated amine and carboxylates interaction. The better mechanical properties of chitosan-alginate films can minimize pressure damage of food coated during packaging and transportation. Meanwhile, film resistance to pH change is shown in Figure 2. Alginate-chitosan film shows higher stability to pH change than alginate or chitosan films, in line with previous study [11].

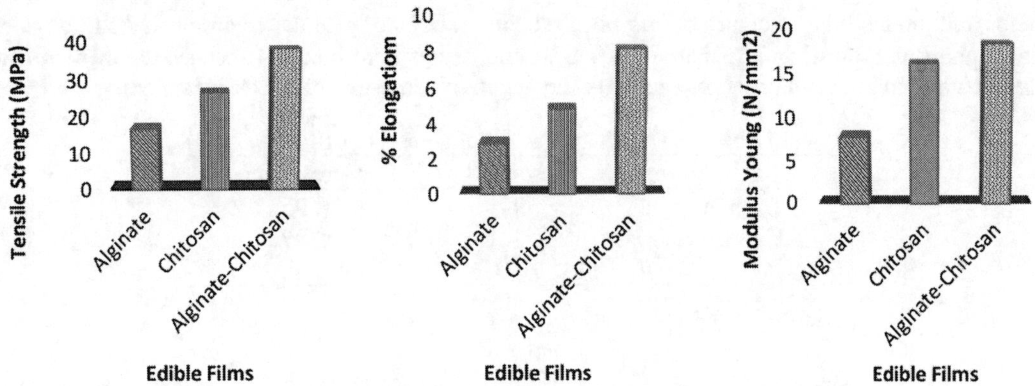

Figure 1. The mechanical properties of films.

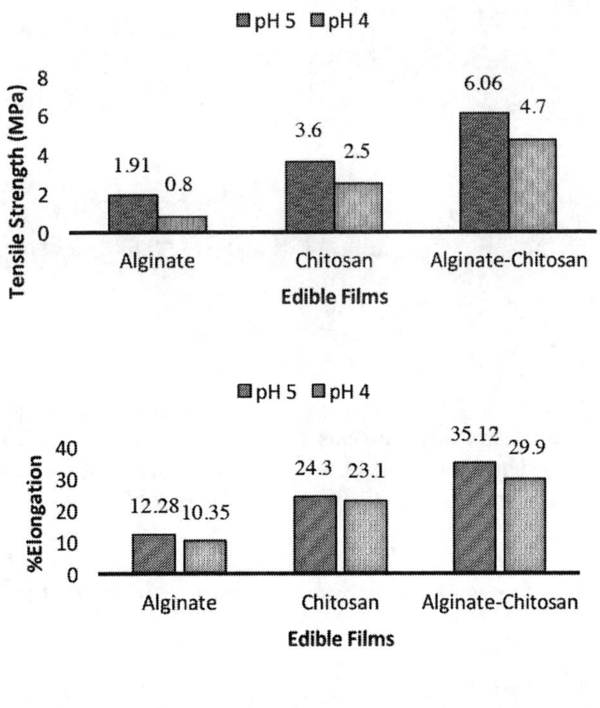

Figure 2. The pH effect on the mechanical properties of films.

The water vapor sorption capacity reached the maximum amount after its saturation which is 8 hours, as shown in Table 2. The value shows that the alginate-chitosan films have greater water vapor sorption and resistance than single films (alginate and chitosan). It is influenced by the hydrophilic and hydrophobic balance of the film, crosslinking, the degree of ionization and its interaction with the counter ion. Sorption on alginate-chitosan films is probably due to the presence of hydrophilic COO^- ions in the film. The film's hydrophobicity increases due to the formation of amine and carboxylate interaction and it is expected to have greater resistance to water-insoluble resistance.

The antibacterial activity of films is compared in Figure 3. Alginate-chitosan films shows the highest antibacterial activity resist *E. coli* and *S. aureus* compare to other films, except gentamicin as a positive control. The polymer structure of alginate-chitosan which has a positively charged of amine group can interact with the negative charge of a molecule such as proteins from microbes that cause leakage of proteins and intracellular structure of microbes. The interaction of protonated amine and carboxylate in alginate-chitosan is ionic bonds that are not as strong as covalent bonds. It causes the availability of protonated amino group in order to react with the negatively charged of bacterial protein.

Table 2. Water vapor sorption capacity of films.

Film	% sorption	% resistance
Alginate	94	95
Chitosan	98	99
Alginate-Chitosan	104	105

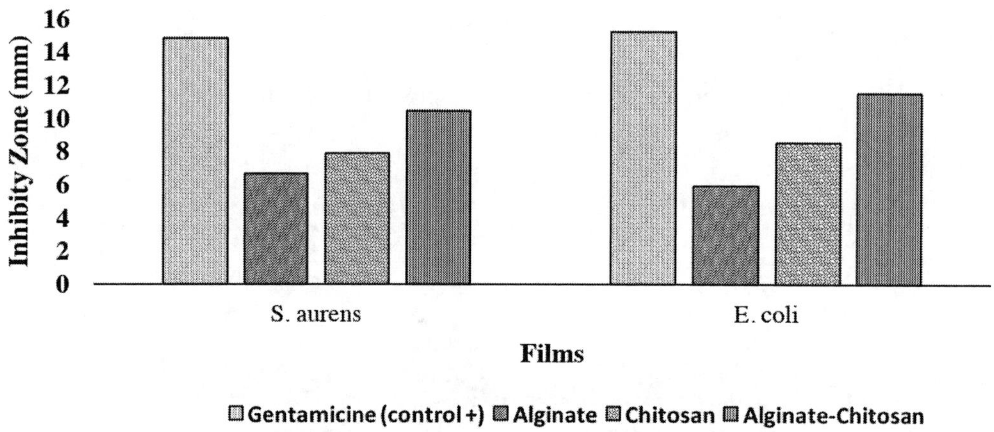

Figure 3. The antibacterial activity of films.

4. Conclusion

The interaction of alginate and chitosan in the film occur through ionic interaction of protonated amine and carboxylate. The alginate-chitosan film has better mechanical strength and antibacterial activity resist *E. coli* and *S. aureus* than both alginate and chitosan film as the native membrane. Therefore, alginate-chitosan film has the potential application for safe and efficient fruit coating.

Acknowledgments
The authors are grateful University of Mataram for the research funding of DIPA BLU (PNBP).

References

[1] Galgano F *et al* 2015 *Ital. J Food Sci.* **27** 1.
[2] Butler B L *et al* 1996 *J. Food Sci.* **61** 5 953.
[3] Aloui H and Khwaldia K 2016 *Compr. Rev. Food Sci. Food Saf.* **00** 1.
[4] Zheng L Y and Zhu J F 2003 *Carbohydr. Polym.* **54** 527.
[5] Sucharitha K V *et al* 2018 *Int. Food Res. J.* **25** 1 93.

[6] Menezes J and Athmaselvi K A 2016 *Int. Agrophysics* **19** 551.
[7] Bhagath Y B and Manjula K 2019 *Int. Food Res. J.* **26** 2 377.
[8] Hermanto D *et al* 2019 *Ind. J. Chem.* **19** 3 786.
[9] Vargas M *et al* 2008 *Crit. Rev. Food Sci. and Nutr.* **48** 6 496.
[10] Hermanto D *et al* 2019 *J. Chem. Sci. Appl.* **22** 1 11.
[11] Bueno C Z *et al* 2015 *J. Supercrit. Fluids* **10** 1 1.
[12] Kaya M *et al* 2018 *RSC Adv.* **8** 3941.
[13] Han J *et al* 2010 *Int. J. Biol. Macromol.* **46** 2 199.
[14] Lee J W *et al* 1999 *J. Appl. Polym. Sci.* **73** 113.
[15] Kulig D *et al* 2016 *Polym.* **8** 5 1.
[16] Sankalia M G *et al* 2007 *Eur. J. Pharm. Biopharm.* **65** 2 215.
[17] Hermanto D *et al* 2019 *J. Math. Fund. Sci.* **51** 3 309.

IOP Conf. Series: Materials Science and Engineering **833** (2020) 012074 doi:10.1088/1757-899X/833/1/012074

Modification of Gunung Kidul Natural Zeolite and Its Application in the Corrosion Rate Reduction of Steel Plates in Acid Media

Mohammad Abdul Khafid[*1], Aditya Pandu Wicaksono[1], Firstyananda Wahyu Andita[2], and Asifa Ihya Nurdina[2]

[1]Department of Environmental Engineering, Faculty of Mineral Technology, Universitas Pembangunan Nasional Veteran Yogyakarta, DI. Yogyakarta, Indonesia
[2]Department of Chemical Engineering, Faculty of Industrial Engineering, Universitas Pembangunan Nasional Veteran Yogyakarta, DI Yogyakarta, Indonesia

*Corresponding author: abdulkhafid925@gmail.com

Abstract. The use of metals in the industry plays a very important role, where nearly 90% of industrial equipment comes from metallic materials. Corrosion is an electrochemical process between metals with its environment that can damage the metal quality. Gunung Kidul natural zeolite can be modified into a mixture of alkyd paint as a coating material to reduce the corrosion rate of steel plates effectively. In this research, the zeolite was activated by NaOH 1 N for 120 minutes. The zeolite was characterised using Surface Area Analyzer (SAA) and nitrogen adsorption analysis to investigate the surface area of zeolite by Brunauer-Emmet-Teller (BET) modelling. The lining was done by coating the steel plates by alkyd-zeolite paints and inserted into a 100 mL H_2SO_4 1N, 2N, and 3N for 20 days. The results of this study showed that the corrosion rate of steel plate was 0.1374 miles/year (uncoated), 0.0407 miles/year (coated by alkyd paint), and 0.0095 miles/year (coated by alkyd paint-zeolite). Using activated zeolite as a mixture of alkyd paint is environmentally friendly and effective in lowering the corrosion rate of mild steel plates.

1. Introduction

The use of metal materials for industrial and infrastructure activities plays a very important role because metals are raw materials in infrastructure development, as well as in equipment for everyday needs. Some of the technical properties are owned by metals such as tensile strength, resistant to temperature, electrical and thermal conductivity [1]. However, steel is easily corroded and causes damage to material. Corrosion is a process of material destruction in the reaction to the chemical and electrochemical environment. Therefore, their resistance all depends on the corrosion resistance [2]. Degradation of steel or metal strength due to corrosion often occurs in the construction of concrete structures. Generally, the corrosion occurs to construction in areas with the condition of sour sulphate land such as swamp areas that are still affected by the tide of sea or coastal areas [3]. For an example of iron metal corrosion in moist air, the reaction process is as follows:

Anode $2Fe_{(S)} \rightarrow 2Fe^{2+}_{(aq)} + 4e$

Cathode $\underline{O_{2\,(g)} + 4H^{+}_{(aq)} + 4e \rightarrow 2H_2O_{(l)}}$ +

Redox $2Fe_{(s)} + O_{2\,(g)} + 4H^{+}_{(aq)} \rightarrow 2Fe^{2+} + 2H_2O_{(l)}$ (1)

The methods of controlling corrosion in metals are metal coating, cathode protection, and addition. One of the common methods of repair to control corrosion is coating. There are two coating materials, namely liquid coating and concrete coating. The liquid coating is plating on the surface of the steel with chemicals, thus the steel can be protected from corrosion [4]. However, the use of these chemical compounds is expensive and quite dangerous for human beings and the environment. Hence, it needs environmentally friendly raw materials from natural resources that can lower the corrosion rate of steel plate. One of the quality compounds to reduce the corrosion rate of steel plate is zeolite [5].

Zeolite natural resources can be obtained in Indonesia, especially in West Java (District Bayah and Cipatuja), Central Java (District Wonosari, Gunung Kidul Regency, Yogyakarta), and East Java (Pacitan Regency) which has a zeolite reserve of more than two hundred million tonnes [6]. Zeolite is a mineral containing silicate aluminium compounds with the condition of a three-dementing skeleton structure formed by the tetrahedral composition of the AlO_4^{5-} and the SiO_4^{4-}. Zeolite's pores are filled by metal ions, dominated by alkali and alkali earth metals (Na, Ca, K, Mg, and Fe), and water molecules that can move freely in the cavity [7,8]. Zeolite appears as a corrosion-resistant material with porous and crystal coatings with uniform cross-section of 0.3 nm to 1.2 nm. Zeolite can be utilized as a catalyst and adsorbent in petrochemical and petroleum refining processes. Low silica zeolite has been used for many years as a hydrophilic coating material and antimicrobial for the industrial needs [8]. The process of cation and mixing exchange of alkyd paint, the permissible of zeolite solids are made homogeneous first through the process of physics with the grinding. Alkyd is a paint with a polyester base with a composition of ethane glycol, flax oil, and phthalate anhydride [9]. This paint dries through oxidative polymerization of oil so that it has a high oil content.

The process of enhancing the diffusion of reactants on catalysts and the additional mesoporous system is required in crystal micro pores. Thus, it is expected that molecules that have large size can enter into the pore system, to be processed and leave the pores system back [10]. The process of natural zeolite activation can be done chemically through addition of alkaline substance (NaOH), which aims to clean the surface of the zeolite pores, eliminating metal oxide as an impurity, and rearranging the location of atoms that can interchangeable. The activation of zeolite causes the occurrence of ion which causes increased surface area of zeolite due to impurities elimination that cover the pores of the zeolite. Increased surface area is expected to increase the ability of zeolite in the absorption process [11]. Zeolite is a very useful catalyst that demonstrates some important qualities that are better compared to amorphous catalysts. The amorphous catalysts are almost always made in powder form to provide a large surface area so that the number of catalytic sides is greater. The existence of the cavity in zeolite provides a very wide internal surface area so that it can hold 100 times more molecules than the amorphous catalysts with the same amount [12].

This research aims to determine the effect of activated zeolite addition in alkyd paint as a coating material on steel plates and study the corrosion rate of the plate in acid media. The corrosion rate MPY (miles per year) on the steel plate is calculated and discussed. The result is also compared with some samples to get optimum conditions of the coating process and corrosion rate.

2. Methods

2.1. Raw Material and Instrumentation

In this research, natural zeolite was obtained from Wonosari, Gunung Kidul Regency, Yogyakarta, Indonesia which has been separated through the 80-mesh sieve. The instruments used in this study were pH-meters (HI 9126), a Nabertherm oven, digital analytical balances (Matrix type Esj 210-4B), 80-mesh sieve, magnetic stirrer, desiccator, stopwatch, and Surface Area Analyzer (SAA) Quadrasorb Station 2 version 7.01. The chemicals used were H_2SO_4, NaOH, alkyd paint, steel plates, and *oil* thinner.

2.2. Zeolite Activation

Firstly, zeolite was cleaned from other inorganic substances (Na, K, Ca, Mg and Fe), then grinded to obtain finer zeolite powder. Next, the zeolite powder was passed on an 80-mesh. The powder passed the sieve was used in the next step. The activation process was done by soaking 250 grams of zeolite in a NaOH 1 N for 2 hours at room temperature. The zeolite was then washed using distilled water until the color of the solution changes to almost clear. After that, the zeolite was filtered off, dried at room temperature, and then heated in the oven at 180°C until a constant mass was gained [13].

2.3. Zeolite Characterization

To determine the surface area of zeolite, Surface Analyzer (SAA) with Quadrasorb Station 2 version 7.01 was used. Firstly, the gas-absorbed (degassing) gases was removed by heating in a vacuum condition at 300 °C for 6 hours. Phase analysis of the isothermal adsorption was carried out on two samples (activated and inactivated zeolites) using nitrogen gas for 5 hours and an outgases temperature of 300 °C. Based on a partial pressure nitrogen, the adsorption test was done in the P/Po range of 0.01 – 0.30. The surface area of the activated zeolite was analysed by Brunauer-Emmet-Teller (BET) modelling [14].

2.4. Corrosion Testing

The steel plate was firstly weighted, then it was immersed in alkyd paint (four samples). Next, zeolite (50, 60, and 70 grams) was mixed with the paint alkyd and stirred until homogenous. Thus, there are five plate samples with different sample conditions: steel plate that was uncoated, steel plate that was coated with alkyd paint, and steel plates that were coated with alkyd paints + zeolite (50, 60, 70 grams). The five samples of steel plates were immersed in the 100 mL H_2SO_4 (1N, 2N, 3N) for 20 days. Data collection of each sample was collected three times. After the immersion, the samples were dried first and then weighed to determine the mass reduction of steel plates (every 5 day to up to 20 days). Steel plate mass reduction results were used to calculate the rate of corrosion. The corrosion rate was evaluated using the method of weight loss of steel plate as shown in Equation (2), in which W is the difference in metal weight or weight loss (g), D is density (g/cm^3), A is the surface area of rusty metal (inch^2) and T is corrosion time (Hour) [15].

$$M.P.Y = (534.W) / (D.A.T) \tag{2}$$

3. Result and Discussion

3.1. SAA Results (Surface Area Analyzer)

To determine the surface area of zeolite specifically, a Surface Area Analyzer (SAA) with Branauer-Emmett-Teller (BET) modelling was applied. This method was done by using adsorption of N_2 gas to determine the surface area, the pores volume, and the pores diameter of a zeolite. This technique was done by measuring the amount of gas that was given by a solid in the gas pressure variation in an

IOP Conf. Series: Materials Science and Engineering **833** (2020) 012074 doi:10.1088/1757-899X/833/1/012074

isotherm state. The zeolite isotherm adsorption profile, treated with alkaline treatment and before alkaline treatment, is shown in Figure 1, respectively.

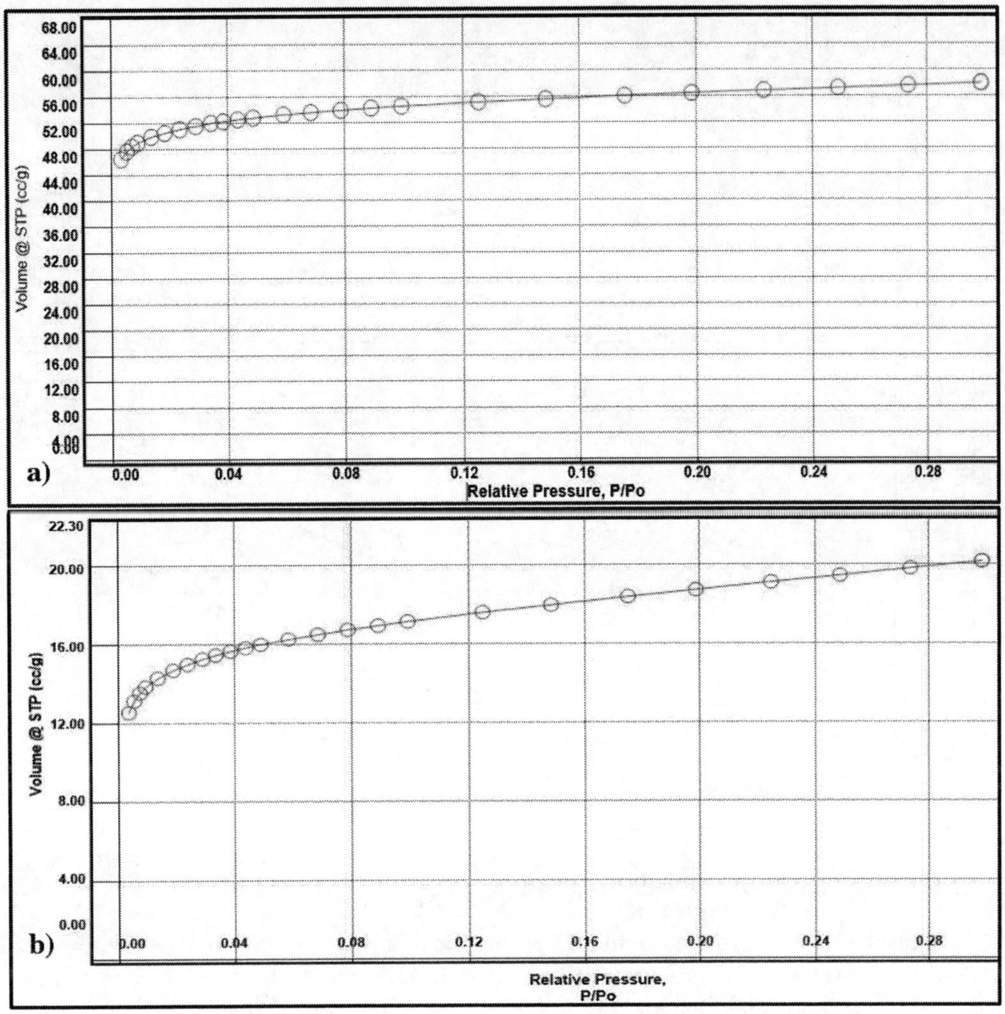

Figure 1. Graph of isotherm adsorption, a) activated zeolite and b) un-activated zeolite.

Based on Figure 1, the amount of nitrogen adsorption volumes on the activated and inactivated zeolite are considerably fluctuated and the relative pressure (P/PO), which tends to be low with the vulnerable value, is slowly increases in the middle and rapidly rises again at the pressure (P/PO) value of 0.01-0.30. The first increment process is caused by an adsorbed gas molecules interacting with an area that is energized on the surface of the solids. The filling of nitrogen gas had been formed into monolayer coating. The result of the isotherm data of these two samples is the type 1 isotherm, which is a type of physio-sorption of gas in the microporous solids and isothermal chemical adsorption.

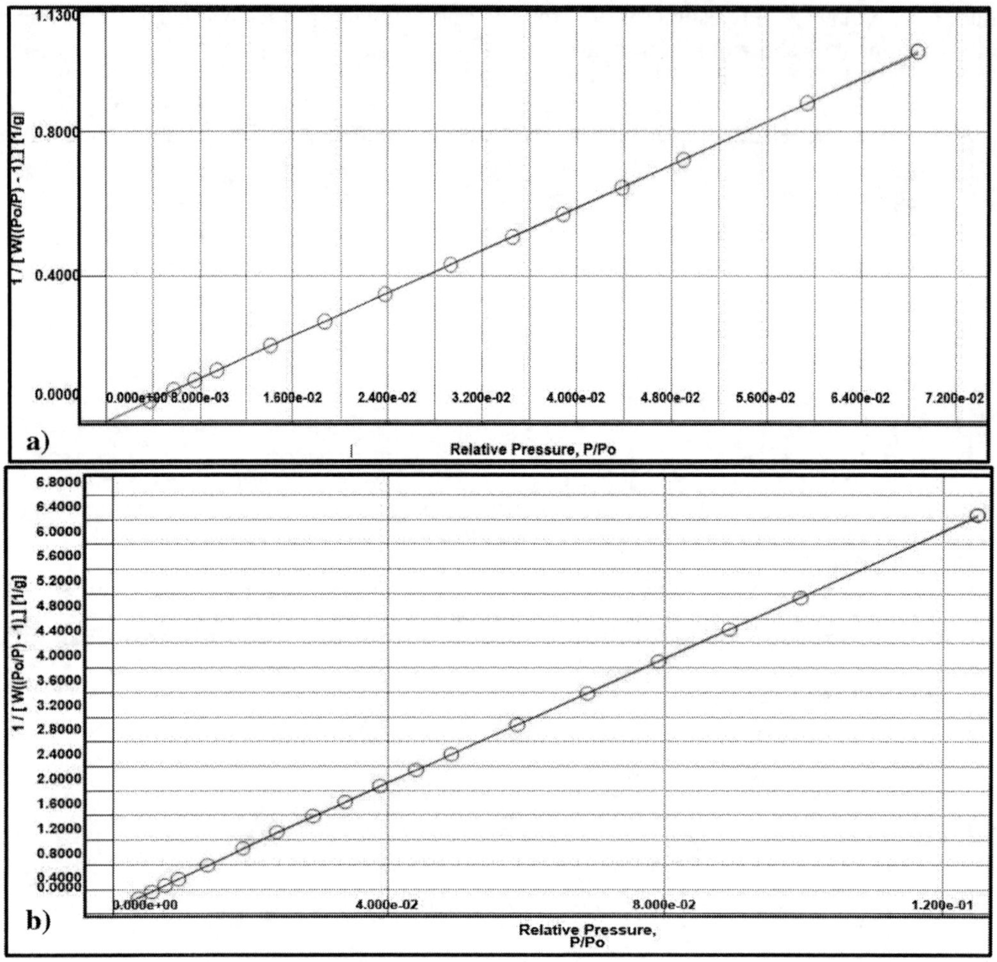

Figure 2. Graph multi-point BET plots, a) activated zeolite and b) inactivated zeolite.

BET test results indicate that the inactivated zeolite has relatively different pore volume and pressure than that of the activated zeolite. Figures 2 shows that the activated zeolite has a greater pressure with fewer pore volumes than that of inactivated zeolite. The result of calculation by BET method resulting in surface area values of 235.484 m^2/g (activated zeolite) and 68.279 m^2/g (inactivated zeolite). Activated zeolite has a larger surface area compared to the inactivated zeolite. This is due to the physical–chemical process of cation exchange and eliminating impurities, both organic and inorganic in the Gunung Kidul natural zeolite, resulting in pores enlargement and expanding surfaces. The internal surface area of zeolite can reach tens even hundreds of times greater than the surface of the exterior. This large surface area is very beneficial to the utilization of zeolite either as an adsorbent or as a heterogeneous catalyst. This result also support the fact that the alkaline treatment of Gunung Kidul natural zeolite can increase the surface area of zeolite which eventually increasing the capacity of ion exchanges and the absorption of zeolite [16].

3.2. Results of Corrosion rate testing

The results of the corrosion rate calculation using equations (2) for all samples shows varying results. The uncoated sample indicate that on the day five of corrosion test in H_2SO_4 3N, the corrosion rate of 0.1784 miles/years reaches the highest. Then it went much higher on day twenty with a value of 0.2082 miles/year. In general, there is a periodic decline every five days of calculation regarding the corrosion rate. The value of the corrosion rate is influenced by the condition of the steel plate media, the acid media increases the corrosion rates. Meanwhile, the corrosion rates on the steel plate coated with alkyd paint in H_2SO_4 3N shows a smaller value of 0.0645 miles/years. This is due to the presence of a coating of alkyd paint resin as a corrosion inhibitor that has protection or barrier properties. The corrosion rate results of uncoated and coated steel plates are shown in Figure 3.

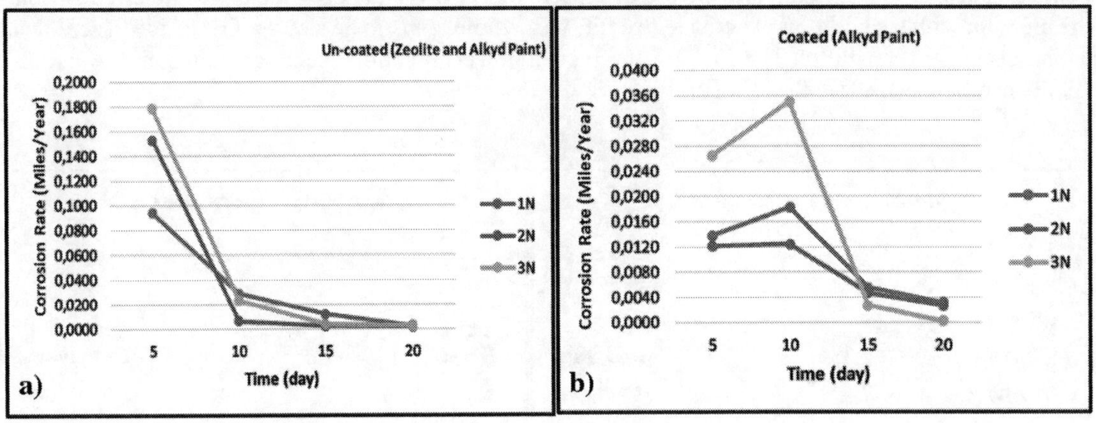

Figure 3. Graph of corrosion rate on steel plates a) uncoated and b) coated of alkyd paint.

Table 1. Comparison of corrosion rates on steel plates in H_2SO_4 (1N, 2N, 3N).

	Corrosion Rate (Uncoated)	(MPY)			Corrosion Rate (MPY) (Coated alkyd paint)	
Day	1N	2N	3N	1N	2N	3N
5	0.0938	0.1531	0.1784	0.0138	0.0121	0.0265
10	0.0289	0.0063	0.0234	0.0182	0.0124	0.0351
15	0.0118	0.0026	0.0040	0.0055	0.0047	0.0026
20	0.0029	0.0016	0.0024	0.0032	0.0026	0.0002

The corrosion rate shows smaller value for steel plates coated by alkyd + zeolite. The steel plate in H_2SO_4 1N shows the lowest corrosion rate of 0.0086 miles/year with addition of 70 g zeolite, whereas steel plate in H_2SO_4 2N shows the lowest corrosion rate of 0.0011 miles/year with addition of 60 g zeolite. The lowest corrosion rate in a higher acidity environment (H_2SO_4 3N) is 0.0018 miles/year, obtained by addition of 70 g zeolite. These results indicated that the optimum inhibition can be achieved in the immersion period from day 1 to day 10 using a combination of alkyd paint and zeolite powder of 60 - 70 grams. Mass loss suggests that the bonding between the zeolite and alkyd paints can prevent direct contact with the corrosive environment (H_2SO_4), thus the steel plate can have more resistance to corrosion [17].

Based on Figure 4, the effect of activated zeolite addition against inhibitors efficiency in the corrosive medium is noticeable from reduction in corrosion rate occurred in day 5. The efficiency percentage of was increased by almost 90% by addition of 70 g zeolite. However, in the addition of 50 g zeolites, the efficiency value tends to decrease. The corrosion test was done three times to minimize data errors and get an overall average value of each sample tested for obtaining valid data. The effect of corrosion rate decreasing for steel plates is significantly influenced by voids and cracks in the protective coating of zeolite [18]. The process of zeolite activation aims to minimize the presence of impurities and form a better homogeneous and compact layer, which is a way to obtain an efficient and effective corrosion-resistant coating, structures, and the morphology of the Gunung Kidul natural zeolite. The process of zeolite activation has the ability in formation of zeolite coating and the growth of zeolite crystal. The results show that the large surface area of zeolite causes smaller zeolite size, hence it becomes more homogeneous. This eventually affects the increased barrier layer to corrosion. The rate of corrosion in the medium with the addition of activated zeolite as a whole tends to be slower, this is due to zeolite and cation exchanger capability. The H_2SO_4 solution is a powerful cation exchange, in which Na^+ cation in zeolite can be replaced by the H^+ [19].

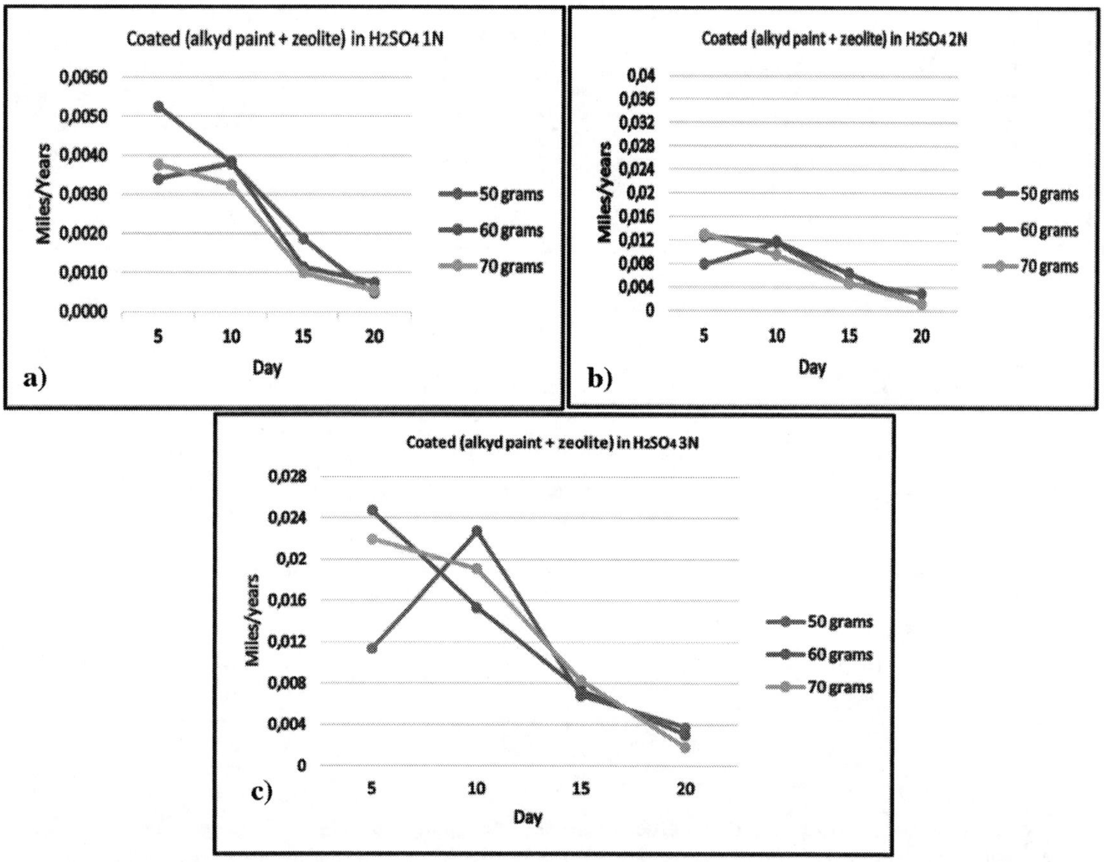

Figure 4. Graph of corrosion rate on steel plates in a) H_2SO_4 1N, b) H_2SO_4 2N, and c) H_2SO_4 3N.

4. Conclusion

The modification of Gunung Kidul natural zeolite by NaOH can be applied as a corrosion inhibitor. Through a surface area analysis with BET modelling, the zeolite has a surface area value of 235.484 m^2/g (activated zeolite) and 68.279 m^2/g (inactivated zeolite). The mixture of 70 g zeolite with 150 mL alkyd paints in 20-day immersion of 100 mL H_2SO_4 (1N, 2N, 3N) produces the suitable composition in lowering the corrosion rate with a total value of 0.0086 miles/year, 0.0285 miles/year, and 0.0511 miles/year, respectively.

Acknowledgments
This research was supported by the Chemical Engineering Laboratory, Faculty of Industrial Engineering and Geotechnical Environmental Laboratory, Department of Environmental Faculty of Mineral Technology, Universitas Pembangunan Nasional "Veteran" Yogyakarta.

References
[1] McCafferty E 2010 *Introduction to Corrosion Science* (New York: Springer Science+Business Media) p 13
[2] Jones D A 1996 *Principles and Prevention of Corrosion Second Edition* (United State of America: Prentice Hall Inc) p 6
[3] Shukla H S, Haldar N and Udaybhanu G 2012 *E-Journal of Chemistry* **9** 149–60
[4] P Marcus and J Oudar 1995 *Corrosion Mechanisms in Theory and Practice* (New York: Marcel Dekker Inc) p 261-265
[5] Haryono G, Sugiarto B, Farid H, and Tanoto Y 2010 *Proc. Seminar Nasional Teknik Kimia "Kejuangaan" 5 maret 2010 (Yogyakarta)* Vol 1 (Yogyakarta: Teknik Kimia, Universitas Pembangunan Nasional "Veteran" Yogyakarta)
[6] Schneider K 1974 *Use of Local Minerals in the Treatment of Radioactive Waste Technical Report Series No 136* (Vienna: IAEA) p 11-13
[7] Arryanto Y, and Amini S, and Lu MGQ 2002 *J. Zeolit Indones (Journal Indones. Zeolites)* Chapter 1 pp 1–4.
[8] Breck D W 1974 *Zeolite Molecular Sieves Structure Chemistry and Use* (New York: John Wıley and Sons Inc) p 25
[9] Anton J Hartomo 1992 *Paint Coating on Metals* (Bandung: Ganesa) p 30
[10] Othmer K 1981 *Encyclopedia of Chemical Technology (3th ed)* vol 15 (New York: John Wiley and Sons Inc)
[11] Trisunaryanti W, Triyono, Rizki C, Saptoadi H, Zainal Z A, Syamsiro M, Yoshikawa K 2013 *IOSR Journal of Applied Chemistry* **3** 29–34
[12] Smith J V 1988 *Chemical Reviews* **88** 149–82
[13] Rosyid M, Nawangsih E and Dewita 2012 *Proc. Seminar Penelitian dan Pengelolaan Perangkat Nuklir* ISSN 1410-8178
[14] Wei Z, Xia T, Liu M, Cao Q, Xu Y, Zhu K and Zhu X 2015 *Frontiers of Chemical Science and Engineering* **9** 450–60
[15] S K Chowdhary 2001 *J. Of Metallurgy and material science* **43** 265
[16] Muhriz M A, Subagio and Pradoyo 2011 *J. Sains dan Matematika* **19** 11-17
[17] Ertan A and Çakicioğlu-Özkan F 2005 *Adsorption* **11** 151–6
[18] Mitra A, Wang Z, Cao T, Wang H, Huang L and Yan Y 2002 *Journal of The Electrochemical Society* **149**
[19] Dyer A 1988 *An Introduction to Zeolite Molecular Sieves* (Chichester: John Wiley and Sons Ltd)

Hydrothermal synthesis of titanium dioxide nanotube with methylamine for photodegradation of Congo red

C Y Leong[1], P W Koh[1], Y S Lo[1] and S L Lee[1,2*]

[1]Department of Chemistry, Faculty of Science, Universiti Teknologi Malaysia, 81300 Johor Bahru, Malaysia
[2] Centre for Sustainable Nanomaterials, Ibnu Sina Institute for Scientific and Industrial Research, Universiti Teknologi Malaysia, 81300 Johor Bahru, Malaysia

*Corresponding author: sllee@ibnusina.utm.my

Abstract. Titanium dioxide (TiO_2) nanotube photocatalyst is highly desired for the photodegradation of dye in wastewater treatment. A series of titanium dioxide nanotube photocatalysts were successfully synthesized using methylamine as N-ligand via hydrothermal treatment at different hydrothermal temperatures and durations. The effect of these two parameters on the photocatalytic activity of synthesized materials were investigated. TEM micrographs and XRD analysis depicted methylamine assisted the transformation of anatase TiO_2 nanoparticles to nanotube via the exfoliation of TiO_2 crystallite into layered sheet and promoted the curling of layered sheet. Hydrothermal temperature up to 180°C was able to fully transform the morphology of anatase TiO_2 nanoparticles into nanotube. The reaction duration was further modified. Fluorescence analysis showed that 24 h hydrothermal duration gave the slowest electron-hole recombination rate. DR-UV-Vis analysis indicated that the synthesized samples were active under UV region. The photocatalytic performance of the synthesized materials was tested in the photodegradation of Congo red under UV irradiation. The results suggested that among the materials synthesized, TiO_2 nanotube synthesized at 180°C, under 24 h hydrothermal duration appeared to be the most superior photocatalyst which gave the highest photocatalytic activity of 77%. Possible mechanism of the TiO_2 nanotube formation with methylamine as N-ligand is presented.

1. Introduction

Titanium dioxide (TiO_2) and TiO_2-based materials have been intensively investigated due to their excellent catalytic performance [1-5]. However, TiO_2 is suffering from its low surface area, hence leads to limited activity. Since last decade, TiO_2 nanotubes (NTs) have attracted much attraction as compared to the commonly used TiO_2 nanoparticles due to its excellent unique features, including high ion-exchangeability, large specific surface area and good photocatalytic activity [6]. The current developed fabrication methods of TiO_2 NT include the template-assisted method, sol-gel method, hydrothermal method, and electrochemical anodic oxidation [7-8]. Among these fabrication methods, hydrothermal fabrication is believed to be the most promising due to its environmentally friendly properties, simple setup and cheap apparatus is used. For the other methods, for instance, electrochemical anodic oxidation method, usage of extremely toxic aqueous hydrochloric acid solutions is needed for the anodization of titanium foil [9].

In this study, hydrothermal treatment method was used to synthesize TiO_2 nanotube (TNT). Although hydrothermal synthesis involves simple setup, every single step in the process from choice of

TiO_2 precursor to hydrothermal conditions and lastly the post-treatment process of TiO_2 NT is very important in determining the crystallography and morphology as well as the physical-chemical properties of the TiO_2 NT final products. The hydrothermal conditions include the temperature, hydrothermal time and concentration of reactant [3]. According to the study by Weng [10], increasing the hydrothermal time contributes to the lengthening of TNTs. When the hydrothermal duration exceeds 24 h, there is no further increase in the length of TNTs. From the other studies, however, it has been stated that the morphology of the synthesized products is dependent on the hydrothermal duration [11]. It seems that the hydrothermal duration plays an important role in affecting the morphology of TNTs formed which drives the author to further study on the effect of hydrothermal duration and temperature on the formation of TNTs in this study.

During the hydrothermal synthesis of TNTs, agglomeration of individual morphological form titanate into secondary particles occurs. Secondary particles include nanotubular bundles, nanofibers, nanofibers which are linked hierarchically and others [2]. Even though there were extensive studies on the effect of synthesizing condition on the geometry of single nanotube, there are few studies reported on the principles of controlling the agglomeration of nanotubes and treatment of the shape and geometry of secondary particles [12]. In addition, based on the study by Peng *et al.* [13], the $-NH_2$ functional group of DETA captured the Ti^{4+} in the solution hence facilitate crystal grow of TiO_2 to be formed uniformly on the carbon nanotube. Methylamine, which is also one of the N-atom bearing ligand, is expected to behave the same as DETA, holding the Ti^{4+} ion and could assist the titanate layer to curl up and thus promote the formation of TiO_2 NT in this research. Its effectiveness in helping the intermediate TiO_2 nanosheet to curl up, forming nanotubular structure is yet to be known. The mechanism for the formation of TiO_2 NT with methylamine as a N-bearing ligand was proposed in this research.

2. Experimental

2.1. Synthesis of titanium dioxide nanotube

TiO_2 NT was synthesized using hydrothermal method with the presence of methylamine. Commercial anatase titanium dioxide served as the Ti precursor while methylamine hydrochloride was used as N-ligand.

In a typical synthesis, methylamine hydrochloride was added to the 7 M NaOH aqueous solution, followed by the addition of TiO_2 into the reaction mixture based on the molar ratio TiO_2:CH_3NH_2 of 1:1. The reaction mixture was allowed to stir vigorously (1000 rpm) for one hour. Later, reaction mixture was transferred to a 100 mL Teflon-lined stainless-steel autoclave and kept in an electric oven at 130°C for 24 h. It should be noted that in order to ensure the desired structure to be formed successfully, any disturbance during the reaction, such as opening of the oven door, had to be prevented. After 24 h of hydrothermal duration, the autoclave was taken out and left to cool to room temperature. The white precipitate was obtained by centrifugation, washed thoroughly with sulphuric acid and double distilled water, then dried at 70°C overnight.

Experiment was repeated with different ratio titanium dioxide to methylamine of 1:1, 1:3, 1:5 and 1:8. The samples were notated as xM-TiO_2 NT-y-z, where x, y, and z represent the ratio of methylamine to TiO_2 and M represent methylamine, hydrothermal temperature and duration, respectively. The optimized ratio was then used to synthesize TiO_2 NT at different temperatures (130°C and 180°C) and different hydrothermal durations (6 h, 12 h and 24 h).

2.2. Characterization

The physical-chemical properties of synthesized TiO_2 NT were studied via characterization with Transmission Electron Spectrometer (TEM), X-ray diffractometer (XRD), Ultraviolet-Visible Diffuse Reflectance Spectrometer (DR-UV-Vis), and spectrofluorometer.

TEM (JEOL, JEM-2100F, 200 kV) was used to investigate the morphologic, crystallographic and compositional information on the TiO_2 samples. Meanwhile, the crystallinity and the crystal structure of the synthesized TiO2 samples were examined via XRD (Bruker Advance D8 with Cu Kα radiation;

λ = 1.5406 Å; 40 kV, 40 mA). UV-Vis diffuse reflectance spectra of the prepared samples were evaluated using a UV-Vis spectrometer (Perkin-Elmer Lambda Lambda 35) equipped with a diffuse reflectance attachment with $BaSO_4$ coated 76 mm integration sphere as a reference material. The reflection in percentage was measured and presented by Kubalka-Munk function. Meanwhile, the electron-hole recombination rate was calculated using a spectrofluorometer (JASCO, FG-8500).

2.3. Photocatalytic testing

The synthesized TiO_2 anatase was tested for its photodegradation activity on Congo red under dark condition where the UV radiation is absent and normal condition where the UV radiation is present.

The photocatalytic activity of each of the synthesized TiO_2 NT was calculated by using the formula as follow:

$$\text{Photodegradation activity} = \frac{C_0 - C_t}{C_0} \times 100\% \quad (1)$$

where, C_0 = Concentration of Congo red before reaction (ppm)
C_t = Concentration of Congo red after reaction (ppm)

3. Results and Discussion

3.1. Surface morphologies

Figure 1 shows the TEM micrographs of the synthesized samples. As shown, with the addition of methylamine, most of the structure of TiO_2 nanosheet had been fully transformed into nanotubes. Both the as-synthesized TiO_2 NT-130°C-24 h and 3M/TiO_2 NT-130°C-24 h had interplanar spacing of 0.35 nm, represented the (1 0 1) lattice plane of anatase TiO_2 [14]. Meanwhile, 3M/TiO_2 NT-130°C-24 h had one additional d-spacing of 0.64 nm, represented (001) lattice plane of nanotube [15].

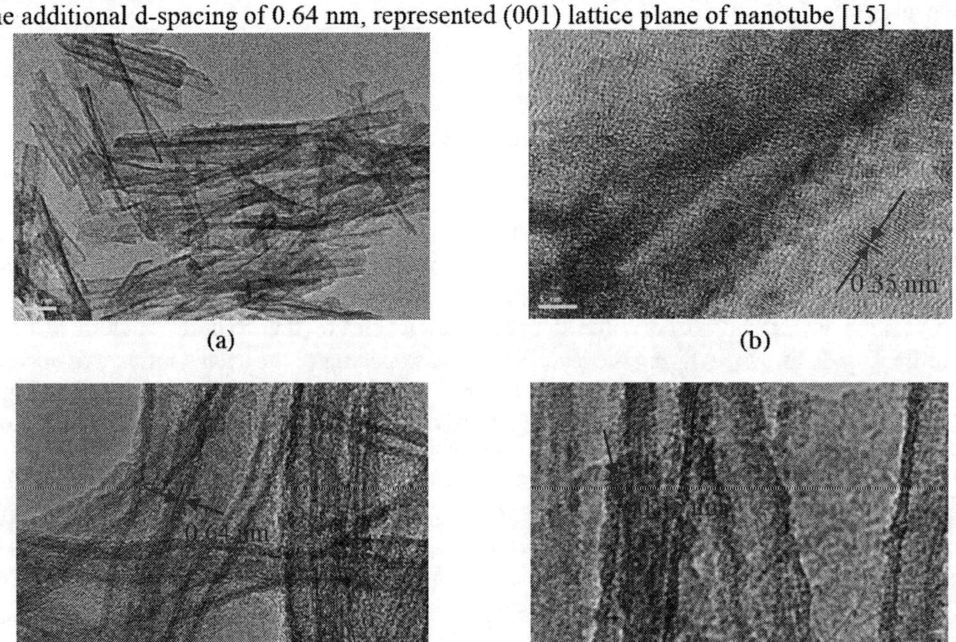

Figure 1. (a) TEM images and (b) HRTEM image of as-synthesized TiO_2 NT-130°C-24 h; (c) and (d) HRTEM image of 3M/TiO_2 NT-130 °C-24 h.

3.2. Surface morphologies

Figure 2(a) illustrates the XRD patterns of the methylamine modified titanium dioxide nanotube samples. As illustrated, the characteristic of anatase phase of titania could be observed clearly in all the samples at 2θ value of 25.35°, which attributed to the lattice plane (1 0 1) [16]. The intensity of the peak however, decreased as higher ratio of methylamine ligand was added to the titania. Peaks assigned to the tetragonal titania anatase phase were also observed at 37.90°, 53.90°, 54.05°, 55.10°, 62.85°, 68.85°, 70.25° and 75.20°, with lattice plane (0 0 4), (1 0 5), (1 0 5), (2 1 1), (2 0 4), (1 1 6), (2 2 0), and (2 1 5) respectively. As compared to the as-synthesized TiO_2 nanotube, samples modified with methylamine gave three feature peaks of $Na_2Ti_3O_7$ titanate at ~9.4°, 28.20°, and 48.15°, with lattice planes (0 0 2), (1 1 1), and (2 0 0), respectively. The calculated crystallite size of the synthesized samples decreased with the increasing molar ratio of methylamine and increased in crystallite size when the molar ratio reached 8. Possible explanation could be due to the promotion of crystal growth by eliminating the inter-crystallite pores.

Figure 2(b) shows the XRD patterns of materials synthesized at different hydrothermal temperatures, namely 130°C and 180°C. It can be seen that when the hydrothermal temperature was increased from 130°C to 180°C, the intensity of peaks corresponded to anatase phase of titania reduced and peaks corresponded to $Na_2Ti_3O_7$ type of titanate (JCPDS 31-1329) were observed with increasing intensity. Thus, it could be deduced that increase of hydrothermal temperature up to 180°C was able to fully transform the spherical morphology of anatase TiO_2 particle to nanotube shape. Meanwhile, the modification of reaction condition by varying the hydrothermal duration had insignificant changes within the samples. Increasing hydrothermal duration would not distort the nature of the synthesized materials.

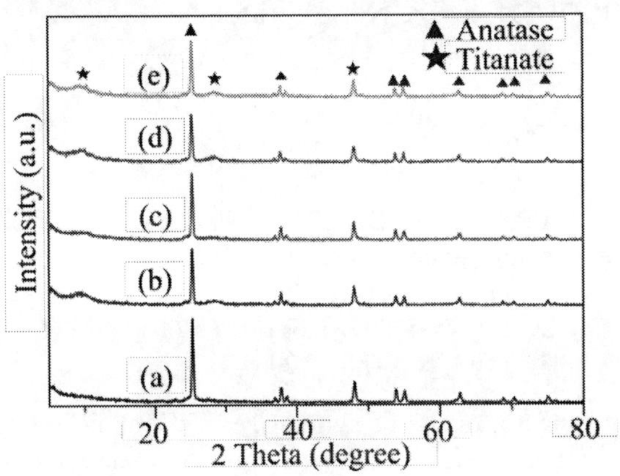

Figure 2(a). XRD patterns of (a) as-synthesized TiO_2 NT-130°C-24 h, (b) 1M/TiO_2 NT-130°C-24 h, (c) 3M/TiO_2 NT-130°C-24 h, (d) 5M/TiO_2 NT-130°C-24 h, (e) 8M/TiO_2 NT-130°C-24 h

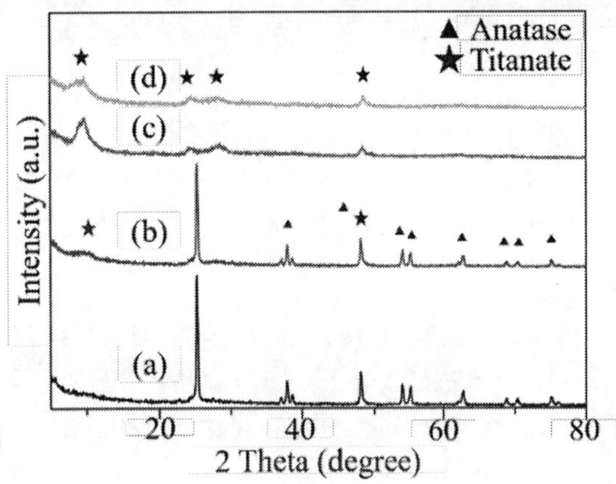

Figure 2(b). XRD patterns of (a) as-synthesized TiO_2 NT-130°C-24 h, (b) 3M/TiO_2 NT-130°C-24 h, (c) as-synthesized TiO_2 NT-180°C-24 h, and (d) 3M/TiO_2 NT-180°C-24 h

3.3. Optical properties

Figure 3(a) illustrates the DR-UV-Vis spectra of the synthesized TiO_2 nanotubes. There were two absorption peaks located at 270 – 280 nm, which were contributed by amorphous Ti species in all the synthesized samples [17]. Also, an absorption peak at 320 nm was observed, which was associated to anatase TiO_2 [18]. The intensity of the absorption band at 270 nm decreased with increasing xM/TiO_2 ratio, while it was accompanied by an increased in the intensity of adsorption band at 320 nm. This proves that with an increasing amount of methylamine, amorphous TiO_2 could be reduced, transforming

IOP Conf. Series: Materials Science and Engineering **833** (2020) 012075 doi:10.1088/1757-899X/833/1/012075

into crystalline anatase phase of TiO_2. Addition of methylamine increased the band gap energy of as-synthesized TiO_2 due to quantum size effect [19].

Figure 3(b) elucidates the DR UV-Vis spectra of the samples synthesized at different hydrothermal temperature, namely 130°C and 180°C. Increasing the hydrothermal temperature would increase the intensity of absorption band at 270 nm which assigned to amorphous Ti species and the absorption band at 320 nm which denoted the anatase TiO_2 was fully disappeared. It could be deduced that hydrothermal temperature of 180°C was able to fully transform the morphology of anatase TiO_2 particle into nanotube shape.

As for the modification of reaction condition based on hydrothermal duration (6 h, 12 h, 24 h), hydrothermal duration of 6 h and 12 h were too short for the complete formation of TiO_2 nanotube. As a result, mixture of unreacted nanoparticles, nanosheet and nanotube were present within the sample. The current findings strongly suggested that hydrothermal duration of 24 h was important for the complete transformation of spherical anatase TiO_2 nanoparticles into nanotube.

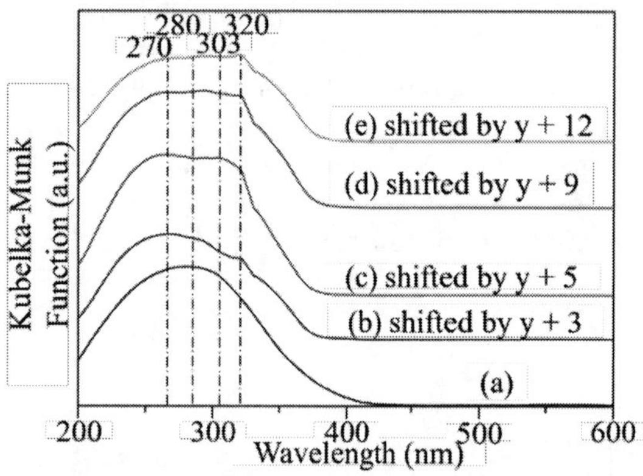

Figure 3(a). DR-UV-Vis spectra of (a) as-synthesized TiO_2 NT-130°C-24 h, (b) 1M/TiO_2 NT-130°C-24 h, (c) 3M/TiO_2 NT-130°C-24 h, (d) 5M/TiO_2 NT-130°C-24 h, and (e) 8M/TiO_2 NT-130°C-24 h

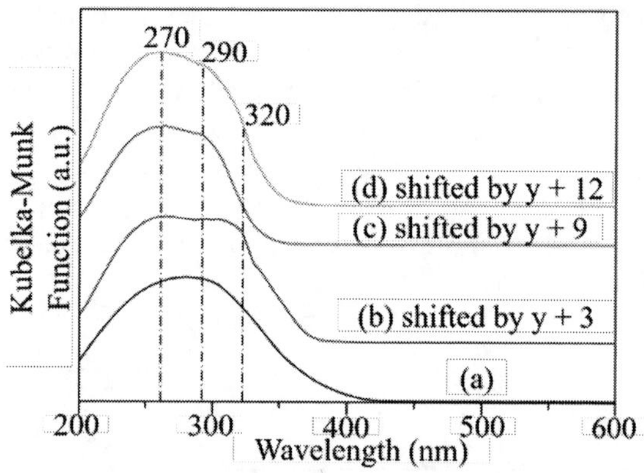

Figure 3(b). DR UV-Vis spectra of (a) as-synthesized TiO_2 NT-130°C-24 h, (b) 3M/TiO_2 NT-130°C-24 h, (c) as-synthesized TiO_2 NT-180°C-24 h, and (d) 3M/TiO_2 NT-180°C-24 h

As for the fluorescence analysis, the fluorescence spectra of the as-synthesized TiO_2 and the best photocatalyst (3M/TiO_2 NT-130°C-24 h) were shown in Figure 3(c). With the addition of methylamine, the intensity of the peak decreased, denoting a decrease in the electron-hole recombination rate. A possible explanation to this is due to an increase in the separation effectiveness and longer lifetime of the photoinduced charge carriers [20]. It could be suggested that addition of methylamine as N-ligand suppressed the electron-hole recombination rate by increasing their lifetime.

As the hydrothermal temperature increased up to 180°C, the electron-hole recombination rate was greatly suppressed as evidence in Figure 3(d). Possible explanation could be the crystalline nature of the nanotube enable smooth, fast and efficient electron transfer to the surface hence retarded the electron-hole recombination rate [21]. Hydrothermal duration of 24 h was optimum for the complete transformation of nanoparticles to nanotube. The fluorescence intensity of 3M/TiO_2 NT-180°C-24 h was the lowest as compared to those synthesized at 6 h and 12 h, denoting the slowest recombination rate.

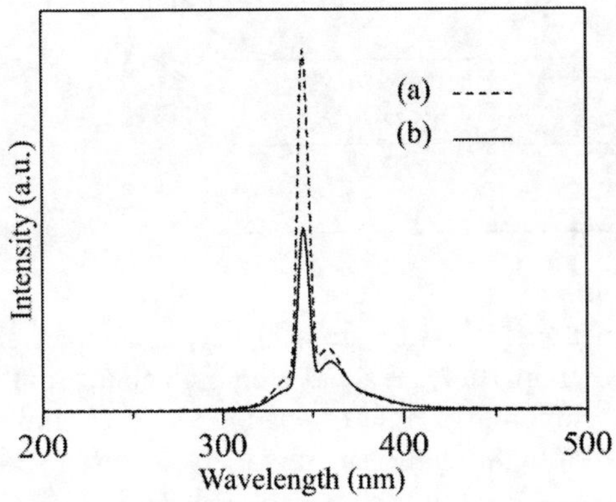

Figure 3(c). Fluorescence spectra of (a) as-synthesized TiO_2 NT-130°C-24 h and (b) 3M/TiO_2 NT-130°C-24 h

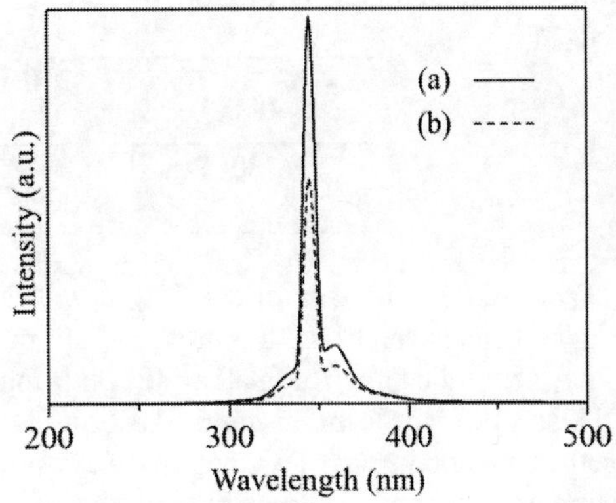

Figure 3(d). Fluorescence spectra of (a) 3M/TiO_2 NT-130°C-24 h and (b) 3M/TiO_2 NT-180°C-24 h

3.4. Photocatalytic activity

The photocatalytic performance of the synthesized samples was evaluated via photodegradation of Congo red dye under dark condition and with UV irradiation. The obtained absorbance at 498 nm was used to calculate the photodegradation efficiency of the synthesized samples. As illustrated in Table 1, the photocatalytic activities of the xM/TiO_2 (x = 1, 3, 5, 8) were higher than the as-synthesized TiO_2. The addition of methylamine increased the photocatalytic activity of TiO_2 by enhancing nanotubes formation. Methylamine might had assisted the scrolling of the intermediate nanosheet, forming nanotube. Sample 3M/TiO_2 had the best photocatalytic activity of 69.4% which could be attributed to the lowest electron-hole recombination rate.

Table 1. Photodegradation of Congo red over samples xM/TiO_2-130°C-24 h.

x	**Photocatalytic activity (%)**
1	68.2
3	69.4
5	60.5
8	58.5

Table 2 elucidates the photocatalytic activity of samples that were synthesized at different hydrothermal durations, namely 6 h and 24 h, under constant hydrothermal temperature at 180°C. It is demonstrated clearly in Table 2 that 3M/TiO_2 NT-180°C-24 h had a higher photocatalytic activity as compared to the sample synthesized at 6 h, which could be due to the reduction in electron-hole recombination rate of nanotube. The photocatalytic activity of the samples decreased drastically with hydrothermal duration, around 19%. The increase in photocatalytic activity of TiO_2 NT might be attributed to the higher amount of nanotube formation from mixture of unreacted particle, nanosheet and nanotube. Apparently, hydrothermal duration of 6 h was too short for the TiO_2 NT formation. Larger amount of nanotube could be formed at 24 h hydrothermal duration with the presence of methylamine that facilitated in scrolling up the edge of intermediate nanosheet.

Table 2. Photodegradation of Congo red over samples 3M/TiO_2 NT -130°C-24 h and 3M/TiO_2 NT -180°C-24 h.

Temperature	Duration (h)	Photocatalytic Activity (%)
180°C	24	77.1
	6	58.7

3.5. Proposed mechanism for the TiO_2 nanotube formation

2D TiO_2 nanosheet was the intermediate product of transformation from 3D TiO_2 nanoparticles to 1D TiO_2 nanotube under high alkaline condition. TiO_2 nanoparticles reacted with high concentration of NaOH solution to form a highly disordered phase containing $H_2Ti_3O_7$ nanocrystals. Under the hydrothermal conditions, TiO_2 nanoparticles are stripped into thin lamellar fragments. After the process of stripping develops to certain extend, the large amounts of Ti-O bonds are broken to form $H_2Ti_3O_7$. It should be noted that the transformation process was gradual and it started from the surface region to the centre region. Thus, the Ti-O bonds on the surface were broken in the early stage. During the breaking process of Ti-O bonds, the O atoms were twisted or rotated, creating an angle between Ti and O atoms. This led to the formation of the lepidocrocite-type layers, which were suggested to be the construction units in any layered titanate. It was reported that the surface energy of TiO_2 nanosheet and TiO_2 nanotube was 1 J/m^2 and 500×100^{-18} J/m^2, respectively [22]. Due to the stable lower surface energy, it drives TiO_2 to fold the sheet-like structure into tube-like structure. However, it was found out that the hydrothermal temperature of 130°C in the current study was not enough to transform all the nanosheet curl up into nanotube structure. Presence of nitrogen atom within the methylamine could have held the O^- of the Ti-O bond and assisted the scrolling of nanosheet at the edges, which resulted in higher degree of nanotube formation.

4. Conclusion

The synthesis of titanium dioxide nanotube using methylamine as N-ligand via hydrothermal method was conducted successfully. Effect of different temperature and hydrothermal duration on the properties and photocatalytic activity were evaluated. At higher hydrothermal temperature, up to 180 °C, there was a reduction in the electron-hole recombination rate as all the anatase TiO_2 nanoparticles had transformed into nanotube. On the other hand, hydrothermal duration of 24 h was needed to allow complete transformation of spherical anatase TiO_2 nanoparticles to nanotube shape. Sample 3M/TiO_2 NT-180°C-24 h appeared to be the best photocatalyst due to its low electron-hole recombination rate, giving photocatalytic activity of 77.1% in photodegradation of Congo red organic dye. Evidently, methylamine was vital in assisting the complete TiO_2 nanotube formation and, hydrothermal temperature and hydrothermal duration were optimum at 180 °C and 24 h, respectively.

Acknowledgement

The authors gratefully acknowledge the Ministry of Higher Education, Malaysia (MOHE) for the Fundamental Research Grant Scheme (vote no.: R.J130000.7854.5F238) and Universiti Teknologi Malaysia (UTM) for the Research University Grant (Vote no.: Q.J130000.3554.07G57).

References

[1] Koh P W, Hatta M H M, Ong S T, Yuliati L, Lee S L 2017 *Journal of Photochemistry & Photobiology, A: Chemistry* **332** p 215–223

[2] Ekhsan J M, Lee S L, Nur H 2014 *Applied Catalysis A: General* **471** p 142-8

[3] Koh P W, Yuliati L, Lee S L 2019 *Iranian Journal of Science and Technology, Transactions A: Science* **43** p 95–103

[4] Lee S L, Hamdan H 2008 *Journal of Non-Crystalline Solids* **354** (33) p 3939-43.

[5] Ooi Y K, Yuliati L, Hartanto D, Nur H, Lee S L. 2016 *Microporous and Mesoporous Materials* **225** p 411-20.

[6] Muniyappan S, Solaiyammal T, Karthigeyan A and Murugakoothan P 2017 *Modern Electronic Materials* **3**(4) p 174-8
[7] Ge Y, Wu J, Li H, Zhang K and Li X 2016 *Journal of Nanoscience and Nanotechnology* **16**(8) p 8748-59
[8] Sahrin N T, Nawaz R, Chong F K, Lee S L, Wirzal M D H 2020 *Nanomaterials* **10** p 128-145
[9] Liu N, Chen X, Zhang J and Schwank J W 2014 *Catalysis Today* **225** p 34-51
[10] Weng L-Q, Song S-H, Hodgson S, Baker A and Yu J 2006 *Journal of the European Ceramic Society* **26**(8) p 1405-1409
[11] Li Q, Liu R, Zou B, Cui T and Liu B 2014 *Physica B: Condensed Matter* **445** p 42-7
[12] Bavykin D V, Kulak A N and Walsh F C 2011 *Langmuir* **27**(9) p 5644-9
[13] Peng H, Yang X, Zhang P, Zhang Y, Liu C, Liu D and Gui J 2017 *RSC Adv.* **7**(79) p 50216-24
[14] Eskandarloo H, Hashempour M, Vicenzo A, Franz S, Badiei A, Behnajady M A and Bestetti M 2016 *Applied Catalysis B: Environmental* **185** p 119-132
[15] Wang G, Liu Z Y, Wu J N and Lu Q 2012 *Materials Letters* **71** p 120-2
[16] Li Y, Ma Q, Han J, Ji L, Wang J, Chen J and Wang Y 2014 *Applied Surface Science* **297** p 103-8
[17] Xiong G, Cao Y, Guo Z, Jia Q, Tian F and Liu L 2016 *Phys Chem Chem Phys* **18**(1) p 190-6
[18] Xiong G, Jia Q, Liu L and Guo Z 2017 *RSC Advances* **7**(39) p 24046-54
[19] Yu J C, Ho W, Yu J, Hark S K and Lu K 2003 *Langmuir* **19**(9) p 3889-96
[20] Li H, Xia Z, Chen J, Lei L and Xing J 2015 *Applied Catalysis B: Environmental* **168-169** p 105-113
[21] Sun W-T, Yu Y, Pan H-Y, Gao X-F and Chen Q 2008 **130**(4) p 1124-25
[22] Bavykin D V, Parmon V N, Lapkin A A and Walsh F C 2004 *Journal of Materials Chemistry* **14**(22) p 3370-77

Green Synthesis of ZnO Nanoparticles by Using Banana Peel Extract as Capping agent and Its Bacterial Activity

Siti Marfu'ah, Santias Megasani Rohma, Faisal Fanani, Emi Nurul Hidayati, Diah Wahyu Nitasari, Thutug Rahardiant Primadi, Endang Ciptawati, Sumari Sumari, Fauziatul Fajaroh*

Department of Chemistry, State University of Malang, Malang, Indonesia

*Corresponding author: fauziatul.fajaroh.fmipa@um.ac.id

Abstract. One of the nanoparticles used in the medical world is ZnO nanoparticles. The purpose of this study was to synthesize ZnO nanoparticles by using peel's extract of kepok, ambon, and tanduk bananas (*Musa paradisiaca L., Musa acuminata Cavendish Subgroup, and Musa x paradisiaca*) as a capping agent. This research was conducted in four main steps, namely: extraction of three kind of bananas peel, phytochemical test on the bananas peel extract, synthesis of ZnO nanoparticles by using the extracts, characterization of ZnO nanoparticles by X-Ray Diffraction (XRD), and antibacterial activity test of ZnO nanoparticles. Based on the phytochemical test it was proven that bananas peel extract contained secondary metabolite such as flavonoids, polyphenols, alkaloids, and saponins in various concentration. The XRD analysis concluded that ZnO nanoparticles were successfully synthesized. Through the agar diffusion test, it can be showed that ZnO nanoparticles are a promising material as antibacterial agents. This statement is based on evidence that the diameter of the clear zone produced ranges from 8 to 10 nm.

1. Introduction

One of the metal oxide nanoparticles which has many benefits for humans is zinc oxide nanoparticles (ZnO NPs). ZnO NPs are used in a variety of applications including antibacterial agents, solar cell manufacturing materials, cosmetic ingredients, and gas adsorbents such as NO_x, CO_2, CO, and NH_3 [1]. Currently, the development of ZnO NPs is not only limited to how it is applied, but also how to synthesize the nanoparticles which produces the desired properties. Many ZnO NPs synthesis methods have been developed, one of them is a chemical method. Unfortunately, this method has disadvantages consider with the environmental impact due to waste from the synthesis process using hazardous chemicals as capping agents (PEG polymers, PVC, and anionic surfactants and others). Moreover, the application of chemical methods is also expensive, requires a long process, and high temperatures [2].

One of environmental-friendly method for synthesis of ZnO NPs nanoparticles is green method. In this method, the use of chemicals as reducing or capping agent is replaced by natural ingredients like plant extracts or mushrooms or algae. Green synthesis of ZnO NPs has been carried out by many researchers including using *Bauhinia tomentosa* leaf extract [3], *Citrus sinensis* extract [4], cavendish banana peel extract [5] *Sargassum sp* [6] and garlic extract [7].

To the best of our knowledge, there is no researchers uses peel extract of Kepok, Ambon, and Tanduk bananas (*Musa paradisiaca* L, *Musa acuminata Cavendish Subgroup,* and *Musa x paradisiaca)* as capping agents in ZnO NPs synthesis. The peels of these three types of bananas are accumulated as waste. Data from Ministry of Agriculture showed that the prospect of banana production until 2020 is about 8000 tons/year [8]. While the waste of banana peels production is about 40% of total production [9]. The use of banana peel extract as a capping agent is possible because the banana peel extract contains secondary metabolite compounds such as flavonoids, polyphenols, alkaloids and saponins with various amounts [3]. Flavonoids are natural compounds that contain 15 carbon atoms arranged in a C6-C3-C6 configuration. Flavonoids are predicted to be the compounds that produce the most nanoparticles in the Green Synthesis process both as capping agents and reducing agents [10]. Polyphenols are secondary metabolites that have many phenol groups in their structure. Alkaloids are secondary metabolites that have one or more nitrogen atoms in their heterocyclic rings, whereas saponins are compounds composed of glycons (sugars) and aglycones (non-sugars) that are rich in hydroxyl groups. Functional group in the secondary metabolites will act as Lewis bases in water. Metal cations that are in the same solution with the compounds will be stabilized and subsequently undergo conversion to hydroxide after passing through the saturation point. Different parts of plant material such as fruit extracts, bark, fruit bark, and roots have been studied for the synthesis of silver, gold, platinum and titanium nanoparticles in various sizes and shapes [2]. According to Ibrahim synthesis of nanoparticles mediated by plant extracts is preferred because it is cheaper, environmental-friendly, and the products are safe for medical purposes [2].

This study aims to carry out phytochemical tests on bananas peel extract, synthesis and characterization of ZnO NPs by using bananas peel extract as a capping agent, followed with antibacterial test. This last goal relates to the ability of ZnO NPs to interfere the permeability of bacterial cell walls which in turn makes bacterial cells undergo a lysis or damage so that the cytoplasmic fluid in cells will come out and cause bacterial death [11]. The results of Manyasree's research [12] show that ZnO NPs is highly reactive to *S. aureus* gram-positive and gram-negative *E. coli* bacteria.

2. Materials and Method

The equipments used in this study were some glass wares, an analytical balance (SHIMADZU) with accuracy of 0.0001 g, hot plate with magnetic stirrer, mortar and pestle, centrifuge, vortex mixer, and XRD (Panalytical Xpert Pro).

The materials used in this study were kepok, ambon, and tanduk bananas (*Musa paradisiaca* L., *Musa acuminata Cavendish Subgroup,* and *Musa x paradisiaca*) peel, $ZnSO_4.7H_2O$ p.a. Merck, NaOH p.a. Merck, Ethanol p.a. Merck, universal pH paper, $HgCl_2$, KI, demineralized water, DMSO p.a, nutrient agar, Merck, *E. coli* and *S. aureus* bacteria. Filter paper (Whatman-42) and universal indicator

Bananas peels were first cleaned using demineralized water and cut into pieces, then weighed in the amount of 150 grams. The banana peels that have been placed in a beaker glass were added with 150 mL of demineralized water and heated at 80 °C for 10 minutes. The banana peels then mashed and macerated, and then filtered off.

Phytochemical tests were carried out by referring to the procedure in the study of Rao et al (2016). Phytochemical tests consist of flavonoid, polyphenol, alkaloid, and saponin tests. The tests are described below.

i. Flavonoid Test
The flavonoid test was carried out by adding concentrated of HCl solution dropwise to the banana peel extract. If the color changes to red, it can be said that the extract contains of flavonoids.

ii. Polyphenol Test
Banana peel extract was added with solution of $FeCl_3$ 1%. A positive reaction gives a strong green, red, purple, blue or black color.

iii. Alkaloid Test
Banana peel extract was added with 0.2 mL of HCl solution and then was added with 1 mL Mayer reagent. A positive test is given by the formation of white deposits to yellowish.

iv. Saponin Test
A 1 mL of banana peel extract was added with 20 mL of hot water then shaken vigorously. Positive result is indicated by the formation of bubbles that are stable for not less than 15 minutes.

Synthesis of ZnO nanoparticles was carried out by mixing 500 mL of 0.0369 M of zinc sulfate solution with 129 mL of banana peel extract and stirred for 10 minutes, then added 2.0 M of NaOH solution dropwise (8 grams of NaOH dissolved into 100 mL of demineralized water) to reach pH 12. The mixture was stirred for 2 hours, then filtered using a Buchner funnel. After that, the solid was rinsed with ethanol until the filtrate had the same color as the ethanol, then the solid was rinsed with demineralized water until the filtrate reached pH 7. The solid was dried in an oven at 60 °C until it reached a constant mass and dried for 48 hours.

ZnO nanoparticles were characterized using X-Ray Diffraction (XRD) to determine crystallinity, phase identification and particle size. This characterization was carried out in the range of 2θ angles at 10°-90° at room temperature, with an X-ray source produced by copper anode material.

The antibacterial test was carried out in 2 stages, namely the stage of making agar medium and the stage of antibacterial testing. The bacteria used were S. aureus as representative of gram-positive bacteria and E. coli as representative of gram-negative bacteria.

i. Agar Medium Preparation
The agar medium was prepared in 24 medium plates made from solid agar nutrient powder. Each plate containing nutrients was wrapped in a cover paper and tied. Next, the plate was sterilized in an autoclave at 121 °C for 15 minutes, then cooled down in a refrigerator.

ii. Antibacterial Testing
0.5 g of ZnO NPs powder was dispersed in 5 mL of DMSO solvents, then vortexed until completely mixed. Furthermore, wells are made in the medium so that previously contaminated with the bacteria used, namely *E. coli* and *S. aureus*. Two drops of ZnO NPs solution was put into a well and the cup was tightly closed then allowed to stand in an incubator at 37° C for 24 hours. The same procedure is performed for each bacterium and each sample. Antibacterial activity was measured based on the diameter of the clear zone produced.

3. Result and Discussion

3.1. The results of phytochemical tests

The results of phytochemical tests on *Musa paradisiaca L., Musa acuminata Cavendish Subgroup, and Musa x paradisiaca* peel extracts are presented in Table 1.

Table 1. The Results of Pythochemical Test

No	Banana	Flavonoid	Polyphenol	Alkaloid	Saponin
1	*Musa paradisiaca L.*	+	+	+	+
2	*Musa acuminata Cavendish Subgroup*	+	++	+	+
3	*Musa x paradisiaca*	+++	+++	++	++

As shown in Table 1, all three types of bananas contain secondary metabolites. The levels of these secondary metabolites are only stated qualitatively, based on the intensity of changes shown by each type of banana in the phytochemical test. *Musa x paradisiaca* peel extracts have the highest levels of secondary metabolites than the other two bananas whether it's the levels of flavonoids, polyphenols, alkaloids, or saponins. This levels difference is predicted to affect the properties of ZnO NPs produced and discussed in the following analysis results.

3.2. X-Ray Diffraction

Characterization using powder XRD aims to determine the synthesis of ZnO nanoparticle s by using banana peel extract as a capping agent. The characterization was based on the standard ZnO nanoparticle Diffractogram JCPDS Card No: 36-1451. The XRD profile of the product is presented in Figure 1. Based on the diffractogram, ZnO NPs have been successfully synthesized using extracts of all 3 types of banana peel varieties, which obtained a value of 2θ angles at 31.78°, 34.42°, 36.25°, 47.53°, 56.57°, 62.87°, 66.42°, 67.93°, and 69.09°. These peaks correspond to JCPDS card No: 36-1451. The peaks marked with the sign (*) on the diffractogram in Figure 1 (b) is the diffraction peak of the compound $Zn(OH)_2$ according to the JCPDS Card No 89-0138 as checked by the application of Match software. This means that $Zn(OH)_2$ also crystallized because of the possibility of temperature process used is not too high and not long enough so that the $Zn(OH)_2$ does not undergo the H_2O molecules releasing into ZnO and remains stable in the form of $Zn(OH)_2$

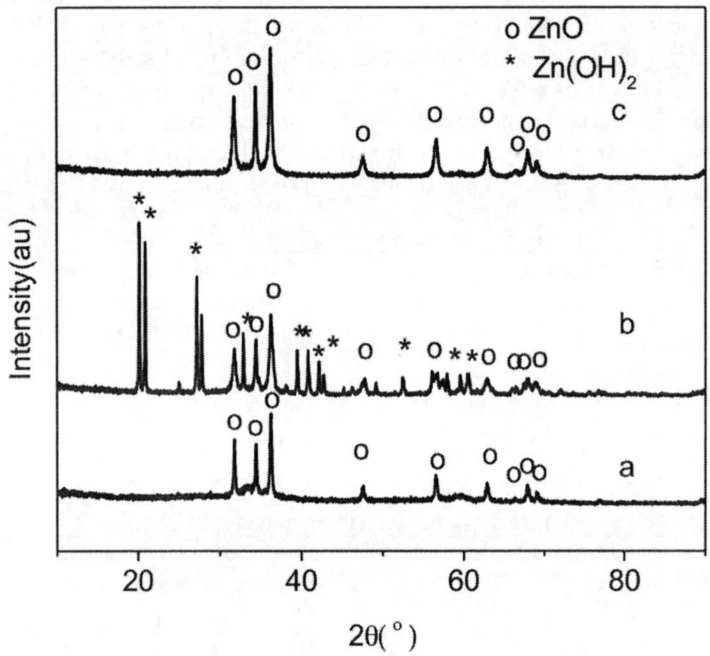

Figure 1. XRD diffractograms of ZnO NPs synthesized from banana peel extract of (a) *Musa paradisiaca L.; (b) Musa acuminata Cavendish Subgroup; (c) Musa x paradisiaca* peel extracts

As shown in Figure 1, ZnO NPs synthesized with Tanduk (*Musa x paradisiaca*) banana showed the lowest peak intensity compared to others due to the levels of secondary metabolites in the Tanduk banana that prevent further crystallization.

3.3. The Crystallite Size

The XRD results data are then processed using the RIETICA application. The crystalite size of each ZnO NPs sample obtained are presented in Table 2. XRD analysis for each type of bananas is only represented by one sample each. Calculation of the crystalite size of each sample is based on the results of the XRD analysis.

Table 2. The Crystallite Size

Banana	The Crystallite Size
Red Kepok (Musa paradisiaca L.)	33.3 nm
Ambon (Musa acuminata Cavendish Subgroup)	26.5 nm
Tanduk (Musa x paradisiaca)	24.5 nm

3.4. Antibacterial Activity

Antibacterial test results expressed as clear zone diameters around the wells in each plate is presented in Table 3 and Table 4. Based on the results of ZnO NPs antibacterial test against *E. coli* and *S. aureus* bacteria, it can be said that all of the three samples have antibacterial activity as shown from the clear zone that produced which is greater than 5 mm. DMSO solution that is used as a negative control does not form a clear zone so that the DMSO (dimethylsulfoxide) has no effect on the antibacterial test. ZnO NPs which were synthesized using capping agent extract of Tanduk banana peel has antibacterial activity > that of Ambon banana peel extract > that of red kepok banana peel extract. The antibacterial test results are directly proportional to the particle size of each ZnO NPs. The smallest particle size of ZnO NPs will be more effective to be used as an antibacterial because the nanoparticles will be easier to penetrate the cell walls of the bacteria and easier to damage the bacterial cells so that cells will undergo lysis quicker and resulting in more damage. All the three ZnO NPs have better antibacterial activity against *E. coli* gram-negative bacteria compared to that of *S. aureus* gram-positive bacteria, because gram-negative bacteria have a thinner layer of peptidoglycan which making it easier to be penetrated by the ZnO nanoparticles. These results are in accordance with those obtained by Diez-Pascual [13].

Table 3. Antibacterial test against *S. aureus*

No	**Banana**	**Average Clear Zone Diameter against *S. aureus* (mm)**	**Conclusion**
1	*Musa paradisiaca L.*	8.16	Antibacterial
2	*Musa acuminata Cavendish Subgroup*	8.20	Antibacterial
3	*Musa x paradisiaca*	9.36	Antibacterial
4	DMSO	0	-

Table 4. Antibacterial test against *E. coli*

No	**Banana**	**Average Clear Zone Diameter against *S. aureus* (mm)**	**Conclusion**
1	*Musa paradisiaca L.*	8.28	Antibacterial
2	*Musa acuminata Cavendish Subgroup*	8.38	Antibacterial
3	*Musa x paradisiaca*	10.73	Antibacterial
4	DMSO	0	-

4. Conclusion

ZnO nanoparticles were successfully synthesized using banana peel extract as a capping agent. The extract from the Tanduk banana peel gives the best results. This is evidenced by the similarity between the diffractogram and the standard diffractogram. In addition, antibacterial test results showed that all three types of bananas have antibacterial activity, where the clear zone produced is more than 5 mm.

References

[1] Basnet P *et al* 2018 *J.Photochem.Photobiol.*B. **183** 201–221
[2] Ibrahim HMM 2015 *J Radiat Res Appl Sc* **8**(3), 265–275.
[3] Sharmila G *et al* 2018 *J.Nanostructure Chem.* **8** 293–299
[4] Luque PA *et al* 2018 *J.* Mater. Sci.Mater. **29** 9764–9770

IOP Conf. Series: Materials Science and Engineering **833** (2020) 012076 doi:10.1088/1757-899X/833/1/012076

[5] Liang WH 2016 *Project Report* Malaysia: Universitas Tunku Abdul Rahman Malaysia
[6] Sari RN *et al* 2018 *Jurnal Pascapanen dan Bioteknologi Kelautan Dan Perikanan* **13** 41
[7] Slman DK *et al* 2018 *J. Pharm. Sci. & Res.* **10** 1590–1596
[8] Kementerian Pertanian 2016 Out Look Komoditas Pertanian Sub Sektor Hortikultura, Jakarta
[9] Anhwange B A 2008 *J Food Technol* **6** 263-266
[10] Marslin G *et al* 2018 *J. Materials* **11** 940.
[11] Ahmed S *et al* 2017 *J. Photochem Photobiol* B. **166** 272–284.
[12] Manyasree D *et al Int. J. App. Pharm.* **10** 224–228.
[13] Diez-Pascual AM *Polymers* 2019 **11** 1790

IOP Conf. Series: Materials Science and Engineering **833** (2020) 012077 doi:10.1088/1757-899X/833/1/012077

Effect of sonication frequency in synthesis of silica aerogel-activated carbon nanocomposite

Siti Muthomimah[1], Widi Rahayu[1], Nazriati[1,*], Adilah Aliyatulmuna[1]

[1]Department of Chemstry, Faculty of Mathematics and Natural Science, Universitas Negeri Malang, Jl. Semarang 5, Malang

*Corresponding Author: nazriati.fmipa@um.ac.id

Abstract. The silica adsorption capacity can be improved by composing it with activated carbon. The study aims to know the effect of sonication frequency in the nanocomposite silica aerogel-activated carbon synthesis using sol-gel method with TMCS (trimethylchlorosilane) and HMDS (hexamethyldisilazane) as modifying agents. Synthesized nanocomposites were further characterized by means of functional groups analysis using FTIR (Fourier Transform Infrared Spectrophotometry) method, morphological analysis using SEM (Scanning Electron Microscopy) method, and surface area analysis using BET (Brunauer-Emmet-Teller) method. The results of functional group analysis using FTIR shows that nanocomposite has a group of Si - OH, Si - O - Si, Si – C, C - H, C = O, C – O, the peaks around of 3300 cm^{-1}, 1051 cm^{-1} - 1180 cm^{-1}, 1253 cm^{-1}, 2960 cm^{-1}, 1649 cm^{-1}, 1417 cm^{-1}; morphological analysis using SEM indicates that nanocomposite synthesis with 42 kHz frequency sonication has a smaller particle size than 35 kHz frequency, where are the surface area analysis using BET method obtained 674.897 m^2 / g at frequency of 42 kHz and 486.909 m^2 / g synthesis at frequency of 35 kHz.

1. Introduction

Silica aerogel is classified as a type of aerogel made from silica with low density property, large surface area and low thermal conductivity [1]. Synthesis of silica aerogel was obtained by using sodium silicate extracted from bagasse ash [2], brown rice husk [3], and rice husk ash [4]. Silica aerogel with a large surface area is utilized as an adsorbent [5]. However, the stability of silica against humid air is not good, thus affecting its adsorption power [6]. Activated carbon as a universal adsorbent has a large surface area, pore volume and large pore diameter. However, the active side of activated carbon limits its adsorption power [7]. The adsorption capacity of silica and activated carbon is improved through both composites [8]. Activated carbon modifies the structure of silica aerogel, limits the contraction of aerogel, and strengthens the frame of silica aerogel [9].

Silica aerogel-activated carbon composites combine the polar-inorganic properties of silica aerogel and non-polar-organic from activated carbon; thus, the composite is able to increase the adsorption ability. Givianrad and co-workers synthesized silica aerogel-activated carbon nanocomposites by using a mixture of TMOS, methanol, MTMS, and NH_4F solution [8]. Mixing was performed by using the help of a magnetic stirrer for 30 minutes. The results of the mixture in the form of gel plus activated carbon were stirred by using a 20000 rpm homogenizer for 30 minutes. Drying gel was stored at room temperature for 5 days. The synthesis product was applied as a heavy metal waste adsorbent with an adsorption capacity reaching 60%. Adsorbents utilizing activated carbon have an adsorption capacity of 50% and adsorbents utilizing silica aerogel have an adsorption capacity of 30%. However, the synthesis of both adsorbents requires a long time since the addition of activated carbon into gel causes the active carbon to not properly stick to silica [5].

Sonication is considered as one of the effective techniques in mixing, reaction processes, and breakdown of materials with the help of high energy at ultrasonic wavelengths. Synthesis of mesoporous silica from rice husk ash along with sonication as a substitute for mixing with magnetic stirrer, indicated results such as large surface area, increased pore volume [4] and large adsorption power [3]. Synthesis by sonication method produces cavitation bubbles which will create nanoscale diameter particles and homogeneous size distribution of nanoparticles [10]. Sonication produces shock waves which will cause cavitation bubbles, separate particle agglomeration, and produce particle dispersion [11]. Frequency indicates the number of shock waves that occur, resulting in the formation of cavitation bubbles, clot separation, particle dispersion, and accelerated synthesis.

Synthesis method with sonication frequency has advantages, such as 1) shorter synthesis, 2) greater particle surface area, 3) larger pore volume, and 4) more homogeneous pore-size. This study aims to determine the effect of sonication frequency on the synthesis of silica aerogel-activated carbon nanocomposites. The synthesis was conducted by the sol-gel method, using bagasse ash as a source of silica.

2. Research Material and Method

The basic ingredients used are bagasse ash, obtained from PG. Krebet Baru II Malang, Malang Regency. The utilized chemicals in this study include: Pure Analysis (pa), such as: NaOH (Merck), Activated Carbon (Merck), Cation exchange resins, TMCS (Merck), HMDS (Merck), NH_4OH (Merck), HCl (Merck), aqua demineralization (Hydrobatt), universal indicator paper (MN), Whatman No. 40 filter paper, and aquadest. The tools used include 42 kHz ultrasonic batch (Branson 1510), 35 kHz ultrasonic batch (Sharp), analytic balance sheet (Durascale Dab-200), magnetic stirrer (Cimarec), oven (Memmert), FTIR (Shimadzu IR-Prestige 21), SEM (FEI-INSPECT-S50), and BET (Quantachrome).

Bagasse ash was sifted and roasted to produce dry bagasse ash. Bagasse ash of 10 grams was added with 60 mL of 2N NaOH in 250 mL Erlenmeyer, then stirred with a magnetic stirrer and heated to boiling for 1 hour. The mixture was cooled to room temperature and filtered with Whatman 40 filter paper and Buchner filter. Sodium silicate was mixed by using a cation exchange resin (pH 2) in a ratio of 1: 3, stirred by using a magnetic stirrer for 30 minutes and was filtered to produce silicate acid filtrate.

The synthesis step was carried out in ultrasonic batches with frequencies of 35 kHz and 42 kHz at 10- and 20-minutes time variations. Silica acid of 10 mL was added by activated carbon as much as 0.03 grams, stirred in an ultrasonic batch for 10 minutes. TMCS was added into silica acid by modifying agent as much as 0.3 mL, stirred for 10 minutes, added with 0.6 mL HMDS, and stirred for 10 minutes. If no gel has been formed, NH_4OH was added 1N dropwise. This method was repeated for variations in frequency and time.

Then the gel was processed at 40 °C for 18 hours and 60 °C for 1 hour. The formed gel was washed by using aquademineral to neutral pH, then the gel was dried at 80 °C for 24 hours to obtain Silica Aerogel-activated Carbon (SAC) nanocomposite.

SAC nanocomposite produced was characterized by: (a) FT-IR (Fourier Transform Infrared Spectrophotometry) to determine the functional groups; (b) SEM (Scanning Electron Microscopy) to determine surface morphology; and (c) surface area analysis using BET (Brunauer-Emmet-Teller) method.

3. Result and Discussion

3.1. Silica Aerogel-activated Carbon (SAC) Nanocomposite

In figure 1 FT-IR spectra show the formation of nanocomposites in the presence of broad absorption of silanol (Si-OH) groups in an area of about 3300 cm^{-1}, siloxane (Si-O-Si) regions around wave numbers 1051 cm^{-1} and 1180 cm^{-1}. Si - C at wave number 1253 cm^{-1}, group C - H at wave number around 2960 cm^{-1}. The absorption area indicates that the Si - OH condensation becomes Si - O and is modified to Si

- CH_3 in the presence of a modifying agent. The functional group is identical to the functional group in the aerogel silica. The C = O group in the absorption region is around 1649 cm^{-1}, C-O at 1417 cm^{-1}, the group is identical to the functional group in activated carbon. This analysis shows the successful synthesis of aerogel silica nanocomposites-activated carbon.

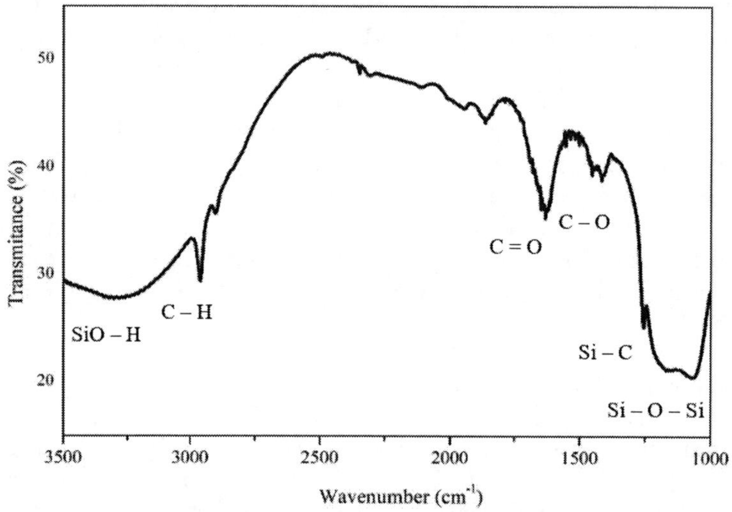

Figure 1. FT-IR Spectra of Silica Aerogel-Activated Carbon Nanocomposite.

3.2. Surface morphology of silica aerogel- activated carbon (SAC) nanocomposite

Figure 2 indicates the surface morphology of SAC nanocomposite which is the result of SEM magnification for 100,000 times at frequencies of 35 kHz and 42 kHz. The surface morphology of SAC nanocomposites presents a round-like cotton shape with homogeneous size distribution.

The particle size of SAC nanocomposite synthesized with sonication at a frequency of 35 kHz is bigger than the size of the nanocomposite particles synthesized at the frequency of 42 kHz, see figure 2 (a) and 2 (b), respectively. The surface morphology of SAC at 42 kHz frequency appears homogeneous and small, demonstrating that the sonication shock wave was able to separate the clumping of particles (*agglomeration*).

3.3. Effect of sonication frequency on nanocomposite characteristics

Physical properties of SAC nanocomposite include surface area and pore volume. The surface area is determined by BET method and pore volume is determined by BJH (Barret-Joyner-Hallenda) method. the data of surface area and pore volume are shown in Table 1.

Sonication frequencies at 35 kHz and 42 kHz affect surface area and pore volume. Frequency indicates the number of shock waves that occur, producing cavitation bubbles which are able to separate lumps and disperse particles. Utilization of greater frequencies results in shorter synthesis times. Overall, the sonication frequencies which produce (from the largest to smallest) SAC nanocomposite surface area are: 42 kHz for 10 minutes, 35 kHz for 20 minutes and 35 kHz for 10 minutes. The data indicates that the sonication method has an energy limit, yet a maximum impact on the continuity of the reaction chemistry. Conversely, if the limit is exceeded, the effectiveness of the chemical reaction will reduce.

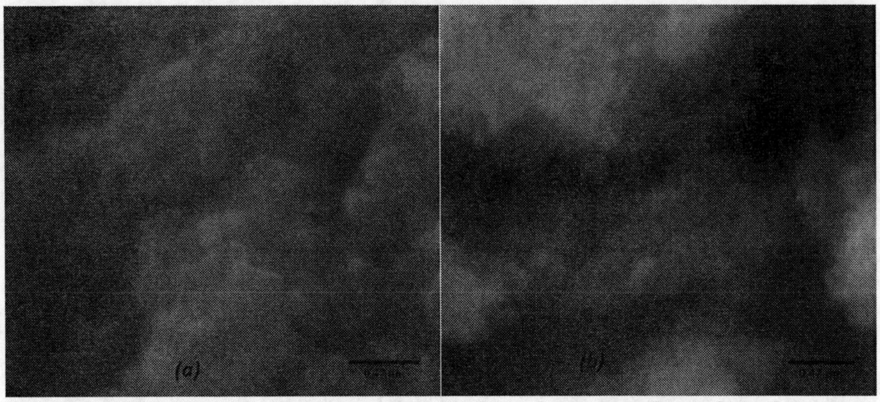

Figure 2. SEM SAC Nanocomposite (a) 35 kHz frequency (b) 42 kHz frequency.

Table 1. Physical properties of SAC nanocomposites.

Frequency (kHz)	Time (minute)	S_{BET} (m^2/g)	Pore volume (cm^3/g)
35	10	486.909	0.377
35	20	507.206	0.431
42	10	674.897	0.634
42	20	485.948	0.397

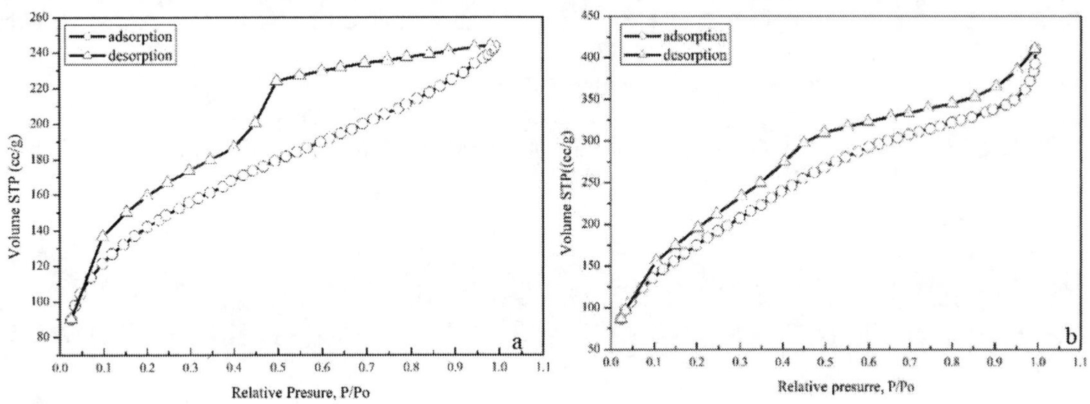

Figure 3. Adsorption-desorption curves of nanocomposites (a) synthesis at a frequency of 35 KHz (b) synthesis at a frequency of 42 KHz.

SAC nanocomposite adsorption-desorption type by using nitrogen gas is presented in Figure 3 (a) and (b). The curve has similarities with typical type IV isotherms according to IUPAC classification, initially the slow gas absorption is caused by stronger intermolecular interaction of adsorbate than the interaction between adsorbate and adsorbent. This type applies to porous material. In addition, particles from the synthesis have mesoporous-sized pore diameters.

4. Conclusion

Synthesis of SAC nanocomposite can be performed by sonication at frequencies of 35 kHz and 42 kHz and synthesis time of 10 minutes and 20 minutes. The results of SAC nanocomposite FTIR indicate potential existence of Si - OH, Si - O – Si, C – H groups (identical functional groups in aerogel silica) and C = O, C - O groups (identical to activated carbon functional groups). SEM image of nanocomposite depicts that synthesis of SAC nanocomposite with sonication frequency of 42 kHz has smaller particle

IOP Conf. Series: Materials Science and Engineering **833** (2020) 012077 doi:10.1088/1757-899X/833/1/012077

size than the results of synthesis with 35 kHz with for 10 minutes. This result proves that sonication shock waves can separate agglomeration. Synthesis of SAC nanocomposite at a frequency of 42 kHz requires a shorter synthesis time than that with a frequency of 35 kHz. Longer synthesis time (20 minutes) indicates smaller surface area of SAC nanocomposite. However, sonication method has energy limits that still have a maximum impact on the sustainability of chemical reactions. If the limit is exceeded, the effectiveness of chemical reactions with sonication frequency will reduce. Sonication frequency influences surface morphology, particle size distribution, surface area, and pore volume. Based on the type IV adsorption-desorption curve, SAC nanocomposite from the synthesis has a mesoporous-sized pore diameter.

5. Acknowledgment

The researchers would like to thank the Directorate of Higher Education for financial support for this research (*Hibah PDUPT*, contract no. 19.3.46/UN32.14.1/LT/2019).

References

[1] Y. Zhao, Y. Li, and R. Zhang 2018 *Ceram. Int.*
[2] N. Nazriati, H. Setyawan, S. Affandi, M. Yuwana, and S. Winardi 2014 *J. Non. Cryst. Solids* **400** 6-11.
[3] S. Sankar, N. Kaur, S. Lee, and D. Y. Kim 2018 *Ceram. Int.* **44** 8720–8724
[4] M. Chareonpanich, A. Nanta-ngern, and J. Limtrakul 2007 *Mater. Lett.* **61** 5153–5156
[5] M. Karnib, A. Kabbani, H. Holail, and Z. Olama 2014 *Energy Procedia* **50** 113–120
[6] A. Mohammadi and J. Moghaddas 2015 *Chem. Eng. Res. Des.* **94** 475–484
[7] P. Valle-Vigón, M. Sevilla, and A. B. Fuertes 2013 *Microporous Mesoporous Mater.* **176** 78–85
[8] M. H. Givianrad, M. Rabani, M. Saber-Tehrani, P. Aberoomand-Azar, and M. Hosseini Sabzevari 2013 *J. Saudi Chem. Soc.* **17** 329–335
[9] Ś. Agnieszka, S. Wojciech, and J. Paulina 2015 *J. Non. Cryst. Solids* 416 **1–3**
[10] T. Kudo, K. Sekiguchi, K. Sankoda, N. Namiki, and S. Nii 2017 *Ultrason. Sonochem.* **37** 16–22
[11] K. S. Suslick, D. J. Casadonte, M. L. H. Green, and M. E. Thompson 1987 *Ultrasonics* **25** 56–59

Preparation of Nanocellulose Bioplastic with a Gradation Color of Red and Yellow

Zulfa Durrotun Nasihin, Masruri MASRURI*, Warsito Warsito, Arie Srihardyastutie

Chemistry Departement, Faculty of Mathematics and Natural Sciences, Brawijaya University, Jl. Veteran 65145 Malang

*Corresponding author: masruri@ub.ac.id

Abstract. A bioplastic derived from nanocellulose isolated from the pine flower (*Pinus* Merkusii) waste has been made. The red and yellow colors of bioplastic were afforded from the extract of dragon fruit and turmeric. Meanwhile, nanocellulose was hydrolyzed by citric acid using concentrations of 10%, 30%, and 60%, and obtained crystallinity index of 61.00%, 57.24%, and 52.82% respectively. Each nanocellulose applied has a crystallite size of 15.09 nm (calculated using XRD data with Scherrer's equation). The result indicated that the increase in both dragon fruit and turmeric extracts concentration increase the red and yellow color intensity of the bioplastic. This study paves the way for further application of nanocellulose in bioplastic.

1. Introduction

The increasing of plastic used in recent time brings a negative impact on human life. The main raw material for plastic comes from petroleum which is increasingly scarce and not sustainable. The European Council has demonstrated the need to reduce dependence on the use of fuel and gas to create sustainable energy. In addition, the use of plastic as food packaging such as plastic teabags can transfer various microparticles and nanoparticles to food, causing adverse effects on human health [1]. Therefore, an alternative raw material is needed to create environmentally, friendly, and biodegradable plastic with the same strength as petroleum-derived plastics. In addition, biodegradability is one of important properties needed in various bioplastic applications such as food packaging to reduce polymer-based waste from oil [2].

Bioplastics have emerged as substitutes for polymeric materials. It can be made using natural ingredients such as starch. Based on previous research, bioplastics can be made using starch, nanocellulose, and plasticizers [3]. Nanocellulose is the most effective material as a biopolymer matrix to improve the performance of bio-based materials [4]. Nanocellulose has the same backbone as cellulose in hydrolysis process [5]. Previously, Rambabu et al. [6] reported pine flower contains 44 wt% cellulose and 22 wt% lignin. Phenolic compounds in pine flower have been widely developed as starting material for industrial chemicals [7-8]. However, cellulose contained in pine flower is still rarely used. Dragon fruit and turmeric are natural dyes used as coloring additives in bioplastics [9]. The addition of dyes to bioplastics makes it more interesting for food packaging. The purpose of this study is determine the color results obtained in bioplastics from pine flower nanocellulose using natural dyes from dragon fruit and turmeric extracts also their further application in bioplastics.

2. Material and Methods

2.1. Materials
The sample used in this study was Merkusii Pine flower waste obtained from Coban Rais area, Batu, East Java, Indonesia. Chemicals used were sodium hydroxide (Merck), sodium hypochlorite 12% w/v (technical grade), citric acid 99-102% (SMART-LAB), polyvinyl alcohol (Merck), cassava, glycerol, turmeric, dragon fruit, and aquades. Equipment used in this study included: glassware, oven (MEMMERT), ultrasonicator (DELTA D150H), magnetic stirrer, hot plate, grinder, and analytical balance (Ohuas Pioneer PA214).

2.2. Procedures

2.2.1. Sample Preparation of Pine Flower. Sample preparation was in accordance with reported procedure [6] with modifications. Pine flowers were soaked in hot water at 100° C for 2 h, washed and dried for 24 h at 50° C. The pine flowers were grounded and soaked in hot water for 4 h then roasted for 24 h at 60° C.

2.2.2. Sample Isolation from Pine Flower. At this stage, cellulose isolation from pine flower was in accordance with references [6,10] with modifications. A total of 50 g of grounded pine flower was taken into a 1000 ml beaker and 500 mL of 6% NaOH was added. The ratio of pine flower powder and NaOH solution was 1:10 (v/v). The mixture was then heated at 70 ° C for 4 h. After extraction, the mixture of NaOH solution and pine flower powder was washed to neutral pH.

A total of 50 g of pine solid resulted from the process of alkalization underwent bleaching process using 6% sodium hypochlorite solution as much as 500 mL. Then it was heated at 70 ° C for 2 h. This process was repeated until white cellulose was produced. The solid formed was filtered and washed with distilled water to obtain neutral pH. The cellulose obtained was then dried in an oven at 30-50 °C until constant weight was obtained. The dried cellulose was blended and filtered.

2.2.3. Synthesis of Nanocellulose. Cellulose isolated from pine flower waste was synthesized to obtain nanocellulose referring to procedures [5,10] with modification. Five grams of cellulose from the isolated pine flower waste was added with 10%, 30%, and 60 % citric acid in 100 ml respectively. Ultrasonication was prepared at the temperature of 45 °C. After the temperature reached 45 °C, the sample was taken into the ultrasonication reactor for 60 mins. The reaction was stopped by the immersion of the rector in a water bath containing cold distilled water followed by drying process using the oven at a temperature of 30-50 °C.

2.2.4. Preparation of Natural Dye. 200 g of fresh turmeric and dragon fruit respectively, was peeled and washed with clean water and followed by drying in an oven at 30 °C for 30 mins. The extract was obtained using juice extractor without additional water. The extract was then filtered and centrifuged at 2000 rpm for 15 mins. Each extract was carried out by varying concentrations of 5%, 15%, and 25% (v/v) [11].

2.2.5. Synthesis of Starch-based Bioplastic using Natural Dye. The starch-based bioplastic synthesis procedure was in accordance with reports by [12,13] with modification. Polyvinyl alcohol (PVA) solution was prepared by mixing 4% PVA, 25% glycerol, and 71% distilled water (w/w) in 500 rpm stirrer at 80 °C for 1 h. The starch solution was made by mixing 3% starch (cassava), 12% glycerol and 85% distilled water (w/w) in a stirrer of 500 rpm at 80 °C for 15 mins. Both solutions were mixed with PVA: starch ratio of 80:20 (w/w) and 500 rpm sterilizer at 80 °C for 10 mins. The resulting mix was then added with 1% nanocellulose, 10% turmeric extract and 10% natural dragon fruit extract with a concentration of 5%, 15%, and 25% (v/v).

2.3. Characterization

2.3.1. X-ray Diffractiometry (XRD) Analysis. The nanocellulose was analyzed by X-ray diffraction (XRD) Merk PanAnalytical type Expert Pro ranging from 10° to 90° with Cu K radiation (1.5406) at room temperature. The crystallinity index (CrI) of the nanocellulose was determined by Seghal's method equation (1) and the crystallite size (D_{hkl}) was determined by Scherrer equation (2) [14,15].

$$CrI = (Itot - Iamorph) / Itot \times 100\% \tag{1}$$

$$Dhkl = \frac{0.9\lambda}{\beta_1 \cos\theta} \tag{2}$$

2.3.2. Fourier Transform Infrared Spectroscopy (FTIR) Analysis. The FTIR spectra of pine flower was recorded using Fourier Transform Infrared (FTIR) Spectrometer SHIMADZU 8400s. The samples were measured in the wavenumber range 4000-400 cm^{-1} using KBr pellet method.

3. Result and Discussion

3.1. Synthesis of Nanocellulose

Citric acid was used for chemical hydrolysis of cellulose from pine flowers as lignocellulosic waste. Nanocellulose hydrolysis produced in the form of yellowish white powder shown Figure 1. The yields obtained from nanocellulose hydrolysis using citric acid with concentrations of 10%, 30%, and 60% were 95.6%, 94.6%, and 93.4%, respectively.

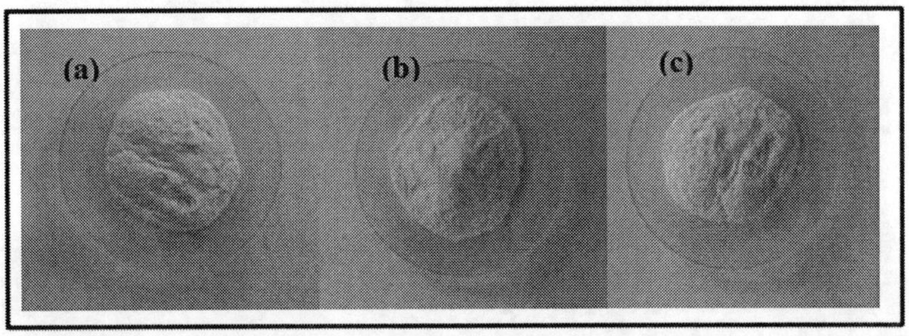

Figure 1. Nanocellulose products. (a) CA 10, (b) CA 30, (c) CA 60.

3.2. Characterization of Nanocellulose using FTIR

The transmission of infrared spectrum on hydrolyzed nanocellulose using citric acid with varying concentrations (CA 10, CA 30, CA 60) is presented in Figure 2. The peaks were interpreted as following: presence of broad peak at 3433 cm^{-1} wave number for O-H stretching, the peak at 2962 cm^{-1} for C-H aliphatic stretching, and the peak at 1692 cm^{-1} of vibrations for C=C lignin [6,16]. The peak at 1160 cm^{-1} can be ascribed to C–O–C asymmetric stretching of lignin, hemicellulose, and cellulose [4].

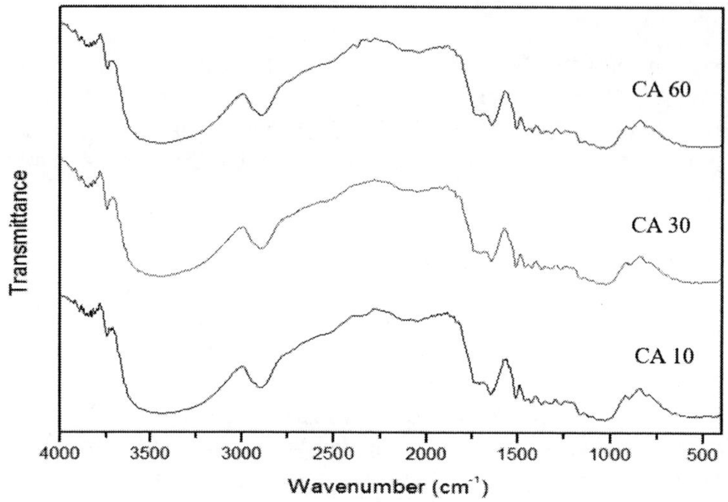

Figure 2. FTIR spectra of nanocellulose.

3.3. Characterization of Nanocellulose using XRD

X-ray diffraction analysis for hydrolyzed nanocellulose using citric acid (CA 10, CA 30, CA 60) is shown in Figure 3. The XRD pattern results showed that in the case of nanocellulose from the pine flower, there were three intense peaks observed at the diffraction angle (2θ) = 15°, 22.5°, and 34°, in accordance to the crystalline cellulose [1].

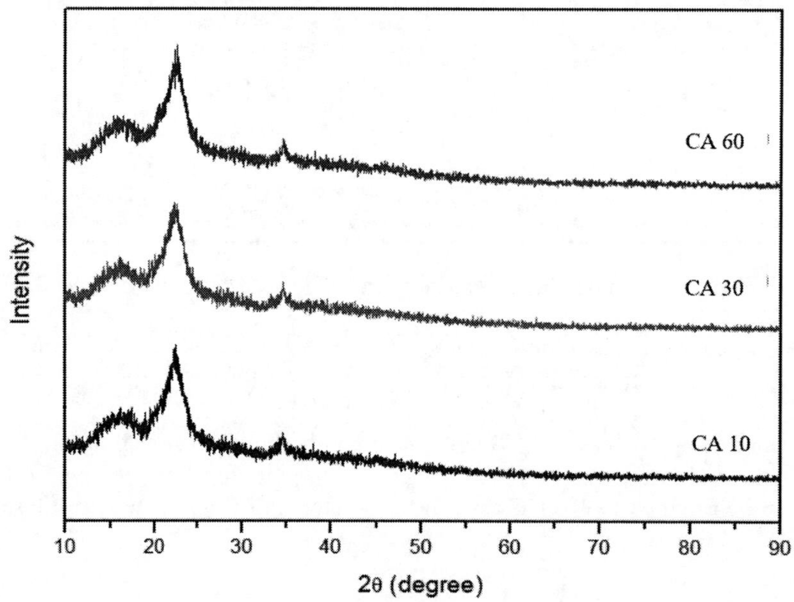

Figure 3. XRD pattern of nanocellulose.

Hydrolysis with acids not only able to remove the amorphous portion of cellulose but also damage the crystalline regions [17]. The increase of citric acid concentration in the hydrolysis process damage the crystalline area of the nanocellulose as shown in Table 1.

Table 1. Result of XRD pattern of nanocellulose.

Sample	Peak 2θ	CI (%)	Crystallite Size
CA 10	22.40	61.00	15.09 nm
CA 30	22.50	57.24	15.09 nm
CA 60	22.55	52.82	15.09 nm

3.4. Synthesis of Starch-based Bioplastics
Bioplastic color gradations were made using extracts of natural dyes which are turmeric and dragon fruit. According to the result in Figure 4, the increase in both dragon fruit and turmeric extracts concentration also increase the red and yellow color intensity of the bioplastic.

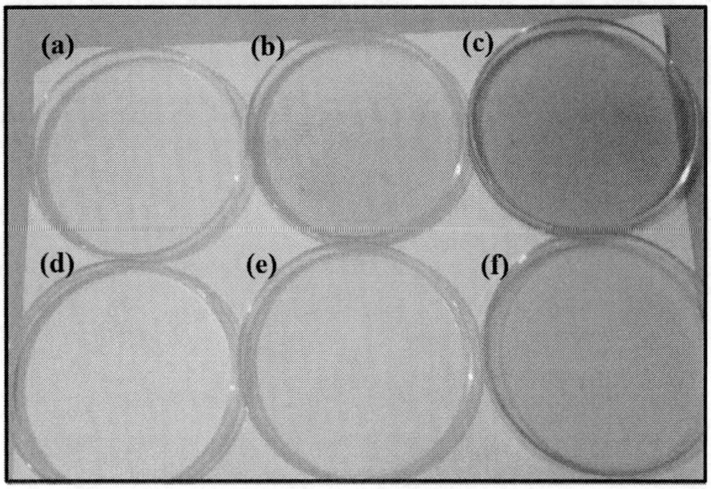

Figure 4. Gradation color of bioplastic using turmeric extract in various concentration (a) 5% (b) 15% (c) 25%, and dragon fruit extract in various concentration (d) 5%, (e) 15%, (f) 25%.

4. Conclusion
In conclusion, bioplastics were produced from pine flower waste and showed promising result. By using natural additive dyes, it is possible to ensure the utilization of pine flower waste for bioplastic as food packaging which is harmless to the environment.

5. Acknowledgement
The research is supported by Penelitian Dasar Unggulan Perguruan Tinggi (2019) from DRPM of Ministry for Research, Technology and Higher Education of Indonesian. The authors gratefully acknowledge the contributions of Mr. Widji Sulistjo and Mr. Hadi Kurniawan.

References
[1] Henrique M A Neto F W P Silvério H A Martins D F Gurgel L V A Barud H and Pasquini D 2015 *Industrial Crops and Products* **76** 128–140
[2] Muñoz-Bonilla A Echeverria C Sonseca Á Arrieta M P and Fernández-García M 2019 *Materials* **12** 641
[3] Agustin M B Ahmmad B Alonzo S M M and Patriana F M 2014 *Journal of Reinforced Plastics and Composites* **33** 2205–13.
[4] Soykeabkaew N Tawichai N Thanomsilp C and Suwantong O 2016 *Walailak Journal of Science and Technology* **14** 353-368

[5] Shaheen T I and Emam H E 2018 *International Journal of Biological Macromolecules* **107** 1599–1606
[6] Rambabu N Panthapulakkal S Sain M and Dalai A K 2015 *Industrial Crop and Product* **10** 1-9
[7] Masruri M and Pamungkas KK 2019 *IOP Conference Series: Materials Science and Engineering* **546** 042023
[8] Masruri M, Amini R W and Rahman M F 2016 *Indonesian Journal of Chemistry* **16** 59-64
[9] Putra E P D Thamrin E S and Saputra H 2019 *IOP Conference Series: Earth Environment Science* **258** 012047
[10] Liu C Li B Du H Lv D Zhang Y Yu G Mu X and Peng H 2016 *Carbohydrate Polymers* **151** 716–724
[11] Maizura M Aminah A and Wan Aida W M 2011 *International Food Research Journal* **18** 529-534
[12] Alashwal B Y Saad B M Gupta A Sharma S and Mishra P 2019 *Journal of King Saud University – Science* **31** 5
[13] Guimarães M Botaro V R Novack K M Teixeira F G and Tonoli G H D 2015 *Industrial Crops and Product* **70** 72–83
[14] Segal L Creely J J Martin J A E and Conrad C M 1959 *Textile Research Journal* **29** 786–794
[15] Jaboyedoff M, Kübler B and Thélin P 1999 *Clay Minerals* **34** 601–617
[16] Habibi Y Lucia L A and Rojas O J 2010 *Chemical Reviews* **110** 3479–3500
[17] Yu H Qin Z Liang B Liu N Zhou Z and Chen L 2013 *Journal of Materials Chemistry A* **1** 3938

Synthesis and Characterization of Copolymer Poly(vinylidene fluoride)/Graphene Nanofiber

[1]Muhammad Nasir*, [1]Triannisa Rahmawati, [1]Indriyati, [1]Fitri Dara, [1]Desak Gede Sri Andayani

[1]Research Unit for Clean Technology, Indonesia Institute of Sciences (LIPI), Cisitu-Sangkuriang, Bandung 40135, Indonesia

*Corresponding author: mnasir71@yahoo.com

Abstract. The copolymer poly(vinylidenefluoride)/graphene (cPVDF/graphene) nanofibers were fabricated using the electrospinning method. The concentration of cPVDF was fixed at 24% (wt.%), graphene 0.01-0.1% (wt.%) in *N, N*-dimethylacetamide (N, N-DMAC). Characterization was conducted to analyse the effect of graphene concentration and electrospinning parameter of cPVDF/graphene using SEM-EDS, ATR-FTIR, TG-DSC, and XRD. Electrospinning parameters (applied voltage, the distance between nozzle and collector, flow rate) effected nanofiber diameter, beads formation, β-crystal phase formation in the flow rate variation, the nanofiber produced is at the size of the nanometer and at optimum conditions, almost no beads are found, as well as homogeneous. SEM images showed that the presence of graphene and applied electrospinning parameter affected morphology and an average diameter of nanofibers. FTIR and XRD pattern indicated that β-crystal phase of cPVDF was dominant than α-phase, which was followed by increasing the crystallinity when graphene added. Graphene also influenced the thermal behaviour of composite nanofibers which was supported by TG-DTG and DSC thermograms. Hydrophobicity of cPVDF nanofibers increased with more graphene content. These composite nanofibers have potential application to solve environmental problems.

1. Introduction

Approximately most of 97.5% hydrosphere in oceans is the combined mass of water on, above, and under the surface, in which only 2.5% of the hydrosphere is freshwater. This freshwater can be found in glaciers, permanent snow, mountainous regions, lakes, rivers, and ground water[1].Along with the development of the world, the need for freshwater increases. Moreover, the increase of population, industry, and water pollution are the concern for the world to demands water treatment technology. There are several ways to overcome this problem, likes constructing the reservoir, use water effectively, and diversion of water needs. On the other hand, escalation water treatment technologies are the target for the environmental engineer, and membrane separation technology has been a concern to solve water crisis[2].

Membrane separation process is a filtration process to separate molecules with different sizes by using pressure as driving force. Polluted water can be treated by using membrane. The polluted water forced by pressure through the semipermeable membrane to separate material based on membrane pore size. This process has several advantages in the applications, such as high efficiency, low cost, can separate in a wide range, can be used for a wide range of application[2,3], and applied in industrial scale such as reverse osmosis (RO), ultrafiltration (UF), microfiltration (MF), and nanofiltration (NF)[3].

A semi-crystalline and piezoelectric poly(vinylidene fluoride) or PVDF is a polymer with –(CH_2CF_2)–repeating unit that widely used in the separation and purification. This hydrophobic polymer generally has five difference polymorphs: α, β, γ, δ, and ε[4,5], but α, β, γ are the most common obtained phase of PVDF[5,6], in which every polymer crystal phase give significant effect to intrinsic properties of PVDF with the degree of crystallinity between 35 to 70%[7]. This crystallinity depends on molecular weight, thermal history, molecular weight distribution, polymerization method, and cooling rates[8]. PVDF has exceptional properties, such as high thermal stability, mechanical strength, high chemical, and excellent aging resistance[9,10] that make this polymer widely used as raw material and suitable for membrane filtration and purification. Although PVDF nature properties have been hydrophobic, in purpose to increase hydrophobicity, another hydrophobic material needed.

Graphene is material with C-C structure that makes this material highly hydrophobic. This material has one atom thick, two-dimensional sheets of carbon atoms[11], hexagonal lattice with sp^2 hybridization[12].The techniques developed to synthesize graphene can be divided into five major methods, like mechanical cleavage, chemical method, epitaxial growth, total organic synthesis, and chemical vapor deposition (CVD)[13]. This material can be used as graphene bulbs, graphene battery, drug carriers, hydrogen storage materials, graphene superconductors, and graphene chips[14]. Because of its outstanding property of graphene, i.e.: high thermal stability, high mechanical properties, flexible, transparent, and hydrophobic, this material used to mix with PVDF to produce more hydrophobic membrane.

There are several methods to produce membrane, which is the conventional method through casting solution and electrospinning. Electrospinning has attracted interest of academia and industry, because of its ability to produce ultrafine fiber[15] with diameter range from few micrometer to tens of nanometers and high versatility and potential for application in many fields[16] than the conventional method. Nevertheless, to reach optimum nanofiber, the electrospinning parameter must be optimized for every polymer solution. This electrospinning parameter encompasses voltage, flow rate, and the distance between collector and nozzle.

In this study, we analyzed the effect of graphene concentration and electrospinning parameter to get optimum nanofiber morphology. Applied voltage, flow rate, the distance between collector and nozzle was varied to get optimum nanofiber using electrospinning method. This research also analyzed the diameter average, diameter distribution, crystallinity, thermal characteristic, and hydrophobicity nanofiber at optimum condition. SEM-EDS, ATR-FTIR, TG-DSC, XRD, and water contact angle were conducted to analyzed nanofiber characteristics.

2. Experimental method

2.1. Materials

Copolymer poly(vinylidene fluoride) (cPVDF) as nanofiber polymer material was purchased from Kynar flex® 2801-00 ARKEMA Inc. USA. *N, N*-dimethylacetamide (N, N-DMAc) as solvent was obtained from Merck. Graphene was supplied by Graphene Supermarket in form of nano powder multilayer flakes. All materials were used without further purification.

2.2. Solution preparation and electrospinning of cPVDF/graphene

The electrospinning solutions with different graphene concentrations were prepared using N,N-DMAc as solvent. cPVDF powder was fixed at 24 wt%, while graphene varied at 0.01; 0.05; and 0.1 wt%. Both of materials were mixed with N,N-DMAc using magnetic stirrer at room temperature until homogenous. During electrospinning, every sample was optimized by varying flow rate (0.5; 1.0; 1.5; and 2.0 μL/min), voltage (16, 19, 22, and 25 kV), and distance between nozzle and collector (12.5; 14.5 cm). The nanofiberproductwascharacterizedusing SEM-EDS toselectfinenanofiber-basedonbeads, branch, homogeneity, and diameter for each graphene concentration. Furthermore, an optimized nanofiber was characterized using ATR-FTIR, XRD, TG-DTG, DSC, and contact angle. All electrospinning process was done at controlled room temperature and humidity.

2.3. Characterization

The surface morphology of nanofiber was studied with scanning electron microscopy-energy dispersive spectroscopy (SEM JEOL JSM IT-300) at 20.0 kV accelerating voltage for scanning the surface, whereas the electron dispersive spectroscopy (EDS) used to monitor the % composition of sample qualitatively. Fiber diameter, fiber diameter distribution, and the average diameter of fiber were calculated using Image J software. 125 points were taken to calculate nanofiber diameters. Thermal characteristics of cPVDF/graphene nanofiber were determined by thermogravimetric-differential thermal gravimetric using TGA 209 Lybra (Netzsch), in which sample was heated from room temperature to 900 °C for 10 °C/min of heating rate under N_2 gas. Fourier transform infrared (FT-IR) in attenuated total reflectance (ATR) mode from Thermo Scientific Nicolet iS5 and x-ray diffraction (XRD) Bruker D8 ADVANCE were used to observe the functional group in the sample, the diffraction peak of nanofiber, and crystallinity. XRD analysis was performed from the 2θ range between 10° to 70° with the rotating anode CuK α tube (λ = 1.5418 Å). Differential scanning calorimetry using DSC 214 Polyma NETZSCH Germany. This instrument performed at room temperature until 275 °C for 10 °C/min of heating rate under N_2 gas. The contact angle measured to quantify sample hydrophobicity and Image J conducted to calculate the contact angle value.

3. Results and discussions

3.1. Nanofiber morphology

SEM-EDS were carried out to investigate the effect of graphene concentration and electrospinning parameter to the nanofiber morphology. The morphology of nanofiber was analyzed to get optimum nanofiber for every specific graphene concentration.

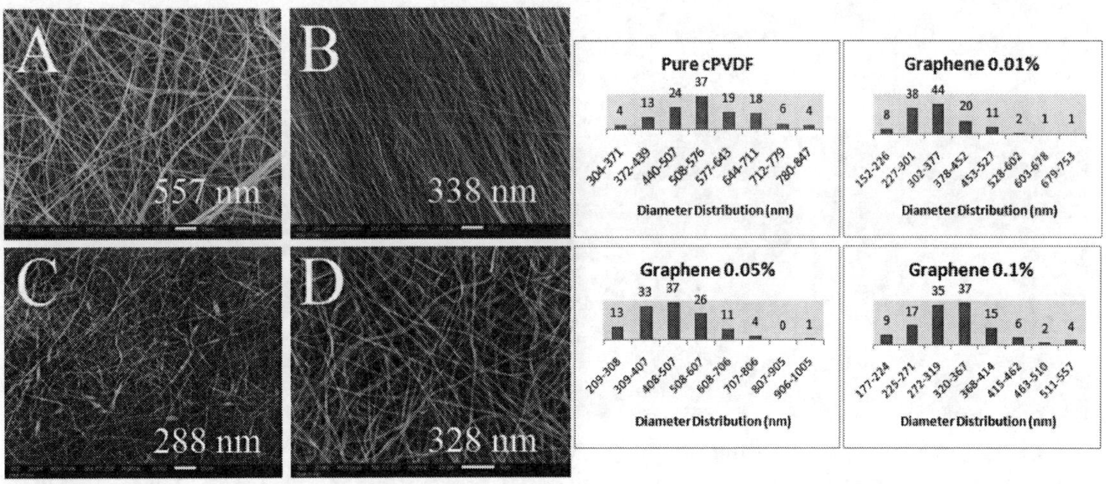

Figure 1. Morphology and diameter distribution of nanofiber using SEM. Image (left) and diameter distribution (right) ofthe effect of graphene concentration on nanofiber morphology. (a) pure cPVDF, (b) graphene 0.01%, (c) graphene 0.05%, and (d) graphene 0.1%. The number in white colour is diameter average.

3.1.1. The effect of graphene concentration on nanofiber morphology. Figure 1 shows typical SEM images of pure cPVDF and cPVDF/graphene at different graphene concentration with same electrospinning parameter condition on the left, and diameter distribution on the right. The electrospinning parameter was 19 kV (voltage), 1 μL/min (flow rate), and 14.5 cm (distance between collector drum and nozzle). At different graphene concentration, electrospun nanofibers can show different nanofiber morphologies, such as branch, bead, nanofiber diameter, porous, and homogeneity. On this cases, all nanofiber in these results had randomly fibrous morphology, except for graphene 0.01% a little bit aligned. As predicted, although electrospinning parameter design in the same condition, graphene concentration significantly affect nanofiber morphology as image resulted. The most significant morphology is evident from Figure 1(c). At this nanofiber condition (graphene 0.05%), highly beaded, branch, and inhomogenious nanofibers were produced. This surface scanning was different compared to pure cPVDF and another nanofiber. However, the calculation of the diameter as shown in Figure 1 on the right side provided that all fibers have a diameter in the nanometer scale. Furthermore, the addition of reduced nanofiber diameter than pure nanofiber. This condition in consequence of the increase of charge density that generates a higher elongation force to the ejected jets[5].

3.1.2. The effect of flow rate to nanofiber morphology. Applied voltage, flow rate, and the distance between collector and nozzle were varied toget optimum nanofiber using electrospinning method. Figure 2 shows typical SEM images of pure cPVDF and cPVDF/graphene 0.05% at different electrospinning flow rate condition (left), and diameter distribution (right). The difference of electrospinning flow rate has changed the morphology of nanofibers, shown inFigure 2(a), 2(b) and 2(c). This nanofibers contains excessively beads, branches, and heterogeneous in size. However, the calculation of the diameter, as shown in the right side of Figure 1, provided that all fibers have a diameter in the nanometer scale. The other variation such as voltage and distance from the nozzle to the collector were also observed. These variations also produce different nanofibers morphologies, with different levels of beads and branches.

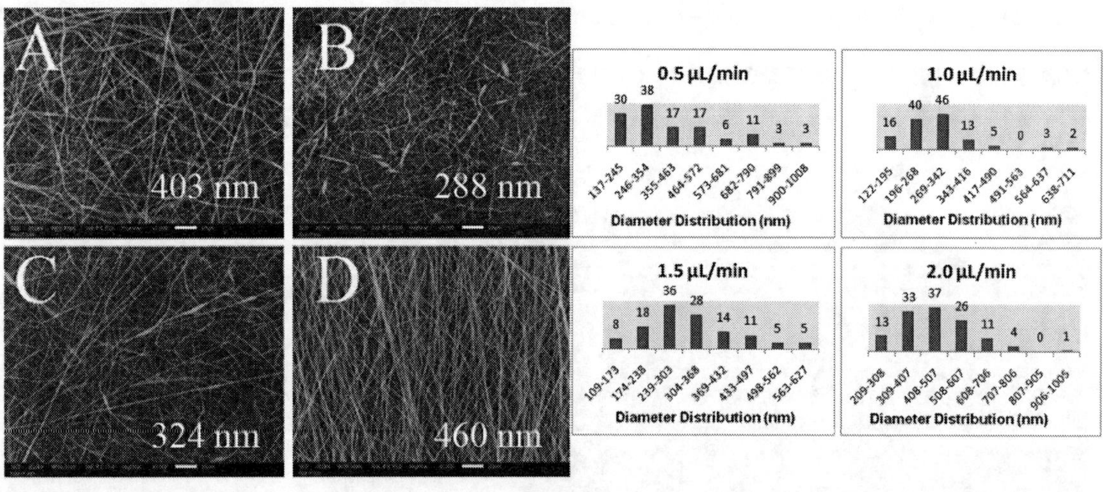

Figure 2. Scanning Electron Microscopy (SEM) images of nanofibers. Image (left) and diameter distribution (right) of electrospinning flow rate variation to nanofiber morphology. (a) pure cPVDF 0.5 μL/min, (b) 1.0 μL/min, (c) 1.5 μL/min, and (d) 2.0 μL/min, when graphene concentration 0.05%. The number in white colour is diameter average.

From observations on the influence of variations in graphene concentration and optimization of the parameters of the electrospinning tool, it was obtained that each polymer solution has optimum conditions for producing nanofibers shown in SEM results on Figure 3. Furthermore, high and low applied voltage made the polymer solution did not be electrospun, whereas high and low flow rate caused beads and abroad range of nanofiber diameter.

IOP Conf. Series: Materials Science and Engineering **833** (2020) 012079 doi:10.1088/1757-899X/833/1/012079

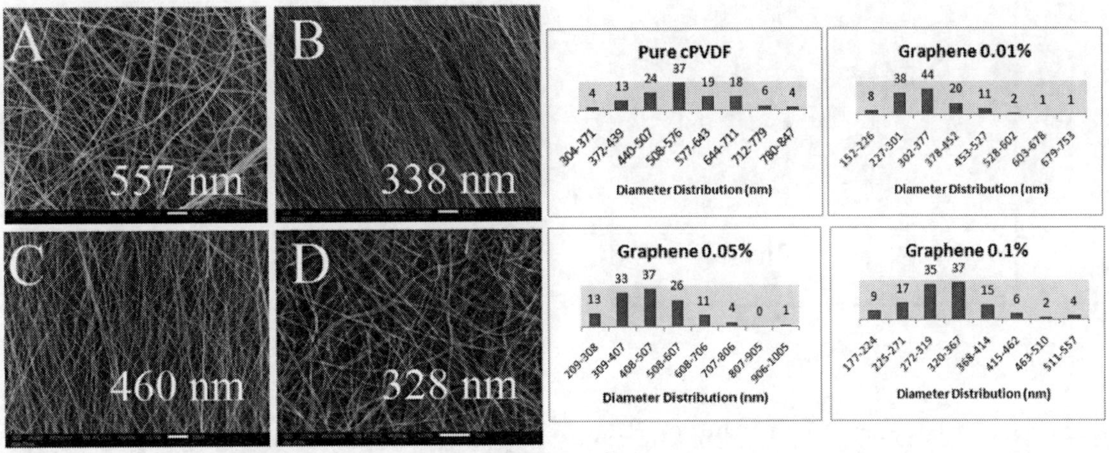

Figure 3. Scanning Electron Microscopy (SEM) images of nanofibers: Image (left) and diameter distribution (right) nanofiber at optimum condition for every graphene concentration. (a) pure cPVDF, (b) graphene 0.01%, (c) graphene 0.05%, and (d) graphene 0.1%. The number in white colour is diameter average.

3.2. Crystallinity

ATR-FTIR, XRD was conducted to measure the crystallinity of nanofibers at the optimum state. Figure 4 shows the results of measurements using ATR-FTIR. The analysis showed that both of pure cPVDF and nanofiber with added graphene, there were three crystalline phases, such as α, β, and γ. In the spectra, α-crystal phase appeared around 612 cm^{-1} and 762 cm^{-1} with low intensity, β-crystal phase around 837 cm^{-1} with high intensity, 1276 cm^{-1} with low intensity, and the crystallization phase of γ-crystal phase around 1429 cm^{-1} with low intensity, and other crystal phase peaks with very low intensity. Thus, it was in line with the literature of FTIR measurements for these three phases[17,18,19]. The band at around 612 cm^{-1} was assigned to a mixed mode of CF_2 bending vibration. The band at around 762 cm^{-1} was related to in-plane bending or rocking vibration. The band at around 837 cm^{-1} was related to a mixed of CH_2 rocking and CF_2 asymmetric stretching. The band at around 1276 cm^{-1} was assigned to C-F stretching vibration[20]. From the result of FTIR spectra, it was known that nanofibers are dominated by β-crystal phase. Crystallinity data from FTIR measurements were confirmed by XRD data. The total crystallinity of nanofibers 37.1%; 37.5%; 45.0%; 55.2% for pure cPVDF; graphene 0.01%; 0.05%; and 0.1% respectively. This measurement indicated that the addition of graphene increase nanofiber crystallinity, in which as much as graphene added to the polymer, theresulted nanofibers more crystalline.The XRD diffractogram as shown in Figure 5.The XRD diffractogram reflected that a value of 2θ around 20.4°was assigned to orthorhombicβ-crystal phase[18],while18.4°, 36.5°and 41.2°were assigned to monoclinicα-crystal phase, but the α-crystal phase intensity was not as strong as β-crystal phase. Thus, the results of the xrd data show that β-crystal phase was the dominant crystal phase in nanofibers and the corresponding FTIR spectra is in good agreement with XRD diffractogram to analyzed the crystalline phase. Theα-crystal phase is phase kinetically favorable, this phase frequently observed in membranes, while β crystal phase thermodynamically more stable and seldom observed in membranes[21,22].

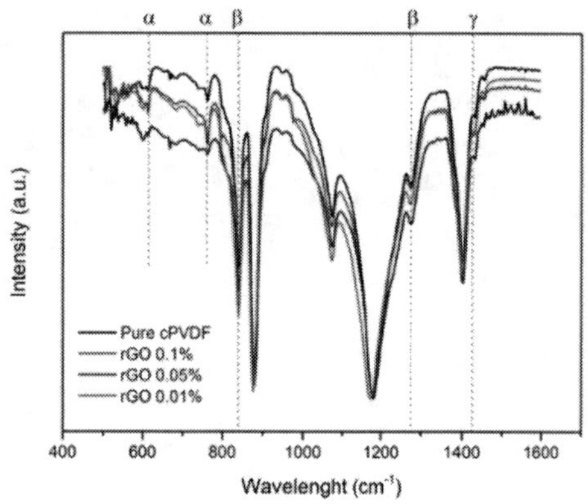

Figure 4. FTIR spectra of pure cPVDF, graphene 0.01%, graphene 0.05%, and graphene 0.1% nanofiber.

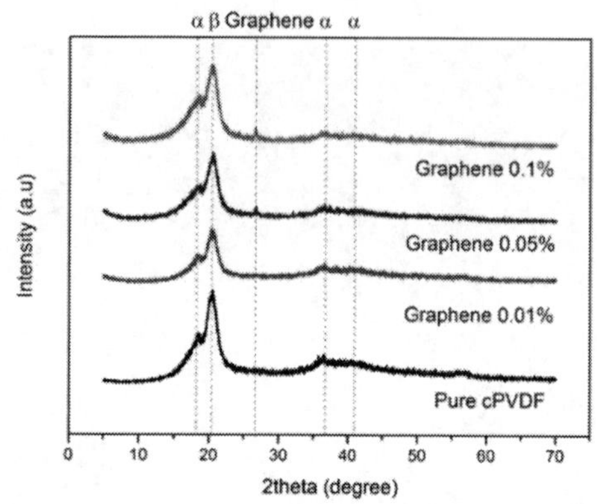

Figure 5. XRD spectra of pure cPVDF, graphene 0.01%, graphene 0.05%, and graphene 0.1% nanofiber.

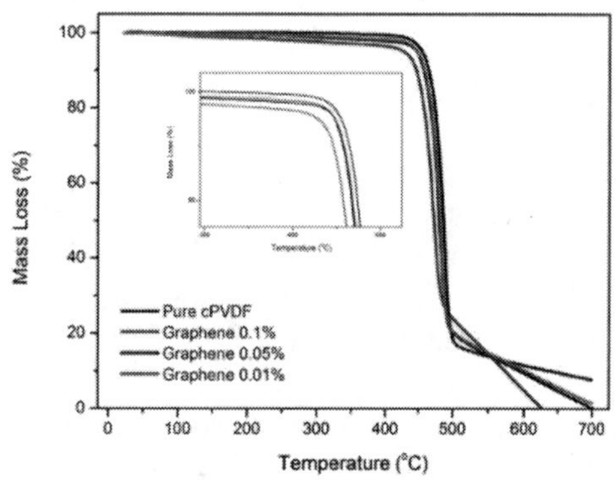

Figure 6. TGA spectra of pure cPVDF, graphene 0.01%, graphene 0.05%, and graphene 0.1% nanofiber.

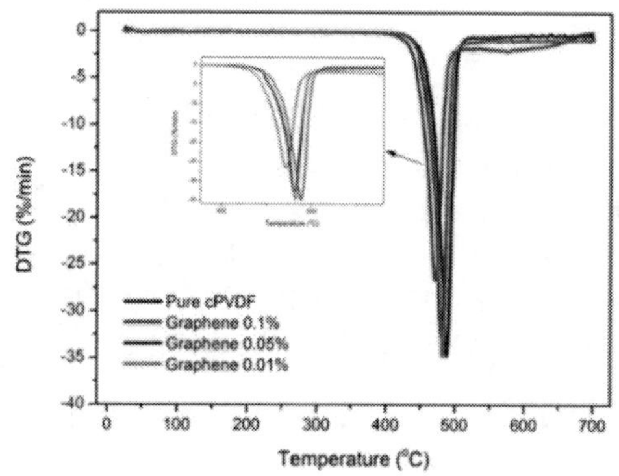

Figure 7. DTG spectra of pure cPVDF, graphene 0.01%, graphene 0.05%, and graphene 0.1% nanofibers.

3.3. Thermal characteristic

The thermal characteristics of the nanofibers were characterized using TG-DTG and DSC. TGA spectra exhibited that decomposition occurred at for all nanofibers, it was started with the loss of moisture or solvent material between the room temperature until below 200°C, followed by decomposition of the polymer. Every composite produced the same trend of decomposition. As Figure 6 showed, pure cPVDF nanofiber consists of volatilization and loss of moisture or solvent bellows 200°C, followed by the decomposition of nanofiber cPVDF between 200-699.7 °C with a mass loss more than 92%. In which at small addition of graphene, the thermal stability decrease, while at more graphene added, the thermal stability increase. These phenomenastill under study. The DTG spectra was attached to confirm TGA spectra (Figure 7). The spectra described that nanofiber has the same tendency on decomposition rate. Furthermore, analysis results using DSC show that the addition of small graphene increases the melting point, whereas the addition of more graphene decreases the melting point. DSC measurement results as in Figure 8. From the results of this analysis, it was known that the average melting point of nanofiber was at a temperature of 144 °C with an endothermic process.

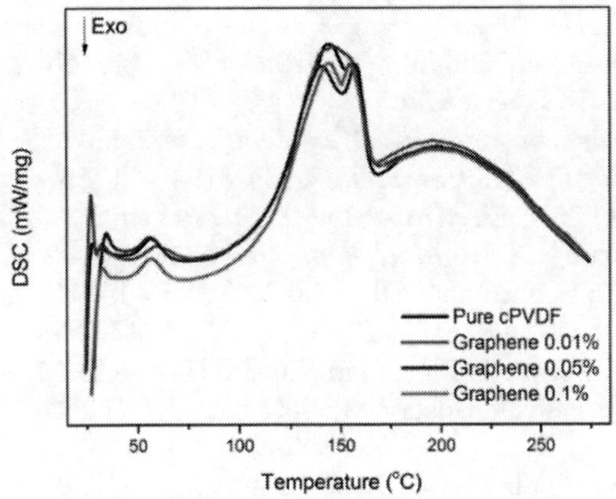

Figure 8. DSC spectra of pure cPVDF, graphene 0.01%, graphene 0.05%, and graphene 0.1% nanofiber.

3.4. Hydrophobicity

Nanofiber water contact angle analysis was carried out to determine the hydrophobicity of nanofiber before and after graphene was added at certain concentrations. Contact angle calculation is done using Image J. software. From the calculation results, it is known that the addition of graphene increases the water contact angle[10]. The higher amount of graphene mixed into the solution, the higher hydrophobicity of the nanofibers produced. The water contact angle valueis 134.2; 134.7; 138.1; and 139.0 for pure cPVDF, graphene 0.01%, graphene 0.05%, and graphene 0.1% respectively.

4. Conclusions

In this study, nanofiber has been fabricated using electrospinning method. The effect of variations in graphene concentration with fixed electrospinning parameters has investigated. The variations in the electrospinning parameters including variations in voltage, flow rates, and the distance between the drum collector and the nozzle had been investigated. It was known that variations in graphene concentration and the electrospinning parameters affect the morphology of the nanofiber. In the optimum conditions, the nanofibers are homogenous, unbranched, and free of beads. The addition of grapheme decreasing the average diameter of the nanofiber, and as the amount of graphene increases, the higher hydrophobicity of the nanofiber produced. Based on the results of the ATR-FTIR and XRD analysis, all variations in graphene concentration are dominated by the β-crystal phase. Thermal analysis found that the reaction was endothermic with 144 °C average of melting point.

Acknowledgements

This research acknowledges Ministry of Research, Technology, and Higher Education of the Republic of Indonesia and Indonesian Institute of Sciences.

References

[1] Della Sala DA 2018 *Elsevier Inc.*
[2] Shi L 2014 University of Delaware
[3] Cui ZF, Jiang Y, and Field RW 2010 *Elsevier Ltd*
[4] Lovinger AJ 1982 *Macromolecules* 15 **1** 40-44
[5] Abolhasani MM, Shirvanimoghaddam K, and Naebe M. 2017 *Compos. Sci. Technol.* **138** 49-56
[6] Tan X, Rodrigue D. 2019 *Polymers (Basel)* 11 **7** 1160
[7] Liu F, Hashim NA, Liu Y, Abed MRM, and Li K 2011 *J. Memb. Sci.* 375 **1-2** 1-27
[8] Roussel S, McElroy KL, and Judovits LH 1992 *Polym. Eng. Sci.* 32 **17** 1300-1308
[9] Kang GD and Cao YM 2014 *J Memb Sci.* **463** 145-165

[10] Yuan X, Li W, Liu H, Han N, and Zhang X 2016 *Compos. Commun.* **2** 5-8
[11] Soldano C, Mahmood A, and Dujardin E 2010 *Carbon N Y.* 48 **8** 2127-2150
[12] Zhen Z, and Zhu H 2018 *Elsevier Inc*
[13] Kim J-KZQ 2013 *Graphene for Transparent Conductors.* **53**
[14] Ren S, Rong P, and Yu Q 2018 *Ceram Int.* 44 **11** 11940-11955
[15] Wang L, and Ryan AJ 2011 *Electrospinning Tissue Regen.* **3** 33
[16] Bhardwaj N, and Kundu SC2010 *Biotechnol. Adv.* 28 **3** 325-347
[17] Shao H, Fang J, Wang H, and Lin T 2015 *RSC Adv.* 5 **19** 14345-14350
[18] Cai X, Lei T, Sun D, and Lin L 2017 *RSC Adv.* 7 **25** 15382-15389
[19] Issa A, Al-Maadeed M, Luyt A, Ponnamma D, and Hassan M 2017 *Carbon.* 3**4**30
[20] Lanceros-Méndez S, Mano JF, Costa AM, and Schmidt VH 2001 *J. Macromol. Sci - Phys.* 40 B **3-4** 517-527
[21] Cui Z, Hassankiadeh NT, Zhuang Y, Drioli E, and Lee YM 2015 *Prog. Polym. Sci.* 5194-126
[22] Correia HMG and Ramos MMD 2005 *Comput. Mater. Sci.* 33 **1-3** 224-229

IOP Conf. Series: Materials Science and Engineering **833** (2020) 012080 doi:10.1088/1757-899X/833/1/012080

Zinc Oxide with Visible Light Photocatalytic Activity Originated from Oxygen Vacancy Defects

Wynona A. Nimpoeno[1,2], Hendrik O. Lintang[1,2], and Leny Yuliati[1,2*]

[1]Ma Chung Research Center for Photosynthetic Pigments, Universitas Ma Chung, Villa Puncak Tidar N-01, Malang 65151, East Java, Indonesia
[2]Department of Chemistry, Faculty of Science and Technology, Universitas Ma Chung, Villa Puncak Tidar N-01, Malang 65151, East Java, Indonesia

Corresponding author: leny.yuliati@machung.ac.id

Abstract. In this study, ZnO material with oxygen defect was successfully synthesized using a hydrothermal method and interestingly, it was found to possess a remarkable photocatalytic activity under the visible light irradiation. The synthesized ZnO material was then characterized using X-ray diffraction (XRD), diffuse reflectance ultraviolet-visible (DR UV-vis), Fourier transform infrared spectroscopy (FTIR), and spectrofluorometer. Both diffraction pattern and absorption spectrum showed the characteristic of ZnO. However, the maximum emission wavelength of the synthesized ZnO was observed at 558 nm, showing that the ZnO was formed with a high concentration of oxygen vacancy. The excitation spectra also indicated that the prepared ZnO could be excited at the visible light region, which was more than 400 nm. The ZnO was evaluated for photocatalytic phenol degradation under UV and visible light. After 15 hours-reaction, the synthesized ZnO showed the ability to degrade 59.8% phenol under UV light and 39.6% phenol under visible light. This result showed a remarkable performance of ZnO under visible light, which was comparable to that performed under UV light. This study proposed that the existence of oxygen vacancy defects successfully induced the photocatalytic activity of ZnO under visible light irradiation through the electron trapping mechanism in the oxygen vacancy state.

1. Introduction

Organic wastes have caused various types of environmental pollution, including water, soil, and air pollution. The ability of organic compounds to be partially dissolved in water and evaporated into the air makes this waste more difficult to handle [1]. Methods to treat organic wastes include adsorption [2], separation [3], and photomineralization [4]. Photomineralization is a greener approach to handle organic waste as it converts organic wastes into harmless water and carbon dioxide.

Recently, many types of researches are focused on improving the photocatalytic activities of metal oxide photocatalysts. Semiconductors such as titanium dioxide (TiO_2) [5-7] and zinc oxide (ZnO) [8–11] are widely used for this photocatalysis application. Efforts to improve their photocatalytic activities include doping [12,13], dye sensitization [14,15], and the addition of another visible light active semiconductor [16,17]. ZnO is a tremendous potential material with its relatively easy synthesis and ability to harvest UV and visible light if doped or sensitized properly.

Defects in ZnO are usually described as intrinsic and extrinsic defects [18]. Extrinsic defects are introduced through impurities addition, such as nitrogen or metal ion doping [19, 20], while intrinsic defects include interstitial, vacancy, and anti-site defects. The existence of these defects could increase photocatalytic activities of ZnO, especially under UV light irradiation [21, 22]. Unfortunately, sunlight consists more of visible light than the UV light. Therefore, in order to utilize the sunlight, improving the activity of ZnO under visible light is very important. In this study, the ZnO was synthesized using a hydrothermal method. The synthesized ZnO was found to possess a remarkable photocatalytic activity under visible light irradiation.

2. Experimental

All chemicals in this research were commercially purchased and used without further treatment.

2.1. ZnO Synthesis

Zinc oxide was synthesized similarly using previously reported method [23]. Zinc acetate dehydrate ($Zn(CO_2CH_3)_2.2H_2O$, Merck, 99,5–101%) and sodium hydroxide (NaOH, Merck, 99%) were dissolved separately in 75 mL of distilled water at the concentration of 0.1 and 0.2 M, respectively. Zinc acetate solution was then added into the sodium hydroxide solution in a dropwise manner while stirring vigorously. The combined solution was then stirred for an hour to promote zinc hydroxide formation. The resulting solution was moved into a Teflon bottle and was heated using an oven at 150 °C for 6 hours to form ZnO. The ZnO suspension obtained was separated from the byproduct solution through decantation. The ZnO solid was washed with distilled water until the filtrate pH reached neutral. It was then dried to obtain ZnO powder.

2.2. Material Characterization

Synthesized ZnO was characterized using X-ray diffraction (XRD; Rigaku, SmartLab), Fourier-transform infrared spectroscopy (FTIR, JASCO FT/IR-6800), diffuse reflectance ultraviolet-visible (DR UV-vis) spectrophotometer (JASCO V-760), and spectrofluorometer (JASCO FP-8500) to study the crystallinity and crystallite size, the functional groups, optical, and fluorescence properties of the prepared samples, respectively. All spectra were recorded at room temperature. As for the FTIR spectra measurement, all the samples were prepared in a pellet form with the addition of potassium bromide (KBr, spectroscopy grade).

2.3. Photocatalytic Test

In the photocatalytic activity test, phenol (C_6H_5OH, Merck, 99–100.5%) was used as the organic pollutant model. The light source employed was a halogen lamp (Dolan-Jenner MI-157, 150W). The photocatalyst (50 mg) was suspended into the phenol solution (25 mL, 50 ppm). The mixture was stirred for 2 hours under the dark condition to achieve the adsorption-desorption equilibrium, followed by irradiation using the halogen lamp for 3 hours. The suspension was filtered to remove the solid photocatalyst using a filter membrane prior to the measurement of phenol content using high-performance liquid chromatography (HPLC; Shimadzu, LC-20AT) with C-18 column (YMC-Triart, S-5 μm, 12 nm). The percentage of phenol degradation was calculated from the percentage ratio of degraded phenol concentration to the initial concentration of phenol.

3. Result and Discussion

3.1. Structural Analysis

The synthesized ZnO powder was characterized using the XRD. The resulting pattern shows diffraction peaks characteristic to zinc oxide at 2θ of 31.65°, 34.30°, 36.13°, 47.42°, 56.47°, 62.73°, 66.23°, 67.81°, and 68.95° associated with the diffractions of (1 0 0), (0 0 2), (1 0 1), (1 0 2), (1 1 0), (1 0 3), (2 0 0), (1 1 2), and (2 0 1) planes, respectively. Based on the diffraction pattern, ZnO crystalline was successfully synthesized in a wurtzite form with a crystallite size of 32.8 nm.

IOP Conf. Series: Materials Science and Engineering **833** (2020) 012080 doi:10.1088/1757-899X/833/1/012080

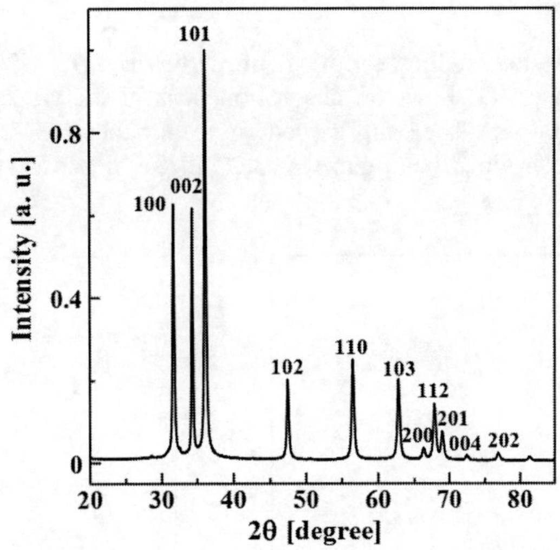

Figure 1. XRD pattern of synthesized ZnO.

3.2. Functional Group Analysis

To clarify further the successful synthesis of ZnO, the changes in functional groups before and after ZnO synthesis were investigated. The spectra of $Zn(OAc)_2$ as the precursor to synthesize the ZnO and the synthesized ZnO were measured using FTIR spectrophotometer and shown in Figure 2. The $Zn(OAc)_2$ gave vibrational peaks assigned to the stretching and bending vibrations of O-H of moisture and atmospheric CO_2, that can be detected at 3106, 2498 and 2348 cm^{-1}. COO^- symmetrical and asymmetric vibrations, C-CH_3 framework, C-C, and acetic ion vibrations can be additionally detected at the wavenumbers of 1563, 1444, 1060, 1018, 846, 690, and 620 cm^{-1} [24]. In contrast, the synthesized ZnO only exhibited peaks at 3400 and 1639 cm^{-1}, corresponding to the stretching and bending vibrations of O-H and H-O-H of water [25], trace leftover of acetic ion at 1639 and 896 cm^{-1} [26], and ZnO vibrational peak at 423 cm^{-1} [27]. The disappearance of acetic ion vibrational peaks indicated the successful ZnO synthesis from the $Zn(OAc)_2$.

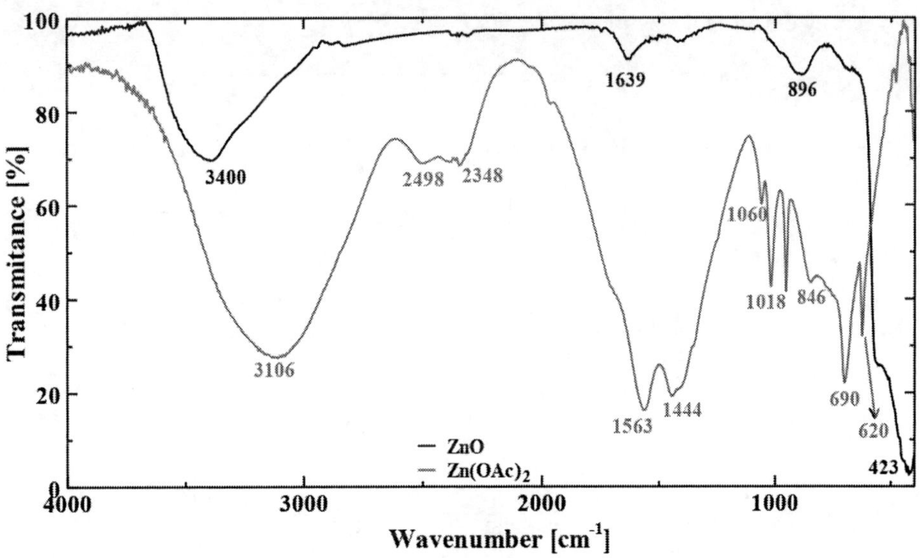

Figure 2. FTIR spectra of $Zn(OAc)_2$ as the ZnO precursor and synthesized ZnO.

IOP Conf. Series: Materials Science and Engineering **833** (2020) 012080 doi:10.1088/1757-899X/833/1/012080

3.3. Optical Properties

The synthesized ZnO was studied for its optical properties by DR UV-vis spectrophotometer. As depicted in Figure 3(a), the ZnO shows an absorption peak at the wavelength of 313 nm, with the absorption edge close to 400 nm. The band gap energy of synthesized ZnO was calculated using the Tauc plot. The obtained determined band gap was 3.22 EV, which was still falling in the UV region (Figure 3(b)).

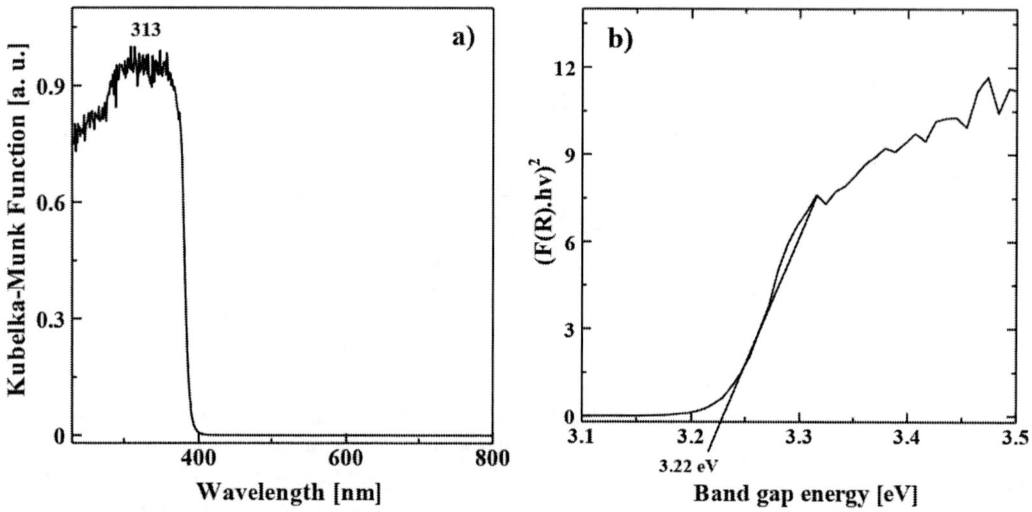

Figure 3. (a) DR UV-vis spectrum and (b) the Tauc plot of synthesized ZnO.

Fig 4 shows the excitation and emission spectra of the synthesized ZnO. The excitation spectrum was measured at the emission wavelength of 558 nm, while emission spectrum was recorded using an excitation wavelength of 372 nm. It has been reported that ZnO prepared by co-precipitation method gave emission peaks at around 420 and 465 nm [28]. Different from this reported sample, the currently synthesized ZnO gave a yellow emission at 558 nm, which was attributed to electron transition from the conduction band (CB) of ZnO to the oxygen vacancy (Vo) [29]. This result indicated the presence of oxygen vacancy defect in the synthesized ZnO.

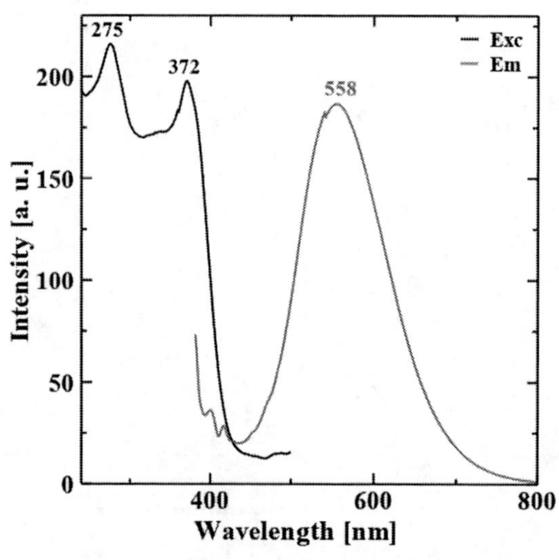

Figure 4. Excitation and emission spectra of synthesized ZnO.

3.4. Photocatalytic Activities
The synthesized ZnO was tested for its photocatalytic activities under UV and visible light irradiation. Figure 5(a) shows the photocatalytic activity of ZnO under UV light irradiation. As shown in Figure 5(a), the amount of left phenol decreased with the increase of the irradiation time. As high as 59.8% of phenol was successfully degraded after 15 hours-reaction. The photocatalytic activity of ZnO under UV light irradiation could be attributed to electron and hole formation, hydroxyl generation, and electron/hole trapping in the oxygen vacancy defect sites [30, 31].

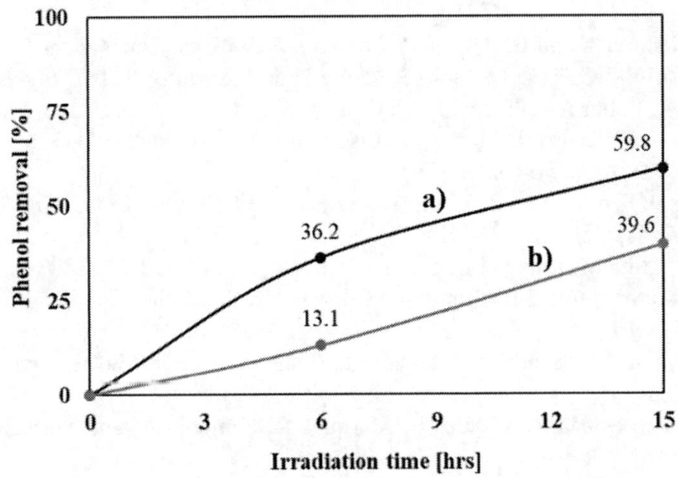

Figure 5. Photocatalytic activity of ZnO under (a) UV and (b) visible light irradiation.

ZnO was also evaluated for its photocatalytic activity under visible light irradiation and the result was shown in Figure 5(b). Contrary to its large band gap of 3.22 eV, the synthesized ZnO showed a remarkable photocatalytic activity under visible light irradiation. After 15 hours-reaction, only falling short of 20% from its photocatalytic activity under UV light irradiation, the ZnO gave 39.6% of phenol degradation. Moreover, comparing the activity of ZnO to other visible-light-active photocatalyst, the currently synthesized ZnO gave a comparable level of photocatalytic activity. For instance, while the ZnO gave 13% phenol degradation after 6-h-reaction, the bulk carbon nitride gave 20% degradation under visible light irradiation [32]. This visible light activity would come from the oxygen vacancy defect as discussed previously. The oxygen vacancy defect would create a new energy level with lower band potential than the conduction band of the ZnO. This lower energy level of the oxygen defects provided the new path for the ZnO to absorb visible light by an electron trapping mechanism in the defect states. The current result was indeed in good agreement with the previous report that the oxygen vacancy led to the enhanced visible light activity of the ZnO [33]. With the presence of the oxygen vacancy defects, the synthesized ZnO was demonstrated to give both UV and visible light activity, which was an important parameter to utilize the whole spectrum of solar light.

4. Conclusion
ZnO was successfully synthesized by the hydrothermal method. XRD patterns showed that ZnO has a crystalline wurtzite structure and a crystallite size of 32.8 nm. FTIR spectra detected the disappearance of acetic acid-related vibration peaks, indicating the formation of ZnO. The fluorescence spectra of the ZnO showed the excitation and emission of oxygen vacancy defect sites in the synthesized ZnO. Photocatalytic tests after 15 hours-reaction showed that the synthesized ZnO has good photocatalytic activity under UV light irradiation (59.8%) and remarkable photocatalytic activity under visible light

irradiation (39.6%). The remarkable photocatalytic activity under visible light irradiation was proposed to be contributed from the oxygen vacancy defect sites.

References

[1] Patnaik P 2018 *Handbook of Environmental Analysis: Chemical Pollutants in Air, Water, Soil, and Solid Wastes* 3rd ed (Boca Raton: CRC Press)
[2] Cheng W, Gao W, Cui X, Ma J and Li R 2016 *J. Taiwan Inst. Chem. Eng.* **62** 192
[3] Boonrod K, Towprayoon S, Bonnet S and Tripetchkul S 2015 *Resources, Conservation and Recycling* **95** 77
[4] Mehta A, Mishraa A and Basu S 2018 *Environ. Sci. Water Res. Technol.* **4** 12 2012
[5] Reddy K, Karthik K, Prasad S, Soni S, Jeong H and Raghu A 2016 *Polyhedron* **120** 16
[6] Suhadolnik L, Pohar A, Likozar B and Ceh M 2016 *Chem. Eng. J.* **303** 292
[7] Lisowski P, Colmenares J, Masek O, Lisowski W, Lisovytskiy D, Kaminska A and Lomot D 2017 *ACS Sustainable Chem. Eng.* **5** 6274
[8] Jia T, Wang W, Long F, Fu Z, Wang H and Zhang Q 2009 *J. Alloys Comp.* **484** 410
[9] Jiang J, Wang H, Chen X, Li S, Xie T, Wang D and Lin Y 2017 *J. Coll. Interf. Sci.* **494** 130
[10] Sampaio M, Lima M, Baptista D, Silva A, Silva C and Faria J 2017 *Chem. Eng. J.* **318** 95
[11] Vaiano V, Matarangolo M, Murcia J, Rojas H, Navío J and Hidalgo M 2018 *Appl. Catal. B: Environ.* **225** 197
[12] Ihara T, Miyoshi M, Iriyama Y, Matsumoto O and Sugihara S 2003 *Appl. Catal. B: Environ.* **42** 403
[13] Ferrari-Lima A, Souza R, Mendes S, Marques R, Gimenes M and Fernandes-Machado N 2015 *Catal. Today* **241** 40
[14] Saikia L, Bhuyan D, Saikia M, Malakar B, Dutta D and Sengupta P 2015 *Appl. Catal. A: Gen.* **490** 42
[15] Molla M, Tateishi I, Furukawa M, Katsumata H, Suzuki T and Kaneco S 2017 *Open J. Inorg. Non-metallic Mater.* **7** 1
[16] Hussin F, Lintang H O, Lee S L and Yuliati L 2016 *Malay. J. Anal. Sci.* **20** 102
[17] Hussin F, Lintang H O, Lee S L and Yuliati L 2018, *Malay. J. Fund. Appl. Sci.* Special Issue on Chromatography and Other Analytical Techniques 159
[18] Ellmer K and Bikowski A 2016 *J. Phys. D: Appl. Phys.* **49** 413002
[19] Hosseini S, Sarsari I, Kameli P and Salamati H 2015 *J. Alloys Comp.* **640** 408
[20] Güy N, Çakar S and Özacar M 2016 *J. Coll. Interf. Sci.* **466** 128
[21] Aggelopoulos C, Dimitropoulos M, Govatsi A, Sygellou L, Tsakiroglou C and Yannopoulos S 2017 *Appl. Catal. B: Environ.* **205** 292
[22] Pascariu P, Homocianu M, Cojocaru C, Samoila P, Airinei A and Suchea M 2019 *Appl. Surf. Sci.* **476** 16
[23] Aneesh P, Vanaja K.A. and Jayaraj M 2007 *Proc. SPIE 6639, Nanophotonic Materials* **IV** 66390J
[24] Phoohinkong W, Foophow T and Pecharapa W 2017 *Adv. Nat. Sci.: Nanosci. Nanotechnol.* **8** 035003
[25] Liu C, Shih K, Gao Y, Li F and Wei L 2012 *J. Soil. Sediment.* **12** 724
[26] Max J and Chapados C 2004 *J. Phys. Chem. A* **108** 3324
[27] Kooti M and Sedeh A 2012 *J. Chem.* **2013** 562028
[28] Hussin F, Lintang H O, Lee S L and Yuliati L 2017 *J. Photochem. Photobiol. A: Chem.* **340** 128
[29] Kayaci F, Vempati S, Donmez I, Biyikli N and Uyar T 2014 *Nanoscale* **6** 10224
[30] Martínez C, Canle L, Fernández M, Santaballa J and Faria J 2011 *Appl. Catal. B: Environ.* **102** 563
[31] Liang Y, Guo N, Li L, Li R, Ji G and Gan S 2016 *New J. Chem.* **40** 1587
[32] Lee S C, Lintang H O and Yuliati L 2012 *Chem. Asian. J.* **7** 2139

[33] Wang J, Wang Z, Huang B, Ma Y, Liu Y, Qin X, Zhang X and Dai Y 2012 *ACS Appl. Mater. Interfaces* **4** 4024

Acknowledgements

We acknowledge the financial support from the Directorate General of Strengthening Research and Development, Ministry of Research, Technology and Higher Education of the Republic of Indonesia via the World Class Research scheme (WCR 2019, No. 041/SP2H/LT/MULTI/L7/2019 and No. 014/MACHUNG/LPPM/SP2H-LIT-MULTI/III/2019).

IOP Conf. Series: Materials Science and Engineering **833** (2020) 012081 doi:10.1088/1757-899X/833/1/012081

Synthesis of ZnO-Ag with Clove Oil Using Ultrasonication Method and Its Antibiofilm Activity Against *Klebsiella pneumoniae*

K Nisyak*[1], Y A Prasetya[1], A Hisbiyah[1], E D Iftitah[2] and A Srihardyastutie[2]

[1] Medical Laboratory Technology Department, STIKES Rumah Sakit Anwar Medika, Sidoarjo Indonesia

[2] Chemistry Department, Brawijaya University, Malang Indonesia

*Corresponding author: nisachemist@gmail.com

Abstract. ZnO-Ag nanocomposite synthesis has been successfully carried out by ultrasonication using clove oil as a reducing agent. Synthesis reactions were carried out in 30, 60, 90, 120, 240, and 300 minutes time variations. ZnO-Ag crystallinity was analyzed using X-Ray Diffractometer (XRD) and the average particle size was calculated using Scherrer equation. The morphology of ZnO-Ag which has the smallest particle size was analyzed by Scanning Electron Microscope - Energy dispersive X-Ray Spectroscopy (SEM-EDX). ZnO-Ag antibiofilm activity against *Klebsiella pneumoniae* was determined by microtiter plate test. Based on research that has been done, the best time for synthesizing ZnO-Ag with clove oil through ultrasonication is 90 minutes, the resulting particle size of 21.46 nm with the composition of 51.4% Ag (cubic), 20% wulfingite (orthorhombic), 22.9% $Zn(OH)_2$ (orthorhombic), and 5.7% ZnO (cubic). ZnO-Ag has strong antibacterial activity against *K. pneumoniae* and can inhibit the growth of *K. pneumoniae* biofilms.

1. Introduction

Biofilms are bacterial cell colonies that attach to inert materials, metals, plastics, and even to human body tissues by synthesizing extracellular polysaccharides, called Extracellular Polymeric Substances (EPS) [1]. Biofilm-forming bacteria can live in conditions of nutrient deficiencies, able to prevent bacterial phagocytosis, inhibit antibody responses, and resistant to antibiotics and disinfectants. The occurrence of infections is related to biofilm forming bacteria, making it difficult in terms of therapy and often causing resistances that lead to death [2]. *Klebsiella pneumoniae* is one of the Gram negative bacteria. It can forms biofilms that cause nosocomial infections which are difficult to cure because they are related to the presence of antibiotic resistance [3]. Biofilms from *K. pneumoniae* cause infections in patients who have immune disorders [4].

The main source of the transmission of *K. pneumoniae* infection is through medical equipment which is less sterile [5]. Several studies have reported the disease outbreaks in hospitals due to *K. pneumoniae* isolates that are resistant to third generation cephalosporins, aminoglycosides, and quinolones [6]. The emergence of various diseases caused by biofilm forming bacteria has encouraged the continued exploration of active materials that can be developed into antibiofilm agents, one of which is nanomaterial from metal composites.

Nowadays the development of materials used to deal with the problem of nosocomial infections in public health services is directed at nanoscale functional materials, one of which is nanocomposite. Nano-sized material has a high surface area and specific reactivity so that it is suitable to be used as a therapeutic agent for a disease in the health field [7]. A nanocomposite is a solid material composed of two or more materials which have different properties from one another and produce different characteristics and properties from the constituent material [8]. Nanocomposites in the health sector can be functionalize as antimicrobials [9]. The material that has been widely studied and has the potential as an antimicrobial is ZnO and silver. ZnO has been shown to have active antimicrobial activity against Gram-positive, Gram-negative, and fungal pathogens [10]. ZnO has unique properties, biocompatible, and low toxicity [11]. Silver shaped nanoparticles have a high surface area and stronger antibacterial activity in damaging bacterial cell walls, inhibiting cell synthesis, as well as disrupting bacterial cell metabolism [12]. Technological innovations to form ZnO-Ag nanocomposites can be functionalized as antibiofilm agents that can inhibit the formation of biofilms, prevent the attachment of biofilms to material, and destroy biofilms.

Development of green synthesis methods for the manufacturing of nano materials using natural product compounds as bioreductors and stabilizing agents continues to be explored. The green synthesis method is considered to be environmentally friendly, effective, and better yield produced [13]. The main requirements for chemical compounds from natural products that can be used as reducing agents are compounds containing electron-rich functional groups such as hydroxyl groups (-OH), carboxyl groups (-COOH), alkene groups, and benzene, for example in flavonoid group compounds [14]. The use of major chemical compounds in essential oils as reducing agents in the synthesis of nanoscale materials has not been widely reported. Research from Azizi et al (2016) reports zerumbone compounds in ginger oil can be used as reducing agents and stabilizers in ZnO-Ag synthesis [15].

In this research, the reducing agent used for ZnO-Ag synthesis is clove oil. Clove oil is one of the typical essential oils from Indonesia which contains eugenol compounds (Figure 1), phenylpropanoid compounds that have benzene rings, methoxy groups, hydroxyl groups, and double bonds. The presence of electron-rich functional groups in eugenol can be a reducing and stabilizing agent in ZnO-Ag synthesis. Important aspects in the green synthesis method are the efficient use of energy, reagents that are not excessive, and short reaction times to produce the best synthesis products, one of which is to utilize ultrasonic waves [16]. Utilization of ultrasonic waves can cause acoustic cavitation, increase product purity, modify the selectivity of starting materials, and break particles to the smallest size [17].

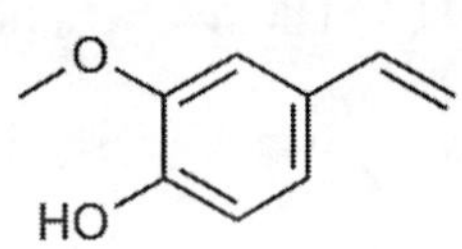

Figure 1. Eugenol from clove oil [18].

Synthesis of ZnO-Ag with clove oil is carried out using the one-pot synthesis method. The synthesis process is carried out in one container with a continuous stage using ultrasonic waves. One of the strengths of the pot synthesis method is that it is economical, efficient, and applicable to industrial scale. The ZnO-Ag produced was characterized by its particle size, crystallinity, and morphology. ZnO-Ag antibiofilm activity against *K. pneumoniae* was tested by microtiter plate test. ZnO-Ag Antibiofilm activities included prevention tests for biofilm adhesion, inhibition of biofilm formation, and biofilm degradation. The profile of ZnO-Ag antibiofilm activity can be the basis for developing safe and effective nanocomposite-based health equipment.

2. Experimental Section

2.1. Material and Instrumentation

The materials used in this research were silver nitrate (Merck), zinc acetate dihydrate (Merck, clove-leaf oil (CV. Nusaroma Depok), aquadest (Brataco Chemical), ethanol (Merck), ammonia (Merck), *K. pneumoniae* isolate from Tropical Disease Center of Airlangga University (ATCC 700603), nutrient broth (Merck), Muller Hinton agar (Merck), and crystal violet (Merck). The instrumentations used in this research were ultrasonicator (40 kHz, 120 W), oven (DGG 9053A), analytical balance (Ohaus px224/E), X-Ray Diffractometer (Panalytical X'Pert Pro, XRD), scanning electron microscope – energy dispersive X-Ray spectroscopy (SEM-EDX, Hitachi FLEXSEM 1000), gas chromatography-mass spectrophotometer (Shimadzu, GC-MS QP2010), and microplate reader.

2.2. Synthesis ZnO-Ag with clove oil

ZnO-Ag was synthesized using modification method according to previous report [19]. A 15 mL of clove-leaf oil was added to a flask containing 50 mL ethanol and sonicated for 30 minutes. Then 0.003 M of zinc acetate dihydrate was added dropwise to the mixture. Ammonia was added until the pH of solution was 9 and sonicated for 3 hours. ZnO-Ag formation begins by adding 0.001 M silver nitrate solution dropwise and sonicated for 30 minutes. The resulting reaction product is centrifugated and the precipitate obtained was washed with ethanol and dried at 110 °C for 2 hours. The reaction with ultrasonication after the addition of silver nitrate was carried out in the time variations of 30, 60, 90, 120, and 240 minutes.

2.3. Characterization of reaction product

ZnO-Ag crystallographic analysis was performed with XRD and morphological analysis with SEM-EDX. The diffractogram obtained was identified by its shape form using Match software version 3.8.3.151 and Origin Pro. The average crystallite size was calculated by using Scherrer's equation as following:

$$I = (0{,}9\lambda) / (\beta\cos\theta) \tag{1}$$

where I stand for grain size (nm), λ for wavelength of the XRD, β for FWHM, and θ for Bragg's angle. The degree of crystallinity was calculated by plotting crystalline area with total area of diffractogram as

$$\text{Degree of crystallinity} = (\text{Area of crystalline} / \text{Total area}) \times 100\% \tag{2}$$

2.4. Antibiofilm Test

A total of 200 µL of ZnO-Ag solution was put into the microplate, then the microplate was closed and incubated at 37 °C for 1 hour. The contents of the microplate were removed and inserted by 200 µL of the bacterial suspension. 200 µL of nutrient broth were put into the microplate, closed and incubated at 37 °C for 24, 48, 72, and 96 hours. After incubation, the microplate is removed and washed with distilled water for three times. The microplate was given a 200 µL of 1% violet crystal and incubated for 15 minutes. Violet crystal dye was washed and rinsed with distilled water for three times. Subsequently 200 µL of ethanol were added to the microplate and incubated for 15 minutes. The absorbance of microplate was measured using microplate reader at λ of 595 nm. The negative control used was DMSO 9.8% solution and positive control used was meropenem as antibiotics. The test was carried out in triplo. The percentage of biofilm inhibition was calculated using this formula:

$$\%\ \text{biofilm inhibition} = ((\text{OD negative control} - \text{OD sample}) / \text{OD negative control}) \times 100\% \tag{3}$$

where OD is optical density from microplate reader.

3. Result and Discussion

3.1. Synthesis ZnO-Ag with clove oil using ultrasonication method

ZnO-Ag synthesis with clove oil consists of two reaction stages, namely the reaction of ZnO formation continued with the addition of silver nitrate solution to form ZnO-Ag. The whole reaction process is carried out in one container so it can be called as one pot synthesis. The content of clove-leaf oil was analyzed by GC-MS to determine its eugenol levels. Based on chromatogram in Figure 2, clove-leaf oil contains three main compounds namely eugenol 71.41%, trans-caryophyllene 25.96%, and alpha-humulene 2.63%. The presence of high levels of eugenol in clove-leaf oil, acts as a bioreductor and stabilizer of the ZnO-Ag synthesis reaction.

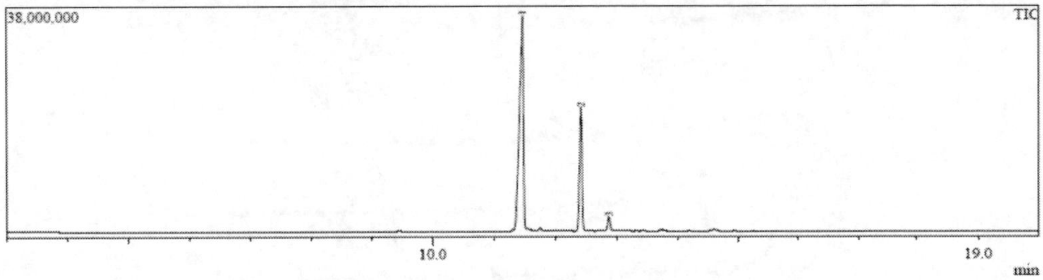

Figure 2. Total Ionic Chromatogram (TIC) of clove-leaf oil.

The use of ultrasonic waves in synthesis reactions encourages physical and chemical changes in the liquid medium and cause acoustic cavitation. The effect of ultrasonic waves in the material synthesis reaction process can produce magnetic particles and break down large crystal aggregates into small sizes to the nanoscale. ZnO formation is done by adding zinc acetate solution to a mixture of ethanol and clove-leaf oil. The formation of ZnO is strongly influenced by the alkaline condition so it needs to be conditioned by the pH of alkaline solution with ammonia. Ammonia and pH play an important role in the formation of ZnO [20]. The formation of ZnO-Ag is done by adding silver to the ZnO that has been formed. The variation of the time reaction is used to determine the effect of time reaction to phase crystal and its particle size.

3.2. Phase identification and crystallinity of ZnO-Ag

ZnO-Ag obtained from ultrasonication reaction was analyzed for type compound, phase identification, particle size, and crystallinity using XRD. In Figure 3, the diffractogram of the all times reactions variations is analyzed for the difference in peaks. At an angle 2θ of 38° and 44°, the difference at each reaction time can be analyzed. The diffraction peaks at 2θ of 38° and 44°, identify the existence of cubic structure Ag according to JCPDS No. 04-0783. At the reaction time of 90 minutes and 300 minutes, the peak at 2θ of 38° is higher than the other reaction times, which shows the high level of Ag in ZnO-Ag.

ZnO-Ag phase identification was done using Match! software from the diffractogram to obtain the results which are shown in Figure 4. Match! software shows that there are differences in the composition of ZnO-Ag elements at each reaction time variation. ZnO-Ag nanocomposites consists of five phases that is hexagonal zinc oxide, orthorhombic $Zn(OH)_2$ (wuelfingite), trigonal (hexagonal axes) $Zn(OH)_2$, hexagonal Ag and cubic Ag. Based on the results of the phase identification, ZnO-Ag with the most AgNP composition were obtained at the reaction time of 90 minutes (51.4%). Whereas, ZnO-Ag with the most ZnO-NP composition was obtained during the reaction time of 30 minutes (33.5%). The side product of this synthesis was $Zn(OH)_2$ which is produced most during the reaction time of 30 minutes. It was consisted of 14.9% trigonal (hexagonal axes) $Zn(OH)_2$ and 17.6% orthorhombic $Zn(OH)_2$ (Wuelfingite). Based on these results, the composition of $Zn(OH)_2$ was inversely proportional to the composition of ZnO, so there was a possibility of an equilibrium reaction between ZnO and Zn $(OH)_2$ during the synthesis process. The best composition of ZnO-Ag is shown at the reaction time of 90 minutes which consists Ag of 51.4% (cubic), wulfingite of 20% (orthorhombic), $Zn(OH)_2$ of 22.9% (orthorhombic), and ZnO of 5.7% (cubic).

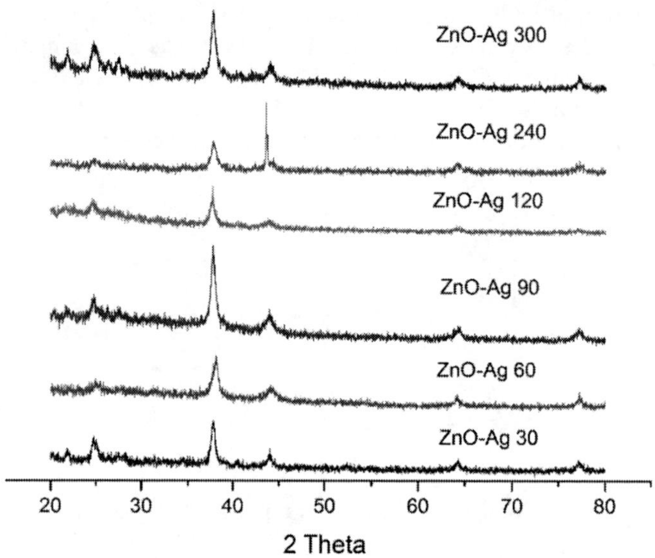

Figure 3. Diffractogram of ZnO-Ag.

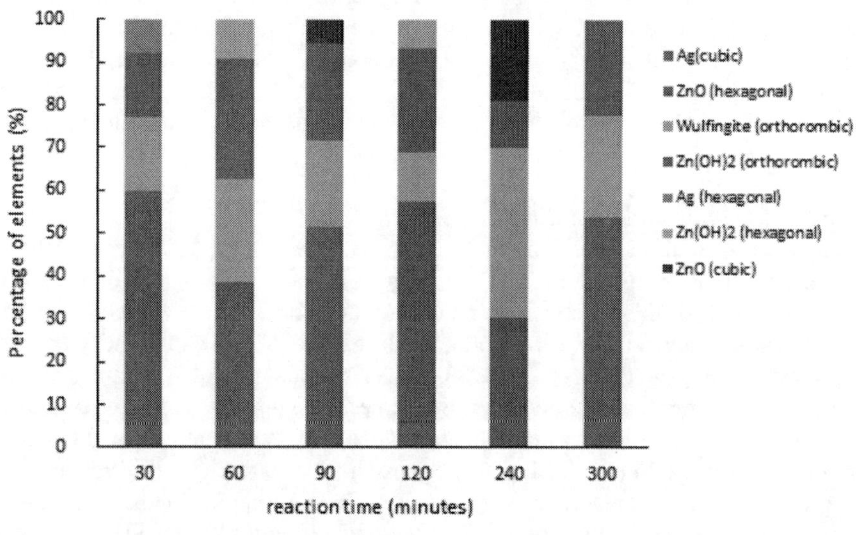

Figure 4. Composition of ZnO-Ag.

The average particle size was determined using the Scherrer equation and degree of crystallinity (%) was calculated using the Origin Pro and Microsoft Excell software. The results of the calculations are shown in Table 1. ZnO-Ag with the smallest average particle size and the highest degree of crystallinity were obtained at a reaction time of 90 minutes, which was 21.46 nm and 24% of degree crystallinity. Based on the data in Table 1, the degree of crystallinity of all ZnO-Ag was categorized to low, because there is no calcination process in the synthesis ZnO-Ag, so not all $Zn(OH)_2$ is converted to ZnO. The size range of ZnO-Ag was from 21.46 nm to 44.67 nm in diameter according to Debye-Scherer's equation.

Table 1. Average particle size and degree crystallinity of ZnO-Ag.

No	Reaction time (minutes)	Average Particle Size (nm)	Crystallinity (%)
1	30	24.98	18
2	60	42.45	18
3	90	21.46	24
4	120	25.54	17
5	240	44.67	19
6	300	22.38	23

3.3. Morphology of ZnO-Ag

The morphology of the ZnO-Ag nanocomposite form was analyzed by SEM to observe the position of Ag and ZnO in the nanocomposite. The morphology of ZnO-Ag synthesized with a reaction time of 90 minutes is shown in Figure 5, in which Ag particles attached to ZnO particles. ZnO looks like fiber, while silver is cuboid. Based on the EDX spectrum in ZnO-Ag, there are Zn (38%), Ag (38%) and O (24%) elements. The presence of ultrasonic waves in the process of formation reaction can act as a source of reaction energy and vibrations that can split particles into small sizes.

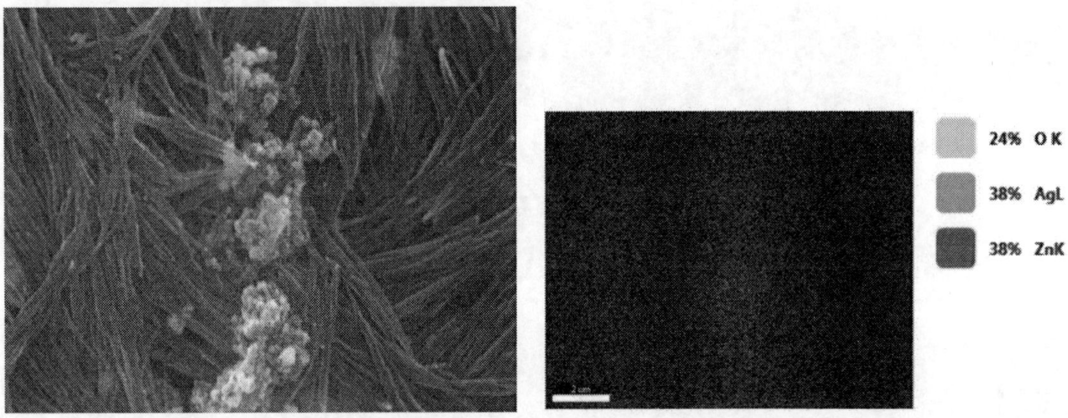

Figure 5. Micrographs SEM of ZnO-Ag with a magnification of 20,000x.

Based on the data analysis using diffractogram of XRD and micrograph of SEM-EDX, ZnO-Ag can be formed properly. Eugenol acts as reducing agent in the formation of ZnO-Ag, the functional group in eugenol can reduce Ag^+ ions to Ag particles. The mechanism for the reaction of ZnO-Ag formation is shown in Figure 6. $Zn(OH)_2$ complexes formed in nanoscopic metabolites occur through electron transfer from hydroxyl groups to the coordinates of transition metal bonds to form coordinating covalent bonds. Complex compounds formed and contained electrons in their allyl groups can give their free electron pairs to Ag^+ ions, so that Ag^+ ions can be reduced and form Ag nanoparticles. In addition, trans-caryophillene and alpha-humulent compounds in clove oil are also thought to reduce Ag^+ ions to form Ag nanoparticles. Furthermore, the formed $Zn(OH)_2$ complex can become ZnO nanoparticles through a thermal decomposition process using ultrasonic waves.

3.4. Antibiofilm activity of ZnO-Ag to K. pneumoniae biofilm

ZnO-Ag antibiofilm activity against *K. pneumoniae* was performed by microtiter plate assay method. ZnO-Ag was made in series of concentrations of 10 – 100 ppm and reacted with biofilm *K.*

IOP Conf. Series: Materials Science and Engineering **833** (2020) 012081 doi:10.1088/1757-899X/833/1/012081

pneumoniae that has been formed. Interactions between ZnO-Ag and *K. pneumoniae* biofilms were observed at incubation times of 24, 48, 72, and 96 hours. ZnO-Ag antibiofilm activity at 24 hours incubation was determined as antibacterial activity, 48 hours incubation time as inhibitor of biofilm attachment, 72 hours incubation time as an effort to prevent biofilm maturation, and 96 hours incubation time as biofilm degradation. The results of ZnO-Ag antibiofilm activity against *K. pneumoniae* are shown in Figure 7.

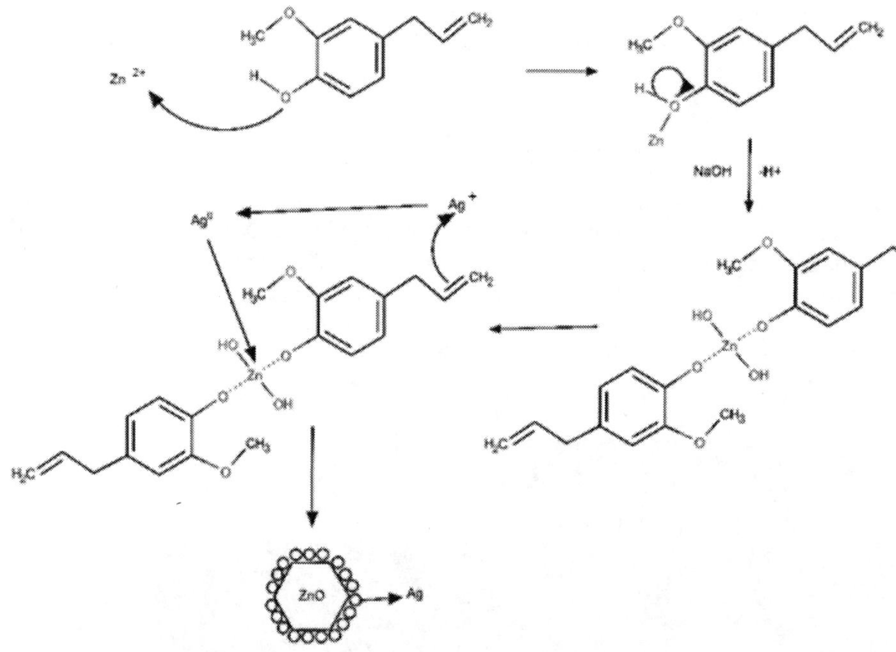

Figure 6. Reaction mechanism of Zn-Ag nanocomposite formation with clove oil.

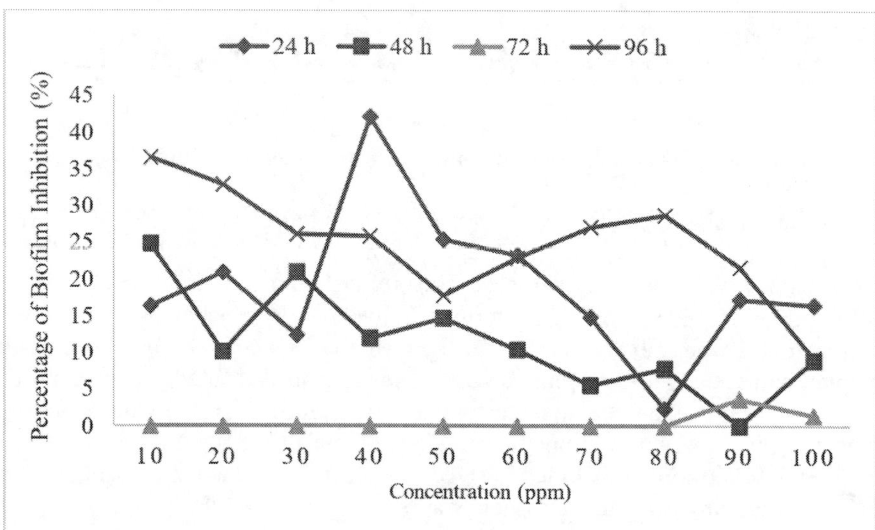

Figure 7. Antibiofilm Activity of ZnO-Ag to *K. pneumoniae* biofilm.

Based on the graph shown in Figure 7, the observation of ZnO-Ag antibiofilm activity against *K. pneumoniae* at 24 hours incubation showed that the concentration of 40 ppm gave the greatest inhibitory value of 42.02%. This value indicates that ZnO-Ag is an antibacterial agent against *K. pneumoniae*. On the observation at 48 hours incubation time, ZnO-Ag showed the best activity at a concentration of 10 ppm which could inhibit the attachment of biofilms to the surface of material by 24.76%. ZnO-Ag can destroy *K. pneumoniae* biofilms at a concentration of 10 ppm with a biofilm inhibition value of 36.54%. The results obtained show that ZnO-Ag concentration is not proportional to its antibiofilm activity. These results can be influenced by the condition of the *K. pneumoniae* biofilm and external influences in the optical density measurement process.

ZnO-Ag nanocomposite antibiofilm activity is influenced by the crystal shape, particle size, and surface area. There are four interactions in ZnO-Ag antibiofilm mechanism that cause bacterial cell death and prevent biofilm attachment. The first interaction is the interaction of ZnO-Ag with bacterial cell membranes by releasing Ag^+ and Zn^{2+} ions on the bacterial cell membrane, then the formation of reactive oxygen species (ROS) which causes damage to the bacterial cell membrane. The interaction of Ag^+ ions with bacterial cell ribosomes inhibits bacterial protein synthesis. ZnO-Ag interacts intracellularly with bacterial cell membranes and causes bacterial cell death [14]. The combination of these interactions is thought the inhibition of biofilms formation and prevention of biofilms attachment to the material. This is because bacterial cells undergo lysis due to interactions with ZnO-Ag.

4. Conclusions

ZnO-Ag has been synthesized via a facile green method using essential oil of clove through ultrasonication method. The main component in clove oil can play a role as a reducing agent in the formation of ZnO-Ag. The best time for synthesis of ZnO-Ag with clove oil through ultrasonication method is 90 minutes. The resulting particle size is 21.46 nm with the composition Ag of 51.4% (cubic), wulfingite of 20% (orthorhombic), $Zn(OH)_2$ of 22.9% (orthorhombic), and ZnO of 5.7% (cubic). ZnO-Ag has strong antibacterial activity against *K. pneumoniae* and can inhibit the growth of *K. pneumoniae* biofilms. For further research, it is recommended to perform calcination procedures to form ZnO so that the crystallinity is good.

References

[1] C. G. Ng, M. F. Loke, K. L. Goh, J. Vadivelu, and B. Ho 2017 *Food Microbiol.* **62** 68–76.
[2] K. Steffy, G. Shanthi, A. S. Maroky, and S. Selvakumar 2018 *J. Trace Elem. Med. Biol.* **50** 229–239.
[3] I. Bhattacharjee, S. K. Chatterjee, and G. Chandra 2010 *Asian Pac. J. Trop. Med.* **3** 547–551.
[4] E. D. Shingadia and G. Syrogiannopoulos 2017 *Pediatr Infect Dis J.* **36** 1002–1005.
[5] E. Cadavid, S. M. Robledo, W. Quiñones, and F. Echeverri 2018 *Antibiotics* **7** 1–14.
[6] K. Seifi, H. Kazemian, H. Heidari, F. Rezagholizadeh, F. Shirvani, and H. Houri 2016 *Jundishapur J. Microbiol.* **9** 2–7.
[7] F. Asghari, Z. Jahanshiri, M. Imani, M. Shams-Ghahfarokhi, and M. Razzaghi-Abyaneh, *Antifungal nanomaterials: Synthesis, properties, and applications*. Elsevier Inc., 2016.
[8] D. Carvalho, T. Sousa, P. V. Morais, and A. P. Piedade 2016 *Appl. Surf. Sci.* **379** 489–496.
[9] M. S. Kumar and A. P. Das 2017 *Adv. Colloid Interface Sci.* **249** 53–65.
[10] A. Sirelkhatim, S. Mahmud, and A. Seeni 2015 *Nano-Micro Lett.* **7** 219–242.
[11] A. Top and H. Çetinkaya 2015 *Mater. Chem. Phys.* **167** 77–87.
[12] R. Salomoni, P. Léo, and M. F. A. Rodrigues 2015 *Portadilia Inicial* **2** 851–857.
[13] J. A. Hernandez-viezcas 2016 *Nanotechnol. Environ. Eng.* **4** 1-29.
[14] I. Matai, A. Sachdev, P. Dubey, S. Uday Kumar, B. Bhushan, and P. Gopinath 2014 *Colloids Surfaces B Biointerfaces* **115** 359–367.
[15] S. Azizi, M. Rosfarizan, R.R. Abdul, M. A. Boroumand, M. Mona, A. Arbakariya, S. W. Zuhainis, and N. Farideh 2016 *Appl. Surf. Sci.* **384** 517–524.
[16] Y. Shao. W. Chunhua, W. Tiantian, Y. Chunhong, C. Shiguo, D. Tian, Y. Xinqian, and X. Yaqin

2018 *Int. J. Biol. Macromol.* **111** 1281–1292.
[17] A. Bayrami, S. Alioghli, S. R. Pouran, A. Habibi-yangjeh, A. Khataee, and R. Singh 2019 *Ultrason. - Sonochemistry*, **55** 57 - 66.
[18] Y. Zhang, Y. Wang, X. Zhu, P. Cao, S. Wei, and Y. Lu 2017 *Microb. Pathog.* **113** 396–402.
[19] T. S. Vijayakumar, S. Karthikeyeni, S. Vasanth, A. Ganesh, G. Bupesh, R. Ramesh, M. Manimegalai, and P. Subramanian 2013 *J. Nanosci.* **2013** 1-7.
[20] R. Mahdavi and S. S. A. Talesh 2017 *Ultrason. - Sonochemistry*, **39** 504-510.

Acknowledgements
The author thanks to the Ministry of Research, Technology and Higher Education who has provided research funding 2019 through *Penelitian Kerjasama Perguruan Tinggi* with contract number 113/SP2H/LT/DRPM/2019. The author also thanks Dr. Elvina Dhiaul Iftitah, M.Sc. and Dr. Arie Srihardyastutie, M.Kes. from Universitas Brawijaya who provided guidances and directions in this research.

IOP Conf. Series: Materials Science and Engineering **833** (2020) 012082 doi:10.1088/1757-899X/833/1/012082

Swelling Effect Observation of The Copper Phthalocyanine Layer on QCM and Its Effect on Surface Roughness and Morphology Changes

Rahmad Oktafiansyah[1,2], D J Djoko Herry Santjojo[1,2], Setyawan P Sakti[1,2], Tyas N Zafirah[1,2], Muhammad Ghufron[1,2], N F Khusnah[2,3,] and Masruroh[1,2*]

[1]Department of Physics, Brawijaya University, Malang, 65145, Indonesia
[2]Collaborative Research Group for Advanced System and Material Technology (ASMAT), Brawijaya University, Malang, Indonesia
[3]Laboratorium Sentral Ilmu Hayati (LSIH), Brawijaya University, Malang, 65145, Indonesia

*Corresponding author: ruroh@ub.ac.id

Abstract. Copper Phthalocyanine (CuPc) is one material often used for the active layer on Quartz Crystal Microbalance (QCM) sensor. One important mechanism which can shorten the recovery time in the sensor is the swelling effect. The swelling effect is studied by observing the morphology of the layer and subsequent changes in QCM frequency during surface contact with three kinds of solution. The solutions used in this study were distilled water, phosphate-buffered saline (PBS) and Tris-HCl. The results of frequency measurements indicate a swelling effect which was shown by the decrease of the frequency change after the injection process. The most significant change in frequency occurs in PBS solutions. This change occurs as PBS has a higher molecular weight and ionic strength than distilled water and Tris-HCl. SEM results show that the swelling of the CuPc layer changes the microstructure layer. A decrease in the roughness of the CuPc layer was also observed due to the presence of buffer solution molecules that diffuse through the layer. Upon diffusion of the solution molecules into the CuPc layer, a continuous decrease in the QCM frequency occurs. Further, the diffusion caused the CuPc microstructure to become more tenuous.

1. Introduction

Quartz Crystal Microbalance (QCM) is a device that is sensitive to changes in the mass of up to 10^{-4} ng, making QCM a vital device and well applied as a gas sensor base. In its usage, the sensitivity and selectivity of the QCM sensor can be increased by coating certain materials on its surface [1]. Copper Phthalocyanine (CuPc) is one of the materials that is often chosen as a thin layer on the surface of QCM because it is chemically and thermally very stable. Moreover, CuPc is proven to increase the number of gas molecules detected by sensors [2,3]. CuPc is composed of Phthalocyanine chains that bind to the Cu cation as its core metal. Generally, CuPc has two structural phases, namely the metastable α-phase and also the more stable β-phase [4]. The structure of CuPc can be conditioned by giving heat treatment to the layer, where the α-phase is found in heat treatment below 200°C, and the β-phase structure is found when heat treatments are above 200°C [5].

One criterion of good coating for gas sensors is the existence of a swelling effect. A high swelling effect can shorten the gas sensor recovery time [6]. Swelling occurs due to the interaction of the thin layer with the solution, causing the solvent molecules to diffuse into the layer creating the thin layer to experience additional volume [7]. The diffusion of liquid into the thin layer tends to balance the tensile and repulsive force between the liquid mixture and the layer thermodynamically. Upon reaching the driving force equilibrium of the liquid with the force that holds the diffusion from the layer, the swelling process will stop [8]. Several factors affect swelling, including layer microstructure, the affinity between solution and layer, the porosity of layers, temperature and ionic strength [9]. Specific buffer solutions i.e. PBS and Tris-HCl, which have ionic strengths of 2.434 M and 0.086 M, are used for swelling observation. The diffusion process in swelling mechanism is categorized in non-Fickian diffusion, making the decrease in diffusion rate proportional to its depth. Observation of the swelling effect of a layer can be done by measuring changes in deposited QCM frequency which are then given a liquid on its surface [9].

In this research, the effect of affinity between the solution of distilled water, PBS, and Tris-HCl with the CuPc layer on the swelling effect was studied by observing frequency changes. The effect of swelling on changes in microstructure (morphology and surface roughness) of CuPc layers is also discussed.

2. Materials and Method

Samples were prepared by superimposing Polystyrene on the surface of a quartz crystal microbalance (QCM) sensor with a frequency of 10 MHz. QCM was obtained from Great Microtama Surabaya. Polystyrene layer with a molecular weight of 192 kDa was made from polystyrene grains obtained from Sigma-Aldrich dissolved in toluene solvent to obtain a solution with a concentration of 6%. Then the solution is homogenized with an ultrasonic cleaner, and coated with a spin coater, then heated in an oven for 60 minutes at 200°C. CuPc deposition was carried out by the thermal evaporation method, in which QCM-coated Polystyrene was used as a substrate, and CuPc Powder (purity 99%) obtained from Sigma-Aldrich was used as a vaporized material. CuPc coating deposition was carried out by setting up 1.5 A current and 28V voltage, with the CuPc deposition time being 5 minutes. After that, the samples were annealed at 220°C in the oven for 60 minutes.

QCM samples deposited with CuPc layers are placed in reaction cells connected with an oscillator to measure the QCM frequency. After the QCM frequency has stabilized, 50 µl of distilled water solution is injected into the QCM. The swelling effect of the CuPc layer was observed by observing changes in the QCM oscillation frequency after being injected with a distilled water solution. Then the same thing was conducted to observe the swelling effect of the CuPc layer from the buffer solution of Phosphate Buffered Saline (PBS), and Tris-Hydrochloride (Tris-HCl) obtained from GIBCO. Figure 1 is an illustration of a set-up tool for observing the swelling effect with QCM. The morphology of the CuPc layer was observed using SEM-EDS (FESEM, FEI, Quanta 650 FEG), in the LSIH, Brawijaya University, and the roughness value of the CuPc layer was measured using the Topography Measurement System (Polytech TMS 1200), in the Material Physics laboratory, Department of Physics, Brawijaya University. Morphological observations and roughness measurements of the CuPc layer were carried out before and after the swelling process for all buffer solutions.

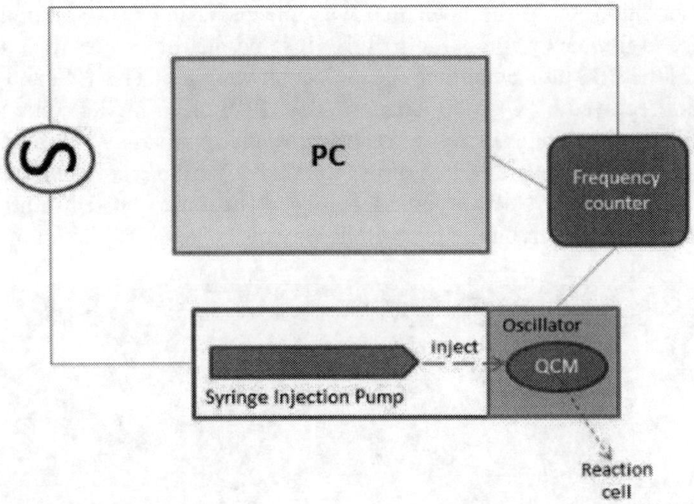

Figure 1. Illustration of an experimental set-up scheme.

3. Result and Discussion

Swelling effect arises as a result of the diffusion of molecules from the solution injected onto the surface of the QCM thin layer. A sudden injection of the solution on the QCM surface causes a mass increase in the QCM sensor, causing the QCM frequency to drop dramatically. The repulsion of the QCM surface layer will increase the QCM frequency on a small scale. Diffusion that occurs in a thin layer of QCM will result in a continuous decrease in the frequency of QCM until saturation conditions are reached, and the solvent molecules are unable to diffuse again.

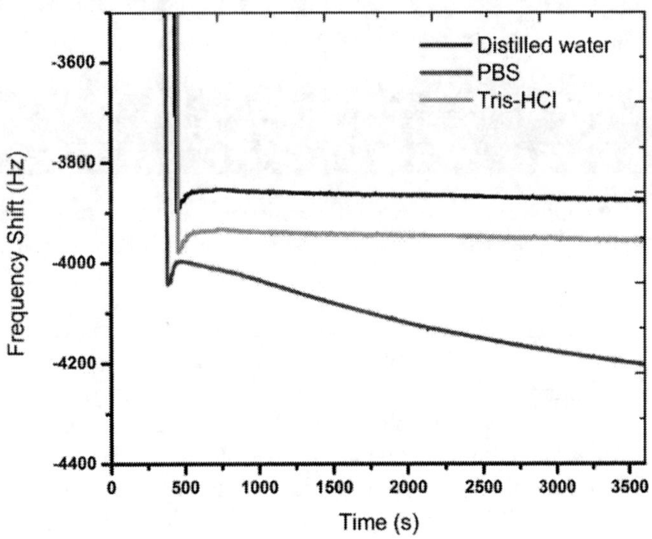

Figure 2. QCM frequency response after injection of the solution on the surface of CuPc.

Figure 2 shows the decrease in the QCM oscillation frequency after the injection process of the solution on the surface of the CuPc layer. The change in QCM frequency indicates the presence of buffer solution molecules that diffuse into the thin layer of CuPc, causing loading on the QCM sensor. After more in-depth observation, the CuPc layer that was injected using PBS solution experienced a more significant decrease in frequency compared to Tris-HCl and distilled water injection, and this is due to the higher weight of the PBS molecular than the molecular weight of Tris-HCl and distilled water. Thus, the mass of loading received by QCM is greater. Also, PBS has a higher ionic strength compared to Tris-HCl and distilled water. Ionic forces cause electrostatic repulsion when buffer molecules interact with the CuPc layer. As the repulsion force of the CuPc layer will be more tenuous, the PBS molecules will be easier to break into the CuPc layer. Molecular diffusion of buffer solution in the CuPc layer results in changes in the microstructure and morphology of the layer, as shown in Figure 1.

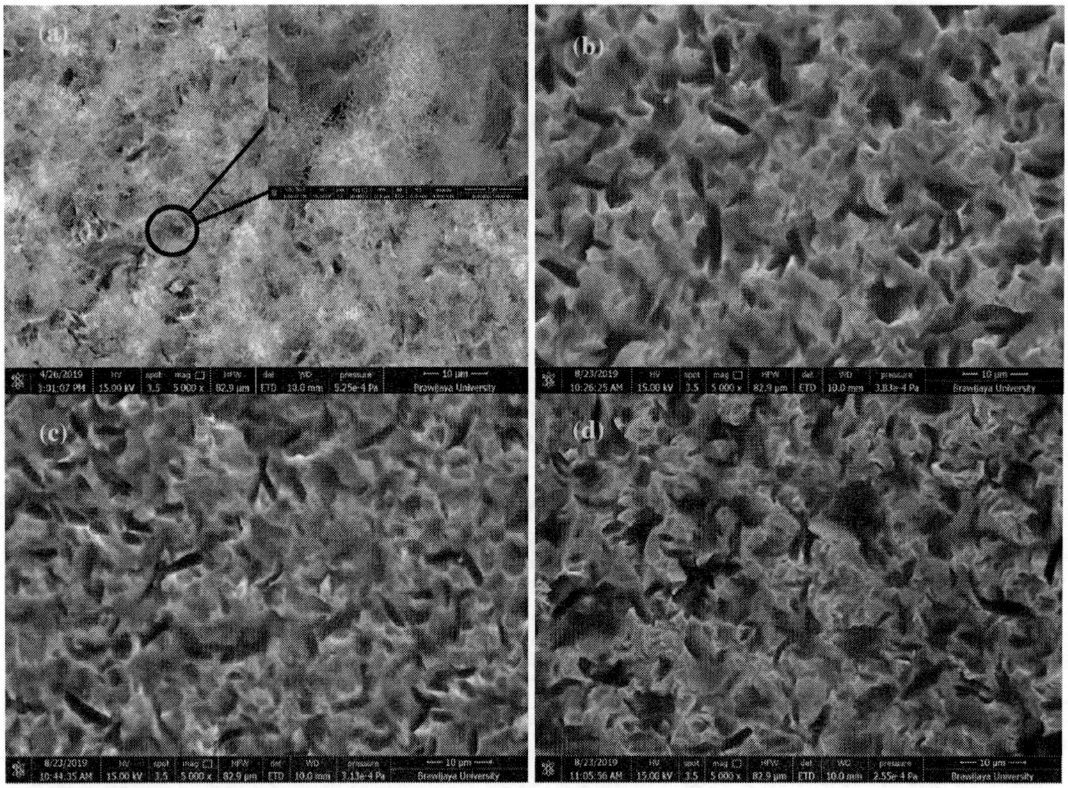

Figure 3. SEM results of CuPc thin layers (a) before injection and after injection with solution (b) Distilled Water, (c) PBS, (d) Tris-HCl.

Figure 3 shows the morphological changes in the CuPc layer due to injection of buffer solution. Figure 3 (a) shows that the CuPc layer has fibrous characteristics that are interconnected with one another. There are no visible interrelated fibers observed after the surface was injected with the buffer solution. However, from Figure 3 (c) and (d), the CuPc fibers are scattered and do not accumulate like the one before injection. The scattered microstructure is possible due to the diffusion of buffer solution molecules that break through the CuPc layer, resulting in the widening of the CuPc layer.

Figure 4 shows that the Na and Cl spectra appear from the results of the EDS analysis of the CuPc layer injected with PBS solution, where Na and Cl are the constituent elements of PBS, thus confirming the existence of a PBS molecule that diffuses into the CuPc layer. The swelling effect that occurs in the CuPc layer results in changes in the properties of the CuPc layer, one of which is the roughness of the CuPc layer. The results of roughness analysis using TMS (Topography Measurement System) are listed in Figure 5.

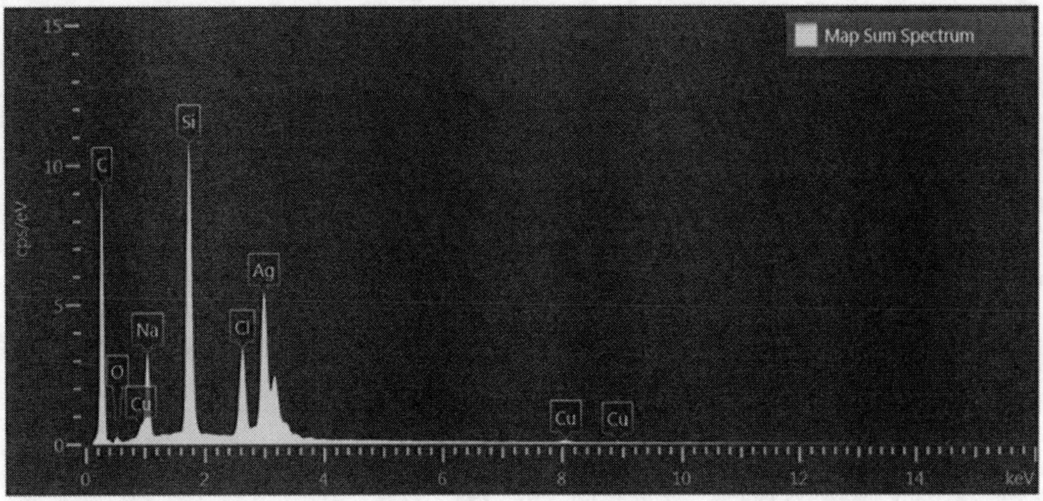

Figure 4. EDS results of CuPc coating after PBS solution injection.

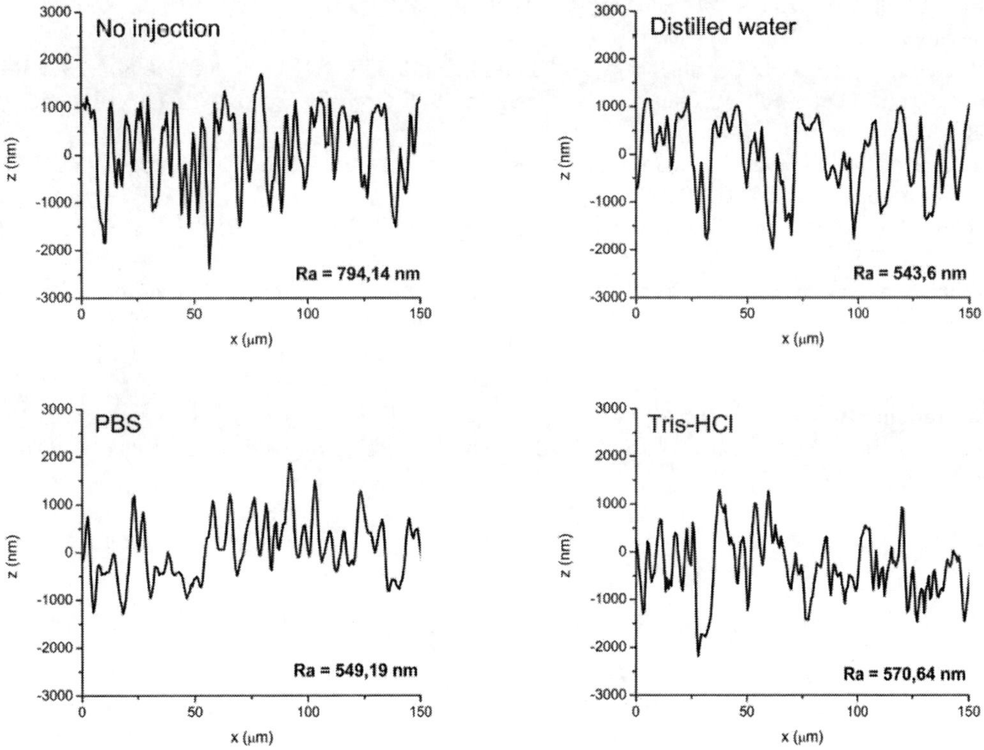

Figure 5. Roughness value of CuPc layers before injection and after injection with Distilled water, PBS and Tris-HCl.

Figure 5 shows a decrease in the level of roughness of the CuPc layer after injection of buffer solution in the CuPc layer. This decrease in roughness is caused by the diffusion of buffer molecules that relax the CuPc layer. The characteristics of the CuPc layer in the form of fibers are interrelated, resulting in a coarser CuPc layer. However, due to the widening of the CuPc layer, the CuPc molecules do not accumulate any more, thus decreasing the layer's roughness. The roughness value of CuPc layers for all buffer solutions shows almost the same value. This similarity shows that the CuPc layer has reached saturation condition, indicating that diffusion occurs at a certain depth. Therefore, swelling that occurs in the CuPc layer is categorized in the case of non-fickian diffusion, where the diffusion rate slows following the deeper layers. Moreover, because of this swelling effect, the CuPc coating can improve QCM's ability to detect gases.

4. Conclusion

The swelling effect occurs in the CuPc layer when it comes in contact with distilled water, PBS, and Tris-HCl solutions as indicated by observations on the changes in QCM frequency. The most significant change in the frequency takes place in the CuPc layer injected with PBS solution, due to the higher PBS molecular weight compared to distilled water and Tris-HCl solutions. The swelling of the CuPc layer causes a decrease in the surface roughness and microstructure changes of the CuPc layer.

References

[1] D V Ivanov, A Yelon 1996 *J. Electrochem. Soc.* **143** 2835–2841
[2] A Arnau, Y Jimenez, R Fernández, R Torres, M Otero, E J Calvo 2006 *J. Electrochem. Soc.* **153** C455
[3] Singh, Sukhwinder, G S S Saini, and S K Tripathi 2014 Sensing Properties of ZnPc Thin Films Studied by Electrical and Optical Techniques Sensors & Actuators *B. Chemical* **203**. Elsevier B.V.: 118–21
[4] El-Nahass, M M, F S Bahabri, a a Al Ghamdi, and S R Al-Harbi 2002. *J. Sol* **25** 2 307–21
[5] Grobosch, M C Schmidt, R Kraus, and M Knupfer 2010 *Organic Electronics* **11** 9. Elsevier B.V.: 1483–88
[6] Rianjanu A, S N Hidayat, T Julian, E A Suyono, A Kusumaatmaja, and K Triyana 2018 *IOP Conf. Ser. Mater. Sci. Eng.* **367** 012020
[7] Shinkai, Tomoaki, Kohzo Ito, and Hideaki Yokoyama 2014 *J Supercritical Fluids* **95**. Elsevier B.V.: 553–59
[8] Sukhorukov, Gleb B, and J Schmitt 1996 *Reversible Swelling of Polyanion/ Polycation Multilayer Films in Solutions of Different Ionic Strength J* **953** 6 948–53
[9] Tyas N Z, D.J.D.H. Santjojo, Masruroh and S.P. Sakti 2019 *Mater. Today Proc* **13** 18-23.

Acknowledgements

This work was partly supported by RISTEKDIKTI for PDUPT Grants through LPPM of Brawijaya University.

Optical Emission Spectroscopy Study of the Electron Temperature and Electron Density Dependence on the Pressure Chamber for the Carbon Deposition Produced by Argon Plasma Sputtering

D R S Pambudi[1], M A Hanif[1], D J D H Santjojo[1,2], M C Padaga[3] and Masruroh[1,2*]

[1]Department of Physics, Brawijaya University, Malang, Indonesia
[2]Collaborative Research Group for Advanced System and Material Technology (ASMAT), Brawijaya University, Malang, Indonesia
[3]Faculty of Animal Husbandry, Brawijaya University, Malang 65145, Indonesia

*Corresponding author: ruroh@ub.ac.id

Abstract. An investigation of the Optical Emission Spectroscopy (OES) to study the effect of the pressure chamber to the chemical species, electron temperature (*Te*) and electron density (*Ne*) was performed on RF 2 MHz plasma sputtering for carbon deposition. A carbon target and argon (Ar) gas were used to deposit carbon films. The carbon deposition was carried out at pressure chamber of 10, 15 and 20 Pa, and the RF voltage, flow rate and substrate temperature were kept at 120 volts, 60 ml/min, and 200°C, respectively. The emission spectrum of Ar plasma was monitored by OES in order to analyze the chemical species present in the plasma during the deposition process. The atomic spectroscopy data center (NIST) database was used to determine the species in the plasma and to calculate the *Te* and *Ne*. The OES emission shows the peak intensity which indicates the Ar I (Ar*) and Ar II (Ar^+ ion). The electron temperature was calculated using the Boltzmann Plot while the electron density was calculated using Stark Broadening. The results show that both *Te* and *Ne* decreased on the increasing of the pressure chamber. The emission line measurements with Boltzmann Plot resulted in the electron temperature dependence of pressure is approximately 0.487±0.005, 0.481±0.006, and 0.474±0.011 eV, respectively. Moreover, the calculation of electron density resulted in approximately $(5.495 \pm 0.407) \times 10^{18}$, $(4.825 \pm 0.384) \times 10^{18}$, and $(4.515 \pm 1.037) \times 10^{18}$ cm^{-3}.

1. Introduction

QCM (Quartz Crystal Microbalance) is a device consisting of quartz crystals coated by electrodes [1]. QCM is very sensitive to the changes in mass and has a piezoelectric principle, i.e., to detect frequency shifts in quartz crystals due to changes in mass on the surface, making it widely used as a sensor [2]. QCM is commonly used as humidity sensor, biosensor, gas sensor, and other uses [3-5]. Coatings are needed on the surface of the QCM that matches its function, enabling the QCM to work optimally. The carbon layer is widely applied to the type of resonant sensor (QCM) as a gas sensor [6,7]. The methods used to form carbon into thin layers include spin coating, deep-coating, plasma chemical vapor deposition (CVD), and plasma sputtering. The plasma sputtering method is widely studied due to its resulting thin layer, homogeneity, and the ability to control the deposition process as needed by the requirements or the desired thin layer [8]. Sputtering is the process of colliding target materials using plasma. The atoms of the target material will be detached and deposited on the substrate, forming a thin layer. The commonly used gas sputter is Argon (Ar), as Argon is an inert gas and has an ideal atomic weight. Argon, when compared to other noble gases like Krypton (Kr), Xenon (Xe), and Radon (Rn), has a lower sputter yield. However, they are hard to find and expensive. In addition, Rn is radioactive [9].

The atomic processes that may occur in plasma are as follows [10]:

$$e + Ar \rightarrow e + Ar^{*} \text{ (Excitation)} \quad (1)$$

$$e + Ar \rightarrow 2e + Ar^{+} \text{ (Ionization)} \quad (2)$$

$$e + Ar^{*} \rightarrow e + Ar \text{ (Deexitation due to collision)} \quad (3)$$

$$e + Ar^{+} \rightarrow Ar^{+*} \text{ (Ar II ion excitation)} \quad (4)$$

Electrons have a crucial role in the plasma process, as most atomic processes require electrons. There are internal and external parameters in the plasma process. External parameters include chamber pressure, RF voltage, gas flow rate, substrate temperature, and deposition time. This work is a preliminary study focusing on the effect of the chamber pressure on the electron temperature and density. The chamber pressure is related to the mean free path, which is the average collision distance. Electron temperature is related to the average kinetic energy of electrons in the plasma, while electron density is related to the degree of ionization from plasma. They are crucial internal plasma parameter determining the deposition quality. The plasma was diagnosed with a spectroscopic technique (OES). The OES results showed a graph of the relationship between intensity vs wavelength. The electron temperature and density can be calculated from the Boltzmann plot [11], and Stark Broadening [12,13]. The temperature and density of these electrons will affect the ion bombardment process in the plasma sputtering target material, which is related to the rate of deposition of thin-film formation. The internal parameters were studied against the variation of the chamber pressure.

2. Methods

2.1 Experimental Method

An RF plasma system with a frequency of 2 MHz was utilized for sputtering deposition of carbon on a QCM substrate. The target material was a carbon disk with a diameter of 25 mm and thickness of 6 mm. The disk was placed 1 cm above the substrate and enclosed with a stainless steel cathode as shown in Figure 1. The carbon deposition process was carried out under the following steps. First, the plasma was generated by introducing argon gas into the system. The argon flow rate was maintained at 60 mL/min during the process. Then the RF voltage was applied across the electrode. The voltage in this study was 120 volts ensuring a stable glow discharge. The temperature of the substrate was kept at 200 ^{0}C, while the chamber pressures were varied by 10, 15, and 20 Pa. Finally, the plasma was set off after 60 minutes of deposition. The experiment was repeated three times for each variation of pressure.

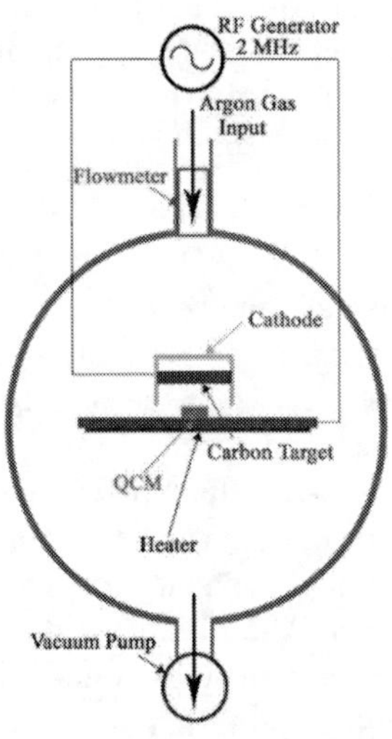

Figure 1. Plasma reactor scheme for carbon deposition.

IOP Conf. Series: Materials Science and Engineering **833** (2020) 012083 doi:10.1088/1757-899X/833/1/012083

2.2 Calculation Method

Intake of the Argon plasma spectrum was performed using OES Aurora 4000 in the wavelength range of 200-1000 nm. Determination of the electron temperature was done by plotting the Boltzmann equation:

$$ln\frac{I\lambda}{gA} = Const - \frac{E_i}{kT} \tag{5}$$

Where I is the emission intensity, λ is the emission wavelength, g is the statistical weight, A is the transition probability, and E_i is the upper level of the transition energy. The related data for the specific wavelength and intensity were obtained from the atomic spectra database (NIST). The NIST database was also used to determine the species of Argon plasma. After obtaining the electron temperature value, the electron density is calculated using the Stark Broadening equation as follows:

$$N_e = (\frac{\Delta\lambda_{\frac{1}{2}}10^{16}}{2\omega}) \tag{6}$$

Where $\Delta\lambda_{\frac{1}{2}}$ is Stark Broadening (FWHM), ω is the electron impact width parameter, and Ne is the electron density.

3. Results and Discussion

The emission spectrum in the plasma sputtering process is measured using OES. Spectrum wavelengths are then used to match with the NIST database to determine the type of Argon and Carbon species. The peak intensity vs wavelength shows the species Ar I and Ar II, while the C species does not appear, as shown in Figure 2. The Ar I species appears at wavelengths of 748.639, 762.044, 810.592 nm, whereas Ar II appeared at wavelengths of 309.429 and 736.036 nm. Ar I is an excited atom or Ar^* [14] and emits photons like the mechanism in the process (1), while Ar II is an Ar atom which ionizes and produces Ar^+ ions as a result in the process (2). Ar I has the highest peaks, indicating that excitation most often occurs in this process. Ar II has a peak that is quite low compared to Ar I as the energy to produce Ar II is high, which enables the ionization process. Ar II has a vital role in plasma sputtering as Ar^+ ions are needed in the ion bombardment process of the target material.

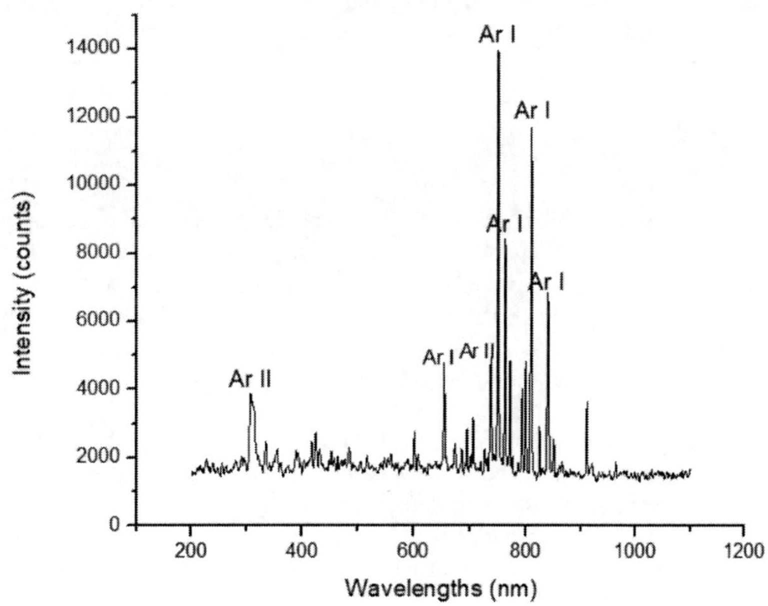

Figure 2. Argon plasma spectrum at 10 Pa pressure.

Electron temperature calculations are performed using the Boltzmann plot. The spectral lines used are the three highest peaks of Ar I, namely 748.639, 762.044, 810.592 nm. The results of the electron temperature and electron density calculation are shown in Table 1.

Table 1. The calculated results of electron temperature and electron density at various pressure.

Pressure (Pa)	Electron temperature (eV)	Electron density (cm^{-3})
10	0.487±0.005	$(5.495 \pm 0.407) \times 10^{18}$
15	0.481±0.006	$(4.825 \pm 0.384) \times 10^{18}$
20	0.474±0.011	$(4.515 \pm 1.037) \times 10^{18}$

The results show that the electron temperature decreases with increasing pressure. Pressure on the chamber affects the mean free path; the greater the pressure, the smaller the value of the mean free path. It means that the increasing of chamber pressure caused electron collision with neutral atoms also increases, resulting in by the mean free path between two successive collisions decrease, which shows that instead of getting energy by electrons from the electric field, more energy is transferred to neutral species as a result of the reduction in Te. Electron density was calculated using stark Broadening. The peaks used for the calculation Ne are selected from the three highest peaks of the Ar I, the same as the electron temperature calculation.

The increase of chamber pressure resulted in a decrease of both the electron temperature and electron density. The decrease is correlated to the distribution of energy in the plasma which is described using two mechanisms, namely Ohmic and stochastic heating. The former is the mechanism for channelling energy from the electric field to electrons. Energy is channelled through the collision of electrons with neutral atoms. In higher pressure, the collision intensity was increased, resulted in a decrease of the electron temperature. The decrease in the electron temperature allows for a decrease in the ionization process of neutral atoms, decreasing the number of electrons and hence the electron density. The latter mechanism was related to the process in the layer around the electrode called the sheath layer. The sheath layer is a layer that has a high electrical potential preventing electrons from passing. Only neutral ions and atoms can pass through the sheath. As the electrons are unable to pass through, electrons approaching the sheath layer will be reflected, thus creating a transfer of momentum from the sheath surface to the electrons. The more electrons reflected the more energy is given which in turn decrease the temperature electron.

4. Conclusion

Studies on the characterization of electron temperature and electron density in Argon plasma in carbon sputtering deposition at the various chamber pressure have been successfully carried out. The argon plasma species appeared in this study are only Ar I and Ar II. The decrease of the electron temperature and electron density due to the increase of chamber pressure were correlated and can be thought as the process controlled by two mechanisms, i.e. ohmic heating and stochastic heating.

References

[1] Vashist S K and Vashist P 2011 *J. Sensors* **2011** 1
[2] Jie H 2006 *Technical Background, Applications and Implementation of Quartz Crystal Microbalance Systems* (Finlandia: University of Jyväskylä)
[3] Li X, Chen X, Chen X, Ding X and Zhao X 2018 *Mater. Chem. Phys.* **207** 135–40
[4] Masruroh, Djoko D J D H, Didik L A, Rahmawati E, Pagaga M, Abdurrouf and Sakti S P 2014 *Appl. Mech. Mater.* **530–531**54–7
[5] Van Quang V, Hung V N, Tuan L A, Phan V N, Huy T Q and Van Quy N 2014 *Thin Solid Films* **568**6–12
[6] Nasir S, Hussein M Z, Zainal Z and Yusof N A 2018 *Materials (Basel).* **11** 1–24
[7] Mazzocco R, Robinson B J, Dickinson J, Boxall C and Kolosov O V. 2012 *Tech. Proc. 2012 NSTI Nanotechnol. Conf. Expo, NSTI-Nanotech* **2012** 173–6

[8] Bunshah R F 1994 *Handbook of Deposition Technologies for Films and Coating - Science, Technology and Applications (2nd Edition)* (New Jersey: Noyes Publications)
[9] Sarkar J 2014 *Sputtering and Thin Film Deposition* vol 2 (USA: Elsevier) 93-170
[10] Wu Y and Wang H 2015 *2nd International Symposium on Plasma Chemistry* (Belgium) 9-10
[11] Gan P P, Huang S, Lu Y R, Wang S Z, Yuan Z and Zhu K 2017 *IPAC 2017: the eighth International Particle Accelerator Conference* (Copenhagen) 2978–81
[12] Xiao D, Cheng C, Shen J, Lan Y, Xie H, Shu X, Meng Y, Li J and Chu P K 2014 *Phys. Plasmas* **21** 1–7
[13] Masruroh, Santjojo D J D H, Abdurrouf, Abdillah M A, Padaga M C and Sakti S P 2019 *IOP Conference Series: Materials Science and Engineering* **515**
[14] Wagatsuma K and Hirokawa K 1985 *Anal. Chem.* **57** 2901–7

Acknowledgements
This work was partly supported by RISTEKDIKTI for Hibah Kompetensi Grants through LPPM University of Brawijaya.

IOP Conf. Series: Materials Science and Engineering **833** (2020) 012084 doi:10.1088/1757-899X/833/1/012084

Effect of Water Depth and Flow Velocity on Microstructure, Tensile Strength and Hardness in Underwater Wet Welding

F. Paundra[1], J. Anindito[1], N. Muhayat[1] and Y. C. N. Saputro[2], Triyono[1*]

[1]Mechanical Engineering Department, Sebelas Maret University, Surakarta

[2]UPTB Solo Techno Park Technical Unit on Regional Development Planning Board Surakarta

*Corressponding author: triyonomesin@uns.ac.id

Abstract. Wet underwater welding is a welding process carried out under water. The depth and water flow velocity greatly affect the welding results. Thus, a further research needs to be done about the effect of the depth and water flow velocity of underwater wet welding. The research aimed to determine the effect of depth and velocity of water flow on underwater wet welding on physical and mechanical properties. The welding process is carried out by using SMAW method at a depth of 2.5 m and 5 m with variations in water flow velocity of 0, 1, and 2 m/s. Land welding is performed as a comparison. X-ray radiography, micro-structure, tensile strength, and Vickers hardness test were performed to determine the type of welding defects. Radiographic test results indicate an incomplete penetration defect (I), spatter (S), porosity (P), undercut (U), concavity (V), and irregular surface (Z). The results of the micro-structure test show that the grain size is fine and coarse as the depth and velocity of the water flow increase. Tensile strength and micro-hardness testing increase along with an increasing depth and velocity of water flow.

1. Introduction

Underwater welding is a welding process carried out underwater. Based on the method, underwater welding can be divided into underwater wet welding and underwater dry welding method [1-3]. Underwater dry welding is a process of underwater welding in dry conditions. This welding is usually undertaken in the chamber [4-5].

Underwater wet welding is performed in the water environment where the welder, workpiece, and electrode are in direct contact with water [2-3]. Underwater welding is widely used in maintenance, offshore construction, platforms, and nuclear power plants [6-7]. This welding has many advantages, including that fact that it is simple and cheap [8]. However, underwater wet welding has a disadvantage as well, in which welding defects are easily occurred [6, 9-10].

Information about underwater wet welding research is still minimal. In previous research, underwater wet welding was carried out on a laboratory scale, where the depth was less than 1 meter without water flow [7, 11-12]; so, the previous research was considered unsuitable to represent the real situation. Therefore, further research was conducted to determine the effect of water flow velocity and depth of underwater wet welding on physical and mechanical properties. The objective of this study is to determine the effect of welding depth and water flow velocity on underwater wet welding towards its micro-structure, hardness and tensile strength.

2. Experimental Methods

In this research, a low carbon steel type SS400 was used as a base metal (BM). The dimension of the welding specimen is 100 x 400 x 4 mm, with a butt joint welding type. Welding specimen are shown in Figure 1.

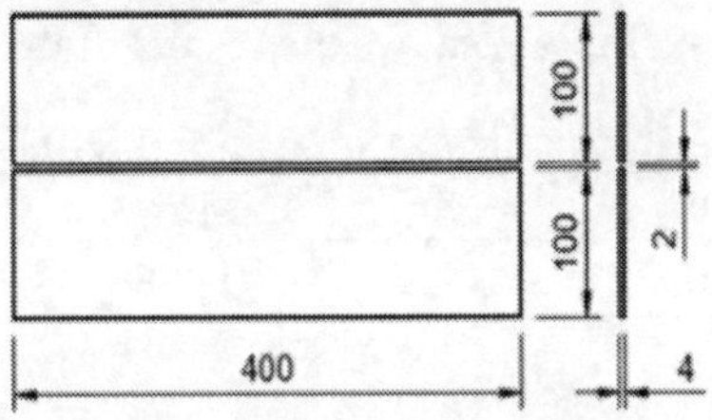

Figure 1. Welding specimen (in mm)

The underwater wet welding process was carried out in Solo Techno Park (STP) by certified welder. The welder was also an instructor at that compound who already has welding experiences. The welding process used was Shielded Metal Arc Welding (SMAW) type DCRP, with a nominal current of 90A and electrode type of E6013 with a diameter of 4mm. Underwater wet welding was carried out at a depth of 2.5 m and 5 m using variations in water flow velocity of 0, 1, and 2 m/s, and land welding as a comparison. The water nozzle was placed with a distance of 15 cm from the welding metal and water spraying was arranged so that the flow velocity was constant. The difference in depth affects the temperature difference. At a depth of 2.5 m, it has a temperature of 25.5° C, and at depth of 5 m it is 24.5° C. Table 1 shows the variables of the research.

Table1. Research Variables

Code	Depth of water (m)	Water flow velocity (m/s)	Temperature (°C)	Total
A	0	0	32	1
B	2.5	0	25.5	1
C	2.5	1	25.5	1
D	2.5	2	25.5	1
E	5	0	24.5	1
F	5	1	24.5	1
G	5	2	24.5	1

Figure 2 shows the underwater wet welding process. Figure 2a is the process of welding without water flow, while 2b shows the welding process of variations in water flow velocity. After the welding process, radiographic X-ray, micro-structure, Vickers hardness and tensile testing were carried out. Tensile testing was carried out using UTM tools with the JIS Z 2201 standard test and micro-hardness testing using Highwood machine with 9.8 N and ASTM E384 standard.

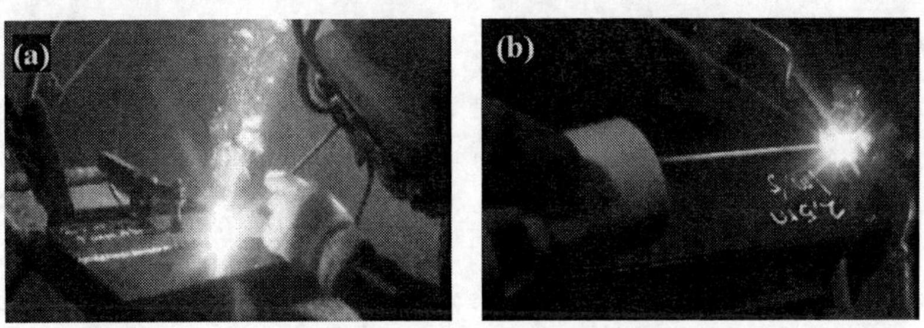

Figure 2. Welding process (a) without water flow, (b) with water flow.

IOP Conf. Series: Materials Science and Engineering **833** (2020) 012084 doi:10.1088/1757-899X/833/1/012084

3. Result and discussion

3.1. Chemical composition

The material used in this research was low carbon steel SS400. Before welding, the material was tested to determine its composition. Table 2 shows the results of the chemical composition test.

Table 2. Chemical composition SS400 (wt.%).

Chemical	Fe	C	Si	Mn	P	Cr	Cu
(%)	96.4	0.0337	0.193	0.288	0.0018	0.0273	0.0136

3.2. X-Ray Radiographic

Welding defect examination was carried out using X-ray radiography. Figure 3 shows the results of X-ray radiographic testing conducted at PT. Robutech Surabaya.

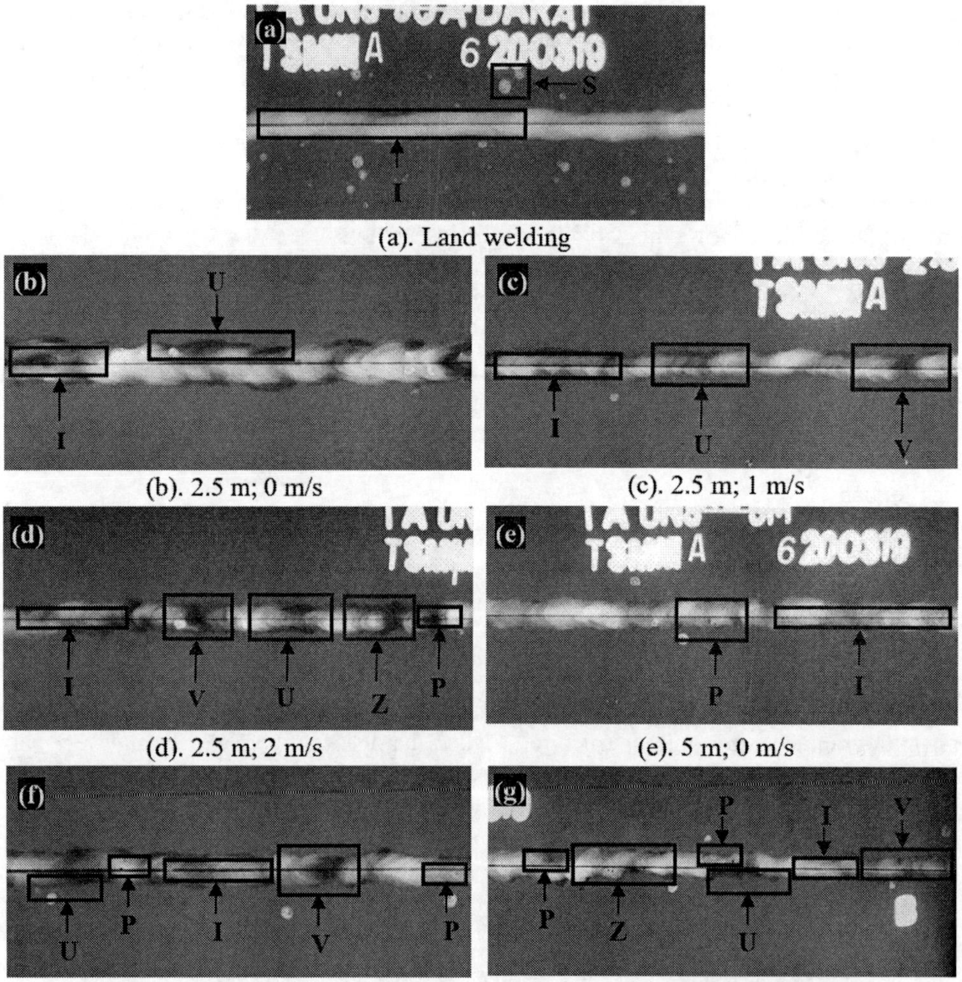

(a). Land welding

(b). 2.5 m; 0 m/s

(c). 2.5 m; 1 m/s

(d). 2.5 m; 2 m/s

(e). 5 m; 0 m/s

(f). 5 m; 1 m/s

(g). 2.5 m; 1 m/s

Figure 3. X-ray Radiographic.

Radiographic test results indicate the presence of various types of welding defects. The types of defects include incomplete penetration (I) that occurs in all welded joints, spatter (S), porosity (P), undercut (U), concavity (V), and irregular surface (Z).

The results of land welding and underwater wet welding show incomplete penetration defects. Incomplete penetration (I) is the lack of penetration in the welding area. This greatly affects the results of welding because it can reduce its mechanical strength. This defect occurs because the welding current is too small, and the electrode distance is too far apart. Spatter defects are welding sparks; this happens because the electrode and base metal are too far apart. The results of the test show that there is more spatter in land welding than in underwater wet welding.

Underwater wet welding is susceptible to porosity. Porosity is a small hole in the weld metal. Porosity in underwater wet welding occurs due to hydrogen inclusion. Hydrogen inclusion is caused by direct contact between water and electricity arcs at high temperatures forming gas particles. Rapid freezing will cause gas particles to be trapped in the metal, which results in porosity[13,14].

Undercut and concavity are conditions where the surface of the weld metal is lower than the base metal, which occurs in underwater wet welding due to lack of filler metal. They are shown in Figure 3 (b, c, d, f, and g), and concavity are presented in Figure 3 (c, d, f, and g). The most common undercut and concavity defects occur at a flow velocity of 2 m/s. The undercut is more common in welding with water flow. The undercut is a groove between weld metal and base metal that is not filled with welded molten metal due to high heat of the welded molten metal. This occurs as a result of an increased welding heat dissipation caused by high-water flow velocity, which results in a decrease of arc temperature and arc shrinkage [15]. High water flow velocity (2 m/s) results in many irregular surface defects. Irregular surface defects are shown in Figure 3 (d and g), which are weld surface conditions that contain notches or abrupt changes in thickness or appearance, in convex and concave surfaces. These conditions occur with variations in size that are too high. Irregular surface occurs due to poor welding arc stability with fluctuations in amperage and volts as water flow velocity increases. The irregular surface also occurs due to the process of liquid metal arc transfer, which is not good due to high water flow and protective bubbles that are less stable.

3.3. Microstructure

Figure 4 (a) is a base metal microstructure of SS400 material. In the picture, it can be observed that the morphology of base metal microstructures are ferrite and pearlite. The ferrite structure has a low hardness and ductile strength, while the pearlite structure has a stronger and harder tensile strength and more brittle than ferrite [16, 17]. Figure 4 (b-h) shows the weld metal area of the welding results with variations in depth and water flow velocity. Weld metal consists of four types of mixed ferrite: grain boundary ferrite (GBF), ferrite with the second phase (FSP), acicular ferrite (AF) and polygonal ferrite (PF). Underwater wet welding with a higher flow of water causes a higher cooling rate[18]. Uneven cooling rates and lower ambient temperatures cause the micro-structure of weld metal regions to be non-uniform and contain less PF. Increased welding heat input results in increasing grain size with a reduced relative amount of FSP and AF and increased GBF and PF. The width of GBF and FSP also increases with the increase in welding heat input. Increased welding heat input means reduced cooling rate during the welding process. FSP has properties with low impact values and low elongation percentage values. AF has high tensile strength and hardness properties and a good impact value compared to PF and GBF.

Figure 4 (b) shows the weld metal area of land welding results containing PF, FSP, and GBF. FSP and GBF have the biggest size compared to welding in water caused by slow cooling speeds. Micro-structure has a lot of PF but there is no AF. Weld metal welding in water with a depth of 2.5 m without flow shows a smaller structure of GBF and FSP as seen in Figure 4 (c) than welding on land and the presence of AF structures started to emerge. Increasing the velocity of water flow up to 1 m/s in welding with a depth of 2.5 m results in the smaller and flatter GBF and FSP structures shown in Figure 4 (d). Many AF structures emerge with a reduced PF structure. The flow velocity of 2 m/s at a depth of 2.5 m causes the GBF structure does not appear but the AF structure becomes more numerous and the PF structure also multiplies.

Underwater wet welding at 5 m depth has a slightly different micro-structure than that of welding at 2.5 m depth. The results of welding without flow when compared to that of 2.5 m depth have a smaller and flatter GBF and FSP micro-structure. A depth of 5 m with a flow velocity of 1 m/s results in FSP and AF which are denser than 2.5 depth. The microstructure of the flow velocity of 2 m/s at a depth of 5 m contains more FSP and AF compared to that of in a depth of 2.5 m and in a depth of 5 m with a flow velocity of 1 m/s or without flow. The amount of PF decreases due to insufficient time to develop. The presence of AF and FSP are the main factors in weld metal areas which have high tensile strength. The more AF structures the weld metal will produce higher tensile strength values[17, 19]. Thus, based on observations of micro-structure, welding with a depth of 5 m and a flow velocity of 2 m/s produces a high tensile strength.

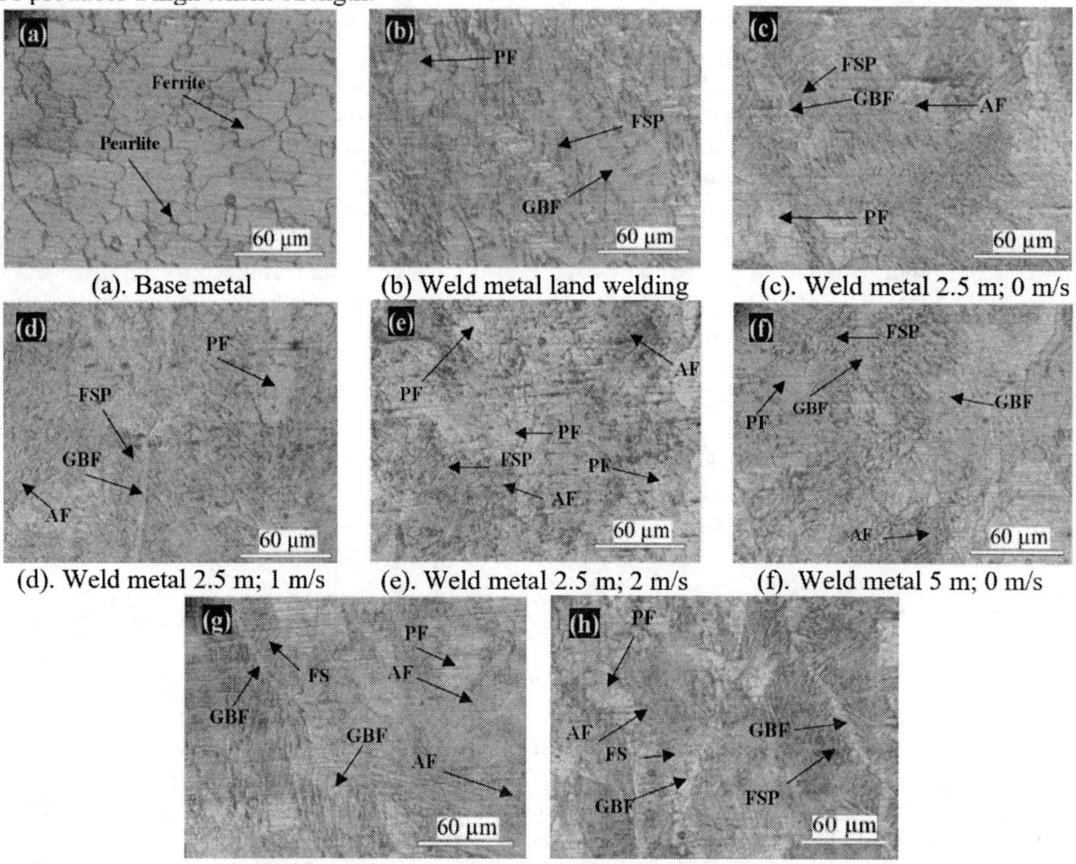

(a). Base metal (b) Weld metal land welding (c). Weld metal 2.5 m; 0 m/s

(d). Weld metal 2.5 m; 1 m/s (e). Weld metal 2.5 m; 2 m/s (f). Weld metal 5 m; 0 m/s

(g). Weld metal 5 m; 1 m/s (h). Weld metal 5 m; 2 m/s

Figure 4. Microstructure of base metal and weld metal.

Figure 5 shows a HAZ area consisting of ferrite and pearlite. The increasing depth and velocity of water flow results in small and smooth micro-structures. The fine grains are found in underwater wet welding with a depth of 5 m and a water flow velocity of 2 m/s as in Fig 5 (g). The base metal has a larger grain size compared to the HAZ area performed on land welding. Based on Figure 5 (a and b) the grain size on land welding has a much rougher size than underwater wet welding. An increase in water flow velocity causes an increase in the cooling rate and a reduction in welding heat input. The cooling rate is very fast as a grain growth and causes grain size to be finer and coarser than welding on land.

The mechanical properties of metals are influenced by micro-structure. The grain size of the weld metal will change according to the cooling rate. An increase in cooling rate will result in fine and coarse grain sizes. The small grain size causes increased tensile strength and hardness[20]. Based on the results of the micro-structure, the welding depth of up to 5 m has a greater effect on changes in the micro-structure compared to that of the influence of flow rates up to 2 m/s. Welding protective bubbles which rise more quickly to the surface result in more frequent metal contact with water which increased the cooling rate.

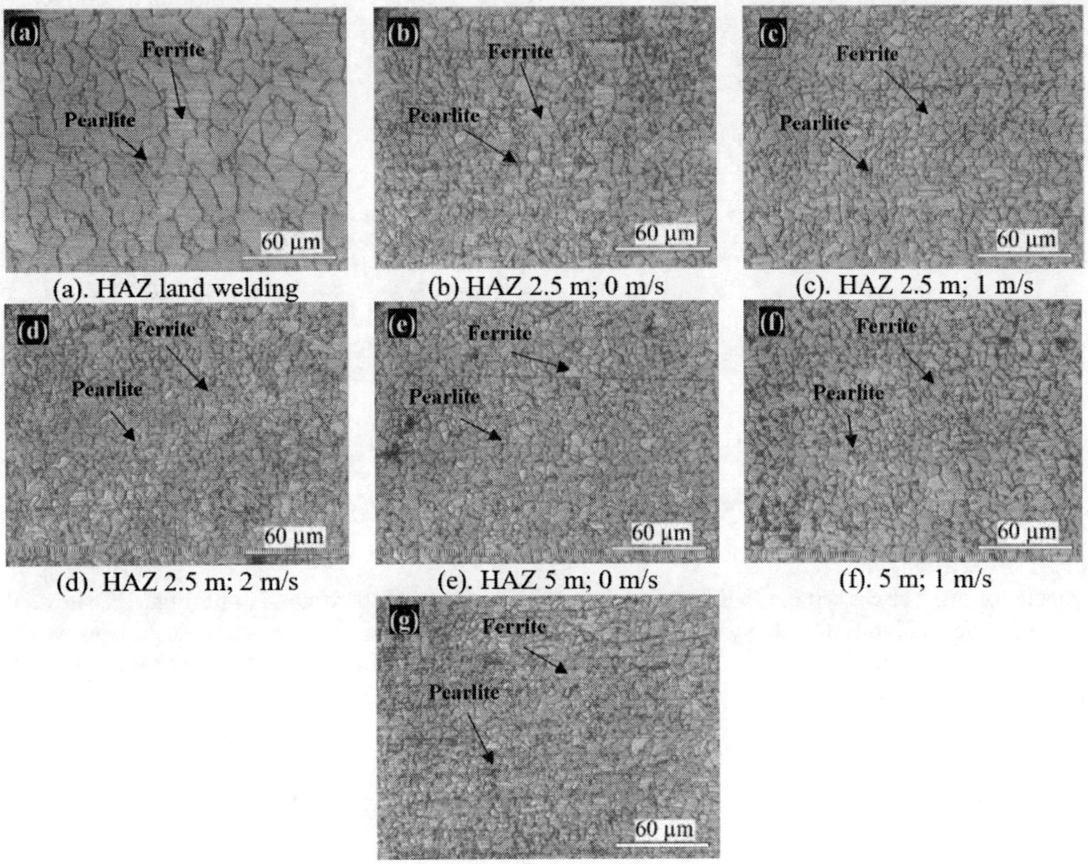

(a). HAZ land welding (b) HAZ 2.5 m; 0 m/s (c). HAZ 2.5 m; 1 m/s

(d). HAZ 2.5 m; 2 m/s (e). HAZ 5 m; 0 m/s (f). 5 m; 1 m/s

(g). HAZ 5 m; 2 m/s

Figure 5. HAZ microstructure.

3.4. Micro-Vickers

Micro-Vickers testing was conducted at three different points, namely weld metal, HAZ and base metal. Figures 6 (a and b) show the results of the micro-Vickers test. The lowest value of hardness is found in welding performed on land with the highest hardness value in the weld metal region of 131.4 HV, HAZ area 139.5 HV, and base metal 108 HV; whereas the highest hardness is observed in underwater wet welding with a depth of 5 m at 2 m/s of water flow velocity, with the highest hardness in the weld metal area of 188.8 HV, HAZ area 233.7 HV, and base metal 110.7 HV. Underwater wet welding with the lowest hardness is found in welding with a depth of 2.5 m without water flow. The highest hardness value in the weld metal area is 141.7 HV, HAZ 157 HV area, and base metal 109.6 HV

Increasing the welding depth from 2.5 m depth to 5 m significantly affects the hardness, as presented in Figures 6. The depth of welding has a greater effect than the increase in water flow velocity on the hardness value in the weld metal and HAZ area. The HAZ area of welding with a depth of 2.5 m at a flow velocity of 2 m/s has the highest hardness of 170.6 HV, while the highest hardness at a depth of 5 m with a flow of 2 m/s is 233.7 HV.

The increase in hardness of the weld joint along with an increase in water depth and speed of water flow is caused by an increase in the cooling rate or due to reduced heat input. This happens because of weld metal and base metal came into direct contact with water and forced convection occurs due to water flow [6,18]. Increasing the welding depth causes the protective bubbles to rise to the surface so that the weld joint is exposed to water directly more frequently. Increased cooling rate or reduced heat input will cause the increasing of AF and FSP growth. Increased AF and FSP contribute to an increase in the hardness of underwater wet welding joints. Reduction of heat input due to rapid cooling produces small and smooth grains that will increase hardness and strength in the HAZ.

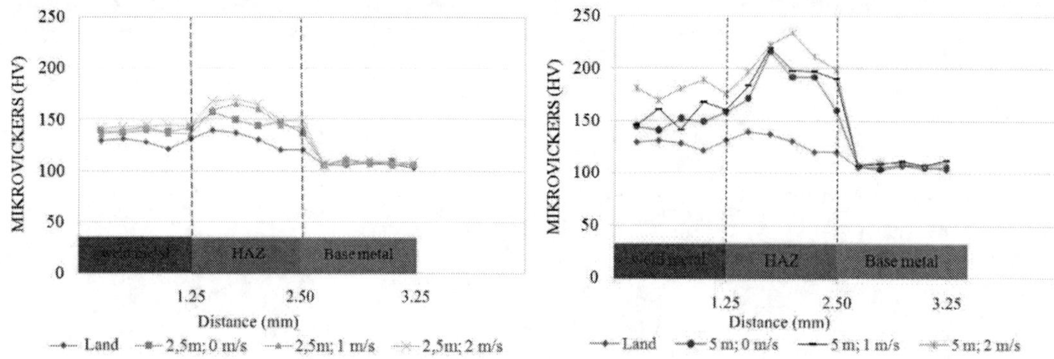

(a) **(b)**

Figure 6. Hardness Distribution: (a) 2.5 m, and (b) 5 m.

3.5. Tensile Strength

Tensile testing was carried out to determine the tensile strength of the welded joint. Figure 7 shows the results of the weld joint tensile test. The graph shows that the value of tensile strength in land welding is below the average of the one in underwater wet welding. This happens because the welding of the land has a deeper but less penetration than that of the underwater wet welding.

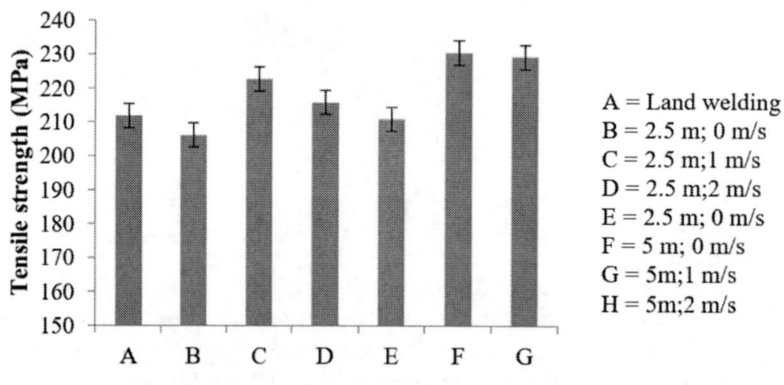

Figure 7. Tensile strength

The tensile strength value of underwater wet welding increases with the addition of variations in depth and water flow velocity. The smallest tensile strength occurs at a depth of 2.5 m without flow, with a tensile strength of 206 N/mm^2. The highest tensile strength value was obtained at a depth of 5 m with a variation of flow velocity of 1 m/s, which is 230.52 N/mm^2. Increased tensile strength in underwater wet welding is due to the increasing number of ferrite with the second phase (FSP) and acicular ferrite (AF) structures. Ferrite with the second phase (FSP) and acicular ferrite (AF) structures are formed due to very fast cooling [19]. Another factor influencing the increase in tensile strength is grain size. In underwater wet welding, the cooling speed increases with the addition of depth and water flow velocity. An increase in cooling speed will result in finer grain sizes.

4. Conclusions

X-ray radiographic test results showed a defect in the welding joints. The defects include incomplete penetration (I) occurred at all welded joints, spatter (S), porosity (P), undercut (U), concavity (V), and irregular surface (Z). Porosity increased with increasing depth and velocity of water flow. Microstructure investigation showed that increasing the depth and flow velocity of water increased the formation of AF and FSP structures. On the other hand, the rapid cooling rate performed the stunted grain growth which results in a fine and coarse grain size. The micro-hardness of the underwater wet weld joint increased as increase of water depth and flow velocity. The welding depth of 5 m with a flow velocity of 2 m/s has the highest Vickers hardness in the weld metal and HAZ, in which the highest hardness weld metal and HAZ area were 188.8 and 233.7 HV respectively. Tensile strength of the underwater wet welding joint increased due to water depth and water flow velocity. The lowest tensile strength was 206.13 N/mm^2 and obtained by specimens welded at a water depth of 2.5 m without flow while the highest tensile strength was 230.52 N/mm^2 and obtained by specimens welded at a water depth of 5 m with a flow velocity of 1 m/s.

References

[1] Kun S, Min Z, Yonghua S, Yu H and Xiaoqin S 2018 *J. Mater. Process. Tech.* **256** 190.
[2] Jyotsna D M 2006 *JNAME.* **3** 39.
[3] Nixon and Billingham 1987 *Pergamon J.* **11**. 3.
[4] Yahya A M, Triyono and Nurul M 2019 *AIP Conf. Proc.* **2097** 030044.
[5] Dariusz F, Grzegorz R and Jerzy L 2014 Problems of underwater welding of higher-strength low alloy steels *Gdańsk University of Technology*. **4** 187.
[6] Hongliang L, Duo L, YaoTian Y, Ning G, Yibo L and JiCai F 2018 *J. Manuf. Processes.* **31** 833.
[7] Jiangfeng W, Qingjie S, Zuchen P, Jie Y and Jicai F 2019 *J. Mater. Process. Tech.* **264** 289.
[8] Ning G, Changsheng X, Wei G, Yongpeng D and Jicai F 2015 *Mater. Des.* **85** 156.
[9] Ning G, Yongpeng D, Jicai F, Wei G and Zongquan D 2015 *J. Mater. Process. Tech.* **225** 133.
[10] Hao C, Ning G, Xianghua S, Yongpeng D and Jicai F 2018 *J. Manuf. Processes.* **21** 103.
[11] Ning G, Xiao X, Hongyun Z, Caiwang T, Jicai F and Zongquan D 2016 *Mater. Des.* **115** 112.
[12] Changsheng X, Ning G, Xin Z, Haiyue J, Hao C and Jicai F 2019 *Mater. Des.* **179** 107899.
[13] Hao C, Ning G, Lu H, Xin Z, Jicai F and Guodong W 2019 *Mater. Des.* **170** 107696.
[14] Robert J 1997 *The effect of water temperature on underbead cracking of underwater wet weldment* (California: Naval Postgraduate School).
[15] Yong Z, Chuanbao J, Bo Z, Jiakun H and Chuansong W 2016 *J. Mater. Process. Tech.* **238** 373.
[16] Wenbin G, Dongpo W, Fangjie C, Xinjie D, Caiyan D and Wei X 2016 *J. Mater. Process. Tech.* **238** 333.
[17] Hao C, Ning G, Xianghua S, Yongpeng D, Jicai F and Guodong W 2018 *J. Manuf. Process.* **31** 103.
[18] Pan Z, Yulin C,Wenlong X, Dehai and Xinqing Z 2016 *Pro. Nat. Sci-Mater.***26** 169.
[19] Liu F C and Nelson T W 2017 *Mater. Sci. Eng.* **710** 280.
[20] Zhao Y, Tong X, Wei X H, Xu S S, Lan S, Wang X L and Zhang Z W 2019 *J. Mater. Process. Tech.* **238** 373.

The Effect of Zirconium Addition on Corrosion Behavior of Zn-Zr Alloys as Biodegradable Orthopedic Implant Application

Rizki Yuni Pratiwi, Achmad Fauzi Trinanda, Moh Waqyan Ghani Fahmi, Sotya Astutiningsih, Ahmad Zakiyuddin

Dept. of Metallurgy and Materials Engineering, University of Indonesia

*Corresponding email: ahmadzakiyuddin@ui.ac.id

Abstract. In this work, corrosion behaviour of zinc-based alloys with addition of 0.5%, 1%, and 2% of zirconium for biodegradable material as orthopaedic implant were investigated. The potentiodynamic polarization method is carried out to determine the corrosion resistance and corrosion rate of each composition in order to observe the effect of zirconium addition in a Kokubo simulated body fluid solution. The result showed that the addition of 0.5% and 1% of zirconium would decrease the corrosion rate of Zn-xZr alloys corresponding to 0.079 mm/year and 0.116 mm/year whereas the 2% addition would increase the rate to 0.188 mm/year due to the formation of Zr-rich precipitates inside the alloys. The passivation zone on the polarization curve showed the formation of the protected thin layer on the surface of the alloys which caused the corrosion rate to decrease, therefore confirmed the degradable ability of the Zn-xZr alloys. In general, the corrosion rates of Zn-xZr alloys were higher than Fe-based alloys and lower than Mg-based alloys. Moreover, the corrosion rates were much lower than the maximum rate of 0.4 mm/year for biodegradable implants so Zn-xZr alloys were suitable as biodegradable material implant for orthopaedic application.

1. Introduction

Biomaterials are defined as a substance that has been engineered to take a form to replace or assist part of an organ or tissue in living things and will be in direct contact with it [1]. In the body, biomaterials will interact with their supported organs or tissues so they must have good biocompatibility. In the medical world, biomaterials are widely used as orthopaedic, cardiovascular, ophthalmic, dental implants, wound healing, and drug delivery system. Orthopaedic implants are medical devices used to replace or assist the healing process of bones, or to replace articulating surfaces of a joints [2]. The main function of orthopaedic implants such as joint prostheses, plates, and screws is to provide strength and stability, directs the process of bone or joint growth by interacting directly with the damaged tissue [3].

Based on World Health Organization data, it is estimated that there will be an increase in the need for orthopaedic implants for the next few years. Based on research conducted by the World Health Organization, there is an increase in world average life expectancy between 2000 and 2016 that was 5.5 years to 72 years (74.2 for women and 69.8 for men), the fastest increase since the 1960s. In 2019, the average age of the world's life expectancy is 72.5 years. With the increasing of life expectancy, people worldwide will live longer. In 2050, the world population above 60 years is estimated to reach 2 billion people, of which 80% come from low and middle income countries. As the world's population is getting older, there will be an increase in musculoskeletal diseases, such as osteoporosis and osteoarthritis as well as an increase in demands for orthopaedic implants. Therefore, the development of biomaterial implant will greatly influence the application of these orthopaedic implants.

These days, biomaterials used as orthopaedic implants are still dominated by non-degradable metals, such as stainless steel, titanium alloys, and Co-Cr alloys. Biodegradable materials are begun to develop rapidly following disadvantages of the non-degradable metals, such as inflammatory reactions [4], the required of second surgery for removal of the implant [5], bone degradation [6], and so on. Meanwhile, biodegradable material does not require further surgery because it will dissolved in the recipient body. Therefore, essential metals in the body are needed to be developed as biodegradable material. Until now, only three alloys have been developed as biodegradable material, namely Mg-alloys, Fe-alloys, and Zn-alloys.

Magnesium is the first material developed as a biodegradable material and has been used commercially as an orthopaedic implant. Magnesium is harmless to the body and plays a role in several biological functions of the body, such as enzyme reactions, digestive functions and nerves, metabolism, synthetic proteins, andthe immune system [7]. Magnesium has mechanical properties that are similar to the mechanical properties of human bones [8]. However, pure magnesium has a high rate of degradation and can produce hydrogen gas which can harm the tissue around the implantation sites [9][10].

Iron is the second metalin development as a biodegradable material which has reached in vivo stage development in animals. Iron is one of the essential elements for oxygen transfer into body tissues and an important component in many types of enzymes[8]. Iron alloys have good mechanical properties so they are suitable for orthopaedic implants. However, iron alloys have very low degradation rates and require a long time for dissolving into the body. Thus, they can produce similar side effects to permanent non-degradable implants [11][12].

Zinc is the third metal developed as biodegradable material. In human body, zinc is the second most transition metal element after iron that present in all organs, tissues, and secretions of body fluids with the most amount found in muscles and skeleton (86%), skin (6%), liver (5%), brain (1.5%), and other tissues that are evenly distributed. As one of the essential elements in the body, zinc has various roles in human biological functions, specifically to support the immune system, enzyme synthesis, wound healing and protein formation, and bone formation [13][14]. Zinc is known to have a degradation ratebetween magnesium alloys and iron alloys. With a lower degradation rate, zinc alloys will produce less hydrogen gas which are expected to be able to dissolve in accordance with bone healing time (approximately 6 months) [15]. In addition, zinc alloys are also easier to be manufactured than magnesium alloys. However, zinc has relatively low mechanical properties, so the combination with other alloy elements is needed.

Zirconium is one of the transition metal elements that has been widely used in the biomaterial world, especially for dental implants and alloying element in magnesium based biodegradable material. Inside the body, zirconium has no crucial role in biological functions. Moreover, zirconium shows no toxic effects on body tissues. The addition of zirconium is known to be able to improve the mechanical properties of magnesium alloys through the mechanism of precipitate strengthening [16]. In this study, we will further investigate the effect of zirconium addition on the corrosion behavior of zinc alloys.

2. Materials and experimental details

2.1. Material preparation

Zn-xZr alloys with different content of Zr (0.5, 1, and 2 wt.%) as well as the pure Zn reference sample were prepared by melting high purity Zn (99.99%) and high purity Zr (99.99%) ingots using a graphite crucible under air atmosphere. The melts were stirred and kept for one hour in 550°C to ensure that zirconium are completely dissolved, and then poured into graphite mold and cooled at room temperature. The melted alloys then homogenized at the temperature of 380°C for 8 to 10 hours. After that, the samples were prepared for the microstructure analysis by cutting it for 5 mm in thickness. The surface of the alloys were mounted, grinded, polished and etched with 2 ml HNO_3 and 98 ml distilled water to prepare for surface analyzing using an optical microscope carl zeiss primotech and scanning electron microscope FEI inspect 50 to determine the shape of the grain and intermetallic distribution of the alloys.

2.2. Electrochemical measurements
In this study, the corrosion properties of Zn-xZr alloys would be measured in a simulated body fluid using potentiodynamic polarization tests on Autolab (PGSTAT-302N) with a standard set-up of three electrodes. The three electrodes used in the tests were platinum plate as counter electrode, silver silver chloride (SSC) as reference electrode, and the alloys as the working electrodes. Before the tests, the Open Circuit Potentials (OCP) of the speciments would be monitored for 3600 s. The OCP was measured to see whether the alloys would be stable inside the fluids. The tests were recorded at scanning rate of 1 mV/s, starting from -250 mV to +500 mV at the temperature of 37°C in simulated body fluid. From the tests, the corrosion rate would be calculated based on ASTM G102-89.

3. Results and discussion

3.1. Microstructure of Zn-xZr alloys
Fig.1 shows the surface of the samples after potentiodynamic polarization test while Fig. 2 shows the microstructures of the alloys under scanning electron microscope. The characterization of the microstructures shows the addition of the higher value of zirconium that led to the formation of the interconnected networks of intermetallic compounds. Consequently, the formation of these intermetallic compounds would increase the corrosion rates, although it also displaced the corrosion potentials of the Zn-xZr alloys to more noble values.

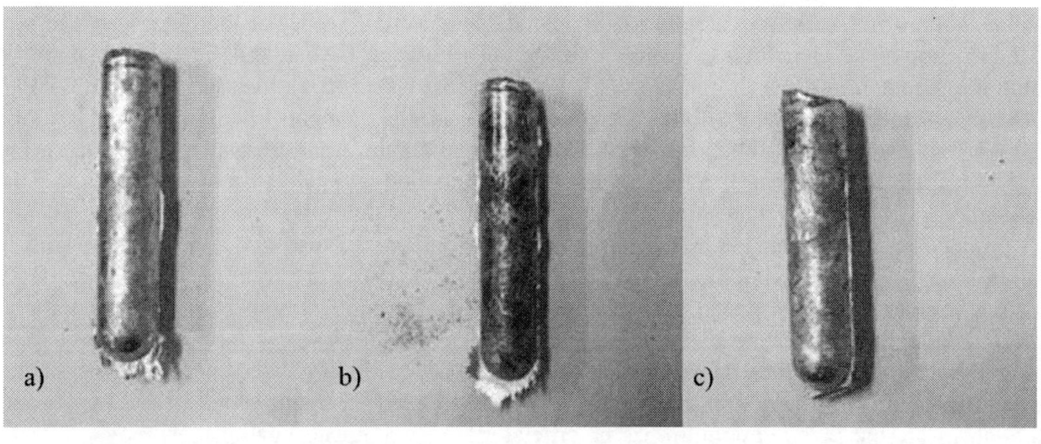

Figure 1. Visual inspection of (a)Zn-0.5Zr, (b) Zn-1Zr, (e) Zn-2Zr after potentiodynamic polarization test.

3.2. Corrosion behaviour of the Zn-xZr alloys
The potentiodynamic polarization curves of the Zn alloys are shown in Fig.3. The corresponding electrochemical performance parameters are listed in Table 1. The corrosion potential (E_{corr}) and corrosion current density (I_{corr}) are -0.875 V vs SSC in saturated KCl and 9.88 μA/cm^2 for pure zinc, -0.965 V and 5.29 μA/cm^2 for Zn-0.5Zr, -0.984 V and 7.73 μA/cm^2 for Zn-1Zr, and -0.893 V and 12.50 μA/cm^{2}for Zn-2Zr, respectively. The corrosion potentials of zinc alloys are lower than that of pure zinc which suggest that the Zn-xZr alloys are more prone to corrosion than pure zinc. Meanwhile, with the exception of Zn-2Zr, the current density of the Zn-xZr alloys are also relatively lower than the pure zinc.The corrosion current density would be used to determine the corrosion rate of the alloys. The corrosion rate (Vcorr) was calculated based on ASTM G102 in which pure zinc will corrode with the rate of 0.148 mm/year, 0.079 mm/year for Zn-0.5Zr, 0.116 mm/year for Zn-1Zr, and 0.188 mm/year for Zn-2Zr, respectively. The reduction of alloy corrosion rate can be linked to the refinement of the

microstructures caused by the addition of zirconium. It has been studied that the smaller grain size creates more grain boundaries and, also reduces the corrosion rates of the alloys. On the other hand, the addition of zirconium beyond the previously mentioned value led to a deterioration of the corrosion behaviour of the Zn-2Zr alloy with a higher current density, which indicates that Zn-2Zr has the lowest corrosion resistance between the zinc alloys examinedin this study.

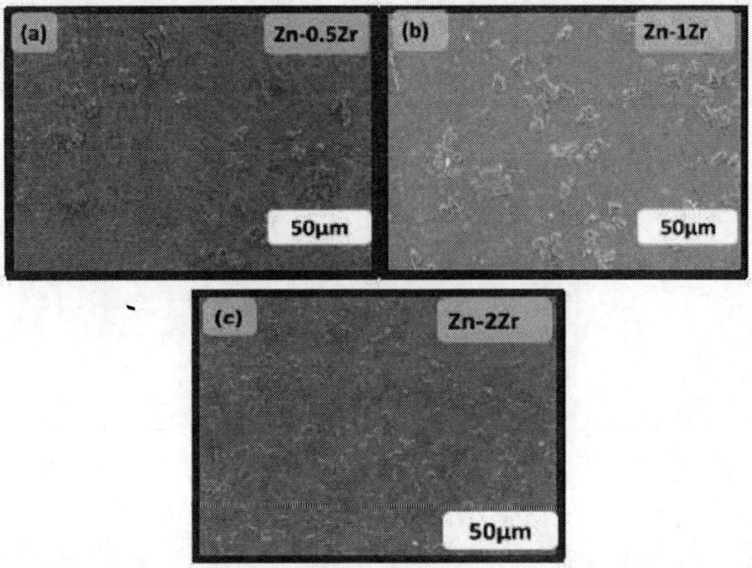

Figure 2. Microstructures of (a) Zn-0.5Zr, (b) Zn-1Zr, (c) Zn-2Zr with SEM analysis.

During anodic polarization, a passive film is formed on the surface and provide a protective barrier that keeps the corrosion current at low level. Figure 3 shows that passive films were formed in the alloy which indicated by the passivation potential (E_{pp}) and passivation current density (ι_{pp}) in the polarization curve. Passivation potential and passivation current density are the point in which the passivation zone started to form on the surface of each alloy. On the other hand, the end of the passivation zone is signed by the breakdown potential (E_b). This is the potential in which the passive film breaks to the surface of alloys. The details of the breakdown potentials are also shown in Table 1. As shown in Fig. 3, Zn-1Zr would be the first to form the passivation zone, followed by Zn-0.5Zr, Zn-2Zr, and lastly the pure zinc. On the other side, the breakdown potential of pure zinc is higher than that of Zn-0.5Zr and Zn-1Zr alloys, suggesting that the two alloys are more sensitive to local corrosion by destroying the protective layers whilst Zn-2Zr is more likely to avoid the local corrosion due to passivation phenomena. The formation of the passivation zone also shows the resistancy of the alloys to corrode where Zn-2Zr shows the highest resistance among the alloys with its high E_{pp} and E_b.

As biomaterial, it is required for an orthopedic implant to hold 95% of its area during to the first six weeks of implantation. It is needed to ensure that the material used as the orthopedic implant will not degrade completely before the healing process of the bone which usually takes six to 12 weeks is finished. If the material in used breaks down too fast, the sudden lost of bone support will disrupt the healing process and may impair the formation of the bone as well. Therefore, it is studied that the maximum corrosion rate allowed for biomaterial used for implant is 0.4 mm/year[17]. For that reason, the alloys studied here will be suitable to use as biodegradable implant. However, Zn-0.5Zr will not be effective because its corrosion rate is even lower than that of pure iron. On the other hand, Zn-2Zr will be a better choice with its higher corrosion rate. In comparison, Zn-1Zr has a slightly lower corrosion rate than Zn-2Zr, but it also has low corrosion resistance which is proven by its low OCP and the short range of potentials in which the passivation zone is formed. Therefore, in practice, Zn-1Zr will corrode faster than Zn-2Zr as shown in Fig. 1 where Zn-1Zr has the more blackened area than the Zn-2Zr alloys.

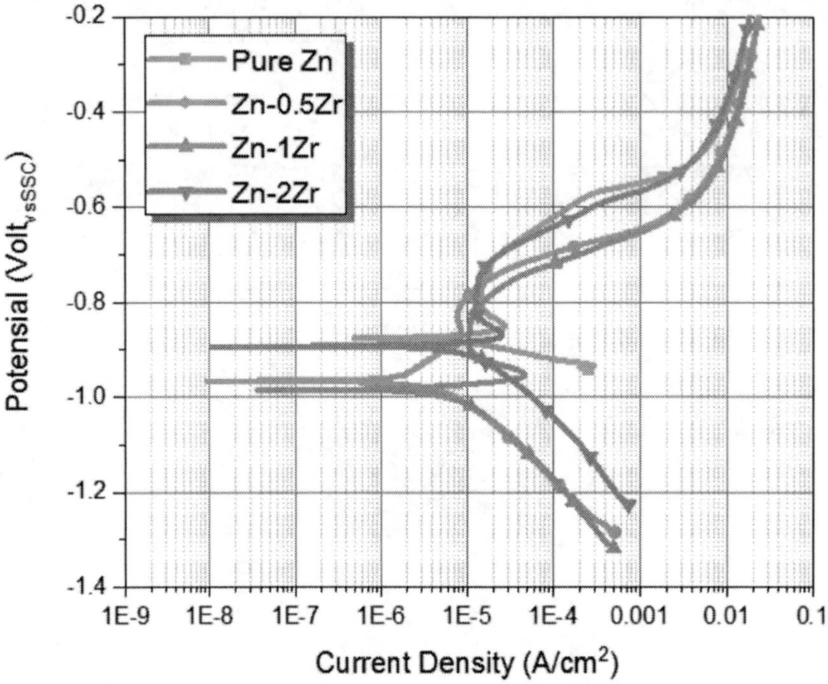

Figure 3. Polarization curve of Zn-xZr alloys with pure zinc as reference.

Table 1. Value measured from potentiodynamic polarization tests of Zn-xZr alloys and pure Zn.

Alloys	Ecorr(V_{vsSSC})[a]	icorr($\mu A/cm^2$)[b]	CR (mm/y)[c]	E_{pp} (V_{vsSSC})[d]	ipass ($\mu A/cm^2$)[e]	E_b (vsSSC)[f]
Zn	-0.875	9.88	0.148	-0.84841	27.40	-0.71241
Zn-0.5Zr	-0.965	5.29	0.079	-0.86951	8.55	-0.80451
Zn-1Zr	-0.984	7.73	0.11	-0.95145	46.79	-0.82045
Zn-2Zr	-0.893	12.50	0.188	-0.86788	25.53	-0.71088

[a] Corrosion potential
[b] Corrosion current density
[c] Corrosion rate
[d] Potential in which the passivation zone starts to form
[e] Current density in which the passivation zone starts to form
[f] Breakdown potential

4. Conclusions

Based on the research, it can be concluded that the addition of zirconium will decrease the corrosion rate of Zn-xZr alloys until a certain value of zirconium where the intermallic compounds will be formed and increase the corrosion rate of the alloys. The formation of passivation zone shows that the protected layer covering the surface of Zn-xZr alloys will also play a part to their biodegradable capability. In general, Zn-xZr alloys are suitable to be used as biodegradable implant for orthopedic application with Zn-2Zr as the most effective alloy among others.

Acknowledgment

The authors would like to thank the Direktorat Riset dan Pengabdian Masyarakat Universitas Indonesia (DRPM- UI) for the financial support under the grant of Hibah PITTA UI 2018 No. 2535/UN2.R3.1/HKP.05.00/2018.

References

[1] Chen Q and Thouas G. A 2015 *Mater. Sci. Eng. R Rep* **87** 1–57
[2] Rouhi G and Amani M A Brief Introduction into Orthopaedic Implants: Screws, Plates, and Nails p. 20
[3] Agarwal R and García A. J 2018 *Adv. Drug Deliv. Rev.* **94** 53–62
[4] Shen C et al. 2016 *RSC Adv.* **6** 86410–86419
[5] Qin H et al. 2015 *Biomaterials* **53** 211–220
[6] Rahim M, Ullah S, and Mueller P 2018 *Metals* **8** 532
[7] Homayun B and Afshar A 2014 *J. Alloys Compd.* **607** 1–10
[8] Vojtěch D, Kubásek J, Čapek J, and Pospíšilová I 2014 Magnesium, Zinc and Iron Alloys for Medical Applications in Biodegradable Implants p. 6
[9] Galib R. H and Sharif A 2016 Development of Zn-Mg Alloys as a Degradable Biomaterial vol. 1 no. 1 8
[10] Li H et al. 2015 Development of biodegradable Zn-1X binary alloys with nutrient alloying elements Mg, Ca and Sr *Sci. Rep.* **5**
[11] Zheng Y, Gu X, and Witte F 2014 *Mater. Sci. Eng. R Rep.* **77** 1–34
[12] Kafri A, Ovadia S, Goldman J, Drelich J, and Aghion E 2018 *Metals* **8** 153
[13] Li H, Zheng Y, and Qin L 2014 *Prog. Nat. Sci. Mater. Int.* **24** 414–422
[14] Katarivas Levy G, Goldman J, and Aghion E 2017 *Metals* **7** 402
[15] Tong X et al. 2018 *Acta Biomater.* **82** 197–204
[16] Wątroba M, Bednarczyk W, Kawałko J, and Bała P 2018 *Mater. Charact.* **142** 187–194
[17] Vojtěch D, Kubásek J, and Čapek J 2015 *Mater. Tehnol.* **49** 877–882

IOP Conf. Series: Materials Science and Engineering **833** (2020) 012086 doi:10.1088/1757-899X/833/1/012086

The Effect of SrO Impregnation on The Characteristics of Cobalt Ferrite ($CoFe_2O_4$) Nanoparticles Synthesized by Coprecipitation

Thutug Rahardiant Primadi, Aman Santoso, Fauziatul Fajaroh*

Department of Chemistry, Universitas Negeri Malang

*Corresponding author: fauziatul.fajaroh.fmipa@um.ac.id

Abstract. Cobalt Ferrite ($CoFe_2O_4$) nanoparticles are magnetic materials that have greater hardness than Fe_3O_4. $CoFe_2O_4$ is widely applied in many fields, including as a heterogeneous catalyst in various chemical reactions such as photocatalysis, phenol degradation, and transesterification reactions. This is supported by the thermal and chemical stability possessed by $CoFe_2O_4$. Coating or impregnation of non-magnetic material on the surface of the ferrite, on one side will reduce its magnetism and possibly its surface area. But on the other hand, the positive impact of coating or impregnation is the increase in particle monodispersity. Weakening the magnetism of the particles will reduce the interaction of attraction between particles, so that the particles are getting more monodisperse. This will optimize its potential as a heterogeneous catalyst. In this research impregnation of strontium oxide (SrO) would be carried out on the surface of $CoFe_2O_4$. After that, the characters were compared with $CoFe_2O_4$ before being impregnated. The $CoFe_2O_4$ synthesis was carried out through the coprecipitation method, followed by wet impregnation with SrO. The results showed that the impregnation of SrO on $CoFe_2O_4$ increased the zeta potential of the particles which indicated the stability of the particles in the dispersion.

1. Introduction

Cobalt Ferrite ($CoFe_2O_4$) is a spinel ferrite crystal classified AB_2O_4 type. Element A is a divalent metal which occupies a tetrahedral position, while B is a trivalent metal that occupies an octahedral position [1]. $CoFe_2O_4$ is a magnetic compound with a magnetic value slightly lower than Fe_3O_4, $CoFe_2O_4$ is a hard metal while Fe_3O_4 is a soft metal [2]. This causes $CoFe_2O_4$ have good thermal and chemical stability [3]. Based on this stability makes $CoFe_2O_4$ to be useful material with wide applications, such as material for electronic, sensor, drug delivery, and catalyst [4]. This material is often used as a heterogeneous catalyst in several chemical reactions, including photocatalysis, phenol degradation of acetylene decomposition in carbon nanotube production (CNT), biomedical applications [5].

The homogeneous catalyst usually makes reactions going quickly and efficient, but the catalyst cannot be recovered for reuse and also corrosive for environment. That is different from heterogeneous catalyst which is easy to separate, recoverable, and environment-friendly [6]. Catalyst properties will be maximized if the size of material is very small (especially in nanometers orders). Because of the smaller the particle, the wider the specific surface area. The wider the surface area of the catalyst, the wider the active side which is in contact with substrate [7].

The factor that must be avoided in order to obtain uniform particle distribution and small particle size is material agglomeration. For this reason, it is necessary to add materials that can reduce or prevent particle agglomeration in the synthesis process. One of the metal oxides that can be used as a deterrent for agglomeration in ferrite is Strontium Oxide (SrO). SrO also has a dual function to assist the catalytic process, especially the transesterification reaction [5] [8].

In this research, we intend to synthesize and characterize material of $SrO/CoFe_2O_4$ nanocomposite. The stages of this research include: (1) synthesis of SrO by precipitation; (2) synthesis of $CoFe_2O_4$ by coprecipitation; (3) Impregnation of SrO into $CoFe_2O_4$ by wet-impregnation; (4) Characterization of the synthesized material by XRD, BET, SEM, EDX, Zeta Potential, and VSM.

2. Materials and Method

Materials used are Strontium Chloride hexahydrate ($SrCl_2.6H_2O$) *Merck*, Ammonium Carbonate (NH_4CO_3) *Merck*, Iron (III) Chloride hexahydrate ($FeCl_3.6H_2O$) *Merck*, and Cobalt Chloride hexahydrate ($CoCl_2.6H_2O$) *SAP*, and Sodium Hydroxide (NaOH) *Merck*.

2.1. Synthesis of Strontium Oxide (SrO)

Strontium Oxide (SrO) was synthesized by precipitation method from Strontium Chloride ($SrCl_2$) solution and Ammonium carbonate (NH_4CO_3) solution. This process was going on stirring at 80 °C for 2 hours. The result of this process was smooth white powder; It's Strontium Carbonate ($SrCO_3$). The powder was then washed to remove the impurities with demineralized water until pH of filtrate became neutral and free from chloride ions. After that washing process, the powder was continued washed using ethanol and acetone. SrO was formed by calcination of $SrCO_3$ at 1000 °C for 4 hours. The product was brittle white powder, which is the specific characterization of SrO. Then this material was cooled at room temperature and stored in a desiccator.

2.2. Synthesis of Cobalt Ferrite ($CoFe_2O_4$)

Nano-magnetic Cobalt Ferrite ($CoFe_2O_4$) was synthesized by co-precipitation reaction of an aqueous solution of Iron (III) Chloride and Cobalt Chloride by an alkaline solution. A mixture of aqueous solution of $CoCl_2$ and $FeCl_3$ in mole ratio of 1:2 was mixed with solution of Chloride Acid. An alkaline solution is added drop by drop to this solution. In this research, using Sodium Hydroxide (NaOH) as an alkaline solution.

This Co-precipitation process was carried out in a reflux system equipped with a magnetic stirrer. The set of equipment used is presented in Figure 1. This process took place at 70 °C for about 4 hours. The result was black sediment at the bottom of the flask. Then the black sediment was separated from the solution by filtering. The material was washed with demineralized water until pH of the filtrate is neutral and free from chloride ions. After that, the material was dried at 150 °C for 5 hours and calcined at 850 °C for 4 hours. The end result was black Cobalt Ferrite ($CoFe_2O_4$).

2.3. Impregnation SrO on $CoFe_2O_4$

The impregnation process was conducted by using wet-impregnation method. Strontium Oxide (SrO) and Cobalt Ferrite ($CoFe_2O_4$) were mixed with a weight ratio of 5:1. This mixture was dispersed in pure methanol and stirred in a reflux system at 80 °C for about 8 hours. The resulting deposit was filtered and dried at 120 °C for 6 hours. The end result was a composite of $SrO/CoFe_2O_4$.

2.4. Characterization of $SrO/CoFe_2O_4$

Characterization of the particles was consisting of several tests for material characteristics. X-ray diffraction (XRD) patterns were recorded in the 2θ range of 10°–90° with Cu Kα radiation source (λ = 0.1542 nm, 40 mA, 45 kV) by using Panalytical Xpert Pro Diffractometer. Crystallinity size of the material can be calculated from XRD result using *Debye-Scherrer Equation*.

$$D = \frac{K\lambda}{\beta cos\theta} \quad (1)$$

where λ is the wavelength of X-ray beam, β is full width at half maximum (FWHM) and θ is Bragg scattering angle and K (=0.89) is shape factor.

The specific surface area was determined by N_2 adsorption at −196 °C using Nova 1200 Quantchrome by the BET methods. The morphology of the prepared nanocomposites was determined by scanning electron microscopy (SEM) images with Type Inspect S50, FEI. Zeta Potential was done by Malvern Zetasizer using methanol as dispersant at room temperature. Analysis with Vibrating sample magnetometer (VSM) was done by using Oxford VSM 1,2 H with magnetic fields in the range − 8000 to 8000 Oe at room temperature.

Figure 1. Co-precipitation method.

3. Result and Discussion

3.1. XRD Result

The synthesized SrO was started through $SrCO_3$ synthesis by coprecipitation method through reaction:

$$SrCl_{2(aq)} + (NH_4)_2CO_{3(aq)} \rightarrow SrCO_{3(s)} + NH_4Cl_{(s)}$$

Then SrO formed by calcination of $SrCO_3$ at 1000 °C following this reaction:

$$SrCO_{3(s)} \rightarrow SrO_{(s)} + CO_{2(g)}$$

Meanwhile, the synthesis process of cobalt ferrite ($CoFe_2O_4$) was through co-precipitation method using an alkaline (NaOH) solution as precipitant. The XRD results shown in Figure 2. The diffraction peaks of $CoFe_2O_4$ suitable with JCPDS Card No. 96-591-0064 for $CoFe_2O_4$ indicated by the crystal planes of (220), (311), (222), (400), (422), (333), (440), (531), (620), (553), and (444). Furthermore, through the refining method using Rietvield for $CoFe_2O_4$ diffractogram as shown in Figure 3, there is a Good of Fitness (GOF) value of 3.213, and obtained data that the synthesis material has molecular weight of 235,73 g/mol with density of 66.324 g/cm^3 that suitable for $CoFe_2O_4$.

Another curve in Figure 2 show XRD spectrum of SrO/$CoFe_2O_4$ which was the result of SrO impregnation on the surface of $CoFe_2O_4$ by wet impregnation using methanol as solvent. The successful indicator of the impregnation can be seen in Figure 4, the XRD spectrum appears to be a combination of typical peaks of SrO corresponding to JCPDS Card no. 96-152-9601 at 2θ 25.3° and typical peaks of $CoFe_2O_4$ with JCPDS Card No. 96-591-0064. This result is similar with result of Falcao [9].

Crystallite size calculated by the *Debye-Scherrer* equation of the material before and after impregnation is 24 nm and 44 nm. This result showed that SrO covers $CoFe_2O_4$ nanoparticle. Based on this size, this composite still can be classified as nanoparticle because its lower than 100 nm. So it can be concluded that co-precipitation method is an effective preparation method of nanomaterial.

3.2. BET Result

BET characterization gave a specific surface area of $CoFe_2O_4$ and SrO/$CoFe_2O_4$ of about 24.317 m²/g and 9.223 m²/g, respectively. This indicated that surface of $CoFe_2O_4$ has been covered with large amounts of SrO powder. This result is match with crystallinity size analysis using *Debye-Scherrer Equation,* i.e. the nanomaterial became bigger after impregnation, so the specific area of nanomaterial became tighter than before impregnation. But the composite with specific surface area of 9.223 m^2/g is still potential as nanocatalyst.

3.3. SEM and EDX Result

Based on SEM photographs as shown in Figure 5, the surface morphology of $CoFe_2O_4$ is similar to a boulder polydisperse (Figure 5a). This is different from the photo of $SrO/CoFe_2O_4$ (Figure 5b) which looks much smaller and monodisperse, although taken with the same magnification.

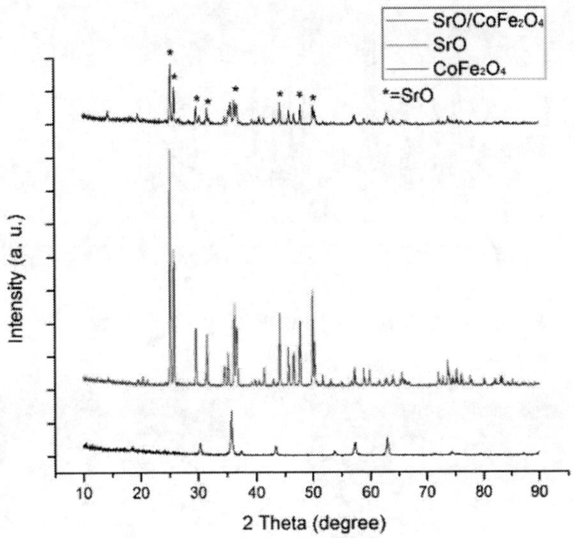

Figure 2. XRD Diffractograms.

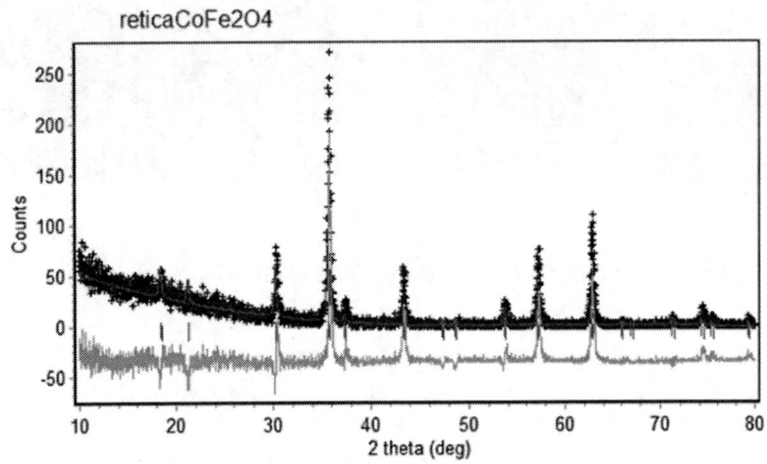

Figure 3. Rietvield of $CoFe_2O_4$ Diffractogram.

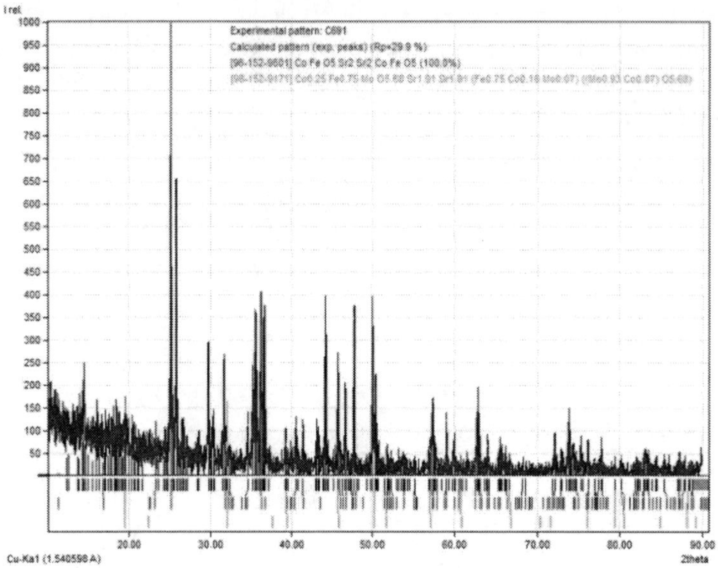

Figure 4. Match of $SrO/CoFe_2O_4$ Diffractogram.

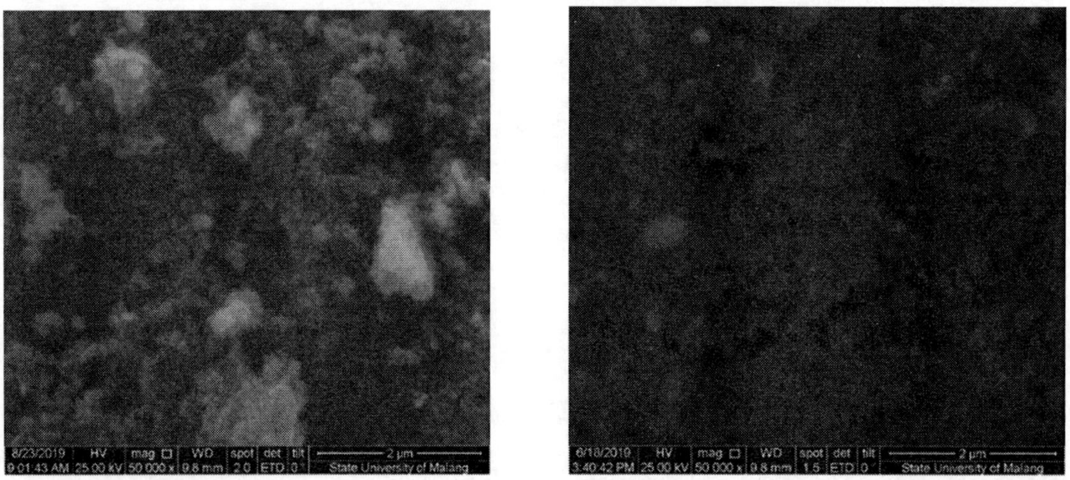

(a) (b)

Figure 5. SEM of $CoFe_2O_4$ (a) before impregnation and (b) after impregnation using SrO.

From the $CoFe_2O_4$ EDX result shown in Figure 6 that it is proven that material consists of Cobalt, Iron, and Oxygen with correct atom and weight ratio. In Figure 7 we can see the presence of SrO in composite, but we can't find a correct number result. It's very difficult to get the correct number in composite material, because it was very difficult to get homogenous impregnation material.

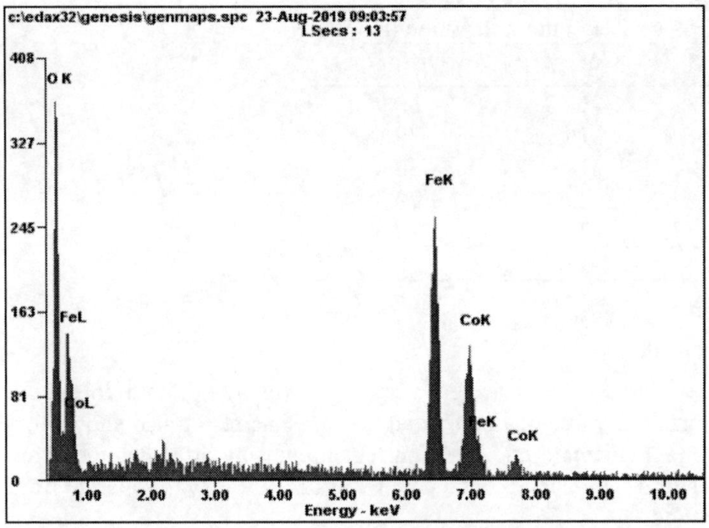

Element	*Wt%*	*At%*
OK	26.27	55.91
FeK	46.89	28.59
CoK	26.84	15.51
Matrix	Correction	ZAF

Figure 6. EDX of $CoFe_2O_4$.

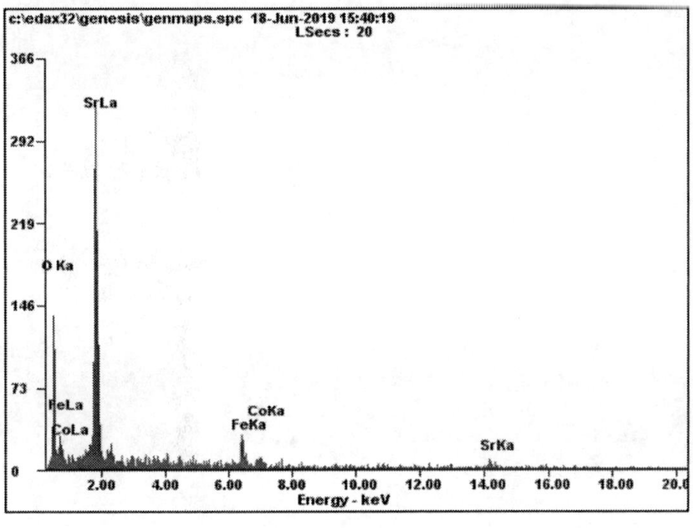

Element	*Wt%*	*At%*
OK	35.64	73.52
FeK	07.39	04.37
CoK	03.57	02.00
SrK	53.41	20.12
Matrix	Correction	ZAF

Figure 7. EDX of $SrO/CoFe_2O_4$ composite.

3.4. Results of Zeta Potential Analysis

Zeta potential analysis is potentially useful for checking the stability component in disperse condition and to know the surface charge of material. If the zeta potential value of a material in a dispersion outside the range of -30 mV to 30 mV, the dispersion is stable. But if zeta potential value in a range between -30 mV and 30 mV, the dispersion of the material is not stable and tend to aggregate and then flocculate [10]. In relation with biodiesel catalytic reaction, zeta potential parameter of a catalyst affects the conversion value of biodiesel. the more positive the zeta potential of a material, the more potential it is as a Lewis acid which is the more potential for starting the formation of free radicals needed in the transesterification reaction to produce biodiesel.

The zeta potential result value show in Table 1, the average value for $CoFe_2O_4$ is -18.5 mV. After impregnation process using SrO, the value becomes 30.3 mV. This shows that the presence of SrO makes the material more stable in dispersion and has the potential as a transesterification catalyst.

Table 1. Zeta Potential Results.

No. Experiment	Zeta Potential Value (mV)	
	$CoFe_2O_4$	$SrO/CoFe_2O_4$
1	-12.8	30.9
2	-16.6	29.8
3	-18.4	30.4
4	-23.6	30.8
5	-21.0	29.6
Average	**-18.5**	**30.3**

3.5. Results of VSM Analysis

Figure 8 shows the magnetic moment of $CoFe_2O_4$ is relatively very high, which is around 76 emu/g. This value is almost same with magnetization value of Fe_3O_4 (magnetite). This result also shows that $CoFe_2O_4$ can be classified as ferro-paramagnetic material. After impregnation using SrO, magnetization value goes down until 12 emu/g. But it still be concluded that the material can be easily to separate from the system using an external magnetic field.

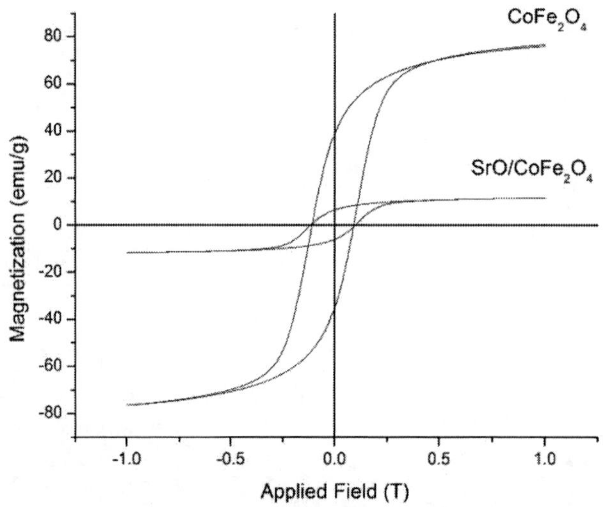

Figure 8. VSM Result of $CoFe_2O_4$ and $SrO/CoFe_2O_4$.

4. Conclusion

Based on the results of this study, it can be concluded that the $SrO/CoFe_2O_4$ nanocomposite has been synthesized by the coprecipitation method and followed by the impregnation of SrO on the surface of $CoFe_2O_4$. Based on XRD analysis results that provide information on crystal size, BET analysis results that show large surface area, SEM results that inform particle morphology, zeta potential analysis results that show the dispersion of synthesized particles in reaction media is stable, and VSM results that show magnetism material, it can be concluded that $SrO/CoFe_2O_4$ material has the potential as a catalyst in biodiesel synthesis.

References

[1] Houshiar M, Zebhi F, Razi Z J, Alidoust A and Askari Z 2014 *J. Magn. Magn. Mater.* **371** 43–8
[2] Horng L, Chern G, Chen M C, Kang P C and Lee D S 2004 *J. Magn. Magn. Mater.* **270** 389–96

IOP Conf. Series: Materials Science and Engineering **833** (2020) 012086 doi:10.1088/1757-899X/833/1/012086

[3] Chitu L, Jergel M, Majkova E, Luby S, Capek I, Satka A, Ivan J, Kovac J and Timko M 2007 *Mater. Sci. Eng. C* **27** 1415–7

[4] Amiri S and Shokrollahi H 2013 *Mater. Sci. Eng. C* **33** 1–8

[5] Semwal S, Arora A K, Badoni R P and Tuli D K 2011 *Bioresour. Technol.* **102** 2151–61

[6] Yan S, DiMaggio C, Mohan S, Kim M, Salley S O and Ng K Y S 2010 *Top. Catal.* **53** 721–36

[7] Ross J R H 2012 Catalyst Preparation *Heterogeneous Catalysis* (Elsevier) pp 65–96

[8] Gryglewicz S 1999 *Bioresour. Technol.* **70** 249–53

[9] Falcão M, Garcia M, de Moura C, Nicolodi S and de Moura E 2017 *J. Brazil. Chem. Soc.*

[10] Joseph E and Singhvi G 2019 Multifunctional nanocrystals for cancer therapy: a potential nanocarrier *Nanomaterials for Drug Delivery and Therapy* (Elsevier) pp 91–116

IOP Conf. Series: Materials Science and Engineering **833** (2020) 012087 doi:10.1088/1757-899X/833/1/012087

Performance of Additive in Metal Separation from Nickel Ore

Angella Natalia Ghea Puspita, Adji Kawigraha*, Nur Vita Permatasari, Santi Ari Respati[1]

[1]Center of Mineral Resource Technology (PTSPM), Agency for Assessment and Application Technology (BPPT), Serpong, Banten, Indonesia.

*Corresponding author: adji.kawigraha@bppt.co.id

Abstract. The potential resources and reserves of nickel ore in Indonesia is the third largest in the world. The potential quite large, but the nickel content found in nature is very small. To increase the added value of nickel ore, it is necessary to process/ refine nickel ore. In addition, coal ratio parameter and process temperature in the reduction process, additive is also used as parameter for the composite reduction process. This research used 6 combination saprolite ore composite based on result of previous research. The objective of this research is to know the performance of additive in separation of metal from nickel ore based on result of XRD analysis. XRD analysis is used to know chemical compound saprolite ore composite after reduction process. This results from XRD analysis shown that additive CaCO3 has the best performance additive with higher Fe and Tetrataenite content with percentage of coal ratio of 30% in temperature of 1300 ^{0}C. For additive Ca_2SO_4 percentage coal ratio 15% have higher Fe and NiFeO content in temperature of 1200 ^{0}C and 1300 ^{0}C.

1. Introduction

Nickel ore is the potential resources and reserves in Indonesia and the third largest in the world after New Caledonia and Philippines [1]. The potential for resources and nickel ore reserves are quite large, but the nickel content contained in nature very small. Laterite nickel ore divided into 2 types namely, Saprolite ore that has small Fe and large Ni content (around 1.5-2.5%), and limonite ore that has large Fe and small Ni content (around 0.8-1.5%) [2]. To increase the added value of nickel ore, it is necessary to process/refine nickel ore. The processing/refining technology of nickel ore is carried out through the Pyro metallurgy and Hydro metallurgy technique. The hydro metallurgy technique involves chemical reactions with other additional loads, while the Pyro metallurgy technique involves high temperatures and large energy [3].

The composite use to leads a better process, because it does not require cooking and sintering plants and minimize the use of gas. The uses of composite pellets raise the idea of using low grade iron ore to produce pig iron [4]. The process of mixing saprolite ore, coal, additive and bentonites; pelletizer, and drying are the process of making composites. In pyro metallurgy technique, reduction process is an important role in making nickel. Many parameters are involved in the reduction process. The availability of carbon along with processing time and temperature is the main concern in the reduction process [5], based on the result of previous experiment. In addition to the coal ratio parameter and process temperature in the reduction process [6], additive is also used as parameter for the composite reduction process.

The reduction process of saprolite ore composite is carried out in tube furnace with a temperature increase rate of 10 ^{0}C/minute. Composites are placed in the graphite crucible, that's suitable for heating samples in a tube furnace. Saprolite ore composites after reduction process is analyzed using X-Ray-Difference Diffraction (XRD). The XRD analysis uses to know the chemical compound. In this previous research of reduction process saprolite ore composite in Tube Furnace [7][8][9], 4 parameters are used, namely coal ratio, temperature, duration time and type additive with each 3 level. The results show that coal ratio, temperature and type additive affect nickel percentage content [7]. From the research of optimization of saprolite ore composite reduction process using Artificial Neural Network (ANN) and Adaptive Neuro- Fuzzy Inference System (ANFIS) method, the results show that the optimal combination is $SB_{15}Ca_{10}P_2$ with a percentage coal ratio of 15%, a temperature of 1200 ^{0}C, process time of 3 hours and additive type of Ca_2SO_4 [8] [9]. From the research by [7] with Response Surface Methodology (RSM) method, it shows that the optimal combination is $SB_{30}Ca_{10}P_2$ with percentage coal ratio of 30%, a temperature of 1300 ^{0}C, process time of 2 hours and type of additive of Ca_2SO_4.

The composite combination used in this research are $SB_{15}Ca_{10}P_2$ and $SB_{30}Ca_{10}P_2$ with temperature comparison of 1200 ^{0}C and 1300 ^{0}C and analyzed using XRD. To know additive performance, another type of additive, $CaCO_3$, is used in this research [7], because the composite combination with a type of additive $CaCO_3$ have higher percentage in Ni content in XRF result analysis. The result of XRF analysis in the previous research [7] is shown in figure 1. To compare the performance additive in separation metal of nickel ore, composite combination $SB_{15}Ca_{10}P_2$ is used. $SB_{30}Ca_{10}P_2$ with a temperature of 1200 ^{0}C and $SB_{15}Cc_{10}P_2$, $SB_{30}Ca_{10}P_2$ with a temperature of 1300 ^{0}C are used in this research. Saprolite ore composite combination reduction process in Tube Furnace is analyzed in XRD analysis. The objective of this research is to know the performance of additive in metal separation from nickel ore based on the result of XRD analysis.

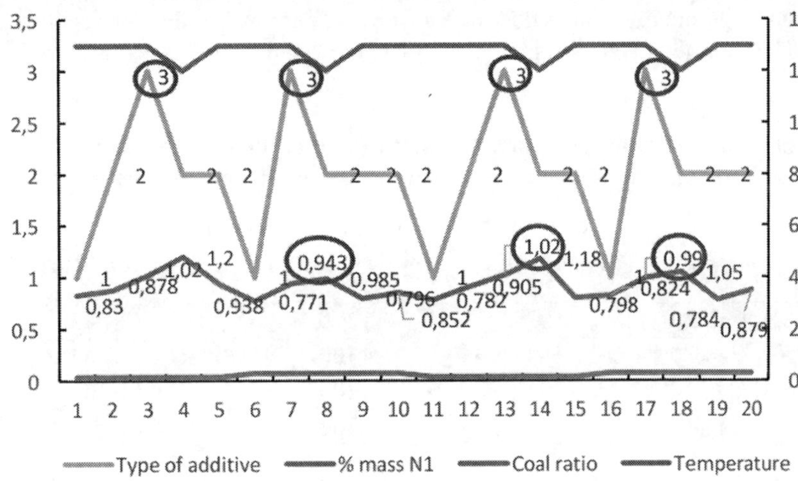

Figure 1. Type of additive with % mass Ni value.

2. Material and Methods

2.1. Materials

The type of materials used in this research is saprolite ore from Pomala, Southeast Sulawesi, Indonesia. Saprolite ore contains small Fe and large Ni content (around 1.5-2.5%). The sampling of material for the manufacture of saprolite ore composite is carried out directly from mining products in Pomala. Samples of mining results in Pomala undergo a grinding process to get a finer composition, then carried out by sampling using the Coning and Quartering method. Saprolite ore composite are produced by mixing percentage of saprolite ore, coal, additive and bentonite. Process of mixing, pelletizer and drying are the process of making saprolite ore composite. The composites produced in diameter 1.2-2 centimeters are used in figure 2. Saprolite ore composite that used in this research are $SB_{15}Ca_{10}P_2$ and $SB_{30}Ca_{10}P_2$ with each weighting 500 grams detail in table 1.

Figure 2. Saprolite ore composite.

Table 1. Saprolite ore composite.

Composite	Saprolite ore (gram)	Coal (gram)	Bentonite (gram)	Additive (Ca2SO4) (gram)	Total weight (gram)
$SB_{15}Ca_{10}P_2$	374	66	50	10	**500**
$SB_{30}Ca_{10}P_2$	308	132	50	10	**500**

The reduction process of saprolite ore in Tube Furnace through pyro metallurgy technique need high temperature and large energy. The reduction has several parameters that are used to get the optimal combination. The parameter that used in this research are % coal ratio, temperature, and type of additive.

2.2. *Methods*

Combination saprolite ore with the parameter that used in reduction process in Tube Furnace results Saprolite ore combination. Saprolite ore combination used in this research are in table 2.

Table 2. Saprolite ore combination.

Composite	% coal ratio	Temperature (^0C)	Type of additive
$SB_{15}Ca_{10}P_2$	15	1200	Ca_2SO_4
$SB_{15}Ca_{10}P_2$	15	1300	Ca_2SO_4
$SB_{30}Ca_{10}P_2$	30	1200	Ca_2SO_4
$SB_{30}Ca_{10}P_2$	30	1300	Ca_2SO_4
$SB_{15}Cc_{10}P_2$	15	1300	$CaCO_3$
$SB_{30}Cc_{10}P_2$	30	1300	$CaCO_3$

Saprolite ore composite combination with reduction process in tube furnace detail in Table 2. After reduction process, the composite will be analyzed using XRD.

2.3. *Characterization*

After reduction process, saprolite ore composite will be analysed using XRD. The XRD analysis uses to know the chemical compound. Furthermore, XRD analysis used to determine the different mineralogical phases present in the reduced ore [10]. The result of XRD analysis than analyzed between composite $SB_{30}Ca_{10}P_2$ and $SB_{30}Cc_{10}P_2$ at temperature 1300 ^0C, composite $SB_{15}Ca_{10}P_2$ and $SB_{15}Cc_{10}P_2$ at temperature 1300 ^0C, composite $SB_{15}Cc_{10}P_2$ and $SB_{15}Cc_{10}P_2$ at temperature 1300 ^0C, composite $SB_{15}Cc_{10}P_2$ and $SB_{15}Cc_{10}P_2$ at temperature 1300 ^0C, and composite $SB_{15}Ca_{10}P_2$ and $SB_{30}Ca_{10}P_2$ at temperature 1200 ^0C.

3. Result and Discussion

3.1. Results

Saprolite ore composite combination in Table 2 was reduced through the reduction process in tube furnace. The results of the saprolite ore composite were reduced through the reduction process in tube furnace and analyzed using XRD in Figure 3-7. From Figure 3 between composite composite $SB_{30}Ca_{10}P_2$ and $SB_{30}Cc_{10}P_2$ at temperature 1300 ^{0}C, composite $SB_{30}Cc_{10}P_2$ have high Fe content and similarity tetrataenite content compare at composite $SB_{15}Cc_{10}P_2$. That concludes the type of additive $CaCO_3$ is better than Ca_2SO_4 for percentage coal ratio of 30%. Figure 4 XRD analysis between composite $SB_{15}Ca_{10}P_2$ and $SB_{15}Cc_{10}P_2$ at temperature 1300 ^{0}C shows that $SB_{15}Cc_{10}P_2$ has higher Fe and Trevorite content compared to composite $SB_{15}Ca_{10}P_2$. This concludes the percentage coal ratio of 15%, the additive type of $CaCO_3$ is better than Ca_2SO_4. The percentage coal ratio 15% and 30%, type of additive $CaCO_3$ is better than Ca_2SO_4 because of the higher content of Fe, tetratenite and trevorite (NiFeO).

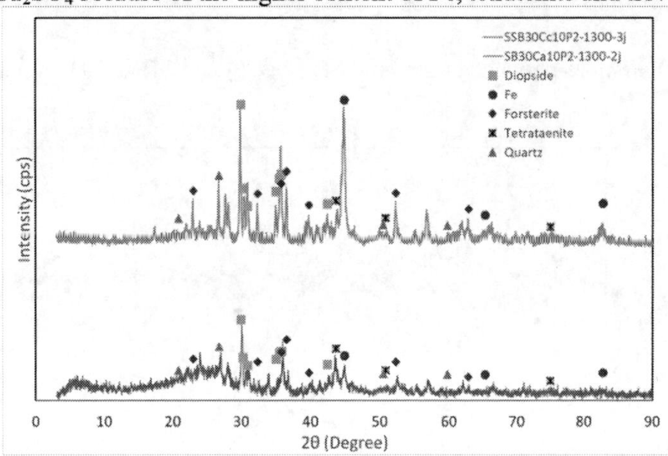

Figure 3. XRD analysis between composite $SB_{30}Ca_{10}P_2$ and $SB_{30}Cc_{10}P_2$ at temperature 1300 ^{0}C.

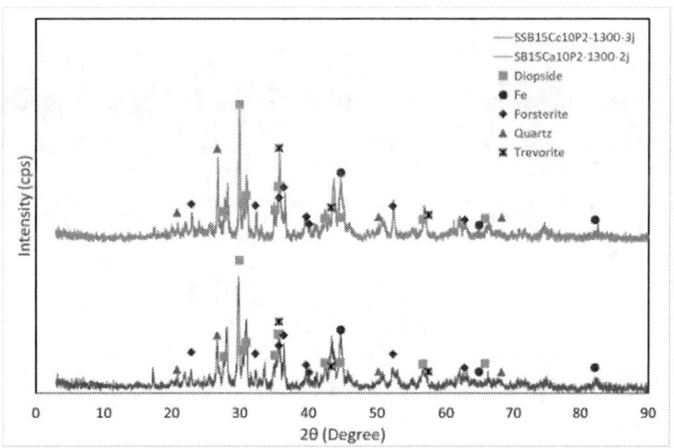

Figure 4. XRD analysis between composite $SB_{15}Ca_{10}P_2$ and $SB_{15}Cc_{10}P_2$ at temperature 1300 ^{0}C.

IOP Conf. Series: Materials Science and Engineering **833** (2020) 012087 doi:10.1088/1757-899X/833/1/012087

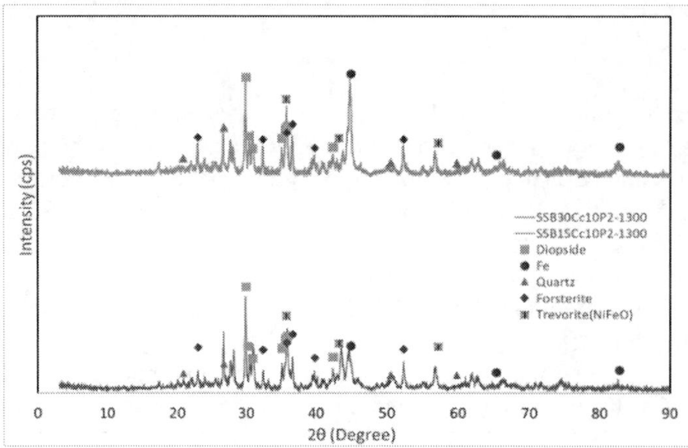

Figure 5. XRD analysis between composite $SB_{15}Cc_{10}P_2$ and $SB_{30}Cc_{10}P_2$ at temperature 1300 ^{0}C.

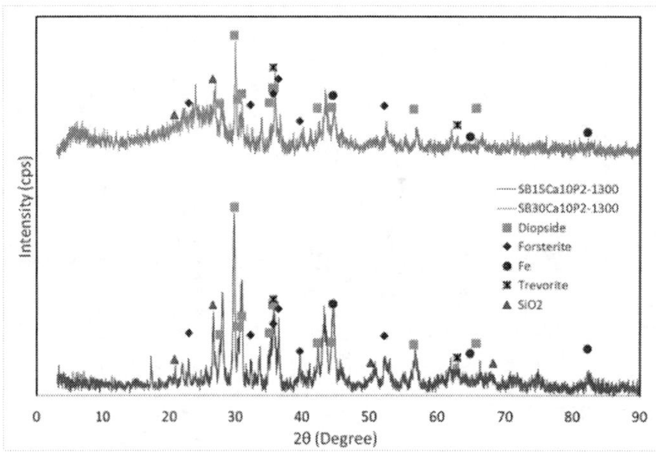

Figure 6. XRD analysis between composite $SB_{15}Ca_{10}P_2$ and $SB_{30}Ca_{10}P_2$ at temperature 1300 ^{0}C.

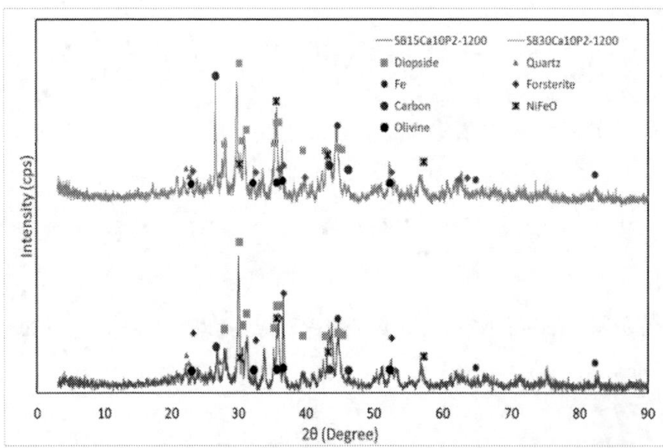

Figure 7. XRD analysis between composite $SB_{15}Ca_{10}P_2$ and $SB_{30}Ca_{10}P_2$ at temperature 1200 ^{0}C.

Figure 5 XRD analysis between composite $SB_{15}Cc_{10}P_2$ and $SB_{30}Cc_{10}P_2$ at temperature 1300 ^{0}C shows that the type of additive $CaCO_3$ at temperature of 1300 ^{0}C, composite combination $SB_{30}Cc_{10}P_2$ is better than composite combination of $SB_{15}Cc_{10}P_2$, because it has Fe and trevorite (NiFeO) content. The result from Figure 3, 4, and 5, conclude that $SB_{30}Cc_{10}P_2$ is the best performance additive at temperature of 1300 ^{0}C with type of additive $CaCO_3$.

XRD analysis in figure 6 between composite $SB_{15}Cc_{10}P_2$ and $SB_{30}Cc_{10}P_2$ at temperature 1300 ^{0}C, shows that type of additive Ca_2SO_4 in coal ratio 15% has higher Fe and NIFeO content compared with coal ratio 30%. The result of XRD analysis is similar to figure 7 between composite $SB_{15}Ca_{10}P_2$ and $SB_{30}Ca_{10}P_2$ at temperature 1200 ^{0}C coal ratio 15% has higher Fe and NiFeO content compared with coal ratio 30%, but in temperature 1200 ^{0}C carbon content still appears and hasn't been oxidized perfectly yet. In type of additive Ca_2SO_4, coal ratio 15% is better than coal ratio 30% in temperature 1200 ^{0}C and 1300 ^{0}C.

Based on result XRD analysis from figure 3-7, in temperature 1300 ^{0}C, with coal ratio 15% and 30%, type of additive $CaCO_3$ is better than because it contains more Fe, tetratenite and trevorite (NiFeO) content than type of additive Ca_2SO_4. In type of additive $CaCO_3$, in temperature 1300 ^{0}C, coal ratio 30% is better than coal ratio 15%. But the different result with the additive type of Ca_2SO_4, in temperature of 1200 ^{0}C and 1300 ^{0}C, coal ratio 15% is better than coal ratio 30%.

3.2. *Discussion*

Reduction process was affected by temperature and percentage coal ratio. The higher coal ratio, it can faster reduce. At higher temperature, if the percentage coal ratio small, the reduction process is difficult to detect. Based on the result of this research, XRD analysis shows that type of additive in $CaC0_3$ and Ca_2SO_4, percentage coal ratio 30% contain higher Fe, fosterite, tetratenite and trevorite (NiFeO) content in temperature 1300 ^{0}C. But the different result in temperature 1200 ^{0}C with a type of additive Ca_2SO_4, percentage coal ratio 15% is better than percentage coal ratio 30% because has higher Fe and NiFeO content, but carbon content still appears and hasn't been oxidized perfectly. It can conclude type of additive also affected in the reduction process in Tube Furnace, in addition to temperature and the percentage coal ratio and answer the objective of this research is to know the performance of additive in separation of metal from nickel ore based on the result of XRD analysis.

4. Conclusion

Composite and reduction process are important in making nickel. Coal ratio, process temperature, and additive are parameter of the composite reduction process in this research, based on the result of previous research [7][8][9] that uses the reduction process of saprolite ore composite in tube furnace. The reduction process of saprolite ore composite was carried out in tube furnace and analyzed using XRD to know the chemical compound. In this research, saprolite ore composite combination used are $SB_{15}Ca_{10}P_2$, $SB_{30}Ca_{10}P_2$, $SB_{15}Cc_{10}P_2$, and $SB_{30}Cc_{10}P_2$ at temperature of 1200 ^{0}C and 1300 ^{0}C. The results show that additive $CaC0_3$ has the best performance in separation of metal from nickel ore with higher Fe and tetrataenite content, percentage coal ratio of 30% and temperature of 1300 ^{0}C. For Ca_2SO_4 additive, percentage coal ratio of 15% is better than percentage coal ratio of 30% and have higher Fe and NiFeO content at temperature of 1200 ^{0}C and 1300 ^{0}C.

Acknowledgment

This research is part of Research on the activities of Nickel Mineral Processing and Refining Technology Innovation, Center of Mineral Resource Technology (PTPSM), Agency for Assessment and Application Technology (BPPT).

IOP Conf. Series: Materials Science and Engineering **833** (2020) 012087 doi:10.1088/1757-899X/833/1/012087

References

[1] Team of Geological Resource Center 2014 *Mineral balance 2014*
[2] Chen G, Shiau J, Liu S, Hwang W 2016 *Material Transactions* **57 (9)** 1560-1566
[3] Rochani S, Saleh N 2013 *Mineral and Coal technology research and development center*
[4] Kawigraha A, Wahyudi J, Harjanto S, Pramusanto 2013 *Applied Mechanic and Materials* **281 (1)** 490-495
[5] Wahyuadi et al 2014 *Advance Material Research* **893 (1)** 95-99
[6] Yongli L, Sun T, Zou A, Xu C 2012 *Internation journal of Mining Science Technology* **22(3)** 323-328
[7] Puspita A 2019 *Universitas Indonesia library* Depok, Indonesia
[8] Puspita A, Surjandari I, Zulkarnain, Kawigraha A, Permatasari N 2019 *Procedia Computer Science* **161** 424-432
[9] Puspita A, Surjandari I, Zulkarnain, Kawigraha A, Permatasari N 2019 *16th International Conference on Service System and Service Management* 8887655
[10] A Shofi et al 2019 *IOP Conf. Ser.: Mater. Sci. Eng.* **541** 012002

IOP Conf. Series: Materials Science and Engineering **833** (2020) 012088 doi:10.1088/1757-899X/833/1/012088

Effect of Sonication Time on Characteristics of Synthesized Silica Aerogel Activated Carbon Nanocomposite Based on Bagasse Ash

Widi Rahayu[1], Siti Muthomimah[1], Sumari Sumari[1], Nazriati Nazriati[1]

[1]Department of Chemistry, Faculty of Mathematics and Natural Sciences, Universitas Negeri Malang, Jl Semarang 5 Malang 65145, Indonesia

Corresponding author : nazriati.fmipa@um.ac.id

Abstract. The purpose of this research was to identify the effect of sonication time on the characteristics of silica aerogel activated carbon (SA-AC) nanocomposite produced based on bagasse ash. SA-AC nanocomposites was synthesized by sol-gel method at ambient pressure. The silicic acid solution was mixed with activated carbon and the surface modifying agent (TMCS and HDMS). The mixture was then put in an ultrasonic bath at the varying sonication time of 5, 10, 15, and 20 seconds on frequency of 42 kHz. SA-AC nanocomposites were characterized by using FTIR, SEM, and BET. The results showed that SA-AC is formed, indicated by the wave number FTIR spectra of Si-O-Si (1076.28 cm^{-1}), Si-C (1253.73 cm^{-1}), C-H (2960.73 cm^{-1}), C-O (1087.85 cm^{-1}), C=C (2113.98 cm^{-1}), and C=O (1942.32 cm^{-1}). SA-AC nanocomposites synthesized by sonication assistence has a smaller agglomerate than that of without sonication. The largest surface area of SA-AC nanocomposite of 677.719 m^2/g and the biggest pore volume of 0.6338 m^3/g is obtained when the nanocomposite is synthesized at sonication time of 10 seconds

1. Introduction

Silica aerogel is one of the adsorbents with large specific surface area, low density, hydrophobes, and high thermal stability. Therefore, silica aerogel is more effective at absorbing heavy metals compared to xerogel and hydrogel silica. Nazriati et al. [1], reveals that synthesized silica aerogel has a surface area of 540 to1114 m^2/g, pore volume of 0.75 to 2.16 cm^3/g, pore size of 3.39 to 7.38 nm and contact angle of 140 °C. The silica aerogel was synthesized by the sol gel method from baggase ash silica derived from sugar factory waste. Besides silica aerogel, activated carbon also has a large surface area that can be used as an adsorbent. However, activated carbon does not have enough functional groups to adsorb heavy metals economically. Therefore, several methods for modifying the surface of activated carbon are introduced chemically or physically [2],[3]. One of them is by making a nanocomposite[4]. By combining the two adsorbents, it is hoped that the new nanocomposite adsorbents can be produced with greater adsorption capacity. The adsorption capacity of nanocomposite is higher than silica aerogel and activated carbon [5][6]. The capacity of nanocomposite of activated carbon and silica aerogel can eliminate more than 60% Cd metal on the optimum experimental conditions [7].

Other method is needed to increase the surface area of nanocomposites. One method that might be used is nanocomposite synthesis using ultrasonic, or better known as the sonication method. Sankar et al. [8] reported that there were the effect of sonication time on the characteristics of silica nanoparticles derived from brown rice husk. The results obtained an increasing in the specific surface area of nanoparticles. The synthesis of nanoparticles used sonication time variations 0, 10, 30, and 50 seconds. Where the specific surface area without ultrasonic (0 seconds) 201.45 m^2/g increased significantly to 271.22 m^2/g at 50 seconds sonication time. The synthesized nanoparticles had an increasing surface area with increasing sonication time. Sonication method in synthesis material is produces cavitation waves which make the particles size smaller, so it has a large surface area [9][10][11][12]. Accordingly, synthesis of nanocomposite of SA-AC using ultrasonic assistant is needed to increase the surface area of the resulting product.

The sonication method can be used to replace conventional stirring using a magnetic stirrer in the synthesis of activated carbon silica nanocomposites with a sol gel method. The cavitation energy of the ultrasound in making the nanocomposite is also need to know the effect on its specific surface area. This method is one of the easiest, cheapest, and the most effective tools to provide energy for environmentally friendly reactions called Green Technique and having relatively faster synthesis times [13],[14].

2. Materials and Method

The materials used in this study were bagasse ash, NaOH (p.a Merck), HCl 37% (p.a Merck), activated carbon, cation exchange resin, NH_4OH (p.a Merck), trimethylchlorosilane (TMCS) p.a, hexamethyldisilazane (HMDS) p.a, and aquades. Equipments used include glassware, hot plates, magnetic stirrer (Cimarec), oven (Memmert), vacuum pump (Ulvac Sinkukiko DA-20D), analytical balance (*Durascale Dab-200*), Ultrasonic (*BranSonic*), FTIR (*Shimadzu IR-Prestige 21*), SEM (*FEI-INSPECT-S50*), dan BET (*Nova Touch LX4 Quantachrome*).

Synthesis of SA-AC nanocomposites was started by extracting sodium silicate from bagasse ash using NaOH 2 N solution for 1 hour. The sodium silicate solution that has been obtained was then mixed with cation exchange resin to produce a silicic acid solution. A 10 mL silicic acid solution was mixed with 0.03 g of activated carbon. The mixture was put in an ultrasonic bath with sonication time variations of 5, 10, 15, and 20 seconds with a frequency of 42 kHz. Subsequently, the surface modifying agent was added in the form of trimethylchlorosilane (TMCS) as much as 0.3 mL and hexamethyldisilazane (HMDS) as much as 0.6 mL. The addition of modifying agent was carried out by droping while stirred in an ultrasonic bath with sonication time variations of 5, 10, 15, and 20 seconds. The formed hydrogel was then aged at 40 °C in 18 hours and 60 °C in 1 hour for tissue strengthening. The hydrogel was then washed using demineralized water until the neutral pH and it was filtered. The hydrogel obtained was drained at 80 °C in 24 hours to get the SA-AC nanocomposite.

The SA-AC nanocomposite produced was analyzed to determine its characteristics. The analysis was used several instruments such as functional group analysis using the Fourier Transform Infra Red (FTIR) instrument, nanocomposite surface morphology analysis using Scanning Electron Microscopy (SEM) instruments, and specific surface area analysis using the Brunauer-Emmet-Teller (BET) instrument.

3. Results and Discussion

FTIR spectra of SA-AC nanocomposites with each sonication time of 5, 10, 15, and 20 seconds is then compared with standard wave number to determine the functional groups contained in the synthesized SA-AC nanocomposite. The SA-AC nanocomposite FTIR spectra at every sonication time of 5.10, 15, and 20 seconds can be seen in Figure 1.

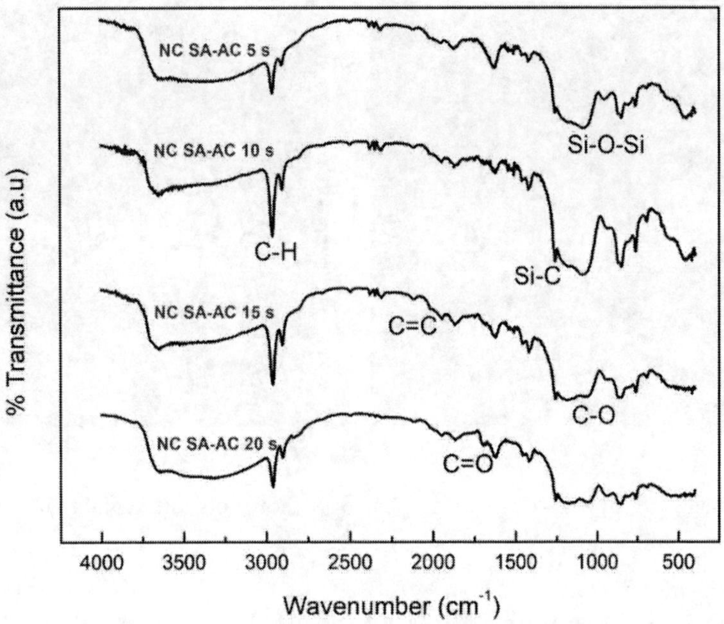

Figure 1. FTIR Nanocomposite Spectra of SA-AC at sonication time of 5, 10, 15, and 20 seconds.

Based on the spectra in Figure 1, there are several typical absorption bands. The FTIR spectra of the nanocomposite SA-AC at sonication time of 5, 10, 15, and 20 seconds show an absorption band at wave number of 1076.28 cm^{-1} indicating the presence of Si-O-Si vibration. It shows that condensation of silanol (Si-OH) into siloxan (Si-O-Si) has occurred. While the absorption band at wave number of 1253.73 cm^{-1} indicates the presence of Si-C vibration in the Si-CH_3 functional group of SA-AC nanocomposite. It indicates that the nanocomposite has been successfully modified due to a modifying agent. In addition, the successful modification of nanocomposites is shown by the absorption band at the wave number of 2960.73 cm^{-1} which indicates the C-H vibrations of CH_3.

The difference between FTIR spectra of SA-AC nanocomposite and silica aerogel is shown by the absorption band at wave numbers of 1942.32 cm^{-1} and 2113.98 cm^{-1} indicating the vibrations of C = O and C = C. In addition, there is an absorption band on waves around 1087.85 cm^{-1} which indicates the presence of C-O vibrations. Vibration at wave numbers of 1087.85 cm^{-1}, 1942.32 cm^{-1}, and 2113.98 cm^{-1} is also present in the active carbon spectrum. It is related to the functional groups of C-O, C = O, and C = C that are typical functional groups possessed by activated carbon. In addition, the SA-AC nanocomposite absorption band synthesized at sonication time of 5, 10, 15, and 20 seconds has the similarity wave number of peaks. It shows that SA-AC nanocomposites were successfully formed at sonication time of 5, 10, 15, and 20 seconds.

Morphology of SA-AC nanocomposites was analized using SEM as presented in Figure 2. It shows that the morphology of SA-AC nanocomposites synthesized by sonication assistence has smaller agglomerates than that of without sonication. It can be related to the cavitation process generated from ultrasonic waves that propagate in the liquid medium. The energy caused by the collapse of cavitation will reduce the size of the agglomeration of material formed. The shock waves of sonication can separate agglomeration resulting in a smaller particle size. Characterization of the surface area used adsorption and desorption N_2 with BET method and the measuring pore volume used BJH method. The data of BET and BJH method were presented in Table 1.

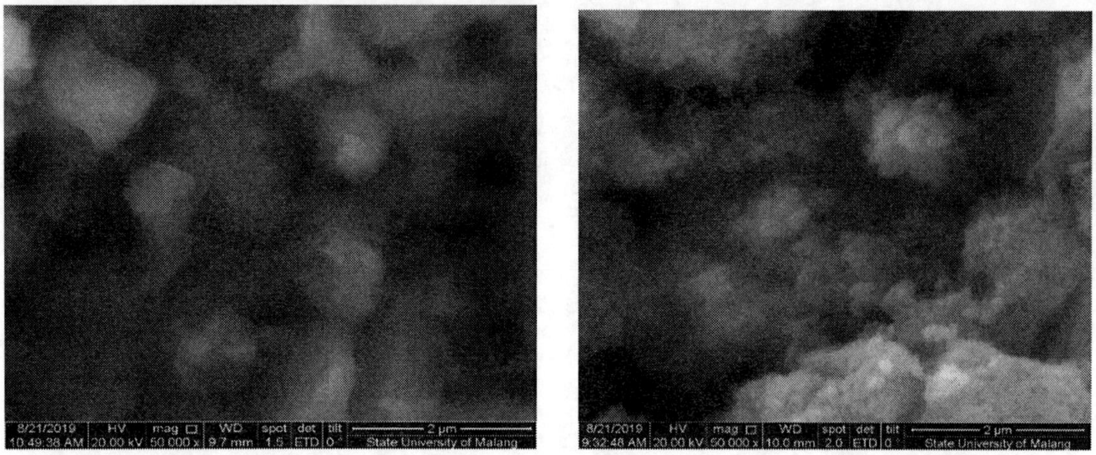

(a) **(b)**

Figure 2. SEM images of (a) SA-AC nanocomposites without sonication and (b) Nanocomposites with sonication.

Table 1. BET Analysis of SA-AC Nanocomposite.

Sample	S_{BET} **(m²/g)**	**Pore volume (cm³/g)**
NC SA-AC TU	509.139	0.5280
NC SA-AC 5 s	509.940	0.5675
NC SA-AC 10 s	677.719	0.6338
NC SA-AC 15 s	435.682	0.4705
NC SA-AC 20 s	485.948	0.3969

Data at Table 1 shows that the surface area of the SA-AC nanocomposite synthesized at sonication time of 10 seconds has the largest surface area of 677.719 m^2/g and pore volume of 0.6338 m^3/g. The surface area of the nanocomposite is increasing when it is synthesized at optimal time of sonication. Ultrasonic waves can prevent agglomeration or clumping of particles [15]. It is also in line with Suslick's statement [16] that the sonication induces particle size and tends to be more homogeneous and becoming smaller which is eventually leads to the size of the stable nanoparticles and avoid clumping. At the optimum sonication time, the surface area becomes larger because the particle size is getting smaller. After passing the optimum sonication time, however, the surface area of nanocomposite becomes smaller because of aglomeration.

The phenomena of adsorption and desorption type of N_2 gas in the BET method can be seen in Figure 3. All nanocomposite samples synthesized at sonication time of 5, 10, 15, and 20 seconds showed the type IV of isotherm. It indicates that the SA-AC nanocomposites has pores in the range of 2 to 50 nm and the adsorbent was a mesoporous material.

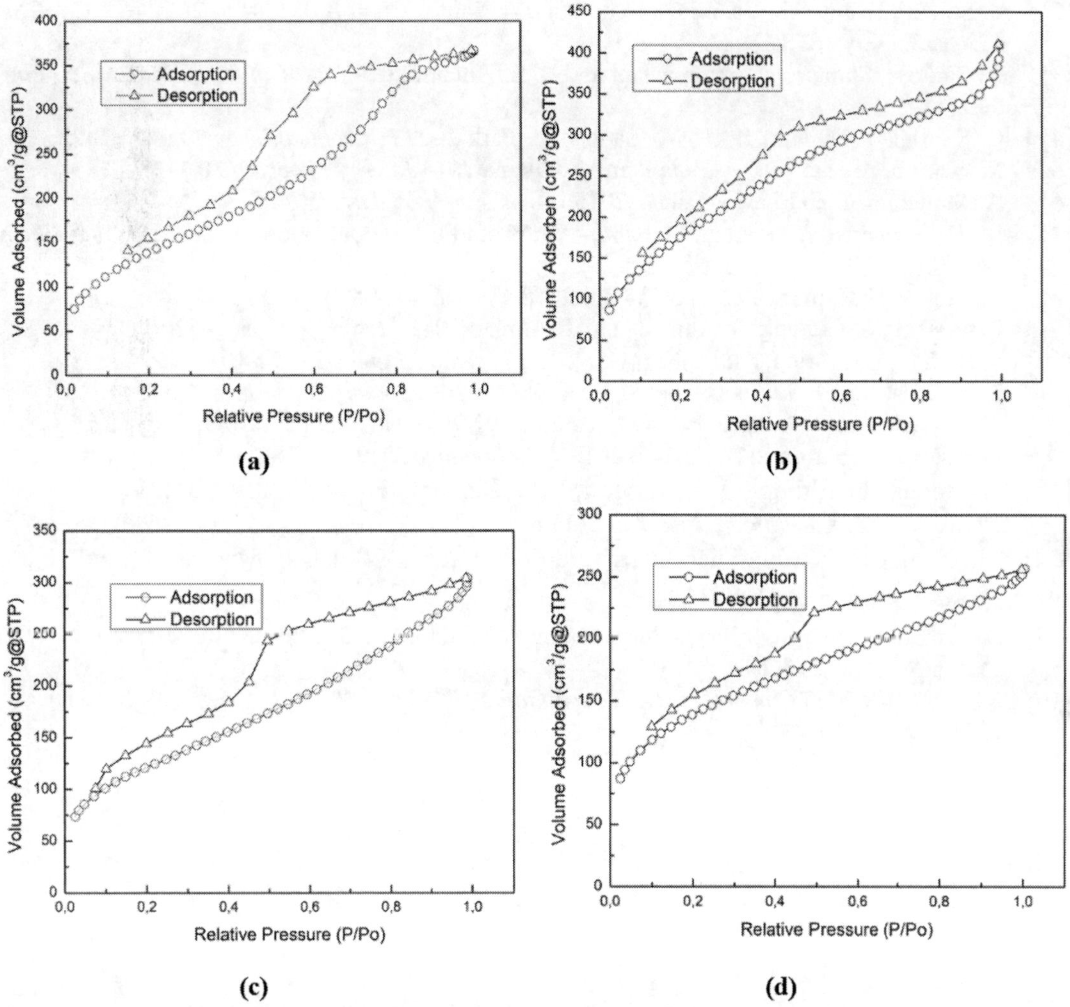

Figure 3. The Adsorption Isotherm BET of composites at various sonication time of (a) 5 seconds (b) 10 seconds (c) 15 seconds (d) 20 seconds.

4. Conclusions

Silica aerogel-activated carbon nanocomposites have been successfully synthesized using ultrasonic assistant. The nanocomposite synthesized by varying the sonication time of 5, 10, 15, and 20 seconds at a frequency of 42 kHz produced mesoporous materials. The surface area of SA-AC nanocomposite synthesized at sonication time of 10 seconds has the largest surface area of 677.719 m^2/g and the biggest pore volume of 0.6338 m^3/g. The longer sonication time resulted on the smaller surface area as well as pore volume of SA-AC nanocomposites.

5. Aknowledgment

The authors are highly thankful to the Directorate of Higher Education for financial support for this research (*Hibah PDUPT*, contract no. 19.3.46/UN32.14.1/LT/2019).

References

[1] N. Nazriati, H. Setyawan, S. Affandi, M. Yuwana, and S. Winardi 2014 *J. Non. Cryst. Solids* **400** 6

[2] S. Sumari, F. Fajaroh, Yahmin, N. Sholihah, A. Santoso, and A. Budianto 2019 *IOP Conf. Ser. Mater. Sci. Eng.* **515(1)**.
[3] A. Santoso, Sumari, R. Joharmawan, and L. B. Hutami 2019 *IOP Conf. Ser. Mater. Sci. Eng.* **509 (1)**.
[4] C. K. Ahn, Y. M. Kim, S. H. Woo, and J. M. Park 2009 *Hydrometallurgy* **99 (3-4)** 209.
[5] M. Karnib, A. Kabbani, H. Holail, and Z. Olama 2014 *Energy Procedia* **50** 113.
[6] A. Mohammadi and J. Moghaddas 2015 *Chem. Eng. Res. Des.* **94** 475.
[7] M. H. Givianrad, M. Rabani, M. Saber-tehrani, and P. Aberoomand-azar 2013 *J. Saudi Chem. Soc.* **17(3)** 329.
[8] S. Sankar, N. Kaur, S. Lee, and Y. Kim 2018 *Ceram. Int.* **44(7)** 8720.
[9] I. A. Wani, A. Ganguly, J. Ahmed, and T. Ahmad 2011 *Mater. Lett.* 65(3) 520.
[10] S. Sompech, A. Srion, and A. Nuntiya 2012 *Procedia Engineering* **32** 1012.
[11] D. N. F. Muche, F. L. Souza, and R. H. R. Castro 2018 *React. Chem. Eng.* **3(3)** 244.
[12] N. M. Bahadur,F. Chowdhury, Md. Obaidullah, Md. S. Hossain,R. Rashid,Y. Akter,T. Furusawa, M. Sato, and N. Suzuki. 2019 *J. Nanomater* **2019**.
[13] Navinchandra Shimpi, Ananda Mali, D.P. Hansora, Satyendra Mishra 2015 *Nanoscience and Nanoengineering* **3(1)** 8.
[14] Zhixiong Chen, Amin Shahsavar. Abdullah A.A.A.Al-Rashed, Masoud Afrand 2020 *Powder Technology* **360** 1134
[15] S. Sumari, F. Fajaroh, I. Bagus Suryadharma, A. Santoso, and A. Budianto 2019 *IOP Conf. Ser. Mater. Sci. Eng.* **515(1)**.
[16] K. S. Suslick 2003 *Comprehensive Coordination Chemistry II* (USA: Elsevier) Chapter 1.41 pp. 731–739.

Inexpensive solar cell using TiO_2/coffee composite as photon absorbing material

D Y Rahman[1], F D Utami[1], D O Margaretta[1], N Amalia[1], R Sulistyowati[2] and M Abdullah[1].

[1]Department of Physics, Faculty of Mathematics and Natural Sciences, Bandung Institute of Technology, Jalan Ganeca 10 Bandung 40132, Indonesia.
[2]Department of Physics Education, Faculty of Teacher Training and Education Science, Universitas PGRI Palembang, Jalan Ahmad Yani, Lrg. Gotong Royong 9/10 Ulu, Palembang, Indonesia.

E-mail : duiyantorahmanssi@gmail.com and mikrajuddin@gmail.com

Abstract. Solar cell using TiO_2/coffee composite as photon absorbing material has been successfully fabricated. A suspension of TiO_2/coffee was dropped on Fluorine Tin Oxide substrate which was previously heated on a hotplate with temperature of 100 °C. The fraction of coffee was varied from 10, 20, 30, 40, and 50 %, compared to the mass of TiO_2. PVA-LiOH electrolyte polymer was functionalized as the hole transport medium. The highest efficiency of 0.76 % was achieved with 40% of coffee fraction. The solar cell is promising to be developed in the future due to the easy method and cheap materials used.

1. Introduction

Currently, the hope of the world in possession of cheap solar cell devices is increasing continuously. This condition feeds the research into the manufacture of solar cells comprising inexpensive materials, a convenient and necessarily easy process and ready to be adopted to mass production. Great efforts have been done to produce these types of solar cells starting from the first generation of bulk silicon solar cells [1,2], followed by the second generation of thin-film solar cells [3,4], and the third generation of organic solar cells [5,6]. The thin-film solar cell appeared as the solution to the high cost of silicon production since it consumed a thick layer and high purity of silicon [7,8]. The organic solar cell was then developed to largely overcome the shortcomings of its previous generation which was caused by the high cost of encapsulation and furthermore, the toxic materials involved [9]. Each development has been carried out mainly for obtaining much easier method of fabrication, much cheaper materials used, and much more comparable efficiency than the present applied solar cell especially silicon based solar cell.

Dye-sensitized solar cell (DSSC) is a type of organic solar cell that has been mostly developed by researchers over the last decade. The structure of this solar cell is comprised of a dye as photon absorber and titanium dioxide (TiO_2) as transport medium of electron injected by the dye to move towards transparent conducting electrode [10,11]. Since the efficiency of the solar cell is still relatively small compared to the most commercial-silicon-based solar cells, it is in need to find a new structure of solar cell made with much cheaper materials and easier method.

Further development of solar cell using much cheaper materials and much easier fabrication method than DSSC is a challenge. A new structure of solar cell replacing most of all materials used in DSSC has been proposed. The solar cell is comprised of TiO_2 as the photon absorbing material, metal particles as the mean of transporting the electron out to the front electrode (FTO), PVA-LiOH electrolyte polymer as the medium for transporting positive charge, and aluminium as counter electrode [12,13]. Technical anatase TiO_2, which also can be used as photocatalytic and antibacterial activity material [14-18], is very cheap. Another materials such as PVA-LiOH and aluminium sheet are also inexpensive, and the processes involved in fabricating this solar cell module are simple. Since the lifetime of electron-hole pair in TiO_2 is quite short, which is typically in order of *ns*, the deposition of copper particle among TiO_2 particle is aimed to dissociate electron-hole pairs freely and inhibit electron-hole recombination through producing high Schottky barrier at the metal-semiconductor junction that accommodated electron capture [12,13].

Since TiO_2 has limited absorption range of solar spectrum, further development was carried out using graphite powder to have a wider absorption of solar spectrum and ultimately increased the current of the cell [19]. Although graphite is commonly used as counter electrode [20-22], the presence of graphite powder has successfully improved the performance of the solar cell through higher current resulted [19].

Due to the limited availability of graphite as mined materials, using another carbon-based material replacing graphite is challenging, especially from renewable materials. Coffee powder is a promising material to be combined with TiO_2 in absorbing the solar spectrum since it has a narrow bandgap of 2.59 eV [23], capable of absorbing the solar spectrum until the visible region. In addition, it is a renewable material, reasonably cheap, and massively planted in many places around the world.

In the present study, combined materials of large bandgap titanium dioxide (TiO_2) with material of small bandgap (coffee) in order to get higher current is reported. The function of TiO_2 as a large bandgap material-in absorbing high energy of photons and producing high V_{oc} will be supported by the role of coffee in absorbing low energy of photons and increasing I_{sc} of the solar cell. Optimum composition is studied by varying the coffee fraction. High I_{sc} and V_{oc} would simultaneously improve the efficiency of the solar cell. The effect of using coffee in various different fractions are observed in I-V characterization.

2. Material and Experiment

2.1. Materials

Anatase TiO_2 technical grade was purchased from Bratachem Indonesia. Coffee powder was obtained from local production in Bandung. Fluorine tin doped oxide (FTO) was procured from Solaronix Switzerland. LiOH was purchased from Chemical Kanto Japan. Polyvinyl alcohol (PVA) was purchased from Bratachem Indonesia. Aqua mineral water was acquired from local stall in Bandung. The solar cell performance was measured using current-voltage (IV) counter (Keithley 617) under xenon lamp illumination at intensity of 300 W/m^2. The intensity of illumination was measured using the Newport 841-PE energy/power meter model.

2.2. Preparation of solar cell.

A 5 g of TiO_2 powder was stirred homogeneously in 12 mL of mineral water (Aqua, Indonesia) for 15 minutes. Subsequently, a coffee powder was added to the suspension with the amount of 10% up to 50% to the amount of TiO_2 powder. The suspension of TiO_2 and coffee was then stirred for 30 minutes. Meanwhile, FTO (fluorine doped tin oxide) transparent electrode was heated on a hot plate at 100° C for 30 minutes. The suspension was then dropped onto the FTO surface using a small spatula surface and was distributed evenly on the FTO surface. To make the TiO_2/coffee particles well attached on the surface of FTO, the film was then heated further at 200 °C for 2 hours.

Separately, 0.38 g of LiOH was dissolved in 20 ml mineral water for 15 minutes. A 1.8 g of polyvinyl alcohol (PVA) was then added to the solution and heated at 100 °C for 60 minutes to form an electrolyte polymer gel. The solar cell module was constructed by sandwiching the electrolyte polymer between film and aluminium electrode. The detailed description of the solar cell structure is illustrated at Figure 1.

The efficiency (η) and the fill factor of the solar cell were calculated based on the equation below :

$$\eta = \frac{I_{SC} V_{OC} FF}{P_{in}} \times 100\% \quad (1)$$

$$FF = \frac{I_m V_m}{I_{SC} V_{OC}} \quad (2)$$

Where I_{sc} is the short circuit current, V_{oc} is the open voltage, FF is the fill factor , P_{in} is the power of light, I_m and V_m are the current and voltage that produce the maximum power.

For each coffee fraction, four samples were measured to obtain the I-V characteristics. Since the efficiencies obtained from I-V characterization for each coffee fractions are almost the same, only one data was listed to represent the four data obtained.

3. Result and Discussion

Figure 1 describes the proposed mechanism of photo-current of the developed solar cell. After absorbing the photons, electrons in the valence band of TiO_2 and coffee will be excited to the conduction band where accumulated electrons located on the conduction band of TiO_2 will directly flow to the conduction band of coffee due to the lower band gap of coffee particles and finally arrived at FTO. The electrons located on the conduction band of coffee will directly move to FTO. As the ion minerals filled the voids of particles of absorber materials act as holes' scavengers, the holes will be easily captured and converted into positive ions. The movement of ions to the electrolyte polymer will be easy since they are located deep inside the film.

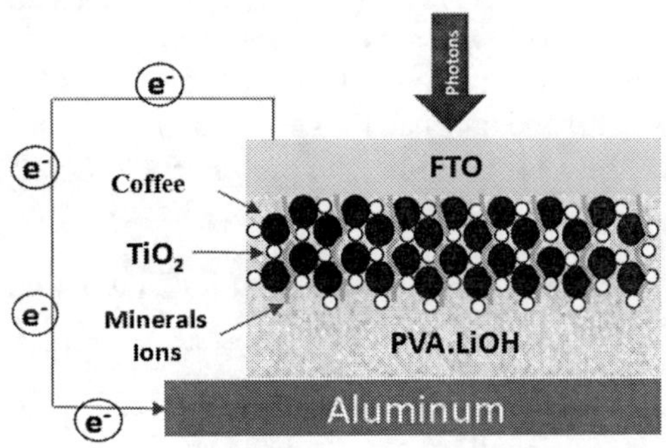

Figure 1. Schematic diagram of TiO_2/coffee solar cell.

Figure 2 shows the I-V curve of TiO_2 based solar cell containing different fractions of coffee powder. The corresponding parameters obtained from each fraction are summarized in Table 1. It is clearly shown that the presence of coffee powder in the TiO_2 solar cell has increased the I_{sc}, compared to the I_{sc} of TiO_2-based solar cell (0.99 mA). An I_{sc} of 1.39 mA was observed when the fraction of coffee was 10% and continued to increase up to coffee fraction of 40% (1.7 mA). However, when the coffee fraction surpassed 40%, the I_{sc} decreased to 1.35 mA for the coffee fraction of 50%.

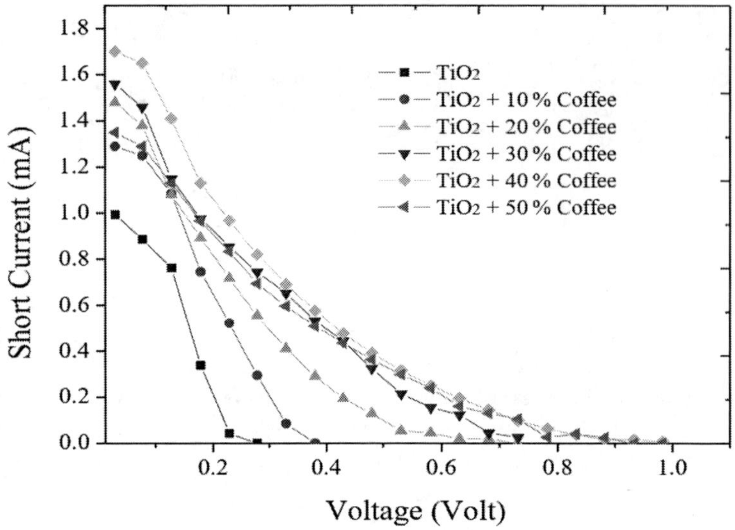

Figure 2. I-V curve of TiO_2 based solar cell with variation of coffee fractions.

The increase of I_{sc} after inserting small bandgap materials was theoretically calculated by Soga [24]. I_{sc} strongly depends on the absorption characteristics of the active material [24], in which I_{sc} is higher when larger number of electron-hole pairs can be generated or more photons can be absorbed. Therefore, it is highly suggested that the active material in the solar cell should have a small bandgap to increase I_{sc}. The increase of I_{sc} was also followed by the increase of V_{oc} from 0.23 V, without coffee, to 0.78 V, when the fraction of coffee was 40 %. The increase of I_{sc} and V_{oc} boosted the efficiency of the solar cell from 0.32 %, without coffee, to 0.76 %, when the fraction of coffee was 40 %. It is predicted that this is the optimum amount of a combination of small and large bandgap materials in achieving highest efficiency. The function of small bandgap materials in producing high I_{sc} and large bandgap materials in achieving high V_{oc} is immensely confirmed in the present study [25, 26].

Table 1. The parameters obtained from I-V measurement for TiO_2 based solar cell containing different fractions of coffee powder.

Coffee fractions (%)	I_{sc} (mA)	V_{oc}(V)	FF	Efficiency (%)
0	0.99	0.23	0.35	0.32
10	1.39	0.33	0.28	0.46
20	1.48	0.68	0.15	0.54
30	1.56	0.78	0.18	0.71
40	1.70	1.02	0.13	0.76
50	1.35	0.90	0.16	0.65

Several researches have reported the same attempt of employing small and large bandgap materials in a solar cell. Peet *et al* succeded in improving the cell efficiency by using combined small and large bandgap materials [27]. An efficiency of 10.6 % was succesfully achieved by You *et al.*, who combined small and large bandgap materials in a polymer solar cell [28]. Employing small bandgap material to support main absorbing material was also carried out by Kim *et al.*, and the work succeeded in improving the efficiency of a plastic-based solar cell [28]. Although the efficiency of the developed solar cell in this study is much lower than those of solar cells, it is possible to be further developed in the future since it used much cheaper materials and easier method.

4. Conclusion

We have successfully fabricated inexpensive solar cell using TiO_2/coffee composite as photon absorbing material. Combination of small bandgap material (coffee) and large bandgap material (TiO_2) in producing high I_{sc} and V_{oc} has remarkably boosted the efficiency of the solar cell. The highest efficiency of 0.76 % with I_{sc} of 1.70 mA and V_{oc} of 1.02 V was achieved when the fraction of coffee of 40 % was used. Since easy method and cheap materials were applied in this work, this solar cell is promising to be developed in the future.

5. References

[1] Chapin D M, Fuller C S and Pearson G L 1954 *J. Appl. Phys.* **25**(5) 676-677
[2] Green M A, Emery K, Hishikawa Y, and Warta W 2010 *Prog. Photovoltaics Res. Appl.* **18** 346-352
[3] Contreras M A, Egaas B, Ramanathan K, Hiltner J, Swartzlander A, Hasoon F, and Noufi R 1999 *Prog. Photovoltaics Res. Appl.* **7**(4) 311-316
[4] Green M A 2007 *J. Mater. Sci-Mater. El.* **18**(1) 15-19
[5] Ameri T, Dennler G, Lungenschmied C, and Brabec C J 2009 *Energ. Environ. Sci.* **2**(4) 347-363
[6] Hoppe H and Sariciftci N S 2004 *J. Mater. Res.* **19**(7) 1924-1945
[7] Chopra K L, Paulson P D and Dutta V 2004 *Prog. Photovoltaics Res. Appl.* **12**(2-3) 69-92
[8] Glunz S W 2007 *Adv. OptoElectron.* 97370
[9] Green M A 2001 *Prog. Photovoltaics Res. Appl.* **9**(2), 123-135
[10] O'regan B and Grätzel M 1991 *Nature* **353**(6346) 737
[11] Mathew S, Yella A, Gao P, Humphry-Baker R, Curchod B, Ashari-Astani N, Tavernelli I, Rothlisberger U, Nazeeruddin, and Gratzel M 2014 *Nat. Chem.* **6** 242
[12] Saehana S, Arifin P, Khairurrijal, and Abdullah M 2012 *J. Appl. Phys.* **111** 123109
[13] Rokhmat M, Wibowo E, Sutisna, Yuliza E, Khairurrijal, and Abdullah M 2015 *Adv. Mater. Res.* **1112** 245–250
[14] Utami F D, Rahman D Y, Sustini E, and Abdullah M 2019 *J. Phys. Conf. Ser.* **1171**(1) 012030.
[15] Wibowo E, Rokhmat M, Rahman D Y, Murniati R and Abdullah M 2017 *Procedia Eng.* **170** 78-83.
[16] Utami F D, Rahman D Y, Sutisna, Kamirul, Margaretta D O, and Abdullah M 2019 *J. Phys. Conf. Ser.* **1204**(1) 012086
[17] Margaretta D O, Permadi K W, Rahman D Y, Utami F D, Viridi S, and Abdullah M 2019 *J. Phys. Conf. Ser.* **1204**(1) 012051
[18] Margaretta D O, Permadi K W, Rahman D Y, Utami F D, Viridi S, and Abdullah M 2019 *IOP Conf. Ser.: Mater. Sci. Eng.* **599**(1) 012005
[19] Rahman D Y, Rokhmat M, Yuliza E, Sustini E, and Abdullah M 2016 *Int. J. Energy Environ. Eng.* **7**(2016) 289–296
[20] Rahman D Y, Utami F D, Margaretta D O, Sustini E, and Abdullah M 2019 *IOP Conf. Ser.: Mater. Sci. Eng.* **599**(1) 012004
[21] Rahman D Y, Utami F D, Margaretta D O, Sustini E, and Abdullah M 2019 *J. Phys. Conf. Ser.* **1204**(1) 012084
[22] Rahman D Y, Sustini E and Abdullah M 2017 *Jurnal Matematika dan Sains* **22**(1) 33-36
[23] Syafinar R, Gomesh N, Irwanto M, Fareq M, and Irwan Y M 2015 *Energy Procedia* **79** 799-807
[24] Soga T 2007 *Nanostructured materials for solar energy conversion* (Nagoya: Elsevier)
[25] Koster L J A, Mihailetchi V D, and Blom P W 2007 *SPIE Newsroom* 9-10
[26] Kroon R, Lenes M, Hummelen J C, Blom P W, and De Boer B 2008 *Polym. Rev.* **48** 531-582
[27] Peet J, Kim J Y, Coates N E, Ma W L, Moses D, Heeger A J, and Bazan G 2007 *Nat. Mater.* **6**(7)

497
[28] You J, Dou L, Yoshimura K, Kato T, Ohya K, Moriarty T, and Yang Y 2013 *Nat. Comms.* **4** 1446-1446
[29] Kim J Y, Lee K, Coates N E, Moses D, Nguyen T Q, Dante M, and Heeger A J 2007 *Science* **317**(5835) 222-225

Acknowledgements
This work was supported by fellowship grant from Indonesia Endowment Fund for Education (LPDP). The PMDSU fellowships from the Ministry of Research, Technology and Higher Education, Republic of Indonesia F.D.U. is highly acknowledged.This work was also supported by KK Research Grant from Bandung Institute of Technology 2019 and WCR Research Grant from the Ministry of Research, Technology and Higher Education, Republic of Indonesia 2019.

IOP Conf. Series: Materials Science and Engineering **833** (2020) 012090 doi:10.1088/1757-899X/833/1/012090

Preparation of Silver-Incorporated *Rhynchospora corymbosa* (L.) Cellulose via *in-situ* Green Reduction and Its Antibacterial Study

Fitri Rosdiana,[1]Aswin Falahudin[1], Risky Hadi Wibowo,[2]SalprimaYudha S.,[1,*] Irfan Gustian,[1]

[1]Department of Chemistry, Faculty of Mathematics and Natural Sciences, University of Bengkulu, Jalan W.R, Supratman, Kandang Limun, Bengkulu 38371A, Indonesia
[2]Department of Biology, Faculty of Mathematics and Natural Sciences, University of Bengkulu, Jalan W.R, Supratman, Kandang Limun, Bengkulu 38371A, Indonesia

*Corresponding author: salprima@unib.ac.id; sp.yudha.s@gmail.com

Abstract. Silver incorporated *R. corymbosa (L.)* cellulose was obtained through the direct attachment (insertion) of silver ions to natural cellulose, followed by the green reduction of ions to metallic silver, using aqueous extract of fresh *R. corymbosa* (L.). In addition, Fourier transform infra-red (FTIR) analysis of the isolated cellulose showed some peaks at 3330 cm^{-1}, 2890 cm^{-1}, 1320 cm^{-1}, 1030 cm^{-1}, 895 cm^{-1} corresponding to OH stretching, methylene ($-CH_2-$) stretching, OH bonding, C-O bonding, and 1,4 β-glycoside, respectively. The silver incorporated *R. corymbosa* (L.) cellulose was prepared by impregnating silver ions obtained from silver nitrate solution, followed by ion reduction to the metallic silver *in-situ*. Furthermore, another interesting outcome of this study is seen in the products' ability to show inhibition zones in the growth medium of *Escherichia coli*.

1. Introduction

There has been rapid development in the adoption of green technologies during the preparation of sustainable materials. This is possibly implemented through the physical methods, by direct deposition of silver nanoparticles on cellulose matrices, or chemical methods, which include the *in-situ* formation of silver nanoparticles within the mixture of cellulose-Ag salts, and also covalent binding [1]. Furthermore, some of the current development approaches are described using the following examples: the adoption of spuned-cellulose fiber was used in the preparation of silver/cellulose composites, obtained by the dispersion of cellulose in silver nitrate solution, followed by heating at 80 °C to form silver nanoparticles, without the addition of chemical reductors. The reaction product had greater effectiveness as an antibacterial material against *Staphilococusaeureus* than *Eschericia coli* [2].

Paper–cellulose was used as a supporting material in the preparation of Au-Ag-cellulose nanocomposites, which was performed using tri-sodium citrate dihydrate as a reductor, in the presence of silver enhancer, and the product had better activity against *E.coli* (AATCC 100) [3]. The synthesis silver-cellulose nanocomposites involved the *in-situ* bioaccumulation of cellulose matrices in bio-flocculants, followed by a reaction using some variation of silver ions. The reduction process was successfully conducted using the bio-flocculants identified by the reaction color and supported by some characterization analysis, and the product possesses good antibacterial activity against *E. coli* [4].

The dispersion of antibacterial silver nanoparticle in deacetylated cellulose was obtained using $NaBH_4$ as a reductor, while $KMnO_4$ as an oxidator was used to reduce the particle size, and the resulting activity was markedly higher than the un-oxidized composites [5]. In addition, other studies have demonstrated the insertion of silver and silica nanoparticles in the matrices of cellulose pulp, in a 50% water solution of N-methylmorpholine-N-oxide, and the material produced possesses effective antibacterial activity, due to the synergetic work between both components [6]. Meanwhile, synthetic method of producing cellulose-supported silver nanoparticles was also developed through the use of paper cellulose as the main support, and adding a small amount of stabilizers, e.g., ✔-cellulose, carboxymethyl cellulose or amino cellulose, to decrease the incidence of aggregation. $NaBH_4$ was adopted as a synthetic reductor, and the material produced conferred effective antibacterial activity, especially against methicilin-resistant *S. aureus* (MRSA) [7].

It is important to develop new sustainable materials with higher biocompatibility in the production of antibacterial agents, using simple treatment. This report is, therefore, aimed at providing new information regarding the use of *R. corymbosa* as sustainable cellulose in the preparation of silver incorporated cellulose and its possible application as an antibacterial agent. Furthermore, green reduction of silver ions was performed to form metallic silver, using the aqueous extract of fresh *R. corymbosa*.

2. Material and Methods

2.1. Materials

Silver nitrate ($AgNO_3$) was purchased from Merck and used without further purification for the generation of metallic silver. Also, *R.corymbosa*(L.) was collected from Bengkulu coastal area, while *Escherichia coli* (ATCC 8739) was used as an antibacterial assay representative to investigate the prepared composites, and demineralized water was used in most of the experimental procedures.

2.2. Procedure

R.corymbosa (L.) cellulose was extracted using the modification of a known technique [8-10], where *R. corymbosa* (L.) powder (10 gram) was placed in a round bottom flask, followed by the addition of 120 mL NaOH (1 M). Therefore, the mixture was refluxed for 3 hours, followed by the immersion of solid materials in 100 mL of 5% H_2O_2, at 80 °C for 30 minutes. The product was then soaked in 100 mL sulfuric acid solution (1 N), stirred vigorously for 2 hours, subsequently decanted, and washed using demineralized water every 24 h for 4 days. Finally, the solid materials obtained were then dried at 60 °C for 6 h.

Aqueous extract of fresh *R. corymbosa* (L.) was obtained by heating the fresh plant powder (10 gram) in 100 mL demineralized water, and the filtrate obtained using Whatman paper (Number 3) was freshly adopted in the reduction of silver ions. On the other hand, *R.corymbosa* (L.) cellulose obtained from the previous step was immersed in $AgNO_3$ solution (0.01 M) for 24 hours, and the as-prepared aqueous extract of fresh *R. corymbosa* (L.) was added dropwise and stirred at room temperature. This reaction mixture was preserved for 24 hours, followed by the separation of solid materials from the reaction flask, and washing using demineralized water.

The bacterial target used was *Escherichia coli* ATCC 8739, of which one loop was cultured in trypticase soy broth (TSB) and incubated with a shaker for 24 hours at 27 °C. Furthermore, about 500 μL of *E. coli* suspension was inoculated into 100 mL of melted trypticase soy agar (TSA) medium and poured into sterile plates. After solidification, the separated sterile paper discs were sprinkled with *R. corymbosa* (L) powder, and silver-*R. corymbosa* (L) cellulose, while 20 μL of 15 mg silver-sulfadiazine cream, diluted in 1 mL of demineralized water was dropped into the sterile paper disk and adopted as a positive control. Conversely, demineralized water (20 μL) was applied as a negative control, and the plates were incubated for 24 hours at 300°C. These assays were conducted in duplicates, and the

IOP Conf. Series: Materials Science and Engineering **833** (2020) 012090 doi:10.1088/1757-899X/833/1/012090

materials' antimicrobial activity was indicated by formation of a clear zone around the bacterial colonies.

3. Results and Discussions

The use of simple methods in the extraction of cellulose, following the modified procedure creates a white powder form as shown in Figure 1. This isolation was initiated by grinding the dried pieces of *R. corymbosa* (L.) to produce powder samples as shown in Figure 1(b), followed by treatment with NaOH, H_2O_2, and H_2SO_4 to give the product shown in Figure 1(c). Specifically, the addition of NaOH is to remove lignin and other compounds, followed by a bleaching treatment with hydrogen peroxide, and then the sample preparation was terminated by hydrolyzing the powder with sulfuric acid, producing a final product yield of 22.6 %.

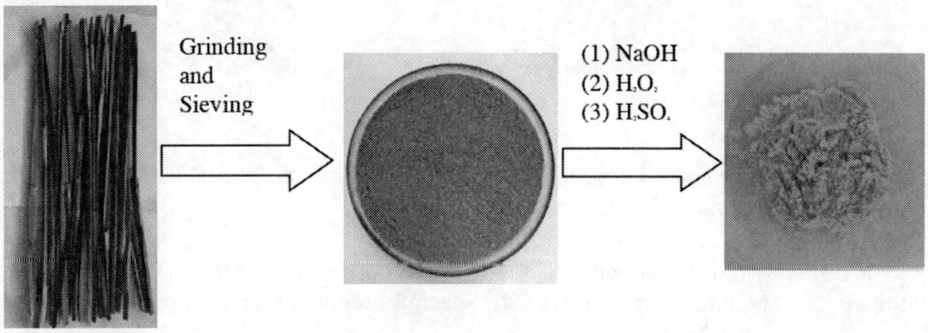

(a) (b) (c)

Figure 1. Cellulose isolation processes of *R. corymbosa* (L.), (a) dried *R. corymbosa* (L.) plant, (b) powder of *R. corymbosa* (L.), (c) final product.

The fresh pieces of *R. corymbosa* (L.) were warmed up in hot water for 30 minutes, followed by filtration, and the clean extract was added to silver nitrate ($AgNO_3$) solution. Interestingly, the mixture demonstrates a specific peak at 449 nm, using UV-Vis spectrophotometry analysis, as shown in Figure 2, therefore providing preliminary evidence on the reduction and stabilizing capability of the metallic silver particles produced in the corresponding solution.

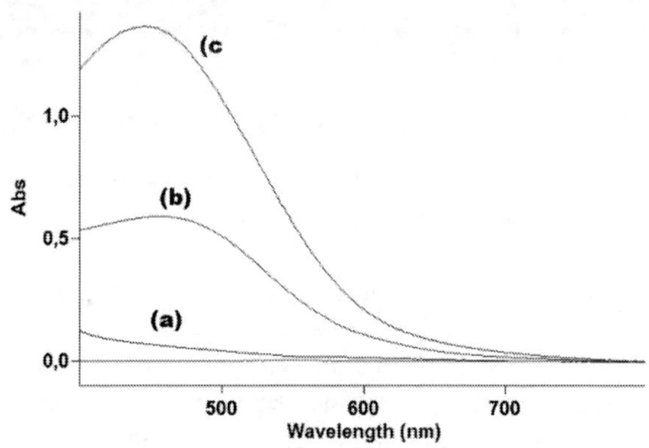

Figure 2. UV–vis spectra of (a) fresh *R. corymbosa* (L.) extract solution, (b) $AgNO_3$ - *R. corymbosa* (L.) extract reaction mixture after 2 h, (c) $AgNO_3$- *R. corymbosa* (L.) extract reaction mixture after 24 h, peak observed at 449 nm.

Based on these results in Figure 1 and 2, further experiments were performed to obtain *in-situ* incorporated silver-cellulose through a green approach. The final powder product shown in Figure 1(c) was treated with $AgNO_3$ solution for 24 h to provide enough time for the attachment of silver ions to the cellulose surface as well as for entry into the matrix [11]. This was then treated with the fresh extract of *R. corymbosa* (L.), causing a color change from clear solution to brown after 24 hours, and a red-brown solid material was obtained as shown in Figure 3 after further treatment indicating the change from silver ions to metallic silver, as seen in Figure 2. Figure 3(b) shows the color of silver-incorporated cellulose powder, which was different from the initial cellulose shown in Figure 3(a).

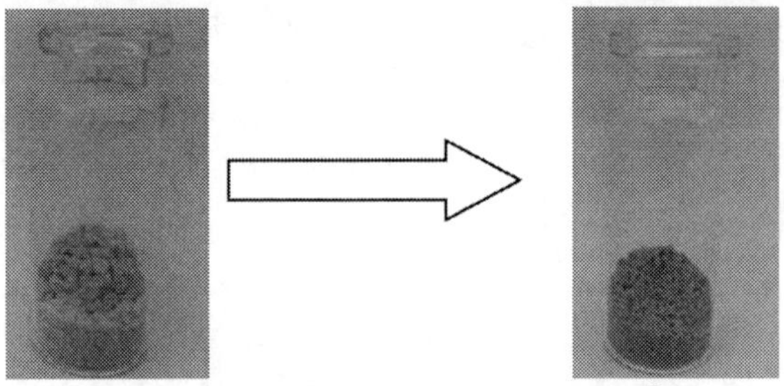

(a) (b)

Figure 3. (a) *R. corymbosa* (L.) cellulose powder, (b) *R. corymbosa* (L.) cellulose treated with $AgNO_3$ and fresh aqueous extract of *R. corymbosa* (L.)

The identification of functional group changes required an investigation on both powders obtained using the Fourier transform infra-red (FTIR) spectrophotometer, and the results are shown in Figure 4.

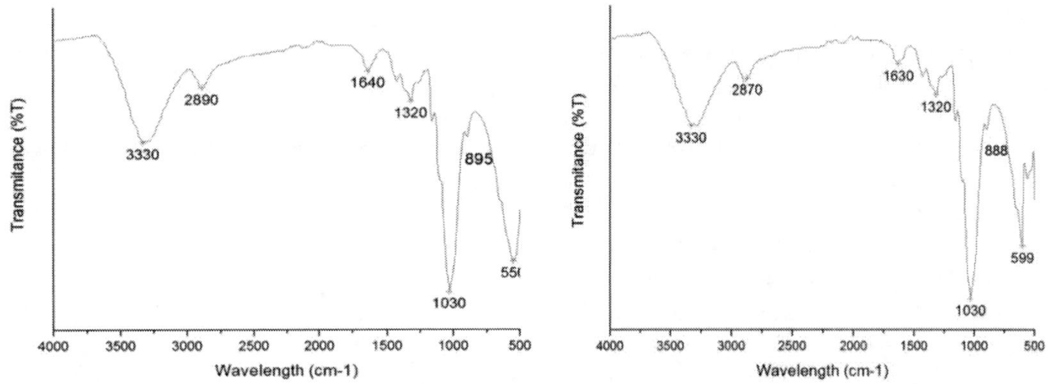

(a) (b)

Figure 4. FTIR spectra of (a) *R. corymbosa* (L.) cellulose powder, (b) *R. corymbosa* (L.) cellulose treated with $AgNO_3$ and fresh extract of *R. corymbosa* (L.)

Based on Figure 4, the *R. corymbosa* (L) cellulose powder gives peaks at 3330 cm^{-1}, 2890 cm^{-1}, 1320 cm^{-1}, 1030 cm^{-1}, 895 cm^{-1} and silver-fiber *R. corymbosa* (L) peaks at 3330 cm^{-1}, 2870 cm^{-1}, 1320 cm^{-1}, 1030 cm^{-1}, 888 cm^{-1}, which corresponds to OH stretching, methylene stretching, O-H bending, C-O bonding, and 1,4 beta-glycoside. This FTIR analysis demonstrates the appearance of a new peak at 599 cm^{-1} in *R. corymbosa* (L.) cellulose treated with silver nitrate in the presence of the fresh extract of *R. corymbosa* (L.) extract. Despite the absence of exact evidence, the new peak was assumed to have resulted from the interaction between silver and the cellulose functional group.

The experimental product was applied as an antibacterial agent against the gram-negative bacteria *E. coli*, and the results are shown in Figure 5 and Table 1 (two repetitions). The results showed an average inhibition zone of 3.97 mm for silver incorporated- *R. corymbosa* (L) cellulose, which indicates a lesser activity than the control material (silver sulfadiazine), measuring 5.35 mm. This clearly indicates the

IOP Conf. Series: Materials Science and Engineering **833** (2020) 012090 doi:10.1088/1757-899X/833/1/012090

propensity to use the current material as an alternative antibacterial agent, as well as the potential for further development in aspects of wound healing and water filtration [12].

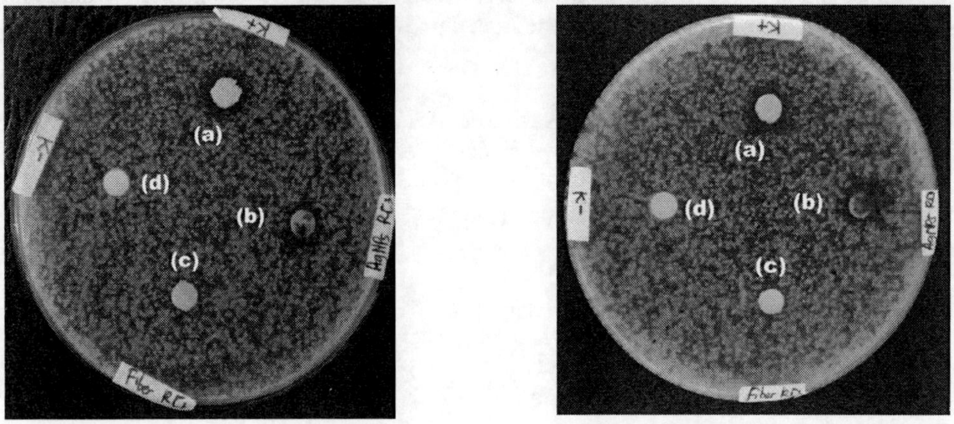

Figure 5. Antibacterial assay of filter paper (a) immersed in silver sulfadiazin cream as control positive, (b) sprinkled with silver incorporated *R. corymbosa* (L) cellulose powder, (c) sprinkled with fiber *R. corymbosa* (L) powder, (d) immersed in demineralized water

Table 1. Results of antibacterial activity toward *E coli* growth represented by zone of inhibition of as-prepared materials and standard antibiotic by Diffusion Disc Methodology

No	Material	Inhibition Zone		Average (mm)
		Repetition I (mm)	Repetition II (mm)	
1	Filter paper immersed in silver sulfadiazin cream as control positive	5.50	5.20	5.35
2	Filter paper sprinkled with silver incorporated *R. corymbosa* (L) cellulose powder,	4.03	3.90	3.97
3	Filter paper sprinkled with fiber *R. corymbosa* (L) powder,	0	0	0
4	Filter paper immersed in demineralized water	0	0	0

These findings provide evidence stipulating the ability for *R. corymbosa* (L) plant grown in coastal area to valorize as an alternative cellulose source for further research. Interestingly, the metallic silver synthesized from $AgNO_3$ salt in this study required the use of aqueous extract sourced from the same plant, which possibly attaches to the cellulose matrices.

4. Conclusion

Silver-incorporated *R. corymbosa* (L.) cellulose was successfully produced using *R. corymbosa* (L) as a source of cellulose, while the fresh aqueous extract was adopted as a reducing agent in the conversion of $AgNO_3$ to Ag (0). In addition, FTIR analysis demonstrated the occurrence of a new peak on the cellulose-Ag composite at 599 cm^{-1}, and the final product displayed characteristics of a growth inhibitor against *E. coli* (ATCC 8739). Furthermore, product effectiveness is shown by the presence of an inhibition zone, which was almost equal to that of the control, silver-sulfadiazine (volume 20 µL; concentration 15 mg/mL).

5. Acknowledgement

The research fund was obtained from Direktorat Riset dan Pengabdian Masyarakat (DRPM), Direktorat Jenderal Penguatan Riset dan Pengembangan, Kementerian Riset, Teknologi, dan Pendidikan Tinggi, Republik Indonesia, under the scheme of Penelitian Tim Magister Fiscal Year 2019.

References

[1] Xu Y, Li S, Yue, X and Lu W 2018 *BioResources* **13**(1) 2150 – 2170.
[2] Li R, He M, Li T and Zhang L 2015 *Carbohydrate Polymers* **115** 269 –275
[3] Tsai T.-T, Huang T.-H, Chang C.-J, Ho N, Y.-J, Tseng, Y.-T and Chen C.-F 2017 *Scientific Reports* 7 3155 DOI:10.1038/s41598-017-03357-w
[4] Muthulakshmi L, Rajini N, Rajalu A.V, Siengchin S and Kathiresan T 2017 *Intl. J. Biological Macromolecules* 103, 1113 – 1120
[5] Kalwar K, Bhutto M A, Dali L dan Shan D 2017 *Mater. Res. Express* 4 105405 https://doi.org/10.1088/2053-1591/aa925b
[6] Smiechowicz E, Niekraszewicz B, Kulpinski P and Dzitko K, 2018 *Cellulose* **25** 3499 – 3517.
[7] Alahmadi N S, Betts J W, Heinze T, Kelly S M, Koschella A and Wadhawan J D 2018 *RSC Adv.*, **8** 3646 – 3656.
[8] Hamisan A F, Abd-Aziz S, Kamaruddin K, Shah U K, Shahab N and Hassan M A 2009 *J. Agricultural Res.* 4, 250 – 256.
[9[Ahmad Z, Roziaizan N N, Rahman R, Mohamad A F and Wan Ismail W I N 2016 *MATEC Web of Conferences* 47(Mcc), hal. 05013. doi: 10.1051/matecconf/20164705013.
[10] Yuningsih L M, Mulyadi D and Aripandi I 2017, *American J. Mat. Sci.* **7**(3) 59 – 63.
[11] Basta A H, El-Saied H, Hasanin M S and Deftar M M, 2018 *Intl. J. Biological Macromolecules* 107 Part A 1364 – 1372.
[12] Praveena S M, Han L S, Than L T L and Aris A Z 2016 *J. Experimental Nanosci.* **11**(17) 1307 – 1319.

Impedance Spectrum of QCM Sensor Coated With 18-Crown-6-Ether Solved in THF, Chloroform and Toluene

Mira Setiana, Tyas N. Zafirah, Masruroh, Istiroyah, Setyawan P. Sakti*

Department of Physics, Faculty of Mathematics and Natural Sciences, Brawijaya University, Jl. Veteran, Malang 65145, Indonesia

*Corresponding author: sakti@ub.ac.id

Abstract. Crown ether is a form of ethylene oxide cyclo-oligomer. Crown ether has a strong ability to form a complex with cations. The crown ether has been used as part of the electrochemical sensor and optical sensor because of its selectivity to several cations. There was no report on the use of the 18-crown-6-ether as a single coating layer on the QCM sensor. This work investigates the loading effect of the 18-crown-6-ether as a selective coating for QCM sensors as a pre-requirement for a selective coating layer. The impedance spectrum of the QCM sensor coated with 18-crown-6-ether at different concentrations solved in three different solvents was measured. The effects of solvents and 18-crown-6-ether concentrations to the calculated thickness and the effect of the QCM sensor impedance was investigated. The solvents used are Toluene, Tetrahydrofuran (THF), and Chloroform. The results show that the film behaves like a rubber material, which is shown by the impedance value of the sensor at the series resonance frequency. It can be concluded that the thickness of the 18-crown-6-ether layer must be maintained less than one μm to avoid damping on the QCM sensor.

1. Introduction

The coating layer is an important part of the Quartz Crystal Microbalance (QCM) sensor for these chemical sensors and biosensors. The coating material functions as a selective material or immobilization matrix. The mechanical [1] and functional properties of the material used must be considered so as not to disturb the working frequency of the QCM sensor. In many applications, a combination of mechanical and functional properties can be obtained in the form of a monolayer layer, a thin film layer, and also a thick film layer. In most applications, especially in the sensor field, the monolayer layer and thin-film layer are used [2–4]. However, thick film coatings are also used [5–8]

Thin films in nanometer thickness or monolayer films are used based on the basic assumptions of rigid layer behavior. Therefore, the effect of the mechanical properties of the coating layer on sensor performance is neglected, and the Sauerbrey equation can be used [9]. The mass deposited on the sensor surface can then be calculated using a linear relationship between changes in frequency and deposited mass. Several non-Sauerbrey approaches have been reported in the use of QCM sensors for chemical sensors. Among other things, the mechanical effect of the damping of the coating material to the electrical impedance of the QCM sensor needs to be controlled to minimize the unexpected behavior of the QCM sensor. The solvent effect to the microstructure of the film [10] of the

The 18-Crown-6-Ether (crown ether) is known as a selective material. The crown ether is used in the form of polymer or attached to other supporting materials [11–13]. The use of crown ethers solely as a coating material was not reported. A study has been conducted on the effect of the thickness of the crown ether coating on the electrical impedance of the QCM sensor. The research was carried out to investigate the thickness of the crown ether layer to the electrical impedance of the QCM sensor.

2. Experimental Method

2.1. Material

The QCM sensor used is one that has a silver electrode, which is made of AT-cut with a fundamental resonant frequency of 10 MHz. The crown ether used was 18-crown-6-ether purchased from Sigma-Aldrich (India). The solvents used are toluene, THF, and chloroform from the same factory.

2.2. Methods

The crown ether solution was made from 18-crown-6-ether dissolved in 1 mL of each solvent. The solvent was toluene, chloroform, and tetrahydrofuran (THF). The solution concentration was 1%, 2%, 3%, 4% and 5% mass for each solvent. Furthermore, 50μL of this solution was deposited on the QCM electrode sensor using the spin coating method at a rotation speed of 300 rpm. The solvent in the coating film was left evaporated at room temperature.

Before the coating process, the electrical impedance spectrum of the QCM sensor was measured using the Bode 100 Vector Network Analyzer from OmicronLab. The impedance spectrum of the coated sensor with the crown ether was taken after the evaporation process of the film material solvent took more than 24 h. The resonance frequency and the minimum impedance of the sensor before and after the coating were taken from the impedance spectrum. The film thickness was calculated based on the resonance frequency shift using the Sauerbrey equation [14]. The calculation was done by assuming that the crown ether behaves as a glassy material. Therefore the film thickness (h) can be calculated using the following equation:

$$h = -\frac{\Delta f\ A\sqrt{\rho_Q \mu_Q}}{2 f_0^2\ \rho_c} \tag{1}$$

where Δf is frequency shift of the series resonance frequency of the QCM sensor, A is the cross-section area of the QCM electrodes, μ_Q is shear modulus of quartz crystal (2.947 x 1011 $gcm^{-2}s^{-2}$), ρ_Q is mass density of quartz crystal (2.648 g/cm^3), f_0 is initial frequency prior to the existence of mass adsorbed, and ρ_c is the mass density of crown ether. The impedance measurement was done in ambient air condition at room temperature. Room humidity varied around 50% to 60%.

3. Result and Discussion

3.1. Impedance Spectrum and Value

The impedance spectrum of the Quartz Crystal Microbalance before coating showed the initial condition of the sensor. The minimum impedance of the sensor was found at 8.4±1.7Ω , and the series resonance frequency of 10.005.000±2000 Hz. The frequency at the series resonance frequency, where phase was zero and the frequency at the minimum impedance was equal. Figure 1 shows the impedance spectrum of the sensor before coating and after coated with crown ether.

The impedance spectrum of the coated sensor shows an increase in the minimum impedance and a decrease in the maximum impedance at series resonance. The impedance curve shifts to the left and formed similar curves, except the lost in a sharp phase change around the series resonance and parallel resonance frequency.

The impedance spectrum of the sensor after coating shows a change in the curve behavior with the increasing crown ether concentration being deposited using a spin coating. Figure 2 shows the minimum impedance of the sensor coated with crown ether at different concentrations and different solvents. The minimum impedance value of the sensor increases with the increasing concentration of the crown ether. The impedance value at a concentration of less than 3% has the same value, although the solvent was different. However, for the higher concentration, the minimum impedance was varied and affected by the solvent. It seems that the evaporation rate which affects the microstructure of the film [10] leads to an impedance change of the sensor. The microstructure effect becomes noticeable as the film thickness increased.

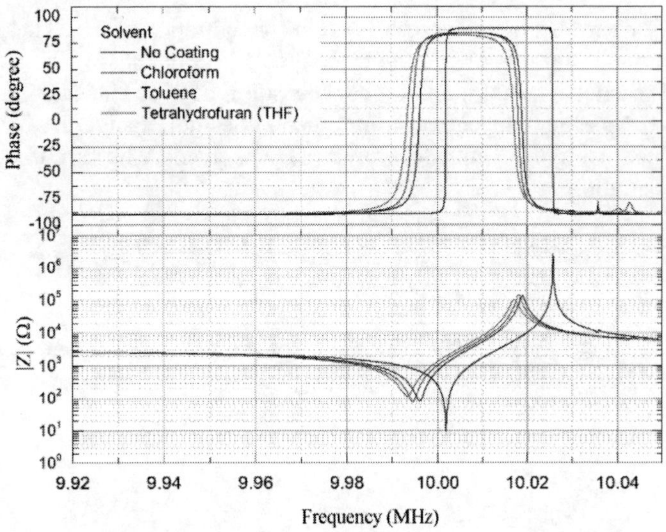

Figure 1. Impedance spectrum of the sensor before coating and after coating with crown ether at 6% concentration.

Although the minimum impedance of the sensor with crown ether coating increases with the crown ether concentration up to the concentration of 6%, the conditions that indicate a glassy characteristic are observed. The impedance and phase curves shifted to the left. The phase shift where the minimum impedance of the sensor exists is equaled to zero. It means that the frequency of the minimum impedance is at the series resonance frequency. There is no different in the frequency at the minimum impedance and zero phases.

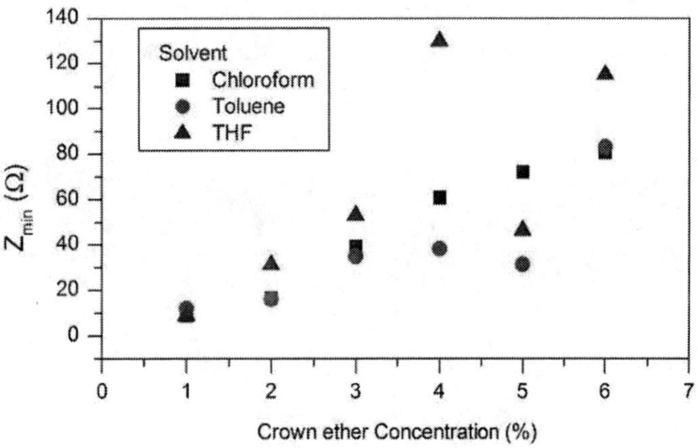

Figure 2. Sensor minimum impedance coated with crown ether using the spin coating.

3.2. Coating thickness
Based on the impedance analyses that showed a glassy property of the coating layer, we can apply the Sauerbrey equation to calculate the coating thickness. It was known that when using a spin coating method, the concentration and rotation speed determined the film thickness. At constant speed, the bigger concentration of crown ether should result in a thicker film. Based on the Sauerbrey equation, the film thickness was calculated using the calculated deposited crown ether mass, sensor electrode diameter, and crown ether mass density.

The effect of crown ether concentration on film thickness is shown in Figure 3. The behavior of the calculated thickness of the crown ether depends on the solvent. The thickness of the crown ether film solved in chloroform shows a linear increase up to a concentration of 5%. The coating thickness reaches its maximum at around 300 nm. While the calculated film thickness originated from the crown ether solved in toluene reaches its saturation at a concentration of 3%. The calculated thickness is less than 300 nm. Higher crown ether concentration in toluene does not change the film thickness. The highest thickness is obtained using the Tetrahydrofuran (THF) solvent. The film thickness reaches 780 nm at crown ether concentration of 4%

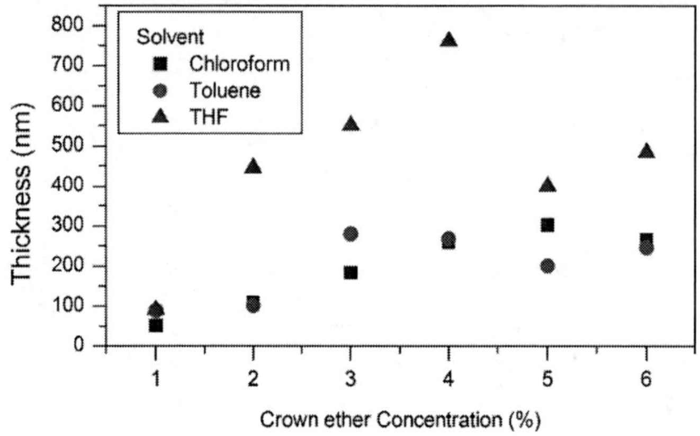

Figure 3. The calculated thickness of the crown ether film based on the Sauerbrey equation.

3.3. Coating thickness and impedance

The effect of thickness on the impedance of the QCM sensor after being coated with Crown ether is shown in Figure 4. The minimum impedance of the sensor is related to the crown ether thickness. As the thickness of the crown ether layer increases, the minimum impedance of the sensor rises. The minimum impedance value of the sensor was significantly affected by the crown ether thickness. Pearson correlation between the film thickness and minimum impedance is 0.7681. It indicated a strong correlation between the minimum impedance and the film thickness.

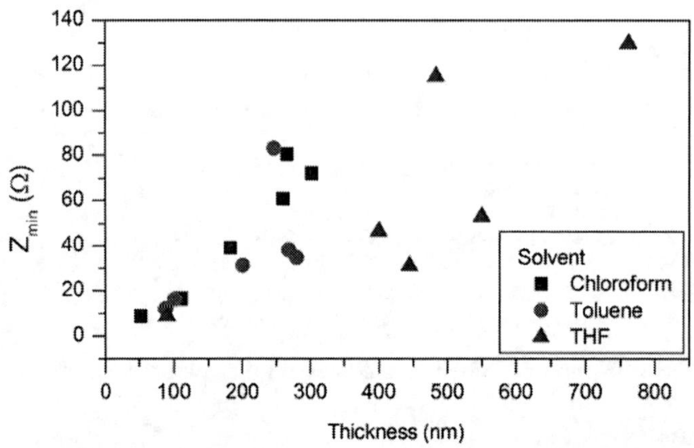

Figure 4. The impedance value of the sensor coated with crown ether with different thicknesses.

Compare to the effect of the film thickness on the QCM sensor minimum impedance which was coated using polystyrene, the increase of minimum impedance caused by crown ether film is much bigger. Using polystyrene coating with similar order of thickness [14], the change of minimum impedance was negligible. The polystyrene film is known as a rigid film [15]; therefore, the effect of the film thickness to the minimum impedance is very small. The change of the minimum impedance of the sensor coated using crown ether tends to mimic a rubbery film.

4. Conclusion

The impedance spectrum of the QCM sensor coated with an 18-crown-6-ether layer deposited using a spin coating method shows high electrical damping caused by a thin layer of 18-crown-6-ether film. Increasing crown ether concentration in solvent results in thicker film calculated using the Sauerbrey equation, but there is a maximum film thickness that can be achieved. The maximum film thickness depends on the solvent. The crown ether solved in tetrahydrofuran results in the thickest film comparing to the film formed by solved crown ether in chloroform and toluene. The minimum impedance value relates to the film thickness, in which higher film thickness results in bigger minimum impedance value. At the film thickness close to 1 µm, the minimum impedance of the sensor coated with 18-crown-6-ether goes up more than 100Ω.

Acknowledgments

This work was funded by the Brawijaya University Professorship Research Grant scheme.

References

[1] Wolff O and Johannsmann D 2000 *J. Appl. Phys.* **87** 4182–8
[2] Mateescu A, Wang Y, Dostalek J and Jonas U 2012 *Membranes (Basel).* **2** 49–69
[3] Llanes D, Hempel U and Ocampo A 2005 *Rev. EIA* 9–19
[4] Zeng Y, Lin S, Gu D and Li X 2018 *Nanomaterials* **8** 1–16
[5] Brignell J E, White N M and Cranny A W J 1988 *IEE Proc. I Solid State Electron Devices* **135** 77–84
[6] Prudenziati M and Morten B 1986 *Sensors and Actuators* **10** 65 82
[7] Belavic D, Hrovat M, Holc J, Zarnik M S, Kosec M and Pavlin M 2007 *J. Electroceramics* **19** 363–8
[8] Morten B, Prudenziati M and Taroni A 1983 *Sensors and Actuators* **4** 237–45
[9] Sauerbrey G 1959 *Zeitschrift für Phys.* **155** 206–22
[10] Kim, G., & Libera, M. 1998 *Macromolecules* **31** 2569–2577.
[11] Price G J and Drake P L 2006 *Sensors Actuators, B Chem.* **114** 466–72
[12] Faridbod F, Ganjali M, Dinarvand R, Norouzi P and Riahi S 2008 *Sensors* **8** 1645–703
[13] Zhang Z, Dou Q, Gao H, Bai B, Zhang Y, Hu D, Yetisen A K, Butt H, Yang X, Li C and Dai Q 2018 *Adv. Healthc. Mater.* **7** 1–7
[13] Sakti S P, Masruroh, Kamasi D D and Khusnah N F 2019 *Mater. Today Proc.* **13** 53–8
[14] Guo D, Li J N, Xie G X, Wang Y Y and Luo J B 2014 *Langmuir* **30** 7206–12

Effect of sulfur and sodium sulfate on phase transformation and microstructure on carbothermic reduction of Indonesian ilmenite

Agung Setiawan[1*], Sri Harjanto[1]

[1]Department of Metallurgy and Materials Engineering
Faculty of Engineering, Universitas Indonesia, Kampus Baru UI Depok, 16424, Indonesia

*Coreresponding author: agung.setiawan72@ui.ac.id

Abstract. Utilization of ilmenite and pulverized biomass of palm kernel shell is a beneficial solution to deal with the depletion of high-grade titanium minerals and manage of palm oil waste. The purposes of this research are to analysis the phase transformation and microstructure with additive addition. Sulfur and sodium sulfate have used an additive with an additional dosage of 0%, 1.5%, 3.0%, and 4.5%, and the reductant was the pulverized palm kernel shell biomass. The ilmenite concentrate was reduced at 1200 °C for 60 minutes in inert condition (N_2 atmosphere). Testing methods performed are Optical microscope (OM), X-Ray diffraction (XRD), Field-Emission Scanning Electron Microscopy (FE-SEM), and Energy Dispersive X-ray spectroscopy (EDX), calculating of the average particle size, and metal recovery. The results showed that the active additive used was sodium sulfate compared to sulfur. At the addition of 4.5 wt.% sodium sulfate, it was obtained phases of titanium suboxide, Ti_6O_{11} and Ti_3O_5, with very high intensity. Recovery and grades of the reduced sample with 4.5 wt.% sodium sulfate are reached 70.91% and 94.20% Fe, respectively. Moreover, the average particle size of metallic iron increased from 11.96 μm to 38.36 μm with the addition of 4.5wt.% sodium sulfate.

1. Introduction

The primary raw materials to make TiO_2 pigment and Ti metal are ilmenite ($FeTiO_3$) and rutile (TiO_2). TiO_2 pigment has the best refractive index than other chemicals, and it also has good thermal stability and excellent resistance to chemical attack and ultraviolet degradation [1]. Because of these unique characteristics, TiO_2 pigment is widely applied in several industries for instance coating, papers, plastics, rubber, paints, textiles, pharmaceutical, and food [1, 2]. On the other hand, titanium metal has high chemical durability, thermal conductivity, and high strength, as a result, it is applied to thermal insulation, aerospace, marine ships, and military armor [3, 4]. High-grade titanium ores have decreased with the fast growth of the titanium processing industry. Thus, ilmenite, as an alternative raw material, must be utilized but also is complex to separate or reduce because of the impurities.

There have been some of the studies on carbothermic reduction of ilmenite with the different additives. Huang et al. [5] have reported that the particle size of the metallic iron of the sample escalated with increasing of Fe-Si to ilmenite. Also, because the silicon in this process supplies heat, the maximum temperature of the mass loss rate is reduced so that it can enhance the rapid ilmenite reduction with carbon. In another study, Liao et al. [6] examined the Fe-Si effect to improve carbothermic reduction ilmenite placer. The addition of Fe-Si powder acts as the agent of nucleation and aggregation on reduced ilmenite.

Song et al. [7] found an addition of 2% $Na_2B_4O_7$ on carbothermic reduction of ilmenite can improve the particle size of iron up to 46.1 μm and the metallization ratio achieved 89.5%. Chen et al. [8] tested some additives such as CaF_2, Na_2SO_4, $Na_2B_4O_7$, NaCl, and Na_2CO_3. They reported that the metallization ratio of pre-oxidized ilmenite escalated up to 94% with the addition of 3.0% additives. Lv et al. [9] carried out the effect of sodium sulfate on pre-oxidized ilmenite reduction at a reaction temperature of 1350, 1400, and 1450 °C for 30 minutes. They reported that the metallization degree of samples would decrease during the dosage of sodium sulfate goes up to 12%, but the iron particle continuously increased with adding sodium sulfate.

In this work, ilmenite concentrate was reduced in the presence of sulfur and sodium sulfate and pulverized biomass. The addition of additive during the carbothermic reduction of ilmenite increases carbon reactivity and enlarges the size of the iron particle and titanium-rich slag [8]. This study aims to investigate the phase transformation and microstructure with sulfur and sodium sulfate as an additive. The effect of the addition dosage of additives, recovery, and the average particle size of metallic iron are discussed.

2. Material and Methods

2.1. Materials

The Indonesian ilmenite concentrate was obtained from PT. Monokem Surya, Indonesia. The mineralogical analysis and chemical composition of the ilmenite concentrates were investigated by XRD and XRF. Figure 1 showed the diffractogram of Indonesian ilmenite concentrate, which implies that the significant phases in the Indonesian ilmenite concentrates are $FeTiO_3$, TiO_2, and Fe_3O_4. The chemical composition of the ilmenite concentrates as measured by XRF is shown in Table 1. The Indonesian ilmenite concentrate is composed of impurities, namely Cr_2O_3, Al_2O_3, and MnO.

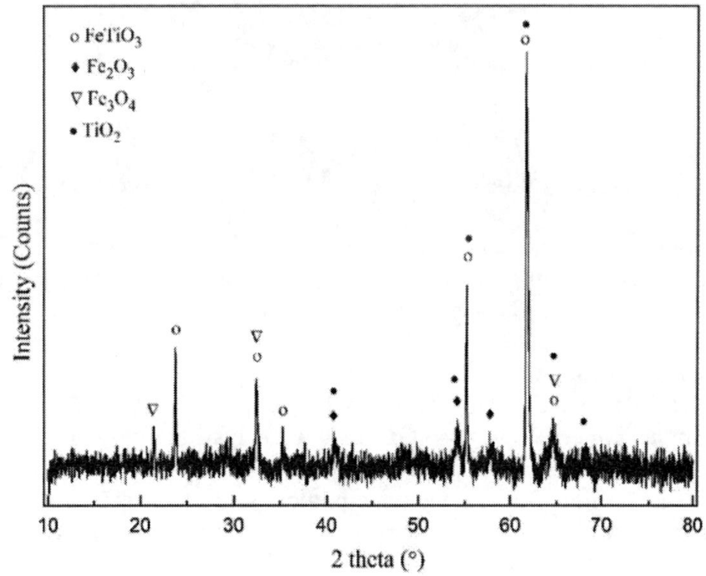

Figure 1. Diffractogram of Indonesian ilmenite concentrate.

Table 1. Elements composition of Indonesian ilmenite concentrates (wt.%).

Compositions	TiO_2	Fe_2O_3	Cr_2O_3	Al_2O_3	MnO
Content	60.27	35.01	2.11	1.11	1.50

The reductant was the pulverized palm kernel shell biomass (particle size < 250 μm). Table 2 showed the ultimate and proximate analysis of the reductant. Sodium sulfate and sulfur were used as an additive and carboxymethyl cellulose was used as a binder.

Table 2. Elemental composition of pulverized palm kernel shell biomass.

Proximate Analysis	Quantity (%)	Ultimate Analysis	Quantity (%)
Moisture in air-dried sample	6.06	Carbon	80.59
Ash	3.60	Hydrogen	3.05
Volatile Matter	11.12	Nitrogen	0.79
Fixed Carbon	79.22	Total Sulfur	0.08
		Oxygen	11.89

2.2. Experimental Method

The ilmenite concentrate was homogenously mixed with pulverized biomass and additives using magnetic stirrer for 15 min. The molar ratio of C to O (in iron oxide) in the mixture was set to 1.3. The additive addition levels were set to 0%, 1.5%, 3%, and 4.5% (wt.), respectively. Next, mixtures with masses of ten-gram were mixed to homogeneity with the binder (0.5%) and distilled water (0.6%). Then, the mixture samples were compressed into a cylindrical shape (diameter: 20 mm and height: 10 mm) at a pressure of 6 ton, as shown in Figure 2. In the next step, the mixture samples were dried at 100 °C for 5 hours.

Figure 2. Image of mixture sample after compressing.

Reduction experiments were conducted in a horizontal tube electric furnace. Nitrogen was flowed into the alumina tube to release air at least for 15 min before placing the samples into the furnace until the end of reduction experiment. The samples were placed in the alumina crucible boat with samples and put into the hot zone of the furnace. The samples were reduced at temperature 1200 °C with-holding time 60 min and heating rate 5 °C/min.

The reduction samples were examined by XRD with a Cu Kα source to determine phase compositions. The optical microscopy was performed to analysis the growth of the metal phase of the reduction sample. The surface morphology and chemical analysis of the reduction samples were investigated by FE-SEM with EDX. The average particle size of metallic iron was measured by the free software ImageJ using images (more than 5 pictures) from OM and FE-SEM (Backscattered electrons). The OM and Back-Scattered Electron (BSE) images with magnification 50, 100, and 1000 times of the reduced samples were used as the calculation samples. In addition, the process of determining the area of an iron particle was based on its brightness level and uses a threshold mode.

The recovery was defined as the following equation,

$$\text{Recovery M (\%)} = \frac{c\,(f-t)}{f\,(c-t)} \tag{1}$$

where *c*, *f*, and *t* are the grade in the concentrate, the grade in the feed, and the grade in the slag, respectively. The grade in concentrated, feed, and slag was identified by EDX.

3. Result and Discussion

3.1. Phase transitions during reduction

The XRD experiments were carried out with a scanning rate from 15° to 80° (2θ) with scan speed 2°/minute. Diffractogram of the reduction samples with different dosage of added sulfur was shown in Figure 3. The main phases of samples without additives were metallic iron, magnetite, pseudo-brookite, TiO_2, and $FeTiO_3$. However, several new phases appeared in the samples, such as FeO and Ti_6O_{11}. It is indicated that pulverized biomass as a reductant can reduce TiO_2 to titanium suboxide. At 1.5 wt.% sulfur, the intensities of FeO and Fe_2O_3 disappeared utterly, and the intensity of pseudo-brookite became strong. The XRD examination implied that a relatively high manganese content in ilmenite will react with sulfur to form manganese sulfide.

Furthermore, the diffractogram of reduction sample with dosage of 3.0 wt.% sulfur showed that the formation of Ti_6O_{11} with high intensity followed by the formation of Ti_3O_5 with low intensity and the intensity of Fe_3O_4 disappeared utterly. The Ti_3O_5 phase became stronger when the sulfur dosage added was 4.5 wt.%. Moreover, the intensity of metallic iron significantly increased with an increasing dosage of sulfur until 3 wt.% and decreased with a dosage of 4.5 wt.% sulfur. It is due to the chemical reaction between metallic iron and sulfur to form FeS and FeS_2. These phases are agreed with the diffractogram of 4.5% sulfur, as shown in Figure 3.

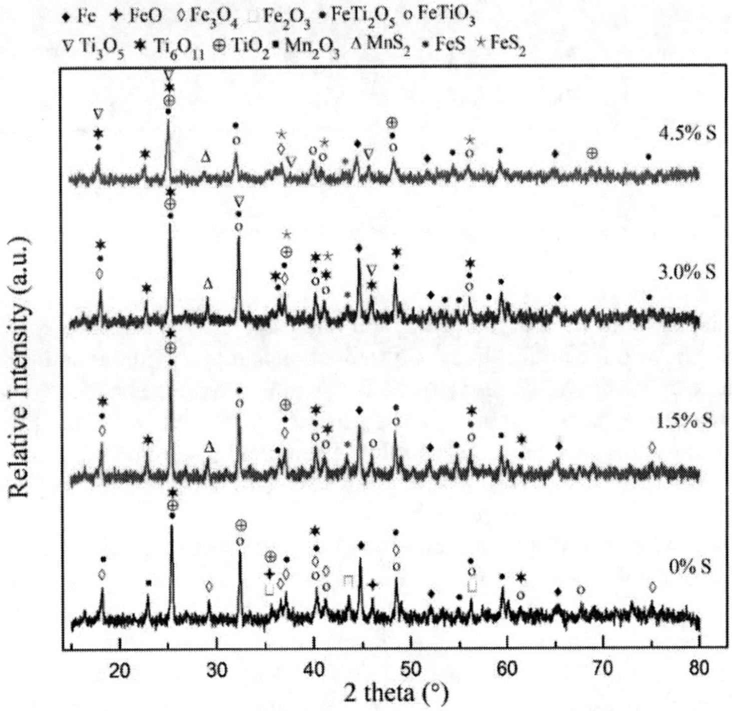

Figure 3. Diffractogram of reduction samples with different dosage of added sulfur.

The diffractogram of the reduction samples with different dosages of added sodium sulfate was shown in Figure 4. The sample with the addition of 1.5 wt.% sodium sulfate showed that the formation of pseudo-brookite with high intensity followed by the formation of Ti_6O_{11} with low intensity, but the FeO phase is still formed. The magnetite phase, moreover, still formed in the sample with the addition of sodium sulfate. On the other hand, the Ti_3O_5 phase grew stronger with the addition of 4.5 wt.%

IOP Conf. Series: Materials Science and Engineering **833** (2020) 012092 doi:10.1088/1757-899X/833/1/012092

sodium sulfate compared to the addition of sulfur and TiO_2 formed with high intensity. It explains that sodium sulfate can improve the reduction ability of ilmenite. Lv et al. also reported that at temperature 1000-1100 °C CO and CO_2 will promote which are caused by the chemical reaction between sodium sulfate and carbon [9].

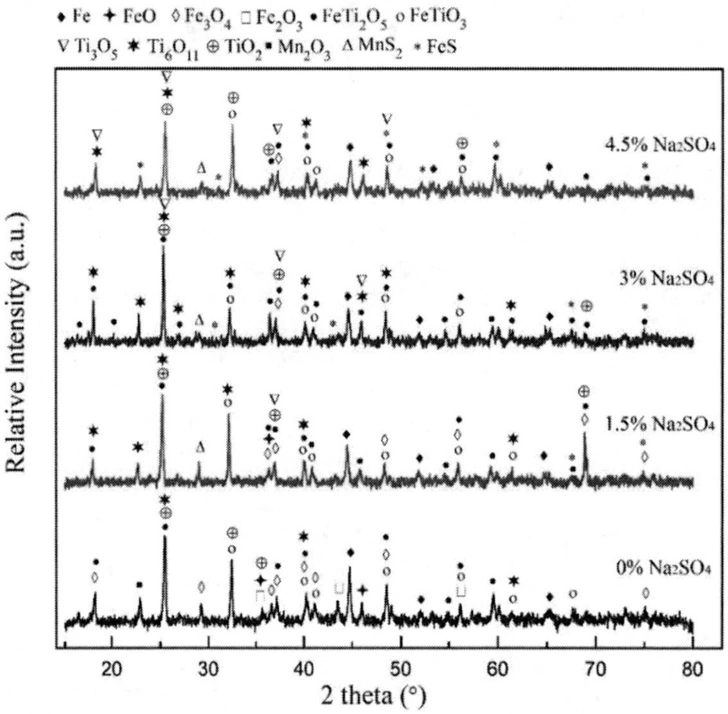

Figure 4. Diffractogram of reduction samples with different dosages of added sodium sulfate.

3.2. Morphological analysis

The growth of the metallic iron of the reduction samples enlarged gradually with increasing the additives. Figure 5 showed the optical images of reduction samples without additives and the addition of 4.5 wt.% additives. The bright area is metallic iron and a gray area is a slag containing titanium. In this image, the area of the metallic iron grew significantly with additional additives. The addition of sodium sulfate has a particle size larger than sulfur. The metallic iron in the reduction sample without additive has spherical structures, as in Figure 5a and 5d, but in the reduction sample with 4.5 wt.% additive has lump structures, as in Figure 5b, 5c, 5e, and 5f.

Figure 6 showed the effect of different additives on the average particle size of metallic iron. The average particle size of the metallic iron was markedly when addition 4.5 wt.% sodium sulfate than sulfur. The average particle size in samples without additives, 4.5 wt.% sulfur, and 4.5 wt.% sodium sulfate was 11.96 μm, 18.98 μm, and 38.36 μm, respectively. Sodium sulfate, which reacts during the process, will break down into Na_2O, Na_2S, and S. Where Na_2O will react with silica minerals to form several sodium silicates phases that have a low melting point. Besides, S will react with Fe to form FeS, which has a low melting point as well. Thus, the additive will dissolve metal particles that have an impact on the growth of metal particles. It has an impact on the evolution and aggregation of large metal particles, which are confirmed on the results of microstructure analysis [10].

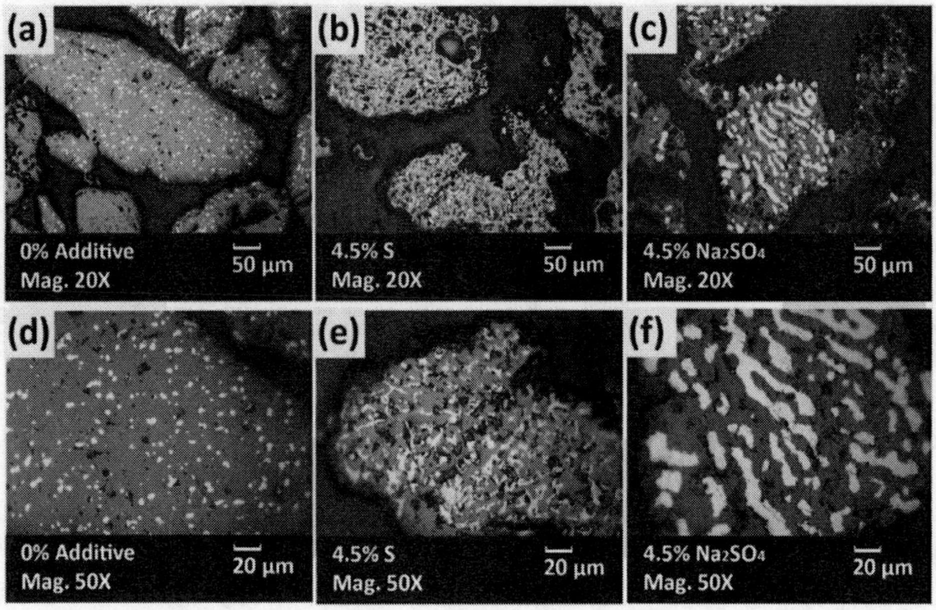

Figure 5. Optical images of reduction samples. (a) and (d) micrograph of reduction sample without additive using magnification 20 and 50 times, respectively. (b) and (e) micrograph of reduction sample with 4.5% sulfur using magnification 20 and 50 times, respectively. (c) and (f) micrograph of reduction sample with 4.5% Na_2SO_4 using magnification 20 and 50 times, respectively. The bright area is metallic iron.

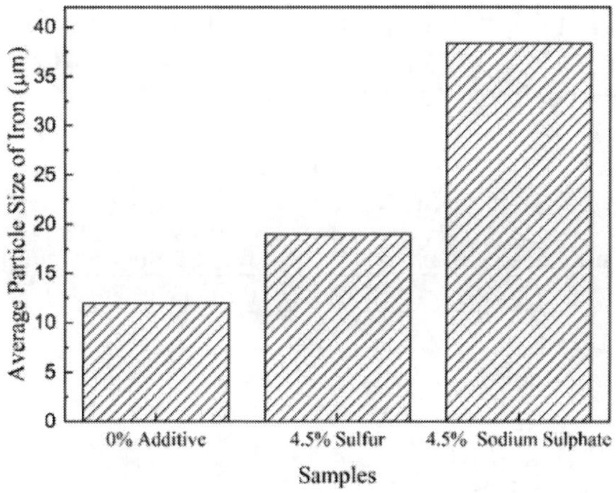

Figure 6. The effect of different additives on the average particle size of metallic iron.

The surface morphology and elemental mapping of the reduction samples without and with 4.5 wt.% additives are shown in Figure 7. Figure 7a showed that the metallic iron particle in the reduction sample without additives gradually grow to form spherical structures. Otherwise, in reduction samples with 4.5 wt.% additives, the metallic iron particle formed lump structures, as in Figure 7b and 7c. Besides, elemental mapping, as seen in Figure 7, showed the dispersion of metallic iron, Ti, C, S, and O. In the reduction sample with 4.5 wt.% sulfur, sulfur distributed in the metallic iron to form FeS and FeS_2, as in Figure 7b. It is following by the XRD results in Figure 3. On the contrary, in the reduction sample with 4.5 wt.% sodium sulfate, the distribution of sulfur was very little detected, as in Figure 7c.

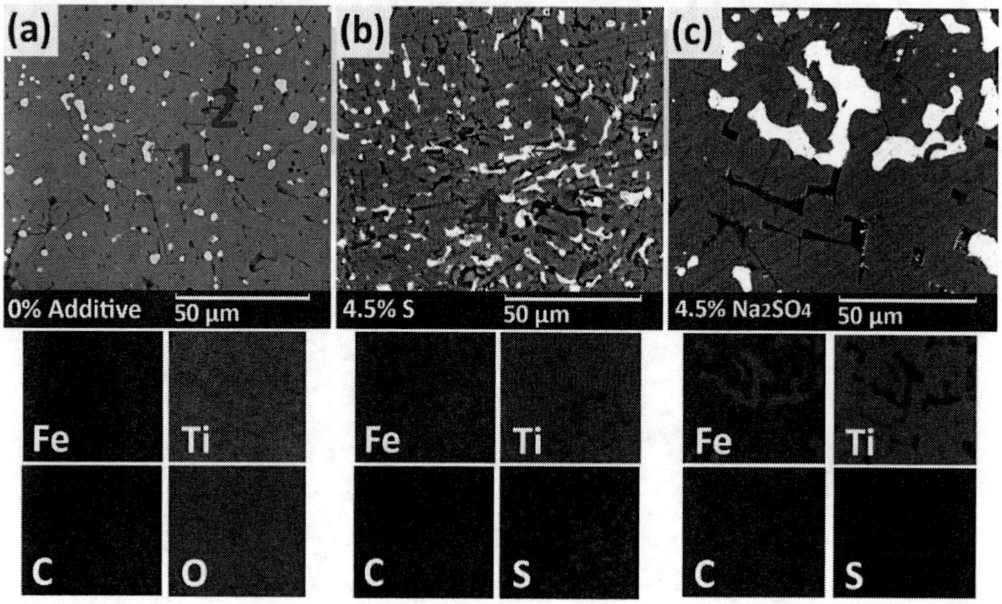

Figure 7. Morphology and elemental mapping in the reduction samples with different additives. The additive content is 0%, 4.5% sulfur, and 4.5% Na_2SO_4 from (a) to (c), respectively. The bright area is metallic iron.

Table 3 showed the EDX analysis results of the reduction samples from Fig. 7. Conforming to the EDX result of the point marked in Figure 7 (point 1, 3, and 5), present in Table 3, it can be found that the bright area is mainly metallic iron. in Figure 7c, the reduction sample with 4.5 wt.% sodium sulfate comprises approximately 93.54% Fe, which can be classified as iron phase. Also, the gray area contains 51.86% Ti, 23.00% O, and some impurities such as Al, Si, Cr, Mn, Mg, Na, C, and S. Furthermore, the reduction sample with 4.5 wt.% sulfur, in bright area, contains 52.51% Fe and 33.05% S, it can be concluded that as iron sulfide phase. It is consistent with the XRD results, as shown in Figure 3.

Table 3. EDX analysis results of the reduction samples.

Element	Point (wt.%)					
	1	2	3	4	5	6
Iron (Fe)	92.90	15.77	52.51	26.09	93.54	15.95
Titanium (Ti)	3.49	52.08	5.62	38.50	2.86	51.86
Aluminum (Al)	-	1.45	0.32	1.83	0.33	1.40
Silicon (Si)	0.26	0.33	0.36	0.68	0.15	0.46
Chromium (Cr)	0.27	0.08	0.17	0.36	0.21	1.05
Manganese (Mn)	0.46	2.46	0.68	1.28	0.08	1.70
Magnesium (Mg)	0.42	0.44	0.09	0.82	0.09	2.82
Carbon (C)	0.82	0.74	2.28	1.57	1.38	1.14
Sodium (Na)	-	-	-	-	0.18	0.30
Sulfur (S)	-	-	33.05	5.90	0.21	0.31
Oxygen (O)	1.38	26.65	4.72	22.55	0.97	23.00

3.3. Recovery of reduction sample

The grade and recovery of Fe with different dosage of sulfur, from 0%, 1.5%, 3.0%, and 4.5% of sulfur, are shown in Figure 8a. The Fe recovery of the reduction sample with the addition of sulfur, from 0% to 4.5% of sulfur, is 72.64%, 82.65%, 69.48%, and 77.33%, respectively. Furthermore, the

Fe grade of the reduction sample with the addition of sulfur, from 0% to 4.5% of sulfur, is 76.52%, 55.15%, 90.07%, and 91.21%, respectively. Moreover, the grade and recovery of Fe with different dosages of sodium sulfate, from 0%, 1.5%, 3.0%, and 4.5% of sodium sulfate, are shown in Figure 8b. The Fe recovery of the reduction sample with the addition of sodium sulfate, from 0% to 4.5% of sodium sulfate, is 72.64%, 72.99%, 70.46%, and 70.91%, respectively. Furthermore, the Fe grade of the reduction sample with the addition of sodium sulfate, from 0% to 4.5% of sodium sulfate, is 76.52%, 90.40%, 93.58%, and 94.20%, respectively.

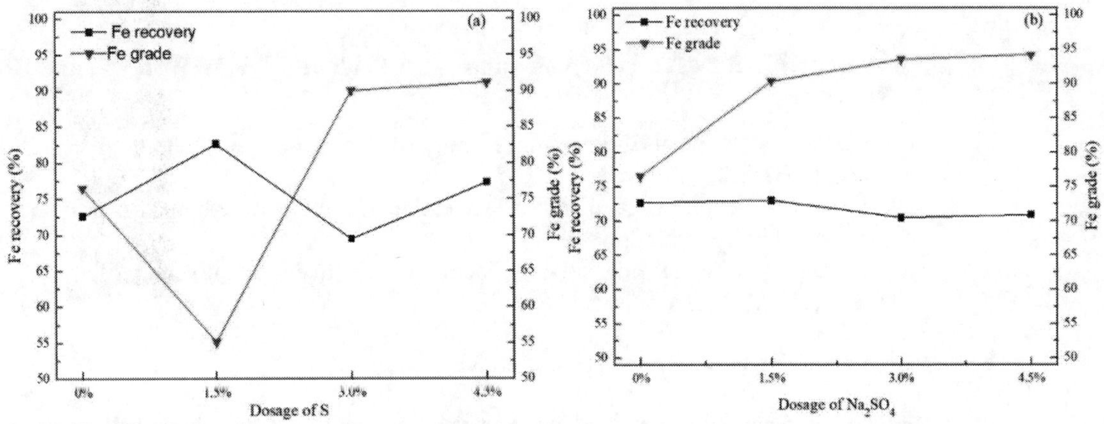

Figure 8. Recovery and grade of Fe with different dosages of added sulfur (a) and sodium sulfate (b).

4. Conclusion

The order of carbothermic reduction of Indonesian ilmenite can be explained as follow: first, the $FeTiO_3$, TiO_2, and Fe_2O_3 phase in ilmenite are reduced to Fe, FeO, Fe_3O_4, $FeTi_2O_5$, $FeTiO_3$, TiO_2, and Ti_6O_{11}. Secondly, $FeTiO_3$ and $FeTi_2O_5$ are reduced to Fe, TiO_2, Ti_6O_{11}, and Ti_3O_5. These results are beneficial to separate (e.g., magnetic separator) the iron and slag containing titanium slag. The addition of sodium sulfate can naturally improve the aggregation and growth of metallic iron particles and separate the slag and metallic iron than the addition of sulfur.

Reference

[1] Gázquez M J, Bolívar J P, Garcia-Tenorio R, and Vaca F 2014 *Mater. Sci. Appl.* **05** 441
[2] Zhang W, Zhu Z, and Cheng C Y 2011 *Hydrometallurgy* **108** 177
[3] Salimon A, Brechet Y, Ashby M F, and Greer A L 2005 *J. Mater. Sci.* **40** 5793
[4] Setoyama D, Matsunaga J, Muta H, Uno M, and Yamanaka S 2004 *J. Alloys Compd.* **381** 215
[5] Huang R, Lv X, Bai C, Zhang K, and Qiu G 2013 *Steel Res. Int.* **84** 892
[6] Liao X, Peng J, Zhang L, Hu T, and Li J 2017 J. *Alloys Compd.* **708** 1110
[7] Song B, Lv X, Miao H H, Han K, Zhang K, and Huang R 2016 *ISIJ Int.* **56** 2140
[8] Chen D S, Song B, Wang L N, Qi T, Wang Y, and Wang W J 2011 *Miner. Eng.* **24** 864
[9] Lv W, Bai C, Lv X, Hu K, Lv X, Xiang J, and Song B 2018 *Powder Technol.* **340** 354
[10] Lv W, Lv X, Xiang J, Wang J, Lv X, Bai C, and Song B 2017 Int. J. *Miner. Process.* **169** 176

Acknowledgments

The authors would like to acknowledge the Ministry of Research, Technology, and Higher Education, the Republic of Indonesia through the PMDSU grant for financial support in the research, and PT. Monokem Surya, Indonesia, for providing ilmenite concentrate.

IOP Conf. Series: Materials Science and Engineering **833** (2020) 012093 doi:10.1088/1757-899X/833/1/012093

Natural Dyes Extraction Intended for Coloring Process in Fashion Indutries

P D Sutrisna[1,*], R P Hadi[1], J J Valentina[1], H R Priyantini[1], P W Waluyo[2], and I M Ronyastra[3]

[1]Department of Chemical Engineering, Faculty of Engineering, University of Surabaya (UBAYA), Indonesia
[2]Department of Fashion Design and Lifestyle, Faculty of Creative Industry, University of Surabaya (UBAYA), Indonesia
[3]Department of Industrial Engineering, Faculty of Engineering, University of Surabaya (UBAYA), Indonesia

*Corresponding author: pudod@staff.ubaya.ac.id

Abstract. Recently, small scale fashion industries are growing quite rapidly in Indonesia. This is due to the continuous encouragement and facilitation by the Government of Indonesia to improve workforce in creative industry. As fashion industries depend on the availability of dyes as one of their resource, it is very important to develop a new and sustainable supply of dyes for dyeing process in fashion industries. Hence, research on the extraction of dyes from sustainable and natural resources for fashion industries are considered very important and urgent. Dyes for fashion and textile are naturally occurred in plants or parts of plants. The extraction of dyes from these sources can be conducted by using simple mixing between solvent and the plants inside a soxhlet extractor or reflux apparatus. In this study, the extraction processes of natural dyes from two materials, i.e. sappanwood, and mango leaf employing combination of maceration, soxhlet and reflux techniques were reported. The experiments were conducted to analyse the effects of extraction time and solvent to materials ratios on the yield of dyes. The experimental results showed that for all materials the optimum ratio of material and solvent was 1 : 6. In addition, the reflux followed by maceration could produce dyes that can be utilised for fabric coloring that suits the industrial needs.

1. Introduction

The textile industries have relied on synthetic color for decades since it is easy to apply to fabrics. However, the wastewater of synthetic dyes usually damages the environment, especially the body of water where it would dispose to. This phenomenon happens because of the heavy metals used for additional material to fix the color called mordant [1]. Though treated before disposal, the wastewater treatment is limited only to reduce the concentration of organic materials instead of to completely remove it.

As a result, the textile industries start getting back to natural dyes which attained through the extraction of natural sources. Two natural sources of these natural dyes that are often used in industries are *Caesalpinia sappan* barks which contain brazilin that gives red color and *Mangifera indica* leaves which contain mangiferin that gives brown color. A lot of research has been conducted to define the best method and process condition upon the extraction of those sources [2], so the standardized quality of natural dyes obtained with maximum quantity.

The extraction of *Mangifera indica* leaves using batch extraction method has been done by Chavan to determine the effects of solvent used, particle size, and temperature upon the extract percentage. The research used methanol, ethanol, and acetone as its solvent, a varied range of temperature and size of particles. The results show that methanol is the best solvent for extraction. In addition, the extract percentage increased by increasing temperature and reducing the size of particles [2]. Another study conducted by Chavan upon the different species of *Mangifera indica* by using Hapoos leaves. It gives extract percentage of 48.66%(b/b) and yield of 2.43%(b/b) while Kesar leaves gives extract percentage of 34.95%(b/b) and yield of 1.75%(b/b) [3]. On the other hand, the extraction upon Rajapuri leaves conducted by Shinde gives extract percentage of 60.71%(b/b) and yield of 3.035%(b/b) [4]. These studies exhibit that the species take part in determining the value of mangiferin extracted.

So much for *Mangifera indica*, the extraction of *Caesalpinia sappan* barks by reflux method using water and ethanol as solvent was also studied. The usage of water as a solvent gives 15.6±7.5%(b/b) yield while ethanol gives 20.5±4.7%(b/b) yield. On the other hand, Sinsawasdi proved that maceration method is more effective compared to the reflux method while using water as the solvent since Brazilin is thermolabile that easily degrades at high temperature [5].

The effectivity of solvent used in the extraction of *Caesalpinia sappan* L. has been studied by Hung. The solvents compared in this study were methanol, ethanol, and water whose value of each yield were 18.8±5.4%(b/b), 20.5±4.7%(b/b), and 15.6±4.7%(b/b), respectively. It shows that ethanol is the best solvent among them [6].

Despite the previous studies on *Mangifera indica* leaves and *Caesalpinia sappan* barks, there was no research reported about the efficient method and process condition using the best solvent for both materials. Thus, this research tends to fill that gap by comparing some methods with different process condition for both natural sources.

2. Materials and methods

2.1. *Mangifera indica leaves*

Mangifera indica leaves being used were specifically chosen from the species *Gadung* gathered directly from the trees planted in Surabaya, Indonesia. The leaves were washed then sun dried and grinded using kitchen blender until powder were attained. The powder were screened to get -40 mesh + 70 mesh particle size and then were extracted using water and methanol as solvent. The methods employed for this extraction were maceration, reflux, and soxhlet. For each variable, samples were taken periodically then were filtered using filter paper, and the absorbances were measured using HP-8453 Double Beam UV-Vis Spectrophotometer using wavelength of 361 nm. Maceration method was coupled with mixing and was carried out until the value of the absorbance became stable with sample taken for every two hours. While reflux method was employed with time variations of 2, 4, and 6 hours. For each reflux experiment, a little amount of sample was taken every half an hour and then analysed using spectrophotometer. In addition, for soxhlet method the experiment was carried out until as little solvent as possible was left with sample taken every two hours.

2.2. *Caesalpinia sappan barks*

Caesalpinia sappan L. was purchased as powder then screened to get -40 mesh + 70 mesh particle size and extracted using water and ethanol as solvent. The methods employed for this extraction were maceration, reflux, and soxhlet. For each variable, samples were taken periodically then were filtered using filter paper, and the absorbances were measured using HP-8453 Double Beam UV-Vis Spectrophotometer (wavelength of 543 nm for water extract and 448 nm for ethanolic extract). Maceration method was coupled with mixing with time variations of 2, 4, and 6 hours. During the maceration processes, a little amount of samples was taken every half an hour and then analysed using spectrophotometer. While reflux method was carried until the value of absorbance became stable with sample taken for every two hours, and Soxhlet method was carried out until as little solvent as possible was left with sample taken for every two hours.

For both materials, the yields of extraction process were measured by taking the ratio between the mass of product after drying and the mass of initial material. The LC-HRMS analysis was conducted to analyse the qualitative existence of mangiferin in the extract of mango leaf and also brazilin in the extract of sappan barks.

3. Results and Discussion

3.1. *Mangifera indica*

The extraction of dyes from mango leaf in this study has been conducted by three different processes, such as reflux at different times, soxhlet, and maceration. The yields of dyes from the experiment are presented in Table 1.

Table 1. Yield of dyes from the extraction of mango leaf

Material	Method	Yield of water as solvent (wt. %)	Yield of methanol as solvent (wt. %)
Mango leaf	2 hours reflux	2.18	2.60
	4 hours reflux	2.52	2.73
	6 hours reflux	2.78	2.67
	Soxhlet	1.05	6.00
	Maceration	1.53	2.43
		Yield of water as solvent (wt. %)	Yield of ethanol as solvent (wt. %)
Sappan barks	Maceration for 2 hours	0.409	0.2545
	Maceration for 4 hours	0.6263	0.2525
	Maceration for 6 hours	0.4211	0.5076
	Reflux	0.8475	2.0202
	Soxhlet	1.4614	4.9628

The yields data in Table 1 presents the highest yield achieved by the soxhlet method with the using of methanol as solvent. The reason for high yield when methanol applied was because of the high solubility in mangiferin [1], so methanol would be easier to dissolve mangiferin than water. On the other hand, the reflux method produced higher yield than maceration due to its material mixed with the solvent and those will face the high temperature. The reason for high temperature excellence is because of the increased of kinetic energy in the solvent will contribute to the lower solvent's viscosity, then ease the diffusional process through the leaves' matrix and the higher intermolecular forces within the solvent causes higher local temperature in leaves' matrix that increase the mangiferin's solubility. This process also supported that the materials do not degrade in high temperature [2]. In addition, it was also observed that the maximum time for *Mangifera indica* extraction using reflux method and water for its solvent happens in 60 minutes.

The existence of mangiferin inside the produced dyes was then analysed qualitatively by using LC-HRMS technique. The qualitative presence of mangiferin can be detected by comparing the mass (m) to charge ratio (z) from the equipment with the literature value [7]. The typical LC-HRMS patterns and the values of m/z of the sample are depicted in Figure 1 and Table 2.

3.2. *Caesalpinia sappan L.*

The extraction of natural dyes from sappan barks was conducted by employing water and ethanol as solvents. The presence of thermolabile chemical such as brazilin in sappan wood that produces red color provide an information that extraction process at low temperature will enhance the yield and stability of the dyes. At a relatively high temperature, the oxidation of brazilin into brazilein is possible that can produce brown colour, instead of red colour as expected. The yields of extraction of sappan barks are shown in Table 1.

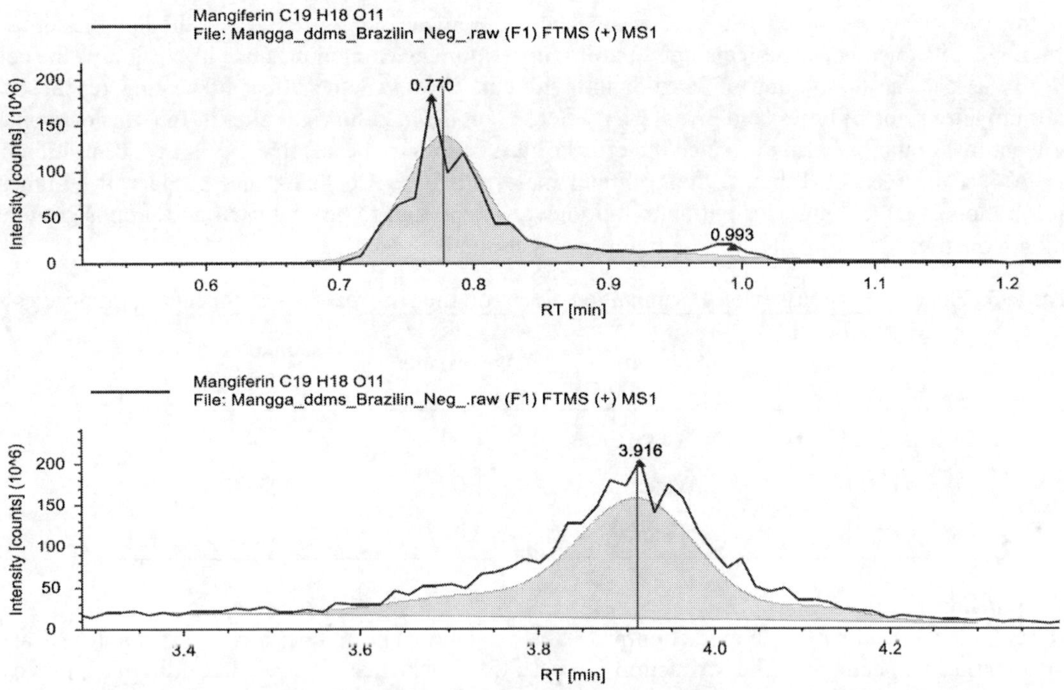

Figure 1. Typical pattern of LC-HRMS analysis of dyes from mango leaf

Table 2. m/z values of dyes from mango leaf

Retention time (minutes)	Area	m/z
0.778	12288712.0399	423.0920
3.912	31460199.2237	423.0914
4.986	2737568.1677	423.0919
5.366	2828636.7901	423.0917
16.738	15665.8189	423.0916

The yields data in Table 1 show that maceration method gives lower yields compared to reflux and soxhlet methods, since reflux and soxhlet employs heating process to boil the solvent while maceration does not. Heating helps rupturing cell wall making it easier for the solvent to dissolve brazillin. In addition, soxhlet method gives higher absorbance compared to reflux method on both solvent. In soxhlet method, the solvent used to extract the solute is always fresh with little or no solute, since this method utilize the vapor phase of solvent. As a result, the driving force of extraction, which is the difference of solute concentration between materials and solvent is always high, producing rapid mass transfer at any time. On the other hand, extracted solute and solvent is being mixed up in reflux method decreasing solute concentration difference resulting on lower mass transfer compared to Soxhlet method.

Although heating boost extraction rate, the prolonged exposure of heat leads to degradation of brazilin, giving red-brownish color instead of red, as brazilin is thermolabile resulting the degradation of brazilin to brazilein, which is brown-colored, in high temperature. Nonetheless, since water boiling point is higher than ethanol, the degradation of brazilin while being extracted using water occurs earlier than the one being extracted using ethanol. The experiment shows that even though brazilin is thermolabile, its tolerance is high enough to utilize heat upon the extraction, since heat ruptures the cell wall and assists the dissolution of brazilin into solvent. However, the effect of heating reaches its maximum after some of times, and gives reversed effect on the brazilin right after it. In terms of sappan to solvent mass ratio, we also evaluated the effect of this ratio into the absorbance and concentration of dyes. As can be seen in Table 3, the optimum mass ratio was 1:6 that could produce the highest concentration of red colour. This optimum ratio is very important to be evaluated and proposed to the fashion industries that currently still use traditional technique.

Table 3. The effect of mass ratio of sappan to solvent on the absorbance and concentration of dyes

Mass ratio of sappan to solvent ratio	Solvent	Absorbance	Concentration (ppm)
1 : 4	Water	1.042	1884.4
1 : 6	Water	1.071	1940.2
1 : 8	Water	0.687	1249.6

4. Conclusion

This research has conducted the extraction processes of yellow and red colours from mango leaf and sappan barks, respectively. The extraction process for each material required different process conditions to optimize the yield of dyes produced. The extraction of mango leaf suggested the high temperature extraction process to produce high yield of dyes. On the other hand, the extraction of sappan barks required a low temperature extraction process due to the presence of brazilin, which has a thermolabile property. In order to be utilised in industrial scale, the utilisation of water is much preferred than organic solvent like methanol and ethanol, hence the optimised ratio of 1 mass of sappan wood and 6 mass of water as solvent was also suggested from the results of this study.

Acknowledgement

The authors gratefully acknowledge the funding support from The Ministry of Research and Higher Education of The Government of Indonesia under the scheme of National Research Competitive of Applied Research 2019 under contract No. 004/SP2H/LT/MULTI/L7/2019.

References

[1] Borges M E, Tejera R L, Díaz L, Esparza P, and Ibáñez E 2012 *Food Chem.* **132** 1855-1860.
[2] Setyopratomo P, Fatmawati A, Sutrisna P D, Savitri E, and Allaf K 2012 *Asia Pac. J. Chem. Eng.* **7** 726-732.
[3] Shinde S S and Chavan A R 2014 *Int. J. Sci. Eng. Res.* **5** 928-934.
[4] Shinde S, Thorat L and Mulay A 2015 *Int. J. Innovative Emerg. Res. Eng.* **2** 22-25.
[5] Sinsawasdi, V K 2012 *Sappanwodd Water Extract: Evaluation of Color Properties, Functional Properties, and Toxicity* (Florida : University of Florida).
[6] Hung T M, Dang N H, and Dat N T 2014 *Biol. Res.* **47** 20.
[7] https://pubchem.ncbi.nlm.nih.gov/compound/Mangiferin#section=Spectral-Information

The Fabrication of Zn-Zr Alloy through Centrifugal Casting and Box Furnace Method for Implant Application

Achmad Fauzi Trinanda*, Moh Waqyan Ghani Fahmi, Rizki Yuni Pratiwi, Ahmad Zakiyuddin*, Sotya Astutiningsih

Universitas Indonesia

*Corresponding author: ahmadzakiyuddin@ui.ac.id

Abstract. Recently, zinc (Zn) is extensively observed as biomaterial or biodegradable implant for medical application. A biomaterial is any substance that has been engineered to interact with biological systems for a medical purpose. Zn is chosen because it has moderate degradation compared to Mg and Fe. Unfortunately, Zn has limitation on its mechanical properties that still low. The Alloying element is needed to improve the mechanical properties of Zn. Alloying element, such as zirconium (Zr) can improve the mechanical properties of Zn. In this study the addition of Zr with the composition variations of 0%, 0.5%, 1%, and 2% were used. Two kinds of alloying method used are centrifugal casting and box furnace with a temperature of 550°C. The aims of this study are to know the effect of each fabrication method to the mechanical properties and the result of Zn-Zr alloy microstructures. The results show that the microstructure of the alloy from centrifugal casting is more homogenous than the box furnace method. The mechanical properties of centrifugal casting are also higher than the box furnace method. The hardness value obtained from centrifugal casting of 0, 0.5, 1, 1.5 and 2% of Zr are 35.162 HV, 41.988 HV, 42.324 HV, 57.112 HV, respectively.

1. Introduction

The number of traffic accidents in Indonesia has been increased. This traffic accident becomes the main reason of some bone fracture chases in Indonesia. It is noted that 50 percents of bone fracture chases are caused by the traffic accident. Not only the traffic accidents, another accident such as the work accident also contributed to the number of bone fracture chases [1]. To overcome these bone fracture chases, doctors are using implant for the replacement of bone structure, while the injured bone is still in the healing process. The implant materials are commonly made of stainless steel and titanium alloy [2].

Stainless steel and titanium alloy has broadly used as implant materials [3]. These materials are chosen to be implant material because of their excellent corrosion resistance in the biological environment. The combination has an exception degree of biocompatibility which it shares with only a handful of other materials. The stainless steel and titanium alloy are good mechanical properties to be used as implant material [4]. However, these types of materials are not biodegradable, makes them need to be extracted after the healing process. This implant extraction process or second operation makes the healing process not effective because of the longer recovery time, even when the bones injure are fully-healed.

So that, to overcome this conventional implant problem, researchers are developing biodegradable implant made of biodegradable materials or metals such as Iron, Magnesium and Zinc alloy[5].Biodegradable metals, also known as bioresorbable or bioabsorbable metals, are compatible with human tissues and degrade to nontoxic by-products through oxidation and dissolution of the metal [6]. Degradation rate of the metals takes the important role to the biodegradable implant. The degradation rate

IOP Conf. Series: Materials Science and Engineering **833** (2020) 012094 doi:10.1088/1757-899X/833/1/012094

of these metals has been extensively studied. Zinc has moderate degradation rate compared to the iron and magnesium alloy [7,8]. This characteristic makes zinc a promising material for biodegradable implant.

Unfortunately, the mechanical properties of zinc are still lower than the bones. It is noted that the ultimate tensile strength of zinc about 30 MPa, while the bone's strength about 150 MPa [9].This property is not enough for implant application. Addition of alloying element into the zinc can improve the ultimate tensile strength of zinc. For example, addition of 3wt% copper into zinc can improve the mechanical properties of zinc up to 257 MPa, with 47% of elongation and 0.372 $\mu A/cm^2$of degradation rate. However, the overdose of copper content inside the human body can cause some neurodegenerative diseases such as Alzheimer, Menkes, and Wilson [10].

Therefore, another alloying element, zirconium is used in this research. Zirconium as alloying element for zinc is still rarely observed by the researchers. Zirconium is a biocompatible and bioneutral element. The Zirconium also has low systemic toxicity. Based on the Zn-Zr phase diagram, with a little addition of zirconium into zinc can form Zn-rich intermetallic phase ($Zn_{22}Zr$). The number of dissolved zirconium inside Mg-Zn-Zr alloy is 50 µg/day and is still allowable in human body [11-13]. These potentials make zirconium a promising alloying element to improve zinc mechanical properties for implant application.

2. Experimental Method

This research is started with the making of samples through the centrifugal and box furnace casting. The pure zinc and zirconium materials were obtained from Alfa Aesar. The zinc was mixed with the zirconium with variation of zirconium of 0.5, 1.0, 1.5, and 2.0 wt%. Each variable was heated at 550 ^{O}C for 15 minutes and poured into the mold until solidified for the box furnace casted samples. The molten metal that will be casted through centrifugal casting method was rotated in the mold with speed of 500 rpm using Neutrodyn Easyti machine as shown in Figure 1. The casted samples then homogenized at 380 ^{O}C and followed by the quenching process using medium water to make homogeneous grain size. \

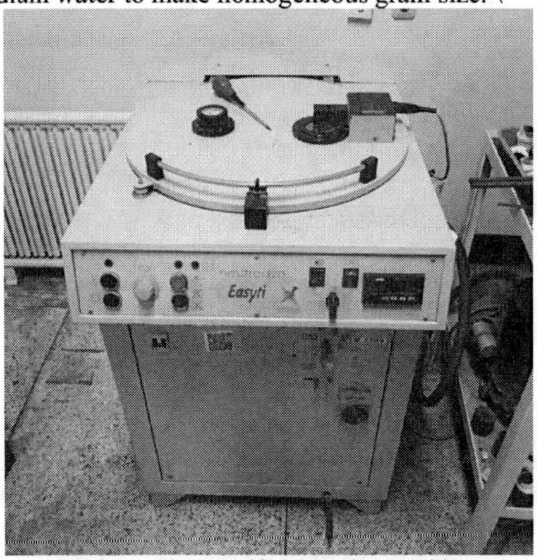

Figure 1. The centrifugal casting machine.

Then, the homogenized samples are prepared for the Vickers hardness test and SEM-EDS examination. The samples are cut for easier handling during the preparation process. The samples are grinded by using sand paper with grit number of 240, 320, 600, 800, 1000, 1200 and 2000. Then, followed by the polishing process used polishing machine and alumina paste to get fine surface. The polished samples etched with the mixture of 2 ml of HNO_3 and 98 ml of H_2O then examined under the Vickers hardness test and SEM-EDS.

3. Results and Discussion

3.1. Microstructure

After the addition of zirconium into the zinc, it is found that the microstructure of zinc has changed. The SEM examination in Figure 2 using secondary electron shows that after the addition of zirconium there is formation of some particles. These particles are intermetallic phase which is formed after the addition of zirconium.

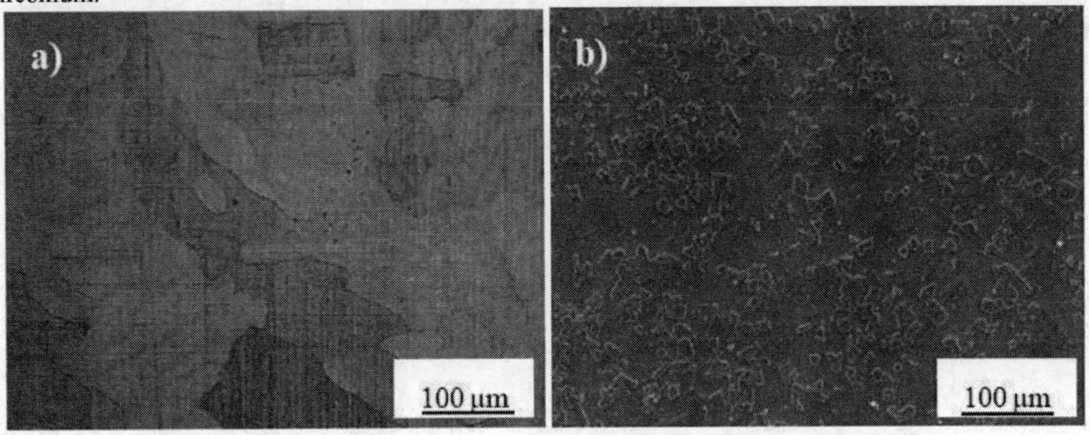

Figure 2. The microstructure of zinc; a) pure zinc, b) zinc with zirconium addition.

Figure 3 shows the differences of microstructures resulted from two different processes, the centrifugal casting and box furnace casting. It is found that the centrifugal casting gives more homogeneous microstructure than the box furnace casting method. The intermetallic particles are well-distributed all around the zinc matrix. This condition happened because of the centrifugal force in centrifugal casting that makes the intermetallic particles well-distributed.

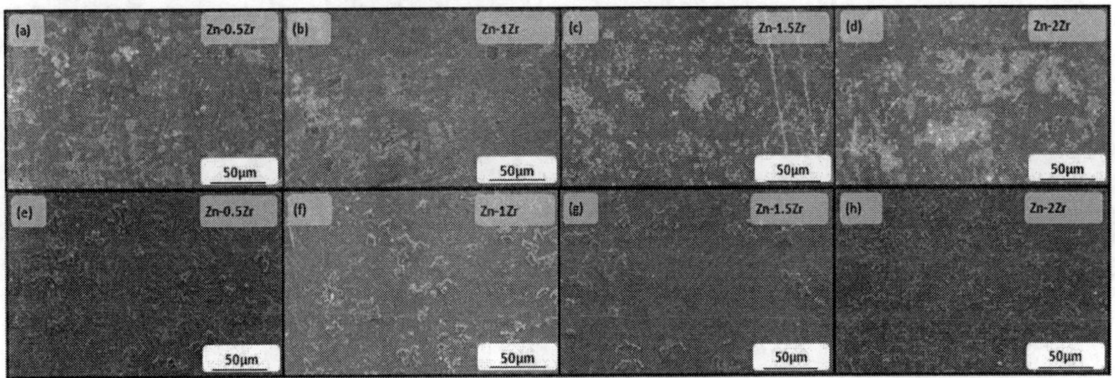

Figure 3. The microstructures of (a-d) box furnace casted and (e-h) centrifugal casted alloy.

The centrifugal casting is also cleaner than the box furnace casting, it has a closed process thus there is no time for the molten metal to interact with the environment. There is no white region in the centrifugal casted sample's microstructure, but some white regions inside the box furnace casted sample's microstructure is found. These white regions are the oxide that is formed when the casting process was done by the box furnace method.

$$2\ Zn + O_2 \rightarrow 2\ ZnO \qquad (1)$$

In contrast with the centrifugal casting, the box furnace casting method is an open process, which molten metal can interact with the atmosphere and then the molten zinc reacted with oxygen, formed oxides as the equation 1.

3.2. Vickers Hardness

The addition of zirconium into zinc can improve the hardness of zinc. It is found that the hardness increased with the increment of zirconium content inside zinc. The optimum strength is reached by the addition of 2 wt% zirconium with Vickers hardness number of 57 VHN (Figure 4).

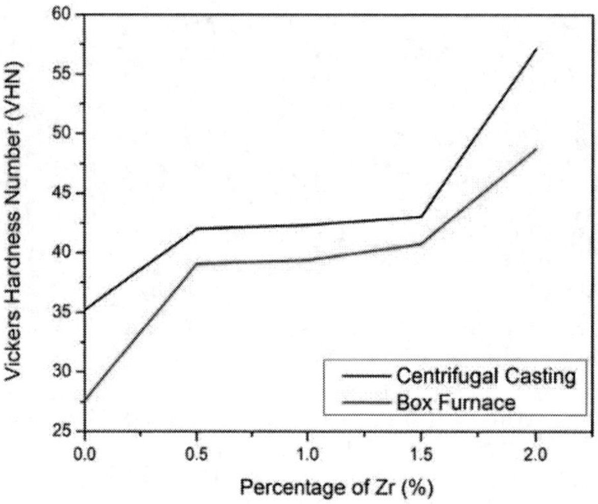

Figure 4. The Vickers hardness number of centrifugal casted and box furnace casted alloy.

The increment of hardness is caused by the existence of intermetallic particles. These particles have good adhesion with the zinc matrix, thus in can hinder the dislocations movement and strengthen the material. The addition of zirconium also makes the grain size of the alloy is finer. Finer grain size can strengthen the material because the number of grain boundaries that can hinder the dislocations movement also increase. The addition of zirconium inside the zinc made the material becomes stronger.

3.3. Energy Dispersive Spectroscopy (EDS)

The energy dispersive spectroscopy is used to investigate the composition inside some particles that are formed after the alloying process. Figure 5 shows the EDS result of Zn-2Zr which is made through centrifugal casting.

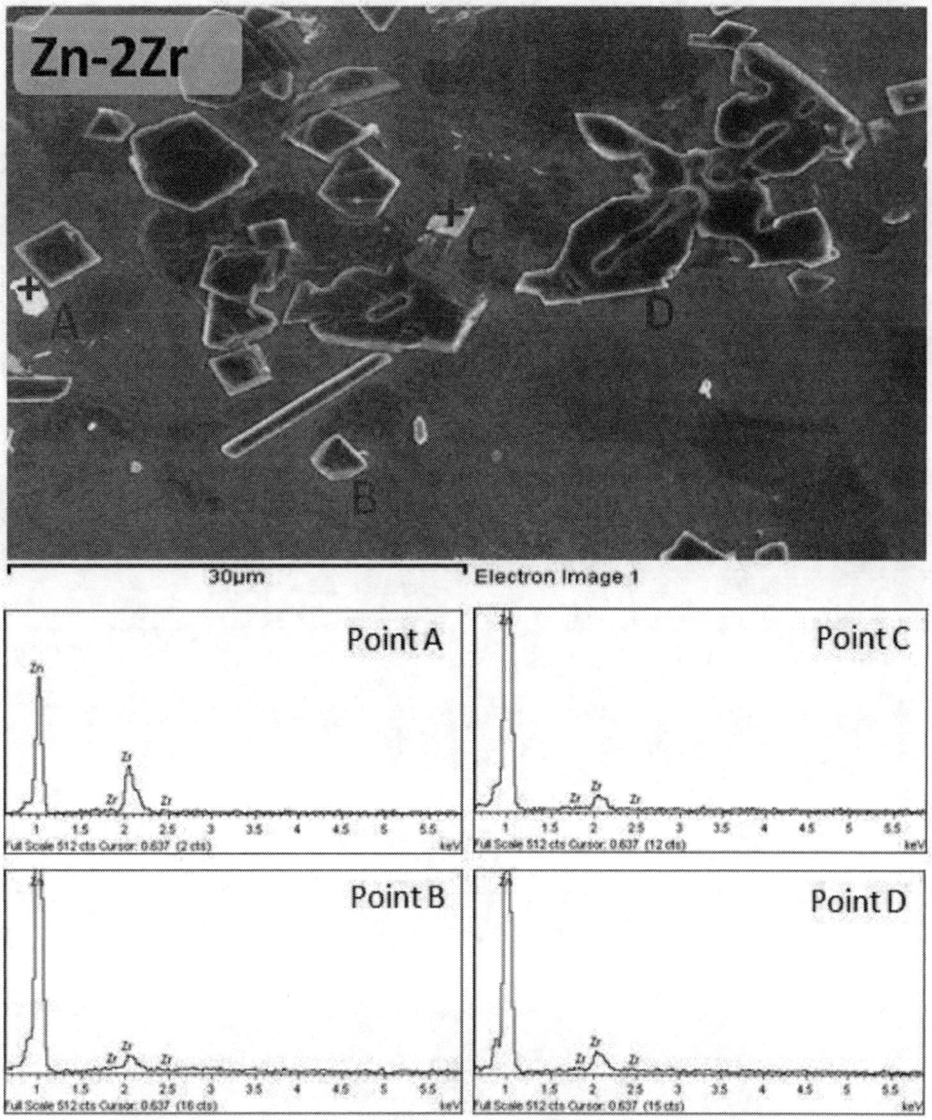

Figure 5. The EDS spectroscopy of centrifugal casted Zn-2Zr alloy.

The EDS spectroscopy of centrifugal casted alloy confirmed that the particles are formed after the addition of zirconium. The spectroscopy shows only zinc and zirconium inside the particles. This indicates that the particles formed are $Zn_{22}Zr$. The solubility of zirconium inside zinc is very limited, so the addition of zirconium inside zinc will lead to the formation of intermetallic phases [13].

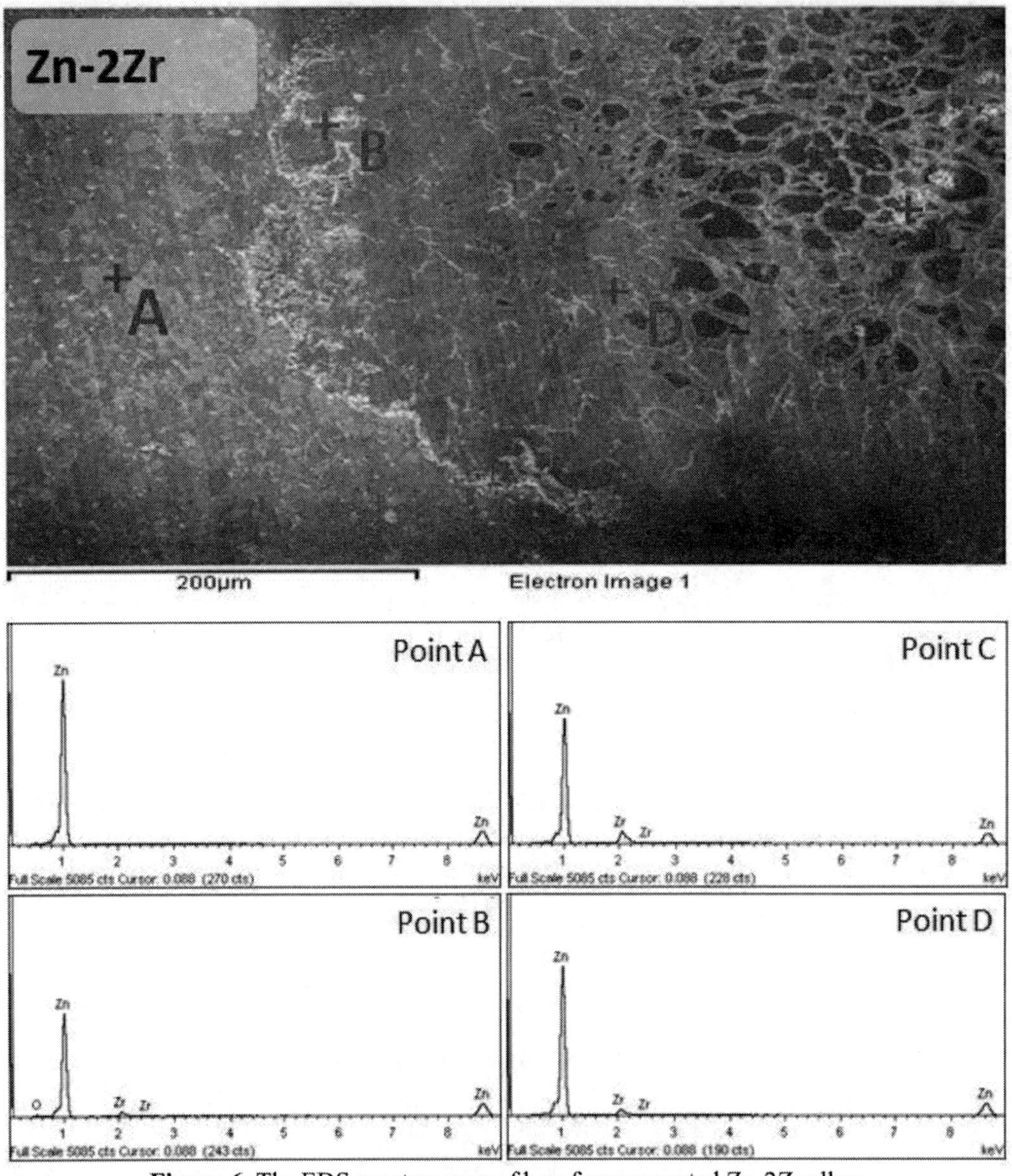

Figure 6. The EDS spectroscopy of box furnace casted Zn-2Zr alloy.

Figure 6 shows the spectroscopy of box furnace casted alloy. The spectroscopy of box furnace casting shows the existence of oxygen inside the alloy. This result indicates the formation of oxides with the box furnace method. As mentioned before, the box furnace method is an open process. This process let the molten zinc reacted with oxygen and formed oxides.

4. Conclusions

The fabrication of Zn-Zr alloy has been done successfully. The microstructure of zinc changed after the addition of zirconium with the formation of intermetallic phases. These intermetallic phases or particles are formed because the solubility of zirconium inside zinc is very low. Thus, the addition of zirconium inside zinc will lead to the formation of intermetallic phases. The addition of zirconium also gives finer grain size compared than the zinc without any addition of zirconium. The addition of zirconium can improve the hardness of zinc. The optimum hardness is reached by the centrifugal casted alloy with addition of 2 percent zirconium with Vickers hardness number of 57 VHN.

The centrifugal casting method gives more homogenous microstructure of Zn-Zr alloy. The intermetallic phases that are formed in this process are well-distributed due to the centrifugal force. The alloy from this process also gives higher Vickers hardness number than the box furnace casting because the intermetallic phases that act as reinforcement and grain refinement are well-distributed, thus it can strengthen the material. The EDS spectroscopy also shows that the centrifugal casting is cleaner than the box furnace casting. There is no oxygen spectroscopy inside centrifugal casted alloy, this is indicating there is no oxides formed with this process.

Acknowledgement
This publication is supported by Hibah PITTA-B Number 2535/UN2.R3.1/HKP.05.00/2018 from Universitas Indonesia.

References

[1] Amrc.org.hk 2019 [online] Available at: https://www.amrc.org.hk/sites/default/files/Indonesia_1.pdf [Accessed 10 Nov. 2019]
[2] Davis J R 2003 Handbook of Materials for Medical devices ASM International
[3] Shrivastava S 2004 Medical Device Materials Proceedings of the Materials & Processes for Medical Devices Conference ASM International
[4] AZoM.com 2019 *Stainless Steel and Titanium in Surgical Implants* [online] Available at: https://www.azom.com/article.aspx?ArticleID=7156 [Accessed 10 Nov. 2019]
[5] Li H F, Xie X H, Zheng Y F, Cong Y, Zhou F Y, Qiu K J, Wang X, Chen S H, Huang L, Tian L and Qin L 2015 *Scientific Reports* **5**
[6] Murni N, Dambatta M, Yeap S, Froemming G and Hermawan H 2015 *Materials Science and Engineering: C* **49** 560–6
[7] Liu L, Gebresellasie K, Collins B, Zhang H, Xu Z, Sankar J, Lee Y-C and Yun Y 2018 *Applied Sciences* **8** 1459
[8] Su Y, Cockerill I, Wang Y, Qin Y-X, Chang L, Zheng Y and Zhu D 2019 *Trends in Biotechnology* **37** 428–41
[9] Vojtěch D, Kubásek J, Šerák J and Novák P 2011 *Acta Biomaterialia* **7** 3515–22
[10] Tang Z, Huang H, Niu J, Zhang L, Zhang H, Pei J, Tan J and Yuan G 2017 *Materials & Design* **117** 84–94
[11] Dutkiewicz J 1992 *Journal of Phase Equilibria* **13** 430–3
[12] Hong D, Saha P, Chou D-T, Lee B, Collins B E, Tan Z, Dong Z and Kumta P N 2013 *Acta Biomaterialia* **9** 8534–47
[13] Wątroba M, Bednarczyk W, Kawałko J and Bała P 2018 *Materials Characterization* **142** 187–94

IOP Conf. Series: Materials Science and Engineering **833** (2020) 012095 doi:10.1088/1757-899X/833/1/012095

Influence of Activation Time on Ibuprofen Adsorption Using Zinc Oxide from Gelatin Templating Method

Maria Ulfa[1,*] and Umi Wahidatul Latifah[1]

[1]Study Program of Chemistry Education, Faculty of Teacher Training and Education, Sebelas Maret University, Jl. Ir. Sutami 36A, 57126 Surakarta, Central Java Indonesia

*Corresponding author: ulfa.maria2015@gmail.com

Abstract. The present work aimed to investigate the performance of activated zinc oxide for ibuprofen adsorption. The zinc oxide was synthesized by the templating method using F127 and gelatine as a soft template and sulfuric acid as a catalyst. The zinc oxide was activated by an HCl solution 1M for 1, 8, and 24 hours. The raw and activated zinc oxides were characterized by XRD and by infrared spectroscopy. The HCl treatment increases both of the numbers of Zn-O-Zn groups and peak of Zn-O from diffractogram but decreases hydroxyl group content. Ibuprofen adsorption studies of kinetics and isotherms were carried out at room temperature with solid-liquid ratio 3:5(v/v) and water-hexane solution. The adsorption properties were correlated to the activated conditions. The isotherms of adsorption were better reproduced by Langmuir–Freundlich models using activated zinc oxide by HCl at 8 hours. The best performance of ibuprofen adsorption resulted from activated zinc oxide at 8h using water-hexane 1:9 as ibuprofen solvent. The activation time and a high ratio of water-hexane both influence the adsorption of ibuprofen performance.

1. Introduction

Contamination of the water environment by natural compound waste as well as pharmaceuticals and personal care products (PPCP) is widely recognized as one of the serious environmental problems. PPCP is a substance used to treat and prevent diseases in humans and animals. The most used compound in this group is NSAIDs, including ibuprofen, diclofenac, and naproxen. The substance can enter the environment by the excretion of a compound that does not react, followed by the disposal of waste. In the long run, serious health problems can occur in humans who have consumed water that has been contaminated with waste pharmaceuticals, including ibuprofen [1]. Non-steroidal anti-inflammatory drugs (NSAIDs) are over-the-counter medications and commonly prescribed by physicians. This class of drugs has analgesic, anti-inflammatory, and anti-pyretic properties so that they are widely used. Ibuprofen is a potential hazard to human health due to this numerous pharmaceutical residue in aquatic area with [2].

Previous studies have discussed some of the methods that are used to eliminate the residual NSAIDs that contaminate water. Some of these methods are biological degradation, wet catalytic oxidation, and ozonation. Adsorption is one of the alternative methods; it is simple, of low cost, and potentially effective for eliminating residual NSAIDs, including ibuprofen, from environmental waters [3].

There are lots of materials used to eliminate the residual NSAIDs in waters, such as carbon active, biosorbentofchitin and lignin, nanoparticles silica magnetic or graphene oxide, metal-organic frameworks, etc. Some carbon-active modifications of waste husk rice, zeolite, fly ash, and coconut shell can also be used as an adsorbent for eliminating residues of NSAIDs with low cost [4].

Photocatalytic semiconductors, such as zinc oxide, if activated to become the adsorbent, also show great adsorption ability. In maximizing the process of adsorption, adjustments were carried out, one of them by adding activator on metal oxides that aim to reach the volume of pores and the spacious surface of the adsorbent [5,6].

Zinc oxide (ZnO) could activated in several ways, such as using photons or light so Zinc oxide can be a photocatalyst, addition of activated carbon, zeolite, or certain solutions so that zinc oxide can be used as a good adsorbent [7]. Based on previous research, zinc oxide, which has been activated, can be used as an adsorbent for wastewater to adsorb toxic dyes, pollutants [8][9].

Zinc oxide is the material of metal oxides that are usually visibly white and not soluble in water and alcohol. Zinc oxide is one of the semiconductor inorganics that are not toxic and can provide mobility and height and whose thermal stability is good. Zinc oxide has a band distance of eV with a bond energy of 60 meV at room temperature, the most stable structure of zinc oxide is wurtzite [10][11].

Activation of zinc oxide is done so that zinc oxide can be used as an adsorbent. Through activation, the surface morphology of zinc oxide experiences changes, namely an increase in the size of the micro-pore and in the adsorption of substances. The length of time of activation affects the size of the pores of micro, making an impact on power adsorption, where the increasing size of the pores so the power adsorbed increasingly large so that will be a lot of substance that is able in adsorption application[12][13]

In this study, the formation of zinc oxide was investigated using Pluronic F127 with gelatin as co-template via the hydrothermal method. Pluronic F127 is a soft organic template with a relatively high dissolution rate and low mechanical strength. The zinc oxide was characterized using the X-ray diffraction method (XRD) and infrared analysis. The potential use of synthesized zinc oxide as an adsorbent for drug removal was investigated using ibuprofen as a probe molecule.

2. Materials and Methods

2.1. Materials

In the present research study, all of the starting materials obtained commercially from the authentic companies and used without further purification. Herein, zinc sulfate ($ZnSO_4$), ethanol (EtOH), F127 and gelatin as starter materials and sulfuric acid as an activation solution.

2.2. Synthesis of ZnO

EtOH is dropped into the F127 solution with a burette (1 drop of EtOH per 20 seconds) at room temperature). After the F127 has been dropped, the mixture is left in the stirrer for 4 hours 150 rpm. Closed condition (crepe plastic). When the droplet is attempted, the air is prevented from entering by making crepe plastic drop holes and covering them out with a mask. Add gelatin then stir for 60 minutes.

$ZnSO_4$ was dropped on the solubility of F127 with a burette (1 drop per 10 minutes) at room temperature and room conditions. After everything has been dropped, the mixture is stirred for 20 hours of 150 rpm closed condition (crepe plastic).

During the droplet, the air is prevented from entering by making holes in the crepe plastic and covering the mouth; the mixture is put in the hydrothermal reactor and then the oven at 100°C for 24 hours and then allowed to cool. The white solid is filtered and washed with distilled water so that pH=7. The solid is then dried at 100°C for 24 hours and then calcined at 550°C for 12hours; then the solid is stored in a clear, closed bottle.

2.3. Activation of ZnO

Samples that have been synthesized, then activated using HCl solution with a variation of the time of activation are different. Sample ZnO added with HCl solution in the state is stirred using magnetic stirrer with heating for 1, 8, and 24 hours. Then the samples were washed using distilled water until a pH between 6 and 7.

Sample ZnO was subsequently filtered and then dried in an oven at a temperature of 110°C. Samples ZnO that been washed with water were dried and allowed to stand at the temperature of space in the state closed, thus the obtained samples of ZnO were activated.

2.4. Characterization

Sample ZnO was activated using a solution of HCl with the variation of time, then characterized using a Fourier Transform Infrared (FTIR) spectrophotometer brand Shimadzu 2100 resolution of $0.5cm^{-1}$ X-ray diffraction (XRD). This characterization aims to determine the presence of functional groups in the ZnO sample compound as well as the crystalline from the sample. Subsequently, the samples that have been characterized were used to ibuprofen adsorption.

2.5. Ibuprofen Adsorption

To find out the adsorption of ibuprofen with activated ZnO adsorbents, the solution of ibuprofen was prepared in a solution of Water: Hexane in a ratio of 10:0; 9:1; 5:5; 1:9; 0:10 concentrates 100 ppm. The concentration of ibuprofen was monitored with a UV-Vis spectrophotometer at the wavelength of 291.5 nm. A total of 60 mg of nano-ZnO was added to the solution of ibuprofen and stirred at a temperature chamber for 60 minutes. A total of 3 mL solution of ibuprofen is taken every interval of 5 minutes for 1 hour, and then the absorbance is measured using a spectrophotometer UV-Vi spectrophotometer Hitachi U-2000 Japan.

Calibration was done by preparing a standard solution of ibuprofen with a concentration of 5 to 100 ppm. The amount of ibuprofen adsorbed by ZnO was calculated using the equation. C_0 is the concentration of the initial ibuprofen, Ct is the concentration equilibrium of ibuprofen, v is the volume of solution that is tested absorbance, and w is the number of ZnO used for adsorption.

$$qe = (Co - Ct)\left(\frac{v}{w}\right) \tag{1}$$

3. Result and Discussion

3.1. Characterization of Zinc Oxide

The XRD pattern on the zinc oxide with different activation time is shown in Figure 1b. The X-Ray Diffraction pattern of the ZnO which is activated for 1 h (labelled as ZnO-1h) shows three strong diffraction peaks at of 2θ (18.862°, 26.398°, 29.394°, 35.864°), ZnO-8h show strong peaks at of 2θ (18.801°, 26.337°,29.424°, 35.762°) and ZnO-8h shows strong peak at 2θ (20.504°, 26.435°,31.026°, 35.901°). All the peak of sample nothing shows differences significantly which is described that crystallinity of all sample has been stable in different of activation time. All XRD of sample be indexed to the known structure of the ZnO and hexagonal phase compare to the ZnO hexagonal wurtzite phase that shown in Figure 1(c) [14]. The best diffractogram results are ZnO with a 24-hour activation time. The diffractogram results for the activation time under 24 hours show a lower peak when compared with the 24 hours activation time, as shown in Figure 1.

The FTIR spectra of all sample of ZnO show the similar peak as functional group content. in Figure 2, shows that the general peak just of ZnO-24h that means that zinc hydroxyl groups have a huge degree of an ion characteristic. FTIR of ZnO-1h and ZnO-8h not shown. The broad peak centered at 3450 cm^{-1}, which corresponded to the stretching vibration of the hydroxyl group, was observed in all carbon materials. The presence of surface hydroxyl reflects the number of carbon terminal bonded to an oxygen atom that may be due to the formation of surface defects. The peak at wavelength 1300 cm^{-1} shown there is Zn-O-Zn which is mean that zinc oxide synthesis has been successfully done in this research.

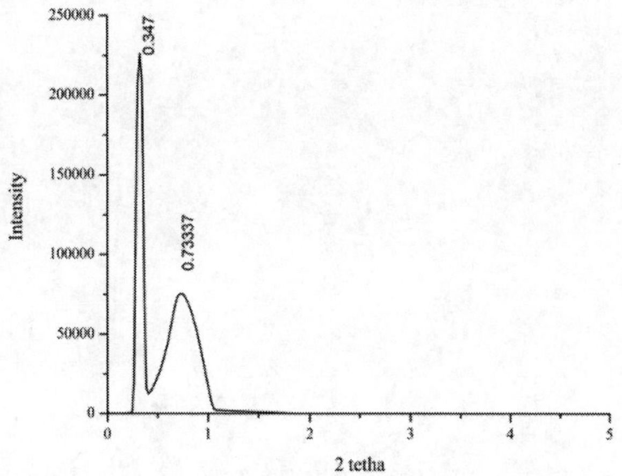

(a) XRD pattern of ZnO with 24 hours activation time at 0-5".

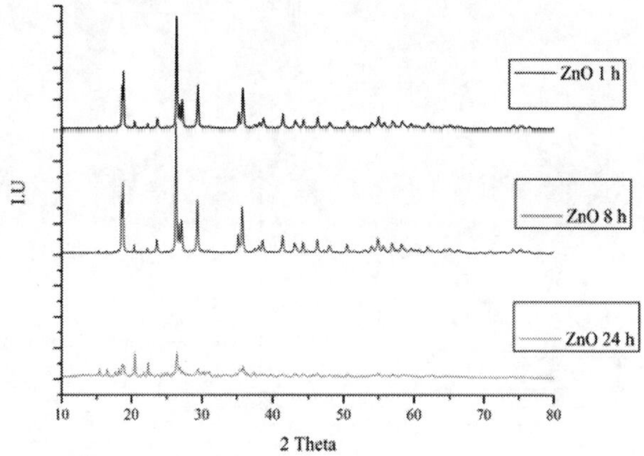

(b) XRD patterns of ZnO with 1, 8 and 24 hours activation time at 10-80".

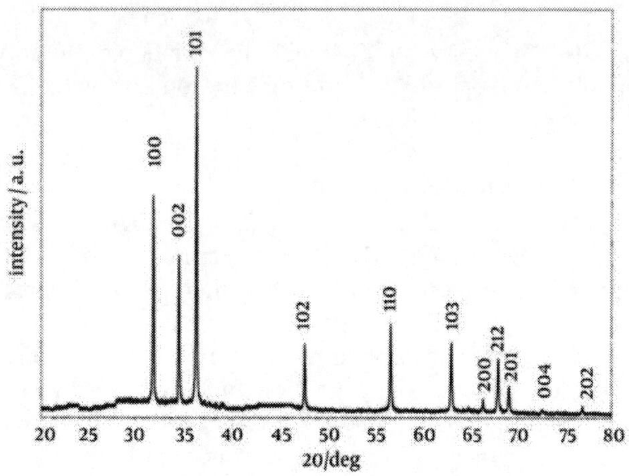

(c) The XRD pattern of zinc oxide hexagonal phase.

Figure 1. (a) dan (b) Diffractogram result of Zinc Oxide with XRD at 2 Theta with 24 hours activation time and (c) The XRD pattern of zinc oxide hexagonal phase.

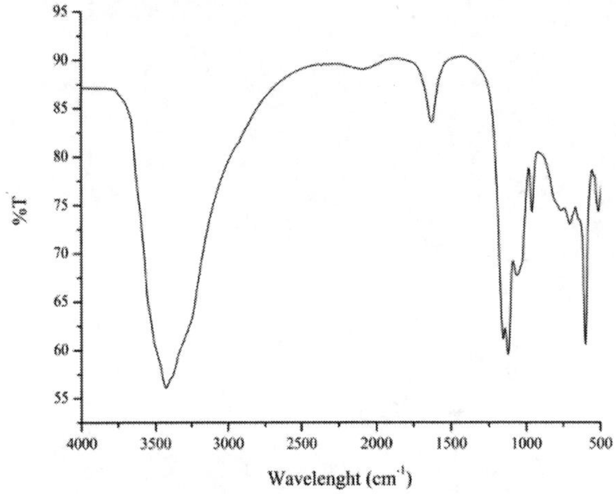

Figure 2. Characterization results of Zinc Oxide 24 hours activation time with FTIR.

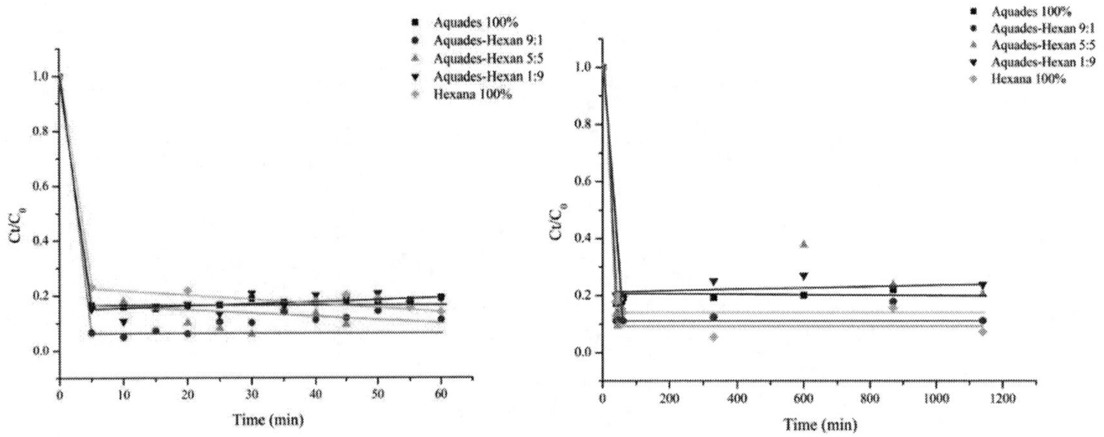

(a) (b)

Figure 3. (a) Ibuprofen Adsorption curve on ZnO with the variation of the solvent ratio between water and hexane during 60 minutes (b) Ibuprofen Adsorption curve on ZnO with the variation of solvent ratio between water and hexane during 1440 minutes.

3.2. Adsorption studies of ibuprofen on Zinc Oxide

The adsorption profile of ibuprofen at ZnO 24 hours, as shown in Figure 3, is activated by HCl obtained from the variation of water to the hexane ratio. High adsorption of ibuprofen was observed on Zinc Oxide using solvent variation 1:9 water to hexane ratio with 77.2% of ibuprofen adsorbed from initial concentration.

The best performance of ibuprofen adsorption resulted from activated zinc oxide at 24 hours using water-hexane 1:9 as ibuprofen solvent. The activation time and a high ratio of water-hexane both influence the adsorption of ibuprofen performance. The effect of the activation time is that the activation time will open the surface of ZnO, thereby increasing the micro-pore size and thus increasing the direct absorption of ibuprofen.

4. Conclusion

The best performance of ibuprofen adsorption results from activated zinc oxide at 24 h using water-hexane 1: 9 as ibuprofen solvent. The length of time will increase the micropore size of ZnO so that it can adsorb more ibuprofen. The best Ct / C0 value of each variation of the existing solvent is pure water, hexane 10%, 50%, 90% and 100% are 0.1661; 0.0726; 0.1552; 0.1689 and 0.1678, respectively. Not only the activation time but also a high ratio of water-hexane influences the adsorption of ibuprofen performance.

Acknowledgment
Authors acknowledge the Research Collaboration Indonesian Program (PPPKI) of Sebelas Maret University 2019 under contract Number 1362/UN27.21/PN/2019 for Maria Ulfa.

References

[1] W. Phasuphan, N. Praphairaksit, And A. Imyim 2019 *J. Mol. Liq.* **294** 111554.
[2] A. P. Zahra and Novita Carolia 2017 *J. Majority* **6** 153–158.
[3] J. C. Lancheros, C. A. Madera-Parra, A. Caselles-Osorio, W. A. Torres-López, And X. M. Vargas-Ramírez 2019 *J. Ecol. Eng.* **135** 89–97
[4] F. H. Pusceddu *Et Al.* 2018 *J. Environ. Pollut.* **232** 274–283
[5] A. A. Abdullah, T. A. Saleh, S. A. Ganiyu, M. N. Siddiqui, And K. R. Alhooshani 2017 *J. Anal. Appl. Pyrolysis* **128** 246–256
[6] J. Saini, V. K. Garg, R. K. Gupta, And N. Kataria 2017 *J. Environ. Chem. Eng.* **5** 884–892
[7] C. Yang, S. Yang, H. Fan, Y. Wang, And J. Shangguan 2019 *J. Colloid Interface Sci.* **555** 548–557
[8] N. Ghasemi And S. Rohani 2019 *J. Mol. Liq.* **285** 252–269
[9] S. Chaudhary, Y. Kaur, A. Umar, And G. R. Chaudhary 2016 *J. Mol. Liq.* **224** 1294–1304
[10] T. E. P. Alves, C. Kolodziej, C. Burda, And A. Franco 2018 *J. Mater. Des.* **146** 125–133
[11] K. M. Fang, Z. Z. Wang, M. Zhang, A. J. Wang, Z. Y. Meng, And J. J. Feng 2013 *J. Colloid Interface Sci.* **402** 68–74
[12] L. Huber *Et Al.* 2019 *J. Microporous Mesoporous Mater.* **276** 239–250
[13] M. Sharma *Et Al.* 2019 *Chem. Eng. J.*, **358** 540–551
[14] S. Karim Fatah 2018 *Diyala J. Pure Sci.* **14** 40–47

IOP Conf. Series: Materials Science and Engineering **833** (2020) 012096 doi:10.1088/1757-899X/833/1/012096

Ibuprofen Adsorption Study by Langmuir, Freundlich, Temkin and Dubinin-Radushkevich Models Using Nano Zinc Oxide from Mild Hydrothermal Condition

Maria Ulfa[1] and Yuli Iswanti[1]

[1]Study Program of Chemistry Education, Faculty of Teacher Training and Education, Sebelas Maret University, Jl. Ir. Sutami 36A, 57126 Surakarta, Central Java Indonesia

[a]Corresponding author: ulfa.maria2015@gmail.com

Abstract. Ibuprofen in the aquatic area as inflammatory drug residue is dangerous for environmental sustainability. The removal of ibuprofen at the aquatic area is carried out on nano zinc oxide (NZO) as an adsorbent. The NZO adsorbent was synthesized by gelatin self-assembly as a green method and characterized by FT-IR and XRD techniques. The residual ibuprofen model was prepared by dissolution of ibuprofen in water-hexane at ratio 10:0; 1:9; 5:5, 9:1 and 0:10. The optimized batch of experimental parameters were 100 ppm for 55 min at room temperature. Adsorption data was investigated by Langmuir, Freundlich, Temkin and Dubinin-Radushkevich models. The result shows that Temkin model is the best model for describing the adsorption mechanism between ibuprofen as adsorbate and zinc oxide as an adsorbent. The adsorption capacity of 9:1 is the highest on that ibuprofen adsorption indicated that the composition of both solvents is favorable for ibuprofen dissolution due to polar-nonpolar interaction which is reaching 92% removal of ibuprofen. In the future, zinc oxide as adsorbent may be used for the removal of ibuprofen from any aquatic core area due to the economic and environmental reason.

1. Introduction

The presence of drugs in water is dangerous to humans and the environment. Ibuprofen is the drug which most often used by humans and found mostly in water. If the concentration in water is > 1 μg / L, it will endanger existing microorganisms, which can also affect the human health [1]. The presence of this drug in water can disrupt the balance of the ecosystem and the residue can modify the genetic information of natural bacteria so that resistance occurs [2]. Several ways have been used to reduce the concentration of ibuprofen in water such as ultrasonic irradiation [3], ozonation [4], chlorination [5], biodegradation [6], and adsorption [7]. Adsorption is the easiest and cheapest way to reduce the concentration of ibuprofen in water. However, the adsorption regeneration process is required when adsorption is needed. In addition, pollutants that have been adsorbed cannot be degraded into other compounds that are not harmful. Heterogeneous photocatalysis with activated semiconductors such as zinc oxide (ZnO) is very possible for the process of decomposition of hazardous substances (pollutants) [8]. By using nano-sized zinc oxide (nano-ZnO), it is expected that the adsorption process of ibuprofen can run optimally.

The adsorption data analysis in this study uses the Langmuir, Freundlich, Temkin, and Dubinin-Radushkevich models. The Langmuir model assumes uniform energies of adsorption onto the surface and no transmigration of adsorbate in the plane of the surface. The Freundlich model commonly used to describe the adsorption characteristics of the heterogeneous surface. The Temkin model assumes that the heat of adsorption (a function of temperature) of all molecules in the layer would decrease linearly rather than logarithmic with coverage by ignoring the extremely low and large value of concentrations. Dubinin–Radushkevich isotherm is generally applied to express the adsorption mechanism with a Gaussian energy distribution onto a heterogeneous surface [9]

Based on the description above, a study of ibuprofen adsorption using nano-ZnO was carried out with adsorption data analysis using the Langmuir, Freundlich, Temkin, and Dubinin-Radushkevich models.

2. Materials and Method

2.1. Material

The chemical in this research are analytical grade. The material are zinc sulfate ($ZnSO_4$), ethanol (EtOH), F127, and gelatin as precursor and sulfuric acid as an activation solution.

2.2. Synthesis of ZnO

The ZnO synthesis was started by making 190 mL of ethanol solution with a ratio of ethanol and water 1: 3 then added to the 31-grams of $ZnSO_4$ base material. Then react with 3.6 grams of F127. Ethanol was dropped into the F127 solution with a burette (1 drop per 20 seconds) at room temperature. After all, ethanol has been dropped, the mixture is stirred for 2 hours 250 rpm closed condition (crepe plastic). When the drip is attempted, no air enters by making a drip hole in the crepe and covering the mouth with a mask. Then, the mixture was added with 0.0346 grams of gelatin and then sterilized for 60 minutes (solution A).

After that, $ZnSO_4$ solution was added to solution A with a burette (1 drop per 10 minutes) at room temperature stirrer conditions. After everything is dropped, the mixture is stirred for 20 hours 150 rpm closed condition (crepe plastic). During the dripping effort, there is no air entering by making holes in the plastic crepe and covering the mouth with a mask.

The mixture is put into a hydrothermal reactor while roasted at 100^{0}C for 24 hours and then cooled. One other part is left at room temperature. After that the mixture is filtered, the solid is dried in an oven at 100^{0}C for 24 hours then calcined at 550^{0}C for 12 hours. Then the solids were stored in a clear, closed bottle.

2.3. Characterization of ZnO

The resulting sample will then be characterized using the Fourier Transform Infrared (FTIR) instrument brand Shimadzu 2100 resolution 0.5cm^{-1} and X-Ray Diffraction (XRD). This characterization aims to determine the size and morphology of the particles. Then the sample was applied in ibuprofen adsorption.

2.4. Adsorption of Ibuprofen on nano-ZnO

To find out the adsorption on nanomaterial, ibuprofen solution was prepared in a solution of water: hexane in a ratio of 10:0; 9:1; 5:5; 1:9; and 0:10 concentrates of 100 ppm. The concentration of ibuprofen was monitored with a UV-Vis spectrophotometer at a wavelength of 291.5 nm. Calibration was done by preparing a standard solution of ibuprofen with a concentration of 5 to 100 ppm. 60 mg of nano-ZnO was added to the ibuprofen solution and sterilized at room temperature for 60 minutes. A total of 3 mL of ibuprofen solution was taken at 5 minutes intervals and then absorbance was measured using a Hitachi Japan UV-Vis U-2000 spectrophotometer. The amount of ibuprofen absorbed by nano-ZnO is calculated using the equation

$$qe = (Co - Ce)\left(\frac{v}{w}\right) \quad (1)$$

Co is the initial concentration of ibuprofen, *Ce* is the equilibrium concentration of ibuprofen, v is the volume of the solution tested for its absorption, and w is the amount of nano-ZnO used for adsorption.

Changes in the concentration of adsorbate by the adsorption process in accordance with the mechanism of adsorption can be studied through the determination of the adsorption isotherm. The adsorption data analysis in this study uses the Langmuir, Freundlich, Temkin, and Dubinin-Radushkevich models [10]

Langmuir equation: $\frac{1}{\mathrm{qe}} = \frac{1}{\mathrm{qmax}} + \frac{1}{\mathrm{qmax\ Kl}} \times \frac{1}{\mathrm{Ce}}$

Freundlich equation: $\mathrm{Log\ qe} = \frac{1}{\mathrm{n}}\mathrm{Log\ Ce} + \mathrm{Log\ Kf}$

Temkin equation: $q_e = \frac{RT}{b} \ln A + \frac{RT}{b} \ln Ce$

Dubinin-Radushkevich equation: $\text{Ln } q_e = \text{Ln } q_m - \beta\varepsilon^2$

Where qmax, the monolayer capacity of the adsorbent (mg/g); KL, the Langmuir constant (L/mg) and related to the free energy of adsorption; qm, the theoretical saturation capacity (mg/g); and ε, the Polanyi potential, which is equal to RT Ln (1 + (1/Ce)), where R (J/mol K) is the gas constant and T (K) is the absolute temperature; β, a constant related to the mean free energy of adsorption per mole of the adsorbate (mol^2/kJ^2).

3. Result and Discussion

3.1. Characterization of nano-ZnO

Analysis of zinc oxide using X-Ray Diffraction is shown in Figure.1 Characterization result of nano-ZnO with 0-800. The XRD pattern on ZnO synthesized showed the diffraction peaks at 2θ= 270. When nano-ZnO were produced at low concentration of gelatin, high order geometric nanoparticle was formed based on relatively small full width at half maximum intensity of (002) plane. The graphitization degree of the nano-ZnO changed with the concentration of gelatin evident by the decrease in the intensity and FWHM of the peaks which indicates the influence of gelatin in controlling the physical and chemical structure of the nanoparticle.

The synthesized of ZnO were also characterized using infrared analysis. Figure 2 showed that the broad peak centered at 3750 cm^{-1}, which is corresponded to the stretching vibration of the hydroxyl group were observed in all nanomaterials based on [11] that compounds containing hydroxyl and carboxyl functional groups play a role in the formation of nanoparticles. The presence of surface hydroxyl reflects the number of terminals bonded to oxygen atom that may due to the formation of surface defects. The C=O group showed at 1600 cm^{-1} and the C-C group showed at 600 cm^{-1}. The vibration of ZnO can showed at 600 cm^{-1} which is corresponded to the literature that the vibration of metal with O group from ligan will show at 600-400 cm^{-1} [12]

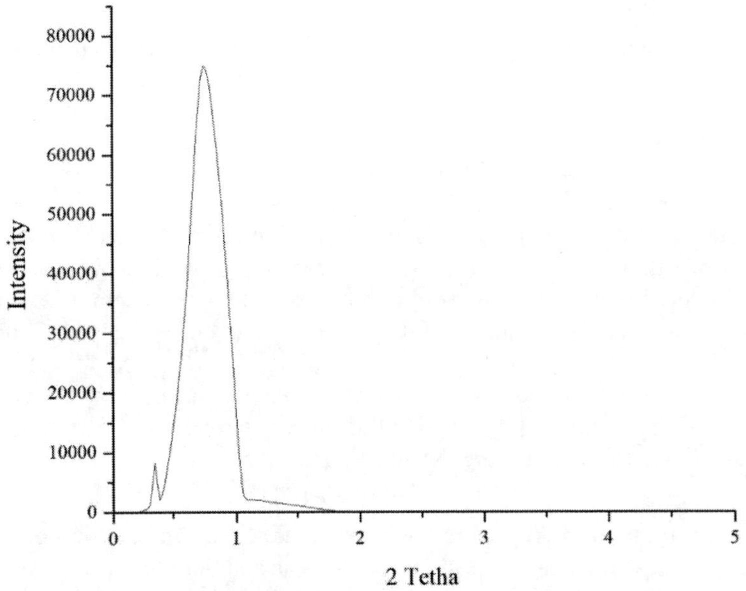

Figure 1. Characterization result of nano-ZnO in the scanning range 0-80^0.

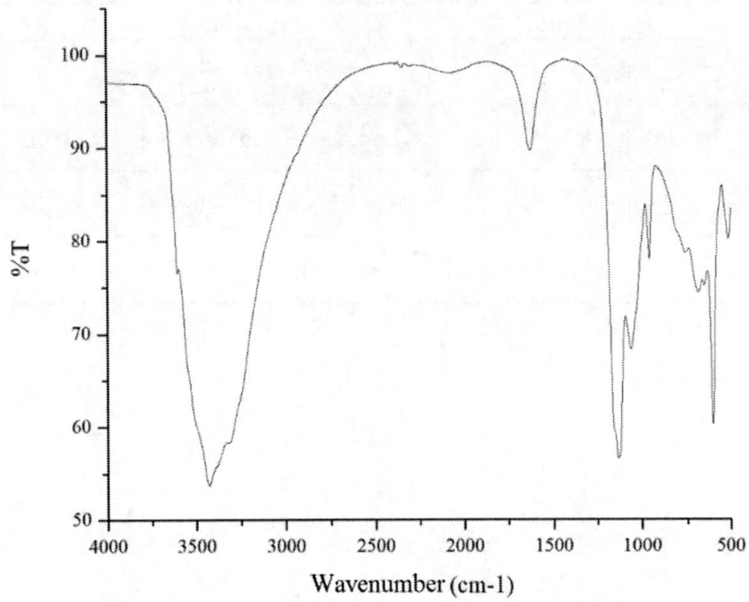

Figure 2. Characterization result of nano-ZnO with FTIR.

3.2. Adsorption studies of ibuprofen on nano-ZnO

The use of nano-ZnO as an adsorbent in the drug removal process has been carried out with ibuprofen as a molecule probe. Figure 3 shows the adsorption of ibuprofen with nano-ZnO using a variation of water solvent:hexane. The best absorption of ibuprofen is shown by variations of water solvents: hexane of 9:1. Adsorption experiences equilibrium in the 330 minutes. The adsorption of ibuprofen is significantly influenced by the ratio of solvent which is the component of hexane in water was increased not only the solubility of ibuprofen, but also the adsorption of ibuprofen molecules onto ZnO.

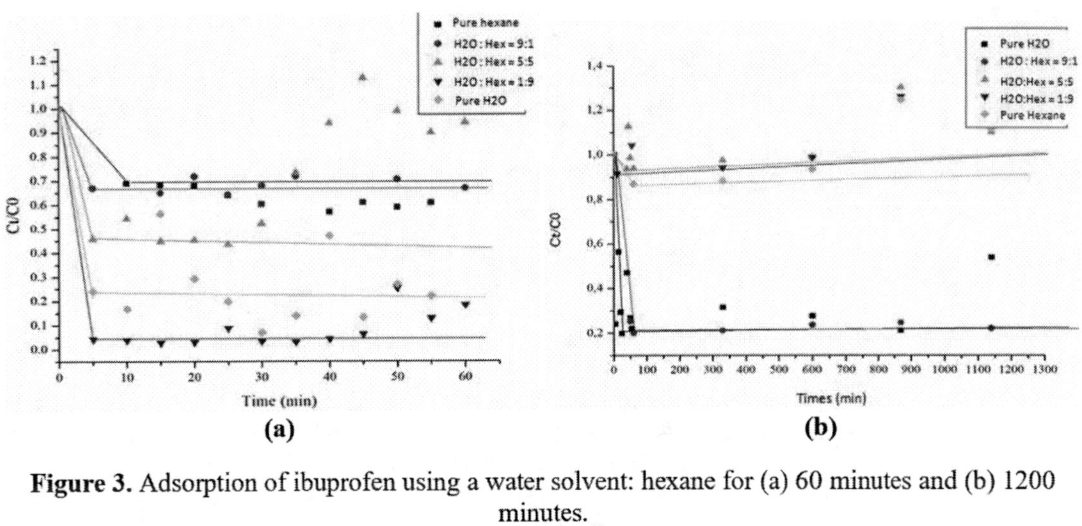

(a) (b)

Figure 3. Adsorption of ibuprofen using a water solvent: hexane for (a) 60 minutes and (b) 1200 minutes.

Table 1. Isotherm analysis on ibuprofen adsorption using the Langmuir, Freundlich, Temkin, and Dubinin-Radushkevich models.

No	Model	Equation	R^2
1	Langmuir	$:\frac{1}{qe} = \frac{1}{qmax} + \frac{1}{qmax\ Kl} x \frac{1}{Ce}$	0.5688
2	Freundlich	$\text{Log } qe = \frac{1}{n} \text{Log } Ce + \text{Log } Kf$	0.908
3	Temkin	$q_e = \frac{RT}{b} \ln A + \frac{RT}{b} \ln Ce$	0.997
4	Dubinin-Radushkevich	$\text{Ln } q_e = \text{Ln } q_m - \beta\varepsilon^2$	0.8709

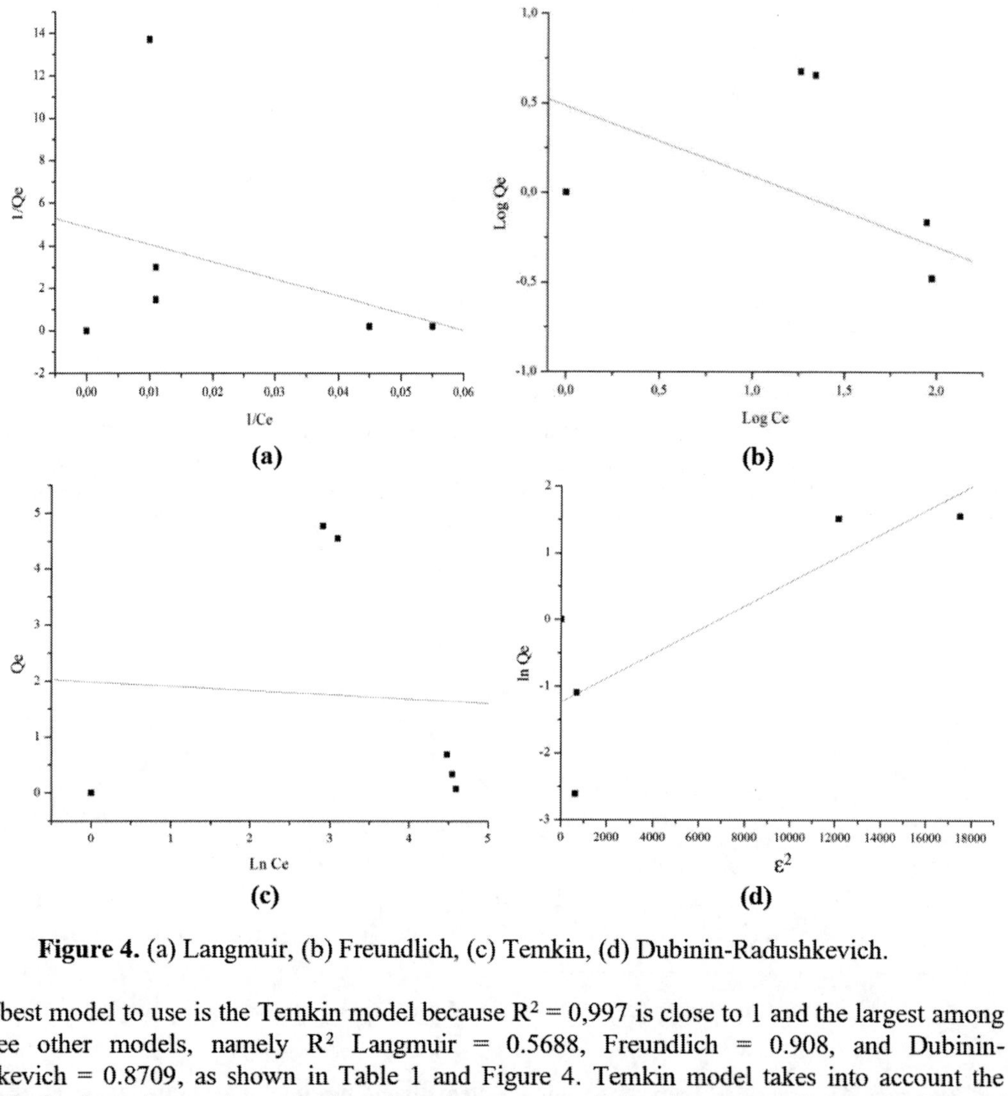

Figure 4. (a) Langmuir, (b) Freundlich, (c) Temkin, (d) Dubinin-Radushkevich.

The best model to use is the Temkin model because $R^2 = 0{,}997$ is close to 1 and the largest among the three other models, namely R^2 Langmuir = 0.5688, Freundlich = 0.908, and Dubinin-Radushkevich = 0.8709, as shown in Table 1 and Figure 4. Temkin model takes into account the effect of adsorbate interactions on the adsorption process. It is also assumed that the heat of adsorption of all molecules in the layer decreases linearly as a result of increase surface coverage. This model confirms that the adsorption of ibuprofen onto zinc oxide follows a chemisorption process because the application of this model is for chemical adsorption [13]

4. Conclusion

Nano-ZnO has been successfully synthesized using the hydrotermal method. The use of nano-ZnO as an adsorbent in ibuprofen adsorption is highly dependent on the surface area and shape of ZnO.

Effective ibuprofen adsorption is performed with a water: hexane solvent with ratio of 9:1 and reaches an equilibrium point at 330 minutes. Isotherm analysis on ibuprofen adsorption is more effective using the Temkin model than the Langmuir, Freundlich, and Dubinin-Radushkevich models. So, the adsorption of ibuprofen onto nano-ZnO follows chemisorption process. And it could be conclude that nano-ZnO is a potential and active adsorbent for removal ibuprofen from the water.

Acknowledgment
Authors acknowledge the Research Collaboration Indonesian Program (PPPKI) of Sebelas Maret University 2019 under contract Number 1362/UN27.21/PN/2019 for Maria Ulfa.

References

[1] Martin, J., M. D. M. Orta, S. M. Carraso, J. L. Santos, I. Aparicio, and E. Alonso 2019 *J. App. Clay Sci.* **171** 29–37
[2] Anggraini, D. I. and I. Pujilestari 2017 *J. Sci. App. Chem.* **20** 105-109
[3] F. Méndez-Arriaga, R.A. Torres-Palma, C. Petrier, S. Esplugas, J. Gimenez, and C. Pulgarin 2008 *J. Water Res.* **42** 4243–4248
[4] S. Esplugas, D.M. Bila, L. Gustavo, T. Krause, and M. Dezotti 2007 *J. Hazard Mater* **149** 631–664
[5] G.R. Boyd, S. Zhang, and D.A. Grim 2005 *J. Water Res.* **39** 668–676
[6] A. Joss, S. Zabczynski, A. Göbel, B. Hoffmann, D. Löffler, C.S. McArdell, T.A. Ternes, A. Thomsen, and H. Siegrist 2006 *J. Water Res.* **40** 1686–1696
[7] I.A. Katsoyiannis, S. Canonica, and U. von Gunten 2017 *J. Water Res.* **45** 3811–3822
[8] Choinaa, J., A. Bagabas, Ch. Fischer, G.-U. Flechsig, H. Kosslicka, A. Alshammarib, and A. Schulz 2015 *J. Cat. Today* **241** 47–54
[9] Dada, A. O., A. P. Olalekan, A. M. Olatunya, and O. Dada 2012 *J. App. Chem.* **3** 3-45
[10] Balarak, D., F. K. Mostafapour, H. Azarpira, and A. Joghataei 2017 *JPRI* **20** 1-9
[11] Sari, R. N., Nurhasni, and M. A. Yaqin 2017 *JPHPI* **20** 238-254
[12] Mintari, N., Suhartana, Sriatun 2015 *J. Sci. App. Chem.* **18** 29-33
[13] D. Ringot, B. Lerzy, K, Chaplain, J. P. Bonhoure, E. Auclair, and Y. Larondelle 2007 *J. Bioresource Tech.* **98** 1812-1821

The Influence of Hollow Cathode Geometry and N_2-H_2 Gas Mixture on the 2 MHz RF-DC Plasma Species and Density

J M Windajanti[1], Abdurrouf[1], D J Djoko H Santjojo[1] and M A Pamungkas[1]

[1]Physics Department, Faculty of Mathematics and Sciences, Brawijaya University, Jalan Veteran Malang, East Java, Indonesia

*Corresponding author: abdurrouf@ub.ac.id

Abstract. RF-DC plasma combined with the hollow cathode effect is used to generate plasma inside a hollow cavity. A high-density plasma was formed inside a vacuum chamber using a 2 MHz RF generator. A hollow cathode was utilized to focus the power at the voltage of 150 V with DC bias of -500 V. The plasma system was specially designed for titanium nitriding process at low temperature using N_2-H_2 gas mixture. The N_2-H_2 gas with a flow rate of 90-10 mL/minute was conducted at pressure 20 to 150 Pa. The cylindrical and rectangular hollow cathodes are used to make the discharge containing attractive atoms or molecules inside the hollow cavity. The trapped electrons could trigger the secondary electron emission that induced high-density plasma. The geometry components such as gap size, diameter, and cavity depth combined with variations of chamber pressure have been studied to produce high-density plasma. The species inside of N_2-H_2 plasma were detected by OES based on light emission. The high-intensity peaks of N_2* molecules from the second positive system and N_2^+ ions from the first negative system at the wavelength range of 300 to 500 nm were detected to obtain effective plasma species for the nitriding process.

1. Introduction

The plasma glow discharge consists of largely neutral atoms and molecules. Although the positive charge is equally the same as the negative charge, some ionized gas may exist in the plasma [1]. The plasma discharge is generated from a low pressure gas contained in a vacuum chamber. When a potential difference is given between the two electrodes, plasma discharge will be formed after a 'gas breakdown' so that the gas ionization process occurs to release electrons and positive ions. When positive ions are forced to the cathode, a bombardment of positive ions with the cathode can trigger the release of advanced electrons called 'secondary electron emission'. Moreover, when these electrons collide with gas particles in the plasma, it will cause excitation and ionization of the gas particles. Excitation of gas atoms or molecules will be followed by the radiative decay to a lower level while it is emitting light with special characteristics (i.e. the appearance of violet colors on nitrogen plasma). The collision events that occur in the ionization process create new ion and electron pairs that make self-sustained on plasma glow discharge [2]. In radio frequency (RF) discharges, the power and the electrons are coupled efficiently so the discharges have more efficient ionization than the DC discharge. The current in an RF discharge is generally a displacement current, results from the oscillatory motion of the electrons when they comply with the electric field [3]. An addition of DC bias voltage could attract the positive ions onto the cathode table to obtain the high-density plasma [4].

A plasma glow discharge is used in various fields of application, such as in the process of surface hardening, thin film deposition, and etching in materials technology, in the microelectronics industry, and also as a laser and lighting [5], [6]. The nitriding process using plasma glow discharge is a common method for increasing wear and corrosion resistance in surface hardening processes without affecting bulk material and also avoiding fatigue in the material due to high temperature processes. The advantage of this process is energy savings and no anti-pollution equipment requirements [7]. In the nitriding process, the bombardment of positive ions onto nitrided specimens placed on the cathode plate is a cathodic process that requires a high rate of nitrogen ionization. Plasma nitriding is faster than conventional nitriding because the surface of the specimen is saturated with nitrogen. Plasma process parameters, i.e. chamber pressure, DC bias and the RF Power control the composition and the depth of the nitride layer [4]. The effectiveness of plasma nitriding is always being developed by reducing the temperature and processing time [5].

With the cathode geometry like a cavity, a hollow cathode discharge will be formed where the plasma is trapped by the electrode wall. The loss of electrons in the plasma will decrease because the electrons are trapped by the cathode geometry, and result in the higher plasma density (10^{11} to $>10^{13}$ cm^{-3}) inside the hollow tube. The decrease is due to the occurrence of the pendulum effect which makes electrons to be reflected by the sheath on both sides of the cathode wall which induces a long electron trajectory to produce a large number of high-energy secondary electrons. Also, the electrons have only one direction to escape from the plasma, which is along the axis of the hollow cathode [8]. The negative glow of the plasma can expand to occupy the interior volume of the cathode using an appropriate combination of a given hollow cathode diameter or separation of the cathode surface with the chamber gas pressure under the applied power. The increasing of the discharge current depends on the hollow cathode geometry. An inter-cathode distance of the rectangular hollow or diameter of the cylindrical hollow is the important parameter that influences the discharge behavior. The decreased of the inter-cathode distance or the diameter could increase plasma density due to the emission of high energy electrons by the hollow cathode and then enhance the gas ionization [9].

The present work aims to investigate the influence of both pressure and hollow cathode geometry on the luminous intensity emitted by the plasma. The emission is originated from the positive column of glow discharge plasma that generated using pure N_2 gas and N_2-H_2 gas mixture for constants discharge voltage of RF and DC bias generator.

2. Experimental Method

2.1. High-density RF-DC plasma system

A high-density RF-DC plasma nitriding system is utilized to generate a plasma of N_2-H_2 gas. In this plasma system there is no need for a mechanical matching box, but rather by adjusting the frequency of 2 MHz. The input and output power are automatically matched with the response time limited to only one millisecond. This fast power control provides full use of plasma with a pressure range of more than 50 Pa. With a neutral electric condition of the vacuum chamber, the RF and DC power can be controlled independently. The RF plasma is produced by the dipole electrodes, and then the DC bias is directly applied to attract the ions towards the cathode.

The RF-DC plasma system composed by a vacuum chamber which equipped with the RF and DC bias plasma generator, vacuum pump, and gas supply system. In the vacuum chamber interior, the main part components are a pair of RF dipole rod electrodes and a stainless steels cathode plate. An addition of a hollow cathode device was placed on the cathode plate. The gas pressure was controlled by Pirani vacuum gauge control (Ulvac GP-2G) and the RF and DC bias voltage by the voltmeter. The nitrogen and hydrogen gas was injected into the chamber via a flowmeter through an inlet at the top side of the chamber. The schematic design of the RF-DC plasma system is depicted in Figure 1.

In this investigation process, the hollow cathode device is first placed on the cathode plate, followed by a vacuum process until it reaches a bottom pressure of 4 Pa. After that, nitrogen gas is introduced into the chamber as a plasma carrier gas until the pressure rises up to 100 Pa, and then the plasma was ignited. On the use of the N_2-H_2 gas mixture, the hydrogen gas would be entered after the plasma was generated by using pure nitrogen gas. Table 1 shows the processing parameter.

IOP Conf. Series: Materials Science and Engineering **833** (2020) 012097 doi:10.1088/1757-899X/833/1/012097

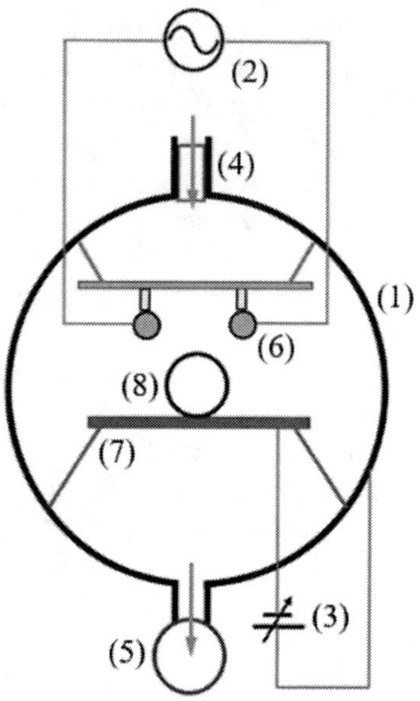

Figure 1. The schematic design of the RF-DC plasma system, consist of: (1) vacuum chamber; (2) 2 MHz RF generator; (3) DC bias generator; (4) N_2-H_2 gas inlet with flowmeter; (5) vacuum pump; (6) anode dipole rod; (7) cathode plate; and (8) hollow cathode device.

Table 1. The operating conditions of the high-density plasma system.

Processing Parameter	Set-up Value
Base pressure vacuum of chamber	4 Pa
RF frequency	2 MHz
RF voltage	150 V
DC bias voltage	-500 V
Gas flow rate	N_2 gas: 100 mL/min; N_2-H_2 gas: 90-10 mL/min
Pressure of gas mixture	Variance at 20-150 Pa

2.2. Quantitative plasma diagnosis

An identification of ion density of N_2-H_2 plasma inside the hollow cathode device in the high-density RF-DC plasma nitriding system was carried out using a quantitative plasma diagnosis. The various plasma species and their characteristics were identified by the optical emission spectroscopy (OES Aurora-4000 GE UV-NIR) with the typical setup as illustrated in Figure 2.

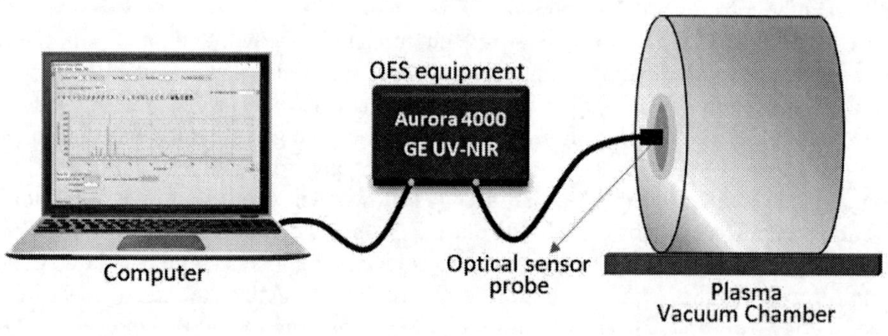

Figure 2. A typical setup of optical emission spectroscopy (OES).

IOP Conf. Series: Materials Science and Engineering **833** (2020) 012097 doi:10.1088/1757-899X/833/1/012097

The existence of plasma reactive species strongly depends on the control parameters such as applied voltage and gas flow rate. The emission spectra of N_2-H_2 gas were recorded in the wavelength range from 200 nm to 1100 nm. The emission light is detected through the optically transparent silica window located in front of the chamber by a photosensor amplifier. Knowing the nitrogen plasma species is important for the adjustment of process parameters in order to the optimization process of the specific applications using the plasma treatment.

2.3. Hollow cathode geometry

A device of the hollow cathode is inserted into the chamber and placed on the cathode plate. There are two models of hollow tube used in this study, i.e. hollow cathode with a cylindrical and rectangular cross-section. The 80 mm length of cylindrical hollow has two kinds of a diameter of 50.8 and 38.1 mm, respectively. The same length of rectangular hollow has a gap height of 25 and 20 mm, respectively. A photograph of the hollow cathode depicted in Figure 3.

Figure 3. The geometry of hollow cathode tube.

3. Result and Discussion

Photons emitted by radiative decay after the electron boost the plasma species into upper electronic states can be obtained and analyzed by recording the OES spectrum because it has a characteristic wavelength in the optical region of V-UV spectra. The density of plasma ions can be measured by this measurement technique, by analogy that the specific wavelength of the light emission intensity from excited states of a specific wavelength is proportional to the species concentration in that excited states [10].

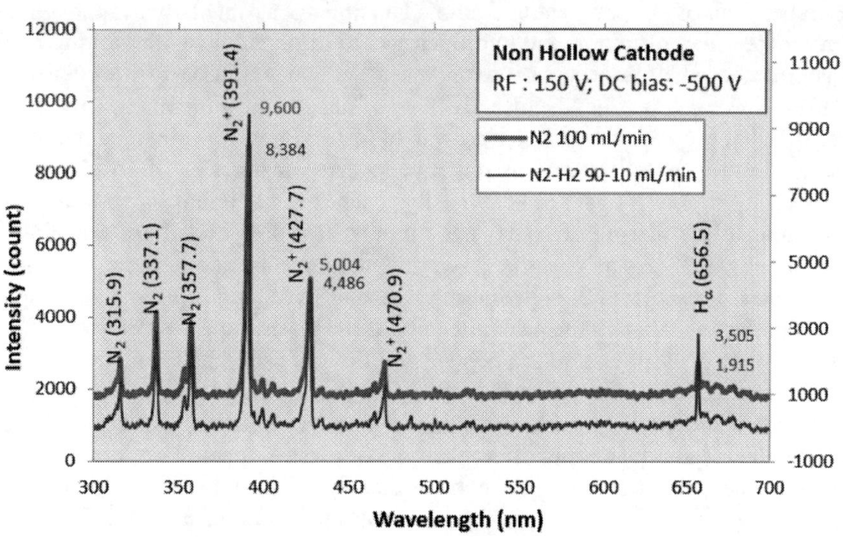

Figure 4. OES Spectra of N_2 Plasma and N_2-H_2 Plasma generated without hollow cathode chamber at 100 Pa.

The glow discharge was generated by using the RF- DC plasma generator with the power of 100 W and the RF frequency of 2 MHz. At the time of observation, the voltage used to generate plasma was 150 V and -500 V for each RF and DC bias generator. The reactive species of the nitrogen molecules and hydrogen atoms spectra were identified in the wavelength range from 300 to 700 nm on the recorded optical emission spectra from 200 to 1100 nm. The result of nitrogen peaks was determined from the glow discharge from pure nitrogen gas with the flow rate of 100 mL/min in comparison with those from the N_2-H_2 gas mixture with the flowrate of each one of 90-10 mL/min as seen in Figure 4. From the recorded of OES spectra, it showed that there were several strong nitrogen molecular lines of N_2^* (activated nitrogen molecules) and N_2^+ (nitrogen ion molecules) [4]. There were three peaks with a high-intensity line of N_2^* that come from the transition of $C^3\Pi_u \rightarrow B^3\Pi_g$ at the second positive system. There were also three peaks with a high intensity of N_2^+ in which the strongest emission is the lines of nitrogen ion at first negative system (1NS) with the wavelength of 391.4 nm from the transition of $B^2\Sigma_u^+ \rightarrow X^2\Sigma_g^+$ electronic state [11], [12]. Several dominant peaks of nitrogen molecule obtained from the recording of OES on discharge generated by this RF-DC generator can be observed in Table 2. It was also observed the hydrogen peak at a wavelength of 656.5 nm which is the Balmer series transition from the 3rd to 2nd energy level called H_α peak.

Table 2. The nitrogen gas species in glow discharge RF-DC plasma.

Nitrogen Species	Wavelength (nm)	Upper Energy Level	Lower Energy Level
N_2^*	315.9	$C^3\Pi$ [1]	$B^3\Pi$ [0]
N_2^*	337.1	$C^3\Pi$ [0]	$B^3\Pi$ [0]
N_2^*	357.7	$C^3\Pi$ [1]	$B^3\Pi$ [1]
N_2^+	391.4	$B^2\Sigma$ [0]	$X^2\Sigma$ [0]
N_2^+	427.7	$B^2\Sigma$ [0]	$X^2\Sigma$ [1]
N_2^+	470.9	$B^2\Sigma$ [0]	$X^2\Sigma$ [2]
H_α	656.5	Balmer series (n = 3 to n = 2)	

Without a hollow cathode, the plasma OES spectra of N_2^+ ions at 391.4 and 427.7 nm produced with a mixture of N_2-H_2 gas has a higher intensity than the pure N_2 gas as shown in Figure 4. There is also an obtained of higher peak of H_α species at 656.5 nm. The other peak intensity is almost the same. Further investigation on the effect of mixing the hydrogen gas to nitrogen gas with varying concentrations of both gases is still needed. It is necessary to increase the concentration of nitrogen ions because the presence of these ions greatly affect the effectiveness of plasma nitriding process used.

When the chamber pressure was increased from 20 to 150 Pa, it was obtained that the peak intensity of each species of nitrogen molecules was sensitive to these changes as illustrated in Figure 5. In the plasma generated from pure N_2 gas, peak intensity reaches a maximum value at a pressure of 80 Pa processes, whereas in the plasma of N_2-H_2 gas mixture at a pressure of 100 Pa. Even with a small concentration, mixing of H_2 gas in N_2 gas is very influential on plasma generation. Plasma becomes very dim at chamber pressures below 60 Pa, while in the use of pure N_2 gas plasma still appears to hold up to a pressure of 15 Pa even though the peak intensity of the plasma species decreases.

In theory, the glow discharge structure depends on the gas used, the chamber gas pressure, and the material and geometric shape of the electrodes [13]. In this investigation, the effect of the addition of the hollow cathode device to the plasma structure formed was obtained. Based on the OES spectrum, the addition of hollow cathode will not affect the type of plasma species are formed. The difference in shape and size of the hollow cathode geometry affects only the intensity of each molecule peak of nitrogen and hydrogen species, as well as the effect on the optimum pressure that was produced the highest intensity.

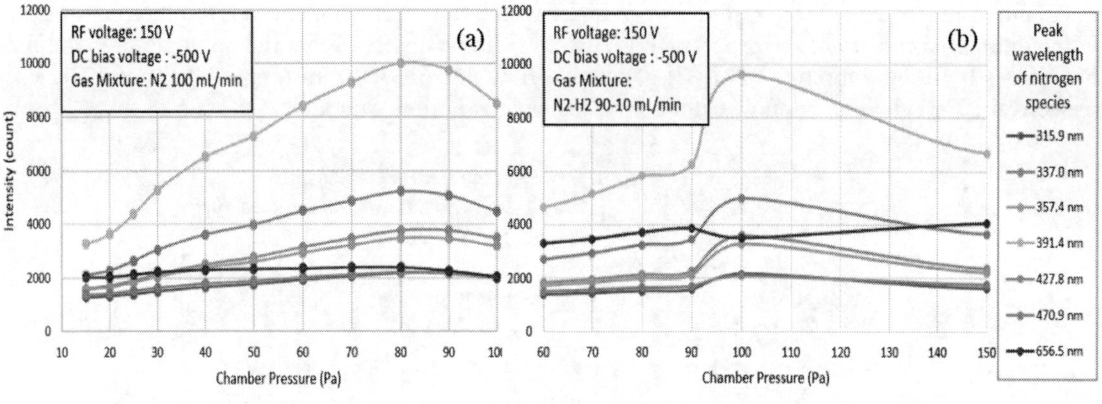

Figure 5. Effect of chamber pressure to the intensity of nitrogen plasma species in pure nitrogen (a) and nitrogen-hydrogen gas mixture (b) plasma.

The use of cylindrical hollow cathode did not increase the intensity of nitrogen ions in the plasma. However, in the use of rectangular hollow cathodes, there appears to be an increase in the intensity of nitrogen ions. The uses of cylindrical hollow cathode slightly decreased the intensity the peak intensity of N_2^+ at 391.4 and 427.7 nm but it increased the N_2^* peak at 337.1 and 357.7 nm. In the cylindrical hollow cathode, the adding of hydrogen gas slightly reduced the intensity of N_2^+ peak at 391.4 and 427.7 nm and also the N_2^* peak at 315.9, 337.1 and 357.7 nm. However, there was a significant increase in the peak intensity of H_α species at 656.5 nm, as seen in Figure 6.

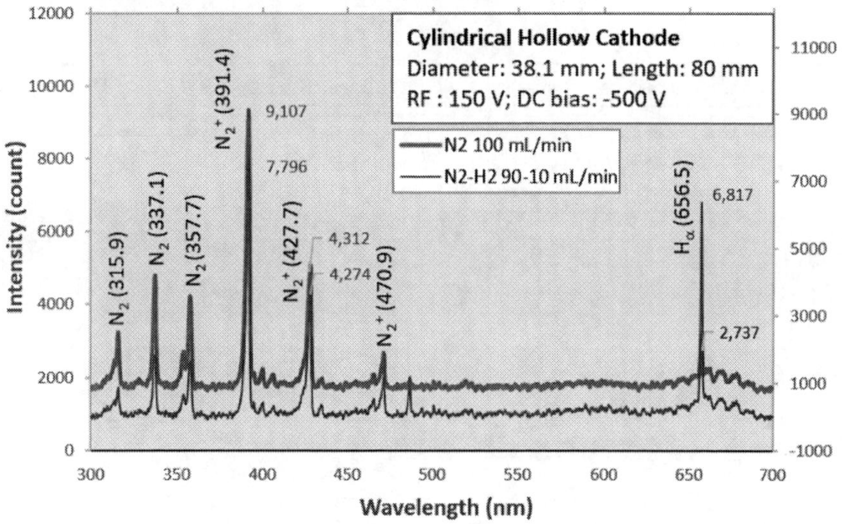

Figure 6. OES spectra of N_2 plasma and N_2-H_2 plasma at cylindrical hollow cathode at 30 Pa.

Meanwhile, the use of rectangular hollow cathodes can increase the intensity of nitrogen ions in the plasma. Compare to plasma without hollow cathode, the uses of rectangular hollow cathode increase the intensity of almost all nitrogen species. The N_2^* peak at 315.9, 337.1 and 357.7 nm are enhanced, but the key point was the increase of N_2^+ peak intensity at 391.4 and 427.7 nm. On the other hand, the intensity of N_2^* peak and N_2^+ peak was almost the same for the adding of hydrogen gas with the slight increase of the peak intensity of H_α species at 656.5 nm.

An interesting result in the use of hollow cathodes with different cross-sections is the enormous effect on the optimum pressure of the process. In Figure 8, it can be observed that the optimum pressure for the process by using a mixture of N_2-H_2 gas with a flow rate of 90-10 mL/min is 30 Pa and 100 Pa, respectively, for cylindrical hollow cathode and rectangular hollow cathode.

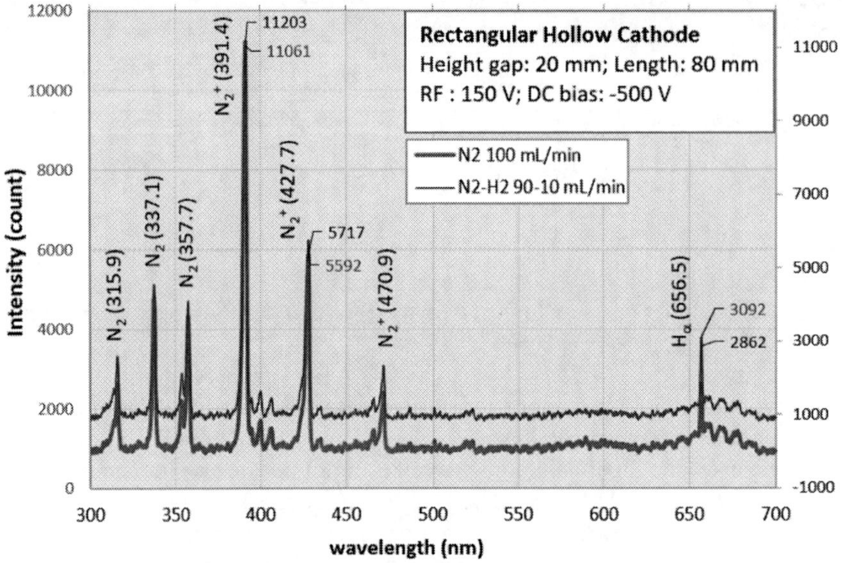

Figure 7. OES spectra of N_2 plasma and N_2-H_2 plasma at rectangular hollow cathode at 100 Pa.

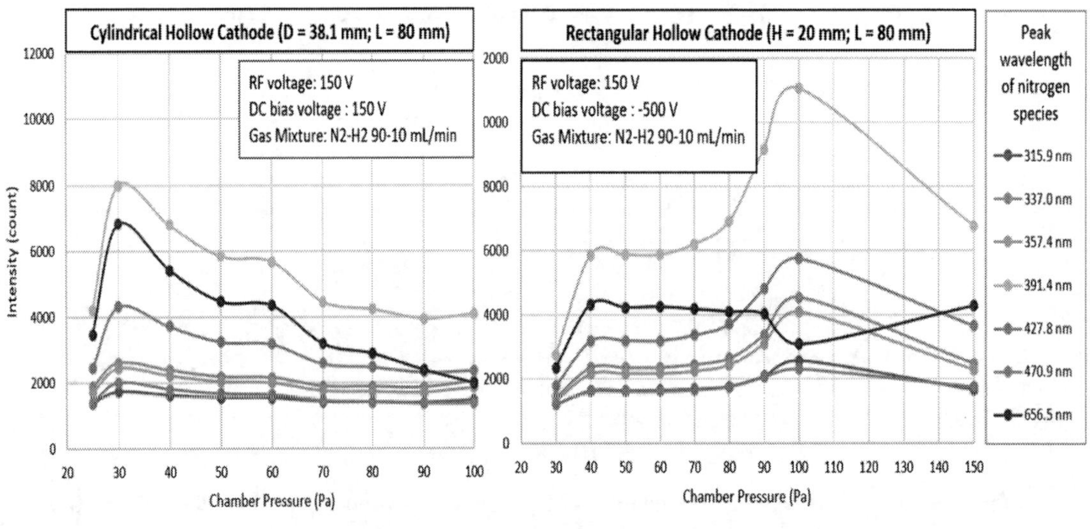

Figure 8. Effect of chamber pressure to the plasma species intensity of N_2-H_2 gas mixture in the cylindrical hollow cathode (a) and in the rectangular hollow cathode (b).

4. Conclusion

In conclusion, the plasma glows discharge structure is influenced by the type of gas used to generate plasma, the amount of pressure in the chamber, and also the geometrical shape of the additional hollow cathode device. The addition of hydrogen gas as an N_2-H_2 gas mixture with a flow rate of 90-10 mL/min could change the intensity of nitrogen species inside the glow discharge plasma. The use of a rectangular hollow cathode increases the intensity of nitrogen ion species at the high-pressure process. The plasma species peak intensity was increased when the hollow gap reduces. On the other hand, the cylindrical hollow cathode is effective for plasma processing at low pressure. And, reducing the hollow diameter could increase the plasma species intensity.

References

[1] Bellan P M 2006 *Fundamentals of Plasma Physics* (Cambridge: Cambridge University Press)
[2] Bogaerts A 1999 *J. Anal. At. Spectrom.* **14** 1375–84
[3] Söderström D 2008 Modelling and Applications of the Hollow Cathode Plasma Digital *Comprehensive Summaries of Uppsala Dissertations from the Faculty of Science and Technology* **433** 1-87
[4] Aizawa T 2018 Low Temperature Plasma Nitriding of Austenitic Stainless Steels *Stainless Steels and Alloys* (IntechOpen.78365) chapter 3 pp 31–50
[5] Bogaerts A, Neyts E, Gijbels R, and van der Mullen J 2002 *Spectrochim. Acta, Part B*, **57** 609–58
[6] Winchester M R and Payling R 2004 *Spectrochim. Acta Part B At. Spectrosc.* **59** 607–66
[7] Matsuda F, Nakata K, and Makishi T 1987 *Trans. JWRI* **16** 139–44
[8] Muhl S and Pérez A 2015 *Thin Solid Films* **579** 174-98 *(Accepted manuscript* TSF 34146)
[9] Brunatto S F, Louis J, and Muzart R 2008 *J. of Braz. Soc. of Mech. Sci. & Eng.* **XXX** (2) 145–151
[10] Qayyum A, Zeb S, Naveed M A, Ghauri S A, Zakaullah M, and Waheed A 2005 *J. Appl. Phys.* **98** 1-9
[11] Baek E J, Joh H M, Kim S J, and Chung T H 2016 *Phys. Plasmas* **23** 1-9
[12] Wagatsuma K and Hirokawa K 1985 *Anal. Chem.* **57** 2901–07
[13] Awsi S K 2013 *American J. of Modern Physics* **2** 276–81

Acknowledgements

The work was financially supported by Brawijaya University through "Program Hibah Penelitian Guru Besar dan Doktor Tahun Anggaran 2019" with the contract number: 28/UN10.10/PN/2019.

AUTHOR INDEX